North American Terrestrial Vegetation

North American Terrestrial Vegetation, Second Edition, is a major contribution to botanical and ecological literature and provides comprehensive coverage of the major vegetation types of North America from the arctic tundra of Alaska to the tropical forests of Central America. Each chapter gives the reader a feeling for the composition, architecture, environment, and conservation status of each ecosystem. In addition to vegetation descriptions, information is also included on the abiotic environment, paleoecology, productivity, nutrient cycling, autecological behavior of dominant species, environmental issues, management problems, the role of natural disturbance, and critical areas for future research.

This is an outstanding new edition of a well-received text, and it is essential reading for students and researchers in plant science, ecology, and conservation.

FEATURES OF THE SECOND EDITION

- There are new chapters on freshwater wetlands, coastal marine wetlands, temperate Mexico, the Caribbean Islands, and the Hawaiian Islands.
- The book has been updated throughout with new text and figures and with new literature reviewed.
- Every chapter includes information on habitat loss and restoration-preservation programs that are now mitigating these losses.
- Richly illustrated with many new diagrams, photographs, and tables.
- Includes a complete, cross-referenced topics, species, and place names and regions index

FROM THE REVIEWS OF THE FIRST EDITION

"Far more than a mere compilation of vegetation types, this book synthesizes our current history of the understanding of the history, dynamics, and physical setting of the continent's plant cover . . . a text and a reference for field scientists at all levels . . . it is unsurpassed as a guide to the vegetation blanketing North America and to the processes structuring that vegetation."

– Ecology

" . . . will be the standard reference on North American plant communities for quite some time."

– BioScience

Michael G. Barbour is Professor of Plant Ecology at the University of California, Davis. He is co-author of *Terrestrial Plant Ecology, Second Edition* (Benjamin/Cummings, 1987) and *Terrestrial Vegetation of California, Second Edition* (California Native Plant Society, 1988).

William Dwight Billings (1910–1997) has provided inspiration, advice, example, argument, and encouragement to generations of students and professionals in plant ecology. Professor Billings's love of nature and dedication to the study of ecology have been passed on to 52 men and women who received their Ph.D.'s under his direction.

Generalized vegetation formations of North America.
A=ice; B=arctic tundra; C=taiga; D=Pacific coastal–Cascadian forests; E=Palouse prairies; F=Intermountain deserts, shrub-steppes, woodlands, and forests; G=Californian forests and alpine vegetation; H=Californian grasslands, chaparral, and woodlands; I=Mojave and Sonoran deserts; J=Rocky Mountain forests and alpine vegetation; K=central prairies and plains; L=mixed deciduous forests; M=Chihuahuan deserts and woodlands; N=Appalachian forests; O=piedmont oak-pine forests; P=coastal plain forests, bogs, swamps, and strand; Q=tropical forests. Boundaries and formation names according to W. D. Billings; art by Christina Weber-Johnson.

North American Terrestrial Vegetation
Second Edition

Edited by

Michael G. Barbour
William Dwight Billings

CAMBRIDGE
UNIVERSITY PRESS

PUBLISHED BY THE PRESS SYNDICATE OF THE UNIVERSITY OF CAMBRIDGE
The Pitt Building, Trumpington Street, Cambridge, United Kingdom

CAMBRIDGE UNIVERSITY PRESS
The Edinburgh Building, Cambridge CB2 2RU, United Kingdom   www.cup.cam.ac.uk
40 West 20th Street, New York, NY 10011-4211, USA   www.cup.org
10 Stamford Road, Oakleigh, Melbourne 3166, Australia
Ruiz de Alarcón 13, 28014 Madrid, Spain

First published 1988
Second edition 2000

Printed in the United States of America

Typeface, 9/11 Palatino pt.   *System* DeskTopPro$_{/UX}$ [RF]

A catalog record for this book is available from the British Library.

Library of Congress Cataloging-in-Publication Data
North American terrestrial vegetation / edited by Michael G. Barbour,
William Dwight Billings. – 2nd ed.
p.   cm.
Includes bibliographical references (p.   ) and index.
ISBN 0-521-55027-0 (hardbound)
1. Plant communities – North America.   2. Plant ecology – North
America.   3. Phytogeography – North America.   I. Barbour, Michael G.
II. Billings, W. D. (William Dwight), 1910–1997.
QK110.N854   1999
581.7'22'097 – dc21   97-29061
                                                                              CIP

ISBN 0 521 55027 0 hardback
ISBN 0 521 55986 3 paperback

Contents

Contributors

Michael G. Barbour
Environmental Horticulture Department
University of California
Davis, California

William Dwight Billings (deceased)
Botany Department
Duke University
Durham, North Carolina

Lawrence C. Bliss
Botany Department
University of Washington
Seattle, Washington

Norman L. Christensen, Dean
Nicholas School of the Environment
Duke University
Durham, North Carolina

Julio Figueroa Colón
International Institute of Tropical Forestry
USDA Forest Service

Hazel R. Delcourt
Ecology and Evolutionary Biology Department
University of Tennessee
Knoxville, Tennessee

Paul A. Delcourt
Ecology and Evolutionary Biology and the Geological Sciences Departments
University of Tennessee
Knoxville, Tennessee

Deborah L. Elliott-Fisk
Wildlife, Fish, and Conservation Biology
University of California
Davis, California

Jerry F. Franklin
College of Forest Resources
University of Washington
Seattle, Washington

Charles B. Halpern
College of Forest Resources
University of Washington
Seattle, Washington

Gary S. Hartshorn
Organization for Tropical Studies
Duke University
Durham, North Carolina

Jon E. Keeley
Biological Resources Division
Sequoia-Kings Canyon Field Station
Three Rivers, California

Lloyd L. Loope
Geological Survey, Biological Resources Division
Pacific Island Ecosystem Research Center
Haleakala Field Station
Makawao Maui, Hawaii

Ariel E. Lugo
International Institute of Tropical Forestry
USDA Forest Service

Isolda Luna
Herbario, Facultad de Ciencias
Universidad Nacional Autonoma de México
Mexico DF, Mexico

James A. MacMahon, Dean
College of Biological Sciences
Utah State University
Logan, Utah

Karen L. McKee
National Wetlands Research Center
Lafayette, Louisiana

Irving A. Mendelssohn
Wetland Institute
Louisiana State University
Baton Rouge, Louisiana

Richard A. Minnich
Earth Sciences Department
University of California
Riverside, California

Robert K. Peet
Biology Department
University of North Carolina
Chapel Hill, North Carolina

Curtis J. Richardson
Nicholas School of the Environment
Duke University
Durham, North Carolina

Paul G. Risser, President
Oregon State University
Corvallis, Oregon

Frederick N. Scatena
Institute of Tropical Forestry
Rio Piedras, Puerto Rico

Phillip L. Sims
Southern Plains Range Research Station
U.S. Department of Agriculture
Woodward, Oklahoma

Victor Manuel Toledo
Centro de Ecologia
Universidad Nacional Autonoma de México
Mexico DF, Mexico

Alejandro Velázquez
Laboratorio de Biogeografia
Facultad de Ciencias, Universidad Nacional Autonoma de México
Mexico DF, Mexico

Neil E. West
Range Science Department
Utah State University
Logan, Utah

James A. Young
Agricultural Research Service
U.S. Department of Agriculture
Reno, Nevada

Preface to the First Edition

Rather naively, back in the fall of 1982, when this book first began to be organized, we expected to have a 300-page publication in hand three years later. The complexity of the subject and the intricate schedules of fourteen contributing authors and two editors lengthened the preparation period by 100% and increased the size of the book by 50%. We were helped in our early discussions about the book's content, format, and contributors by the hard work and encouragement of Richard Ziemacki, then editor at the New York office of Cambridge University Press. We thank him now for his vision when ours was still in the "what if..." stage.

We reached final decisions about book objectives, format, and the list of contributors by mid-1983. But the distribution of fine-scale topics was in a state of dynamic equilibrium well into 1986. Book writing, it seems, is a seral procedure, with many analogies to succession in vegetation. We trust that the climax presented to the reader is an accurate and challenging summary of what is currently understood about North American vegetation.

Chapters focus on the major plant formations of North America, but they also include information on many other, more local, vegetation types. The authors have devoted enough attention to each vegetation type discussed to give the reader details on vegetation structure, response to disturbance, community/environment relations, nutrient cycling and productivity, and autecological behavior of dominant species. We have selected contributors who are active researchers in extensive regions or vegetation types, and we have invited them to flavor their reviews with their own research biases (many chapters include previously unpublished data, analyses, or models). The same standard topical outline was given to each author to follow, but each region or vegetation type has a peculiarly

skewed literature that reflects the environmental factors long considered to be important by regional ecologists. In one area, those factors may be fire and soil moisture; in another, wind storms and soil nutrients; in another, paleoecological events of the Pleistocene or Holocene; in another, human actions that have introduced grazing animals or weeds or modified the scale of natural disturbances. Consequently, any two chapters may have little in common beyond starting with a frontispiece map and introduction and concluding with suggestions for future research. Although the sequences of topics vary from chapter to chapter, each chapter includes sections of paleobotany, the modern environment, human-caused vegetation changes, successional patterns following disturbance, and a quantitative description of major vegetation types.

There is some overlap, in terms of vegetation types discussed, between pairs of chapters. Petran chaparral, for example, is described in Chapter 3 (Rocky Mountains), Chapter 6 (chaparral), and Chapter 7 (intermountain deserts, shrub steppes, and woodlands); the New Jersey pine barrens are considered in both Chapter 10 (deciduous forest) and Chapter 11 (southeastern coastal plain); the forest-tundra ecotone is discussed in both Chapter 1 (arctic tundra and polar desert biome) and Chapter 2 (boreal forest). In the interest of efficient use of space, only one chapter of such a pair describes an overlapping vegetation unit in detail, but the units are such logical inclusions within each chapter that the remaining degree of overlap is important to the book's overall objective and to the reader's understanding of vegetation gradients.

Our intended audience includes knowledgeable laypeople, advanced undergraduates, graduate students, and professional ecologists in both basic and applied fields. Every chapter addresses management questions and problems, and every chapter includes considerable information on basic ecological topics, quite apart from immediate management considerations. Common names of dominant plants are sometimes used, but text, tables, and figures emphasize scientific names. The metric system is adopted exclusively. Figures and tables are added when they can efficiently convey a sense of the vegetation better than can the text. Space limitations prevent us from illustrating every vegetation type included in the text, and some types simply do not lend themselves well to black-and-white photography. Some of the graphs in the text were drawn by Christina Weber-Johnson.

There is a rich heritage of texts on North American vegetation, and we stand on the shoulders of their authors in order to see just a bit farther. Our predecessors include Clements, Daubenmire, Harshberger, Knapp, Küchler, and Vankat; their works span the past 75 years. Most recently, Chabot and Mooney have edited a volume on the physiological ecology of North American vegetation, and we consider their work to be an appropriate complementary companion to this book.

Many colleagues have served as reviewers for portions of this book, and we take this opportunity now to thank them. We also congratulate the authors on the care and effort with which they have created their chapters, and we anticipate that these chapters will stand as the best summaries for some years to come. Quite simply, we believe that the authors have made this text a balanced, modern, detailed reference work that has had no equal in the past and that certainly has no current equal.

Finally, we deeply appreciate the help of the following people for their work in indexing: Julie Barbour, Frederika Bowcutt, Robert Boyd, Sonia Cook, Ronnie James, Steven Jennings, Scott Martens, Tisa Owen, Wayne Owen, Oren Pollack, Robert Rhode, George Robinson, Diane Ryerson, Jane Sakauye, Judy Sindel-Dorsey, Glenn Turner, and Debbie Woodward. Mrs. Judi Steinig was invaluable to us in the production aspects of this book; we are indebted to her for the exceptional care she consistently applied to our project.

MGB. and WDB.

Preface to the Second Edition

When Dwight Billings and I finished checking the last proofs of *North American Terrestrial Vegetation* in 1988, we never expected to visit the process again — let alone within 10 years. However, that book became so well and widely received that the publisher asked us to write a second edition. We undertook the project with the proviso that our first effort might be expanded as well as updated.

The expansion has been in (1) breadth of vegetation types and (2) an additional theme that all chapters embrace. This second edition contains new chapters on freshwater wetlands, coastal marine wetlands, temperate Mexico, the Caribbean Islands, and the Hawaiian Islands. Every chapter, whether new or among the original 13, now includes information on habitat loss and the restoration and preservation programs that are mitigating these losses. First edition chapters have all been variously revised, some completely, but all have new tables, illustrations, chapter sections, and references. Some chapters have been written by new authors or co-authors, in this way injecting different perspectives.

In early January of 1997, near the very end of manuscript preparation, Dwight Billings passed away. His academic life has recently been well chronicled by Boyd Strain (*Bulletin of the Ecological Society of America*, April, 1997 issue), so there is no need to review here his enormous impact on American plant ecology. Rather, those of us who have participated in this edition simply want to say that we miss him and hope that our work will serve as a testimonial to his influence in our lives. Fittingly, we have chosen a painting of Dwight's for the book's jacket. It is a pastel he did at the age of 36, of quaking aspen (*Populus tremuloides*) in the Sierra Nevada. Because aspen is the tree species with the widest distribution in North America, the painting provides a fitting ecological – as well as personal – motif for our book to wear.

We dedicate this volume to Shirley Miller Billings, Dwight's wife of 38 years. As Boyd Strain so well recalls in his obituary, Shirley and Dwight were a research *team*.

Finally, we thank our many reviewers for their careful reading of various chapters and their helpful comments. We also thank Ms. Judi Steinig, who once again served as production editor.

Chapter
1

Arctic Tundra and Polar Desert Biome

L. C. BLISS

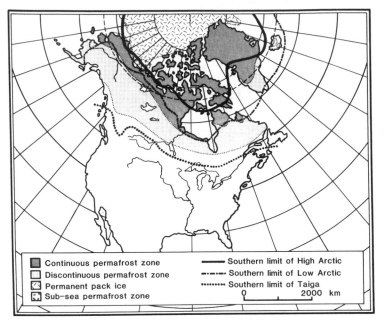

Continuous permafrost zone   —— Southern limit of High Arctic
Discontinuous permafrost zone   —·—·— Southern limit of Low Arctic
Permanent pack ice   ········· Southern limit of Taiga
Sub–sea permafrost zone   0   2000   km

Figure 1.1. Major subdivisions of the North American and Greenland arctic.

INTRODUCTION

The Arctic is often viewed as a monotonous landscape with a limited number of vascular plants and an abundance of cryptogams. In reality, the arctic tundra and polar desert biome is as diverse in its vegetation types and soils as are the grassland biomes and the coniferous biomes of the western mountains and the taiga. The Arctic constitutes about 20% of North America (Fig. 1.1, general map of biomes), about 2.5 M km² in Canada, 2.0 M km² in Greenland, and 0.3 M km² in Alaska. Over this large area there are considerable variations in climate, ice cover, soils, sizes of the flora (cryptogam and vascular), and plant communities.

As used here, "Arctic" refers to those areas beyond the climatic limit of the boreal forest and treeline. There are often small pockets of trees, usually *Picea glauca*, in protected mesohabitats (south-facing slopes, river terraces) beyond the treeline, but the uplands are dominated by arctic tundra vegetation. Throughout these cold-dominated landscapes, soils are permanently frozen (permafrost) with only the upper portion (20–60 cm, except 100–200 cm along rivers), the active layer, thawing in summer. The Circumpolar Arctic is divided into the Low and High Arctic (Fig. 1.1) based on many ecological characteristics (Table 1.1). Tundras predominate in the Low Arctic, and polar semideserts and polar deserts dominate the High Arctic. "Tundra" is a generic term that includes vegetation types that range from tall shrub (2–5 m

high) to dwarf shrub heath (5–20 cm high) and graminoid-moss. These landscapes have a total plant cover of 80–100%, including an abundance of cryptogams in most sites.

PHYSICAL ENVIRONMENT

Climate

The macroclimate of the Arctic is characterized by continuous darkness in midwinter, with a nearly continuous cover of snow and ice for 8–10 mo and by continuous light in summer, with a short growing season of 1.5–4 mo. In winter a large semipermanent high-pressure system of intensely cold dry air occurs over the Yukon Territory and the western Northwest Territories (N.W.T.). The average southern position of these anticyclonic systems parallels the southern boundary of the taiga in winter. However, outbreaks of cold arctic air periodically extend into the midwestern and southern states during severe storms. In summer, the arctic high-pressure system is weaker, with its southern limit paralleling the arctic–boreal forest boundary over 50% of the time (Bryson 1966). Low-pressure systems in the Gulf of Alaska and between Baffin Island and Greenland bring increased precipitation to the Arctic in summer, when these storm systems move across Alaska, into the Canadian Arctic Archipelago and Greenland. Winter penetration of low-pressure systems into the Arctic seldom occurs

Table 1.1. *Comparison of environmental and biotic characteristics of the Low and High Arctic in North America*

Characteristics	Low Arctic	High Arctic
Environmental		
Length of growing season (months)	3–4	1.5–2.5
Mean July Temperature (°C)	8–12	3–6
Mean July soil temperature at		
− 10 cm (°C)	5–8	2–5
Annual degree days above 0°C	600–1400	150–600
Active-layer depth (cm)		
fine-textured soils	30–50	30–50
coarse-textured soils	100–300	70–150
Botanical/vegetational		
Total plant cover (%)		
Tundra	80–100	80–100
Polar semidesert	20–80	20–80
Polar desert	1–5	1–5
Vascular plant flora (species)	700	350
Bryophytes	Common	Abundant
Lichens	Fruticose and foliose growth forms common	Frutcose growth form minor, crustose and foliose common
Growth-form types	Woody and graminoid common	Graminoid, cushion, and rosette common
Plant height (cm)		
Shrubs	10–500	5–100
Forbs	5–30	2–10
Sedges	10–50	5–20
Shoot:root ratios (alive)		
Shrubs	1:1	1:1
Forbs	1:1–2	1:0.5–1
Sedges	1:3–5	1:2–3

except in southern Baffin Island and southern Greenland, regions with much higher annual precipitation (Table 1.2).

Whether the positioning of the Arctic Front (leading edge of polar air) in summer along the treeline results from changes in surface albedo, evapotranspiration, and surface roughness of vegetation, as suggested by Hare (1968), or from an assemblage of climatic variables, as suggested by Bryson (1966), is not known. However, both the vegetation limits of forest and tundra and the airmass systems are responding to the basic solar radiation controls of net radiation (Hare and Ritchie 1972). Annual net radiation averages 1000–1200 MJ m^{-2} yr^{-1} in the northern boreal forest, 750–800 MJ m^{-2} yr^{-1} across Canada, and about 670 MJ m^{-2} yr^{-1} across Alaska at treeline. In the High Arctic it averages 200–400 MJ m^{-2} yr^{-1} over much of the land and drops to near zero in northern Ellesmere Island and northern Greenland. A similar pattern of reduced summer temperature, mean annual temperature, and precipitation occurs northward (Fig.

1.2). These dramatic shifts in summer climate and length of the growing season explain much of the change in vegetation from the Low Arctic to the High Arctic. Winter snow cover averages 20–50 cm over many upland areas, whereas in the Low Arctic, lee slopes and depressions may have snowbanks 1–5+ m. In the High Arctic, with greatly reduced winter precipitation, many uplands have only 10–20 cm snow and some areas even less. Deep snowbanks (1–3+ m) occur in protected slopes and in depressions. With a lower sun angle and lower summer temperatures many of these snowbanks do not melt until late July or August, whereas in the Low Arctic nearly all snow is melted by mid-June. The time of snowmelt greatly influences the distribution of species, especially north of 74° latitude. There is also a very strong correlation between shrub height and average snow depth, especially in the Low Arctic. This explains why tall shrubs (2–5 m) are confined to river bottoms, steep banks, and drainages in uplands.

Global climate change is a major subject of cur-

Table 1.2. *Climatic data for selected stations in the Low and High Arctic*

Station and latitude	Temperature (0° C)					Precipitation (mm)			
	Mean monthy			Mean annual	Degree-days (>0° C)	Mean monthly			Mean annual
	June	July	August			June	July	August	
Low Arctic									
Frobisher Bay, N.W.T., 64°	0.4	3.9	3.6	−12.7	610	38	53	58	425
Baker Lake, N.W.T., 65°	3.2	10.8	9.8	−12.2	1251	16	36	34	213
Tuktoyaktuk, N.W.T., 70°	4.7	10.3	8.8	−10.4	903	13	22	29	130
Kotzebue, AK, 67°	8.2	12.4	8.5	−4.3	1462	22	44	37	246
Umiat, AK, 69 °	9.1	11.7	7.3	−11.7	993	43	24	20	119
Angmagssalik, Greenland, 65°	5.8	7.4	6.6	−0.4	793	44	35	62	770
Godthaab, Greenland, 64°	5.7	7.6	6.9	−0.7	809	46	59	69	515
Umanak, Greenland, 71°	4.8	7.8	7.0	−4.0	682	12	12	12	201
High Arctic									
Barter Island, AK, 70°	1.8	4.6	3.7	−11.6	368	10	18	25	124
Barrow, AK, 71°	1.3	3.8	2.2	−12.1	288	8	22	20	100
Cambridge Bay, N.W.T., 69°	1.5	8.1	7.0	−14.8	579	13	22	26	137
Sacks Harbour, N.W.T., 72°	2.2	5.5	4.4	−13.6	458	8	18	22	102
Resolute, N.W.T., 75°	−0.3	4.3	2.8	−16.4	222	13	26	30	136
Isachsen, N.W.T., 79°	−0.9	3.3	1.1	−19.1	161	8	21	22	102
Eureka, N.W.T., 80°	2.2	5.5	3.6	−19.3	318	4	13	9	58
Alert, N.W.T., 83°	−0.6	3.9	0.9	−18.1	167	13	18	27	156
Scoresbysund, Greenland, 70°	2.4	4.7	3.7	−6.7	333	26	38	33	248
Nord, Greenland, 81°	−0.4	4.2	1.6	−11.1	202	5	12	19	204

Source: Data from various sources

rent research, especially in relation to boreal and arctic regions. Plant and animal communities and ecosystems have changed greatly in the past and no doubt will respond significantly to future climate changes (Warrick, Shugart, Antonovsky, Tarrant, and Tucker 1986). Arctic ecosystems are likely to be the most altered with climate warming (Maxwell and Barrie 1989; Maxwell 1992, 1995); there is a potential for a 2–5° C increase in summer temperatures. Variations in surface ocean temperatures are believed to play a major role in governing terrestrial climate (Lowe et al. 1994). Recent ice-core data from Greenland (Taylor et al. 1993) indicate that these climate changes have been abrupt in the past (as brief as one to two decades).

Data for 65 arctic stations from 1881–1981 (Kelley and Jones 1981) indicate that 1881–1920 was a cool period, followed by warming until the early 1950s. Climate warming has continued in Alaska and the western Canadian Arctic in the 1980s and 1990s, whereas areas in the eastern Canadian Arctic are cooling (Chapman and Walsh 1993; Maxwell 1995). Other indications of warming are the earlier melt of snow, especially at Barrow compared with the central Canadian Arctic (Foster 1989), the 2° C rise in permafrost temperature in the past few decades in northern Alaska (Lachenbruck and Marshall 1986), and the more recent data from the Toolik Lake, Alaska, site. The mean July–August water

temperature of Toolik Lake has risen about 3° C from 1975 through 1991 (Hobbie et al. 1994) and air temperatures at Barrow by AK ≃ 1.5° C since 1971 (Oechel, Vourlitis, Hastings, and Bochkarev 1995). However, on Devon Island N.W.T., where lakes thawed completely in the early 1970s and 1980s, only 30–50% of the ice has melted in 1991–1996. The overall indications are that there is a warming trend that is within the range of historic variability yet that also appears to be a response to increased levels of CO_2.

Permafrost

"Permafrost" refers to those areas of soil, rock, and sea floor where the temperature remains below 0° C for 2+ yr. In coarse gravels and frozen rock there is little ice; a dry permafrost predominates. Ice, however, is generally an important component of permafrost, reaching 80–100% by volume in massive ice wedges and ice lenses. Permafrost underlies all of the Arctic and extends into the boreal forest, tapering off as discontinuous permafrost where mean annual temperature in the atmosphere ranges from −8.5° C at the northern limit of discontinuous permafrost to about −1° C at its southern limit. The depth of the permafrost depends on mean annual temperature, thermal conductivity of soil and rock, proximity to the sea, and topographic position. Permafrost is

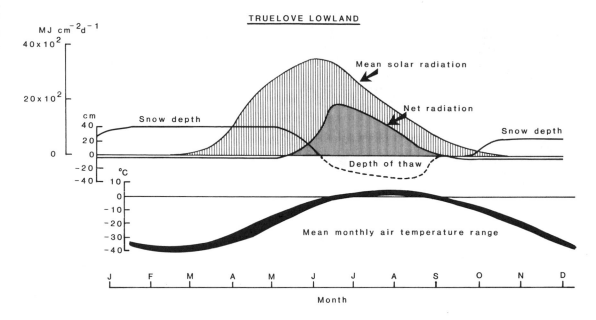

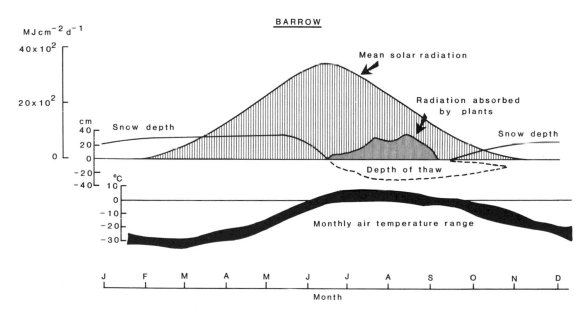

Figure 1.2. Diagram of mean maximum and mean minimum temperatures, snow depth, active-layer depth, and solar radiation at Truelove Lowland (from NATV, 1st edition) and Barrow (From Chapin, F. S. III and G. R.

Shaver 1985, Arctic, pp. 16–40 in B. F. Chabot and H. A. Mooney (ed.), Physiological ecology of North America plant communities, Chapman & Hall, New York; with kind permission of Kluwer Academic Publishers.

about 320 m thick at Barrow, 500–600 m in the Mackenzie River Delta region, and 400–650+ m throughout the High Arctic. Mean annual temperature near the top of the permafrost table has risen 2° C in parts of the Arctic in the past 30–100 yr, an indication of climate warming.

Permafrost and the dynamic processes of the annual freeze–thaw cycle result in many surface features. Ice-cored hills (pingos) (Fig. 1.3), and small ice

mounds (palsas) occur in the Low Arctic but are more limited farther north. The most common features, which are not restricted to the Arctic, are sorted and nonsorted circles and stripes, polygons, earth hummocks, and solifluction steps. The circles, stripes, and polygons may be 10–25 cm across each unit, but many polygons are 3–10 m, even 50–100 m in diameter. Raised and depressed-center polygons, sorted and nonsorted circles, soil hummocks, and

Figure 1.3. Pingo (ice-cored hill) east of the Mackenzie River Delta. These form in old lake basins when permafrost invades formerly unfrozen soils.

terraces all influence the distribution of plants and the development of soils (Fig. 1.4A–C). Large soil polygons (>1 m) are common in lowlands, whereas stone nets and stripes are typical of uplands. Polygonal patterns occur on lands that are level or have gentle slopes (1–3°); elongated polygons, stripes, and solifluction steps are common on slopes >3–5° (Washburn 1980). Cushion plants often predominate on the step (tread), with taller herbs and shrubs predominating on the riser where winter snow cover is deeper and continuous.

Not all patterned ground features result from ice formation. In the High Arctic, many small polygonal patterns (10–30 cm) result from desiccation cracks (Fig. 1.4D) that form as the silty to sandy soils dry each summer. Vegetated soil hummocks probably result from plant establishment, desiccation crack formation, and the slow accumulation of organic matter as plants grow on the hummocks and they become more defined (see Fig. 1.4C). Needle ice is another important feature in finer-textured soils. This ice forms in many soils, especially in the fall, and the resulting lifting of surface soils greatly inhibits seedling establishment in some sites.

Soils

Plant communities and soils are interrelated in the Arctic, as they are in all biomes. Because of reduced plant cover, a short growing season, and the short time since ice retreat (<12,000 yr) or emergence from the sea, the soils are less well developed than in temperate regions. The presence of permafrost, which limits the vertical movement of water and the churning of some soils (cryoturbation), further restricts soil and plant development. Consequently, chemical decomposition, release of

nutrients, and synthesis of minerals from weathering of clay all progress very slowly.

The early concepts that arctic soils form under processes quite different from those in temperate regions have been replaced by the realization that the same soil-forming processes occur but at greatly reduced rates. The process of podzolization is limited in the Arctic to well-drained soils with a deep active layer. Where dwarf shrub heath species and dwarf birch predominate, weakly developed podzols (Spodosols) are found. Soils of uplands and dry ridges that are less well developed are arctic brown soils (Inceptisols). Cushion plant and heath shrub communities occur in these areas. The most common group of soils in the Low Arctic includes the tundra soil (Inceptisols) of cottongrass–dwarf shrub heath and some sedge communities of imperfectly drained habitats. These soils form under the process of gleization. Poorly drained lowlands, where soils remain saturated all summer, accumulate both sedge and moss peats. These bog and half-bog soils (Histosols) are dominated by sedge-moss or grass-moss plant communities (Fig. 1.5).

The arctic podzols (Spodosols) and arctic browns (Inceptisols) show some translocation of humus and iron, with iron-enriched B_2 horizons and weakly eluviated A_2 horizons in the podzols. The surface layers tend to be quite acidic (pH 6–4) and low in available nutrients but quite well drained above the perma frost layer. Inceptisols of imperfectly drained lands (arctic tundra soils) are acidic (pH 6.5–4.5), contain B horizons that have subangular to angular structures, are grayish in color with iron oxide mottles, and are low in available nutrients. Histosols of poorly drained lands are acidic (pH 6.5–5.0) and are similar to the arctic tundra soils in having limited translocation of minerals into the B horizon.

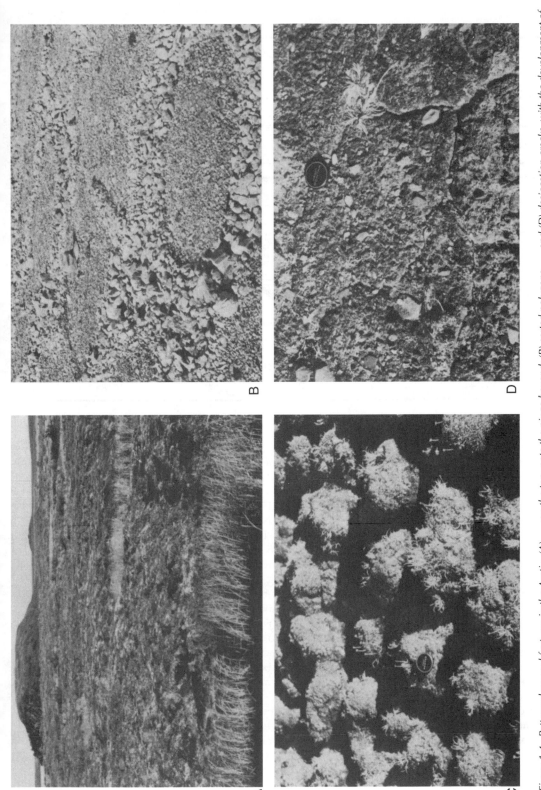

Figure 1.4. Patterned ground features in the Arctic: (A) raised-center polygons in the Mackenzie River delta area with sedges in the troughs and heath shrubs and lichens on the top; note the pingo beyond; (B) sorted polygons on Cornwallis Island within a polar desert landscape; (C) soil hummocks with Dryas integrifolia, Banks Island; and (D) desiccation cracks, with the development of mosses and "safe sites" for vascular plants to grow, King Christian Island.

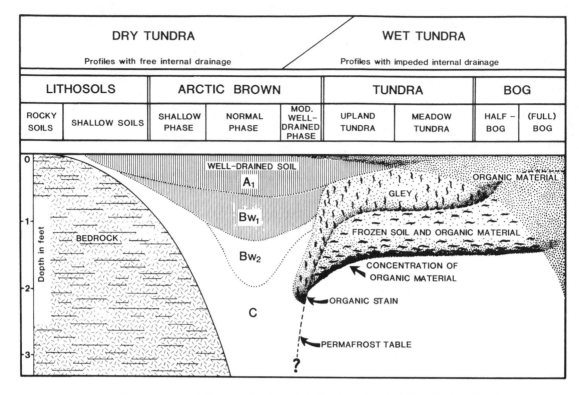

DRY TUNDRA				WET TUNDRA			
Profiles with free internal drainage				Profiles with impeded internal drainage			
LITHOSOLS		ARCTIC BROWN			TUNDRA		BOG
ROCKY SOILS	SHALLOW SOILS	SHALLOW PHASE	NORMAL PHASE	MOD. WELL- DRAINED PHASE	UPLAND TUNDRA	MEADOW TUNDRA	HALF - BOG
							(FULL) BOG

Figure 1.5. Generalized diagram of major soils in the Arctic (modified from Tedrow 1977).

Within the High Arctic, soil-forming processes are further reduced. The very limited lowlands of sedge-moss or grass-moss communities have thin peats, 2–20 cm thick overlying gleyed (arctic tundra) soils. Many slopes covered with cushion plant–cryptogam communities overlie well-drained soils of the arctic brown group (Inceptisols). Within the barren polar deserts, soils develop under the process of gleization or calcification. These soils are generally basic (pH 7.5–8.5), contain very little organic matter, and are base saturated, but they are very deficient in nitrogen and phosphorus. There are some areas where free carbonates are released from sedimentary rocks or from recently uplifted marine sediments along coastal areas. In these habitats, calcium and magnesium salts effervesce on the soil surface during the brief warm and drier periods in summer. Halophytic species predominate in these sites, as they often do in the extremely depauperate salt marshes in some coastal areas. For a detailed discussion of arctic soils, see Tedrow (1977).

PALEOBOTANY

On a geological basis, arctic ecosystems are relatively young. The fossil record from Cook Inlet, Alaska, indicates that tropical, subtropical, and warm temperate forests predominated in Eocene time ($\approx$ 55 M yr BP) (Wolfe 1977). In Miocene time ($\approx$ 22 M yr BP) there were mixed coniferous-deciduous forests on Devon Island at what is now 75 ° N (Whitlock and Dawson 1990), and comparable forests occupied uplands in central Alaska ($\approx$ 18–15 M yr BP) (Wolfe 1969). By Pliocene time ($\approx$ 6–3 M yr BP), coniferous forests still predominated, along with insects typical today of southern British Columbia and northern Washington (Hopkins, Matthews, Wolfe, and Silberman 1971). The Beaufort Formation, extending from Banks Island (71° N) to Meighen Island (80° N) and estimated to be of middle Miocene to Pliocene age, contains fossils of mixed coniferous-deciduous forests (Hills, Klovan, and Sweet 1974). An amazing group of fossil seed plants, mosses, and invertebrates from Meighen Island indicates a forest-tundra vegetation in late Miocene to early Pliocene time on a land surface that is now polar desert. Tundra probably occurred farther north and at higher elevations on Axel Heiberg and Ellesmere islands and in northern Greenland. Much of the arctic flora and fauna is believed to have evolved in the highlands of Central Asia (Hoffman and Taber 1967; Yurtsev 1972) and to a more limited degree in the central and northern Rocky Mountains (Billings 1974;

Packer 1974). From these centers of origin, plants and animals spread north and across Beringia to enrich the biota on both continents.

A circumpolar flora of perhaps 1500 species developed prior to the onset of Pleistocene glaciations (Löve and Löve 1974). Glacial advances and retreats, with accompanying climatic changes, reduced both the flora and fauna of these northern lands. The result is a truly Circumpolar Arctic vascular plant flora of 1000–1100 species, with about 700 of these species occurring in the North American Arctic.

The evolution of arctic ecosystems in Beringia (the land connection between Alaska and Siberia during the Late Pleistocene) and the northern Yukon Territory has been discussed by Hopkins et al. (1982) and Ritchie (1984). The pollen record from the Bering Strait region from the Duvanny Yar Interval (30,000–14,000 yr BP) was interpreted as a fairly uniform herbaceous tundra in which upland sedges, grasses, and *Artemisia* predominated (Colinvaux 1964, 1967; Matthews 1974). It was assumed that this vegetation was similar to that of the steppe tundra of eastern Siberia and that it occupied much of northern Alaska and the Yukon Territory. More recent studies by Ager (1982), Anderson and Brubaker (1986, 1994), Anderson, Reanier, and Brubaker (1988) in Alaska, and Cwynar and Ritchie (1980), Ritchie (1984, 1987), and Ritchie, Cwynar, and Spear (1983) in the Yukon indicate that more mesic graminoid communities typified the western Alaskan lowlands and that more xeric herbaceous communities dominated eastern Alaska and the Yukon Territory. Ericaceous heath shrubs probably have been underestimated because of their low pollen production. Spruce became extinct, and balsam poplar and aspen became more restricted during that period. Willow scrub was confined to river bottoms, as today.

Although many fossils remains of large mammals have been found in central Alaska, most have been reworked by rivers, so that accurate carbon dates are difficult to determine. Earlier studies by Guthrie (1972) indicated that an arctic grassland must have predominated to support these mammals. However, more recent studies and summaries of various research indicate little basis for this belief (Ritchie 1982, 1984; Ritchie and Cwynar 1982; Schweger 1982). Relict examples of this "mammoth steppe" exist in Alaska today on steep south-facing river bluffs. Dominants include *Artemisia frigida*, *Agropyron spicatum*, and *Calamagrostis purpurascens* (Wesser and Armbruster 1991).

The Birch Zone (14,000–12,000 yr BP) was a warmer period during which a rapid rise in the sea level drowned the Bering land bridge. The increase in birch pollen indicates a rise in shrub tundra along with the entrance of cottongrass and cottonwood pollen. The large mammal fauna changed with the loss of woolly mammoth at about 14,000 yr, bison at about 13,000 yr, and horse and wapiti at about 10,000 yr BP. Ritchie (1984) has presented new data from the Bluefish Cave site in the northern Yukon indicating that large changes occurred in vegetation from herb tundra (15,000–12,000 yr BP) to closed and open woodlands at low elevations and to herb tundra in the uplands (9,000 BP to the present). The Late Pleistocene fauna of woolly mammoth, horse, Dall sheep caribou, bison, wapiti, and musk-ox shifted to caribou and moose with major changes in vegetation. This is one of the very few records in the north that conclusively documents simultaneous changes in mammal and vegetation patterns. The extinction and reduction of large mammal species over a 4000 yr period probably resulted from a combination of climate change that induced major changes in vegetation and from overhunting by humans (Martin 1974, 1982). Grayson (1991) presents evidence that the extinction of 35 genera of mammals occurred over several thousand years and that the overkill hypothesis alone does not adequately explain the large number of extinct mammals.

Studies indicate that a warmer period prevailed across northern Alaska, the central Yukon, and the adjacent N.W.T. from 10,000–6500 yr BP (Ritchie 1984, 1987; Anderson et al. 1988; Cwynar and Spear 1991). Balsam poplar formed gallery forests and localized groves. Open stands of white spruce dominated uplands from 9400–5000 yr when climate cooled and shrub and tussock tundra again became dominant.

VEGETATION

Given the diversity of arctic landscapes (mountains, Precambrian Shield, low-elevation lands) and their great latitudinal extent (55° along Hudson Bay to 83° at northern Ellesmere Island and Greenland), with associated climatic changes, it is not surprising that there are diverse patterns of plant communities. Although arctic vegetation has a general structure and pattern of herbaceous species and (often) low scattered shrubs, not all of it has the same appearance.

The major types include (1) tall shrub tundra (*Salix, Alnus, Betula* 2–5 m high) along river terraces, stream banks, steep slopes, and lake shores; (2) low shrub tundra (*Salix, Betula* 40–60 cm high) on slopes and uplands beyond the forest tundra; (3) dwarf shrub heath tundra (5–20 cm high) and cottongrass–dwarf shrub heath tundra (tussock tundra) on rolling terrain with soils of intermediate drainage; (4) graminoid-moss tundra (20–40 cm high) on poorly drained soils; and (5) cushion plant–herb–cryptogam polar semidesert (2–5 cm high) on wind-exposed slopes and ridges with limited snow cover. This last vegetation type is more typical of the High Arctic.

There are general reductions in total plant cover, plant height, and woody and vascular plant species and increases in lichens and mosses from the Low Arctic to the High Arctic. At the final limit of plant growth in the polar desert barrens, the ultimate restriction to plant growth appears to be soils which are saturated in the spring but of which the upper 1–2 cm bakes hard later in some summers, preventing seedling establishment. These soils lack nitrogen and phosphorus and have few safe sites in which vascular plants can become established because of the lack of cryptogamic crusts that provide sites with more water and nutrients. These bare soils also favor an abundance of needle ice in autumn. Plant limits are not caused by temperature alone, because vascular plants grow to within a few meters of glaciers and ice caps at elevations of 650–720 m on Ellesmere and Devon islands (Bliss et al. 1994).

A vegetation classification system that follows Russian concepts of the Arctic and an associated circumpolar vegetation map are still under development (Walker et al. 1995).

Forest Tundra

Throughout northern Alaska and Canada there is a relatively narrow zone, 10–50 km wide, of forest tundra. This vegetation consists of scattered clumps of trees in more protected sites where snow accumulates within a matrix of low shrub tundra. An added factor in the role of climate (position of the Arctic Front) is the role of soils. In Canada, where the forest tundra lies farther north, the transition zone is narrower and the soils tend to be more nutrient-rich and of finer texture. Where the transition zone lies south of the climatic potential for forest, the zone is wider and the soils are nutrient-poor and more droughty (Timoney, LaRoi, Zoltai, and Robinson 1992, 1993; Timoney 1995).

In many places, the transition from forest tundra to shrub tundra consists only in loss of stunted trees. Because of this patterning, earlier Russian ecologists considered shrub tundra a part of the Subarctic rather than the Arctic (Aleksandrova 1980), for they believed that fire had eliminated the trees in many places. Repeated fires within these Subarctic communities have little effect on the understory vascular plants, for they easily resprout.

However, lichens and mosses are slow to return. In the Mackenzie River delta region, it takes about 200+ yr for an extensive lichen cover to develop on the fine-textured soils (Black and Bliss 1978). On sandy soils to the southeast, where shrubs are minor, a lichen cover developed in 60–120+ yr in the Abitan Lake region (Maikawa and Kershaw 1976). Studies conducted near Inuvik, N.W.T., indicate that there is a period of only 5–8 yr following fire during which seedlings of *Picea mariana* have a chance to become established, because of a short period of seed viability, rapid seed release from surviving cones, and seed destruction by rodents and insects (Black and Bliss 1980). Massive fires within the forest tundra could drive the treeline 50–100 km south should they occur during a cool climatic period, for the seeds germinate only when temperatures are >15° C. Chapter 2 describes this vegetation and the role of fire in more detail.

Low Arctic Tundra

The general appearance of most landscapes within the Low Arctic is that of a grassland in which low to dwarf shrubs are common, except in the wetter habitats. The vascular flora in an area of 100–200 km² may total 100–150 species, although any single plant community may have only 10–50 species of vascular plants and 20–30 species of mosses and lichens.

Tall shrub tundra. The rivers that flow across the tundra have sandbars and gravel bars, islands, and terraces with well-drained soils that are relatively warm, have a deep active layer (1–1.5 m), and contain higher nutrient levels than do adjacent uplands. These habitats and steeper slopes above lakes and rivers that have a deep snow cover in winter (2–5 m) are generally covered with various mixtures of *Salix, Betula,* and *Alnus.* In arctic Alaska, the Yukon, and the northwestern Northwest Territories, *Salix alaxensis* predominates on the coarser-textured alluvium, dune sands, and slopes (Fig. 1.6). Associated shrubs, 1–2 m high, of lesser importance and of varying abundance from site to site, include *S. arbusculoides, S. glauca* ssp. *richardsonii, S. pulchra,* and *Alnus crispa* (Table 1.3) (Bliss and Cantlon 1957; Drew and Shanks 1965; Johnson, Viereck, Johnson, and Melchior 1966; Gill 1973; Komarkova and Webber 1980). There is often a rich understory of grasses and forbs in these shrub communities.

Depauperate outliers of the *Salix alaxensis, S. lanata* ssp. *richardsonii, S. pulchra* shrub community, with sedges and forbs, occur along rivers and steep slopes on southern Banks and Victoria islands (Kuc

Figure 1.6. Tall shrub tundra dominated by Salix alaxensis *with lesser amounts of* S. glauca *ssp.* desertorum, S. arbusculoides, *and herbaceous plants including* Lupinus arcticus, Hedysarum mackenzii, Deschampsia caespitosa, *and* Agropyron sericeum, *on river gravels along the Colville River, Umiat, Alaska.*

1974) within the southern High Arctic. *Salix alaxensis* is usually only 0.5–1.5 m tall, and the stands look similar to those in the upper river drainages on the north slope of the Brooks Range.

Low shrub tundra. Plant communities dominated by varying combinations of dwarf birch, low willows, heath species, scattered forbs, and graminoids are common on rolling uplands beyond the forest tundra in Alaska and northwestern Canada (Hanson 1953; Corns 1974). Open forest and forest tundra may have occupied these lands in northwestern Canada in the past, but fires and changing climate have forced the treeline south (Ritchie and Hare 1971; Ritchie 1977).

The open canopy of shrubs is 40–60 cm high and is dominated by *Betula nana* ssp. *exilis, Salix glauca* ssp. *acutifolia, S. planifolia* ssp. *pulchra,* and *S. lanata* ssp. *richardsonii,* with the combinations of species varying from place to place (Fig. 1.7). Ground cover includes *Carex lugens* and *C. bigelovii* in Alaska (*C. bigelowii* alone in the Northwest Territories), *Eriophorum vaginatum,* and numerous forbs, grasses, and heath shrubs (10–20 cm high). Varying combinations of *Vaccinium uliginosum, V. vitis-idaea* ssp. *minus, Empetrum nigrum* ssp. *hermaphroditum, Ledum palustre* ssp. *decumbens, Arctos (Arctostaphylos) alpina, A. rubra, Rubus chamaemorus,* and dwarf species of *Salix* occur along with an abundant ground cover of lichens and mosses (Table 1.3). Where snow lies late into June, the heaths are often dominated by *Cassiope tetragona.* The most common mosses include *Aulacomnium turgidum, Hylocomium splendens,* and *Polytrichum juniperinum.* Important fruticose lichens are *Cetraria nivalis, C. cucullata, Cladonia gracilis, Cladina mitis, C. rangiferina,* and *Thamnolia vermicularis.*

In much of the northeastern mainland of Canada, low shrub tundra is minor, probably because of limited winter snow and abrasive winter winds (Savile 1972). In the Chesterfield Inlet of Hudson Bay, northern Quebec, and southern Baffin Island, the dominant shrubs are *Betula glanulosa, Salix glauca* ssp. *callicarpaea,* and the aforementioned heath species, graminoids, and cryptogams (Polunin 1948).

In Greenland, low shrub tundras occur in the inner fjord regions that are warmer in summer. These shrub communities, dominated by *Salix glauca* ssp. *callicarpaea, Betula nana,* and *B. glandulosa* predominate on steep slopes and along streams and rivers from Disko Island (69°N) and Sønder Stromfjord (67° N) southward. The shrubs (>1 m in height) are associated with herbs and dwarf heath shrubs (Böcher 1954, 1959; Hansen 1969).

Dwarf shrub heath tundra. As used here, "heath" refers to species within the Ericaceae, Empetraceae, and Diapensiaceae. Some authors have broadened the concept to include *Dryas* and dwarf species of *Salix.* A characteristic feature of these heath plants is the evergreen leaf, although deciduous-leaved species also occur (*Vaccinium uliginosum, Arctostaphylos alpina, A. rubra*). Heath-dominated communities occur on well-drained soils of river terraces, slopes, and uplands where winter snows are at least 20–30 cm deep. Heath tundra occupies relatively small areas, a few hundred meters square, rather than the many hectares or square kilometers of other tundra vegetation types.

In western Alaska (Hanson 1953; Churchill 1955; Johnson et al. 1966), the northern Yukon (Hettinger et al. 1973), and the Mackenzie River

Table 1.3. *Prominence values (cover × square root of frequency) for plant communities in the Low Arctic at Umiat, Alaska. Sampling area for each stand was 10 m⁻².*

	Plant communities				
Species	Tall shrub	Low shrub	Cottongrass-heath	Cushion plant	Graminoid-moss
Eriophorum vaginatum	—	—	90	—	—
Arctagrostis latifolia	29	27	19	—	27
Carex lugens	—	30	27	—	—
Carex glacialis	—	—	—	27	—
Luzula confusa	—	—	19	—	—
Dryas integrifolia	—	—	—	90	—
Ledum palustre ssp. *decumbens*	9	30	30	—	—
Vaccinium vitis-idaea	9	29	70	—	—
Vaccinium uliginosum	14	23	—	—	—
Arctostaphylos alpina	—	21	27	19	—
Rhododendron lapponicum	—	—	—	9	—
Cassiope tetragona	—	14	29	9	—
Empetrum nigrum ssp. *hermaphroditum*	—	21	27	—	—
Rubus chamaemorus	—	—	25	—	—
Salix planifolia ssp. *pulchra*	80	14	9	—	—
Salix glauca ssp. *acutifolia*	9	29	—	25	—
Betula nana ssp. *exilis*	9	30	30	—	—
Alnus crispa	375	23	—	—	—
Pedicularis lanata	—	—	—	25	—
Lupinus arcticus	—	—	—	29	—
Saussurea angustifolia	—	14	—	—	—
Saxifraga punctata	—	14	19	—	—
Saxifraga cernua	14	—	—	9	29
Polemonium acutiflorum	—	14	—	—	—
Petasites frigidus	9	—	19	—	—
Polygonum bistorta	—	23	16	—	—
Pyrola grandiflora	21	25	—	—	—
Stellaria longipes	—	—	—	—	27
Caltha palustris	—	—	—	—	23
Carex aquatilis	—	—	—	—	375
Eriophorum angustifolium	—	—	—	—	21
Cardamine pratensis	—	—	—	—	16
Hedysarum alpinum	—	—	—	19	29
Mosses	29	190	625	30	90
Lichens	—	90	375	—	—
Total vascular species	13	20	17	17	9

Source: Churchill (1955)

Delta region (Corns 1974), communities of heath species (10–20 cm high) are quite common. The species occur in various combinations including *Ledum palustre* ssp. *decumbens, Vaccinium uliginosum, V. vitis-idaea, Empetrum nigrum* ssp. *hermaphroditum, Loiseleuria procumbens, Rhododendron lapponicum, R. kamschaticum,* and *Cassiope tetragona. Betula nana* ssp. *exilis* and one or more species of dwarf *Salix* are commonly associated, along with abundant lichens and mosses.

In the Keewatin District, N.W.T, west of Hudson Bay, dwarf shrub heath predominates on north exposures, fellfields, and gravel summits. Here, the heaths are less rich floristically, with *Ledum palustre* ssp. *decumbens, V. uliginosum, V. vitis-idaea, Empetrum nigrum* ssp. *hermaphroditum,* and *Cassiope te-*

tragona dominating (Larsen 1965, 1972). Where snow is deep, *Betula glandulosa* and highly deformed *Picea mariana* (60–90 cm high) occur near the treeline at Ennadai.

At Chesterfield Inlet along the west coast of Hudson Bay, at Wakeham Bay in northern Quebec, and at Lake Harbour, Baffin Island, heath vegetation is common. The dominant species include *V. uliginosum, Ledum palustre* ssp. *decumbens, Empetrum nigrum* ssp. *hermaphroditum, Phyllodoce caerulea,* and the sedge *Carex bigelovii.* Where snow lies into July, *Cassiope tetragona* dominates, with *Salix reticulata, S. herbacea, Dryas integrifolia, Carex misandra, Luzula nivalis,* and numerous lichens and mosses (Polunin 1948).

In southeast and west Greenland, heath vege-

Figure 1.7. Low shrub tundra dominated by Salix planifolia *ssp.* pulchra,
Salix glauca *ssp.* acutifolia, Betula
nana *ssp.* exilis, Vaccinium uliginosum, V. vitis idaea, Empetrum nigrum *ssp.* hermaphroditum, Ledum
palustre *ssp.* decumbens, *and* Carex
lugens *in the Brooks Range.*

Figure 1.8. Cottongrass–dwarf shrub
heath tundra in the Caribou Hills,
northeast of Inuvik, N.W.T.

tation is found on steep, moist slopes, usually in inland valleys, which have a warmer and drier continental climate as compared with the moist maritime climate along the coast and outer fjords (Böcher 1954). Species dominating in various combinations include *Cassiope tetragona, Vaccinium uliginosum* ssp. *microphyllum, V. vitis-idaea* ssp. *minus, Ledum palustre* ssp. *decumbens, Rhododendron lapponicum, Phyllodoce caerulea, Empetrum nigrum* ssp. *hermaphroditum,* and *Loiseleuria procumbens* (Sørensen 1943; Böcher 1954, 1959, 1963; Böcher and Laegaard 1962; Hansen 1969; Daniels 1982). *Betula nana* is a component of the more southern heaths. *Ledum, Phyllodocae,* and *Loiseleuria* drop out in the Melville Bugt region (72–75° N). At Thule (78° N) the dominant species of the limited heaths include *Cassiope tetragona, Vaccinium uliginosum* ssp. *microphyllum,*

and *Salix arctica* (Sørensen 1943), as they do on Ellesmere in the High Arctic.

Cottongrass–Dwarf shrub heath tundra. Large areas of rolling uplands between the mountains and the wet coastal plain in Alaska and the Yukon Territory are dominated by tussocks of *Eriophorum vaginatum,* dwarf shrubs, lichens, and mosses (Hanson 1953; Churchill 1955; Britton 1957; Johnson et al. 1966; Hettinger et al 1973; Wein and Bliss 1974; Komarkova and Webber 1980). This vegetation type is of limited occurrence in the Northwest Territories, except in the Mackenzie River Delta region (Corns 1974), where it is common (Fig. 1.8), and it is absent from most of the eastern Canadian Arctic, with the exception of southern Baffin Island and northern Quebec (Polunin 1948).

Historically, these landscapes have been called cottongrass tussock tundra or tussock–heath tundra because of the conspicuousness of the cottongrass tussocks. However, the phytomass, net annual production, and cover of heath and low shrub species often are greater than for cottongrass. The Alaskan and western Canadian cottongrass is *E. vaginatum* ssp. *vaginatum*, whereas the subspecies *spissum* occurs in the eastern Arctic.

The predominant heath species are *V. vitis-idaea* ssp. *minus*, *V. uliginosum* ssp. *alpinum*, *Ledum palustre* ssp. *decumbens*, and *Empetrum nigrum* ssp. *hermaphroditum*. Scattered low shrubs of *Betula nana* ssp. *exilis* and *Salix pulchra* are common, along with *Carex bigelovii*, *C. lugens*, and several species of forbs (Table 1.3). The cryptogam layer is well developed and includes the mosses *Dicranum elongatum*, *Aulacomnium turgidum*, *A. palustre*, *Rhacomitrium lanuginosum*, *Hylocomium splendens*, *Tomenthypnum nitens*, and several species of *Sphagnum*. The most common lichens include *Cetraria cucullata*, *C. nivalis*, *Cladina rangiferina*, *C. mitis*, *Cladonia arbuscula*, *Dactylina arctica*, and *Thamnolia vermicularis* (Bliss 1956; Johnson et al. 1966).

Graminoid-moss tundra. The concept of arctic tundra is often associated with treeless wetlands in which species of *Carex*, *Eriophorum*, and sometimes the grasses *Arctagrostis*, *Dupontia*, *Alopecurus*, and *Arctophila* predominate, along with an abundance of bryophytes, but few lichens. In northern and western Alaska, arctic wetland meadows are of minor extent in the mountain valleys, but they increase in importance in the Foothill Province and dominate the Coastal Plain Province (Churchill 1955; Britton 1957; Webber 1978; Komarkova and Webber 1980). Arctic wetlands are again minor in the Yukon mountains, but they increase in importance on the coastal plain and eastward to the Mackenzie River region (Hettinger, Janz, and Wein 1973; Corns 1974).

The dominant sedges are *Carex aquatilis*, *C. rariflora*, *C. rotundata*, *C. membranacea*, *Eriophorum angustifolium*, *E. scheuchzeri*, and *E. russeolum* (Table 1.3). The grasses *Arctagrostis latifolia*, *Dupontia fisheri*, *Alopercurus alpinus*, and *Arctophila fulva* occur along a gradient from drier sites to saturated soils and standing water. In shallow waters of lakes and ponds (50–60 cm deep), *Menyanthes trifoliata*, *Equisetum variegatum*, and *Arctophila fulva* predominate, with *Potentilla palustris* and *Hippuris vulgaris* in water 20–30 cm deep. The various species of *Carex* and *Eriophorum* and *Dupontia fisheri* occur with little (10–20 cm deep) or no standing water. *Carex aquatilis*, *Eriophorum angustifolium*, and *E.*

scheuchzeri are common (Fig. 1.9) in low-center polygons and in the troughs of high-center polygons (Britton 1957; Hettinger et al. 1973; Corns 1974). The nearly continuous cover of mosses includes species of *Aulacomnium*, *Ditrichum*, *Calliergon*, *Drepanocladus*, *Sphagnum*, *Hylocomium splendens*, and *Tomenthypnum nitens* (Britton 1957; Johnson et al. 1966).

Studies of plant succession in the thaw-lake cycle of northern Alaska have shown that high-center polygons form in drained lake basins from the melting of ice wedges and postdrainage thermokarst erosion. Pioneer plants include the moss *Psilopilum cavifolium* on dry peaty sites and the graminoids *Arctophila fulva* in wet sites, and *Eriophorum scheuchzeri* and *Dupontia fisheri* in the moist swales. *Eriophorum angustifolium* and *Carex aquatilis* enter somewhat later, but in time they predominate (Britton 1957; Billings and Peterson 1980).

Sedge-dominated wet meadows are found in the Ennadai area (Larsen 1965), Chesterfield Inlet, northern Quebec, and Lake Harbour (Polunin 1948). Common species include *Eriophorum angustifolium*, *E. scheuchzeri*, *Carex rariflora*, *C. membranacea*, and *C. stans* (Polunin 1948).

In Greenland, graminoid-moss tundra is present, but as with large areas of eastern Canada, the marshes are less common. Important species include *Carex rariflora*, *C. vaginata*, *C. holostoma*, *Eriophorum angustifolium*, and *E. triste* (Böcher 1954, 1959).

Other graminoid vegetation. With thousands of kilometers of arctic shoreline, one might assume that salt marshes, coastal dune complexes, and other coastal vegetation would be common. Such is not the case, for favorable habitats are limited. Factors that restrict arctic salt marshes are the limited areas of fine sands and silts, the annual reworking of shorelines by sea ice, a modest tidal amplitude, the low salinity of coastal waters, a very short growing season, and low soil temperatures. Coastal salt marshes have been described from subarctic Alaska (Vince and Snow 1984), northern Alaska (Jefferies 1977; Taylor 1981), Tuktoyaktuk (Jeffries 1977) and Hudson Bay, N.W.T. (Kershaw 1976; Bazely and Jefferies 1986a), the west-central part of southern Greenland (Søorensen 1943; Vestergaard 1978), and Devon, Ellesmere, and Baffin islands (Polunin 1948; Jefferies 1977; Muc, Freedman, and Svoboda 1989; Bliss and Gold 1994). See also Chapter 13 of this volume.

Many of these marshes have only a 5–20% plant cover, with plant heights of 1–5 cm. Floristically, most salt marshes have only three to five herba-

Figure 1.9. *Graminoid-moss tundra in the High Arctic dominated by* Dupontia fisheri *on Melville Island.*

ceous species, none of which is woody or belongs to the Chenopodiaceae, a family of plants common in temperate and tropical latitudes. The most common species are *Puccinellia phryganodes* on mud, with scattered clumps of *Carex ursina, C. ramenskii, C. subspathacea, Stellaria humifusa,* and *Cochlearia officinalis.* In many ways it is quite amazing that there are any species at all adapted to living in these harsh coastal environments, in contrast with the extensive and highly productive salt marshes of temperate regions.

The *Puccinellia* marshes along Hudson Bay are heavily grazed by lesser snow geese. Grazing in summer increases the aboveground net primary production of these marshes by 40–100% because the geese produce so many droppings that accelerate nitrogen cycling (Bazely and Jefferies 1986b; Jefferies 1988) in what are nitrogen-limited systems.

Beach and dune grasslands are minor features in the Arctic. *Elymus arenarius* ssp. *mollis* often dominates, with small amounts of the semisucculents *Honckenya peploides, Mertensia maritima,* and *Cochlearia officinalis* ssp. *groenlandica,* along with *Festuca rubra* and *Matricaria ambigua* (Barbour and Christensen 1993).

Small areas of grassland are found in the inner fjords of east-central Greenland, dominated by *Calamagrostis purpurascens, Arctagrostis latifolia, Poa arctica,* and *P. glauca* (Seidenfaden and Sørensen 1937; Oosting 1948). In west Greenland, larger areas (1–5 ha) of grassland occur that are dominated by *Calamagrostis neglecta, Poa pratensis,* and *P. arctica* (Böcher 1959). Similar small grasslands are found in the eastern Canadian Arctic of Baffin Island (Polunin 1948) and northern Keewatin (Larsen 1972).

Low Arctic Semidesert: Cushion Plant–Cryptogam

From the Rocky Mountains of Montana north to the Yukon Territory and Alaska, windswept slopes and ridges are often dominated by cushions or mats of *Dryas integrifolia* and *D. octopetala.* This vegetation type, often called *Dryas* fellfield or *Dryas* tundra, is limited in areal extent within the western Low Arctic but increases in importance in the large areas of barren rock and lag gravel surfaces in the eastern Arctic. In the Mackenzie District and Keewatin District, N.W.T., where acidic soils derived from granites predominate, *Dryas* is less prominent. Other cushion plants include *Silene acaulis, Saxifraga oppositifolia,* and *S. tricuspidata.* The lichens *Alectoria nitidula, A. ochroleuca, Cetraria cucullata,* and *C. nivalis,* along with the moss *Rhacomitrium lanuginosum,* are common in the Repulse Bay, Peely Lake, and Snow Bunting Lake areas (Larsen 1971, 1972).

Within the ecotone between forest and tundra in the Campbell–Dolomite uplands near Inuvik, N.W.T., mats of *Dryas integrifolia* and *Cladonia stellaris* cover the limestone and dolomite rocks. Where small pockets of soil occur, open stands of *Picea glauca* predominate (Ritchie 1977). In the Brooks Range (Spetzman 1959) and northward along exposed ridges within the Foothill Province, *Dryas*-lichen vegetation is common. *Dryas integrifolia* or *D. octopetala* and their hybrids often comprise 80–90% of the vascular plant cover. Associated species include *Silene acaulis, Carex rupestris, C. capillaris, Kobresia myosuroides, Anemone parviflora,* and *Polygonum viviparum.* Cushions of *Dryas* are often 0.5–1.0 m across, and they seldom reach a height greater than 2–5 cm. The ground cover of herbs, lichens, and mosses is generally sparse. As a result, there is often much bare rock and soil, although crustose lichens are often common.

High Arctic Tundra

In contrast with the Low Arctic, the High Arctic is characterized by herbaceous rather than woody species, and there is generally much less plant cover, especially of vascular plants. Lichens and mosses contribute a much larger percentage of total cover and biomass than in the Low Arctic. Vast areas are dominated by lichens and mosses, with only a 5–25% cover of flowering plants. Equally large areas of polar deserts have almost no cryptogams (0–3% cover) and only a 0.1–4% cover of vascular plants (Table 1.1). Vascular plants are much smaller in size (3–10 cm) than those of the same species (10–30 cm) in the Low Arctic, and their root systems are also smaller.

Based on vegetation types, the High Arctic can be divided into small areas of tundra (tall shrub, dwarf shrub heath, cottongrass tussock, and graminoid-moss), vast areas of cushion plant–cryptogam and cryptogam-herb polar semidesert, herb barrens, and very limited snowflush herb-moss vegetation of the polar deserts. With the exception of a few locations, woody species other than dwarf willows and semiwoody *Dryas* and *Cassiope* are very minor components.

Graminoid-moss tundra. Of the common vegetation types in the Low Arctic, only the graminoid-moss tundra is ecologically important farther north. It occupies 5–40% of the lands on the southern islands, with the exception of Baffin Island, which is mountainous. In the Queen Elizabeth Islands, <2% of the area contains this vegetation, yet it provides the major grazing habitat for musk-ox and breeding grounds for waterfowl and shore birds.

In the eastern and southern islands, *Carex stans* (*C. aquatilis*) and *C. membranacea* are the dominants, with lesser amounts of *Eriophorum scheuchzeri, E. triste, Dupontia fisheri,* and *Alopecurus alpinus.* Small clumps of *Salix arctica* and *Dryas integrifolia* are restricted to moss hummocks or rocky areas within the wet meadows – the best-drained and best-aerated microsites. Plant communities of this vegetation type occur where drainage is impeded along river terraces, small valleys, and coastal lowlands (Beschel 1970; Bird 1975; Muc 1977; Thompson 1980; Sheard and Geale 1983; Freedman et al. 1983; Bliss and Svoboda 1984; Muc et al. 1989; Schaefer and Messier 1994). In the northwestern is-

lands, including northern Melville Island, the grass *Dupontia fisheri* dominates (Fig. 1.9), with lesser amounts of *Juncus biglumis* and *Eriophorum triste* in wetlands. Where there are shallow ponds, *Pleuropogon sabinei* occurs – the ecological equivalent of *Arctophila fulva* to the south (Savile 1961; Bliss and Svoboda 1984).

Bryophytes are abundant in the various wetland plant communities, including *Orthothecium chryseum, Campylium arcticum, Tomenthypnum nitens, Drepanocladus revolvens, Ditrichum flexicaule,* and *Cinclidium arcticum.* Cyanobacteria are abundant, including species of *Nostoc* and *Oscillatoria.* Lichens are minor in these wetlands.

Dwarf shrub heath tundra. Compared with the Low Arctic, these heaths have few species, and they are almost always in snowbed sites that melt by early July. The heaths of central and northern Baffin Island are richer floristically than elsewhere. *Cassiope tetragona* is the dominant and characteristic species and is commonly associated with *Vaccinium uliginosum* ssp. *microphyllum, Salix herbacea, S. arctica, Carex bigelovii, Luzula nivalis,* and *L. confusa.* Important lichens and mosses include *Cladina mitis, Dactylina arctica, Aulacomnium turgidum, Drepanocladus uncinatus, Hylocomium splendens,* and *Rhacomitrium lanuginosum* (Polunin 1948).

Heath vegetation is sparse in western Greenland in terms of species and areal extent. *Cassiope* dominates, with lesser amounts of *Salix arctica, Dryas integrifolia,* and *Luzula confusa* (Sørensen 1943). Small areas of heath with *Cassiope, Dryas,* and *Salix* occur in northern Greenland at 81–83; dg N (Holmen, 1957).

Farther north and west in the Queen Elizabeth Islands, the depauperate heaths are dominated by *Cassiope, Dryas,* and *Luzula.* Only in the warmer eastern High Arctic is *Vaccinium uliginosum* ssp. *microphyllum* present (Fig. 1.10) (Beschel 1970; Brassard and Longton 1970; Bliss, Kerik, and Peterson 1977; Reznicek and Svoboda 1982; Muc et al. 1989). Lichens and mosses are also common in these heaths.

Other tundra vegetation types. Cottongrass tussock and tall shrub tundra occupy small areas on Banks and Victoria islands, the northern limits for these common vegetation types of the Low Arctic. Tussock tundra is dominated by *Eriophorum vaginatum* ssp. *vaginatum,* with scattered plants of *Vaccinium vitis-idaea* ssp. *minus,* the only heath species, and a few other vascular species and cryptogams. Along some of the rivers and near the lakes of these same southern islands, there are small willow thickets (0.5–1.5 m), with *Salix alaxensis* and *S. pul-*

Figure 1.10. Dwarf shrub heath tundra dominated by Cassiope tetragona *with lesser amounts of* Dryas integrifolia, Vaccinium uliginosum *ssp.* microphyllum, *and* Luzula parviflora. *Note the wet sedge-moss meadows beyond, Truelove Lowland, Devon Island.*

chra (Kuc 1974) and in southeastern Victoria Island *S. lanata* ssp. *richardsonii* (Schaefer and Messier 1994).

High Arctic Semidesert

Arctic vegetation and plant communities included within this landscape unit have generally been considered polar desert (Aleksandrova 1980; Andreyev and Aleksandrova 1981). However, in terms of plant cover, plant and animal biomass, species richness, and soil development, the High Arctic semidesert is so different from barren polar deserts that the two have been separated for North America (Bliss 1975, 1981). Areas of polar semidesert cover about 50–55% of the southern islands and about 25% of the northern islands. The two major vegetation types include cushion plant–cryptogam and cryptogam–herb.

Cushion plant–cryptogam vegetation. Large areas of the southern islands and the Boothia and Melville peninsulas are covered with large mats of *Dryas integrifolia*. Associated species include *Salix arctica, Saxifraga oppositifolia, S. caespitosa, S. cernua, Draba corymbosa, Papaver radicatum*, and two to three species each of *Minuartia* and *Stellaria*. Graminoids are always present as scattered plants, including *Carex rupestris, C. nardina, Luzula confusa*, and *Alopecurus alpinus* (Table 1.4). Lichens and mosses provide 30–60% of the total plant cover; vascular plants provide 5–25% of the cover (Fig. 1.11).

This vegetation is common on rolling uplands, gravelly raised beaches, and river terraces that are drier and warmer than surrounding lands in summer. Plant communities of this type have been described from Victoria Island (Schaefer and Messier

1994), Devon Island (Svoboda 1977), Ellesmere Island (Brassard and Longton 1970; Freedman et al. 1983; Muc et al. 1989), Bathurst Island (Sheard and Geale 1983; Bliss, Svoboda, and Bliss 1984), Cornwallis and Somerset islands (Bliss et al. 1984), and South Hampton Island (Reznicek and Svoboda 1982). These landscapes are the major habitats for Peary's caribou, collared lemming, ptarmigan, and several species of passerine birds.

Cryptogam–herb vegetation. This vegetation type is common on the western Queen Elizabeth Islands, where the previous vegetation type is minor or totally lacking. Cryptogam–herb vegetation covers low rolling uplands with sandy to silty and clay loam soils. Vascular plants contribute 5–20% cover and cryptogams 50–80%.

Alopecurus alpinus, Luzula confusa, and *L. nivalis* are the common graminoids, along with three to five species of *Saxifraga, Papaver radicatum, Draba corymbosa, Cerastium alpinum*, and *Juncus albescens* (Table 1.4). Abundant bryophytes include the mosses *Rhacomitrium lanuginosum, R. sudeticum, Aulacomnium turgidum, Polytrichum juniperinum, Pogonatum alpinum, Ditrichum flexicaule, Dicranoweisia crispa, Tomenthypnum nitens*, and *Schistidium holmenianum*, and in wetter soils, the liverwort *Gymnomitrion corallioides*. Common lichens include crustose *Lecanora epibryon, Lepraria neglecta*, and *Dermatocarpon hepaticum*, and fruticose *Cladonia gracilis, Parmelia omphalodes, Cetraria cucullata, C. delisei, C. nivalis*, and *Dactylina ramulosa*. There is often a black cryptogamic crust of lichens, mosses, cyanobacteria, and fungi on soils (Bliss and Svoboda 1984). Seeds of vascular plants germinate at random on a variety of microsite surfaces, but their survival to adult plants is strongly favored by moss mats or desiccation cracks where mosses are also

Table 1.4. *Prominence values (cover × square root of frequency) for plant communities in the High Arctic. Sampling area for each stand was 40 m⁻².*

	Plant communities				
Species	Cushion plant 31S	Cryptogam-herb 4M	Graminoid-steppe 27ER	Snowflush 12C	Herb barrens 19C
Dryas integrifolia	28	—	—	—	—
Saxifraga oppositifolia	8	—	—	4	—
Saxifraga caespitosa	—	11	—	1	—
Saxifraga hieracifolia	—	9	—	—	—
Saxifraga cernua	—	10	—	1	—
Saxifraga flagellaris	—	17	—	9	—
Phippsia algida	—	—	—	—	—
Papaver radicatum	—	14	—	6	2
Cerastium alpinum	—	4	—	—	—
Oxyria digyna	—	7	—	—	—
Ranunculus sulphureus	—	4	—	—	—
Minuartia rubella	—	3	—	—	—
Stellaria crassipes	—	6	—	—	—
Draba corymbosa	—	2	—	5	9
Draba subcapitata	—	1	—	—	—
Festuca brachyphylla	—	14	—	—	—
Alopecurus alpinus	—	37	20	—	—
Puccinellia angustata	—	—	—	—	6
Luzula confusa	—	6	81	—	—
Luzula nivalis	—	5	2	—	—
Mosses	—	116	76	66	—
Lichens	22	41	24	14	—
Total vascular species	6	19	4	14	5

Note: Sampling area for each stand was 40 m².
Source: Data from Bliss and Svoboda (1984) and Bliss et al. (1984).

Figure 1.11. Cushion plant–cryptogam vegetation of the polar semideserts. Dryas-*dominated community with lesser amounts of* Saxifraga oppositifolia, *and* Salix arctica *with* Cassiope tetragona *in the small depressions, Thomson River, Banks Island.*

important (Sohlberg and Bliss 1984). Plant communities within this vegetation type have been described from Axel Heiberg Island (Beschel 1970), Melville, Cameron, King Christian, and Ellef Ringnes islands (Bliss and Svoboda 1984), Laugheed Island (Edlund 1980), and Prince Patrick Island (Bird 1975).

Graminoid steppe. Large areas of fine sand to clay loam soils are dominated by *Luzula confusa* and *Alopecurus alpinus* on Ellef Ringnes and King Christian islands which is seen in Figure 1.12. Other areas on Melville, Laugheed, King Christian, and Ellef Ringnes islands are dominated by *A. alpinus*

Figure 1.12. Luzula confusa *and* Alopecurus alpinus *dominate this graminoid steppe on Ellef Ringnes Island.*

on black soils derived from shales of the Lower Cretaceous age. There are very few herbs or cryptogams associated with *Alopecurus* (Bliss and Svoboda 1984). This great reduction in species richness may result from soil churning, which is so common in these dark soils, as well as low nutrient status. *Alopercurus* is rhizomatous and can tolerate soil movement.

High Arctic Polar Desert

Herb barrens. Landscapes that are almost totally devoid of plants occupy thousands of square kilometers, especially in the Queen Elizabeth Islands (Fig. 1.13A). These landscapes occur from near sea level (10–20 m) to upland plateaus at over 200–300 m. The basic controls appear to be the geologic substrate (which influences soil development), the very low amount of soil nutrients, formation of needle ice in autumn, surface soil drying (1–2 cm) in many summers, and a very short growing season (1–1.5 mo). Soil texture ranges from medium-grained sands to silt loams. In many areas there is a thin veneer of rocks that covers much of the surface, reducing the area of safe sites for plant establishment. In other areas there is at least 30–70% open soil.

Woody or semiwoody species are very rare in these barren landscapes, where rosette (*Draba*), cushion (*Saxifraga*), and mat-forming (*Puccinellia*) species predominate. In a study of 23 barren sites on six islands, a total of 17 vascular species and 14 species of cryptogams were found, with a mean species richness of 9 per site (60 m²). Vascular plant cover averaged 1.8% and cryptogams 0.7%; the remaining 97.5% was bare soil, frost-shattered rocks, and pebbles (Fig. 1.13B). Most vascular plants were found in desiccation cracks or adjacent to stones within the polygonal patterns (sorted nets and stripes). Cryptogams were seldom present in close association with vascular plants, in contrast to polar semidesert and snowflush habitats within this barren landscape, where moss mats and cryptogamic crusts are preferred sites for vascular plants. Of the vascular plants recorded, *Draba corymbosa*, *D. subcapitata*, *Papaver radicatum*, *Minuartia rubella*, *Saxifraga oppositifolia*, and *Puccinellia angustata* were most commonly present (Table 1.4). The most important cryptogams were *Hypnum bambergeri*, *Tortula ruralis*, *Thamnolia subuliformis*, *Dermatocarpon hepaticum*, and *Lecanora epibryon* (Bliss et al. 1984). The lack of lichens on rocks and the very depauperate flora and plant cover suggest that these lands, especially uplands, may have become vegetated only since the Little Ice Age (130–430 yr BP) (Svoboda 1982).

Not all uplands in the High Arctic constitute true polar deserts. The uplands above Alexandra Fiord, Ellesmere Island, are much richer in vascular (37), bryophyte (22), and lichen (23) species than other comparable sites (Batten and Svoboda 1994). *Saxifraga oppositifolia*, *Luzula confusa*, and *L. nivalis* dominated at 520–920 m elevation with three sites having limited amounts of *Salix arctica*. Large mats of *S. arctica* were sampled at 525–530 m on the western plateau (Batten and Svoboda 1994; Bliss, Henry, Svoboda, and Bliss 1994) as well as populations of *Cassiope tetragona* and *Dryas integrifolia* above 500 m. Cryptogamic crusts indicating higher surface moisture levels in some sites and higher temperatures due to dark rock surfaces receiving reflected radiation off the nearby glaciers (Labine 1994) are important factors that account for the presence of woody species and a richer flora than is typical of most polar deserts (Bliss et al. 1984; Bliss et al. 1994).

Figure 1.13. (A) Aerial view of a polar desert landscape on Prince Patrick Island. The darker areas are snowflush communities with an abundance of mosses and a few vascular plants. (B) Polar desert barrens on Ellef Ringnes Island, with 1–2% cover of Papaver radicatum and Puccinella angustata.

Snowflush community. Within the polar barrens, the only habitats that have significant plant cover are those below large snowbanks and snowfields. The meltwaters, present part of the summer, enable the development of cryptogamic crusts and moss mats in which flowering plants become established. Although snowflush communities occupy only 3–5% of the landscape, they are very conspicuous features when present. Species richness was found to be much greater in these habitats (30 species of vascular plants and 27 species of cryptogams) at 12 sites (60 m²) on three islands (Bliss et al. 1984). Plant cover was also much greater in these habitats (vascular plants 9.5%, bryophytes 18.5%, and lichens 8.2%) than in the herb barrens (Table 1.4).

Plant composition can be quite variable, with graminoid-dominated meadows in some sites (*Eriophorum triste, Alopecurus alpinus*) and herbs (*Saxifraga oppositifolia, S. cernua, Papaver radicatum,*

Draba corymbosa, Minuartia rubella, and *Stellaria longipes*) in others. The polygonal troughs and stripes contain *Orthothecium chryseum, Ditrichum flexicaule,* and *Drepanocladus revolvens,* and nearly all vascular plants are restricted to these moss mats or cryptogamic crusts (Fig. 1.14). These microsites always contain cyanobacteria and thus have the ability to fix nitrogen. Few species other than *Phippsia algida, Alopecurus alpinus, Papaver radicatum, Cerastium alpinum,* and *Stellaria longipes* occur in the large areas of bare soil.

PLANT COMMUNITY DYNAMICS

Succession–Low Arctic

Primary succession. Plant succession has received less attention in the arctic literature, for it is not a conspicuous feature. Natural disturbances (fire, massive erosion, permafrost melt, glacier retreat)

Figure 1.14. *Snowflush habitats within the polar desert barrens. Snowflush site dominated by mosses, with a few scattered vascular plants of* Alopecurus alpinus, Luzula nivalis, *and* Draba alpina *on Cornwallis Island.*

or human-induced disturbances (villages, mining, petroleum exploration, and development) are much less common features than in temperate regions. The flora is small, and the number of pioneer species (r-selected) is greatly reduced. Arctic environments, limited in heat, nutrients, often water, and length of growing season, are not characterized by classical succession where species replacement and directional change predominate, as in temperate regions. Svoboda and Henry (1987) recognized that with poleward increase in environmental severity, succession shifts from directional change and species replacement to species establishment and survival with nonreplacement of species.

Directional change with species replacement (primary succession) is characteristic of succession along major rivers, with their meandering channels that drain the Low Arctic. Plant communities dominated by herbs (legumes, composites, grasses) establish as seedlings and stranded plants on river alluvium. With increased deposition of sands and silts, *Salix alaxensis* becomes established, often the result of clumps stranded with decreased spring flooding. These willow clumps grow in size and accelerate the deposition of fines and the establishment of more forbs and grasses (often 20–25 species). With time, *Salix arbusculoides, S. glauca* ssp. *desertorum, S. niphoclada* var. *farrae, S. pulchra, S. lanata* ssp. *richardsonii,* and *Alnus crispa* establish, and provide greater shade and ground cover. Herbaceous species decrease in numbers (10–15 species, moss cover increases (20–40%), and the active layer decreases in thickness (>1.0 m in late July) (Bliss and Cantlon 1957) (Fig. 1.15).

On older gravel bars, decadent stands of *Salix alaxensis* predominate, with an understory of willows, a significant decrease in herbaceous species

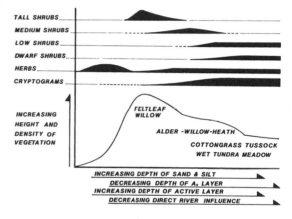

Figure 1.15. *Pattern of plant succession along the Colville River, northern Alaska. Note the changes in plant cover, organic matter, and depth of the active layer (from Bliss and Cantlon 1957).*

(2–8 species), and an increase in moss cover (70–100%). An organic layer develops, and the active layer decreases further (0.5–0.7 m in late July). Farther back on the first terrace there is a shift from decadent *Salix alaxensis* to greenleaf willows, an increase in heath species, and a further reduction in herbaceous species. Associated changes include an increase in the organic layer and a decrease in the active layer (20–30 cm, late July). These transitional areas gradually shift to open stands of *Alnus crispa* with a well-developed ground cover of dwarf heath shrubs and sedges (Bliss and Cantlon 1957). A similar pattern of succession from tall willow species to low shrub tundra with abundant heath species and cottongrass tussocks occurs in the outer areas of the Mackenzie River delta. Portions of the successional pattern were described by Gill (1973).

Successional patterns have also been described for sand dunes along the Meade River, near Atkasuk, Alaska. Sandbars along the river are stabilized by *S. alaxensis*, which over time are replaced by the low willows *S. glauca* and *S. lanata* (Peterson 1978). *Elymus mollis* is also a pioneer species of both riparian and upland dunes. On upland sites, *Carex bigelovii* and *Arctostaphylos rubra* dominate along with *Dryas integrifolia* over time. In less exposed sites, communities of lichens and dwarf heath shrubs replace *Dryas* (Peterson and Billings 1978, 1980). Long-stabilized dunes support tussock tundra with its associated forbs, cryptogams, and dwarf heath shrubs (Komarkova and Webber 1980).

A cyclic pattern of succession occurs with the development and eventual drainage of thaw lakes in the coastal plain of Alaska (Fig. 1.16). The cycle is driven by physical rather than biological processes in contrast to most examples of succession. The thaw-lake cycle of succession was studied by Britton (1957), who proposed the concept, and later by Billings and Peterson (1980), who studied the physical and biological changes associated with the artificial drainage of four shallow lakes in 1950 and succession on older, naturally drained lake basins. Soon after drainage, the surface consists of relatively low, high-center polygons with wide troughs above the ice wedges (Fig. 1.16). The troughs are invaded by mosses: *Dupontia fisheri, Eriophorum scheuchzeri,* and *Arctophila fulva.* In a matter of a century or two, *Carex aquatilis* and *Eriophorum angustifolium* become established, and a mixed graminoid-moss vegetation develops. With time, 1000 yr or so, the ice wedges increase in size, the polygonal pattern becomes more prominent, and the polygon centers with their high ice content (ice lenses) melt, forming miniponds in which *Carex aquatilis* dominates (Fig. 1.16). As the ice lenses and wedges continue to melt, the low-center polygons coalesce, forming thaw ponds. As these increase in size and water depth, *Arctophila fulva* becomes dominant (Fig. 1.16). The thaw ponds continue to increase in area and become oriented NE–SW because of differential erosion from prevailing NE–SW winds (Black 1969). In this successional sequence, the decreased availability of phosphorus decreased the pH and depth of the active layer, and the accumulation of organic matter played important roles (Billings and Peterson 1992).

Secondary Succession

Fire. Fire is common to many biomes, but until the 1980s and 1990s, tundra fires received little attention. Fires are more common on the Seward

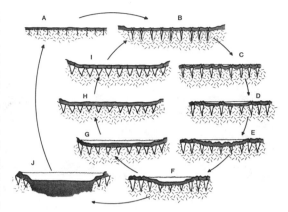

Figure 1.16. Sequence of events in the thaw-lake cycle, coastal plain, Alaska (from Billings and Peterson 1980).

Peninsula in western Alaska than on the North Slope, in the Yukon Territory, or in the Northwest Territories (Racine 1981; Racine, Johnson, and Viéreck 1987; Wein 1976; Wein and Bliss 1973a; Wein and Shilts 1976). Most fires burn 1–10 ha, but some have covered >1000 km². These fires, mostly within low shrub and dwarf heath shrub tundras, remove litter and biomass, darken the soil surface, increase the depth of summer thaw 30–50%, and remove nutrients. As a consequence, there is often a rapid resprouting of shrubs and the seeding in of 1) native grasses *Arctagrostis latifolia, Calamagrostis canadensis, Poa arctica,* and 2) the sedges *Eriophorum vaginatum* and *Carex bigelovii.* The grasses are normally minor components of tussock and shrub tundra, but they seed abundantly following a fire, with the result that there can be rapid recovery of plant cover and net plant production, presumably because of a large seed bank (McGraw 1980), although seedling survival is often low. There is a slower recovery of most heath species, lichens, and some mosses.

Studies following the large fires in 1977 in western Alaska found that vascular plant cover returned to prefire levels in 6–10 yr. Much of the regrowth resulted from established *Eriophorum vaginatum* and resprouting of low and dwarf shrubs (Racine et al. 1987). Thirteen years after a burn along the Elliott Highway in central Alaska, net production had recovered to preburn levels in a tussock–dwarf shrub heath tundra (Fetcher, Beatly, Mullinax, and Winkler 1984).

Near the treeline, fires may eliminate tree seedling establishment unless the following 3–6 yr are unusually warm and moist (Black and Bliss 1980). Consequently, tundra advances south. Russian scientists have referred to this as "pyrogenic tundra" or "subarctic tundra." Fires rarely, if ever, occur in the High Arctic, because of a lack of fuel load and summers that are relatively cool and moist.

Studies of nutrient regimes 1–2 yr after a burn within cottongrass–dwarf heath shrub tundra show a higher nutrient content in new plant shoots than in controls. Nutrient release from the burn, deeper thawed and warmer soils, and increased microbial activity all contribute to a more rapid uptake of nutrients by graminoid and heath species (Wein and Bliss 1973a). However, 6 yr after the Inuvik, N.W.T. fire, nitrogen and phosphorus were still being lost in large amounts from the tundra and forest-tundra ecosystems by erosion and decomposition (McLean, Woodley, Weber, and Wein 1983).

Human disturbance. The development of winter roads, well pads, supply camps, and borrow pits related to petroleum exploration and development have led to disturbed soil surfaces and the opportunity to study secondary succession. Bladed surfaces and vehicle tracks reduce surface peats and often expose mineral soil. Soils are warmed, the active layer increases, and nutrient availability and uptake by surviving and invading species often increase (Bliss and Wein 1972; Hernandez 1973; Gersper and Challinor 1975). Unless underground plant parts are totally destroyed, recovery of plant cover and plant production is generally rapid, especially in sites where native grasses can establish. Following both fire and surface soil disturbances, succession is usually nondirectional, although the density of some species (heath, lichen, and moss) is often reduced for a number of years.

In the Low Arctic of the Mackenzie River delta region and the North Slope of Alaska, where petroleum exploration began in the 1960s, human-induced disturbance and succession in surface soil are most evident. Six years following exposure of peats and mineral soils, plant cover was usually 30–50% versus >100% cover in low shrub–heath and cottongrass–heath tundra. In wet sedge-moss tundra, *Carex aquatilis* and *C. rariflora* provided 95% cover within 1 yr because of rhizomatous growth (Hernandez 1973). At Eagle Creek, Alaska, *Eriophorum vaginatum* provided nearly 100% cover from seedlings 10 yr after disturbance (Shaver, Gartner, Chapin, and Linkins 1983). Shrubs of heath, birch, and willow are slow to reestablish unless rootstalks remain.

Where permafrost is ice-rich, soil slumpage and erosion gullies may form. This is generally a minor problem unless most or all of the organic mat is destroyed. A major summary of disturbance and recovery of Alaskan arctic tundra vegetation has been prepared by Walker et al. (1987).

Succession–High Arctic

Primary succession. Recognition of plant succession in the High Arctic is more difficult than in the Low Arctic because of little surface disturbance, a smaller flora with even fewer ruderal species, and an environment in which abiotic rather than biotic factors dominate. There are scattered polar oases in which the vegetation resembles the mires (wet sedge-moss meadows) of the Low Arctic. The Truelove Lowland, Devon Island, is such an oasis, having a series of raised beaches that extend across the lowland and that result from isostatic rebound of the land, with mires occupying the wetter habitats. Along the coast there is a distinct patterning of plant communities and associated edaphic factors from the shoreline inland, indicating that primary succession occurs. This pattern includes directional change with species replacement (Fig. 1.17).

Marine algae wash into shallow lagoons and eventually are blown onto the coarse gravel beaches during fall storms. These algae are invaded by cyanobacteria, which fix large amounts of nitrogen, and by green algae. With time *Puccinellia phryganodes* and several other salt-tolerant species establish. In topography grades that slope up by means of concave areas, meadows of *Dupontia fisheri* followed by *Carex stans* predominate where meltwater keeps the soils moist all summer. Farther upslope, where water is more limited, *Salix arctica* hummocks, with lesser amounts of *Juncus biglumis* and *Alopecurus alpinus*, predominate. This community grades into cushion plant communities of *Saxifraga oppositifolia*, *Salix arctica*, and lichens on the rocky uplands. In drier, convex areas, the smaller mats of marine algae are invaded by cyanobacteria and *Puccinellia*. Here succession proceeds through rosette–herb hummocks that are dominated by several species of *Draba*, *Cochlearia officinalis* ssp. *groenlandica*, *Cerastium alpinum*, and an abundance of crustose lichens and mosses, especially *Ditrichum flexicaule*. These communities then grade upslope into the cushion plant–lichen communities (Bliss and Gold 1994). Reduced levels of salinity and of nitrogen fixation upslope from the shoreline and an increased depth of organic soils, with their ability to remain wet all summer, are the major edaphic factors in the conversion of meadows dominated successively by *Puccinellia*, *Dupontia*, and *Carex* (Fig. 1.17).

These two toposequences become end points in wet sedge-moss meadows in concave sites that hold water and cushion plant–lichen communities in well-drained sites. These sites closely resemble the older communities (>3000 yr) that dominate

Characteristic	Cushion plant-lichen	Salix hummock	Dupontia meadow	Carex meadow	Dupontia meadow shoreline	Puccinellia marsh
Soil depth (cm)	2.1	4.0	11.4	18.1	11.4	3.9
	0-12	2-5	0-29	4-25	3-17	0-12
Soil moisture (%)	------	229	292	358	323	293
Conductivity (dS m^{-1})	------	0.45	0.60	0.46	0.97	3.64
pH	6.4	5.9	5.5	5.6	5.6	6.0
NO$_3$ (mg l^{-1})	------	1.2	2.0	1.2	1.6	1.6
NH$_4$ (mg l^{-1})	------	8.7	10.2	24.8	10.4	10.9
Net annual plant production (g m^{-2})	37.5	------	59.6	100.4	------	17.0
# Vascualar plant species	8	25	24	27	17	6
Acetylene reduction (μmol C$_2$H$_4$ m^{-2} h^{-1})	18	38	50	59	------	215

Figure 1.17. Successional pattern of plant communities with supporting environmental and biological data along the topographic sequence from a brackish lagoon to a rock *outcrop along Rocky Point, Devon Island, N.W.T. (Illustration © 1995 by Phyllis A. Woolwine).*

the lowland. There, *Dryas* hummocks (rather than *Salix* hummocks) dominate in the transition to the xeric, gravelly beach ridges associated with cushion plant species and lichens, and *Carex* meadows dominate the intervening basins.

Secondary succession. The limited examples of secondary succession in the Canadian Arctic Archipelago show that the pattern is nondirectional and nonreplacement. Bladed surfaces and vehicle tracks, the result of petroleum exploration, are the major sources of disturbance. *Carex stans, Eriophorum angustifolium,* and *Alopecurus alpinus* reestablish by means of rhizomes in wet and mesic sedge-moss meadows, with limited establishment by seedlings (Bliss and Grulke 1988). In upland sites, grasses such as *Poa arctica, Alopecurus,* and *Arctagrostis latifolia* and rosette species of *Draba, Papaver,* and *Cerastium* establish mostly by seedlings (Babb and Bliss 1974; Forbes 1992). A recent study by Forbes (1994) shows the indicator value of bry-

ophytes rather than vascular plants to date past surface disturbances. The time required for recovery of vascular plant cover is probably >50 yr, but it is probably less for mosses in moist to wet sites.

STANDING CROP AND PLANT PRODUCTION

It is well known that arctic plants are small in stature, that they grow slowly, and that a standing crop of plant and animal material accumulates slowly. Decomposition is also very slow. However, in the Arctic there is great variation in carbon allocation and in the production of dry matter among plant communities. As used here, "standing crop" refers to living and dead plant material, "phytomass" refers only to living material, and "plant production" refers to net production over time, usually per year.

Table 1.5 summarizes data on major arctic vegetation types. These data show the dramatic shifts

Table 1.5. *Summary data from various sources on annual plant production, root: shoot ratios, and leaf area index (LAI) for various Arctic vegetation types*

| Vegetation type | Plant production (g m^{-2}) | | Root:shoot | | LAI |
	Vascular	Cryptogam	Phytomass	Net production	
Low Arctic tundra					
Tall shrub	250–400	5–25	1:2	1:2	1–2
Low shrub	125–175	25–50	1:2	1:2	1–2
Cottongrass-herb	150–200	25–100	3:1	1:1	0.5–1
Wet sedge-moss	150–200	5–25	20:1	3:1	0.5–1
High Arctic tundra					
Wet sedge-moss	100–175	10–40	20:1	3:1	0.5–1
High Arctic polar semidesert					
Cushion plant-cryptogam	5–25	2.5	1:1	1:5	0.1–0.2
Cryptogam-herb	10–20	5–30	2:1	1:2	0.1–0.2
High Arctic polar desert					
Herb barrens	0.1–2	T^a–1	1:3	1:2	—^{b}

^{a}T = trace.
^{b}Dash = no data

in plant production, root:shoot ratios, and the relative role of cryptogams in major Low and High Arctic systems.

Tundra

Standing crop and net annual production are greatest in tall shrub tundra dominated by *Salix alaxensis, S. lanata* ssp. *richardsonii,* and *S. glauca* (Komarkova and Webber 1980). Total net annual production is nearer 250–400 g m^{-2} if it is assumed that root production is 50–100% of aboveground production. These values will be higher south of the Meade River (ca. 70°15' N) in the warmer sector of the Foothills Province. Low shrub tundra dominated by *Betula nana* ssp. *exilis, S. glauca, S. pulchra,* various species of heath, and *Carex bigelovii* or *C. lugens* has an aboveground standing crop similar to that of cottongrass–dwarf shrub heath and of wet sedge-moss tundras (Table 1.6). Total annual production for these three types is similar – 150–300 g m^{-2} if root production is included. Based on root:shoot ratios for annual production of 1:2 for low shrub, 2:1 for cottongrass–heath, and 3:1 for wet sedge-moss tundras, estimates of total production of vascular plants are approximately 125–175 g m^{-2}, 150–200 g m^{-2}, and 150–200 g m^{-2}, respectively. Cryptogams play a significant role in the low shrub, dwarf shrub heath, some cottongrass–heath, and the wet sedge-moss community types (Tables 1.5, 1.6). In the latter communities, biomass root:shoot ratios are generally 15:1 to 21:1, and net annual production ratios are 2:1 to 3:1.

Note that plant production in the wet sedge-moss communities in the High Arctic of Devon Island is similar to that at Barrow, Alaska (Table 1.5).

From very limited data, low shrub and graminoid-dominated communities have a leaf area index (LAI) of 0.6–2. Chlorophyll content and LAI are generally well correlated with aboveground vascular plant production (Wielgolaski, Bliss, Svoboda, and Doyle 1981).

Polar Semideserts and Polar Deserts

There are significant shifts in the magnitude of standing crop and net annual production of plant communities in the High Arctic (Tables 1.5 and 1.6). This reflects the shorter growing season, reduced degree days, and colder soils with less available nitrogen and phosphorus. The cushion plant–cryptogam vegetation has a large aboveground standing crop because of the large amount of dead material associated with *Dryas integrifolia* and *Salix arctica*. Elsewhere, the cryptogam–herb vegetation has a considerable standing crop and biomass of bryophytes and lichens, but vascular plants contribute a relatively small percentage. Belowground biomass and annual production are minor contributors to carbon allocation; root:shoot ratios are generally 1:1 to 1:2 for phytomass and annual production, indicating that root development is generally much more restricted in these cold soil environments that range from moist to dry. Net annual production is generally 25–50 g

Table 1.6. *Standing crop (g m⁻²) for select Arctic vegetation types*

| | Vascular plants | | | | |
Location	Aboveground	Belowground	Litter	Cryptogams	Total
Low Arctic tundra					
Atkasook, AK[1]					
Tall shrub	1496	—[a]	142	82	—
Low shrub	342	—	427	137	—
Dwarf shrub heath	421	—	423	403	—
Wet sedge-moss	257	—	24	499	—
Cottongrass-heath	418		323	577	—
Eagle Creek, AK[2]					
Cottongrass-heath	419	6563	77	89	7148
Dempster Highway, YK[2]					
Heath-cottongrass	168	6724	39	302	7233
High Arctic tundra					
Barrow, AK[3]					
Wet sedge-moss	83	4366	67	45	4561
Devon Island, N.W.T.[4]					
Wet sedge-moss	273	2230	—[b]	908	3411
Barrow, AK[3]					
Herb–cushion plant	75	3376	41	31	3524
High Arctic polar semidesert					
Devon Island, N.W.T.[4]					
Cushion plant–cryptogam	387	57	—[b]	79	523
Victoria Island, N.W.T.[5]					
Cushion plant–cryptogam	1078	300	—[b]	327	1705
Melville Island, N.W.T.[6]					
Cryptogam-herb	127	21	138	782	920
King Christian Island, N.W.T.[6]					
Cryptogam-herb	41	23	—[b]	2146	2210
High Arctic polar desert					
Devon Island, N.W.T.[7]					
Herb barrens	11	1	12	2	26
Somerset Island, N.W.T.[7]					
Herb barrens	8	1	15	1	25

[a] Dash = no data.
[b] Litter included in aboveground.
Sources: 1. Komarkova and Webber (1980); 2. Wein and Bliss (1974); 3. Webber (1978); 4. Bliss (1977); 5. Bliss and Svoboda (unpublished); 6. Bliss and Svoboda (1984); 7. Bliss et al. (1984).

m⁻². The LAI (based on two examples of the cushion plant–cryptogam vegetation) is only 0.2 and it may be as low as 0.1 in the cryptogam–herb vegetation.

In the polar barrens, plant production and standing crop are extremely low. Cryptogams are nearly absent in contrast with tundra and polar semideserts, where they are almost always large contributors of carbon. In 18 stands sampled on five islands, annual plant production averaged only 0.8 g m⁻² (Bliss et al. 1984).

Controlling Parameters

Why are phytomass and annual production so low in these diverse arctic landscapes? In the polar de-

serts, the reasons are a short growing season (20–45 d), extremely low levels of nitrogen and phosphorus, and low soil and air temperatures. Plant cover and plant production are considerably higher in snowflush sites where cryptogam crusts provide higher levels of nitrogen, greatly reduced soil churning and needle ice formation, and more consistent surface soil moisture.

In the polar semideserts, there is considerably greater cover of mosses and lichens. These surfaces hold more moisture in summer, provide safe sites for vascular plant establishment and development, and contain cyanobacteria that fix limited amounts of nitrogen (Sohlberg and Bliss 1984). In these High Arctic ecosystems, with their limited vascular plant cover, low air and soil temperatures are further limitations. Soil temperatures 5–10 cm below the surface, where most roots occur, are generally 2–4° C. At higher temperatures, root growth increases dramatically (Bell and Bliss 1978).

Plant production is significantly higher in wet sedge-moss meadows. Soil nitrogen is also higher because cyanobacteria are more abundant. These habitats also have cold soils with a shallower active layer (20–30 cm) compared with 40–60 cm beneath cushion plant–lichen or herb–cryptogam communities and the polar desert barrens.

In the Low Arctic, these same parameters are central to the control of plant growth, although the environmental extremes are not as severe. Greater production of the shrub complexes, the cotton-grass–dwarf shrub heath, and the wet sedge-moss communities reflect their higher nutrient content. The growing season is longer, and the number of degree days is higher (Table 1.1). Simulation modeling within the cottongrass–dwarf shrub heath tundra clearly shows the interaction of nitrogen and phosphorus, soil water, and the thermal regime in controlling plant growth. When fertilizer is added, growth is stimulated in the graminoids, but growth is depressed in heath species (Miller et al. 1984). Similar results have been reported using fertilizer experiments at Toolik Lake, Alaska (Shaver and Chapin 1986), on cottongrass–dwarf shrub heath and in the High Arctic on *Carex stans* wet meadows and *Dryas integrifolia* on raised beach ridges (Babb and Whitfield 1977).

LIFE HISTORY PATTERNS AND PHYSIOLOGICAL ADAPTATIONS

Arctic plants grow in some of the most extreme environments in the world. Consequently, they have been intensively studied for four decades. These plants are subjected to low air and soil temperatures, to low nutrient availability, and often to extremes in water availability. Plants of some species grow in cold, water-saturated soils with low oxygen concentrations; others grow in soils that are saturated after snowmelt but that may dry considerably by midsummer. These environmental constraints are further compounded by a short growing season that limits carbon gain and sexual reproduction.

Reproduction and Plant Establishment

Arctic plants are long-lived; there are very few annuals ($\approx$10 with arctic-alpine distribution) or short-lived perennials. Within plant communities of the Low Arctic, many perennials depend on tillering (*Carex, Dupontia, Eriophorum, Luzula, Petasites*) and

branch layering (*Betula, Empetrum, Salix, Vaccinium*), yet these same species flower, fruit, and produce viable seeds. Seedlings are not common in undisturbed tundra, but with surface disturbance (fire, vehicle tracks, animal burrows), seedling establishment of graminoids is common (Bliss and Wein 1972; Hernandez 1973; Chapin and Chapin 1981; Chester and Shaver 1982; McGraw and Shaver 1982). Seed dormancy mechanisms are generally absent or minor in both Low Arctic (Bliss 1971) and High Arctic species (Bell and Bliss 1980). The optimum temperature for germination is surprisingly high (15–30° C) for most species (Billings and Mooney 1968; Bell and Bliss 1980).

In the High Arctic, seed germination is generally low, with seedling survival more successful on bryophyte or desiccation-crack microsites (Bell and Bliss 1980; Sohlberg and Bliss 1984). These microsites are moister and have a higher nutrient content and less needle ice than do bare soil or lichen surfaces. Seed accumulation within snowbanks and desiccation cracks indicate the importance of microsites as safe sites in the establishment and maintenance of species. Seedling mortality is much greater in summer than in winter, indicating the importance of soil moisture in seedling survival.

Estimates of seed bank sizes (germinable seeds) range from 150 seeds m^{-2} at Barrow (Leck 1980) in a sedge-moss tundra, to 430–820 seeds m^{-2} in upland shrub and lowland wet sedge tundra in northern Alaska (Roach 1983), and to 3376 seeds m^{-2} in cottongrass tussock tundra at Eagle Creek in central Alaska (McGraw 1980). In another study, which followed a gradient from a *Dryas* fellfield (85 seeds) to a dwarf shrub heath (3142 seeds), and to a late snowbed willow community (597 seeds m^{-2}), a major correlation of seed bank size was with aboveground plant production (Fox 1983).

In the High Arctic, Freedman, Hill, Svoboda, and Henry (1982) reported germinable seed bank sizes of 1 seed m^{-2} for a wet sedge meadow, 131 seeds m^{-2} for a *Dryas* cushion plant community, 6026–152 seeds m^{-2} along a 3 m gradient and from a lemming burrow, and 7810 seeds m^{-2} near a fox den on Ellesmere Island. Estimated seed bank sizes were 570 seeds m^{-2} in a graminoid barren and 1600 seeds m^{-2} in a cryptogam–herb community on King Christian Island (Sohlberg and Bliss 1984). McGraw and Vavrek (1989) have reviewed much of the literature on arctic seed banks and report a similar abundance of seedlings from disturbed and undisturbed soils.

The abundance of flowering and the presence of seed banks and seedlings confirm that arctic plants allocate rather large amounts of carbon to sexual reproduction, even though the cold, short, and sometimes dry summer environments are not conducive to seedling establishment.

Plant Growth and Phenological Development

The general patterns of plant growth and development differ little from those in temperate regions. There is, however, a telescoping of these events into a shorter time. Early season growth of deciduous forbs and shrubs is rapid because of their near-synchronous leaf initiation and leaf expansion from preformed buds of the previous year (Sorensen 1941; Bliss 1956; Johnson and Tieszen 1976; Tieszen 1978). In contrast, graminoids continue to grow until shorter days and lower temperatures suppress leaf and shoot elongation. Many species carry some green leaves over the winter and initiate growth as soon as the snow melts in June. In the High Arctic, of 27 species examined, only *Oxyria digyna* failed to retain wintergreen or evergreen leaves over winter. Eight of these species were grown under controlled summer environmental conditions (continuous light, alternating "day" and "night" temperature conditions, and high humidity), and all entered dormancy in spite of favorable growing conditions (Bell and Bliss 1977), indicating a periodic growth pattern. Field studies of *Luzula confusa* showed that leaf senescence began in 45–50 d and that root growth dropped to near zero in 49 d. Under controlled environment conditions, the growing period was 55 d (Addison and Bliss 1984). Nonperiodic growth patterns have been observed in snowbed forbs in Greenland (Sorensen 1941) and in Alaska (Murray and Miller 1982). Species may flower and fruit synchronously in one microenvironment and be asynchronous in other environments (Bliss 1956; Billings 1974; Murray and Miller 1982).

Root growth often begins as soon as the active layer thickens and soils warm above 0° C, but lateral root growth is often more active in the latter half of the summer (Shaver and Billings 1975; Bell and Bliss 1978). The temporal partitioning of root and shoot growth may reflect adjustments in plant growth to reduce the demand for limited carbon and nutrient resources and to take advantage of the most favorable environmental conditions aboveground (early) and belowground (late). This partitioning is well documented for *Salix arctica* on Devon Island, where leaf growth and flowering began 0–5 d after snowmelt, shoot growth 3–6 d later, and root growth 5–10 d after shoot growth began

(Dawson 1987). Root elongation in at least two sedges (*Eriophorum angustifolium*, *Carex aquatilis*) is strongly influenced by both temperature and photoperiod. At 2–4° C the growth rates are little different, but at soil temperatures of 14–16° C growth rates nearly double for *Carex* when the photoperiod increases from 15 to 24 hr. In *Eriophorum* roots there is no growth at the 15 hr photoperiod, 4.7 mm d^{-1} at 18 hr, and 7.5 mm d^{-1} at 24 hr (Shaver and Billings 1977). The mechanisms that control this photoperiodic response are not known. Both shoots and roots are capable of growth at low temperatures (0–5° C) (Bell and Bliss 1978; Billings, Peterson, and Shaver, 1978), but experimental studies in field minigreenhouses and plant growth chambers show that growth is enhanced in different growth forms (graminoids, forbs, deciduous shrubs) by higher temperatures (10–20° C) (Chapin and Shaver 1985b).

Plant growth and phenological development are more rapid in warmer microenvironments both for vascular plants (Sørensen 1941; Bliss 1956; Warren Wilson 1959) and bryophytes (Oechel and Sveinbjörnsson 1978). In general, the height of arctic plants conforms to the height aboveground where the air is warmest in summer and to the mean depth of snow cover in winter. This is especially true for shrubs, cushion- and mat-forming plants such as *Dryas*, *Saxifraga caespitosa*, and *S. oppositifolia*, the tussock form of *Eriophorum vaginatum*, and the small rosette species of *Draba*, *Saxifraga*, *Papaver*, and *Minuartia*. Leaf temperatures within the boundary layer are often 5–10° C higher than ambient and may reach 25–30° C above ambient in *Dryas* leaves on clear, calm days. The parabola-shaped flowers of some arctic species serve as basking sites for insects and raise the temperature of the developing ovary 5–10° C, which enhances pollination and speeds seed development (Kevan 1972).

The growth rate of arctic graminoids on a daily basis is often as high as that for temperate graminoids, although arctic temperatures at plant height are 10–15° C rather than the 20–35° C in grasslands to the south. Total plant production is less in the Low Arctic than in comparable temperate grasslands and herblands because of the shorter growing season and a lower range of daily soil and air temperatures.

Physiological Characteristics of Plant Growth Forms

It is now known that there are strong correlations between the physiological characteristics of taxonomically diverse species with the same growth form and their adaptation to certain habitats. I shall discuss deciduous shrubs, evergreen shrubs, graminoids, tussock graminoids, and cushion plants.

Deciduous shrubs require large carbon gains each summer to produce an entirely new leaf surface. Consequently, this group of species has high photosynthetic rates (10–30 mg CO_2 dm^{-2} hr^{-1}) and relatively high leaf conductances (Tieszen 1978; Limbach et al. 1982) (Table 1.7). Root growth, root respiration, and rates of phosphorus and nitrogen uptake are relatively high in *Salix pulchra* and *Betula nana*, as compared to evergreen species. These shrubs also have higher root and shoot growth rates, which places greater demands on nutrient uptake. Mycorrhizae further facilitate nutrient uptake.

There are few dioecious species in the Arctic and thus the role of *Salix arctica*, a deciduous, long-lived woody shrub that has a wide arctic-alpine distribution is of interest. The sex ratio is female-biased (1.9:1), with females more abundant in wet, nutrient-rich, cold-soil habitats (6:1) whereas male plants predominate (1.5:1) in xeric habitats and in those with lower nutrients and higher soil temperatures. In wet sedge-moss communities, female plants maintained higher rates of transpiration and higher leaf-water potentials than did male plants in cold soils (<4° C). Rates of photosynthesis were lower than in males as water stress increased (lower leaf-water potential). In contrast, male plants maintained higher rates of transpiration, lower leaf-water potentials, and lower tissue-osmotic potentials than did female plants in more xeric habitats (Dawson and Bliss 1989a, 1989b, 1993). Developing leaves of both sexes had lower rates of photosynthesis and transpiration but higher rates of respiration and higher levels of nitrogen than mature leaves (Dawson and Bliss 1993). Reproductive shoots also had higher rates of photosynthesis, transpiration, and respiration and higher levels of nonstructural carbohydrates but lower tolerance to water deficits than vegetative shoots. These data help to explain both the sexual segregation of this species and to characterize the ecophysiological breadth within and across highly variable environments.

The evergreen-leaf heath species are conservative plants in a number of ways (Table 1.7). Their photosynthetic rates are low, leaf growth and shoot growth are slow, their leaves function for 2–5 yr, they have low rates of water conductance, the plants maintain small root systems and limited annual root production, they are conservative in use of nutrients, they have low rates of nutrient up-

take, and they show limited response to added nutrients. Respiration rates are relatively low, in part because growth rates are low and leaves are maintained for more than 1 yr (Tieszen 1978; Chapin and Tryon 1982; Limbach, Oechel, and Lowell 1982). These plants have a further energy drain on their production of tannins, alkaloids, terpenes, and anthraquinones, especially in *Empetrum hermaphroditum, Ledum palustre* ssp. *decumbens*, and *Cassiope tetragona* (Batzli and Jung 1980). These heath species store more of their carbon and nutrient reserves in leaves and shoots and apparently have been selected for more effective antiherbivore defenses. Compared with deciduous shrubs, they are less palatable, less digestible, and more toxic to large and small mammals and insects (Batzli 1983; White 1983; MacLean and Jensen 1984; Jefferies, Klein, and Shaver 1994).

Single-stemmed graminoids such as the sedges (*Carex aquatilis, C. stans*), the cottongrasses (*Eriophorum angustifolium, E. vaginatum*, and *E. scheuchzeri*), and the grasses (*Dupontia fisheri* and *Alopecurus alpinus*) are in various combinations or in monoculture the dominant growth form over large areas of the Arctic. These species carry some green stem and leaf tissue over winter, they grow rapidly as soon as snow melts and soils warm in June, and they continue growth until dieback in August and September. They have relatively high photosynthetic rates (9.5–12.6 µmol CO_2 m^{-2} s^{-1}) and high rates of leaf conductance, and they are adapted to living in cold soils that are often water saturated (Mayo et al. 1977; Tieszen 1978). As a group, these species produce per year two to three times more root biomass than shoot biomass.

Nutrients are taken up in smaller quantities by single-stemmed graminoids, in part because of the lack of mycorrhizae. There is considerable retranslocation of nutrients out of senescing leaves, an adaptation of plants growing in nutrient-poor soil systems (Chapin, Van Cleve, and Tieszen 1975). Leaves of *Dupontia* and *Carex aquatilis* and *Carex acquatilis* are mature in terms of high photosynthetic rates for only 10 and 27 d respectively (Tieszen 1978). In the High Arctic, the two or three leaves produced per shoot apparently function efficiently for a longer time (Mayo et al. 1977).

The tussock graminoid *Eriophorum vaginatum* has been studied in more detail than any other arctic species (Miller et al. 1984). It is more physiologically conservative than single-stemmed graminoids (Table 1.7). Net photosynthesis averages only 1.3–1.9 µmol CO_2 dm^{-2} s^{-1}, but root and stem respiration rates are relatively high compared with those for other species (Limbach et al. 1982).

Table 1.7. *General physiological characteristics of different plant growth forms in the Arctic*

	Plant growth form				
Characteristic	Deciduous shrubs	Evergreen shrubs	Single-stem graminoids	Tussock graminoids	Cushion plants
Resource allocation					
Reproduction	Low	Low	Low	Low	Low
Shoots	High	Medium	Medium	Medium	Low
Belowground	Low	Low	High	High	Low
Photosynthetic rate	High	Low	Medium	Low	Low
Respiration rate	High	Low	Medium	High	High
Leaf conductance	High	Low	High	High	Low
Leaf longevity (yr)	1	2–5	1+	1+	2–3
Plant or shoot age (yr)	50–100+	20–50+	5–7	50–100+	20–100+
Carbohydrate reserves	Medium	Low	High	Medium	Medium
Habitat preference	Snow protected, moist	Late snow moist	Wet to dry	Moist	Dry

Leaves of cottongrass typically live for 1^+ yr, but all roots are new each year, a very large energy drain, because root production is three to five times that for shoots. Most roots (75%) are found in the elevated and organic-rich tussock (Chapin, Van Cleve, and Chapin 1979), although some roots grow into the mineral soil, where they grow to within 1 cm of the permafrost table at a temperature near 0° C (Bliss 1956; Chapin et al. 1979). The freeze–thaw cycle that occurs at the top of the permafrost table (bottom of the active layer) makes phosphorus and cations more available, and water movement along this surface may also facilitate water uptake. This species also retranslocates nutrients into the tussock each fall, conserving the limited nutrient pool. The fact that most roots remain in this organic layer enables maximum nutrient uptake each year. Rates of nutrient uptake are as high in *Eriophorum* as they are in *Carex aquatilis* (Chapin and Tryon 1982). Tussocks are long-lived, ranging in age from 122–187 yr (Mark, Fetcher, Shaver, and Chapin 1985).

The graminoids *Carex aquatilis*, *C. stans* in wet meadows, *Eriophorum vaginatum* in soils of intermediate drainage, and *Puccinellia phryganodes* and *C. subspathacea* in coastal brackish marshes are preferred food of caribou, musk-ox, and lesser snow geese because of their nutritional quality and digestibility (Jefferies et al. 1994).

Four contrasting community types that form a toposequence have been studied near Toolik Lake, Alaska (Shaver and Chapin 1991). The communities include dwarf heath shrub (evergreen), low shrub (deciduous), tussock–dwarf heath, and wet sedge tundra. Although there were orders of magnitude differences among the communities in biomass, net annual production, nutrient require-ments, and allocation patterns, rates of biomass turnover and patterns of nutrient use differed only moderately. They concluded that stem processes, especially element storage, plays a central role in both the woody species with large underground stems and the sedges with large rhizomes. Thus the large-stem biomass of woody species, with their slow turnover rates and high capacity for nutrient storage, dilutes the importance of leaf turnover rates with respect to whole-plant element use and rates of recycling.

Plants with the cushion growth form predominate in the High Arctic. This growth form is near the ultimate in physiological conservatism. *Dryas integrifolia*, as an example, occupies windy sites that are also droughty in summer and that have limited snow cover in winter. This is also a long-lived species, 100–150 yr (Svoboda 1977). *Dryas* has low photosynthetic rates (1.3–1.9 µmol CO_2 m^{-2} s^{-1}), low leaf conductances, and very low growth rates of shoots and roots. Leaves are evergreen and function for 2 yr but remain on the plant for many years before decomposing (Mayo et al. 1977; Svoboda 1977). Each clump is a self-contained unit that conserves the very limited nutrient pool at the base of the shoots. Most active roots are to this zone. Respiration rates are high for both roots and shoots. Root:shoot ratios are low, 0.4:0.6, as they are for rosette species in the High Arctic. This is in strong contrast to the patterns for sedges and grasses of wet sites throughout the Arctic for which ratios are as large as 10–20:1.

General Physiological Characteristics

Although arctic plants live in some of the world's most severe environments, adaptations to some stresses are not especially well developed. Seeds of most species germinate on only warm microsites (15–20° C), and in general germination is a slow process, especially in the High Arctic. On the other hand, arctic plants initiate spring growth with a burst, some when there is still a thin snow cover. Plants in warmer microsites grow larger and often flower and fruit more abundantly than do plants in such nutrient-rich sites as animal burrows and bird cliffs.

Arctic plants use only the C_3 carbon photosynthetic pathway, and their maximum rates of photosynthesis are very similar to those for plants of the same growth form in temperate regions. However, the temperature optima for photosynthesis are on the lower end of the range for temperate region species. Most species can maintain positive carbon accumulation over a wide temperature range; some to -2–$4°$ C. In general, arctic plants have a high Q_{10} of 2–3 for both photosynthesis and respiration. Arctic plants have higher rates of mitochondrial activity, shoot respiration, and phosphate absorption than temperate plants do. These high reaction rates may result from higher rates of enzyme activity and probably are necessary to maintain rapid growth rates at low temperatures.

Although many species in the Low Arctic do not experience water stress, many in the High Arctic are periodically drought-stressed. Cushion plants, some grasses, and rosette species in sites with bare and well-drained soils commonly have leaf-water potentials of -2.0 to -3.0 MPa and sometimes less than -4.0 MPa, especially in summers with less precipitation (Mayo et al. 1977; Grulke and Bliss 1988). Leaf-water potentials of plants in wet sites seldom are -1.0 to 1.5 MPa. In cold soils, root resistance to water movement can be an important factor. Arctic plants control water loss through osmoregulation, tissue elasticity, and stomatal control, as do plants of temperate regions. Water relations of arctic plants are summarized by Oberbauer and Dawson (1992).

Nutrients are limiting in arctic plants and ecosystems, as they are in many other systems. Nitrogen and phosphorus are most limiting, and thus they have received the most attention. Tundra plants, like temperate plants, have their temperature optimum at 40° C for nutrient absorption (Chapin and Bloom 1976). Although phosphorus and ammonium absorption is temperature dependent, tundra plants are less temperature sensitive at 0.5–10.0° C than are temperate species. Nutrient uptake increases sharply with increased nutrient concentration (Nadelhoffer et al. 1992), and a high capacity for nutrient uptake enables species to take advantage of nutrient flushes in spring during snow and active layer melt. This uptake has been termed "luxury consumption" (Shaver and Chapin 1980).

Nutrient uptake rates can vary 20-fold, with deciduous shrubs having the highest annual requirements, followed by graminoids, with evergreen shrubs having the lowest requirements (Chapin and Tryon 1982). Numerous studies have shown that inorganic nitrogen in the form of ammonium is preferred over nitrate by arctic vascular plants (McCown 1978; Chapin, Shaver, and Kedrowski 1986), and new information indicates that organic nitrogen in the form of amino acids is also readily absorbed, especially glycine (Chapin, Moilanen, and Kielland 1992; Kielland 1994). Kielland has shown that ectomycorrhizal species have higher ratios of amino acid to ammonium absorption than do nonmycorrhizal species and that uptake rates are higher in deciduous than in evergreen heath shrub species. From various studies it has been shown that rates of mineralization and nitrogen fixation are inadequate to account for the annual rates of nitrogen absorption by both Low and High Arctic species (Chapin and Bledsoe 1992; Kielland and Chapin 1992; Kielland 1994). Nitrogen fixation by legumes appears to be less important as a source of nitrogen than are cyanobacteria in the High Arctic (Chapin and Bledsoe 1992; Chapin, Bliss, and Bledsoe 1991).

Chapin et al. (1991) have shown that rates of nitrogen fixation can vary more than 10-fold in diverse habitats on the Truelove Lowland, Devon Island. Rates are highest in grass-dominated saline coastal marshes (215 μmol $C_2 H_4$ m^{-2} h^{-1}) and very low in cushion plant–lichen communities on gravelly raised beach ridges (18 μmol $C_2 H_4$ m^{-2} h^{-1}). This is another significant difference from much of the Low Arctic, where cyanobacteria appear to be less important except in wet sedge-moss communities.

Recent studies indicate that mycorrhizae may play a more central role in nutrient uptake in arctic plants than previously indicated. Mycorrhizal associations occur in all major plant growth forms at Eagle Summit, Alaska (Miller 1982), and in all growth forms in the High Arctic (Bledsoe, Klein, Bliss 1990; Kohn and Stasovski 1990). Woody species throughout the Arctic support ectomycorrhizae, including many populations of *Salix arctica* and heath species. Endomycorrhizae with well-developed vercicular-arbuscular mycorrhizae (VAM) structures are present in nonwoody species in the Low Arctic, but VAM associations become less and less frequent with increasing latitude where soils are colder and the growing season is shorter (Bledsoe et al. 1990). The less well-developed mycorrhizal structures in these more northern species may nevertheless play an important role in the uptake of organic as well as inorganic nitrogen, but this research remains to be done.

INDUSTRIAL DEVELOPMENT

Resource development in the Arctic has been slow for several reasons. The climate is severe, making work in winter very difficult. Remoteness from markets and environmental regulations greatly increase the costs of exploration and development. The consequences are that only large pools of hydrocarbons and large ore bodies have been developed.

The passage of the National Environmental Policy Act (NEPA) in 1970 required the development of an Environmental Impact Statement (EIS) for each project in the United States. Canada established guidelines for northern pipelines in 1970 and more recently for mining, but laws comparable to NEPA have not been passed.

Mining

There are large deposits of coal, copper, and other minerals in arctic Alaska, but their development awaits new mining technology and transportation systems. A large zinc mine operates near Kotzebue, Alaska. Lead and zinc mines are located on northern Baffin Island and on Little Cornwallis Island in the Canadian Arctic Archipelago. The high-grade ores are stockpiled in winter and shipped by freighter in summer. The presence of ice-rich permafrost requires special precautions in constructing and maintaining roads, buildings, and tailing dumps. New diking techniques have been developed to prevent leaching and thermal erosion where there are massive ice lenses and ice wedges (Bliss and Peterson 1973; Walker 1995). The lack of smelters in the North American Arctic has prevented the many environmental problems associated with these developments in the Russian Arctic and Subarctic (Vilchek and Bykova 1992).

Oil and Gas

Petroleum exploration began in arctic Alaska in the 1940s and in mainland arctic Canada in the 1960s. Oil was discovered at Prudhoe Bay in early 1968 and one year later in the Mackenzie River delta region. Additional oil fields were discovered and developed near Prudhoe Bay in the 1970s and 1980s. The Prudhoe Bay field has peaked in its production, and there is growing political pressure to

open the Arctic Wildlife Range for drilling production wells. Large gas fields and limited oil fields have been found in the northern Canadian arctic islands, but so far only oil has been shipped from Cameron Island, N.W.T., each year. Exploratory gas wells were drilled only in winter in the Canadian High Arctic. Wells were drilled on ice platforms in the sea by pumping sea water onto the ice. Drilling was limited to the three to four coldest months.

Most of the early terrain problems associated with summer road construction, seismic activity, and off-road vehicles have been minimized through using winter snow and ice roads (Adam and Hernandez 1977) and restricting the use of ground-level seismic activity to winter (Bliss and Peterson 1973; Walker and Walker 1991; Walker 1995). Recent studies of winter seismic lines in the Alaskan Arctic Wildlife Range show that soil compression results in a deeper summer active layer, some collection of surface water in depressed tracks, and broken shrubs (Felix, Reynolds, Jorgenson, and DuBois 1992). However, these are minor impacts compared with those of the 1960s.

A major review of the Alyeska pipeline and the petroleum exploration in the Mackenzie River delta region indicated that these large projects have been constructed and maintained in an environmentally acceptable manner (Alexander and Van Cleve 1983), although some problems have occurred. As a result, maintenance costs of the pipeline and the associated Dalton Highway have been less than originally estimated. The enforcement of environmental regulations has been a central factor.

Within the massive Prudhoe–Kuparuk oil fields and their support facilities, a detailed study from 1968 to 1983 showed that the cumulative effects of roads, gravel pads for wells, supply camps, and storage yards have resulted in some blocked drainages. Walker et al. (1987) reported flooding in 9% of the mapped areas and 3% of the total field (500 km²). There have been continual problems with seepage of drilling fluids from retention ponds on the large well pads. It was assumed that the permafrost table would rise into these gravels and prevent loss of fluids, but leakage has occurred. Plastic liners have been installed recently to retain these fluids, which are then removed by tanker truck and reinjected into wells.

The pipeline is elevated through ice-rich permafrost terrain, and the ammonia heat exchangers have been effective in keeping the support columns frozen in place. Culverts and bridges along water courses have enabled maintenance of fish populations, although the original surveys were inadequate to estimate the number of culverts necessary. Survey studies following pipeline operation have shown that moose and male caribou cross freely under the pipeline but that calves and females avoid the pipeline, road, and adjacent rangeland. The passive pipeline is believed to be a lesser problem than are road dust and the road traffic that parallel the pipeline (Alexander and Van Cleve 1983; Walker and Everett 1987; Walker et al. 1987; Bliss 1990).

Revegetation in northern Alaska and the Mackenzie delta region has been quite successful, due to the use of native grasses often mixed with more southern species (Younkin 1976; Younkin and Martens 1987; Densmore, Nieland, Zasada, and Masters 1987; Densmore 1992; Johnson 1987). Over 2–4 yr, the native species begin to take over (Younkin and Martens 1987). With the very limited flora in the High Arctic, there are few species capable of re-vegetating disturbed lands. Over time, one or two native grasses establish in drier sites (Bliss and Grulke 1988; Forbes 1992, 1994). Experimental studies of oil and diesel fuel spills in Alaska and Canada have shown that wet sedge vegetation shows considerable recovery in 1–3 yr but that upland cottongrass tussock and shrub tundra requires 10–20 yr (Wein and Bliss 1973b; Freedman and Hutchinson 1976; articles in Arctic 31:1978; Linkens and Fetcher 1983; Racine 1994).

Future impacts by global climate change could, of course, be enormous (Chapin et al. 1992). Oechel et al. (1995) have published evidence that climate change is already affecting vegetation at Barrow, Alaska. Air temperature during the growing season significantly increased during the 1980s and 1990s, the water table has lowered, and respiration by plants and soil organisms has increased to the point that the ecosystem releases more CO_2 to the atmosphere than it fixes – that is, the past traditional role of Arctic tundra as a CO_2 sink is reversing into a role as CO_2 source.

CONSERVATION

Resource exploration in recent years, along with a growing tourist potential, has helped circumpolar nations realize that setting aside lands now is an imperative. Consequently, large landscape units have been and are being established for conservation purposes in all arctic countries.

The United States and Canada have led the world in establishing national parks and preserves. Although the first parks were in forest-alpine mountain settings, more recently established parks have been in the Arctic. Pressure to establish these national parks and national preserves has resulted

from the need to set aside large areas necessary to maintain large arctic mammals and terrestrial and marine birds, and their habitats. Denali National Park was established in 1917, and Katmai National Park was established as a national monument in 1918 to preserve vast arctic-alpine areas, although each lies within the northern coniferous forest. During the 1970s a major push was made to preserve large portions of arctic Alaska, with the establishment of Kobuk Valley and Gates of the Arctic national parks, the Gates of the Arctic, the Noatak and Bering Land Bridge national preserves, and the Arctic National Wildlife Range (1960). These parks and preserves are intended to maintain populations of caribou, grizzly, polar bear, wolves, and waterfowl.

Canada has established five arctic national parks, the largest one being on northern Ellesmere Island (Table 1.8). In addition, the Polar Bear Park Reserve on Bathurst Island and the Cape Henrietta-Marie and Cape Churchill Wildlife Area Wilderness on Hudson Bay provide protection for a diversity of wildlife and arctic vegetation. There are at least 11 migratory bird sanctuaries, mostly in the eastern arctic. Additional wildlife sanctuaries are the Thelon Game Sanctuary and the Hershel Island Territorial Park (Anon. 1994).

The entire northeastern part of Greenland has been set aside as the Northeast Greenland National Park. This is the largest arctic reserve to protect vegetation, rugged mountains with their icecap and glaciers, and populations of musk-ox, caribou, wolves, and numerous terrestrial and marine bird species.

AREAS FOR FUTURE RESEARCH

Plant communities within the Low Arctic have been well studied, with the exception of those in the Brooks Range, the mountains of the northern Yukon, and the barren lands of the Keewatin District and the northeastern Mackenzie District, N.W.T. Shrub communities of riparian habitats deserve more emphasis in terms of successional patterns, nutrient relationships, and their roles as wildlife habitats. This is as true now as it was a decade ago in our first edition.

There have been few detailed studies of plant communities in the High Arctic, with vast areas having received little botanical research. The historical, environmental, and biological reasons for the barrenness of the polar deserts and the limited productivity of the semideserts need detailed study.

In recent years more emphasis has been placed on nutrient budgets and nutrient cycling within

Table 1.8. *Arctic national parks and preserves in North America: their size and major features*

Region and name	Size (ha)	Features
Canada		
Aulavik, Banks Island	1,227,500	Deep valleys, musk-ox, Peary caribou
Auyuittuq, Baffin Island	2,147,110	Massive icecap, fiords, birds
Ellesmere Island	3,950,000	Icecap, fiords, wildlife
Ivvavik, Yukon Territory	1,016,840	Caribou migration, wildlife
North Baffin, Baffin Island	2,220,000	Vegetation, wildlife, fiords
Polar Bear Pass, Bathurst Island	262,400	Vegetation, wildlife
Thelon Game Sanctuary	2,396,000	Wildlife, vegetation
Queen Maud Gulf	6,278,200	Migratory birds, vegetation
Greenland		
Northeast Greenland	97,200,000	Icecap, fiords, wildlife, vegetation
Melville Bay	1,050,000	Fiords, wildlife
Alaska		
Gates of the Arctic	2,458,700	Mountains, vegetation, wildlife
Katmai	1,384,000	Volcano, wildlife
Kobuk Valley	275,400	Vegetation, wildlife, archaeology
Denali	2,023,428	Mountains, vegetation, wildlife
Bearing Land Bridge	1,127,033	Vegetation, wildlife, waterfowl
Noatak	2,660,598	Wildlife, vegetation, archaeology

Source: World Book Encyclopedia (1995).

various plant communities. Further detailed studies on nutrient cycling and the role of mycorrhizae are needed if we are to better understand the implications of climate change in the Arctic. Studies on fluxes of carbon dioxide and methane from the vast areas of peats are under way but will no doubt intensify in the near future, as will more detailed studies on the implications of increased CO_2 and temperature as they relate to mineralization of peats.

Little attention has been given to species richness and species diversity, but because of an increased interest in conservation, more studies are needed. Finally, should the huge coal reserves in Alaska and petroleum reserves in western Canada be exploited, there will be need for a second generation of expertise on rehabilitation of these lands.

REFERENCES

Adam, K. M., and H. Hernandez. 1977. Snow and ice roads: ability to support traffic and effects on vegetation. Arctic 30:13–27.

Addison, P. A., and L. C. Bliss. 1984. Adaptations of *Luzula confusa* to the polar semi-desert environment. Arctic 37:21–132.

Ager, T. A. 1982. Vegetational history of western Alaska during the Wisconsin Glacial Interval and the Holocene, pp. 75–93 in D. M. Hopkins, J. V. Matthews, C. E. Schweger, and S. B. Young (eds.), Paleoecology of Beringia. Academic Press, New York.

Aleksandrova, V. D. 1980. The Arctic and Antarctic: their division into geobotanical areas. Cambridge University Press, Cambridge.

Alexander, V., and K. Van Cleve. 1983. The Alaska pipeline: a success story. Ann. Rev. Ecol. Syst. 14: 443–463.

Anderson, P. M., and L. B. Brubaker. 1986. Modern pollen assemblages from northern Alaska. Rev. Paleobot. Palynol. 46:273–291.

Anderson, P. M., and L. B. Brubaker. 1994. Vegetation history of northcentral Alaska: a mapped summary of Late-Quaternary pollen data. Quater. Sci. Rev. 13: 71–92.

Anderson, P. M., R. E. Reanier, and L. B. Brubaker. 1988. Late Quaternary vegetational history of the Black River region in northeastern Alaska. Can. Jour. Earth Sci. 25:84–94.

Andreyev, V. N., and V. D. Aleksandrova. 1981. Geobotanical division of the Soviet arctic, pp. 25–37 in L. C. Bliss, D. W. Heal, and J. J. Moore (eds.), Tundra ecosystems: a comparative analysis. Cambridge University Press, Cambridge.

Anon. 1994. The state of protected areas in the circumpolar arctic. CAFF Habitat Conservation Rept.#1. CAFF International Secretariat, Can. Wildlife Serv., Ottawa.

Babb, T. A., and L. C. Bliss. 1974. Effects of physical disturbance on arctic vegetation in the Queen Elizabeth Islands. J. Appl. Ecol. 11:549–562.

Babb, T. A., and D. W. A. Whitefield. 1977. Mineral nutrient cycling and limitation of plant growth in the Truelove Lowland ecosystem, pp. 587–606 in L. C. Bliss (ed.), Truelove Lowland, Devon Island, Canada: a high arctic ecosystem. University of Alberta Press, Edmonton.

Barbour, M. G., and N. L. Christensen. 1993. Vegetation, pp. 97–131 in Flora of North America Editorial Committee, Flora of North America, Vol. 1. Oxford Univ. Press, New York, New York.

Batten, D. S., and J. Svoboda. 1994. Plant communities on the uplands in the vicinity of the Alexandra

Fiord lowland, pp. 97–110 in J. Svoboda and B. Freedman (eds.), Ecology of a polar oasis, Alexandra Fiord, Ellesmere Island, Canada. Captus Univ. Pub., Toronto.

Batzli, G. O. 1983. Responses of arctic rodent populations to nutritional factors. Oikos 40:396–406.

Batzli, G. O., and H. G. Jung. 1980. Nutritional ecology of microtene rodents: resource utilization near Atkasook, Alaska. Arct. Alp. Res. 12:483–499.

Bazely, D. R., and R. L. Jefferies. 1986a. Changes in the composition and standing crop of salt marsh communities in response to removal of a grazer. J. Ecol. 74:693–706.

Bazely, D. R., and R. L. Jefferies. 1986b. Goose faeces: a source of nitrogen for plant growth in a grazed salt marsh. J. Appl. Ecol. 22:693–703.

Bell, K. L., and L. C. Bliss. 1977. Overwinter phenology of plants in a polar semi-desert. Arctic 30:118–121.

Bell, K. L., and L. C. Bliss. 1978. Root growth in a polar semidesert environment. Can. J. Bot., 56:2470–2490.

Bell, K. L., and L. C. Bliss. 1980. Plant reproduction in a high arctic environment. Arct. Alp. Res. 12:1–10.

Beschel, R. E. 1970. The diversity of tundra vegetation, pp. 85–92 in W. A. Fuller and P. G. Kevan (eds.), Productivity and conservation in northern circumpolar lands. IUCN new series 16. Morges, Switzerland.

Billings, W. D. 1974. Adaptations and origins of alpine plants. Arct. Alp. Res. 6:129–142.

Billings, W. D., and H. A. Mooney. 1968. The ecology of arctic and alpine plants. Biol. Rev. 43:481–530.

Billings, W. D., and K. M. Peterson. 1980. Vegetation change and ice-wedge polygons through the thaw-lake cycle in arctic Alaska. Arct. Alp. Res. 12:413–432.

Billings, W. D., and K. M. Peterson. 1992. Some possible effects of climate warming on arctic tundra ecosystems of the Alaskan North Slope, pp. 233–243 in R. L. Peters and T. E. Lovejoy (eds.), Global warming and biological diversity. Yale University Press, New Haven.

Billings, W. D., K. M. Peterson, and G. R. Shaver. 1978. Growth, turnover and respiration rates of roots and tillers in tundra graminoids, pp. 415–434 in L. L. Tieszen (ed.), Vegetation and production ecology of an Alaskan arctic tundra. Ecological Studies, vol. 29. Springer-Verlag, New York.

Bird, C. D. 1975. The lichen, bryophyte, and vascular plant flora and vegetation of the Landing Lake area, Prince Patrick Island, Arctic Canada. Can. J. Bot. 53:719–744.

Black, R. A., and L. C. Bliss. 1978. Recovery sequence of Picea mariana–Vaccinium uliginosum forests after fire near Inuvik, Northwest Territories, Canada. Can. J. Bot. 56:2020–2030.

Black, R. A., and L. C. Bliss. 1980. Reproductive ecology of Picea mariana (Mill.) BSP, at tree line near Inuvik, Northwest Territories, Canada. Ecol. Monogr. 50:331–354.

Black, R. F. 1969. Thaw depressions and thaw lakes – a review. Builetyn Peryglacjalny 19:131–150.

Bledsoe, C., P. Klein, and L. C. Bliss. 1990. A survey of mycorrhizal plants on Truelove Lowland, Devon Island, N.W.T., Canada. Can. J. Bot. 68:1848–1856.

Bliss, L. C. 1956. A comparison of plant development in

microenvironments of arctic and alpine tundras. Ecol. Monogr. 26:303–337.

Bliss, L. C. 1971. Arctic and alpine plant life cycles. Ann. Rev. Ecol. Syst. 2:405–438.

Bliss, L. C. 1975. Tundra grasslands, herblands, and shrublands and the role of herbivores. Geoscience and Man 10:51–79.

Bliss, L. C. 1977. General summary, Truelove Lowland ecosystems, pp. 657–675 in L. C. Bliss (ed.), Truelove Lowland, Devon Island, Canada: a High Arctic ecosystem. University of Alberta Press, Edmonton.

Bliss, L. C. 1981. North American and Scandinavian tundras and polar deserts, pp. 8–24 in L. C. Bliss, O. W. Heal, and J. J. Moore (eds.), Tundra ecosystems: a comparative analysis. Cambridge University Press, Cambridge.

Bliss, L. C. 1990. Arctic ecosystems: Patterns of change in response to disturbance, pp. 347–366 in G. M. Woodwell (ed.), The earth in transition: patterns and and processes of biotic impoverishment. Cambridge University Press, Cambridge.

Bliss, L. C., and J. E. Cantlon. 1957. Succession on river alluvium in northern Alaska. Amer. Midl. Nat. 58:452–569.

Bliss, L. C., and W. G. Gold. 1994. The patterning of plant communities and edaphic factors along a high arctic coastline: implications for succession. Can. J. Bot. 72:1095–1107.

Bliss, L. C., and N. E. Grulke. 1988. Revegetation in the High Arctic: its role in reclamation of surface disturbance, pp. 43–55 in P. Kershaw (ed.), Northern environmental disturbances. Boreal Inst. Northern Studies, University of Alberta Press, Edmonton.

Bliss, L. C., G. H. R. Henry, J. Svoboda, and D. I. Bliss. 1994. Patterns of plant distribution within two polar desert landscapes. Arct. Alp. Res. 26:46–55.

Bliss, L. C., J. Kerik, and W. Peterson. 1977. Primary production of dwarf shrub heath communities, Truelove Lowland, pp. 217–244 in L. C. Bliss (ed.), Truelove Lowland, Devon Island, Canada: a High Arctic ecosystem. University of Alberta Press, Edmonton.

Bliss, L. C., and E. B. Peterson. 1973. The ecological impact of northern petroleum development, pp. 505–537 in J. Malaurie (ed.), Arctic oil and gas problems and possibilities. Contributions du Centre D'etudes Arctiques XII, Paris.

Bliss, L. C., and J. Svoboda. 1984. Plant communities and plant production in the Western Queen Elizabeth Islands. Holarctic Ecol. 7:325–344.

Bliss, L. C., J. Svoboda, and D. I. Bliss. 1984. Polar deserts, their plant cover and plant production in the Canadian High Arctic. Holarctic Ecol. 7:305–324.

Bliss, L. C., and R. W. Wein. 1972. Plant community responses to disturbances in the western Canadian Arctic. Can. J. Bot. 50:1097–1109.

Böcher, T. W. 1954. Oceanic and continental vegetational complexes in southwest Greenland. Medd. om Grønland 148(1):1–336.

Böcher, T. W. 1959. Floristic and ecological studies in Middle West Greenland. Medd. om Grønland 156(5):1–68.

Böcher, T. W. 1963. Phytogeography of Middle West Greenland. Medd. om Grønland 148(3):1–289.

Böcher, T. W., and S. Laegaard. 1962. Botanical studies along the Arfersiorfik Fiord, west Greenland. Botanisk Tidsskrift 58:168–190.

Brassard, G. R., and R. E. Longton. 1970. The flora and

vegetation of Van Huen Pass, northwestern Ellesmere Island. Can. Field-Nat. 84:357–364.

Britton, M. E. 1957. Vegetation of the arctic tundra, pp. 26–61 in H. P. Hansen (ed.), Arctic biology. Oregon State University Press, Corvallis.

Bryson, R. A. 1966. Airmasses, streamlines and the boreal forest. Geogr. Bull. 8:228–269.

Chapin, D. M., and C. S. Bledsoe. 1992. Nitrogen fixation in arctic plant communities, pp. 301–319 in F. S. Chapin III, R. L. Jefferies, J. F. Reynolds, G. R. Shaver, and J. Svoboda (eds.), Arctic ecosystems in a changing climate: an ecological perspective. Academic Press, New York.

Chapin, F. S., III et al. (eds.). 1992. Arctic ecosystems in a changing climate: an ecological perspective. Academic Press, New York.

Chapin, D. M., L. C. Bliss, and L. J. Bledsoe. 1991. Environmental regulation of nitrogen fixation in a high arctic lowland ecosystem. Can. J. Bot. 69:2744–2755.

Chapin, F. S. III, and A. J. Bloom. 1976. Phosphate absorption: adaptation of tundra graminoids to a low-temperature, low-phosphorus environment. Oikos 26:111–121.

Chapin, F. S. III, and M. C. Chapin. 1981. Revegetation of an arctic disturbed site by native tundra species. J. Appl. Ecol. 17:449–456.

Chapin, F. S. III, L. Moilanen, and K. Kielland. 1992. Preferential use of organic nitrogen for growth by a nonmycorrhizal arctic sedge. Nature 631:150–153.

Chapin, F. S. III, and G. R. Shaver. 1985a. Arctic, pp. 16–40 in B. F. Chabot and H. A. Mooney (ed.), Physiological ecology of North America plant communities. Chapman & Hall, New York.

Chapin, F. S. III, and G. R. Shaver. 1985b. Individualistic growth response of tundra plant species to environmental manipulations in the field. Ecology 66:564–576.

Chapin, F. S. III, G. R. Shaver, and R. A. Kedrowski. 1986. Environmental controls over carbon, nitrogen, and phosphorus chemical fractions in *Eriophorum vaginatum* L. in Alaskan tussock tundra. J. Ecol. 74:167–195.

Chapin, F. S. III, and P. R. Tryon. 1982. Phosphate absorption and root respiration of different plant growth forms from northern Alaska. Holarctic Ecol. 5:164–171.

Chapin, F. S. III, K. Van Cleve, and M. C. Chapin. 1979. Soil temperature and nutrient cycling in the tussock growth form of *Eriophorum vaginatum*. J. Ecol. 67:169–189.

Chapin, F. S. III, K. Van Cleve, and L. L. Tieszen. 1975. Seasonal nutrient dynamics of tundra vegetation at Barrow, Alaska. Arct. Alp. Res. 7:209–226.

Chapman, W. L., and J. E. Walsh. 1993. Recent variations of sea ice and air temperature in high latitudes. Bull. Amer. Met. Soc. 74:33–47.

Chester, A. L., and G. R. Shaver. 1982. Reproductive effort in cottongrass tussock tundra. Holarctic Ecol. 5:200–206.

Churchill, E. D. 1955. Phytosocialogical and environmental characteristics of some plant communities in the Umiat region of Alaska. Ecology, 36:606–627.

Colinvaux, P. A. 1964. The environment of the Bering Land Bridge. Ecol. Monogr. 34:297–329.

Colinvaux, P. A. 1967. Quaternary vegetation history of arctic Alaska, pp. 207–231 in D. M. Hopkins (ed.), The Bering land bridge. Stanford University Press, Stanford, Cal.

Corns, I. G. W. 1974. Arctic plant communities east of the Mackenzie Delta. Can. J. Bot. 52:1730–1745.

Cwynar, L. C., and J. C. Ritchie. 1980. Arctic steppe-tundra: A Yukon perspective. Science 208:1375–1377.

Cwynar, L. C., and R. W. Spear. 1991. Reversion of forest to tundra in the central Yukon. Ecology 72:202–212.

Daniels, F. J. A. 1982. Vegetation of the Angmagssalik District, Southeast Greenland. IV. Shrub, dwarf shrub and terricolous lichens. Medd. om Grønland, Biosci. 10:1–78.

Dawson, T. E. 1987. Comparative ecophysiological adaptations in arctic and alpine populations of a deciduous shrub, *Salix arctica*. Ph.D. Dissertation, University of Washington, Seattle.

Dawson, T. E., and L. C. Bliss. 1989a. Patterns of water use and tissue water relations in the dioecious shrub, *Salix arctica*: the physiological basis for habitat partitioning between the sexes. Oecologia 79:332–343.

Dawson, T. E., and L. C. Bliss. 1989b. Interspecific variation in the water relations of *Salix arctica*, an arctic-alpine dwarf willow. Oecologia 79:322–331.

Dawson, T. E., and L. C. Bliss. 1993. Plants as mosaics: leaf-, ramet-, and gender-level variation in the physiology of the dwarf willow, *Salix arctica*. Functional Ecol. 7:293–304.

Densmore, R. V. 1992. Succession on an Alaskan disturbance with and without assisted revegetation of grass. Arct. Alp. Res. 24:238–243.

Densmore, R. V., B. J. Nieland, J. C. Zasada, and M. A. Masters. 1987. Planting willow for moose habitat restoration on the North Slope of Alaska, USA. Arct. Alp. Res. 19:537–543.

Drew, J. V., and R. E. Shanks. 1965. Landscape relationships of soils and vegetation in forest tundra, Upper Firth River Valley, Alaska-Canada. Ecol. Monogr. 35:285–306.

Edlund, S. A. 1980. Vegetation of Lougheed Island, District of Franklin. Geol. Surv. Can. 80-1A:329–333.

Felix, N. A., M. K. Reynolds, J. C. Jorgenson, and K. E. DuBois. 1992. Resistance and resilience of tundra plant communities to disturbance of winter seismic vehicles. Arct. Alp. Res. 24:69–77.

Fetcher, N., T. F. Beatly, B. Mullinax, and D. S. Winkler. 1984. Changes in arctic tussocks tundra thirteen years after fire. Ecology 65:1332–1333.

Forbes, B. C. 1992. Tundra disturbance studies. I. Long-term effects of vehicles on species richness and biomass. Environ. Conserv. 19:48–58.

Forbes, B. C. 1994. The importance of bryophytes in the classification of human-disturbed high arctic vegetation. J. Veg. Sci. 5:877–884.

Foster, J. L. 1989. The significance of the late snow disappearance on the arctic tundra as a possible indicator of climate change. Arct. Alp. Res. 21:60–70.

Fox, J. R. 1983. Germinable seed banks of interior Alaska tundra. Arct. Alp. Res. 15:405–411.

Freedman, B., N. Hill, J. Svoboda, and G. Henry. 1982. Seed banks and seedling occurrence in a high arctic oasis at Alexandra Fiord, Ellesmere Island, Canada. Can. J. Bot. 60:2112–2118.

Freedman, B., J. Svoboda, C. Labine, M. Muc, G. Henry, M. Nams, J. Stewart, and E. Woodley. 1983. Physical and ecological characteristics of Alexandra

Fiord, a high arctic oasis on Ellesmere Island, Canada, pp. 301–304 in Permafrost: Fourth international conference proceedings. National Academy Press, Washington, D.C.

Freedman, W., and T. C. Hutchinson. 1976. Physical and biological effects of experimental crude oil spills on Low Arctic tundra in the vicinity of Tuktoyaktuk, NWT, Canada. Can. J. Bot. 54:2219–2230.

Gersper, P. L., and J. L. Challinor. 1975. Vehicle perturbation effects upon a tundra soil-plant system: 1. Effects on morphological and physical environmental properties of the soils. Soil Sci. Soc. Amer. Proc. 39:737–744.

Gill, D. 1973. Ecological modifications caused by the removal of tree and shrub canopies in the Mackenzie Delta. Arctic 26:95–111.

Grayson, D. K. 1991. Late Pleistocene mammalian extinctions in North America: taxonomy, chronology, and explanations. Jour. World Prehistory 5:193–231.

Grulke, N. E., and L. C. Bliss. 1988. Comparative life history characteristics of two high arctic grasses, Northwest Territories. Ecology 69:484–496.

Guthrie, R. D. 1972. Mammals of the mammoth steppe as paleoenvironmental indicators, pp. 307–326 in D. H. Hopkins, J. V. Matthews, C. E. Schweger, and S. B. Young (eds.), Paleoecology of Beringia. Academic Press, New York.

Hansen, K. 1969. Analyses of soil profiles in dwarf-shrub vegetation in south Greenland. Medd. om Grønland 178(5):1–33.

Hanson, H. C. 1953. Vegetation types in northwestern Alaska and comparisons with communities in other arctic regions. Ecology 34:111–140.

Hare, F. K. 1968. The Arctic. Q. J. Royal Meteor. Soc. 94: 439–459.

Hare, F. K., and J. C. Ritchie. 1972. The boreal bioclimates. Geogr. Rev. 62:333–365.

Hernandez, H. 1973. Natural plant recolonization of surficial disturbances, Tuktoyaktuk Peninsula region, Northwest Territories. Can. J. Bot. 51:2177–2196.

Hettinger, L., A. Janz, and R. W. Wein. 1973. Vegetation of the northern Yukon Territory. Arctic Gas Biol. Rept. Serv. Vol. 1, Canadian Arctic Gas Study Ltd., Calgary.

Hills, L. V., J. E. Klovan, and A. R. Sweet. 1974. *Juglans eocinerea* n. sp., Beaufort Formation (Tertiary), southwestern Banks Island, Arctic Canada. Can. J. Bot. 52:65–90.

Hobbie, J. E. et al. 1994. Long-term measurements at the arctic LTER site, pp. 391–409 in T. M. Powell and J. H. Steele (eds.), Ecological time series. Springer-Verlag, New York.

Hoffman, R. S., and R. D. Taber. 1967. Origin and history of holarctic tundra ecosystems, with special reference to vertebrate faunas, pp. 143–170 in W. H. Osburn and H. E. Wright, Jr. (eds.), Arctic and alpine environments. Indiana University Press, Bloomington.

Holmen, K. 1957. The vascular plants of Peary Land, north Greenland. Medd. om Grønland 124(9):1–149.

Hopkins, D. M., J. V. Matthews, J. A. Wolfe, and M. L. Silberman. 1971. A Pliocene flora and insect fauna from the Bering Strait region. Palaeogeogr. Palaeoclimatol. Palaeoecol. 9:211–213.

Hopkins, D. M., J. V. Matthews, C. E. Schwegner, and

S. B. Young (eds.). 1982. Paleoecology of Beringia. Academic Press, New York.

Jefferies, R. L. 1977. The vegetation of salt marshes at some coastal sites in arctic North America. J. Ecol. 65:661–672.

Jefferies, R. L. 1988. Pattern and process in arctic coastal vegetation in response to foraging by lesser snow geese, pp. 281–300 in M. J. A. Werger, J. J. M. van der Aart, H. J. During, and J. T. A. Verhoeven (eds.), Plant form and vegetation structure, adaptation, plasticity, and relation to herbivory. SPB. Academic Press, The Hague.

Jefferies, R. L., D. R. Klein, and G. R. Shaver, 1994. Vertebrate herbivores and northern plant communities: reciprocal influences and responses. Oikos 71:193–206.

Johnson, A. W., L. A. Viereck, R. E. Johnson, and H. Melchior. 1966. Vegetation and flora, pp. 277–354 in N. J. Wilimovsky and J. N. Wolfe (eds.), Environment of the Cape Thompson region, Alaska. U.S.A.E.C., Div. Tech. Info., Washington, D.C.

Johnson, D. A., and L. L. Tieszen. 1976. Aboveground biomass allocation, leaf growth, and photosynthesis patterns in tundra plant forms in Arctic Alaska. Oecologia 24:159–173.

Johnson, L. 1987. Management of northern gravel sites for successful reclamation: a review. Arct. Alp. Res. 19:530–536.

Kelley, P. M., and P. D. Jones. 1981. Annual temperatures in the arctic, 1881–1981. Climate Monitor 10: 122–124.

Kershaw, K. A. 1976. The vegetation zonation of the East Pen Island salt marshes, Hudson Bay. Can. J. Bot. 54:5–13.

Kevan, P. 1972. Insect pollination of high arctic flowers. J. Ecol. 60:831–847.

Kielland, K. 1994. Amino acid absorption by arctic plants: implications for plant nutrition and nitrogen cycling. Ecology 75:2373–2383.

Kielland, K., and F. S. Chapin, III. 1992. Nutrient absorption and accumulation in arctic plants, pp. 321–335 in F. S. Chapin, III, R. L. Jefferies, J. F. Reynolds, G. R. Shaver, and J. Svoboda (eds.), Arctic ecosystems in a changing climate, an ecophysiological perspective. Academic Press, New York.

Kohn, L. M., and E. Stasovski. 1990. The mycorrhizal status of plants at Alexandra Fiord, Ellesmere Island, Canada, a high arctic site. Mycologia 82:23–35.

Komarkova, V., and P. J. Webber. 1980. Two low arctic vegetation maps near Atkasook, Alaska. Arct. Alp. Res. 12:447–472.

Kuc, M. 1974. Noteworthy vascular plants collected in southwestern Banks Island, N.W.T. Arctic 26:146–150.

Labine, C. 1994. Meteorology and climatology of the Alexandra Fiord lowland, pp. 23–39 in J. Svoboda and B. Freedman (eds.), Ecology of a polar oasis, Alexandra Fiord, Ellesmere Island, Canada. Captus Univ. Pub., Toronto.

Lachenbruck, A. H., and B. V. Marshall. 1986. Changing climate: geothermal evidence from permafrost in the Alaskan Arctic. Science 234:689–696.

Larsen, J. A. 1965. The vegetation of the Ennadai Lake area, N.W.T.: studies in subarctic and arctic bioclimatology. Ecol. Monogr. 35:37–59.

Larsen, J. A. 1971. Vegetation of Fort Reliance, Northwest Territories. Can. Field-Nat. 85:147–178.

Larsen, J. A. 1972. The vegetation of northern Keewatin. Can. Field-Nat. 86:45–72.

Leck, M. A. 1980. Germination in Barrow, Alaska tundra soil cores. Arct. Alp. Res. 12:373–349.

Limbach, W. E., W. C. Oechel, and W. Lowell. 1982. Photosynthetic and respiratory responses to temperature and light of three tundra growth forms. Holarctic Ecol. 5:150–157.

Linkens, A. E., and N. Fetcher. 1983. Effect of surface-applied Prudhoe Bay crude oil on vegetation and soil processes in tussock tundra, pp. 723–728 in Permafrost: Fourth International Conf. Proc. National Academy Press. Washington, D.C.

Löve, A., and D. Löve. 1974. Origin and evaluation of the arctic and alpine floras, pp. 571–603 in J. D. Ives and R. G. Barry (eds.), Arctic and alpine environments. Methuen, London.

Lowe, J. J., A. Ammann, H. H. Birks, S. Bjorck, G. R. Coope, L. Cwynar, J. L. de Beaulieu, R. J. Mott, D. M. Petect, and M. J. C. Walker. 1994. Climatic changes in areas adjacent to the North Atlantic during the last glacial-interglacial transition (14–9 ka BP): a contribution to IGCP-253. J. Quat. Sci. 9:185–198.

McCown, B. H. 1978. The interactions of organic nutrients, soil nitrogen, and plant growth and survival in the arctic environment, pp. 435–456 in L. L. Tieszen (ed.), Vegetation and production ecology of an Alaskan Arctic tundra. Springer-Verlag, New York.

McGraw, J. B. 1980. Seed bank size and distribution of seeds in cottongrass tussock tundra, Eagle Creek, Alaska. Can. J. Bot. 58:1607–1611.

McGraw, J. B., and G. L. Shaver. 1982. Seedling density and seedling survival in Alaskan cottongrass tussock tundra. Holarctic Ecol. 5:212–217.

McGraw, J. B., and M. C. Vavrek. 1989. The role of buried viable seeds in arctic and alpine plant communities, pp. 91–106 in M. A. Leck, V. T. Parker, R. L. Simpson (eds.), Ecology of soil seed banks. Academic Press, New York.

MacLean, D. A., S. J. Woodley, M. G. Weber, and R. W. Wein. 1983. Fire and nutrient cycling, pp. 111–132 in R. W. Wein and D. A. MacLean (eds.), The role of fire in northern circumpolar ecosystems. Wiley, New York.

MacLean, S. F., and T. J. Jensen. 1984. Food-plant selection by insect herbivores: the role of plant growth form. Oikos 41:211–221.

Maikawa, E., and K. A. Kershaw. 1976. Studies on lichen-dominated systems. XIX. The postfire recovery sequence of black spruce–lichen woodland in the Abitan Lake Region, N.W.T. Can. J. Bot. 54:2679–2689.

Mark, A. F., N. Fetcher, G. R. Shaver, and F. S. Chapin. 1985. Estimated ages of mature tussocks of Eriophorum vaginatum along a latitudinal gradient in central Alaska, U.S.A. Arct. Alp. Res. 17:1–5.

Martin, P. S. 1974. Paleolithic plays on the American stage, pp. 669–700 in J. D. Ives and R. G. Barry (eds.), Arctic and alpine environments. Methuen, London.

Martin, P. S. 1982. The pattern and meaning of Holarctic mammoth extinction, pp. 399–408 in D. M. Hopkins, J. V. Matthews, C. E. Schweger, and S. B. Young (eds.), Paleoecology of Beringia. Academic Press, New York.

Matthews, J. V. 1974. A preliminary list of insect fossils from the Beaufort Formation, Meighen Island, District of Franklin. Geol. Surv. Can. 74–1A:203–206.

Maxwell, J. B. 1992. Arctic climate: potential for change under climate warming, pp. 11–34 in F. S. Chapin III, R. L. Jefferies, J. F. Reynolds, G. R. Shaver, and J. Svoboda (eds), Arctic ecosystems in a changing climate: an ecophysiological perspective. Academic Press, New York.

Maxwell, J. B. 1995. Recent climate patterns in the Arctic, in W. C. Oechel, T. Callaghan, T. Gilmanov, J. I. Molten, B. Maxwell, U. Molau, and B. Sveinbjornsson (eds.), Global change and Arctic terrestrial ecosystems. Springer-Verlag, New York.

Maxwell, J. B., and L. A. Barrie. 1989. Atmospheric and climatic change in the Arctic and Antarctic. Ambio 18:42–49.

Mayo, J. M., A. P. Hartgerink, D. G. Despain, D. G. Thompson, R. G. Thompson, E. M. van Zinderin Bakker, and S. D. Nelson. 1977. Gas exchange studies of Carex and Dryas, Truelove Lowland, Devon Island, Canada: a high arctic ecosystem. University of Alberta Press, Edmonton.

Miller, O. K. 1982. Mycorrhizae, mycorrhizal fungi and fungal biomass in subalpine tundra at Eagle Summit, Alaska. Holarctic Ecol. 5:125–134.

Miller, P. C., P. M. Miller, M. Blake-Jacobson, F. S. Chapin, K. R. Everett, D. W. Hilbert, J. Kummerow, A. E. Linkins, G. M. Marion, W. C. Oechel, S. W. Roberts, and L. Stuart. 1984. Plant–soil processes in Eriophorum vaginatum tussock tundra in Alaska: systems modeling approach. Ecol. Monogr. 54:361–405.

Muc, M. 1977. Ecology and primary production of the Truelove Lowland sedge-moss meadow communities, pp. 157–184 in L. C. Bliss (ed.), Truelove Lowland, Devon Island, Canada: a high arctic ecosystem. University of Alberta Press, Edmonton.

Muc, M., B. Freedman, and J. Svoboda. 1989. Vascular plant communities of a polar oasis at Alexandra Fiord, Ellesmere Island. Can. J. Bot. 67:1126–1136.

Murray, C., and P. C. Miller. 1982. Phenological observations of major plant growth forms and species in montane and Eriophorum vaginatum tussock tundra in central Alaska. Holarctic Ecol. 5:109–116.

Nadelhoffer, K. J., A. E. Giblin, G. R. Shaver, and A. E. Linkens. 1992. Microbial processes and plant nutrient availability in Arctic soils, pp. 281–300 in F. S. Chapin III, R. L. Jefferies, J. F. Reynolds, G. R. Shaver, and J. Svoboda (eds.), Arctic ecosystems in a changing climate: an ecological perspective. Academic Press, New York.

Oberbauer, S. F., and T. E. Dawson. 1992. Water relations of arctic vascular plants, pp. 259–279 in F. S. Chapin III, R. L. Jefferies, J. F. Reynolds, G. R. Shaver, and J. Svoboda (eds.), Arctic ecosystems in a changing climate: an ecological perspective. Academic Press, New York.

Oechel, W. C., and B. Sveinbjörnsson. 1978. Primary production processes in arctic bryophytes at Barrow, Alaska, pp. 269–298 in L. L. Tieszen (ed.), Vegetation and production ecology of an Alaskan arctic tundra. Springer-Verlag, New York.

Oechel, W. C., G. L. Vourlitis, S. J. Hastings, and S. A. Bochkarev. 1995. Change in Arctic CO_2 flux over two decades: effects of climate change at Barrow, Alaska. Ecol. Applic. 5:846–855.

Oosting, H. J. 1948. Ecological notes on the flora of east

Greenland and Jan Jayen, pp. 225–269 in L. A. Boyd et al. (eds.), The coast of northeast Greenland. American Geographical Society special publication 30, Washington, D.C.

Packer, J. G. 1974. Differentiation and dispersal in alpine floras. Arct. Alp. Res. 6:117–128.

Peterson, K. M. 1978. Vegetational successions and other ecosystem changes in two arctic tundras. Ph.D. dissertation, Duke University, Durham.

Peterson, K. M., and W. D. Billings. 1978. Geomorphic processes and vegetational change along the Meade River sand bluffs, northern Alaska. Arctic 31:7–23

Peterson, K. M., and W. D. Billings. 1980. Tundra vegetational patterns and succession in relation to microtopography near Atkasook, Alaska. Arct. Alp. Res. 12:473–482.

Polunin, N. 1948. Botany of the Canadian eastern Arctic. Part III. Vegetation and ecology. National Museum of Canada Bulletin 104, Ottawa.

Racine, C. H. 1981. Tundra fire effects on soils and three plant communities along a hill-slope gradient in the Seward Peninsula, Alaska. Arctic 34:71–85.

Racine, C. H. 1994. Long-term recovery of vegetation on two experimental crude oil spills in interior Alaska black spruce taiga. Can. J. Bot. 72:1171–1177.

Racine, C. H., L. A. Johnson, and L. A. Viereck. 1987. Patterns of vegetation recovery after tundra fires in northwestern Alaska, U.S.A. Arct. Alp. Res. 19:461–469.

Reznicek, S. A., and J. Svoboda. 1982. Tundra communities along a microenvironmental gradient at Coral Harbour, Southampton Island, N.W.T. Naturaliste Canadien 109:585–595.

Ritchie, J. C. 1977. The modern and Late Quaternary vegetation of the Campbell-Dolomite uplands near Inuvik, N.W.T., Canada. Ecol. Monogr. 47:401–423.

Ritchie, J. C. 1982. The modern and Late-Quaternary vegetation of the Doll Creek area, northern Yukon, Canada. New Phytol. 90:563–603.

Ritchie, J. C. 1984. Past and present vegetation of the far northwest of Canada. Toronto University Press, Toronto.

Ritchie, J. C. 1987. Postglacial vegetation of Canada. Cambridge University Press, Cambridge.

Ritchie, J. C., and L. C. Cwynar. 1982. The Late Quaternary vegetation of the North Yukon, pp. 113–126 in D. M. Hopkins, J. V. Matthews, C. E. Schweger, and S. B. Young (eds.), Paleoecology of Beringia. Academic Press, New York.

Ritchie, J. C., L. C. Cwynar, and R. W. Spear. 1983. Evidence from northwest Canada for an early Holocene Milankovitch thermal maximum. Nature. 305:126–128.

Ritchie, J. C., and F. K. Hare. 1971. Late Quaternary vegetation and climate near the arctic tree line of northwestern North America. Quat. Res. 1:331–342.

Roach, D. A. 1983. Buried seed and standing vegetation in two adjacent tundra habitats, northern Alaska. Oecologia 60:359–364.

Savile, D. B. O. 1961. The botany of the northwestern Queen Elizabeth Islands. Can. J. Bot. 39:909–942.

Savile, D. B. O. 1972. Arctic adaptations in plants. Canadian Department of Agriculture Monograph 6, Ottawa.

Seidenfaden, G., and T. Sørensen. 1937. The vascular plants of northeast Greenland from 74°30' to 79°00' N. Lat. Medd. om Grønland 101(4):1–215.

Schaefer, J. A., and F. Messier. 1994. Composition and spatial structure of plant communities on southeastern Victoria Island, arctic Canada. Can. J. Bot. 72:1264–1272.

Schweger, C. E. 1982. Late Pleistocene vegetation of Eastern Beringia: Pollen analysis of dated alluvium, pp. 95–112 in D. M. Hopkins, J. V. Matthews, C. E. Schweger, and S. B. Young (eds.), Paleoecology of Beringia. Academic Press, New York.

Shaver, G. R., and W. D. Billings. 1975. Root production and root turnover in a west tundra ecosystem, Barrow, Alaska. Ecology, 56:401–409.

Shaver, G. R., and W. D. Billings. 1977. Effects of daylength and temperature on root elongation in tundra graminoids. Oecologia 28:57–65.

Shaver, G. R., and F. S. Chapin, III. 1980. Response to fertilization by various plant growth forms in an Alaskan tundra: Nutrient accumulation and growth. Ecology 61:662–675.

Shaver, G. R., and F. S. Chapin, III. 1986. Effect of NPK fertilization on production and biomass of Alaskan tussock tundra. Arct. Alp. Res. 18:261–268.

Shaver, G. R., and F. S. Chapin, III. 1991. Production: Biomass relationships and element cycling in contrasting arctic vegetation types. Ecol. Monogr. 61:1–31.

Shaver, G. R., B. L. Gartner, F. S. Chapin, and A. E. Linkins. 1983. Revegetation of arctic disturbed sites by native tundra plants, pp. 1133–1138 in Permafrost: Fourth international conference proceedings. National Academy Press, Washington, D.C.

Sheard, J. W., and D. W. Geale. 1983. Vegetation studies of Polar Bear Pass, Bathurst Island, N.W.T. I. Classification of plant communities. Can. J. Bot. 61:1618–1636.

Sohlberg, E., and L. C. Bliss. 1984. Microscale pattern of vascular plant distribution in two high arctic communities. Can. J. Bot., 62:2033–2042.

Sørensen, T. 1941. Temperature relations and phenology of the northeast Greenland flowering plants. Medd. om Grønland 125:1–305.

Sørensen, T. 1943. The flora of Melville Bugt. Medd. om Grønland 124(5):1–70.

Spetzman, L. A. 1959. Vegetation of the Arctic Slope of Alaska. U.S. Geological Survey professional paper 302-B. Washington, D.C.

Svoboda, J. 1977. Ecology and primary production of raised beach communities, Truelove Lowland, pp. 185–216 in L. C. Bliss (ed.), Truelove Lowland, Devon Island, Canada: a high arctic ecosystem. University of Alberta Press, Edmonton.

Svoboda, J. 1982. Due to the Little Ice Age climatic impact, most of the vegetation cover in the Canadian High Arctic is of recent origin. A hypothesis. Proceedings 33rd Alaskan Science Conference 33:206.

Svoboda, J., and G. H. R. Henry. 1987. Succession in marginal arctic environments. Arct. Alp. Res. 19:373–384.

Taylor, K. C. et al. 1993. The 'flickering switch' of late Pleistocene climate change. Nature 351:432–436.

Taylor, R. J. 1981. Shoreline vegetation of the arctic Alaskan coast. Arctic 34:37–42.

Tedrow, J. C. F. 1977. Soils of the polar landscape. Rutgers University Press, New Brunswick, N.J.

Thompson, D. C. 1980. A classification of the vegetation of Boothia Peninsula and the northern district of Keewatin, N.W.T. Arctic 33:73–99.

Tieszen, L. L. 1978. Photosynthesis in the principal Barrow, Alaska species: a summary of field and laboratory responses, pp. 247–268 in L. L. Tieszen (ed.), Vegetation and production ecology of an Alaskan arctic tundra. Springer-Verlag, New York.

Timoney, K. P. 1995. Tree and tundra cover anomalies in the subarctic forest-tundra of northwestern Canada. Arctic 48:13–21.

Timoney, K. P., G. H. LaRoi, S. C. Zoltai, and A. L. Robinson. 1992. The high subarctic forest-tundra of northwestern Canada: Position, width, and vegetation gradients in relation to climate. Arctic 45:1–9.

Timoney, K. P., G. H. LaRoi, S. C. Zoltai, and A. L. Robinson. 1993. Vegetation communities and plant distributions and their relationships with parent materials in the forest-tundra of northwestern Canada. Ecography 16:174–188.

Vestergaard, P. 1978. Studies in vegetation and soil of coastal salt marshes in the Disko area, west Greenland. Medd. om Grønland 204(2):1–51.

Vilchek, G. E., and O. Y. Bykova. 1992. The origin of regional ecological problems within the northern Tyumen Oblast, Russia. Arct. Alp. Res. 24:99–107.

Vince, S. W., and A. A. Snow. 1984. Plant zonation in an Alaskan salt marsh. J. Ecol. 72:651–667.

Walker, D. A. 1995. Disturbance and recovery of arctic Alaskan vegetation, in J. F. Reynolds and J. D. Tenhunen (eds.), Landscape function: implications for ecosystem response to disturbance, a case study in Arctic tundra. Ecological Study Series. Springer-Verlag, New York.

Walker, D. A., D. Cate, J. Brown, and C. Racine. 1987. Disturbance and recovery of arctic Alaskan tundra terrain: A review of recent investigations. U.S. Army Corps of Engineers, Cold Regions Research and Eng. Lab. Rept. 87–11. Hanover, N.H.

Walker, D. A., and K. R. Everett. 1987. Road dust and its environmental impact on Alaskan taiga and tundra. Arct. Alp. Res. 19:479–489.

Walker, D. A., and M. D. Walker. 1991. History and pattern of disturbance in Alaskan arctic terrestrial ecosystems: a hierarchical approach to analysing landscape change. J. Appl. Ecol. 28:244–276.

Walker, D. A., P. J. Webber, E. F. Binnian, K. R. Everett, N. D. Lederer, E. A. Nordstrand, and M. D. Walker. 1987. Cumulative impacts of oil fields on northern Alaskan landscapes. Science 238:757–761.

Walker, D. A. et al. 1995. Toward a new arctic vegetation map: a review of existing maps. J. Veg. Sci. 6: 427–436.

Warren Wilson, J. 1959. Notes on wind and its effects on arctic-alpine vegetation. J. Ecol. 47:415–427.

Warrick, R. A., H. H. Shugart, M. J. Antonovsky, J. R. Tarrant, and C. J. Tucker. 1986. The effects of increased CO_2 and climate change in terrestrial ecosystems, pp. 363–392 in B. Bolin, B. Doos, J. Jager,

and R. A. Warrick (eds.), The greenhouse effect, climate change, and ecosystems. Chichester, England.

Washburn, A. L. 1980. Geocryology: a survey of periglacial processes and environments. Wiley, New York.

Webber, P. J. 1978. Spatial and temporal variation: I. the vegetation and its production, Barrow, Alaska, pp. 37–112 in L. L. Tieszen (ed.), Vegetation and production ecology of an Alaskan arctic tundra. Springer-Verlag, New York.

Wein, R. W. 1976. Frequency and characteristics of arctic tundra fires. Arctic 29:213–222.

Wein, R. W., and L. C. Bliss. 1973a. Changes in arctic Eriophorum tussock communities following fire. Ecology 54:845–852.

Wein, R. W., and L. C. Bliss. 1973b. Experimental crude oil spills on arctic plant communities. J. Appl. Ecol. 10:671–682.

Wein, R. W., and L. C. Bliss. 1974. Primary production in arctic cottongrass tussock tundra communities. Arct. Alp. Res. 6:261–274.

Wein, R. W., and W. W. Shilts. 1976. Tundra fires in the District of Keewatin. Can. Geol. Surv. 76-1A:511–515.

Wesser, S. D., and W. S. Armbruster. 1991. Species distribution controls across a forest-steppe transition. Ecol. Mon. 61:323–342.

White, R. G. 1983. Foraging patterns and their multiplier effects on productivity of northern ungulates. Oikos 40:377–384.

Whitlock, C., and M. R. Dawson. 1990. Pollen and vertebrates of the early Neogene Houghton Formation, Devon Island, Arctic Canada. Arctic 43:324–330.

Wielgolaski, F. E., L. C. Bliss, J. Svoboda, and G. Doyle. 1981. Primary production of tundra, pp. 187–225 in L. C. Bliss, O. W. Heal, and J. J. Moore (eds.), Tundra ecosystems: a comparative analysis. Cambridge University Press, Cambridge.

Wolfe, J. A. 1969. Neogene floristic and vegetational history of the Pacific Northwest. Madroño 20:83–110

Wolfe, J. A. 1977. Paleogene floras from the Gulf of Alaska region. U.S. Geol. Surv. Prof. Paper 997. Washington, D.C.

Younkin, W. E. (ed.). 1976. Revegetation in the northern Mackenzie River Valley region. Arct. Gas Biol. Ser. Vol. 38. Arctic Gas Study Ltd., Calgary.

Younkin, W. E., and H. Martens. 1987. Long-term success of seeded species and their influence on native species invasion at abandoned rig site A-201, Caribou Hills, NWT, Canada. Can. J. Bot. 19:566–571.

Yurtsev, B. A. 1972. Phytogeography of northeastern Asia and the problems of transberingian floristic interrelations, pp. 19–54 in A. Graham (ed.), Floristics and paleofloristics of Asia and eastern North America. Elsevier, New York.

Chapter
2

The Taiga and Boreal Forest

DEBORAH L. ELLIOTT-FISK

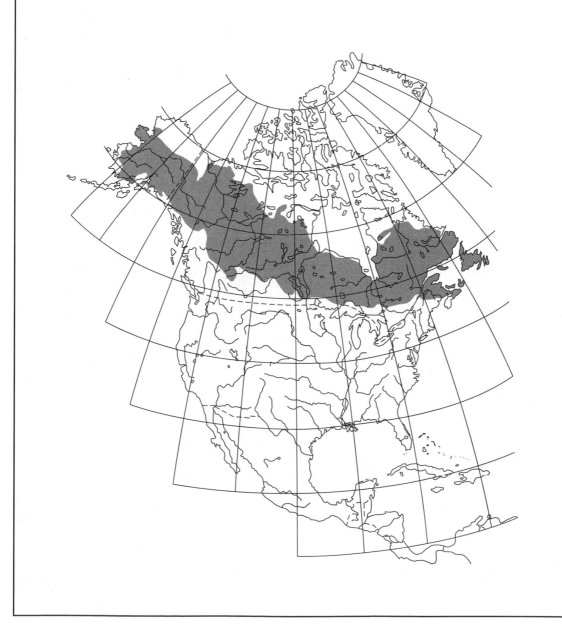

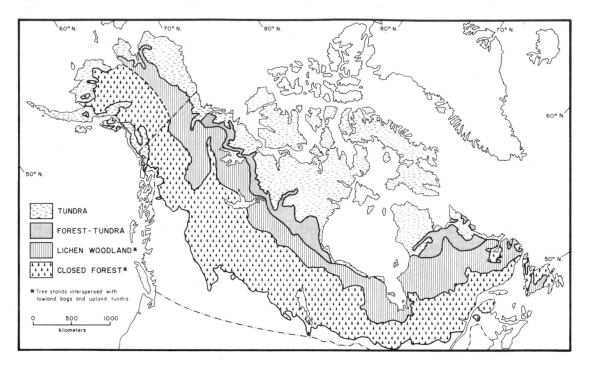

Figure 2.1. The boreal forest of North America can be broadly subdivided into three formations: closed forest, lichen woodland, and forest-tundra. The northern treeline separates the lichen woodland from the forest-tundra eco-tone (or occasionally, from the tundra proper). The northern tree limit is most often found at the northern edge of the forest-tundra ecotone immediately south of the Low Arctic tundra.

INTRODUCTION

The boreal forest of North America is a continuous vegetation belt at high latitudes stretching across the continent from the Atlantic shoreline of central Labrador westward across Canada to the mountains and interior and central coastal plains of Alaska. Larsen (1980) has referred to this major vegetation formation as the North American boreal continuum. This northern coniferous forest, or taiga, is part of the larger Northern Hemisphere circumpolar boreal forest belt.

In North America (Fig. 2.1), the boreal forest spans more than 10° of latitude in places, with transitions to the Arctic tundra to the north (Chapter 1), the subalpine forest of mountainous Alberta and British Columbia to the west (Chapter 3), the prairie grasslands to the southern interior (Chapter 9), and the Great Lakes–St. Lawrence mixed forest to the southeast (Chapter 10) (Rowe 1972). This large area contains many ecotonal communities, with boreal species ranging into the Arctic tundra, tundra species ranging southward into the boreal forest, and so forth. The forest-tundra ecotone in particular is discussed later in this chapter in regard to its physiognomy, phytosociology, and history, and it has been the subject of a book by Lar-

sen (1989). Finally, this region includes extensive wetlands, called bogs. Bog vegetation is described in this chapter and in Chapter 12.

Delimitation of the boreal forest has confused many scientists because of the often broad transitions into other vegetation formations. In western Canada, the Rocky Mountain forests mingle with the boreal forest, with some species, such as *Picea glauca* (white spruce), *Pinus contorta* (lodgepole pine), and *Populus tremuloides* (quaking or trembling aspen), being components of both. Even more confusion stems from the many and differing definitions of treeline and tree limit (Hustich 1966, 1979; Elliott 1979b; Elliott-Fisk 1983; Larsen, 1989).

Although the boreal forest is composed of diverse plant communities, its physiognomy is quite uniform throughout its large geographical range (Larsen 1980; Archibold 1995). This is largely because of the dominance of conifers in the formation. Despite the large extent of the forest, its diversity is low compared with more temperate and tropical communities. *Pica glauca* and *P. mariana* (black spruce) are almost ubiquitous species, with *Larix laricina* (larch or tamarack), *Abies balsamea* (balsam fir), *Pinus banksiana* (jack pine), *P. contorta*, *Populus tremuloides*, *P. balsamifera* (balsam poplar), and *Betula papyrifera* (white or paper birch) as com-

Table 2.1. *Characteristics of the abiotic environment*

Parameter	West	Central	East
Geology	Precambrian shield: basement complex of granitoid-gneissic rocks rimmed by sedimentary Paleozoic rocks (outcropping along Hudson Bay and in the mountainous region of the west)		
Surficial deposits	Till and outwash predominate; lacustral and fluviatile sediments locally important; occasional bedrock outcrops and unglaciated regions		
Topography	Mountainous (elev. > 5000 m)	Gentle plain (elev. < 300 m)	Dissected, tilted plateau (elev. 600–1000 + m)
Glaciation	Extensive coverage by Cordilleran and western edge of Laurentide ice sheets; some ice-free area in Richardson and Mackenzie mountains	Inundated by the Laurentide ice sheet	Mostly inundated by the Laurentide ice sheet; coastal and possibly alpine nunataks
Soils	Shallow spodosols/podzols (forest and woodland), entisols/regosols, and histosols/fibrisols (tundra)		
Climate	Humid, cool microthermal (see Table 2.2 for details)		

mon associates. Intermediate forms, subspecies, varieties, and ecotypes are present (Larsen 1980). Many shrubs and herbs are also characteristic of boreal communities and are discussed later. In particular, lichens and feather mosses are well-known understory components, playing important roles in successional processes and tree regeneration.

The North American boreal forest covers such a large geographical area that it is important to discuss the variations in the physical environment of the region. In particular, climate, surficial geology, glacial history, topography, and soils influence forest structure (Table 2.1).

Climate

Climate encompasses the major group of limiting factors for the boreal forest. Although the forest spans large ranges of latitude and longitude, the regional climate may be classified as cool, humid microthermal (Hare 1950), with very cold winters of 7–9 mo allowing persistence of snow cover during all but the brief, relatively cool summer season. Its physiological processes (translocation, water absorption, rates of photosynthesis, rate of respiration) (Warren Wilson 1967; Oechel and Lawrence 1985) are often temperature-controlled and temperature-limited, though soil drought and waterlogging may play roles in limiting growth at certain sites.

The northern and southern limits of the boreal forest appear to be thermally established, though the causal relationships are as yet largely unknown (Vowinckel et al. 1975; Black and Bliss 1980; Larsen 1980). According to Larsen (1980, 1989), the northern boundary of the forest corresponds roughly to the July isotherm of 13° C, with departures due to montane or marine influences. The southern forest limit

in central and eastern Canada approximates the 18° C July isotherm, though in the west the forest border shifts to slightly cooler regions with higher precipitation.

Radiation has been well investigated by Ritchie and Hare (1971) as a control over boreal forest limits. Net radiation, in particular, as a measure of available energy, may partly control the position of the northern forest border. Other climatic influences and climate summaries for the North American boreal forest have been presented by Hare (1950, 1954), Bryson (1966), Barry (1967), Larsen (1971, 1980, 1989), Hare and Ritchie (1972), Rowe (1972), Haag and Bliss (1974), Hare and Hay (1974), Miller and Auclair (1974), Streten (1974), Lettau and Lettau (1975), Vowinckel and associates (1975), and Wein and MacLean (1983). Slaughter and Viereck (1986) provide an outstanding overview of Alaska's taiga climates.

Climatological gradients through the boreal forest reflect latitude (with temperature and net radiation generally decreasing to the north), air-mass trajectories and sources (influencing cloud cover and precipitation patterns), topography (with an altitudinal decrease in temperature), and maritime/continental position. Table 2.2 presents select climatological data for several sites west to east across the boreal forest.

Topoclimate (mesoclimate) and microclimate variations may be more important than regional gradients in controlling forest existence. Topoclimates are especially important in mountainous regions, such as Labrador, northwestern Canada, and interior Alaska, where altitudinal forest limits may be established. The presence of a mountain range may also restrict the forest to sites south of its true climatic latitudinal limit (Elliot 1979b; Elli-

Table 2.2. *Climatological data for northern boreal forest sites in Alaska and western, central, and eastern Canada*

Parameter	Fairbanks, Alaska	Doll Creek, Yukon	Ennadai Lake, NWT	Napaktok Bay, Labrador
Mean daily temperature, January (° C)	−24	−35	−31.5	−20
Mean daily temperature, July (° C)	15	14	13	7.5
Mean annual temperature (° C)	−3	−10	−9.3	−5
Mean annual precipitation (mm)	290	240	290	500
Mean growing-season length (d)	100	120	100	75
Annual growing degree-days (5.6° C base)	1200	600	1000	550
Mean annual net radiation (kly yr^{-1})[b]	22	20	19	18
Net radiation, July (1y yr^{-1})	—[a]	—[a]	225	200

[a]Dash indicates data not available.
[b]Kilolangleys per year.
Sources: Canada (1957, 1978), Wilson (1971), Hare and Hay (1974), Ritchie (1982), and Elliott-Fisk (1983).

ott and Short 1979; Elliott-Fisk 1983). A forest canopy efficiently modifies the microclimate by acting as a windbreak and energy trap (Larsen 1965, 1980) that provides an ameliorated environment for understory species. However, the microclimate is so modified as the stand develops that stand degeneration (tree exclusion) may occur in certain settings.

Geology and Glacial History

Bedrock and structural geology through the boreal forest region does not strongly affect forest existence or composition. Bedrock varies from Precambrian granitic and gneissic rocks of the Canadian Shield in eastern and central Canada to Paleozoic and Cretaceous sedimentary rocks in the region southwest of Hudson Bay, the south-central prairie–forest border, western Canada, and Alaska. Hustich (1949) and others have noted differences in forest communities dominated by either *Picea glauca* or *P. mariana* on acidic versus basic parent materials in eastern Canada and elsewhere, but it appears that surficial deposits may account for more of these differences than bedrock geology does. As discussed briefly later, *P. mariana* is better able to tolerate shallow, poorly drained soils than is *P. glauca*, which does best on alluvial substrates. Therefore, a patterning is often seen in a forest region, with communities dominated by white spruce on the alluvial deposits along river valleys, and black spruce communities on upland bedrock sites but also on the lowland peats.

Almost all the region now occupied by the boreal forest of North America experienced intensive glaciation as recently as the late Pleistocene (maximum 18,000 yr BP). Forest communities were generally forced south with the expansion of the Laurentide and Cordilleran ice sheets, though refugia existed along the Labrador coast, the Old Crow region of the Yukon Territory, in interior Alaska, and elsewhere. It is unlikely that many of these displaced communities remained intact, because boreal and arctic species were faced with competition by taxa from more temperate regions.

Soils

Soil and vegetation types are highly correlated, with boreal forest communities being most frequently associated with podzol (Spodosol) soils (Canadian system of soil classification and U.S. Seventh Approximation System, respectively). These soils are the result of podzolization, which is a consequence of low temperatures and excess precipitation above that expended on evapotranspiration. In this process, iron and aluminum (with organic materials) are leached from the upper A horizon and illuviated (deposited) in the lower B horizon. Nutrients are removed from the upper soil, which then becomes low in important bases such as calcium. With low temperatures, soil microorganisms may be unable to decompose organic matter effectively, resulting in acidic soil conditions and low nitrogen and mineral levels (Larsen 1980).

Soil and permafrost development may thus inhibit what is thought of as the normal succession of seral stages in more temperate or tropical settings. Instead, conditions become more inimical to tree growth, nutrient cycling is limited, and permafrost (frozen ground) may form, with the active layer becoming shallow as the forest canopy closes. The site may become unfavorable for trees.

Although podzols are the dominant soils of the taiga, other soil types are found, with variations in parent material, plant cover, and other factors. Podzols are most easily derived from coarse-grained granitic rocks or sandy deposits, with regosols (entisols), fibrisols (Histosols), and gray-and brown-wooded soils developing in other settings (Larsen 1980). Soil diversity is high in mountainous regions of Alaska and the Yukon and in part of Labrador and northern Quebec (Larsen 1989). Viereck et al. (1986) note that the taiga soils of interior Alaska are uniformly immature and largely Pergelic Cryaquepts (Inceptisol) and Typic Cryofluvents (Entisol) in lowland sites and Aquic Cryothents (Entisol) and Alfic Cryochrepts (Inceptisol) in upland sites. Bog and meadow soils (Histosols) are common in wet sites and in topographic depressions throughout the boreal forest (Larsen 1982).

Minor soil changes are seen latitudinally through the boreal forest, with podzol generally shallower to the north. A transition to gray-wooded and gray-brown podzols soils and, more southerly, brown forest soils of the temperate deciduous forest proper is seen at the southern limit of the boreal forest (boreal forest–eastern deciduous forest ecotone). In this ameliorated environment, organic decomposition is more rapid, and the deciduous litter is higher in bases, with more rapid recycling of nutrients. As a result, the upper soil horizons tend to be enriched in nutrients rather than leached of nutrients (Tedrow 1970, 1977; Larsen 1980).

Integrated Environmental Factors and Processes

The ecological complexity of boreal forest ecosystems is well recognized. It is believed that we can improve our understanding of these complex systems through the use of both conceptual and simulation models. Shugart (1984) and colleagues (Shugart et al. 1992) have pushed our knowledge forward through model construction, testing, and verification. The qualitative, conceptual model of Bonan and Shugart (1989) is especially useful for illustrating the interrelationships among climate, solar radiation, soil moisture, soil temperature, permafrost, the forest floor organic layer, nutrient availability, forest fires, insect outbreaks, and forest structure. This model helps to explain the mosaic of vegetation types that exist across the northern forests and provides a conceptual framework of further mechanistic research on boreal ecosystems. Bonan (1989a, 1989b, 1992) has quantified this conceptual model into a simulation model in which the physical environmental conditions, as well as forest structure and vegetation patterns, are simulated for boreal ecosystems in interior Alaska.

MAJOR VEGETATION TYPES

Although the boreal forest formation is often depicted as a monotonous coniferous forest of uniform composition and physiognomy, in reality it is a complex mosaic of different plant communities, though almost all are dominated by (principally coniferous) trees. Bogs and meadows of varying sizes are found throughout the region. Changes in dominant tree, shrub, and herb taxa can be seen latitudinally through the forest, though there are no dramatic discontinuities (Rowe 1972; Larsen 1980, 1982).

The term "taiga" is broadly used in ecological literature and may be defined as coniferous northern forest with no admixture of nonconiferous species except *Betula* and *Populus* (Ritchie 1962; Larsen 1980). The forest canopy becomes more open and generally shorter northward, with a transition from closed forest to open lichen woodland to scattered trees of the forest-tundra ecotone.

Rowe (1972) has divided what he terms the "boreal forest region" of Canada into 35 sections. Neiland and Viereck (1978) have prepared a detailed and very workable subdivision of the forest types and ecosystems of Alaska, with basic divisions into bottomland, lowland, and upland forests. Alaska's boreal forest is perhaps more diverse in both types and patterns than that over all of Canada, because of the complex topography of the state. Neiland and Viereck point out that the mosaic of forest communities is a response to (1) topography, microtopography, climate, microclimate, river flooding, permafrost occurrence and depth, organic matter, and fires, (2) chance, and (3) variations in reproductive abilities, productivity, and distribution patterns of the various species.

Boreal Forest Regions

Larsen's (1980) division of the North American boreal forest zone (continuum) into seven regions is most easily used and is outlined briefly here. The Alaskan taiga (region 1) is a mixture of upland and lowland forests dominated by *Picea glauca*, *P. mariana*, *Betula papyrifera*, and *Populus tremuloides* (Fig.

Figure 2.2. Boreal forest along the Klutina River west of the Wrangell Mountains in interior Alaska (elev. 500 m). The closed forest here is dominated by Picea glauca on the uplands and P. glauca, Populus balsamifera, P. tremuloides, and Betula papyrifera on the terrace slopes and lowlands. The understory vegetation varies with the degree of canopy closure, with mosses dominating in closed sites and woody shrubs (Vaccinium, Rosa, and Salix species) under canopy openings.

2.2). Fire and permafrost contribute to successional processes and vegetation patterning.

The boreal forests of the Cordillera, or northern Rocky Mountains, occupy large sections of the Mackenzie and Richardson mountains of the Yukon Territory and the northwestern Mackenzie District of the Northwest Territories (region 2). Ritchie and his colleagues (Ritchie 1962, 1977, 1982; Ritchie and Cwynar 1982; Spear 1982; MacDonald 1983) have contributed a great deal to our understanding of forest dynamics in these regions. The forest reaches its northernmost location here along the Mackenzie River Delta, extending almost to the Arctic Ocean (Larsen 1980).

In the southwestern Mackenzie District and northern Alberta, the interior (region 3) forest is composed predominantly of *Picea mariana, Picea glauca, Pinus banksiana, P. contorta var. latifolia, Populus tremuloides,* and *P. balsamifera,* with occasional stands of *Larix laricina* and *Abies balsamea.* This forest extends from the Cordilleran foothills eastward to the western edge of the Canadian Shield.

On the Canadian Shield proper (region 4), both west and east of Hudson Bay, the forest is relatively uniform in appearance, with topographic and microclimatic factors affecting vegetation structure at certain sites. The relative importance of the tree species changes from south to north, with richer forests in the south, *Picea mariana, P. glauca,* and *Pinus banksiana* increasing in dominance northward and *P. mariana* being dominant

near the northern forest border. Boreal forest outliers to the south exist in northern Minnesota, Wisconsin, and Michigan, with communities transitional to the northern conifer-hardwood forest (Curtis 1959), but disturbance by agriculture and logging has severely altered many of these stands (Larsen 1980).

In the Gaspé–Maritime region of eastern Canada (region 5), a wide variety of forest communities is found, as best illustrated by the delimitation (Rowe 1972) of 18 major divisions of the boreal forest there. Many of these forests are closely related to those of the Great Lakes–St. Lawrence region, as is apparent from their understory vegetation. Again, the boreal forest of eastern Canada is predominantly coniferous, but with *Abies balsamea* playing a more dominant role there than elsewhere, especially on well-drained uplands. In contrast, monospecific stands of *Picea mariana* are rare (Larsen 1980).

The forest of Labrador–Ungava (region 6) has been intensively studied by Hustich (1949, 1950, 1954), Payette and colleagues (Payette and Gagnon 1979; Payette and Filion 1984; Lavoie and Payette 1994, 1996), and others. Although *Picea mariana* and *P. glauca* are often dominants, *Abies balsamea, Larix laricina, Betula paprifera, P. balsamifera, Pinus banksiana,* and *Thuja occidentalis* (white cedar) are also found. The complex topography of this dissected, faulted, and glaciated plateau has led to the development of a mixture of community types partitioned along moisture and geologic gradients. Black spruce communities become increasingly important to the north, with understory vegetation changing in species frequency and composition whereas the arborescent stratum remains essentially unchanged (Larsen 1980). Elevational treelines are also encountered, reaching progressively lower elevations northward.

Much research has been concentrated along the northward forest border (region 7), which is also variously termed the continental Arctic treeline, the northern tree limit, the northern forest-tundra ecotone, and so forth. The forest usually extends northward along river lowlands, with tundra found to the south along upland surfaces (Larsen 1980, 1989). *Picea mariana* again dominates, but *P. glauca* is present on well-drained sites and *Larix laricina* in boggy situations. Occasionally, all three species are found together in outlier stands.

Closed forest The boreal forest can be broadly divided into three structural units (formations): closed forest, lichen woodland, and forest-tundra ecotone. These are briefly discussed, along with shrublands, whose ecology is as yet poorly understood.

Over large areas, closed forest communities dominate the southern boreal forest zone on a variety of topographic sites, soils, and lithologies. Though the species composition of these stands may change, the structural appearance of the forest is almost unchanged throughout.

The characteristic forest community here is the spruce–feathermoss forest, with either *Picea mariana* or *P. glauca* being dominant (Fig 2.3). La Roi (1967; La Roi and Stringer 1967) has classified these forests as follows: (1) western, black spruce–feathermoss forest (mixed wood); (2) eastern, black spruce–feathermoss forest and white spruce–fir–feathermoss forest (mixed wood).

The transition from western to eastern forest occurs along a longitudinal axis from western Lake Superior to Lake Winnipeg to Churchill (southwestern Hudson Bay). The semipermanent trough of the upper (westerly) circumpolar vortex is situated over Hudson Bay and this transitional area. These western and eastern regions experience different temperature, precipitation, and radiation regimes, especially as witnessed through changes in humidity and potential evapotranspiration.

La Roi and Stringer (1976) classified 60 boreal spruce-fir forest stands across North America. Their stands occur in 24 of Rowe's (1972) 38 boreal sections. Both the vascular flora and bryophyte assemblages proved useful in this classification, with the complex patterns of both gradual and abrupt

Figure 2.3. *Eastern white spruce–feathermoss forest at Napaktok Bay, Labrador. The tree stratum here is composed solely of* Picea glauca. *This stand is at the northern tree limit; yet the population is vigorously reproducing (see Fig. 2.6). The bryophyte stratum interfingers with an irregular, patchy dwarf shrub stratum of principally* Betula glandulosa. *The maximum age of trees in this stand is* ~400 yr.

changes in species composition giving a geographic basis for the two primary types of stands (Figs. 2.4 and 2.5).

The black spruce–feathermoss forests have a uniform tree stratum that is moderately dense, with an understory of almost continuous bryophyte cover. In contrast, the tree stratum of the white spruce (mixed) forests is more irregular and open, with strata of broadleaf shrubs and herb–dwarf shrubs that are both dense and species-rich, with a patchy bryophyte stratum. The white spruce–fir forest is 50% richer in number of bryophyte species than is the black spruce forest, even though the latter has a much higher bryophyte cover (La Roi and Stringer 1976). The vascular flora is also richer on the average (Table 2.3). The higher overall diversity may be a function of the higher productivity and greater mean age of the white spruce–fir stands or of a greater habitat diversity. The microclimates of these stands should be studied, because the physiognomy of white spruce–feathermoss forests across Canada appears to be much more diverse than that of the closed black spruce–feathermoss stands.

Hylocomium splendens and *Pleurozium schreberi* are the characteristic cover dominants in bryophyte strata of the spruce–feathermoss forests *H. splendens* dominates in the mixed woods, *P. schreberi* in the black spruce forests. Though the two species

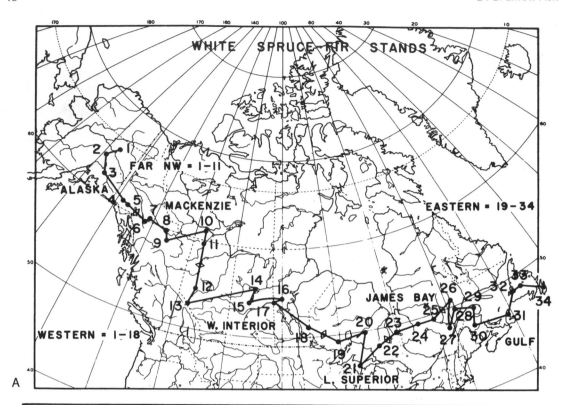

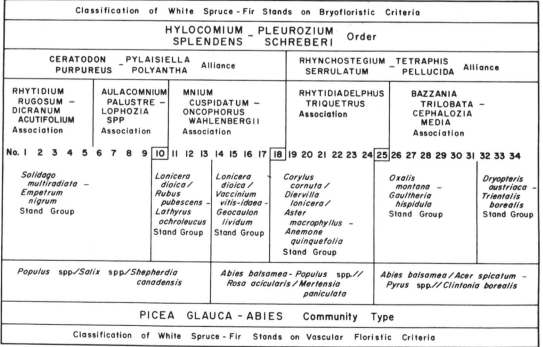

Figure 2.4. Picea glauca–Abies *stands in the North American boreal forest (A) have been classified by La Roi and* Stringer (1976) (part B) using both vascular and bryophyte floristic criteria.

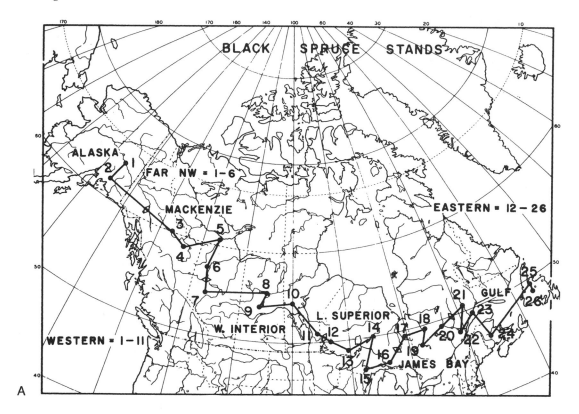

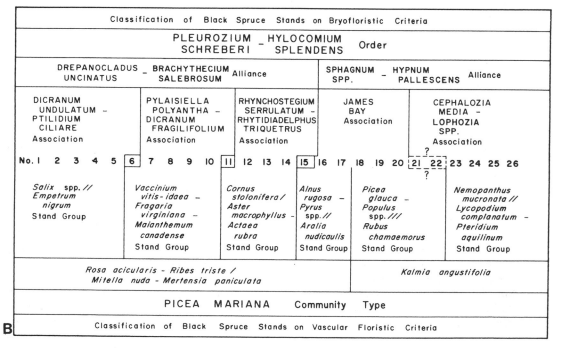

Figure 2.5. *Picea mariana stands in the North American boreal forest (A) have been classified by La Roi and Stringer (1976) (part B) using both vascular and bryophyte floristic criteria.*

Table 2.3. *Vascular flora of spruce-feathermoss forests*

Stratum	Presence class	White spruce-fir	X̄ spp.	Black spruce	X̄ spp.
Tree	61–100%	*Picea glauca* *Betula papyrifera* *Abies balsamea* *Picea mariana*	4.8	*Picea mariana* *Betula papyrifera*	4.3
	41–60%	*Populus tremuloides* *Populus balsamifera*		*Picea glauca* *Abies balsamea* *Populus tremuloides*	
Low tree & tall shrub	41–60%	*Alnus crispa* *Pyrus decora*	3.3	*Alnus crispa*	2.2
Medium & low shrub	61–100%	*Viburnum edule* *Rosa acicularis* *Ribes triste*	6.4	*Ledum groenlandicum* *Rosa acicularis*	6.4
	41–60%	*Rubus idaeus*		*Viburnum edule*	
Herb-dwarf	61–100%	*Linnaea borealis* *Cornus canadensis* *Pyrola secunda* *Mitella nuda* *Maianthemum canadense* *Rubus pubescens* *Moneses uniflora*	28.4	*Cornus canadensis* *Linnaea borealis* *Maianthemum canadense* *Pyrola secunda* *Gaultheria hispidula* *Coptis groenlandica* *Geocaulon lividum* *Vaccinium myrtilloides*	22.9
	41–60%	*Trientalis borealis* *Goodyera repens* *Lycopodium annotinum* *Aralia nudicaulis* *Clintonia borealis* *Epilobium angustifolium* *Mertensia paniculata* *Coptis groenlandica* *Dryopteris austriaca* *Viola renifolia* *Gymnocarpium dryopteris* *Pyrola asarifolia* *Streptopus roseus*		*Clintonia borealis* *Trientalis borealis* *Vaccinium angustifolium* *Goodyera repens* *Epilobium angustifolium* *Equisetum sylvaticum* *Mitella nuda* *Petasites palmatus* *Rubus pubescens* *Equisetum arvense* *Vaccinium vitis-idaea*	
Total for stand			42.9		35.8

Source: La Roi (1967).

can be considered ubiquitous, their distributions may reflect different microclimatic conditions in the two types of communities.

Although La Roi and Stringer's community analysis shows discrete clusters of *Picea mariana* versus *P. glauca* communities in the southern (closed) boreal forest, Larsen (1972, 1974, 1980) has found that *P. mariana* forest stands are almost uniform in structure throughout their latitudinal range. This is most likely a function of similar habitat conditions, with the trees exerting a considerable influence on their own environment. The growth and closure of the forest canopy result in warmer winters, cooler summers under calm conditions, warmer summers in windy sites, higher moisture, encroachment of the permafrost table, and poor drainage.

To the north, lichen woodlands dominate the landscape. The feathermosses require a high pre-cipitation input, with summer humidity especially critical; thus, their importance decreases northward as climates become cooler and more xeric, with lichens then dominating the lower forest strata (La Roi and Stringer 1976). The partitioning of tree, shrub, and herb taxa along moisture gradients is obvious in the southern closed forest (Table 2.4) and contributes to community development (Larsen 1980, 1982).

Lichen woodland. The lichen woodlands of northern Canada in particular have intrigued scientists for decades. These woodlands have been variously regarded as a successional sere following fire in the progression to closed boreal forest or muskeg, as a fire climax, and as a true climax community. These woodlands dominate the northern region of the boreal forest zone (Hare 1959; Ritchie 1962; Kershaw 1977; Larsen 1980; Elliott-Fisk 1983; Larsen 1989).

Table 2.4. *Distribution of selected tree, shrub, and herb species along moisture gradients in the southern boreal forest*

Strata	Very dry/dry	Fresh	Moist	Very moist	Wet
Trees	———————— *Pinus banksiana* ————————				
			———————— *Larix laricina* ————————		
	———————————— *Picea glauca* ————————————				
	———————— *Betula papyrifera* ————————				
	—————————— *Populus tremuloides* ——————————				
			—————— *Populus balsamifera* ——————		
			———— *Abies balsamea* ————		
			———— *Picea mariana* ————————		
Tall shrubs	*Alnus crispa*	*Amelanchier alnifolia*	*Acer spicatum*	*Acer negundo*	*Alnus rugosa*
Medium shrubs	*Juniperus communis*	*Vaccinium myrtilloides*	*Lonicera dioica*	*Ledum groenlandicum*	*Betula glandulosa*
Herbs	*Arctostaphylos uva-ursi*	*Viola rugulosa*	*Petasites palmatus*	*Geocaulon lividum*	*Petasites sagittatus*
	Lycopodium complanatum	*Lycopodium obscurum*	*Mertensia paniculata*	*Equisetum scirpoides*	*Equisetum arvense*
	Potentilla tridentata	*Pyrola asarifolia*	*Vaccinium vitis-idaea*	*Mitella nuda*	*Rubus chamaemorus*

Source: Larsen (1980).

The transition from southern closed spruce–feathermoss forest to lichen woodland is sharp in many areas (such as northern Manitoba and Saskatchewan) but transitional in others, with stands of lichen woodland alternating with closed forest and muskeg according to topographic position. This zonal pattern is especially disrupted in Alaska and the Yukon Territory because of the mountainous terrain (Larsen 1980) but also east of Hudson Bay in the maritime climates and the pronounced ridge and valley topography of northern and central Quebec and Labrador (Marr 1948; Hustich 1949, 1950; Payette 1975; Elliott-Fisk 1983; Payette 1983).

Kershaw (1977) has classified lichen woodlands into two broad groups (1) western *Stereocaulon paschale* woodland and (2) eastern *Cladonia stellaris* woodland (Table 2.5). Although knowledge of the environmental controls of these formations is incomplete, the transition of woodland types occurs southwest of Hudson Bay (in the same area as the western-eastern spruce–feathermoss forest transition) and may be a function of summer moisture regime (Kershaw 1977). Edaphic conditions may be important as well, but they are intimately tied to microclimate. Undoubtedly, the mosaic of vegetation types seen in many areas is a function of soil characteristics, as well as the fire history of the region.

Picea marina and *P. glauca* are the dominant trees of the lichen woodlands, with *P. mariana* increasing in importance to the north. Although tree density is obviously less in these woodlands than in the closed forest, tree stature is often equal to, or greater than, that in the closed forests. Branching also occurs to ground level because of higher light levels in the open canopy.

Forest-tundra ecotone. The location of the forest-tundra ecotone reflects the general circulation of the atmosphere, with irregularities caused by differences in topography, glacial history, bedrock, fire history, and topoclimatic and microclimatic influences. The ecotone is found north of the treeline, which is defined as the boundary separating arboreal and nonarboreal vegetation, with forest/woodland communities dominating at least 50% of the terrestrial landscape at and south of the treeline (Larsen 1974; Elliott-Fisk 1983; Larsen 1989). Physiological controls over the treeline have been discussed by Billings and Mooney (1968) and Oechel and Lawrence (1985). Tranquillini (1979) has given a physiological definition for the treeline, stating that it is the limit beyond which forest existence is no longer possible because damage halts development somewhere in the life cycle from seed set to maturity. Unsuccessful or irregular sexual regeneration has been well documented for these northernmost tree stands (Elliott 1979a, 1979b, 1979c; Black and Bliss 1980; Elliott-Fisk 1983; Lavoie and Payette 1994) (Fig. 2.6).

North of the treeline, woodland patches, clones, and individual trees are scattered over the tundra landscape (Fig. 2.7), eventually reaching the northern tree limit (limit of tree growth). Occasionally,

Table 2.5. *Arboreal density and understory cover and composition of western and eastern lichen woodlands*

Parameter	Western	Eastern	Parameter	Western	Eastern
Mean density:			Club moss and moss		
(trees ha^{-1})			*Pleurozium* spp.	19.3	0
Larix laricina	16	10	*Lycopodium annotinum*	0.1	0.6
Picea glauca	39	124	*Dicranum* spp.	0.1	2.3
Picea mariana	445	422	*Hylocomium splendens*	0	1.0
Total	500	556	*Polytrichum* spp.	5.8	1.8
Composition and			Total	5.7	30.4
mean cover of:			Lichens		
Shrub layer			*Cetraria* spp.	0.4	0.5
Betula glandulosa	9.5	5.5	*Cladonia alpestris*	—	78.1
Empetrum nigrum	5.0	1.7	*C. mitis*	—	2.1
Ledum groenlandicum	1.9	4.3	*C. rangiferina*	—	—
Vaccinium angustifolium	0	4.0	Other *Cladonia* spp.	16.9	9.3
V. uliginosum	0.6	1.0	*Stereocaulon paschale*	15.8	5.0
V. vitis idaea	9.6	0.5	Total	34.0	97.4
Total	27.2	17.0	Grand total		
Grass layer			(shrub and ground		
Festuca ovina	0	0.3	layer)	91.6	120.4
Total	0	0.3			

Sources: Maini (1966) and Rencz and Auclair (1978).

tree species may be found assuming a shrub growth form even north of this, where they form the northern tree species limit (Elliott-Fisk 1983). The diversity of plant communities in the ecotone is high and is related to environmental gradients and site history, as is well illustrated by Ritchie's (1982) ordination of communities in the Doll Creek region of the northern Yukon Territory (Fig. 2.8).

The width of the forest-tundra ecotone varies across North America, being narrower at both its western and eastern ends and up to 235 km in width in central Canada. In Quebec, the ecotone may exceed 300 km in width, with isolated woodland patches found in river lowlands where the rest of the landscape is covered by tundra. The decrease in elevation from Schefferville north toward Ungava Bay has effectively widened the vegetation zones by lessening the apparent latitudinal decrease in temperature northward (Hare 1950; Elliott-Fisk 1983). The western Cordillera may serve to compress latitudinal temperature variation, effectively anchoring air-mass trajectories (Sorenson 1977). The dissection of the landscape has also decreased the width of the ecotone, with severe climatic conditions on ridge tops blocking migration routes. Historical fluctuations in the geographical location of the ecotone, as reflective of climate changes, have been greatest in the relatively flat interior plains of Canada and least along its western and eastern margins.

The physiognomy of the ecotone vegetation is extremely variable across North America, though species compositions remain similar. In certain areas, such as southwestern Keewatin, small clonal spruce may extend up river valleys as outliers, perhaps 50 km from their nearest neighbors. Trees here are typically dwarfed (<2m tall) and only occasionally krummholzed (Fig. 2.9). Marr (1948) has documented the dramatic changes in the forms of tree growth both away from the coast and to the north for the Great Whale River region east of Hudson Bay. Other areas of the ecotone have very large woodland patches that often have an understory of true boreal species, as related to the history of the site (Larsen 1965, 1972; Elliott 1979a, 1979b; Elliott and Short 1979; Elliott-Fisk 1983).

Larsen (1965, 1972, 1974, 1980, 1982, 1989) has investigated the plant communities of the forest-tundra ecotone in western and central Canada. Many of the species found here have large geographic ranges in both the northern boreal forest and Low Arctic tundra. However, the majority of ecotonal communities are depauperate, because neither true arctic nor boreal species can occupy this zone of variable climatic (frontal) conditions. Larsen found that only the more ubiquitous species occurred in sufficient abundance to appear regularly with high frequencies in transects. The physiological characteristics of the plants that account for this are unknown.

Shrublands

Shrublands of varying composition are found throughout the boreal forest. These communities may in some instances be successional to forest or

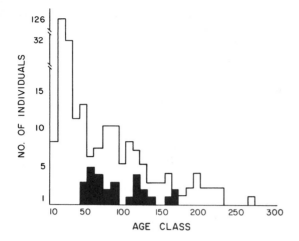

Figure 2.6. The age structures of isolated stands of Picea glauca *in the forest-tundra ecotone at Ennadai Lake, N.W.T. (black bars) and Napaktok Bay, Labrador (open bars) show distinct differences. The Ennadai Lake stand is not in equilibrium (note lack of juveniles), and it is maintaining itself through layering, whereas the Napaktok Bay stand is in equilibrium and is maintaining itself through seed production and seedling establishment.*

Figure 2.7. This mixed stand of Picea glauca, P. mariana, *and* Larix laricina *is found in a ravine in the forest-tundra ecotone at Ennadai Lake, N.W.T. Many typical boreal forest taxa are found in the understory, and paleoecological data suggest that this stand is a relict of a formerly more extensive forest cover. Source: Elliott (1979a), permission of Arctic and Alpine Research and Regents, University of Colorado.*

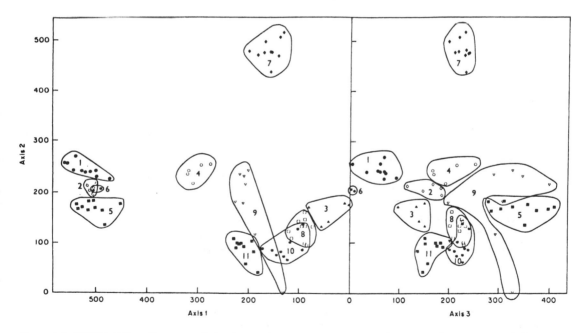

Figure 2.8. DECORANA ordination analysis of vegetation plots from the Doll Creek area of the northern Yukon Territory illustrates the diversity of plant communities in the forest-tundra ecotone (Ritchie 1982). The subjectively determined vegetation types are (1) tundra, (2) black spruce woodlands on moraine surfaces, (3) snowpatch, (4) limestone cliff and crag, (5) black spruce-larch on bottomland mires, (6) open white spruce forest on steep shale-sandstone surfaces, (7) tundra, (8) tundra, (9) white spruce woodlands on south- and southwest-facing limestone slopes, (10) white spruce woodland on north-facing limestone slopes, and (11) spruce-larch woodland at treeline on north-facing limestone slopes.*

Figure 2.9. Dwarf clones of Picea mariana, as shown here in the Northwest Territories, are frequently seen scattered through the forest-tundra ecotone. This pattern is most pronounced over flat terrain that was inhabited by forest/woodland in the Hypsithermal. Alnus crispa shrubs are found in the same stratum, with an understory of various low ericaceous shrubs and lichens. An Eriophorum species bog is visible between spruce clones in the foreground and along the lake shore.

woodland; at other sites they appear to be the climax vegetation.

Dense thickets 2–6 m tall of *Alnus crisps* (green alder), *Betula glandulosa* (dwarf birch), and various *Salix* species (willow) have been documented across Canada and Alaska (Fig. 2.10). These stands are most frequently seen near the northern elevational treeline and frequently are classified as arctic tundra rather than boreal in nature. Neiland and Viereck (1978), Oswald and Senyk (1977), Short (1978a, 1978b), Elliott and Short (1979), and Larsen (1989) have described these communities at upland sites in Alaska, the Yukon Territory, and Labrador–Ungava. Some of these stands show little or no sign of tree invasions, though site conditions appear amicable for tree growth. Further research on these shrublands and their history is needed (see also Chapter 1).

Bogs and Other Wetlands

Various wetland plant communities are found in the boreal forest region. Bogs are the most common of these and have been defined by Curtis (1959) as plant communities of specialized shrubs and herbs growing on a wet, acidic, peat substrate. *Sphagnum* moss species dominate the bryophyte layer. Trees are found as a minor component of bog vegetation in many areas.

Bogs can be contrasted with fens, which form under more alkaline conditions on wet peat. Heath forms on wet to moist sandy soil with a shallow layer of peat, and moor is a community of upland heath on dry or damp but not wet peat. Whereas marsh communities form on wet or periodically wet soils not including peat, meadows are wet or periodically wet on sedge peat or muck. Conifer swamps have a greater tree cover of typically *Picea mariana* or *Larix laricina*, with a wet, mossy understory (Larsen 1982). Heinselman (1963, 1970) and others have devised more elaborate schemes for classifying these northern wetlands on the basis of topography, hydrology, water chemistry, and landscape evolution (Larsen 1982). Characteristic and to a large extent circumglobal forms of plant life and plant families are found on each wetland type. The reader is referred to Larsen (1982) for extensive floristic summaries of various boreal wetland vegetation communities in North America, and to Chapter 12 of this book.

Few studies have been conducted on northern bogs, largely because of their poor accessibility. These communities are widespread in regions that have undergone continental glaciation. Bogs have formed in kettle lakes and shallow ponds, on stable lake margins, in marshland, along stream margins, and over dense shrublands. Many of them are apparently long-lived, whereas others are ephemeral in sites that are frequently disturbed (Larsen 1982).

Bog taxa are predominantly members of the Orchidaceae or Ericaceae, though Cyperaceae, Asteraceae, and Poaceae species are not uncommon. Bog ericads and sedges generally develop on a layer of *Sphagnum* species, with black spruce a nearly ubiquitous component of treed bogs. *Larix laricina* is frequently found in bogs having an open tree canopy. In southern Alaska, *Picea mariana* is largely replaced by coastal forest species such as *Tsuga heterophylla*, *Thuja plicata*, and *Chamaecyparis nootkatensis*, and also by *Pinus contorta*, which has a taproot and may survive hygric conditions longer than other conifers. All these taxa are ombrotropic and adapted to at least seasonally saturated, acidic, nutrient-poor, organic soils (Larsen 1982). Input of nutrients into the system occurs principally through wet or dry atmospheric deposition.

Bog vegetation is uniform in physiognomy throughout the boreal forest (Larsen 1982). The component species are evergreen xeromorphs because of the short growing season, the fluctuating groundwater table, and the nutrient-deficient substrate. Nutrients are largely retained in the biomass.

Trees of several species may be found around the margins of bogs and their meadows, and in many instances they appear to be invading the bog vegetation. However, in many cases it is not an invasion. Many of these trees are dwarfed adults several decades in age that have grown slowly be-

Figure 2.10. Extensive, tall shrublands are associated with forest/woodland vegetation in certain regions of the northern boreal forest. Dark, dense thickets of Alnus crispa, Be-*tula glandulosa, and* Salix planifolia *are shown here just north of Hebron Fiord along the central Labrador coast.*

cause of the low availability of nutrients and the saturation of the bog soils (Lawrence 1958). Long-term survival under these conditions is tenuous unless the tree can extend its roots into a more favorable substrate or unless the climate changes and the site becomes more mesic. Many bogs serve as valuable recorders of climate and other environmental changes; consequently, bog deposits have been investigated by paleoecologists interested in both succession and climatic history in various regions of the boreal forest.

COMMUNITY DYNAMICS

Succession

Researchers have focused on the question of whether or not the closed spruce forest is the ultimate (climax) stage in boreal succession (Larsen 1980). Various workers have argued the question pro (Raup and Denny 1950; Lutz 1956; Kershaw 1977; Oswald and Senyk 1977; Black and Bliss 1980; Larsen 1980) and con (Rowe 1961; Strang 1973; Viereck 1973; Black and Bliss 1978; Larsen 1980) with regard to particular forest communities at cer-

tain sites. In regard to the entire boreal forest, the best answer undoubtedly is "perhaps" (Larsen 1980), because site conditions exert a considerable influence on the successional process.

In this light, it is important to understand that the boreal forest is a disturbance forest, as argued by Rowe (1961), who asserts that there are no species in the Canadian western boreal forest possessing in full the silvical characteristics appropriate to participation in a self-perpetuating climax. As such, the boreal forest is a disturbance forest, usually maintained in youth and health by frequent fires to which all species, with the probable exception of fir, are in various ways adapted.

Larsen (1980) has argued that traditional concepts of forest succession are inapplicable here. Instead of succession resulting in more mesophytic conditions (as occurs in the most temperate regions), tree survival becomes more difficult as the canopy closes. Some of the degradative environmental changes accompanying succession in the boreal forest are (1) an increase in thickness of the organic mat, (2) a decrease in nutrient availability, (3) a decrease in summer soil temperatures, (4) an increase in the level of the permafrost table

Table 2.6. *A rating of regenerative successes of conifers following fire*[a]

Parameter	Pine	Black spruce	Larch	White spruce	Balsam fir
Seed retention on tree	10	6	2	3	1
Early seed production	5	4	2	3	1
Seed mobility	2	5	5	3	1
Seedling frost-hardiness	5	3	4	2	1
Seedling growth rate	5	2	4	3	1
Seedling palatability	3	5	2	4	1
Seedling response to full sun	4	3	5	2	1
Totals	34	28	24	20	7

[a] Relative scale ranges from low (1) to high (10).
Source: Rowe and Scotter (1973).

(i.e., a decrease in the depth of the active layer), (5) a decrease in soil drainage, resulting in waterlogging, anaerobic conditions and gleization, and (6) an increase in frost heave and thrust (Larsen 1980; Wein and MacLean 1983). Patchiness of the microenvironments (substrate, fire flooding, logging, and other disturbances) contributes to the mosaic of seral communities seen in the region and has led Larsen (1980) to state that succession does not progress to a single regional climax. The climax could be a closed forest on eskers, floodplains, and other well-drained sites, or it could be an open muskeg or bog in lowlands and other poorly drained sites. Carleton and Maycock (1978), in their work in the James Bay region of Ontario, found through ordination studies that species did not show any tendency to progress toward a single climax type, though certain species tended to associate in seral stages.

Succession in bogs and other wetland areas in the boreal forest region is poorly understood. As Heinselman (1963, 1970) stated, the only trend in bog succession is toward landscape diversity, not mesophytism. Topography, the biochemistry of the site, climate, and disturbance (including fire) all alter the successional process such that there is a wide variety of bog/peatland communities. Nevertheless, Gates (1942) has outlined a general pattern of boreal bog succession, with *Chamaedaphne* species pioneering succession around the edge of a depression in to open water; a mat soon forms from roots and rhizomes and grows over the pool, and organic matter is deposited on the lake bed; this surface is then invaded by shrubs and moss (generally *Sphagnum*), forming a dense thicket conducive to the establishment of black spruce and larch. *Carex* species typically pioneer secondary succession on peat lands following disturbance (Larsen 1982). Fire, drought, and other disturbances alter the moisture and nutritional status of

the peatland, adding to the diversity of peatland plant communities throughout the boreal forest. Furthermore, Wein (1983) discusses the role of fire on organic terrains and notes that even though the frequency of fire is rare, under drought conditions fires can burn deep into organic soils, typically resulting in low energy but prolonged glowing combustion. These fires are of particular interest economically because they destroy valuable peat depositions and are very costly to suppress. On the other hand, Wein (1983) notes that fires on organic substrates can shift successional trends and enhance the establishment of tree seedlings.

Fire has been documented as the primary disturbance factor disrupting successional processes in the boreal forest (Rowe and Scotter 1973; Viereck 1973; Kershaw 1977; Larsen 1980; Wein and MacLean 1983). Many tree species of the boreal forest are adapted to fire, as shown by their morphologic and reproductive characteristics (Table 2.6) (Rowe and Scotter 1973).

Populus balsamifera (Fig. 2.11) is the most resistant of all the boreal forest trees to destruction by fire, a function of its thick bark (>10 cm near the ground) and regeneration by root suckers as well as seeds (Larsen 1980). Neiland and Viereck (1978) have rated balsam popular alluvial forests, along with those composed of *Populus trichocarpa* (black cottonwood), as the most productive of all Alaskan taiga communities.

Although the conifers are not able to produce stump sprouts or root suckers, all but *Abies* retain some of their seeds on the tree. Thus a seed source (especially for species that are serotinous) is readily available following a fire. Layering may also be adaptive. Layering has been documented for *Picea mariana* in many regions, resulting in stand stability (maintenance) in the absence of fire or major climatic deterioration. Black and Bliss (1980), in their studies of black spruce communities at the

Figure 2.11. *Populus balsamifera is the most fire-resistant tree of the boreal forest. Here individuals 20 m or greater in height are important components of the vegetation at the southern margin of the boreal forest in Elk Island National Park, Alberta.*

northern treeline and tree limit near Inuvik (Northwest Territories), found layering able to replace the loss of dominant canopy trees during periods of fire exclusion. Lavoie and Payette (1996) also point to structural changes (e.g., changes in tree life forms and tree densities) in the northernmost forests in subarctic Quebec in reference to climatic warming with fire exclusion in this century. If sexual reproduction is rare or infrequent, however, as is often the case near the northern treeline (Larsen 1965; Elliott 1979a, 1979c; Black and Bliss 1980; Elliott-Fisk 1983), fire can destroy stands that are normally maintained through layering, and a rapid retreat of the northern forest border can occur. This is believed to be the explanation for very rapid, large-scale retreats of the northern Canadian treeline with neoglacial climatic deterioration.

In reference to vegetation, the question of the seral versus climax status of lichen woodland needs to be addressed. There are compositional (if not structural) variations in lichen woodlands across North America, with the most important being understory dominance by *Stereocaulon paschale* in the west versus *Cladonia stellaris* in the east. Lichens can dominate the ground vegetation rapidly if an area is cleared by fire, especially if the substrate is such that water is in short supply (Kershaw 1977). Under these conditions, tree cover is inhibited long enough for a complete cover of lichens to develop. The lichen mat will then serve to decrease summer soil temperatures, further retarding tree growth and, for at least a period, conserving the open nature of the woodland. Brown and Mikola (1974) have also found that *C. stellaris* inhibits the growth of pine and spruce seedlings, though the mechanisms involved in this process are not known (Kershaw 1977).

Although our record of historical fires, even over the past 100 years, is incomplete for the North American taiga, it is clear that fire suppression, largely in the past 50–60 yr, has altered fire recurrence intervals (Lynham and stocks 1991), especially around human settlements.

If fires are excluded from a woodland site for at least 200 years, a closed-canopy spruce–feathermoss forest is likely to develop, especially in the west (Kershaw 1977). Layering is prevalent in lichen woodlands in these open-light situations; hence, trees eventually are able to dominate the site because lichens are largely intolerant of shade. However, fires are rarely excluded from any given area for 200 years; thus, lichen woodlands are perpetuated. As Auclair (1983) notes, lichendominated systems are predisposed to fire due to the continuous nature of these lichen fuels on the ground surface, their high surface-to-volume ratio, and the rapid descication of these lower plants.

Fox (1983) has compiled a summary of the postfire succession of small mammal and bird communities in the north American boreal forest (and tundra). Bird assemblages exhibit a broad increase in species diversity, species richness, numbers and consuming biomass through succession while biomass decreases, but the trend was not monotonic (Fox 1983). In contrast, small mammal assemblages, numbers, biomass, and consuming biomass are higher in the herb stage of succession and lower in the shrub and sapling stages than in mature forests; species diversity of small mammals is high and increasing up to the shrub stage and lower in the sapling stage than in mature forest (Fox 1983). Fox notes that these trends reflect changes in ground vegetation (which provides food and cover for the small mammals) and in horizontal and vertical vegetation structure as the shrub and tree canopy develops, which is important for the bird assemblages.

Productivity

The productivity of boreal forest communities is a function of latitude, maritime/continental position, substrate, topographic position, and seral status. Net primary productivity varies between 400 and 2000 g m^{-2} yr^{-1}, with an average value of 800 g m^{-2} yr^{-1}, relatively low compared with those for other forests (Whittaker and Likens 1975). Decreased productivity in the continental north is well documented by smaller annual growth increments, a reduction in aboveground biomass, and the lower stature of trees (Larsen 1965, 1972; Zasada, Van Cleve, Werner, McQueen, and Nyland 1978), with a decreased length of the growing season (Fig. 2.12). (In Alaska, however, the most productive forests are found north of the Alaskan Range, because cooler summers due to marine influence exist south of the range). Tree biomass ranges from 9590 to 163,360 kg ha^{-1} for spruce communities and from 21,500 to 51,200 kg ha^{-1} for aspen communities (Oechel and Lawrence 1985). Bonan (1992) provides further stand-specific data across Alaska and Canada as a basis for simulation modeling of environmental factors and ecological processes controlling the structure and function of boreal forest vegetation, making a mechanistic classification of forest stand types possible. Furthermore, his analyses indicate that, although boreal forests are not floristically complex, they are complex in reference to their environmental factors and ecological processes (Bonan 1992), with forest dynamics being very sensitive to the interactions among site conditions, mosses, and trees.

Tremendous differences in biomass production exist along a gradient from uplands to lowland river bottoms. Uplands are dominated by evergreens, which tend to be nutrient-conservative (retaining foliage for several years) and hence often are adapted to nutrient-poor conditions. As a thick organic (peat) mat builds up during succession, the thickness of the 0 horizon increases, with nutrients largely found in the 0 and A horizons. Organics, largely in an undecomposed state, are stored on the forest floor. Decomposition activity is slow because of low air and soil temperatures, a decreased thickness of the active layer, and impeded drainage. The shallow root systems of *Picea* and other boreal conifers (such as *Larix*) are well adapted to these conditions up to a point.

Nutrient availability in the soil generally decreases over time, though the forest biomass may continue to increase. In early seral stages, most of

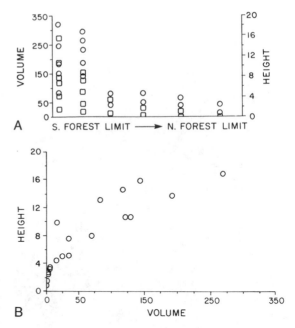

Figure 2.12. (A) The decreasing productivity of the boreal forest with increasing latitude is reflected by decreases in stand volume (m^3 ha^{-1}, squares) and in mean tree height (m, circles). (B) Stand volume and tree height are positively correlated ($r^2 = 0.781$ at the 0.01 confidence level). Data from Black and Bliss (1978).

the biomass ($\approx$ 99 %) is found in the overstory, whereas understory vegetation (especially the mosses) plays a dominant role in later stages (tree biomass <50%). The partitioning of nutrients in the system then changes. For example, nitrogen availability decreases in the soil, but nitrogen accumulates in the phytomass. Living feathermosses act as a nutrient "sponge," so that any nitrogenous or ionic nutrient that is deposited on the moss surface is absorbed by the moss and immobilized (Oechel and Lawrence 1985). This may, however, limit growth on some sites, because not all species are well adapted to low nitrogen levels (Zasada et al. 1978). The upland forest community may then operate as a relatively closed system unless the succession progresses to an even more nutrient-poor state, that of peatland. The partitioning of bio-mass between the overstory, forest floor, and rhizosphere is shown in Table 2.7 for select forest communities in Alaska (Zasada et al. 1978).

A peatland (bog or muskeg) may develop in such a way that nutrients are largely unavailable for tree growth. This successional pathway does not usually lead to increased nutrient availability and ameliorated conditions, in contrast to the situation in more temperate regions, unless thermokarst collapse and erosion occur (Luken and Bil-

Table 2.7. *Distribution of biomass for selected boreal forest communities in Alaska.*

Community dominant	Stand age (yr)	Biomass (g m^{-2}, dry wt.)		
		Overstory	Forest floor	Roots
Alnus incana	20	4790	2078	2473
Populus tremuloides	50	4872	3728	1804
Betula papyrifera	60	11,894	6877	4429
Populus balsamifera	60	18,031	2222	6678
Picea mariana	61	2416	13,325	1040
Betula papyrifera	120	14,763	4374	5467
Picea mariana	130	10,398	11,923	5169
Picea glauca	165	24,945	19,786	12,401

Source: Zasada et al. (1978).

lings 1983). Furthermore, Pare and Bergeron (1995) point to the role of insect outbreaks and windthrow in some boreal forest stands as further factors in forest decline after the aggradation phase of forest succession.

Picea mariana communities in northernmost latitudes and some maritime settings are then the least productive forest communities of the boreal region, partly because of their tolerance of poor site conditions. *Picea glauca*, which is more demanding but may be competitively excluded from poorer sites, is more productive.

In contrast, lowland floodplain communities are very productive. Hardwoods such as *Populus balsamifera* dominate these sites and tend to be less nutrient-conservative than the conifers. These floodplain communities function more as open systems, because nutrients are periodically added to the system by flooding. Flooding also erodes the mat of insulating organic matter; hence, soil temperatures are higher at these sites. The high groundwater table not only promotes maintenance of a deep active layer but also provides additional nutrients to the rooting zone. However, water infiltration promotes loss of soluble constituents through leaching, with these coarse soils exhibiting a low, capacity for cation exchange. Organic matter turns over relatively rapidly in these systems (Zasada et al. 1978). Such communities are often kept in a state of disequilibrium, with lower seral states being maintained by disturbance. Nonetheless, nutrient retention in the biomass increases through the successional cycle.

AUTECOLOGY OF TREE SPECIES

The trees of the boreal forest are relatively diverse in terms of their floristic affinities and histories, life histories, and ecologies. The common trees of bo-

real North America number approximately 18 species. The cold-hardy Pinaceae dominate this list, with 5 genera and 13 species. *Picea glauca* and *P. mariana* are the most widespread members of this family, ranging from Alaska to Labrador and from the northern to southern regions of the boreal forest. The deciduous guild in the boreal forest is dominated by the Salicaceae. The genus *Populus* has four common species in the boreal forest, with one of these, *P. tremuloides*, being the most wide-ranging tree species in all of North America.

Important ecological attributes of the eight most common dominant tree species of the boreal forest are shown in Tables 2.8 and 2.9. Unfortunately, many studies of these species have been done in only the southern portions of their ranges because of their greater economic significance in these areas. Inferences can generally be made for the northern populations, and research on these populations is increasing. Refer to Fowells (1965) and the U.S. Department of Agriculture (1974) for information on all the common tree species of the boreal forest.

The boundaries of the range of any species are determined by the genetically controlled tolerance ranges of its populations and the biotic and abiotic factors of the environments of its individuals, including competition and drought stress. At both the northern and southern limits of the boreal forest, trees are exposed to both severe and highly fluctuating environmental conditions (Vowinckel, Oechel, and Boll 1975; Larsen 1980). Researchers generally agree that plants survive at the limits of their ecological amplitude (range) through conservative allocation of their resources, especially energy reserves.

The ecophysiology of boreal tree species is poorly understood. Although no one has studied carbon balance (photosynthetic rates) for all the

Table 2.8. *Ecological attributes of common boreal tree species*

	Balsam fir	Paper birch	Larch	White spruce	Black spruce	Jack pine	Balsam poplar	Aspen
Evergreen	Yes	No	No	Yes	Yes	Yes	No	No
Deciduous	No	Yes	Yes	No	No	No	Yes	Yes
Needleleaf	Yes	No	Yes	Yes	Yes	Yes	No	No
Broadleaf	No	Yes	No	No	No	No	Yes	Yes
Growth rate	Moderate	Rapid	Rapid	Moderate	Moderate	Rapid	Rapid	Rapid
Max. age (yr)	200	200	350	600	350	230	200	200
Plasticity	Moderate	High	Low	High	High	High	High	Low
Disturbance tolerance	Low	High	Moderate	Low	High	High	High	Moderate
Competitive ability	Moderate	Moderate	Low	High	High	Moderate	Moderate	Low
Frost-free period (d)	80	75	75	60	60	80	75	80
Site requirements								
Substrate preference	Acidic loam	None	Acidic	Basic loam	Acidic	Acidic	Alluvium	Basic
Permafrost table	Moderate	Low	High	Low	High	Low	Low	Low
Soil moisture	Mesic	Xeric-mesic	Mesic-hygric	Xeric-mesic	Xeric-hygric	Xeric-mesic	Mesic	Mesic
Shade tolerance	High	Low	Low	High	High	Low	Low	Low
Nutrient requirements	Moderate	High	Low	Moderate	Low	Moderate	Moderate	High
Seral status	Late	Early	Subclimax	Late	Early-late	Early	Early	Early

Table 2.9. *Reproductive attributes of common boreal tree species*

	Balsam fir	Paper birch	Larch	White spruce	Black spruce	Jack pine	Balsam poplar	Aspen
Vegetative growth								
Layering	Yes	No	Yes	Yes	Yes	No	No	No
Root suckers	No	No	No	No	No	No	Yes	Yes
Crown/stump sprouts	No	Yes	No	No	No	No	Yes	Yes
Sexual reproduction								
Age of maturation (yr)	15	15	12–15	13	10	5–10	15–20	20
Period of seed set (yr)	2	1	3	3	3	3	1	1
Seed crop frequency (yr)	2–4	1	2–6	2–6	4	3–4	3–4	4–5
Germination requirements								
Light	Low	High	High	Low–high	High	Moderate–high	High	High
Temperature (°C)	21	?	18–21	15	15	21	?	?
Moisture	Moderate	High	Moderate	Moderate	Moderate	Moderate	High	Moderate
Cone retention	No	No	A few yr	No	Several yr	Many yr	No	No
Dispersal distance	Low	High	Low	Moderate	Low	Low	High	High
Optimal seedbed	Organic/mineral	Mineral	Organic/mineral	Mineral	Organic/mineral	Mineral	Mineral	Mineral

tree species (or their understory flora) at a particular site, species-specific measurements have been taken at a few sites. The photosynthetic period has been shown to be longer for evergreen than for deciduous taxa because of leaf senescence in the late summer for the deciduous plants (Oechel and Lawrence 1985; Prudhomme 1983). Maximum photosynthetic rates (P_{max}) are also consistently higher for deciduous hardwoods versus conifers, though the maximum rates reached during the growing season are similar (Oechel and Lawrence 1985). Temperature does not generally limit photosynthesis during the growing season. Photosynthesis remains positive from temperatures less than 0° C to those greater than 40° C. The temperature optimum within this range for black spruce in central Quebec is approximately 15° C (Vowinckel et al. 1975). Light intensity has been proposed as a possible factor limiting summer photosynthesis in areas with frequent cloud cover, though the low light compensation and light saturation points of many boreal forest species allow daily carbon gains under low light intensities (Vowinckel et al. 1975); Skre, Oechel, and Miller 1983; Oechel and Lawrence 1985).

Chapin (Chapin, Tyron, and Van Cleve 1983, 1986) has studied the growth rates of seedlings of Alaskan taiga tree species under controlled conditions. His research depicts poplar as the most productive taiga tree, followed by aspen, birch, and alder, with all conifer species (larch, white spruce, and black spruce) growing more slowly. Furthermore, Chapin shows that there is a strong positive correlation between the growth potential of these seedlings and the productivity of 60-yr-old individuals of the same species in natural forest stands (Chapin 1986). The species also differ in their age-specific growth rates, with aspen declining at 60 yr whereas in contrast relative growth rates of white spruce remain high for at least 350 yr (Chapin et al. 1983). These growth rates are thus linked to successional processes.

The study of water stress on boreal tree taxa has been confined to *Picea mariana*. Field measurements of water potential indicate little water stress, with minimum water potentials of −1.5 MPa at Schefferville, −2.1 MPa in the Mackenzie Delta (Black and Bliss 1980; Oechel and Lawrence 1985). However, measurements of water stress by Black and Bliss (1980) on greenhouse-grown seedlings show a 50% reduction in net photosynthesis under a water potential of −1.5 MPa, and a decrease in leaf-water potential to −2.5 MPa, further reducing net photosynthesis to the compensation point. In the field, trees are able to maintain a constant turgor of −1 MPa by osmoregulation, allowing continuation of growth and cell expansion (Black and Bliss 1980).

Long-lived trees must first allocate resources to trunk and foliage maintenance ("self-maintenance") and may therefore exhibit sporadic sexual reproduction to marginal environments. Studies have shown that allocation of resources to seed production decreases the growth of the individual (Harper 1977). The successful marginal trees will then be of species that also have the ability to reproduce vegetatively (through layering, root suckers, or stump/crown sprouts), because environ-

mental conditions necessary for the development of vegetative tissues are not as exacting as those for the development of sexual organs and propagules (Elliott 1979c; Elliott-Fisk 1983). Whether the processes of clonal growth are competitive with those involved in producing seeds is not known (Harper 1977); nonetheless, vegetative reproduction (clonal growth) is a valuable adaptive attribute in the boreal forest because of natural cyclic disturbances by fire, flooding, windthrow, and rises in the permafrost table.

All the dominant tree species of the boreal forest are able to reproduce vegetatively to some extent, with the exception of *Pinus banksiana* (Table 2.9). Species in the genus *Pinus* rarely layer, but members of this taxon in the boreal forest are adapted to fire because they possess serotinous or semi-serotinous cones that retain a viable seed store on the tree for several years. *Picea mariana* also has this valuable trait.

Deciduous trees are adapted to coping with site disturbances by the production of large crops of light wind-dispersed seeds. Populations may be maintained on site through rapid root suckering and stump/crown spreading. Many of these species are "r-strategists," though some are able to remain the climax vegetation on special sites. Deciduous trees also frequently occur along floodplains.

True mixed (evergreen and deciduous) forests occur only in special situations, and thus a comparison of their competitive abilities is warranted. Trees of many of the deciduous species (such as *Larix laricina*) are able to outgrow the evergreens at the seedling and sapling stages, thus attaining dominance in the canopy. Chapin (1986) has shown that all the broadleaf deciduous tree species, as well as the deciduous larch, outgrow the evergreen conifers at the seedling stage and as 60-yr-old adults. The eventual demise of these deciduous trees in many stands may be a function of the evergreen's photosynthetic capabilities. The evergreens have an advantage where the growing season is either short or unpredictable, and they are able to retain their needles for as many as 15 yr (Black and Bliss 1980).

Insect pests and biological agents may damage or kill individual trees and occasionally entire stands. Defoliation by budworms (*Choristoneura fumiferana* and *C. pinus*), tent caterpillars (*Malacosoma disstria*), and the larch sawfly (*Pristiphora erichsonii*) is an episodic problem in certain forest regions. Spruce beetle (*Dendroctonus rufipennis*) infestation can lead to the death of mature and overmature white spruce, but this is easily controlled by salvage cutting. Fungal pests are of minor importance, as is dwarf mistletoe (*Arceuthobium american-*

um and *A. pusillum*), which triggers the formation of witches' brooms in spruce and pine (Moody and Cerezke 1984).

Holling (1992) shows how species-specific insect, foliage and tree life, and growth periodicities can become coupled and result in the loss of a tree stand (or forest patch), and then how this loss can spatially shift to larger mesoscales with dispersal of forest insects. The disturbance/death of a tree species becomes self-propagating, effecting the spatial distribution of age structure for a species and thus the forest species composition as well. The competitive edge that one species (e.g., spruce or fir or pine) may have over another will then shift (Holling 1992). Insect outbreaks that spread can thus be modeled much like forest fire through the taiga and can shift up the hierarchical scale from the local patch to the landscape scale (Holling 1992).

Rowe (1983) further discusses the autecology of boreal forest trees, as well as shrubs, herbs, and lower plants, in reference to their fire strategies as fire-invaders, -evaders, -avoiders, -resisters, and -endurers. He classifies *Pinus banksiana*, *P. contorta*, and *Picea mariana* as fire-evaders because they hold their seed banks largely in serotinous to semiserotinous cones at some height off the ground. Rowe also classifies adults of *Pinus banksiana* and *P. contorta* as fire-resisters because of the thick bark of the pines. *Abies balsamea* and *Picea glauca* are regarded as fire-avoiders that arrive late in succession and become dominants where fire cycles are long. Fire-endurers that re-sprout following a fire include *Populus tremuloides*.

From Table 2.8 it appears that tree species of the boreal forest have relatively narrow site requirements. It should be remembered that these are *optimal* requirements. The majority of the species are able to tolerate other conditions, though populations on these sites may be less vigorous and eventually become competitively excluded. Many of these species are also phenotypically plastic, with ecotypes and local genetic populations (Halliday and Brown 1943; Elliott 1979c). Hybridization, which is frequently introgressive, has also been documented between many congenerics (Gordon 1979; Elliott 1979c).

Picea Glauca and Picea Mariana

The genus *Picea* A. Diertr. is composed of approximately 40 species of evergreen trees ranging through cool, temperate regions of the Northern Hemisphere. *Picea glauca* is geographically the most widespread member of this genus and of the Coniferales as a whole in North America (Nien-

staedt and Teich 1972; Elliott 1979c). However, *P. mariana* is the most frequently encountered tree of the boreal forest, because it is able to occupy a wider range of sites than *P. glauca*.

The phenology of *Picea* has been outlined by Zasada and Gregory (1969), Rowe (1970), and Owens and Molder (1976, 1977; Owens, Molder, and Langer 1977). These monoecious trees have distinct male and female cones, though bisexual cones are occasionally found (Zasada et al. 1978; Elliott 1979a, 1979b). Three years are required for completion of the sexual reproductive cycle from the initiation of axillary bud scale to seed maturation.

Trees of *P. mariana* reach sexual maturity slightly earlier than those of *P. glauca*; so black spruce populations may be able to invade new habitats more rapidly. Although seed crop intervals are usually more frequent for *P. mariana* in northern latitudes (Zasada 1971), *P. glauca* produces more cones (and seeds) in a typical crop. Seedfall peaks in late summer and autumn for *P. glauca*, whereas *P. mariana* releases most of its seeds immediately following a fire. Some seeds are released in both species throughout the year, though this is minimal. Because fire frequency is highest in the north in June and July (with frontal activity), *P. mariana* releases most of its seeds immediately following a fire. Trees of *P. mariana* retain their seed in semiserotinous cones on the tree, an appropriate strategy, because a thin seed coat and small endosperm make the embryo vulnerable to destruction on the forest floor. Rowe and Scotter (1973) stated that *P. mariana* is the most successful regenerator of all boreal trees following a fire; this species appears to be the most opportunistic of the boreal trees.

Although the sizes and morphologies of the seeds of these two spruces differ only slightly, *P. glauca* has proved to be a better disperser of its seed, a useful strategy, because it is susceptible to destruction by fire (Zasada and Gregory 1969; Elliott 1979c). Squirrels are heavy predators on *P. glauca* cones and are effective agents for seed dispersal, though they may reduce the available seed crop (Rowe 1952; Elliott 1979b).

Seedling establishment is optimum for both species on mesic sites, though seedlings of *P. mariana* can tolerate more moisture than those of *P. glauca*. Growth of both species, however, is poorest on wet sites (Maini 1966). Although some early research showed that *P. mariana* seedlings are also able to grow a little faster than those of *P. glauca*, controlled growth experiments generally show more rapid growth of *P. glauca* (Chapin 1986), which also has adult individual trees that are longer-lived. *P. mariana* clones, however, may outlive *P. glauca* individuals or clones.

Layering is common in both spruces, though *P. mariana* layers vigorously in almost all environments, whereas *P. glauca* layers only occasionally and then only in open, marginal sites (Elliott 1979b, 1979c; Elliott-Fisk 1983). Layering allows populations to persist during times of relatively minor (but perhaps prolonged) climatic deterioration (Elliott-Fisk 1983; Lavoie and Payette 1994, 1996).

Both *P. mariana* and *P. glauca* are phenotypically plastic, with numerous growth forms found in different settings. White spruce is typically either an upright tree or simply a dwarfed (short) tree; it does not usually take on a twisted or mat krummholz form. Black spruce, on the other hand, assumes many growth forms, both in individuals and in clones. This has triggered discussion of which of the species is hardier. Hustich (1953) states that *P. glauca* is hardier, keeping an erect form on sites where *P. mariana* is prostrate. Yet *P. mariana* has a much easier time propagating vegetatively through adventitious rooting than does *P. glauca*. These decumbent forms seem well adapted to the environment, with their height equaling that of the average snow cover (thus, shoots are not exposed to desiccating, abrasive winter winds and low temperatures).

Occasionally there is confusion in identifying particular specimens as either *P. glauca* or *P. mariana*, especially in marginal locations where cone production is infrequent. This may be due to introgressive hybridization of the two species (Dugle and Bols 1971; Elliott 1979c; Larsen 1965, 1980; Little and Pauley 1958). Species of *Picea* are known to hybridize readily (Owens and Molder 1977). Hybrids have been reported from Minnesota (the Rosendahl spruce of Little and Pauley 1958), Manitoba, Ontario, British Columbia, and the Northwest Territories (Larsen 1965; Roche 1969; Dugle and Bols 1971; Parker and McLachlan 1978). Unfortunately, the techniques used in identifying putative hybrids have not been consistent. Von Rudloff and Holst (1968), for example, identified the Rosendahl spruce on the basis of leaf oil terpenes, but other studies have focused on morphology. Some individuals that have been studied have exhibited intermediate morphologies for some characters but not others, and this may be a function of whether the individual is a true F_1 hybrid or whether the hybrid population has back-crossed with a parental type (Abercrombie et al. 1973; Parker and McLachlan 1978). Further work is needed, especially on introgressive hybridization of these species at the northern tree limit, an area where stress may induce rare and unusual reproductive events (Anderson 1948; Elliott 1979c). In such an area, species that would otherwise maintain quite separate identities occasionally hybridize, with introgres-

sion resulting in an intermingling of morphological traits of each species.

Although both *P. glauca* and *P. mariana* are wide-ranging through the boreal forest, occasionally even being found as co-dominants in a stand, the species have distinct site preferences that have no doubt promoted their maintenance as separate species. *Picea glauca* is the more site-demanding of the two species, strongly preferring well-drained, basic mineral soils and being intolerant of high permafrost tables. *Picea mariana*, on the other hand, is more tolerant of poor environmental conditions and is able to survive in poorly drained areas and on various substrates (though, like *P. glauca*, it does best on well-drained sites). Black spruce is frequently found on poorly drained acidic soils (i.e., peat) that are unsuitable for most tree species. It is not as nutrient-demanding as white spruce, which according to Fowells (1965) is one of the most exacting conifers in terms of nutrient requirements. The very shallow root system of *P. mariana* allows it to exist in areas where there is a high permafrost table, but this also makes it susceptible to windthrow; its prostrate growth form at exposed treelimit sites may compensate for the shallow root system (Elliot 1979c). Black spruce is not well adapted to coastal fog (Payette and Gagnon 1979) or to salt-laden winds. This is thought to be the explanation for its limited occurrence in maritime settings, in both Labrador–Ungava and Alaska, where *P. glauca* dominates and is the northern treeline species (Elliott and Short 1979).

The two species form mixed stands, sometimes in association with *Larix laricina*, on well-drained sites near the northern forest border (Larsen 1965; Elliott 1979b). *Picea glauca* is frequently the larger, older tree, but because both species are shade-tolerant climax species, neither is excluded. Stands such as this are hypothesized to have escaped fire for several hundred to perhaps several thousand years (Elliott 1979a, 1979c; Elliott-Fisk 1983), in agreement with the finding by Johnson and Rowe (1975) that the frequency of fires is lower north of the forest border, outside of frequent frontal-zone activity. It is certainly well established, however, that fire is a natural process at the northern forest border, in the forest-tundra ecotone, and in the tundra itself (Auclair 1983).

HISTORICAL ASPECTS AND MANAGEMENT IMPLICATIONS

Paleoecology

The history of the boreal forest is complex and is not understood in its details because of the scarcity of Tertiary fossiliferous materials from boreal associations and the magnitude and frequency of late Cenozoic glacials and interglacials. Our best picture of historical forest dynamics is from late glacial and particularly Holocene deposits, as these deposits are typically better preserved, more abundant, and more readily dated than older materials. We must also remember that each species has its own unique floristic history. Variations in genetic plasticity, dispersal strategy, and competitive ability allow floras to evolve through time, experiencing environmental changes, such that present communities may no longer resemble those of even the recent past.

Both the boreal and Arctic floras of North America are believed to be derived from the Arcto-Tertiary geoflora. Conifers and deciduous hardwoods were dominant in this geoflora, which ranged as much as 20° north of the present boreal and Low Arctic vegetation zones (Larsen 1980). Wolfe (1971, 1972, 1978) and others have debated the circumpolar extent of this geoflora, contesting that it did not exist in all northern regions (e.g., Alaska). This is certainly possible, because North American geofloras are known to have intermingled, much as our present floras do, giving the flora of a particular region multiple origins.

The earth's Cenozoic climatic history (spanning the last 70 m yr) is well known on the global scale. The first part of this period saw a general warming trend (with more tropical conditions globally) that lasted to the Eocene-Oligocene boundary about 38 m yr ago, followed by a rapid cooling triggered by the opening of Drake Passage between Antarctica and Australia. This allowed the southern circumpolar current (Westerly Wind Drift) to develop, thermally isolating the high latitudes, and resulted in the spread of cool-temperate taxa globally and, possibly, the evolution of polar and subpolar floras. A gradual cooling trend continued through the Oligocene to the Miocene, when temperatures once again dropped in the higher latitudes. This led to the development in the middle to late Miocene of the East Antarctic Ice Sheet and to mountain glaciation in the Northern Hemisphere. By late Pliocene time (about 3 m yr ago) near the end of the Tertiary, ice sheets had also developed in the Northern Hemisphere, especially adjacent to the North Atlantic Ocean. Boreal forest species and communities continued to evolve throughout this time, with many of the modern tree species coming into existence by the late Pliocene.

Late Tertiary cooling resulted in elimination of many temperate species from the northern floras, leaving evergreen conifers such as *Picea* and *Pinus* dominant (Larsen 1980). Changes in understory vegetation must have accompanied these changes

in the forest canopy, though our knowledge of these is poor.

Research on the Quaternary history of the boreal forest has concentrated on the past 15,000 yr, focusing on (1) the presence of possible glacial refugia of boreal species, (2) the southernmost extent of boreal forest taxa during glacials, (3) rates of northward migration of the forest/woodland accompanying deglaciation, (4) fluctuations of the northern forest border, and (5) vegetation dynamics related to succession, climate change, and anthropogenic disturbance.

It is important to note that many boreal species occurred in associations different from those that exist today. Paleoecological data show that each species behaved individualistically, as proposed long ago by Shreve (1915) and Gleason (1926), because of differences in phenotypic plasticity, reproductive strategy, and competitive ability (Hulten 1937; Larsen 1980; Davis 1983; Webb, Cushing, and Wright 1983). Thus, many of the present boreal forest associations are recent in the geologic sense, having existed for only the past 3000–6000 yr (Ritchie and Yarranton 1978a). This youth may account for the relative impoverishment of the boreal flora and the small number of associations and communities present. Many species can be considered generalists, not occupying tight ecological niches but instead occurring in a wide array of sites (Larsen 1980).

Phytogeographers have long thought that boreal forest species or entire associations may have persisted in isolated glacial refugia during the Wisconsin or previous glacials. Matthews (1974) and Rampton (1971) have reported that their studies support the existence of local spruce communities in Alaska during the mid-Wisconsin glacial. Limited areas of spruce woodland are also documented for east-central Alaska during the late Wisconsin glacial, when the elevational treeline was 400–600 m lower than today (Colinvaux 1967; Hopkins 1967; Matthews 1970, 1979; Pewe 1975). Ager (1975) and Hopkins and associates (1981) have presented pollen and macrofossil evidence for a refugium in interior Alaska, with *Picea, Betula papyrifera, Larix laricina,* and an *Alnus* species present. This refugium was eliminated from Alaska by low temperatures during the last glacial; however, it is probable that some *Alnus* species, *Populus balsamifera,* and possibly *P. tremuloides* survived in unglaciated Alaska during the Wisconsin glacial, perhaps in thermal belts on hill slopes (Hopkins, Smith, and Matthews 1981; Ager 1982). The existence of refugia on nunataks (unglaciated highlands and lowlands) in eastern North America is more controversial (Ives 1974; Short 1978a).

The glacial and postglacial history of the vegetation in Alaska is extremely complex because of topographic variability, multiple glaciations of mountainous areas, and repeated land-bridge connections with Siberia (Ager 1983). In contrast, most of the Canadian Shield and surrounding planated rock formations were covered by the Laurentide and Cordilleran ice sheets. Far-southern positions of boreal species have been found in the southeastern United States (Watts 1970) some 1200 km south of their present limits. Various tundra formations occupied much of the area immediately south of the ice sheets, regions now vegetated by various hardwood and mixed evergreen-deciduous forest associations. Vegetation changes in subarctic and temperate North America have been exceedingly rapid in the last geologic epoch.

With the deterioration of the late glacial ice sheets, tundra and forest components migrated rapidly northward, colonizing what is now the area occupied by the boreal forest. Nichols (1975, 1976) found that in central Canada the early Holocene forest species moved about 480 km north in 2000 yr. If we analyze the Holocene history of the entire North American boreal forest, we see a general transgressive pattern of deglaciation, followed by progressive establishment of a herbaceous tundra, shrub tundra, and spruce-dominated woodland or forest. *Picea* generally appeared in southern sites between 14,000 and 10,000 BP and in northern sites by 8000–6500 BP (Ritchie 1976). These site histories have not been summarized for the entire boreal forest of North America because there were great differences in the timing of these vegetation changes, depending on the latitudinal location of the site, among other factors. Some regional summaries, such as that for select localities in Alaska by Ager (1983), are available.

Sites at the margins of boreal species ranges have recorded the most detailed history of Holocene changes, which were principally a function of climate but also were influenced by soil and community development. Along the southern boreal forest border, shifts in grassland, open parkland, and the southern transitional boreal forest have been detected for the Holocene. This southern border has remained in the same position for the past 2000 yr (Ritchie and Yarranton 1978a, 1978b), though anthropogenic changes have occurred in many regions. Shifts in the northern boreal forest border have been more thoroughly documented because the northern treeline has been correlated with several atmospheric parameters, and changes in the treeline position have then been used to record climatic changes and aid in their explanation (Bryson 1966; Sorenson and Knox 1974; Elliott

1979c; Elliott-Fisk 1983). Tables 2.10 and 2.11 illustrate climatically induced fluctuations of the entire northern Canadian treeline, with the accompanying vegetation changes shown for Labrador–Ungava (Elliott-Fisk 1983). Although Holocene climatic changes have been both directional and synchronous across Canada, deteriorations/ameliorations perhaps (1) lagged from west to east, (2) were buffered by local geographic factors, or (3) were not registered in the fossil record owing to the inherent persistence of the vegetation at some sites (Elliott-Fisk 1983). Some portions of the north-central and northwestern Canadian treelines also are relictual (from hypsithermal warmer climates), with low sexual regeneration capacities, and these are susceptible to destruction by further climatic deterioration or anthropogenic disruption (Elliott-Fisk 1983). Furthermore, Lavoie and Payette (1996) show the late Holocene persistence of the northern forest limit in Quebec during the past 2000–3000 yr. Although they state that their spruce macrofossil and peat evidence shows stability of the northern tree limit in this maritime climate during the warm medieval times (ca. 2000 yr BP) and cold periods (ca. 1300 yr BP and the Little Ice Age) of the late Holocene, they further state that this demonstrates that mechanisms allowing forest limit advance or retreat are not easily triggered by climatic change, as has been documented by other researchers. They therefore question the use of ecological models to show future forest migration in response to climatic warming and CO_2 increase, a topic of much current research (Shugart, Leemans, and Bonan 1992; Lenihan 1993; Sirois, Bonan, and Shugart 1994; Martin 1996).

Management Implications

Although human disruption of the boreal forest ecosystem typically results in a succession of seral communities (like those described previously that are associated with fire), particular disturbances can greatly alter not only the vegetation and the associated fauna but also the substrate. The water budget for a site is readily altered by mining, hydropower development, and destruction of the surface vegetation. In northern regions, this can lead to destruction of frozen ground (permafrost) and rapid degradation of the landscape. In certain marginal settings, trees may be unable to regenerate following their removal, reducing the vegetation to a more or less permanent tundra subclimax (Larsen 1965, 1980; Strang 1973).

The majority of the boreal forest, however, is suitable for sustained harvest; 25% of Canada's land (approximately 10^9 ha) is covered by forest

Table 2.10. *Reconstructed Holocene treeline migrations for Canada*

Years BP	Western Canada	Central Canada	Eastern Canada
1000	RETREAT (major cooling) dwarf birch/heath tundra	RETREAT (cooling) forest-tundra	(further cooling?)
2000		ADVANCE (minor warming) forest	CONTRACTION (cooling) decreased productivity
3000		?(minor warming)?	
4000		RETREAT (cool) tundra ?RETREAT (onset of cooling)?	ADVANCE (warming) spruce woodland
5000	RETREAT (cooling) tall shrub tundra		
6000		ADVANCE (warming) forest	ADVANCE (warming) tall shrub tundra
7000	ADVANCE (warming) closed spruce/birch	(cool tundra)	(peat initiation)
8000	forest	DEGLACIATION	DELAY IN MIGRATION
9000			(cold, dry)? low (herb) tundra
10,000	ADVANCE (warming) forest-tundra		
11,000			DEGLACIATION (?)
12,000	(cool) low tundra		
13,000	DEGLACIATION		

Source: Elliot-Fisk (1983).

Table 2.11. *Holocene vegetation change in Labrador–Ungava. Years are yr BP.*

Site[a]	Low tundra	Shrub tundra	Spruce woodland	Contraction
Hebron Lake	10,200	6900/6400	absent	none
Napaktok Lake/Bay	8700	5500	4800/4300	2200
Ublik Pond	10,500	6500	4300	300/2400
Umiakoviarusek Lake bog	absent	absent	2650	1000
Nain Pond	8000(?)	6500	4500	2500
Pyramid Hills Pond	7000	6500	4500	2500
Kogaluk Plateau Lake	9000	6700	4400	3000(?)

[a] Sites arranged from north to south and from coast to interior.
Source: Short and Ellott-Fisk (unpublished).

suitable for regular harvesting (Larsen 1980). About 20% (approximately 220 × 10[6] ha) of this is being harvested (Love and Overend 1978).

Harvest is typically accomplished by clear-cutting. This form of logging can have seriously negative consequences for surface and ground water quality, soil erosion, and wildlife. Some of the largest clearcuts in the world have been created in Canada late in the 20th century. The largest single clearcut – 270,000 ha – may be in what had

been the Gordon Cousins Forest of central Ontario. Provincial governments have sold timber rights to vast, cumulative areas in the last two decades. For example an area the size of Ohio has been sold in Manitoba and an area the size of Great Britain has been sold in Alberta. British Columbia's annual harvest has been averaging 60 million cubic meters in the 1990s, a volume which ecologists estimate is 30% above the sustainable yield. Canada's boreal forest has been called "the Brazil of the north" by critics of these provincial logging practices (Devall 1993). Reforestation and management of boreal ecosystems are the primary concerns of several governmental agencies, with the Institute of Northern Forestry of the USDA Forest Service (Fairbanks, Alaska) a prime example. A great deal of attention has been paid to genetic improvement of breeding stock and management of plantations to reduce rotation time between harvests. The following recent studies illustrate the types of research being done to better address adaptive management of the North American taiga.

Forest genetics have long been a concern of silviculturalists, with the careful selection of genotypes for plantation and reforestation efforts. Wang, Lechowicz, and Potvin (1994) state that future selections should be made on the basis of genotypes responsive to climatic changes, specifically elevated CO_2, increased daily mean temperatures, and nitrogen enrichment. In their study of *Picea mariana*, families are selected for rapid growth under present conditions that will also do well under future conditions, especially on sites regenerating by seeding. In this regard, the authors state that Holocene historical data show that the natural migration of existing stock will be too slow to regenerate productive forests with trees suited to the changed environment, and that the indigenous provenances may not give maximum yields in the new climatic regime (Wang et al. 1994). They belive that the rapid development of improved stock and replanting on a massive scale may become a necessity to maintain a forest with reasonable timber yields.

Wein (1990) concurs that the northernmost trees will not be able to migrate as quickly as the climatic forest-tundra boundary is predicted to shift (potentially 300–900 km northward) due to climatic change over the next 30–50 yr, a doubling of CO_2, and a temperature increase of 1.5–4.5° C in Canada's northern forests. These climate conditions will stress some tree species and lead to an increase in fire frequency and intensity, increasing the fuel load over the short term. Wein (1990) believes that wildfire could reinforce the greenhouse effect by releasing CO_2 with combustion.

However, it should be noted that a forest succession model designed to simulate the dynamics of subarctic spruce-lichen woodland in northeastern Canada (Sirois et al. 1994) suggests that under a climate-warming scenario, physiognomic response in the woodland would be more pronounced in areas subject to moderately frequent forest fires in contrast to those areas where fire frequency is very low. This model takes into account not simply tree growth but seed regeneration, sim-

ulating the regeneration dynamics more realistically than most other gap models in which successional processes are driven by resource constraints on tree growth.

The implications of various silvicultural practices on wildlife species are continuing to be assessed as a contribution toward better ecosystem management practices. For example, Sturtevant, Bissonette, and Long (1996) have modeled the population dynamics of the American marten (*Martes americana*), an associate of late-successional–old-growth forests, for various stand-level prescriptions and scaling up to a landscape management area. In Newfoundland, the marten has been listed as threatened and in a crisis situation, with only a single recognized viable population. Research shows that a system of old-growth forest reserves for the marten and associated species cannot be sustained indefinitely due to the natural dynamics/disturbance events of the taiga ecosystem. Staging timber harvest to ensure the persistence of marten habitat on the landscape is not feasible, although it is recommended that leaving coarse woody debris following harvest will help the martens, for it will provide habitat for their prey (e.g., voles) (Sturtevant et al. 1996).

Martin (1996) concurs that, in light of future climate change, a system of forest preserves will not protect boreal ecosystem biodiversity. As zonal and mesoclimates shift, ecosystems will respond and shift beyond the edges of the targeted preserved lands. Welsh and Venier (1996) further add that conservation strategies need to address all geographic scales (from the individual plant to the landscape), and they propose using both a "binocular" and a "satellite" perspective in planning processes. Birds are used in their case study to illustrate the complexity of vegetation mosaics for avifauna in relation to resident birds, short-distance migrants, neotropical migrants, and nomadic species. They propose using a vegetation classification scheme as a reference for birds and all wildlife species. Furthermore, they state that reserves should be a part of conservation strategies.

On a more general basis, Harvey, Leduc, and Bergeron (1995) recommend the use of forest models and competition indices to facilitate the design of forest plantations, vegetation management regimes, and wildlife habitat decisions. They present a multivariate analysis for boreal forest communities in northwestern Quebec that allows for site stratification and classification, using many environmental parameters. Bondrup-Nielsen (1995) states that the timber industry must develop environmentally sound management practices to ensure that wildlife populations do not become further endangered. Process and pattern are key and must mimic natural landscape conditions such that a diverse mosaic of stand types are retained on the landscape. Geographic information systems are proposed as a tool to aid management decision-making and landscape design and preservation. Fitzsimmons (1995) carries this a step further in suggesting that the emphasis of management should be shifted from the vegetation to the atmosphere in light of present and future changes in air pollution and climate. The reduction of CO_2 emissions is strongly advocated.

Acid deposition by both wet and dry mechanisms has been proposed as a potential threat to northern forests. Acid rain is the product of long-distance transport and transformation of sulfur and nitrogen oxides from combustion and fossil fuels (Shugart 1984). At the present time, only the southeastern portions of the boreal forest (Ontario, southern Quebec, and southern Labrador), with an average annual precipitation pH of 4.5–5.0, appear to be vulnerable to acid deposition. They are (at least seasonally) downwind of major sulfur sources in the industrialized Northeast and Great Lakes region (Turk 1983). With the dominant wind circulation from the west over North America, most of the boreal forest in Canada and Alaska is not susceptible to regional acid deposition (Barrie 1982; Turk 1983; Shugart 1984).

The future impact of acid deposition on the North American boreal forest should be carefully monitored (Hall 1995). However, at present, the link between acid deposition and forest health is still unclear for the following reasons:

1. Site-specific studies near large smelters show acute effects immediately downwind, but chronic exposure to multiple pollutants from multiple sources in a regional airshed has not been documented (Shugart 1984).
2. The region lacks a historical sampling network and data base on the acidification of natural waters and soils; however, a network of precipitation sampling sites for monitoring acid deposition was established in 1970–1980 (Barrie 1982).
3. The pH of unpolluted precipitation can vary between 4.5 and 5.6, making trends in acidification difficult to detect (Turk 1983).
4. The presence of a winter background of sulfur in Arctic air masses makes it difficult to detect anthropogenic sulfur; this is coupled with a general lack of careful evaluation of natural versus anthropogenic sources of acidity (Barrie 1982; Turk 1983).
5. Field studies have not documented the impact

of SO_2 and other regional-scale pollutants on the dynamics of forest stands (Shugart 1984).
6. The results of natural soil formation (leaching of nutrients, release of aluminum, and acidification of soil and water) are the same as those attributed to acid deposition (Krug and Frink 1983).
7. The ability of hydrogen ions to remove nutrient cations from soils more acidic than pH 5 is low (Krug and Frink 1983).
8. The proportion of bases to acids in precipitation is usually greater than that in strongly acidic forest soils (Krug and Frink 1983).
9. Changes in land use and subsequent vegetation recovery and soil recovery following disturbance play an important role in an area's susceptibility to acidification (Krug and Frink 1983).

It is safe to state that the boreal forest ecosystems present many management challenges because of the ecological complexities of these systems. Despite this, many natural resource managers see boreal forests as being much easier to manage than many other natural vegetation types, for the boreal ecosystem is adapted to widely fluctuating environmental conditions and thus can better resist exploitation than can ecosystems adapted to more stable environments or narrow ecological regimes (Larsen 1980).

AREAS FOR FUTURE RESEARCH

The boreal forest remains relatively unexplored and poorly understood. Physical factors exert important influences on the natural vegetation, yet the physiological tolerances of even the most dominant species are not well known. Environmental conditions in a stand can change markedly through the successional process, such that conditions that favor maintenance of mature trees are unfavorable for the establishment of tree seedlings. Bioclimatological studies are desperately needed to decipher (1) the extension of forest along river bottoms, (2) the ecology of understory lichen and bryophyte species and associations, (3) the physiology of ecotonal shrubs and herbs that occupy both boreal and Arctic vegetation formations, and (4) the existence of shrub lands in sites apparently suitable for forest vegetation.

Though many researchers have tried to link the boreal forest to a particular climatic type, this has not yet been done successfully. Atmospheric scientists need to work with boreal ecologists to link the vegetation to large-scale upper-level dynamics, as well as to surface dynamics, of the atmosphere. Atmospheric chemistry needs to be fully linked to boreal biogeochemical cycling so that we can evaluate the impact of acid deposition on northern ecosystems.

The genetic and phenotypic plasticity of dominant boreal species needs to be investigated if we are to fully understand the evolution and ecology of the vegetation. How can a given tree species (such as *Picea mariana* or *Populus tremuloides*) occupy such a large and apparently diverse geographical area? Future documentation of genotype variation, ecotypes, and introgressive hybridization may more clearly explain this. Such information will also aid future reforestation efforts and the conservation of genetic heterogeneity.

The physiology of boreal trees also needs to be investigated, especially in regard to their short life spans. In temperate subalpine regions, many conifers live to great ages in stressful habitats, yet trees in the equally stressful forest-tundra ecotone are <300 yr old. Many boreal stands can maintain populations for thousands of years during (minor) climatic deterioration. Although several good ecological studies of treeline vegetation have been done, the subject of the dynamics of long-term stands deserves a great deal more effort and will inform our understanding of forest response to future climate change.

The boreal and boreal–Arctic shrublands are vegetation types that have largely been uninvestigated. Shrub communities are important colonizers following deglaciation and prior to forest establishment at many sites. The current distribution and persistence of these communities, in what appear to be some of the best environments for tree establishment, remain mysteries. Little is known about productivity and nutrient cycling in these systems. These stands provide valuable wildlife habitat and browse. In addition, the belowground component in all boreal forest communities warrants further research.

The structure of boreal forest vegetation is well documented. Its history is not completely understood, though great advances are being made on the late Quaternary history of the forest. We must now turn our research efforts toward physiological ecology, bioclimatology, biogeochemistry, wildlife–habitat relations, and ecosystem processes, with integrated studies of plant, animal, physical components, and ecological functions. Well-designed experimental studies will continue to be very useful in improving our basic mechanistic understanding of boreal forest ecosystems and in improving the accuracy of ecological models. Increasing human disruption and destruction of this ecosystem are imminent in the next century, and many biodiversity elements could be lost. Preservation of undisturbed communities is necessary to ensure ecological stability of the landscape. The immediate future is surely going to bring intensive development and change to "northern rim" countries such as Canada (Brownson 1995).

REFERENCES

Abercrombie, M., C. J. Hickman, and M. L. Johnson. 1973. A dictionary of biology. Penguin, Middlesex, England.

Ager, T. A. 1975. Late Quaternary environmental history of the Tanana Valley, Alaska. Ohio State University, Institute of Polar Studies Report 54. Columbus.

Ager T. A. 1982. Vegetational history of western Alaska during the Wisconsin glacial interval and Holocene, pp. 75–93 in D. M. Hopkins, J. V. Matthews, Jr., C. E. Schweger, and S. B. Young (eds.), Paleoecology of Beringia. Academic Press, New York.

Ager, T. A. 1983. Holocene vegetation history of Alaska, pp. 128–141 in H. E. Wright, Jr. (ed.), The Holocene, vol. 2, Late-Quaternary environments of the United States. University of Minnesota Press, Minneapolis.

Anderson, E. 1948. Hybridization of the habitat. Evolution 2:1–9.

Archibold, O. W. 1995. Ecology of world vegetation. Chapman & Hall, New York.

Auclair, A. N. D. 1983. The role of fire in lichen-dominated tundra and forest-tundra, pp. 235–256 in R. W. Wein and D. A. MacLean (eds.), The role of fire in northern circumpolar ecosystems. SCOPE 18. Wiley, New York.

Barrie, L. A. 1982. Environment Canada's long-range transport of atmospheric pollutants program: atmospheric studies, pp. 141–161 in F. M. D'itri (ed.), Acid precipitation: effects on ecological systems. Ann Arbor Science, Ann Arbor, Michigan.

Barry, R. G. 1967. Seasonal location of the Arctic front over North America. Geogr. Bull 9:79–95.

Billings, W. D., and H. A. Mooney. 1968. The ecology of Arctic and alpine plants. Biol. Rev. 43:481–530.

Black, R. A. 1977. Reproductive biology of *Picea mariana* (Mill.) B.S.P. at treeline. Ph.D. dissertation, University of Alberta, Edmonton.

Black, R. A., and L. C. Bliss. 1978. Recovery sequence of *Picea mariana/Vaccinium uliginosum* forests after burning near Inuvik, Northwest Territories, Canada. Can. J. Bot. 56:2020–2030.

Black, R. A., and L. C. Bliss. 1980. Reproductive ecology of *Picea mariana* (Mill.) B.S.P. at treeline near Inuvik, Northwest Territories, Canada. Ecol. Monogr. 50:331–354.

Bonan, G. B. 1989a. A computer model of the solar radiation, soil moisture, and soil thermal regimes in boreal forests. Ecol. Modell. 45:275–306.

Bonan, G. B. 1989b. Environmental factors and ecological processes controlling vegetation patterns in boreal forests. Land. Ecol. 3:111–130.

Bonan, G. B. 1992. A simulation analysis of environmental factors and ecological processes in North American boreal forests, pp. 404–427 in H. H. Shugart, R. Leemans, and G. B. Bonan (eds.), A systems analysis of the global boreal forest. Cambridge University Press, Cambridge.

Bonan, G. B., and H. H. Shugart. 1989. Environmental factors and ecological processes in boreal forests. Annu. Rev. Ecol. Syst. 20: 1–28.

Bondrup-Nielsen, S. 1995. Forestry and the boreal forest: maintaining inherent landscape patterns. Water, Air Soil Poll. 82: 71–76.

Brown, R. T., and P. Mikola. 1974 The influence of fruticose soil lichens upon the mycorrhizal and seedling growth of forest trees. Acta For. Fennica 141:5–22.

Brownson, J. M. J. 1995. In cold margins. Northern Rim Press, Missoula, MT.

Bryson, R. A. 1966. Air masses, streamlines, and the boreal forest. Geogr. Bull. 8:228–260.

Canada. 1957. Atlas of Canada. Department of Mines and Technical Surveys, Geographical Branch, Ottawa.

Canada. 1978. Miscellaneous climatic data obtained from Atmospheric Environment Service, Fisheries and Environment Canada, Ottawa.

Carleton, T. J., and P. F. Maycock. 1978 Dynamics of boreal forest south of James Bay. Can. J. Bot. 56: 1157–1173.

Chapin, F. S. III. 1986. Controls over growth and nutrient use by taiga forest trees, pp. 96–111 in K. Van Cleve, F. S. Chapin III, P. W. Flanagan, L. A. Viereck, and C. T. Dyrness (eds.), Forest ecosystems in the Alaskan taiga. Springer-Verlag, New York.

Chapin, F. S. III, P. R. Tyron, and K. Van Cleve. 1983. Influence of phosphorus supply on the growth and biomass distribution of Alaskan taiga seedlings. Can. J. For. Res. 13: 1092–1098.

Colinvaux, P. A. 1967. Quaternary vegetational history of Arctic Alaska, pp. 207–231 in D. M. Hopkins (ed.), The Bering land bridge. Stanford University Press, Stanford, Cal.

Curtis, J. T. 1959. The vegetation of Wisconsin. University of Wisconsin Press, Madison.

Cwynar, L. C., and J. C. Ritchie. 1980. Arctic steppe-tundra: a Yukon perspective. Science 208: 1375–1377.

Davis, M. B. 1983. Holocene vegetational history of the eastern United States, pp. 166–181 in H. E. Wright, Jr. (ed.), The Holocene, vol. 2, Late-Quaternary environments of the United States. University of Minnesota Press, Minneapolis.

Devall, B. 1993. Clearcut: the tragedy of industrial forestry. Sierra Club Books and Earth Island Press, San Francisco.

Dugle, J. R., and N. Bols. 1971. Variation in *Picea glauca* and *P. mariana* in Manitoba and adjacent areas. Publication AECL-3681. Atomic Energy Commission, Canada Limited, Whiteshell Nuclear Research Establishment, Pinawa, Manitoba.

Elliott, D. L. 1979a. The current regenerative capacity of the northern Canadian trees, Keewatin, N.W.T., Canada: some preliminary observations. Arct. Alp. Res. 11:243–251.

Elliott, D. L. 1979b. The occurrence of bisexual strobiles on black spruce [*Picea mariana* (mill.) B.S.P.] in the forest–tundra ecotone: Keewatin, N.W.T. Can. J. Forest Res. 9:284–286.

Elliott, D. L. 1979c. The stability of the northern Canadian tree limit: current regenerative capacity. Ph.D. dissertation, University of Colorado, Boulder.

Elliott, D. L., and S. K. Short. 1979. The northern limit

of trees in Labrador: a discussion. Arctic 32:201–206.

Elliott-Fisk, D. L. 1983. The stability of the northern Canadian tree limit. Ann. Assoc. Amer. Geogr. 73:560–576.

Fitzsimmons, M. J. 1995. Conserving the boreal forest by shifting the emphasis of management action from vegetation to the atmosphere. Water, Air Soil Poll. 82: 25–34.

Fowells, H. A. 1965. Silvics of forest trees of the United States. USDA Forest Service agricultural handbook 271.

Fox, J. F. 1983. Post-fire succession of small-mammal and bird communities, pp. 155–180 in R. W. Wein and D. A. MacLean (eds.), The role of fire in northern circumpolar ecosystems. Wiley, New York.

Gates, F. C. 1942. The bogs of northern lower Michigan. Ecol. Monogr. 12:216–254.

Gleason, H. A. 1926. The individualistic concept of the plant association, Bull. Torrey Botanical Club 53:7–26.

Gordon, A. G. 1979. The taxonomy and genetics of *Picea rubens* and its relationship to *Picea mariana*. Can. J. Bot. 54:781–813.

Haag, R. W., and L. C. Bliss. 1974. Functional effects of vegetation on the radiant energy budget of boreal forest. Can. Geotech. J. 11:374–379.

Hall, J. P. 1995. Forest health monitoring in Canada: how healthy is the boreal forest? Water, Air and Soil Pollution 82:77–85.

Halliday, W. E. D., and A. W. A. Brown. 1943. The distribution of some important forest trees in Canada. Ecology 24:353–373.

Hare, F. K. 1950. Climate and zonal divisions of the boreal forest formation in eastern Canada. Geogr. Rev. 40:615–635.

Hare, F. K. 1954. The boreal conifer zone. Geogr. Studies 1:4–18.

Hare, F. K. 1959. A photo-reconnaissance survey of Labrador–Ungava. Geographical Survey of Canada Memoir 6:1–83.

Hare, F. K., and J. E. Hay 1974. The climate of Canada and Alaska, pp. 49–192 in R. A Bryson and F. K. Hare (eds.), World survey of climatology, vol. 2, Climates of North America. Elsevier, Amsterdam.

Hare, F. K., and J. C. Ritchie. 1972. The boreal bioclimates. Geogr. Rev 62:333–365.

Harper, H. L. 1977. Population biology of plants. Academic Press, New York.

Harvey, B. D., A. Leduc, and Y. Bergeron. 1995. Early postharvest succession in relation to site type in the southern boreal forest of Quebec. Can. J. For. Res. 25: 1658–1672.

Heinselman, M. L. 1963. Forest sites, bog processes, and peatland types in the glacial Lake Agassiz region, Minnesota. Ecol. Monogr. 33:327–374.

Heinselman, M. L. 1970. Landscape evolution, peatland types, and the environment in the Lake Agassiz peatlands natural area, Minnesota. Ecol. Monogr. 40:235–261.

Holling, C. S. 1992. The role of forest insects in structuring the boreal landscape, pp. 170–191 in H. H. Shugart, R. Leemans, and G. B. Bonan (eds.), A systems analysis of the global boreal forest. Cambridge University Press, Cambridge.

Hopkins, D. M. 1967. The Bering land bridge. Stanford University Press, Stanford, Cal.

Hopkins, D. M., P. A. Smith, and J. V. Matthews, Jr.

1981. Dated wood from Alaska: implications for forest refugia in Beringia. Quat. Res. 15:217–249.

Hulten, E. 1937. Outline of the history of arctic and boreal biota during the Quaternary period. Stockholm.

Hustich, I. 1949. On the forest geography of the Labrador peninsula. A preliminary synthesis. Acta Geogr. 10:3–63.

Hustich, I. 1950. Notes on the forest on the east coast of Hudson Bay and James Bay. Acta Geogr. 11:1–83.

Hustich, I. 1953. The boreal limits of conifers. Arctic 6: 149–162.

Hustich, I. 1954. On forests and tree growth in the Knob Lake area, Quebec–Labrador peninsula. Acta Geogr. 13:1–60.

Hustich, I. 1966. On the forest-tundra and the northern tree-lines. Ann. Univ. Turku [Series A2] 36:7–47.

Hustich, I. 1979. Ecological concepts and biogeographical zonation in the north: the need for a generally accepted terminology. Holarctic Ecol. 2:208–217.

Ives, J. D. 1974. Biological refugia and the nunatak hypothesis, pp. 605–636 in J. D. Ives and R. G. Barry (eds.), Arctic and alpine environments. Methuen, London.

Johnson, E. A., and J. S. Rowe. 1975. Fire in the subarctic wintering ground of the Beverly Caribou herd. Amer. Midl. Nat. 94:1–14.

Kershaw, K. A. 1977. Studies on lichen-dominated systems. An examination of some aspects of the northern boreal lichen woodlands in Canada. Can. J. Bot. 55:393–410.

Krug, E. C., and C. R. Frink. 1983. Acid rain on acid soil: a new perspective. Science 221:520–525.

La Roi, G. H. 1967. Ecological studies in the boreal spruce-fir forests of the North American taiga. Ecol. Monogr. 37:229–253.

La Roi, G. H., and M. H. L. Stringer. 1976. Ecological studies in the boreal spruce-fir forests of the North American taiga. II. Analysis of the bryophyte flora. Can. J. Bot. 54:619–643.

Larsen, J. A. 1965. The vegetation of the Ennadai Lake area, N.W.T.: studies in subarctic and arctic bioclimatology. Ecol. Monogr. 35:37–59.

Larsen, J. A. 1971. Vegetation of Fort Reliance, Northwest Territories. Can. Field Nat. 85:147–178.

Larsen, J. A. 1972. Vegetation and terrain (environment): Canadian boreal forest and tundra. University of Wisconsin Report UW-G1128, Madison.

Larsen, J. A. 1974. Ecology of the northern continental forest border, pp. 341–369 in J. D. Ives and R. G. Barry (eds), Arctic and alpine environments. Methuen, London.

Larsen, J. A. 1980. The boreal ecosystem. Academic Press, New York.

Larsen, J. A. 1982. Ecology of the northern lowland bogs and conifer forests. Academic Press, New York.

Larsen, J. A. 1989. The northern forest border in Canada and Alaska. Springer-Verlag, New York.

Lavoie, C., and S. Payette. 1994. Recent fluctuations of the lichen-spruce forest limit in subarctic Quebec. Journal of Ecology 82:725–734.

Lavoie, C., and S. Payette. 1996. The long-term stability of the boreal forest limit in subarctic Quebec. Ecology 77:1226–1233.

Lawrence, D. B. 1958. Glaciers and vegetation in southeast Alaska. Amer. Sci. 46:89–122.

Lenihan, J. M. 1993. Ecological response surfaces for North American boreal tree species and their use in forest classification. J. Veg. Sci. 4: 667–680.

Lettau, H., and K. Lettau. 1975. Regional climatonomy of tundra and boreal forest in Canada, pp. 210–221 in G. Weller and S. Bowling, (eds.), Climate of the Arctic. University of Alaska, Fairbanks.

Little, E. L., and S. S. Pauley. 1958. A natural hybrid between black and white spruce in Minnesota. Amer. Midl. Nat. 60:202–211.

Love, P., and R. Overend. 1978. Tree power: an assessment of the energy potential of forest biomass in Canada. Energy, Mines and Resources Canada report ER 78-1.

Luken, J. O., and W. D. Billings. 1983. Changes in bryophyte production associated with a thermokarst erosion cycle in a subarctic bog. Lindbergia 9:163–168.

Lutz, H. J. 1956. Ecological effects of forest fires in the interior of Alaska. USDA technical bulletin 1133.

Lynham, T. J., and B. J. Stocks. 1991. The natural fire regime of an unprotected section of the boreal forest in Canada, pp. 99–109 in Proceedings, tall timbers fire ecology conference. Tall Timbers Research Station, Tallahassee.

MacDonald, G. M. 1983. Holocene vegetation history of the upper Natla River area, Northwest Territories, Canada. Arct. Alp. Res. 15:157–168.

Maini, J. S. 1966. Phytoecological study of sylvotundra at Small Tree Lake, N.W.T. Arctic 19:220–243.

Marr, J. W. 1948. Ecology of the forest-tundra ecotone on the east coast of Hudson Bay. Ecol. Monogr. 18: 117–144.

Martin, P. H. 1996. Will forest preserves protect temperature and boreal biodiversity from climate change? For. Ecol. Mgt. 85: 335–341.

Matthews, J. V., Jr. 1970. Quaternary environmental history of interior Alaska: pollen samples from organic colluvium and peats. Arct. Alp. Res. 2:241–251.

Matthews, J. V., Jr. 1974. Wisconsin environment of interior Alaska: pollen and macrofossil analysis of a 27-meter core from the Isabella Basin (Fairbanks, Alaska). Can. J. Earth Sci. 11:828–841.

Matthews, J. V., Jr. 1979. Beringia during the late Pleistocene: arctic-steppe or discontinuous herb-tundra? A review of the paleontological evidence. Geological Survey of Canada open-file report 649. GSC. Ottawa.

Miller, W. S., and A. N. Auclair. 1974. Factor analytic models of bioclimate for Canadian forest regions. Can J. For. Res. 4:536–548.

Moody, B. H., and H. F. Cerezke. 1984. Forest insect and disease conditions in Alberta, Saskatchewan, Manitoba, and the Northwest Territories in 1983 and predictions for 1984. Canadian Forestry Service, Northern Forest Research Centre, information report NOR-X-261.

Neiland, B., and L. A. Viereck. 1978. Forest types and ecosystems, pp. 109–136 in North American forest lands at latitudes north of 60 degrees. Proceedings of a symposium, September 19–22, 1977, University of Alaska, Fairbanks.

Nichols, H. 1975. Palynological and paleoclimatic study of the late Quaternary displacement of the boreal forest-tundra ecotone in Keewatin and Mackenzie, N.W.T., Canada. Institute of Arctic and Alpine Research, occasional paper 15. University of Colorado, Boulder.

Nichols, H. 1976. Historical aspects of the northern Canadian treeline. Arctic 29:38–47.

Nienstaedt, H., and A. Teich. 1972. The genetics of white spruce. USDA Forest Service research paper WO-15.

Oechel, W. C., and W. T. Lawrence. 1985. Taiga, pp. 66–94 in B. F. Chabot and H. A. Mooney (eds.), Physiological ecology of North American plant communities. Chapman & Hall, New York.

Oswald, E. T., and J. P. Senyk. 1977. Ecoregions of Yukon Territory. Canadian Forest Service, Pacific Forest Research Centre, Victoria, British Columbia.

Owens, J. N., and M. Molder. 1976. Bud development in Sitka spruce. II. Cone differentiation and early development. Can. J. Bot. 54:766–779.

Owens, J. N., and M. Molder. 1977. Bud development in *Picea glauca*. II. Cone differentiation and early development. Can. J. Bot. 55:2746–2760.

Owens, J. N., M. Molder, and H. Langer. 1977. Bud development in *Picea glauca*. I. Annual growth cycle of vegetation buds and shoot elongation as they relate to date and temperature sums. Can. J. Bot. 55: 2728–2745.

Pare, D., and Y. Bergeron. 1995. Above-ground biomass accumulation along a 230-year chronosequence in the southern portion of the Canadian boreal forest. J. Ecol. 83:1001–1007.

Parker, W. H., and D. G. McLachlan. 1978. Morphological variation in white and black spruce: investigation of natural hybridization between *Picea glauca* and *P. mariana*. Can. J. Bot. 56:2512–2520.

Payette, S. 1975. La limite septentrionale des forets sur las cote orientale de la baise d'Hudson. Naturaliste Canadien 102:317–329.

Payette, S. 1983. The forest tundra and present tree-lines of the northern Quebec–Labrador peninsula, pp. 3–23 in P. Morisset and S. Payette (eds.), Tree-line ecology, Nordicana No. 47. University of Laval, Sainte-Foy.

Payette, S., and L. Filion. 1984. White spruce expansion at tree line and recent climatic change. Can. J. For. Res. 15:241–251.

Payette, S., and R. Gagnon. 1979. Tree-line dynamics in Ungava Peninsula, northern Quebec. Holarctic Ecol. 2:239–248.

Peinado, M., J. L. Aguirre, and M. de la Cruz. 1998. A Phytosociological survey of the Boreal Forest (*Vaccinio-Piceetea*) in North America. Plant Ecology 137: 151–202.

Pewe, T. L. 1975. Quaternary geology of Alaska. U.S. Geological Survey professional paper 835.

Prudhomme, T. I. 1983. Carbon allocation to antiherbivore compounds in a deciduous and an evergreen subarctic shrub species. Oikos 40:344–356.

Rampton, V. 1971. Late Quaternary vegetational and climatic history of the Snag–Klutlan area, southwestern Yukon Territory, Canada. Geol. Soc. Amer. Bull. 82:959–978.

Raup, H. M., and C. S. Denny. 1950. Photo interpretation of the terrain along the southern part of the Alaska Highway. U.S. Geol. Surv. Bull. 963-D: 95–135.

Rencz, A. N., and A. N. D. Auclair. 1978. Biomass distribution in a subarctic *Picea mariana–Cladonia alpestris* woodland. Can. J. For. Res. 8:168–176.

Ritchie, J. C. 1962. A geobotanical survey of northern Manitoba. Arctic Institute of North America technical paper 9. AINA, Montreal.

Ritchie, J. C. 1976. The late Quaternary vegetational history of the western interior of Canada. Can. J. Bot. 54:1793–1818.

Ritchie, J. C. 1977. The modern and late Quaternary vegetation of the Campbell–dolomite uplands near Inuvik, N.W.T., Canada. Ecol. Mongr. 47:401–423.

Ritchie, J. C. 1982. The modern and late-Quaternary vegetation of the Doll Creek area, north Yukon, Canada. New Phytol. 90:563–603.

Ritchie, J. C., and L. C. Cwynar. 1982. The late-Quaternary vegetation of the north Yukon, pp. 113–126 in D. M. Hopkins, J. V. Matthews, Jr., C. E. Schweger, and S. B. Young (eds.), Paleoecology of Beringia. Academic Press, New York.

Ritchie, J. C., and F. K. Hare. 1971. Late Quaternary vegetation and climate near the arctic treeline of northwestern North America. Quat. Res. 1:331–342.

Ritchie, J. C., and G. A. Yarranton. 1978a. Patterns of change in the late-Quaternary vegetation of the western interior of Canada. Can. J. Bot. 56:2177–2183.

Ritchie, J. C., and G. A. Yarranton. 1978b. The late-Quaternary history of the boreal forest of central Canada, based on standard pollen stratigraphy and principal components analysis. J. Ecol. 66: 199–212.

Roche, L. 1969. A genecological study of the genus *Picea* in British Columbia. New Phytol. 68:505–554.

Rowe, J. S., 1952. Squirrel damage to white spruce. Canada Department of Resource Development, Division of Forest Research silvic leaflet 61.

Rowe, J. S. 1961. Critique of some vegetational concepts as applied to forests of northwestern Alberta. Can. J. Bot. 39:1007–1017.

Rowe, J. S. 1970. Spruce and fire in northwest Canada and Alaska, pp. 245–254 in Proceedings of the annual Tall Timbers fire ecology conference. Tall Timbers Research Station, Tallahassee, Fla.

Rowe, J. S. 1972. Forest regions of Canada. Department of the Environment, Canadian Forestry Service, pub. no. 1300.

Rowe, J. S. 1983. Concepts of fire effects on plant individuals and species, pp. 135–154 in R. W. Wein and D. A. MacLean (eds.), The role of fire in northern circumpolar ecosystems. SCOPE 18. Wiley, New York.

Rowe, J. S., and G. W. Scotter. 1973. Fire in the boreal forest. Quat. Res. 3:444–464.

Short, S. K. 1978a. Holocene palynology in Labrador–Ungava: climatic history and culture change in the central coast. Ph.D. dissertation, University of Colorado, Boulder.

Short, S. K. 1978b. Palynology: a Holocene environmental perspective for archaeology in Labrador–Ungava. Arctic Anthropol. 15:9–35.

Shreve, F. 1915. The vegetation of a desert mountain range as conditioned by climatic factors. Publication 217. Carnegie Institution of Washington, Washington, D. C.

Shugart, H. H. 1984. A theory of forest dynamics: the ecological implications of forest succession models. Springer-Verlag, New York.

Shugart, H. H., R. Leemans, and G. B. Bonan. 1992. A systems analysis of the global boreal forest. Cambridge University Press, Cambridge.

Sirois, L., G. B. Bonan, and H. H. Shugart. 1994. Development of a simulation model of the forest-tundra transition zone of northeastern Canada. Can. J. For. Res. 24: 697–706.

Skre, O., W. C. Oechel, and P. M. Miller. 1983. Moss leaf-water content and solar radiation at the moss surface in a mature black spruce forest in central Alaska. Can. J. For. Res. 13:860–868.

Slaughter, C. W., and L. A. Viereck. 1986. Climatic characteristics of the taiga in interior Alaska, pp. 9–21 in K. Van Cleve, F. S. Chapin III, P. W. Flanagan, L. A. Viereck, and C. T. Dryness (eds.), Forest ecosystems in the Alaskan taiga. Springer-Verlag, New York.

Sorenson, C. J. 1977. Reconstructed Holocene bioclimates. Ann. Assoc. Amer. Geogr. 67:214–222.

Sorenson, C. J., and J. C. Knox. 1974. Paleosols and paleoclimates related to late Holocene forest/tundra border migrations: Mackenzie and Keewatin, N.W.T., Canada, pp. 187–203 in International conference on prehistory and paleoecology of western North American Arctic and subarctic. Archeological Association, University of Calgary, Calgary, Alberta.

Spear, R. W. 1982. The late Quaternary distribution of spruce (*Picea glauca, P. mariana*) in the Mackenzie Delta, Northwest Territories, Canada, p. 45 in Abstracts of the eleventh annual Arctic Workshop, Boulder, Colorado.

Strang, R. M. 1973. Succession in unburned sub-Arctic woodlands. Can. J. For. Res. 3:140–142.

Streten, N. A. 1974. Some features of the summer climate of interior Alaska. Arctic 27:273–286.

Sturtevant, B. R., J. A. Bissonette, and J. N. Long. 1996. Temporal and spatial dynamics of boreal forest structure in western Newfoundland: silvicultural implications for marten habitat management. For. Ecol. Mgt. 87: 13–25.

Tedrow, J. C. F. 1970. Soils of the subarctic regions, pp. 189–206 in Ecology of the subarctic regime, Proceedings of the Helsinki symposium. UNESCO, New York.

Tedrow, J. C. F. 1977. Soils of the polar landscapes. Rutgers University Press, New Brunswick, N.J.

Tranquillini, W. 1979. Physiological ecology of the alpine timberline. Springer-Verlag, New York.

Turk, J. T. 1983. An evaluation of trends in the acidity of precipitation and the related acidification of surface water in North America. U.S. Geological Survey water-supply paper 2249.

U.S. Department of Agriculture. 1974. Seeds of woody plants in the United States. USDA agricultural handbook 450.

Viereck, L. A. 1973. Wildfire in the taiga of Alaska. Quat. Res. 3:465–495.

Viereck, L. A., K. Van Cleve, and C. T. Dyrness. 1986. Forest ecosystem distribution in the taiga environment, pp. 22–43 in K. Van Cleve, F. S. Chapin III, P. W. Flanagan, L. A. Viereck, and C. T. Dyrness (eds.), Forest ecosystems in the Alaskan taiga. Springer-Verlag, New York.

Von Rudloff, E., and M. J. Holst. 1968. Chemosystematic studies in the genus *Picea* (Pinaceae). III. The leaf oil of a *Pica glauca x mariana* (Rosendahl spruce). Can. J. Bot. 46:1–4.

Vowinckel, T., W. C. Oechel, and W. G. Boll. 1975. The effect of climate on the photosynthesis of *Picea mariana* at the sub-arctic tree line. I. Field measurements. Can. J. Bot. 53:604–620.

Wang, Z. M., M. J. Lechowicz, and C. Potvin. 1994. Early selection of black spruce seedlings and global change: which genotypes should we favor? Ecol. Applica. 4:604–616.

Warren Wilson, J. 1967. Ecological data on dry-matter production by plants and plant communities, pp. 77–127 in E. F. Bradley and O. T. Denmead (eds.), The collection and processing of field data. Wiley, New York.

Watts, W. A. 1970. The full glacial vegetation of northwestern Georgia. Ecology 51:631–642.

Webb III, T., E. J. Cushing, and H. E. Wright, Jr. 1983. Holocene changes in the vegetation of the Midwest, pp. 142–165 in H. E. Wright, Jr. (ed.), The Holocene, vol. 2, Late-Quaternary environments of the United States. University of Minnesota Press, Minneapolis.

Wein, R. W. 1983. Fire behavior and ecological effects in organic terrain, pp. 81–95 in R. W. Wein and D. A. MacLean (eds.), The role of fire in northern circumpolar ecosystems. SCOPE 18. Wiley, New York.

Wein, R. W. 1990. The importance of wildfire to climate change – hypotheses for the taiga, pp. 185–190 in J. G. Goldammer and M. J. Jenkins (eds.), Fire in ecosystem dynamics, Mediterranean and northern perspectives, Proceedings of the third international symposium on fire ecology. SPB Academic Publ., The Hague.

Wein, R. W., and D. A. MacLean. 1983. The role of fire in northern circumpolar ecosystems. SCOPE 18. Wiley, New York.

Welsh, D. A., and L. A. Venier. 1996. Binoculars and satellites: developing a conservation framework for boreal forest wildlife at varying scales. For. Ecol. Mgt. 85: 53–65.

Whittaker, R. H., and G. E. Likens, 1975. The biosphere and man, pp. 305–328 in H. Leith and R. H. Whittaker (eds.), The primary production of the biosphere. Springer-Verlag, New York.

Wilson, C. V. 1971. The climate of Quebec. I. Climatic Atlas. Climatological Studies 11, Canadian Meterological Service, Ottawa.

Wolfe, J. A. 1971. Tertiary climatic fluctuations and methods of analysis of Tertiary floras. Palaeogeogr., Palaeoclimatol., Palaeoecol. 9:27–57.

Wolfe, J. A. 1972. An interpretation of Alaskan Tertiary floras, pp. 225–233 in A. Graham (ed.), Floristics and paleofloristics of Asia and eastern North America. Elsevier, New York.

Wolfe, J. A. 1978. A paleobotanical interpretation of Tertiary climates in the northern hemisphere. Amer. Sci 66:694–703.

Zasada, J. C. 1971. Natural regeneration of interior Alaska forests – seed, seedbed, and vegetative reproduction considerations, pp. 231–246, in Proceedings, fire in the northern environment, a symposium. Pacific Northwest Forest and Range Experiment Station, Portland, Oregon.

Zasada, J. C., and R. A. Gregory. 1969. Regeneration of white spruce with reference to interior Alaska: a literature review. U.S. Forest Service research paper PNW-79. Portland, Oregon.

Zasada, J. C., K. Van Cleve, R. A. Werner, J. A. McQueen, and E. Nyland. 1978. Forest biology and management in high-latitude North American forests, pp. 137–195 in North American forest lands at latitudes north of 60 degrees. Proceedings of a symposium, September 19–22, 1977, University of Alaska, Fairbanks.

Chapter 3

Forests and Meadows
of the Rocky Mountains

ROBERT K. PEET

INTRODUCTION

The east slope of the Rocky Mountains can be an impressive sight to the traveler who has just crossed the flat expanses of the Great Plains only to encounter the Colorado Front Range, which rises abruptly from 1500 m at the base of the foothills to peaks reaching 4400 m and gives the impression of an impassable wall. The slopes of these mountains are accented by a distinct dark band of forest bordered both above and below by lighter-colored grassland vegetation, the abrupt topography making the vertical zonation of vegetation particularly dramatic. Closer examination shows the forests to be dominated almost exclusively by conifers and to have both physiognomic and floristic similarities with forests of the boreal region.

The Rocky Mountains consist of massive chains of mountains that form the backbone of the North American continent. In the broad sense, Rocky Mountain forests can be defined as extending from north of 65°N latitude in the boreal regions of Alaska and the Yukon (see Chapter 1) south to the towering peaks of the Mexican volcanic belt (see Chapter 15) at 19°N latitude (Fig. 3.1). Given the length of the cordillera, the surprise is not that latitudinal variation exists in Rocky Mountain forests but that major vegetation types, often representing elevation zones, remain relatively constant over great distances.

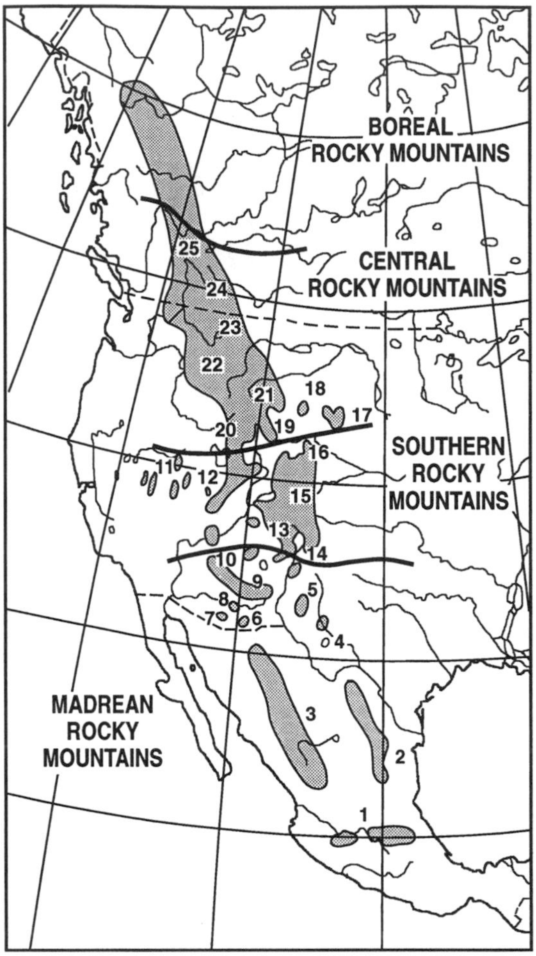

Floristic Regions and Latitudinal Trends

The Rocky Mountains have been variously divided into floristic regions by a series of authors (e.g., Daubenmire 1943; Arno and Hammerly 1984; Axelrod and Raven 1985; Peet 1988; Billings 1990). Although no one classification is ideal, the four-region scheme proposed by Billings (1990) captures the major natural discontinuities in species distributions (Fig. 3.1).

The Boreal Rocky Mountain Region (the Far Northern Rockies of Daubenmire 1943 and Peet 1988) is transitional in composition to the true boreal forest. The northern limit can be defined by the northern range limits of *Abies lasiocarpa* and *Pinus contorta* (common names of major tree species are listed in Table 3.1) in the central Yukon, the last of the major cordilleran tree species to drop out, though forested mountains continue significantly farther north. The southern limit of the Boreal Rockies coincides with the divide between the Peace and Fraser river drainages, corresponding roughly with the northern limit of two other major

Botanical nomenclature follows Kartesz 1994.

Figure 3.1. Distribution of Rocky Mountain vegetation in North America. Approximate boundaries of the four major floristic provinces are indicated by solid lines. Major mountain ranges and locations mentioned in the text include: (1) Trans-Mexican volcanic belt, (2) Sierra Madre Oriental, (3) Sierra Madre Occidental, (4) Davis Mountains, (5) Sierra Blanca, (6) Chiricahua Mountains, (7) Santa Catalina Mountains, (8) Pinaleño Mountains, (9) Mogollon Plateau, (10) San Francisco Peaks, (11) Great Basin ranges, (12) Wasatch Range, (13) San Juan Mountains, (14) Sangre de Cristo Mountains, (15) Front Range, (16) Medicine Bow Mountains, (17) Black Hills, (18) Bighorn Mountains, (19) Wind River Range, (20) Teton Mountains, (21) Yellowstone National Park, (22) Bitterroot Mountains, (23) Glacier National Park, (24) Banff National Park, (25) Jasper National Park.

cordilleran species, *Picea engelmannii* and *Pseudotsuga menziesii*, which are replaced in the Boreal Rockies largely by *Picea glauca*. Much of this region is remote and not readily accessible, with the consequence that little has been written on its vegetation.

The Central Rocky Mountain Region extends

Table 3.1. *Geographic distributions and common names of tree species in the four Rocky Mountain floristic regions*

Species	Boreal	Central	Southern	Madrean
Conifers				
Abies concolor concolor (white fir)			x	x
Abies grandis (grand fir)		x		
Abies lasiocarpa var. *arizonica* (corkbark fir)			+	x
Abies lasiocarpa var. *lasiocarpa* (subalpine fir)	x	x	x	
Cupressus arizonica (Arizona cypress)				x
Juniperus communis (common juniper)	x	x	x	+
Juniperus deppeana (alligator juniper)			+	x
Juniperus flaccida (drooping juniper)				x
Juniperus monosperma (one-seed juniper)			x	x
Juniperus osteosperma (Utah juniper)		x	x	
Juniperus scopulorum (Rocky Mountain juniper)		x	x	x
Larix laricina (tamarack)	+			
Larix yallii (subalpine larch)		+		
Larix occidentalis (western larch)		x		
Picea engelmannii (Engelmann spruce)		x	x	x
Picea glauca (white spruce)	x	+		
Picea mariana (black spruce)	x			
Picea pungens (blue spruce)		+	x	+
Pinus albicaulis (whitebark pine)		x		
Pinus aristata (bristlecone pine)			x	
Pinus arizonica var. *arizonica* (Arizona pine)				x
Pinus cembroides (Mexican pinyon)				x
Pinus contorta var. *latifolia* (lodgepole pine)	x	x	x	
Pinus discolor (pinyon)				x
Pinus edulis (pinyon)			x	+
Pinus engelmannii (Apache pine)				x
Pinus flexilis (limber pine)		x	x	
Pinus leiophylla (Chihuahua pine)				x
Pinus longaeva (bristlecone pine)			x	
Pinus monophylla (singleleaf pine)			x	x
Pinus monticola (western white pine)		x		
Pinus ponderosa var. *ponderosa* (ponderosa pine)		x		
Pinus ponderosa var. *scopulorum* (ponderosa pine)		x	x	x
Pinus strobiformis (Mexican white pine)			+	x
Pseudotsuga menziesii (Douglas fir)		x	x	x
Taxus brevifolia (Pacific yew)		+		
Thuja plicata (western redcedar)		x		
Tsuga heterophylla (western hemlock)		x		
Tsuga mertensiana (mountain hemlock)		x		
Selected angiosperms				
Acer glabrum (Rocky Mountain maple)	x	x	x	x
Acer grandidentatum (bigtooth maple)			+	+
Alnus incana ssp. *tenuifolia* (thinleaf alder)	x	x	x	+
Alnus oblongifolia (Arizona alder)				+
Alnus rubra (red alder)		+		
Alnus viridis ssp. *sinuata* (Sitka alder)	x	x		
Arbutus arizonica (Arizona madrone)				x
Arbutus xalapensis (Texas madrone)				x
Betula occidentalis (water birch)		x	x	
Betula papyrifera (paper birch)	x	x		
Frangula purshiana (cascara buckthorn)		x		
Ostrya species (hop hornbeam)			+	+
Populus angustifolia (narrowleaf cottonwood)		+	x	x
Populus balsamifera (balsam poplar)		x	+	
Populus balsamifera ssp. *trichocarpa* (black cottonwood)		x	x	+
Populus deltoides ssp. *monilifera* (plains cottonwood)		x	x	
Populus fremontii (Fremont cottonwood)			+	+
Populus tremuloides (quaking aspen)	x	x	x	x

Table 3.1. *(cont.)*

Species	Boreal	Central	Southern	Madrean
Prunus serotina (black cherry)				x
Quercus chrysolepis (canyon live oak)				+
Quercus emoryi (Emory oak)				x
Quercus gambelii (Gambel oak)			x	x
Robinia neomexicana (New Mexican locust)			x	x
Salix scouleriana (Scouler willow)	x	x	x	+
Sorbus scopulina (mountain-ash)	x	x	x	+
Sorbus stichensis (Sitka mountain-ash)	x	x		

Note: Not included are the numerous southern Rocky Mountain species found only in Mexico. A minor presence on the margin of the region is indicated by a plus (+) sign.
Sources: Distributions from Little (1971, 1976); nomenclature follows Kartesz (1994).

from north of Jasper National Park in Alberta through the Wind River and Bighorn mountains of Wyoming and combines areas sometimes separated physiographically as the Central and Northern Rockies (e.g., Hunt 1967). This floristic region contains numerous species typical of the Cascade Mountains. As one example, some sheltered valley bottoms on the west slope contain forests with *Tsuga heterophylla*, *Thuja plicata*, and even *Taxus brevifolia*, species otherwise not found in the Rockies. Similarly, *Tsuga mertensiana* and *Larix lyallii* are subalpine species of the Far West that occur in the Rockies only in this area of more moderate climate. Perhaps the species with the range that best defines the Central Rockies is *Pinus albicaulis*, which extends from the Wind River Range to nearly the northern edge of the Central Rockies.

The Southern Rocky Mountain Region (the Central Rockies of Daubenmire 1943; the Southern and Central Rocky Mountain Physiographic Provinces of Thornbury 1965) was recognized by Axelrod and Raven (1985) and Billings (1990) as extending south from the Medicine Bow, Laramie, and Wasach mountains through the Sangre de Cristo Mountains of northern New Mexico and the San Francisco Peaks and Mogollon Plateau of Arizona. Among the conifers, only the bristlecone pines (*Pinus aristata*, *P. longaeva*) appear to be restricted to this region, but several species, including *Pinus edulis* and *Juniperus monosperma*, have more southerly ranges that reach their northern limits in the region.

The Madrean Rocky Mountain Region (Axelrod and Raven 1985; the Southern Rockies of Daubenmire 1943), in contrast, has numerous distinctive species. Pines and oaks are particularly well represented (Perry 1991; Nixon 1993) and probably are components of a distinct flora that evolved in the Sierra Madre of Mexico (Axelrod 1958, 1979) and, following glacial retreat, expanded northward. The

Madrean Region is sufficiently large and rich in endemics that few species are in common between the northern and southern extremes. *Pinus cembroides* does occur throughout, albeit with striking latitudinal changes in stand structure (see Woodin and Lindsey 1954; Segura and Snook 1992). The flora of these northern mountain regions of Mexico is very poorly known (Bye 1995), but as additional information becomes available on the floristics and vegetation, the Madrean Region will likely be divided into multiple distinct floristic regions.

Daubenmire (1943, 1975) hypothesized the location of the Central Rocky Mountain Region to be largely a consequence of a major storm track along which Pacific air penetrates to the Rockies between northwestern Oregon and southwestern British Columbia. Mitchell (1976) has shown that from June through September the Pacific air mass does typically penetrate to the Rockies, the southern boundary of this area of intrusion running northeast from northern California into western Montana. Little Pacific air penetrates south of this line, where warmer continental air dominates. The Cascadian species in the Rockies are likely limited on the south by summer heat and drought and on the north by extremes of cold.

The border between the Southern Rocky Mountain and Madrean floristic regions is less well defined. I follow Axelrod and Raven (1985) in defining the Madrean Region to start with the ranges of southern New Mexico and Arizona, but Mitchell (1976) delimited an area southeast of a line from southwestern Arizona to Salt Lake and across to the Front Range as characterized by summer rains (the "Arizona monsoon") generated by the Bermuda high. This region's precipitation is concentrated during the summer months, in contrast to a region to the west, running from the Mojave Desert to central Idaho, in which winter rains predominate. Building on Mitchell's work, Neilson and

Wullstein (1983) proposed that the northern limit of *Quercus gambelii*, one of the most northerly madrean species, occurs where its lower elevation limit, set by drought, intersects its upper elevation limit, set by cold temperature. This species appears typical in that a combination of summer drought and winter cold likely defines the northern limit of most madrean species. Because the winter cold is less extreme on the west slopes of the Rockies, which are protected from southward flows of arctic air, madrean species extend nearly to Salt Lake on the west, but in Colorado they are nearly absent from the east slope. This leaves areas like the San Juan Mountains of southwestern Colorado poorly defined floristically, which is consistent with their being the northern limit of a small number of conspicuous, clearly Madrean taxa such as *Pinus strobiformis* and *Abies lasiocarpa* var. *arizonica* (Jamieson, Romme, and Somers 1996). Nonetheless, the real dominance of madrean species is not reached until nearly the Mexican border.

Elevation Zonation and Environmental Gradients

Because of their apparent geographical constancy, vegetation zones defined by elevation provide the oldest and simplest means of classifying Rocky Mountain vegetation. Such zones have been identified on the basis of either climate or dominant species. Ramaley (1907, 1908), familiar with the efforts of Merriam (1890) and Schimper (1898) to delimit life zones, recognized four major Rocky Mountain climatic zones – Foothill, Montane, Subalpine, and Alpine – corresponding roughly to Merriam's Transition, Canadian, Hudsonian, and Arctic–Alpine zones. In contrast, Daubenmire (1943) used dominant species to designate zones. Between the basal plains and the alpine he recognized (1) the oak–mountain mahogany (*Quercus–Cercocarpus*) zone, (2) the juniper-pinyon (*Juniperus–Pinus edulis*) zone, (3) the ponderosa pine (*Pinus ponderosa*) zone, (4) the Douglas fir (*Pseudotsuga menziesii*) zone, and (5) the spruce-fir (*Picea engelmannii–Abies lasiocarpa*) zone. The first and second correspond to the Foothill zone of Ramaley, whereas the third and fourth have often been designated as the Lower and Upper Montane, respectively (Marr 1961). The spruce-fir zone is synonymous with the Subalpine zone of Ramaley.

The vegetation zones recognized by Daubenmire and others portray vegetation on an idealized mountain. In actuality, not all zones are necessarily present in a region, nor do they occur at consistent elevations. A northerly aspect, high moisture availability, and increased latitude all tend to lower the elevation at which vegetation representative of a zone occurs.

The primary difficulty with an elevation zonation approach to vegetation classification is that vegetation usually is not composed of discrete bands. Rather, composition varies continuously along environmental gradients, with the consequence that no two investigators are likely to recognize the same zones. In addition, climatic factors do not map directly onto elevation, and edaphic and disturbance factors also influence vegetation composition. In this chapter I take two complementary approaches: I recognize 11 major groups of communities as foci for discussion, but I also use environmental gradients to summarize major vegetation patterns.

The single most important vegetation gradient in the Rockies undoubtedly corresponds to elevation. Elevation is a complex gradient that combines several environmental variables important for plant growth. For example, with increasing elevation temperature generally drops, precipitation increases (though perhaps with a secondary decline at highest elevations), solar radiation and particularly ultraviolet radiation increase, wind increases, and snow depth and duration may increase (Whittaker and Niering 1965; Greenland et al. 1985).

Local topographic variation makes climatic information difficult to obtain or summarize for the mountainous regions of western North America (Baker 1944). Most weather stations are located at low elevations, and few long-term records are available from sites above the foothills. One notable exception to the dearth of climatic data is a result of work at the Mountain Research Station of the University of Colorado by Marr and colleagues (Marr 1961, 1967; Marr, Clark, Osburn, and Paddock 1968; Marr, Johnson, Osburn, and Knorr 1968; Barry 1972, 1973). They collected long-term data from weather stations situated at 2195 m (lower montane), 2580 m (upper montane), 3050 m (subalpine), and 3750 m (alpine). Supplemented with a National Oceanic and Atmospheric Administration (NOAA) station in the nearby foothills (1603 m), these stations provide a detailed view of a typical elevation gradient in climate (Fig. 3.2). Along this transect, the mean annual temperature drops from 8.8°C at 1603 m to −3.3°C at 3750 m. The mean daily minimum for January decreases from −7.8°C at 2195 m to −16.1°C at 3750 m, and the mean July maximum drops from 30.7°C at 1603 m to 19.4°C at 3750 m. Monthly precipitation is greatest in May and increases with elevation from 395 mm yr^{-1} at the base of the foothills to 1050 mm yr^{-1} in the alpine zone.

Outliers of Rocky Mountain vegetation can be

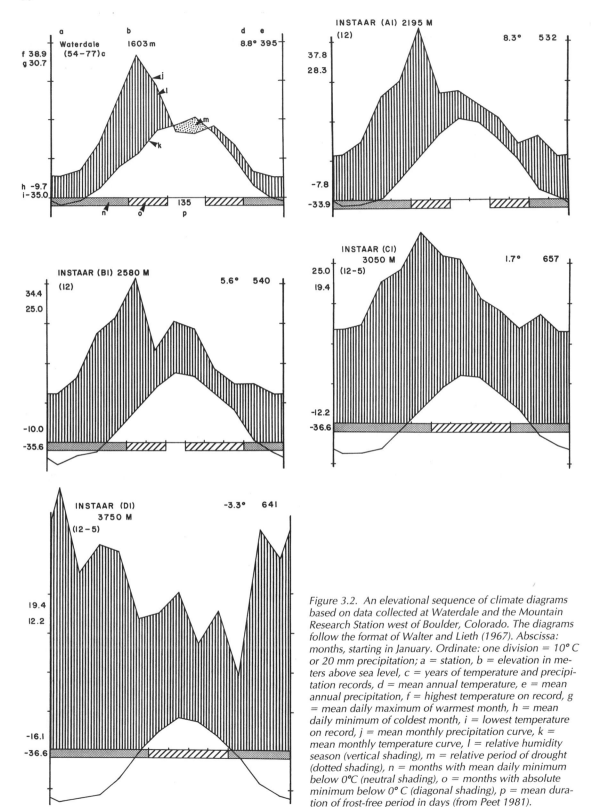

Figure 3.2. An elevational sequence of climate diagrams based on data collected at Waterdale and the Mountain Research Station west of Boulder, Colorado. The diagrams follow the format of Walter and Lieth (1967). Abscissa: months, starting in January. Ordinate: one division = 10° C or 20 mm precipitation; a = station, b = elevation in meters above sea level, c = years of temperature and precipitation records, d = mean annual temperature, e = mean annual precipitation, f = highest temperature on record, g = mean daily maximum of warmest month, h = mean daily minimum of coldest month, i = lowest temperature on record, j = mean monthly precipitation curve, k = mean monthly temperature curve, l = relative humidity season (vertical shading), m = relative period of drought (dotted shading), n = months with mean daily minimum below 0°C (neutral shading), o = months with absolute minimum below 0° C (diagonal shading), p = mean duration of frost-free period in days (from Peet 1981).

found at middle and high elevations on small desert mountain ranges west of the Southern Rockies and between the Southern Rockies and the main body of the Madrean Region in the Sierra Madre Occidental. The mass of such a range is often more important than its elevation in determining vegetation, a phenomenon Lowe (1961) referred to as the Merriam effect. Smaller ranges tend to be more arid and warmer, with vegetation zones shifted upward accordingly (Gehlbach 1981; Gottfried, Ffolliott, and DeBano 1995). Whether these climatic shifts are due to the greater oregraphic precipitation on large ranges, wind speed, or the buffering of temperatures remains unclear

The topographic-moisture gradient is a second major complex gradient determining vegetation composition. Sites situated so as to receive high levels of incident solar radiation, such as south-facing slopes and ridge tops, are warmer and drier than north-facing slopes and sheltered valley bottoms. In addition, upper slopes and ridges lose water to down-slope flow, whereas lower slopes and bottoms often have a net gain from runoff. The topographic moisture gradient combines these factors and others that influence temperature and moisture within a given elevation belt (Whittaker 1967, 1973; Peet 1981).

When placed as orthogonal axes, the elevation and topographic moisture gradients provide a useful frame of reference for studying mountain vegetation. An idealized representation for the Central Rockies is shown in Figure 3.3. The interaction of the two gradients is clearly evident in the predominantly diagonal orientation of the vegetation zones: A given vegetation type tends to be found at higher elevations on drier sites. This trend is particularly marked in sheltered valleys where high moisture content, low incident radiation, and drainage of cold air combine to produce an environment and vegetation more characteristic of that nearly 500 m higher on open slopes.

Soil provides a third complex environmental variable important for interpretation of vegetation composition. Much of the Rocky Mountain region has young soils derived from Precambrian granites and chemically similar gneiss and schist. In areas dominated by these rocks, the major soil factors influencing vegetation are texture and depth, both of which in turn have direct impacts on water availability (Smith 1985). Trees almost always dominate sites with thin or rocky soils, but where deep, fine-textured soils occur, grasses and forbs can form a dense sod that inhibits tree regeneration.

The conspicuous importance of elevation and moisture appears to have obscured the importance

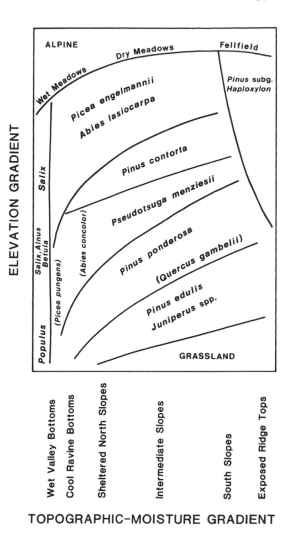

Figure 3.3. Distribution of major vegetation zones of the central Rocky Mountains in relation to elevation and topographic-moisture gradients. Parenthetical species are not consistent dominants.

of substrate variation for most Rocky Mountain vegetation research. Limestone and granitic soils, when in close proximity to each other, provide a clear exception. Such dissimilar ranges as the Bighorn Mountains of Wyoming (Despain 1973) and the Santa Catalina Mountains of Arizona (Whittaker and Niering 1968) have much more xeric vegetation at a given site on limestone as contrasted with granitic soils, though this may be more a consequence of differences in soil texture (Wentworth 1981). The influence of soil chemistry on Rocky Mountain vegetation remains largely unstudied. In one of the few Rocky Mountain vegetation studies to emphasize the importance of parent material, Despain (1973) demonstrated that vegetation in the

Bighorn Mountains of north-central Wyoming is strongly dependent on whether the soil parent material is granite or limestone. Subsequently, Despain (1983, 1990) documented contrasting patterns of vegetation for rhyolite- and andesite-derived soils in Yellowstone National Park, where rhyolite-derived soils are more sterile and the resulting vegetation is more xeromorphic in character than on otherwise equivalent but chemically more base-rich andesite-derived soils. In other studies, highly sterile, acidic soils derived from hydrothermally altered rocks have been shown to support conifer woodland in regions where the primary vegetation is semidesert shrubland (Billings, 1950; Salisbury 1964; Schlesinger, DeLucia, and Billings 1989).

DISTURBANCE, SUCCESSION, AND ECOSYSTEM DEVELOPMENT

Disturbance Events

Nearly all Rocky Mountain forests are in some stage of recovery from prior disturbance. Fire, wind, insects, disease, ungulate browsing, avalanches, landslides, extreme weather, volcanism, and, of course, humans all have major impacts on the landscape. As a consequence, the vegetation is perhaps best thought of not as a uniform, stable cover but, rather, as a disturbance-derived mosaic, with the character of each tessera frequently changing and the borders being periodically redefined.

For most of the Rocky Mountain landscape, fire has historically been the most important as well as the most conspicuous form of natural disturbance. Charcoal can be found in the soil of virtually any forest. Mean fire intervals have been calculated in a sufficient number of localities for general patterns to be hypothesized. However, the available data must be interpreted with caution, as fire intervals vary with local conditions within a forest type, and forests cover such large areas that multiple studies are needed for general characterization.

Current fire regimes give little indication of earlier fire regimes. Over the past few centuries fire regimes have varied considerably in response to modest changes in climate (Johnson and Larsen 1991; Johnson and Wowchuck 1993). Moreover, human influence on fire regimes has been far from constant. Although lightning likely caused many presettlement fires, current evidence suggests that native peoples were responsible for the bulk of the fires (Arno 1985; Gruell 1985). Barrett and Arno (1982) were able to document the role of aboriginal fires in the Bitterroot Mountains of Montana by determining fire frequencies in matched areas, some

heavily used by native people and others remote. Mean fire intervals in the remote areas were roughly twice as long as those in the heavily used areas. How fire frequencies in Rocky Mountain forests may have varied through the Holocene in response to changes in population densities and cultural practices of the native peoples remains largely unknown (see Kay 1994). The great majority of the Rocky Mountain region has experienced a major reduction in fire frequency since the beginning of the century, owing to cessation of aboriginal burning, fire suppression, fuel reduction through grazing, public "education," and the continuing dissection of the landscape into smaller units separated by artificial firebreaks such as roads (see Reed, Johnson-Barnard, and Baker 1996). In contrast, for much of the region, the activities of early European visitors who engaged in prospecting and land clearing, to say nothing of simple arson, led to a major increase in fire frequency during the later part of the nineteenth century (Veblen and Lorenz 1986). Nonetheless, portions of the ponderosa pine forest showed a decline in fire frequency during this same period, owing variously to increased grazing by domestic stock and the subjugation and declining populations of the native peoples who originally used fire as a common management practice (Dieterich 1980; Arno 1985; Gruell 1985; Savage and Swetnam 1990).

The Rocky Mountain forests that had the highest fire frequencies in the original landscape were the low-elevation woodlands dominated by *Pinus ponderosa* var. *scopulorum* and close relatives (e.g., *P. ponderosa* var. *arizonica*, *P. engelmannii*, *P. durangensis*) with their abundance of flammable grasses and forbs and proximity to the fire-maintained grasslands of the plains. A combination of lightning fires and aboriginal burning led in some areas to nearly annual fires (Dieterich 1980, 1983), with a mean fire interval of 5–14 yr being typical of the region as a whole (Weaver 1951; Arno 1980; Gruell 1985; Fisher, Jenkins, and Fisher 1987; Fulé and Covington 1996). Such frequent fires were mostly of low intensity, burning accumulated dead grass and surface litter but rarely causing major damage to the thick-barked dominant trees. The low-elevation *Pseudotsuga* woodlands of Wyoming and Montana had fire intervals slightly longer than those of the more southern *Pinus ponderosa* woodlands. Arno and Gruell (1983) reported pre-1910 intervals of 35–40 yr, and Gruell (1985) suggested 20–40 yr to be typical. The low-elevation pinyon-juniper woodlands also burned regularly with low-intensity surface fires but with lower frequency than did the *P. pon-*

derosa woodlands owing to lower rates of fuel accumulation (Arno 1985).

Pinus contorta and *Picea–Abies* forests had longer fire intervals than *Pinus ponderosa* woodlands did, but the fires that did occur were often severe stand-replacing crown fires (Wellner 1970). Romme and colleagues (Romme 1982; Romme and Knight 1982; Romme and Despain 1989) report that the *Pinus contorta*–dominated subalpine forests of Yellowstone National Park burn with a 300–400 yr mean fire interval but that large areas burn within a few relatively short intervals, alternating with relatively long fire-free periods during which fuel accumulates. This pattern, consistent with the 1988 Yellowstone fire, appears typical for moderate-to-high-elevation forests throughout at least the Central Rocky Mountain region (Christensen et al. 1989; Romme and Despain 1989). Nonetheless, not all fires are catastrophic; throughout the Rockies, *Pinus contorta* and *Picea–Abies* forests in proximity to frequently burned communities such as *Pinus ponderosa* or *Pseudotsuga* woodlands have experienced low-intensity surface fires at a shorter fire interval, typically 15–50 yr (Houston 1973; Heinselman 1981).

Stand ages in the Rockies south of Yellowstone largely lend support to a 200–400 yr mean fire interval for subalpine *Pinus contorta* and *Picea–Abies* forests (Peet 1981; Romme and Knight 1981), but an interval of 50–150 yr is likely more typical of lower-elevation *Pinus contorta* forests (Clements 1910; Loope and Gruell 1973; Peet 1981). Studies conducted farther north in Jasper Park, Alberta (Tande 1979), and in Montana (Arno 1980) also suggest a pattern of a few widespread fires but with a shorter (65–100 yr) mean fire interval. Hawkes (1980), working in Kananaskis Park, Alberta, found a similar fire interval (90 yr) for low-elevation forests (1830 m) but a somewhat longer interval (153 yr) for high-elevation forests (>1830 m).

Forms of forest disturbance other than fire can also be characterized by mean return intervals, but only for fire has a sufficient number of studies determined return intervals to allow a clear pattern to emerge. For example, windstorms destroy substantial patches of subalpine forest, with major damage usually being concentrated in old stands that have a high incidence of trunk rot. No regional determinations of windthrow frequency are currently available for Rocky Mountain forests, probably because the areas of damage tend to be more limited in extent than is the case for fire (see Peet 1981; Alexander 1987; Veblen, Hadley, Reid, and Rebertus 1989).

Insect outbreaks destroy large areas of forest. In the 1940s, an outbreak of spruce beetle (*Dendroctonus rufipennis*) killed virtually all the *Picea* and most of the *Abies* trees of greater than 10 cm diameter on the White River Plateau of northwestern Colorado, destroying an estimated timber volume of 10^7 m^3 of *Picea* alone (Miller 1970; Alexander 1974). Mountain pine beetle causes extensive damage to such species as *Pinus contorta* and *P. ponderosa* (Amman 1977, 1978; Romme, Knight, and Yavitt 1986); during the period 1979–1983 alone it infested nearly 2M ha in the western United States (Schmidt and Amman 1992). Similarly, budworm epidemics kill large numbers of trees (McKnight 1968); an epidemic in the 1980s destroyed vast areas of *Pseudotsuga* and *Abies concolor* forest in Colorado. Both pine beetles and budworms cause sufficient damage to have been labeled the most damaging pests of western forests (see Schmidt and Amman 1992; Lynch and Swetnam 1992), and both have been cycling through periodic epidemics since long before European settlement (Baker and Veblen 1990; Swetnam and Lynch 1993; Wilson and Tkacz 1996).

Fungal pathogens such as *Armillaria mellea* and *Phellinus weirii* are less well known but can be significant. James and associates (1984) reported that about 35% of the annual tree mortality in two national forests in Idaho was associated with root diseases caused by such fungi. Furthermore, fire suppression management during the past 50–100 yr may be largely responsible for conditions that have allowed major outbreaks of several insects (e.g., western spruce budworm, Douglas fir tussock moth, mountain pine beetle) and root diseases (e.g., *Armillaria*). Insects and fungal pathogens typically are far less devastating than fire. Indeed, loss of 50% or more of the mature trees in a *Pinus contorta* stand over a 2 yr period, though causing an immediate drop in production, can be compensated for within 5 yr by increased production on the part of the remaining trees (Romme et al. 1986).

Disturbance types of seemingly limited importance, such as avalanches and volcanism, also play major roles in specific regions. For example, in a study of the influence of avalanches on vegetation in the central third of Glacier National Park, Butler (1979) identified more than 800 avalanche paths. Because avalanches occur annually or nearly so in many such tracks, the variances in velocity and distance of flow of avalanches within a track become critical variables determining the impact on vegetation (Johnson, Hogg, and Carlson 1985). As might be expected, the most common woody plants in tracks with high avalanche frequencies are short-lived species with flexible stems, such as *Alnus* spp., *Acer glabrum*, *Betula pumila*, and *Salix* spp. (Butler 1979; Malanson and Butler 1984; Johnson et al. 1985).

Grazing and browsing by ungulates, both native and domestic, have had profound impacts on Rocky Mountain vegetation, at least south of the Canadian border. Overgrazing by domestic stock was widespread by the mid-1800s. For example, Knight (1994) has described the grazing boom in Wyoming that became widespread with the completion of the Union Pacific Railroad and that did not subside until climate and market conditions produced a dramatic crash in 1887. A thriving livestock trade existed in New Mexico already in the early 1700s, with severe overgrazing reported in some areas by the early 1800s (Denevan 1967).

The unambiguous impact of domestic stock for a long time distracted attention from the impact of native ungulates on the vegetation. However, dramatic increases in densities of elk (*Cervus elaphus*), especially since 1970, in Yellowstone and Rocky Mountain national parks have focused attention on the natural role of such ungulates in the original landscape. These population increases have dramatically decreased the extent of aspen woodland as well as riparian willow shrubland (Baker, Monroe, and Hessl 1997), by as much as 95% in Yellowstone (Kay 1994). The ultimate impact of the still increasing elk herds remains unclear. Whether elk populations were kept low prior to the 1800s by hunting pressure from Native Americans or were in equilibrium with then healthy populations of carnivores is also unresolved (see Kay 1994; Romme, Turner, Wallace, and Walker 1995; Baker et al. 1997).

Stand Development

Just as community composition varies across environmental gradients such as elevation and moisture, the course of stand development following a major disturbance such as fire is a function of gradient position. Three scenarios characteristic of low, middle, and high elevations serve to illustrate the range of stand development patterns normally encountered

Stands in the middle portions of the elevation and moisture gradients typically exhibit a developmental sequence widely encountered in montane and boreal conifer forests. Four-stage sequences have been envisaged by several workers (e.g., Bloomberg 1950; Bormann and Likens 1979; Oliver 1980; Peet 1981; Peet and Christensen 1987), with the details varying but the main outline being consistent. The first (establishment) stage is one of little competition and extensive tree establishment. During the second (self-thinning) stage, there is in-

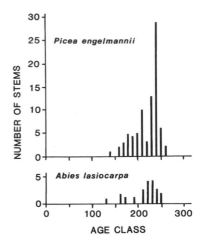

Figure 3.4. Age distributions in 10-yr age classes for Picea engelmannii and Abies lasiocarpa in a 260-yr-old, even-aged stand in Rocky Mountain National Park. Based on stems >2.5 cm diameter at 2 cm above the ground, cored at the ground surface.

tense competition among the established trees such that little, if any, new establishment is possible. This is primarily a period of decreasing tree density and increasing tree size. In the third (breakup) stage, mortality finally exceeds the ability of established trees to fill canopy gaps, with the result that resources again become available for new establishment. The fourth (equilibrium) stage is a form of steady state in which tree mortality is balanced by tree recruitment. This basic sequence can be found in forests dominated by *Pinus contorta*, *Picea engelmannii*, *Pseudotsuga*, and various other species.

The synchrony of tree establishment following disturbance can be highly variable and depends on site conditions and the intensity of competition from herbaceous species. On middle-elevation sites with abundant seed and limited competition from herbaceous species, establishment can be confined to a 10 yr window. More typically, one encounters a 40–70 yr range of establishment dates. The establishment window can be significantly wider on sites where severe environmental conditions inhibit seedling growth (Franklin and Hemstrom 1981; Peet 1981) or where especially fertile, moist soils allow rapidly growing herbs to overtop and out-compete tree seedlings. Age distributions (determined by increment cores at ground level) for *Picea* and *Abies* on a typical site in Rocky Mountain National Park are shown in Figure 3.4. In this stand *Picea* largely became established in the 80 yr following an extensive forest fire 260 yr prior to measurement, and very little establishment has taken place since that period. Notice also that *Abies*,

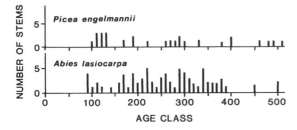

Figure 3.5. Age distributions in 10 yr age classes for Picea engelmannii and Abies lasiocarpa in a near steady-state stand in Rocky Mountain National Park. Based on stems >2.5 cm diameter at 2 cm above the ground, cored at the ground surface.

which is primarily confined to the subcanopy of this stand, became established during the same interval as *Picea*. *Pseudotsuga* often plays the same role in stands dominated by *Pinus contorta* and *Pinus ponderosa*, appearing to be younger than the primary canopy species owing to its smaller size but actually having become established during nearly the same interval.

Only infrequently do stands reach the steady-state stage of development in Rocky Mountain forests, owing to the high disturbance frequency. Usually the steady-state stage is recognized by a reverse-J or negative-exponential size distribution of the dominant species (Leak 1965; Parker and Peet 1984), though this can also occur in broadly even-aged stands. Figure 3.5 shows age distributions for an old stand in Rocky Mountain National Park that comes as close to a steady state as can normally be found. Even this stand fails to have a true negative-exponential age distribution and probably represents the second generation of trees following a disturbance, with most of the trees having become established during a 250 yr interval of canopy breakup.

Tree establishment on extreme, high-elevation sites or xeric sites is often very slow as a consequence of the harsh environment and sometimes the more rapid establishment of a highly competitive herbaceous stratum (Stahelin 1943; Bollinger 1973; Noble and Ronco 1978; Peet 1981). Establishment can be sufficiently slow that by the time the canopy is approaching closure, trees are dying from senescence-related causes. Consequently, the second stage of stand development with its intense tree competition and lack of recruitment is bypassed, as is the third stage with pronounced canopy breakup and renewed establishment. Instead, one finds a stage in which both stand density and tree size slowly increase toward steady-state conditions (Fig. 3.12b shows a 70-yr-old stand of such a type in which canopy closure is nowhere near

complete and establishment is likely to continue for another century or more).

Although stand development on extreme, high-elevation sites appears slow and gradual, the actual establishment events are often episodic inasmuch as only in unusually favorable years are new seedlings able to store sufficient reserves to survive their first winter (Rebertus, Burns, and Veblen 1991; Baker 1992; Kearney 1982; Tomback, Sund, and Hoffman 1993). Similarly, unusual conditions are often required for tree seedlings to become successfully established in competition with dense grasses and forbs. For example, several ecologists working in the Pacific Northwest have reported tree invasions of subalpine meadows to occur only during extended periods of drought and associated long snow-free periods (Franklin, Moir, Douglas, and Wiberg 1971; Agee and Smith 1984). Dunwiddie (1977) has described expansion of trees into subalpine meadows of the Wind River Mountains of Wyoming as being correlated with periods of grazing by domestic stock.

In open, low-elevation communities, such as *Pinus ponderosa* woodlands, episodic establishment can be particularly important. On these sites, fires generally are not detrimental to the canopy trees but occur frequently (though less so than prior to 1900) and at low intensity, killing young woody plants and removing accumulated dead grass and litter. Here, tree establishment occurs in pulses associated with the co-occurrence of a good seed year, favorable weather (e.g., ample spring and summer moisture), and absence of fire (or perhaps occurrence of an earlier fire that removed competition). Such strict requirements can result in long intervals between pulses of establishment (Potter and Green 1964; Peet 1981; White 1985; cf. Cooper 1960). Low-elevation riparian forests dominated by cottonwood (e.g., *Populus angustifolia*) also tend to be episodic in establishment, though the establishment is mostly concentrated in years of high spring discharge that scour the bottoms followed by moist summers (Baker 1990).

All intermediates between the three development patterns described can be found. For example, stand dynamics can vary continuously along environmental gradients such as that reported from Montana (Tesch 1981), where on a mesic, north-slope *Pseudotsuga* site, even-aged development occurred, but on the neighboring south-facing slope, recruitment was slow, and an uneven-aged stand developed with shade-intolerant pioneer species like *Pinus contorta* continuing to reproduce beneath the open *Pseudotsuga* canopy. Multiple disturbance events can lead to intermediate forms of stand development, such as when tree regeneration resumes in the middle of the self-thinning phase owing to wind (Veblen et al. 1989) or insect (Hadley and Veblen 1993) damage opening the canopy. Soils of exceptionally high fertility often support a lush herb and shrub layer that inhibits tree regeneration (Graves 1995), leading to a forest structure that is less even-aged and less dense than is otherwise encountered. It is probably this last phenomenon that is responsible for the treeline in Glacier National Park being poorly defined on fertile soils (see Malanson and Butler 1994).

Biomass, Production, and Species Diversity

As with stand development, three characteristic and contrasting patterns of successional change in biomass, production, and diversity are associated with low-, middle-, and high-elevation forests (Peet 1978b, 1981; also see Aplet, Laven, and Smith 1988). Following a major disturbance on a middle-elevation site, biomass is low but increases steadily to an asymptote in the self-thinning phase of forest development (Fig. 3.6a). With canopy breakup, biomass declines rapidly, only to increase again with the renewal of regeneration. Production similarly starts low but increases rapidly to reach a peak early in the self-thinning phase, shortly after canopy closure. Production then appears to decline slowly until canopy breakup frees resources and again stimulates production. Species diversity (vascular plant species per 1000 m²) tends to track resource availability. Early in the sequence diversity is quite high, but diversity drops dramatically with the onset of self-thinning and stays low until canopy breakup again provides resources for understory growth (see Table 3.4 for specific examples). Epidemics such as those of pine beetles simply mimic natural thinning, albeit at an accelerated rate, so that no long-term reduction in production is observed (Romme et al. 1986). Similarly, diversity increases following the release from canopy tree competition that accompanies beetle outbreaks, windthrow, and fire (Peet 1978; Christensen et al. 1989).

On exposed ridges and extreme, high-elevation sites where tree establishment is slow, biomass and production do not show the pronounced cycles characteristic of middle-elevation stands but, rather, they increase asymptotically (Fig. 3.6b). Species diversity is less variable but still tracks available resources and thus shows an initial high followed by a modest decline to steady-state values. In low-elevation woodlands, which are characterized by episodic regeneration, production is

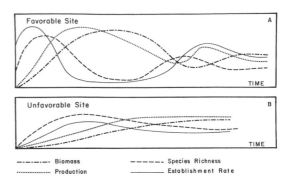

Figure 3.6. Generalized development patterns for forests of the Colorado Front Range. The favorable site (a) is typical of forests dominated by Pinus contorta, Pseudotsuga menziesii, and Picea engelmannii. Time scales vary with site, the secondary low in biomass occurring at ~ 300 yr for Pinus contorta and ~ 450 yr for Picea engelmannii. The unfavorable site (b) is typical for extreme, high-elevation Picea-Abies forests and xeric Pinus flexilis forests (from Peet 1981).

relatively constant, and biomass tracks the oscillations in numbers of large trees.

Biomass and production also vary with site and regional climate, but detailed data are few and only major patterns are clear (Table 3.2). An elevation sequence in the Santa Catalina Mountains of Arizona (first 10 stands in Table 3.2) shows that almost the entire range of biomass and production normally found in Rocky Mountain forests can be found across a single, environmentally heterogeneous landscape. Indeed, Weaver and Forcella (1977), studying data from some 50 forest stands in Montana, Idaho, and eastern Oregon, reported a range of biomass values contained within the range Whittaker and Niering (1975) found in the Santa Catalina Mountains.

The final 12 stands in Table 3.2 illustrate variation within forests dominated by a single species, *Pinus contorta*. In these stands, the biomass ranges from 5200 to 24,500 gm⁻² and production from 220–840 gm⁻² yr⁻¹. After study of a subset of these data, including two very different but adjacent stands (18 and 19), Pearson and associates (1984) concluded that *Pinus contorta* biomass is at least as tightly linked to stand density as to site conditions.

Species diversity also can vary greatly within a small but environmentally heterogeneous area. In Rocky Mountain National Park and the adjacent foothills, species richness (per 1000 m²) takes the form of a cup-shaped response surface on a plot of elevation versus site moisture gradient. Sites near the middle of moisture and elevation gradients generally have low numbers of species, with the bottom of the "cup" falling near an average of 13

Table 3.2. *Selected aboveground living biomass and production estimates for Rocky Mountain forests*

Dominant species[a]	Elevation (m)	Age (yr)	Density (ha⁻¹)	Ba[b] (m² ha⁻¹)	Biomass[c] (t ha⁻¹)	LAI[d] (m² m⁻²)	Production[e] (t ha⁻¹ yr⁻¹)	Location	Ref.[f]
Abies lasiocarpa	2720	106	590	57.8	357	14.7	8.6	Arizona	1
Abies concolor	2340	124	1510	58.6	361	15.5	11.1	Arizona	1
Pseudotsuga menziesii *Abies concolor*	2640	321	400	118.1	790	16.7	10.7	Arizona	1
Pseudotsuga menziesii	2650	252	340	70.5	438	15.5	8.3	Arizona	1
Pinus ponderosa *Pinus strobiformis*	2740	93	2700	39.4	161	7.6	6.1	Arizona	1
Pinus ponderosa	2470	142	1100	46.3	250	5.9	5.7	Arizona	1
Pinus ponderosa *Quercus hypoleucoides*	2180	150	1280	34.9	163	4.7	4.9	Arizona	1
Pinus leiophylla *Quercus arizonica*	2040	101	2780	26.0	114	3.7	4.3	Arizona	1
Pinus cembroides *Juniperus deppeana*	2040	115	570	4.3	19	2.0	0.65	Arizona	1
Quercus oblongifolia *Quercus emoryi*	1310	117	190	4.0	11	1.8	0.72	Arizona	1
Pinus edulis *Juniperus osteosperma*	2135	270	1111	—	119	—	5.6	Arizona	2
Populus tremuloides	3200	80	3070	31.8	157	—	11.7	New Mexico	3
Juniperus occidentalis	1356	>350	246	—	21	2.0	1.1	Oregon	4
Thuja plicata *Pinus monticola (3)*	—	105	798	56.8	290	—	8.7	Idaho	5
Tsuga heterophylla *Pinus monticola*	—	>250	105	49.8	316	—	5.5	Idaho	5
Pinus monticola	—	103	710	63.7	446	—	13.4	Idaho	5
Abies grandis *Larix occidentalis*	—	105	1127	53.5	290	—	9.2	Idaho	5
Pinus monticola *Abies grandis (5)*	—	103	314	66.4	504	—	13.8	Idaho	5
Picea engelmannii *Abies lasiocarpa (20)*	~3050	>250	908	40.2	125	—	1.8–5.2	Colorado	6
Pinus contorta	2800	110	2217	42	142	7.3	—	Wyoming	7
Pinus contorta	2800	110	14640	50	101	7.1	—	Wyoming	7
Pinus contorta	2800	110	9700	55	124	8.8	—	Wyoming	7
Pinus contorta	2900	110	1850	64	144	9.9	—	Wyoming	7
Pinus contorta	3050	75	1280	26	96	9.0	—	Wyoming	7
Pinus contorta	2950	240	420	37	132	4.5	—	Wyoming	7
Pinus contorta	—	72	8600	—	122	4.5	4.3	Colorado	8
Pinus contorta	—	71	1650	—	52	4.5	2.2	Colorado	8
Pinus contorta	—	77	3800	—	275	14.0	8.4	Colorado	8
Pinus contorta	1400	100	2520	52.3	245	—	—	Alberta	9
Pinus contorta	1400	100	717	34.9	194	—	—	Alberta	9
Pinus contorta	1400	100	12256	35.9	92	—	—	Alberta	9

[a] Numbers of stands averaged together shown in parentheses.
[b] Basal area.
[c] Aboveground biomass of trees.
[d] Leaf area index, two-sided.
[e] Aboveground primary production of trees.
[f] References: 1. Whittaker and Niering (1975); 2. Darling (1966); 3. Gosz (1980); 4. Gholz (1980); 5. Hanley (1976); 6. Arthur (1991); 7. Pearson et al. (1984); 8. Moir (1972); Moir and Francis (1972); 9. Johnstone (1971).

species per 1000 m² in dry *Pinus contorta* forests. At both the high- and low-elevation transitions from forest to grassland, diversity is much higher, typically averaging 40 species per 1000 m². The highest-diversity sites are mesic ravine forests at moderately low elevations, where deciduous trees often dominate and diversity averages up to 60 species per 1000 m² (Peet 1978b).

The complex relationship between diversity and environment in the Rocky Mountain National Park landscape illustrates the futility of modeling diversity as a simple function of elevation, moisture, or successional development; strong interactions are involved, and diversity can be understood only as a multidimensional phenomenon. What is clear is that dense conifer forests depress diversity, and that this depression can be ameliorated by increased levels of such resources as moisture, light (Daubenmire and Daubenmire 1968; Moral 1972; Peet 1978b; McCune and Antos 1981), and fertility (Allen et al. 1991).

MAJOR VEGETATION TYPES

The combined influences of elevation, moisture, soil chemistry, and latitude make for complex regional vegetation patterns. The necessary addition of communities recovering from various forms of disturbance yields, at a minimum, a five-dimensional vegetation model. Such continuous, multidimensional variation is difficult to present graphically or in words. In Figure 3.7 several dimensions of variation are displayed through a series of two-dimensional gradient diagrams. Each diagram has a vertical gradient of elevation and a horizontal gradient of moisture and exposure. In two of these diagrams, major seral species are indicated by shading. The latitudinal gradient can be deduced by comparison of the diagrams. To simplify discussion, I recognize 11 major vegetation types in the Rocky Mountain forest region, the characteristics of, and variation within, are discussed individually in the following sections.

Riparian and Canyon Forests

Low-elevation, stream-side forests of the Rockies are both the first forests encountered by the traveler and the least typical of the region as a whole. Broadleaf deciduous trees and shrubs, mostly cottonwoods (e.g., *Populus angustifolia*) and willows (*Salix spp.*), line the major streams of the foothills and adjacent semiarid lowlands (Fig. 3.8) and appear as a series of fingers of mesophytic forest in an otherwise semiarid landscape of low grass or desert scrub. The dominant species of these low-

elevation forests shift geographically, with *Populus balsamifera* ssp. *trichocarpa* occurring from the Central Rockies to coastal Alaska, *Populus deltoides* ssp. *monilifera* dominant on the eastern fringe of the Southern Rockies, and *Populus fremontii* dominant in the western portion of the Southern Rockies and much of the northern Madrean Region.

Riparian communities, like Rocky Mountain forests in general, exhibit variation with elevation. With increasing elevation in the Southern Rockies, forests of the wide-leaved *Populus deltoides* ssp. *monilifera* or *P. fremontii* give way to forests of the narrower-leaved *Populus angustifolia*. Still higher, *Alnus incana* ssp. *tenuifolia* and *Betula occidentalis*, along with numerous species of *Salix*, replace *Populus* on alluvial flats and moist meadow edges. *Salix* thickets continue to dominate many stream sides and valley bottoms up to timberline. Where middle-elevation streams pass through sheltered valleys or canyons, the dominant deciduous species are often replaced by conifers such as *Picea pungens*, *Pseudotsuga*, and *Abies concolor* (Pace and Layser 1977, Peet 1978a; Romme and Knight 1981). In the Central Rockies *Alnus viridis* ssp. *sinuata* is the dominant streamside shrub of middle elevations, *Picea engelmannii* being the only regularly co-occurring tree.

The similarities of the foothill riparian forests to the deciduous forests of eastern North America do not stop with the deciduous habit of the dominant trees. Numerous species of herbs typical of eastern forests occur in a narrow band along the foothills of the east slope, where they are confined to mesophytic riparian habitats. These include *Aralia nudicaulis*, *Carex sprengelii*, *Maianthemum racemosum*, and *Ranunculus abortivus*. Other species typical of eastern forests, such as *Maianthemum stellatum* and *Ratibida pinnata*, are widespread in the Central Rockies but are largely confined to similar riparian habitats or to the deciduous *Populus tremuloides* woodlands.

In the Madrean Region, the low-elevation riparian vegetation can be divided into a lower *Platanus wrightii*, *Juglans major*, and *Fraxinus* community and a higher *Alnus oblongifolia* community. Here cool, moist canyons take on added importance as habitat for rare or relict mesophytic species. Trees typical of eastern forests, such as *Ostrya virginiana*, *Cercis canadensis*, and *Quercus muehlenbergii*, occur as isolated populations in a few moist canyon bottoms and lower slopes, and some predominantly madrean species, such as *Cupressus arizonica*, are mostly confined to such canyons at their northern limits. Except for the uncommonly high importance of *Cupressus*, Niering and Lowe's (1984) description of Bear Canyon in the Santa Catalina

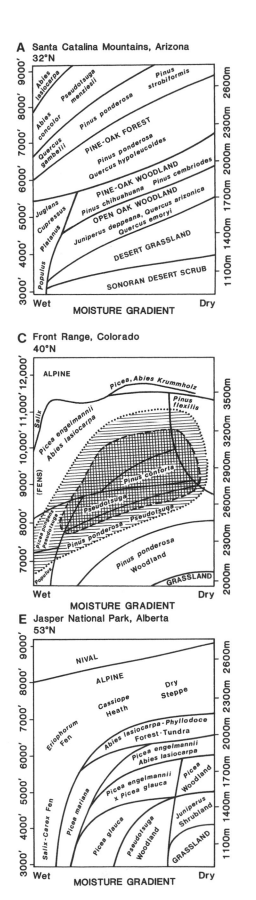

A Santa Catalina Mountains, Arizona
32°N

B Sangre de Cristo Mountains, New Mexico
36°N

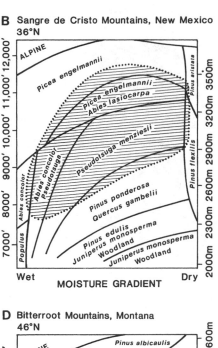

C Front Range, Colorado
40°N

D Bitterroot Mountains, Montana
46°N

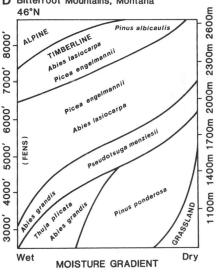

E Jasper National Park, Alberta
53°N

Figure 3.7. Gradient mosaic diagrams illustrating variation in vegetation composition with elevation and topographic position for seven sites along a latitudinal sequence. In B and C, the shading down to the left indicates the range of Populus tremuloides as an important postdisturbance species, whereas shading down to the right indicates the range of Pinus contorta as an important postdisturbance species. Successional P. contorta and P. tremuloides communities may also occur at locations D and E, but they are not indicated because of insufficient data on the exact distributions. A: Santa Catalina Mountains, Arizona (from Whittaker and Niering 1965). B: Southern Sangre de Cristo Mountains near Santa Fe, New Mexico (redrawn from Peet 1978a). C: Northern Front Range, northern Colorado (redrawn from Peet 1978a, 1981). D: Bitterroot Mountains, central western Montana (adapted from Habeck 1972, and Arno 1979) E: Jasper National Park, Alberta (adapted from La Roi and Hnatiuk 1980).

Figure 3.8. Riparian Populus angustifolia (cottonwood) forest bordering the Wind River in western Wyoming. Riparian forests regularly extend down into the grass- and shrub-
dominated foothills and plains.

Mountains of southern Arizona serves well to represent this type. They reported the common tree
species (>2.5 cm dbh) to be, in descending order
of abundance: *Cupressus arizonica, Alnus oblongifolia, Quercus rugosa, Fraxinus velutina, Q. hypoleucoides, Q. oblongifolia, Platanus wrightii, Q. emoryi,
Pinus cembroides,* and *Juglans major.*

Canyon and riparian vegetation has been
greatly modified by human activities, probably
more so than any of the other major vegetation
types of the Rocky Mountain region. River courses,
already narrow and of limited extent, provide a
natural route for roads and railroads through the
rugged terrain. These bottomlands also contain the
best lands for agriculture in terms of level terrain,
fertility, and water availability. Finally, the aesthetic attraction of streamsides has led to a concentration of dwellings. Just as pervasive but less
widely appreciated is the impact of water removal
and water control structures such as dams. Elimination of high-water events has meant that the
chronically early succession character of streamsides is rapidly being lost and streams are stabiliz-
ing in both location and vegetation. Open cottonwood woodlands are growing up to intricate

forest, and the nearly treeless scour zone vegetation is rapidly being narrowed (see Baker 1988;
Knight 1994; Baker and Walford 1995).

Pygmy Conifer Woodland

Over much of the Southern and Madrean Rocky
Mountain regions, pygmy conifer woodlands (Fig.
3.9A) form the transition from grassland or desert
on the basal plains to montane conifer forests. The
dominant junipers (e.g., *Juniperus osteosperma, J.
scopulorum*) and pinyon pines (*Pinus* section *Cembroides*) generally have a low stature seldom exceeding 7 m, rounded crowns, and multiple stems
that give the trees a shrub-like appearance. The
lowest, driest sites are dominated by *Juniperus osteosperma*, with pinyon assuming dominance at
higher elevations. *Juniperus scopulorum* is more frequent in the pinyon zone and the adjacent lower
reaches of the ponderosa pine woodlands. Pygmy
conifer woodlands are best developed in the semiarid lands west of the main cordillera and are,
therefore, more fully described in the chapter on
cold deserts (Chapter 7).

On the east slope of the Rockies from the Texas

Figure 3.9. Lower-treeline forests. A: Pygmy conifer wood-land of Pinus edulis and Juniperus monosperma east of Taos, New Mexico. B: Quercus gambelii chaparral with scattered Pinus ponderosa at the mountain front near Colorado Springs, Colorado.

border north to central Colorado, *Pinus edulis* and *Juniperus monosperma* are the dominant trees of foothill woodlands. Southward, *Pinus cembroides* and *Juniperus deppeana* assume dominance, though with a progressively greater intermixing of madrean forest elements. *Pinus cembroides* continues south to the Mexican volcanic belt (Chapter 15), but throughout the woodlands of Mexico pinyon shares dominance with various oak species, except on the most highly disturbed sites (Robert 1977; Passini 1982; Segura and Snook 1992). In north-central Colorado, both *Juniperus monosperma* and *Pinus edulis* are absent from the east slope except for a disjunct population of pinyon northwest of Fort Collins, likely established in the last few hundred years (Wright 1952; Betancourt, Seguster, Mitton, and Anderson 1991). In a region stretching from Idaho on the west to the Black Hills of South Dakota on the east, and north into Alberta, *Juniperus scopulorum* dominates low-elevation woodlands. This expansion of *J. scopulorum* from its typical forest margin habitat in the south might be a consequence of the absence of competing pinyon and juniper species, though other environmental and biotic factors cannot yet be discounted. In some locations in the Central Rockies, *Pinus flexilis* occurs with *J. scopulorum*, filling the niche of the co-dominant pine (Daubenmire 1943; Peet 1978a; Arno 1979). West of the main Rocky Mountain massif, from the Mogollon Mesa of Arizona north to the Wyoming border, *Juniperus osteosperma* and *Pinus edulis* dominate vast areas. Westward, in the Great Basin region of Nevada and western Utah, *Pinus monophylla* replaces *P. edulis* as the dominant pine. As a general rule, *Juniperus monosperma* is associated with winter moisture, whereas *J. osteosperma* and *J. deppeana* are commonly associated with summer rains, and *J. scopulorum* is the least tolerant of drought (Gottfried, Swetnam, Allen, Betancourt, and Chung-Mac Coubrey 1995). On the east side of the cordillera in the Bighorn Mountains of Wyoming, *J. osteosperma* occurs in association with limestone substrate (Despain 1973), except where grazing and fire suppression have allowed a recent expansion in habitat (Waugh 1986).

Fires in pinyon-juniper stands are infrequent owing to the low fuel levels. In particular, droughty conditions and competition from tree roots limit grass and forb growth. Moreover, seedling establishment is infrequent and unpredictable owing to the droughty habitat. As a consequence, most pinyon-juniper stands, at least on sites with a long history of woodland occupation, are uneven-aged, and some contain trees up to 300–400 yr old (Swetnam and Brown 1992), Gottfried, Swetnam, et al. 1995). Nonetheless, major fires oc-

cur in these landscapes, most typically when a particularly dry year follows a moist year in which grass and other fuels accumulated (Gottfried, Swetnam, et al. 1995). The sensitivity of pinyon to fire together with low fuel in most pinyon stands probably meant that the original presettlement landscapes tended to alternate between areas with frequent fires and low tree density and those with higher density and low fire frequency.

Numerous sources of information, including old photographs, age-structure analyses, and personal observations of long-time residents, clearly document a major historic expansion of juniper woodlands into adjacent grasslands and, to a lesser extent, an increase in density of the existing woodlands (Springfield 1976; Tausch, West, and Nabi, 1981; Gruell 1983; Waugh 1986; Veblen and Lorenz 1991; Knight 1994; Floyd-Hanna, Spencer, and Romme 1996). Several studies, such as those of Burkhardt and Tisdale (1969, 1976), have shown older *Juniperus* to be associated with rocky sites, whereas the expansion has been largely onto finer-textured soils. This, together with the correlation between heavy grazing and juniper expansion (Johnsen 1962; West, Rea, and Tausch 1975; Waugh 1986), suggests a scenario wherein the introduction of intense grazing by domestic cattle and sheep reduced the cover and vigor of the dominant grasses. This both allowed establishment of *Juniperus* and reduced the frequency of grass fires owing to fuel reduction (Savage and Swetnam 1990). With increased establishment opportunities and reduced fire frequency and intensity, *Juniperus* quickly spread out of its traditional rocky refugia (Young and Evans 1981). Little is known of the fire frequency needed to prevent juniper encroachment, although Burkhart and Tisdale (1976) do estimate that a fire interval of 30–40 yr is sufficient to keep *Juniperus* from invading the adjacent *Artemisia tridentata* steppe in southwestern Idaho. Nonetheless, other explanations are plausible for the *Juniperus* expansion.

Aboriginal exploitation of this pygmy woodland for fuel may have reduced the importance of juniper and pinyon in at least some areas (Samuels and Betancourt 1982) so that part of the current expansion of these species might simply be a result of recovery following this earlier land use. Bahre (1991) presents compelling evidence that pinyon-juniper woodlands in general are rebounding from extensive Euro-American harvest in the 19th century.

Also, a marked shift in climate toward decreased severity of drought that occurred around 1900 has been shown to have had a significant impact on establishment of grasses in the semiarid

Southwest (Neilson 1986). This climatic shift could also have been responsible for the coincident increase in woody plant establishment.

Ponderosa Pine Woodland

The vegetation type most Americans associate with western mountains is *Pinus ponderosa* woodland. Novels, films, and television programs have romanticized this landscape of tall but sparse trees growing over grassy rangeland (Fig. 3.10). Although best developed in the southwestern states, *P. ponderosa* woodland extends from the Sierra Madre Occidental of Mexico north into the dry interior valleys of southern British Columbia. East of the Continental Divide, from the Colorado–Wyoming border northward, the low-elevation *P. ponderosa* woodland occurs only along the eastern fringe of the Rocky Mountain region. Thus, the more interior Medicine Bow, Wind River, and Teton ranges of Wyoming have little, if any, *P. ponderosa*, whereas outlying ranges like the Laramie and Bighorn, plus the Black Hills of South Dakota and numerous rocky scarps of the western Great Plains, have extensive *P. ponderosa* forests and woodlands (Wells 1965; Alexander and Edminster 1981; Knight 1994).

Near its upper elevation limit, *P. ponderosa* increases in density to form well-developed forests, though often with the pine being only successional to *Pseudotsuga* (Peet 1981). In the Madrean Region, low-elevation *P. ponderosa* woodland grades downward into either pygmy conifer woodland or encinal (an oak-dominated, chaparral-like community). Where pygmy conifers are absent, as along the Front Range in northern Colorado, the forest of *P. ponderosa* becomes progressively more open with decreasing elevation, until only scattered individuals remain in the most rocky areas.

The lush grass understory of a *P. ponderosa* woodland is highly flammable during the dry summer months. As a consequence, both lightning and aboriginal fires were common in the years before European settlement. Studies of multiple fire scars on old pines have shown that, in many areas of the Southwest, fires occurred almost annually during the eighteenth and nineteenth centuries (Cooper 1960; Dieterich 1980, 1983; Dieterich and Swetnam 1984). Farther north along the east slope of the Colorado Front Range, where the landscape is more dissected, the fire return time was longer, generally 25–40 yr (Rowdabaugh 1978; Laven, Omi, Wyant, and Pinkerton 1980).

The frequent occurrence of fire not only kept the understory of *P. ponderosa* woodlands free of invading tree species but also limited pine regeneration. In addition, the semiarid climate of the pinelands led to low establishment rates (Pearson 1923), and the pronounced year-to-year variation in seed production (Schubert 1974) further limited most regeneration to only occasional mast years. The age structures of old *P. ponderosa* stands still reflect the episodic nature of regeneration (Cooper 1960; Potter and Green 1964; Larsen and Schubert 1969; Schubert 1974; White 1985).

Original *Pinus ponderosa* forests of the Southwest have been reported to have had a patchy appearance, most trees occurring in small, family-like groups with considerable grass and forb development between, but not within, the groups. Cooper (1960, 1961) described this two-phase mosaic and reported that the patches were even-aged and represented synchronous regeneration following the death of an earlier patch of trees. Specifically, the earlier occupants supplied fuel that produced a particularly hot fire, which in turn provided openings in the grass for tree invasion. Subsequently, White (1985) showed thatsome sites with such a patch-like appearance have uneven-aged within-patch structure. He suggested that there was periodic regeneration of trees in the patches, usually following a fire that burned the accumulated fuel beneath the living trees and thus provided regeneration sites safe from competition with grasses. The possibility exists that Cooper and White were describing essentially the same phenomenon, with differences in fire intensities being responsible for the death or survival of the original patch dominants.

Following suppression of wildfires and the introduction of domestic cattle, *Pinus ponderosa* woodlands underwent dramatic transformation. The first change was establishment of numerous new trees, which caused a shift from a woodland to a forest physiognomy, a change documented essentially throughout the range of the species (e.g., Cooper 1960; Veblen and Lorenz 1986; Covington and Moore 1994; Morgan 1994). This increased regeneration could have been the result of either reduced grass density, owing to grazing (Savage and Swetnam 1990), or of fire suppression that allowed greater postestablishment survival (Marr 1961). These dense stands of young trees, lacking recurrent fires, readily built up high fuel loads so that fires became catastrophic, killing many of the trees in a forest (Weaver 1959; Cooper 1960; Kallander 1969).

The understory vegetation of natural, fire-maintained ponderosa pine woodlands is dominated by grasses and contains a rich assemblage of forbs. Although superficially consistent throughout and poorly documented, the composition of the

Figure 3.10. Ponderosa pine forest. A: Pinus ponderosa with Quercus gambelii *understory near Santa Fe, New Mexico. With decreasing latitude, low-elevation pine stands gain madrean elements, especially oaks. B:* Pinus ponderosa *woodland with a shrub layer of* Artemisia tridentata *and a herb stratum dominated by* Muhlenbergia montana *in Rocky Mountain National Park, Colorado. Beyond the trees, open park vegetation is visible.*

ponderosa woodland understory varies dramatically with site conditions and geography (Peet 1981; Pase 1982). Modern fire suppression and the associated increase in tree density have greatly altered and simplified the composition of this once species-rich herbaceous layer over most of the range of the species (see Peet 1981; Fulé and Covington 1996).

Madrean Pine-Oak Woodland

With decreasing latitude and more moderate climate, *Pinus ponderosa* woodlands are replaced by a diverse assemblage of pines and oaks, as well as other broad-leaved, sclerophyllous angiosperms. The northern limit of this formation can be defined approximately by the Santa Catalina and Chiricahua mountains of Arizona on the west and the Davis Mountains of Texas on the east. These mountain ranges are, in effect, the northernmost of an archipelago of desert mountain ranges that provide floristic stepping-stones to the species-rich madrean pine-oak woodlands and forests of the Sierra Madre Occidental. The northern limit of madrean influence is reached with the *Pinus ponderosa* forests of the southern Sangre de Cristo Mountains of northern New Mexico and southern Colorado, where *Quercus gambelii* is an important understory shrub species (Fig. 3.10a) (Peet 1978a).

At elevations between roughly 1750 and 2200 m, the mountains of southern Arizona and New Mexico (e.g., Santa Catalina Mountains, Chiricahua Mountains) support a pine-oak woodland composed of such species as *Pinus leiophylla, P. cembroides, P. arizonica var. arizonica, Quercus hypoleucoides, Q. arizonica, Q. emoryi, Q. rugosa, Juniperus deppeana,* and *Arctostaphylos pungens* (Whittaker and Niering 1965; Niering and Lowe 1984; Dick-Peddie 1993; Barton 1994), several of which are important over extensive areas in Mexico (see Chapter 15). Pine-oak woodlands at 2250 m in the Davis Mountains of Texas are not as diverse as in the Santa Catalina woodlands but contain such species as *Pinus edulis, P. ponderosa, P. strobiformis, Juniperus deppeana, Quercus grisea, Q. hypoleucoides, Q. gravesii, Q. gambelii, Populus tremuloides,* and *Arbutus texana* (Hinckley 1944). South of the Davis Mountains near the Texas–Mexico border, the Chisos Mountains have a more diverse madrean flora, with numerous broadleaf tree species (Whitson 1965). Similar but significantly more species-rich vegetation has been described from the mountains of Chihuahua (Shreve 1939), Coahuila (Muller 1947), and the Sierra Madre Occidental south to the central Mexican Highlands (Brown 1982; Felger and Johnson 1995; Felger and Wilson 1995). The vegetation

grades from grasslands at the lowest elevations through open savanna to woodlands, the most common species at the grassland transition over much of the region being *Q. chihuahuensis* but with *Quercus emoryi, Q. oblongifolia,* and *Q. arizonica* more important toward the north (Brown 1982). Higher in the mountains the woodlands become a mixture of pine and oak, with the pine generally dominating the upper canopy and the oaks the understory. On the western slope of the Sierra Madre, with its more moderate climate, there is a considerable admixture of tropical elements, including pines (e.g., *Pinus engelmannii, P. durangensis,* and *P. oocarpa*), various tropical epiphytes (e.g., *Tillandsia* spp., *Encyclia microbulbon., Oncipidium cebolleta*), and various species of woody plants usually associated with the drought-deciduous tropical forests farther south (Brown 1982; Felger and Johnson 1995).

The latitudinal gradient in madrean woodlands is perhaps most dramatically illustrated by Leopold's report (1950; also see Perry 1991, Nixon 1993) that approximately 112 species of *Quercus* and 39 species of *Pinus* can be found in the pine-oak woodlands that form the dominant vegetation over much of the Mexican Plateau. Moreover, the high diversity of the madrean forests and woodlands occurs not only at the regional scale but also at the scale of the individual stand. Fulé and Covington (1996) report locations in the Sierra Madre Occidental where they found 20 or more overstory species within a few hectares, with seven or eight pines and a similar number of oaks.

Leopold (1937), Marshall (1962), and Fulé and Covington (1994, 1996) all comment on how the Mexican tradition of burning the montane pine-oak communities maintained a natural structure as well as high biodiversity, at least through the 1940s. Fire return intervals of roughly 4 yr were not uncommon (Fulé and Covington 1994). However, fire suppression is now widespread, and many areas not exploited by timber extraction are now growing up to dense stands of substantially lower diversity, as happened in many areas north of the international border a half century earlier.

Southward, with the progressively more moderate winters, broadleaf sclerophyls largely replace the conifers. The resultant encinal communities, dominated by genera such as *Rhus, Ceanothus, Quercus, Cercocarpus, Arctostaphylos,* and *Arbutus,* have clear floristic affinities with the chaparral of California, despite a very different climatic regime (Muller 1939; Brown 1982; Axelrod and Raven 1985; Rundel and Vankat 1889; Vankat 1989). They can be found at the lower limit of the forest on the west slope of the Sierra Madre Oriental, as well as in ranges farther west and even on the shallow soils of exposed ridge tops in the Santa Catalina and Chiricahua mountains of southern Arizona (Whittaker and Niering 1965).

Species-poor encinal communities, dominated almost exclusively by *Quercus gambelii*, with various amounts of *Rhus* and *Cercocarpus*, can be found as far north as Colorado Springs, Colorado, on the east slope (Fig. 3.9b), and north to the Wyoming and Idaho borders on the west (Ream 1964). On the east slope beyond the range of *Quercus*, an attenuated form of the encinal community composed of low shrubs such as *Rhus trilobata*, *Cercocarpus montanus*, and *Purshia tridentata* occurs between the grasslands of the plains and the foothill *Pinus ponderosa* woodlands as far north as the Laramie Mountains in Wyoming (Ramaley 1931; Peet 1978a).

In Utah and western Colorado, *Quercus gambelii* dominates extensive areas along the lower slopes of the mountains. The sites appear similar to those typically dominated by either *Pinus edulis* or *Pinus ponderosa* elsewhere in the Rockies, but no explanation for the dominance of oak over pine has received widespread acceptance (Harper et al. 1985). The successional status of oak varies with site conditions. Although stable stands have been described (e.g., Brown 1958), in some locations *Quercus* is clearly successional to *Pinus ponderosa* (Dixon 1935; Cronquist, Holmgren, Holmgren, and Reveal 1972), whereas in others it is successional to *Pinus edulis* (Floyd 1982), *Acer grandidentatum* (Nixon 1967), or *Abies concolor* and *Pseudotsuga* (Harper, Wagstaff, and Kunzler 1985).

Shrub-dominated communities having floristic affinities with the *Quercus gambelii* encinal are widely distributed in the Rockies, particularly on the western slope. Floyd-Hanna et al. (1996) report such communities in the San Juan Mountains to have an amazingly high diversity of woody plants including such taxa as *Quercus gambelii*, *Purshia tridentata*, *Prunus virginiana*, *Fendlera rupicola*, *Cercocarpus montanus*, *Symphoricarpos oreophilus*, *Ribes* spp., and *Rhus trilobata*. They suggest such communities to be particularly common on north-facing slopes where snow accumulates and slides frequently damage the inflexible trees, whereas shrubs simply bend and spring back. In contrast, other researchers (Brown 1958; Erdman 1970) find much shrubland where pinyon-juniper woodland has been subjected to frequent fires. Normally, the relatively low fuel of pinyon pine woodland inhibits fire and protects the pine, but once the pines are removed by catastrophic or unusually frequent fire, the more readily burned shrublands assume dominance and maintain themselves by their fuel

production and intrinsic flammability (see Floyd-Hanna et al. 1996).

Douglas Fir Forest

Pseudotsuga menziesii (Douglas fir) is found on appropriate sites throughout the montane zone of the Rocky Mountains, ranging from central British Columbia deep into Mexico. Only in northern British Columbia and the Yukon is *Pseudotsuga* absent, though the Rocky Mountain montane forests of which it is a dominant are essentially absent north of Jasper National Park, Alberta (Stringer and La Roi 1970). Over its range, this species is the potential climax dominant for a broad range of sites. In addition, *Pseudotsuga* functions as an important seral species in many additional areas. *Pseudotsuga* is often associated with shade-intolerant seral species such as *Pinus contorta*, *Pinus ponderosa*, and, in the Northwest, *Larix occidentalis*. All these species regenerate well following fire, and all except *Pinus contorta* are tolerant of repeated, low-intensity surface fires.

In the Front Range of Colorado, *Pseudotsuga* is the dominant tree species of north-facing slopes and steep ravines from the lower treeline at 1650 m up to 2700 m. However, on open slopes it is more restricted and is confined mostly to sites between 2300 and 2800 m. Northward, in western Montana, *Pseudotsuga* is more broadly distributed, ranging from either the lower treeline or *Pinus ponderosa* woodland upward to *Picea-Abies* forest.

On mesic, fertile slopes of the Southern Rocky Mountain Region, *Pseudotsuga* often co-dominates with or is successional to *Abies concolor* (Moir and Ludwig 1979; Jamieson et al. 1996). (*A. concolor* extends north along the east slope to central Colorado, whereas in Utah, where minimum temperatures are not as extreme, it ranges north to the Idaho border.) *Picea pungens* can similarly co-dominate with *Pseudotsuga* in the Southern Rockies and northern Madrean Region, but it is largely confined to the frequently saturated soils of moist canyon bottoms. In the Madrean Region, *Pseudotsuga* is more a species of high peaks, north slopes, and mesic sites. In the Huachuca (Brady and Bonham 1976) and Santa Catalina Mountains of Arizona, it can dominate on peaks above 2450 m (Niering and Lowe 1984). In northeastern Sonora and adjacent Chihuahua, mixed conifer forests of *Pseudotsuga*, *Pinus*, and *Abies* occur above 2150 m, whereas farther south in Sonora they are very limited in distribution and occur only as small patches on north-facing slopes and in canyons (Felger and Johnson 1995).

Cascadian Forests

Several tree species that dominate extensive areas in the Cascade Mountains of Oregon, Washington, and British Columbia reach eastward into the Rockies only in ranges near the Columbian Plateau and adjacent Canada. These forests are particularly well developed in the Idaho panhandle and extend as far east as the western slope of Glacier National Park, Montana (Pfister, Kovalchik, Arno, and Presby 1977; Antos and Habeck 1981; Habeck 1987; Cooper, Neiman, and Robert 1991). This is an area where Pacific air penetrates the Cascades, bringing heavy rains and cool but moderate temperatures to the western slopes of the Rockies. Near the eastern limits of *Tsuga heterophylla* (western hemlock) and *Thuja plicata* (western red cedar), outbreaks of Arctic air periodically damage trees not in sheltered localities. Among the Cascadian species, *Tsuga heterophylla*, *Thuja plicata*, *Abies grandis*, *Taxus brevifolia*, *Tsuga mertensiana*, and *Larix lyallii* (at treeline) are locally climax species of the Northern Rockies. Important understory species with similar distribution patterns include *Menziesia ferruginea*, *Oplopanax horridus*, *Philadelphus lewisii*, *Rhododendron albiflorum*, *Sorbus sitchensis*, *Vaccinium membranaceum*, and *Xerophyllum tenax* (Aller 1960; Pfister et al. 1977).

On mesic sites at moderately low elevations (ca. 800–1600 m), particularly in sheltered valley bottoms, luxuriant forests of *Thuja plicata* and *Tsuga heterophylla* dominate and appear to be essentially equivalent to forests in the Cascade Mountains to the west (see Chapter 4). Basal areas of 100 m² ha⁻¹ are not uncommon, and values in excess of 200 have been reported (Daubenmire and Daubenmire 1968). *Tsuga* is ubiquitous in those forests, and like *Abies* in the *Picea-Abies* forests, it clearly dominates the seedling and sapling strata. The relative abundance of seedlings suggests that *Tsuga* will slowly replace *Thuja* (Daubenmire and Daubenmire 1968; Habeck 1968), though the evidence is equivocal. On wet bottomlands and on warmer mesic slopes near the lower limit of *Tsuga*, *Thuja* typically assumes dominance. Although generally a species of wet sites, *Thuja* can occupy any aspect or slope position in portions of northern Idaho between 500 and 1600 m. At the upper limits of *Tsuga heterophylla* on moist sites, it is replaced by *Tsuga mertensiana* or, more commonly, by *Abies lasiocarpa* (Cooper et al. 1991).

Between the *Tsuga heterophylla–Thuja plicata* forests on moist sites and the *Pseudotsuga* forests on drier sites, *Abies grandis* is the potential climax dominant. In addition, *Abies grandis* dominates between the warm, low-elevation *Pseudotsuga* forests

and the *Picea-Abies* forests of cool, high-elevation sites.

Like most Rocky Mountain forest types, *Tsuga*, *Thuja*, and *Abies grandis* forests are periodically subject to severe crown fires (Habeck 1968; Antos and Habeck 1981). The principal successional species are *Pinus contorta*, *P. monticola*, and *Larix occidentalis*. On moister sites, *Abies grandis* can occur as a successional species, as can even *Tsuga* and *Thuja* (Daubenmire and Daubenmire 1968; Antos and Habeck 1981; Cooper et al. 1991). *Larix occidentalis* is a special case in that not only is it especially shade-intolerant, it is arguably the most fire-tolerant tree species of the Rockies. Not uncommonly, *Larix* co-dominates with *Pinus ponderosa* in open grassy woodlands maintained in an open state by low-intensity ground fires (Arno and Fischer 1995). Although technically these forests could be viewed as successional, the frequency of fire in the presettlement landscape was sufficient to assure long-term compositional stability.

Montane Seral Forests

The ubiquity of disturbance in Rocky Mountain forests, particularly by fire, has resulted in seral species dominating large proportions of the landscape. The two best known seral species are *Pinus contorta* (lodgepole pine) and *Populus tremuloides* (quaking aspen). Both are widespread and can form steady-state forests under certain conditions, but they are much more important as postfire invaders (Fig. 3.11). *Pinus contorta* dominates many postfire forests from the northern end of the Rockies in the Yukon Territory of Canada south to the Sangre de Cristo Mountains of southern Colorado. The species is absent, however, from the Madrean Region and most of the more isolated and xeric ranges of the Great Basin. *Populus tremuloides* is the most widely distributed tree species in North America and is the only important deciduous tree species in the uplands of the Rocky Mountains north of the Madrean Region. It can be found on postfire sites as well as along forest margins from near the Arctic treeline south into Mexico (although south of 32°N latitude it occurs only as isolated, apparently relict populations). Trees of both species grow rapidly following fire and from extensive, even-aged stands.

Pinus contorta is often viewed as the archetypal postfire species. The pattern described long ago by Clements (1910) and others (e.g., Mason 1915) is that following fire the serotinous cones of *P. contorta* release large quantities of seeds that produce a dense, even-aged stand. Such a forest undergoes initial rapid growth, followed by slower growth

Figure 3.11. Middle-elevation seral forests. A: Populus tremuloides *on a middle-elevation open slope in the southern Sangre de Cristo Mountains near Santa Fe, New Mexico. The location is south of the range of* Pinus contorta *on a site which that species might otherwise dominate. B: A senescent, even-aged* Pinus contorta *stand in Yellowstone National Park, Wyoming.*

and natural thinning until the next fire comes along to reset the cycle. Though correct in broad outline, that scenario has had to be revised as knowledge of regeneration and age structure has increased (e.g., Veblen 1986a, 1986b; Johnson and Fryer 1989). Often, seedling establishment is not particularly rapid, with the consequence that a typical lodgepole stand has an initial cohort with a broad range of establishment ages, upward of 30–50 yr (Peet 1981; Veblen and Lorenz 1986) and sometimes reaching 100 yr. The rate of postfire seedling establishment is influenced by distance from seed source, rate of recovery of competing herbaceous vegetation, soil texture, and availability of resources such as water and nutrients (see Peet 1981; Knight 1994). In those situations in which numerous seedlings do become established within a few years following fire, "dog-hair" stands can develop with little size hierarchy being apparent; the result often is slow growth for all stems.

Although *Pinus contorta* does produce serotinous cones, serotiny is genetically controlled and its prevalence in the population varies geographically and with stand history. In areas with extensive, gentle topography over which fires can spread readily with little interruption, serotiny is the rule (Lotan 1975). However, in areas of rugged topography where bare rocky ridges break up the landscape keeping fires small, and where some patches of pines can be certain to escape fire, many trees are not serotinous. In intermediate habitats in Montana, where the average return time for fire approaches the longevity of the species, the level of serotiny has been shown to correlate with stand history; stands originating following fire have a high percentage of serotinous trees, whereas stands originating following blowdown or insect damage have a high percentage of trees with nonserotinous cones (Muir and Lotan 1985). Tinker, Romme, Hargrove, Gardner, and Turner (1994) confirmed and expanded on this result by observing that in Yellowstone Park the percentage of serotiny is rather constant within small-scale patches (<1 km) corresponding to consistent disturbance history, and at large scales (>10 km) where disturbances would average out, but is quite variable at scales in-between, corresponding to observations with different disturbance history. In addition, Mutch (1970) has proposed that selection has favored flammability in the shade-intolerant *P. contorta*, with high flammability leading to catastrophic fires that kill invading understory climax species, thus ensuring continued success for the serotinous phenotypes of *P. contorta*. The increasing prevalence of stands that have developed following logging rather than fire is resulting in a shift in the degree of serotiny, synchrony of establishment, and even the degree of dominance by *Pinus contorta* (versus *Abies* and *Pseudotsuga*; Muir 1993).

The ecology of *Populus tremuloides* contrasts greatly with that of *Pinus contorta*. Aspen regenerates largely from root sprouts instead of seeds (Larson 1944; Brown and DeByle 1987; Romme et al. 1997). This allows it to establish on sites with fertile soils and the associated lush herb growth that cannot be invaded by most conifers. Until recently, nearly all forest stands between 2000 and 3200 m in the Colorado Front Range contained small *Populus* sprouts, even where mature trees were no longer found (although the recent explosion of the elk population in some areas appears to have changed this situation). Cottam (1954) found that seedling establishment is almost nonexistent in most situations. He suggested that nearly all the aspen in the extensive woodlands of western Colorado and eastern Utah are of sprout origin and that the clones largely date from an earlier period of greater and more seasonally even precipitation. To illustrate the clonal nature of the aspen woodlands, he described how distinct clones could be identified in the spring or fall by leaf phenology and coloration.

The Yellowstone fires of 1988 represent a conspicuous exception to the general lack of seed regeneration in that aspen seedlings were abundant the year following the fire, and these individuals have continued to persist (Kay 1993; Romme et al. 1997). These fires were exceptional in their intensity and extent and caused ground surfaces to be more highly scarified and free of herbaceous competition than normal. Moreover, this was also a mast year for aspen.

Following fire, aspen sprouts rapidly from previously established root systems, often developing complete canopy coverage in only 3–5 yr (Jones and Trujillo 1975; Brown and DeByle 1987). Extremely hot fires can damage or kill aspen roots, but intense fires are uncommon in *Populus* stands, which typically have a well-developed understory of mesophytic forbs and grasses in contrast to most conifer stands, which have sparse understories and considerable accumulated woody litter (DeByle et al. 1987). The undergrowth is to some extent a consequence of the fine-textured and more fertile soils of sites that *Populus* occupies, but it must also be viewed as a consequence of the biology of the species. This is particularly evident south of the range of *Pinus contorta*, where sites that in the north would be typically dominated by *Pinus contorta* with little undergrowth are occupied by *Populus* with a typically luxuriant herbaceous understory. Apparently, the rapidly decomposing wood and relatively nutrient-rich deciduous leaves of *Populus* result in low levels of woody fuel accumulation and rapid nutrient cycling, all of which encourages growth of mesophytic understory species (Vitousek, Gosz, Grier, Melillo, and Reiners 1982; Parker and Parker 1983). Thus, *Populus* maintains its root-sprouting potential by maintaining a forest understory that burns with low intensity. At least for some regions, the possibility then exists that *Pinus contorta* and *Populus tremuloides* represent alternative stable states for the same site (Peterson 1984; Wilson and Agnew 1992).

The ecological relationships between *Pinus contorta* and *Populus tremuloides* have been the source of considerable discussion and confusion. *Populus* has been variously described as growing on moister, drier, finer-textured, rockier, and less acidic sites than *Pinus contorta*. The relative dominance of *Pinus contorta* on coarse, granitic soils and *Populus* on finer, more calcareous soils is well documented (Langenheim 1962; Patten 1963; Reed 1971; Despain 1973). The distributional pattern is further clarified in the Sangre de Cristo Mountains of northern New Mexico, south of the range of *Pinus contorta*, where *Populus* occupies a broad range of habitat types. However, as one moves northward, *Pinus contorta* occupies an increasing portion of the range of habitats, so that in the Front Range the distribution of *Populus tremuloides* takes on the appearance of a doughnut when viewed in a graph of elevation versus site moisture (Fig.3.7B–C). *Pinus contorta* appears to be the better competitor in the middle of the gradient mosaic, whereas *Populus* appears to have the broader range of tolerance, which allows it to win at the periphery. Both *Pinus contorta* and *Populus tremuloides* stands typically succeed to stands of more shade-tolerant species. *Abies lasiocarpa* is the most important invader of high-elevation successional stands, although *Picea engelmannii* can also be important. At lower elevations, *Pseudotsuga* is typically the most important invader.

Despite their similarity as invading species, the actual patterns of growth and establishment are typically quite different in *Pinus contorta* and *Populus tremuloides* stands. In *Pinus contorta* stands, the understory herb layer is usually sparse, offering little competition for new tree seedlings. Thus, seedling establishment is frequent. However, intense competition from the established trees, and low soil fertility, result in slow growth rates for new seedlings. As a consequence, a typical 200-yr-old *P. contorta* stand has numerous stunted seedlings of *Pseudotsuga* or *Abies* and, to a lesser extent, of *Picea*. These seedlings can remain in the understory indefinitely, "waiting" for the death of nearby canopy trees. It is common to find such *Abies* seedlings < 1 m high but > 100 or even 150 yr of age. Following a catastrophic windstorm or pine beetle outbreak in which the canopy is largely destroyed, a seemingly even-aged *Abies* stand can develop from the released seedlings.

In contrast, in *Populus tremuloides* stands there is usually a well-developed herb layer that interferes with seedling establishment. Consequently, tree seedlings are much less common than in *P. contorta* stands, and the transition to conifer dominance is often much slower. However, once established, conifer seedlings in *P. tremuloides* stands appear to grow rather steadily, either because *Populus* does not offer as much competition for resources or because the associated soil is not as deficient in nutrients (see Vitousek et al. 1982).

Although usually viewed as seral species, both *Pinus contorta* and *Populus tremuloides* can form stable, self-maintaining stands. In the Colorado Front Range, occasional old, stable *P. contorta* stands do occur (Fig. 3.11) (Whipple and Dix 1979; Peet 1981), though not as commonly as sometimes suggested (see Moir 1969). Northward they appear rather frequently (Despain 1973, 1983; La Roi and Hnatiuk 1980), though usually in situations in which there is no nearby seed source for other potential climax species like *Pseudotsuga* or *Abies lasiocarpa*, or where soils are distinctly infertile, as on rhyolite in Yellowstone Park (Despain 1990). In the southern Rockies stable *Populus tremuloides* stands are best developed on the western slopes of isolated Great Basin ranges, where *Populus* often forms a low-elevation belt transitional from steppe or shrubland to forest (Langenheim 1962; Morgan 1969; Reed 1971; Harniss and Harper 1982; Mueggler 1988), but such stands also occur on the more fertile, finer-textured soils of the east slope. Similar fringing aspen woodlands can be found along the east slope in Montana and southern Alberta (Lynch 1955; Anderson and Baily 1980).

Populus tremuloides is frequently observed to be decreasing in abundance and vigor in such well-known areas as Rocky Mountain and Yellowstone national parks. The basis for this decline has been the focus of considerable debate in the literature (reviewed in Kay 1990; Romme et al 1995; Knight 1994; Baker et al. 1997). Although multiple factors doubtless contribute to this decline, it is evident from permanent plot and exclosure studies that increased elk (*Cervus elaphus*) density has been the critical factor. The nature of the interaction between elk and *P. tremuloides* in the original, pre-settlement landscape is largely unknown. What is clear is that when elk densities were low to nonexistent owing to hunting pressure, there was abundant (although perhaps episodic) establishment of new *Populus* stems. Since at least the 1930s (and since the 1880s in Yellowstone) aspen has been declining and elk density has been increasing. Recent decades have seen an explosion of elk density within Rocky Mountain National Park and environs and a concurrent near-absolute disappearance of *Populus* regeneration and a significant increase in tree mortality, owing to elk browse and bark consumption.

Populus tremuloides and *Pinus contorta* are not the only important seral species of the Rockies. *Larix occidentalis* is a widespread postfire species in the forests of northern Idaho and adjacent portions of Montana, British Columbia, Washington, and Oregon. The species is shade-intolerant and requires a mineral seedbed, but in contrast to other seral species of the region its mature trees with their thick bark, lack of resins, and open-branching pattern are strongly fire-resistant (Schmidt, Shearer, and Roe 1976). Surface fires of moderate intensity that did not kill *Larix* occurred at 10–30 yr intervals, whereas destructive crown fires occurred at intervals of around 140 yr (Davis 1980; Arno and Fischer 1995). *Pinus monticola*, another seral species of the central Rockies, is most characteristic of the mesic sites where *Tsuga heterophylla* or *Thuja plicata* is the potential climax species. Several additional species play important seral roles, even though they are best known as climax species in the central Rocky Mountain region: *Pseudotsuga menziesii*, *Pinus flexilis*, and *Pinus ponderosa* all act as successful species on sites more mesic than those on which they are typically climax (Peet 1981). In Wyoming, both *Pseudotsuga* and *Pinus contorta* are common postfire successional species; *Pinus contorta* is generally the dominant on coarse, acidic, infertile soils, whereas *Pseudotsuga* is more likely to dominate on fine-textured and more fertile soils or those derived from limestone.

Spruce-Fir Forest

Spruce-fir (*Picea-Abies*) forest characterizes the subalpine portion of the Rocky Mountains from the Yukon Territory south to the Mexican border (Fig. 3.12). These forests represent a southern extension and modification of the boreal conifer forests to which they are similar both floristically and structurally. The primary tree species of the high mountains are *Picea engelmannii* (Engelmann spruce) and *Abies lasiocarpa* (subalpine fir), both of which are genetically similar to and sometimes hybridize with their boreal counterparts, *Picea glauca* and *Abies balsamea*. Indeed, as in the case for *Pinus contorta* and *P. banksiana*, the mountain species of each

pair represents a genetically heterogeneous species from which the more homogeneous boreal species likely derived during one or more of the numerous Pleistocene range expansions (Boivin 1959; Taylor 1959; Parker, Maze, and Bradfield 1981; Parker, Maze, Bennett, Cleveland, and W. G. McLachlan 1984; Critchfield 1985).

Picea engelmannii is the dominant high-elevation spruce south of about 54°N latitude, whereas *Picea glauca* replaces *P. engelmannii* in the Boreal Rockies and occurs at lower elevations throughout the Canadian Rockies and parts of northern Montana. Hybrids not uncommonly link populations of the two species (Moss 1955; Daubenmire 1974). In the Black Hills of western South Dakota, *Picea glauca* occurs some 1000 km disjunct from the main part of its range. Otherwise, *Picea engelmannii* and *Abies lasiocarpa* are the dominant subalpine species of the Central and Southern Rockies.

On isolated peaks and ranges, particularly in semiarid portions of the Madrean Region, the occurrence of *Picea* and *Abies* appears to reflect nearly random, postglacial events of colonization and extinction. The Chiricahua Mountains of southeastern Arizona only contain *Picea engelmannii* (Sawyer and Kinraide 1980), whereas the Santa Catalina Mountains 160 km to the west only contain *A. lasiocarpa* (Whittaker and Niering 1965), and both are present in the Penaleño Mountains (Stromberg and Patten 1991). A similar situation exists in the Great Basin of western Utah and eastern Nevada (Loope 1969; Harper, Freeman, Ostler, and Klikoff 1978; Wells 1983; Arno and Hammerly 1984).

Spruce-fir forests are poorly developed in the mountains of Mexico. Probably neither *Picea engelmannii* nor *Abies lasiocarpa* var. *arizonica* occurs south of the U.S. border (Little 1971, and see also Chapter 15). Spruce is ecologically unimportant in Mexican forests, occurring only as two narrow endemics on high northern peaks: *Picea chihuahuana* in the Sierra Madre Occidental (Gordon 1968) and *Picea mexicana* in the Sierra Madre Oriental (Martinez 1961) [= *P. engelmannii* var. *mexicana* of Taylor and Patterson (1980)]. *Abies* is represented by eight species, with *A. religiosa* of the high southern peaks and *A. durangensis* of the northwest being the most important. Nowhere do these forests cover a major portion of the Mexican landscape (Leopold 1950; Lauer 1973; Rzedowski 1981).

Spruce-fir forests are remarkably consistent in composition and structure over the length of the Rockies. On open slopes, the understory often consists of a nearly continuous cover of low *Vaccinium* (e.g., *V. scoparium*, *V. myrtillus*) and often a dense layer of moss. Typically, few herbaceous species are present, often not more than 10 in a

Figure 3.12. Subalpine Picea engelmannii–Abies lasiocarpa forests, Rocky Mountain National Park, Colorado. A:
Typical dense stand in a site sheltered from fire behind a
cirque lake. B: High-elevation (~3400 m) Picea-Abies forest exhibiting slow recovery from a fire that occurred ~70
yr earlier.

Table 3.3. *Successional sequences (1→4) for* Picea engelmannii–Abies lasiocarpa *forests*

Site and stage	Basal area (m² ha⁻¹)	Relative importance		Species number (per 1000 m²)
		Picea	Abies	
Wet				
1	9.8	27.6	72.4	49.0
2	73.9	66.9	33.1	25.5
3	51.9	65.8	34.2	34.2
4	49.7	59.5	40.5	34.8
Mesic				
1	12.2	37.0	67.3	23.7
2	41.2	76.5	22.5	10.3
3	53.8	63.1	34.0	15.7
4	40.5	56.6	40.8	28.7
Xeric				
1	12.2	60.4	39.6	35.0
2	29.7	52.1	36.9	22.3
3	45.3	59.2	39.7	8.7
4	37.5	56.1	32.5	15.7

Note: Changes in average basal area and relative importance (average of relative density and relative basal area) of *Picea* and *Abies* are shown for three contrasting site conditions in the Colorado Front Range. Stage 1 contains even-aged stands (as judged by their bell-shaped diameter distributions) with an average diameter for *Picea* <12.5 cm. Stages 2 and 3 contain similar stands, with average diameters 12–25 and >25 cm, respectively. Stage 4 contains stands with diameter distributions indicative of mature, steady-state forests.
Source: Peet (1981).

1000 m² plot. Among the more widespread species are *Arnica cordifolia, Carex geyerii, C. rossii, Orthila secunda, Pedicularis racemosa,* and *Polemonium pulcherrimum* ssp. *delicatum. Abies* regeneration provides the bulk of additional woody understory vegetation. On moist sites with seeps or adjacent to running water, a lush herbaceous stratum can be expected (e.g., *Senecio, Mertensia, Erigeron, Cardamine, Saxifraga, Veratrum*), together with taller shrubs (e.g., *Lonicera involucrata, Viburnum edule*). Forests such as these occur from the Canadian Rockies south at least to central New Mexico (see Oosting and Reed 1952; Pfister et al. 1977; Peet 1978a; Moir and Ludwig 1979; Allen, Peet, and Baker 1991). The major latitudinal pattern is higher diversity of spruce-fir communities at higher latitudes. In Arizona, Nevada, and New Mexico, *Picea-Abies* forest covers only a narrow range of sites, and consequently only a few species combinations are found in any one locality. In contrast, in the Northern Rockies, *Picea-Abies* forests dominate a large portion of the landscape, and they are expressed in numerous variations (Pfister et al. 1977).

Because forest structure and dynamics vary with site conditions and disturbance history, it is not possible to provide a simple structural characterization. Table 3.3. illustrates some of this variation with averages for sets of stands along suc-cessional gradients in wet, mesic, and dry *Picea-Abies* forests in the Colorado Front Range. For these stands, peak basal areas range from 45 to 74 m² ha⁻¹ and species numbers from 10 to 50 per 1000 m².

Picea and *Abies* have somewhat different ecological characteristics. Perhaps the most striking difference is that *Picea* is more tolerant of extreme environmental conditions than is *Abies.* For example, in the Front Range *Picea* is the dominant species on very wet or boggy sites, and on drier sites (Peet 1981). *Abies* is numerically dominant on more mesic sites, although *Picea* usually dominates in basal area. In the Sangre de Cristo Range, *Picea* dominates the upper slopes, with *Abies* being largely absent from the upper 300 m of forest (although this may reflect physiological characteristics of *Abies lasiocarpa* var. *arizonica*, which replaces the typical form of *Abies* in the Sangre de Cristo Mountains and southward). On the high peaks of the Medicine Bow Mountains and other ranges to the north, however, *Abies* can dominate at treeline as it does in eastern North America.

The relative ecological roles of *Picea* and *Abies* in mixed stands is an unresolved puzzle of long standing. The most common pattern is for *Picea* to have the greater number of large trees and the greater basal area but for the greater portion of seedlings and saplings to be *Abies*. For the Rocky Mountains as a whole, Hobson and Foster (1910) found 50–90% of the small stems to be *Abies* but 75% of the mature timber to be *Picea*. Numerous subsequent authors have reported similar patterns. One interpretation is that the mixed stands are successional and that *Abies* will, in the absence of large-scale disturbance, largely replace *Picea* (Hansen 1940; Bloomberg 1950; Daubenmire and Daubenmire 1968; Loope and Gruell 1973; Peet 1981). A smaller number of researchers expect *Picea* to increase relative to *Abies* owing to greater longevity (Miller 1970; Alexander 1974; Schmidt and Hinds 1974). Indeed, individuals of *Picea engelmannii* > 500 yr of age are not uncommon (Peet 1981; Rebertus, Veblen, Roovers, and Mast 1992) and trees live to exceed 800 yr (Brown, Shepperd, Brown, Meta, and MO Cain 1995), whereas *Abies* > 350 yr are infrequent (Peet 1981). Still others favor the hypothesis that these forests are near equilibrium and that the structure will remain relatively constant owing to the greater longevity of *Picea* (Oosting and Reed 1952; Veblen 1986a; Aplet et al. 1988; Rebertus et al. 1992). It is probable that versions of each interpretation are true for the appropriate site conditions and landscape context (Peet 1981).

In the debate concerning the relative roles of *Picea* and *Abies*, the successional status of stands has largely been ignored. *Picea* appears to be a far more successful species at establishing on mineral soil following fire (Day 1972; Alexander 1974; Whipple and Dix 1979; Peet 1981). *Abies*, in contrast, is the better of the two at establishing in the shade and on organic substrate (Knapp and Smith 1982). Following fire on mesic sites in the Front Range, *Picea* usually dominates. In young stands with dense canopies, *Picea* regeneration is virtually absent and *Abies* regeneration occurs mostly as stunted seedlings. With thinning of the canopy, be it artificial, wind-induced, or as a result of insect outbreaks, *Abies* seedlings are released to assume greater dominance (although *Picea* regeneration also increases). Gradually, over a period of perhaps 500 yr, *Abies* largely replaces the canopy *Picea*. Alternatively, peculiarities of climate and seed rain can lead to an initial dominance of *Abies*. In this situation, *Picea* may slowly increase in importance but cannot be expected to achieve dominance.

Subalpine White Pine Forests

Dry ridges and exposed southern slopes of the subalpine zone do not support dense forests of tall, conical *Picea* and *Abies*. Instead, forests of shorter, round-crowned, more widely spaced trees of the white pine group (*Pinus* subgenus *Haploxylon*) occur (Fig. 3.13). These sites typically possess immature, skeletal soils (Arno and Hoff 1989). The white pines, because of their tolerance for extreme environmental conditions, can colonize these high-elevation sites. On the less extreme of these sites, *Picea* and *Abies* can subsequently become established in the understory and eventually grow up through, and out-compete, the white pines (Beasley and Klemmedson 1980; Peet 1981; Veblen 1986b; Baker 1992). The typical pattern is one of a broad environmental range over which white pines are potentially dominant but with a more restricted core region where they are potential climax species. Because of its particularly broad elevation range, *Pinus flexilis* can be an important postfire seral species, replacing *Pinus contorta* in this role on drier sites in the southern Rockies.

In the central Rockies, south through the Wind River Range of Wyoming, the dominant high-elevation white pine is *Pinus albicaulis* (whitebark pine). In the Southern Rockies, from central Colorado south through the Sangre de Cristo Mountains of northern New Mexico and westward to the San Francisco Peaks in northeastern Arizona, *Pinus aristata* (bristlecone pine) dominates these sites. Northwestward on the high peaks of the Great Basin, the closely related *Pinus longaeva* replaces *P. aristata*. The bristlecone pines are famous for their great longevity, *Pinus aristata* sometimes reaching 2500 yr (Brunstein and Yamaguchi 1992) and *P. longaeva* reaching 5000 yr (Hiebert and Hamrick 1984). The xeric, high-elevation sites of the Madrean Region have a greater range of associated species, but throughout the northern portion of the region north to the San Juan Mountains *Pinus strobiformis* is the usual dominant. In its extreme form, *P. strobiformis* is highly distinctive, but over much of the northern portion of its range the species appears to intergrade genetically with *P. flexilis*. Like *P. flexilis*, *P. strobiformis* grows on dry, exposed ridges, though not often at treeline, perhaps because the mountains do not typically reach treeline elevations in the areas where it occurs.

Pinus flexilis ranges across most of both the Southern and Central Rockies, including the higher mountains of the Great Basin. Where *Pinus aristata* or *Pinus albicaulis* occurs, *P. flexilis* is mostly confined to dry, exposed ridges at lower elevations, though sometimes dominance is shared in subalpine sites. In the Central Rockies in particular, *P. flexilis* is largely confined to thin-soil, droughty sites at low and middle elevations (Pfister et al. 1977). In some of the dry Great Basin ranges such as the Ruby Mountains of Nevada, *Pinus flexilis*, *P. albicaulis*, and *P. longaeva* are the only important

Figure 3.13. Timberline forests. A: Wind-shaped Pinus aristata *forming the treeline at 3600 m in the Spanish Peaks, Colorado. B: The upper boundary of dense* Picea-Abies *forest in Rocky Mountain National Park, Colorado. On this* *relatively sheltered site, the trees are well formed. Regeneration is nearly absent as a consequence of snow accumulation in the glades.*

montane and subalpine conifers (Loope 1969; Lewis 1971; Arno and Hammerly 1984). Moreover, there appears to be an east-west gradient in conifer forest composition and diversity in these ranges, with species number dropping off to the west and the dominance on *P. longaeva* increasing in compensation (Hiebert and Hamrick 1984). Along the

Front Range, between the geographical ranges of the southern *P. aristata* and the northern *P. albicaulis*, *P. flexilis* shows what appears to be competitive release, its range having expanded upward to dominate exposed ridges near treeline (Peet 1978a; Allen et al. 1991).

The understories of the various white pine forests show only gradual compositional change with latitude. *Pinus albicaulis* stands described from Montana (Weaver and Dale 1974) and Wyoming (Reed 1976) are essentially equivalent in structure and species composition to *P. flexilis* stands in the Colorado Front Range (Peet 1981), and to a lesser extent to *Pinus aristata* stands in northern New Mexico (Peet 1978a). In nearly all cases they are open, with considerable mineral soil being visible. The only conspicuous exceptions occur over limestone, where grassy understories are sometimes present (see Arno and Hoff 1989; Allen and Peet 1990).

White pines burn less frequently than other Rocky Mountain conifers, owing to lower fuel accumulations and the often nearly bare ground beneath. Nonetheless, they are susceptible to fires that start at lower elevations and burn upslope; catastrophic crown fires were generally responsible for reestablishment of these pines on sites where succession had been leading to dominance by *Picea-Abies*. Typical presettlement fire return intervals ranged from 30–300 yr depending on site conditions (Morgan, Buntin, Keane, and Arena 1995). As a consequence of dependence on fire for establishment on many sites, together with twentieth-century fire suppression, the white pine forests of the Rocky Mountains are steadily decreasing in extent (Morgan et al. 1995). Moreover, insects and disease cause significant mortality. *Pinus albicaulis* is particularly vulnerable and has suffered major diebacks from mountain pine beetle (*Dendroctonus ponderosae*) and the introduced white pine blister rust (*Cronartium ribicola*) (Tomback, Clary, Koehler, Hoff, and Arno 1995). The pine beetle is particularly destructive where large populations of *Pinus contorta* are present to support epidemic-sized beetle populations, and it was responsible for a major dieback of *Pinus albicaulis* in the early twentieth century. Blister rust is more common in the moister parts of the range where *Ribes*, the alternate host, flourishes (Arno and Hoff 1989). For example, in the relatively moist forests of northwestern Montana, *Pinus albicaulis* has been particularly hard hit by rust outbreaks of recent decades, resulting in >90% mortality (Keane and Morgan 1995). Fortunately, a few individuals appear to be resistant to the disease (Hoff, Hagle, and Krebill 1995; Tombeck et al. 1995)

Treeline Vegetation

At the northern end of the Rockies, in the southern Yukon, the upper limit of forest growth is at about 1400 m on favorable sites. With decreasing latitude, the treeline rises steadily at a rate of about 100 m per degree of latitude to an elevation of over 3600 m in northern New Mexico. Farther south, the timberline does not rise as rapidly, the highest treelines occurring near 4000 m in the tropical mountains of the Mexican volcanic belt (Fig. 3.14). The rate of descent of the treeline with increasing latitude is similar to the −83 m per degree of latitude reported for eastern North America by Cogbill and White (1991), and the continent-wide value of −110 m per degree of latitude reported by Daubenmire (1954).

Two species, *Picea engelmannii* and *Abies lasiocarpa*, are the dominant treeline species throughout much of the Rocky Mountain region (Figure 3.13B). Nonetheless, treeline vegetation patterns can be extraordinarily complex. Environmental factors such as wind, snow depth, and snow duration assume special importance at high elevations (Stevens and Fox 1991). Snow accumulation and duration can be critical at the treeline. Where no snow accumulates, winter desiccation can be severe, and tree establishment and growth can fail altogether; where accumulation is sufficient for snow to persist through much of the growing season, tree establishment can be greatly inhibited. Once established, trees can trap blowing snow, and their shade can slow snowmelt.

One of the most intriguing forms of interaction between wind and snow is the formation of "ribbon forests" (Fig. 3.15B). These alternating parallel strips of forest and intervening "snow glades" of moist alpine meadow have been described from the subalpine regions of Wyoming, Montana (Billings 1969), and Colorado (Buckner 1977; Holtmeier 1978, 1982). The trees apparently function as snow fences, with snow accumulation inhibiting seedling establishment in a band on the lee side of the ribbon. Adjacent to the drift, however, tree growth is improved because of continual summer water from snowmelt, plus increased protection from desiccation during the winter. In many cases, the forest ribbons appear to have established along solifluction terraces, where seedlings would have had little protection from blowing snow and ice (Holtmeier 1985).

Where wind is strong, abrasion and winter desiccation can cause trees to assume stunted, often cushion-like growth forms (Fig. 3.15A). On extreme sites, the shrub-like trees occur only where sheltered behind rocks or undulations in the topography (Holtmeier 1985). In similar but less exposed sites, tree islands can form that often consist of a clone of trees produced vegetatively from one parent through layering (i.e., rooting of branches pressed against the ground by the snow). In other cases, multiple seedlings become established in the

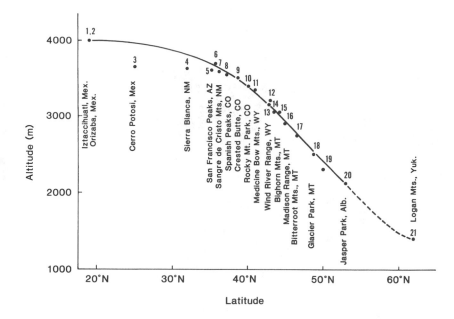

Figure 3.14. Treeline as a function of elevation and latitude. Elevations in parentheses following the references indicate uppermost limits for trees, in contrast to the prevailing treeline indicated in the figure. The rate of decline of the treeline between 40°N and 55°N latitude is approximately 100 m elevation per degree of latitude. 1, Beaman 1962 (4100 m); 2, Lauer and Klaus 1975; 3, Beaman and Andresen 1966; 4, Dye and Moir 1977; 5, Rominger and Paulik 1983; 6, Peet 1978a (3800 m); 7, Baker 1983; 8, Peet 1978a (3650 m); 9, Langenheim 1962; 10, Peet 1978a, 1981 (3550 m); 11, Oosting and Reed 1952, Peet 1978a (3500 m); 12, Griggs 1938; 13, Reed 1976; 14, Griggs 1938; 15, Despain 1973; 16, Patten 1963; 17, Arno and Habeck 1972, Habeck 1972; 18, Kessell 1979 (2600 m); 19, Wardle 1965; 20, Griggs 1938; 21, Oswald and Senyk 1977.

shelter of a single original plant. In either case, the appearance is that of a clump of shrubs. Where the wind is somewhat less strong, emergent trees rise from the shrub island, often with flag-shaped crowns (Holtmeier 1978, 1985). Marr (1977) and Benedict (1984) have documented centuries of movement of tree islands at rates of roughly 2 cm yr^{-1} caused by wind abrading the windward side while vegetative regeneration extends the plant on the lee.

Treeline forests of the Rockies contain frequent areas of standing dead trees with little to no tree regeneration (Fig. 3.12B). These relics of forests that burned centuries ago owe their occurrence to the extremely low rates of dead wood decomposition and tree seedling establishment associated with the extreme climatic conditions at treeline (Stahelin 1943; Peet 1981). Forest fires that started low in the *Picea-Abies* forests or in montane *Pinus contorta* forests burned uphill until they reached the forest boundary. Consequently, in many areas treeline must be viewed as dynamic; fire periodically destroys the forest, which only slowly creeps back up the mountain, taking centuries to regain lost ground (Peet 1981; Arno and Hammerly 1984). Moreover, climatic cooling has in some cases low-

ered the zone of natural forest regeneration, leaving at least some occurrences of timberline forest as remnants from periods of more favorable conditions for tree establishment (Ives and Hansen-Bristow 1983; Hansen-Bristow and Ives 1984; Carrar, Trimble, and Rubin 1991). The occurrence of significant tree regeneration in some old, high-elevation burns has been presented as evidence against the hypothesis of recent climatic depression of treeline (e.g., Shankman 1984; Daly and Shankman 1985). Such observations, which represent cases of slow recovery, do not refute observations of climatic depression on other Rocky Mountain sites. Similar downward shifts in treeline with climatic cooling in recent centuries are well documented throughout the Northern Hemisphere (e.g., Kullman, 1987; Arseneault and Payette 1992), and researchers have begun to look for expansion of tree regeneration into meadows and tundra as indicators of global warming (e.g., Butler 1986).

The availability of satellite imagery and image analysis software has led to increasingly sophisticated models of the position of alpine treeline (e.g., Brown 1994; Baker and Weisberg 1995; Walsh, Butler, Allen, and Mclanson 1994; Allen 1995). For example, Brown modeled the distribution of treeline

Figure 3.15. Influences of wind and snow at treeline. A: Islands of predominantly Abies lasiocarpa krummholz with conspicuous layering in Rocky Mountain National Park, Colorado. B: Ribbon forest in Montana (from Billings 1969). The interaction of snow, wind, and fire has resulted in bands of forest perpendicular to the prevailing wind.

in Glacier National Park and found that elevation, potential solar radiation, winter snow accumulation, and soil moisture were critical factors, along with aspects of geomorphology and past disturbance. Although the significance of each of these factors had long been recognized, Brown's models allowed a quantitative evaluation of their individual importance and interactions. One outcome was that Brown found strikingly different patterns in the residuals from his models when they were examined for different watersheds on the west slope of Glacier National Park, which led him to conclude that treeline varies within a region with topographic orientation and setting of the basin. Specifically, north-facing basins have lower treelines, and south-facing basins contain greater areas of forest than would be expected from site conditions alone. Whether the basin-specific differences result from mesoscale climatic differences associated with differences in solar radiation or from region-specific differences in wind patterns remains to be clarified.

Picea and *Abies* are not the only treeline dominants in the Rockies. The important roles of *Pinus aristata*, *P. flexilis*, and *P. albicaulis* on dry or exposed ridges have already been described. Other pines dominate at treeline in the high mountains of central Mexico, the most important being *Pinus hartwegii*, one of the only two tree species to reach above 4000 m elevation in North America (*Juniperus standleyi* reaches 4100 m in Guatemala). Most of these pines do not assume the krummholz shape characteristic of treeline *Picea* and *Abies*, though krummholz forms of *P. albicaulis* and *P. flexilis* do occur. However, the low density of understory vegetation in the white pine stands results in low levels of fuel and thus lower fire frequencies than is typical for other treeline forests. *Pinus hartwegii* stands, in contrast, are often grassy and savanna-like in appearance.

Two additional tree species characteristic of treeline habitats only in the Northern Region are *Larix lyallii* and *Tsuga mertensiana*. Both of these species occur only where the climate is moderated by incursion of Pacific air. *Larix*, which is restricted to sites between central Idaho and Banff National Park, grows largely on moist talus slopes (often north-facing) of cirque basins where snow lies much of the summer and the soils are kept moist by seepage. These sites are often above the normal *Picea-Abies* treeline where no other tree would be expected to grow (Arno and Habeck 1972; Arno, Worrall, and Carlson 1995). *Tsuga mertensiana* is of even more limited range in the Rockies, its distribution being confined primarily to the Cascades and the Sierra Nevada (see Chapters 4 and 5). Habeck (1967) reported *T. mertensiana* in Montana on a few moist ridges along the western border of the state, though it is more common at the headwaters of the Saint Joe River in northern Idaho and in the Selkirk Mountains of British Columbia's Glacier National Park (Shaw 1909; Arno and Hammerly 1984).

Meadows and Parks

Scattered through the forest lands of the Rockies are treeless areas of various sizes dominated by grasses, sedges, forbs, and sage (*Artemisia*) (Fig. 3.16). The larger examples are locally called parks and include such well-known examples as South Park and Estes Park in Colorado and Cinnabar Park in Wyoming. Smaller, mesic-to-moist examples are usually called meadows. Although the reasons for the treeless state of mountain parks and meadows have often been discussed in the ecological literature, little agreement exists, probably because no one explanation fits all examples, and often a variety of factors and their interactions are involved.

Soil texture is one of the most critical factors for explaining the existence of treeless areas (Dunnewald 1930; Ives 1942; Daubenmire 1943; Peet 1981; Veblen and Lorenz 1986). Parks and meadows often occupy rounded valley bottoms with predominantly fine-textured alluvial or colluvial soils. This is in marked contrast to the coarse, rocky material of adjacent forested slopes (Fig. 3.16A). Dense fibrous roots of the dominant grasses can form such a thick sod in the fine-textured soils that tree seedlings are rarely able to penetrate. Where trees such as *Pinus ponderosa* or *Juniperus scopulorum* do occur in these grasslands, they are often located immediately adjacent to large rocks that provide entry routes for roots to penetrate to the deeper soil horizons (Robbins and Dodds 1908; Peet 1981). Similar mechanisms have been invoked to explain the occurrence of disjunct populations of Rocky Mountain trees on the rocky scarps of the Great Plains (Wells 1965). However, this explanation does not preclude gradual invasion of parkland by trees, the control of which must be attributed to some other factor, such as periodic fire, drought, or grazing (Veblen and Lorenz 1986).

Soil texture alone cannot explain the complex pattern of grassland occurrence. Drought can be an important contributing factor; south-facing slopes in the *Pinus ponderosa* and lower *Pseudotsuga* zones of the Front Range, particularly where soil is thin, frequently support park-like vegetation dominated by *Muhlenbergia montana* or other graminoids (Peet 1981). Daubenmire (1968) has reported similar thin, droughty soils in northern Idaho to be dominated by a *Festuca idahoensis–Pseudoroegneria spicatum* steppe at an elevation that otherwise would support mesic *Tsuga*, *Thuja*, and *Abies* forest.

High-elevation meadows of the Rockies have been attributed to an even greater range of possible causal mechanisms. Subalpine valley bottoms often are treeless, here giving the appearance of alpine vegetation extending as narrow fingers sometimes kilometers down into the *Picea-Abies* zone (Fig. 3.16b). Although soil texture likely plays a role on drier sites at high elevations, it probably is not as

Figure 3.16. Treeless areas in forested landscapes, Rocky Mountain National Park, Colorado. A: Two types of parks can be seen in the photograph. The valley bottom with fine-textured alluvial and colluvial soils has an extensive treeless area. In addition, treeless areas can be seen on the south slope above and behind the Pinus flexilis in the fore-ground. The mountain across the valley is largely dominated by even-aged Pinus contorta. B: Subalpine valley bottom dominated by Salix shrubs (mostly S. brachycarpa), forbs, and grasses. The treeless condition is due to a combination of saturated soils, high accumulation of snow, cold-air drainage, and fine-textured soils.

frequently the pivotal factor in this elevation zone as excess soil moisture is. In addition, cold-air drainage and frost pockets, high snow accumulation, slow postfire regrowth, lake filling, paludification, beavers, and avalanches all can contribute to the local dominance of grass over trees.

Excess soil moisture is the explanation most often suggested for montane and subalpine meadows. Sites with soils saturated or nearly so occur along streams or valley bottoms, on lake margins, at the seepy bases of slopes, behind clayey morainal deposits, and even on open slopes where an impermeable substrate keeps water at the surface. Ives (1942) suggested that valley bottoms often undergo paludification, with peat and silt accumulation resulting in lateral spread of meadow or bog into the adjacent forest. Beavers *(Castor)* have often accelerated the paludification process. For millennia, beavers have repeatedly dammed streams, only to have the impoundments fill with peat and silt and the old dams become buried beneath the bog surface. Today, many of these old beaver meadows give the appearance of a terraced landscape.

Grassy balds similar to those of the Southern Appalachians (see Billings and Mark 1957 and Chapter 10) have been described from high open slopes and mountaintops, particularly in Montana (Korterba and Habeck 1971; Root and Habeck 1972). Drought resulting from thin soils and high winds provides the most likely explanation (Daubenmire 1968; Root and Habeck 1972). Nonetheless, the limited ability of woody plants to invade established grass turf doubtless helps maintain these balds as it does those of the southeastern United States.

Slow recovery from disturbance is critically important in the maintenance of subalpine meadows and balds (Stahelin 1943). On sites where tree regeneration is slow and forests require up to 500 yr to recover from fire, postfire meadow communities can cover a significant portion of the landscape. Meadows also can dominate avalanche tracks, where the frequency of disturbances precludes development of mature forest (Butler 1979). In both cases, graminoids and forbs, once established, competitively inhibit the establishment of trees and thus further reduce the rate of forest recovery (Robbins 1918; Daubenmire 1943; Bollinger 1973; Peet 1981).

Some subalpine treeless areas seemingly defy simple explanation. Cinnabar Park, a 44-ha upland meadow at 2926 m in the Medicine Bow Mountains of Wyoming, has been shown to have been treeless for a long period. Despite extensive study (Miles and Singleton 1975; Vale 1978; Doering and Reider 1992), no explanation for its origin or maintenance has yet received wide acceptance. Similarly, Fish Creek Park, Wyoming, has been extensively studied, but the mechanism behind its origin remains obscure. This park is known from pollen sediments to have originated within the last 3500 yr, so it is neither constrained by edaphic conditions nor a relict of past vegetation. A combination of tree removal by fire, slow invasion, and climate change seems to offer the best explanation (Lynch 1995).

Relatively little has been written on the composition of the meadow and park vegetation of the Rockies (see Windell et al. 1986). The enormous geographic range coupled with the range of site conditions makes summary difficult. Montane parks of the Colorado Front Range are often dominated by *Festuca kingii* or *Muhlenbergia montana* (Peet 1981). In the San Juan Mountains of southwestern Colorado, *Festuca thurberi* and various bunch grasses dominate the parks (Jamieson et al. 1996). Daubenmire (1968) reports *Festuca idahoensis* and *Pseudoroegneria spicatum* to dominate the parks of the montane zone in Idaho.

The wetter meadows are effectively a group of fen communities and are extremely variable in composition, in part because of their small size and often insular setting, and because small changes in amount of water or water chemistry can correspond to major changes in species dominance.

Working in the wet meadows on the west slope of the Front Range, Wilson (1969) reported two meadow sequences. For typically ponded or floodplain habitats, mostly at middle elevations, he described a sequence with *Carex utriculata* dominant on the wettest sites, *Carex aquatilis* dominant on moderately wet sites, and *Calamagrostis canadensis* dominant on the moist sites. In some valley-bottom sites, *Salix* thickets or carrs dominate in place of graminoids, but the alternation of dominance between the growth forms is not well understood and in part reflects past browsing by ungulates and beavers. A second series, more characteristic of higher elevations and often occurring on seepage slopes, has *Eleocharis quinqueflora* dominant on the wettest sites, a mixed community with species such as *Caltha leptosepala, Deschampsia caespitosa, Carex illota,* and *Agrostis humilis* on the moderately wet sites, and again a diverse community with *Phleum alpinum, Poa reflexa, Erigeron peregrinus, Bistorta bistortoides,* and others on mesic sites (cf. Rydberg 1915; Reed 1917; Robbins 1918; Langenheim 1962). A more complete elaboration of these communities can be found in Windell et al. (1986).

Some high-elevation seepage slopes and occasional shallow-lake margins develop a mat of *Sphagnum* and assume a character not unlike bogs of the boreal zone (see Chapters 2 and 12). In the Front Range, the heath component of these includes *Vaccinium myrtillus, V. caespitosum, Kalmia polifolia,* and *Gaultheria humifusa* (Peet 1981). Such fen-like wetlands are relatively uncommon in the Southern Rockies but increase in frequency northward.

AREAS FOR FUTURE RESEARCH

Rocky Mountain vegetation, stretching from Alaska and the Yukon to northern Central America, is both varied and complex. The depth of our current understanding of this vegetation is equally varied, but additional research is needed in virtually all areas of ecological inquiry.

The first stage in ecological analysis should almost always be descriptive. Few modern descriptive studies have been published for Rocky Mountain areas north of Jasper Park, Alberta, or south of the Mexico–United States border. Studies are particularly needed for Mexico because of the floristic complexity of the region and because the high intensity of land use and growing population are rapidly altering the remnants of natural vegetation. As a consequence of the land classification activities of the USDA Forest Service and The Nature Conservancy, considerable information is available for the Southern and Central Rockies (e.g., Pfister et al. 1977; Steele, Pfister, Ryker, and Kittams 1981; Steele, Cooper, Ondor, Roberts, and Pfister 1983), though there remains ample opportunity for synthetic analysis of these and other data sets. Only a few studies have organized compositional information relative to environmental or geographical gradients, and even fewer have carefully examined the interactions between vegetation and soil chemistry (e.g., Allen et al. 1991; Despain 1990).

Rocky Mountain forests are disturbance forests, with climax stands being less common than seral communities. Quantitative assessment of disturbance regimes is limited, and the impact of various forms of disturbances on tree population dynamics and various ecosystem processes needs further study. Reliable estimates of mean return intervals in either presettlement or contemporary forests exist only for fire, and even here the data are scant. The implications of changes in disturbance regimes through human activities, such as landscape fragmentation, carnivore removal, and fire suppression, need more careful study.

Few in-depth, integrated studies of ecosystem processes, including such phenomena as primary production, nutrient cycling, and water fluxes, have been undertaken. The most notable excep-

tions have been studies in the southern Sangre de Cristo Mountains of New Mexico (Gosz 1980) and the Medicine Bow Mountains of Wyoming (Fahey 1983; Pearson et al. 1984; Fahey et al. 1985; Knight and Fahey 1985). In no case are we yet able to superimpose isopleths for rates of ecological processes on a gradient representation of vegetation.

Landscape-scale studies of the Rocky Mountains are in their infancy. Nonetheless, the efforts of Turner, Romme, and their colleagues to understand the implications of landscape position for forest recovery from the Yellowstone fires of 1988, and of Walsh, Brown, Baker, and their respective colleagues to understand how vegetation patterns at timberline vary with landscape context clearly demonstrate the importance of incorporating landscape-level processes in future studies of Rocky Mountain vegetation.

REFERENCES

Agree, J. K., and L. Smith. 1984. Subalpine tree establishment in the Olympic Mountains, Washington. Ecology 65:810–819.

Alexander, R. R. 1974. Silviculture of subalpine forests in the central and southern Rocky Mountains: the status of our knowledge. USDA Forest Service, Res. Pap. RM-121.

Alexander, R. R. 1987. Ecology, silviculture and management of Engelmann spruce–subalpine fir type in the central and southern Rocky Mountains. USDA Handbook 659.

Alexander, R. R., and C. B. Edminster. 1981. Management of ponderosa pine in even-aged stands in the Black Hills. USDA Forest Service, Res. Pap. RM-228.

Allen, T. R. 1995 Relationships between spatial pattern and environment at alpine treeline ecotone, Glacier National Park, Montana. Ph.D. dissertation, University of North Carolina, Chapel Hill.

Allen, R. B., and R. K. Peet. 1990. Gradient analysis of forests of the Sangre de Cristo Range, Colorado. Can. J. Bot. 68:193–201.

Allen, R. B., R. K. Peet, and W. L. Baker. 1991. Gradient analysis of latitudinal variation in Southern Rocky Mountain forests. J. Biogeog. 18:123–139.

Aller, A. R. 1960. The composition of the Lake McDonald forest, Glacier National Park. Ecology 41:29–33.

Amman, G. D. 1977. The role of mountain pine beetle in lodgepole pine ecosystems: impact on succession, pp. 3–18 in W. J. Mattson (ed.), The role of arthropods in forest ecosystems. Springer-Verlag, New York.

Amman, G. D. 1978. The biology, ecology, and causes of outbreaks of mountain pine beetle in lodgepole pine forests, pp. 39–53 in M. A. Berryman, G. D. Amman, and R. W. Stark (eds.), Theory and practice of mountain pine beetle management in lodgepole pine forests. University of Idaho Experiment Station, Moscow.

Anderson, H. G., and A. W. Bailey. 1980. Effects of annual burning on grassland in the aspen parkland of east-central Alberta. Can. J. Bot. 58:985–996.

Antos, J. A., and J. R. Habeck, 1981. Successional development in *Abies grandis* forests in the Swan Valley, western Montana. Northwest Sci. 55:26–39.

Aplet, G. H., R. D. Laven, and F. W. Smith 1988. Patterns of community dynamics in Colorado Engelmann spruce–subalpine fir forests. Ecology 69:312–319.

Aplet, G., F. Smith, and R. Laven. 1989. Stemwood biomass and production during spruce–fir stand development. J. Ecol. 77:70–77.

Arno, S. F. 1979. Forest regions of Montana. USDA Forest Service, Res. Pap. IN-218.

Arno, S. F. 1980. Forest fire history in the northern Rockies. J. Forestry 78:460–465.

Arno, S. F. 1985. Ecological effects and management implications of Indian fires. USDA Forest Service, Gen. Tech. Report IN-182, pp. 81–86.

Arno, S. F., and W. C. Fischer. 1995. *Larix occidentalis* – Fire ecology and fire management, pp. 130–135 in Ecology and management of *Larix* Forests: A look ahead. USDA Forest Service, Gen. Tech. Report INT-319.

Arno, S. F., and G. E. Gruell. 1983. Fire history of the forest-grassland ecotone in southwestern Montana. J. Range Mgt. 36:332–336.

Arno, S. F., and J. R. Habeck. 1972. Ecology of alpine larch (*Larix lyallii* Parl.) in the Pacific Northwest. Ecol. Monogr. 42:417–450.

Arno, S. F., and R. P. Hammerly. 1984. Timberline: mountain and arctic forest frontiers. The Mountaineers, Seattle.

Arno, S. F., and R. J. Hoff. 1989. Silvics of Whitebark Pine (*Pinus albicuulis*). USDA Forest Service, Gen. Tech. Report INT-253.

Arno, S. F., J. Worrall, and C. E. Carlson. 1995. *Larix lyallii*: colonist of tree-line and talus sites, pp. 72–76 in Ecology and management of *Larix* Forests: A look ahead. USDA Forest Service, Gen. Tech. Report INT-319.

Arseneault, D., and S. Payette. 1992. A postfire shift from lichen–spruce to lichen–tundra vegetation at treeline. Ecology 73:1067–1081.

Arthur, M. A. 1991. Vegetation, pp. 76–92 in J. Baron (ed.). Biogeochemistry of a subalpine ecosystem: Loch Vale Ecosystem. Springer-Verlag, New York.

Axelrod, D. I. 1958. Evolution of the Madro-Tertiary geoflora. Bot. Rev. 24:433–509.

Axelrod, D. I. 1979. Age and origin of Sonoran Desert vegetation. Cal. Acad. Sci., Occ. Pap. 132:1–74.

Axelrod, D. I., and P. H. Raven. 1985. Origins of the Cordilleran flora. J. Biogeog. 12:21–47.

Bahre, C. J., Jr. 1991. Legacy of change. University of Arizona Press, Tucson.

Baker, F. S. 1944. Mountain climates of the western United States. Ecol. Monogr. 14:223–254.

Baker, W. L. 1983. Alpine vegetation of Wheeler Peak, New Mexico, U.S.A.: Gradient analysis, classification, and biogeography. Arct. Alp. Res. 15:223–240.

Baker, W. L. 1988. Size-class structure of contiguous riparian woodlands along a Rocky Mountain river. Physical Geog. 9:1–14.

Baker, W. L. 1990. Climatic and hydrologic effects on regeneration of *Populus angustifolia* James along the Animas River, Colorado. J. Biogeog. 17:59–73.

Baker, W. L. 1992. Structure, disturbance, and change in the bristlecone pine forests of Colorado, U.S.A. Arct. Alp. Res. 24:17–26.

Baker, W. L., and G. M. Walford. 1995. Multiple stable states and models of riparian vegetation succession on the Animas River, Colorado. Ann. Assoc. Am. Geogr. 85:320–338.

Baker, W. L., J. J. Honaker, and P. J. Weisberg. 1995. Using aerial photography and GIS to map the forest-tundra ecotone in Rocky Mountain National Park, Colorado, for global change research. Photogram. Eng. Rem. Sens. 61:313–320.

Baker, W. L., J. A. Munroe, and A. E. Hessl. 1997. The effects of elk on aspen in the winter range in Rocky Mountain National Park, Colorado, USA. Ecography. 20:155–165

Baker, W. L., and T. T. Veblen. 1990. Spruce beetles and fire in the nineteenth-century subalpine forests of western Colorado, U.S.A. Arct. Alp. Res. 22:65–80.

Baker, W. L., and P. J. Weisberg. 1995. Landscape analysis of the forest – tundra ecotone in Rocky Mountain National Park, Colorado. Prof. Geogr. 47:361–375.

Barrett, S. W., and S. F. Arno. 1982. Indian fires as an ecological influence in the northern Rockies. J. For. 10:647–651.

Barry, R. G. 1972. Climatic environment of the east slope of the Front Range, Colorado. Inst. Arct. Alp. Res., Occ. Pap. 2.

Barry, R. G. 1973. A climatological transect on the east slope of the Front Range, Colorado. Arct. Alp. Res. 5:89–110.

Barton, A. M. 1994. Gradient analysis of relationships among fire, environment, and vegetation in a southwestern USA mountain range. Bull. Torrey Bot. Club 121:251–265.

Beaman, J. H. 1962. The timberlines of Iztaccihautl and Popocatepetl, México. Ecology 43:377–385.

Beaman, J. H., and J. W. Andresen. 1966. The vegetation, floristics, and phytogeography of the summit of Cerro Potosi, Mexico. Amer. Midl. Natur. 75:1–33.

Beasley, R. S., and J. O. Klemmedson. 1980. Ecological relationships of bristlecone pine. Amer. Midl. Nat. 104:242–252.

Benedict, J. B. 1984. Rates of tree-island migration, Colorado Rocky Mountains, USA. Ecology 65:820–823.

Betancourt, J. L., W. S. Scguster, J. B. Mitton, and R. S. Anderson. 1991. Fossil and genetic history of a pinyon pine (*Pinus edulis*) isolate. Ecology 72:1685–1697.

Billings, W. D. 1950. Vegetation and plant growth as affected by chemically altered rocks in the western Great Basin. Ecology 31:62–74.

Billings, W. D. 1969. Vegetational pattern near alpine timberline as affected by fire–snowdrift interactions. Vegetatio 19:192–207.

Billings, W. D. 1990. The mountain forests of North America and their environments, pp. 47–86 in C. B. Osmund, L. F. Pitelka, and G. M. Hidy (eds.), Plant biology of the basin and range. Springer-Verlag, New York.

Billings, W. D., and A. F. Mark. 1957. Factors involved in the persistence of montane treeless balds. Ecology 38:140–142.

Bloomberg, W. G. 1950. Fire and spruce. Forest Chron. 26:157–161.

Boivin, B. 1959. *Abies balsamea* (Linné) Miller et ses variations. Naturaliste Canadien 86:219–223.

Bollinger, W. H. 1973. The vegetation patterns after fire at the alpine forest-tundra ecotone in the Colorado Front Range. Ph.D. dissertation, University of Colorado, Boulder.

Bormann, F. H., and G. E. Likens. 1979. Pattern and process in a forested ecosystem. Springer-Verlag, New York.

Brady, W., and C. D. Bonham. 1976. Vegetation patterns on an altitudinal gradient, Huachuca Mountains, Arizona. Southwestern Nat. 21:55–66.

Brown, D. E. 1982. Madrean evergreen woodland. Desert Plants 4:59–65.

Brown, D. G. 1994. Predicting vegetation types at treeline using topography and biophysical disturbance variables. J. Veg. Sci. 5:641–656.

Brown, H. E. 1958. Gambel oak in west central Colorado. Ecology 39:317–327.

Brown, J. K., and N. V. DeByle. 1987. Fire damage, mortality, and suckering in aspen. Can. J. For. Res. 17:1100–1109.

Brown, P. M., W. D. Shepperd, C. C. Brown, S. A. Mata, and D. L. McCain. 1995. Oldest known Engelmann spruce. USDA Forest Service, Res. Note RM-534.

Brunstein, F. C., and D. K. Yamaguchi. 1992. The oldest known Rocky Mountain bristlecone pines (*Pinus aristata* Englem.). Arct. Alp. Res. 24:253–256.

Buckner, D. L. 1977. Ribbon forest development and maintenance in the central Rocky Mountains of Colorado. Ph.D. dissertation, University of Colorado, Boulder.

Buckhardt, J. W., and E. W. Tisdale. 1969. Causes of juniper invasion in southwestern Idaho. Ecology 57: 472–484.

Burkhardt, J. W., and E. W. Tisdale. 1976. Causes of juniper invasion in southwestern Idaho. Ecology 57: 472–484.

Butler, D. R. 1979. Snow avalanche path terrain and vegetation, Glacier National Park, Montana. Arct. Alp. Res. 11:17–32.

Butler, D. R. 1986. Conifer invasion of subalpine meadows, central Lemhi Mountains, Idaho. Northwest Sci. 60:166–173.

Bye, R. 1995. Prominence of the Sierra Madre Occidental in the biological diversity of Mexico, pp. 19–27 in Biodiversity and management of the Madrean Archipelago: The sky islands of southwestern United States and northwestern Mexico. USDA Forest Service, Gen. Tech. Report RM-264.

Carrar, P. E., D. A. Trimble, and M. Rubin. 1991. Holocene treeline fluctuations in the northern San Juan Mountains, Colorado, U.S.A., as indicated by radiocarbon-dated conifer wood. Arct. Alp. Res. 23:233–246.

Christensen, N. L., J. K. Agee, P. F. Brussard, J. Hughes, D. H. Knight, G. W. Minshall, J. M. Peek, S. J. Pyne, F. J. Swanson, J. W. Thomas, S. Wells, S. E. Williams, and H. A. Wright. 1989. Interpreting the Yellowstone fires of 1988. BioScience. 39:678–685.

Clements, F. E. 1910. The life history of lodgepole burn forests. USDA Forest Service, Bulletin 79.

Cogbill, C. V., and P. S. White 1991. The altitude-elevation relationship for spruce-fir forest and treeline along the Appalachian Mountain chain. Vegetatio 94:153–175.

Cooper, C. F. 1960. Changes in vegetation, structure, and growth of southwestern pine forests since white settlement. Ecol. Monogr. 30:129–164.

Cooper, C. F. 1961. Pattern in ponderosa pine forests. Ecology 42:493–499.

Cooper, S. V., K. E. Neiman, and D. W. Robert. 1991. Forest habitat types of northern Idaho: a second approximation. USDA Forest Service, Gen. Tech. Report INT-236.

Cottam, W. P. 1954. Prevernal leafing of aspen in Utah Mountains. J. Arnold Arb. 35:239–250.

Covington, W. W., and M. M. Moore. 1994. Southwestern ponderosa forest structure: changes since Euro-American settlement. J. For. 92:39–47.

Critchfield, W. B. 1985. The late Quaternary history of lodgepole and jack pines. Can. J. For. Res. 15:749–772.

Cronquist, A. A., A. H. Holmgren, N. H. Holmgren, and J. L. Reveal. 1972. Intermountain flora – vascular plants of the intermountain west, U.S.A., Vol. 1. Hafner, New York.

Daly, C., and D. Shankman. 1985. Seedling establishment by conifers above tree limit on Niwot Ridge, Front Range, Colorado, USA. Arct. Alp. Res. 17:389–400.

Darling, M. S. 1966. Structure and productivity of a pinyon-juniper woodland in northern Arizona. Ph.D. dissertation, Duke University, Durham.

Daubenmire, R. 1943. Vegetation zonation in the Rocky Mountains. Bot. Rev. 9:325–393.

Daubenmire, R. 1954. Alpine timberlines in the Americas and their interpretation. Butler Univ., Bot. Stud. 11:119–136.

Daubenmire, R. 1968. Soil moisture in relation to vegetation distribution in the mountains of northern Idaho. Ecology 49:431–438.

Daubenmire, R. 1974. Taxonomic and ecologic relationships between *Picea glauca* and *Picea engelmannii*. Can. J. Bot. 52:1545–1560.

Daubenmire, R. 1975. Floristic plant geography of eastern Washington and northern Idaho. J. Biogeog. 2:1–18.

Daubenmire, R., and J. B. Daubenmire. 1968. Forest vegetation of eastern Washington and northern Idaho. Washington Agr. Exp. Station, Tech. Bull. 60.

Davis, K. M. 1980. Fire history of a western larch/Douglas-fir forest type in northwestern Montana. USDA Forest Service, Gen. Tech. Report RM-81, pp. 69–74.

Day, R. J. 1972. Stand structure, succession, and use of southern Alberta's Rocky Mountain forest. Ecology 53:474–478.

DeByle, N. V., W. C. Fischer, and C. Bevins. 1987. Wildfire occurrence in aspen stands in the interior western United States. West. J. Appl. For. 2:73–76.

Denevan, W. M. 1967. Livestock numbers in nineteenth-century New Mexico, and the problem of gullying in the southwest. Ann. Assoc. Amer. Geog. 57:691–703.

Despain, D. G. 1973. Vegetation of the Big Horn Mountains, Wyoming, in relation to substrate and climate. Ecol. Monogr. 43:329–355.

Despain, D. G. 1983. Nonpyrogenous climax lodgepole pine communities in Yellowstone National Park. Ecology 64:231–234.

Despain, D. G. 1990. Yellowstone vegetation: consequences of environment and history in a natural setting. Roberts Rinehart, Boulder, Colorado.

Diaz, H. F., R. G. Berry, and G. Kiladis. 1982. Climatic

characteristics of Pike's Peak, Colorado (1874–1888) and comparisons with other Colorado stations. Mountain Res. Dev. 2:359–371.

Dick-Peddie, W. A. 1993. New Mexico vegetation: past present and future. University of New Mexico Press, Albuquerque.

Dieterich, J. H. 1980. Chimney Spring forest fire history. USDA Forest Service, Res. Pap. RM-220.

Dieterich, J. H. 1983. Fire history of southwestern mixed conifers: a case study. Forest Ecol. Mgt. 6:13–31.

Dieterich, J. H., and T. W. Swetnam. 1984. Dendrochronology of a fire-scarred ponderosa pine. For. Sci. 30: 238–247.

Dixon, H. 1935. Ecological studies on the high plateaus of Utah. Bot. Gaz. 97:272–320.

Doering, W. R., and R. G. Reider, 1992. Soils of Cinnabar park, Medicine Bow Mountains, Wyoming, USA: indicators of park origin and persistence. Arct. Alp. Res. 24:27–39.

Dunnewald, T. J. 1930. Grass and timber soils distribution in the Big Horn Mountains. J. Amer. Agron. 22: 577–586.

Dunwiddie, P. W. 1977. Recent tree invasion of subalpine meadows in the Wind River Mountains, Wyoming. Arct. Alp. Res. 9:393–399.

Dye, A. J., and W. H. Moir. 1977. Spruce-fir forest at its southern distribution in the Rocky Mountains, New Mexico. Amer. Midl. Nat. 97:133–146.

Erdman, J. A. 1970. Piñon-juniper succession after natural fires on residual soils of Mesa Verde Colorado. Brigham Young University Sci. Bull., Biol. Ser. 2.

Fahey, T. J. 1983. Nutrient dynamics of aboveground detritus in lodgepole pine (*Pinus contorta* ssp. *latifolia*) ecosystems, southeastern Wyoming. Ecol. Monogr. 53:51–72.

Fahey, T. J., J. B. Yavitt, J. A. Pearson, and D. H. Knight. 1985. The nitrogen cycle in lodgepole pine forests, southeastern Wyoming. Biogeochemistry 1: 257–275.

Felger, R. S., and M. B. Johnson. 1995. Trees of the Northern Sierra Madre Occidental and sky islands of southwestern North America, pp. 71–83 in Biodiversity and management of the Madrean Archipelago: the sky islands of southwestern United States and northwestern Mexico. USDA Forest Service, Gen. Tech. Report RM-264.

Felger, R. S., and M. F. Wilson. 1995. Northern Sierra Madre Occidental and its Apachian outliers: a neglected center of biodiversity, pp. 36–59 in Biodiversity and management of the Madrean Archipelago: The sky islands of southwestern United States and northwestern Mexico. USDA Forest Service, Gen. Tech. Report RM-264.

Fisher, R. F., M. J. Jenkins, and W. F. Fisher. 1987. Fire and the prairie-forest mosaic of Devil's Tower National Monument. Amer. Midl. Nat. 117:250–257.

Floyd, M. E. 1982. The interaction of pinyon pine and gambel oak in plant succession near Dolores, Colorado. Southwestern Nat. 27:143–147.

Floyd-Hanna, L., A. W. Spencer, and W. H. Romme. 1996. Biotic communities of the semiarid foothills and valleys, pp. 143–158 in R. Blair (ed.), The western San Juan Mountains: their geology, ecology, and human history. University Press of Colorado, Niwot.

Franklin, J. F., and M. A. Hemstrom. 1981. Aspects of succession in the coniferous forests of the Pacific Northwest, pp. 212–229 in D. C. West, H. H.

Shugart, and D. B. Botkin (eds.), Forest succession: concepts and application. Springer-Verlag, New York.

Franklin, J. F., W. H. Moir, G. W. Douglas, and C. Wiberg. 1971. Invasion of subalpine meadows by trees in the Cascade Range, Washington and Oregon. Arct. Alp. Res. 3:215–224.

Fulé, P. Z., and W. W. Covington. 1994. Fire regime disruption and pine-oak forest structure in the Sierra Madre Occidental, Durango, Mexico. Restoration Ecol. 2:261–272.

Fulé, P. Z., and W. W. Covington. 1996. Conservation of pine-oak forests in northern Mexico, pp. 80–88 in Conference on adaptive ecosystem restoration and management: restoration of cordilleran conifer landscapes of North America. USDA Forest Service, Gen. Tech. Report RM-278.

Gehlbach, F. R. 1981. Mountain islands and desert seas: a natural history of the U.S.–Mexican borderlands. Texas A & M University Press, College Station.

Gholz, H. L. 1980. Structure and productivity of *Juniperus occidentalis* in central Oregon. Amer. Midl. Nat. 103:251–261.

Gordon, A. G. 1968. Ecology of *Picea chihuahuana* Martinez. Ecology 49:880–896.

Gosz, J. R. 1980. Biomass distribution and production budget for a nonaggrading forest ecosystem. Ecology 61:507–514.

Gottfried, G. J., P. F. Ffolliott, and L. F. DeBano. 1995. Forests and woodlands of the Sky Islands: stand characteristics and silvicultural prescriptions, pp. 152–164 in Biodiversity and management of the Madrean Archipelago: the Sky Islands of southwestern United States and northwestern Mexico. USDA Forest Service, Gen. Tech. Report RM-264.

Gottfried, G. J., T. W. Swetnam, C. D. Allen, J. L. Betancourt, and A. L. Chung-MacCoubrey. 1995. Pinyon-juniper woodlands, pp. 95–132 in Ecology, diversity, and sustainability of the Middle Rio Grande Basin. USDA Forest Service, Gen. Tech. Report RM-268.

Graves, J. H. 1995. Resource availability and the importance of herbs in forest dynamics. Ph.D. dissertation, University of North Carolina, Chapel Hill.

Greenland, D., J. Burban, J. Key, L. Klinger, J. Moorehouse, S. Oaks, and D. Shankman, 1985. The bioclimates of the Colorado Front Range. Mountain Res. Dev. 5:251–262.

Griggs, R. F. 1938. Timberlines in the northern Rocky Mountains. Ecology 19:548–564.

Gruell, G. E. 1983. Fire and vegetative trends in the northern Rockies: interpretations from 1871–1982 photographs. USDA Forest Service, Gen. Tech. Report INT-158.

Gruell, G. E. 1985. Fire on the early western landscape: an annotated record of wildland fires 1776–1900. Northwest Sci. 59:97–107.

Habeck, J. R. 1967. Mountain hemlock communities in Western Montana. Northwest Sci. 41:169–177.

Habeck, J. R. 1968. Forest succession in the Glacier Park cedar-hemlock forests. Ecology 49:872–880.

Habeck, J. R. 1972. Fire ecology investigations in Selway-Bitterroot Wilderness: historical considerations and current observations. University of Montana Publication R1-72-001.

Habeck, J. R. 1987. Present-day vegetation in the northern Rocky Mountains. Ann. Missouri Bot. Gard. 74: 804–840.

Hadley, K. S., and T. T. Veblen. 1993. Stand response to western spruce budworm and Douglas fir bark beetle outbreaks, Colorado Front Range. Can. J. For. Res. 23:479–491.

Hanley, D. P. 1976. Tree biomass and production estimated for three habitat types in northern Idaho. College of Forestry, Wildlife and Range Sciences, University of Idaho, Bulletin 14.

Hansen, H. P. 1940. Ring growth and dominance in a spruce-fir association in southern Wyoming. Amer. Midl. Nat. 23:442–448.

Hansen-Bristow, K. J., and J. D. Ives. 1984. Changes in the forest-alpine tundra ecotone: Colorado Front Range. Phys. Geogr. 5:186–197.

Harniss, R. O., and K. T. Harper. 1982. Tree dynamics in seral and stable aspen stands of central Utah. USDA Forest Service, Res. Pap. INT-297.

Harper, K. T., D. C. Freeman, W. K. Ostler, and L. G. Klikoff. 1978. The flora of Great Basin mountain ranges: diversity, sources, and dispersal ecology. Great Basin Nat. Mem. 81–103.

Harper, K. T., F. J. Wagstaff, and L. M. Kunzler. 1985. Biology and management of the Gambel oak vegetation type: a literature review. USDA Forest Service, Gen. Tech. Report INT-179.

Hawkes, B. C. 1980. Fire history of Kananaskis Provincial Park – mean fire return intervals. USDA Forest Service, Gen. Tech. Report RM-81, pp. 42–45.

Heinselman, M. L. 1981. Fire intensity and frequency as factors in the distribution and structure of northern ecosystems, pp. 7–57 in H. A. Mooney, T. M. Bonnicksen, N. L. Christensen, J. E. Lotan, and W. A. Reiners (eds.), Fire regimes and ecosystems properties. USDA Forest Service, Gen. Tech. Report WO-26.

Hiebert, R. D., and J. L. Hamrick. 1984. An ecological study of bristlecone pine (*Pinus longaeva*) in Utah and eastern Nevada. Great Basin Nat. 44:487–494.

Hinckley, L. C. 1944. The vegetation of the Mount Livermore area in Texas. Amer. Midl. Nat. 32:236–250.

Hobson, E. R., and J. H. Foster. 1910. Engelmann spruce in the Rocky Mountains. USDA Forest Service, Circular 170.

Hoff, R. J., S. K. Hagle, and R. G. Krebill. 1995. Genetic consequences and research challenges of blister rust in whitebark pine forests, pp. 118–126 in Proceedings – International workshop on subalpine stone pines and their environment: the status of our knowledge. USDA Forest Service, Gen. Tech. Report INT-309.

Holtmeier, F. K. 1978. Die bodennahen Winde in den Hochlagen der Indian Peaks Section (Colorado Front Range). Müenstersche Geographische Arbeiten 3:7–47.

Holtmeier, F. K. 1982. "Ribbon-forest" and "Hecken", streifenartige Verbreitungsmuster des Baumwuchses an der oberen Waldgrenze in den Rocky Mountains. Erdkunde 36:142–153.

Holtmeier, F. K. 1985. Climatic stress influencing the physiognomy of trees at the polar and mountain timberline. Eidgenöessische Anstalt für das forstliche Versuchswesen, Berichte 270:31–40.

Houston, D. B. 1973. Wildfires in northern Yellowstone National Park. Ecology 54:1111–1117.

Hunt, C. B. 1967. Natural regions of the United States and Canada. Freeman, San Francisco, Calif.

Ives, J. D., and K. J. Hansen-Bristow. 1983. Stability and instability of natural and modified upper timberline landscapes in the Colorado Rocky Mountains, USA. Mountain Res. Dev. 3:149–155.

Ives, R. L. 1942. A typical subalpine environment. Ecology 23:89–96.

James, R. L., C. A. Stewart, and R. E. Williams. 1984. Estimating root disease losses in northern Rocky Mountain national forests. Can. J. For. Res. 14:652–655.

Jamieson, D. W., W. H. Romme, and P. Somers. 1996. Biotic communities of the cool mountains, pp. 159–173 in R. Blair (ed.), The western San Juan Mountains: their geology, ecology, and human history. University Press of Colorado, Niwot.

Johnsen, T. N., Jr. 1962. One-seed juniper invasion of North Arizona grasslands. Ecol. Monogr. 32:187–207.

Johnson, E. A., L. Hogg, and C. S. Carlson. 1985. Snow avalanche frequency and velocity for the Kananaskis Valley in the Canadian Rockies. Cold Regions Sci. Technol. 10:141–151.

Johnson, E. A., and G. I. Fryer. 1989. Population dynamics in lodgepole pine – Engelmann spruce forests. Ecology 70:1335–1345.

Johnson, E. A., and C. P. S. Larsen. 1991. Climatically induced change in fire frequency in the southern Canadian Rockies. Ecology 72:194–201.

Johnson, E. A., and D. R. Wowchuck. 1993. Wildfires in the southern Canadian Rocky Mountains and their relationship to mid-tropospheric anomalies. Can. J. For. Sci. 23:1213–1222.

Johnstone, W. D. 1971. Total standing crop and tree component distributions in three stands of 100-year-old lodgepole pine, pp. 81–89 in H. E. Young (ed.), Forest biomass studies. University of Maine, Orono.

Jones, J. R., and D. P. Trujillo. 1975. Development of some young aspen stands in Arizona. USDA Forest Service, Res. Pap. RM-151.

Kallander, H. 1969. Controlled burning on the Fort Apache Indian Reservation, Arizona. Tall Timbers Fire Ecol. Conf. Proc. 9:241–249.

Kartesz, J. T. 1994. A synonymized checklist of the vascular flora of the United States, Canada, and Greenland. Timber Press, Portland, Ore.

Kay, C. E. 1990. Yellowstone's northern elk herd: a critical evaluation of the "natural regulation" paradigm. Ph.D. dissertation, Utah State University, Logan.

Kay, C. E. 1993. Aspen seedlings in recently burned areas of Grand Teton and Yellowstone National parks. Northwest Sci. 67:94–104.

Kay, C. E. 1994. Aboriginal overkill: the role of Native Americans in structuring western ecosystems. Human Nature 5(4):359–398.

Keane, R. E., and P. Morgan. 1995. Decline of whitebark pine in the Bob Marshall Wilderness Complex of Montana, U.S.A., pp. 245–253 in Proceedings – International workshop on subalpine stone pines and their environment: the status of our knowledge. USDA Forest Service, Gen. Tech. Report INT-309.

Kearney, M. S. 1982. Recent seedling establishment at timberline in Jasper National Park, Alberta. Can. J. Bot. 60:2283–2287.

Kessell, S. R. 1979. Gradient modeling: resource and fire management. Springer-Verlag, New York.

Knapp, A. K., and W. K. Smith. 1982. Factors influencing understory seedling establishment of Engelmann spruce (*Picea engelmannii*) and subalpine fir (*Abies lasiocarpa*) in southeast Wyoming. Can. J. Bot. 60:2753–2761.

Knight, D. H. 1994. Mountains and plains: The ecology of Wyoming landscapes. Yale University Press, New Haven.

Knight, D. H., and T. J. Fahey. 1985. Water and nutrient outflow from contrasting lodgepole pine forests in Wyoming. Ecol. Monogr. 55:29–48.

Koterba, W. D., and J. R. Habeck. 1971. Grasslands of the North Fork Valley, Glacier National Park, Montana. Can. J. Bot. 49:1627–1636.

Kullman, L. 1987. Little Ice Age decline of a cold marginal *Pinus sylvestris* forest in the Swedish Scandes. J. Biogeography 14:1–8.

Langenheim, J. H. 1962. Vegetation and environmental patterns in the Crested Butte area, Gunnison County, Colorado. Ecol. Monogr. 32:249–285.

La Roi, G. H., and R. J. Hnatiuk. 1980. The *Pinus contorta* forests of Banff and Jasper national parks: a study in comparative synecology and syntaxonomy. Ecol. Monogr. 50:1–29.

Larsen, M. M., and G. H. Schubert. 1969. Root competition between ponderosa pine seedlings and grass. USDA Forest Service, Res. Pap. RM-54.

Larson, G. C. 1944. More on seedlings of western aspen. J. For. 42:452.

Lauer, W. 1973. The altitudinal belts of the vegetation in the central Mexican highlands and their climatic conditions. Arct. Alp. Res. 5:A99–A113.

Lauer, W., and D. Klaus. 1975. Geoecological investigations on the timberline of Pico de Orizaba, México. Arct. Alp. Res. 7:315–330.

Laven, R. D., P. N. Omi, J. G. Wyant, and A. S. Pinkerton. 1980. Interpretation of fire scar data from a ponderosa pine ecosystem in the central Rocky Mountains, Colorado, pp. 46–49 in M. A. Stokes and J. H. Dieterich (eds.), Proceedings of the fire history workshop, October 20–24, 1980, Tucson, Arizona. USDA Forest Service, Gen. Tech. Report RM-81.

Leak, W. B. 1965. The J-shaped probability distribution. For. Sci. 11:405–409.

Leopold, A. S. 1950. Vegetation zones of Mexico. Ecology 31:507–518.

Leopold, A. 1937. Conservationist in Mexico. Am. Forests 37:447–502.

Lewis, M. E. 1971. Flora and major plant communities of the Ruby–East Humboldt Mountains. USDA Forest Service, Humboldt National Forest.

Little, E. L., Jr. 1971. Atlas of United States trees. Vol. 1. Conifers and important hardwoods. USDA, Misc. Publ. 1146.

Little, E. L., Jr. 1976. Atlas of United States trees. Vol. 3. Minor western hardwoods. USDA, Misc. Publ. 1314.

Loope, L. L. 1969. Subalpine and alpine vegetation of northeastern Nevada. Ph.D. dissertation, Duke University, Durham.

Loope, L. L., and G. E. Gruell, 1973. The ecological role of fire in the Jackson Hole area, northwestern Wyoming. Quat. Res. 3:425–443.

Lotan, J. E. 1975. The role of cone serotiny in lodgepole pine forests, pp. 471–495 in D. M. Baumgartner (ed.), Management of lodgepole pine ecosystems. Washington State University Cooperative Extension Service, Pullman.

Lowe, C. H. 1961. Biotic communities of the Sub-

Mogollon Region of the inland Southwest. J. Arizona Acad. Sci. 2:40–49.

Lynch, A. M., and T. W. Swetnam. 1992. Old-growth mixed-conifer and western spruce budworm in the southern Rocky Mountains, pp. 66–80 in Old-growth forests in the Southwest and Rocky Mountain regions: proceedings of a workshop. USDA Forest Service, Gen. Tech. Report RM-213.

Lynch, D. 1955. Ecology of the aspen groveland in Glacier County, Montana. Ecol. Monogr. 25:321–344.

Lynch, E. A. 1995. Origin of a park-forest vegetation mosaic in the Wind River Range, Wyoming. Ph.D. dissertation, University of Minnesota, Minneapolis.

McCune, B., and J. A. Antos. 1981. Diversity relationships of forest layers in the Swan Valley, Montana. Bull. Torrey Bot. Club 108:354–361.

McKnight, M. E. 1968. A literature review of the spruce, western, and 2-year cycle budworms. USDA Forest Service, Res. Pap. RM-44.

Malanson, G. P., and D. R. Butler. 1984. Transverse pattern of vegetation on avalanche paths in northern Rocky Mountains, Montana. Great Basin Nat. 44:453–458.

Malanson, G. P., and D. R. Butler. 1994. Tree-tundra competitive hierarchies, soil fertility gradients, and the elevation of treeline in Glacier National Park, Montana. Physical Geog. 15:166–180.

Marr, J. W. 1961. Ecosystems of the east slope of the Front Range in Colorado. University of Colorado Stud., Biol. 8.

Marr, J. W. 1967. Data on mountain environments. I. Front Range, Colorado, sixteen sites, 1952–1953. Univ. Colorado Stud., Biol. 27.

Marr, J. W. 1977. The development and movement of tree islands near the upper limit of tree growth in the southern Rocky Mountains. Ecology 58:1159–1164.

Marr, J. W., J. M. Clark, W. S. Osburn, and M. W. Paddock. 1968. Data on mountain environments. III. Front Range, Colorado, four climax regions, 1959–1964. University of Colorado Stud., Biol. 29.

Marr, J. W., A. W. Johnson, W. S. Osburn, and O. A. Knorr. 1968. Data on mountain environments. II. Front Range, Colorado, four climax regions, 1953–1958. University of Colorado Stud., Biol. 28.

Marshall, J. T., Jr. 1962. Land use and native birds of Arizona. J. Arizona Acad. Sci. 2:75–77.

Martinez, M. 1961. Una nueva especie de *Picea* en México. Anales del Instituto de Biologia, Universidad Nacional México 32:137–142.

Mason, D. T. 1915. The life history of lodgepole pine in the Rocky Mountains. USDA, Bulletin 154.

Merriam, C. H. 1890. Life zones and crop zones of the United States. USDA Division of Biological Survey, Bulletin 10:9–79.

Miles, S. R., and P. C. Singleton, 1975. Vegetative history of Cinnebar Park in Medicine Bow National Forest, Wyoming. Soil Science Society of America, Proceedings 39:1204–1208.

Miller, P. C. 1970. Age distributions of spruce and fir in beetle-killed forests on the White River Plateau, Colorado. Amer. Midl. Nat. 83:206–212.

Mitchell, V. L. 1976. The regionalization of climate in the western United States. J. Appl. Meteorol. 15:920–927.

Moir, W. H. 1969. The lodgepole pine zone in Colorado. Amer. Midl. Nat. 81:87–98.

Moir, W. H. 1972. Litter, foliage, branch, and stem production in contrasting lodgepole pine habitats of the Colorado Front Range, pp. 189–198 in J. F. Franklin, L. J. Dempster, and RH. Waring (eds.), Research on coniferous forest ecosystems – a symposium, USDA Forest Service, Portland, Ore.

Moir, W. H., and R. Francis. 1972. Foliage biomass and surface area in three *Pinus contorta* plots in Colorado. For. Sci. 18:41–45.

Moir, W. H., and J. A. Ludwig. 1979. A classification of spruce-fir and mixed conifer habitat types of Arizona and New Mexico. USDA Forest Service, Res. Pap. RM-207.

Moral, R. del. 1972. Diversity patterns in forest vegetation of the Wenatchee Mountains, Washington. Bull. Torrey Bot. Club 99:57–64.

Morgan, M. D. 1969. Ecology of aspen in Gunnison County, Colorado. Amer. Midl. Nat. 82:204–228.

Morgan, P. 1994. Dynamics of ponderosa and jeffrey pine forests, pp. 47–73 in Flammulated, boreal, and great gray owls in the United States. USDA Forest Service, Gen. Tech. Report RM-253.

Morgan, P., S. C. Buntin, R. E. Keane, and S. F. Arena. 1995, pp. 136–141 in Proceedings – International workshop on subalpine stone pines and their environment: the status of our knowledge. USDA Forest Service, Gen. Tech. Report INT-309.

Moss, E. H. 1955. The vegetation of Alberta. Bot. Rev. 21:493–567.

Mueggler, W. F. 1988. Aspen community types of the Intermountain Region. USDA Forest Service, Gen. Tech. Report INT-250.

Muir, P. S. 1993. Disturbance effects on structure and tree species composition of *Pinus contorta* forests in western Montana. Can. J. For. Res. 23:1617–1625.

Muir, P. S., and J. E. Lotan, 1985. Disturbance history and serotiny of *Pinus contorta* in western Montana. Ecology 66:1658–1668.

Muller, C. H. 1939. Relations in the vegetation and climatic types in Nuevo Leon, México. Amer. Midl. Nat. 21:687–729.

Muller, C. H. 1947. Vegetation and climate of Coahuila, Mexico. Madroño 9:33–57.

Mutch, R. W. 1970. Wildland fires and ecosystems – a hypothesis. Ecology 51:1046–1051.

National Oceanic and Atmospheric Administration (NOAA). Environmental Data Service. Climatological data, Colorado. NOAA, Asheville, N.C.

Neilson, R. P. 1986. High-resolution climatic analysis and Southwest biogeography. Science 232:27–34.

Neilson, R. P., and L. H. Wullstein. 1983. Biogeography of two southwestern American oaks in relation to atmospheric dynamics. J. Biogeography 10: 275–297.

Niering, W. A., and C. H. Lowe. 1984. Vegetation of the Santa Catalina Mountains: community types and dynamics. Vegetatio 58:3–28.

Nixon, E. S. 1967. A comparative study of the mountain brush vegetation in Utah. Great Basin Nat. 27:59–66.

Nixon, K. C. 1993. The genus *Quercus* in Mexico, pp. 447–458 in Ramamoorthy, T. P., R. Bye, A. Lot, and J. Fa (eds.), Biological diversity of Mexico: Origins and distribution. Oxford University Press, New York.

Noble, D. L., and F. Ronco. 1978. Seedfall and establishment of Engelmann spruce and subalpine fir in clearcut openings in Colorado. USDA Forest Service, Res. Pap. RM-200.

Oliver, C. D. 1980. Forest development in North America following major disturbances. Forest Ecol. Mgt. 3:153–168.

Oosting, H. J., and J. F. Reed. 1952. Virgin spruce-fir forest of the Medicine Bow Mountains, Wyoming. Ecol. Monogr. 22:69–91.

Oswald, E. T., and J. P. Senyk. 1977. Ecoregions of Yukon Territories. Canadian Forest Service, Publication BC-X-164.

Parker, A. J., and K. C. Parker. 1983. Comparative successional roles of trembling aspen and lodgepole pine in the southern Rocky Mountains. Great Basin Nat. 43:447–455.

Parker, A. J., and R. K. Peet. 1984. Size and age structure of conifer forests. Ecology 65:1685–1689.

Parker, W. H., J. Maze, F. E. Bennett, T. A. Cleveland, and W. G. McLachlan. 1984. Needle flavinoid variation in *Abies balsamea* and *A. lasiocarpa* from western Canada. Taxon 33:1–12.

Parker, W. H., J. Maze, and G. E. Bradfield. 1981. Implications of morphological and anatomical variation in *Abies balsamea* and *A. lasiocarpa* (Pinaceae) from western Canada. Am. J. Bot. 68:843–854.

Pase, C. P. 1982. Rocky Mountain (Petran) and Madrean montane conifer forests. Desert Plants 4:43–48.

Pase, C. P., and C. E. Layser. 1977. Classification of riparian habitat in the southwest. USDA Forest Service, Gen. Tech. Report RM-43:5–9.

Passini, M. 1982. Les forêts de *Pinus cembroides* au Méxique: Etude phytogéographique et écologique. Etudes Mesoamericaines II–5. Mission Archeologique et éthnologique Francaise au Méxique. Paris.

Patten, D. T. 1963. Vegetational pattern in relation to environments in the Madison Range, Montana. Ecol. Monogr. 33:375–406.

Pearson, G. A. 1923. Natural reproduction of western yellow pine in the Southwest. USDA, Bull. 1105.

Pearson, J. A., T. J. Fahey, and D. H. Knight. 1984. Biomass and leaf area in contrasting lodgepole pine forests. Can. J. For. Res. 14:259–265.

Peet, R. K. 1978a. Latitudinal variation in southern Rocky Mountain forests. J. Biogeog. 5:275–289.

Peet, R. K. 1978b. Forest vegetation of the Colorado Front Range: patterns of species diversity. Vegetatio 37:65–78.

Peet, R. K. 1981. Forest vegetation of the Colorado Front Range: composition and dynamics. Vegetatio 45:3–75.

Peet, R. K. 1988. Forests of the Rocky Mountains, pp. 63–101 in M. G. Barbour and W. D. Billings (eds.), North American Terrestrial Vegetation. Cambridge University Press, Cambridge.

Peet, R. K., and N. L. Christensen. 1987. Competition and tree death. BioScience 37:586–595.

Perry, J. P., Jr. 1991. The pines of Mexico and Central America. Timber Press, Portland.

Peterson, C. H. 1984. Does a rigorous criterion for environment identity preclude the existence of multiple stable points? Am. Nat. 124:127–133.

Pfister, R. D., B. L. Kovalchik, S. F. Arno, and R. C. Presby. 1977. Forest habitat types of Montana. USDA Forest Service, Gen. Tech. Report INT–34.

Potter, L. D., and D. L. Green. 1964. Ecology of ponderosa pine in western North Dakota. Ecology 45:10–23.

Ramaley, F. 1907. Plant zones in the Rocky Mountains of Colorado. Science 26:642–643.

Ramaley, F. 1908. Botany of northeastern Larimer

County, Colorado. University of Colorado Stud. 5: 119–131.

Ramaley, F. 1931. Vegetation of chaparral-covered foot-hills, southwest of Denver, Colorado. University of Colorado Stud. 18:231–237.

Ream, R. R. 1964. The vegetation of the Wasatch Mountains, Utah and Idaho. Ph.D. dissertation, University of Wisconsin, Madison.

Rebertus, A. J., B. R. Burns, and T. T. Veblen. 1991. Stand dynamics of *Pinus flexilis*-dominated subalpine forests in the Colorado Front Range. J. Veg. Sci. 2:445–458.

Rebertus, A. J., T. T. Veblen, L.M., N. Roovers, and J. N. Mast. 1992. Structure and dynamics of old-growth Engelmann spruce–subalpine fir in Colorado, pp. 139–151 in Old-growth forests in the Southwest and Rocky Mountain regions. USDA Forest Service, Gen. Tech. Report RM-213.

Reed, E. L. 1917. Meadow vegetation in the montane region of northern Colorado. Bull. Torrey Bot. Club 44:97–109.

Reed, R. A., J. Johnson-Barnard, and W. L. Baker. 1996. Contribution of roads to forest fragmentation in the Rocky Mountains. Cons. Biol. 10:1098–1106.

Reed, R. M. 1971. Aspen forests of the Wind River Mountains, Wyoming. Amer. Midl. Nat. 86:327–343.

Reed, R. M. 1976. Coniferous forest habitat types of the Wind River Mountains, Wyoming. Amer. Midl. Nat. 95:159–173.

Robbins, W. W. 1918. Successions of vegetation in Boulder Park, Colorado. Bot. Gaz. 65:493–525.

Robbins, W. W., and G. S. Dodds. 1908. Distributions of conifers on the mesas. University of Colorado Stud. 6:37–49.

Robert, M. 1977. Aspects phytogéographiques et écologiques des forêts de *Pinus cembroides*. I. Les forêts de l'est et du nord-est du Méxique. Bull. Société Botanique de France 124:197–216.

Rominger, J. M., and L. A. Paulik. 1983. A floristic inventory of the plant communities of the San Francisco Peaks Research Natural Area. USDA Forest Service, Gen. Tech. Report RM-96.

Romme, W. H. 1982. Fire and landscape diversity in subalpine forests of Yellowstone National Park. Ecol. Monogr. 52:199–221.

Romme, W. H., and D. G. Despain. D. G. 1989. Historical perspective on the Yellowstone fires of 1988. BioScience 39:695–699.

Romme, W. H., and D. H. Knight. 1981. Fire frequency and subalpine forest succession along a topographic gradient in Wyoming. Ecology 62:319–326.

Romme, W. H., and D. H. Knight. 1982. Landscape diversity: the concept applied to Yellowstone National Park. BioScience 32:664–670.

Romme, W. H., D. H. Knight, and J. B. Yavitt. 1986. Mountain pine beetle outbreaks in the Rocky Mountains: regulators of primary productivity? Am. Nat. 127:484–494.

Romme, W. H., M. G. Turner, R. H. Gardner, W. H. Hargrove, G. A. Tuskan, D. G. Despain, and R. A. Renkin. 1997. A rare episode of sexual reproduction in aspen. (*Populus tremuloides* Michx.) following the 1988 Yellowstone fires. Natural Areas J. 17:17–25.

Romme, W. H., M. G. Turner, L. L. Wallace, and J. S. Walker. 1995. Aspen, elk, and fire in northern Yellowstone National Park. Ecology 76:2097–2106.

Root, R. A., and J. R. Habeck. 1972. A study of high elevation grassland communities in western Montana. Amer. Midl. Nat. 87:109–121.

Rowdabaugh, K. M. 1978. The role of fire in the ponderosa pine-mixed conifer ecosystems. Master's thesis, Colorado State University, Fort Collins.

Rundel, P. W., and J. L. Vankat. 1989. Chaparral communities and ecosystems, pp 127–139 in S. C. Keeley (ed.), The California chaparral: paradigms re-examined. Publication No. 34, Science Series, National History Museum, Los Angeles, CA.

Rydberg, P. A. 1915. Phytogeographical notes on the Rocky Mountain region. V. Grasslands of the subalpine and montane zones. Bull. Torrey Bot. Club 42: 629–642.

Rzedowski, J. 1981. Vegetacion de México. Limusa, México, D. F.

Salisbury, F. B. 1964. Soil formation and vegetation on hydrothermally altered rock material in Utah. Ecology 45:1–9.

Samuels, M. L., and J. L. Betancourt. 1982. Modeling the long-term effects of fuelwood harvests on pinyon-juniper woodlands. Environ. Mgt. 6:505–515.

Savage, M., and T. W. Swetnam. 1990. Early 19th-century fire decline following sheep pasturing in a Navajo ponderosa pine forest. Ecology 71:2374–2378.

Sawyer, D. A., and T. B. Kinraide. 1980. The forest vegetation at higher altitudes in the Chiricahua Mountains, Arizona. Amer. Midl. Nat. 104:224–241.

Schimper, A. F. W. 1898. Pflanzen-Geographie and physiologischer Grundlage. Fischer, Jena.

Schlesinger, W. H., E. H. DeLucia, and W. D. Billings. 1989. Nutrient-use efficiency of woody plants on contrasting soils in the western Great Basin, Nevada. Ecology 70:105–113.

Schmidt, J. M., and G. D. Amman. 1992. *Dendroctonus* beetles and old-growth forests in the Rockies, pp. 51–59 in Old-growth forests in the Southwest and Rocky Mountain Regions: Proceedings of a workshop. USDA Forest Service, Gen. Tech. Report RM-213.

Schmidt, J. M., and T. E. Hinds. 1974. Development of spruce-fir stands following spruce beetle outbreaks. USDA Forest Service, Res. Pap. RM-131.

Schmidt, W. C., R. C. Shearer, and A. L. Roe. 1976. Ecology and silviculture of western larch forests. USDA Forest Service, Tech. Bull. 1520.

Schubert, G. H. 1974. Silviculture of southwestern ponderosa pine: the status of our knowledge. USDA Forest Service, Res. Pap. RM-123.

Segura, G., and L. C. Snook. 1992. Stand dynamics and regeneration patterns of a pinyon pine forest in eastcentral Mexico. For. Ecol. Mgt. 47:175–194.

Shankman, D. 1984. Tree regeneration following fire as evidence of timberline stability in the Colorado Front Range, U.S.A. Arct. Alp. Res. 16:413–417.

Shaw, C. H. 1909. The causes of timberlines on mountains: the role of snow. Plant World 12:169–181.

Shreve, F. 1939. Observations on the vegetation of Chihuahua. Madroño. 5:1–13.

Smith, W. K. 1985. Western montane forests, pp. 95–126 in B. F. Chabot and H. A. Mooney (eds.), Physiological ecology of North American plant communities. Chapman & Hall, New York.

Springfield, H. W. 1976. Characteristics and management of southwestern pinyon-juniper ranges: the status of our knowledge. USDA Forest Service, Res. Pap. RM-160.

Stahelin, P. 1943. Factors influencing the natural restocking of high altitude burns by coniferous trees in the central Rocky Mountains. Ecology 24:19–30.

Steele, R., R. D. Pfister, R. A. Ryker, and J. A. Kittams. 1981. Forest habitat types of central Idaho. USDA Forest Service, Gen. Tech. Report INT-114.

Steele, R., S. V. Cooper, D. M. Ondov, D. W. Roberts, and R. D. Pfister. 1983. Forest habitat types of eastern Idaho–western Wyoming. USDA Forest Service, Gen. Tech. Report INT-144.

Stevens, G. C., and J. F. Fox. 1991. The causes of treeline. Ann. Rev. Ecol. Syst. 22:177–191.

Stringer, P. W., and G. H. La Roi. 1970. The Douglas fir forests of Banff and Jasper national parks, Canada. Can. J. Bot. 48:1703–1726.

Stromberg, J. C., and D. T. Patten. 1991. Dynamics of the spruce-fir forests on the Pinaleño Mountains, Graham Co., Arizona. Southwestern Nat. 36:37–48.

Swetnam, T. W., and P. M. Brown. 1992. Oldest known conifers in southwestern United States: temporal and spatial patterns of maximum age, pp. 24–38 in Old-growth forests in the Southwest and Rocky Mountain regions. USDA Forest Service, Gen. Tech. Report RM-213.

Swetnam, T. W., and A. M. Lynch. 1993. Multicentury, regional-scale patterns of western spruce budworm outbreaks. Ecol. Monogr. 63:399–424.

Tande, G. F. 1979. Fire history and vegetation pattern of coniferous forests in Jasper National Park, Alberta. Can. J. Bot. 57:1912–1931.

Tausch, R. J., N. E. West, and A. A. Nabi. 1981. Tree age and dominance in Great Basin pinyon-juniper woodlands. J. Range Mgt. 34:259–264.

Taylor, R. J., and T. F. Patterson. 1980. Biosystematics of Mexican spruce species and populations. Taxon. 29: 421–440.

Taylor, T. M. C. 1959. The taxonomic relationship between *Picea glauca* (Moench) Voss and *Picea engelmannii* Parry. Madroño 15:111–115.

Tesch, S. D. 1981. Comparative stand development in an old-growth Douglas fir (*Pseudotsuga menziesii* var. *glauca*) forest in western Montana. Can. J. For. Res. 11:82–89.

Thornbury, W. D. 1965. Regional geomorphology of the United States. Wiley, New York.

Tinker, D. B., W. H. Romme, W. W. Hargrove, R. H. Gardner, and M. G. Turner. 1994. Landscape-scale heterogeneity in lodgepole pine serotiny. Can. J. For. Res. 24:897–903.

Tomback, D. F., S. K., Sund, and L. A. Hoffman. 1993. Post-fire regeneration of *Pinus albicaulis*: height-age relationships, age structure, and microsite characteristics. Can. J. For. Res. 23:113–119.

Tomback, D. F., J. K. Clary, J. Koehler, R. J. Hoff, and S. F. Arno. 1995, The effects of blister rust on a post-fire generation of whitebark pine: the Sundance Burn of northern Idaho (U.S.A.). Cons. Biol. 9:654–664.

Vale, T. R. 1978. Tree invasion of Cinnabar Park in Wyoming. Am. Midl. Nat. 100:277–284.

Vankat, J. L. 1989. Water stress in chaparral shrubs in summer-rain versus summer-drought climates – whither the mediterranean-type climate paradigm?, pp 117–124 in S. C. Keeley (ed.), The California chaparral: paradigms re-examined. Publication No.

34, Science Series, National History Museum, Los Angeles, CA.

Veblen, T. T. 1986a. Treefalls and the coexistence of conifers in subalpine forests in the Central Rockies. Ecology 67:644–649.

Veblen, T. T. 1986b. Size and age structure of subalpine forests in the Colorado Front Range. Bull. Torrey Bot. Club 113:225–240.

Veblen, T. T., K. S. Hadley, M. S. Reid, and A. J. Rebertus. 1989. Blowdown and stand development in a Colorado subalpine forest. Can. J. For. Res. 19:1218–1225.

Veblen, T. T., and D. C. Lorenz, 1986. Anthropogenic disturbance and recovery patterns in montane forests, Colorado Front Range. Phys. Geogr. 7:1–24.

Veblen, T. T., and D. C. Lorenz, 1991. The Colorado Front Range: a century of ecological change. University of Utah Press, Logan.

Vitousek, P. M., J. R. Gosz, C. C. Grier, J. M. Melillo and W. A. Reiners. 1982. A comparative analysis of potential nitrification and nitrate mobility in forest ecosystems. Ecol. Monogr. 52:155–177.

Walsh, S. S., D. R. Butler, T. R. Allen, and G. P. Malanson. 1994. Influence of snow patterns and snow avalanches on the alpine treeline ecotone. J. Veg. Sci. 5:657–672.

Walter, H., and H. Lieth. 1967. Klimadiagramm-Weltatlas. Gustav Fischer Verlag, Jena.

Wardle, P. 1965. A comparison of alpine timberlines in New Zealand and North America. New Zealand J. Bot. 3:113–135.

Waugh, W. J. 1986. Verification, distribution, demography, and causality of *Juniperus osteosperma* encroachment on a Big Horn Basin, Wyoming site. Ph.D. dissertation, University of Wyoming, Laramie.

Weaver, H. 1951. Fire as an ecological factor in the southwestern ponderosa pine forests. J. For. 49:93–98.

Weaver, H. 1959. Ecological changes in the ponderosa pine forest of the Warm Springs Indian Reservation in Oregon. J. For. 57:15–20.

Weaver, T., and D. Dale. 1974. *Pinus albicaulis* in central Montana: environment, vegetation and production. Amer. Midl. Nat. 92:222–230.

Weaver, T., and F. Forcella. 1977. Biomass of fifty conifer forests and nutrient exports associated with their harvest. Great Basin Nat. 37:395–401.

Wellner, C. A. 1970. Fire history in the northern Rocky Mountains, pp. 42–64 in The role of fire in the intermountain West. University of Montana, Missoula.

Wells, P. V. 1965. Scarp woodlands, transported grassland soils, and concept of grassland climate in the Great Plains region. Science 148:246–249.

Wells, P. V. 1983. Paleobiogeography of montane islands in the Great Basin since the last glaciopluvial. Ecol. Monogr. 53:341–382.

Wentworth, T. R. 1981. Vegetation on limestone and granite in the Mule Mountains, Arizona. Ecology 62:469–482.

West, N. E., K. H. Rea, and R. J. Tausch. 1975. Basic synecological relationships in juniper pinyon woodlands, pp. 41–53 in G. F. Gifford and F. E. Busby (eds.), The pinyon-juniper ecosystem: a symposium. Utah Agricultural Experiment Station, Utah State University, Logan.

Whipple, S. A., and R. L. Dix. 1979. Age structure and successional dynamics of a Colorado subalpine forest. Amer. Midl. Nat. 101:142–158.

White, A. S. 1985. Presettlement regeneration patterns in a southwestern ponderosa pine stand. Ecology 66:589–594.

Whitson, P. D. 1965. Phytocoenology of Boot Canyon Woodland, Chisos Mountains, Big Bend National Park, Texas. Master's thesis, Baylor University, Waco.

Whittaker, R. H. 1967. Gradient analysis of vegetation. Biol. Rev. 42:207–264.

Whittaker, R. H. 1973. Direct gradient analysis: techniques, pp. 9–31 in R. H. Whittaker (ed.), Ordination and classification of communities. Handbook of Vegetation Science 5. Junk, The Hague.

Whittaker, R. H., and W. A. Niering. 1965. Vegetation of the Santa Catalina Mountains, Arizona: a gradient analysis of the south slope. Ecology 46:429–452.

Whittaker, R. H., and W. A. Niering. 1968. Vegetation of the Santa Catalina Mountains, Arizona. IV. Limestone and acid soils. J. Ecol. 56:523–544.

Whittaker, R. H., and W. A. Niering. 1975. Vegetation of the Santa Catalina Mountains, Arizona. V. Biomass, production, and diversity along the elevation gradient. Ecology 56:771–790.

Wilson, H. C. 1969. Ecology and successional patterns of wet meadows. Rocky Mountain National Park, Colorado. Ph.D. dissertation, University of Utah, Salt Lake City.

Wilson, J. B., and A. D. Q. Agnew. 1992. Positive feedback switches in plant communities. Adv. Ecol. Res. 23:264–336.

Wilson, J. L., and B. M. Tkacz. 1996. Historical perspectives on forest insects and pathogens in the Southwest: implications for restoration of ponderosa pine and mixed conifer forests, pp. 26–31 in Conference on adaptive ecosystem restoration and management: restoration of Cordilleran conifer landscapes of North America. USDA Forest Service, Gen. Tech. Report RM-278.

Windell, J. T., B. E. Willard, D. J. Cooper, S. Q. Foster, C. F. Knud-Hansen, L. P. Rink, and G. N. Kiladis. 1986. An ecological characterization of Rocky Mountain montane and subalpine wetlands. U.S. Fish and Wildlife Service, Biol. Report 86.

Woodin, H. E., and A. A. Lindsey. 1954. Juniper-pinyon east of the continental divide, as analyzed by the line-strip method. Ecology 35:473–489.

Wright, C. W. 1952. An ecological description of an isolated pinyon pine grove. Master's thesis, University of Colorado, Boulder.

Young, J. A., and R. Evans. 1981. Demography and fire history of a western juniper stand. J. Range Mgt. 34:501–506.

Chapter
4

Pacific Northwest Forests

JERRY F. FRANKLIN   CHARLES B. HALPERN

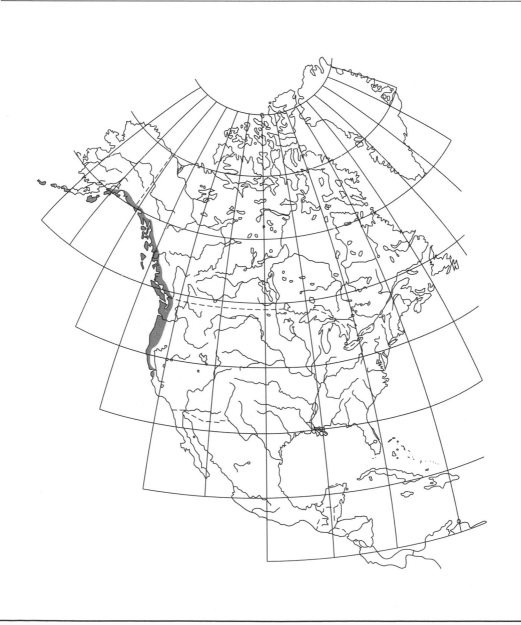

INTRODUCTION

The forests and associated vegetation of coastal northwestern North America cover nearly 20 degrees of latitude, extending from the Gulf of Alaska to northern California but lying within 60–120 km of the Pacific Ocean (Frontispiece). Mild, moist maritime conditions characterize the region, producing expanses of forest dominated by massive evergreen conifers, including *Pseudotsuga menziesii* (Douglas fir), *Tsuga heterophylla* (western hemlock), *Thuja plicata* (western red cedar), *Picea sitchensis* (Sitka spruce), and *Sequoia sempervirens* (coast redwood). The outstanding ecological features of these northwestern forests and the contrasts that they offer with forests in other temperate regions of the world include the following:

1. Dominance by numerous coniferous species, species that are the largest and longest-lived representatives of their genera (Table 4.1). Hardwoods

typically dominate other mesic, temperate forest regions of the world, but in the Pacific Northwest forests the ratio of hardwood to coniferous trees has been estimated at 1:1000 (Kuchler 1946). Hardwood species are few in number and generally are confined to specialized habitats (*Quercus qarryana*, Oregon white oak, and *Populus trichocarpa*, black cottonwood) or to early successional stages (*Alnus rubra*, red alder). *Lithocarpus densiflorus* (tan oak) is an evergreen hardwood with broad environmental and successional amplitude within the Klamath Mountain region.

2. Forests that exhibit the greatest biomass accumulations and some of the highest productivity levels of any in the world, temperate or tropical.

3. A climate of wet, mild winters and relatively warm, dry summers. These conditions favor evergreen life forms and needle-leaved conifers by permitting extensive photosynthesis outside of the growing season and reducing net photosynthesis

Table 4.1. *Ages and dimensions typically attained by some tree species on productive sites in the Pacific Northwest and their relative shade tolerances and fire sensitivities[a].*

Species	Age (yr)	Diameter (cm)	Height (m)	Shade tolerance[b]	Fire sensitivity
Abies amabilis	400+	90–110	45–55	VTOL	HIGH
Abies concolor	300+	100–150	40–55	TOL	INTER
Abies grandis	300+	75–125	40–60	TOL	INTER
Abies lasiocarpa	250+	50–60	25–35	TOL	HIGH
Abies magnifica var. *shastensis*	300+	100–125	40–50	INTER	INTER
Abies procera	400+	100–150	45–70	INTOL	HIGH
Chamaecyparis lawsoniana	500+	120–180	60	TOL	INTER
Chamaecyparis nootkatensis	1000+	100–150	30–40	TOL	INTER
Larix occidentalis	700+	140	50	INTOL	LOW
Libocedrus decurrens	500+	90–120	45	INTER	INTER
Picea engelmannii	400+	100+	45–50	TOL	HIGH
Picea sitchensis	500+	180–230	70–75	TOL	HIGH
Pinus contorta	250+	50	25–35	INTOL	INTER
Pinus lambertiana	400+	100–125	45–55	INTER	INTER
Pinus monticola	400+	110	60	INTER	INTER
Pinus ponderosa	600+	75–125	30–60	INTOL	LOW
Pseudotsuga menziesii	750+	150–220	70–80	INTOL	LOW
Sequoia sempervirens	1250+	150–380	75–100	TOL	LOW
Thuja plicata	1000+	150–300	60+	TOL	INTER
Tsuga heterophylla	400+	90–120	50–65	VTOL	HIGH
Tsuga mertensiana	500+	75–100	35+	TOL	HIGH
Acer macrophyllum	300+	50	15	TOL	INTER
Alnus rubra	100	55–75	30–40	INTOL	HIGH
Castanopsis chrysophylla	150	30+	20+	INTER	INTER
Lithocarpus densiflorus	180	25–125	15–30	TOL	INTER
Populus trichocarpa	200+	75–90	25–35	INTOL	HIGH
Prunus emarginata	50	15–30	15	INTOL	HIGH
Quercus garryana	500	60–90	15–25	INTOL	LOW

[a] Developed from a variety of sources including Burns and Honkala (1990). Maximum ages and sizes for species are generally much greater than those indicated here.
[b] Tolerance scale: VTOL = very tolerant of shade; TOL = tolerant; INTER = intermediate shade tolerance (greater in youth, lesser at maturity); INTOL = intolerant.

during the summer months (Waring and Franklin 1979).

4. Strong climatic gradients associated with latitude, longitude (distance from coast or from major mountain ranges), and elevation, as well as strong localized gradients associated with slope, aspect, and topographic position. Vegetation gradients follow climatic gradients closely: thus, primary and secondary environmental correlates in ordinations are indices of moisture and temperature.

5. Disturbance regimes that, within the heart of the northwestern forests, are dominated by infrequent catastrophic events, such as wildfires, at intervals of several hundred years (Agee 1993). This contrasts with the patterns of frequent, noncatastrophic fires that dominate disturbance regimes in many other forested regions of western North America, including California and the Rocky Mountains (see Chapters 3 and 5).

6. Extended forest-alpine transition zones with attractive and diverse parklands of forest, tree patches, and meadowlands. Deep, persistent winter snowpacks within a relatively warm subalpine zone are believed to be key factors in the formation and maintenance of these parklands.

It is our intent in this chapter to provide a broad overview of the forests of the coastal Pacific Northwest and some of the associated nonforested communities. Although we incorporate information from the entire latitudinal range, we emphasize the more southerly forests of western Washington and Oregon and limit our discussions of forests of coastal British Columbia and southeastern Alaska. Our primary focus is on the composition, distribution, and general successional patterns of the major (but not all) forest types, emphasizing mature and old forests with well-developed understories. Our approach, and consequently our terminology, reflect the strong phytosociological tradition in the Pacific Northwest, which Professor Rexford Daubenmire strongly influenced. We use several terms in our descriptions that merit definition: "succession" refers to a directional change in community structure or composition (or both), either observed or inferred; "climax" refers to the ability of a species or community to persist in the absence of catastrophic disturbance; and "series" refers to the group of community types (or plant associations) that have the same reproducing tree species and canopy dominants at climax. The series grouping has become the most commonly used level in the phytosociological hierarchy (above habitat type or association) because of its importance in the western regional ecology programs of the USDA Forest Service. Plant nomenclature follows Hitchcock and Cronquist (1973).

Additional sections of this chapter describe the major sources and patterns of disturbance, some of the important ecosystem-level attributes and processes in these forests, and the paleoecological history of the region. We conclude with discussions of current issues in forest conservation and emerging areas of ecological research.

ENVIRONMENTAL CONDITIONS

The Pacific Northwest is characterized by north/south-trending mountain ranges, including the Cascade Range, the Coastal Ranges of Oregon, the Olympic Mountains, and the British Columbia Coastal Range (McKee 1972). Lowland areas, such as the Willamette Valley and Puget Trough, separate the coastal and Cascadian mountain systems for most of their length. The major topographic and climatic divide of the region is the Cascade Range, which is bisected by only three river systems: the Fraser, the Columbia, and the Klamath. Geologic conditions are highly varied, with sedimentary rock types typical of the Oregon Coastal Ranges and metamorphic rocks dominating much of the northern Cascade Range and Olympic Mountains. Volcanic rocks of Miocene, Pliocene, and Pleistocene age are typical of the southern two-thirds of the Cascade Range. Glaciation has been an important process at higher elevations in the Cascade Range and Olympic Mountains, as well as in the Coastal Ranges of British Columbia and southeastern Alaska. Continental glaciation extended a short distance south of Puget Sound.

Forest soils are highly varied, reflecting the diverse parent materials and topography of the region (Franklin and Dyrness 1988). Haplumbrepts, Haplohumults, Haplorthods, Xerochrepts, Cryorthods, and Vitrandepts are most characteristic. Deposition of parent materials by alluvial, colluvial, glacial, or eolian action is an important soil-forming process. Soils in the Cascade Range, for example, are much deeper than might be expected, because of periodic aerial deposits of volcanic ejecta.

A maritime climate characterizes the Pacific Northwest (Fig. 4.1). In coastal regions, temperatures are mild, with prolonged cloudy periods, muted extremes, and narrow diurnal fluctuations (6–10°C). Winters are mild, with precipitation of 800–3000 mm, of which 75–85% occurs between October 1 and March 31, mostly as rain, or as snow at higher elevations. Summers are cool to warm, depending on location, and can be relatively dry. Most precipitation is the result of low-pressure systems that approach from the Pacific Ocean.

There are major variations in the climate of the region associated with latitude, elevation, and po-

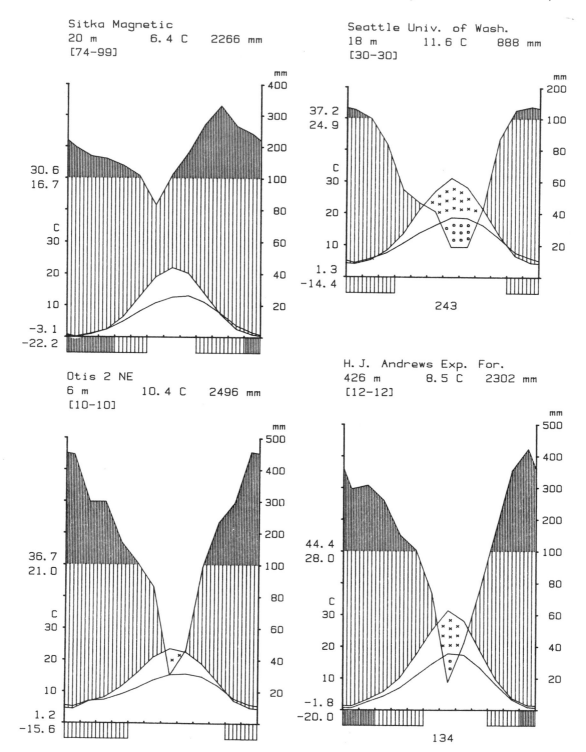

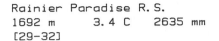

Rainier Paradise R. S.
1692 m   3. 4 C   2635 mm
[29-32]

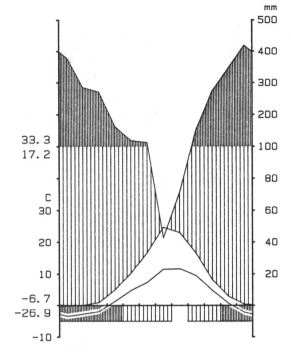

Figure 4.1. Climatic diagrams illustrating temperature and moisture regimes for selected locations in the Pacific Northwest using the conventions of Walter (1979). The Sitka, Alaska, and Otis, Oregon (Cascade Head Experimental Forest), stations are near opposite latitudinal ends of the Picea sitchensis *zone. Seattle, Washington, and H. J. Andrews Experimental Forest, Oregon, are in the* Tsuga heterophylla *zone in the Puget Trough and western Oregon Cascade Range, respectively. Longmire and Paradise, Washington, are located in the lower* Abies amabilis *and* Tsuga mertensiana *zones, respectively, within Mount Rainier National Park (diagrams produced by Bradley Smith from U.S. Weather Bureau data).*

sitions relative to mountain ranges (Fig. 4.1). There is a general latitudinal increase in precipitation and decrease in temperature. Mountain masses block maritime air masses, creating rainshadows in their lee; for example, the coastal mountains are responsible for the drier and less muted climates of the Willamette Valley and Puget Trough. Mountain ranges also produce local increases in precipitation (because of orographic effects) and in the proportion of precipitation that falls as snow.

MAJOR TEMPERATE FORESTS

Pseudotsuga menziesii – Tsuga heterophylla Forests

This is the major forest complex of the Pacific Northwest, encompassing seral forests dominated by *Pseudotsuga menziesii* and massive old-growth forests of *Pseudotsuga, Tsuga heterophylla, Thuja plicata,* and other species (Fig. 4.2). These forests occur from sea level up to elevations of 700–1000 m in the Coast Ranges, Olympic Mountains, and Cascade Range north of latitude 43°15' N. They are equivalent to the *Tsuga heterophylla* zone defined by Franklin and Dyrness (1988) (Table 4.2). Many detailed studies of community composition and environmental relationships have been conducted: Spilsbury and Smith (1947), McMinn (1960), Dyrness et al. (1974), Zobel et al. (1976), Halverson et al. (1986), Hemstrom and Logan (1986), Topik et al. (1986), Hemstrom et al. (1987), Franklin et al. (1988), Henderson et al. (1989), Moir (1989), Henderson et al. (1992), and Diaz and Mellen (1996).

Pseudotsuga menziesii–Tsuga heterophylla forests occupy a wide range of environments and are highly variable in composition and structure, depending on local conditions, especially the moisture regime. Major forest trees are *Pseudotsuga menziesii, Tsuga heterophylla,* and *Thuja plicata; Abies grandis* (grand fir), *Picea sitchensis,* and *Pinus monticola* (western white pine) occur sporadically.

Rainier Longmire
842 m   7. 3 C   2094 mm
[30-30]

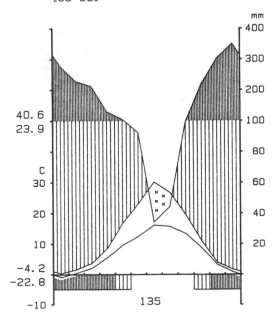

Figure 4.2. Typical old-growth forest stand of Pseudotsuga menziesii *and* Tsuga heterophylla *in the H. J. Experimental* Forest, western Oregon Cascade Range (photo courtesy USDA Forest Service).

In central Oregon, *Calocedrus decurrens* (incense cedar), *Pinus lambertiana* (sugar pine), and even *Pinus ponderosa* may be encountered. *Abies amabilis* (Pacific silver fir) is common near the upper altitudinal limits of the *Pseudotsuga* forests and at lower elevations in the Olympic Mountains, northern Cascade Range, and British Columbia Coastal Ranges. *Taxus brevifolia* (western yew) is ubiquitous but is always a subordinate tree (Busing et al. 1995). Hardwoods are not common except on recently disturbed sites or in specialized habitats, such as riparian zones. Typical species are *Alnus rubra, Acer macrophyllum* (bigleaf maple), and *Prunus emarginata* (bitter cherry). *Populus trichocarpa*

and *Fraxinus latifolia* (Oregon ash) also occur along waterways. *Arbutus menziesii* (Pacific madrone), *Castanopsis chrysophylla* (golden chinkapin), and *Quercus garryana* are found on warmer, drier sites, often with shallow or lithosolic soils. More information on the ecology of these species can be found in Burns and Honkala (1990).

In forest communities, composition, structure, and productivity vary markedly along moisture gradients, as has been demonstrated by numerous studies (see earlier citations). Less productive communities dominated by *Gaultheria shallon* or *Holodiscus discolor* characterize the hot, dry end of the gradient; *Pseudotsuga menziesii* may even be a ma-

Table 4.2. *Major series and the corresponding vegetation zones and Society of American Foresters (SAF) cover types in the Pacific Northwest.*

Type	Series[a]	Zone	Major SAF types
Pseudotsuga menziesii– Tsuga heterophylla	*T. hererophylla*	*T. heterophylla*	Pacific Douglas fir–western hemlock, western hemlock, western red cedar–western hemlock, western red cedar
Picea sitchensis–Tsuga heterophylla	*T. heterophylla, P. sitchensis*	*P. sitchensis*	Western hemlock–Sitka spruce
Sequoia sempervirens	*Lithocarpus densiflorus, T. heterophylla, S. sempervirens*	*P. sitchensis*	Redwood
Sierra-type mixed conifer	*Abies concolor, P. menziesii*	Mixed conifer	Sierra Nevada mixed conifer, white fir, Pacific ponderosa pine–Douglas fir
Klamath Mountain mixed evergreen	*L. densiflorus, P. menziesii, A. concolor*		Douglas fir–tan oak–Pacific madrone, Pacific ponderosa pine–Douglas fir
Abies amabilis– T. heterophylla	*A. amabilis*	*A. amabilis*	Coastal true fir–hemlock
Abies magnifica var. *shastensis*	*A. magnifica* var. *shastensis, A. concolor*	*A. magnifica* var. *shastensis*	Red fir
Tsuga mertensiana	*A. amabilis, T. mertensiana*	*T. mertensiana*	Mountain hemlock, coastal true fir–hemlock

[a]"Series" refers to a group of community types or plant associations that have the same climax tree species; it is a category widely used in the USDA Forest Service Area Ecology Program.
Sources: Eyre (1980), Franklin and Dyrness (1988).

jor reproducing tree species on such sites (Means 1980). At the moist end of the gradient, herbaceous understories dominated by *Polystichum munitum* and *Oxalis oregana* are characteristic. Intermediate, mesic sites have understories dominated by a variety of evergreen shrubs, subshrubs, and herbs, of which *Acer circinatum, Berberis nervosa,* and (in Oregon) *Rhododendron macrophyllum* are most common.

Although the details of community compositions vary with locale, the basic pattern – *Gaultheria* on dry sites and *Polystichum* on wet sites – is repeated throughout the *Pseudotsuga menziesii–Tsuga heterophylla* type. This pattern is illustrated in Table 4.4, using data from five old-growth associations on the western slope of the Oregon Cascade Range at about 45° N latitude (Dyrness et al. 1974). The same types of understories are found in much younger forests, including stands solely dominated by *Pseudotsuga menziesii.*

Successional patterns in these forests are well documented for the period that precedes closure of the tree canopy (e.g., Dyrness 1973; Halpern 1988, 1989; Schoonmaker and McKee 1988; Halpern and Franklin 1990; Clark 1991; Halpern et al. 1992, 1997; Halpern and Spies 1995). In contrast, only limited research has been devoted to the patterns and correlates of understory development during later successional stages (but see Long and Turner 1975; Spies 1991; Halpern and Spies 1995). Large-scale disturbances (e.g., clear-cut logging and broadcast burning) initiate a complex but generally predictable series of species and growth-form replacements characterized by gradual shifts in the abundance and diversity of ruderal, colonizing species (invaders) and surviving understory plants (residuals) (see Table 4.3). The most dynamic changes in composition occur soon after disturbance: winter annuals (e.g., *Senecio sylvaticus, Epilobium paniculatum*) attain early dominance but are

Table 4.3. *The abundance and dispersal/establishment characteristics of 11 groups of plant species defined from long-term studies of succession in* Pseudotsuga menziesii–Tsuga heterophylla *forests of the H. J. Andrews Experimental Forest, western Oregon.*

Species group[a]	Growth-form(s)[b]	Modes of dispersal/ establishment[c]	Peak abundance Phase[d]	Peak abundance Magnitude[e]	Duration of elevated abundance[f]	Representative species
Invading species groups						
I1	H	W,(S?)	early	major	short, moderate	*Senecio sylvaticus, Epilobium paniculatum, Conyza canadensis*
I2	H	W	early-middle	major	long	*Epilobium angustifolium*
I3	H	W,(S?)	early-middle	minor	—	*Agoseris* spp., *Cirsium* spp., *Gnaphalium microcephalum, Lactuca serriola*
I4	H, S	W,S,A	middle	minor	—	*Anaphalis margaritacea, Rubus leucodermis, Collomia heterophylla, Vicia americana, Bromus* spp.
I5	S	S	middle	major	long	*Ceanothus velutinus, Ceanothus sanguineus*
I6	H, S, T	S,V,W,A	late	major, minor	—	*Pteridium aquilinum, Rubus parviflorus, Salix scouleriana, Prunus emarginata*
Residual species groups						
R1	H	V,(S,A)	early	major	long	*Rubus ursinus*
R2	H	V,(S,W)	middle	major	moderate-long	*Trientalis latifolia, Whipplea modesta, Hieracium albiflorum*
R3	H, S, T	V,(A,W,S)	middle-late, late	major	moderate-long	*Acer circinatum, Polystichum munitum, Gaultheria shallon, Berberis nervosa, Corylus cornuta, Tsuga heterophylla*
R4	H, S, T	V,(A)	—	minor	—	*Coptis laciniata, Viola sempervirens, Vaccinium parvifolium, Castanopsis chrysophylla, Oxalis oregana, Rubus nivalis, Acer macrophyllum, Cornus nuttallii*
R5	H, T	V,(A,W)	—	minor	—	*Chimaphila umbellata, Thuja plicata, Goodyera oblongifolia, Synthyris reniformis, Taxus brevifolia*

[a] Species groups: I = invaders, i.e., species generally absent from the mature or old-growth forest understory; R = residual forest species. Groups are described more fully in Halpern (1989).
[b] Growth forms: H = herbs, subshrubs, and low shrubs; S = tall shrubs; and T = understory trees.
[c] Modes of dispersal/establishment: W = wind-dispersed seed; A = animal-dispersed seed; S = germination from the soil seedbank; V = vegetative (resprouting from roots, rhizomes, or stem bases). Modes in parentheses represent less common pathways.
[d] Phase of peak abundance (period after logging/burning); dash (—) = no peak in abundance, early = 0–4 yr, middle = 5–10 yr, late = >10 yr.
[e] Magnitude of peak abundance: major = >5% cover in any year, minor = consistently <5% cover.
[f] Duration of elevated abundance (number of years with >5% cover); dash (—) = 0 yr, short = 1–2 yr, moderate = 3–5 yr, long = >5 yr.
Source: Adapted from Halpern (1989).

rapidly replaced by invasive perennial herbs (e.g., *Cirsium* spp., *Gnaphalium microcephalum, Epilobium angustifolium*) and subordinate forest herbs and subshrubs that are released by removal of the overstory (e.g., *Rubus ursinus, Trientalis latifolia, Whipplea modesta*). These species are gradually over-

topped by taller woody colonists (e.g., *Ceanothus* spp., *Rubus parviflorus*) (Table 4.2).

The principal colonists of these seral forest communities have diverse (often multiple) mechanisms for dispersal and establishment (Table 4.2): wind dispersed seeds or spores (*Senecio sylvaticus, Epi-*

Table 4.4. *Constancy and cover percentages of important species in five plant associations representing dry to wet sites (left to right) of the* Tsuga heterophylla *zone of the western Oregon Cascade Range.*

Stratum/species	PSME/HODI[a] Con.	Cover	TSHE/RHMA/ GASH Con.	Cover	TSHE/RHMA/ BENE Con.	Cover	TSHE/POMU Con.	Cover	TSHE/POMU/ OXOR Con.	Cover
Overstory tree										
Tsuga heterophylla	—[b]	—	76	20	100	43	100	44	100	29
Pseudotsuga menziesii	100	41	100	45	100	45	100	42	100	38
Thuja plicata	—	—	47	3	72	13	80	16	75	13
Libocedrus decurrens	50	6	12	1	—	—	—	—	—	—
Pinus lambertiana	50	1	12	2	—	—	—	—	—	—
Acer macrophyllum	25	2	12	1	—	—	47	2	62	7
Arbutus menziesii	38	2	—	—	—	—	—	—	—	—
Understory tree										
Tsuga heterophylla	25	T	100	8	100	8	100	9	100	11
Taxus brevifolia	38	4	59	6	78	7	47	4	50	1
Cornus nuttallii	50	2	53	3	50	2	33	1	38	1
Castanopsis chrysophylla	50	1	82	2	78	2	13	T	25	T
Pseudotsuga menziesii	100	8	—	—	—	—	—	—	—	—
Thuja plicata	—	—	29	1	56	2	53	3	38	2
Libocedrus decurrens	38	3	—	—	—	—	—	—	—	—
Pinus lambertiana	38	T	6	T	—	—	—	—	—	—
Acer macrophyllum	12	T	6	T	11	T	13	T	—	—
Arbutus menziesii	12	T	—	—	—	—	—	—	—	—
Shrub										
Acer circinatum	88	19	88	21	83	9	87	2	88	6
Rhododendron macrophyllum	12	T	100	40	89	13	67	1	12	T
Holodiscus discolor	88	5	—	—	—	—	—	—	—	—
Corylus cornuta var. *californica*	88	7	12	T	11	T	33	T	12	T
Vaccinium parvifolium	62	1	71	2	83	1	87	2	100	3
Beberis nervosa	100	16	100	14	100	11	100	8	100	13
Gaultheria shallon	62	7	100	40	89	4	53	2	75	4
Symphoricarpos mollis	88	2	—	—	17	2	—	—	—	—
Herb										
Rubus ursinus	75	1	65	1	83	2	67	1	75	1
Achlys triphylla	50	T	29	T	33	T	47	1	75	2
Viola sempervirens	12	T	65	1	83	2	73	2	62	1
Trillium ovatum	—	—	29	T	56	1	87	1	62	1
Polystichum munitum	100	4	65	1	94	4	100	26	100	27
Linnaea borealis	75	3	82	5	100	13	80	13	50	11
Vancouveria hexandra	25	T	6	T	39	T	27	T	88	4
Galium triflorum	38	T	18	T	11	T	60	1	38	1
Trientalis latifolia	100	1	29	T	44	T	27	T	12	T
Lathyrus polyphyllus	38	3	—	—	—	—	—	—	—	—
Madia gracilis	50	1	—	—	—	—	—	—	—	—
Collomia heterophylla	38	1	—	—	—	—	—	—	—	—
Hieracium abliflorum	62	1	12	T	28	T	20	T	25	T
Synthyris reniformis	75	4	12	T	17	T	—	—	—	—
Xerophyllum tenax	25	2	53	2	50	2	7	T	—	—
Iris tenax	62	1	6	T	—	—	—	—	—	—
Festuca occidentalis	88	1	6	T	—	—	7	T	—	—
Whipplea modesta	100	8	29	1	17	1	20	T	—	—
Chimaphila umbellata	88	1	82	2	83	4	53	T	38	1
Coptis laciniata	—	—	53	1	89	4	73	3	25	1
Tiarella trifoliata	—	—	12	T	28	T	73	4	12	T
Disporum hookeri	—	—	18	T	17	T	27	T	50	1
Asarum caudatum	—	—	—	—	—	—	27	T	25	1
Blechnum spicant	—	—	—	—	—	—	27	1	38	1
Oxalis oregana	—	—	—	—	—	—	7	T	100	38

constancy = con.

[a]PSME/HODI = *Pseudotsuga menziesii–Holodiscus discolor*; TSHE/RHMA/GASH = *Tsuga heterophylla–Rhododendron macrophyllum–Gaultheria shallon*; TSHE/RHMA/BENE = *Tsuga heterophylla–Rhododendron macrophyllum–Berberis nervosa*; TSHE/POMU = *Tsuga heterophylla–Polystichum munitum*; TSHE/POMU/OXOR = *Tsuga heterophylla–Polystichum munitum–Oxalis oregana*.

[b]T = trace (<0.5% cover); a dash (—) = absent.

Source: Based on Table 8 of Franklin and Dyrness (1988).

lobium spp., *Pteridium aquilinum*, *Salix* spp.); animal dispersed fleshy fruits (*Rubus* and *Ribes* spp.); germination from a buried seedbank (*Ceanothus*, *Rubus*, and *Ribes* spp.); or vegetative release of plants present in forest gaps or other disturbed microsites (e.g., *Pteridium aquilinum*, *Rubus parviflorus*). Species establishment occurs early in succession (largely within 2–3 yr), after which populations spread largely by vegetative means. As a group, invaders can dominate early seral communities for more than a decade, but their relative importance and persistence vary with disturbance intensity (e.g., *Ceanothus* shows greatest development on intensely burned sites); disturbance history (as it influences seedbank composition and viability); climatic conditions; and rates of tree regeneration and canopy closure (Halpern 1988, 1989; Halpern and Franklin 1990).

Although logging and burning markedly reduce the cover and biomass of most forest understory species, a majority of taxa survive disturbance (albeit at greatly reduced densities) (Halpern 1989; Halpern et al. 1992; Halpern and Spies 1995). Moreover, the woody plants that typically dominate these understories (e.g., *Acer circinatum* and *Berberis nervosa* on mesic sites; *Rhododendron macrophyllum*, *Gaultheria shallon*, *Corylus cornuta* on drier sites) are fairly tolerant of disturbance, having the ability to resprout from root collars or deeply buried rhizomes. Gradual but continuous growth and/or clonal spread of these and other species contributes to rapid recovery of understory composition and abundance, often well before tree canopies begin to close (Table 4.2). However, rates and magnitudes of recovery vary among plant communities (reflecting the relative disturbance tolerances and growth rates of the dominant species) and with the intensity of the initiating disturbance (Halpern 1988). Long-term (30+ yr) observations of permanent successional plots on the H. J. Andrews Experimental Forest (western Cascade Range, Oregon) suggest that the diversity of forest understory species may remain depressed for decades on severely burned sites, and some taxa may experience local extinction (e.g., *Chimaphila* spp., *Goodyera oblongifolia*, *Pyrola* spp., *Taxus brevifolia*). Many of the species that appear most susceptible to catastrophic disturbance are found most commonly, and with greatest abundance, in late seral forests of the region, suggesting that long periods are required between catastrophic disturbances for populations to recover (Halpern and Spies 1995).

Pseudotsuga menziesii and *Tsuga heterophylla* are the major seral and climax species, respectively. *Pseudotsuga* typically dominates young forests, often forming pure stands because of its relatively large and hardy seedlings and rapid growth rate. *Tsuga heterophylla* or *Thuja plicata* may establish early in succession or later under the canopy of *Pseudotsuga*. Stand basal areas shift toward *Tsuga–Thuja* dominance after 400–600 yr. Individual *Pseudotsuga menziesii* may persist for more than 1000 yr. *Thuja plicata* is a minor climax species; low but continuous reproduction and high survival contribute to long-term expansion and persistence of populations within many stands on moist to wet habitats. *Pseudotsuga menziesii* can be a climax species on dry sites within the *Pseudotsuga–Tsuga* forest complex (Means 1980); such habitats actually belong to the *Pseudotsuga* series, not the *Tsuga heterophylla* series. Several other species, such as *Acer macrophyllum* and *Calocedrus decurrens*, appear capable of playing minor climax roles. *Alnus rubra* is the most conspicuous of the seral hardwoods, although the short-lived *Prunus emarginata* may have a wider ecological amplitude.

There are many variants of the widespread *Pseudotsuga–Tsuga* forests. A notable contrast is the dominance of many understories by *Rhododendron macrophyllum* in Oregon forests and its virtual absence in Washington. Forests dominated by *Thuja plicata*, with understories of *Oplopanax horridum* and *Athyrium filix-femina*, or even *Lysichitum americanum* and *Carex obnupta*, occur on very wet sites. *Chamaecyparis lawsoniana* (Port Orford cedar) is a localized endemic often associated with these forests in southwestern Oregon (Zobel et al. 1985). The Puget lowlands have a number of notable features, including prairies, Mima mounds, and common occurrence of *Pinus contorta* and *P. monticola* (del Moral and Deardorff 1976; Franklin and Dyrness 1988). Talus communities of *Acer circinatum* and *A. macrophyllum* are often associated with the forests and are particularly noticeable in the Columbia Gorge.

Picea sitchensis–Tsuga heterophylla Forests

These forests characterize a relatively narrow band adjacent to the Pacific Ocean, extending from northern California to the Gulf of Alaska. Maritime influences are maximum, with cool, muted temperatures, high precipitation, and frequent fogs. *Picea-Tsuga* forests typically extend only a few kilometers inland but can extend further up river valleys or where the coastal plain is unusually broad. These forests are found from sea level to several hundred meters elevation or, in Alaska, essentially to timberline.

Picea sitchensis and *Tsuga heterophylla* are the major tree species (Hemstrom and Logan 1986; Franklin and Dyrness 1988; Henderson et al. 1989). *Tsuga* may dominate numerically by a factor of two or more, but *Picea* distinguishes these forests from types found in less humid, inland regions. In areas immediately adjacent to the ocean, *Picea sitchensis* may form nearly pure forests or co-dominate with *Pinus contorta* (lodgepole pine), because of the spruce's high tolerance for salt spray. Typical forest associates vary with latitude: *Pseudotsuga menziesii*, *Thuja plicata*, *Abies grandis*, *Acer macrophyllum*, and *Abies amabilis* are typical over much of the range. *Alnus rubra* has come to occupy large areas following logging; prolific seeding and rapid early growth make it an aggressive competitor. *Sequoia sempervirens* occurs in the south. *Tsuga mertensiana* (mountain hemlock) and *Chamaecyparis nootkatensis* (Alaska yellow cedar) are associates in southeastern Alaska and parts of British Columbia. *Pinus contorta* occurs on swampy or boggy habitats, in ocean-front areas, and on sand dunes.

Understories in mature forests typically include substantial shrub and herb coverage and a well-developed moss layer. One widespread community type in coastal Washington and Oregon is the *Picea sitchensis–Oxalis oregana* type (Hemstrom and Logan 1986; Henderson et al. 1989). Typical herbs include *Polystichum munitum*, *Oxalis oregana*, *Rubus pedatus*, *Blechnum spicant*, *Tiarella trifoliata*, *Montia sibirica*, and *Maianthemum dilatatum*. Typical shrubs include *Acer circinatum*, *Rubus spectabilis*, *Vaccinium parvifolium*, *Menziesia ferruginea*, and *Vaccinium alaskaense*. Wetter habitats may incorporate an even greater variety of herbs, as well as *Oplopanax horridum* and abundant *Rubus spectabilis*. The understory composition is similar in southeast Alaska, with well-developed ericaceous shrub layers but an absence of *Oxalis oregana* (Alaback 1982). The *Picea sitchensis–Gaultheria shallon* community type is characteristic of sand dunes and areas of salt spray (Hemstrom and Logan 1986). Its shrubby understory of *Gaultheria shallon* and depauperate herb layer contrast sharply with the other types.

Revegetation is very rapid following disturbances on the highly productive habitats characteristically occupied by *Picea sitchensis–Tsuga heterophylla* forests. *Alnus rubra* is an aggressive colonizer and often forms pure or nearly pure stands following a disturbance. *Alnus* is typically replaced by coniferous species such as *Tsuga*, *Picea*, and *Thuja*. *Alnus* stands may develop dense shrubby understories, however, which retard conifer regeneration; *Rubus spectabilis* is a typical dominant (Carlton 1988; Tappeiner et al. 1991; Tappeiner and Zasada 1993). Conifers may also regen-

nerate directly following disturbances, resulting in dense, even-aged stands. The high leaf areas in young and fully stocked stands can prevent development of understory plants; reestablishment of a significant shrub and herb community may take several centuries, a matter of concern where such forests are used as wintering grounds by herbivores (Alaback 1982). *Tsuga* is the major climax species based on size-class distributions. Under old-growth conditions, canopies often are sufficiently open for reproduction of *Picea* (Taylor 1990).

Alluvial rain forests of the western Olympic Peninsula are outstanding examples of *Picea–Tsuga* forest (Franklin and Dyrness 1988; Kirk and Franklin 1992) (Fig. 4.3). Forests have relatively low tree densities, with open canopies. Epiphytes, mostly cryptogams, are conspicuous. *Selaginella oregana* is particularly abundant on *Acer macrophyllum*. Branches covered with such vascular and non-vascular epiphytes may sprout roots in a process of air-layering (Nadkarni 1984). Rotting logs are important seedbeds for tree reproduction throughout the *Picea sitchensis* zone, but nurse-log phenomena are especially conspicuous in the alluvial forests (McKee et al. 1982). Competition from understory plants is the major factor restricting tree seedlings to rotten-wood seedbeds (Harmon 1986). Finally, *Cervis canadensis* var. *roosevelti* (Roosevelt elk) significantly influence the composition and structure of the forest understory by grazing; such preferred species as *Oplopanax horridum* and *Rubus spectabilis* are found only in protected microsites (Schreiner et al. 1996).

Sequoia *sempervirens* Forests

The coastal forests of northern California and southern Oregon are distinguished by the presence of *Sequoia sempervirens*. This is the world's tallest tree (112 m); with its long life span and massive growth form, it results in forests that have the greatest biomass accumulations known. The primary belt of *S. sempervirens* is about 16 km in width, with its western boundary often several kilometers inland from the ocean, and its eastern limits 35+ km or more inland (Zinke 1977). A wide variety of environments and community types with *Sequoia* are represented along the coast to inland gradient (Waring and Major 1964; Zinke 1977).

Associates of *Sequoia sempervirens* vary substantially with local environmental conditions (Waring and Major 1964). *Pseudotsuga menziesii* and *Lithocarpus densiflorus* are present on a wide variety of

Figure 4.3. Alluvial Picea sitchensis-Tsuga heterophylla *forest in the Hoh River valley of Olympic National Park, Washington (photo courtesy USDA Forest Service).*

sites. *Tsuga heterophylla, Umbellularia californica,* and *Alnus rubra* are common on moist habitats. *Picea sitchensis* is associated near the ocean. Dry-site species include *Arbutus menziesii, Calocedrus decurrens,* and occasionally *Pinus attentuata.*

The major climax species in *Sequoia sempervirens* forests include *Lithocarpus densiflorus, Tsuga heterophylla,* and *Sequoia sempervirens.* Regeneration of *Sequoia* was once believed to be dependent on disturbance, specifically fire or flood (Stone and Vasey 1968), but age-structure analyses of cutover stands have shown that *Sequoia* can reproduce in sufficient numbers to perpetuate its population, even in the absence of catastrophic disturbance (Viers 1982).

Waring and Major (1964) have provided detailed information on understory compositions in various *Sequoia* forest types. Forests on alluvial flats and moist lower slopes have the lushest understories: *Polystichum munitum, Oxalis oregana, Disporum smithii, Anemone deltoidea, Tiarella unifoliata, Trillium ovatum, Asarum caudatum, Viola glabella,* and *Hierochloe occidentalis* are characteristic. Stands in middle- and upper-slope positions are often characterized by evergreen shrubs, such as *Gaultheria shallon, Rhododendron macrophyllum,* and *Vaccinium ovatum,* and by reduced diversity and coverage of herbs.

Klamath Mountains
Mixed Evergreen Forests

Mixed forests of conifers and evergreen hardwoods characterize the majority of the forested landscape within the Klamath Mountains of northwestern California and southwestern Oregon. They are bounded by mesic coastal forests on the west, subalpine forests of *Abies* species at high elevations, and dry *Quercus* woodlands and *Pinus* forests in the interior valleys to the east. The Klamath Mountains contain regions of complex and strongly contrasting geology, including ultrabasic rock types, such as serpentine; consequently, soil

Figure 4.4. *Mixed evergreen forests in southwestern Oregon consist of a mixture of evergreen hardwood and conifer species that include* Arbutus menziesii *and* Pseu-dotsuga menziesii *in this stand (photo courtesy of USDA Forest Service).*

conditions are highly variable. There are also sharp gradients in moisture and temperature regimes between the coastal and inland valleys. Complex vegetational patterns are therefore characteristic of the region (Whittaker 1960). Furthermore, the region contains a large number of endemic or relictual species that increase the vegetational complexity.

The mixed evergreen forests typically include conifers, primarily *Pseudotsuga menziesii*, and one or more evergreen hardwoods, such as *Lithocarpus densiflorus, Quercus chrysolepis, Arbutus menziesii,* and *Castanopsis chrysophylla* (Sawyer et al. 1977; Franklin and Dyrness 1988; Tappeiner and McDonald 1984; Atzet et al. 1992, 1996) (Fig. 4.4). Stands are often two-storied, with *Pseudotsuga* forming a canopy up to 65 m in height emergent above a hardwood canopy up to 35 m in height (Thornburgh 1982). Three series comprise most of this formation on normal soils: *Lithocarpus densiflorus, Pseudotsuga menziesii,* and *Abies concolor* (Thornburgh 1982; Atzet et al. 1992, 1996). The *Pinus jeffreyi* series is most characteristic on ultrabasic parent material.

Pseudotsuga-Lithocarpus communities are widespread on the windward or coastal side of the Klamath Mountains, regions with relatively high summer humidities (Atzet et al. 1992, 1996). Twenty-one *Lithocarpus densiflorus* community types have been recognized in the Siskiyou National Forest (Atzet and Wheeler 1984), and 11 of 16 habitat types listed for northwestern Californian mixed evergreen forests are also *Pseudotsuga-Lithocarpus* (Thornburgh 1982). Characteristic tree species are *Pseudotsuga* and *Pinus lambertiana* in the overstory and *Pseudotsuga, Lithocarpus, Arbutus menziesii, Castanopsis chrysophylla, Quercus chrysolepis,* and *Abies concolor* in the lower tree layers (Table 4.5). Dense evergreen ericaceous shrub layers occur in the *Lithocarpus* series, with dominants such as *Rhododendron macrophyllum, Vaccinium ovatum, Gaultheria shallon,* and *Berberis nervosa* (Table 4.5). *Polystichum munitum* is the most common herb.

The *Pseudotsuga* series characterizes drier environments on the interior slopes of the Klamath Mountains (Atzet et al. 1992, 1996). *Pseudotsuga, Pinus lambertiana,* and *Pinus ponderosa* are overstory dominants, and *Castanopsis chrysophylla, Quercus chrysolepis, Arbutus menziesii,* and *Quercus kelloggii* form a lower tree canopy. *Lithocarpus densiflorus*

Table 4.5. *Percentages of constancy/cover of selected species in stands belonging to the four major series in the Siskiyou National Forest in southwestern Oregon. Community types generally correspond with moist to dry sites (left to right).*

Layer/species	Community type[a]			
	ABCO	LIDE	PSME	PIJE
Tree upper story				
Pseudotsuga menziesii	87/40	99/59	94/48	36/4
Abies concolor	83/34	3/9	9/9	T/T[b]
Pinus lambertiana	32/6	37/10	34/14	10/7
Abies magnifica var. shastensis	21/13	—	1/1	—
Calocedrus decurrens	17/18	4/12	15/13	44/5
Pinus ponderosa	18/10	8/8	35/22	—
Pinus jeffreyi	1/6	1/5	7/15	95/18
Chamaecyparis lawsoniana	4/4	6/22	3/8	—
Picea breweriana	1/4	—	—	—
Tsuga heterophylla	T/8	—	1/20	—
Pinus monticola	2/2	2/15	—	15/4
Tree lower story				
Abies concolor	100/29	13/6	28/5	10/2
Quercus sadleriana	21/16	4/17	8/18	—
Pseudotsuga menziesii	53/7	70/7	95/27	49/4
Calocedrus decurrens	25/4	11/6	22/11	62/8
Abies magnifica var. shastensis	17/5	—	1/1	—
Pinus lambertiana	13/2	28/3	33/5	5/7
Pinus ponderosa	5/2	1/3	21/10	—
Lithocarpus densiflorus	11/7	100/50	28/13	—
Arbutus menziesii	17/10	48/11	63/12	3/1
Castanopsis chrysophylla	36/8	29/16	27/16	—
Quercus chrysolepis	19/9	46/13	49/14	8/5
Quercus kelloggii	2/7	5/4	30/10	10/3
Chamaecyparis lawsoniana	8/6	11/7	4/4	8/3
Pinus jeffreyi	T/1	—	6/8	95/8
Pinus monticola	2/3	3/8	4/8	23/9
Umbellularia californica	—	13/18	3/12	26/10
Shrub				
Rosa gymnocarpa	77/4	45/3	56/3	3/1
Symphoricarpos mollis	66/5	22/3	44/7	—
Berberis nervosa	65/16	62/14	38/15	—
Gaultheria shallon	2/29	38/42	—	—
Vaccinium ovatum	—	32/36	—	—
Rhododendron macrophyllum	1/40	30/36	—	—
Rhus diversiloba	1/2	25/9	38/13	—
Lonicera hispidula	3/4	17/3	25/3	—
Berberis piperiana	6/2	6/3	25/5	3/1
Holodiscus discolor	35/4	6/6	29/6	8/9
Arctostaphylos nevadensis	5/7	2/7	5/22	31/18
Arctostaphylos viscida	—	1/12	7/17	54/14
Herb				
Grasses	100/1	100/1	100/7	100/39
Achillea millifolium	5/2	T/1	7/5	36/3
Chimaphila umbellata	56/5	31/4	27/4	—
Goodyera oblongifolia	52/1	55/1	43/1	3/1
Anemone deltoidea	50/2	3/1	7/2	—
Disporum hookeri	44/2	29/1	34/2	—
Achlys triphylla	41/16	33/6	21/7	—
Smilacina stellata	39/4	1/2	9/5	5/1
Linnaea borealis	33/17	20/11	15/13	—
Polystichum munitum	20/3	55/4	36/5	3/1
Xerophyllum tenax	8/7	25/6	37/9	26/8
Pteridium aquilinum	21/3	39/3	34/4	—

Table 4.5. *(cont.)*

Layer/species	Community type[a]			
	ABCO	LIDE	PSME	PIJE
Whipplea modesta	27/7	36/6	29/8	8/1
Galium aparine	2/2	11/1	27/4	—
Hieracium albiflorum	38/2	20/2	39/2	3/1
Trientalis latifolia	60/3	25/2	37/3	10/1

[a] *Abies concolor* series (ABCO, 331 plots in 24 community types), *Lithocarpus densiflorus* series (LIDE, 251 plots in 21 community types), *Pseudotsuga menziesii* series (PSME, 188 plots in 20 community types), and *Pinus jeffreyi* series (PIJE, 39 plots in 4 community types).
[b] T = trace; dash (—) = absent.
Source: Based primarily on Atzet and Wheeler (1984).

and *Abies concolor* are associates in some *Pseudotsuga* community types, but they are considerably less important than in their respective series (Table 4.5). *Rhus diversiloba* and other deciduous species are typical components of the shrub layer along with *Berberis* species, and grasses are a conspicuous element of the herbaceous layer.

The *Abies concolor* series generally occurs at elevations above the *Lithocarpus* series on coastal slopes; it is also found on moister and cooler habitats than the *Pseudotsuga* series that is found further inland. The *Abies concolor* community types are really conifer types rather than mixed evergreen types and are analogous to some of the Sierran mixed conifer types. They occur in the Klamath Mountains in close association with mixed evergreen types and have some distinctive compositional features (Sawyer et al. 1977). *Picea breweriana* is associated with this series (Atzet et al. 1996). The most common evergreen hardwood associate is *Quercus sadleriana* (Table 4.5).

Forest communities found on ultrabasic rock types, with their poorly developed and chemically unique soils, vary greatly with temperature and moisture regime (Whittaker 1960). *Chamaecyparis lawsoniana* is a characteristic species on wet sites (Zobel et al. 1985; Jimerson and Daniel 1994). The *Pinus jeffreyi* series is confined to ultrabasic types and occurs across a broad elevational range. Open woodlands of *Pinus jeffreyi* are characteristic, with *Pinus monticola* commonly associated at higher elevations and *Umbellularia californica* on moist sites (Table 4.5). *Arctostaphylos* species and grasses are characteristic of the understory.

Successional relationships in mixed evergreen forests are complex. Fire, grazing, and logging have been important in creating the current community mosaic, which includes many multiaged stands. Major climax species (and community series) are *Pseudotsuga*, *Abies concolor*, and *Lithocarpus*

(Atzet et al. 1992, 1996). There are many minor climax associates, including *Arbutus*, *Castanopsis*, *Quercus chrysolepis*, *Pinus lambertiana*, and *Chamaecyparis lawsoniana* on various habitats (Thornburgh 1982). The successional status of *Calocedrus* is problematic on many sites; it appears at least as shade-tolerant as *Pseudotsuga* and is represented in smaller size classes within stands that may represent either older suppressed trees or relatively recent reproduction.

On *Lithocarpus* habitat types, two-stories stands of *Pseudotsuga* and *Lithocarpus* are the expected climax. Even-aged stands of *Lithocarpus* and *Pseudotsuga*, either pure or in mixture, typically develop following disturbances. *Lithocarpus* has a distinct advantage because it has the ability to sprout from the root crown. *Pseudotsuga*, however is less likely to be killed by ground fire and has a faster growth rate that allows it to overtop *Lithocarpus* in 15–30 yr if they establish simultaneously (Thornburgh 1982). Where pure stands of *Lithocarpus* are established, canopy gaps develop after 60–100 yr that permit *Pseudotsuga* establishment. Brushfields of *Ceanothus*, *Arctostaphylos*, and other species may develop in mixed evergreen habitats following fire or other disturbances and can retard establishment of conifers for several decades.

More southerly versions of mixed evergreen forest, generally with associated conifers other than *Pseudotsuga*, are discussed in Chapter 5.

Sierran-Type Mixed Conifer Forests

Mixed stands of *Pseudotsuga menziesii*, *Pinus lambertiana*, *Abies concolor*, *Pinus ponderosa*, and *Calocedrus decurrens* characterize the montane forests of the southern Cascade Range and eastern Siskiyou Mountains (Franklin and Dyrness 1988; Atzet and McCrimmon 1990). These forests are northern extensions of the Sierran mixed conifer type described in Chapter 5. The transition between the Sierran mixed conifer forest and *Pseudotsuga menziesii–Tsuga heterophylla* forest is gradual, with southern species extending into northern Oregon (e.g., *Pinus lambertiana* and *Calocedrus decurrens* into the Mount Hood region), and northern species (e.g., *Tsuga heterophylla* and *Thuja plicata*) extending well into southern Oregon. The McKenzie River drainage, at about 45° N, is often identified as a major transitional point.

There is considerable variability in the Sierran mixed conifer forest associated with latitude, elevation, and local site conditions (Griffin 1967; Franklin and Dyrness 1988; Atzet and McCrimmon 1990). *Pinus ponderosa* and *Quercus garryana* are most common at lower elevations, especially adjacent to major valleys. *Quercus chrysolepis* may be a major associate of *Pseudotsuga menziesii* on somewhat moister sites, producing stands resembling the mixed evergreen forests of the Klamath Mountains. Highly varied forests of *Pseudotsuga, Abies, Pinus lambertiana, Calocedrus decurrens*, and *Quercus kelloggii* characterize middle elevations. Stands at higher elevations, adjacent to montane *Abies magnifica* var. *shastensis* forests, are often totally dominated by *Abies concolor* (Atzet and McCrimmon 1990). *Pinus jeffreyi* and *Pinus ponderosa* forests occur at high elevations on the eastern slope of the southern Cascade Range (Rundel et al. 1977), but *Pinus jeffreyi* is absent from the western slope mixed conifer forest in the Cascade Range, in contrast to the Sierra Nevada.

The *Pseudotsuga menziesii* and *Abies concolor* series characterize most of the Sierran-type mixed conifer forest. The *Pseudotsuga* series typifies hotter, drier habitats. Shrubby understories are common. The *Abies concolor* series dominates the bulk of the mixed conifer region, including modal and moist sites. Understories associated with the *Abies* series vary from shrubby types (with shrubs of either northern or southern origin) to those with a rich array of herbs (e.g., *Abies concolor–Linnaea borealis* community) (Franklin and Dyrness 1988).

Successional relationships in the mixed conifer forests resemble those in the mixed evergreen types. Fire, grazing, and logging have been important in creating the current mosaic of communities and the characteristic multiaged stands. The major climax species are *Abies concolor* and *Pseudotsuga menziesii*. There are many minor associates, because often there are sufficient openings or gaps in the forest to allow for regeneration of less shade-tolerant species such as *Pinus lambertiana* and *Quercus kelloggii*. *Calocedrus decurrens* appears to be a climax species on many sites and has a bimodal

Figure 4.5. *Mixed forest of* Abies amabilis, Abies procera, *and* Tsuga heterophylla *in the Cascade Range of Oregon, Wildcat Mountain Research Natural Area (photo courtesy USDA Forest Service).*

distribution. It typically occurs in very dry forests and, with *Abies concolor*, on moist to wet sites (often invading meadows) at relatively high elevations.

UPPER MONTANE/SUBALPINE FORESTS AND PARKLANDS

Abies amabilis Forests

These forests characterize montane regions from the central Oregon Cascades north through the mountains of southern British Columbia and in the Olympic Mountains (Krajina 1965; Fonda and Bliss 1969; Franklin and Dyrness 1988). They support a mixture of temperate-zone and subalpine species (Fig. 4.5). Many detailed studies of these forest communities exist – for example, Franklin (1966), Brockway et al. (1983), Hemstrom et al. (1987), Franklin et al. (1988), Henderson et al. (1989, 1992), and Moir (1989). Cooler temperatures and permanent winter snowpacks distinguish the environment of the *Abies amabilis* zone from that of the lower-elevation *Tsuga heterophylla* zone.

Forest compositions are highly variable, depending on stand age, history, and locale (Franklin 1965). *Abies amabilis, Tsuga heterophylla, Abies procera* (noble fir), *Pseudotsuga menziesii, Thuja plicata*, and *Pinus monticola* are typical tree species. In the High Cascades and more continental environments, *Pinus contorta, Abies lasiocarpa* (subalpine fir), *Picea engelmannii* (Engelmann spruce), *Abies grandis*, and *Larix occidentalis* (western larch) may occur. *Tsuga mertensiana* and *Chamaecyparis nootkatensis* can be common at higher elevations and in more northerly latitudes.

Understories are typically dominated by ericaceous genera, such as *Vaccinium, Menziesia, Gaultheria, Chimaphila, Rhododendron*, and *Pyrola. Cornus canadensis, Clintonia uniflora, Rubus pedatus, Rubus lasiococcus, Linnaea borealis*, and *Xerophyllum tenax* are common species in the herb layer. The four major groups of communities are (1) *Vaccinium alaskaense*, (2) herb-rich communities, (3) *Vaccinium membranaceum–Xerophyllum tenax*, and (4) *Gaultheria shallon* (Table 4.6). *Abies amabilis–Vaccinium alaskaense* communities are widely distributed, occurring from northern Oregon to southern British Columbia and in the Olympic Mountains of Washington (Franklin 1966; Henderson et al. 1989, 1992). The herb-rich communities typically have high diversity and coverage of ferns and dicotyledonous herbs; shrubs may be abundant, as in the *Abies amabilis–Oplopanax horridum* type, or essentially absent. Community types with understories dominated by *Vaccinium membranaceum* and/or *Xerophyllum tenax*, or by *Gaultheria shallon* typify, respectively, the snowier and drier *Abies amabilis* types.

In many respects, patterns of early successional change within these forests are similar to those within the *Tsuga heterophylla* zone. A combination of colonizing and residual forest species contribute to the early seral community, with open-site species gradually replaced by residual forest herbs and shrubs (Shlisky 1996; White et al. 1996). Many of the wind-dispersed, invasive taxa that dominate early seral stages at lower elevations play similar roles here (e.g., *Epilobium angustifolium, Pteridium aquilinum*). Some invaders appear more widespread and persistent in these generally cooler, moister forest environments (e.g., *Anaphalis margaritacea, Salix* spp.), whereas others appear restricted to particular site conditions or plant associations (e.g., *Ceanothus velutinus* on burned sites and in drier habitats). As in the *Tsuga heterophylla* zone, the dominant forest understory species are

fairly tolerant of disturbance and through vegetative reproduction contribute to gradual recovery of understory composition. If ground disturbances are moderate (e.g., partial timber harvests or light-intensity wildfires), shrubfields dominated by *Vaccinium* spp. or other herbaceous communities (e.g., composed of *Xerophyllum tenax*) may persist for decades, inhibiting the establishment and growth of trees. As observed in forests at lower elevations, there are groups of forest herbs and subshrubs that are particularly sensitive to disturbance (e.g., *Chimaphila umbellatum, Clintonia uniflora, Goodyera oblongifolia, Pyrola* spp.); these taxa are largely absent from early seral communities but are found with increasing frequency and abundance in mature and older forests (Shlisky 1996).

Succession typically leads from more diverse mixed forests toward old-growth stands of *Abies amabilis* and *Tsuga heterophylla. Abies procera, Pseudotsuga menziesii*, and *Pinus monticola* are typical seral species. Any species may regenerate directly following a disturbance, although the heavy-seeded, fire-sensitive *Abies amabilis* is often the last to invade (Schmidt 1957). *Abies amabilis* is the major climax species, as shown by size- and age-class analyses (e.g., Franklin et al. 1988). Although *Tsuga heterophylla* often dominates stands initially, it reproduces poorly within the *Abies amabilis* zone, perhaps because of its small, fragile seedlings that are buried under snow-compressed accumulations of litter (Thornburgh 1969).

Dispersed throughout these upper montane forests is a diverse assemblage of meadows and other nonforested habitats. Although, in total, they occupy a small portion of the landscape, they support a large proportion of the regional flora (Hickman 1976). Despite their botanical and aesthetic appeal, there have been relatively few studies of the phytosociology and ecology of nonforested communities (but see Roach 1958; Hickman 1976; Vale 1981; Halpern et al. 1984; Hemstrom et al. 1987; Miller and Halpern *in press*). Dense shrub communities of *Alnus sinuata* are common on steep, north-facing slopes – sites that experience deep snow accumulations and recurrent avalanches (Franklin and Dyrness 1988) (Fig. 4.6). Succession to forest is prevented by periodic disturbance: A common tree species that can establish and reproduce in this environment is *Chamaecyparis nootkatensis* (Alaska yellow cedar), which has flexible stems that layer when in contact with the soil (Antos and Zobel 1986). More open shrubfields dominated by *Acer circinatum* and other woody taxa (e.g., *Sambucus* spp., *Ribes* spp., *Rhamnus purshiana*) characterize talus slopes and lava flows throughout the central western Cascade region

Table 4.6. *Constancy (%) and average basal area (m² ha⁻¹) of trees and constancy and cover (%) of shrubs and herbs in five associations found in the* Abies amabilis *zone of Mount Rainier National Park, Washington. Associations are arranged from wettest to driest (left to right).*

Growth form/species	ABAM/OPHO[a] Con.	BA	ABAM/TIUN Con.	BA	ABAM/MEFE Con.	BA	ABAM/VAAL Con.	BA	ABAM/GASH Con.	BA
Trees										
Abies amabilis	94	25.6	100	34.4	100	46.1	98	23.4	79	2.5
Tsuga heterophylla	94	40.2	75	20.1	79	21.1	98	33.4	100	40.3
Pseudotsuga menziesii	35	28.0	31	6.4	21	8.0	49	15.9	79	14.9
Chamaecyparis nootkatensis	6	2.7	25	3.9	84	14.1	28	3.2	29	1.0
Thuja plicata	47	16.6	—	—	11	0.1	38	4.9	86	7.5
Tsuga mertensiana	—[b]	—	16	1.3	68	9.3	15	1.0	—	—
Abies lasiocarpa	—	—	1	0.5	—	—	—	—	—	—
Shrubs										
Vaccinium alaskaense	100	11	33	T	54	7	97	28	77	9
Vaccinium membranaceum	28	T	80	2	96	12	83	3	85	2
Vaccinium ovalifolium	83	4	68	2	96	8	71	7	46	1
Menziesia ferruginea	56	2	20	1	100	16	65	1	54	1
Berberis nervosa	11	1	6	T	4	T	38	1	100	7
Rubus spectabilis	72	2	40	T	10	T	7	T	7	T
Oplopanax horridum	100	12	33	T	5	T	14	T	—	—
Acer circinatum	22	1	30	3	—	—	28	2	—	—
Ribes lacustre	44	1	13	T	—	—	5	T	—	—
Vaccinium parvifolium	61	1	7	T	—	—	66	33	100	5
Gaultheria shallon	11	T	—	—	—	—	12	T	100	25
Rhododendron albiflorum	—	—	—	—	63	3	—	—	—	—
Herbs										
Rubus pedatus	94	7	72	1	79	4	87	4	40	T
Clintonia uniflora	94	3	74	3	63	2	78	2	15	T
Rubus lasiococcus	44	1	85	5	96	2	68	1	23	T
Linnaea borealis	72	5	13	T	8	T	64	4	92	3
Cornus canadensis	72	3	13	T	10	T	46	1	69	1
Chimaphila umbellata	6	T	20	T	5	T	53	1	100	2
Blechnum spicant	67	5	20	1	15	T	68	4	20	T
Pyrola secunda	50	T	68	1	79	T	72	T	31	T
Viola sempervirens	61	1	78	2	54	T	26	T	46	T
Trillium ovatum	94	T	78	T	42	T	32	T	7	T
Achlys triphylla	92	14	84	14	25	1	38	1	—	—
Streptopus roseus	47	3	72	7	46	1	37	1	—	—
Valeriana sitchensis	22	1	72	4	29	T	5	T	—	—
Athyrium filix-femina	83	3	33	1	5	T	12	T	—	—
Gymnocarpium dryopteris	45	7	50	3	17	T	7	T	—	—
Smilacina stellata	50	2	59	4	13	T	23	T	—	—
Viola glabella	56	1	53	2	13	T	2	T	—	—
Streptopus streptopoides	28	T	48	4	17	T	9	T	—	—
Tiarella unifoliata	83	6	87	15	58	1	47	T	—	—
Adenocaulon bicolor	78	11	—	—	8	T	5	T	—	—
Tiarella trifoliata	67	5	53	1	—	—	9	T	—	—
Maianthemum dilatatum	50	1	20	T	—	—	12	T	—	—
Xerophyllum tenax	—	—	—	—	54	3	29	1	100	15

[a] ABAM/OPHO = *Abies amabilis–Oplopanax horridum*, ABAM/TIUN = *Abies amabilis–Tiarella unifoliata;* ABAM/MEFE = *Abies amabilis–Menziesia ferruginea;* ABAM/VAAL = *Abies amabilis–Vaccinium alaskaense;* ABAM/GASH = *Abies amabilis–Gaultheria shallon.*
[b] T = trace (<0.5% cover); dash (—) = absent.
Source: Franklin et al. (1988).

Figure 4.6. Shrub communities dominated by Alnus sinuata typify snow avalanche tracks and other disturbed habitats within the Abies amabilis zone, H. J. Andrews Experimental Forest, Oregon (photo courtesy of USDA Forest Service).

(Roach 1958; Hemstrom et al. 1987; Franklin and Dyrness 1988).

A more diverse set of herb- and graminoid-dominated communities occupy various positions on the landscape: small hydric basins; broad open areas of flat and often poorly drained topography; and steep, south-facing slopes and ridgetops (Table 4.7). Hydric montane basins support a mosaic of wetland or mire communities that correlate with depth and seasonal movement of the water table (Halpern et al. 1984). Species-poor *Carex sitchensis* communities occupy areas of deep, persistent standing water. Herb-rich and moss-covered *Eleocharis pauciflora* communities characterize sites with shallow surface water. Communities dominated by *Deschampsia caespitosa* occur where soils are moist but not inundated over the course of the growing season; they show their greatest development at higher elevations, in the broad, poorly drained flats

and basins of the High Cascades Province. Hydric meadow communities with similar physiognomies and many of the same species and genera can be found in montane and subalpine settings much farther south in the southern Cascades and the Sierra Nevada (e.g., Seyer 1979; Ratliff 1982; Benedict 1983; Halpern 1986).

Extensive mosaics of forest and meadow are also common on steep, south- and west-facing slopes of the western Cascades (Hickman 1976; Halpern et al. 1984). Gradients in vegetation composition parallel those of topographic position reflecting variation in soil depth and available moisture (Table 4.7). Along lower slopes, deep, loamy soils support tall, lush, herb-rich communities dominated by *Veratrum* spp., *Senecio triangularis*, and *Heracleum lanatum*. These grades upslope into dense shrub- and forb-dominated communities whose dominants, *Rubus parviflorus* and *Pteridium aquilinum*, can be found in the understory of bordering forests. Farther upslope in drier, shallower soils, these are replaced by more open, short-statured communities dominated by graminoids, *Carex pensylvanica* and *Bromus* spp. Upper slopes and ridgetops support highly diverse and showy lithosolic or "rock-garden" communities characterized by an array of early-summer ephemerals (*Gilia capitata*) or drought-tolerant forbs and subshrubs (e.g., *Eriophyllum lanatum*, *Penstemon procerus*, *Eriogonum* spp.).

Various factors have contributed to the creation and maintenance of these montane forest openings: high water tables in hydric basins; shallow, rocky soils along steep, upper slopes; and catastrophic or recurrent disturbance (wildfire, landslides, and soil creep; snow movement; historic sheep grazing; and the burrowing activities of small mammals). Hickman (1968, 1976) has speculated that some meadow types have persisted for more than 1000 yr as topo-edaphic climax communities. Elsewhere, grazing by sheep during the first half of the century has maintained or enlarged meadow openings and significantly altered species composition (Kuhns 1917; Halpern et al. 1984; Miller and Halpern *in press*). Studies in the Three Sisters Biosphere Reserve, Oregon, and elsewhere in the western Cascades indicate that rates and patterns of succession from meadow to forest vary across the landscape in response to local environment (hydrology, topography, soils); meadow composition; and climatic and disturbance history (Vale 1981; Halpern et al. 1984; Miller and Halpern *in press*). Tree invasion of hydric meadows (primarily *Abies lasiocarpa* and *Picea engelmannii*) has been slow but continuous, restricted to elevated microsites and existing tree islands. Recent invasion of

Table 4.7. *The constancy (% of plots sampled)/mean cover (%) of species that characterize selected meadow communities in the montane zone, central Oregon Cascade Range.*

	Communities[a]									
	Hydric sites			Mesic sites					Xeric sites	
Growth-form/*species*	CASI	ELPA	DECA–HYAN– MUFI	DECA– TRLO	VERAT– SETR	RUPA	PTAQ– ELGL	CAPE– BROMU	ERLA– GICA	PEPR– SAGR
Trees										
Picea engelmannii	—[b]	50/T	18/T	—	—	—	—	—	—	—
Pinus contorta	—	33/T	18/T	—	—	—	—	—	—	—
Tsuga mertensiana	—	33/T	—	—	—	—	—	—	—	—
Abies lasiocarpa	—	33/T	—	—	—	13/T	50/4	—	—	—
Abies amabilis	—	—	—	—	—	25/1	—	—	—	—
Shrubs/subshrubs										
Salix myrtillifolia	44/10	17/T	27/T	—	—	—	—	—	—	—
Salix geyeriana	—	33/2	9/T	—	—	—	—	—	—	—
Vaccinium occidentale	56/12	100/8	82/3	11/T	—	—	—	—	—	—
Salix commutata	33/1	33/T	45/1	11/T	—	—	—	—	—	—
Rubus parviflorus	—	—	—	—	—	100/59	50/2	—	25/T	—
Amelanchier alnifolia	—	17/T	—	—	—	—	—	17/T	50/T	67/1
Comandra umbellata	—	—	—	—	—	—	—	33/T	50/T	100/2
Eriogonum umbellatum	—	—	—	—	—	—	—	—	75/4	67/1
Erigonum compositum	—	—	—	—	—	—	—	—	75/2	33/T
Graminoids										
Carex sitchensis	100/76	67/11	82/7	—	50/7	—	—	—	—	—
Carex rostrata	56/12	83/2	55/11	11/T	—	—	—	—	—	—
Eleocharis pauciflora	56/6	100/58	91/15	22/1	—	—	—	—	—	—
Deschampsia caespitosa	56/2	67/1	100/36	100/65	50/1	—	—	—	—	—
Carex luzulina	33/2	67/7	64/19	33/1	50/T	—	—	—	—	—
Muhlenbergia filiformis	11/T	83/4	82/9	33/3	25/T	—	—	—	—	—
Elymus glaucus	—	—	—	11/T	75/16	100/15	100/20	100/7	25/1	33/T
Carex pachystachya	—	—	—	22/1	75/2	75/1	100/3	67/T	50/T	33/T
Carex pensylvanica	—	—	—	11/T	—	38/4	50/8	100/23	50/3	100/8
Bromus spp.[c]	—	—	—	—	—	75/8	100/13	100/l9	100/5	67/T
Forbs/Ferns										
Dodecatheon jeffreyi	78/4	100/8	82/7	11/T	—	—	—	—	—	—
Hypericum anagalloides	67/1	83/4	91/9	—	—	—	—	—	—	—
Pedicularis groenlandica	44/T	100/2	45/3	11/T	—	—	—	—	—	—
Caltha biflora	44/3	67/7	55/3	11/T	—	—	—	—	—	—
Polygonum bistortoides	56/2	67/1	45/2	—	50/T	—	—	—	—	—
Ranunculus gormanii	33/1	83/3	64/12	33/T	—	—	—	—	—	—
Microseris boreale	33/T	83/T	55/8	11/1	—	—	—	—	—	—
Trifolium longipes	11/T	17/T	55/2	100/25	—	—	—	—	—	—
Aster foliaceus	11/T	—	36/T	89/8	100/6	38/1	100/1	67/3	—	33/T
Potentilla drummondii	11/T	33/T	18/1	78/7	50/1	—	—	—	—	—
Senecio triangularis	11/T	33/T	9/T	—	100/15	13/T	—	—	—	—
Viola glabella	11/T	—	—	—	100/18	13/T	13/T	—	—	—
Veratrum spp.[d]	—	—	—	11/T	100/51	13/T	—	—	—	—
Heracleum lanatum	—	—	—	—	50/28	38/T	50/T	—	—	—
Rudbeckia occidentalis	—	—	—	—	50/4	50/3	—	17/T	—	—
Pteridium aquilinum	—	—	—	—	25/4	88/15	100/80	17/1	—	33/1
Lathyrus nevadensis	—	—	—	—	75/1	100/4	100/4	67/3	75/1	33/T
Cirsium callilepes	—	—	—	—	—	100/2	100/1	83/T	—	—
Erigeron aliceae	11/T	—	—	—	—	25/1	100/T	83/22	—	33/1
Achillea millefolium	—	—	—	11/4	—	13/T	—	67/1	100/6	33/1
Gilia capitata	—	—	—	—	—	13/T	—	—	100/14	33/T
Sanicula graveolens	—	—	—	—	—	—	50/1	33/T	75/1	100/5
Castilleja hispida	—	—	—	—	—	—	—	17/T	100/2	—
Eriophyllum lanatum	—	—	—	—	—	—	—	—	75/16	—
Penstemon procerus	—	—	—	—	—	—	—	—	50/1	100/41

[a] CASI = *Carex sitchensis;* ELPA= *Eleocharis pauciflora;* DECA = *Deschampsia caespitosa;* VERAT–SETR = *Veratrum* spp.–*Senecio triangularis;* RUPA = *Rubus parviflorus;* PTAQ–ELGL = *Pteridium aquilinum–Elymus glaucus;* CAPE–BROMU = *Carex pensylvanica–Bromus* spp.; ERLA–GICA = *Eriophyllum lanatum–Gilia capitata;* PEPR–SAGR = *Penstemon procerus–Sanicula graveolens.*
[b] T = trace (<0.5% cover); dash (—) = absent.
[c] *Bromus carinatus* and *B. sitchensis.*
[d] *Veratrum californicum* and *V. viride.*
Source: Adapted from Halpern et al. (1984).

mesic hillslopes by *Abies grandis* appears linked to cessation of sheep grazing in the 1940s, although it also coincides with a regional climatic shift to cooler and wetter springs and summers (Miller and Halpern *in press*).

Abies magnifica var. *shastensis* Forests

Forests characterized by *Abies magnifica* var. *shastensis* (Shasta red fir) dominate the upper portion of the montane zone in the southern Cascade Range (to about 44° N latitude) and the Klamath Mountains. These forests typically occur between 1600 and 2000 m elevation in the Cascades and 1800 and 2200 m in the Klamath Mountains. These represent northern extensions of the Sierra Nevada *Abies magnifica* forests (see Chapter 5).

Typical associated tree species are *Abies concolor* at lower elevations and *Tsuga mertensiana* at higher elevations in the zone. *Pinus monticola* and *P. contorta* are other common associates. Many other tree species may occur sporadically. The Klamath endemic, *Picea breweriana*, is noteworthy and may be a local dominant.

The composition and density of understory vegetation in *Abies magnifica* var. *shastensis* stands vary widely. Many stands have relatively depauperate understories; under dense canopies the understory may consist solely of small ericads and orchids, many of which are nearly or completely achlorophyllous – the "*Pirola-Corallorrhiza* union" of Oosting and Billings (1943). On moist sites, Shasta red fir forests typically have luxuriant herbaceous understories incorporating grasses, sedges, and forbs, many of which are also found in associated mountain meadows. Tall ericaceous shrubs, such as *Vaccinium ovalifolium*, may also be present. On colder, more stressful sites, mixed forests of *Abies magnifica* var. *shastensis* and *Tsuga mertensiana* are characterized by depauperate understories of *Vaccinium scoparium*.

Abies magnifica var. *shastensis* appears to be the major climax species. *Abies concolor* is a climax associate on more productive sites and replaces red fir at lower elevations. *Tsuga mertensiana* is a climax associate on colder sites. *Pinus contorta* and *P. monticola* are primarily seral species.

Tsuga mertensiana Forests

These forests occupy the coldest and snowiest forest zone in the Pacific Northwest. It is a true subalpine environment, with deep, persistent winter snowpacks: 400–1400 cm of snowfall are typical, accumulating in snowpacks up to 7.5 m in depth. The sharp increase in snow accumulation reflects the typical elevation of the freezing isotherm during the winter (Brooke et al. 1970). Temperatures are cool, but much warmer during winter months than in comparable zones in more continental regions.

The *Tsuga mertensiana* zone extends throughout the Cascade Range, the British Columbia Coastal Ranges, the Olympic Mountains, the Klamath Mountains, and two-thirds the length of the Sierra Nevada (see Chapter 5). These forests occur between 1250 and 1850 m in central Washington and 1700–2000 m in southern Oregon. Throughout much of its distribution, the zone can be divided into a lower subzone of closed forest and an upper parkland subzone. Parklands are considered in the following section.

Tsuga mertensiana provides the continuity for this latitudinally extensive forest formation. Forests of essentially pure *Tsuga* are best developed in the central and southern Oregon Cascade Range, where extensive, undulating topography occurs at subalpine elevations. *Abies amabilis* co-dominates stands from central Oregon to central British Columbia and in the Olympic Mountains, except at the highest elevations. Other common associates are *Pinus contorta, P. monticola, P. albicaulis, Picea engelmannii,* and *Abies lasiocarpa. Chamaecyparis nootkatensis* is a distinctive associate from northern Oregon northward. *Abies procera* and *Abies magnifica* var. *shastensis* occur with *Tsuga* north and south, respectively, of the McKenzie River in Oregon.

Communities vary dramatically in their diversity. Forests with *Abies amabilis* as an associate typically have greater species and structural diversity (Dyrness et al. 1974; Franklin et al. 1988; Henderson et al. 1989). Communities dominated by understories of *Xerophyllum tenax* or *Vaccinium membranaceum* or both are common in northern Oregon and southern Washington. Well-developed tall shrub layers composed of species such as *Menziesia ferruginea, Vaccinium ovalifolium, Vaccinium alaskaense,* and *Rhododendron albiflorum* are more common to the north, as are forests with relatively lush understories of dicotyledonous herbs, ferns, and shrubs such as *Oplopanax horridum*.

Tsuga mertensiana stands in the southern Cascade Range are typically depauperate (Fig. 4.7). *Tsuga mertensiana–Vaccinium scoparium* is the most common community type. Stands near the forest line often have a nearly monospecific understory of *Luzula hitchcockii*.

Succession can be very slow following wildfire or other disturbance in the *Tsuga mertensiana* zone. Early successional communities usually are dominated by surviving taxa such as *Xerophyllum tenax* or *Vaccinium*. Reestablishment of closed forest may

Figure 4.7. Pure stands of Tsuga mertensiana *are very common in the High Cascades of Oregon, but often have very depauperate understories typified by* Vaccinium sco-parium; *Torrey-Charlton Research Natural Area (photo courtesy USDA Forest Service).*

take a century or more (Hemstrom and Franklin 1982; Franklin et al. 1988). Successional sequences of tree species vary geographically, with *Pinus* species common in the southern half of the zone. *Abies lasiocarpa*, a major climax species in the Rocky Mountains, is apparently seral when associated with *Tsuga mertensiana* (Franklin and Mitchell 1967).

Abies amabilis appears to be the major climax species wherever it occurs within the closed forest subzone of the *Tsuga mertensiana* zone; consequently, the *Abies amabilis* series dominates much of the zone. *Tsuga* reproduction is typically much less abundant than that of *Abies amabilis. Tsuga* and *Chamaecyparis nootkatensis* can be minor climax associates. In the southern Cascade Range and Klamath Mountains, where more shade-tolerant tree species are absent, *Tsuga* is the major climax species, and the *Tsuga mertensiana* series is characteristic.

Subalpine Parkland

The subalpine meadow-forest mosaic, or parkland, is one of the most distinctive features of the mountains of the Pacific Northwest (Fig. 4.8). Deep, late-lying snowpacks are believed to be responsible for the 300- to 400-m-wide elevational ecotone that defines the upper subzone of the *Tsuga mertensiana* zone. The variety and richness of the meadow flora and communities make the parkland attractive to scientists and laypeople alike (Brooke et al. 1970; Douglas 1970, 1972; Kuramoto and Bliss 1970; Henderson 1973; Halpern et al. 1984).

Subalpine parklands attain maximal development in the Olympic Mountains and Cascade Range of Washington. Forest patches and tree groups present within the parkland may have a composition different from that of the closed forest subzone (Franklin and Dyrness 1988). The non-forested communities are numerous and varied

Figure 4.8. An attractive parkland, or mosaic of meadow and tree communities, typifies the upper portion of the Tsuga mertensiana zone in the Olympic Mountains and Cascade Range; Butter Creek Research Natural Area, Mount Rainier National Park, Washington (photo courtesy USDA Forest Service).

Table 4.8. *Mean prominence values (average percentage cover multiplied by square root of frequency) for selected shrubs and herbs in subalpine meadow communities in the northern Washington Cascade Range.*

Species	Communities[a]					
	CAME–PHEM	VASI–VEVI	CASP	CANI	LUPE	SATO
Cassiope mertensiana	441	—[b]	—	11	11	T
Phyllodoce empetriformis	386	—	—	2	15	1
Vaccinium deliciosum	92	—	T	T	21	1
Luetkea pectinata	73	—	2	21	502	—
Lycopodium sitchense	16	—	—	T	16	—
Deschampsia atropurpurea	9	T	T	15	44	1
Polygonum bistortoides	1	8	30	T	9	—
Valeriana sitchensis	5	305	15	—	47	T
Carex spectabilis	1	52	782	6	41	2
Veratrum viride	—	290	2	—	T	—
Lupinus latifolius	4	59	42	—	—	—
Carex nigricans	2	—	16	803	13	1
Epilobium alpinum	—	T	T	32	10	3
Hieracium gracile	3	—	1	3	36	4
Luzula wahlenbergii	—	—	—	2	10	47
Potentilla flabellifolia	—	4	7	1	13	—
Castilleja parviflora	T	—	—	T	16	4
Anemone occidentalis	T	—	2	—	11	T
Saxifraga tolmei	—	—	—	T	—	78

[a]CAME–PHEM = *Cassiope mertensiana–Phyllodoce empetriformis;* VASI–VEVI = *Valeriana sitchensis–Veratrum viride;* CASP = *Carex spectabilis;* CANI = *Carex nigricans;* LUPE = *Luetkea pectinata;* SATO = *Saxifraga tolmei.*
[b]T = trace amounts; dash (—) = absent.
Source: Douglas (1972).

and largely reflect snowpack duration (e.g., Douglas 1972; Evans and Fonda 1990) (Fig. 4.8). Meadows can typically be assigned to one of five broad groups (Henderson 1973; Douglas 1970, 1972): (1) *Phyllodoce–Cassiope–Vaccinium*, the heath shrub or heather-huckleberry group; (2) *Valeriana sitchensis–Carex spectabilis*, the lush herbaceous group; (3) *Carex nigricans*, the dwarf sedge group; (4) the pioneer and low herbaceous group, and (5) *Festuca viridula*, the grass or dry grass group.

Exemplary data are provided for communities belonging to each of these types in Table 4.8. Heather-huckleberry communities are typically dominated by some combination of *Phyllodoce empetriformis*, *Cassiope mertensiana*, and *Vaccinium deliciosum*. These communities are closely related to heather types of the lower alpine zone. Lush herbaceous types are typically dominated by showy herbs greater than 1 m tall, such as *Valeriana sitchensis* and *Veratrum viride*, as well as the robust *Carex spectabilis*. *Carex nigricans* occupies sites with late-lying snowpacks and cold, wet soils. These communities are characterized by nearly pure, dense mats of this short sedge. A series of diverse, poorly developed pioneer communities compose

the early pioneer group. *Saxifraga tolmei, Luetkea pectinata,* and *Antennaria lanata* are typical dominants, but total plant cover is low. Grassy meadows dominated by *Festuca viridula* or *F. idahoensis* characterize interior subalpine parklands but also occur in rainshadow regions of the coastal mountains. These communities typically contain significant development of forbs (e.g., *Lupinus latifolius, Potentilla flabellifolia, Polygonum bistortoides, Ligusticum grayi, Anemone occidentalis,* and *Aster ledophyllus*).

Subalpine parklands to the south (Oregon Cascade Range) support many similar meadow assemblages, although community distributions and compositions vary somewhat in response to the warmer, drier climate; the deep, well-drained pumice soils that characterize the Oregon High Cascades; and the presence of taxa with more southerly affinities (see Table 4.9). As in subalpine parklands to the north, community composition reflects the depth and duration of snowpack, as well as the availability of soil moisture late in the growing season (Campbell 1973; Halpern et al. 1984). South-facing slopes and other landforms that emerge from snow early in the summer support

Table 4.9. The constancy (% of plots sampled)/mean cover (%) of species that characterize selected meadow communities in the subalpine zone, central Oregon Cascade Range. Communities represent warmer/drier to colder/wetter sites (left to right).

Growth-form/species	FEVI	PONE	PHEM–CAME	CASP–LULA	DECA–CASP	CANI	CASC
Trees							
Abies lasiocarpa	38/T[b]	—	—	—	—	—	—
Tsuga mertensiana	13/T	—	50/1	50/T	—	—	20/T
Shrubs/subshrubs							
Eriogonum umbellatum	50/1	67/4	25/1	—	—	—	—
Eriogonum pyrolaefolium	38/T	67/6	50/1	50/T	—	—	—
Phyllodoce empetriformis	—	—	100/39	100/T	50/T	50/1	20/T
Cassiope mertensiana	—	—	75/30	—	—	—	—
Luetkea pectinata	—	—	75/5	100/T	—	10/T	—
Vaccinium occidentale	—	—	—	—	50/T	1/T	20/T
Salix commutata	—	—	—	—	—	50/8	40/1
Kalmia microphylla	—	—	—	—	—	30/6	20/1
Graminoids							
Festuca viridula	100/47	50/3	—	—	—	—	—
Carex pensylvanica	75/6	—	—	—	—	—	—
Sitanion hystrix	25/1	83/3	—	50/T	—	—	—
Juncus parryi	38/T	50/1	100/5	100/T	25/T	40/T	20/T
Danthonia californica	25/1	33/3	25/T	50/T	50/6	20/1	40/T
Carex spectabilis	13/1	—	50/T	100/51	75/35	30/6	—
Carex nigricans	—	17/T	50/3	100/2	50/5	100/39	60/9
Deschampsia caespitosa	—	—	—	—	100/26	40/3	80/24
Muhlenbergia filiformis	—	—	—	—	50/1	10/1	80/10
Carex scopulorum	13/1	—	—	—	25/3	60/9	100/45
Eleocharis pauciflora	—	—	—	—	—	30/4	100/16
Forbs							
Lomatium martindalei	13/T	83/2	25/T	—	—	—	—
Spraguea umbellata	38/T	83/2	25/T	50/1	—	—	—
Calochortus subalpinus	75/2	17/T	—	—	—	10/T	—
Lupinus latifolius	75/9	17/T	25/T	100/33	—	20/T	20/T
Polygonum newberryi	63/3	100/11	75/2	100/2	—	30/T	—
Microseris alpestris	75/2	100/1	75/2	100/2	—	30/1	—
Ligusticum grayi	38/T	—	75/6	100/1	75/5	30/T	60/T
Aster alpigenus	25/T	33/T	100/5	—	50/9	80/19	80/21
Potentilla flabellifolia	—	17/T	50/1	50/1	100/19	60/7	20/1
Trifolium longipes	—	17/T	—	—	100/25	10/T	80/5
Veronica wormskjoldii	—	—	—	—	100/T	—	40/T
Senecio triangularis	—	—	—	—	100/T	—	20/T
Dodecatheon jeffreyi	—	—	—	—	75/1	—	60/1
Ranunculus alismaefolius	—	—	—	—	100/24	20/5	80/7

[a]FEVI = *Festuca viridula;* PONE = *Polygonum newberryi;* PHEM–CAME = *Phyllodoce empetriformis–Cassiope mertensiana;* CASP–LULA = *Carex spectabilis–Lupinus latifolius;* DECA–CASP = *Deschampsia caespitosa–Carex spectabilis;* CANI = *Carex nigricans;* CASC = *Carex scopulorum.*
[b]T = trace (<0.5% cover); dash (—) = absent.
Source: Adapted from Halpern et al. (1984).

Festuca viridula communities similar to those of interior parklands to the north. These commonly grade downslope into broad basins or "pumice plains" that support sparsely vegetated (<50% cover) communities of *Polygonum newberryi, Eriogonum* spp., and other showy forbs. Typically pros-trate, drought tolerant, and/or deeply rooted, these taxa are well suited to the coarse-textured soils, high soil surface temperatures, and drying winds that are common in these exposed environments. Heath-shrub communities (*Phyllodoce empetriformis–Cassiope mertensiana*), best developed in the northern Cascades, occupy gentle slopes, benches, and forest openings of varying exposures. They commonly intergrade with dense ground layers of *Vaccinium* spp. at the margins of forest and meadow. The subshrub *Luetkea pectinata* is a common associate. On benches and in basins, where snowmelt is delayed or soil moisture is available late into the growing season, are lush, graminoid-dominated communities of *Carex spectabilis* and *Deschampsia caespitosa*. The latter are very similar in physiognomy and composition to those of the upper montane zone. As in the northern Cascades, basins with late-lying snow and cold wet soils support dense and depauperate mats of the short-statured sedge, *Carex nigricans*. Where there is standing water or perennial surface flows, stream channels, or seeps, hydric communities develop, dominated by *Carex scopulorum*. Many graminoids and forbs found in hydric basins of the montane zone (e.g., *Eleocharis pauciflora*, *Muhlenbergia filiformis*, *Dodecatheon jeffreyi*) find their elevational limits here.

Throughout the subalpine zone, succession from meadow to forest is thought to be controlled by climate, largely through the effects of deep, persistent snowpacks on tree establishment. Widespread invasion of subalpine meadows by trees (primarily *Abies lasiocarpa* and *Tsuga mertensiana*) earlier in the century has been attributed to regional warming (e.g., Franklin et al. 1971; Heikkinen 1984; Rocheforte et al. 1994; Taylor 1995), although this pattern is limited to colder sites with deeper snowpacks (primarily within *Phyllodoce–Cassiope* communities). In contrast, recent invasion of more interior environments (e.g., the eastern portions of Mount Rainier and Olympic National Parks) or of warmer, drier, south-facing slopes (Three Sisters Biosphere Reserve) has occurred when climatic conditions have been cooler and wetter (Woodward et al. 1995; Rocheforte and Peterson 1996; Miller and Halpern *in press*). Unlike patterns in the montane zone, historical changes in grazing regime appear to have played a relatively minor role in the dynamics of subalpine parklands (Miller and Halpern *in press*).

DISTURBANCE PATTERNS

Wildfire and wind are the major natural disturbances in the forests of the Pacific Northwest. The tendency toward infrequent catastrophic disturbances in these forests contrasts with patterns of more frequent, noncatastrophic wildfires typical of coniferous forests in the Sierra Nevada and Rocky Mountains (Agee 1993). Insect outbreaks are also of lesser importance in the northwestern coastal

forest types. There are some broad gradients in disturbance patterns within the region. Wind increases in relative importance from interior to coastal regions. Fire increases in frequency from north to south.

Large, intense, and infrequent forest fires appear to be the most important natural agent of forest destruction in the Pacific Northwest. Evidence includes the extensive acreages of comparable forest age classes and records of forest fires dating from the early 1800s. An analysis of fire patterns in Mount Rainier National Park, Washington (Hemstrom and Franklin 1982) identified 16 important fire events since the year 1230 and a natural fire rotation of 434 yr. The largest episode occurred in 1230 and affected >47% of the park area. The Yacholt Burn in southwestern Washington (initial fire in 1902) and Tillamook Burn in coastal Oregon (initial fire in 1933) are historical demonstrations that fires can cover thousands of hectares in a very short time when conditions are appropriate. Trees frequently survive within such burns as individuals, groups and small stands, and so revegetation, including reestablishment of trees, can be rapid. Such burned sites have a tendency to reburn in the century following the initial fire, however, and repeatedly burned sites reforest slowly because of reduced seed supplies (fewer survivors), increased competing vegetation, and a more severe physical environment for tree seedlings. Large-scale deforestation can drastically modify macroclimate, as well as microclimate, by reducing effective precipitation in regions where condensation in tree crowns is a significant process (Harr 1982). Repeated wildfires are believed to be one factor contributing to the wide range in age classes typical of old-growth stands of *Pseudotsuga* (Franklin and Hemstrom 1981).

The frequencies of wildfires, both light and catastrophic, vary both locally and regionally (Agee 1993). Natural fire rotation is shorter, for example, in the central Oregon Cascade Range than at Mount Rainier. Noncatastrophic wildfires do occur, creating gaps and thinning forest canopies; these can result in very complex forest structures (Stewart 1986). More frequent, lower-intensity wildfires apparently increase with decreasing latitude, and fire regimes south of the Willamette–Umpqua River divide begin to approach those of the Sierra Nevada. Partial burns are important factors in maintaining *Pseudotsuga menziesii* as a part of coastal *Sequoia sempervirens* stands (Viers 1982).

Wind is an important catastrophic agent. Forests in coastal *Picea sitchensis–Tsuga heterophylla* forests in southeastern Alaska and coastal British Columbia are typically initiated by major windstorms. Catastrophic windstorms occur in other areas as well. Henderson et al. (1989) have reported that windthrow, rather than fire, has been the most important agent of disturbance on the western Olympic Peninsula. A particularly destructive storm occurred on the peninsula on January 29, 1921 (Morgenroth 1991). Ruth and Yoder (1953) have documented catastrophic windstorms on the Oregon coast. The Columbus Day (October 12, 1962) windstorm affected both the Coast and Cascade Ranges of Oregon and Washington, destroying several billion board-feet of timber.

Wind also acts as a chronic agent of disturbance, creating small- to moderate-sized gaps within intact forest patches (e.g., Spies et al. 1990; Taylor 1990; Lertzman 1992; Gray 1995; Van Pelt 1995). Wind often interacts with fungal diseases, affecting roots, butts, and lower boles as a part of the gap-creating process. Gaps are important features of late-successional forests (e.g., Spies and Franklin 1996) and typically persist for many decades, providing for habitat and species diversity (e.g., Lertzman 1992; Van Pelt 1995). There are strong regional patterns in wind-related mortality: It accounts for approximately 80% of the total mortality in coastal *Picea sitchensis–Tsuga heterophylla* forests, 40% in *Pseudotsuga–Tsuga heterophylla* forests of the Cascade Range, and less than 15% in *Pinus ponderosa* forests east of the Cascade Range (J. F. Franklin, unpublished data).

Pathogens can create significant disturbances in some situations, but generally they are not as important as they are in many other western coniferous forests (Franklin et al. 1987). Outbreaks of bark beetles typically kill only individuals or groups of trees. Outbreaks of defoliators are uncommon and rarely appear to threaten stands. Several diseases may create patch-wise mortality (e.g., *Phellinus wierii*) but leave stands basically intact. Several introduced pathogens (a root rot on *Chamaecyparis lawsoniana*, an aphid on several species of *Abies*, and blister rust on five-needled pines) have seriously affected individual species but, again, have rarely eliminated entire stands. Thus, pathogens appear to be chronic, rather than catastrophic, agents of disturbance.

Geomorphic processes are also key elements of the disturbance regime in Pacific Northwest forests; of primary importance are the interactions between forests and fluvial processes (Swanson and Lienkaemper 1982), landslides and earthflows (Miles and Swanson 1986), and volcanic eruptions. The May 18, 1980, eruption of Mount St. Helens, Washington, provided an exceptional laboratory for studying the varied effects of disturbance intensity, pre-eruption plant composition, erosion,

and snow on forest recovery (e.g., Antos and Zobel 1985a, 1985b; Franklin et al. 1985; Halpern et al. 1990; Franklin et al. 1995).

The types and intensities of disturbances have varying effects on the paths and rates of succession and on ecosystem structures and processes. For example, fire in the forests of the Pacific Northwest initially produces large numbers of standing dead trees and tends to kill from below (i.e., killing smaller and less fire-resistant individuals), thereby favoring regeneration of *Pseudotsuga menziesii* or other pioneer species. Windthrow, on the other hand, generates downed logs, rather than snags, and tends to eliminate larger individuals, leaving most of the seedlings and saplings of shade-tolerant species untouched. Therefore, windthrow accelerates succession toward the climax tree species. Because snags and logs fulfill different wildlife functions and decay at different rates (snags decompose three to four times as fast as logs of comparable size in the Pacific Northwest forests), wildfire and windstorms also have significantly different impacts on wildlife populations and nutrient and energy cycling (Harmon et al. 1986).

Clear-cutting has been the most common current agent of disturbance in Pacific Northwest forests, and it differs markedly from wildfire in its effects (Kohm and Franklin 1997). Considerable research has been conducted on succession following clear-cutting (e.g., Dyrness, 1973 Halpern 1988, 1989; Schoonmaker and McKee 1988; Halpern and Franklin 1990), and on the effects on erosion and nutrient losses (e.g., Sollins and McCorison 1981; Feller and Kimmins 1984; Martin and Harr 1989). Rates and paths of succession are altered by planting and other cultural practices, such as elimination of nonarboreal species. Furthermore, logging usually removes most snags and logs, eliminating their potential functional roles.

Generalized successional relationships have been discussed in the sections on individual forest types. Additional interpretations can be made for individual tree species utilizing relative shade tolerances (Table 4.1). Two cautionary notes are essential, however. First, shade tolerance is a physiological feature of each species, whereas successional role is dependent on a community context. For example, *Pinus ponderosa* is a shade-intolerant tree species, but it may play either a seral or climax role, depending on whether or not more shade-tolerant tree species are capable of growing on a specific site. In general, shade tolerance is required for climax status on sites that can develop closed canopies. Second, few climax tree species require that a seral species precede to ameliorate site conditions. Hence, most climax species can also function as pioneers, although they may be unable to compete with faster-growing seral species early in succession. As an example, *Tsuga heterophylla* can dominate young stands following fire or clear-cutting over much of the Pacific Northwest, but *Pseudotsuga menziesii* is more common.

Much attention is currently focused on relations between forest types and successional stages and use by vertebrates (Brown 1985). Both early (before tree canopy closure) and late (old-growth) stages in succession have been identified as periods of special interest because of higher levels of diversity and/or of special-interest species than are found in young forest stands. Several bird and mammal species find optimum habitats in old-growth *Pseudotsuga* forests, for example, and require special management consideration (Franklin et al. 1981; Harris 1984; Ruggiero et al. 1991). One key to the special role of old-growth forests is their structural complexity compared with younger forest ecosystems.

ECOSYSTEM CHARACTERISTICS

Biomass

The outstanding structural feature of forests in the Pacific Northwest is the huge biomass accumulation typically present (Franklin and Dyrness 1988), which results from long-lived species capable of growing to very large sizes and from high productivity. Values for above-ground live biomass are typically in the range of 500–1000 Mg ha^{-1} (Table 4.10), exceeding values for temperate deciduous forests and tropical rain forests by factors of 2 to 4 (Franklin and Waring 1981). *Sequoia sempervirens* forests hold the world record for maximum biomass, with a basal area of 343 m^2 ha^{-1} and a stem biomass of 3461 Mg ha^{-1} (Fujimori 1977); with addition of branch, leaf, and root biomass, the estimate of standing crop would probably approach Fujimori's (1972) estimate of 4525 Mg ha^{-1}. Although maximum values for *Pseudotsuga menziesii* and *Abies procera* are less than half those for *Sequoia* (Fujimori et al. 1976), they still greatly exceed maxima for any other forests in the world.

Foliage is a particularly important biomass component. In addition to the environment, foliage quantity (Waring 1983) and foliage morphology (Leverenz and Hinckley 1990; Brooks et al. 1994) are largely responsible for the generally high net productivities of Pacific Northwest forests. On well-developed soils with adequate moisture and drainage, very high quantities of individual tree and stand foliage are found. Nine old-growth *Pseudotsuga men-*

ziesii trees averaged 2850 m^2 total needle surface area per tree (Massman 1982). Stands may support foliage masses in excess of 3 kg m^{-2} or projected foliage areas of 12 m^2 m^{-2}. Leaf surface areas are strongly related to site water balance (Grier and Running 1977; Gholz 1982) and to temperature regimes (Waring et al. 1978). In addition, shade-tolerant conifers have a number of morphological features (e.g., a highly plastic shoot and needle structure) and physiological traits (acclimation to sun and shade conditions) that allow them to function within environments of both high and low light.

Except for some of the faster-growing deciduous hardwoods (e.g., *Alnus rubra* or *Populus tricho-carpa*), the majority of the tree species in these forests take a relatively long time to accumulate, on both an individual tree and stand basis, large quantities of foliage. Foliage losses due to disturbances, even during relatively mild winters, can be high (Grier 1988). Leaf biomass and surface area can require decades to recover to former levels following moderate to severe disturbances, a very slow rate compared with forests of the eastern United States, where equilibria may be reached in less than a decade. Because foliage drives system productivity, its temporal and spatial dynamics are critical in understanding many ecosystem processes.

Coarse woody debris, primarily standing dead trees and downed logs, is increasingly recognized as an important organic structure in forests and streams, particularly in the Pacific Northwest (Franklin et al. 1981; Maser and Trappe 1984; Maser et al. 1988). This woody debris plays significant roles in energy and nutrient cycling, geomorphic processes, and provision of habitat for terrestrial and aquatic organisms (Harmon 1986; Harmon et al. 1986). Large masses of such material are typically present in natural forests of all ages, because few catastrophes consume or remove much wood from trees that are killed. Tonnages in old-growth stands average 75–100 Mg ha^{-1} and may range to more than 500 Mg ha^{-1} (Franklin and Waring 1981; Spies et al. 1988).

Productivity and Nutrient Cycling

Productivity of forest stands in the Pacific Northwest is generally comparable to that in other temperate forest regions (Table 4.11). Biomass in young stands accumulates at 15–25 Mg ha^{-1} yr^{-1} in fully stocked stands on better than average sites. Mature and old-growth stands have lower net productivities. On the best sites, particularly in *Sequoia* and *Tsuga–Picea* stands, annual net productivity is as high as at any place on earth (Fujimori

Table 4.10. *Aboveground total biomass, leaf biomass, and projected leaf area (one side only) for four forest types in the Pacific Northwest.*

Forest type and age class	Number of stands	Aboveground biomass		Leaf mass (Mg ha^{-1})	Projected leaf area (m^{2}m^{-2})
		Average (Mg ha^{-1})	Range (Mg ha^{-1})		
Pseudotsuga menziesii (70–170 yr)	10	604	422–792	19	9.7
Pseudotsuga menziesii- Tsuga heterophylla	19	868	317–1423	23	11.7
Picea sitchensis- Tsuga heterophylla	4	1163	916–1492	21	13.2
Abies procera	1	880	—	18	10

Source: Franklin and Waring (1981).

Table 4.11. *Above-ground net primary production estimates for forests west of the Cascade Range crest in Oregon and Washington.*

Forest type	Stand age (yr)	Biomass (Mg ha^{-1})	Net primary production (Mg ha^{-1} yr^{-1})	Source
Abies amabilis	23	49	6.4[a]	Grier et al. (1981)
Tsuga heterophylla	26	192	36.2	Fujimori (1971)
Pseudotsuga menziesii	40	248, 467	7.3, 13.7[b]	Keyes and Grier (1981)
Pseudotsuga menziesii- Tsuga heterophylla	100	661	12.7	Fujimori et al. (1976)
Tsuga heterophylla- Picea sitchensis	110	871	10.3	Fujimori et al. (1976)
Abies procera- Pseudotsuga menziesii	115	880	13.0	Fujimori et al. (1976)
Pseudotsuga menziesii	125	449	6.2	Gholz (1982)
Picea sitchensis- Tsuga heterophylla	130	1080, 1492	15, 13	Gholz (1982)
Pseudotsuga menziesii- Tsuga heterophylla	150	527, 865	9.5, 10.5	Gholz (1982)
Abies amabilis	180	446	4.6[a]	Grier et al. (1981)
Pseudotsuga menziesii- Tsuga heterophylla	450	718	10.8	Grier and Logan (1977)

[a]Total net primary production was 18.3 and 16.8 Mg ha^{-1}yr^{-1} in young and old stands, respectively; below-ground production accounted for 65 and 73%, respectively, of those totals.
[b]Total net primary production was 15.4 and 17.8 Mg ha^{-1} on high- and low-quality sites; below-ground production accounted for 65 and 73%, respectively, of those totals.

1971, 1977). A record net annual production of 36.2 Mg ha^{-1} yr^{-1} occurred in a 26-yr-old *Tsuga heterophylla* stand on the Oregon coast.

Such high productivities are exceptional, however. Nonetheless, although the peak productivity of other mesic, temperate forests (e.g., eastern deciduous and southeastern coniferous forests) is comparable with, or exceeds that of Pacific Northwest forests, what makes coastal northwestern forests unique is their ability to sustain near-maximum rates of net primary production over much longer periods. Greater biomass accumulation in northwestern forests reflects sustained height growth and longevity of the dominant tree species. This growth is aided by the trees' ability to accumulate and maintain a large amount of foliage. Trees of northwestern species continue to grow substantially in diameter and height, and stands continue to increase in biomass, long after forests in other temperate regions have reached equilibrium. This is well illustrated by comparing growth of *Pinus taeda* (loblolly pine) in the Southeast and *Pseudotsuga* in the Northwest. Wood production from a single 100-yr rotation of *Pseudotsuga* is about 22% greater than from two 50-yr rotations of pine (Worthington 1954). Recent studies of height growth patterns of several northwestern conifers show that height growth may be sustained into their second and third centuries.

Patterns of nutrient cycling have been described for several northwestern forest types (Sollins et al. 1980; Edmonds 1982). Typical and important features of these cycles are very large nutrient pools and the "tightness" of the forests, as indicated by low nutrient losses. Winter decomposition is important because of mild temperatures and moisture limitations on decomposition processes during the relatively dry summer season. Nitrogen is generally considered the limiting nutrient on most sites; studies of sources and losses of nitrogen are focal areas of research. Numerous sources for nitrogen fixation have been identified, including (1) shrubs and trees with nitrogen-fixing symbionts such as *Alnus, Ceanothus*, and *Purshia tridentata* (Tarrant et al. 1967; Conard et al. 1985); (2) canopy lichens with blue-green algal associates (Carroll 1980); (3) microbial organisms in rotting wood (Harmon et al. 1986); (4) organisms living in the rhizosphere; and (5) free-living organisms associated with decaying leaf litter.

Research on hydrologic cycling has focused on effects of forest cutting on stream flow and water quality (e.g., Rothacher 1970; Brown et al. 1973; Harr 1986; Martin and Harr 1989). Forest removal typically results in increased water yields, particularly during summer low-flow periods, as a consequence of reduced transpirational losses. Condensation of fog or cloud moisture in tree canopies results in substantial amounts of fog drip in some coastal and mountain forests, however, and forest cutting may reduce water yields under such circumstances. In one study in the Oregon Cascade Range, for example, fog drip added 30%, or 88 cm, of precipitation to the 216 cm received in the open (Harr 1982). The deep crowns and large surface areas of needles, lichens, twigs, and branches (Pike et al. 1977) found in old-growth forests make them particularly effective as condensing and precipitating surfaces for moisture, nutrients, and pollutants.

PALEOECOLOGICAL CONSIDERATIONS

Paleoecological research on the vegetation of the Pacific Northwest includes studies of fossil floras (Chaney 1956; Axelrod 1976), pollen profiles (Hansen 1947; Heusser 1960; Baker 1983; Heusser 1983), glacial records (Burke and Birkeland 1983; Porter et al. 1983; Waitt and Thorson 1983), volcanic-ash depositions (Mullineaux 1974; Sarna-Wojcicki et al. 1983), and tree-ring records (Sigafoos and Hendricks 1972; Brubaker and Cook 1983; Graumlich and Brubaker 1986; Ettl and Peterson 1995a, 1995b). These approaches have been particularly successful in reconstructing Quaternary vegetational history. Some interesting linkages between native peoples and vegetation development have been discovered, such as between *Thuja* expansion and evolution of a woodworking technology (Hebda and Mathewes 1984).

Daubenmire (1978) has summarized vegetational development up to the Quaternary. The Arcto-Tertiary geoflora was an important ancestral formation. This flora comprised a widespread and complex temperate forest in the warm period at the close of the Eocene. The mixed hardwood and coniferous forests included *Abies, Chamaecyparis, Calocedrus, Picea, Pinus, Pseudotsuga*, and *Tsuga*. Cooling and the rise of mountain ranges during the Oligocene, Miocene, and Pliocene resulted in development of a xerophytic flora, northward expansion and incorporation of some elements of the Madro-Tertiary geoflora, and loss of most of the hardwood tree species and genera.

Both continental and alpine glaciations were important during the Pleistocene. The continental ice sheet occupied the Puget Trough to a few kilometers south of Olympia, Washington, and affected additional areas by creating outwash plains, channels, lakes through damming of river valleys (Waitt and Thorson 1983). The maximum extent of the continental ice sheet during the Fraser Glaciation was achieved at 22,000–18,000 BP. Alpine glaciation was extensive in both the Olympic Mountains and Cascade Range (Burke and Birkeland 1983; Porter et al. 1983). The histories of the glaciations are complex, and the patterns and extents of recent glaciations vary substantially among mountain ranges (Burke and Birkeland 1983). There have been numerous studies of glacial fluctuations in the Pacific Northwest post–1800 CE.

Reconstructions of the more recent climate and vegetation histories are based primarily on pollen records and plant macrofossils (Tsukada et al. 1981; Leopold et al. 1982; Thompson et al. 1993; Whitlock 1993). Deglaciation occurred between 14,000 and 10,000 yr BP and resulted in the formation of new plant communities and modifications of existing communities. *Pinus contorta* was abundant in deglaciated areas and in mixed forest communities of low and high elevation conifers, where conditions were suited to forest development.

The early Holocene (10,000–6000 yr BP) was a period of warming characterized by warmer summers, intensified summer drought, and cooler winters than at present. Forests are believed to have been more open and representative of earlier successional stages, characterized by species such as *Pseudotsuga, Alnus rubra,* and *Pteridium aquilinum.*

Modern vegetation patterns were established in the late Holocene (about 6000 yr BP to present). Representation of genera such as *Thuja* and *Tsuga* increased, suggesting cooler, moist conditions and reduced frequency of fire. Glaciological evidence (Burke and Birkeland 1983) and computer simulations also provide evidence for climatic cooling and an increase in moisture during the second half of the Holocene (Barnosky et al. 1987; Thompson et al. 1993). Old-growth forests similar in characteristics to those found today probably developed at the beginning of this period (Brubaker 1991).

CONSERVATION ISSUES

Management of the forests of the Pacific Northwest has been the subject of increasingly intense debate during the last several decades (e.g., Norse 1990; Yaffee 1994; Kohm and Franklin 1997). Major foci have been preservation of primeval forests, maintenance of biological diversity (particularly species associated with old-growth forests), protection of forest-associated aquatic ecosystems, and forest practices that provide alternatives to clear-cutting.

Scientific research on forest ecosystems and their component species has been central to these debates. Old-growth coniferous forests in the Pacific Northwest are increasingly understood as being functionally distinct and structurally complex ecosystems (e.g., Franklin et al. 1981; Franklin and Spies 1991; Spies and Franklin 1991; Franklin 1992). These forests also have high levels of biological diversity, providing habitat for many species with specialized requirements (e.g., Ruggiero et al. 1991; Forest Ecosystem Management Assessment Team 1993).

Relevant research has also considered the ecological effects of patterns at larger spatial scales, from watersheds to landscapes (Franklin 1992). Fragmentation of forests and associated edge effects are receiving increasing attention (e.g., Franklin and Forman 1987; Chen et al. 1992, 1993, 1995; Spies et al. 1994). Riparian and aquatic ecosystems associated with forests have also been the focus of these larger-scale assessments (e.g., Forest Ecosystem Management Assessment Team 1993; Scientific Panel for Sustainable Forestry Practices in Clayoquot Sound 1995; Kohm and Franklin 1997).

Major changes in forest policy are emerging from these debates, and ecologists have played a major role in shaping solutions (e.g., Forest Ecosystem Management Assessment Team 1993; Yaffee 1994; Scientific Panel for Sustainable Forest Practices in Clayoquot Sound 1995; Tuchmann et al. 1996). As one example, 77%, or 7.6 out of 9.8 M ha, of federal forestland in the Pacific Northwest has been withdrawn from timber harvest (Tuchmann et al. 1996); included in this withdrawal are 3 M ha of late successional reserves and more than 1 M ha of riparian reserves.

Other major conservation efforts include additions to the extensive system of ecological reserves (Research Natural Areas) that have been established on federal, state, and private lands. A major obligation of the scientific community is to use these reserved areas for research whenever possible.

AREAS FOR FUTURE RESEARCH

There are many areas of exciting ecological research under way in the forests of the Pacific Northwest, and numerous possibilities for contributing to or benefiting from existing studies and databases. The USDA Forest Service Area Ecology Program, headquartered in Portland, Oregon, has developed an extensive database on forested and other wildland communities found on the national forests. There are outstanding opportunities to use these data for major syntheses of forest community patterns within the region (e.g., Ohmann 1995). Detailed studies of patterns, rates, and mechanisms of succession are needed, including consideration of small-scale disturbances, effects of forest management, and the autecology and population biology of understory plants.

Studies of ecosystem-level processes have only begun, and there are many areas of research that merit further attention. Below-ground and canopy processes, nitrogen dynamics, forest-stream interactions, and the effects of herbivores are all important but underexplored areas of research. The development of structural complexity in forest stands and its implications for biological diversity and ecosystem processes also require investigation. Creation of the Wind River Canopy Crane Research Facility near Stevenson, Washington, has provided a new opportunity for studies of the canopies of 2 ha of old-growth *Pseudotsuga–Tsuga–Thuja* forest.

Long-term studies are essential for understanding population, community, and ecosystem processes. A substantial network of permanent sample plots and benchmark watersheds has been established within the region, with both academic and Forest Service scientists playing central roles.

Long-term research efforts are under way at the H. J. Andrews Experimental Forest in western Oregon (a site supported by the National Science Foundation as part of its Long-Term Ecological Research Program), the Hoh River valley of Olympic National Park in Washington, and the Wind River Canopy Crane Research Facility. It is the collective responsibility of the scientific community to develop, maintain, and use these long-term databases.

In conclusion, the forests of the coastal Pacific Northwest are an extraordinary resource for ecological study by scientists and students. They provide examples of the levels of productivity and massiveness that can be achieved by forest ecosystems, the importance of dead trees and coarse woody debris, and the intense interactions that can occur between forest and stream ecosystems. They provide an interesting and varied contrast with the moist temperate forests found in many other parts of the world. Finally, they inspire with their majesty, beauty, and variety.

REFERENCES

Agee, J. K. 1993. Fire ecology of Pacific Northwest forests. Island Press, Washington, D.C.

Alaback, P. B. 1982. Forest community structural changes during secondary succession in southeast Alaska, pp. 70–79 in J. E. Means (ed.), Forest succession and stand development research in the northwest. Oregon State University Forest Research Laboratory, Corvallis.

Antos, J. A., and D. B. Zobel. 1985a. Recovery of forest understories buried by tephra from Mount St. Helens. Vegetatio 64:103–111.

Antos, J. A., and D. B. Zobel. 1985b. Upward movement of underground plant parts into deposits of tephra from Mount St. Helens. Can. J. Bot. 63:2091–2096.

Antos, J. A., and D. B. Zobel. 1986. Habitat relationships of *Chamaecyparis nootkatensis* in southern Washington, Oregon, and California. Can. J. Bot. 64:1898–1909.

Atzet, T., and D. L. Wheeler. 1984. Preliminary plant associations of the Siskiyou Mountain Province. USDA Forest Service, Pacific Northwest Region, Portland.

Atzet, T. A., and L. A. McCrimmon. 1990. Preliminary plant associations of the southern Cascade Mountain Province. USDA Forest Service, Siskiyou National Forest, Grants Pass.

Atzet, T., D. L. Wheeler, B. Smith, J. Franklin, G. Riegel, and D. Thornburgh. 1992. Vegetation, pp. 92–113 in S. D. Hobbs, S. D. Tesch, P. W. Owston et al. (eds.), Reforestation practices in southwest Oregon and northern California. Oregon State University Forest Research Laboratory, Corvallis.

Atzet, T., D. E. White, L. A. McCrimmon, P. A. Martinez, P. R. Fong, V. D. Randall. 1996. Field guide to the forested plant associations of southwestern Oregon. USDA Forest Service Pacific Northwest Region technical paper R6-NR-ECOL-TP-17–96, Portland.

Axelrod, D. I. 1976. History of the coniferous forests, California and Nevada. University of California Press, Berkeley.

Baker, R. G. 1983. Holocene vegetational history of the western United States, pp. 109–127 in H. E. Wright, Jr. (ed.), Late-Quaternary environments of the United States, Vol. 2, The Holocene. University of Minnesota Press, Minneapolis.

Barnosky, C. W., P. M. Anderson, and P. J. Bartlein. 1987. The Northwestern U.S. during deglaciation: vegetation and climate implications, pp. 289–331 in W. F. Ruddiman and H. E. Wright, Jr. (eds.), North America and adjacent oceans during the last deglaciation. Geological Society of America, The Geology of North America Series, Boulder.

Benedict, N. B. 1983. Plant associations of subalpine meadows, Sequoia National Park, California. Arc. Alp. Res. 15:383–396.

Brockway, D. G., C. Topik, M. A. Hemstrom, and W. H. Emmingham. 1983. Plant association and management guide for the Pacific silver fir zone, Gifford Pinchot National Forest. USDA Forest Service Pacific Northwest Region R6-Ecol-130a-1983, Portland.

Brooke, R. C., E. B. Peterson, and V. J. Krajina. 1970. The subalpine mountain hemlock zone. Ecol. West. North Am. 2:147–349.

Brooks, J. R., T. M. Hinckley, and D. G. Sprugel. 1994. Acclimation responses of mature *Abies amabilis* sun foliage to shading. Oecologia 100:316–324.

Brown, E. R. (ed.). 1985. Management of wildlife and fish habitats in forests of western Oregon and Washington, Part I, chapter narratives. USDA Forest Service, Pacific Northwest Region, Portland.

Brown, G. W., A. R. Gahler, and R. B. Marston. 1973. Nutrient losses after clearcut logging and slash burning in the Oregon Coast Range. Water Resources Res. 9:1450–1452.

Brubaker, L. B. 1991. Climate change and the origin of old-growth Douglas-fir forests in the Puget Sound lowland, pp. 17–24 in L. F. Ruggiero, K. B. Aubry, A. B. Carey, and M. H. Huff (tech. coords.), Wildlife and vegetation of unmanaged Douglas-fir forests. USDA Forest Service general technical report PNW-GTR-285, Portland.

Brubaker, L. B., and E. R. Cook. 1983. Tree-ring studies of Holocene environments, pp. 222–238 in H. E. Wright, Jr. (ed.), Late-Quaternary environments of the United States, vol. 2, The Holocene. University of Minnesota Press, Minneapolis.

Burke, R. M., and P. W. Birkeland. 1983. Holocene glaciation in the mountain ranges of the western United States, pp. 3–11 in H. E. Wright, Jr. (ed.), Late-Quaternary environments of the United States, Vol. 2, The Holocene. University of Minnesota Press, Minneapolis.

Burns, R. M., and B. H. Honkala. 1990. Silvics of North America. Vol. I, Conifers. Vol. II, Hardwoods. USDA agricultural handbook 654, Washington, D.C.

Busing, R. T., C. B. Halpern, and T. A. Spies. 1995. Ecology of Pacific Yew (*Taxus brevifolia*) in western Oregon and Washington. Cons. Biol. 9:1199–1207.

Campbell, A. G. 1973. Vegetation ecology of Hunt's Cove, Mt. Jefferson, Oregon. M. S. thesis, Oregon State University, Corvallis.

Carlton, G. C. 1988. The structure and dynamics of red alder communities in the central coast range of western Oregon. M. S. thesis, Oregon State University, Corvallis.

Carroll, G. C. 1980. Forest canopies: complex and independent subsystems, pp. 87–107 in R. H. Waring (ed.), Forests: fresh perspectives from ecosystem research. Oregon State University Press, Corvallis.

Chaney, R. W. 1956. The ancient forests of Oregon. University of Oregon Press, Eugene.

Chen, J., J. F. Franklin, and T. A. Spies. 1992. Vegetation responses to edge environments in old-growth Douglas fir forests. Ecol. Appl. 2:387–396.

Chen, J., J. F. Franklin, and T. A. Spies. 1993. Contrasting microclimates among clearcut, edge, and interior of old-growth Douglas fir forest. Agric. For. Meteorol. 63:219–237.

Chen, J., J. F. Franklin, and T. A. Spies. 1995. Growing-season microclimatic gradients from clear-cut edges into old-growth Douglas fir forests. Ecol. Appl. 5:74–86.

Clark, D. L. 1991. Factors determining species composition of post-disturbance vegetation following logging and burning of an old-growth Douglas fir forest. Master's thesis, Oregon State University, Corvallis.

Conard, S. G., A. E. Jaramillo, K. Cromack, Jr., and S. Rose (comps.). 1985. The role of the genus *Ceanothus* in western forest ecosystems. USDA Forest Service general technical report PNW-182, Portland.

Daubenmire, R. 1978. Plant geography. Academic Press, New York.

del Moral, R., and D. C. Deardorff. 1976. Vegetation of the Mima Mounds, Washington State. Ecology 57:520–530.

Diaz, N. M., and T. K. Mellen. 1996. Riparian ecological types, Gifford Pinchot and Mt. Hood National Forests, Columbia River Gorge National Scenic Area. USDA Forest Service Pacific Northwest Region R6-NR-TP-10-96, Portland.

Douglas, G. W. 1970. A vegetation study in the subalpine zone of the western north Cascades, Washington. M.S. thesis, University of Washington, Seattle.

Douglas, G. W. 1972. Subalpine plant communities of the western north Cascades, Washington. Arc. Alp. Res. 4:147–166.

Dyrness, C. T. 1973. Early stages of plant succession following logging and burning in the western Cascades of Oregon. Ecology 54:57–69.

Dyrness, C. T., J. F. Franklin, and W. H. Moir. 1974. A preliminary classification of forest communities in the central portion of the western Cascades in Oregon. U.S. International Biological Program Coniferous Forest Biome Bull. 4:1–123.

Edmonds, R. L. (ed.). 1982. Analysis of coniferous forest ecosystems in the western United States. Hutchinson Ross, Stroudsburg, Pa.

Ettl, G. J., and D. L. Peterson. 1995a. Growth response of subalpine fir (*Abies lasiocarpa*) to climate in the Olympic Mountains, Washington. Global Change Biol. 1:213–230.

Ettl, G. J., and D. L. Peterson. 1995b. Extreme climate and variation in tree growth: individualistic response in subalpine fir (*Abies lasiocarpa*). Global Change Biol. 1:231–241.

Evans, R. D., and R. W. Fonda. 1990. The influence of snow on subalpine meadow community pattern, North Cascades, Washington. Can. J. Bot. 68:212–220.

Eyre, F. H. (ed.). 1980. Forest cover types of the United States and Canada. Society of American Foresters, Washington, D.C.

Feller, M. C., and J. P. Kimmins. 1984. Effects of clear-cutting and slash burning on streamwater chemistry and watershed nutrient budgets in southwestern British Columbia. Water Resources Res. 20:29–40.

Fonda, R. W., and L. C. Bliss. 1969. Forest vegetation of the montane and subalpine zones, Olympic Mountains, Washington. Ecol. Monogr. 39:271–301.

Forest Ecosystem Management Assessment Team. 1993. Forest ecosystem management: An ecological, economic, and social assessment, 1993–793–071. Joint publication of the USDA Forest Service; USDC National Oceanic and Atmospheric Administration and National Marine Fisheries Service; USDI Bureau of Land Management, Fish and Wildlife Service, and National Park Service; and US Environmental Protection Agency, Washington, D.C.

Franklin, J. F. 1965. Tentative ecological provinces within the true fir-hemlock forest areas of the Pacific Northwest. USDA Forest Service Research paper PNW-22, Portland.

Franklin, J. F. 1966. Vegetation and soils in the subalpine forests of the southern Washington Cascade Range. Ph.D. dissertation, Washington State University, Pullman.

Franklin, J. F. 1992. Scientific basis for new perspectives in forests and streams, pp. 25–72 in Robert J. Naiman (ed.), Watershed management: balancing sustainability and environment change. Springer-Verlag, New York.

Franklin, J. F., K. Cromack, Jr., W. Denison, A. McKee, C. Maser, J. Sedell, F. Swanson, and G. Juday. 1981. Ecological characteristics of old-growth Douglas fir forests. USDA Forest Service general technical report PNW-118, Portland.

Franklin, J. F., and C. T. Dyrness. 1988. Natural vegetation of Oregon and Washington. Oregon State University Press, Corvallis.

Franklin, J. F., and R. T. T. Forman. 1987. Creating landscape patterns by forest cutting: ecological consequences and principles. Landscape Ecol. 1:5–18.

Franklin, J. F., P. M. Frenzen, and F. J. Swanson. 1995. Re-creation of ecosystems at Mount St. Helens: contrasts in artificial and natural approaches, pp. 287–333 in J. Cairns, Jr. (ed.), Rehabilitating damaged ecosystems, 2nd ed. CRC Press, Boca Raton.

Franklin, J. F., and M. A. Hemstrom. 1981. Aspects of succession in the coniferous forests of the Pacific Northwest, pp. 219–229 in D. C. West, H. H. Shugart, and D. B. Botkin (eds.), Forest succession: concepts and application. Springer-Verlag, New York.

Franklin, J. F., M. Klopsch, K. Luchessa, and M. Harmon. 1986. Tree mortality in some mature and old-growth forests in the Cascade Range of Oregon and Washington. Unpublished manuscript, Forestry Sciences Laboratory, Corvallis.

Franklin, J. F., J. A. MacMahon, F. J. Swanson, and J. R. Sedell. 1985. Ecosystem responses of Mount St. Helens. National Geogr. Res. 1:198–216.

Franklin, J. F., and R. G. Mitchell. 1967. Successional status of subalpine fir in the Cascade Range. USDA Forest Service research paper PNW-46, Portland.

Franklin, J. F., W. H. Moir, G. W. Douglas, and C. Wiberg. 1971. Invasion of subalpine meadows by trees in the Cascade Range, Washington and Oregon. Arc. Alp. Res. 3:215–224.

Franklin, J. F., W. H. Moir, M. A. Hemstrom, S. E. Greene, and B. G. Smith. 1988. The forest communities of Mount Rainier National Park. USDI National Park Service Scientific Monograph Series No. 19, Washington, D.C.

Franklin, J. F., H. H. Shugart, and M. E. Harmon. 1987. Tree death as an ecological process. BioScience 37: 550–556.

Franklin, J. F., and T. A. Spies. 1991. Composition, function, and structure of old-growth Douglas fir forests, pp. 71–80 in L. F. Ruggiero, K. B. Aubry, A. B. Carey, and M. H. Huff (tech. coords.), Wildlife and vegetation of unmanaged Douglas fir forests. USDA Forest Service general technical report PNW-GTR-285, Portland.

Franklin, J. F., and R. H. Waring. 1981. Distinctive features of the northwestern coniferous forest: development, structure, and function, pp. 59–86 in R. H. Waring (ed.), Forests: fresh perspectives from ecosystem research. Oregon State University Press, Corvallis.

Fujimori, T. 1971. Primary productivity of a young *Tsuga heterophylla* stand and some speculations about biomass of forest communities on the Oregon coast. USDA Forest Service research paper PNW-123, Portland.

Fujimori, T. 1972. Discussion about the large forest biomasses on the Pacific Northwest in U.S.A. J. Jpn. For. Soc. 54:230–233.

Fujimori, T. 1977. Stem biomass and structure of a mature *Sequoia sempervirens* stand on the Pacific coast of northern California. J. Jpn. For. Soc. 59:435–441.

Fujimori, T., S. Kawanabe, H. Saito, C. C. Grier, and T. Shidei. 1976. Biomass and primary production in forests of three major vegetation zones of the northwestern United States. J. Jpn. For. Soc. 58:360–373.

Gholz, H. L. 1982. Environmental limits on aboveground net primary production, leaf area, and biomass in vegetation zones of the Pacific Northwest. Ecology 63:469–481.

Graumlich, L. J., and L. B. Brubaker. 1986. Reconstructions of annual temperature (1590–1979) for Longmire, Washington, derived from tree-rings. Quat. Res. 25:223–234.

Gray, A. N. 1995. Tree seedling establishment on heterogeneous microsites in Douglas fir forest canopy gaps. Ph.D. dissertation, Oregon State University, Corvallis.

Grier, C. C. 1988. Foliage loss due to snow, wind, and winter drying damage: Its effects on leaf biomass of some western conifer forests. Can. J. For. Res. 18: 1097–1102.

Grier, C. C., and R. S. Logan. 1977. Old-growth *Pseudotsuga menziesii* communities of a western Oregon watershed: biomass distribution and production budgets. Ecol. Monogr. 47:373–400.

Grier, C. C., and S. W. Running. 1977. Leaf area of mature northwestern coniferous forests: relation to site water balance. Ecology 58:893–899.

Grier, C. C., K. A. Vogt, M. R. Keyes, and R. L. Edmonds. 1981. Biomass distribution and above- and below-ground production in young and mature *Abies amabilis* zone ecosystems of the Washington Cascades. Can. J. For. Res. 11:155–167.

Griffin, J. R. 1967. Soil moisture and vegetation patterns in northern California forests. USDA Forest Service research paper PSW-46, Berkeley.

Halpern, C. B. 1986. Montane meadow plant associations of Sequoia National Park, California. Madroño 33:1–23.

Halpern, C. B. 1988. Early successional pathways and the resistance and resilience of forest communities. Ecology 69:1703–1715.

Halpern, C. B. 1989. Early successional patterns of forest species: Interactions of life history traits and disturbance. Ecology 70:704–720.

Halpern, C. B., J. A. Antos, M. A. Geyer, and A. M. Olson. 1997. Species replacement during early secondary succession: The abrupt decline of a winter annual. Ecology 78:621–631.

Halpern, C. B., and J. F. Franklin. 1990. Physiognomic development of *Pseudotsuga* forests in relation to initial structure and disturbance intensity. J. Veg. Sci. 1:475–482.

Halpern, C. B., J. F. Franklin, and A. McKee. 1992. Changes in plant species diversity after harvest of Douglas fir forests. Northwest Env. J. 8:205–207.

Halpern, C. B., P. M. Frenzen, J. E. Means, and J. F. Franklin. 1990. Plant succession in areas of scorched and blown-down forest after the 1980 eruption of Mount St. Helens, Washington. J. Veg. Sci. 1:181–194.

Halpern, C. B., B. G. Smith, and J. F. Franklin. 1984. Composition, structure, and distribution of the ecosystems of the Three Sisters Biosphere Reserve/Wilderness Area. Final report to the U.S. Department of Agriculture. USDA Forest Service, Pacific Northwest Research Station, Forestry Sciences Laboratory, Corvallis.

Halpern, C. B., and T. A. Spies. 1995. Plant species diversity in natural and managed forests of the Pacific Northwest. Ecol. Appl. 5:913–934.

Halverson, N. M., C. Topik, and R. VanSickle. 1986. Plant association and management guide for the western hemlock zone, Mt. Hood National Forest. USDA Forest Service Pacific Northwest Region R6-ECOL-232A-1986, Portland.

Hansen, H. P. 1947. Postglacial forest succession, climate, and chronology in the Pacific Northwest. Trans. Amer. Phil. Soc. New Series 37:1–130.

Harmon, M. E. 1986. Logs as sites of tree regeneration in *Picea sitchensis–Tsuga heterophylla* forests of Washington and Oregon. Ph.D. dissertation, Oregon State University, Corvallis.

Harmon, M. E., J. F. Franklin, F. J. Swanson, et al. 1986. Ecology of coarse woody debris in temperate ecosystems. Adv. Ecol. Res. 15:133–302.

Harr, R. D. 1982. Fog drip in the Bull Run municipal watershed. Water Resources Bull. 18:785–789.

Harr, R. D. 1986. Effects of clearcutting on rain-on-snow runoff in western Oregon: a new look at old studies. Water Resources Res. 22:1095–1100.

Harris, L. D. 1984. The fragmented forest: island biogeography theory and the preservation of biotic diversity. University of Chicago Press, Chicago.

Hebda, R. J., and R. W. Mathewes. 1984. Holocene history of cedar and native Indian cultures of the North American Pacific coast. Science 225:711–713.

Heikkinen, O. 1984. Forest expansion in the subalpine zone during the past hundred years, Mount Baker, Washington, U.S.A. Erdkunde 38:194–202.

Hemstrom, M. A., and J. F. Franklin. 1982. Fire and other disturbances of the forests in Mount Rainier National Park. Quat. Res. 18:32–51.

Hemstrom, M. A., and S. E. Logan. 1986. Plant association and management guide, Siuslaw National Forest. USDA Forest Service Pacific Northwest Region R6-ECOL-220-1986a, Portland.

Hemstrom, M. A., S. E. Logan, and W. Pavlat. 1987. Plant association and management guide, Willamette National Forest. USDA Forest Service Pacific Northwest Region R6-ECOL-257-B-86, Portland.

Henderson, J. A. 1973. Composition, distribution, and succession of subalpine meadows in Mount. Rainier National Park, Washington. Ph.D. dissertation, Oregon State University, Corvallis.

Henderson, J. A., R. D. Lesher, D. H. Peter, and D. C. Shaw. 1992. Field guide to the forested plant associations of the Mt. Baker–Snoqualmie National Forest. USDA Forest Service Pacific Northwest Region technical paper R6-ECOL-TP-028-91, Portland.

Henderson, J. A., D. H. Peter, R. D. Lesher, and D. C. Shaw. 1989. Forested plant associations of the Olympic National Forest. USDA Forest Service Pacific Northwest Region technical paper R6-ECOL-TP-001-88, Portland.

Heusser, C. J. 1960. Late-Pleistocene environments of North Pacific North America. American Geographical Society special publication 35, New York.

Heusser, C. J. 1983. Vegetational history of the northwestern United States including Alaska, pp. 239–258 in S. C. Porter (ed.), Late-Quaternary environments of the United States, Vol. 1, The late Pleistocene. University of Minnesota Press, Minneapolis.

Hickman, J. C. 1968. Disjunction and endemism in the flora of the central western Cascades of Oregon: An historical and ecological approach to plant distributions. Ph.D. dissertation, University of Oregon, Eugene.

Hickman, J. C. 1976. Non-forest vegetation of the central western Cascade Mountains of Oregon. Northwest Sci. 50:145–155.

Hitchcock, C. L., and A. Cronquist. 1973. Flora of the Pacific Northwest. University of Washington Press, Seattle.

Jimerson, T. M., and S. L. Daniel. 1994. A field guide to Port Orford cedar plant associations in northwest California. USDA Forest Service Pacific Southwest Region technical paper R5-ECOL-TP-002, Berkeley.

Keyes, M. R., and C. C. Grier. 1981. Above-and belowground net production in 40-year-old Douglas fir stands on low and high productivity sites. Can. J. For. Res. 11:599–605.

Kirk, R., and J. F. Franklin. 1992. The Olympic rain forest. University of Washington Press, Seattle.

Kohm, K. A., and J. F. Franklin (eds.). 1997. Creating a forestry for the 21st century: The science of ecosystem management. Island Press, Washington, D.C.

Krajina, V. J. 1965. Biogeoclimatic zones and classification of British Columbia, pp. 1–17 in V. J. Krajina (ed.), Ecology of western North America, Vol. 2. University of British Columbia Department of Botany, Vancouver.

Kuchler, A. W. 1946. The broadleaf deciduous forests of the Pacific Northwest. Ann. Assoc. Amer. Geogr. 36: 122–147.

Kuhns, J. C. 1917. G-Reconnaissance-Cascade, report to the forest supervisor. Supervisor's Office, Willamette National Forest, Eugene.

Kuramoto, R. T., and L. C. Bliss. 1970. Ecology of subalpine meadows in the Olympic Mountains, Washington. Ecol. Monogr. 40:317–347.

Leopold, E. B., R. Nickman, J. I. Hedges, and R. Ertel,

Jr. 1982. Pollen and lignin records of Late-Quaternary vegetation, Lake Washington. Science 218:1305–1307.

Lertzman, K. P. 1992. Patterns of gap-phase replacement in a subalpine, old-growth forest. Ecology 73: 657–669.

Leverenz, J. W., and T. M. Hinckley. 1990. The effect of shoot structure on leaf area index and productivity of evergreen conifer stands. Tree Physiol. 6:135–149.

Long, J. N., and J. Turner. 1975. Aboveground biomass of understory and overstory in an age sequence of four Douglas fir stands. J. Appl. Ecol. 12:179–188.

Martin, C. W., and R. D. Harr. 1989. Logging of mature Douglas fir in western Oregon has little effect on nutrient output budgets. Can. J. For. Res. 19:35–43.

Maser, C., R. F. Tarrant, J. M. Trappe, and J. F. Franklin. 1988. From the forest to the sea: A story of fallen trees. USDA Forest Service general technical report PNW-GTR-229, Portland.

McKee A., G. La Roi, and J. F. Franklin. 1982. Structure, composition, and reproductive behavior of terrace forests, South Fork Hoh River, Olympic National Park, pp. 19–29 in E. E. Starkey, J. F. Franklin, and J. W. Matthews (eds.), Ecological research in national parks of the Pacific Northwest. Oregon State University Forest Research Laboratory, Corvallis.

McKee, B. 1972. Cascadia. The geologic evolution of the Pacific Northwest. McGraw-Hill, New York.

McMinn, R. G. 1960. Water relations and forest distribution in the Douglas fir region on Vancouver Island. Canadian Department of Agriculture publication 1091, Victoria, B. C.

Maser, C., and J. M. Trappe. 1984. The seen and unseen world of the fallen tree. USDA Forest Service general technical report PNW-164, Portland.

Massman, W. J. 1982. Foliage distribution in old-growth coniferous tree canopies. Can. J. For. Res. 12:10–17.

Means, J. E. 1980. Dry coniferous forests in the western Oregon Cascades. Ph.D. dissertation, Oregon State University, Corvallis.

Miles, D. W. R., and F. J. Swanson. 1986. Vegetation composition on recent landslides in the Cascade Mountains of western Oregon. Can. J. For. Res. 16: 739–744.

Miller, E. A., and C. B. Halpern. 1998. Effects of environment and grazing disturbance on tree establishment in meadows of the central Cascade Range, Oregon, USA J. Veg. Sci. 9:265–282.

Morgenroth, C. 1991. Footprints in the Olympics: An autobiography. Ye Galleon Press, Fairfield, Washington.

Mullineaux, D. R. 1974. Pumice and other pyroclastic deposits in Mount Rainier National Park, Washington. U.S. Geol. Surv. Bull. 1326:1–83.

Nadkarni, N. M. 1984. Biomass and mineral capital of epiphytes in an *Acer macrophyllum* community of a temperate moist coniferous forest, Olympic Peninsula, Washington State. Can. J. Bot. 62:2223–2228.

Norse, E. A. 1990. Ancient forests of the Pacific Northwest. Island Press, Washington, D.C.

Ohmann, J. L. 1995. Regional gradient analysis and spatial pattern of woody plant communities in Oregon. Ph.D. dissertation, Oregon State University, Corvallis.

Oosting, H. J., and W. D. Billings. 1943. The red fir forest of the Sierra Nevada: *Abietum magnificae*. Ecol. Monogr. 13:259–274.

Pike, L. H., R. A. Rydell, and W. C. Denison. 1977. A 400-year-old Douglas fir tree and its epiphytes: biomass, surface area, and their distributions. Can. J. For. Res. 7:680–699.

Porter, S. C., K. L. Pierce, and T. D. Hamilton. 1983. Late Wisconsin mountain glaciation in the western United States, pp. 71–111 in S. C. Porter (ed.), Late-Quaternary environments in the United States, Vol. 1, The late Pleistocene. University of Minnesota Press, Minneapolis.

Ratliff, R. D. 1982. A meadow site classification for the Sierra Nevada, California. USDA Forest Service Pacific Southwest Forest and Range Experiment Station, general technical report PSW-60, Berkeley.

Roach, A. W. 1958. Phytosociology of the Nash Crater lava flows, Linn County, Oregon. Ecol. Monogr. 22: 169–193.

Rocheforte, R. M., R. L. Little, A. Woodward, and D. L. Peterson. 1994. Changes in subalpine tree distribution in western North America: A review of climatic and other causal factors. The Holocene 4:89–100.

Rocheforte, R. M., and D. L. Peterson. 1996. Temporal and spatial distribution of trees in subalpine meadows of Mount Rainier National Park, Washington. Arc. Alp. Res. 28: 52–59.

Rothacher, J. 1970. Increases in water yield following clear-cut logging in the Pacific Northwest. Water Resources Res. 6:653–658.

Ruggiero, L. F., K. B. Aubry, A. B. Carey, and M. H. Huff (tech. coords.). 1991. Wildlife and vegetation of unmanaged Douglas fir forests. USDA Forest Service general technical report PNW-GTR-285, Portland.

Rundel, P. W., D. J. Parsons, and D. T. Gordon. 1977. Montane and subalpine vegetation of the Sierra Nevada and Cascade Ranges, pp. 559–599 in M. G. Barbour and J. Major (eds.), Terrestrial vegetation of California. Wiley, New York.

Ruth, R. H., and R. A. Yoder. 1953. Reducing wind damage in the forests of the Oregon Coast Range. USDA Forest Service, Pacific Northwest Forest and Range Experiment Station research paper 7, Portland.

Sarna-Wojcicki, A. M., D. E. Champion, and J. O. Davis. 1983. Holocene volcanism in the coterminous United States and the role of silicic volcanic ash layers in correlation of latest-Pleistocene and Holocene deposits, pp. 52–77 in H. E. Wright, Jr. (ed.), Late-Quaternary environments of the United States, Vol. 2, The Holocene. University of Minnesota Press, Minneapolis.

Sawyer, J. O., D. A. Thornburgh, and J. R. Griffin. 1977. Mixed evergreen forest, pp. 359–381 in M. G. Barbour and J. Major (eds.), Terrestrial vegetation of California. Wiley, New York.

Schmidt, R. L. 1957. The silvics and plant geography of the genus *Abies* in the coastal forests of British Columbia. British Columbia Forest Service technical publication T46.

Schoonmaker, P., and A. McKee. 1988. Species composition and diversity during secondary succession of coniferous forests in the western Cascade Mountains of Oregon. For. Sci. 34:960–979.

Scientific Panel for Sustainable Forest Practices in Clay-

oquot Sound. 1995. Report 5, Sustainable ecosystem management in Clayoquot Sound: Planning and practices. Cortex Consultants, Victoria, B.C.

Schreiner, E. G., K. A. Krueger, P. J. Happe, and D. B. Houston. 1996. Understory patch dynamics and ungulate herbivory in old-growth forests of Olympic National Park, Washington. Can. J. For. Res. 26:255–265.

Seyer, S. C. 1979. Vegetation ecology of a mountain mire, Crater Lake National Park, Oregon. M.S. thesis, Oregon State University, Corvallis.

Shlisky, A. 1996. Early successional plant communities, Pacific silver fir zone, Gifford Pinchot National Forest. Draft report, USDA Forest Service, Pacific Northwest Region technical paper R6-NR-ECOL-TP-17-96, Portland.

Sigafoos, R. S., and E. L. Hendricks. 1972. Recent activity of glaciers of Mount Rainier, Washington. U.S. Geological Survey professional paper 387-B.

Sollins, P., C. C. Grier, F. M. McCorison, K. Cromack, Jr., R. Fogel, and R. L. Fredriksen. 1980. The internal element cycles of an old-growth Douglas fir ecosystem in western Oregon. Ecol. Monogr. 50:261–285.

Sollins, P., and F. M. McCorison. 1981. Nitrogen and carbon solution chemistry of an old-growth coniferous forest watershed before and after clearcutting. Water Resources Res. 17:1409–1418.

Spies, T. A. 1991. Plant species diversity and occurrence in young, mature, and old-growth Douglas fir stands in western Oregon and Washington, pp. 111–121 in L. F. Ruggiero, K. B. Aubry, A. B. Carey, and M. H. Huff (tech. coords.), Wildlife and vegetation of unmanaged Douglas fir forests. USDA Forest Service general technical report PNW-GTR-285, Portland.

Spies, T. A., and J. F. Franklin. 1991. The structure of natural young, mature, and old-growth Douglas fir forests in Oregon and Washington, pp. 91–110 in L. F. Ruggiero, K. B. Aubry, A. B. Carey, and M. H. Huff (tech. coords.), Wildlife and vegetation of unmanaged Douglas fir forests. USDA Forest Service general technical report PNW-GTR-285, Portland.

Spies, T. A., and J. F. Franklin. 1996. The diversity and maintenance of old-growth forests, pp. 296–314 in R. C. Szaro and D. W. Johnston (eds.), Biodiversity in managed landscapes: theory and practice. Oxford University Press, New York.

Spies, T. A., J. F. Franklin, and M. Klopsch. 1990. Canopy gaps in Douglas fir forests of the Cascade Mountains. Can. J. For. Res. 20:649–658.

Spies, T. A., J. F. Franklin, and T. B. Thomas. 1988. Coarse woody debris in Douglas fir forests of western Oregon and Washington. Ecology 69:1689–1702.

Spies, T. A., W. J. Ripple, and G. A. Bradshaw. 1994. Dynamics and pattern of a managed coniferous forest landscape in Oregon. Ecol. Appl. 4:555–568.

Spilsbury, R. H., and D. S. Smith. 1947. Forest site types of the Pacific Northwest. British Columbia Forest Service publication T30.

Stewart, G. H. 1986. Population dynamics of a montane conifer forest, western Cascade Range, Oregon, USA. Ecology 67:534–544.

Stone, E. C., and R. B. Vasey. 1968. Preservation of coast redwood on alluvial flats. Science 159:157–161.

Swanson, F. J., and G. W. Lienkaemper. 1982. Interactions among fluvial processes, forest vegetation, and aquatic ecosystems, South Fork Hoh River,

Olympic National Park, pp. 30–34 in E. E. Starkey, J. F. Franklin, and J. W. Matthews (eds.), Ecological research in national parks of the Pacific Northwest. Oregon State University Forest Research Laboratory, Corvallis.

Tappeiner, J. C., II, and P. M. McDonald. 1984. Development of tan oak understories in conifer stands. Can. J. For. Res. 14:271–277.

Tappeiner, J., J. Zasada, P. Ryan, and M. Newton. 1991. Salmonberry clonal and population structure: the basis for a persistent cover. Ecology 73:609–618.

Tappeiner, J. C., and J. C. Zasada. 1993. Establishment of salmonberry, salal, vine maple, and bigleaf maple seedlings in the coastal forests of Oregon. Can. J. For. Res. 23: 1775–1780.

Tarrant, R. F., J. M. Trappe, and J. F. Franklin (eds.). 1967. Biology of alder. USDA Forest Service, Pacific Northwest Forest and Range Experiment Station, Portland.

Taylor, A. H. 1990. Disturbance and persistence of Sitka spruce (*Picea sitchensis* (Bong) Carr.) in coastal forests of the Pacific Northwest, North America. J. Biogeogr. 17:47–58.

Taylor, A. H. 1995. Forest expansion and climate change in the mountain hemlock (*Tsuga mertensiana*) zone, Lassen National Park, California. Arc. Alp. Res. 27:207–216.

Thompson, R. S., C. Whitlock, P. J. Bartlein, S. P. Harrison, and W. G. Spaulding. 1993. Climatic changes in the western United States since 18000 BP, pp. 468–513 in H. E. Wright, J. E. Kutzbach, T. Webb III, W. F. Ruddiman, F. A. Street-Perrott, and P. J. Bartlein (eds.), Global climates since the last ice age. University of Minnesota Press, Minneapolis.

Thornburgh, D. A. 1969. Dynamics of the true fir-hemlock forests of the west slope of the Washington Cascade Range. Ph.D. dissertation, University of Washington, Seattle.

Thornburgh, D. A. 1982. Succession in the mixed evergreen forests of northwestern California, pp. 87–91 in J. E. Means (ed.), Forest succession and stand development research in the northwest. Oregon State University Forest Research Laboratory, Corvallis.

Topik, C., N. M. Halverson, and D. G. Brockway. 1986. Plant association and management guide for the western hemlock zone, Gifford Pinchot National Forest. USDA Forest Service Pacific Northwest Region technical paper R6-ECOL-230A-1986, Portland.

Tsukada, M., S. Sugita, and D. M. Hibbert. 1981. Paleoecology in the Pacific Northwest. I. Late Quaternary vegetation and climate. Internationale Vereinegung fur theoretische und angemandte Limnologie 21:703–737.

Tuchmann, E. G., K. P. Connaughton, L. E. Freedman, and C. B. Moriwaki. 1996. The Northwest Forest Plan. A report to the President and Congress. USDA Forest Service Pacific Northwest Research Station, Portland.

Vale, T. R. 1981. Tree invasion of montane meadows in Oregon. Amer. Midl. Nat. 105:61–69.

Van Pelt, R. 1995. Understory tree response to canopy gaps in old-growth Douglas fir forests of the Pacific Northwest. Ph.D. dissertation, University of Washington, Seattle.

Viers, S. D., Jr. 1982. Coast redwood forest: stand dynamics, successional status, and the role of fire, pp. 119–141 in J. E. Means (ed.), Forest succession and stand development research in the northwest. Oregon State University Forest Research Laboratory, Corvallis.

Waitt, R. B., Jr., and R. M. Thorson. 1983. The Cordilleran ice sheet in Washington, Idaho, and Montana, pp. 53–70 in S. C. Porter (ed.), Late-Quaternary environments of the United States, Vol. 1, The late Pleistocene. University of Minnesota Press, Minneapolis.

Walter, H. 1979. Vegetation of the earth and ecological systems of the geo-biosphere. Springer-Verlag, Berlin.

Waring, R. H. 1983. Estimating forest growth and efficiency in relation to canopy leaf area. Adv. Ecol. Res. 13:327–354.

Waring, R. H., W. H. Emmingham, H. L. Gholz, and C. C. Grier. 1978. Variation in maximum leaf area of coniferous forests in Oregon and its ecological significance. For. Sci. 24:131–140.

Waring, R. H., and J. F. Franklin. 1979. Evergreen coniferous forests of the Pacific Northwest. Science 204: 1380–1386.

Waring, R. H., and J. Major. 1964. Some vegetation of the California coastal redwood region in relation to gradients of moisture, nutrients, light, and temperature. Ecol. Monogr. 34:167–215.

White, J. D., J. C. Haglund, and T. K. Mellen. 1996. Early seral plant communities, Pacific silver fir zone, Mt. Hood National Forest. USDA Forest Service, Pacific Northwest Region R6-NR-TP-16-96, Portland.

Whitlock, C. 1993. Vegetational and climatic history of the Pacific Northwest during the last 20,000 years: implications for understanding present-day biodiversity. Northwest Env. J. 8:5–28.

Whittaker, R. H. 1960. Vegetation of the Siskiyou Mountains, Oregon and California. Ecol. Monogr. 30:279–338.

Woodward, A., E. G. Schreiner, and D. G. Silsbee. 1995. Climate, geography, and tree establishment in subalpine meadows of the Olympic Mountains, Washington, U.S.A. Arc. Alp. Res. 27:217–225.

Worthington, N. 1954. The loblolly pine of the south versus the Douglas fir of the Pacific Northwest. Pulp Paper 28:87–90.

Yaffee, S. L. 1994. The wisdom of the spotted owl. Island Press, Washington, D.C.

Zinke, P. J. 1977. The redwood forest and associated north coast forests, pp. 679–698 in M. G. Barbour and J. Major (eds.), Terrestrial vegetation of California. Wiley, New York.

Zobel, D. B., A. McKee, G. M. Hawk, and C. T. Dyrness. 1976. Relationships of environment to composition, structure, and diversity of forest communities of the central western Cascades of Oregon. Ecol. Monogr. 46:135–156.

Zobel, D. B., L. F. Roth, and G. M. Hawk. 1985. Ecology and management of Port Orford cedar. USDA Forest Service general technical report PNW-184, Portland.

Chapter
5

Californian Upland Forests and Woodlands

MICHAEL G. BARBOUR   RICHARD A. MINNICH

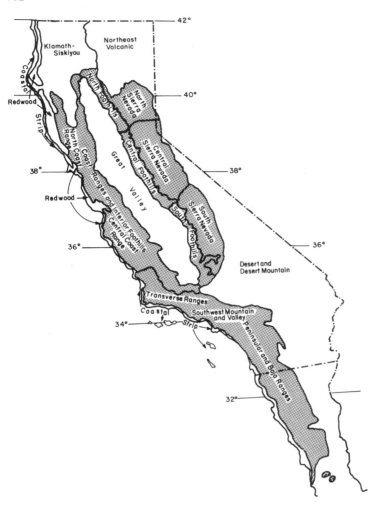

Figure 5.1. Regions of California and Baja California discussed in this chapter. Major landscape province names and boundaries are taken from Mason (1970) and Nelson (1922).

INTRODUCTION

"Californian" is used here to indicate the area within the Californian Floristic Province as defined and discussed by Stebbins and Major (1965), Raven and Axelrod (1978), and Hickman (1993). However, the vegetation of the entire province is not discussed in this chapter. Omissions include forests of northernmost California, lowland riparian forests, and upland vegetation not dominated by trees. In the northern part of the state, the Californian province increasingly blends with an Oregonian province, and several vegetation types in that area are described in Chapter 4. Lowland riparian forests and other freshwater wetlands are included in Chapter 12; coastal wetlands in Chapter 13; chaparral in Chapter 6; coastal and central valley grasslands in Chapter 9; alpine tundra in Chapter 14; and semiarid pinyon-juniper woodlands in Chapter 7.

Woodlands and forests are defined by UNESCO

(1973) and Paysen, Derby, and Conrad (1982) as being dominated by trees at least 5 m tall and having a tree canopy covering at least 25% of the ground. The part of California covered by such vegetation and discussed in this chapter totals approximately 170,000 km², or 42% of the state's area (Fig. 5.1). In addition, the Californian floristic province extends south into Baja California, largely along the western foothills and flanks of the Sierra Juarez and Sierra San Pedro Mártir. The woodlands and forests within this 27,000 km² area are also discussed here. According to the USDA's ECOMAP (McNab and Avers 1994), those 200,000 km² fit within the Humid Temperate Domain, the two Divisions of Mediterranean and of Mediterranean Regime Mountain, and the two Provinces of California Coastal Chaparral Forest and of Shrub and Sierran Steppe–Mixed Forest–Coniferous Forest–Alpine Meadow.

There have been several efforts to classify and map California vegetation. Statewide maps include

Kuchler's (1977) "potential natural formations," a Vegetation Type Map (VTM) Survey of nearly half the state conducted in 1928–1940 (Wieslander 1935; Jensen 1947; Wieslander and Jensen 1946), and several Forest Service projects: 300 quadrangles and 4.5 m ha of upland vegetation mapped by a soil-vegetation program that recognized units as small as 4 ha (Colwell 1988), a CALVEG map of existing vegetation by Matyas and Parker (1980), a follow-up map a decade later (Anon. 1991), and a map of landscape types (Goudey and Smith 1994). Classifications include lists and keys to habitats (Mayer and Laudenslayer 1988), ecosystems (Barry 1989; McNab and Avers 1994), habitats (Cheatham and Haller 1975; Holland 1986), major vegetation units (Munz and Keck 1959; Thorne 1976); forests (Eyre 1980); and more local vegetation types called series (Sawyer and Keeler-Wolf 1995). These attempts have variously divided California into as few as 29 and as many as 375 natural vegetational units.

For the sake of simplicity, in Table 5.1 we divided the woodlands and forests into just 7 major categories and then into 37 prominent series recently defined by Sawyer and Keeler-Wolf (1995). A series is a regional vegetation type, the most continuous layer of which is dominated by a single species, a regular mix of two or more species, or by a genus of ecologically related species. Each series carries the common name of the dominant. A series is only relatively homogeneous in terms of physiognomy and flora; it may contain many local variations, equivalent to the traditional associations of phytosociologists. Sawyer and Keeler-Wolf described more than 200 series for all of California, and we can only guess that there might be an order of magnitude larger number of associations waiting to be documented.

Eight taxa suffice to characterize and differentiate the seven categories in Table 5.1: the foothill oaks *Quercus agrifolia*, *Q. chrysolepis*, and *Q. douglasii*; the lower montane yellow pines *Pinus ponderosa* and *P. jeffreyi*; and the upper montane pines *Pinus albicaulis*, *P. contorta* var. *murrayana*, and *P. flexilis*. Tree nomenclature follows Eyre (1980), and all other nomenclature in this chapter follows Hickman (1993).

Mediterranean Climate and the Californian Floristic Province

Aschmann (1973, 1985) has defined mediterranean climate as having a cool, wet winter in which >65% of the annual 275–900 mm of precipitation falls, and the mean temperature of which is <15° C, yet frost is restricted to <3% of annual hours. Mediterranean climates occupy western or south-

Table 5.1. *Major categories and prominent series of vegetation discussed in this chapter. The series within each major category largely share the characteristic taxa listed.*

Major category	Prominent series/most characteristic taxa
Northern oak woodlands	blue oak, interior live oak, Oregon white oak, mixed oak, and foothill pine series/*Quercus douglasii*
Southern and coast oak woodlands	Engelmann oak, California walnut, and coast live oak series/*Quercus agrifolia*
Mixed evergreen forest	Douglas fir–tan oak, Coulter pine, Coulter pine–canyon live oak, black oak, Santa Lucia fir, tan oak, California buckeye, bigcone Douglas fir, bigcone Douglas fir–canyon live oak, and California bay series/*Quercus chrysolepis*
Midmontane forest	mixed conifer, white fir, giant sequoia, Jeffrey pine, Jeffrey pine–ponderosa pine, ponderosa pine, Douglas fir, Douglas fir–ponderosa pine, and incense cedar series/*Pinus ponderosa* and *P. jeffreyi*
Upper montane forest	lodgepole pine–red fir, aspen, western white pine, mountain juniper, and Jeffrey pine series/*Pinus contorta* var. *murrayana*
Subalpine woodland	whitebark pine, limber pine, lodgepole pine, mountain hemlock, and mixed subalpine forest series/*Pinus albicaulis* and *P. flexilis*
Sierran east-side and Baja California montane forests	Jeffrey pine aspen, Washoe pine, aspen, lodgepole pine, red fir, and white fir series/*Pinus jeffreyi*

western edges of five continents at 28–42°N or S latitude: North America, South America, the Mediterranean rim of Europe and Africa, the tip of South Africa, and southern parts of Australia. Vegetation and ecosystems within each region have been summarized and compared by many biogeographers, most recently Barbour and Minnich (1990) and Archibold (1995), as well as by Keeley in this volume.

According to Aschmann's definition, mediterranean climate is not completely synonymous with the entire Californian Floristic Province. It is restricted to low-elevation, noncoastal areas of California situated to the west of the Cascade–Sierra and Peninsular Range crests. A marine influence moderates mediterranean climate in northwestern California, dampening annual and daily temperature oscillations and increasing the amount of summer and annual rainfall to the point that they exceed mediterranean limits. Above elevations of approximately 1000 m in the north and 2000 m in the south, precipitation is too high and winter temperatures too low to fall within mediterranean limits, even though the seasonality of precipitation remains mediterranean. In upper montane forests, the dominant form of precipitation shifts from rain to snow, and some of the deepest snowpacks of the state develop there (Barbour, Berg, Kittel, and Kunz 1991).

A mediterranean climate was not present in California until the late Tertiary, probably not until more recently than 5 m yr ago (Axelrod, in press). Space limitations for this chapter prohibit a full discussion of the warm-temperate climates and vegetation that preceded the Pleiocene, but we recommend four reviews: Axelrod (1976, 1988); Raven and Axelrod (1978) and Graham (1993).

MAJOR VEGETATION TYPES

Northern Oak Woodlands: Blue Oak

This complex of woodlands forms a nearly continuous ring around the Central Valley of California, generally between 100 and 1200 m elevation (Fig. 5.2). Most of the range is well away from the coast, except for some outliers in Monterey and San Luis Obispo counties. Recent state-wide censuses by Greenwood, Marose, and Stenback (1993) and Bolsinger (1988) reported that these woodlands cover about 5% of the state and that only 3% of those 5 million acres (2 m ha) has any sort of reserve status. Conversion of woodland to pasture, agriculture, and urban areas has been extensive.

This is essentially two-layered vegetation (Fig. 5.3). An overstory canopy, 5–15 m tall, is 10–60% closed, and blue oak (*Quercus douglasii*) – with an importance percentage value >67 (Brooks 1971; Vankat and Major 1978) – is dominant. Associated evergreen trees include coast and interior live oaks (*Q. agrifolia* toward the coast, *Q. wislizenii* toward the interior), and foothill pine (*Pinus sabiniana*). Associated deciduous oaks include valley oak (*Q. lobata*) at lower elevations with shallow water tables, black oak (*Q. kelloggii*) at higher elevations or on mesic slopes, and Oregon white oak (*Q. garryana*) at similar mesic sites in the North Coast Range. A somewhat shorter, more spreading deciduous tree, *Aesculus californica*, occurs as scattered individuals or in small clumps. Sapling and tree densities, combined for all species, usually total fewer than 200 ha^{-1} on gentle slopes but increase in density and cover on steep, north-facing slopes to 550 ha^{-1} (Borchert 1994; Griffin 1988). Dense stands of 1000

ha^{-1} exist. Mature trees in such dense stands are only 10–15 cm dbh, but more open woodland trees are 15–30 cm dbh and have a combined basal area of 5–9 m^{2}ha^{-1}.

Shrubs 1–2 m tall are regularly present, but cover is insignificant, typically <5%. Common genera include *Arctostaphylos, Ceanothus, Cercis, Heteromeles, Rhamnus*, and *Toxicodendron*. Only *Cercis* and *Toxicodendron* are deciduous. The understory is almost continuous cover by (mainly) introduced grasses and forbs of the modern California annual grassland series (e.g., the genera *Avena, Brassica, Bromus, Centauria, Cynosurus, Erodium, Eschscholzia, Lolium, Lupinus*, and *Vulpia*).

Some of the most detailed sampling summaries of blue oak woodland have been compiled by Allen and Holzmann (1991) and by Allen, Holzmann, and Evett (1991) for northern California and by Borchert, Cunha, Krosse, and Lawrence (1993) for the central coast area. In both cases, the authors were able to identify more than a dozen associations or their rough equivalent.

The northern oak woodlands occur on moderately rich, loamy, well-drained soils with neutral or slightly basic pH, largely in the Inceptisol, Alfisol, and Mollisol orders. Topography is often gently rolling to steep (10–30% slope). Based on our own survey of weather stations within these woodlands, we estimate that mean annual temperature is 16° C, mean annual amplitude (warmest month minus coldest month) is 19° C, and mean annual precipitation is 530 mm (range = 280–1000). The macroclimate for northern oak woodland is not significantly different from that of adjacent grassland and chaparral. The three vegetation types occur in a complex mosaic, often with narrow ecotones, reflecting differences in slope, aspect, soil depth, soil texture, and frequency of fire more than differences in macroclimate. In general, northern oak woodland lies between more xerophytic chaparral and grassland downslope or on locally arid sites and more mesic mixed evergreen forest upslope or on locally mesic sites.

Stand dynamics. There has been considerable examination of the age structure of blue oak stands and research into the causes for its typical nonclimax nature. White (1966), for example, made a detailed study of 38 ha of blue oak woodland in the Central Coast Range, Monterey County. Two blue oak age groups were evident: 60–100 yr old and 150–260 yr old. Successful establishment had been declining for the past 90 yr, and there had been no establishment in the last 30 yr. Establishment appeared to be episodic, with the most recent flush having occurred in the 1870s. Vankat and Major

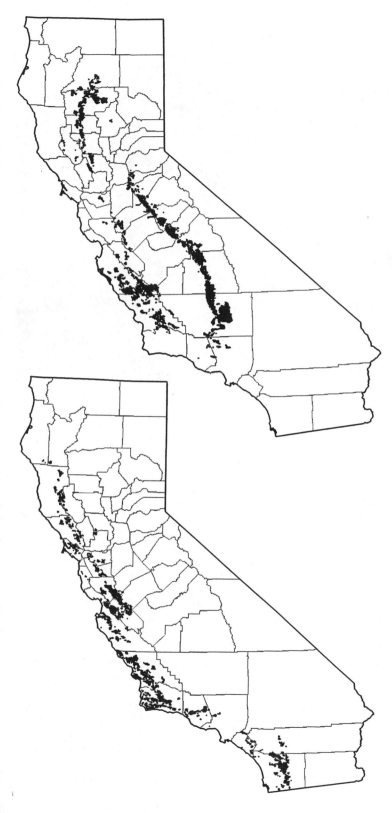

Figure 5.2. Map of the distribution of (top) northern oak woodlands and of (bottom) coast/southern oak woodlands (redrawn from Greenwood et al. 1993).

Figure 5.3. Aspect of blue oak woodland.

(1978), working from historical records and photographs in the Sequoia–Kings Canyon area, also concluded that there had been a flush of blue oak establishment in the 1870s. Bartolome and his associates (McClaran 1986; Muick and Bartolome 1987; McClaran and Bartolome 1990;) demonstrated a widespread pattern of suppressed establishment throughout blue oak's range. Evergreen oak populations sympatric with blue oak were typically not so depressed.

The causes of regeneration loss in the twentieth century has been variously ascribed to herbivory by rodents, deer, and cattle; to competition with weedy annual plants; to soil compaction by domesticated livestock; and to changes in the wildfire regime. In a series of papers, Griffin (1971, 1976, 1988, 1980a) concluded that the major causes of deciduous oak seedling mortality in Monterey County were inability of the root to penetrate compact soil, summer drought away from tree canopies on southern exposures, and browsing by pocket gophers, rodents, deer, and cattle. Thousands of seedlings were marked and placed in a variety of exclosures. Only those on north-facing slopes, in partial shade, and protected from all grazing animals survived for 6 yr. Griffin imag-

ined that the flush of establishment in the 1870s could have coincided with low herbivore population numbers and optimal fall germination conditions, followed by mild summers. He concluded that rodent, deer, and livestock populations currently are all too high to permit significant establishment. Borchert, Davis, Michaelsen, and Oyler (1989) later repeated many of those experiments farther south in San Luis Obispo County and reached similar conclusions. Griffin's conclusion that shade was critical for seedling survival has been recently supported by Callaway's (1992) field observations that blue oak survival is enhanced in the shade of shrubs. Shrub cover also reduced aboveground herbivory.

Manipulative experiments by Gordon, Welker, Menke, and Rice (1989), Gordon and Rice (1993), and Welker and Menke (1990) illustrated the competitive effect of the root systems of introduced annuals on soil water availability. Blue oak seedling growth, photosynthesis, and survivorship were all negatively related to the extent of soil water depletion.

Stand structure may, of course, reflect episodic disturbances in the form of ground or crown fires. Fire is a natural part of the landscape of the mediterranean climate. It is typically started by lightning strikes in late summer or fall that ignite fuel on the ground. Given appropriate conditions of wind, fuel, and humidity, a low-intensity ground fire will move slowly across the landscape. High levels of wind and fuel may carry flames into tree canopies, producing a fast-moving, high-intensity crown fire. One would expect both kinds of fire to visit oak woodlands, yet there has been surprisingly little research either on documenting the periodicity of fire or on assessing the effect of fire on oak woodlands. It is known that all oak tree species are capable of sprouting in California, though *Q. lobata* and *Q. douglasii* lose that ability once they reach a certain mass or age (Griffin 1980b; Plumb 1980).

Recovery from ground fires can be rapid, as documented by Hagerty (1994) for a blue oak woodland in Sequoia National Park. Fireline intensities were estimated to have been 300–1100 kWm^{-1}, flame lengths were 1–4 m, and scorch height was 50 cm. Within 2 yr epicormic and basal sprouting had regenerated the preburn crown volume. Only 6% of the blue oaks and 11% of the interior live oaks had been killed. Most canopy leaves were killed. This was considered to be a hot ground fire.

Perhaps the best data we have on oak woodland fire history are for a single locale. McClaran (1988; see also a review by Standiford, Klein, and Garrison 1996) cut down 181 fire-scarred blue oaks at a field station in the northern Sierra Nevada foothills. The fire scar record extended between 1968 and 1681 (277 yr) and revealed that the average fire return period prior to European settlement was 25 yr, and it was shortened thereafter to 7 yr (excluding a period of complete fire suppression after 1950). This is an unexpectedly short interval, because anecdotal guesses by decades of California ecologists have tended to put the fire return interval at >50 yr. Standiford et al. (1996) believe that poor regeneration of blue oak in this century is largely a result of changes in fire frequency, rather than a result of grazing by livestock or other animals.

Ecophysiology. In terms of water relations, blue oak is much more xerophytic than any associated sympatric oak species. At Hastings Natural History Reservation in Monterey County, blue oak occurs with four other oaks. Johannes, Knops, and Koenig (1994) showed that *Quercus douglassi* and *Q. kelloggii* sustained the most negative water potentials (−3.4 MPa at midday), dividing xeric sites by elevation: blue oak at lower elevations than

black oak. *Q. lobata* and *Q. chrysolepis* sustained the least negative water potentials (−2.1 MPa at midday), dividing mesic sites with shallow water tables by elevation; valley oak at lower elevations than coast live oak. *Q. chrysolepis* (canyon live oak) was intermediate (−2.5 MPa at midday). Griffin (1973) recorded midday water potentials in blue oak at the same site as low as −4.0 MPa midday and −1.1 MPa predawn. At the same time, nearby valley oak and coast live oak averaged only −0.4 MPa predawn. Rundel (1980) and Baker, Rundel, and Parsons (1981) showed even more negative values for Sierran foothill populations of blue oak, averaging −2.0 MPa predawn.

Rundel's (1980) review of oak ecology concluded that we have insufficient data to define or compare photosynthetic rates, productivity, or patterns of biomass allocation of California oak trees. That is still true today as we write this second edition, 18 yr later. We cannot yet assume with confidence that the generally held tenets about photosynthetic differences between evergreen and deciduous species (Mooney 1972; Chabot and Hicks 1982; Larcher 1995) apply specifically to California oaks. California oaks provide an excellent test for such hypotheses, and they even offer hybrids between evergreen and deciduous species (e.g., *Q.* x *morehus*, a common hybrid of *Q. kelloggii* and *Q. wislizenii*) (Tucker 1980).

Consequently, we are still uncertain as to the environmental factors that limit the distribution of blue oak. Major (1988), Myatt (1980), and Vankat (1982) have documented gradients in precipitation, mean annual temperature, and annual amplitude of temperature with elevation that correspond with the upper limit of blue oak woodland (Fig. 5.4). There is also some evidence (Dunn 1980; Baker et al. 1981) that soil C:N ratios and contents of N, P, Ca, and organic matter rise as one moves upslope out of blue oak woodland. Within blue oak woodland, there is conflicting evidence for a positive or negative impact of blue oak on the growth and nutritional content of associated understory species (Holland 1980; Holland and Morton 1980; Kay and Leonard 1980; Callaway, Nadkarni, and Mahall 1991). The most recent work, by Callaway and colleagues, revealed significant competition for water by the fine roots of trees in the upper 50 cm of soil and the possibility that the roots were also exuding allelopathic substances that further inhibited understory grasses. As is true for most California vegetation, even the most basic ecosystem information such as productivity, standing crop, nutrient cycling details, and efficiencies in nutrient use is unknown.

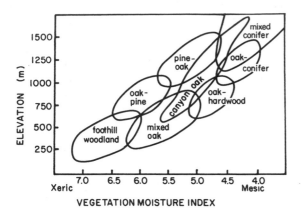

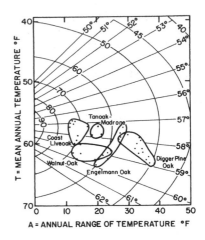

Figure 5.4. Vegetation-environment relationships among blue oak woodland (foothill woodland), mixed evergreen forest (mixed oak, pine-oak, canyon live oak, oak-conifer), and midmontane conifer forest (mixed conifer) in the Stanislaus River region of the Sierra Nevada (redrawn from Myatt 1980).

Figure 5.5. Thermal relations of five woodland and forest types of vegetation in California. "Digger pine-oak" is a synonym for northern oak woodland as we use the term in this chapter. "Tan oak–madrone" is a mixed evergreen type of forest to be discussed in the next section. Dots represent meteorological stations within each vegetation type. Radii are lines of warmth, or effective temperature; arcs represent temperateness or equability, reflecting departures from 100, a yearly constant of 14°C (redrawn from Axelrod 1988).

Southern and Coastal Oak Woodlands: Coast Live Oak

These woodlands extend from eastern Mendocino County at 40° N latitude in the North Coast Range, through the Central Coast Range, and into the Transverse Ranges, where they occupy north-facing and coast-facing slopes and ravines below 1200 m elevation. They also occur in interior valleys and on gentle foothill slopes of the Peninsular Ranges, mainly at 150–1400 m elevation, continuing south to the Sierra San Pedro Mártir, 30° N, on western slopes below 2000 m (Fig. 5.2). They occupy 2% of the state's area (Greenwood et al. 1993) and seldom extend farther than 100 km from the Pacific Ocean.

Coast live oak woodland has a physiognomy similar to that of blue oak woodland. The overstory is 9–22 m tall and incompletely closed, and the understory herbaceous layer approaches 80% cover. Shrubs (*Heteromeles arbutifolia, Rhamnus californica,* and *Salvia leucophylla*) are regularly present but contribute very little cover. In the Coast Range, *Quercus agrifolia* is the major dominant, and its basal area is 2–4 times greater than for northern oak woodlands (Allen, Holzman, and Evett 1991). In the Transverse, Peninsular, and Baja California ranges, stands are more open, and coast live oak is associated with (and sometimes subordinate to) two deciduous species: California walnut (*Juglans californica,* especially from Orange County to Santa Barbara County) and mesa or Engelmann oak (*Q. engelmannii,* especially in an 80-km-wide belt run-

ning north–south about 30 km from the coast, from Los Angeles County to San Diego County). Synonyms for these deciduous woodlands include walnut-oak woodland and Engelmann oak woodland.

Axelrod's (1988) analysis of climatic relationships among California oak woodlands suggests that the southern types differ from each other and from blue oak woodland in terms of warmth (ET, effective temperature) and equability (M, a measure of temperature amplitude during the year). As shown in Figure 5.5, equability declines in the following order: coast live oak, walnut woodland, Engelmann oak, and blue oak. The mean annual temperature is still 16° C, as in blue oak woodland, but the maritime influence and summer fog reduce annual fluctuations to only 10° C. Mean annual precipitation declines from 530 mm in the north to 260 mm along the slopes of the Baja California mountains (Hastings and Humphrey 1969; Minnich, in press).

Coast live oak woodland has been poorly quantified. The absence of data on historic stands is especially unfortunate in southern California, where walnut and Engelmann woodlands have been largely disturbed, modified, or supplanted because of human activity. The best regional description of these southern, interior woodlands comes from work in the Santa Ana Mountains and from Vegetation Type Map (VTM) surveys conducted by Wieslander in Riverside and San Diego counties in

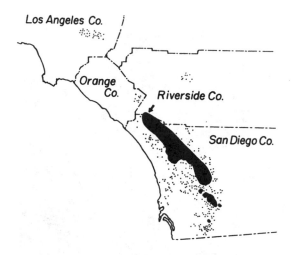

Figure 5.6. *Range of* Quercus engelmannii *in mainland California. The range also continues south into the Sierra Juarez of Baja California, to about 31°N latitude, and offshore to the west onto Santa Catalina Island (redrawn from Scott (1991).*

Table 5.2. *Average composition of three California walnut stands in southern California. Cover is relative (in percentages), basal area as in m² ha⁻¹, and stems or individuals are expressed as number per hectare. N = number of stands containing that species.*

Species	N	Cover	Basal area	Stems	Individuals
Juglans californica	3	78	23.1	2692	2113
Quercus agrifolia	1	13	12.7	463	413
Heteromeles arbutifolia	3	7	1.7	646	437
Prunus ilicifolia	2	1	0.2	80	70
Sambucus mexicana	1	<1	<0.1	30	27
Rhamnus crocea	1	<1	<0.1	10	10

Source: Keeley, 1990.

the 1930s (summarized by Griffin 1988). Only 10% of the VTM plots located in oak woodland contained *Q. engelmannii*; the rest were dominated by *Q. agrifolia*. Some Engelmann oak stands were park-like, with only 27 trees per ha and 10–50 % tree cover; 90% of the trees in such stands were *Q. engelmannii*. Other stands were denser, with 50–150 trees per ha and up to 100% cover. These dense stands either had equal densities of Engelmann and coast live oaks or exhibited a mix of Engelmann, coast, and black oaks in a 6:2:1 ratio. The Engelmann oak diameter distribution is much like that for blue oaks, most trees being in the 20–30 cm dbh class; few young trees were present. Elements of adjacent grassland or coastal sage scrub vegetation regularly form an understory.

In general, *Q. agrifolia* is more abundant on cooler, steeper slopes, and *Q. engelmannii* on warmer, gentler slopes. *Q. engelmannii* is relatively intolerant of frost and occupies warm slopes below 1300 m elevation that recieve >370 mm annual precipitation, and it also extends down into riparian drainages where it becomes associated with *Platanus* and *Salix* (Scott 1991). The pattern continues into Baja California, where *Q. engelmannii* occupies dry slopes below 1200 m, while *Q. agrifolia* ranges up to 2000 m (Wiggins 1980). Occasional trees may attain an age of 350 yr, but typical overstory individuals are only 50–80 yr old (Scott 1991). Both Engelmann and coast live oaks resprout following fire or cutting, *Q. engelmannii* doing so more vigorously (Plumb 1980).

In his 1991 review of Engelmann oak woodland,

Scott wrote that it had the smallest range of any oak in the entire southwestern part of the United States (only 35,400 ha), and ironically a distribution that fell within one of the country's fastest-growing urban landscapes (Fig. 5.6). *Quercus engelmannii* may be considered the extreme northwestern outlier of a large group of subtropical white oak species.

Much less is known about oak-walnut woodlands. Recent surveys of three stands by Keeley (1990) showed that coast live oak was not a consistent associate. One stand in Ventura County exhibited significant cover by walnut, oak, and Christmas berry, whereas two stands in Los Angeles County had walnut, Christmas berry, and holly-leaf cherry in the canopy (Table 5.2). Most walnut trees were shorter than 8 m and less than 100 yr of age. Stands were quite dense, about 2000 individuals per ha, but the small trunks yielded only 11–38 m² ha⁻¹ basal area. The age structures suggested that there has been episodic establishment during this century, reflecting mast years in wet periods and low years in drought periods. One of the stands had many multiple-stemmed trees, indicating that past fire had swept through the forest frequently; the other two stands had few multiple-stemmed trees.

Mixed Evergreen Forests: Canyon Live Oak

This complex of nearly a dozen series occurs as an ecotone between oak woodland below and montane conifer forest above. Consequently, the list of characteristic species can be long, taking in species that range considerably above and below the mixed evergreen forest. The floristic composition and distribution limits of the component series are

Figure 5.7. Mixed evergreen forest on Mt. Palomar, 1500 m elevation. Overstory species include Pinus coulteri, Pseudotsuga macrocarpa, Quercus chrysolepis, *and* Q. kelloggii.

well known, but quantitative descriptions are few. The lack of stand data is surprising, given the forest's wide distribution, covering 3–4% of the state's area (Barbour and Major 1988) and in addition extending south into Baja California and north into Oregon (see Chapter 4). Cooper (1922) included it in his "broad sclerophyll forest formation," and synonyms for some of its series include Coulter pine forest, bigcone Douglas fir forest, mixed hardwood forest, Douglas fir–hardwood forest, tan oak–madrone forest, and Santa Lucia fir forest. To the north, this complex grades into Douglas fir forest and coast redwood forest, both described in Chapter 4.

Kuchler's (1977) brief description of the forest's structure is a good beginning for our discussion: "Low to medium tall, broad-leaved evergreen forest with an admixture of broad-leaved deciduous and needle-leaved evergreen trees; the latter may be towering above the canopy (p. 925)." A review by Sawyer, Thornburg, and Griffin (1988) identified three trees as common throughout the forest's range – *Acer macrophyllum, Quercus chrysolepis*, and *Umbellularia californica* – but we choose to emphasize the oak in this chapter as the most common thread. The maple and bay are more representative

of mesic drainages and are not as widespread as the oak.

The coniferous overstory, when present, is generally scattered and 30–65 m in height. Beneath it is a more completely closed canopy, 15–30 m tall, of broadleaf evergreen trees and often some deciduous trees. Both canopies together contribute 50–100% cover (Fig. 5.7). The combined cover of understory shrubs, mosses, and perennial herbs ranges from 5–25%. The annual grasses and forbs of oak woodland understories do not continue into mixed evergreen forest. In some hardwood series, the community is essentially one-layered, the ground covered with a thick mat of undecomposed leaf litter and with shrubs and herbs largely absent.

As defined in this chapter, mixed evergreen forest extends in a broken ring around the Central Valley, facing the valley on northern Sierran slopes at 600–1200 m elevation but expanding its zone to 300–1500 m elevation in the Coast Ranges. To the north, its range reaches through the Klamath Mountains and well into Oregon (Franklin and Dyrness (1973). In the Transverse, Peninsular, and Baja ranges, the forest is 900–1400 m in elevation (Wright 1968; Minnich 1976, 1987; Vogl 1976; Wiggins 1980).

Climatic data for 11 mixed evergreen forest stations (Thorne 1988; Major 1967, 1988; Talley 1974; Plumb 1980, Wainwright and Barbour 1984) show that mean annual temperature is 14° C, significantly cooler than oak woodlands, and that mean annual precipitation is 870 mm, about 140% that of oak woodlands. Although Axelrod's thermal scheme (refer back to Fig. 5.5) showed that the tan oak–madrone type of mixed evergreen forest (equivalent to the tan oak series in Table 5.2) did not differ significantly from the cluster of oak woodland phases, Myatt (1980) inferred that the various Sierran types of mixed evergreen forest did fall out distinctly between oak woodlands and mixed conifer forest on an elevational/moisture gradient (refer back to Fig. 5.4). Mixed evergreen forest corresponds to the mesomediterranean bioclimatic level, ecologically equivalent to sclerophyllous forests of *Pinus halepensis, P. brutia, Quercus ilex, Q. rotundifolia*, and *Q. calliprinos* around the Mediterranean Sea, whereas the warmer, lower foothill woodland corresponds to the thermomediterranean woodlands of *Argania spinosa, Juniperus phoenicea, J. oxycedrus, Quercus suber*, and *Tetraclinis articulata* (Quezel and Barbero 1989).

Northern forests with Douglas fir. In the North Coast Ranges, the western portion of the Klamath Region, and the northwestern slopes of the Sierra Nevada, mixed evergreen forests have the consistent presence – but variable cover – of *Psuedotsuga menziesii*. Other conifers, such as *Abies concolor, Calocedrus decurrens, Pinus lambetiana, P. ponderosa, Taxus brevifolia*, and *Sequoia sempervirens*, may be associated in local areas. The broadleaf evergreen component consists of seven widespread species: *Arbutus menziesii, Chrosolepis chrysophylla, Lithocarpus densiflorus, Quercus agrifolia, Q. chrysolepis, Q. wislizenii*, and *Umbellularia californica* (Keeler-Wolf 1988; Bingham and Sawyer 1991; Hunter 1995).

Bingham and Sawyer (1991) surveyed 62 stands in the North Coast Ranges and were able to divide them into young (40–100 yr), mature (100–200 yr), and old-growth (>200 yr) phases. The authors summarized successional patterns. By the time stands were into their third century, they had a two-tiered canopy structure, with conifers above the hardwoods. Emergent conifers had an average dbh exceeding 90 cm, a density of 82 trees per ha, and a basal area of 75 m² per ha; the understory hardwoods had an average dbh <40 cm, a density of 75 trees per ha, and a basal area of 5 m² per ha. The old-growth stands were also unique in having high amounts of herb and moss cover (20–30%).

Another detailed study, at Annadel State Park in Sonoma County (Wainwright and Barbour 1984), showed that *Pseudotsuga menziesii* had a commanding importance percentage (based on the sum of relative density and basal area) of 40%. The only other conifer, *Sequoia sempervirens*, had the lowest importance value of 2%, for a combined conifer importance percentage of 42%. The broadleaf evergreens (*Umbellularia californica, Heteromeles arbutifolia, Arbutus menziesii*, and *Quercus agrifolia*) had a combined importance percentage almost equal to that of the conifers (39%), and the deciduous trees (*Quercus kelloggii* and *Q. garryana*) had a much lower combined importance percentage (19%).

Giant chinquapin (*Chyrsopsis chrysophylla*) is an uncommon hardwood evergreen associate of this forest, except in the Klamath Mountains area. In that area, chinquapin can be the dominant hardwood on north-facing slopes 1000–1400 m elevation that receive an annual precipitation >1500 mm (Keeler-Wolf 1988). Of 19 tree species encountered there, *Pseudotsuga menziesii* had the highest importance percentage, 35%. Other conifers contributed another 32%. Chinquapin had 21%, and other broadleaf evergreens contributed another 8%. Deciduous trees again had a low combined importance percentage (4%).

Douglas fir–mixed evergreen forest occurs along the west flank of the northern Sierra Nevada at about 500 m elevation. Gudmunds and Barbour (1984) quantified nine old-growth stands in the Yuba River drainage (39° 15'N). *Calocedrus decurrens* and *Pseudotsuga menziesii* were the leading conifers; *Arbutus menziesii, Cornus nuttallii, Lithocarpus densiflorus, Quercus chrysolepis*, and *Q. kelloggii* were the leading broadleaf species. Species composition and community architecture were similar to forests in the Coast Ranges at the same latitude. In both cases, as total stand basal area increased, the contribution of broadleaf species declined (e.g., forests with <35 m² basal area per ha had more than half of that total contributed by broadleaf species, whereas forests with 50–60 m² basal area per ha had a very low broadleaf contribution). Evergreens accounted for the decline in broadleaf basal area because deciduous hardwoods showed no statistically significant trend and their contribution was typically low.

In a similar comparative study, Gray (1978) sampled vegetation on Snow Mountain in Lake County and at a similar latitude in the Sierra Nevada. We interpreted his Sierran stands 1–8 and his Snow Mountain stands 5–8 to be equivalent to Douglas fir–mixed evergreen forest. In both cases, *Pinus ponderosa* and *Pseudotsuga menziesii* were the leading conifer dominants, contributing a combined importance percentage of 20%. *Quercus chry-*

solepis was a consistent leading broadleaf ever-green in both regions (averaging 7%), whereas the oaks *Quercus wislizenii* and *Q. kelloggii* were re-gionally distinct, the former important in the Coast Range (7%) and the latter in the Sierra Nevada (7%).

The recurrence and role of fire in Douglas fir–mixed evergreen forest are not well known. Based on buried fire scars, fires of low to moderate inten-sity revisit the same hectare of forest in the Siski-you Mountains every 11–37 yr (Agee 1991; Atzet and Martin 1992). Intense fires apparently kill Douglas fir, and if the return period is short enough, Douglas fir will be removed from the for-est (Thornburgh 1982; Sawyer et al. 1988). More recently, John Hunter examined forest dynamics in the North Coast Ranges (Hunter and Parker 1993; Hunter 1995) and concluded that Douglas fir is shorter-lived in these forests (300–500 yr) than in the Pacific Northwest (500–1000 yr). The hard-woods have a life span of 200–300 yr. If we start with a crown fire at time zero, all Douglas firs pres-ent are killed, but all hardwoods resprout. Douglas fir gradually comes back as seedlings capable not only of surviving in shade but of producing adults that grow up through hardwood crowns. Among the hardwoods, *Arbutus menziesii, Quercus garry-ana,* and *Q. wislizenii* are the least shade-tolerant, and they decline as Douglas fir canopy cover in-creases. *Chrysolepis chrysophylla* and *Quercus agri-folia* are moderately shade-tolerant, and *Lithocarpus densiflorus* and *Umbellularia californica* are the most shade-tolerant; they remain as a dense understory. Hunter concluded that a recurring crown fire is likely to happen before overstory Douglas firs die, and therefore gap-phase dynamics do not charac-terize this forest. Softwoods simply grow up through hardwoods, there is a thinning of hard-woods and the disappearance of the least shade-tolerant species, then a crown fire sets the stand back to hardwood sprouts.

Southern forests with Coulter pine. *Pinus coulteri* is scattered from Mount Diablo (38°N) south through the Coast Range, then as large patches into the Transverse Range and down the Peninsular Range into Baja California. Within this area it becomes an important, sometimes dominant element in the mixed evergreen forest. Douglas fir is either absent or replaced with bigcone Douglas fir (*Pseudotsuga macrocarpa*).

In the Santa Lucia Range, Coulter pine is a reg-ular but minor part of forests on steep slopes at 1200–1600 m elevation (Talley and Griffin 1980). Extending down to lower elevations of 250–1500 m, *Abies bracteata* can become an added element,

Table 5.3. *Basal area (m² per ha) for all trees encountered in 45 plots within a Santa Lucia fir forest. Summit plots range from 1280 to 1560 m elevation; slope plots range from 1160 to 1420 m; ravines extend down to 240 m.*

Species	Summit $N = 10$	Slope $N = 6$	Ravine $N = 29$
Quercus chrysolepis	12	25	13
Abies bracteata	6	13	12
Pinus lambertiana	1	1	<1
Pinus coulteri	1		<1
Lithocarpus densiflora		2	<1
Pinus ponderosa		1	<1
Arbutus menziesii		<1	1
Calocedrus decurrens			1
Umbellularia californica			1
Sequoia sempervirens			1
Acer macrophyllum			<1
Aesculus californica			<1
Quercus agrifolia			<1
Quercus wislizenii			<1
Platanus racemosa			<1
Alnus rhombifolia			<1

Source: Talley 1974 and his personal communication.

but the other associated species do not change (Sawyer et al. 1988; Table 5.3). Tree cover is 40–65%, and herb and shrub covers are each 5%. These steep, rocky sites are relatively fire-free (Griffin 1978). *Abies bracteata* has a very limited range of 1800 km² in the Central Coast Range (Griffin and Critchfield 1972), so it is today a minor element of mixed evergreen forest as a whole. Historically, it has been an associate of the forest for >13 million years (D. I. Axelrod, personal communication).

In the Transverse and Peninsular ranges, the Coulter pine phase is best developed between 1200 and 1800 m elevation (Minnich 1976, 1987; Thorne 1976, 1988, and it is associated with *Pinus ponderosa* (upper elevations), *Quercus chrysolepis, Q. kelloggii,* and an open shrub stratum. Vogl (1976), Vale (1979), and Minnich (1988) have pointed out the close spatial relationship between chaparral and Coulter pine forest and have theorized that the tree is an obligate fire type. Borchert (1985), however, has shown that Coulter pine cones are serotinous only where the tree is associated with chaparral, *Quercus chrysolpis,* or *Cupressus sargentii* but not where associated with *Q. agrifolia* forest, which burns less frequently. Fire-return periods before the time of fire suppression averaged 65 yr, similar to those of nearby chaparral.

A southern California endemic, bigcone Doug-las fir grows with canyon live oak in compact, dis-crete stands on steep north-facing slopes, in ra-vines, and on bedrock surfaces (Bolton and Vogl

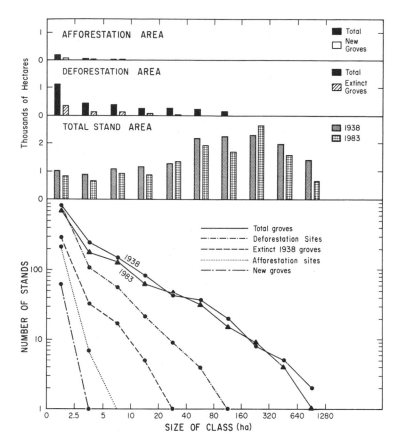

Figure 5.8. Size frequency distributions of Pseudotsuga macrocarpa stands in the San Bernardino Mountains during the period of 1938–1983. The top graph shows regrowth of burned stands or the establishment of new stands; the second graph shows diminution and extinction of burned stands; the third graph shows cumulative extant area, by stand size class, for 1938 and 1983; and the bottom graph details size distribution for lost and recovered stands. During this period, there was a net loss of 2500 ha of stand area out of a total of 15,000 ha in 1983, or −17% (from Minnich 1988 and his unpublished data).

1969; Haston, Davis, and Michaelson 1988; Fig. 5.7). At lowest elevations of 700 m, *Pseudotsuga macrocarpa* trees are scattered individuals 15–30 m tall above a closed canopy of *Quercus chrysolepis* (Minnich 1976, 1980). At about 1500 m elevation, the oak thins and the conifer becomes increasingly abundant so that total tree canopy cover (85%) is about equally divided between oak and conifer. The upper-elevation limit for the species is 2200 m.

As with *Abies bracteata*, bigcone Douglas fir forest appears to be more fire-free than surrounding mixed evergreen phases such as canyon live oak and Coulter pine (Minnich 1980). A study in the Santa Ana Mountains by Littrell and McDonald (1974), for example, showed a stable age structure for oak and conifer in mature stands. However, fire-suppression policies of this century may be altering wildfire intensities and destabilizing bigcone Douglas fir distribution. Minnich (1988 and unpublished data) has reconstructed a half-century of the dynamics of bigcone Douglas fir stands in the San Bernardino Mountains by examining repeat aerial photographs between 1938 and 1983. The fire return period during this time averaged 65 yr. Many of the stands that burned became extinct; these

were all the smallest stands (Fig. 5.8, "deforestation area"). In the same period, although some burned stands recovered and increased in area and more than 100 new stands became established, there was a net decline of >2000 ha (see Fig. 5.8, "total stand area").

Although all three species – *Quercus chrysolepis*, *Pinus coulteri*, and *Pseudotsuga macrocarpa* – occur sympatrically in the Transverse and Peninsular ranges, the two conifers tend to dominate different sites. Coulter pine and canyon live oak occur on more xeric, frequently disturbed sites adjacent to chaparral, and bigcone Douglas fir and canyon live oak occur on more mesic, protected, fire-free sites (Minnich 1980). Hanes (1976) called mixed evergreen forest with *P. macrocarpa* a "high elevation riparian woodland" to emphasize its mesic habitat in the San Gabriel Mountains.

Nonconiferous phases. Within the range of mixed evergreen forest, climax stands may apparently be dominated exclusively by the broadleaf trees *Aesculus californica*, *Arbutus menziesii*, *Lithocarpus densiflora*, *Quercus agrifolia*, *Q. chrysolepis*, *Q. kelloggii*, *Q. wislizenii*, and *Umbellularia californica*. These appear

Table 5.4. *Nonconiferous mixed evergreen forest. Sequoia-Kings Canyon National Park, 1000 m elevation. C = relative cover, D = relative density, BA = relative basal area, IV = importance percentage. Summary of three stands.*

Tree species	C	D	BA	IV
Quercus kelloggii	77	60	70	69
Quercus douglasii	13	29	16	19
Aesculus californica	7	6	10	8
Quercus wislizenii	2	6	4	4
Totals:				
Absolute cover	77%			
Absolute density	600 trees per ha			
Absolute basal area	17 m² per ha			
Absolute shrub cover	10%			

Source: Vankat 1970.

to be related to unique edaphic or topographic situations such as canyons, steep slopes, or poor soils, and they can be nearly pure stands of *Q. chrysolepis* or *Q. kelloggii* (Table 5.4). The USDA Forest Service map by Matyas and Parker (1980) and a more recent map (Anon. 1991) both show that a considerable area of the west flank of the Sierra Nevada – 170,000 ha – is in these two oak phases, which extend upslope well into the montane conifer zone (Bolsinger 1988; McDonald and Tappeiner 1996; Myatt 1980). In contrast to lower-elevation foothill woodland, a considerable volume of biomass in montane hardwood stands is in trees >70 cm dbh.

Canyon live oak forest, as mentioned earlier, is also common at all elevations within the mixed evergreen forest in southern California and northern Baja California.

Midmontane Forest

The midmontane forest is the most extensive forest in California, covering 13–14% of the state's area (Barbour and Major 1988). It is most commonly called the mixed conifer forest, but various phases of it are given equal ranking by some ecologists. Such phases have been called ponderosa pine forest, yellow pine forest, white fir forest, big tree forest, Transition Life Zone (Merriam 1898), and upper-mediterranean level (Quezel and Barbero 1989), sometimes with north-south or east-west prefixes attached. *Pinus ponderosa* is the biological thread that holds the phases of this forest together, much as *Quercus douglasii* was the matrix for northern oak woodland and *Q. chrysolepis* for mixed evergreen forest. An additional five taxa associate with ponderosa pine and form such a complicated fine-grained pattern of dominance that no single

species can be identified as a regional dominant. In more xeric or colder portions of the forest's range, especially in the eastern and southern Sierra Nevada and in the Transverse and Peninsular ranges, *P. jeffreyi* replaces ponderosa.

The mixed conifer forest is a four-layered community, though the cover contributed by each layer can be quite variable, and in the most xeric sites the herbaceous element is not significant. Overstory trees are conifers 30–60 m tall, commonly > 1 m dbh, with interlocking crowns that total 50–80% cover (Fig. 5.9). Exceptionally, *Pinus lambertiana* and *Sequoiadendron giganteum* crowns approach 80 m tall. Early accounts of the physiognomy of the forest prior to fire-suppression policies often emphasized its openness. Open areas were almost park-like, acre-size clusters of large trees alternating with openings, and with relatively few trees of intermediate height. Other early descriptions mention open areas as alternating with dense forest, and these convey a sense that patches of dark and impenetrable forest were more common than patches of open forest (see for example SNEP Science Team, vol. I, 1996, p. 63). In general, however, there is agreement that evidence exists for a change in shrub and tree cover and for a shift in the balance of species over the past 100 yr, as described more fully in a later section on fire ecology.

A subdominant tree canopy 5–15 m tall is present, but individual members are so scattered that total cover is relatively low. Members include winter-deciduous species (*Acer macrophyllum*, *Cornus nuttallii*, *Corylus cornuta*, and *Quercus kelloggii*) and broad-leaved evergreens (*Quercus chrysolepis* and tree forms of *Cercocarpus ledifolius*).

A shrub canopy <2 m tall is regularly present, and patches contribute 10–30% cover. Deciduous and evergreen species are common in the genera *Arctostaphylos*, *Ceanothus*, *Chamaebatia*, *Castanopsis*, *Lithocarpus*, *Prunus*, *Quercus*, *Ribes*, *Symporicarpus*, and *Vaccinium*.

Herb cover may reach 20% (Vankat and Major 1978) but is more commonly 5–10%. Perennial forbs predominate, some of which are hemiparasitic, and species richness is relatively high. The more common genera include *Adenocaulon*, *Clintonia*, *Disporum*, *Galium*, *Iris*, *Lupinus*, *Osmorhiza*, *Pteridium*, *Pyrola*, *Smilacina*, and *Viola*.

Physical factors. The elevational limits of mixed conifer forest reflect latitude, proximity to humid air, slope aspect, and soil depth. Considering its entire range, the type may extend from 800–2600 m in elevation. The forest is absent from the western Transverse Ranges and the Central Coast Range, where peaks are below 1400 m (for exam-

Figure 5.9. Aspect of the midmontane mixed conifer forest in the central Sierra Nevada, Placer County, 1550 m elevation. Prominent species include Abies concolor, Calocedrus decurrens, Pinus lambertiana, P. ponderosa, Pseudotsuga menziesii, and Quercus kelloggii.

ple, the highest peaks in the Santa Lucia Range of Monterey County support a depauperate mixed evergreen forest dominated by *Quercus chrysolepis*, *Pinus lambertiana*, and *Abies bracteata*).

Judging by Major's (1967, 1988) analyses of climatic gradients along the Sierra Nevada's west flank and our own analyses of gradients in southern California, the lower elevational limit corresponds to a mean annual temperature of 13° C and an annual precipitation of 850 mm. Well within the forest, the mean annual temperature is 11° C, and annual precipitation exceeds 1000 mm. Temperature and lapse rates along the west flank of the Sierra Nevada are relatively similar, north to south, averaging about −0.5° C and +40 mm per 100 m rise. Our own analyses of precipitation lapse rates for the Transverse and Peninsular ranges show considerably shallower slopes: only +11 to +5 mm precipitation per 100 m rise. The percentage of precipitation that falls as snow also increases with elevation. In the northern Sierra Nevada, about 33% falls as snow within the mixed conifer forest zone (Major 1988).

Parker (1994) has shown that the mean elevational displacement of all tree species that occur between Lassen National Volcanic Park and Yosemite National Park is −172 m per increasing degree of latitude. This rate of change is much steeper than the −100 m per degree for Asian temperate mountains, −91 m for the Appalachians, and −77 m for the Rocky Mountains (see Chapters 3 and 10).

Soils at lower elevations and on moderate slopes are typically loamy Alfisols or Ultisols, often distinctively red in color, with a pH of 5–6, and having a depth of 80–180 cm. Common soil series include Aiken, Josephine, Mariposa, and Sites (all Ultisols) and Atwell, Boomer, Cohasset, Holland, and Musick (all Alfisols). On steeper terrain, with less soil development, Inceptisols are common (Chaix, Corbett, Hugo, Masterson, McCarthy, Neuns, Sheetiron) (CDFR 1979; Laacke 1979).

In the pages that follow, we describe only four major phases of midmontane-forests: mixed conifer, white fir, Jeffrey pine, and Sierra big tree. This is a simplification of the detailed understanding that regional ecologists are developing about a diversity of series and associations. Fites (1993) has recognized 20 associations in the northern Sierra Nevada and southern Cascade Range alone, and Sawyer and Keeler-Wolf (1995) include more from other parts of the forest's range.

Phases

Mixed conifer. The mixed conifer forest exhibits shared or shifting dominance by *Abies concolor*

Table 5.5. *Importance percentage of all conifer species in an old-growth forest at Placer County Big Trees State Park, 1600 m elevation. Trees were sampled into two categories: 3–40 cm dbh and >40 cm dbh.* Abies concolor *trees 40 cm dbh are approximately 125 yr old. The last two columns summarize average tree diameter, breast height (dbh, in centimeters) and density (trees per hectare).*

Species	3–40 cm	>40 cm	dbh	Density
Abies concolor var. *lowiana*	41	12	8	258
Calocedrus deccurens	12	9	15	67
Pinus lambertiana	12	23	23	87
Pinus ponderosa	16	30	17	114
Pseudotsuga menziesii	19	26	19	122

Source: Barbour et al. 1987.

var. *lowiana* (northern California white fir) or *A. c.* var. *concolor* (southern California white fir (Hamrick and Libby 1972; Vasek 1985), *Calocedrus decurrens* (incense cedar), *Pinus ponderosa* (ponderosa pine, mainly in northern California) or *P. jeffreyi* (Jeffrey pine, mainly in southern California), *P. lambertiana* (sugar pine), and *Pseudotsuga menziesii* (Douglas fir, limited to northern California). An old-growth stand at Placer County Big Trees State Park is a classic example of this phase (Table 5.5). The overstory canopy is weighted toward ponderosa pine, Douglas fir, and sugar pine, whereas the understory canopy is dominated by white fir. The absolute density of white fir trees of all diameters is 258 per ha, yet because their average dbh is only 8 cm, their contribution to basal area is modest.

In southern California, the mixed conifer forest contains almost equal contributions of Jeffrey pine and ponderosa pine. Gemmill's (1980) data for the San Bernardino Mountains, for example, show a ratio of densities for ponderosa pine–Jeffrey pine–white fir–incense cedar–sugar pine as approximately 7-4-4-1-1 (total absolute density = 197 trees per ha). Thorne (1988), Vogl (1976), Borchert and Hibberd (1984), and Kuchler (1977) all agreed that 2100 m elevation in southern California mountains is a transition point between dominance by ponderosa and Jeffrey pines. A lower zone, 1400–2100 m, is a pondersoa pine variant, and a higher zone, 2100–2700 m, is a Jeffrey pine variant of a "yellow pine forest" (Thorne 1988). Haller (1962) examined the traits of the two yellow pines and of putative hybrids. Hybrids are regularly found but not in high frequencies. The hybrid frequency is kept low by seasonal differences in dates of pollen maturity, by the reduced viability of hybrid seed, and by failure of hybrids to become established.

Figure 5.10. *Aspect of mixed conifer forest in the Sierra San Pedro Mártir. Prominent species include* Pinus jeffreyi, P. lambertiana, *and* Abies concolor *var.* concolor.

Haller found the elevational span for hybrids to be rather narrow, about 275 m around a mean elevation of 1970 m.

Ponderosa pine is completely supplanted by Jeffrey pine south of the United States–Mexico border (Minnich 1987). Midmontane conifer forest occurs at elevations between 1500 and 2900 m in the Sierra Juarez and Sierra San Pedro Mártir at the southern limit of the Peninsular Ranges (Passini, Delgadillo, and Salazar 1989; Minnich, Barbour, Burk, and Fernau 1995, and Minnich, Barbour, Burk, and Sosa-Ramirez 1998). The 44,000 ha of forests in the Mártir have never experienced logging or fire-suppression management, and thus they represent the last Californian old-growth mixed conifer forest at a landscape scale. A zone of monotypic Jeffrey pine forest occupies the lower fringe of the midmontane, 1500–2100 m. Above this is a mixed conifer forest with *Abies concolor* var. *concolor, Calocedrus decurrens, Cupressus montana, Pinus jeffreyi, P. lambertiana, P. contorta* ssp. *murrayana,* and *Populus tremuloides.* The physiognomy is open and park-like (Fig. 5.10); overstory cover ranges between 25 and 45%, and tree density ranges between 65 and 145 per ha (with saplings, <3 cm dbh, having 2–3 times that density). Shrub and herb cover is only 5–10%. Tree size and age distri-

butions indicate multiage stands with steady-state dynamics (Fig. 5.11). The authors argue that these forests are not open because of a comparatively arid environment. Annual precipitation, seasonality of occurrence, and soil hydrology patterns are well within southern Californian ranges. Where local topography prevents ground fires from encroaching, tree density approaches that of Sierran forests: One mixed conifer stand, dominated by white fir and situated on a north-facing slope which had been unburned for at least the last century, contained 387 trees per ha.

White fir. With increasing elevation in the midmontane zone, white fir becomes dominant over a broad spectrum of habitats. Many regional ecologists recognize the existence of a climax white fir forest (Thorne 1976; Rundel, Parsons, and Gordon 1988; Matyas and Parker 1980; Parker 1982; Vankat 1982). Some elements of the upper montane forest penetrate down into this white fir forest: *Abies magnifica* in the more mesic northern part of California, and *Pinus contorta* ssp. *murrayana* in the more xeric southern part.

Tree canopy cover is generally >75%, but herb and shrub covers are lower than those of mixed conifer forest. Perhaps the most detailed compari-

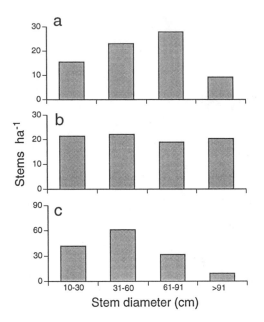

Figure 5.11. Stem densities for all species by size class in Sierra San Pedro Mártir forests: (A) monotypic Jeffrey pine forest (average of nine stands); (B) mixed conifer forest dominated by Jeffrey pine (seven stands); (C) mixed conifer forest dominated by white fir (five stands) (from Minnich et al. 1995, reprinted by permission of Blackwell Science, Inc.).

Table 5.6. Importance value of trees in 27 samples of mixed conifer and white fir phases of the mid-montane forest, southern Sierra Nevada, 36°30' N Latitude, 1500–2200 m elevation.

Species	Mixed conifer	White fir
Calocedrus decurrens	30	24
Quercus kelloggii	25	5
Pinus ponderosa	16	<1
Abies concolor var. lowiana	16	45
Pinus lambertiana	10	12
Quercus chrysolepis	3	1
Cornus nuttallii	<1	4
Abies magnifica		4
Corylus cornuta		2
Pinus jeffreyi		<1
Populus balsamifera ssp. trichocarpa		<1
Torreya californica		<1

Source: Vankat and Major 1978.

son of white fir and mixed conifer stands in one area is that by Vankat and Major (1978) (see also Table 5.6 this chapter). The importance percentage of Abies concolor was three times as great in the white fir phase as in the mixed conifer, whereas Pinus ponderosa's fell to <1. A coefficient of community or similarity index between the two phases in Table 5.6, weighted by importance percentage, is 58.

Farther north, white fir stands in Sierra County are common on mesic sites between 1500 and 2000 m (Conard and Radosevich 1982). These stands show an even more pronounced dominance by white fir, with an importance percentage of 93%. The western slopes of the Klamath Mountains are also dominated by white fir, with Douglas fir as the most abundant associate (Sawyer and Thornburgh 1988). Ponderosa pine, sugar pine, chinquapin, and Pacific yew (Taxus brevifolia) are also present. More interior, eastern slopes have higher importance by ponderosa pine and sugar pine, "producing forests similar to Sierra Nevada mixed conifer." It is in such an interior place that the richest collection of conifers in all of California exists: the Sugar Creek Research Natural Area, within Klamath National Forest in Siskiyou County (Keeler-Wolf 1990). As many as 10 conifer species co-occur in homogeneous stands between

1645 and 1829 m elation. The mix is rich in part because it includes midmontane, upper montane, and subalpine elements; it also includes southern outliers of species that are otherwise absent from the state. Midmontane conifers include Abies concolor, Calocedrus decurrens, Pinus lambertiana, P. ponderosa, Pseudotsuga menziesii, and Taxus brevifolia. Upper montane conifers include Abies magnifica var. shastensis, Pinus jeffreyi, and P. monticola. Subalpine conifers include Pinus albicaulis, P. balfouriana, and Tsuga mertensiana. Outlier relicts of extra-Californian species include Abies lasiocarpa, A. procera, Chamaecyparis lawsoniana, C. nootkatensis, Picea breweriana, and P. engelmannii. The overstory reaches to 60 m, cover is >60%, tree density is >1000 per ha, and basal area approaches 72 m² per ha. Shrub and herb diversity is also high.

On ultrabasic parent materials, the white fir–mixed conifer Klamath forest becomes a woodland of scattered Abies concolor, Chamaecyparus lawsoniana, Pinus jeffreyi, P. monticola, and Psuedotsuga menziezii over a continuous shrub canopy. Shrub taxa include Quercus vaccinifolia, Arctostaphylos nevadensis, A. patula, and Quercus vaccinifolia, among many others. Herb cover is low, Xerophyllum tenax and bunchgrasses being the more common (Sawyer and Thornburgh 1988).

As summarized by Vankat (1970) and more recently by SNEP (1996), there is evidence from many studies throughout California that tree density has increased and shrub cover has decreased in the white fir phase over the past century. The changes are usually ascribed to lack of ground fires, because young white firs are easily killed by such fires, in comparison to pines. It may also be possible that young white firs are more shade-tolerant than pines (Lanini and Radosevich 1986).

Sierra big tree. The Sierra big tree, Sierra redwood, or giant sequoia (*Sequoiadendron giganteum*) is distributed in discrete groves from Placer County to the southern boundary of Tulare County. Rundel (1972a) identified 75 groves, whereas a more recent survey by Willard (1992) condenses them into 65 groves. Individual trees may attain a biomass of 6000 tons, 5 m dbh, 100 m in height, and 3200 yr of age. Cones remain green and closed so long as they are attached to the parent tree. If killed by heated air from a ground fire (or if removed from the tree by animals), the scales open and seeds are released; thus Sierra big tree is a closed-cone conifer. Its establishment requirements for bare mineral soil and high light levels lead some ecologists to label it a pioneer species (Stephenson 1992).

Apart from the imposing presence of *Sequoiadendron* (importance percentage about 33%) and a richer understory of shrubs and perennial herbs, the floristic and physiognomic characteristics of big-tree groves are not significantly different from the white fir phase (e.g., ponderosa pine's importance percentage is <1% and white fir's is >30%).

The groves range from 825 to 2680 m in elevation and are situated in particularly mesic microenvironments (Fig. 5.12). Grove boundaries appear to have been stable for the past 500 yr (Rundel 1971, 1972b). Because of the brittleness of the wood and high cost of handling such large boles, logging had removed only 34% of big-tree acreage up to the time that most stands were placed in public ownership (Hartesveldt, Harvey, Shellhammer, and Stecker 1975). Today, 90% of big-tree acreage is in public ownership, though that does not necessarily mean they are protected. Only in 1990, for example, did the USDA Forest Service agree to set aside all big-tree groves but one in Sequoia National Forest from timber harvest (McDonald 1996).

Jeffrey pine. As previously mentioned, Jeffrey pine gradually supplants ponderosa pine with decreasing latitude. In Baja California it is found throughout the midmontane forest zone; in most of California it is in the upper part of the midmontane; and in northern California it is typically restricted to the upper montane. In the southern Sierra, for example (Fig. 5.12), the two species coexist at about 1800 m. Above this elevation Jeffrey pine not only is a major associate in the midmontane forest, but it extends well into the upper montane forest (Vankat 1982). Jeffrey pine appears to respond more to local edaphic or microenvironmen-

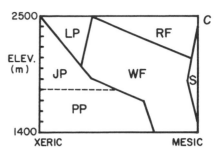

Figure 5.12. Diagrammatic relationships between several phases of the midmontane conifer forest in the southern Sierra Nevada, Sequoia–Kings Canyon National Park. PP = ponderosa pine mixed conifer phase, WF = white fir phase, S = Sierra big tree, JP = Jeffrey pine, LP = lodgepole pine, RF = red fir. The latter three are upper-montane phases that penetrate into the midmontane forest (from Rundel et al. 1988).

tal factors than to regional climates or lapse rates. It is often on more arid south- or west-facing slopes, on shallow soils, or on serpentinite substrates.

Haller (1959, 1962), Wright (1968), Yeaton, Yeaton, and Horenstein (1980), and Yeaton (1978, 1981, 1983a, 1983b) have all examined the elevational replacement of montane pines in some detail. With respect to the low-elevation replacement of *Pinus sabiniana* by *P. ponderosa*, both Yeaton (1981) and Griffin (1965) implicated competition for soil moisture at the sapling stage. Their reasoning is deductive rather than based on experimental manipulation, but it is persuasive. Gray pine's habitat range becomes more and more restricted with increasing elevation, to the most open habitats on serpentine. In mixed stands (700–800 m elevation), saplings of ponderosa pine do have the faster growth rates, implying a more favorable carbon balance, but we have seen no data on photosynthetic rates for the two pines. Both are diploxylon pines, a taxonomic group to which Yeaton (1981) ascribed the following traits: colonizing arid, rather open habitats; reproducing at a young age; reproducing frequently; producing small seeds; and in general possessing r-selected traits. Consequently, the two diploxylon pines are more likely to compete for similar habitats than is a haploxylon–diploxylon mix.

Jeffrey pine is another diploxylon pine, and this may explain its rather narrow ecotone with ponderosa pine. At the same elevation where these two pines displace each other, so too do two haploxylon white pines: sugar pine gives way to western white pine (*Pinus monticola*). Here again, Yeaton (1981) hypothesized that competition for soil moisture is the driving mechanism. To test his hypothesis, we need comparative data on water-use efficiency, photosynthetic rates, and carbon allocation patterns.

Fire ecology. Fire plays an integral role in the dynamics and structure of Californian mixed conifer forest (Kilgore 1981; Rundel et al. 1988; Swetnam 1993; Wright and Bailey 1982). The seed release, germination, and seedling establishment phases of many woody species in the midmontane zone are enhanced by the effects of ground fires. Furthermore, the architecture of the forest and the balance in relative abundances of tree species were maintained by low-intensity surface fires. *Abies concolor* has increased in importance dramatically coincident with the emplacement of fire-suppression management in the early part of the twentieth century. White fir seedling density and sapling mortality are inversely related to fire intensity and fire-return period.

The classic early surveys of the newly formed forest reserves in California by Leiberg, Sudworth, and others at the turn of the century described more open forests than those of today – forests that were frequently visited by surface fires (see reviews by McKelvey and Johnston 1992 and others in SNEP Vol. II 1996, and photographic evidence gathered by Gruell 1998). Using fire scar dendrochronology methods, presuppression fire intervals estimated for mixed conifer forest range from 4–20 yr, depending on vegetation type and local topography (Show and Kotok 1924; Wagener 1961; Weaver 1974; McBride and Laven 1976; Kilgore and Taylor 1979; Finney and Martin 1989; Swetnam 1993). Because fire-free intervals seem to have been short, it was deduced that intensities were low, leaving pole-size and larger trees unharmed (Kilgore 1981). Fire areas cannot be easily determined by fire scar data, but the present model of presuppression surface fire assumes that fires were typically small (1–800 ha), patchy, and irregularly shaped.

Unfortunately, data on nineteenth-century forests are few, so much of the literature on vegetation change caused by fire suppression is anecdotal. One of the more detailed studies recently resurveyed 68 forest plots in the San Bernardino Mountains initially sampled by a California VTM survey in 1929–1934 (Minnich et al. 1995). The co-authors found that tree densities ("tree" defined as having dbh >10 cm) had increased overall in the 60 yr from 116 to 207 per ha, or by 79%. Density increase in the smallest size class (10–30 cm dbh) was 200–1000%, whereas increase in the larger size classes (>61 cm) were insignificant or even declined because more individuals left the classes through death than entered them

through growth. Thickening rates were also proportional to mean annual precipitation: Forests with annual precipitation of 1000 mm showed the greatest overall increase in density, whereas those with annual precipitation of 400 mm showed no statistically significant change. The most mesic conifer stands previously dominated by ponderosa pine had shifted to dominance by incense cedar and white fir, with about 250 trees per ha. If the presuppression fire return period of 15–30 yr (McBride and Laven 1976) had been in effect during the 60 yr of demographic change, a majority of the 68 plots would have been burned; instead, only 6 had been burned, giving a postsuppression fire return period of 700 yr. A similar estimate (644 yr) has recently been generated for modern Sierra Nevada mixed conifer forest by McKelvey and Busse (1996).

Additional details on the fire ecology of mixed conifer forest come from the Sierra San Pedro Mártir, mentioned earlier as being the last landscape-scale piece of old-growth midmontane vegetation that still experiences a natural wildfire regime (Minnich and Franco-Vizcaino 1997). Minnich et al. (1998) reconstructed fire history there from repeat fire scars and aerial photography for the past 65 yr. The fire scar data indicate an average fire return period of 26 yr and reveal no decline in the twentieth century (in contrast to studies elsewhere in the West that have all experienced fire-suppression management). Repeat aerial photography gave additional information on fire intensity, tree mortality, and fire patch size. Fire patch size (Table 5.7) had a skewed distribution, the majority of the 865 fires being <16 ha in size, but the 19 largest ones accounting for almost half of 143,000 ha burned. Less than 3% of the area experienced stand-replacing crown fires in the same period.

Estimated fire return period from the photographic evidence was much longer than that generated from fire scar evidence: 52 yr. Such a long fire-free interval should permit considerable fuel accumulation, predicting that surface fires will be hotter than the commonly accepted California model postulates. In fact, Minnich and his associates were able to document that such fires killed about 67% of pole-size trees and 20% of overstory trees. If the behavior of fire in the Sierra San Pedro Mártir is representative of mixed conifer forest throughout the Californias, then the current surface-fire model has two flaws in its reconstruction of presuppression fires: The fires had a much longer fire return period, and they burned much hotter, having a significant impact on the survival of mature trees.

Despite ambiguities in reconstructing presuppression fire behavior, there is wide agreement

Table 5.7. *Number of fires and area burned in the period 1925–1991 in mixed conifer forest of the Sierra San Pedro Mártir, as recorded in repeat aerial photographs. The total area of the forest is only 44,000 ha and the total area burned = 143,355 ha, so some areas burned more than once. The estimated fire-return period is 52 yr. Total number of fires = 865.*

	Fire size class (hectares)									
	<16	16–50	50–100	100–200	200–400	400–800	800–1600	1600–3200	3200–6400	>6400
Number	436	166	83	61	58	20	22	8	9	2
Area	2511	4687	6174	8706	16,738	11,127	23,717	18,145	35,241	16,311

Source: Minnich et al. 1998.

that a century of fire-suppression management has resulted in a large increase of dead and living fuel, increasing the probability of surface fires becoming crown fires of greater areal extent and intensity. That is, "pyrodiversity" has declined from a once wide spectrum over space, time, and intensity. The likely consequence is that biodiversity has declined as well (Martin and Sapsis 1992). Fires can be agents of diversity. They can also reduce stress and mortality caused by competition. A succession of drought years in the late 1980s and early 1990s produced historically high die-offs of white fir and yellow pine throughout California in dramatic contrast to unchanged low mortality in the less dense Baja California forests. Savage (1994), for example, studied this episode of mortality in an 18,000 m² of scattered plots in the San Jacinto Mountains. She reported 42% mortality in ponderosa pine, 50% in Jeffrey pine, 49% in Coulter pine, and <10% in white fir–sugar pine–incense cedar (combined). This episode created approximately 180 snags per ha, reducing stand density from 550 per ha to 370 per ha. The number of snags was two orders of magnitude greater than those created at the same time in the Sierra San Pedro Mártir forests, which were about one-third as dense, probably because of continuing exposure to natural wildfires (Minnich et al., 1998).

As a move toward reinstating fire in the landscape, the USDA Forest Service, the National Park Service, the California Department of Forestry, and the California Department of Parks and Recreation have begun programs of prescribed burning and have adopted "let-burn" policies for natural fires in certain circumstances. Unfortunately, the area burned annually is still only a small fraction of that required to adequately reduce fuel, raise biodiversity, and improve forest health.

Air pollution. Air pollutants that affect vegetation in the midmontane zone are primarily those released in automobile exhaust and secondarily those that result from exhaust fumes interacting with oxygen in the presence of sunlight. Major phytotoxicants are ozone, nitrogen oxides, hydrocarbons such as peroxy acetyl nitrate (PAN), and sulfur dioxide.

Ozone appears to be the most important phytotoxicant in the complex. During summer months, temperature inversions are common, and the result is an accumulation of ozone along mountain slopes up to the inversion boundary layer, commonly 1700 m elevation. Current federal EPA standards are that ozone should not exceed a concentration of 0.12 ppm for more than 1 hr per yr. There is evidence now that mixed conifer forests in the Transverse, Peninsular, and Sierra Nevada Ranges exceed this limit (Cahill, Carroll, Campbell, and Gill 1996; Miller 1973, 1996; McBride, Semion, and Miller 1975; Williams, Brady, and Willison 1977; Pronos and Vogler 1981; Williams 1983). California's state agency limit (0.09 ppm) is a better ecological target, according to SNEP (Vol. 1, 1996).

Ponderosa pine and Jeffrey pine are the species most sensitive to ozone. Symptoms include mottling on the needles, premature needle drop, reduction in net photosynthesis, and reduction in needle size (Miller 1996; Patterson and Rundel 1995; Taylor 1980). Suppressed understory individuals are more sensitive than mature overstory trees. White fir, incense cedar, and sugar pine are relatively insensitive to ozone and rarely exhibit symptoms.

Ozone-affected trees were first detected in the 1950s in mixed conifer forest of the Transverse and Peninsular ranges, on slopes facing the Los Angeles basin. Tree decline and death there were significant enough to require salvage harvests in the 1960s and 1970s. Ozone damage has since been documented along the west slope of the Sierra Nevada as far north as Lake Tahoe (Fig. 5.13). In 1993, the proportion of ponderosa or Jeffrey pine trees showing symptoms was 100% at Barton Flats in the Transverse Range, 84% at Sequoia–Kings Canyon

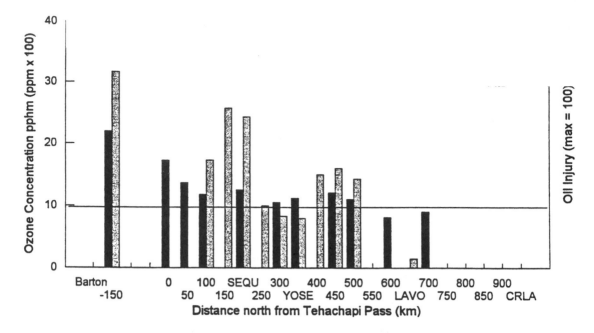

Figure 5.13. Maximum ozone concentrations (black bars) and ozone injury indices (gray bars) along a south-to-north transect from Barton Flats in southern California along the west flank of the Sierra Nevada to Crater Lake, Oregon, in the Cascade Range. SEQU = Sequoia–Kings Canyon Na- tional Park; YOSE = Yosemite National Park; LAVO = Lassen Volcanic National Park. Maximum ozone concentrations are means of the 30 highest daily exposures per year, for 3 yr (from Cahill et al. 1996).

Figure 5.14. Lodgepole pine forest, San Gabriel Mountains in the Transverse Ranges, 2440 m elevation. Openings are occupied by Ceanothus cordulatus and Castanopsis sem- pervirens. (Photo courtesy of R. F. Thorne.)

National Park in the southern Sierra Nevada, 53% in the central Sierra east of Sacramento, and down to 19% in Lassen Volcanic National Park in the southern Cascade Range (Miller 1996). Air circulation patterns in the San Joaquin Valley allow pollutants to accumulate in a north-to-south gradient that increases from the latitude of Sacramento to Tehachapi Pass. The Sacramento Valley to the north does not have such a closed airshed.

Oxides of nitrogen and sulfur, deposited on mountain slopes either as acid rain or in the form of dry deposition, have the potential for significant environmental impact because granitic substrates – common in California – have little buffering capacity. McColl (1981) concluded that most acid rain in California was caused by NO_x rather than SO_x. He showed that germination, seedling mortality, growth, and needle longevity of ponderosa pine and Douglas fir were all adversely affected in growth chamber simulations with artificially applied mists. Bradford, Page, and Straughan (1981) surveyed 170 Sierran lakes and found no statistically significant change over the previous 15 yr, leading them to conclude that montane ecosystems were sufficiently distant to allow dilution to levels that are not yet having an ecological effect.

Upper Montane Forests

One species that provides a thread of continuity through the upper montane forests, from the Klamath Mountains and Cascade Range in the north to the Sierra San Pedro Mártir in the south, is lodgepole pine (*Pinus contorta* ssp. *murrayana*). The forests correspond to the Canadian Life Zone of Merriam (1898) and to the mountain-Mediterranean altitudinal level of Quezel and Barbero (1989). Some ecologists (Hanes 1976; Thorne 1982, 1988) refer to all lodgepole pine forests as subalpine, but in a global review of subalpine forests, MacMahon and Andersen (1982) defined subalpine as the habitat of woodlands of *Pinus albicaulis* (in the north) or *P. flexilis* (in the south). We adopt their view here, placing closed forests of lodgepole pine and such associates as red fir (*Abies magnifica*) and quaking aspen (*Populus tremuloides*) in the upper montane.

Upper montane forests cover about 3% of the state's area. Elevational limits are 1800–2400 m in northern California, 2200–3000 m in the southern Sierra Nevada, and 2400–2800 m in southern California.

Soils are shallow, rocky Inceptisols or Entisols. Weather data are modest, but they suggest that mean annual precipitation is 820 mm (though northernmost sites may receive 1000–1600 mm).

The total amount of precipitation, thus, does not significantly differ between midmontane and upper montane forests; what does change is the form of precipitation and the mean annual temperature. About 70–90% of all precipitation falls as snow (compared to 33% in the midmontane), leading to snowpack depths of 2.5–4.0 m and snow duration nearly 200 d per year. Mean annual temperature is 5° C, well below the 11° C of midmontane forest.

Lodgepole pine phase. Pacific lodgepole pine of California differs from the Rocky Mountain *P. contorta* ssp. *latifolia* in being open-cone. The Sierran lodgepole forest is not a fire type community (Parker 1986a). More has been published about stand structure, population dynamics, and autecology for the Rocky Mountain variety than for the Pacific taxon.

The bimodal habitat distribution of lodgepole pine in California is well known: It can occur on arid, windswept sites on shallow soils in the subalpine zone, and it dominates relatively wet sites at the edges of upper montane meadows or lakes that also receive cold-air drainage (Rundel et al. 1988). In these wetter sites, red fir often dominates the understory, suggesting a succession to red fir dominance. Stands of *Populus tremuloides* or *P. balsamifera* ssp. *trichocarpa* may alternate with lodgepole stands in similar habitats.

Lodgepole pine forests are moderately dense, with 55–80% cover, and are of modest stature, the trees typically <300 yr old, <20 m tall, and <70 cm dbh (Fig. 5.14). Tree density is high, as many as 2000 per ha, but basal area is modest, about 60 m² per ha. Lodgepole pine is a thorough dominant of such forests, with an importance percentage >67% (Parker 1986a; Vankat and Major 1978).

Shrub cover is typically insignificant but may in places reach >15%. Prominent genera include *Arctostaphylos*, *Ceanothus*, *Castanopsis*, and *Ribes*. Except where lodgepole encroaches on meadows, herb cover is also insignificant.

Vankat and Major (1978) reviewed the evidence for lodgepole invasion of Sierran meadows and concluded that the most recent wave of increase began about 1900. They ascribed the increase to elimination of sheep grazing but added that there may be an overall cycle relating to fire that ultimately regulates lodgepole movement. Helms (1987), on the other hand, was able to relate lodgepole pine dynamics solely to annual precipitation, dry years permitting invasion into meadows. Other biologists theorize that the dynamics of lodgepole stands are driven by episodic infections of the needle miner *Coleotechnites milleri*. Many stands, however, do have an age structure that implies a climax state (Eyre 1980; Parker 1986a).

Lodgepole pine forests experience variable fire patterns, including low-intensity surface fires, moderately intense understory fires, and canopy fires (Baumgartner 1985; Agee 1993). Fire scar data are few, and scarring from insect predations may confuse the historical reconstruction (Agee 1993); furthermore, years with late snowmelt often pass without any xylem ring having been laid down. Fire return periods for southern Oregon stands appear to be in the range of 60–80 yr, but stands free of fire for 350 yr are known (Agee 1993; Stuart 1984).

Red fir phase. In northern California, the most mesic sites within the upper montane zone are dominated by red fir. Lodgepole pine is here a consistent but minor element. *Abies magnifica* is among the largest species of the genus *Abies*. The physiognomy of this forest contrasts with that of lodgepole pine forest. Mature overstory trees are commonly 30–45 m tall, >1 m dbh, and >300 yr old. Stands are somewhat open and park-like, the impressive columnar boles seemingly evenly spaced (Fig. 5.15). Overstory cover averages 60%; herb and shrub cover <5%. Tree regeneration is notably patchy.

The range of red fir is largely restricted to northern California, but does extend into the southern Cascades of Oregon. It is not genetically homogeneous throughout this range but, rather, exists in three forms: the species itself, as Shasta red fir (*A.m.* var. *shastensis*), and as a series of hybrids in the southern Cascades between *A. magnifica* and noble fir (*Abies procera*). Shasta red fir has a disjunct distribution, primarily restricted to the southern part of the Sierra Nevada and to the southern Cascade–Klamath–North Coast Range area (Hallin 1957). Significant genetic differences distinguish the two nodes (Constance Millar, USDA Forest Service, Albany, CA, personal communication). It appears to be completely sympatric and interfertile with red fir, and in terms of stand structure Barbour and Woodward (1985) found no significant differences between stands dominated by either taxon.

In pioneering work more than five decades ago, Oosting and Billings (1943) described the *Abietum magnificae* association. Their work was based on rather limited sampling data, and red fir vegetation is still less intensively understood than the mixed conifer forest of the midmontane zone. Barbour and Woodward (1985) summarized data from 84 widely distributed stands (Table 5.8). Associated trees (not listed in the table) included white fir, western white pine (*Pinus monticola*), Jeffrey pine, sugar pine, lodgepole pine, and mountain hemlock

Figure 5.15. *Understory view of red fir forest, central Sierra Nevada, Placer County, 2000 m elevation. Wolf lichen* (Letharia vulpina) *is prominent on the trunks. A dense patch of red fir saplings occupies the opening. Note human figure for scale.*

(*Tsuga mertensiana*), but their contribution to community composition was modest, as shown by the huge average importance percentage for red fir of 84%. Basal area for red fir stands (83 m^2 per ha) is the highest for any California montane type. Diversity from stand to stand is indicated by the broad range of values in Table 5.8. Additional diversity, not shown in that table, comes from the presence and abundance of understory species. Potter (1994) described 11 different associations within red fir forest just for the central and southern Sierra Nevada, Jimmerson (1994) has described 16 more from the Klamath area, and Sawyer and Keeler-Wolf (1995) list another dozen.

Stand dynamics. Stand dynamics and the role of disturbance in red fir forest are less completely understood than classification. Fire history in these forests has not been intensively or extensively studied (Agee 1993; Andrews 1993; Potter et al.

Table 5.8. *Red fir contribution to 86 stands distributed from the North Coast Range through the Sierra Nevada. BA = basal area; abs = absolute; rel = relative (%); IP = importance percentage. Commonly associated trees in these stands included* Abies concolor var. lowiana, Pinus monticola, *and* P. jeffreyi.

| | Trees per hectare | | BA in m^2 ha^{-1} | | Cover (%) | | |
	Abs	Rel	Abs	Rel	Abs	Rel	IP
Range	70–1832	8–100	8–160	20–100	16–90	39–100	22–100
Mean	418	82	69	83	55	89	84

Source: Barbour and Woodward 1985.

1992). Fire occurs in red fir forests, as is evident from external fire scars and agency fire records, but only three published investigations exist regarding fire return intervals and population responses. Pitcher (1987) completed a study of three stands in the Mineral King area of Sequoia National Park, Taylor and Halpern (1991) and Taylor (1993) examined two other stands at Swain Mountain Experimental Forest in the southernmost portion of the Cascade Range, and Chappell and Agee (1996) studied three stands in Crater Lake National Park, Oregon. Fewer than 100 scars, distributed over a combined sample of only 7 ha, represent the total published information we have for a forest type that covers 1.2 million ha. The data suggest that surface fires recur in red fir forests every 26–65 yr and that regeneration of red fir may or may not be linked with such fires; also that occasional stand-replacing crown fires do occur but at an unknown frequency.

Donald Potter, zone ecologist for the USDA Forest Service, has accumulated unpublished scar data for tree stumps on 68 clearcut old-growth red fir stands throughout the central and southern Sierra Nevada. Each stand sample was based on the first 10 scarred stumps he encountered. His work suggests a 53-yr fire return period. He has no information on fir population response as correlated with fire scars. Barbour and Antos (unpublished data) examined six other Sierran old-growth clearcuts in a more exhaustive manner. Each clearcut was sampled with an 80 × 80 m plot, within which all stumps remaining after harvest were aged and searched for fire scars. Fewer than 10% of the stumps encountered had fire scars, and most were scarred only once. In 6 of the 9 stands there was a concordance among stumps showing the same fire date; that subset of data suggests fire-return periods between 30 and 195 yr. There was no consistent demographic response to this subset of dates; increased regeneration, depressed regeneration, and no change were all seen in the years immediately postfire.

The age structure of red fir stands is also poorly known, quite apart from any presumed relationship with the timing of disturbance. Studies by Pitcher, Taylor, and Barbour and Woodward (cited earlier), as well as additional ones by Potter (1997), Potter et al. (1992), and Regalia (1978), relied heavily on age-diameter regressions that have only modest r^2 values. These studies concluded that old-growth stands were many-aged, indicating continuous – rather than episodic – regeneration, that there was a steep drop-off in survivorship between 50 and 150 yr, and that a long tail of low densities of old individuals approaches 600 yr of age. Starting from a stand-replacing disturbance, approximately 200 yr are required for old-growth characteristics to appear.

Barbour and Antos (unpublished data) determined actual age distributions by counting the rings of all stumps within 6400 m^2 clear-cut plots. In three cases they were also able to sample the plots just prior to harvest, and in those cases they censused all individuals, including cohorts as young as 1-yr-old seedlings. The decline of density with age in those populations (Fig. 5.16) follows an L-shaped curve, typical for a climax type of population, and the real-age distributions are not significantly different from those generated by diameter-age approximations.

Early succession in red fir following stand-replacing events has recently been summarized for more than 100 locations in the Sierra Nevada by Barbour et al. (1998). Regeneration of red fir in clearcuts of age 4–32 yr was most negatively correlated with summer insolation and most positively correlated with date of snowmelt. On average, there was a 12 yr delay in the successful invasion of red fir into a clearcut, during which time there was increasing cover by brush. This delay suggests that survival of red fir seedlings requires shade, a suggestion that has been supported by field studies which show that red fir seedlings survive optimally where solar radiation is less than

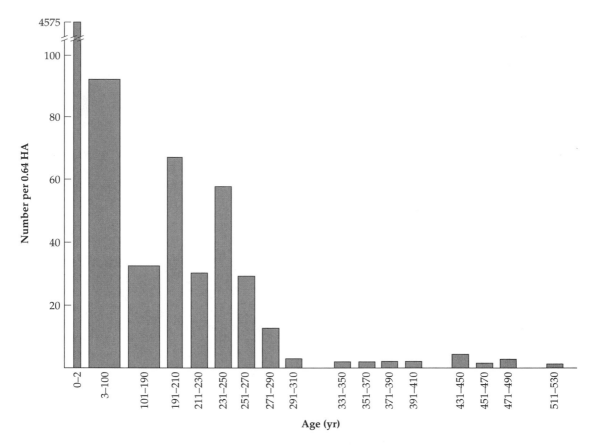

Figure 5.16. Age structure of an old-growth red fir forest. Every individual, from 1-yr-old seedlings to the largest trees, was aged by ring count. The sample area was 6400 m², and it is located near Yuba Pass in Sierra County, Tahoe National Forest.

one-third full sun (Ustin, Woodward, Barbour, and Hatfield 1984; Selter, Pitts, and Barbour 1986). A similar delay does not occur in the southern Cascades (see, for example, Gordon 1970; Laacke and Tomascheski 1986; Oliver 1985), perhaps because summer temperatures and aridity are moderate in comparison to those in the Sierra Nevada.

The ecotone between white fir and red fir.   A major ecotone between midmontane and upper montane forests exists at approximately 2000 m elevation in northern California (Kunz 1988; Rundel et al. 1988). At the ecotone, the midmontane forest is dominated by white fir and the upper montane is dominated by red fir. The transition from one to the other occurs over an elevational span of about 250 m. Hybrids – if they exist – are undescribed in the literature.

The two forests are very different in terms of biomass, productivity, population dynamics, and microenvironmental traits (Parker 1984, 1986b,

1991; Pitcher 1987; Westman 1987), and they are floristically different as well. All conifer species, for example, have similar range limits: ponderosa pine is supplanted by Jeffrey pine; sugar pine is replaced by western white pine; Douglas fir and incense cedar reach their upper limits and lodgepole pine reaches its lower limits. Understory broadleaf trees also reach their upper limits in this ecotone. A turnover pattern in herbaceous species parallels that of trees (Mellmann-Brown and Barbour 1995).

In a series of articles, Barbour and associates demonstrated that the location of the ecotone was not related to differing ecophysiology of red and white fir seedlings and saplings during the growing season but instead seemed related to differing tolerance of snowpack (Barbour, Pavlick, and Antos 1990; Barbour et al. 1991; Pavlik and Barbour 1991). The elevation of the ecotone corresponds to the mean freezing level during storms from December to March, and thus there is a steep gradient of increasing snowpack at the ecotone. Manipulative snowpack experiments remain to be done, however, so at this time the ecological role of snowpack on red and white fir seedlings and saplings is unknown.

Working in the southern Sierra Nevada, Royce (1997) has been successful in modeling the location of the ecotone on the basis of a summer drought index that takes into consideration the timing of the melt of the winter snowpack, slope aspect, and storage capacity of soil. Red fir does not occur in southern California; lodgepole pine is the upper montane dominant there. Snowfall may, nevertheless, play some role in the location of the upper montane's lower boundary. The long-term mean snowline elevation in southern California is at 2300 m (Minnich 1986), very close to the 2400 m elevation of the upper montane's lower boundary.

Aspen parkland phase. Quaking aspen is the most widely distributed tree species in North America (Perala 1990). It is represented in the Californias by the variety *Populus tremuloides* var. *aurea*, according to a monograph by Barry (1971). In the Sierra Nevada it occurs from elevations as low as 1500 m at 40°N to near timberline >3000 m at 36°N. It is nearly absent from southern California, occurring in a stand at 2200 m in the San Gorgonio Wilderness Area of the San Bernardino Mountains (Minnich 1976; Thorne 1988) and in a stand west of Lake Arrowhead in the San Bernardino Mountains (D. I. Axelrod, personal communication). It reappears in the Sierra San Pedro Mártir above 2100 m (Wiggins 1980; Minnich 1987).

Quaking aspen stands can be spatially adjacent to upper montane forest, subalpine forest, dry *Wyethia* meadows, and wet *Carex-Salix* meadows. Along the east slope of the Sierra Nevada they may also be adjacent to Jeffrey pine forest and sagebrush scrub. In a study of Murphy Meadows in Placer County, elevation 2450 m, Barry (1971) found that ecotones were often abrupt and related to soil moisture, temperature, and pH. Aspen parkland patches are typically 1–10 ha in area, and the hundreds of trees included in a patch belong to one-to-several clones. Genetic differences detectable among Sierran clones include bole color (white to yellow-green), time of spring bud break, radial growth rate, coloration of fall leaves, and time of leaf drop.

Quaking aspen is absent from montane riparian habitats, which are instead dominated by *Populus balsamifera* ssp. *trichocarpa*, *P. fremontii*, and *Alnus incana* ssp. *tenuifolia* (see, for example, a study of Truckee River riparian vegetation by Caicco, Swan-

Figure 5.17. Aspen parkland, central Sierra Nevada, Placer County, 2400 m elevation.

son, and Macoubrie in 1993 and a study of eastside riparian by Stromberg and Patten in 1992).

We have found no published quantitative studies of Californian aspen stands. Anecdotal information and agency reports (e.g., Potter et al. 1992) suggest that overstory trees may commonly reach 65 cm dbh and 20 m in height, and that canopy cover averages 60% (Fig. 5.17). Red fir saplings can be common in the otherwise herbaceous understory, implying that some stands are seral. Aspen trees are not resistant to fire, but the roots remain alive after fire and are capable of sending up new shoots (Perala 1990).

Mixed Subalpine Woodland

This vegetation qualifies more as a woodland than a forest. Although exceptional individuals reach 25 m in height, and some stands in locally protected sites approach the density of upper montane forests, typical subalpine woodland has a canopy height of 10–15 m and consists of clusters of widely spaced individuals that contribute <40% cover (Fig. 5.18). Dominant conifers include whitebark pine (*Pinus albicaulis*), limber pine (*P. flexilis*), mountain hemlock (*Tsuga mertensiana*), lodgepole pine, western white pine, and foxtail pine (*P. balfouriana*). Western white pine and foxtail pine rarely form krummholz, both persisting as single-stemmed trees to the treeline (Ryerson 1984). The Society of American Foresters (SAF) (Eyre 1980) has adopted the name "mixed subalpine" (SAF

Figure 5.18. Aspect of subalpine woodland, about 2800 m elevation, along the slopes of Slide Mountain in the central Sierra Nevada. Prominent trees include Pinus albicaulis, P. contorta ssp. murrayana. P. monticola, and Tsuga mertensiana.

256) to emphasize the pattern of shared or shifting dominance exhibited by these six taxa. Pure stands of different individual taxa may occur in adjacent patches, or several taxa may share dominance in a single stand (Potter et al. 1992).

In addition, there are north-south patterns. Only lodgepole pine extends throughout the Californian subalpine. A northern trio – white bark pine, western white pine, and mountain hemlock – extends through all or most of northern California but not into southern California. Limber pine extends from the central Sierra Nevada (38° N) south into southern California. Northern stands thus tend to be dominated by *Pinus albicaulis* and *Tsuga mertensiana*, southern stands by *Pinus flexilis*.

Pinus balfouriana has a unique disjunct distribution pattern among the subalpine conifers. It is present in the Klamath region and in the southern Sierra Nevada and not in-between. Populations in the two regions appear to be worthy of subspecific rank (Mastroguiseppi and Mastroguiseppi 1980), though the ecology of the trees and the floristic composition of the community seem remarkably uniform throughout California (Rourke 1988; Ryerson 1983; Ryerson, personal communication).

Klamath populations are *P.b.* ssp. *balfouriana*, and Sierran populations are *P.b.* ssp. *austrina*.

Secondary species commonly associated with the six listed include quaking aspen, Jeffrey pine, white fir, red fir, Sierra juniper (*Juniperus occidentalis* var. *australis*), and singleleaf pinyon (*Pinus monophylla*). Total overstory cover is variable and depends on the dominant. Pine-dominated stands average 30–50% cover and hemlock-dominated stands average 75% cover (Potter et al. 1992). Understory shrub and herb cover is a modest 5–10%.

Stand structure and autecology. Most of the published data on stand structure comes from the Sierra Nevada. Table 5.9 is a summary of 31 southern Sierran stands described by several authors. *Tsuga mertensiana* was not represented in those stands, but tree species richness otherwise was high. Average absolute basal area and tree density for all taxa in the stands were 50 m² per ha and 300 trees per ha, respectively, values about 70% those of the upper montane red fir forest.

The most detailed published study for hemlock-dominated vegetation included 45 stands west of Lake Tahoe (Nachlinger and Berg 1988). A phyto-

Table 5.9. *Summary of 31 subalpine woodland stands, south-central Sierra Nevada. PR = presence (%), BA = relative basal area, D = relative density, IP = importance percentage.*

Species	PR	BA	D	IP
Pinus balfouriana ssp. *austrina*	81	48	52	50
Pinus albicaulis	32	20	13	17
Pinus flexilis	35	11	13	12
Abies magnifica var. *shastensis*	26	9	7	8
Pinus contorta ssp. *murrayana*	35	8	7	7
Pinus monticola	16	4	3	3
Abies concolor var. *lowiana*	13	1	3	4
Populus balsamifera ssp. *trichocarpa*	3	<1	2	1
Juniperus occidentalis var. *australis*	6	<1	<1	<1
Pinus monophylla	3	<1	<1	<1
Pinus jeffreyi	3	<1	<1	<1

Sources: Lepper 1974, Rundel et al. 1988, Ryerson 1983, and Vankat and Major 1978.

sociological analysis produced 4 associations and 12 subassociations. Relative density for hemlock was 49%, western white pine 23%, whitebark pine 18%, and red fir 10%. The authors reported that the distribution of size classes of hemlock trees fell into four types: unimodal, bimodal, inverse-J, and even (Fig. 5.19). About 30 of the stands were unimodal or bimodal, indicating episodic periods of establishment. They were on north-facing aspects with moderately steep slopes. Virtually all sites with evidence of avalanche disturbance exhibited this kind of age/size distribution. Half of the remaining sites had even distributions; these were all at lower elevations and on steeper slopes. They were sites with the longest growing seasons, and they had more western white pine and red fir than other stands did. Sites with inverse-J distributions were at the highest elevations, with the shortest growing seasons. Hemlock had the lowest densities in these latter stands, whereas white bark pine was most abundant. Three years of observation revealed that mean April 1 snowpack depth was 513 cm and mean snow-free date was July 24. There was no difference in snowpack depth and time of melt between hemlock stands and adjacent red fir or lodgepole pine stands. However, the authors added, "The presence of large, late-lying snow cornices above hemlock stands and not above . . . adjacent subalpine vegetation . . . probably provides additional moisture in the growing season."

Taylor (1976) sampled 26 stands from the Carson Pass area that were dominated by either *Pinus albicaulis, Tsuga mertensiana,* or both. They were comparatively open stands, with low tree density, average basal area of only 26 m² per ha, and canopy cover of 39%. Lodgepole pine, western

white pine, and red fir were common associates. Sawyer and Thornburgh (1988) analyzed 68 relevés of hemlock woodland in the Klamath Mountains and divided them into three associations. The most mesic had an understory characterized by *Phyllodoce empetriformis*; drier associations were characterized by either *Quercus vaccinifolia* or *Pyrola picta*. Herb cover and richness were very low in the latter association.

Lepper (1974) sampled three subalpine stands in southern California mountains. On the average, *Pinus flexilis* contributed 43% relative density, *Abies concolor* 24%, *P. contorta* 18%, *Juniperus occidentalis* 6%, *P. jeffreyi* 4%, and *P. monophylla* 1%. She found that these stands fit well within a regional western association that she called the *Pinus flexilis–Ribes cereum–Artemisia tridentata* community. Commonly associated species were *Koeleria cristata, Populus tremuloides, Sitanion hystrix,* and *Symporicarpos vaccinoides*. Lepper also examined stand age structure based on increment cores. The maximum limber pine age in her Sierran–southern California range of stands averaged 337 yr (extremes were 85–1176 yr). Age distributions were generally very flat, showing uniform, sporadic establishment over a long period. Seedling densities were quite low, about 15 per ha. Establishment appeared to be positively correlated with the depth of the winter snowpack but negatively correlated with the duration of the snowpack into the growing season. Successful establishment may also require seed dispersal and burial by animals, such as Clark's nutcracker (Tomback 1986). Limber pine has a maximum net photosynthesis rate of 2.8 mg CO_2 g^{-1} hr^{-1} at 15° C (Lepper 1974, 1980). Temperatures above 25°C depressed net photosynthesis significantly, as did soil moisture from −1.3 to −2.0 MPa. Growth rates did not correlate positively with importance percentage; that is, limber pine grew fastest in sites where it had a low importance and where recruitment of young plants was also low.

Ryerson (1983, 1984) investigated 15 southern Sierran stands that contained *Pinus balfouriana*. She found diameter growth rates to be similar to those of limber and whitebark pines (7–10 cm per century) and to be about half the rates of associated species from lower elevation zones, such as red fir. Trees younger than 50 yr did not reproduce sexually. Senescence (associated with significant amounts of heart rot) begins by 1000 yr of age, and the maximum tree age she encountered was 3300 yr.

Physical factors. In the Klamath region, the subalpine woodland lies between 1900 and 2700 m elevation; in the northern Sierra Nevada it is 2300–

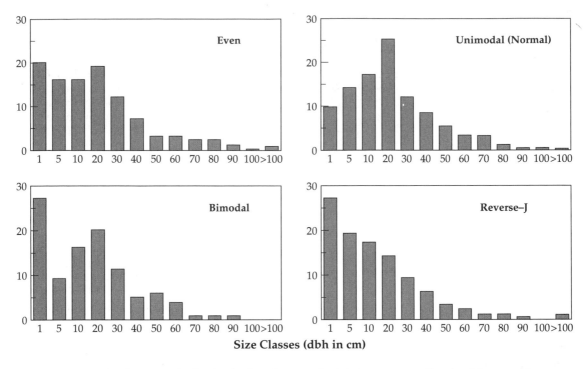

Figure 5.19. Size class histograms for four kinds of stands of mountain hemlock in the Lake Tahoe area. The horizontal axis is in centimeters of diameter breast height; the vertical axis is in percentages (from Nachlinger and Berg 1988).

2900 m; in the southern Sierra Nevada it is 3000–3400 m; in the Transverse and Peninsular ranges it extends from 2800 m to the tops of the tallest peaks at 3500 m; and it is absent from peaks in Baja California, which reach to only 3100 m (Ferlatte 1974; Minnich 1976; Eyre 1980; Vankat 1982; Major 1988; Sawyer and Thornburgh 1988; Thorne 1988, 1982).

Annual precipitation on the west flank of the Sierra Nevada in this zone is 750–1400 mm, with 85% or more of it falling as snow (Major 1967, 1988; Taylor 1976; Eyre 1980; Rundel et al. 1988); along the east flank it is only 350–750 mm (Major 1967; Lepper 1974). Snow depth in late March may average 2 m (Klikoff 1965), and trees may be absent where snow accumulation is <1 m (Taylor 1976). The growing season is 2 mo long, but hard frosts may occur at any time. On the basis of only four sites described by Lepper (1974), Major (1967, 1988), and Taylor (1976), mean annual temperature is <4° C.

Soils are shallow, rocky, coarse-textured Inceptisols, judging from studies by Klikoff (1965) and Lepper (1974). Ultrabasic parent materials, especially peridotite, are common at high elevations in the Klamath area. In the eastern Klamath area (e.g., Mt. Eddy) *Pinus balfouriana* replaces *Tsuga mertensiana* on such sites, herb species richness is high,

and some Great Basin taxa are present (Sawyer and Thornburgh 1988).

The physical factors thought to correlate with timberline in California have barely been explored. Klikoff (1965) showed that soil moisture (water potential) beneath whitebark pine forest and krummholz was not significantly different from that beneath upland *Carex* meadow, and he concluded that snow depth and duration are more important in determining treeline. Mooney, Wright, and Strain (1964) found bristlecone pine timberline in the White Mountains to be where annual photosynthesis was only twice respiration. In the central Sierra Nevada, the elevational distance between the limit of the tree growth form and the limit of krummholz is about 150 m. Based largely on field observations, Clausen (1965) developed a hypothesis that krummholz forms of trees there are genetically dwarfed. To our knowledge, this theory has not been tested experimentally. Most ecologists ascribe dwarfing to winter desiccation and ice blast above the snowpack.

Pollen studies in sediments beneath high elevation Sierran lakes have shown that subalpine conifers first appeared about 10,000 yr ago. These early Holocene woodlands were more open than at present and contained a subcanopy of montane chaparral shrubs. Anderson (1990) concluded that the climate then was more arid than at present. More mesic species, such as mountain hemlock and red fir, became abundant only about 6000 yr ago. The elevation of timberline has fluctuated up and down several times since then, suggesting that subalpine vegetation has never been stable, right up to the present.

Evidence of more recent climate change is revealed in tree ring records. Graumlich (1991, 1993) reconstructed temperature and precipitation patterns for the past 1000 yr using rings in foxtail pine and Sierra juniper growing in the central Sierra Nevada. Her conclusions were that temperature fluctuations were on centennial or longer scales of time (e.g., a warm period 1100–1375 A.D.; a cold period 1450–1850), whereas precipitation fluctuations were at a decadal scale. Twentieth-century precipitation appears to have been above average, a conclusion supported by reviews in SNEP (1996). There is no evidence yet of CO_2 enrichment – or of recent putative climate change – affecting Sierran subalpine conifers.

East-side Sierran Forests

The eastern, desert-facing slopes of the Sierra Nevada are physically quite different from the western slopes. The elevation drops more precipitously, soils are more skeletal, and forest cover is less continuous. In addition, the eastern escarpment is within a rainshadow, as prevailing winds bring precipitation from the west. Occasional monsoonal summer thunderstorms coming from the Gulf of California reach the east side as far north as the Sierra Nevada, but their contribution is modest. Even as far south as the Sierra San Pedro Mártir, summer precipitation accounts for only 15% of annual precipitation (Minnich et al., 1998).

Forest zonation is also different, being less marked into discrete elevational phases. Subalpine and upper montane tree species tend to associate in the same stands, and the upper montane and midmontane forests are both thoroughly dominated by *Pinus jeffreyi*. We can, for convenience, divide the east slope into just two zones: subalpine–upper montane and midmontane.

In the northern Sierra Nevada, the subalpine–upper montane zone lies between 2500 and 2900 m elevation; in the southern Sierra Nevada it is between 2800 and 3400 m (Rundel et al. 1988). *Abies magnifica* is a prominent species, but it does not often form pure stands of large individuals as it does on the west slope. *Pinus contorta* is a common associate.

The midmontane zone extends from 2000–2500

m in the north and from 2600–2800 m in the south. Jeffrey pine is the typical dominant. Old-growth forests are open and park-like (Fig. 5.20), with <200 trees per ha and canopy cover <60% (Table 5.10). On the better sites, the largest overstory trees may reach 40 m in height and 120 cm dbh. Apart from occasional intrusions by black oak, trembling aspen, or black cottonwood (along streams), deciduous understory trees are rare. Shrub cover is variable but may average 20%, and the list of common species shows a cold desert flavor: *Artemisia tridentata*, *Chrysothamnus parryi*, *Haplopappus bloomeri*, and *Purshia tridentata* join *Arctostaphylos patula*, *Ceanothus prostratus*, *C. velutinus*, and *Cercocarpus ledifolius*.

Herb species richness is relatively high, including mainly perennial herb and bunchgrass growth forms. Among the most abundant herbs is *Wyethia mollis*. In addition to being an understory element, it can form monospecific patches of considerable size and duration that seem to inhibit invasion by other taxa. Some of these patches are created following stand-replacing fires, and succession is stalled at a dry meadow stage. In Tahoe National Forest alone, *W. mollis* dominates 5200 ha on the eastern slope of the Sierra Nevada (Parker and Yoder-Williams 1989). Aquatic extracts of *W. mollis* foliage have been shown to inhibit the germination and early growth of Jeffrey pine in both growth chamber and natural field situations (Parker and Yoder-Williams 1989). The allelopathic effect is similar to that more recently reported for *Cirsium vulgare* on ponderosa pine (Randall and Rejmanek 1993).

Ponderosa pine replaces Jeffrey pine on a north-to-south gradient, Jeffrey pine finally being absent altogether along the Sierran-Cascade east side, north of the town of Canby, Modoc County, 41°28' N (Griffin and Critchfield 1972; Smith 1994). Between that latitude and Carson Pass (38°42'N), the two coexist, and then ponderosa pine is absent except for fewer than a dozen scattered locations to the south. Leach (1994) studied six of those stands, which occurred in riparian habitats between Rock Creek and Independence Creek facing the Owens Valley. The most commonly associated species included *Artemisia ludoviciana*, *A. tridentata*, *Betula occidentalis*, *Chrysothamnus parryi*, *Populus balsamifera* ssp. *trichocarpa*, *Rhamnus californica*, *Rosa woodsii*, and *Salix lutea* – species that typify midelevation riparian vegetation on the western side of the Owens Valley (Harris 1988). Harris found that age distributions indicated sporadic regeneration episodes linked to short-term flood disturbance. One significant period of regeneration corresponds with a period of historically high water flows between

Figure 5.20. Jeffrey pine forest, east side of the central Sierra Nevada, Mono County, 2400 m elevation.

Table 5.10. *Composition of east-side old-growth Jeffrey pine forest. Data represent means of six stands between 36 and 40° N latitude. C = relative cover, D = relative density, BA = relative basal area, IP = importance percentage. Total absolute cover = 58%, density = 872 trees per hectare, and BA = 47 m² ha⁻¹.*

Species	C	D	BA	IP
Pinus jeffreyi	52	21	53	42
Calocedrus decurrens	22	33	28	28
Abies concolor var. *lowiana*	16	42	13	24
Populus balsamifera ssp. *trichocarpa*	9	12	4	8
Pinus contorta ssp. *murrayana*	2	2	<1	1
Pinus ponderosa	<1	1	<1	<1
Juniperus occidentalis var. *australis*	<1	<1	<1	<1
Quercus kelloggii	<1	<1	2	<1
Pinus lambertiana	<1	<1	<1	<1

Source: Rundel et al. 1988.

1854 and 1916. These east-side Sierran populations were not on hydrothermally altered substrate, the habitat of ponderosa pine outliers farther east in Nevada, as described by Billings (1950), DeLucia, Schlesinger, and Billings (1988) and DeLucia and Schlesinger (1990).

A third east-side yellow pine, *P. washoensis*, is one of the most narrowly distributed pines of North America. The species is limited to three locations: Mt. Rose, Babbitt Peak, and the Warner Mountains of northeastern California. Allozyme analysis of the three populations revealed a high degree of similarity, suggesting that they were once continuous and became separated only in the last 12,000 yr (Niebling and Conkle 1990). Washoe pine is most closely related to north plateau and Rocky Mountain races of *P. ponderosa* and has less in common with *P. jeffreyi* (Critchfield 1984; Niebling and Conkle 1990).

All three yellow pines co-exist in the midmontane forest of the Warner Mountains. Both Washoe and Jeffrey pines reach their northern limits here. White fir typically dominates this forest (importance percentage = 87%, yellow pines = 8%); total canopy cover ranges from 40–60%, tree density = 585 per ha, and basal area = 74 m² ha⁻¹ (Riegel, Thornburgh, and Sawyer 1990; Schierenbeck and Jensen 1994). The elevational range is 1670–2195 m. Above this forest is a mixed *Pinus albicaulis–P. contorta* subalpine–upper montane woodland that extends to the highest peaks at 3016 m.

Below the Jeffrey pine forest lies the pinyon-

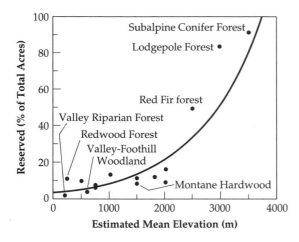

Figure 5.21. Relationship among elevation, range, and preserve area for selected woodlands and forests. Low elevation types such as foothill woodland and mixed conifer have the most extensive ranges, are the most impacted, and yet have the lowest fraction of their range in protected status (based on data provided by Jensen et al. 1993 and by B. Pavlik (in a personal communication).

juniper woodland, an extensive vegetation type described in Chapter 7.

CONVERSATION ISSUES

California is not particularly unique among western states in terms of how different its modern landscape is from the landscape in existence before Euroamerican contact. Forest cover has been thinned, the ratio of late successional to early successional hectares has grown smaller, wildfires have been suppressed, ecotones displaced, physiognomy changed, weedy species introduced, overgrazing by domesticated livestock permitted, and wildland vegetation replaced by pasture and city.

As merely one example of the kind of data available to quantify landscape change and habitat loss, the final report of the Sierra Nevada Ecosystem Project (SNEP 1996) concluded that old-growth forests of mid- and upper-montane regions today cover only 8% of the Sierran landscape, whereas they would have covered four times that area in the pre-Euroamerican landscape. Furthermore, the average size of old-growth stands is much smaller than that of early successional phases; climax stands of woodland and forest now are often fragile, non-functional, extinction-prone islands in a matrix of sharply different successional stages. And finally, those vegetation types most endangered are least protected (Jensen, Torn, and Harte 1993; Davis and Stoms 1996; Fig. 5.21). Oak woodland, east-side Jeffrey pine forest, and mixed conifer forest have <10% of their range located within the boundaries of preserves.

Ecosystem ramifications that ripple out from vegetation change are becoming clear. For example, significant declines in neotropical bird migrants have now become correlated with fire suppression in midmontane forests (Beedy 1981; Kilgore 1971; Marshall 1988; SNEP 1996). We can certainly expect similar trends to exist in less-well-documented groups such as microbes, insects, and herbs.

Although the California landscape has fared no better than that of any other state, Californian conservation policy has sometimes been in a leadership position. One bioregion that has benefited from that leadership is the Sierra Nevada region. Spurred by the *Sacramento Bee*'s Pulitzer Prize–winning series on environmental problems in the Sierra Nevada (Knudson 1991), the California Resources Agency held a major conference of scientists, citizens, agency personnel, and representatives of user groups to discuss ideas about sustainable management of Sierran ecosystems (Wheeler 1992). A follow-up group of 37 individuals then met in 1993–1994 to develop a series of "critical research questions" (Parrish and Erman 1994). The short list of questions was both sophisticated and pointed. Some examples are the following:

What indicators can we use to assess ecosystem health and biodiversity?
What is the appropriate balance between reserve area and that used for resource extraction?
How can ecological resources be assessed for real-world cost-benefit models?
How can the analysis and flow of information be integrated among agencies of various scale so as best to resolve management issues?
What kind of incentives can be used to facilitate effective planning and management?

Most recently, a 3 yr, multimillion dollar, interdisciplinary study of the Sierra Nevada was completed by a core team of 20 scientists and more than 40 associates (SNEP 1996). This report grappled with a wide range of issues, including the identification of biological hot spots, evaluation of options for the conservation of resources including old-growth forests, summarizing what is known about air pollution, an assessment of human well-being in selected Sierran communities, imagining the kind of governmental units and structure that would be more bioregional and less political, making projections of human population growth, and explaining the costs and benefits of grazing, logging, fire suppression, and tourism. This comprehensive study could well serve as a model for other ecosystems in other states.

During the 1990s, California also created a formal interagency structure for the discussion of regional environmental issues that transcend the usual political units. Representatives of more than 30 governmental units now convene at monthly staff meetings and at quarterly public meetings. At their best, these meetings have led to creative solutions for bioregional problems; at a minimum, they have educated high-level agency staff and the general public about the intrinsic importance of biotic diversity.

Baja California has also been active in developing proactive conservation policy. A research and graduate student training center in environmental science and conservation biology is growing at the Universidad Autonoma de Baja California (Ensenada). Mexico has been a strong participant in the United Nations program for biosphere reserves. Since the program's creation in 1974, more than 300 reserves in 82 countries have been formally entered into the UNESCO network (Franco-Vizcaino and Cueva 1995). Mexico has 12 and may soon acquire a thirteenth at Parque Nacional de San Pedro Mártir, a 63,000 ha mixed conifer forest ecosystem described earlier in this chapter. Among many features, it offers a natural fire regime, an extensive old-growth forest, 15 endemic taxa of plants, 20 birds, 8 mammals, and 1 fish, and it is a potential site for the reintroduction of the California condor. At this writing, the future of that montane ecosystem is uncertain. Should it not become a biosphere preserve, large sections at the perimeter may be logged: A local *ejido* has been provisionally authorized to cut 8000 m³ (1 m bd ft) of timber per year for 5 yr (Minnich et al. 1994). Such an extensive harvest could have enormous ecological consequences.

Preservation, of course, is only one among several conservation options. The enhancement or restoration of degraded ecosystems potentially would affect a much larger area than the 1% of the state that is in preserve status. Recently, Bruce Pavlik of Mills College and Peggy Fiedler of San Francisco State University have been calling for the creation of a California center for ecological restoration (personal communication). They point out that restoration activities in the state have been typically "science-lite": ad hoc and haphazard, not grounded in ecological theory. They also conclude that few plant species and communities have been experimented with (most are hydric with low species richness) and that self-sustainability of the restored sites has yet to be demonstrated. A restoration center would have propagation facilities, a research mission, demonstration areas, and public education outreach activities, and would be a locus for professional training. It could be supported with a combination of mitigation funds from developers, foundation grants, public agency budgets, and entrance fees paid by lay visitors.

AREAS FOR FUTURE RESEARCH

Significant contributions to our understanding of Californian vegetation can still be made at all levels: basic stand descriptions, vegetation classification, autecology and demography of key species, stand dynamics, energy and nutrient transfer rates, and predictive models of vegetation change caused by environmental stress, climate change, or prescribed fire.

With respect to vegetation sampling and classification, California has an opportunity to begin an intensive application of traditional phytosociological methods and association nomenclature. Although the state's ecological literature has only a modest past history in the use of Braun-Blanquet relevé techniques, recent regional surveys by Rivas-Martinez (1997) are providing a basis and a stimulus for others to follow. Braun-Blanquet methodology is used more widely in the world than any other scheme of vegetation description; is it not about time for American vegetation scientists to realize their isolation and join their global colleagues?

With regard to autecological studies, we have particularly little information about such widespread dominants as *Abies magnifica, Pinus jeffreyi, Populus balsamifera* ssp. *trichocarpa, P. tremuloides, Quercus agrifolia, Q. chyrsolepis,* and all the subalpine conifers. We do not know how pairs of closely related taxa share habitats or displace each other: Why, for example, does a series of five-needle conifer species segregate along an elevational gradient rather than occurring sympatrically? What environmental and biotic factors determine the major midmontane–upper montane ecotone at 2000 m elevation? What is the role of wintertime conditions in determining species and vegetation limits?

With regard to ecosystem studies, it is fair to state that we have a rudimentary grasp of energy and nutrient transfer dynamics only for mixed conifer forest and northern oak woodland and virtually zero knowledge for any other upland forest or woodland type. We know very little about the periodicity of disturbance and patterns of recovery for mixed evergreen forest, upper montane forest, subalpine woodland, quaking aspen parkland, Jeffrey pine forest, and montane riparian forest. It is not surprising, then, that current management practices for sustainable use of these vegetation types are far from successful.

Not only have we yet been unable to understand how most Californian upland vegetation types function today – we have also failed to reconstruct (as accurate models) the pre-European contact ecosystems. We therefore do not know the ecological impacts of displacement of Native Americans, logging, grazing, and fire suppression activities since the time of contact. Anthropologists, for example, have not yet been able to translate cultural gathering and harvesting practices to the landscape level. We need a new kind of science, which Kat Anderson calls "ethnoecology," to take us beyond ethnobotany (Anderson and Moratto 1996; Blackborn and Anderson 1993).

Our lack of ecohistorical curiosity cannot be attributed merely to the absence of solid historical information. Why is it that the thousands of VTM vegetation plot records, taken early enough in the twentieth century to serve as a record of prefiresuppression vegetation patterns, have been chronically underused even a decade after Dr. Barbara Allan-Diaz at the University of California, Berkeley, has advertised their ready availability? Our record in this state of forest and woodland studies that employ repeat sampling of permanently marked plots is almost nil.

Finally, conservation and restoration work in forests and woodlands needs increased attention. Although Californian environmental protection programs have often led the nation, they have focused too often on preserving the pristine at the expense of forsaking the damaged. As a recent book on California vegetation oriented to the lay person concludes, "We must continue to protect remaining undisturbed areas, certainly; but we must also begin a process of restoration and enhancement of much larger areas in California, those beyond the pristine preserves" (Barbour, Pavlik, Drysdale, and Lindstrom 1993).

REFERENCES

Agee, J. K. 1991. Fire history along an elevational gradient in the Siskiyou Mountains, Oregon. Northwest Sci. 65:188–189.

Agee, J. K. 1993. Fire ecology of Pacific northwest forests. Island Press, Washington, D.C.

Allen, B. H. and B. A. Holzman. 1991. Blue oak communities in California. Madroño 38:80–95.

Allen, B. H., B. A. Holzman, and R. R. Evett. 1991. A classification system for California's hardwood rangelands. Hilgardia 59:1–45.

Anderson, M. K., and M. J. Moratto. 1996. Native American land-use practices and ecological impacts, Chapter 9 in SNEP Science Team (eds.), Status of the Sierra Nevada, Vol. II, Centers for Water and Wildland Resources, Report No. 36. University of California, Davis.

Anderson, R. S. 1990. Holocene forest development and paleoclimates within the central Sierra Nevada, California. J. Ecol. 78:470–489.

Andrews, R. S. 1993. Ecological support team workshop proceedings for the California spotted owl environmental impact statement. USDA Forest Service, unpublished report, Albany, Calif.

Anony. 1991. California vegetation. California Teale Data Center, Sacramento. [map, 1:1 million]

Archibold, O. W. 1995. Ecology of world vegetation. Chapman and Hall, New York.

Aschmann, H. H. 1973. Distribution and peculiarity of Mediterranean ecosystems, pp. 11–19 in F. di Castri and H. A. Mooney (eds.), Mediterranean type ecosystems: origins and structure, Springer-Verlag, New York.

Aschmann, H. H. 1985. A more restrictive definition of Mediterranean climates. Bull. Soc. Bot. Fr., Actual. Bot. 1984d (2,3,4):21–30.

Atzet, T., and R. E. Martin. 1992. Natural disturbance regimes in the Klamath Province, pp. 40–48 in H. M. Kerner (ed.), Proceedings of the symposium on biodiversity of northwestern California, Wildland Resources Center, Report 29. University of California, Berkeley.

Axelrod, D. I. 1976. History of the coniferous forests, California and Nevada. Univ. Calif. Pub. Bot. 70:1–62.

Axelrod, D. I. 1988. Outline history of California vegetation, pp. 139–192 in M. G. Barbour and J. Major (eds.), Terrestrial vegetation of California, 2nd ed. California Native Plant Society, Spec. Pub. 9. Sacramento.

Axelrod, D. I. In press. Interior Late Tertiary paleoenvironments, eastern Washington to Mexico. University of California Publications in Geology, Berkeley, Calif.

Baker, G. A., P. W. Rundel, and D. J. Parsons. 1981. Ecological relationships of *Quercus douglasii* (Fagaceae) in the foothill zone of Sequoia National Park, California. Madroño 28:1–12.

Barbour, M. G., and J. Major (eds.). 1988. Terrestrial vegetation of California, expanded edition. California Native Plant Society, Special Publication No. 9. Sacramento.

Barbour, M. G., and R. A. Minnich. 1990. The myth of chaparral convergence. Israel J. Bot. 39:453–463.

Barbour, M. G., N. Berg, T. Kittel, and M. Kunz. 1991. Snowpack and the distribution of a major vegetation ecotone in the Sierra Nevada of California. J. Biogeog. 18:141–149.

Barbour, M. G., J. H. Burk, and W. D. Pitts. 1987. Terrestrial plant ecology, 2nd ed. Benjamin/Cummings, Menlo Park, Calif.

Barbour, M. G., R. F. Fernau, J. M. Rey, N. Jurjavcic, and E. B. Royce. 1998. Tree regeneration and early succession following clearcuts in red fir forests of the Sierra Nevada, California. J. Forest Ecol. and Manage. 104:101–111.

Barbour, M. G., B. M. Pavlik, and J. A. Antos. 1990. Seedling growth and survival of red and white fir in a Sierra Nevada ecotone. Amer. J. Bot. 77:927–938.

Barbour, M., B. Pavlik, F. Drysdale, and S. Lindstrom. 1993. California's changing landscapes: diversity and conservation of California vegetation. California Native Plant Society, Sacramento.

Barbour, M. G., and R. A. Woodward. 1985. The Shasta

red fir forest of California. Can. J. For. Res. 15:570–576.

Barry, W. J. 1971. The ecology of *Populus tremuloides*, a monographic approach. Ph.D. dissertation, University of California, Davis.

Barry, W. J. 1989. A hierarchical vegetation classification system with emphasis on California plant communities. California Resources Agency, Dept. Parks and Recreation, Sacramento.

Baumgartner, D. M. (ed.). 1985. Lodgepole pine: the species and its management. Washington State University Press, Pullman.

Beedy, E. C. 1981. Bird communities and forest structure in the Sierra Nevada of California. Condor 83: 97–105.

Billings, W. D. 1950. Vegetation and plant growth as affected by chemically altered rocks in the western Great Basin. Ecology 31:62–74.

Bingham, B. B., and J. O. Sawyer, Jr. 1991. Distinctive features and definitions of young, mature, and old-growth Douglas fir/hardwood forests, pp. 363–377 in L. F. Ruggiero et al. (eds.), Wildlife and vegetation of unmanaged Douglas fir forests. USDA Forest Service, PNW-GTR-285, Portland, Ore.

Blackborn, T. C., and M. K. Anderson (eds.). 1993. Before the wilderness: environmental management by Native Californians. Ballena Press, Menlo Park, Calif.

Bolsinger, C. L. 1988. The hardwoods of California timberlands, woodlands, and savannas. USDA Forest Service, PNW RB-148, Portland, Ore.

Bolton, R. B. Jr., and R. J. Vogl. 1969 Ecological requirements of *Pseudotsuga marcrocarpa* (Vasey) Mayr in the Santa Ana Mountains, California. J. For. 67:112–116.

Borchert, M. 1985. Serotiny and cone-habit variation in populations of *Pinus coulteri* (Pinaceae) in the southern coast ranges of California. Madroño 32:29–48.

Borchert, M. I. 1994. Blue oak woodland, p. 11 in T. N. Shiflet (ed.), Rangeland cover types of the United States. Society for Range Management, Denver, Colo.

Borchert, M. I., N. D. Cunha, P. C. Krosse, and M. L. Lawrence. 1993. Blue oak plant communities of southern San Luis Obispo and northern Santa Barbara counties, California. USDA Forest Service, PSW-GTR-139, Albany, Calif.

Borchert, M. I., F. W. Davis, J. Michaelsen, and L. D. Oyler. 1989. Interactions of factors affecting seedling recruitment of blue oak (*Quercus douglasii*) in California. Ecology 70:389–404.

Borchert, M., and M. Hibberd. 1984. Gradient analysis of a north slope montane forest in the western Transverse Ranges of southern California. Madroño 31:129–139.

Bradford, G. G., A. L. Page, and I. R. Straughan. 1981. Are Sierra lakes becoming acid? Calif. Agr. (May–June):6–7.

Brooks, W. H. 1971. A quantitative ecological study of the vegetation in selected stands of the grass-oak woodland in Sequoia National Park, California. Unpublished progress report to Superintendent of Sequoia–Kings Canyon National Park, Three Rivers, Calif.

Cahill, T. A., J. J. Carroll, D. Campbell, and T. E. Gill. 1996. Air quality, chapter 48, in SNEP Science Team (eds.), Status of the Sierra Nevada, Vol. II.

Wildland Resources Center Report No. 36, University of California, Davis.

Caicco, S., M. Swanson, and M. Macoubrie. 1993. Truckee River riparian vegetation and fluvial geomorphology study, final report. USDI Fish and Wildlife Service, Region One, Reno, Nev.

Callaway, R. M. 1992. Effect of shrubs on recruitment of *Quercus douglasii* and *Q. lobata* in California. Ecology 73:2118–2128.

Callaway, R. M., N. M. Nadkarni, and B. E. Mahall. 1991. Facilitation and interference of *Quercus douglasii* on understory productivity in central California. Ecology 72:1484–1499.

CDFR (California Department of Forestry). 1979. California's forest resources. Sacramento.

Chabot, B. F., and D. J. Hicks. 1982. The ecology of leaf life spans. Ann. Rev. Ecol. Syst. 13:229–259.

Chappell, C. B., and J. K. Agee. 1996. Fire severity and seedling establishment in *Abies magnifica* forests, southern Cascades, Oregon. Ecol. Applic. 6:628–640.

Cheatham, N. H., and J. R. Haller. 1975. An annotated list of California habitat types. Unpublished report, University of California, Berkeley.

Clausen, J. 1965. Population studies of alpine and subalpine races of conifers and willows in the California high Sierra Nevada. Evolution 19:56–68.

Colwell, W. L. Jr. 1988. The status of vegetation mapping in California today, pp. 195–200 in M. G. Barbour and J. Major (eds.), Terrestrial vegetation of California, 2nd ed. California Native Plant Society, Special Pub. 9, Sacramento.

Cooper, W. S. 1992. The broad-sclerophyll vegetation of California. Carnegie Institution of Washington, No. 319, Washington, D.C.

Conard, S. G., and S. R. Radosevich. 1982. Post-fire succession in white fir (*Abies concolor*) vegetation of the northern Sierra Nevada. Madroño 29:42–56.

Critchfield, W. B. 1984. Crossability and relationships of Washoe pine. Madroño 31:144–170.

Davis, F. W., and D. M. Stoms. Sierran vegetation: a gap analysis, Chapter 23 in SNEP Science Team, Status of the Sierra Nevada, Vol. II, Centers for Water and Wildland Resources, Report No. 36, University of California, Davis.

DeLucia, E. H., and W. H. Schlesinger. 1990. Ecophysiology of Great Basin and Sierra Nevada vegetation on contrasting soils, pp. 143–178 in C. B. Osmond, L. F. Pitelka, and G. M. Hidy (eds.), Plant biology of the Basin and Range, Springer-Verlag, Berlin.

DeLucia, E. H., W. H. Schlesinger, and W. D. Billings. 1988. Water relations and the maintenance of Sierran conifers on hydrothermally altered rock. Ecology 69:303–311.

Dunn, P. H. 1980. Nutrient-microbial considerations in oak management, pp. 148–160 in T. R. Plumb (ed.), Ecology, management, and utilization of California oaks. USDA Forest Service PSW-44, Berkeley.

Eyre, F. H. (ed.). 1980. Forest cover types of the United States and Canada. Society of American Foresters, Washington, D.C.

Ferlatte, W. J. 1974. A flora of the Trinity Alps of Northern California. University of California Press, Berkeley.

Finney, M. A., and R. E. Martin. 1989. Fire history in a *Sequoia sempervirens* forest at Salt Point State Park, California. Can. J. For. Res. 19:1451–1457.

Fites, J. A. 1993. Ecological guide to mixed conifer plant

associations. USDA Forest Service, R5-ECOL-TP-001, Albany, Calif.

Franco-Vizcaino, E., and H. Cueva. 1995. Strategies for conservation of the Sierra San Pedro Mártir. UCMexus, University of California, Riverside.

Franklin, J. F., and C. T. Dyrness. 1973. Natural vegetation of Oregon and Washington. USDA Forest Service Gen. Tech. Rep. PNW-118, Portland, Ore.

Gemmill, B. 1980. Radial growth of California black oak in the San Bernardino Mountains, pp. 128–135 in T. R. Plumb (ed.), Ecology, management, and utilization of California oaks. USDA Forest Service, PSW-44, Berkeley, Calif.

Gordon, D. R., and K. J. Rice. 1993. Competitive effects of grassland annuals on soil water and blue oak (*Quercus douglasii*) seedlings. Ecology 74:68–82.

Gordon, D. R., J. M. Welker, J. W. Menke, and K. J. Rice. 1989. Competition for soil water between annual plants and blue oak (*Quercus douglasii*) seedlings. Oecologia 79:533–541.

Gordon, D. T. 1970. Natural regeneration of white and red fir. USDA Forest Service Res. Pap. PSW-58, Berkeley, Calif.

Goudey, C. B., and D. W. Smith. 1994. Ecological units of California: subsections. USDA Forest Service, PSW-map, 1:1 million, Albany, Calif.

Graham, A. 1993. History of the vegetation: Cretaceous (Maastrichtian) – Tertiary, pp. 57–70 in Flora of North America Editorial Committee, Flora of North America, Vol. 1. Oxford University Press, New York.

Graumlich, L. J. 1991. Subalpine tree growth, climate, and increasing CO_2: an assessment of recent growth trends. Ecology 72:1–11.

Graumlich, L. J. 1993. A 1000-year record of temperature and precipitation in the Sierra Nevada. Quat. Res. 39:249–255.

Gray, J. T. 1978. The vegetation of two California mountain slopes. Madroño 25:177–185.

Greenwood, G. B., R. K. Marose, and J. M. Stenback. 1993. Extent and ownership of California's hardwood rangelands. California Dept. Forestry and Fire Protection, Sacramento.

Griffin, J. R. 1965. Digger pine seedling response to serpentinite and non-serpentinite soil. Ecology 46:801–807.

Griffin, J. R. 1971. Oak regeneration in the upper Carmel Valley, California. Ecology 52:862–868.

Griffin, J. R. 1973. Xylem sap tension in three woodland oaks of central California. Ecology 54:152–159.

Griffin, J. R. 1976. Regeneration in *Quercus lobata* savannas, Santa Lucia Mountains, California. Amer. Midl. Nat. 95:422–435.

Griffin, J. R. 1978. The marble-cone fire ten months later. Fremontia 6:8–14.

Griffin, J. R. 1980a. Animal damage to valley oak acorns and seedlings, Carmel Valley, California, pp. 242–245 in T. R. Plumb (ed.), Ecology, management, and utilization of California oaks. USDA Forest Service PSW-44, Berkeley.

Griffin, J. R. 1980b. Sprouting in fire-damaged valley oaks, Chews Ridge, California, pp. 216–219 in T. R. Plumb (ed.), Ecology, management, and utilization of California oaks. USDA Forest Service PSW-44, Berkeley.

Griffin, J. R. 1988. Oak woodland, pp. 383–415 in M. G. Barbour and J. Major (eds.), Terrestrial vegetation of California, 2nd ed. California Native Plant Society, Spec. Pub. 9, Sacramento.

Griffin, J. R., and W. B. Critchfield. 1972. The distribution of forest trees in California. USDA Forest Service Research Paper PSW-82, Berkeley.

Gruell, G. 1998. Sierra Nevada forests, past and present: 145 years of photographic records. DANR Communication Services, University of California, Davis. In press.

Gudmunds, K. N., and M. G. Barbour. 1984. Mixed evergreen forest stands in the northern Sierra Nevada, pp. 32–37 in T. R. Plumb and N. H. Pillsbury (eds.), Proceedings of the symposium on multiple-use management of California's hardwood resources. USDA Forest Service, Gen. Tech. Rep. PSW-100, Berkeley.

Haggerty, P. K. 1994. Damage and recovery in southern Sierra Nevada foothill oak woodland after a severe ground fire. Madroño 41:185–198.

Haller, J. R. 1959. Factors affecting the distribution of ponderosa and Jeffrey pines in California. Madroño 15:65–71.

Haller, J. R. 1962. Variation and hybridization in ponderosa and Jeffrey pines. Univ. Calif. Pub. Bot. 34:123–166.

Hallin, W. E. 1957. Silvical characteristics of California red fir and Shasta red fir. USDA Forest Service Technical Paper 16, Berkeley.

Hamrick, J. L., and W. J. Libby. 1972. Variation and selection in western montane species. I. White fir. Silvae Genet. 21:29–35.

Hanes, T. L. 1976. Vegetation types of the San Gabriel Mountains, pp. 65–76 in J. Lating (ed.), Plant communities of southern California. California Native Plant Society, Spec. Pub. 2, Berkeley.

Harris, R. R. 1988. Associations between stream valley geomorphology and riparian vegetation as a basis for landscape analysis in the eastern Sierra Nevada, California, USA. Env. Mgt. 12:219–228.

Hartesveldt, R. J., H. T. Harvey, H. S. Shellhammer, and R. E. Stecker. 1975. The giant sequoia of the Sierra Nevada. USDI National Park Service, Washington, D.C.

Haston, L. L., F. W. Davis, and J. Michaelson. 1988. Climatic response functions for bigcone spruce. Phys. Geog. 9:81–97.

Hastings, J. R., and R. R. Humphrey (eds.). 1969. Climatological data and statistics for Baja California. University of Arizona Institute of Atmospheric Physics, Technical Report 18. Tempe, Ariz.

Helms, J. A. 1987. Invasion of *Pinus contorta* var. *murrayana* (Pinaceae) into mountain meadows at Yosemite National Park, California. Madroño 34:91–97.

Hickman, J. C. (ed.). 1993. The Jepson manual: higher plants of California. University of California Press, Berkeley.

Holland, R. F. 1986. Preliminary descriptions of the terrestrial natural communities of California. Unpublished report, California Resources Agency, Dept. Fish and Game, Sacramento.

Holland, V. L. 1980. Effect of blue oak on rangeland forage production in central California, pp. 314–318 in T. R. Plumb (ed.), Ecology, management, and utilization of California oaks. USDA Forest Service PSW-44, Berkeley.

Holland, V. L., and J. Morton. 1980. Effect of blue oak on nutritional quality of rangeland forage in central

California, pp. 319–322 in T. R. Plumb (ed.), Ecology, management, and utilization of California oaks. USDA Forest Service PSW-44, Berkeley.

Hunter, J. C. 1995. Architecture, understory light environments, and stand dynamics in northern California's mixed evergreen forests. Ph.D. dissertation, University of California, Davis.

Hunter, J. C. and V. T. Parker. 1993. The disturbance regime of an old-growth forest in coastal California. J. Veg. Sci. 4:19–24.

Jensen, H. A. 1947. A system for classifying vegetation in California. California Fish and Game 33:199–266.

Jensen, D. B., M. S. Torn, and J. Harte. 1993. In our own hands: a strategy for conserving California's biological diversity. University of California Press, Berkeley.

Jimmerson, T. M. 1994. A field guide to Port Orford cedar plant associations in northwest California. USDA Forest Service, R5-ECOL-TP-002, Albany, Calif.

Jimmerson, T. M. 1993. Preliminary plant associations of the Klamath Province, Six Rivers and Klamath National Forests. USDA Forest Service, Six Rivers National Forest. Unpublished Report, Eureka.

Jimmerson, T. M. et al. 1995. A field guide to serpentine plant associations and sensitive plants in northwestern California. USDA Forest Service, R5-ECOL-TP-006, Albany.

Johannes, M., H. Knops, and W. D. Koenig. 1994. Water use strategies of five sympatric species of Quercus in central coastal California. Madroño 41:290–301.

Kay, B. L., and O. A. Leonard. 1980. Effect of blue oak removal on herbaceous forage production in the north Sierra foothills, pp. 323–328 in T. R. Plumb (ed.), Ecology, management, and utilization of California oaks. USDA Forest Service, PSW-44, Berkeley.

Keeler-Wolf, T. 1988. The role of Chrysolepis chrysophylla (Fagaceae) in the Psuedotusga-hardwood forest of the Klamath Mountains of California. Madroño 35: 285–308.

Keeler-Wolf, T. 1990. Ecological surveys of Forest Service Research National Areas in California. USDA Forest Service Gen. Tech. Rep. PSW-125, Berkeley.

Keeley, J. E. 1990. Demographic structure of California black walnut (Juglans californica; Juglandaceae) woodlands in southern California. Madroño 37:237–248.

Kilgore, B. M. 1971. Fire in ecosystem distribution and structure: western forests and scrublands, pp. 58–59 in H. A. Mooney (ed.), Fire regimes and ecosystem properties. USDA Forest Service, Gen. Tech. Rep. WO-26, Washington, D.C.

Kilgore, B. M. 1981. Fire in ecosystem distribution and structure: western forests and scrublands, pp. 58–59 in H. A. Mooney (ed.), Fire regimes and ecosystem properties. USDA Forest Service, Gen. Tech. Rep. WO-26.

Kilgore, B. M., and D. Taylor. 1979. Fire history in a sequoia–mixed conifer forest. Ecology 60:129–142.

Klikoff, L. 1965. Microenvironmental influence on vegetational pattern near timberline in the central Sierra Nevada. Ecol. Monogr. 35:187–211.

Knudson, T. 1991. The Sierra in peril. Sacramento Bee (5-part series), Sacramento, Calif.

Kuchler, A. W. 1977. The map of the natural vegetation, pp. 909–938 in M. G. Barbour and J. Major (eds.), Terrestrial vegetation of California. Wiley, New York.

Kunz, M. E. 1988. Patterns of distribution, growth, and seedling survival in sympatric red fir (Abies magnifica) and white fir (A. concolor) in the Sierra Nevada. Ph.D. dissertation, University of California, Davis.

Laacke, R. J. 1979. California forest soils. Univ. Calif., Div. Agricultural Sciences, Pub. 4094, Berkeley.

Laacke, R. J., and J. H. Tomascheski. 1986. Shelterwood regeneration of true fir: conclusions after 8 years. USDA Forest Service, Res. Pap. PSW-184, Berkeley.

Lanini, W. T., and S. R. Radosevich. 1986. Responses of three conifer species to site preparation and shrub control. For. Sci. 32:61–77.

Larcher, W. 1995. Physiological plant ecology, 3rd ed. Springer-Verlag, New York.

Leach, S. E. 1994. Ecological relationships of disjunct populations of Pinus ponderosa Laws. in the Owens Valley region of eastern California. Master's thesis, University of California, Davis.

Lepper, M. G. 1974. Pinus flexilis James, and its environmental relationships. Ph.D. dissertation, University of California, Davis.

Lepper, M. G. 1980. Carbon dioxide exchange in Pinus flexilis and P. strobiformis (Pinaceae). Madroño 27:17–24.

Littrell, E. E., and P. M. McDonald. 1974. Within-stand dynamics of bigcone Douglas fir in southern California. USDA Forest Service, unpublished report, Berkeley.

MacMahon, J. A., and D. C. Anderson. 1982. Subalpine forests: a world perspective with emphasis on western North America. Prog. Phy. Geogr. 6:368–425.

Major, J. 1967. Potential evapotranspiration and plant distribution in western states with emphasis on California, pp. 93–126 in R. H. Shaw (ed.), Ground level climatology. AAAS, Washington D.C.

Major, J. 1988. California climate in relation to vegetation, pp. 11–74 in M. G. Barbour and J. Major (eds.), Terrestrial vegetation of California, exp. ed. California Native Plant Society, Spec. Publ. No. 9, Sacramento.

Marshall, J. T. 1988. Birds lost from a giant sequoia forest during fifty years. Condor 90:359–372.

Martin, R. E., and D. B. Sapsis. 1992. Fires as agents of biodiversity: pyrodiversity promotes biodiversity, pp. 150–157 in H. M. Kerner (ed.), Proceedings of the symposium on biodiversity of northwestern California. Wildland Resources Center, Report No. 29. University of California, Berkeley.

Mason, H. L. 1970. The scenic, scientific, and educational values of the natural landscape of California. California Dept. of Parks and Recreation, Sacramento.

Mastroguiseppe, R. J., and J. D. Mastroguiseppe. 1980. A study of Pinus balfouriana Grev. and Balf. (Pinaceae). Syst. Bot. 5:86–104.

Matyas, W. J., and I. Parker. 1980. CALVEG, mosaic of existing vegetation of California [map]. USDA Forest Service, Regional Ecology Group, San Francisco.

Mayer, K., and W. Laudenslayer. 1988. A guide to wildlife habitats of California. The California Resources Agency, Department of Forestry and Fire Protection, Sacramento.

McBride, J. R., and R. D. Laven. 1976. Fire scars as an

indicator of fire frequency in the San Bernardino Mountains, California. J. For. 74:439–442.

McBride, J. R., V. P. Semion, and P. R. Miller. 1975. Impact of air pollution on the growth of ponderosa pine. Calif. Agr. 29(12):8–9.

McClaran, M. P. 1986. Age structure of *Quercus douglasii* in relation to livestock grazing and fire. Ph.D. dissertation, University of California, Berkeley.

McClaran, M. P. 1988. Comparison of fire history estimates between open-scarred and intact *Quercus douglasii*. Amer. Midl. Nat. 120:432–435.

McClaran, M. P., and J. W. Bartolome. 1990. Comparison of actual and predicted blue oak age structures. J. Range Mgt. 43:61–63.

McColl, J. G. 1981. Effects of acid rain on plants and soils in California. Calif. Air Resources Board, Contract A8-136-31, unpublished final report.

McDonald, J. E. 1996. The sequoia forest plan settlement agreement, pp. 126–128 in SNEP Science Team (eds.), Status of the Sierra Nevada, vol. III, Wildland Resources Center Report 38, University of California, Davis.

McDonald, P. M., and J. C. Tappeiner. 1996. Silviculture-ecology of forest-zone hardwoods in the Sierra Nevada, pp. 621–636 in SNEP Science Team (eds.), Status of the Sierra Nevada, vol. III, Wildland Resources Center Report 38. University of California, Davis.

McKelvey, K. S., and K. K. Busse. 1996. Twentieth-century fire patterns on Forest Service lands, Chapter 41 in SNEP Science Team (eds.), Status of the Sierra Nevada, Vol. II, Wildland Resources Center Report No. 36. University of California, Davis.

McKelvey, K. S., and J. D. Johnston. 1992. Historical perspective on forests of the Sierra Nevada and the Transverse Ranges of southern California: forest conditions at the turn of the century, pp. 225–246 in J. Verner (ed.), The California spotted owl. USDA Forest Service Gen. Tech. Rep. PSW-133. Albany.

McNab, W. H., and P. E. Avers (compilers). 1994. Ecological subregions of the United States: section descriptions. USDA Forest Service, Administrative Publication WO-WSA-5. Washington, D.C.

Mellmann-Brown, S., and M. G. Barbour. 1995. Understory/overstory species patterns through a Sierra Nevada ecotone. Phytocoenologia 25:89–106.

Merriam, C. H. 1898. Life zones and crop zones of the United States. USDA Bull. Biol. Surv. Div. 10:9–79

Miller, P. L. 1973. Oxidant-induced community change in a mixed conifer forest, pp. 101–117 in J. Naegle (ed.), Air pollution damage to vegetation. Advances in chemistry series 122. American Chemical Society, Washington, D.C.

Miller, P. R. 1996. Biological effects of air pollution in the Sierra Nevada, pp. 885–900 in SNEP Science Team (eds.), Status of the Sierra Nevada, Vol. III, Wildland Resources Center. University of California, Davis.

Minnich, R. A. 1976. Vegetation of the San Bernardino Mountains, pp. 99–124 in J. Latting (ed.), Plant communities of southern California. California Native Plant Society, Spec. Pub. 2, Berkeley.

Minnich, R. A. 1980. Wildfire and the geographic relationships between canyon live oak, Coulter pine, and bigcone Douglas fir forests, pp. 55–61 in T. R. Plumb (ed.), Ecology, management, and utilization of California oaks. USDA Forest Service PSW-44, Berkeley.

Minnich, R. A. 1986. Snow levels and amounts in the mountains of southern California. Journal of Hydrology 86:37–58.

Minnich, R. A. 1987. The distribution of forest trees in northern Baja California, Mexico. Madroño 34:98–127.

Minnich, R. A. 1988. The biogeography of fire in the San Bernardino Mountains of California: a historical study. Univ. Calif. Pub. Geog. 27:1–121.

Minnich, R. A. et al. 1998. The el niño Southern oscillation and precipitation and variability in Baja, California, Mexico. Atmosfera (in press).

Minnich, R. A., E. Franco Vizcaino, J. Sosa Ramirez, J. H. Burk, W. J. Barry, and M. G. Barbour. 1994. The potential of the Sierra San Pedro Martir as a biosphere reserve, in A. Gomez-Pompa and R. Dirzo (eds.), Proyecto sobre areas protegidas de Mexico, 1. Areas protegidas establecidas. Report to the Secretary of Social Development, Mexico, DF and to UC Mexus, Univ. Calif. Riverside.

Minnich, R. A., M. G. Barbour, J. H. Burk, and R. F. Fernau. 1995. Sixty years of change in California conifer forests of the San Bernardino Mountains. Conserv. Biol. 9:902–914.

Minnich, R. A., M. G. Barbour, J. H. Burk, and J. Sosa-Ramirez. 1998. Californian conifer forests under unmanaged fire regimes in the Sierra San Pedro Martír, Baja California, México. J. Biogeog (in press).

Minnich, R. A., and E. Franco-Vizcaino. 1997. Mediterranean vegetation of northern Baja California, including the Sierra San Pedro Martir. Fremontia 25:3–21.

Mooney, H. A. 1972. The carbon balance of plants. Ann. Rev. Ecol. Syst. 3:315–346.

Mooney, H. A., R. D. Wright, and B. R. Strain. 1964. Field measurements of the metabolic responses of bristlecone pine and big sagebrush in the White Mountains of California. Amer. Midl. Nat. 72:281–297.

Muick, P. C., and J. W. Bartolome. 1987. An assessment of natural regeneration of oaks in California. Report to California Dept. Forestry and Fire Protection, Sacramento.

Munz, P. A., and D. D. Keck. 1959. A California flora. University of California Press, Berkeley.

Myatt, R. G. 1980. Canyon live oak vegetation in the Sierra Nevada, pp. 86–91 in T. R. Plumb (ed.), Ecology, management, and utilization of California oaks. USDA Forest Service PSW-44, Berkeley.

Nachlinger, J. L., and N. H. Berg. 1988. Snowpack-vegetation dynamics: mountain hemlocks in the Lake Tahoe area, pp. 23–34 in B. Shafer, J. K. Marron, and C. Torendle (eds.), Proceedings of the Western Snow Conference, Colorado State University, Fort Collins. [Copies can be ordered from J. K. Marron at POB 2646, Portland, OR 97208.]

Neibling, C. R., and M. T. Conkle. 1990. Diversity of Washoe pine and comparisons with allozymes of ponderosa pine races. Can. J. For. Res. 20:298–308.

Nelson, E. W. 1922. Lower California and its natural resources. National Academy of Science, XVI (first memoir). Washington, D.C.

Oliver, W. W. 1985. Growth of California red fir advance regeneration after overstory removal and thinning. USDA Forest Service, Res. Pap. PSW-180, Berkeley.

Oosting, H. J., and W. D. Billings. 1943. Abietum magnificae: the red fir forest of the Sierra Nevada. Ecol. Monogr. 13:259–274.

Parker, A. J. 1982. Environmental and compositional ordinations of conifer forests in Yosemite National Park, California. Madroño 29:109–118.

Parker, A. J. 1984. Mixed forests of red fir and white fir in Yosemite National Park, California. Amer. Midl. Nat. 112:15–23.

Parker, A. J. 1986a. Persistence of lodgepole pine forests in the central Sierra Nevada. Ecology 67:1560–1567.

Parker, A. J. 1986b. Environmental and historical factors affecting red and white fir regeneration in ecotonal forests. For. Sci. 32:339–347.

Parker, A. J. 1989. Forest/environment relationships in Yosemite National Park, California, USA. Vegetatio 82:41–54.

Parker, A. J. 1991. Forest/environment relationships in Lassen Volcanic National Park, California, USA. J. Biogeog. 18:543–552.

Parker, A. J. 1994. Latitudinal gradients of coniferous tree species, vegetation, and climate in the Sierran–Cascade axis of northern California. Vegetatio 115: 145–155.

Parker, V. T., and P. Yoder-Williams. 1989. Reproduction, survival, and growth of young *Pinus jeffreyi* by an herbaceous perennial, *Wyethia mllis*. Amer. Midl. Nat. 121:105–111.

Parrish, J. L., and D. C. Erman (eds.). 1994. Critical questions for the Sierra Nevada: recommended research priorities and administration. Centers for Water and Wildland Resources, Report No. 34, University of California, Davis.

Passini, M-F., J. Delgadillo, and M. Salazar. 1989. L'ecosysteme forestier de Basse-Californie: composition floristique, variables ecologiques principales, dynamique. Acta Oecologica 10:275–293.

Patterson, M. T., and P. W. Rundel. 1995. Stand characteristics of ozone-stressed populations of *Pinus jeffreyi* (Pinaceae): extent, development, and physiological consequences of visible injury. Amer. J. Bot. 82:150–158.

Pavlik, B. M., and M. G. Barbour. 1991. Seasonal patterns of growth, water potential, and gas exchange of red and white fir saplings across a montane ecotone. American Midland Naturalist 126:14–29.

Paysen, T. E., J. A. Derby, and C. E. Conrad. 1982. A vegetation classification system for use in California. USDA Forest Service Gen. Tech. Rep. PSW-63, Berkeley.

Perala, D. A. 1990. *Populus tremuloides* Michx., pp. 555–569 in R. M. Burns and B. H. Honkala (eds.), Silvics of North America, vol. 2. USDA Forest Service, Agriculture Handbook 654, Washington, D.C.

Pitcher, D. C. 1987. Fire history and age structure in red fir forests of Sequoia National Park, California. Can. J. For. Res. 17:582–587.

Plumb, T. R. (ed.). 1980. Ecology, management, and utilization of California oaks. USDA Forest Service PSW-44, Berkeley.

Potter, D. A. 1994. Guide to forested communities of the upper montane in the central and southern Sierra Nevada. USDA Forest Service, R5-ECOL-TP-003., Albany, Calif.

Potter, D. A. In press. Ecological classification in upper montane forests of the central and southern Sierra Nevada. USDA Forest Service, Gen. Tech. Rep., Albany, Calif.

Potter, D. A., M. Smith, T. Beck, B. Kermeen, W. Hance, and S. Robertson 1992. Ecological characteristics of old-growth red fir, lodgepole pine, Jeffrey pine, and mixed subalpine forests in California. Old-growth definitions/descriptions for forest cover types, in R. Stewart (ed.), USDA Forest Service, PSW, Albany, Calif.

Pronos, J., and D. R. Vogler. 1981. Assessment of ozone injury to pines in the southern Sierra Nevada, 1979–1980. USDA Forest Service Forest and Pest Management Report 81–20, San Francisco.

Quezel, P., and M. Barbero. 1989. Altitudinal zoning of forest structures in California and around the Mediterranean: a comparative study. Bielefelder Okolog. Beitr. 4:25–43.

Randall, J. M., and M. Rejmanek. 1993. Interference of bull thistle (*Cirsium vulgare*) with growth of ponderosa pine (*Pinus ponderosa*) seedlings in a forest plantation. Can. J. For. Res. 23:1507–1513.

Raven, P. H., and D. I. Axelrod. 1978. Origin and relationships of California flora. Univ. Calif. Pub. Bot. 72:1–134.

Regalia, M. R. 1978. The stand dynamics of Shasta red fir in northwestern California. Master's thesis, Humboldt State University, Arcata, Calif.

Riegel, G. M., D. A. Thornburgh, and J. O. Sawyer. 1990. Forest habitat types of the south Warner Mountains, Modoc County, California. Madroño 37: 88–112.

Rourke, M. D. 1988. The biogeography and ecology of foxtail pine, *Pinus balfouriana* Grev. & Balf., in the Sierra Nevada. Ph.D. dissertation, University of Arizona, Tucson.

Royce, E. B. 1997. Xeric effects on the distribution of conifers in a southern Sierra Nevada ecotone. Ph.D. dissertation, University of California, Davis.

Rivas-Martinez, S. 1997. Syntaxonomical synopsis of the potential natural plant communities of North America, part 1. Itinera Geobatanica 10:5–148.

Rundel, P. W. 1971. Community structure and stability in the giant sequoia groves of the Sierra Nevada, California. Amer. Midl. Nat. 85:478–492.

Rundel, P. W. 1972a. An annotated check list of the groves of *Sequoiadendron giganteum* in the Sierra Nevada, California. Madroño 21:319–328.

Rundel, P. W. 1972b. Habitat restriction in giant sequoia: the environmental control of grove boundaries. Amer. Midl. Nat. 87:81–99.

Rundel, P. W. 1980. Adaptations of mediterranean-climate oaks to environmental stress, pp. 43–54 in T. R. Plumb (ed.), Ecology, management, and utilization of California oaks. USDA Forest Service, PSW-44, Berkeley.

Rundel, P. W., D. J. Parsons, and D. T. Gordon. 1988. Montane and subalpine vegetation of the Sierra Nevada and Cascade Ranges, pp. 559–599 in M. G. Barbour and J. Major (eds.), Terrestrial vegetation of California, 2nd ed. California Native Plant Society, Sacramento.

Ryerson, A. D. 1983. Population structure of *Pinus balfouriana* Grev. & Balf. along the margins of its distribution area in the Sierran and Klamath regions of California. Master's thesis, California State University, Sacramento.

Ryerson, A. D. 1984. Krummholz foxtail pines. Fremontia 11(4):30.

Savage, M. 1994. Anthropogenic and natural disturbance and patterns of mortality in a mixed conifer forest in California. Can. J. For. Res. 24:1149–1159.

Sawyer, J. O., and T. Keeler-Wolf. 1995. A manual of California vegetation. California Native Plant Society, Sacramento.

Sawyer, J. O., and D. A. Thornburgh. 1988. Montane and subalpine vegetation of the Klamath Mountains, pp. 699–732 in M. G. Barbour and J. Major (eds.), Terrestrial vegetation of California. California Native Plant Society, Spec. Pub. No. 9, Sacramento.

Sawyer, J. O., D. A. Thornburgh, and J. R. Griffin. 1988. Mixed evergreen forest, pp. 359–381 in M. G. Barbour and J. Major (eds.), Terrestrial vegetation of California. California Native Plant Society, Spec. Pub. No. 9, Sacramento.

Schierenbeck, K. A., and D. B. Jensen. 1994. Vegetation of the upper Raider and Hornback Creek basins, south Warner Mountains: northwestern limit of *Abies con-color* var. *lowiana*. Madroño 41:53–64.

Scott, T. 1991. The distribution of Engelmann oak (*Quercus engelmannii*) in California. USDA Forest Service, General Technical Report PSW-126, Berkeley.

Selter, C. M., W. D. Pitts, and M. G. Barbour. 1986. Site microenvironment and seedling survival of Shasta red fir. Amer. Midl. Nat. 115:288–300.

Show, S. B., and E. I. Kotok. 1924. The role of fire in the California pine forest. USDA Bulletin 1294, Washington, D.C.

Smith, S. 1994. Ecological guide to eastside pine plant associations. USDA Forest Service, R5-ECOL-TP-004, Albany, Calif.

SNEP Science Team (eds.). 1996. Status of the Sierra Nevada, Final report to Congress of the Sierra Nevada Ecosystem Project, 3 vols. Wildland Resources Center Report No. 36, University of California, Davis.

Standiford, R. B., J. Klein, and B. Garrison. 1996. Sustainability of Sierra Nevada hardwood rangelands, pp. 637–680 in Status of the Sierra Nevada, vol. III, Wildland Resources Center Report No. 38, University of California, Davis.

Stebbins, G. L., and J. Major. 1965. Endemism and speciation in the California flora. Ecolog. Monogr. 35:1–35.

Stephenson, N. L. 1992. Long-term dynamics of giant sequoia populations: implications for managing a pioneer species, pp. 56–63 in P. S. Aune (ed.), Proceedings of the symposium on giant sequoias. USDA Forest Service, Gen. Tech. Rep. PSW-GTR-151, Albany, Calif.

Stromberg, J. C., and D. T. Patten. 1992. Mortality and age of black cottonwood stands along diverted and undiverted streams in the eastern Sierra Nevada, California. Madroño 39:205–223.

Stuart, J. D. 1984. Hazard rating of lodgepole pine stands to mountain pine beetle outbreaks in south-central Oregon. For. Ecol. Mgt. 5:207–214.

Swetnam, T. W. 1993. Fire history and climatic change in giant Sequoia groves. Science 262:885–890.

Talley, S. N. 1974. The ecology of Santa Lucia fir (*Abies bracteata*), a narrow endemic of California. Ph.D. dissertation, Duke University, Durham.

Talley, S. N., and J. R. Griffin. 1980. Fire ecology of a montane pine forest, Junipero Serra Peak, California. Madroño 27:49–60.

Taylor, A. H. 1993. Fire history and structure of red fir (*Abies magnifica*) forests, Swain Mountain Experimental Forest, Cascade Range, northeastern California. Can. J. For. Res. 23:1672–1678.

Taylor, A. H., and C. B. Halpern. 1991. Structure and dynamics of *Abies magnifica* forests in the southern Cascade Range, USA. J. Veg. Sci. 2:189–200.

Taylor, D. W. 1976. Vegetation near timberline at Carson Pass, central Sierra Nevada, California. Ph.D. dissertation, University of California, Davis.

Taylor, O. C. 1980. Photochemical oxidant air pollution effects on a mixed conifer ecosystem. EPA-600/3-80-002, US Environmental Protection Agency, Corvallis, Ore.

Thornburgh, D. A. 1982. Succession in the mixed evergreen forests of northwestern California, pp. 87–91 in J. E. Means (ed.), Forest succession and stand development research in the northwest. Forest Research Laboratory of Oregon State University, Corvallis.

Thorne, R. F. 1976. The vascular plant communities of California, pp. 1–31 in J. Latting (ed.), Plant communities of southern California. California Native Plant Society, Spec. Pub. 2, Berkeley.

Thorne, R. F. 1982. The desert and other transmontane plant communities of southern California. Aliso 102:219–257.

Thorne, R. F. 1988. Montane and subalpine forests of the Transverse and Peninsular Ranges, pp. 537–557 in M. G. Barbour and J. Major (eds.), Terrestrial vegetation of California, 2nd ed. California Native Plant Society, Sacramento.

Tomback, D. F. 1986. Post-fire regeneration of krummholz whitebark pine. Madroño 33:100–110.

Tucker, J. M. 1980. Taxonomy of California oaks, pp. 19–29 in T. R. Plumb (ed.), Ecology, management, and utilization of California oaks. USDA Forest Service, PSW-44, Berkeley.

UNECSCO (United Nations Educational and Scientific Council). 1973. International classification and mapping of vegetation. Paris.

Ustin, S. L., R. A. Woodward, M. G. Barbour, and J. L. Hatfield. 1984. Relationships between sunfleck dynamics and red fir seedling distribution. Ecology 65:1420–1428.

Vale, T. R. 1979. *Pinus coulteri* and wildfire on Mount Diablo, California. Madroño 26:135–139.

Vankat, J. L. 1970. Vegetation change in Sequoia National Park, California. Ph.D. dissertation, University of California, Davis.

Vankat, J. L. 1982. A gradient perspective on the vegetation of Sequoia National Park, California. Madroño 29:200–214.

Vankat, J. L., and J. Major. 1978. Vegetation changes in Sequoia National Park, California. J. Biogeogr. 5:377–402.

Vasek, F. C. 1985. Southern California white fir (Pinaceae). Madroño 32:65–77.

Vogl, R. J. 1976. An introduction to the plant communities of the Santa Ana and San Jacinto Mountains, pp. 77–98 in J. Latting (ed.), Plant communities of southern California. California Native Plant Society, Spec. Pub. 2, Berkeley.

Wagener, W. W. 1961. Past fire incidence in Sierra Nevada forests. J. For. 59:739–748.

Wainwright, T. C., and M. G. Barbour. 1984. Characteristics of mixed evergreen forest in the Sonoma Mountains of California. Madroño 31:219–230.

Weaver, H. 1974. Effects of fire on temperate forests: western United States, pp. 279–319 in T. T. Kozlowski and C. E. Ahlgren (eds.), Fire and ecosystems. Academic Press, New York.

Welker, J. M., and J. W. Menke. 1990. The influence of simulated browsing on tissue water relations, growth, and survival of *Quercus douglasii* (Hook and Arn.) seedlings under slow and rapid rates of soil drought. Func. Ecol. 4:807–817.

Westman, W. E. 1987. Aboveground biomass, surface area, and production relations of red fir (*Abies magnifica*) and white fir (*A. concolor*). Can. J. For. Res. 17:311–319.

Wheeler, D. P. (ed.). 1992. The Sierra Nevada: report of the Sierra Summit Steering Committee. California Resources Agency, Sacramento.

White, K. L. 1966. Structure and composition of foothill woodland in central coastal California. Ecology 47: 229–237.

Wieslander, A. E. 1935. A vegetation type map of California. Madroño 3:140–144.

Wieslander, A. E., and H. A. Jensen. 1946. Forest areas, timber volumes, and vegetation types in California. USDA Forest Service, PSW Release No. 4 with map, Berkeley.

Wiggins, I. L. 1980. Flora of Baja California. Stanford University Press, Stanford.

Willard, D. 1992. The natural giant sequoia (*Sequoiadendron giganteum*) groves of the Sierra Nevada, California – an updated, annotated list, pp. 159–164 in P. S. Aune (ed.), Proceedings of the symposium on giant sequoias. USDA Forest Service, Gen. Tech. Rep. PSW-GTR-151, Albany.

Williams, W. T. 1983. Tree growth and smog disease in the forests of California: case history, ponderosa pine in the southern Sierra Nevada. Environ. Pollut. (Series A) 30:59–75.

Williams, W. T., M. Brady, and S. C. Willison. 1977. Air pollution damage to the forests of the Sierra Nevada Mountains of California. J. Air Pollut. Control Assoc. 27:230–234.

Wright, R. D. 1968. Lower elevational limit of montane trees. II. Environment-keyed responses of three conifer species. Bot. Gaz. 129:219–226.

Wright, H. A., and A. W. Bailey. 1982. Fire ecology, United States and Canada. Wiley, New York.

Yeaton, R. I. 1978. Some ecological aspects of reproduction in the genus *Pinus* L. Bull. Torrey Bot. Club 105:306–311.

Yeaton, R. I. 1981. Seedling characteristics and elevational distributions of pine (Pinaceae) in the Sierra Nevada of central California; a hypothesis. Madroño 28:67–77.

Yeaton, R. I. 1983a. The effect of predation on the elevational replacement of digger pine by ponderosa pine on the western slopes of the Sierra Nevada. Bull. Torrey Bot. Club 110:31–38.

Yeaton, R. I. 1983b. The successional replacement of ponderosa pine by sugar pine in the Sierra Nevada. Bull. Torrey Bot. Club 110:292–297.

Yeaton R. I., R. W. Yeaton, and J. E. Horenstein. 1980. The altitudinal replacement of digger pine by ponderosa pine on the western slopes of the Sierra Nevada. Bull. Torrey Bot. Club 107:487–495.

Chapter
6

Chaparral

JON E. KEELEY

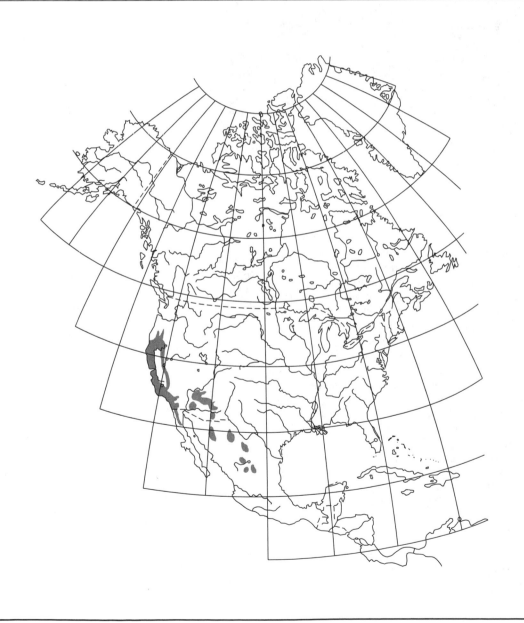

Figure 6.1. View of southern California mixed chaparral.

INTRODUCTION

"Chaparral" is the name applied to the sclero-phyllous shrub vegetation of southwestern North America. This chapter concentrates on the ever-green chaparral centered in California, although related vegetation, including the interior chaparral of Arizona and northeastern Mexico, the winter-deciduous "petran" chaparral of the Rocky Mountains, and the West Coast summer-deciduous "soft chaparral," or coastal sage scrub, are also discussed. For a complete bibliography of chaparral literature, see Keeley (1995a), and for reviews with a more historical perspective, see Mooney and Parsons (1973) and Hanes (1977). Species lists are presented in an earlier version of this chapter in Keeley and Keeley (1988a). For comparisons of chaparral with other mediterranean-climate scler-ophyllous shrub communities, see Mooney (1977b), Miller (1981), Shmida and Barbour (1982), Kruger and associates (1983), Barbour and Minnich (1990), and Keeley (1992a).

California chaparral dominates the foothills

I thank Drs. Ragan Callaway, Frank Davis, Steve Davis, Ron Quinn, and Tom Stohlgren for providing data or artwork.

from the Sierra Nevada to the Pacific Ocean (Wieslander and Gleason 1954). The northern limits are the drier parts of the Rogue River watershed in Oregon (43°N latitude) (Detling 1961), and the southern limits are the San Pedro Mártir Mountains of Baja California (30°N) (Shreve 1936), with isolated montane patches occurring as far south as 27°30'N (Axelrod 1973).

Throughout this region, chaparral characteristically forms a nearly continuous cover of closely spaced shrubs 1.5–4 m tall, with intertwining branches (Fig. 6.1). Herbaceous vegetation is generally lacking, except after fires, which are frequent throughout the range. Chaparral occurs from sea level to 2000 m on rocky, nutrient-deficient soils and is best developed on steep slopes. Because of complex patterns of topographic, edaphic, and climatic variations, chaparral may form a mosaic pattern in which patches of oak woodland, grassland, or coniferous forest appear in sharp juxtaposition. Fire frequency and substrate are important factors determining these patterns. Chaparral is replaced by grassland on frequently burned sites, especially along the more arid borders at low elevations where shrub recovery is more precarious because of drought, and on deeper clay soils and alluvial plains, and by oak woodland on mesic slopes where fires are less frequent and often less intense.

California chaparral is distributed in a region of mediterranean climate: cool, wet winters and hot, dry summers (Fig. 6.2). Rainfall is 200–1000 mm annually, two-thirds of which falls from November to April in storms of several days' duration (Miller and Hajek 1981). Because of the episodic nature of the winter rains, there may be prolonged dry spells, even during the wet season. The annual rainfall variance is significantly greater than in other regions, and extreme droughts are not uncommon (Varney 1925; Major 1977). Significant summer precipitation is rare and arises from convectional storms in the higher elevations or tropical storms in the south. Mean winter temperatures range from less than 0° C at montane sites to greater than 10° C at lower elevations, although coastal sites may occasionally experience subzero temperatures (Collett 1992). Summer temperatures often exceed 40° C but are more moderate along the coast and at the upper elevational limits.

The climate is dominated by the subtropical high-pressure cell that forms over the Pacific Ocean. During the summer, this air mass moves northward and blocks polar fronts from reaching land. During the winter, this high-pressure cell moves toward the equator and allows winter storms to pass onto land. The climate is wettest in the north, where the effect of the Pacific High is least, and it becomes progressively drier to the south.

Winter precipitation patterns are largely controlled by the orographic effect. On the windward side of mountains, air cools adiabatically with increasing elevation, so that temperature decreases and precipitation increases with elevation. On the leeward sides of mountains, air warms as it descends, creating a rainshadow that is hotter and drier than coastal exposures at comparable elevations.

A factor of local importance in southern California is the Santa Ana wind, which is the result of a high-pressure cell in the interior of the United States that drives dry desert air toward the coast. These föehn types of winds may exceed 100 km hr^{-1} and bring high temperature and low humidity. Santa Anas are most common in spring and fall, and some of the most catastrophic wildfires occur under these conditions.

COMMUNITY COMPOSITION

Chaparral is a shrub-dominated vegetation, with other growth forms playing minor or temporal roles. More than 100 evergreen shrub species occur

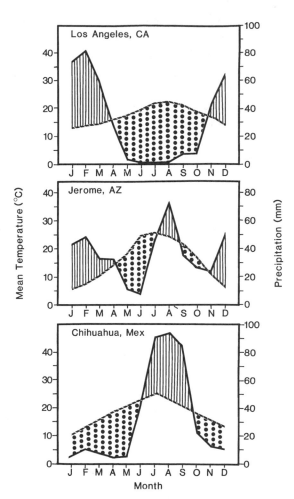

Figure 6.2. Climatic data for three chaparral sites (vertical line area = ppt>evap; dotted area = ppt<evap; solid line = precipitation, dashed line = mean temperature): Los Angeles, California (75 m, 34°05', 118°15'), Jerome, Arizona (1600 m, 34°45', 112°07), and Chihuahua, Mexico (1350m, 28°42', 105°57'). In addition to differences in seasonal distribution of precipitation, the annual variance in precipitation is much greater in California than in interior regions. For example, 40 yr of data for Los Angeles showed 5 mo with a coefficient of variation from 220–350%, whereas the Arizona site had no month with more than 120% (data from U.S. Department of Commerce, Climatic Summary of the United States).

in chaparral (Keeley and Keeley 1988a). At any one site there may be as few as one or more than 20 species, depending on available moisture, slope aspect, slope inclination, distance from the coast, elevation, latitude, and fire history. Broad generalizations about species preferences toward edaphic and topographic features are difficult to make in that they often change with the region. Attempts to ascribe site preferences for most species have

Table 6.1. *Site characteristics and areal coverage (% ground surface cover) for 10 selected stands (A–J) of mature chaparral throughout the range (nomenclature according to Hickman 1993).*

Species	A	B	C	D	E	F	G	H	I	J
Elevation (m)	800	1000	1100	375	300	975	1100	1300	1000	400
Slope aspect	S	E	S	W	N	N	E	level	SE	S
Years since last fire	35	22	95	55	55	118	—	—	>45	—
Bare ground (%)	—	—	7	7	3	5	>47	>37	—	12
Woody species diversity (H')	0.07	1.53	0.79	1.83	1.96	2.96	—	—	1.91	—
Adenostoma fasciculatum	138	49	80	7		5			33	9
Arctostaphylos auriculata										64
A. glandulosa						2			58	
A. glauca			33				17	19		
Ceanothus cuneatus										10
C. greggii		22					9		7	
C. megacarpus				79						
C. spinosus					136					
Cercocarpus betuloides			26	3		11	5			
Heteromeles arbutifolia					5	10				
Quercus berberidifolia		8				93				
Other species	3		1	13	10	12	18	42	5	

Sources: (A) Parsons (1976) in the southern Sierra Nevada Range; (B) Keeley and Johnson (1977) in the interior Peninsular Range; (C) Keeley (1992b) in the southern Sierra Nevada Range; (D) Keeley (1992b) in the coastal Transverse Range; (E) Keeley (1992b) in the coastal Transverse Range; (F) Keeley (1992b) in the coastal Peninsular Range; (G) Vasek and Clovis (1976) in the interior Transverse Range; (H) Vasek and Clovis (1976) in the interior Transverse Range; (I) Schorr (1970) in the interior Peninsular Range; (J) C. Davis (1972) in the northern Central Coast Range.

generally produced weak correlations (Gauss 1964; Wilson and Yogi 1965; Zenan 1967; Hanes 1971; Steward and Webber 1981). On a local scale there is a regular turnover in species from one slope aspect to another and with elevation. For example, the diverse genus *Ceanothus* exhibits a regular turnover of species with elevation, and within an elevational band, species segregate by slope aspect or inclination (Nicholson 1993; Zedler 1995a).

The composition of chaparral stands varies along gradients of moisture, elevation, latitude, and the like, and attempts at classifying chaparral associations have named 28 communities (Holland 1986) and over 50 series (Sawyer and Keeler-Wolf 1995). Sites may be dominated by a single species, or several may co-dominate (Table 6.1). Due to overlapping branches, areal coverage can exceed 100% (ground-surface cover). Typically, bare ground is <10%, but it increases on drier desertic sites (Sites G & H, Table 6.1) and on stressful substrates such as serpentine or gabbro. Demography of mature chaparral has been described in Schlesinger, Gray, and Gilliam (1982a), Keeley (1992b, 1992c), Zammit and Zedler (1992); patterns for three stands are shown in Table 6.2. Sometimes a species that represents a minor part of the cover may be numerically important, as is *Cercocarpus betuloides* at Site 1 (Table 6.2) (nomenclature according to Hickman 1993). In dense chaparral, this spe-

cies spreads by rhizomes, an atypical characteristic for chaparral shrubs; however, it is arborescent on open sites. Sprouting shrubs such as *Adenostoma fasciculatum*, on the other hand, commonly produce only a few stems in dense chaparral but proliferate basal sprouts on more open sites. Nonsprouting species of *Arctostaphylos* and *Ceanothus* usually have a single stem per shrub, and arborescent individuals of the former taxon may dominate a site, even at low densities (Site 2 in Table 6.2).

The most widely distributed chaparral shrub is *Adenostoma fasciculatum* (Fig. 6.3), ranging from Baja to northern California in pure chamise chaparral or in mixed stands (Table 6.1). It often dominates at low elevations and on xeric south-facing slopes, with 60–90% cover. The short needle-like leaves produce a sparse foliage, and soil litter layers are poorly developed. Along its lower elevational limits, *A. fasciculatum* intergrades with subligneous coastal sage subshrubs, particularly *Salvia mellifera*, *S. apiana*, and *Eriogonum fasciculatum*. At higher elevations, *A. fasciculatum* often codominates with one or more species of *Arctostaphylos* or *Ceanothus* (Table 6.1).

Ceanothus and *Arctostaphylos* are large genera (>60 species) and often form pure stands commonly referred to as manzanita chaparral or ceanothus chaparral. Some species are highly restricted, whereas others are nearly as widespread as *Aden-*

Table 6.2. *Demographic structure of three older stands of chaparral in southern California*

Species	Postfire regeneration[a]	Live basal coverage m² basal area ha⁻¹	Shrub density Genets ha⁻¹	% Dead	Ramets /genet	Seedlings & saplings Genets ha⁻¹
Site 1. West-facing slope (19°) in Santa Monica Mountains (375 m), 55 yr old:						
Ceanothus megacarpus	(OS)	34.9	3360	25	1.2	0
Adenostoma fasciculatum	(FR)	4.8	1500	22	6.4	0
Salvia mellifera	(FR)	2.8	1920	45	3.8	0
Cercocarpus betuloides	(OR)	2.1	610	0	65.8	0
Rhamnus crocea	(OR)	0.5	260	23	2.3	80
Additional species (three lianas)	(OR)	0.1	360	31	9.0	0
Site 2. Southeast-facing slope (9°) in San Gabriel Mountains (1000 m), 88 yr old:						
Arctostaphylos glauca	(OS)	41.8	2440	19	4.4	0
Ceanothus crassifolius	(OS)	13.9	3810	82	2.1	0
Adenostoma fasciculatum	(FR)	20.3	6690	33	5.5	0
Quercus berberidifolia	(OR)	2.3	220	0	5.9	0
Additional species (two shrubs, one subshrub)	(OR)	0.1	240	38	1.7	0
Site 3. Northwest-facing slope (37°) in San Gabriel Mountains (975 m), 65 yr old:						
Quercus berberidifola	(OR)	8.9	860	0	7.6	80
Adenostoma fasciculatum	(FR)	7.3	1940	50	4.7	0
Garrya veatchii	(OR?)	6.5	420	7	8.5	0
Heteromeles arbutifolia	(OR)	4.9	330	0	30.5	0
Arctostaphylos glauca	(OS)	2.8	220	14	1.8	0
Ceanothus crassifolius	(OS)	2.3	750	77	1.7	0
Prunus ilicifolia	(OR)	2.3	310	10	29.6	4530
Additional species (four shrubs, two subshrubs, one liana)	(FR & OR)	9.1	480	17	19.1	0

[a] OS = obligate seeder (no resprouting capacity); FR = facultative resprouter (resprouts and establishes seedlings); OR = obligate resprouter (typically does not establish seedlings after fire).
Source: Keeley (1992b, and unpublished data).

ostoma. Most species are endemic to the California chaparral and have suites of characters reflecting a long association with fire. Both *Ceanothus* and *Arctostaphylos* have lignotuberous species that sprout after fire, as well as species lacking the capacity for vegetative regeneration. All species in these two genera produce deeply dormant seeds that require fire cues for germination. The nonsprouting (obligate seedling) species tend to be more abundant on south-facing slopes, ridge tops, and desert exposures. Sprouting species are more important on mesic slopes and at higher elevations (Keeley 1986; Nicholson 1993).

Adenostoma, Arctostaphylos, and *Ceanothus* species predominate in the drier areas of chaparral, but on more mesic sites other broader-leaved evergreen shrubs become important (Site 3, Table 6.2). This association, referred to by Cooper (1922) as broad-sclerophyll chaparral, is more diverse and includes *Quercus berberidifolia* (formerly *Q. dumosa*) (Fig. 6.4), *Q. wislizenii, Heteromeles arbutifolia, Pru-*

nus ilicifolia, Cercocarpus betuloides, Malosma laurina, and species of *Rhamnus, Rhus,* and *Garrya,* and occasionally small winter-deciduous trees such as *Sambucus* species and *Fraxinus dipetala.* Shrubs are generally taller in this chaparral, 3–6+ m. Beneath the canopy, light levels and soil temperatures are much lower than in chamise chaparral, and soil litter layers are much deeper (Keeley 1992c). Most species in this association are long-lived and can become arborescent if left undisturbed (e.g., *Heteromeles* and *Prunus* can reach 11 m or more; Keeley, personal observation), and *Malosma* was considered to be one of the dominant arboreal species of southern California (Hall 1903). Most of these species are common components of woodland communities, where they persist as gap-phase shrubs (Keeley 1990a).

Montane chaparral at the upper elevational limits occupies sites that may be covered by snow for many months. It has a somewhat different physiognomy (see also Chapter 5); the evergreen shrubs

Figure 6.3. The needleleaf Adenostoma fasciculatum (chamise) is the most widely distributed of all California chaparral shrubs, but it is absent from Arizona chaparral.

Figure 6.4. Quercus berberidifolia (formerly Q. dumosa) is a broad-sclerophyll species commonly forming nearly pure stands known as scrub oak chaparral. This species is replaced in Arizona by the very closely related (if not conspecific) Q. turbinella, which is one of the most widely distributed of the interior chaparral species.

have a more rounded, compact shape, with foliage to the ground surface. The association is dominated by vigorous sprouting species capable of dense coppice growth after fire, resulting in nearly impenetrable thickets with more than 100% cover (Wilson and Vogl 1965; Conard and Radosevich 1982). Often montane chaparral is dominated by species that are more typically found as understory or gap-phase coniferous forest shrubs (e.g., *Castanopsis sempervirens, Quercus vaccinifolia, Arctostaphylos patula*, and *Ceanothus integerrimus*), and winter-deciduous shrubs such as species of *Prunus, Ribes, Amelanchior*, and *Symphoricarpos*.

Broad generalizations about regional changes in shrub cover, height, spinescence, and leaf characteristics have been presented in Mooney and Harrison (1972), Parsons (1976), and Rundel and Vankat (1989).

Regional Patterns

The most extensive tracts of continuous, uninterrupted evergreen chaparral is at middle elevations (300–1500 m) in southern California. This area,

sometimes described as the South Coast Region (Sampson 1944), includes the Transverse and Peninsular ranges, which extend from Ventura County to northern Baja California. Here chamise chaparral forms a blanket-like cover over large areas from the coast to the mesas and foothills and into the mountains. Near the coast, chaparral commonly gives way to the summer-deciduous coastal sage scrub, although the evergreen sclerophylls *Rhus integrifolia* and *Malosma laurina* are often associated with coastal sage vegetation. In parts of this region, *Adenostoma fasciculatum* is replaced by *A. sparsifolium*. This latter shrub is distributed from the southern part of the South Coast Ranges through the coastal section of the Transverse Ranges and the interior parts of the Peninsular Ranges into northern Baja California (Marion 1943); it appears to replace *A. fasciculatum* on more mesic and fertile sites (Beatty 1987). Both species co-exist on some sites and niche partitioning occurs by differences in soil microhabitat (Beatty 1987), water relations (Redt-

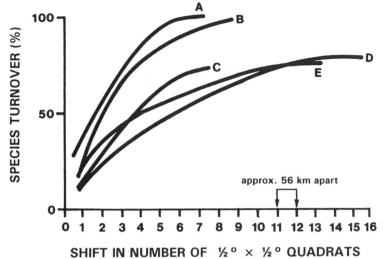

Figure 6.5. Species turnover rates in Ceanothus and Arctostaphylos in 0.5° × 0.5° latitude-longitude quadrats, expressed as a percentage of the species that the sites have in common. Santa Barbara/Santa Cruz to inland transects illustrate coast to interior changes (A, B, C), and coastal transects illustrate latitudinal changes (D, E) (redrawn from Cody 1986).

feldt and Davis 1996), and phenology (Hanes 1965). Epling and Lewis (1942) considered this South Coast Region to hold special significance because of the occurrence of numerous chaparral shrubs that are endemic or that reach their most northern distribution here.

Throughout this region, chaparral occurs on a variety of soils and substrates, including Jurassic, Upper Cretaceous, and Eocene sedimentary rocks and Tertiary volcanics in coastal ranges, and granitic substrates of Precenozoic metamorphic and metavolcanic rock on interior ranges (Minnich and Howard 1984). Particularly common in the Peninsular Ranges are mafic (Mg and Fe rich) soils derived from coarse-grained plutonic rocks known as gabbro (Alexander 1993). Recognizable by their red coloration, these gabbro soils are less favorable for plant growth, resulting in more open chaparral associations. Many herbs, shrubs, and trees are restricted to gabbro throughout their range, or in some cases species disjunct from much farther north (where they are found on various substrates) are restricted to gabbro soils in the south (Oberbauer 1993).

Farther south in northern Baja California, coastal chaparral takes on a somewhat different flavor with an increasing number of drought-deciduous shrubs such as *Fraxinus trifoliata* and *Aesculus parryi* (Peinado, Alcaraz, Aguirre, Delgadillo, and Aguado 1995). Interior plateaus such as the western Sierra Juarez are noteworthy for their extensive stands of *Adenostoma sparsifolium*.

In the central coastal regions of California, chaparral forms a patchwork mosaic with grassland, coastal sage scrub, and broadleaf and coniferous forest (Shreve 1927; F. Davis, Stine, Stomss et al. 1994). Fire is believed to be the determining factor in this mosaic distribution because no consistent pattern of edaphic or topographic factors coincides with the distribution of chaparral (Wells 1962). The region has numerous endemic *Arctostaphylos* and *Ceanothus* species, many of which are edaphic endemics (Wells 1962) or are restricted to coastal areas under marine influence (Griffin 1978). For example, the *Arctostaphylos andersonii* complex in the Santa Cruz Mountains comprises six species, each occupying a habitat characterized by a distinct combination of soil conditions, including water-holding capacity, texture, pH, depth, root penetrability, and fertility. (C. Davis 1972). In both *Arctostaphylos* and *Ceanothus* there is a rapid turnover of species along latitudinal and longitudinal gradients (Fig. 6.5). The steeper gradient is from the coast to the interior, reflecting greater heterogeneity in sediments and climate, suggesting that these taxa represent finely tuned morpho-physiological types adapted to fairly specific ecological conditions (Cody 1986). Both genera have endemics on the Channel Islands that are adapted to a more open, woodland-like chaparral (Bjorndalen 1978), arising from less frequent fires and more intensive grazing by feral animals (Minnich 1980a).

In the north coastal region, chaparral diminishes and is restricted to the driest sites (Clark 1937). *Adenostoma fasciculatum* is common in the drier interior valleys, whereas broad-sclerophyll species dominate the coastal chaparral, and some of them, such as *Prunus ilicifolia*, form small woodlands (Oberlander 1953). Throughout this region, localized outcrops of ultramafic serpentine substrate produce a more open vegetation referred to as serpentine chaparral or "serpentine barrens" (Kruckeberg 1984). Low levels of Ca and high levels of Mg, plus potentially toxic levels of Ni and Cr, in these soils (Koenigs, Williams, and Jones 1982) exclude many species (Whittaker 1960; Kruckeberg 1969, 1984). Serpentine endemics include shrubs such as *Q. durata* var. *durata, Garrya congdoni, Ceanothus jepsonii, C. ferrisae*, various subspecific taxa of *Arctostaphylos*, and herbaceous genera such as *Streptanthus*, where 16 species in one subgenus are restricted to serpentine. In addition, many widespread species have evolved serpentine ecotypes. Mechanisms for tolerating serpentine soils vary with the species; for example, White (1971) found that *Arctostaphylos nevadensis* was selectively able to take up Ca over Mg, whereas *Ceanothus pumilus* was not able to do so but could regulate Mg, Ni, and Cr uptake. The exclusion of serpentine endemics from other sites has been attributed to competition (Kruckeberg 1954), although Tadros (1957) and Wicklow (1964) found that in the case of the fire-following annual *Emmenanthe rosea*, restriction to serpentine sites was due to an inability to establish on more fertile soils that supported greater microbial growth.

Chaparral is absent from the Central Valley of California; however, some claim that this is an artifact of human disturbance (Cooper 1922; Bauer 1930). Above 500 m in the Sierra Nevada foothills, grasslands or xeric woodlands of *Pinus sabiniana* and *Aesculus californica* intergrade into mixed chaparral or nearly pure stands of *Ceanothus cuneatus* or *Adenostoma fasciculatum* (Graves 1932; Rundel and Parsons 1979; Vankat 1982). Montane chaparral is distributed above 1000 m, giving way to coniferous forest above 2000 m. This upper border is dynamic and strongly influenced by fire frequency (Wilken 1967).

On the interior side of the Sierra Nevada, montane chaparral forms a mosaic with coniferous forest, pinyon-juniper woodlands, or scrub vegetation with Great Basin affinities (Skau, Meeuwig, and Townsend 1970). On these sites montane chaparral may replace coniferous forest after wildfires and remain for 50+ yr (Townsend 1966). In the northern Sierra Nevada and Cascade ranges, extending as far north as Bend, Oregon (W. D. Billings, personal communication), montane chaparral is more restricted and forms a mosaic with ponderosa pine forest and a variety of other, more mesic vegetation. It often forms associations with winter-deciduous shrubs, especially on the eastern slopes of the Sierra Nevada and adjacent ranges.

Patterns of Biodiversity

By Northern Hemisphere standards California is unusually diverse, and much of this diversity is in

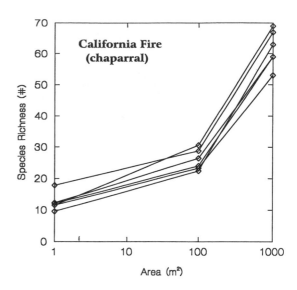

Figure 6.6. Plant species richness at different scales the first year after fire for six chaparral sites in southern California (from Keeley, unpublished data).

postfire chaparral (Jensen, Torn, and Hart 1990; Keeley and Swift 1995). Regionally, shrub diversity is highest in central coastal California, and annual diversity is greatest in the southern coastal ranges (Richerson and Lum 1980). Species richness is, of course, a function of scale. In the first postfire season, 10–20 species m² and 50–70 species per tenth ha are typical for southern California chaparral (Fig. 6.6). At these scales alpha-diversity patterns are comparable with the most diverse mediterranean-climate communities such as South African fynbos or Western Australian heath (Bond 1983). It is curious, however, that species richness in these other shrublands is significantly higher at the scale of 100 m², resulting in straight-line curves on semilog graphs such as in Fig. 6.6 (but cf. Schwilk, Keeler, and Bond 1997). As discussed later, species richness declines markedly after fire, and it appears that the species richness curves likewise change, for richness at 1 m² declines more rapidly than at 100 m² (Keeley, unpublished data).

Thus, at the scale of 100 m², postfire habitat is relatively homogeneous, but with succession this patch size becomes more heterogeneous. For sites <10 yr postfire, F. Davis, Hickson, and Odion (1988) reported an average of 36 species per 100 m² for sites in central California, and the slope of the regression line relating species richness to stand age was −0.93. In mature chaparral, however, site factors play a much greater role than age in determining species diversity; for example, the most diverse site in Table 6.1 was over 100 yrs of age. Topographic and climatic heterogeneity explains a significant portion of diversity patterns in the state (Richerson and Lum 1980), and within chaparral there is considerable evidence for habitat specialization, thus high gamma diversity likely reflects high habitat heterogeneity (Zedler 1995a). Chaparral distributed in small patches (<3 ha) has lower alpha diversity but higher beta and gamma diversity than larger landscape patches, and patchiness enhances the differentiation in species composition among sites (Harrison 1997).

COMMUNITY RESPONSE TO WILDFIRE

The California mediterranean climate is conducive to massive wildfires. Mild, wet winters contribute to a prolonged growing season, which, coupled with moderately fertile soils, result in dense stands of contiguous fuels. Long summer droughts produce highly flammable fuels that are readily ignited by lightning from occasional convection storms. Fire frequency on average is about every two to three decades, but this may be more frequent than in the historical past (Byrne, Michaelson, and Soutar 1977 and unpublished data). Lightning strikes are the natural source of fire ignition, but today humans are responsible for most wildfires (Keeley 1982). Although lightning strikes are common, Minnich, Vizcaino, Sosa-Ramirez, and Chow (1993) reported for northern Baja California that only 2–5% resulted in fires. Lightning-ignited fires increase with elevation, latitude, and distance from the coast, whereas human-ignited fires show the opposite pattern and peak in different months (Parsons 1981; Keeley 1982). Throughout much of its range, chaparral forms a continuous cover over great distances, and as a result, huge wildfires that cover tens of thousands of hectares are common, particularly during Santa Ana wind conditions (Davis and Michaelsen 1995). Minnich (1983) suggested that fires of this size are an artifact of modern-day fire suppression, which results in unnaturally large accumulations of fuel. In support of this theory, he reported that large wildfires are relatively unknown from northern Baja California. One factor responsible for different fire regimes in these two regions is the fact that fires are three times more common in Baja California and are largely anthropogenic in origin. More frequent fires maintain a greater mosaic of fuel loads, and this, coupled with the fact that fires south of the U.S. border are generally not driven by Santa Ana winds (Minnich 1995), results in more localized burns. Although differences in fire regimes seem to have affected landscape patterns, on a smaller scale chaparral communities appear to be little affected (Minnich and Bahre 1995).

It is curious that despite large differences in fire size, Minnich (1995) reports that the average fire rotation period is approximately 70 yr in both Baja California and southern California. A similar estimate of fire recurrence interval was derived for the central Coast Ranges by Greenlee and Langenheim (1980). They did a careful survey of the distribution of lightning-caused fires in conjunction with known patterns of fire behavior and concluded that the natural fire cycle for the inland reaches of Santa Cruz County may have ranged upward to 100 yr and was probably far longer in the coastal and lower-elevation areas. Historical studies by Byrne (1977, unpublished ms.) support the notion that infrequent large catastrophic wildfires were part of the natural landscape prior to modern fire suppression. Simulations based on the Rothermal fire model concluded that for central coastal California, the modern landscape is subjected to much smaller fires and that the prehistoric landscape commonly experienced large-scale catastrophic wildfires (F. Davis and Burrows 1993).

These studies call into question the effectiveness of fire suppression and suggest that, within the chaparral zone, it has had minimal impact on the fire regime. In the largely chaparral-dominated Los Padres National Forest, comparison of early (1911–1950) and later (1951–1991) periods likewise showed fire suppression had no effect on the frequency of very large fires (Moritz 1997). Similar findings were reported for the chaparral belt in the San Bernardino National Forest (Conard and Weise 1998), although in the montane coniferous zone of this forest, fire suppression clearly has reduced fire frequency (Minnich, Barbour, Burk, and Fernau 1995). Undoubtedly, the natural fire regime in chaparral was a stochastic pattern of spatial and temporal heterogeneity in frequency, intensity, and size of fires (Zedler 1995b).

Chaparral fires are generally stand-replacing fires that kill all aboveground vegetation, although fire intensity, described by the total heat release, maximum temperatures, and duration of heating, varies markedly within and between fires. Generalizations about the temperatures that shrub bases or seeds are exposed to during fires are difficult to make, because temperatures vary greatly with depth of burial, stand age and composition, weather conditions, and burning patterns (Sampson 1944; Bentley and Fenner 1958; DeBano, Rice, and Conrad 1979; Anfuso 1982). For example, surface temperatures may remain higher than 500° C for more than 5 min during some fires but not exceed 250° C in others. Temperatures at 2.5 cm depth are more commonly in the range of 50–200° C but often persist for half an hour or longer. Borchert and Odion (1995) illustrate the range of potential variation within a site, where one microsite had a maximum soil surface temperature over 900° C, but within minutes the temperature dropped to below 200° C, whereas another microsite maintained soil surface temperatures above 300° C for over 5 hr. Soil, however, is a good insulator, and at 2 cm depth maximum temperature was approximately 150° C and approximately half that value at 5 cm depth. Variation is due to patterns of fuel accumulation and wind, temperature, and relative humidity during the fire.

Temporary Postfire Flora

In the first year after fire, there is an abundant growth of herbaceous and suffrutescent vegetation (Fig. 6.7), the extent of which varies with site and year (Sampson 1944; Horton and Kraebel 1955; Sweeney 1956; Stocking 1966; Ammirati 1967; S. Keeley, J. E. Keeley, Hutchison, and Johnson 1981; F. Davis, Borchert, and Odion 1989; Rice 1993; Keeley 1997a). This "temporary" vegetation is relatively short-lived, and by the third or fourth year shrubs commonly dominate the site.

The postfire herbaceous flora is often dominated by annuals, and species diversity is typically greatest the first year after fire. Sweeney (1956) studied 10 chaparral burns in northern California and reported 214 herbaceous species, two-thirds of which were annual species. However, Ammirati (1967) reported that herbaceous perennials were more important than annuals on mesic coastal slopes in the north coast ranges, comparable to the pattern discussed below for the southern part of the state. In southern California a study of 90 burned sites recorded 433 species in the first growing season, >60% being annuals. In this study average ground surface cover was 72%, although it ranged from as little as 5% on a few sites to fully one-third of the sites with more than 100% ground surface cover (due to overlapping branches) (Keeley, unpublished data). On average, annuals composed 50% of the cover, although some sites were dominated by herbaceous perennials and others by suffrutescents (Fig. 6.8). Canonical correspondence ordination (CANOCO) analysis of these sites (Fig. 6.9) showed that native annuals increased significantly with increasing distance from the coast and with percentage sand, whereas herbaceous perennials tended to follow an opposite pattern and were more strongly correlated with soil nutrients.

F. Davis et al. (1989) examined microscale vegetation recovery after fire and reported that postburn vegetation was very patchy, with some microsites devoid of seedlings and others with dense patches of seedlings. They also reported that, relative to canopy gaps, prefire soil seed bank density

Figure 6.7. Lush herbaceous growth the first spring after wildfire is in marked contrast to the depauperate herbaceous vegetation under the mature chaparral canopy.

varied with species, some being more concentrated in gaps and others not. However, postfire seedling recruitment was strongly correlated with prefire canopy gaps (Fig. 6.10). Their study also showed that gaps were microsites with reduced soil heating during fire, suggesting that temperatures lethal to seeds played an important role in driving postfire vegetation patterns, and it is noteworthy that these patterns were evident the second year after fire. On a similar scale of 0.1–10 m, Rice (1993) found that fire intensity (measured by the size of the skeletal remains of shrubs) was correlated with vegetation establishment; 38% of her areas of high fire intensity were not vegetated, whereas none of the areas of low fire intensity lacked vegetation. Tyler (1995) also reported an inverse relationship between soil heating during fire and seedling density.

Postfire Shrub Recovery

Rate of shrub recovery varies with elevation, slope aspect, inclination, degree of coastal influence, and patterns of precipitation (Hanes 1971; Keeley and

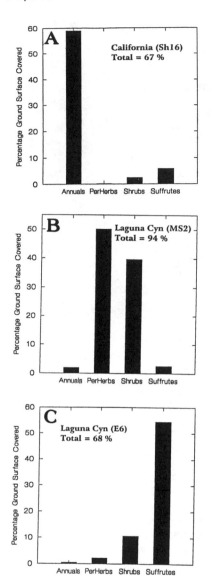

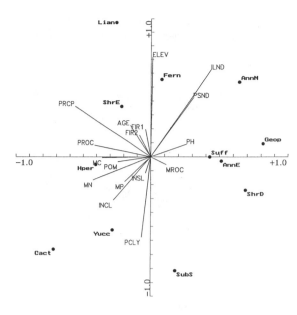

Figure 6.9. CANOCO plot of environmental factors and growth form distribution for the first year postfire flora from 90 sites burned in the same week in autumn 1993 (from Ne'eman and Keeley unpublished data). AnnE = exotic annuals, AnnN = native annuals, Cact = Opuntia spp., Fern = ferns, Geop = geophytes, Hper = herbaceous perennials, ShrE = evergreen shrubs, Lian = lianas; ShrD = deciduous shrubs, SubS = subshrubs, Suff = suffrutescents, and Yucc = Yucca spp.. Environmental parameters are: ELEV = elevation, ILND = distance inland, INCL = slope incline, INSL = calculated annual solar radiation, PRCP = precipitation, FIR1 = fire severity measure 1, FIR2 = fire severity measure 2, AGE = age prior to fire, MROC = rock cover; and soil parameters: MP = phosphorous, MN = nitrogen, MC = carbon, PH = pH, PSND = percentage sand, PCLY = percentage clay, POM = percentage organic matter, PROC = percentage rock.

Figure 6.8. Cover of annuals, herbaceous perennials (per herbs), shrubs, and suffrutescents (suffrutes) in the first postfire year at three chaparral sites burned in the same week (from Keeley, unpublished data).

Keeley 1981; F. Davis et al. 1989). Recovery of shrub biomass is from basal resprouts (Fig. 6.11) and seedling recruitment from a dormant soil-stored seed bank. After a spring or early summer burn, sprouts may arise within a few weeks, whereas after a fall burn, sprout production may be delayed until winter (Biswell 1974). Regardless of the timing of fire, seed germination is delayed until late winter or early spring and is uncommon after the first year.

Resilience of chaparral to fire disturbance is ex-

emplified by the marked tendency for communities to return rapidly to prefire composition (Keeley 1986). Of course, validity of this conclusion is a function of scale and mode of regeneration. Resprouting shrubs may pre-empt the same microsite through repeated fire cycles, whereas microsite variation in seedling recruitment may change fine-scale patterns of community composition. Although not well documented, the potential exists for fire-induced mortality of lignotubers and seeds to alter community composition.

Fire-caused mortality of potentially resprouting shrubs is variable, depending on species. For example, some, such as *Quercus berberidifolia*, *Heteromeles arbutifolia*, and *Malosma laurina*, are seldom killed, whereas others, such as *Adenostoma fasciculatum* and various *Ceanothus* (subgenus *Ceanothus*) species, sometimes suffer extensive mortality. Factors that may be involved include fire intensity, soil moisture, plant size, and physiological condition

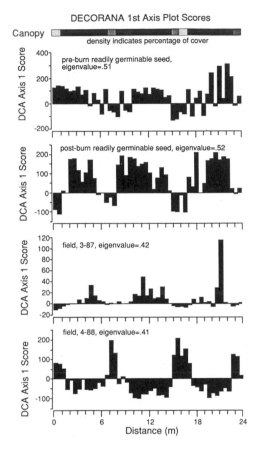

Figure 6.10. Detrended Correspondence Analysis scores for plots based on density of readily germinable seed before and after fire, and postfire seedling recruitment in March 1987 and April 1988, as a function of distance along the transect. Shading on the horizontal bar at the top of the figure indicates canopy coverage and gaps. Fire was in December 1986. (From Davis et al. 1989, Vegetatio 84: 53–67, Fig. 2, 1989 with kind permission of Kluwer Academic Publishers.)

Figure 6.11. Resprout from lignotuber of Adenostoma fasciculatum in the first spring after fire.

(Laude, Jones, and Moon 1961; Plumb 1961; Keeley and Zedler 1978; Tratz 1978; Radosevich and Conard 1980; Stohlgren, Parsons, and Rundel 1984; Rundel, Baker, Parsons, and Stohlgren 1987; Moreno and Oechel 1991b, 1993, 1994; Sparks and Oechel 1993). *Adenostoma faseiculatum* mortality varies from 0–100% but typically is <50% and increases with fire intensity. Accurate estimates of fire intensity are problematical, and thus numerous studies have relied on surrogate measures of fire severity, such as twig diameter on burned skeletons, which is correlated with measures of fire intensity (Moreno and Oechel 1989). Using this surrogate measure of fire intensity, Rundel et al. (1987) found that smaller plants suffered greater mortality and plants burned in late spring suffered

greater mortality than plants burned in late summer. Since they found a similar mortality pattern in unburned but clipped shrubs, they attributed the seasonal effect to a deficit of storage carbohydrates during late spring. Moreno and Oechel (1991b, 1993) likewise found that smaller plants suffered greater mortality; however, other studies have failed to find a relationship between plant size and survivorship (Anfuso 1982; Keeley 1997). Fire intensity not only affects survivorship but resprouting vigor as well (Malanson and O'Leary 1985; Rundel et al. 1987; Moreno and Oechel 1991b, 1993; Borchert and Odion 1995). Mortality of resprouting shrubs is also markedly influenced by fire frequency. For example, fires in adjacent stands burned once, twice, or three times in a 6 yr period generated 1300, 700, and 400 *A. fasciculatum* resprouts per ha, respectively (Haidinger and Keeley 1993), and a similar pattern was reported by Zedler (1995b).

Comparisons of prefire soil seed pools with postfire seedling densities suggest that vast numbers of seeds are killed (Keeley 1977; Bullock 1982; Davey 1982; F. Davis et al. 1989). Seedling density for prolific seeders such as *Adenostoma fasciculatum* or *Ceanothus* spp. may exceed 100 m^{-2} (Sampson 1944; Sweeney 1956; F. Davis et al. 1989; Moreno and Oechel 1991a), although as discussed in the following section there is tremendous inter- and intraspecific variation. Fire intensity affects seed survivorship and consequently seedling recruitment. There is interspecific variation in seed survivorship, thus the potential exists for characteristics of a particular fire having long-term impacts on community composition. For example, by artificially enhancing fire intensity, Moreno and Oechel (1991a) demonstrated that seedling recruitment decreased for *A. fasciculatum* but increased for *C. greggii*. F. Davis and Odion (reported in Borchert and Odion 1995) found that microsites with higher fire intensities diminished *A. fasciculatum* seedling recruitment to a far greater extent than *Arctostaphylos purissima* seedling recruitment, and they attributed this to differences in depth of seed burial. Moreno and Oechel (1991a) suggest that higher temperatures generated by more intense fires play an important role in determining seed mortality during fires, and there is empirical evidence for this effect (Zammit and Zedler 1988). An additional factor is smoke, which has been found not only to induce germination in many chaparral species but also to be lethal at durations of 5–15 min exposure, and there is marked interspecific variation in tolerance of seeds to smoke (Keeley and Fotheringham 1998b).

Seedling mortality is generally high during the first year and is concentrated in the spring (Bullock 1982; Mills 1983; Rundel et al. 1987; Moreno and Oechel 1992). Seedlings are strikingly smaller than resprouts (Sampson 1944; Horton and Kraebel 1955; Keeley and Keeley 1981), however, Bond (1987) and Tyler and D'Antonio (1995) reported that proximity to resprouting shrubs had no negative effect on seedling growth or survivorship; indeed, the latter authors noted that proximity of seedlings to resprouts significantly reduced the incidence of herbivory. However, Tyler and D'Antonio (1995) did find that both postfire survivorship and growth of *Ceanothus impressus* seedlings were positively correlated with increasing distance to near neighbors of all species, and this appeared to be due to reduced competition for water. Moreno and Oechel (1988) likewise reported that high herb density significantly reduced the survivorship of both *Adenostoma fasciculatum* and *C. greggii* seedlings and that survivorship was strongly correlated with soil moisture (Moreno and Oechel 1992).

Herbivory of seedlings and resprouts is important after fire (J. Davis 1967; Howe 1982; Mills 1983, 1986; Mills and Kummerow 1989; Moreno and

Oechel 1991b, 1992) and potentially alters long-term demographic patterns. For example, Bullock (1991) showed that 6 yr after fire, *C. greggii* survivorship was an order of magnitude greater in plots fenced from small mammals. There is also evidence that predator preferences may alter the balance in shrub composition. Mills (1986) reported that the major mammalian herbivores (brush rabbits, *Sylvilagus bachmani*) preferred *C. greggii* over *A. fasciculatum*. This effect, however, was short-lived and 5 yr after fire, with or without predation, there was no significant difference in survivorship between these two species. Quinn (1994), reported an unusual and dramatic postfire shift in species composition from *Ceanothus crassifolius* domination before fire to *Salvia mellifera* domination 10 yr after fire, whereas a plot fenced from herbivores returned to *C. crassifolius*. For the most part, these studies have been done on small controlled burns surrounded by mature chaparral. Wildfires that are orders of magnitude larger may eliminate many potential herbivores, although Tyler (1996) found no difference between herbivory effects at the periphery versus middle of a 40 ha burn.

It has been hypothesized that in the absence of fire, chaparral would be succeeded by other vegetation types. Sampson (1944) suggested that in northern California, grassland would eventually replace chaparral; however, Hedrick (1951) found no evidence of this in a 90 yr stand. He commented that "the most striking feature of this old chamise stand is the lack of evidence that it is dying out or being replaced by herbaceous vegetation." It is true that old plants accumulate deadwood, and many stems die, but these are replaced by additional crown sprouts (Keeley 1992c), indicating that resprouting shrubs rejuvenate themselves in the absence of fire. These observations are at odds with much published dogma on chaparral "senescence," a concept that has been widely criticized (Zedler and Zammit 1989; Keeley 1992b; Zedler 1995b). In addition to Hedrick's study, other investigations have shown no evidence of old chaparral being replaced (Keeley and Zedler 1978; Lloret and Zedler 1991; Keeley 1992b; Sites C & F, Table 6.1; Sites 2 & 3, Table 6.2), and the community appears to be resilient to fire-recurrence intervals of 100+ yr. For most but not all shrub species, mortality in old stands of chaparral is not accompanied by seedling establishment (Table 6.2). When there are unusually long intervals between fires, chaparral may be replaced by sclerophyllous woodland if sufficient seed sources are available (Cooper 1922; Wells 1962). For example, the woodland tree *Quercus agrifolia* readily recruits seedlings under adjacent chaparral shrubs (Calloway and D'Antonio 1991). However, in a study of 12 sites unburned for 55–118 yr, seedlings of several woodland species were reported but saplings were not found, indicating that recruitment was either recent or not successful and that successional replacement of chaparral was not imminent (Keeley 1992c). This conclusion is also supported by landscape level studies that compared vegetation change over 40 yr from aerial photographs (Scheidlinger and Zedler 1980; Callaway and Davis 1993). Although there are significant landscape changes between coastal sage scrub and grassland, as documented in other studies (Keeley 1990b), chaparral transition to other vegetation types is rare (Fig. 6.12).

Postfire Resource Management

Resource agencies often respond to wildfires with emergency revegetation programs designed to reestablish quickly a herbaceous cover sufficient to reduce soil erosion and eliminate the threat of mudslides and flooding. The rationale for this management is that burned soils are more hydrophobic than unburned soils, resulting in increased surface flow of rainwater and increasing soil erosion (DeBano et al. 1979). Emergency seeding is considered essential on sites following exceptionally intense fires because of anticipated negative effects of intensity on natural regeneration. Throughout the state the seed of choice has been the non-native annual ryegrass (*Lolium multiflorum*). There is abundant evidence that this practice fails to substantially reduce threats of mudslides and flooding (Gautier 1983; Conard, Beyers, and Wohlgemuth 1995; Spittler 1995; Booker, Dietrich, and Collins 1995; Keeley 1996) and competitively displaces the native flora (Schultz, Baunchbauch, and Biswell 1955; Corbett and Green 1965; S. Keeley et al. 1981; Griffin 1982; Barro and Conard 1987; Beyers, Conard, and Wakeman 1994; Keeley, Morton, Pedrosa, and Trotter 1995). In response to concern over seeding nonnative species in natural ecosystems, some agencies have discontinued the practice. Other agencies have continued the practice using native species; however, this approach is also fraught with problems (Keeley 1995b).

LIFE HISTORIES OF PLANTS

Although shrubs dominate chaparral, the community comprises a rich diversity of growth forms, many of which are conspicuous only after fire. In addition to evergreen shrubs and trees, there are semi-deciduous subshrubs, slightly ligneous suffrutescents, woody and herbaceous vines, and a rich variety of herbaceous perennials and annuals.

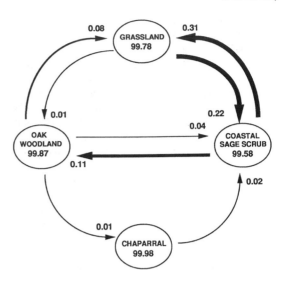

Figure 6.12. Annual transition rates among plant communities in Gaviota State Park, Santa Barbara County, California as determined from vegetation change observed on aerial photographs between 1947 and 1989 (reprinted from Callaway and Davis 1993).

Shrubs

Two modes of seedling recruitment are disturbance-dependent species that restrict recruitment to postfire conditions and disturbance-free species that successfully recruit only in the long-term absence of fire. The former exploit postfire environments for population expansion, and the latter require unusually long fire-free conditions for population expansion. Since fire is a recurrent catastrophic disturbance, species have been classified by their mode of postfire regeneration – that is, obligate seeders, facultative seeders, or obligate resprouters (see Table 6.2). Although these terms are useful, they describe life history response to only one facet of the environment and do not adequately describe reproductive modes for the entire shrub flora.

Disturbance-dependent recruitment. Disturbance-dependent species establish seedlings in the first year after fire, but seedling recruitment is almost nonexistent in subsequent years. Seedlings arise from a long-lived soil seed bank (Keeley 1977; Davey 1982; Parker and Kelly 1989) and germination is cued either by intense heat (Quick 1935; Hadley 1961; Quick and Quick 1961) or by chemical stimulus from smoke or charred wood (Fig. 6.13). The majority of species in the largest woody genera, *Arctostaphylos* and *Ceanothus*, lack the ability to regenerate vegetatively from basal burls or root crowns (Wells 1969).

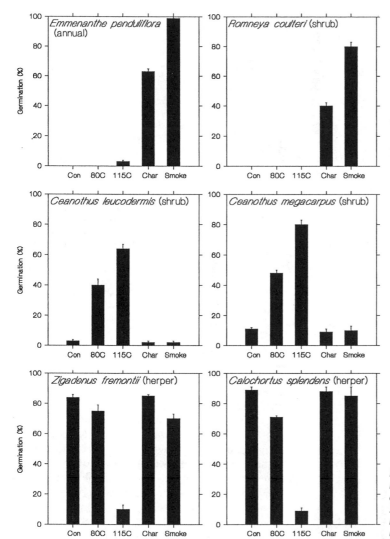

Figure 6.13. Germination of chaparral herbs and shrubs in response to control, 80°C heat for 1 hr, 115°C for 5 min, charred wood, or smoke for 5 min (from Keeley and Fotheringham, unpublished data.

These are often termed obligate-seeding shrubs, and their stems are even-aged, dating back to the last fire (Schlesinger and Gill 1978).

On recently burned sites, obligate-seeding species typically have very high seedling densities but low coverage relative to both facultative-seeding and obligate-resprouting species (Sampson 1944; Horton and Kraebel 1955; Vogl and Schorr 1972; Keeley and Zedler 1978; Keeley and Keeley 1981). Seedling density is typically two orders of magnitude greater than pre-fire shrub density (Fig. 6.14B; Moreno and Oechel 1992).

Reproductive maturity requires 5–15 yr before significant seed crops are produced, and thus fires at intervals more frequent than this can produce localized extinctions (Zedler, Gautier, and McMaster 1983; Zedler 1995a; Fig. 6.15). *Ceanothus* and

Arctostaphylos dispersal in space is very limited (Davey 1982; Evans, Biswell, and Palmquist 1987), but since seeds are dormant, they accumulate in a soil seed bank and thus are temporally dispersed. Despite relatively large seed crops, accumulation in the soil is slow because a large portion of the seed crop is rapidly removed from the soil by predators (Fig. 6.16), which often selectively take the largest viable seeds (Keeley and Hays 1976; Zammit and Zedler 1988; Mills and Kummerow 1989; Kelly and Parker 1990; Quinn 1994). In one study, two *Arctostaphylos* species were found to produce more seeds in a single year than were present in the soil seed bank (Keeley 1977), and 10 yr later there was no statistically significant change in the size of the soil seed banks (Keeley 1987b).

Neither *Ceanothus* nor *Arctostaphylos* produces

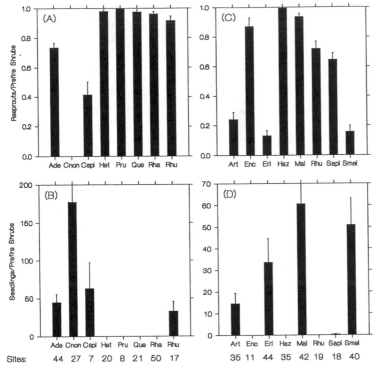

Figure 6.14. Postfire resprouting success (A) and seedling recruitment (B) of chaparral and coastal sage scrub (C and D, resprouts and seedlings, respectively) expressed relative to prefire population size (from Keeley unpublished data). Data from 90 sites burned in southern California, autumn 1993, and number of sites each species was present at is given below; Chaparral shrubs (A&B): Ade = Adenostoma fasciculatum, Cnon = nonsprouting species of Ceanothus, Cspi = Ceanothus spinosus, Het = Heteromeles arbutifolia, Pru = Prunus ilicifolia, Que = Quercus berberidifolia, Rha = Rhamnus crocea, Rhu = Rhus ovata, and coastal sage subshrubs and shrubs (C&D): Art = Artemisia californica, Enc = Encelia californica, Eri = Eriogonum fasciculatum, Haz = Hazardia squarrosa, Mal = Malosma laurina, Rhu = Rhus integrifolia, Sapi = Salvia apiana, Smel = Salvia mellifera.

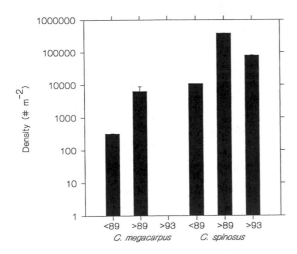

Figure 6.15. Impact of repeat fires (1989 and 1993) on density, prior to the 1989 fire, first year after the 1989 fire, and first year after the 1993 fire for non-sprouting Ceanothus megacarpus and resprouting C. spinosus, at the same site in the Santa Monica Mountains (from Keeley, unpublished data).

significant seed crops annually; rather, crops tend to be biennial (Keeley 1977, 1987c; Keeley and Keeley 1988b). Also, these studies and others (Keeley and Keeley 1977; Kelly and Parker 1990) have shown that congeneric pairs of obligate seeders

and resprouting facultative seeders do not differ significantly in seed output. In addition, Kelly and Parker (1991) reported that species of Arctostaphylos representing these two modes were not significantly different in their seed-to-ovule ratio; however, polyploid species were significantly lower than diploids. Zammit and Zedler (1992) did a detailed 5 yr study of seed production in C. greggii, comparing stands 6, 13, 32, 57, and 82 yr of age on north- and south-facing exposures. Seed production was maximized within two decades after fire and did not decline with shrub age, even on the oldest sites, and they concluded that there was no evidence of senescence in these very ancient shrubs. This pattern also holds for species of Arctostaphylos; in fact shrubs over 90 yr of age may greatly exceed seed production of much younger shrubs (Keeley and Keeley 1977).

As stands age and the canopy closes, there is intense competition and density-dependent thinning, resulting in high mortality (Schlesinger et al. 1982). In stands older than 50 yr, the fate of obligate-seeding species varies with the species and site. On mesic slopes, these shrubs are apparently out-competed by sprouting species (e.g., Site 3, Table 6.2; Fig. 6.17), but they can dominate arid sites 80 to 100 yr after fire (Keeley and Zedler 1978; Keeley 1992b; e.g., Site 2, Table 6.2). Obligate-seeding Ceanothus species may persist on a site for

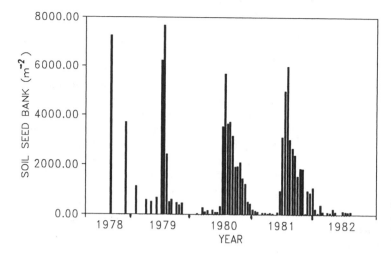

Figure 6.16. Annual fluctuations in the soil seed bank of Ceanothus crassifolius over a 4.5 yr period, n = 10 samples from beneath shrubs. Peak for each year is July (reprinted from Quinn 1994).

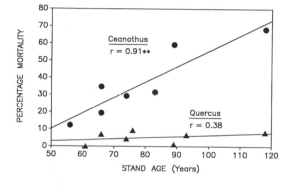

Figure 6.17. Mortality of non-sprouting Ceanothus species and sprouting Quercus species in stands of different ages (** = P <0.01) (from Keeley 1992b).

well over a century without fire, and a viable seed bank for much longer, suggesting selection by a fire regime with unpredictable frequency (Montygierd-Loyba and Keeley 1987; Zammit and Zedler 1992; Keeley and Swift 1995; Zedler 1995b).

Adenostoma fasciculatum and some *Ceanothus* and *Arctostaphylos* species establish seedlings after fire and resprout from a basal lignotuber (Fig. 6.14A, B), although the proportion of resprouting shrubs to seedlings is variable with the species, site, and fire. As with obligate-seeding species, seedling establishment by these facultative seeders is confined to the first postfire year, cued by heat or chemicals. *Adenostoma fasciculatum* is reported to produce two types of seeds: those that germinate readily at maturity and those that require intense heat shock (Stone and Juhren 1951, 1953; Zammit and Zedler 1988). Thus, the former seed type could germinate in the absence of fire. Although successful seedling establishment under the shrub canopy is in fact nonexistent, such seeds do contribute to coloniza-

tion of other types of disturbance. Seedling recruitment by facultative seeders is often not as great as for obligate-seeding shrubs (Fig. 6.14B). These species, however, are far more resilient to recurrent fires; for example, two fires 4 yr apart are sufficient to cause the local extinction of an obligate seeder, whereas a facultative seeder persists (Fig. 6.15).

Disturbance-free recruitment. Heteromeles arbutifolia, Quercus berberidifolia, Prunus ilicifolia, Cercocarpus betuloides, Rhamnus species, and *Rhus integrifolia* seldom establish seedlings after fire (Fig. 6.14B). They persist on burned sites, however, because they are vigorous resprouters (Fig. 6.14A); thus the term "obligate resprouters" applies to these shrubs. In mature chaparral, they produce substantial seed crops that are widely dispersed (Bullock 1978; Horn 1984; Lloret and Zedler 1991; Keeley 1992d). Their seeds are short-lived (<1 yr) and germinate readily with adequate moisture (Keeley 1997b) and thus a dormant seed bank does not accumulate in the soil (Parker and Kelly 1989; Keeley 1991). This, coupled with the observation that these seeds are easily killed by intense heat, accounts for their failure to establish seedlings after fire.

Seedling recruitment is largely restricted to older, more mesic stands of chaparral, although seldom are seedlings of these species very abundant (Keeley 1992b, 1992c). Successful reproduction does occur under some conditions, as illustrated by the age distributions of *Rhamnus crocea* and *Prunus ilicifolia* seedlings and saplings in a 75-yr-old chaparral stand (Fig. 6.18). At this site, seedling recruitment was restricted to beneath the shrub canopy and was absent from gaps. Seedling recruitment is correlated with low light and high litter depth for these and other species (Williams

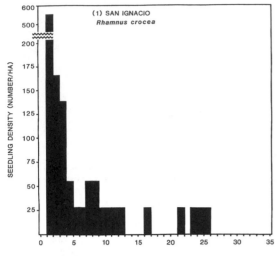

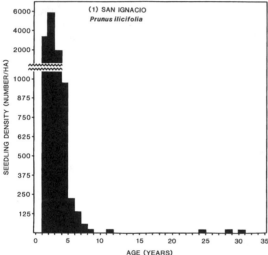

Figure 6.18. Age distribution of seedlings and saplings for two sprouting shrub species in a 75-yr-old stand of southern California chaparral (redrawn from Keeley 1992c).

1991; Keeley 1992c). Significant seedling recruitment in mature chaparral has also been noted for *Cercocarpus betuloides, Heteromeles arbutifolia, Prunus ilicifolia, Quercus wislizenii, Q. durata,* and *Rhus integrifolia* (Gibbens and Schultz 1963; Patric and Hanes 1964; Zedler 1982; Lloret and Zedler 1991; Williams et al. 1991; Keeley 1992c). Despite evidence of successful seedling and sapling recruitment in older stands, it appears that recruitment into the shrub canopy is rare. Most saplings remain stunted beneath the canopy; however, they are capable of resprouting after fire, and therefore fire may ultimately be required in order to make the transition from sapling to adult.

Sprouting shrubs are distinguished from obli-

gate-seeding species by the uneven-aged structure of stems (Keeley 1992c). All sprouting species continually produce new shoots from the root crown throughout their life spans (Fig. 6.19), so that most have one or more stems dating back to resprouts after the last fire, but continue to recruit new stems long after fire (Fig. 6.20). The pattern of stem recruitment varies with species and site; for example in stands >50 yr, the proportion of stems recruited in the previous decade ranged from 20–30% in *Adenostoma fasciculatum* to over 80% in *Cercocarpus betuloides* (Keeley 1992c). Some sprouting species are capable of vegetative spread by roots or rhizomes (e.g., *Fremontodendron decumbens,* Boyd and Serafini 1992; *Cercocarpus betuloides,* Keeley 1992c).

Trees

Evergreen coniferous trees, such as species of *Cupressus* and *Pinus,* often form dense even-aged stands within a matrix of chaparral (Vogl, Armstrong, White, and Cote 1977; Zedler 1977, 1981, 1995b), commonly on serpentine, gabbro, or other unusual substrates (McMillan 1956; Koenigs et al. 1982; Zedler 1995a). These species do not resprout after fire, and seedling establishment is in the first postfire year from a dormant seed bank held in serotinous cones. The serotinous pines produce cones sealed by resins that require high temperatures for opening; they may be sealed for decades and still retain viable seed. Serotinous cypress produce much greater quantities of smaller cones that open upon drying; as branches die or vascular connections are severed. Thus, cypress are more likely to disperse seeds in the absence of fire; however, recruitment within stands is apparently infrequent.

Pines exhibit remarkable variation in cone serotiny. *Pinus attenuata* is strongly serotinous and forms dense even-aged stands dating back to the last fire (Vogl et al. 1977). Flammability is high, and temperatures essentially sterilize the soil beneath the parent plant. Thus, when they are juxtaposed with chaparral, there is often a sharp contrast between the bare ash beds beneath the pines and the dense cover of postfire chaparral stands (Keeley and Zedler 1998). *Pinus coulteri* is strongly serotinous when associated with chaparral but not so in woodland habitats (Borchert 1985). In chaparral, this pine tends to form even-aged stands and exhibits characteristics such as lack of self-pruning that increase chances of stand-replacing fires. However, in forests and woodlands it has quite a different growth form and potentially continuous seedling recruitment. *Pinus torreyana* is not strongly serotinous, but some cones open gradually over many years (McMaster and Zedler 1981).

Figure 6.19. Multistemmed Adenostoma fasciculatum *shrub with shoots of various ages arising from a common root crown.*

Pinus radiata and *P. muricata* are coastal pines that often are juxtaposed with chaparral. They exhibit variable levels of cone serotiny, but because of the unusual delay in dispersal of current cone crops (late winter), some cones are almost certainly closed at the time of most wildfires (Keeley and Zedler 1998).

Pseudotsuga macrocarpa is a nonserotinous conifer but one of the few capable of resprouting after fire. Sprouts arise from epicormic buds along the trunk and main branches of mature trees, but saplings succumb to most fires. Seedling recruitment occurs under the chaparral or oak woodland canopy during fire-free periods (Bolton and Vogl 1969; McDonald and Litterell 1976), and thus populations are uneven-aged. Long fire-free periods are also apparently necessary for successful seedling establishment by *Pseudotsuga macrocarpa* (Minnich 1980b).

There is evidence that for most of these species ranges have become more restricted in modern times because of increased fire frequency (Shantz 1947; Horton 1960; Gause 1966; Zedler 1977). *Cupressus* species, for example, require 40 yr or more to accumulate sufficient aerial seed banks to ensure successful postfire seedling establishment, and

Zedler (1981, 1995b) has shown that postfire seedling establishment of *C. forbesii* is an order of magnitude greater in stands over 50 yr of age at the time of burning than in stands 30 yr of age.

Sclerophyllous hardwood trees such as *Quercus agrifolia, Q wizlizenii, Q. chrysolepis, Arbutus menziesii,* and *Umbellularia californica* form woodlands juxtaposed with chaparral, particularly in ravines and on mesic north-facing slopes. Higher fuel moisture makes such sites less susceptible to complete destruction by wildfires, and after low-intensity fires, all are able to resprout from epicormic buds beneath the bark of stems or from the root crown. Seedling establishment after fire is rare because of the lack of a dormant seed bank, due to production of short-lived, nondormant, heat-sensitive seeds (Keeley 1991, 1997b).

Subshrubs

Subshrubs, which are weakly ligneous small shrubs, are summer-deciduous (e.g., *Salvia mellifera*) or evergreen (e.g., *Eriogonum fasciculatum*) and are most important along the lower-elevation xeric borders (Hanes 1971). They are readily shaded out by the ev-

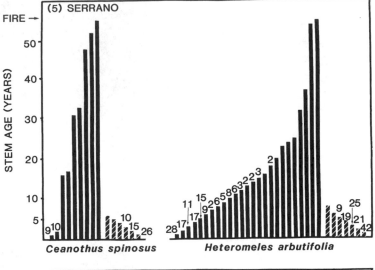

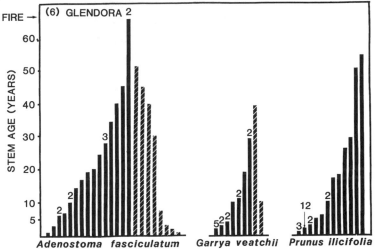

Figure 6.20. Age distribution for all stems on one individual each of five resprouting species in a 55-yr-old (Serrano) or a 66-yr-old (Glendora) stand of chaparral in the Transverse Range of southern California (from Keeley 1992c). Each filled bar represents one live stem, unless topped with a number indicating more than one stem at that age, and each hatched bar represents a dead stem.

ergreen chaparral species (McPherson and Muller 1967; Gray 1983) and thus occupy gaps in the chaparral canopy. Most have light, readily dispersed seeds (Wells 1962) and are capable of recruiting new individuals into gaps, as well as establishing after fire from a dormant seed bank. *Dendromecon rigida* is a relatively short-lived subshrub (<5% survive 10 yr) (Bullock 1989) that is one of only a few myrmecochorous species in chaparral, with a prominent ant-attracting elaisome on each seed (other myrmechochores include the annual *Claytonia perfoliata* and Woody *Fremontodendron* spp. (Boyd 1996; Keeley, personal observations).

Suffrutescents

These dwarf, subligneous species (roughly equivalent to Raunkiaer's chamaephytes) are abundant

in the first year after fire from a soil seed bank. The seeds are dormant until heat shock or smoke or charred wood induces germination. One of these species, *Lotus scoparius, Eriophyllum confertiflorum,* or *Helianthemum scoparium,* typically, dominates a burn in the second or third year after a fire, but as the shrub cover increases, suffrutescent species are eliminated or restricted to gaps (Horton and Kraebel 1955; S. Keeley et al. 1981).

On open rocky sites from central California southward, the rosette-forming *Yucca whipplei* is a conspicuous element. Seeds germinate readily with adequate moisture, and populations are uneven aged. In southern California chaparral, this species is monocarpic; the rosette grows for a decade or more before flowering and dying. On desert slopes of the Transverse Ranges, however, plants produce up to 100 rosettes during the vegetative phase, and flowering may be spread over many decades. In the central Coast Ranges, populations spread by rhizomes (Haines 1941; Keeley, Keeley, and Ikeda 1986). This species neither seeds nor resprouts after fire; however, the lower fuel volume on xeric sites, coupled with the tough, fibrous leaves that protect the apical meristem, result in significant survival during fire. Although many plants initiate flower stalks following fire, flowering is not dependent on such conditions.

Lianas and Vines

Both woody lianas and herbaceous perennial vines grow into the shrub canopy and are particularly abundant in older stands of broad-sclerophyll chaparral. Postfire regeneration modes include facultative seeders and obligate resprouters. *Calystegia macrostegia* germinates in profusion after fire from a soil seed bank. Germination is apparently induced by heat shock. It is noteworthy that this species typically germinates in profusion in early winter, months before most other postfire chaparral species. This liana often dominates sites in the second year, sprawling over regenerating shrub seedlings and resprouts, although it is unclear whether it competitively inhibits regeneration or plays a positive role as a nurse plant. *Marah macrocarpus* is an obligate resprouter after fire, with massive foliage cover generated in the first postfire season, although foliage of this herbaceous perennial dies back rapidly with the onset of summer drought. Seeds are very large and short-lived, and they germinate readily under the canopy, although recruitment is likely limited to years of high rainfall (Schlising 1969).

Perennial Herbs

The herbaceous perennial lifecycle of annual dieback preadapts this growthform to fire because they resprout vigorously from bulbs, corms, and rhizomes in the first postfire year. Seedling recruitment is rare immediately after fire due to the lack of a dormant seed bank (but c.f. Borchert 1989). Seeds are nonrefractory, that is, they do not require heat shock or smoke to induce germination (Fig. 6.13), although many require cold (Sweeney 1956; Keeley 1991). Seedling recruitment is possible anytime between fires but is apparently largely concentrated in the years subsequent to the first postfire season when resources are more generally available and flowering resprouts provide a rain of seeds (e.g., Martin 1995). As the canopy closes in, these herbaceous perennials are restricted to gaps in the canopy, and due to lower light levels, they

Table 6.3. *Growth form distribution of seed germination mode in California chaparral and coastal sage scrub ($\chi^2 = 71.2$, $P < 0.001$). Nonrefractory seeds germinate readily without fire-related cues, although some may be stimulated by brief cold stratification treatment.*

	Annuals	Herbaceous perennials	Woody plants
Nonrefractory	17	28	25
Heatshock	6	1	34
Smoke or charred wood	26	5	14

Source: Keeley and Bond (1997).

flower less frequently (Stone 1951; Stocking 1966; Christensen and Muller 1975a).

Annuals

Annuals make up the most diverse component of the chaparral flora, being abundant in disturbed areas and producing spectacular floral displays in the first spring after fire. Some, such as *Emmenanthe penduliflora* and species of *Phacelia*, *Lupinus*, *Lotus*, *Antirrhinum*, and *Allophyllum*, are "fire annuals" or "pyrophyte endemics," rarely found except on burned sites. Typically, these species have relatively specific site preferences (e.g., slope aspect, soil type, elevation, distance from coast, etc.) (Sweeney 1956; S. Keeley et al. 1981; O'Leary 1988). They have weakly developed dispersal ability (e.g., most Asteraceae) either lack pappus or have deciduous pappus); thus, the long-lived seeds are dispersed more in time (until fire) than in space.

Some native annuals such as *Camissonia micrantha*, *Crassula connata* (formerly *C. erecta*), *Cryptantha* spp., *Filago californica*, *Pterostegia drymarioides*, and others are opportunistic and most abundant on burned sites, but they persist in chaparral canopy gaps (Zammit and Zedler 1994) or in more open communities, arising from repeated disturbances or along xeric margins. In mature chaparral, closely related species may have quite different microsite requirements. For example, Shmida and Whittaker (1981) reported that *Cryptantha muricata* and *Lotus strigosus* were restricted to the open, whereas the congeners *C. intermedia* and *L. salsuginosus* increased nearer to clumps of *Adenostoma fasciculatum*.

Seed Germination and Allelopathy

Patterns of germination in chaparral species are summarized in Table 6.3 and Figure 6.13. Herbaceous perennials (e.g., *Zigadenus* and *Calochortus*) are capable of recruiting in the absence of fire, and their presence on burned sites is due to resprouts. Seeds germinate readily following cold stratification. Postfire recruiters use either heat shock or chemicals to cue germination. Germination induced by heat shock is the mode in shrubs such as *Ceanothus* (Rhamnaceae), suffrutescents such as *Helianthemum scoparium* (Cistaceae), and *Lotus scoparius* (Fabaceae), and in annual *Lupinus* and *Lotus* (Fabaceae). These species have water-impermeable cuticles, and dormant seeds fail to imbibe water. Heat disrupts this barrier, commonly by loosening cells around the strophiolar plug.

Induction of germination in many species has been shown to be due to chemicals leached from charred wood or from smoke (Fig. 6.13). Smoke-induced germination is known from annual and suffrutescent species in the Asteraceae, Boraginaceae, Brassicaceae, Caryophyllaceae, Hydrophyllaceae, Lamiaceae, Loasaceae, Onagraceae, Papaveraceae, Polemoniaceae, and Scrophulariaceae (Keeley and Fotheringham 1998b). Seeds of these plants differ from species that require heat-shock stimulation in that they lack a dense outer palisade cell layer plus exterior cuticle that block imbibition. Rather, smoke-stimulated seeds freely imbibe water in the dormant stage, and thus germination is not triggered by overcoming a barrier to imbibition, as is the case with heat-shock stimulated species. Nitrogen oxides in smoke have been shown to induce germination in some of these species, and potentially nitrogen oxides produced by postfire nitrification may also trigger germination (Keeley and Fotheringham 1997, 1998b). Nitrate was suggested as a potential germination stimulant in these postfire species (Thanos and Rundel 1995), but this ion alone is ineffective (Keeley and Fotheringham 1998a). Germination induced by charred wood (Wicklow 1977; Jones and Schlesinger 1980; Keeley et al. 1985) may represent the same mechanism as smoke-induced germination since the same species respond to both cues (Fig. 6.13). One mechanism tying these phenomena together is suggested by the fact that smoke will induce germination either directly or indirectly by first binding to soil particles and later evolving nitrogen oxides (Keeley and Fotheringham 1997). Some species, e.g. *Dicentra chrysantha*, *Dendromecon rigida*, and *Trichostemma lanatum*, have multiple barriers to germination and require long-term soil storage before they will germinate in response to smoke (Keeley and Fotheringham 1998b).

It has long been known that germination of some chaparral species requires a cold treatment ($<5°$ C) (Keeley 1991). In a great many cases duration of cold seems to be unimportant, and often one day of cold is sufficient to trigger germination (Santillenes and Keeley, unpublished data). Thus, unlike cold stratification, which measures duration of cold and signals the approaching end of winter, this "cold triggering" mechanism signals the beginning of winter. Under field conditions, germination of most native species is normally delayed until midwinter, even in years with substantial autumn rainfall.

The striking contrast between the depauperate herb growth under mature chaparral and the flush of herbs after fire has been hypothesized to be due to allelopathic suppression of germination by the overstory shrubs. After field and laboratory studies, McPherson and Muller (1969) concluded that "nearly all seeds in the soil of mature *A. fasciculatum* stands are prevented from germinating by the toxin (leached from the shrub overstory) which is most abundantly present during the normal germination period." Fire consumes the shrubs and destroys the toxin, thus releasing the herb seeds from inhibition. This conclusion has been criticized by Keeley and Keeley (1989) because of the following:

1. The majority of McPherson and Muller's work focused on the effects of leached inhibitors on growth of non-native seedlings, not on germination of native species.
2. Leachate from *Adenostoma* foliage may inhibit the germination of some species, but apparently has no effect on many others (McPherson and Muller 1969; Christensen and Muller 1975a; Keeley et al. 1985).
3. Temperatures applied to the soils that resulted in enhanced germination were far lower than the temperatures needed to degrade the suspected toxins (Chou and Muller 1972).
4. The concentration of toxins McPherson and associates (1971) found necessary for inhibition were much higher than those found in soil (Kaminsky 1981).
5. Christensen and Muller (1975a) found the concentrations of suspected allelopathic toxins were greatest in soils from recently burned sites.
6. Seeds in soil that had been heat-treated, and then returned to beneath the shrub canopy (and exposure to the putatively allelopathic leachate), germinated readily (Christensen and Muller 1975b).
7. Seeds of many chaparral herbs fail to germinate even if they are never exposed to so-called allelopathic toxins (Keeley 1991).

Kaminsky (1981) hypothesized that toxins produced by soil microbes are responsible for inhibiting herb germination under chaparral, and he demonstrated such an inhibitory effect with lettuce seeds. Pack and Keeley (unpublished data), however, could not duplicate this effect with native seeds, and Christensen and Muller (1975a) reported that fungal and bacterial populations increased after fire.

Field studies show that most seedlings establishing under mature chaparral succumb to either small mammals or competition with the canopy shrubs (McPherson and Muller 1969; Christensen and Muller 1975a, 1975b; Swank and Oechel 1991; Quinn 1994). Studies in which animals have been excluded from plots beneath chaparral show highly significant increases in herb densities, although not comparable in species composition to the postfire flora.

To summarize, in response to the poor conditions of low light, limited water, insufficient nutrients, high predation, and possibly allelopathic toxins under the shrub canopy, many species have evolved mechanisms that ensure seed dormancy until the canopy is removed. Germination cued by intense heat shock or chemicals from smoke or charred wood ensures establishment immediately following fire. Some species are strictly tied to fire, whereas other opportunistic species colonize gaps in the chaparral canopy by a polymorphic seed pool that comprises fire-dependent refractory seeds that remain dormant until fire and nonrefractory seeds capable of germinating in the absence of fire. For most opportunistic species, successful recruitment is restricted to canopy gaps. A hypothesis worth testing is whether or not nonrefractory seeds have evolved a sensitivity to allelopathic compounds leached from the overstory canopy as a means of inducing secondary dormancy until conditions become more suitable.

Nonnative annual grasses and forbs are found throughout chaparral regions. Under a regime of frequent fires, they readily displace the native herb flora if fires are frequent enough, converting chaparral to annual grassland (Cooper 1922; Sampson 1944; Arnold, Burcham, Fenner, and Grab 1951; Hedrick 1951; Wells 1962; Keeley 1990b; Haidinger and Keeley 1993). In the absence of fire, seeds of non-natives have a low residence time in the soil, and thus the presence of these species on burned sites is more often due to colonization after fire. Most, such as species of *Bromus*, *Erodium*, and *Centaurea*, disperse prior to the summer fire season and consequently are less common in first-year burns but are present in subsequent years (Sampson 1944; Horton and Kraebel 1955; S. Keeley et al. 1981). Fall-fruiting species such as *Lactuca serriola* and *Conyza canadensis* are likely to be more common on first-year burns.

SHRUB MORPHOLOGY AND PHENOLOGY

Leaves

The dominant shrubs are evergreen, with small, sclerified, heavily cutinized leaves (Cooper 1922). Leaf longevity is typically 2 yr (Kummerow and Ellis 1989), although shade leaves are often much longer-lived (Mooney, unpublished data). The widespread *Adenostoma fasciculatum* has a linear-terete leaf less than 1 cm in length (0.06 cm²) that is markedly smaller than leaves of other chaparral shrubs (Fishbeck and Kummerow 1977). The isofacial nature of these leaves, with stomata evenly distributed, has led to some confusion in calculations of the leaf area index (Kummerow and Ellis 1989). These leaves are produced individually on new growth and in short-shoot fascicles on old growth (Jow, Bullock, and Kummerow 1980). Juvenile leaves on seedlings and fire-induced basal resprouts (but not on basal sprouts from mature shrubs) are bifacial and deeply lobed (compare Figs. 6.3 and 6.11); occasionally these occur on mature plants under abnormally mesic conditions. Similar leaf heterophylly is known from *Dendromecon rigida* (Stebbins 1959). Other shrub species have broad sclerophyll leaves remarkably convergent in size, shape, and anatomy. Most are simple (<5 cm length and 1–5 cm²), with average thickness ~ 300 μm, plus 5–10 μm cuticle on upper and lower surfaces (Cooper 1922; Fishbeck and Kummerow 1977), and many are sharply serrated (Fig. 6.4). Most *Arctostaphylos* species and *Dendromecon rigida* have isofacial leaves with an upper and lower palisade and stomata on both surfaces (Fig. 6.21). In *Arctostaphylos*, such leaf types are largely restricted to interior species, whereas coastal taxa have stomata restricted to the lower leaf surface, and some have bifacial leaves (Howell 1945). Species in other genera have stomata restricted to the lower leaf surface. Sunken stomata are not uncommon, ranging from slightly sunken stomata in *Heteromeles arbutifolia*, *Rhamnus californica*, and *Yucca whipplei* (Keeley unpublished data), to the extreme case in certain *Ceanothus* species with stomatal crypts that are invaginated to over half the width of the leaf (Nobs 1963). These stomatal crypts are characteristic of the subgenus *Cerastes* (except in the seedling stage) and are absent in subgenus

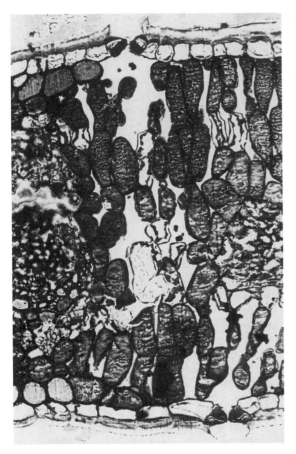

Figure 6.21. Cross section of isofacial leaf of non-sprouting Arctostaphylos bakeri *from northern California chaparral showing distribution of stomata and palisade mesophyll on both the adaxial and abaxial sides; leaf width = 0.33 mm (from Boykin and Keeley, unpublished data).*

steeper angles than shade leaves (H. A. Mooney unpublished data), and other leaf characteristics are highly modifiable, depending on the microenvironment (Cooper 1922; Mortenson 1973; Krause and Kummerow 1977; Hochberg 1980; Ball, Keeley, Mooney, Seaman, and Winner 1983). Such changes in leaf angle can have a profound impact on total leaf absorptance and the heat budget (Ehleringer and Comstock 1989).

Stems and Growth Forms

Chaparral shrubs show a remarkable degree of convergence in growth form. Across landscapes of homogeneous topography, shrubs are of similar height and give the impression of a smooth blanket of vegetation. Detailed studies of plant structure have documented remarkable degrees of convergence in distribution of leaves and stems among unrelated taxa (Mooney et al. 1977; Kummerow et al. 1981). Aridity is a primary factor affecting growth form. Mesic environments result in denser stands of shrubs with arborescent proportions, whereas drier, more open sites result in lower, multistemmed broader shrubs (e.g., Vasek and Ciovis 1976; Keeley 1992b). Growth forms may also change temporally, as in *Arctostaphylos* and *Ceanothus* species, which in older stands grow horizontally to "escape" shading by adjacent plants. Often such branches will root and spread vegetatively by layering, often forming large clones (James 1984).

Fires affect growth form in that many species capable of tree-like proportions resprout after fire, giving rise to a multistemmed shrubby growth form. Resprouting ability is found in all chaparral shrub species, with the obvious exception of obligate-seeding *Ceanothus* and *Arctostaphylos* taxa. The mode of sprouting, however, is variable. *Adenostoma fasciculatum* and sprouting taxa of *Ceanothus* and *Arctostaphylos* initiate a basal lignotuber, or burl, as a normal part of seedling development (Wieslander and Schreiber 1939), although populations of seedlings with and without lignotubers are known for *A. fasciculatum* on mesas north of San Diego and *Arctostaphylos rudis* on mesas east of Lompoc (Keeley, unpublished data). Occasionally, seedlings of *Quercus* spp., *Heteromeles arbutifolia*, *Rhamnus* spp. and others will produce basal burls, apparently in response to intense herbivory under the shrub canopy. Mature shrubs of these species will have large lignotuberous structures that are induced by repeated coppice growth after fires. Others, such as *Cercocarpus betuloides*, sprout from rhizomes a meter or more distal to the main shoot system (Site 1, Table 6.2).

Lignotubers, or basal burls, are uncommon in

Ceanothus. Also, the *Cerastes* have markedly thicker leaves, higher leaf-specific weights, and thicker cuticles than species in the other half of the genus (Barnes 1979).

Leaf orientation is variable among shrub species and depends on environmental conditions. Nearly vertical leaves are prominent in isofacial-leaf species of *Arctostaphylos* and in *Dendromecon rigida*, and many other species have leaf angles greater than 50° (Kummerow, Montenegro, and Krause 1981). In *Arctostaphylos*, leaf angle increases along an aridity gradient (Shaver 1978). In *Ceanothus*, leaf orientation responds to water stress and is seasonally reversible (Comstock and Mahall 1985). If water stress is severe enough, *C. megacarpus* are capable of altering leaf angle by curling (Gill and Mahall 1986), and *Rhus ovata* alter leaf angle by folding (Ehleringer and Comstock 1987). In *Heteromeles arbutifolia* sun leaves have significantly

shrubs outside of mediterranean-climate ecosystems (Keeley 1986). In chaparral shrubs they are often large and commonly exceed the aboveground biomass (Kummerow and Mangan 1981). They are anatomically different among species (Anfuso 1982; Lopez 1983; James 1984), but in all cases they proliferate adventitious buds that are suppressed to various degrees by the dominant stems. The number of epicormic buds is reflected in the fact that after fire, lignotubers typically produce hundreds of new shoots (Kummerow and Ellis 1989), which are fed by carbohydrate stores in the burls (Lopez 1983), and roots (Jones and Laude 1960; Laude et al. 1961). Reserves appear sufficient to sustain the roots for more than 1 yr (Kummerow, Krause, and Jow 1977). The fact that resprouting species tend to be very vulnerable to water stress led DeSouza and associates (1986) to suggest that a primary function of the burl is to provide nutrients that sustain the root system during the summer drought following fire. Storage of inorganic nutrients may also be an important function of lignotubers. After fire, sprouts from burls are more robust because of a much larger pith (Watkins 1939), and often such sprouts will branch and proliferate multiple shoots at the ground level. Sprout production continues in the absence of fire, replacing stems that die (Figs. 6.19 and 6.20).

Stems of most species are ring-porous, with well-developed annual rings, but in others such as *Prunus ilicifolia* and *Malosma laurina*, stems tend toward diffuse-porous, with poorly defined rings (Webber 1936; Watkins 1939; Young 1974; Carlquist 1980, 1989). These shrubs have deep root systems often providing year-round access to water, and this contributes to an unreliable record of annual growth rings (but cf. White 1997). On the other hand, nonsprouting species of *Ceanothus* and *Arctostaphylos* have shallow root systems, and they lay down distinct annual growth rings that have been shown to be a reliable indicator of stand age (Keeley 1993). Wood storage products such as tannins and calcium oxalate are abundant in many species. Older stems of some species e.g., *Adenostoma fasciculatum* tend to rot, whereas others, such as *Quercus* spp., are extremely resistant. Vascicentric tracheids, defined as tracheids present adjacent to vessels, are normally uncommon in dicots but present in stems from 26 families in chaparral (Carlquist 1985). These unusual conducting cells offer a subsidiary conducting system that can supply stems and leaves when the adjacent vessels fail as a result of air embolisms. Carlquist and Hoekman (1985) have developed an index of xeromorphy measured by the density and diameter of xylem vessels as well as the presence of vasicentric

tracheids, and this index appears to be greater for shallow-rooted, obligate-seeding shrubs than for deep-rooted resprouting species. The physiological significance of this index requires careful analysis because embolism induced by water stress is not a direct function of the xylem conduit size but, rather, of the cell wall pore size (Jarbeau, Ewers, and Davis 1995). However, as discussed later, xylem diameter is strongly correlated with freezing-induced embolism.

Stem development in species of *Arctostaphylos* and *Ceanothus (Cerastes)* follows a peculiar pattern. In *Arctostaphylos*, large stripes of bark die, leaving behind only a ribbon of living tissue (Adams 1934). In *Ceanothus*, living tissues grow around these dead stripes, producing a flanged appearance called "longitudinal fissioning" by Jepson (1928). These stripes of dead stem tissue are connected to shaded branches that self-prune (Mahall and Wilson 1986) or to roots in unfavorable microsites, and they are often produced during severe droughts (Parsons et at. 1981). This characteristic of allowing selected strands of vascular tissue to die may have evolved as a way of decreasing the amount of stem cortical surface needed to maintain productive parts of the canopy or root system, and this apparently increases the longevity of these nonsprouting species (C. Davis 1973; Keeley 1975).

Roots

Sprouting shrubs have more deeply penetrating roots than nonsprouting species (Cooper 1922; Hellmers et at. 1955b). Soil depth may limit root penetration, except in highly fractured substrates where roots penetrate bedrock to 9 m or more (Hellmers, Horton, Juhren, and O'Keefe 1955). Roots often penetrate rock fractures and proliferate mats of feeder roots in fine cracks, sometimes producing odd thallus-like structures hundreds of times wider than the thickness. Species differ in ability to penetrate rock fractures, and in one study *Arctostaphylos viscida* was shown to compete far better for water stored in bedrock than did similar aged conifererous trees (Zwieniecki and Newton 1996). During summer drought, this weathered rock mantle is often capable of holding more water within a matric potential range accessible to shrubs than the overlying soils can hold (Newton et al. 1988; Jones and Graham 1993).

Rooting depth may also vary with substrate. Kummerow and associates (1977) excavated a site in southern California on shallow soil overlying bedrock and found that *Adenostoma fasciculatum* roots penetrated to less than 60 cm, and over two-thirds of all root biomass was in the top 20 cm of the soil profile. At another site on deeper soil, roots were distributed deeper for sprouting species *A. fasciculatum* and *Quercus berberidifolia* but not for the nonsprouting *Ceanothus greggii* (Kummerow and Mangan 1981). In northern California, Davis (1972) excavated nonsprouting *Arctostaphylos* species and noted much of the root mass concentrated in the upper 20 cm of the soil profile. Popenoe (1974) found that soil depth was positively correlated with plant height for shrub species in mixed chaparral.

Fine roots (<2.5 mm diameter) are concentrated below the canopy; however, they often overlap those of adjacent plants (Kummerow et al. 1977), suggesting the potential for direct competition for water and nutrients. These feeder roots lack secondary growth and have determinate growth with a life span of one growing season (Kummerow and Ellis 1989). Davis (1972) reported that on sites with a well-developed litter layer, *Arctostaphylos* species proliferate feeder roots near the soil surface, and these roots penetrate the decomposing litter mat. For most shrubs, the radial spread of roots is several times greater than the canopy, although root:shoot biomass ratios are typically less than 1 (Kummerow 1981). Despite the fact that sprouting species maintain their major roots between fires, root:shoot ratios are reportedly similar between sprouting and nonsprouting species.

Root nodules with symbiotic nitrogen-fixing actinomycetes are known for species of *Cercocarpus*) (Vlamis et at. 1964) and *Ceanothus*, but their presence is dependent on various site factors (Vlamis, Schultz, and Biswell 1958; Furman 1959; Hellmers and Kelleher 1959; White 1967; Youngberg and Wollum 1976; Kummerow, Alexander, Neel, and Fishbeck 1978; Ellis and Kummerow 1989) Water stress in particular inhibits nodulation in *Ceanothus* spp. (Pratt, Konopka, Murry, Ewers, and Davis 1977). Ectomycorrhizal associations are common with *Quercus berberidifolia* and *Arctostaphylos glauca*, and vesicular-arbuscular mycorrhizae with *Adenostoma fasciculatum*, *Ceanothus greggii*, and *Rhus ovata* (Kummerow 1981). Ectomycorrhizae have been shown to be important in recovery from water stress in *Q. berberidifolia* (Borth 1986). Root grafting is apparently not common, although it has been observed in *Q. berberidifolia* (Hellmers et al. 1955) and *Prunus ilicifolia* (Bullock 1981).

Vegetative Phenology

Rates of development and growth are controlled by the interaction of low temperatures and irradiance in winter, and low soil moisture and high temperatures in summer. After winter rains have replenished soil moisture, initiation of growth depends on higher temperatures. In general, phenological events begin and end later on north- than on south-facing exposures and are delayed with increasing elevation and latitude. Even at the same site, phenological events are not synchronized. In southern California, mean dates of stem elongation range from March to June, depending on the species, and the growing season ranges from 2 mo for *Adenostoma fasciculatum* to 1 mo for *Rhus ovata* (Kummerow et al. 1981). Such patterns also vary spatially; Bedell and Heady (1959) noted a 3 mo growing season for *A. fasciculatum* in northern California. Some species, such as *Malosma laurina* (Watkins and DeForest 1941) and *Adenostoma sparsifolium* (Hanes 1965), may continue growth during the summer months. In years of severe drought, there may be no new leaf production or stem elongation (Harvey and Mooney 1964).

Secondary stem growth begins earlier and extends later into the season than primary growth (Avila, Lajaro, Araya, Montenegro, and Kummerow 1975), and in *Rhus* species and *Malosma laurina* it may occur year-round. The width of growth rings is significantly lower in drought years (Gray 1982; Keeley 1993) and was shown to be sensitive to the level of late winter and spring precipitation (Guntle 1974).

Fine root growth follows a pattern of growth similar to aboveground growth, with peak biomass levels in midsummer and a massive die-off as soil moisture is depleted (Kummerow et al. 1978).

Reproductive Phenology

Flowers of most shrub species are small and are borne in large showy clusters, but only a small proportion of flowers mature into fruits. Most species are self-incompatible (Raven 1973; Moldenke 1975), however, Fulton and Carpenter (1979) reported self-compatibility in *Arctostaphylos pringlei*, but Brum (1975) found this species to be entirely self-incompatible.

Most species are pollinated by insects, and this trait may have selected for the markedly asynchronous nature of flowering phenology in different species (Mooney 1977b; Steele 1985). *Arctostaphylos* and *Garrya* species flower earliest in the season, prior to the initiation of vegetative growth, followed by *Ceanothus*, *Quercus*, and *Rhus* species. Early flowering in these five genera may be related to the fact that they flower on old growth, from floral buds initiated during the previous year's growing season. *Adenostoma fasciculatum*, *Heteromeles arbutifolia*, and *Malosma laurina* flower later, on new growth after stem elongation is completed

(Bauer 1936; Kummerow et al. 1981; Baker, Rundel, and Parsons, 1982 and *Adenostoma sparsifolium*, with stem growth extending well into summer, is one of the latest flowering species, typically not beginning until August (Hanes 1965).

Annual flower and fruit production patterns are variable. Some species (e.g., *Heteromeles* and *Malosma*) tend to flower (and fruit) more or less annually, whereas others, such as *Arctostaphylos* and *Ceanothus* species, are typically biennial bearers (Keeley 1987a; Keeley and Keeley 1988b; Zammit and Zedler 1992). In these latter two genera, internal competition for photosynthetates during the period of fruit maturation affects the extent of floral bud differentiation for the following year and greatly complicates correlating fruit production with weather patterns.

SHRUB PHYSIOLOGY

Water Relations

Chaparral shrubs vary in their water relations characteristics largely in accordance with species-specific differences in rooting habit (Poole, Roberts, and 1981; S. Davis and Mooney 1986). Shallow-rooted species are able to respond to elevated levels of soil moisture early in the rainy season. During the summer, shallow-rooted nonsprouting species of *Arctostaphylos* and *Ceanothus* (subgenus *Cerastes*) are exposed to extremely negative soil water potentials (e.g., *Ceanothus m.* in Fig. 6.22). At this time they commonly have predawn stem xylem water potentials of -6.5 to -8 MPa versus -1 to -3 MPa during the summer drought for deeper-rooted shrubs such as *Rhus ovata, R. integrifolia, Malosma laurina* (labeled *Rhus l.* in Fig. 6.22), *Heteromeles arbutifolia*, sprouting species of *Arctostaphylos* and *Ceanothus* (subgenus *Ceanothus*, e.g., *Ceanothus s.*, Fig. 6.22), *Prunus ilicifolia*, and *Rhamnus californica* (Poole and Miller 1975, 1981; Dunn, Shrapskire, Song, and Mooney 1976; Burk 1978; Barnes 1979; Miller and Poole 1979; Schlesinger and Gill 1980; Parsons, Rundel, Hedlurd, and Baker 1981; Oechel 1988; S. Davis 1989; Mooney 1989; Rundel 1995).

The critical role rooting depth plays in determining water stress is illustrated by the similarity in water potentials for *Malosma laurina* and *Ceanothus* seedlings in contrast to the marked differences between adults (Fig. 6.22A, B) (Thomas and Davis 1989). Low survivorship of *Malosma* seedlings (Fig. 6.22C) is tied to much greater susceptibility to embolism induced by water stress. *Malosma laurina* exhibits a 50% loss in hydraulic conductivity when branches reach a xylem pressure potential of only

-1.6 MPa, whereas *Ceanothus megacarpus* does not exhibit 50% embolism until -11 MPa (Kolb and Davis 1994; Jarbeau et al 1995). These studies have been extended to include several other species and reveal the same pattern; namely, increasing susceptibility to embolism is correlated with decreased seedling survivorship after fire (Davis 1991; S. Davis, Kolb and Barton 1998). These patterns may be tied to differences in the size of pores in xylem pit membranes, with larger vessels and wider pores contributing to greater xylem efficiency but contributing to greater vulnerability to embolism (Jarbeau et al. 1995). Vulnerability to embolism appears to be typical of deep-rooted sprouting shrubs, and may account for why these species have seedling recruitment restricted to shaded understories of older chaparral.

Ceanothus megacarpus water potentials change markedly during stand development following fire. Comparing stands 6, 13, and 22 yr old, Schlesinger and Gill (1980) showed that throughout the year values were always more negative for plants in the youngest-aged stands and that smaller plants suffered more severe water stress during the summer drought than did larger plants – one small plant even reaching an amazing predawn water potential of -12 MPa (Schlesinger, Gray, Gill, and Mahall 1982a). They contended that water stress is a major factor in stand thinning and that it was reflected in nonrandom mortality that shifted the distribution of shrubs from clumped toward regular, a pattern seen in much older stands as well (Keeley 1992b).

In postfire resprouting shrubs, new sprouts have stem water potentials one-third to one-half as negative as mature shrubs, but these differences disappear by the second year after fire (Radosevich and Conard 1980; Oechel and Hastings 1983; Hastings, Oechel, and Sionit 1989; Stoddard and Davis 1990).

Stem water potentials are more negative during summers following winters of low rainfall (Poole and Miller 1981). For example, during a severe drought in the Sierra Nevada foothills, the shallow-rooted *Arctostaphylos viscida* reached predawn stem xylem water potentials of -7 MPa and showed very little diel variation (indicating little photosynthetic activity), whereas the following summer, after a very wet winter, the lowest predawn potentials were -4 MPa, and there was a large diel change (Parsons et al. 1981). In addition, the monthly pattern of rainfall may be as influential as the seasonal total (Gill 1985). For most species, water potentials are more negative at their lower elevational and southern latitudinal limits than in the center of their distribution (Poole and Miller 1975;

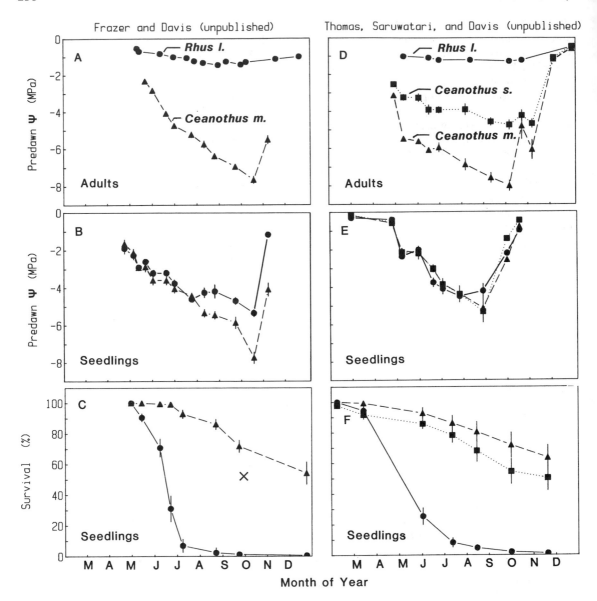

Figure 6.22. Change in predawn water potential in mixed chaparral stands for adult shrubs (A, D) and seedlings (B, E) and postfire seedling survivorship (C, F) for non-sprouting Ceanothus megacarpus *and sprouting* Malosma laurina *(formerly* Rhus laurina) *and sprouting* C. spinosus *(reprinted from Davis 1989).*

Mooney et al. 1977). Root competition may directly induce water stress, as evidenced by the observation that reduction in shrub cover by artificial stand thinning results in a significant reduction in water stress (Knipe 1985), and the ability of grasses to prevent shrub invasion into grasslands by depletion of soil moisture (S. Davis and Mooney 1985; Williams and Hobbs 1989).

For most species there is a midday depression in stomatal conductance that begins earlier in the day as the growing season progresses. Although

deep-rooted shrub species may maintain access to water longer into the summer drought, they exhibit complete stomatal closure under far less water stress than do shallow-rooted species (Poole et al. 1981; Barnes 1979). As a consequence, shallow-rooted *Arctostaphylos* and *Ceanothus* species are likely to maintain more active photosynthesis longer into the drought than many deep-rooted species. Nonetheless, by late summer, these shallow-rooted species frequently exhibit complete stomatal closure for a month or more, whereas deeper-rooted species commonly have a brief period of stomatal conductance each day. Undoubtedly these differences are tied to anatomical characteristics discussed earlier and to seasonal and diel osmotic adjustments (Roberts 1982; Calkin and Pearcy 1984; Bowman and Roberts 1985) and are cued by absccisic acid (Tenhunen, Hanano, Abril, Weiler, and Hartung 1994).

Chaparral communities comprise associations of taxa with distinctive species-specific patterns of annual transpiration that vary with elevation and slope exposure and predict different patterns of productivity for these species. Poole and Miller (1981) hypothesized that communities on different sites should converge at maturity in terms of transpiration per unit of leaf area, and they put that figure at 150–200 mm yr^{-1}. Parker (1984) found that maximum transpiration rates are consistently higher for shallow-rooted obligate-seeding species of *Arctostaphylos* and *Ceanothus* than for deeper-rooted sprouting shrub species. He hypothesized that this characteristic results in more rapid seedling growth rates and hence a better potential for establishment in comparison with seedlings of resprouting species.

Soil Nutrients

Enhanced vegetative and reproductive growth, after fertilizer application suggests that chaparral shrubs are nutrient-limited (Hellmers, Bonner, and Kelleher 1955; Vlamis et al. 1958; Christensen and Muller 1975a; McMaster, Jow, and Kummerow 1982; Gray and Schlesinger 1983). *Adenostoma fasciculatum* is clearly nitrogen-limited, whereas others (e.g., *Ceanothus megacarpus* and *C. greggii*) may not be, although the latter species does respond to phosphorus addition. *Ceanothus* species are nitrogen-fixers, and there is evidence they are capable of enhancing the nitrogen status of surrounding soils (Quick 1944; Hellmers and Kelleher 1959; Vlamis, Schultz, and Biswell 1964; Kummerow et al. 1978). Nitrogen-fixing nodules in *Ceanothus* are deeply buried, and thus some published rates are underestimates. Ellis and Kummerow (1989) reported that young plants derived 10% of their nitrogen from biological fixation by the actinomycete species of *Frankia*. On some chaparral sites, free-living asymbiotic nitrogen fixers may contribute significantly to the nitrogen budget (Ellis 1982; Poth 1982).

Foliar leaching of nitrate with the first fall rains may result in a pulse of nitrogen input to the soil (Christensen 1973). Schlesinger and Hasey (1980) found that this was largely a result of atmospheric deposition, and this, plus foliar-leached ammonium, could exceed the input by symbiotic nitrogen fixation. In many parts of California, deposition of anthropogenic sources of nutrients is becoming an increasingly important part of the nutrient budgets of these ecosystems (Schlesinger et al. 1982b; Bytnerowicz, Miller, and Olszyk 1987; Bytnerowicz and Fenn 1996). Under natural conditions, litter fall is the most important means of returning nutrients to the soil (Gray and Schlesinger 1981). Litter fall is concentrated in summer, and decomposition is relatively rapid (Schlesinger and Hasey 1981). The highest concentrations of soil nutrients tend to be in the upper soil layers (Christensen and Muller 1975a); thus, shallow-rooted shrub species may have a competitive advantage. In *Adenostoma fasciculatum*, most of the nitrogen and phosphorus uptake occurs in winter, prior to growth, and Mooney and Rundel (1979) suggested that this may reduce leaching losses from the soil. These winter uptake patterns, however, are not typical of all species (Shaver 1981; Gray 1983).

Fire has a marked effect on the nutrient status of chaparral soils. Because fire recycles nutrients tied up in plant matter, soil levels of most nutrients increase after fire (Sampson 1944; Christensen and Muller 1975a; Gray and Schlesinger 1981). Fires, however, result in substantial ecosystem losses of K and N through volatilization and runoff (DeBano and Conrad 1978) that may require 60–100 yr to replace (Schlesinger and Gray 1982). The first year after fire, foliage concentrations of important nutrients are very high, although by the second or third year, nutrient levels may be comparable to levels observed for mature vegetation (Sampson 1944; Rundel and Parsons 1980, 1984). On some sites, the postfire proliferation of suffrutescent and annual legumes may add nitrogen through their symbioses with nitrogen-fixing *Rhizobium* bacteria (Poth 1982). Postfire herbaceous species may have different strategies for nitrogen use. Swift (1991) reported that "fire-endemic" species preferentially used the high ammonium levels present immediately after fire but used nitrogen less efficiently than did generalist species that were likely to persist for many years after fire.

Photosynthesis

Shrubs and other growth forms in chaparral are C$_3$ (Mooney, Troughton, and Berry 1974), reflecting the fact that summer drought restricts most photosynthetic activity to the cooler winter and spring months. Nonetheless, photosynthetic activity is possible year-round, although maximum rates are low (5–15 mg CO$_2$ dm^{-2} hr^{-1}, Mooney 1981) due to internal structural limitations of the sclerophyllous leaf (Dunn 1975) and to efficient strategies for use of nitrogen (Field, Merrino, and Mooney 1983; Field and Davis 1989). For most species there is a broad optimum temperature range for photosynthesis between 10° C and 30° C and little capacity for temperature acclimation (Oechel Lawrence, Mustafa, and Martinez 1981). Growth analysis indicates aboveground processes are not highly sensitive to belowground temperature (Larigauderie, Ellis, Mills, and Kummerow 1991).

Peak photosynthetic rates are typically observed only during the spring growing season, and as the season progresses, declining soil moisture reduces stomatal conductance, thus limiting daily CO_2 uptake (Oechel 1982). Surprisingly, in some species the photosynthetic rate is controlled directly by changes in the photoperiod (Comstock and Ehleringer 1986).

Many of the broad-leaf species such as *Heteromeles arbutifolia*, *Prunus ilicifolia*, *Quercus berberidifolia*, and *Rhus* spp. are relatively tolerant of shade (e.g., photosynthesis of *Heteromeles arbutifolia* is saturated at less than one-third full sunlight) (Harrison 1971), whereas others, such as *Ceanothus* and *Arctostaphylos*, are not fully saturated at two-thirds full sunlight (Oechel 1982) and if shaded they have substantially reduced water use efficiency (Mahall and Schlesinger 1982). Efficiency in use of water is higher for species distributed in xeric habitats and lowest in species from mesic habitats, although the latter species have the highest efficiency of nitrogen use (Rundel 1982; Field et al. 1983).

The thermal insensitivity of photosynthesis suggests that low temperatures are likely to play a minor role in limiting wintertime carbon gain, and this suggestion is supported by simulations that Mooney and associates (1975) performed on seasonal patterns of photosynthesis in *Heteromeles arbutifolia*. Their simulation suggests that wintertime depression of photosynthesis is largely a result of limited irradiance.

Across large elevational and climatic gradients, relatively small differences are observed in carbon uptake rates (Oechel et al. 1981). Scaling up potential photosynthetic activity of chaparral patches by using remote sensing techniques is a promising means of examining landscape level patterns (Gamon et al. 1995).

Freezing Tolerance

Freezing damage is occasionally significant, as evidenced by the spectacular freeze of December 1990 (Collett 1992), when widespread dieback was reported throughout the state (Cowden and Waters 1992). One of the most sensitive chaparral shrubs

is *Malosma laurina*, and there is reason to believe its distribution is controlled in part by low temperatures (Misquez 1990). Differential vulnerability to freezing-induced embolism, which is a function of xylem vessel diameter, may explain microhabitat segregation by this and other shrub species (Langan, Ewers, and Davis 1997).

The Myth of Stand Senescence

The concept of chaparral senescence is reminiscent of the Clementsian view of community succession being analogous to development of an organism. Old stands of chaparral have been described as "senescent," "senile," "decadent," and "trashy" – terms that lack clear definition. Without citing any source, Hanes (1971) described older stands of chamise chaparral as unproductive, with little annual growth, and others have suggested that older stands lack diversity and are nutrient-limited, with an overaccumulation of allelopathic toxins. None of these statements has been substantiated.

Apparently the notion that older stands of chaparral are unproductive derives from measurements of browse production for wildlife (Biswell, Taber, Hedrick, and Schultz 1952; Hiehle 1961; Gibbens and Schultz 1963). These studies showed that older chaparral produces very little deer browse. However, these were not valid measures of productivity, because production above 1.5 m, which is normally unavailable to deer, was not included; most new growth in older stands occurs above 2 m. Studies of living biomass accumulation have shown that it remains stable for 60 yr or more (Fig. 6.23). Studies of production that have compared stands a decade or two after fire with stands nearly a century old have repeatedly failed to find any evidence of declining productivity (Keeley and Keeley 1977 unpublished data; Hubbard 1986; Larigauderie, Ellis, Mills, and Kummerow 1991). The idea that older stands are nutrient-limited has been questioned (Schlesinger et al. 1982b) and shown to be erroneous (Rehlaender 1992; Fenn, Poth, Dunn, and Barro 1993). In terms of resilience, stands a century old have seed banks capable of adequate postfire regeneration (Keeley and Zedler 1978; Zammit and Zedler 1994), and some species even require ancient stands for seedling recruitment (Keeley 1992c). Physiological studies of *Adenostoma fasciculatum* and *Ceanothus greggii* show that photosynthetic capacity, on a per leaf basis, does not decrease and even increases in very old stands (Reid 1985).

The primary observations that have led to the suggestion of stand senescence is the inevitable accumulation of dead wood and the occasional mass

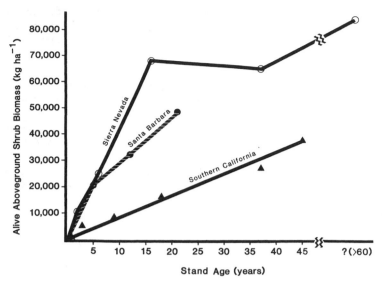

Figure 6.23. Standing living biomass in chaparral stands as a function of age since last fire: southern California mixed chaparral at San Dimas Forest from Specht (1969) (last datum point from Conrad and DeBano 1974); Santa Barbara Ceanothus megacarpus chaparral from Schlesinger and Gill (1980); Sierra Nevada foothills chamise chaparral from Rundel and Parsons (1979), Stohlgren et al. (1984), and Stohlgren (pers. commun. 1984).

dieback of the obligate seeder *Ceanothus*. It is abundantly clear that *Ceanothus* dieback is not due to senescence, for these species are capable of great longevity. Dieback is often associated with drought years (Riggan, Franklin, Brass, and Brooks 1994), which are likely to affect such shallow-rooted shrubs more than others. Sparks and associates (1993) have found differences in allocation of photosynthate between young and old shrubs, which they interpreted as indicators of senescence; however, there are other reasonable interpretations. Additionally, Sparks and Oechel (1993) contend that senescence is suggested by the fact that older stands of *Adenostoma fasciculatum*, with higher biomass of standing dead, sprout less vigorously after fire. However, they failed to rule out fire intensity as a causal factor.

An alternative to the model of stand senescence is one of a shifting balance in the competitive relationships as the community ages, sometimes resulting in successional replacement of disturbance-dependent recruitment species being displaced by disturbance-free species (Keeley 1986; Hilbert and Larigauderie 1988; Zedler and Zammit 1989; Keeley 1992b, 1992c; Zammit and Zedler 1992; Zedler 1995b). For example, on mesic slopes, *Ceanothus* and *Arctostaphylos* species are often displaced by more competitive taxa such as *Quercus, Heteromeles, Prunus*, and the like, which if left undisturbed long enough would form a self-replacing miniature sclerophyll woodland. However, on arid sites, *Ceanothus* and *Arctostaphylos* will persist for a 100 yr or more and continue to replenish the soil seed bank sufficiently to rejuvenate the stand after the inevitable fire.

RELATED PLANT COMMUNITIES

California Coastal Sage Scrub

Often called "soft chaparral," California coastal sage scrub, a largely summer-deciduous vegetation, tolerates more xeric conditions than evergreen chaparral (Mooney 1977a; Westman 1981a). Because of the orographic nature of rainfall, this community is well developed at low elevations near the coast as well as on arid interior slopes, usually below chaparral but occasionally at higher elevations on outcroppings of shallow soil (Bradbury 1978), on fine-textured soils that inhibit downward percolation of winter rains, or on frequently disturbed sites. The dominants include all the subshrubs and suffrutescent species (*Salvia, Eriogonum, Artemisia, Lotus*, and *Mimulus*) that often are associated with chaparral, either in gaps or after fire, plus more restricted species such as *Encelia californica, Hazardia squarrosus, Baccharis pilularis, Viguiera laciniata*, and *Malocothamnus fasciculatus*. Coastal sage scrub is lower (<1.5 m) and more open than chaparral and frequently has some herbaceous understory (Westman 1981a). Various associations have been delineated on the basis of latitudinal changes in species composition from Baja to northern California (Axelrod 1978; Kirkpatrick and Hutchinson 1980; Westman 1983; Malanson 1984; DeSimone and Burk 1992; Davis, Stine, and Stoms 1994) and show alliances with the northern coastal scrub (Heady et al. 1977). Succulents in the Crassulaceae and Cactaceae are important components toward the southern limits, particularly near the coast (Mooney and Harrison 1972). Oftentimes, evergreen sclerophyllous shrubs, such as *Malosma laurina* and *Rhus integrifolia*, are distributed at widely spaced intervals throughout coastal sage, apparently exploiting widely spaced favorable soil microsites.

Most dominants have nonsclerified malacophyllous leaves (3–6 cm²) that initiate growth in early winter and expand until soil moisture, temperature, and photoperiod induce leaf abscission (Nilsen and Muller 1981; Gray 1982). With the onset of drought, these winter leaves are replaced by a few smaller axillary leaves termed brachyblasts (Harrison, Small, and Mooney 1971; Gray and Schlesinger 1981; Westman 1981b), and these two leaf types may differ physiologically (Gigon 1979; Gulmon 1983). In *Salvia* species, a portion of the leaves may curl up during the summer drought but then expand during the following growing season, making these plants technically evergreen (Gill and Mahall 1986). In many species, flowering is on new growth and thus is delayed until summer or fall (Mooney 1977a).

During the spring growing season, maximum leaf conductances, transpiration rates, and photosynthetic rates may be more than double those observed for sclerophyllous shrubs (Harrison et al. 1971; Oechel et al. 1981; Poole et al. 1981). Mooney and Dunn (1970) suggested that the shallow-rooted subshrubs avoid summer drought by losing foliage, and these coastal sage subshrubs have often been termed "drought-avoiders," in contrast to the "drought-tolerators" in evergreen chaparral. Recent studies, however, reveal that the brachyblasts on *Salvia mellifera* can tolerate water potentials as low as -9.0 MPa (Gill and Mahall 1986) and maintain active photosynthesis at -5.0 MPa (Mooney 1982).

Volatilization of aromatic compounds from leaves is a notable feature of the coastal sage dominants *Artemisia californica, Salvia mellifera,* and *S. leucophylla.* Muller and associates (1964) showed that these compounds were potentially allelopathic to herb growth and suggested that this accounted for the typical bare zone of a meter or more that forms between coastal sage and grasslands. Such bare zones also occur between grassland and nonaromatic vegetation such as chamise and scrub oak chaparral. Exclosure experiments in both vegetation types, however, have shown that small mammals are an important factor in the formation of bare zones because of their propensity to forage on grassland species as close to the protective shrub canopy as possible (Bartholomew 1970; Halligan 1974; Bradford 1976; Quinn 1986). Experimental evidence in support of allelopathy has been criticized (Keeley and Keeley 1989); but, the phenom-

enon has not been ruled out. The facts that coastal sage aromatics represent a substantial carbon drain on the plant (Tyson, Dement, and Mooney 1974) and are potentially toxic (Muller and del Moral 1966; Muller and Hague 1967; Halligan 1975) argue strongly for an adaptive role; however, antitranspirant or antiherbivore functions have not been fully explored.

Two modes of postfire recovery in coastal sage scrub include facultative seeders (Fig. 6.14 C,D) and obligate resprouters that flower the first year and recruit seedlings heavily in the second year (Westman, O'Leary, and Malanson 1981; Malanson and O'Leary 1982; Keeley and Keeley 1984). Coastal sage scrub communities have a postfire burst of fire annuals and other herbs comprising the same species as in chaparral (Keeley and Keeley 1984; O'Leary 1988; O'Leary and Westman 1988). Postfire resprouting of coastal sage subshrubs is greatly reduced on inland sites, where recovery is much slower (Westman et al. 1981; Westman 1982). Resprouting is most successful in young stands and least successful in older stands, and this is not driven solely by greater fire intensity, as some coastal sage subshrubs – for example, *Baccharis pilularis* (Hobbs and Mooney 1985) and *Eriogonum fasciculatum, Artemisia californica* (Keeley 1997a) – outgrow the capacity for resprouting. In the absence of fire, coastal sage species are capable of regenerating their canopy from basal sprouts (Malanson and Westman 1985), similar to the pattern observed for sprouting chaparral shrubs (Figs. 6.19 and 6.20), making these subshrubs reasonably resilient to long fire-free periods.

Coastal sage scrub is intermediate between grassland and chaparral in its resilience following frequent fires (Wells 1962; Kirkpatrick and Hutchinson 1980). Fire-recurrence intervals of 5–10 yr may result in chaparral being replaced with coastal sage scrub. More frequent fires, however, will result in transition of sage scrub to grassland (Fig. 6.12) that is dominated by nonnative grasses (Keeley 1990b; Haidinger and Keeley 1993).

Interior Chaparral

Interior regions of western North America have vegetation showing various degrees of similarity to California chaparral. Of these, the Arizona chaparral (Fig. 6.24) is the most similar, even though it is separated by more than 200 km of desert. It is distributed in widely disjunct patches at 1000–2000 m in northern and central portions of the state and intergrades with desert scrub or grassland at the lower margin and with yellow pine forest or pinyon-juniper woodland at the higher elevations

Figure 6.24. *Chaparral from northern Arizona* dominated by Quercus turbinella.

Table 6.4. *Demographic structure of a northern Arizona chaparral stand unburned for 40 years (1450 m, east-facing, 15° incline, Yavapai County)*

Species	Postfire regeneration[a]	Live basal coverage m² basal area ha⁻¹	Shrub density Genets ha⁻¹	% Dead	Ramets /genet	Seedlings & saplings Genets ha⁻¹
Quercus turbinella	(OR)	16.5	1,375	0	88.0	0
Ceanothus greggii	(OS)	2.7	500	0	84.6	0
Quercus pungens	(OR)	1.9	125	0	188.0	0
Rhus trilobata	(OR)	1.3	375	0	36.3	0
Rhamnus crocea	(OR)	0.3	190	0	8.7	0
Gutierrezia sarothrae	(OS)	0.3	12,625	0	7.9	0
Acacia greggii	(OR)	0.1	125	0	14.0	0
Eriogonum fasciculatum	(FS)	0.1	750	0	10.0	0
Three other subshrubs	(OR)	0.7	190	0	39.1	0

[a]Table 6.2 legend.
Source: Welsh and Keeley (unpublished data).

(Carmichael et al. 1978; Brown 1978). The Arizona chaparral shrub flora comprises a subset of the California chaparral, plus shrubs that are closely related if not conspecific with California taxa (e.g. *Quercus turbinella*) (Tucker 1953). Areal coverage is typically from 35–80% ground surface in Arizona chaparral communities (Cable 1957; Carmichael et al. 1978). A typical Arizona site is described in Table 6.4 and, relative to California chaparral (Tables 6.1 and 6.2), it has less basal area and greater bare ground than most California sites. The nearly ubiquitous Californian *Adenostoma fasciculatum* is noticeably absent from Arizona chaparral, and *Q. turbinella* is to Arizona chaparral what *A. fasciculatum* is to California chaparral; it occurs throughout the chaparral region and dominates most sites (Carmichael et al. 1978). Absence of *A. fasciculatum* is likely tied to its optimum growth occurring where coldest monthly temperature ranges from 9–12°C, whereas those California shrub species present in Arizona chaparral have an optimum range from −3 to 3°C (Malanson, Westman, and Yan 1992).

Arizona chaparral sites average 400–650 mm precipitation yr^{-1} (Mooney and Miller 1985), but in contrast to mediterranean-climate California (Fig. 6.2), the summer drought is cut short and 35% of the annual rainfall comes as high-intensity summer thunderstorms (Carmichael et al. 1978). Because summer rains are of short duration and occur at a time of high evaporative loss, precipitation may not be as effectively used as is winter precipitation. Nonetheless, Arizona chaparral shrubs readily use summer rains. Vankat (1989) has documented stem xylem water potential changes from −7 to −1 MPa following summer thunderstorms, although the extent to which these shrubs remain photosynthetically active through the summer is unknown. Swank (1958) noted that most growth is in the spring and that summer growth is unpredictable. Reduced availability of summer precipitation is suggested by the fact that for a specific level of average annual rainfall, Arizona chaparral sites consistently produce sparser and more open communities than does California chaparral (Cable 1957).

Because of the sparser cover, wildfires are less frequent, occurring every 50–100 yr (Cable 1957), although responses to fire by the dominant shrubs are similar to those of California plants. Obligate-seeding species such as *Ceanothus greggii*, *Arctostaphylos pringlei*, and *A. pungens* establish seedlings from soil-stored seeds only after fire (Pase and Pond 1964; Pase 1965; Pase and Lindenmuth 1971). Others, such as *Quercus turbinella* and *Rhamnus crocea*, are obligate resprouters after fire. Despite much overlap in woody floras, there is relatively little similarity in the temporary postfire flora, with the notable exception of *Emmenanthe penduliflora*, which may dominate some postfire sites in Arizona. Although species richness is notably lower than in California, there is a temporary flora that flourishes following fire, but, unlike in California, herbaceous perennials are more important than annuals, likely reflecting the role of summer rain in this interior chaparral (J. Keeley, M. Keeley, Fotheringham, and Zipusch, unpublished data).

A detailed study of stem age structure in Arizona chaparral (Welsh and Keeley, unpublished data) has shown that, as in California chaparral, postfire resprouting species continue to rejuvenate their canopy with sprouts in the absence of fire. Also, as in California, obligate-seeding *Arctostaphylos* were more or less even-aged, dating back to the last fire. Surprisingly, however, the postfire obligate seeder *Ceanothus greggii* proliferated many different-aged stems from the root collar or lower stem, and all four *C. greggii* shrubs that were aged showed continuous recruitment of new stems. Observations suggest that this growth pattern is widespread wherever *Ceanothus* are distributed in open habitats.

On the more western slopes of the Sierra Madre Oriental Range of northeastern Mexico are isolated patches of chaparral-like vegetation (Muller 1939, 1947; Shreve 1942). Muller (1939) suggested that the slightly lower precipitation and greater diel and annual temperature fluctuations in this region accounted for the replacement of thorn scrub by evergreen sclerophyllous vegetation. The strictly summer-rain climate of this region is markedly unlike that of other chaparral regions (Fig. 6.2). This Mexican chaparral is restricted to limestone or shallow rocky soils at 2000–3000 m, with desert scrub below and evergreen forest above. It is distinguished from the surrounding vegetation types by its predominance of shrubs, the importance of evergreen species, greater density of plant cover, and fewer herbs. Important genera include *Quercus*, *Garrya*, *Cercocarpus*, *Rhus*, *Rhamnus*, *Arbutus*, *Ceanothus*, and *Arctostaphylos*, including two California obligate-seeding shrubs, *C. greggii* and *A. pungens*. One striking difference is that this Mexican chaparral has an understory of herbaceous species, including C_4 bunchgrasses (Keeley, personal observations), not commonly observed in California chaparral. Rzedowski (1978) considered this zone to have a high fire frequency, and he noted that the scrub oak species were vigorous sprouters. Also, postfire seedling recruitment has been noted for obligate-seeding *Ceanothus greggii* and *Arctostaphylos pungens* (Valiente, personal communication). Beyond this, little is known of the postfire community, but based on distribution data it is apparent that very few of the hundreds of postfire species in California chaparral extend to northeastern Mexico. Considering the 433 native species found after fire in one southern California study, only 11% are distributed into mainland Mexico, and they include none of the species that contributed significantly to the postfire flora in California (Keeley, unpublished data). Thus, although the shrub overstory is remarkably similar between Mexico and California, the communities are quite different.

Petran "chaparral" is a largely winter-deciduous shrub vegetation at 2000–3000 m in the central Rocky Mountains (Vestal 1917; Daubenmire 1943). Despite its high elevation, annual precipitation is 380–535 mm yr⁻¹, well within the range of values for other chaparral regions (Pase and Brown 1982). Winter temperatures are below freezing, and the summer growing season may be <100 d long. The vegetation has an overall physiognomy similar to evergreen chaparral in its height and thicket-like aspect (Hayward 1948). *Quercus gambelii* is the dominant throughout the range, and it, like the majority of species, is winter-deciduous. Evergreen species, including a few from California chaparral, are minor components, although some, such as *Cercocarpus* species, are sometimes locally abundant (Brooks 1962; J. Davis 1976). A well-developed herb flora is characteristic of the mature vegetation (E. Christensen 1949; Allman 1952). Fires occur, and *Q. gambelii* responds like other scrub oaks by sprouting vigorously from the root-stock (Brown 1958; Kunzler and Harper 1980). All other common shrub species also sprout vigorously after fire, and obligate-seeding species are infrequent (McKell 1950). Most species maintain themselves in the absence of fire through additional sprouts and seedling recruitment, although disturbance-free periods favor some species over others (Allman 1952; Eastmond 1968). Although little has been reported in the way of a temporary postfire flora, one annual species from this region has smoke-stimulated germination (Baldwin, Staszak-Kozinski, and Davidson 1994).

EVOLUTION

Community History

Evergreen Cretaceous vegetation responded to the increasing aridity of the Tertiary Period by the evolution of drought-deciduous and evergreen sclerophyllous taxa. California chaparral sclerophylls owe their origins to physiognomically similar taxa that appeared early in the Tertiary under conditions quite unlike the present Mediterranean climate (Axelrod 1973). In light of the ample summer rains of that period, it is most reasonable that these species evolved on outcroppings of unusually stressful substrates. By analogy with modern vegetation, we can infer the critical factors to have been soil moisture and nutrient stress. Low-nutrient, rocky soils, with high infiltration rates,

produce severe surface soil drought, but they also retain deeper water during droughts. Deep-rooted, woody evergreen sclerophylls can exploit these sites by their enhanced efficiency in nutrient use and ability to remain metabolically active longer into the drought. Drought-deciduous types are favored by more severe droughts, fine-textured soils, or high-nutrient conditions.

By the middle of the Miocene (20 M yr BP), California was an ecotone between the more mesic adapted Arcto-Tertiary geoflora from the north and the xeric adapted Madro-Tertiary from the south. Evergreen sclerophylls were widespread across the Southwest, forming various broadleaf sclerophyllous woodlands and shrublands associated with subtropical species no longer found in this region (Axelrod 1975). Closed-cone pine forests dominated the more mesic coastal regions (Raven and Axelrod 1978), suggesting that fires (and consequently droughts at some time of the year) must have been a predictable feature of the Miocene environment.

At the close of the Tertiary, the climate of California was taking on a mediterranean flavor, possibly a bimodal precipitation regime similar to that in Arizona today (Fig. 6.2), with a greater range of sites exposed to periodic droughts. This, coupled with increased tectonic activity and uplift of mountain ranges, increased the extent of well-drained shallow rocky soils and thus enhanced the spread of evergreen sclerophylls.

The Pleistocene marked the firm establishment of a mediterranean climate in California, accompanied by elimination of summer rains and greater annual temperature extremes. Some Arcto-Tertiary elements persisted on cooler, more mesic sites, and the Madro-Tertiary geoflora lost taxa depending on summer rain and those requiring more equable climates. Pleistocene glacial/interglacial shifts in climate resulted in latitudinal and elevational shifts in elements of both gefloras (Axelrod 1981, 1986), and thus, contemporary chaparral associations are the result of mixing and remixing of these plant assemblages.

The present spatial pattern of chaparral distribution, from the summer-rain region of northern Mexico, through Arizona with bimodal rainfall, to mediterranean-climate California, may be viewed as a model of the temporal development of California chaparral. This model suggests a pattern of evolution of chaparral taxa on islands of poor soils and seasonal drought embedded in a more mesic landscape. As the climate changed, these drought-prone "islands" coalesced into larger patches, with consequent elimination of other vegetation. Despite the fact that some chaparral dominants originated under a summer-rain climate, there is evidence that evolutionary changes have occurred in response to the mediterranean climate. For example, Vankat (in Keeley and Keeley 1988a) has shown that the flowering phenology of *Arctostaphylos pungens* and *Ceanothus greggii* is different in the summer-rain climate of Mexico. In California these genera exhibit flower-bud dormancy through the summer and fall, which may have been selected for by the mediterranean climate. Lignotubers as ontogenetic traits, such as in species of *Adenostoma, Arctostaphylos,* and *Ceanothus* and the coalescence of seed endocarp segments in *Arctostaphylos,* are traits presently absent from Arizona chaparral species, but they are widespread in southern California taxa, and possibly arose in response to a greater predictability of fire in a mediterranean-climate environment. The rapid radiation of annual taxa endemic to postfire environments, an element largely lacking in Arizona and Mexican chaparral, may also reflect the more predictable role of fire in California.

Pleistocene climates were significantly moister than contemporary conditions, with coastal paleoclimates in southern California comparable to modern climates 500 km to the north (Axelrod and Govean 1996). Evidence of this is seen in the elimination of more mesic forests throughout coastal southern and central California during the Holocene (Axelrod 1973; Warter 1976; Heusser 1978). These were largely closed-cone forests, which demonstrates that wildfires have long been important in this region. With the arrival of humans at the close of the Pleistocene, wildfire frequency likely accelerated due to increased ignitions (Wells 1962) and to habitat modification resulting from the elimination of two-thirds of the mammalian genera (Keeley and Swift 1995), as accounted for by the "Pleistocene overkill" model (Martin 1973). The present distribution patterns of most chaparral species date to the xerothermic of recent times (8000–3000 BP) (Raven and Axelrod 1978), a period when severe drought promoted the expansion of chaparral. More recently, between 1100 and 1350 A.D., severe droughts in California, lasting one to two centuries at a time (Stine 1994), further contributed to modern landscape patterns.

Evolution of Chaparral Taxa

The evolutionary history is better known for some chaparral taxa than others. Surprisingly, the ubiquitous *Adenostoma fasciculatum* is largely unknown from the fossil record (Axelrod 1973). It likely evolved on the most xeric sites, and this in itself may account for it not being recorded in the fossil record. Most of the broad-sclerophyll species, in more or less present form, date back to middle Miocene, although as constituents of now extinct plant associations (Axelrod 1975). Many herbaceous genera and *Arctostaphylos* and *Ceanothus* underwent rapid speciation during the Pleistocene, in part because new habitats were created by extensive mountain building and exposure of diverse substrates (Raven and Axelrod 1978) and by an increase in the incidence of wildfires. The three shrubby genera, *Adenostoma, Arctostaphylos,* and *Ceanothus,* are among the most widespread in chaparral and are the taxa that have adapted most closely to fire in their production of lignotubers as a normal ontogenetic character and their timing of seedling recruitment to postfire environments. However, *Adenostoma* has done this without speciating and without extraordinarily high genetic variation (Lardner 1985), whereas the latter two genera have radiated extensively, largely by habitat specialization (Fig. 6.5; Cody 1986; Ball et al. 1983; Zedler 1995a).

Hybridization has played a role in the evolution of both of these large shrub genera. All *Ceanothus* species are diploid ($n = 12$), and hybrids within the subgenera are common (McMinn 1944; Nobs 1963; Phillips 1966; Frazier 1993), but the infrequency of crosses between subgenera *Cerastes* and *Ceanothus* (Hannan 1974) suggest two major clades in the genus *Ceanothus.* Hybridization has been implicated in the evolution of both diploid ($2n = 13$) and tetraploid *Arctostaphylos* taxa, and apparently there are no internal subgeneric barriers (Gottlieb 1968; Schmid, Mallory, and Tucker 1968; Keeley 1976; Ball et al. 1983; Kruckeberg 1977; Ellstrand, Lee, J. Keeley, and S. Keeley 1987; Schierenbeck, Stebbins, and Patterson 1992; Keeley, Massihi, Delgadille, and Hirales 1997). Much evidence points to widespread introgression that has led to a highly reticulate pattern of evolution, and thus the lack of distinct clades presents a challenge to understanding evolutionary radiation within this group. Speciation has led to a proliferation of obligate-seeding species, and the forced genetic mixing in each generation has undoubtedly contributed to the observation that these species tend to form morphologically homogenous populations relative to burl-forming taxa. Populations of the latter often comprise distinctly different clones, treated by Wells (1987) as "forms." An apparently widespread taxonomic problem in this group is the challenge of how to treat subspecific taxa, which appear to be of polyphletic origin; a phenomenon well documented for *A. mewukka* (Schierenbeck et al. 1992) and likely to be true for *A. glandulosa* (Keeley et al. 1997; unpublished data).

In *Arctostaphylos*, tetraploid lignotuberous taxa with foliaceous floral bracts were considered by Wells (1987) to be ancestral, whereas diploid obligate-seeding species with reduced floral bracts were considered to be recently derived (Wells 1969). Determination of derived conditions (apomorphies) is, however, problematical, and recent DNA data (Parker and Vasey, unpublished data) do not fully support the subgeneric alignments suggested by Wells (1992). However, the presence of lignotubers in closely related genera suggest that this is indeed a pleisomorphic (ancestral) character and supports the notion that evolution within the group has proceeded toward loss of the burl. Some patterns within the genus are evident. For example, the Baja California manzanita, *A. peninsularis* is a burl-forming shrub in one part of its range and, apparently due to diminished fire frequency, non-burl-forming shrub in another part of its range (Keeley et al. 1992), and a similar pattern is known for *A. patula*. Two other burl-forming species are known to produce non-burl-forming populations along their arid borders (Keeley, unpublished observations). Wells (1969) suggested that loss of the crown-sprouting trait allowed for a more rapid fine-tuning of adaptation to the relatively recent mediterranean climate. Others have suggested that evolution of the obligate-seeding mode was tied to conditions that created large gaps for seedling establishment after fire, and thus this mode was favored along arid borders or in places of infrequent, intense fires (Keeley and Zedler 1978). Under such conditions, allocation of energy to seeds, as opposed to lignotubers, would be adaptive (Keeley and Keeley 1977). However, the energetic cost of lignotubers may not be high, and elimination of that structure does not preclude maintaining the ability to lay down adventitious buds in root and stem material.

Coupling of Demography, Physiology, and Evolution in Chaparral Shrubs

Adenostoma, *Arctostaphylos*, and *Ceanothus* have adapted their reproductive biology to exploit wildfires for seedling recruitment and population expansion (disturbance-dependent recruitment). In contrast, *Quercus*, *Rhamnus*, *Prunus*, and others have not, and these taxa require long fire-free intervals for seedling recruitment (disturbance-free recruitment). These demographic modes are correlated with character syndromes that reflect physiological and morphological divergence (Keeley, in press; F. Davis et al. 1998)

Disturbance-dependent recruitment derives from the ready availability of resources in postfire environments that have placed high selective value on delaying germination to postfire conditions. Summer droughts in these hot, high light environments, however, have imposed strong selection for physiological tolerance of water stress. As a consequence, these disturbance-dependent species have evolved vascular cells more resistant to embolism and greater osmotic tolerance to extremely low water potentials. Selection for rapid growth rates may have selected against developmental patterns that generate adventitious buds and lignotubers, leading to the obligate-seeding mode, and this mode, with the increased frequency of sexual reproduction, may in turn have allowed for a greater fine-tuning of adaptation to drought. Additionally, with enhanced drought tolerance, there is less selective value to resprouting, an adaptation that can be interpreted as a means of maintaining an established root system with access to year-round moisture. Safe sites for recruitment are rare in time, but when they occur they are spatially extensive, putting little premium on mechanisms that enhance spatial dispersal but a high premium on maintenance of deep dormancy, with germination cues to fire.

Disturbance-free recruitment restricts seedling establishment to cooler, lower-light, moister conditions under the shrub canopy. These shrubs are highly susceptible to drought-induced embolism, and thus they avoid summer drought by maintaining year-round access to water by means of deep, massive root systems. This drought-avoidance strategy works well for adults but makes seedling recruitment in a drought-prone environment precarious. Thus, there has been no effective selection for delaying seed germination to postfire conditions, with the result that seeds are neither dormant nor long-lived (Keeley, 1997b). Additionally, safe sites for seedling recruitment are rare, and thus these taxa all have highly attractive animal-dispersed propagules.

Convergent and parallel evolution is evident in the very similar pattern of disturbance-dependent and disturbance-free recruitment in the Mediterranean macchia of Europe. Here, *Cistus*, *Cytissus*, and others are disturbance-dependent, whereas many taxa such as *Quercus*, *Prunus*, *Rhamnus*, and others require firefree conditions for recruitment. Particularly remarkable is the marked degree of similarity between California and Europe in the character syndromes tied to these demographic modes.

REFERENCES

Adams, J. E. 1934. Some observations on two species of *Arctostaphylos*. Madroño 2:147–152.

Alexander, E. B. 1993. Gabbro and its soils. Fremontia 21:8–10.

Allman, V. P. 1952. A preliminary study of the vegetation in an exclosure in the chaparral of the Wasatch Mountains, Utah. Master's thesis, Brigham Young University, Provo, Utah.

Alpert, P., E. A. Newell, C. Chu, J. Glyphis, S. L. Gulmon, D. Y. Hollinger, N. D. Johnson, H. A. Mooney, and G. Puttick. 1985. Allocation to reproduction in the chaparral shrub, *Diplacus aurantiacus*. Oecologia 66:309–316.

Ammirati, J. F. 1967. The occurrence of annual perennial plants on chaparral burns. Master's thesis, San Francisco State University.

Anfuso, R. F. 1982. Fire temperature relationships of *Adenostoma fasciculatum*. Master's thesis, California State University, Los Angeles.

Arnold, K., L. T. Burcham, R. L. Fenner, and R. F. Grab. 1951. Use of fire in land clearing. Calif. Agr. 5:9–11; 5:4–5, 13, 15; 5:11–12; 5:13–15; 5:6, 15.

Avila, G., M. Lajaro, S. Araya, G. Montenegro, and J. Kummerow. 1975. The seasonal cambium activity of Chilean and Californian shrubs. Amer. J. Bot. 62: 473–478.

Axelrod, D. I. 1973. History of the Mediterranean ecosystem in California, pp. 225–277 in F. de Castri and H. A. Mooney (eds.), Mediterranean ecosystems: origin and structure. Springer-Verlag, New York.

Axelrod, D. I. 1975. Evolution and biogeography of Madrean-Tethyan sclerophyll vegetation. Ann. Missouri Bot. Gard. 62:280–334.

Axelrod, D. I. 1978. The origin of coastal sage vegetation, Alto and Baja California. Amer. J. Bot 65:117–131.

Axelrod, D. I. 1981. Holocene climatic changes in relation to vegetation disjunction and speciation. Amer. Nat. 117:847–870.

Axelrod, D. I. 1986. Cenozoic history of some western American pines. Ann. Missouri Bot. Gard. 73:565–641.

Axelrod, D. I., and F. Govean. 1996. An early Pleistocene closed-cone pine forest at Costa Mesa, southern California. Int. J. Plant Sci. 157:323–329.

Baker, G. A., P. W. Rundel, and D. J. Parsons. 1982. Comparative phenology and growth in three chaparral shrubs. Bot. Gaz. 143:94–100.

Baldwin, I. T., L. Staszak-Kozinski, and R. Davidson. 1994. Up in smoke. I. Smoke-derived germination cues for the postfire annual, *Nicotiana attenuata* Torr. Ex. Watson. J. Chem. Ecol. 20:2345–2371.

Ball, C. T., J. Keeley, H. Mooney, J. Seaman, and W. Winner. 1983. Relationship between form, function, and distribution of two *Arctostaphylos* species (Ericaceae) and their putative hybrids. Oecol. Plant. 4: 153–164.

Barbour, M. G., and R. A. Minnich. 1990. The myth of chaparral convergence. Israel J. Bot. 39:435–463.

Barnes, F. S. 1979. Water relations of four species of *Ceanothus*. Master's thesis, San Jose State University, San Jose, Calif.

Barro, S. C., and S. G. Conard. 1987. Use of ryegrass seeding as an emergency revegetation measure in chaparral ecosystems. USDA Forest Service, Pacific Southwest Forest and Range Experiment Station, General Technical Report PSW – 102.

Bartholomew, B. 1970. Bare zone between California

shrub and grassland communities: the role of animals. Science 170:1210–1212.

Bauer, H. L. 1930. On the flora of the Tehachapi Mountains, California. Bull. South. Calif. Acad. Sci. 29:96–99.

Bauer, H. L. 1936. Moisture relations in the chaparral of the Santa Monica Mountains, California. Ecol. Monogr. 6:409–454.

Beatty, S. W. 1987. Origin and role of soil variability in southern California chaparral. Phys. Geogr. 8:1–17.

Bedell, T. E., and H. F. Heady. 1959. Rate of twig elongation of chamise. J. Range Mgt. 12:116–121.

Bentley, J. R., and R. L. Fenner. 1958. Soil temperatures during fires on California foothills: how to recognize post-fire seedsheds. J. For. 56:738.

Beyers, J. L., S. G. Conard, and C. D. Wakeman. 1994. Impacts of an introduced grass, seeded for erosion control, on postfire community composition and species diversity in southern California chaparral, pp. 594–601 in Proceedings of the 12th international conference on fire and forest meteorology. Society of American Foresters, Bethesda, Md.

Biswell, H. H. 1974. Effects of fire on chaparral, pp. 321–364 in T. T. Kozlowski and C. E. Ahlgren (eds.), Fire and ecosystems. Academic Press, New York.

Biswell, H. H., R. D. Taber, W. W. Hedrick, and A. M. Schultz. 1952. Management of chamise brushlands for game in the north coast region of California. Calif. Fish Game 38:453–484.

Bjorndalen, J. E. 1978. The chaparral vegetation of Santa Cruz Island, California. Norwegian J. Bot. 25:255–269.

Bolton, R. B., and R. J. Vogl. 1969. Ecological requirements of *Pseudotsusa macrocarpa* in the Santa Ana Mountains, California. J. For. 69:112–119.

Bond, W. J. 1983. On alpha diversity and the richness of the Cape flora: a study in southern Cape fynbos, pp. 337–356 in F. J. Kruger, D. T. Mitchell, and J. U. M. Jarvis (eds.), Mediterranean-type ecosystems: the role of nutrients. Springer-Verlag, New York.

Bond, W. J. 1987. Regeneration and its importance in the distribution of woody plants. Ph.D. dissertation, University of California, Los Angeles.

Bond, W. J. 1997. Fire and the evolutionary origins of chaparral, p. 17. MEDECOS VIII Conference on Mediterranean type ecosystems. San Diego, CA (abstracts), October 18–20, 1997, Department of Biology, San Diego State University, San Diego, Calif.

Bond, W. J., and J. J. Midgley. 1995. Kill thy neighbour: an individualistic argument for the evolution of flammability. Oikos 73:79–85.

Booker, F. A., W. E. Dietrich, and L. M. Collins. 1995. The Oakland Hills Fire of 20 October 1991: an evaluation of post-fire response, pp. 163–170 in J. E. Keeley and T. Scott (eds.), Brushfires in California: ecology and resource management. International Association of Wildland Fire, Fairfield, Wash.

Borchert, M. 1985. Serotiny and cone-habit variation in populations of *Pinus coulteri* (Pinaceae) in the southern Coast Ranges of California. Madroño 32: 29–48.

Borchert, M. 1989. Postfire demography of *Thermopsis macrophylla* H. A. var, *agnina* J. T. Howell (Fabaceae), a rare perennial herb in chaparral. Amer. Midl. Nat. 122:120–132.

Borchert, M. I., and D. C. Odion. 1995. Fire intensity and vegetation recovery: a review, pp. 91–100 in

J. E. Keeley and T. Scott (eds.), Brushfires in California wildlands: ecology and resource management. International Association of Wildland Fire, Fairfield, Wash.

Borth, W. B. 1986. Drought tolerance and the mycorrhizal association of *Quercus dumosa* Nutt. Master's thesis, San Diego State University, San Diego, Calif.

Bowman, W. D., and S. W. Roberts. 1985. Seasonal and diurnal adjustments in the water relations of three evergreen chaparral shrubs. Ecology 66:738–742.

Boyd, R. S., 1996. Ant-mediated seed dispersal of the rare chaparral shrub *Fremontodendron decumbens* (Sterculiaceae). Madroño 43:299–315.

Boyd, R. S. and L. L. Serafini. 1992. Reproductive attrition in the rare chaparral shrub *Fremontodendron decumbens* Lloyd (Sterculiaceae). Amer. J. Bot. 79:1264–1272.

Bradbury, D. E. 1978. The evolution and persistence of a local sage/chamise community pattern in southern California. Yearbook of the Assn. Pac. Coast Geogr. 40:39–56.

Bradford, D. F. 1976. Space utilization by rodents in *Adenostoma* chaparral. J. Mammology 57:576–579.

Brooks, A. C. 1962. An ecological study of *Cercocarpus montanus* and adjacent communities in part of the Laramie basin. Master's thesis, University of Wyoming, Laramie.

Brown, D. E. 1978. The vegetation and occurrence of chaparral and woodland flora on isolated mountains within the Sonoran and Mojave deserts in Arizona. J. Ariz. Nev. Acad. Sci. 13:7–12.

Brown, H. W. 1958. Gambel oak in west-central Colorado. Ecology 39:317–327.

Brum, G. D. 1975. Floral biology and pollination strategies of *Arctostaphylos glauca* and *A. pringlei* var. *drupaceae* (Ericaceae). Ph.D. dissertation, University of California, Riverside.

Bullock, S. 1982. Reproductive ecology of *Ceanothus cordulatus*. Master's thesis, California State University, Fresno.

Bullock, S. H. 1978. Plant abundance and distribution in relation to types of seed dispersal in chaparral. Madroño 25:104–105.

Bullock, S. H. 1981. Aggregation of *Prunus ilicifolia* (Rosaceae) during dispersal and its effect on survival and growth. Madroño 28:94–95.

Bullock, S. H. 1989. Life history and seed dispersal of the short-lived chaparral shrub *Dendromecon rigida* (Papaveraceae). Amer. J. Bot. 76:1506–1517.

Bullock, S. H. 1991. Herbivory and the demography of the chaparral shrub *Ceanothus greggii* (Rhamnaceae). Madroño 38:63–72.

Burk, J. H. 1978. Seasonal and diurnal water potentials in selected chaparral shrubs. Amer. Midl. Nat. 99: 244–248.

Byrne, R., J. Michaelsen, and A. Soutar. 1977. Fossil charcoal as measure of wildfire frequency in southern California: a preliminary analysis, pp. 361–367 in H. A. Mooney and C. E. Conrad (eds.), Proceedings of the symposium on environmental consequences of fire and fuel management in mediterranean ecosystems. USDA Forest Service, General Technical Report WO-3.

Bytnerowicz, A., P. R. Miller, and D. M. Olszyk. 1987. Dry deposition of nitrate, ammonium and sulfate to a *Ceanothus crassifolius* canopy and surrogate surfaces. Atmos. Envir. 21:1749–1757.

Bytnerowicz, A., and M. E. Fenn. 1996. Nitrogen deposition in California forests: a review. Envir. Poll. 92: 127–146.

Cable, D. R. 1957. Recovery of chaparral following burning and seeding in central Arizona. USDA Forest Service, Rocky Mountain Forest and Range Experiment Station research note RM-28.

Calkin, H. W., and R. W. Pearcy. 1984. Seasonal progressions of tissue and cell water relations parameters in evergreen and deciduous perennials. Plant Cell Environ. 7:347–352.

Callaway, R. M., and C. M. D'Antonio. 1991. Shrub facilitation of coast live oak establishment in central California. Madroño 38:158–169.

Callaway, F. M., and F. W. Davis. 1993. Vegetation dynamics, fire, and the physical environment in coastal central California. Ecology 74:1567–1578.

Carlquist, S. 1980. Further concepts in ecological wood anatomy, with comments on recent work in wood anatomy and evolution. Aliso 9:499–553.

Carlquist, S. 1985. Vasicentric tracheids as a drought survival mechanism in the flora of southern California and similar regions. Aliso 11:37–86.

Carlquist, S. 1989. Adaptive wood anatomy of chaparral shrubs, pp. 25–35 in S. C. Keeley (ed.), The California chaparral: paradigms reexamined. Natural History Museum of Los Angeles County, Los Angeles, Science Series No. 34.

Carlquist, S., and D. A Hoekmann. 1985. Ecological wood anatomy of the woody southern California flora. Intern. Assoc. Wood Anat. Bull. 6:319–347.

Carmichael, R. S., O. D. Knipe, C. P. Pose, and W. W. Brady. 1978. Arizona chaparral: plant associations and ecology. USDA Forest Service, Rocky Mountain Forest and Range Experiment Station Research Paper RM-202.

Chou, C.-H., and C. H. Muller. 1972. Allelopathic mechanisms of *Arctostaphylos* var. *zacaensis*. Amer. Midl. Nat. 88:324–347.

Christensen, E. M. 1949. The ecology and geographic distribution of oak (*Quercus gambelii*) in Utah. Master's thesis, University of Utah, Salt Lake City.

Christensen, N. L. 1973. Fire and the nitrogen cycle in California chaparral. Science 181:66–68.

Christensen, N. L., and C. H. Muller. 1975a. Effects of fire on factors controlling plant growth in *Adenostoma* chaparral. Ecol. Monogr. 45:29–55.

Christensen, N. L., and C. H. Muller. 1975b. Relative importance of factors controlling germination and seedling survival in *Adenostoma* chaparral. Amer. Midl. Nat. 93:71–78.

Clark, H. W. 1937. Association types in the north Coast Ranges of California. Ecology 18:214–230.

Cody, M. L. 1986. Diversity, rarity, and conservation in Mediterranean-climate regions, pp. 122–152 in M. E. Soulé (ed.), Conservation biology. Sinauer, Sunderland, England.

Collett, R. 1992. Frost report: towns and temperatures. Pac. Hort. 53:6–9.

Comstock, J., and J. R. Ehleringer. 1986. Photoperiod and photosynthetic capacity in *Lotus scoparius*. Pl. Cell Envir. 9:609–612.

Comstock, J. P., and B. E. Mahall. 1985. Drought and changes in leaf orientation for two California chaparral shrubs; *Ceanothus megacarpus* and *Ceanothus crassifolius*. Oecologia 65:531–535.

Conard, S. G., J. L. Beyers, and P. M. Wohlgemuth. 1995. Impacts of postfire grass seeding on chaparral systems – What do we know and where do we go from here?, pp. 149–161 in J. E. Keeley and T. Scott (eds.), Brushfires in California: ecology and resource management. International Association of Wildland Fire, Fairfield, Wash.

Conrad, S. G., and S. R. Radosevich. 1982. Post-fire succession in white fir *Abies concolor* vegetation of the northern Sierra Nevada. Madroño 29:42–56,

Conrad, S. G., and D. R. Weise. In press. Management of fire regime, fuels, and fire effects in southern California chaparral: lessons from the past and thoughts for the future. Tall Timb. Fire Ecol. Conf.

Conrad, C. E., and L. F. DeBano. 1974. Recovery of southern California chaparral. American Society of Civil Engineers national meeting on water resources engineering, meeting reprint 2167.

Cooper, W. S. 1922. The broad-sclerophyll vegetation of California. An ecological study of the chaparral and its related communities. Carnegie Institution of Washington publication 319, Washington, D.C.

Corbett, E. 5., and L. R. Green. 1965. Emergency revegetation to rehabilitate burned watersheds in southern California. USDA Forest Service, Pacific Southwest Forest and Range Experiment Station-Research Paper PSW-22.

Cowden, B., and G. Waters. 1992. Frost report: plants and people. Pac. Hort. 53:10–15.

Daubenmire, R. F. 1943. Vegetational zonation in the Rocky Mountains. Bot. Rev. 9:325–393.

Davey, J. R. 1982. Stand replacement in *Ceanothus crassifolius*. Master's thesis, California State Polytechnic University, Pomona.

Davis, C. B. 1972. Comparative ecology of six members of the *Arctostaphylos andersonii* complex. Ph.D. dissertation, University of California, Davis.

Davis, C. B. 1973. "Bark striping" in *Arctostaphylos* (Ericaceae). Madroño 22:145–149.

Davis, F. W., M. I. Borchert, and D. C. Odion. 1989. Establishment of microscale vegetation pattern in maritime chaparral after fire. Vegetatio 84:53–67.

Davis, F. W., and D. A. Burrows. 1993. Modeling fire regime in Mediterranean landscapes, pp. 247–259, in S. A. Levin, T. M. Powell, and J. H. Steele (eds.), Patch dynamics. Springer-Verlag, New York.

Davis, F. W., D. E. Hickson, and D. C. Odion. 1988. Composition of maritime chaparral related to fire history and soil, Burton Mesa, Santa Barbara County, California. Madroño 35:169–195.

Davis, F. W., and J. Michaelsen. 1995. Sensitivity of fire regime in chaparral ecosystems to climate change, pp. 435–456 in J. Moreno and W. C. Oechel (eds.), Global climate change in Mediterranean-type ecosystems. Springer-Verlag, New York.

Davis, F. W., P. A. Stine, and D. M. Stoms. 1994. Distribution and conservation status of coastal sage scrub in southwestern California. J. Veg. Sci. 5:743–756.

Davis, F. W., P. A. Stine, D. M. Stoms, M. I. Borchert, and A. D. Hollander. 1994. Gap analysis of the actual vegetation of California. 1. The southwestern region. Madroño 42:40–78.

Davis, J. N. 1967. Some effects of deer browsing on chamise sprouts after fire. Amer. Midl. Nat. 77:234–238.

Davis, J. N. 1976. Ecological investigation in *Cercocarpus ledifolius* Null. communities of Utah. Master's thesis, Brigham Young University, Provo, Utah.

Davis, S. D. 1989. Patterns in mixed chaparral stands: differential water status and seedling survival during summer drought, pp. 97–105 in S. C. Keeley (ed.), The California chaparral: paradigms reexamined. Natural History Museum of Los Angeles County, Los Angeles, Science Series No. 34.

Davis, S. D. 1991. Lack of niche differentiation in adult shrubs implicates the importance of the regeneration niche. Trends in Ecol. Evol. 9:272–274.

Davis, S. D., K. J. Kolb, and K. P. Barton. In press. Ecophysiological processes and demographic patterns in the structuring of California chaparral. In P. W. G. Rundel, G. Montenegro, and F. Jaksic (eds.), Landscape Disturbance and Biodiversity in Mediterranean Type Ecosystems. Springer-Verlag, New York, N.Y.

Davis, S. D., and H. A. Mooney. 1985. Comparative water relations of adjacent California shrub and grassland communities. Oecologia 66:522–529.

Davis, S. D., and H. A. Mooney. 1986a. Tissue water relations of four co-occurring chaparral shrubs. Oecologia 70:527–535.

Davis, S. D., and H. A. Mooney. 1986b. Water use patterns of four co-occurring chaparral shrubs. Oecologia 70:172–177.

DeBano, L. F., and C. E. Conrad. 1978. The effect of fire on nutrients in a chaparral ecosystem. Ecology 59:489–497.

DeBano, L. F., R. M. Rice, and C. E. Conrad. 1979. Soil heating in chaparral fires: effects on soil properties, plant nutrients, erosion, and runoff. USDA Forest Service, Pacific Southwest Forest and Range Experiment Station, Research Paper PSW-145.

DeSimone, S. A., and J. H. Burk. 1992. Local variation in floristics and distributional factors in California coastal sage scrub. Madroño 39:170–188.

DeSouza, J., P. A. Silka, and S. D. Davis. 1986. Comparative physiology of burned and unburned *Rhus laurina* after chaparral wildfire. Oecologia 71:63–68.

Detling, L. E. 1961. The chaparral formation of southeastern Oregon with consideration of its postglacial history. Ecology 42:348–357.

Dobzhansky, T. 1953. Natural hybrids of two species of *Arctostaphylos* in the Yosemite region of California. Heredity 7:73–79.

Dunn, E. L. 1975. Environmental stresses and inherent limitations affecting CO_2 exchange in evergreen sclerophylls in mediterranean climates, pp. 159–181 in D. M. Gates and R. B. Schmeri (eds.), Perspectives in biophysical ecology. Springer-Verlag, New York.

Dunn, E. L., F. M. Shropshire, L. C. Song, and H. A. Mooney. 1976. The water factor and convergent evolution in mediterranean-type vegetation, pp. 492–505 in O. L. Lange, L. Kappen, and E.-D. Schultz (eds.), Water and plant life. Springer-Verlag, New York.

Eastmond, R. J. 1968. Vegetational changes in a mountain brush community of Utah during 18 years. Master's thesis, Brigham Young University, Provo, Utah.

Ehleringer, J. R., and J. Comstock. 1987. Leaf absorptance and leaf angle: mechanisms for stress avoidance, pp. 55–76 in J. D. Tenhunen, F. M. Catarino, O. L. Lange, and W. C. Oechel (eds.), Plant response to stress: functional analysis in Mediterranean ecosystems. Springer-Verlag, Berlin.

Ehlcringer, J. R., and J. Comstock. 1989. Stress tolerance and adaptive variation in leaf absorptance and leaf angle, pp. 21–24 in S. C. Keeley (ed.), The California chaparral: paradigms reexamined. Natural History Museum of Los Angeles County, Los Angeles, Science Series No. 34.

Ellis, B. A. 1982. Asymbiotic nitrogen (N₂) fixation and nitrogen content of bulk precipitation in southern California chaparral. Master's thesis, San Diego State University, San Diego, California.

Ellis, B. A., and J. Kummerow. 1989. The importance of N_2 fixation in *Ceanothus* seedlings in early and postfire chaparral, pp. 115–116 in S. C. Keeley (ed.), The California chaparral: paradigms reexamined. Natural History Museum of Los Angeles County, Los Angeles, Science Series No. 34.

Ellstrand, N. C., J. M. Lee, J. E. Keeley, and S. C. Keeley. 1987. Ecological isolation and introgression: biochemical confirmation of introgression in an *Arctostaphylos* (Ericaceae) population. Acta Oecol. 8: 299–308.

Epling, C., and H. Lewis. 1942. The centers of distribution of the chaparral and coastal sage. Amer. Midl. Nat. 27:445–462.

Evans, R. A., H. H. Biswell, and D. E. Palmquist. 1987. Seed dispersal in *Ceanothus cuneatus* and *C. leucodermis* in a Sierran oak-woodland savanna. Madroño 34:283–293.

Fenn, M. E., M. A. Poth, P. H. Dunn, and S. C. Barro. 1993. Microbial N and biomass respiration and N mineralization in soils beneath two chaparral species along a fire-induced age gradient. Soil Biol. Biochem. 25:457–466.

Field, C., J. Merino, and H. A. Mooney. 1983. Compromises between water-use efficiency and nitrogen-use efficiency in five species of California evergreens. Oecologia 60:384–389.

Field, C. B., and S. D. Davis. 1989. Physiological ecology, pp. 154–164 in S. C. Keeley (ed.), The California chaparral: paradigms reexamined. Natural History Museum of Los Angeles County, Los Angeles, Science Series No. 34.

Fishbeck, K., and J. Kummerow. 1977. Comparative wood and leaf anatomy, pp. 148–161 in N. J. W. Thrower and D. E. Bradbury (eds.), Chile California mediterranean scrub atlas: a comparative analysis. Dowden, Hutchinson and Boss, Stroudsburg, Pa.

Frazier, C. K. 1993. An ecological study of hybridization between chaparral shrubs of contrasting life-history strategies. Master's thesis, San Diego State University, San Diego, Calif.

Fulton, R. E., and F. L. Carpenter. 1979. Pollination, reproduction, and fire in *Arctostaphylos*. Oecologia 38: 147–157.

Furman, T. E. 1959. The structure of the root nodules of *Ceanothus sanguineus* and *Ceanothus velutinus*, with special reference to the endophyte. Amer. J. Bot. 46: 698–703.

Gamon, J. A., C. B. Field, M. L. Goulden, K. L. Griffin, A. E. Hartley, G. Joel, J. Peñuelas, and R. Valentini. 1995. Relationships between NDVI, canopy structure, and photosynthesis in three Californian vegetation types. Ecol. Appl. 5:28–41.

Gause, G. W. 1966. Silvical characteristics of bigcone Douglas-fir *Pseudotsuga macrocarpa* (Vasey) Mayr. USDA Forest Service, Pacific Southwest Forest and Range Experiment Station research paper PSW-39.

Gauss, N. M. 1964. Distribution of selected plant species in a portion of the Santa Monica Mountains, California, on the basis of site. Master's thesis, University of California, Los Angeles.

Gautier, C. R. 1983. Sedimentation in burned chaparral watersheds: is emergency revegetation justified? Water Resources Bull. 19:793–802.

Gibbens, R. P., and A. M. Schultz. 1963. Brush manipulation on a deer winter range. Calif. Fish Game 49: 95–118.

Gigon, A. 1979. CO₂-gas exchange, water relations and convergence of mediterranean shrub-types from California and Chile. Oecologia Plantarum 14:129–150.

Gill, D. S. 1985. A quantitative description of the phenology of an evergreen and a deciduous shrub species with reference to temperature and water relations in the Santa Ynez Mountains, Santa Barbara County, California. Master's thesis, University of California, Santa Barbara.

Gill, D. S., and B. E. Mahall. 1986. Quantitative phenology and water relations of an evergreen and a deciduous chaparral shrub. Ecol. Monogr. 56:127–143.

Gottlieb, L. D. 1968. Hybridization between *Arctostaphylos viscida* and *A. canescens* in Oregon. Brittonia 20: 83–93.

Graves, G. W. 1932. Ecological relationships of *Pinus sabiniana*. Bot. Gaz. 94:106–133.

Gray, J. T. 1982. Community structure and productivity in *Ceanothus* chaparral and coastal sage scrub of southern California. Ecol. Monogr. 52:415–435.

Gray, J. T. 1983. Nutrient use by evergreen and deciduous shrubs in southern California. I. Community nutrient cycling and nutrient-use efficiency. J. Ecol. 71:21–41.

Gray, J. T., and W. H. Schlesinger. 1981. Nutrient cycling in mediterranean type ecosystems, pp. 259–285 in P. C. Miller (ed.), Resource use by chaparral and matorral. Springer-Verlag, New York.

Gray, J. T., and W. H. Schlesinger. 1983. Nutrient use by evergreen and deciduous shrubs in southern California. II. Experimental investigations of the relationship between growth, nitrogen uptake and nitrogen availability. J. Ecol. 71:43–56.

Greenlee, J. M., and J. H. Langenheim. 1980. The history of wildfires in the region of Monterey Bay. Unpublished report, California State Department of Parks and Recreation.

Griffin, J. R. 1978. Maritime chaparral and endemic shrubs of the Monterey Bay Region, California. Madroño 25:65–81.

Griffin, J. R. 1982. Pine seedlings, native ground cover, and *Lolium multiflorum* on the marble-cone burn, Santa Lucia Range, California. Madroño 29:177–188.

Gulmon, S. L. 1983. Carbon and nitrogen economy of *Diplacus aurantiacus* a Californian mediterranean climate drought-deciduous shrub, pp. 167–176 in F. Kruger, D. T. Mitchell, and J. Jarvis (eds.), Mediterranean-type ecosystems. The role of nutrients. Springer-Verlag, New York.

Guntle, G. R. 1974. Correlation of annual growth in *Ceanothus crassifolius* Torr. and *Arctostaphylos glauca* Lindl. to annual precipitation in the San Gabriel Mountains. Master's thesis, California State Polytechnic University, Pomona.

Hadley, E. B. 1961. Influence of temperature and other factors on *Ceanothus megacarpus* seed germination. Madroño 16:132–138.

Haidinger, T. L., and J. E. Keeley. 1993. Role of high fire frequency in destruction of mixed chaparral. Madroño 40:141–147.

Haines, L. 1941. Variation in *Yucca whipplei*. Madroño 6: 33–45.

Hall, H. M. 1903. Botanical survey of San Jacinto Mountains. Univ. Calif. Pub. Bot. 1:1–140.

Halligan, J. 1974. Relationship between animal activity and bare areas associated with California sagebrush in annual grassland. J. Range Mgt. 27:358–363.

Halligan, J. P. 1975. Toxic terpenes from Artemisia californica. Ecology 56:999–1003.

Hanes, T. L. 1965. Ecological studies on two closely related chaparral shrubs in southern California. Ecol. Monogr. 35:213–235.

Hanes, T. L. 1971. Succession after fire in the chaparral of southern California. Ecol. Monogr. 41:27–52.

Hanes, T. L. 1977. California chaparral, pp. 417–470 in M. G. Barbour and J. Major (eds.), Terrestrial vegetation of California. Wiley, New York.

Hannan, L. L. 1974. An intersectional hybrid in *Ceanothus*. Madroño 22:402.

Harrison, A. T. 1971. Temperature related effects on photosynthesis in *Heteromeles arbutifolia* M. Roem. Ph.D. dissertation, Stanford University.

Harrison, A. T., E. Small, and H. A. Mooney. 1971. Drought relationships and distribution of two mediterranean-climate California plant communities. Ecology 52:869–875.

Harrison, S. 1997. How natural habitat patchiness affects the distribution of diversity in Californian serpentine chaparral. Ecology 78:1898–1906.

Harvey, R. A., and H. A. Mooney. 1964. Extended dormancy of chaparral shrubs during severe drought. Madroño 17:161–163.

Hastings, S. J., W. C. Oechel, and N. Sionit. 1989. Water relations and photosynthesis of chaparral resprouts and seedlings following fire and hand clearing, pp. 107–113 in S. C. Keeley (ed.), The California chaparral: paradigms reexamined. Natural History Museum of Los Angeles County, Los Angeles, Science Series No. 34.

Hayward, C. L. 1948. Biotic communities of the Wasatch chaparral, Utah. Ecol. Monogr. 18:473–506.

Heady, H. F., T. C. Foin, M. M. Hektner, D. W. Taylor, M. G. Barbour, and W. J. Barry. 1977. Coastal prairie and northern coastal scrub, pp. 733–757 in M. G. Barbour and J. Major (eds.), Terrestrial vegetation of California. Wiley, New York.

Hedrick, D. W. 1951. Studies on the succession and manipulation of chamise brushlands in California. Ph.D. dissertation, Texas A&M College, College Station.

Hellmers, H., J. F. Bonner, and J. M. Kelleher. 1955. Soil fertility: a watershed management problem in the San Gabriel Mountains of southern California. Soil Sci. 80:189–197.

Hellmers, H., J. S. Horton, G. Juhren, and J. O'Keefe. 1955. Root systems of some chaparral plants in southern California. Ecology 36:667–678.

Hellmers, H., and J. M. Kelleher. 1959. *Ceanothus leucodermis* and soil nitrogen in southern California mountains. For. Sci. 5:275–278.

Heusser, L. 1978. Pollen in the Santa Barbara Basin, California: a 12,000-yr record. Geol. Soc. Amer. Bull. 89: 673–678.

Hickman, J. C. (ed.). 1993. The Jepson manual. University of California Press, Los Angeles.

Hiehle, J. L. 1961. Measurement of browse growth and utilization. Calif. Fish Game 50:148–151.

Hilbert, D. W., and A. Larigauderie. 1988. Patterns of

chaparral productivity and decline explained by plant, population, and ecosystem mechanisms, pp. 489–495 in F. di Castri, Ch. Floret, S. Rambal, and J. Roy (eds.), Time scales and water stress. Proceedings of the 5th international conference on Mediterranean ecosystems (MEDECOS V). International Union of Biological Sciences, Paris.

Hobbs, R. J., and H. A. Mooney, 1985. Vegetative regrowth following cutting in the shrub *Baccharis pilularis ssp. consanguinea* (DC) C. B. Wolf. Amer. J. Bot. 72:514–519.

Hochberg, M. C. 1980. Factors affecting leaf size of the chaparral on the California islands, pp. 189–206 in D. M. Power (ed.), The California Islands: proceedings of a multidisciplinary symposium. Santa Barbara Museum of Natural History.

Holland, R. F. 1986. Preliminary descriptions of the terrestrial natural communities of California. California Department of Fish and Game, Nongame Heritage Program, Sacramento.

Horn, S. 1984. Bird dispersal of toyon (*Heteromeles arbutifolia*). Master's thesis, California State University, Hayward.

Horton, J. S. 1960. Vegetation types of the San Bernardino Mountains. USDA Forest Service, Pacific Southwest Forest and Range Experiment Station, Technical Paper 44.

Horton, J. S., and C. J. Kraebel. 1955. Development of vegetation after fire in the chamise chaparral of southern California. Ecology 36:244–262.

Howe, C. F. 1982. Death of chamise *Adenostoma fasciculatum* shrubs after fire as a result of herbivore browsing. Bull. South. Calif. Acad. Sci. 80:138–143.

Howell, J. T. 1945. Concerning stomata on leaves in *Arctostaphylaos*. Wasmann Collector 6:57–65.

Hubbard, T. W. 1986. Stand age and growth dynamics in chamise chaparral. Master's thesis, San Diego State University, San Diego, California.

Jacks, P. M. 1984. The drought tolerance of *Adenostoma fasciculatum* and *Ceanothus crassifolius* seedlings and vegetation change in the San Gabriel chaparral. Master's thesis, San Diego State University.

James, S. M. 1984. Lignotubers and burls—their structure, function and ecological significance in Mediterranean ecosystems. Bot. Rev. 50:225–266.

Jarbeau, J. A., F. W. Ewers, and S. D. Davis. 1995. The mechanism of water-stress-induced embolism in two species of chaparral shrubs. Plant Cell Envir. 18:189–196.

Jensen, D. B., M. Torn, and J. Harte. 1990. In our own hands: a strategy for conserving biological diversity in California. University of California, Berkeley, California Policy Seminar Research Report.

Jepson, W. L. 1928. Biological peculiarities of California flowering plants, part I. Madroño 1:190–192.

Jones, C. S. and W. H. Schlesinger. 1980. *Emmenanthe penduliflora* (Hydrophyllaceae): further consideration of germination response. Madroño 27:122–125.

Jones, D. P., and R. C. Graham. 1993. Water-holding characteristics of weathered granitic rock in chaparral and forest ecosystems. Soil Sci. Soc. Amer. J. 57: 256–261.

Jones, M. D., and H. M. Laude. 1960. Relationships between sprouting in chamise and physiological condition of the plant. J. Range Mgt. 13:210–214.

Jow, W., S. H. Bullock, and J. Kummerow. 1980. Leaf turnover rates of *Adenostoma fasciculatum* (Rosaceae). Amer. J. Bot. 67:256–261.

Kaminsky, R. 1981. The microbial origin of the allelopathic potential of *Adenostoma fasciculatum* H & A. Ecol. Monogr. 51:365–382.

Keeley, J. E. 1975. The longevity of nonsprouting *Ceanothus*. Amer. Midl. Nat. 93:504–507.

Keeley, J. E. 1976. Morphological evidence of hybridization between *Arctostaphylos glauca* and *A. pungens* (Ericaceae). Madroño 23:427–434.

Keeley, J. E. 1977. Seed production, seed populations in soil, and seedling production after fire for two congeneric pairs of sprouting and non-sprouting chaparral shrubs. Ecology 58:820–829.

Keeley, J. E. 1982. Distribution of lightning and man-caused wildfires in California, pp. 431–437 in C. E. Conrad and W. C. Oechel (eds.), Proceedings of the symposium on dynamics and management of mediterranean-type ecosystems. USDA Forest Service, Pacific Southwest Forest and Range Experiment Station, General Technical Report PSW-58.

Keeley, J. E. 1986. Resilience of Mediterranean shrub communities to fire, pp. 95–112 in B. Dell, A. J. M. Hopkins, and B. B. Lamont (eds.), Resilience in mediterranean-type ecosystems. Junk, Dordrecht.

Keeley, J. E. 1987a. Role of fire in the seed germination of woody taxa in California chaparral. Ecology 68: 434–443.

Keeley, J. E. 1987b. Ten years of change in seed banks of the chaparral shrubs, *Arctostaphylos glauca* and *A. glandulosa*. Amer. Midl. Nat. 117:446–448.

Keeley, J. E. 1987c. Fruit production patterns in the chaparral shrub *Ceanothus crassifolius*. Madroño 34: 273–282.

Keeley, J. E. 1990a. Demographic structure of California black walnut (*Juglans californica*; Juglandaceae) woodlands in southern California. Madroño 37:237–248.

Keeley, J. E. 1990b. The California valley grassland, pp. 2–23 in A. A. Schoenherr (ed.), Endangered plant communities of southern California. Southern California Botanists, Fullerton, Spec. Publ. No. 3.

Keeley, J. E. 1991. Seed germination and life history syndromes in the California chaparral. Bot. Rev. 57: 81–116.

Keeley, J. E. 1992a. A California's view of fynbos, pp. 372–388 in R. Cowling (ed.), The ecology of fynbos. Nutrients, fire and diversity. Oxford University Press, Cape Town, South Africa.

Keeley, J. E. 1992b. Demographic structure of California chaparral in the long-term absence of fire. J. Veg. Sci. 3:79–90.

Keeley, J. E. 1992c. Recruitment of seedlings and vegetative sprouts in unburned chaparral. Ecology 73: 1194–1208.

Keeley, J. E. 1992d. Temporal and spatial dispersal syndromes, pp. 251–256 in C. A. Thanos (ed.), MEDE-COS VI. Proceedings of the 6th international conference on Mediterranean climate ecosystems, "Plant–animal interactions in Mediterranean-type ecosystems." University of Athens, Greece.

Keeley, J. E. 1993. Utility of growth rings in the age determination of chaparral shrubs. Madroño 40:1–14.

Keeley, J. E. 1995a. Bibliography on fire ecology and general biology of Mediterranean-type ecosystems. Vol. I: California. International Association of Wildland Fire, Fairfield, Wash.

Keeley, J. E. 1995b. Future of California floristics and systematics: wildfire threats to the California flora. Madroño 42:175–179.

Keeley, J. E. 1996. Postfire vegetation recovery in the Santa Monica Mountains under two alternative management programs. Bull. So. Calif. Acad. Sci. 95:103–119.

Keeley, J. E. 1997a. Postfire ecosystem recovery and management: The October 1993 large fire episode in California, pp. 1–22, in J. Moreno (ed.), Impact of large catastrophic wildfires in mediterranean ecosystems. Backhuys Publishers, The Netherlands.

Keeley, J. E. 1997b. Seed longevity of non-fire recruiting chaparral shrubs. Four seasons 10(3): 36–42.

Keeley, J. E. In press. Coupling of demography, physiology and evolution in chaparral shrubs. In P. W. Rundel, G. Montenegro, and F. Jaksic (eds.), Landscape disturbance and biodiversity in mediterranean type ecosystems. Springer-Verlag, New York, N.Y.

Keeley, J. E., and W. J. Bond. 1997. Convergent seed germination in South African fynbos and California chaparral. Plant Ecol. 133: 153–167.

Keeley, J. E., M. Carrington, and S. Trnka. 1995. Overview of management issues raised by the 1993 wildfires in southern California, pp. 83–89 in J. E. Keeley and T. Scott (eds.), Brushfires in California: ecology and resource management. International Association of Wildland Fire, Fairfield, Wash.

Keeley, J. E., and C. J. Fotheringham. 1997. Trace gas emissions in smoke-induced germination. Science 276:1248–1250.

Keeley, J. E., and C. J. Fotheringham. 1998a. Mechanism of smoke-induced germination in a postfire annual. J. Ecol. 86:27–36.

Keeley, J. E., and C. J. Fotheringham. 1998b. Smoke-induced seed germination in Californian chaparral. Ecology.

Keeley, J. E., and R. L. Hays. 1976. Differential seed predation on two species *Arctostaphylos* (Ericaceae). Oecologia 24:71–81.

Keeley, J. E., and S. C. Keeley. 1977. Energy allocation patterns of sprouting and non-sprouting species of *Arctostaphylos* in the California chaparral. Amer. Midl. Nat. 98:1–10.

Keeley, J. E., and S. C. Keeley. 1981. Postfire regeneration of California chaparral. Amer. J. Bot. 68:524–530.

Keeley, J. E., and S. C. Keeley. 1984. Postfire recovery of California coastal sage scrub. Amer. Midl. Nat. 111: 105–117.

Keeley, J. E., and S. C. Keeley. 1988a. Chaparral, pp. 165–207 in M. G. Barbour and W. D. Billings (eds.), North American terrestrial vegetation. 1st ed. Cambridge University Press, Cambridge.

Keeley, J. E., and S. C. Keeley. 1988b. Temporal and spatial variation in fruit production by California chaparral shrubs, pp. 457–463 in F. di Castri, Ch. Floret, S. Rambal, and J. Roy (eds.), Time scales and water stress. Proceedings of the 5th international conference on Mediterranean ecosystems (MEDECOS V). International Union of Biological Sciences, Paris.

Keeley, J. E., and S. C. Keeley. 1989. Allelopathy and the fire induced herb cycle, pp. 65–72 in S. C. Keeley (ed.), The California chaparral: paradigms reexamined. Natural History Museum of Los Angeles County, Los Angeles, Science Series No. 34.

Keeley, J. E., S. C. Keeley, and D. A. Ikeda. 1986. Seed predation by yucca moths on semelparous, iteroparous and vegetatively reproducing subspecies of

Yucca whipplei (Agavaceae). Amer. Midl. Nat. 115:1–9.

Keeley, J. E., A. Masssihi, J. Delgadillo, S. A. Hirales. 1997. *Arctostaphylos incognita*, a new species and its phenetic relationship to other manzanitas of Baja California. Madroño.

Keeley, J. E., A. Massihi, and R. Goar. 1992. Growth form dichotomy in subspecies of *Arctostaphylos peninsularis* from Baja California. Madroño 39:285–287.

Keeley, J. E., B. A. Morton, A. Pedrosa, and P. Trotter. 1985. The role of allelopathy, heat and charred wood in the germination of chaparral herbs and suffrutescents. J. Ecol. 73:445–458.

Keeley, J. E., and C. C. Swift. 1995. Biodiversity and ecosystem functioning in Mediterranean-climate California, pp. 121–183 in G. W. Davis and D. M. Richardson (eds.), Biodiversity and function in mediterranean-type ecosystems. Springer-Verlag, New York.

Keeley, J. E., and P. H. Zedler. 1978. Reproduction of chaparral shrubs after fire: a comparison of sprouting and seeding strategies. Amer. Midl. Nat. 99:142–161.

Keeley, J. E., and P. H. Zedler. 1998. Evolution of life history patterns in pines, pp. in D. M. Richardson and R. Cowling (eds.), Ecology and biogeography of *Pinus*. Cambridge University Press, Cambridge.

Keeley, J. E., P. H. Zedler, C. A. Zammit, and T. J. Stohlgren. 1989. Fire and demography, pp. 151–153 in S. C. Keeley (ed.), The California chaparral. Paradigms reexamined. Natural History Museum of Los Angeles County, Los Angeles, Science Series No. 34.

Keeley, S. C. 1977. The relationship of precipitation to post-fire succession in the southern California chaparral, pp. 387–390 in H. A. Mooney and C. E. Conrad (eds.), Proceedings of the symposium on environmental consequences of fire and fuel management in mediterranean ecosystems. USDA Forest Service, General Technical Report WO-3.

Keeley, S. C., and A. W. Johnson. 1977. A comparison of the pattern of herb and shrub growth in comparable sites in Chile and California Amer. Midl. Nat. 97:120–132.

Keeley, S. C., J. E. Keeley, S. M. Hutchinson, and A. W. Johnson. 1981. Postfire succession of the herbaceous flora in southern California chaparral. Ecology 62:1608–1621.

Kelly, V. R., and V. T. Parker. 1990. Seed bank survival and dynamics in sprouting and nonsprouting *Arctostaphylos* species. Amer. Midl. Nat. 124:114–123.

Kelly, V. R., and V. T. Parker. 1991. Percentage seed set, sprouting habit and ploidy in *Arctostaphylos* (Ericaceae). Madroño 38:227–232.

Kirkpatrick, J. B., and C. F. Hutchinson. 1980. The environmental relationships of California coastal sage scrub and some of its component communities and species. J. Biogeography 7:23–28.

Knipe, O. D. 1985. Effects of reducing shrub cover on moisture stress in mountain mahogany. J. Soil Water Conserv. 40:445–447.

Koenigs, R. L., W. A. Williams, and M. B. Jones. 1982. Factors affecting vegetation on a serpentine soil. I. Principal components analysis of vegetation data. Hilgardia 50:1–14.

Kolb, K. J., and S. D. Davis. 1994. Drought tolerance and xylem embolism in co-occurring species of coastal sage and chaparral. Ecology 75:648–659.

Krause, D., and J. Kummerow. 1977. Xeromorphic structure and soil moisture in the chaparral. Oecol. Plant. 12:133–148.

Kruckeberg, A. R. 1954. The ecology of serpentine soils. III. Plant species in relation to serpentine soils. Ecology 35:267–274.

Kruckeberg, A. R. 1969. Soil diversity and the distribution of plants, with examples from western North America. Madroño 20:129–154.

Kruckeberg, A. R. 1997. Manzantia (*Arctostaphylos*) hybrids in the Pacific northwest: effects of human and natural disturbance. Syst. Bot. 2:233–250.

Kruckeberg, A. R. 1984. California serpentines: flora, vegetation, geology, soils, and management problems. University of California Press, Los Angeles.

Kruger, F. J., D. T. Michell, and J. U. M. Jarvis (eds.). 1983. Mediterranean-type ecosystems. The role of nutrients. Springer-Verlag, New York.

Kummerow, J. 1981. Structure of roots and root systems, pp. 269–288 in F. di Castri, D. W. Goodall, and R. L. Specht (eds.), Ecosystems of the world. II. Mediterranean-type shrublands. Elsevier Scientific, New York.

Kummerow, J., J. V. Alexander, J. W. Neel, and K. Fishbeck. 1978. Symbiotic nitrogen fixation in *Ceanothus* roots. Amer. J. Bot. 65:63–69.

Kummerow, J., and B. A. Ellis. 1989. Structure and function in chaparral shrubs, pp. 140–150 in S. C. Keeley (ed.), The California chaparral: paradigms reexamined. Natural History Museum of Los Angeles County, Los Angeles, Science Series No. 34.

Kummerow, J., D. Krause, and W. Jow. 1977. Root systems of chaparral shrubs. Oecologia 29:163–177.

Kummerow, J., and R. Mangan. 1981. Root systems in *Quercus dumosa* dominated chaparral in southern California. Oecol. Plant. 2:177–188.

Kummerow, J., G. Montenegro, and D. Krause. 1981. Biomass, phenology and growth, pp. 69–96 in P. C. Miller (ed.), Resource use of chaparral and matorral. Springer-Verlag, New York.

Kunzler, L. M., and K. T. Harper. 1980. Recovery of gambel oak after fire in central Utah. Great Basin Nat. 40:127–130.

Langan, S. J., F. W. Ewers, and S. D. Davis. 1997. Xylem dysfunction caused by water stress and freezing in two species of co-occurring chaparral shrubs. Plant Cell Envir. 20:425–437.

Lardner, M. A. 1985. Genetic and morphological variation in *Adenostoma fasciculatum*. Master's thesis, University of California, Riverside.

Larigauderie, A., B. A. Ellis, J. N. Mills, and J. Kummerow. 1991. The effect of root and shoot temperatures on growth of *Ceanothus greggii* seedlings. Ann. Bot. 67:97–101.

Larigauderie, A., T. W. Hubbard, and J. Kummerow. 1990. Growth dynamics of two chaparral shrub species with time after fire. Madroño 37:225–236.

Laude, H. M., M. B. Jones, and W. F. Moon. 1961. Annual variability in indicators of sprouting potential in chamise. J. Range Mgt. 14:323–326.

Lloret, F., and P. H. Zedler. 1991. Recruitment pattern of *Rhus integrifolia* populations in periods between fire in chaparral. J. Veg. Sci. 2:217–230.

Lopez, E. N. 1983. Contribution of stored nutrients to post-fire regeneration of *Quercus dumosa*. Master's thesis, California State University, Los Angeles.

McDonald, P. M., and E. E. Litterell. 1976. The bigcone Douglas fir–canyon live oak community in southern California. Madroño 23:310–320.

McKell, C. M. 1950. A study of plant succession in the oak brush *Quercus gambelii* zone after fire. Master's thesis, University of Utah, Salt Lake City.

McMaster, G. S., W. Jow, and J. Kummerow. 1982. Response of *Adenostoma fasciculatum* and *Ceanothus greggii* chaparral to nutrient additions. J. Ecol. 70: 745–756.

McMaster, G. S., and P. H. Zedler. 1981. Delayed seed dispersal in *Pinus torreyana* (Torrey pine). Oecologia 51:62–66.

McMillan, C. 1956. Edaphic restriction of *Cupressus* and *Pinus* in the Coast Ranges of central California. Ecol. Monogr. 26:177–212.

McMinn, H. E. 1944. The importance of field hybrids in determining species in the genus *Ceanothus*. Proc. Calif. Acad. Sci. 25:323–356.

McPherson, J. K., C. H. Chou, and C. H. Muller. 1971. Allelopathic constituents of the chaparral shrub *Adenostoma fasciculatum*. Phytochemistry 10:2925–2933.

McPherson, J. K., and C. H. Muller. 1967. Light competition between *Ceanothus* and *Salvia* shrubs. Bull. Torrey Bot. Club 94:41–55.

McPherson, J. K., and C. H. Muller. 1969. Allelopathic effects of *Adenostoma fasciculatum*, "chamise," in the California chaparral. Ecol. Monogr. 39:177–198.

Mahall, B. E., and W. H. Schlesinger. 1982. Effects of irradiance on growth, photosynthesis and water use efficiency of seedlings of the chaparral shrub *Ceanothus megacarpus*. Oecologia 54:291–299.

Mahall, B. E., and C. S. Wilson. 1986. Environmental induction and physiological consequences of natural pruning in the chaparral shrub *Ceanothus megacarpus*. Bot. Gaz. 147:102–109.

Major, J. 1977. California climate in relation to vegetation, pp. 11–74 in M. G. Barbour and J. Major (eds.), Terrestrial vegetation of California. Wiley, New York.

Malanson, G. P. 1984. Fire history and patterns of Venturan subassociations of Califonian coastal sage scrub. Vegetatio 57:121–128.

Malanson, G. P., and J. F. O'Leary. 1982. Post-fire regeneration strategies of California coastal sage shrubs. Oecologia 53:355–358.

Malanson, G. P., and J. F. O'Leary. 1985. Effects of fire and habitat on post-fire regeneration in Mediterranean-type ecosystems: *Ceanothus spinosus* chaparral and California coastal sage scrub. Oecol. Plant. 6: 169–181.

Malanson, G. P., and W. E. Westman. 1985. Postfire succession in Californian coastal sage scrub: the role of continual basal sprouting. Amer. Midl. Nat. 113:309–318.

Malanson, G. P., W. E. Westman, and Y. L. Yan. 1992. Realized versus fundamental niche functions in a model of chaparral response to climatic change. Ecol. Model. 64:261–277.

Marion, L. H. 1943. The distribution of *Adenostoma sparsifolium*. Amer. Midl. Nat. 29:106–116.

Martin, B. D. 1995. Postfire reproduction of *Croton californicus* (Euphorbiaceae) and associated perennials in coastal sage scrub of southern California. Crossosoma 21:41–56.

Martin, P. S. 1973. The discovery of America. Science 179:969–974.

Miller, P. C. (ed.). 1981. Resource use by chaparral and matorral. Springer-Verlag, New York.

Miller, P. C., and E. Hajek. 1981. Resource availability and environmental characteristics of mediterranean type ecosystems, pp. 17–41 in P. C. Miller (ed.), Resource use by chaparral and matorral. Springer-Verlag, New York.

Miller, P. C., and D. K. Poole. 1979. Patterns of water use by shrubs in southern California. For. Sci. 25:84–98.

Mills, J. N. 1983. Herbivory and seedling establishment in post-fire southern California chaparral. Oecologia 60:267–270.

Mills, J. N. 1986. Herbivores and early postfire succession in southern California chaparral. Ecology 67: 1637–1649.

Mills, J. N., and J. Kummerow. 1989. Herbivores, seed predators and chaparral succession, pp. 49–55 in S. C. Keeley (ed.), The California chaparral: paradigms reexamined. Natural History Museum of Los Angeles County, Los Angeles, Science Series No. 34.

Minnich, R. A. 1980a. Vegetation of Santa Cruz and Santa Catalina Island, pp. 123–127 in D. M. Power (ed.), The California Islands – proceedings of a multidisciplinary symposium. Santa Barbara Botanic Garden.

Minnich, R. A. 1980b. Wildfire and the geographic relationships between canyon live oak, Coulter pine, and bigcone Douglas fir forests, pp. 55–61 in T. R. Plumb (ed.), Proceedings of the symposium on ecology, management and utilization of California oaks. USDA Forest Service, Pacific Southwest Forest and Range Experiment Station General Technical Report PSW-44.

Minnich, R. A. 1983. Fire mosaics in southern California and north Baja California. Science 219:1287–1294.

Minnich, R. A. 1995. Fuel-driven fire regimes of the California chaparral, pp. 21–27 in J. E. Keeley and T. Scott (eds.), Brushfires in California wildlands: ecology and resource management. International Association of Wildland Fire, Fairfield, Wash.

Minnich, R. A., M. G. Barbour, J. H. Burk, and R. F. Fernau. 1995. Sixty years of change in Californian conifer forests of the San Bernardino Mountains. Conserv. Biol. 9:902–914.

Minnich, R. A., and C. J. Bahre. 1995. Wildland fire and chaparral succession along the California–Baja California boundary. Int. J. Wildland Fire 5:13–24.

Minnich, R. A., and C. Howard. 1984. Biogeography and prehistory of shrublands, pp. 8–24 in J. J. DeVries (ed.), Shrublands in California: literature review and research needed for management. Contribution 191, Water Resources Center, University of California, Davis.

Minnich, R. A., E. F. Vizcaino, J. Sosa-Ramirez, and Y. Chou. 1993. Lightning detection rates and wildland fire in the mountains of northern Baja California, Mexico. Atmósfera 6:235–253.

Misquez, E. 1990. Frost sensitivity and distribution of *Malosma laurina*. Master's thesis, University of California, Riverside.

Moldenke, A. R. 1975. Niche specialization and species diversity along an altitudinal transect in California. Oecologia 21:219–242.

Montygierd-Loyba, T., and J. E. Keeley. 1987. Demographic structure of *Ceanothus megacarpus* chaparral in the long absence of fire. Ecology 68:211–213.

Mooney, H. A. 1977a. Southern coastal scrub, pp. 471–

478 in M. G. Barbour and J. Major (eds.), Terrestrial vegetation of California. Wiley, New York.

Mooney, H. A. (ed.). 1977b. Convergent evolution of Chile and California – mediterranean climate ecosystems. Dowden, Hutchinson and Boss, Stroudsburg, Pa.

Mooney, H. A. 1981. Primary production in mediterranean-climate regions, pp. 249–255 in F. di Castri, D. W. Goodall, and R. L. Specht (eds.), Ecosystems of the world. II. Mediterranean-type shrublands. Elsevier Scientific, New York.

Mooney, H. A. 1982. Habitat, plant form, and plant water relations in Mediterranean-climate regions, pp. 481–488 in P. Quézel (ed.), Définition et localisation des ecosystèmes Méditerranéens terrestres. NATO, Scientific Affairs Division, Marseille, Ecologia Mediterranea 7(1).

Mooney, H. A. 1989. Chaparral physiological ecology–paradigms revisited, pp. 85–96 in S. C. Keeley (ed.), The California chaparral: paradigms reexamined. Natural History Museum of Los Angeles County, Los Angeles, Science Series No. 34.

Mooney, H. A., and E. L. Dunn. 1970. Photosynthetic systems of Mediterranean climate shrubs and trees in California and Chile. Amer. Nat. 104:447–453.

Mooney, H. A., S. L. Gulmon, D. J. Parsons, and A. T. Harrison. 1974. Morphological changes within the chaparral vegetation type as related to elevational gradients. Madroño 22:281–285.

Mooney, H. A., and A. T. Harrison. 1972. The vegetational gradient on the lower slopes of the Sierra San Pedro Martír in northwest Baja California. Madroño 21:439–445.

Mooney, H. A., A. T. Harrison, and P. A. Morrow. 1975. Environmental limitation photosynthesis on a California evergreen shrub. Oecologia 19:293–301.

Mooney, H. A., J. Kummerow, A. W. Johnson, D. J. Parsons, S. Keeley, A. Hoffman, R. J. Hays, J. Giliberto, and C. Chu. 1977. The producers – their resources and adaptive response, pp. 85–143 in H. A. Mooney (ed.), Convergent evolution of Chile and California-mediterranean climate ecosystems. Dowden, Hutchinson and Boss, Stroudsburg, Pa.

Mooney, H. A., and P. C. Miller. 1985. Chaparral, pp. 213–231 in B. F. Chabot and H. A. Mooney (eds.), Physiological ecology of North American plant communities. Chapman & Hall, New York.

Mooney, H. A., and D. J. Parsons. 1973. Structure and function of the California chaparral: an example from San Dimas, pp. 83–112 in F. di Castri and H. A. Mooney (eds.), Mediterranean ecosystems: origin and structure. Springer-Verlag, New York.

Mooney, H. A., and P. W. Rundel. 1979. Nutrient relations of the evergreen shrub, Adenostoma fasciculatum, in the California chaparral. Bot. Gaz. 140:109–113,

Mooney, H. A., J. Troughton, and J. Berry. 1974. Arid climates and photosynthetic systems. Carnegie Instit. Yearb. 73:793–805.

Moreno, J. M., and W. C. Oechel. 1988. Post-fire establishment of Adenostoma fasciculatum and Ceanothus greggii in a southern California chaparral: influence of herbs and increased soil-nutrients and water, pp. 137–141 in F. di Castri, Ch. Floret, S. Rambal, and J. Roy (eds.), Time scales and water stress. Proceedings of the 5th international conference on Mediterranean ecosystems (MEDECOS V). International Union of Biological Sciences, Paris.

Moreno, J. M., and W. C. Oechel. 1989. A simple method for estimating fire intensity after a burn in California chaparral. Oecol. Plant. 10:57–68.

Moreno, J. M., and W. C. Oechel. 1991a. Fire intensity effects on germination of shrubs and herbs in southern California chaparral. Ecology 72:1993–2004.

Moreno, J. M., and W. C. Oechel. 1991b. Fire intensity and herbivory effects on postfire resprouting of Adenostoma fasciculatum in southern California chaparral. Oecologia 85:429–433.

Moreno, J. M., and W. C. Oechel. 1992. Factors controlling postfire seedling establishment in southern California chaparral. Oecologia 90:50–60.

Moreno, J. M., and W. C. Oechel. 1993. Demography of Adenostoma fasciculatum after fires of different intensities in southern California chaparral. Oecologia 96:95–101.

Moreno, J. M., and W. C. Oechel. 1994. Fire intensity as a determinant factor of postfire plant recovery in southern California chaparral, pp. 26–45 in J. M. Moreno and W. C. Oechel (eds.), The role of fire in Mediterranean-type ecosystems. Springer-Verlag, New York.

Moritz, M. A. 1997. Analyzing extreme disturbance events: fire in Los Padres National Forest. Ecol. Appl. 7:1252–1262.

Mortenson, T. H. 1973. Ecological variation in the leaf anatomy of selected species of Cercocarpus. Aliso 8:19–48.

Muller, C. N. 1939. Relations of the vegetation and climate types in Nuevo Leon, Mexico. Amer. Midl. Nat. 21:687–729.

Muller, C. N. 1947. Vegetation and climate of Coahuila, Mexico. Madroño 9:33–57.

Muller, C. N., and R. del Moral. 1966. Soil toxicity induced by terpenes from Salvia leucophylla. Bull. Torrey Bot. Club 93:130–137.

Muller, C. N., and P. Hague. 1967. Volatile growth inhibitors produced by Salvia leucophylla effect on seedling anatomy. Bull. Torrey Bot. Club 94:182–191.

Muller, C. N., W. H. Muller, and B. L. Haines. 1964. Volatile growth inhibitors produced by aromatic shrubs. Science 143:471–473.

Newton, M., Ortiz-Funez, A., and J. C. Tappeiner. 1988. Pine and manzanita pull water out of rock. Oregon State University Extension Service, Forestry Intensified Research Report 10.

Nicholson, P. 1993. Ecological and historical biogeography of Ceanothus (Rhamnaceae) in the Transverse Ranges of southern California. Ph.D. dissertation, University of California, Los Angeles.

Nilsen, E. T., and W. H. Muller. 1981. Phenology of the drought deciduous shrub Lotus scoparius. Oecologia 53:79–83.

Nobs, M. A. 1963. Experimental studies on species relationships in Ceanothus. Carnegie Institution of Washington Publication 623.

Oberbauer, T. A., 1993. Soils and plants of limited distribution in the Peninsular Ranges. Fremontia 21:3–7.

Oberlander, G. T. 1953. The taxonomy and ecology of the flora of the San Francisco watershed reserve. Ph.D. dissertation, Stanford University, Stanford, Calif.

Oechel, W. C. 1982. Carbon balance studies in chaparral shrubs: implications for biomass production, pp. 158–166 in C. E. Conrad and W. C. Oechel (eds.), Proceedings of the symposium on dynamics and management of mediterranean-type ecosystems. USDA Forest Service, Pacific Southwest Forest and Range Experiment Station General Technical Report PSW-58.

Oechel, W. C. 1988. Minimum non-lethal water potentials in Mediterranean shrub seedlings, pp. 125–131 in F. di Castri, Ch. Floret, S. Rambal, and J. Roy (eds.), Time scales and water stress. Proceedings of the 5th international conference on Mediterranean ecosystems (MEDECOS V). International Union of Biological Sciences, Paris.

Oechel, W. C., and S. J. Hastings. 1983. The effects of fire on photosynthesis in chaparral resprouts, pp. 274–285 in F. J. Kruger, D. T. Mitchell, and J. U. M. Jarvis (eds.), Mediterranean-type ecosystems. The role of nutrients. Springer-Verlag, New York.

Oechel, W. C., W. T. Lawrence, J. Mustafa, and J. Martinez. 1981. Energy and carbon acquisition, pp. 151–183 in P. C. Miller (ed.), Resource use by chaparral and matorral. Springer-Verlag, New York.

O'Leary, J. F. 1988. Habitat differentiation among herbs in postburn California chaparral and coastal sage scrub. Amer. Midl. Nat. 120:41–49.

O'Leary, J. F., and W. E. Westman. 1988. Regional disturbance effects on herb succession patterns in coastal sage scrub. J. Biogeogr. 15:775–786.

Parker, V. T. 1984. Correlations of physiological divergence with reproductive mode in chaparral shrubs. Madroño 31:231–242.

Parker, V. T., and V. R. Kelly. 1989. Seed banks in California chaparral and other Mediterranean climate shrublands, pp. 231–255 in M. A. Leck, V. T. Parker, and R. L. Simpson (eds.), Ecology of soil seed banks. Academic Press, New York.

Parsons, D. J. 1976. Vegetation structure in the Mediterranean climate scrub communities of California and Chile. J. Ecol. 64:435–447.

Parsons, D. J. 1981. The historical role of fire in the foothill communities of Sequoia National Park. Madroño 28:111–120.

Parsons, D. J., P. W. Rundel, R. Hedlund, and G. A. Baker. 1981. Survival of severe drought by a nonsprouting chaparral shrub. Amer. J. Bot. 68:215–220.

Pase, C. P. 1965. Shrub seedling regeneration after controlled burning and herbicidal treatment of dense Pringle manzanita chaparral. USDA Forest Service, Rocky Mountain Forest and Range Experiment Station Research Note RM-56.

Pase, C. P., and D. E. Brown. 1982. Interior chaparral. Desert Plants 4:95–99.

Pase, C. P., and A. W. Lindenmuth, Jr. 1971. Effects of prescribed fire on vegetation and sediment in oak–mountain mahogany chaparral. J. For. 69:800–805.

Pase, C. P., and F. W. Pond. 1964. Vegetation changes following the Mingus Mountain burn. USDA Forest Service, Rocky Mountain Forest and Range Experiment Station Research Note RM-18.

Patric, J. H., and T. L. Hanes. 1964. Chaparral succession in a San Gabriel Mountain area of California. Ecology 45:353–360.

Peinado, M., F. Alcaraz, J. L. Aguirre, J. Delgadillo, and I. Aguado. 1995. Shrubland formations and associations in mediterranean-desert transitional zones of northwestern Baja California. Vegetatio 117:165–179.

Phillips, P. W. 1966. Variation and hybridization in

Ceanothus cuneatus and Ceanothus megacarpus. Master's thesis, University of California, Santa Barbara.

Plumb, T. R. 1961. Sprouting of chaparral by December after a wildfire in July. USDA Forest Service, Pacific Southwest Forest and Range Experiment Station Technical Paper 57.

Poole, D. K., and P. C. Miller. 1975. Water relations of selected species of chaparral and coastal sage communities. Ecology 56:1118–1128.

Poole, D. K., and P. C. Miller. 1981. The distribution of plant water stress and vegetation characteristics in southern California chaparral. Amer. Midl. Nat. 105:32–43.

Poole, D. K., S. W. Roberts, and P. C. Miller. 1981. Water utilization, pp. 123–149 in P. C. Miller (ed.), Resource use by chaparral and matorral. Springer-Verlag, New York.

Popenoe, J. H. 1974. Vegetation patterns on Otay Mountain, California. Master's thesis, San Diego State University.

Poth, M. 1982. Biological dinitrogen fixation in chaparral, pp. 285–290 in C. E. Conrad and W. C. Oechel (eds.), Proceedings of the symposium on dynamics and management of mediterranean-type ecosystems. USDA Forest Service, Pacific Southwest Forest and Range Experiment Station, General Technical Report PSW-58.

Pratt, S. D., A. S. Konopka, M. A. Murry, F. W. Ewers, and S. D. Davis. 1997. Influence of soil moisture on the nodulation of post fire seedlings of *Ceanothus* spp. growing in the Santa Monica Mountains of southern California. Physiol. Plant. 99:673–679.

Quick, C. R. 1935. Notes on the germination of *Ceanothus* seeds. Madroño 3:135–140.

Quick, C. R., 1944. Effects of snowbrush on the growth of sierra gooseberry. J. For. 32:827–932.

Quick, C. R., and A. S. Quick. 1961. Germination of *Ceanothus* seeds. Madroño 16:23–30.

Quinn, R. D. 1986. Mammalian herbivory and resilience in Mediterranean-type ecosystems, pp. 113–128, in B. Dell, A. J. M. Hopkins, and B. B. Lamont (eds.), Resilience in Mediterranean-type ecosystems. Junk, Dordrecht, The Netherlands.

Quinn, R. D. 1994. Animals, fire and vertebrate herbivory in California chaparral and other Mediterranean-type ecosystems, pp. 46–78 in J. M. Moreno and W. C. Oechel (eds.), The role of fire in Mediterranean-type ecosystems. Springer-Verlag, New York.

Radosevich, S. R., and S. G. Conard. 1980. Physiological control of chamise shoot growth after fire. Amer. J. Bot. 67:1442–1447.

Raven, P. H. 1973. The evolution of mediterranean floras, pp. 213–223 in F. di Castri and H. A. Mooney (eds.), Mediterranean ecosystems: origin and structure. Springer-Verlag, New York.

Raven, P. H., and D. I. Axelrod. 1978. Origin and relationships of the California flora. Univ. Calif. Publ. Bot. 72:1–134.

Pratt, S. D., A. S. Konopka, M. A. Murry, F. W. Ewers, and S. D. Davis. 1997. Influence of soil moisture on the nodulation of post fire seedlings of *Ceanothus* spp. growing in the Santa Monica Mountains of southern California. Physiol. Plant. 99:673–679.

Redtfeldt, R. A., and S. D. Davis. 1996. Physiological and morphological evidence of niche segregation between two co-occurring species of *Adenostoma* in California chaparral. Ecoscience 3:290–296.

Reid, C. D. 1985. Possible physiological indicators of senescence in two chaparral shrub species along a fire-induced age sequence. Master's thesis, San Diego State University, San Diego, Calif.

Rehlaender, W. E. 1992. Nutrient status of the chaparral plant-soil system during stand development after fire: the effects of stand age and stubstrate type. Master's thesis, San Diego State University, San Diego, Calif.

Rice, S. K. 1993. Vegetation establishment in post-fire *Adenostoma* chaparral in relation to fine-scale pattern in fire intensity and soil nutrients. J. Veg. Sci. 4: 115–124.

Richerson, P. J., and K. L. Lum. 1980. Patterns of plant species diversity in California: relation to weather and topography. Amer. Nat. 116:504–536.

Riggan, P. J., S. E. Franklin, J. A. Brass, and F. E. Brooks. 1994. Perspectives on fire management in Mediterranean ecosystems of southern California, pp. 140–161 in J. M. Moreno and W. C. Oechel (eds.), The role of fire in Mediterranean-type ecosystems. Springer-Verlag, New York.

Roberts, S. W. 1982. Some recent aspects and problems of chaparral plant water relations, pp. 351–357 in C. E. Conrad and W. C. Oechel (eds.), Proceedings of the symposium on dynamics and management of Mediterranean-type ecosystems. USDA Forest Service, Pacific Southwest Forest and Range Experiment Station, General Technical Report PSW-58.

Roof, J. B. 1978. Studies in *Arctostaphylos* (Ericaceae). Four Seasons 5:2–24.

Rundel, P. W. 1982. Nitrogen use efficiency in Mediterranean-climate shrubs of California and Chile. Oecologia 55:409–413.

Rundel, P. W. 1995. Adaptive significance of some morphological and physiological characteristics in mediterranean plants: facts and fallacies, pp. 119–139 in J. Roy, J. Aronson, and F. di Castri (eds.), Time scales of biological responses to water constraints. SPB Academic Publishing, Amsterdam, The Netherlands.

Rundel, P. W., G. A. Baker, D. J. Parsons, and T. J. Stohlgren. 1987. Postfire demography of resprouting and seedling establishment by *Adenostoma fasciculatum* in the California chaparral, pp. 575–596 in J. D. Tenhunen, F. M. Catarino, O. L. Lange, and W. C. Oechel (eds.), Plant response to stress. Functional analysis in Mediterranean ecosystems. Springer-Verlag, Berlin.

Rundel, P. W., and D. J. Parsons. 1979. Structural changes in chamise *Adenostoma fasciculatum* along a fire-induced age-gradient. J. Range Mgt. 32:462–466 (and unpublished erratum).

Rundel, P. W., and D. J. Parsons. 1980. Nutrient changes in two chaparral shrubs along a fire-induced age gradient. Amer. J. Bot. 67:51–58.

Rundel, P. W., and D. J. Parsons. 1984. Post-fire uptake of nutrients by diverse ephemeral herbs in chamise chaparral. Oecologia 61:285–288.

Rundel, P. W., and J. L. Vankat. 1989. Chaparral communities and ecosystems, pp. 127–139 in S. C. Keeley (ed.), The California chaparral: paradigms reexamined. Natural History Museum of Los Angeles County, Los Angeles, Science Series No. 34.

Rzedowski, J. 1978. Vegetación de México. Editorial Limusa, México City.

Sampson, A. W. 1944. Plant succession on burned chaparral lands in northern California. Agricultural Experiment Station Bulletin 685, University of California, Berkeley.

Sawyer, J. O., and T. Keeler-Wolf. 1995. A manual of California vegetation. California Native Plant Society, Sacramento.

Scheidlinger, C. R., and P. H. Zedler. 1980. Change in vegetation cover of oak stands in southern San Diego County: 1928–1970, pp. 81–85 in T. R. Plumb (ed.), Proceedings of the symposium on the ecology, management, and utilization of California oaks. USDA Forest Service, Pacific Southwest Forest and Range Experiment Station, General Technical Report PSW-44.

Schierenbeck, K. A., G. L. Stebbins, and R. W. Patterson. 1992. Morphological and cytological evidence for polyphyletic allopolyploidy in *Arctostaphylos mewukka* (Ericaceae). Plant Syst. Evol. 179:187–205.

Schlesinger, W. H., and D. S. Gill. 1978. Demographic studies of the chaparral shrub, *Ceanothus megacarpus*, in the Santa Ynez Mountains, California. Ecology 59:1256–1263.

Schlesinger, W. H., and D. S. Gill. 1980. Biomass, production, and changes in the availability of light, water, and nutrients during development of pure stands of the chaparral shrubs, *Ceanothus megacarpus*, after fire. Ecology 61:781–789.

Schlesinger, W. H., and J. T. Gray. 1982. Atmospheric precipitation as a source of nutrients in chaparral ecosystems, pp. 279–284 in C. E. Conrad and W. C. Oechel (eds.), Proceedings of the symposium on dynamics and management of mediterranean-type ecosystems. USDA Forest Service, Pacific Southwest Forest and Range Experiment Station, General Technical Report PSW-58.

Schlesinger, W. H., J. T. Gray, D. S. Gill, and B. E. Mahall. 1982a. *Ceanothus megacarpus* chaparral: a synthesis of ecosystem properties during development and annual growth. Bot. Rev. 48:71–117.

Schlesinger, W. H., J. T. Gray, and F. S. Gilliam. 1982b. Atmospheric deposition processes and their importance as sources of nutrients in a chaparral ecosystem of southern California. Water Resour. Res. 18: 623–629.

Schlesinger, W. H., and M. M. Hasey. 1980. The nutrient content of precipitation, dry fallout, and intercepted aerosols in the chaparral of southern California. Amer. Midl. Nat. 103:114–122.

Schlesinger, W. H., and M. M. Hasey. 1981. Decomposition of chaparral shrub foliage: losses of organic and inorganic constituents from deciduous and evergreen leaves. Ecology 62:762–774.

Schlising, R. A. 1969. Seedling morphology in *Marah* (Cucurbitaceae) related to the Californian Mediterranean climate. Amer. J. Bot. 56:556–560.

Schmid, R., T. E. Mallory, and J. M. Tucker. 1968. Biosystematic evidence for hybridization between *Arctostaphylos nissenana* and *A. viscida*. Brittonia 20:34–43.

Schorr, P. K. 1970. The effects of fire on manzanita chaparral in the San Jacinto Mountains of southern California. Master's thesis, California State University, Los Angeles.

Schultz, A. M., J. L. Baunchbauch, and H. H. Biswell. 1955. Relationship between grass density and brush seedling survival. Ecology 36:226–238.

Schwilk, D. W., J. E. Keeley, and W. Bond. 1997. The intermediate disturbance hypothesis does not explain fire and diversity pattern in fynbos. Plant Ecol. 132: 77–84.

Shantz, H. L. 1947. The use of fire as a tool in the management of brush ranges of California. State of California, Department of Natural Resources, Division of Forests.

Shaver, G. R. 1978. Leaf angle and light absorbance of *Arctostaphylos* species (Ericaceae) along environmental gradients. Madroño 25:133–138.

Shaver, G. R. 1981. Mineral nutrient and nonstructural carbon utilization, pp. 237–257 in P. C. Miller (ed.), Resource use by chaparral and matorral. Springer-Verlag, New York.

Shmida, A., and M. Barbour. 1982. A comparison of two types of mediterranean scrub in Israel and California, pp. 100–106 in C. E. Conrad and W. C. Oechel (eds.), Proceedings of the symposium on dynamics and management of mediterranean type ecosystems. USDA Forest Service, Pacific Southwest Forest and Range Experiment Station, General Technical Report PSW-58.

Shmida, A., and R. H. Whittaker. 1981. Pattern and biological microsite effects in two shrub communities, southern California. Ecology 62:234–251.

Shreve, F. 1927. The vegetation of a coastal mountain range. Ecology 8:37–40.

Shreve, F. 1936. The transition from desert to chaparral in Baja California. Madroño 3:257–264.

Shreve, F. 1942. Grassland and related vegetation in northern Mexico. Madroño 6:190–198.

Skau, C. M., R. O. Meeuwig, and T. W. Townsend. 1970. Ecology of eastside chaparral – a literature review. Agricultural Experimental Station, University of Nevada, Reno.

Sparks, S. R., and W. C. Oechel. 1993. Factors influencing postfire sprouting vigor in the chaparral shrub *Adenostoma fasciculatum*. Madroño 40:224–235.

Sparks, S. R., W. C. Oechel, and Y. Mauffette. 1993. Photosynthate allocation patterns along a fire-induced age sequence in two shrub species from the California chaparral. Int. J. Wildland Fire. 3:21–30.

Specht, T. L. 1969. A comparison of the sclerophyllous vegetation characteristics of mediterranean type climates in France, California, and southern Australia. I: Structure, morphology and succession. Aust. J. Bot. 17:277–292.

Spittler, T. E. 1995. Fire and debris flow potential of winter storms, pp. 113–120 in J. E. Keeley and T. Scott (eds.), Brushfires in California: ecology and resource management. International Association of Wildland Fire, Fairfield, Wash.

Stebbins, G. L., Jr. 1959. Seedling heterophylly in the California flora. Bull. Res. Coun. Israel 7D:248–255.

Steele, R. A. 1985. Timing of flower differentiation and development in the southern California chaparral. Master's thesis, San Diego State University, San Diego, Calif.

Steward, D., and P. J. Webber. 1981. The plant communities and their environments, pp. 43–68 in P. C. Miller (ed.), Resource use by chaparral and matorral. Springer-Verlag, New York.

Stine, S. 1994. Extreme and persistent drought in California and Patagonia during mediaeval time. Nature 369:546–549.

Stocking, S. K. 1966. Influence of fire and sodium calcium borate on chaparral vegetation. Madroño 18: 193–203.

Stoddard, R. J., and S. D. Davis. 1990. Comparative photosynthesis, water relations, and nutrient status of burned, unburned, and clipped *Rhus laurina* after chaparral wildfire. Bull. So. Calif. Acad. Sci. 89:26–38.

Stohlgren, T. J., D. J. Parsons, and P. W. Rundel. 1984. Population structure of *Adenostoma fasciculatum* in mature stands of chamise chaparral in the southern Sierra Nevada, California. Oecologia 64:87–91.

Stone, E. C. 1951. The stimulative effect of fire on the flowering of the golden brodiae (*Broadiae ixioides* Wars. var. *lugens* Jeps.). Ecology 32:534–537.

Stone, E. C., and G. Juhren. 1951. The effect of fire on the germination of the seed of *Rhus ovata* Wars. Amer. J. Bot. 38:368–372.

Stone, E. C., and G. Juhren. 1953. Fire stimulated germination. Calif. Agr. 7:13–14.

Swank, S. E., and W. C. Oechel. 1991. Interactions among the effects of herbivory, competition, and resource limitation on chaparral herbs. Ecology 72: 104–115.

Swank, W. G. 1958. The mule deer in Arizona chaparral and an analysis of other important deer herds. Arizona Game and Fish Department, Wildlife Bulletin 3.

Sweeney, J. R. 1956. Responses of vegetation to fire. A study of the herbaceous vegetation following chaparral fires. Univ. Calif. Pub. Bot. 28:143–216.

Swift, C. 1991. Nitrogen utilization strategies in post-fire chaparral annual species. Ph.D. dissertation, University of California, Los Angeles.

Tadros, T. M. 1957. Evidence of the presence of an edapho-biotic factor in the problem of serpentine tolerance. Ecology 38:14–23.

Tenhunen, J. D., R. Hanano, M. Abril, E. W. Weiler, and W. Hartung. 1994. Above- and belowground environmental influences on leaf conductance of *Ceanothus thyrsiflorus* growing in a chaparral environment: drought response and the role of abscisic acid. Oecologia 99:306–314.

Thanos, C. A., and P. W. Rundel. 1995. Fire-followers in chaparral: nitrogenous compounds trigger seed germination. J. Ecol. 83:207–216.

Thomas, C. M., and S. D. Davis. 1989. Recovery patterns of three chaparral shrub species after wildfire. Oecologia 80:309–320.

Townsend, T. W. 1966. Plant characteristics relating to the desirability of rehabilitating the *Arctostaphylos patula–Ceanothus veluntinus–Ceanothus prostratus* association on the east slope of the Sierra Nevada. Master's thesis, University of Nevada, Reno.

Tratz, W. M. 1978. Postfire vegetational recovery, productivity, and herbivore utilization of a chaparral desert ecotone. Master's thesis, California State University, Los Angeles.

Tucker, J. M. 1953. The relationship between *Quercus dumosa* and *Quercus turbinella*. Madroño 12:49–60.

Tyler, C. M. 1995. Factors contributing to postfire seedling establishment in chaparral: direct and indirect effects of fire. J. Ecol. 83:1009–1020.

Tyler, C. M. 1996. Relative importance of factors contributing to postfire seedling establishment in maritime chaparral. Ecology 77:2182–2195.

Tyler, C. M., and C. M. D'Antonio. 1995. The effects of neighbors on the growth and survival of shrub seedlings following fire. Oecologia 102:255–264.

Tyson, B. J., W. A. Dement, and H. A. Mooney. 1974.

Volatilization of terpenes from *Salvia mellifera*. Nature 252:119–120.

Vale, T. R. 1979. *Pinus coulteri* and wildfire on Mount Diablo, California. Madroño 26:135–140.

Vankat, J. L. 1982. A gradient perspective on the vegetation of Sequoia National Park, California. Madroño 29:200–214.

Vankat, J. L. 1989. Water stress in chaparral shrubs in summer-rain versus summer-drought climates – whither the Mediterranean-type climate paradigm, pp. 117–124 in S. C. Keeley (ed.), The California chaparral. Paradigms reexamined. Natural History Museum of Los Angeles County, Los Angeles, Science Series No. 34.

Varney, B. M. 1925. Seasonal precipitation in California and its variability. Monthly Weather Review 53:208–218.

Vasek, F. C., and J. F. Clovis. 1976. Growth forms in *Arctostaphylos glauca*. Amer. J. Bot. 63:189–195.

Vestal, A. G. 1917. Foothills vegetation in the Colorado Front Range. Bot. Gaz. 64:353–385.

Vlamis, J., A. M. Schultz, and H. H. Biswell. 1958. Nitrogen-fixation by deerbrush. Calif. Agr. 12:11, 15.

Vlamis, J., A. M. Schultz, and H. H. Biswell, 1964. Nitrogen-fixation by root nodules of western mountain mahogany. J. Range Mgt. 17:73–74.

Vogl, R. J., W. P. Armstrong, K. L. White, and K. L. Cote. 1977. The closed-cone pines and cypresses, pp. 295–358 in M. G. Barbour and J. Major (eds.), Terrestrial vegetation of California. Wiley, New York.

Vogl, R. J., and P. K. Schorr. 1972. Fire and manzanita chaparral in the San Jacinto Mountains, California. Ecology 53:1179–1188.

Warter, J. K. 1976. Late Pleistocene plant communities – evidence from the Rancho La Brea tar pits, pp. 32–39 in J. Latting (ed.), Proceedings of the symposium on plant communities of southern California. California Nature Plant Society, Berkeley, Special Publication 2.

Watkins, K. S. 1939. Comparative stem anatomy of dominant chaparral plants of southern California. Master's thesis, University of California, Los Angeles.

Watkins, V. M., and H. DeForest. 1941. Growth in some chaparral shrubs of California. Ecology 22:79–83.

Webber, I. E. 1936. The woods of sclerophyllous and desert shrubs of California. Amer. J. Bot. 33:181–188.

Wells, P. V. 1962. Vegetation in relation to geological substratum and fire in the San Luis Obispo Quadrangle, California. Ecol. Monogr. 32:79–103.

Wells, P. V. 1968. New taxa, combinations and chromosome numbers in *Arctostaphylos*. Madroño 19:193–210.

Wells, P. V. 1969. The relation between mode of reproduction and extent of speciation in woody genera of the California chaparral. Evolution 23:264–267.

Wells, P. V. 1972. The manzanitas of Baja California, including a new species of *Arctostaphylos*. Madroño 21:268–273.

Wells, P. V. 1987. The leafy-bracted, crown-sprouting manzanitas, an ancestral group in *Arctostaphylos*. Four Seasons 7:4–17.

Wells, P. V. 1992. Subgenera and sections of *Arctostaphylos*. Four Seasons 9:64–69.

Went, F. W. 1969. A long-term test of seed longevity. II. Aliso 7:1–12.

West, G. J. 1989. Early historic vegetation change in Alta California: the fossil evidence, pp. 333–348 in D. H. Thomas (ed.), Columbian consequences. Archaeological and historical perspectives on the Spanish borderlands west. Smithsonian Institution Press, Washington, D.C.

Westman, W. E. 1981a. Diversity relations and succession in California coastal sage scrub. Ecology 62:170–184.

Westman, W. E. 1981b. Seasonal dimorphism of foliage in California coastal sage scrub. Oecologia 51:385–388.

Westman, W. E. 1982. Coastal sage scrub succession, pp. 91–99 in C. E. Conrad and W. E. Oechel (eds.), Proceedings of the symposium on dynamics and management of mediterranean-type ecosystems. USDA Forest Service, Pacific Southwest Forest and Range Experiment Station, General Technical Report PSW-58.

Westman, W. E. 1983. Xeric mediterranean-type shrubland association of Alto and Baja California and the community/continiuum debate. Vegetatio 52:3–19.

Westman, W. E., J. F. O'Leary, and G. P. Malanson. 1981. The effects of fire intensity, aspect and substrate on post-fire growth of Californian coastal sage scrub, pp. 151–179 in N. S. Margaris and H. A. Mooney (eds.), Components of productivity of mediterranean regions – basic and applied aspects. Junk, The Hague.

White, C. D. 1967. Absence of nodule formation on *Ceanothus cuneatus* in serpentine soils. Nature 215:875.

White, C. D. 1971. Vegetation-soil chemistry correlations in serpentine ecosystems. Ph.D. dissertation, University of Oregon, Eugene.

White, S. D. 1991. *Quercus wislizenii* forest and shrubland in the San Bernardino Mountains, California. Master's thesis, Humboldt State University, Arcata, Calif.

White, S. D. 1997. *Quercus wislizenii* growth rings, pp. 667–669 in N. H. Pillsbury, J. Verner, W. D. Tietje (eds.), Proceedings of a symposium on oak woodlands: ecology, management, and urban interface issues. USDA Forest Service, Pacific Southwest Research Station, General Technical Report PSW-GTR-160.

Whittaker, R. H. 1960. Vegetation of the Siskiyou Mountains, Oregon and California. Ecol. Monogr. 30:279–338.

Wicklow, D. T. 1964. A biotic factor in serpentine endemism. Master's thesis, San Francisco State University, San Francisco, Calif.

Wicklow; D. T. 1977. Germination response in *Emmenanthe penduliflora* (Hydrophyllaceae). Ecology 58:201–205.

Wieslander, A. E., and C. H. Gleason. 1954. Major brushland areas of the Coastal Ranges and Sierra Cascades Foothills in California. USDA Forest Service, California Forest and Range Experiment Station miscellaneous paper 15.

Wieslander, A. E., and B. O. Schreiber. 1939. Notes on the genus *Arctostaphylos*. Madroño 58:38–47.

Wilken, C. C. 1967. History and fire record of a timberland brush field in the Sierra Nevada of California. Ecology 48:302–304.

Williams, K., S. D. Davis, B. L. Gartner, and S. Karlsson.

1991. Factors limiting the distribution of *Quercus durata* Jeps. in grassland, pp. 70–73 in R. B. Standiford (ed.), Proceedings of the symposium on oak woodlands and hardwood rangeland management. USDA Forest Service, Pacific Southwest Research Station, General Technical Report PSW-126.

Williams, K., and R. J. Hobbs. 1989. Control of shrub establishment by springtime soil water availability in an annual grassland. Oecologia 81:62–66.

Wilson, R. C., and R. J. Vogl. 1965. Manzanita chaparral in the Santa Ana Mountains, California. Madroño 18:47–62.

Young, D. A. 1974. Comparative wood anatomy of *Malosma* and related genera (Anacardiaceae). Aliso 8: 133–146.

Youngberg, C. T., and A. G. Wollum II. 1976. Nitrogen accretion in developing *Ceanothus velutinus* stands. Soil Sci. Soc. Amer. J. 40:109–112.

Zammit, C. A., and P. H. Zedler. 1988. The influence of dominant shrubs, fire, and time since fire on soil seed banks in mixed chaparral. Vegetatio 75:175–187.

Zammit, C. A., and P. H. Zedler. 1992. Size structure and seed production in even-aged populations of *Ceanothus greggii* in mixed chaparral. J. Ecol. 81:499–511.

Zammit, C., and P. H. Zedler. 1994. Organisation of the soil seed bank in mixed chaparral. Vegetatio 111:1–16.

Zedler, P. H. 1977. Life history attributes of plants and the fire cycle: a case study in chaparral dominated by *Cupressus forbesii*, pp. 451–458 in H. A. Mooney and C. E. Conrad (eds.), Proceedings of the symposium on environmental consequences of fire and fuel management in mediterranean ecosystems. USDA Forest Service, General Technical Report WO-3.

Zedler, P. H. 1981. Vegetation change in chaparral and desert communities in San Diego County, California, pp. 406–430 in D. C. West, H. H. Shugart, and D. Botkin (eds.), Forest succession: Concepts and applications. Springer-Verlag, New York.

Zedler, P. H. 1982. Demography and chaparral management in southern California, pp. 123–127 in C. E. Conrad and W. C. Oechel (eds.), Proceedings of the symposium on dynamics and management of mediterranean-type ecosystems. USDA Forest Service, Pacific Southwest Forest and Range Experiment Station, General Technical Report PSW-58.

Zedler, P. H. 1995a. Plant life history and dynamic specialization in the chaparral/coastal sage shrub flora in southern California, pp. 89–115 in M. T. Kalin Arroyo, P. H. Zedler, and M. D. Fox (eds.), Ecology and biogeography of Mediterranean ecosystems in Chile, California, and Australia. Springer-Verlag, New York.

Zedler, P. H. 1995b. Fire frequency in southern California shrublands: biological effects and management options, pp. 101–112 in J. E. Keeley and T. Scott (eds.), Brushfires in California wildlands: ecology and resource management. International Association of Wildland Fire, Fairfield, Wash.

Zedler, P. H. 1995c. Are some plants born to burn? Oikos 10:393–395.

Zedler, P. H., C. R. Gautier, and G. S. McMaster. 1983. Vegetation change in response to extreme events. The effect of a short interval between fires in California chaparral and coastal scrub. Ecology 64:809–818.

Zedler, P. H., and C. A. Zammit. 1989. A population-based critique of concepts of change in the chaparral, pp. 73–83 in S. C. Keeley (ed.), The California chaparral: paradigms reexamined. Natural History Museum of Los Angeles County, Los Angeles, Science Series No. 34.

Zenan, A. J. 1967. Site differences and the microdistributions of chaparral species. Master's thesis, University of California, Los Angeles.

Zwieniecki, M. A., and M. Newton. 1996. Seasonal pattern of water depletion from soil-rock profiles in a mediterranean climate in southwestern Oregon. Can. J. For. Res. 26:1346–1352.

Chapter
7

Intermountain Valleys and Lower Mountain Slopes

NEIL E. WEST   JAMES A. YOUNG

INTRODUCTION

Of concern here is the vegetation that occupies the comparatively lower elevations of the basins, valleys, lower plateaus, foothills, and lower mountain slopes of the intermountain region of western North America. Our primary focus is on what has often been called the "cold" desert biome (Shelford 1963). We dislike that term, however, because this region is cold only in winter and is not to be confused with the permafrost- and glacial-ice-dominated environments of the polar deserts discussed in Chapter 1. Furthermore, only small areas of the intermountain lowlands are deserts in the international sense. Most of the area to be discussed is relatively well vegetated semidesert scrub or shrub-steppe. A popular book about this region, called *The Sagebrush Ocean* (Trimble 1989), reflects these characteristics.

The areas discussed in this chapter, although "lowlands" in the relative sense, are nearly all above 1000 m in elevation. This, plus the latitudinal position of the region, creates temperate climates, as contrasted with the subtropical climates of the warm deserts to the south (see Chapter 8). Thus, West (1983a) titled a major review of the most extensive terrestrial ecosystems in the region *Temperate Deserts and Semideserts*.

The pygmy conifer (pinyon-juniper) woodlands and the mountain mahogany–oak scrub (Wasatch and Arizona "chaparral" types), elevationally above the semideserts, are other major vegetation types also found in semiarid environments of the Intermountain Region. Thus, they are more easily understood by ecologists familiar with semidesert shrublands than by those with experience only in forests outside the region.

Although scrub, steppe, woodlands, and chaparral occupy the bulk of the region, there are other kinds of vegetation present. Küchler's (1970) potential natural vegetation types and reference numbers, for example, include the following:

Sagebrush steppe (*Artemisia-Agropyron*), type 49
Great Basin sagebrush (*Artemisia*), type 32
Saltbush-greasewood (*Atriplex-Sarcobatus*), type 34
Blackbrush (*Coleogyne*), type 33
Galleta–three-awn shrub steppe (*Hilaria-Aristida*), type 51
Juniper-pinyon woodland (*Juniperus-Pinus*), type 21
Mountain mahogany–oak scrub (*Cercocarpus–Quercus*), type 31
Tule marsh–(*Typha* + *Scirpus*), type 42

Other riparian types found along streams do not appear on Küchler's map. These are largely linear shrublands or forests dominated, respectively, by *Salix* (willow) or *Populus* (cottonwood) species where they have not yet been displaced by *Tamarix* (salt cedar) and/or *Eleaganus* (Russian olive). Natural wet meadows are very rare at lower elevations.

Upslope montane forests and meadows of the intermountain region are described in Chapters 3, 4 and 5. The grasslands of the Palouse region (eastern Washington, northern Idaho, northeastern Oregon, and southern British Columbia), the Four Corners region (where New Mexico, Arizona, Colorado, and Utah join); and the Great Plains are described in Chapter 9.

THE REGION IN GENERAL

We start with a brief general discussion of the landforms, geology, climates, and soils of the Intermountain Region. We also need to consider the paleoecological forces that have influenced what we find today. These sections on environmental features are followed by presentations on each of the vegetation types, where more detailed information is brought together.

Landforms and Geology

The region is bounded by high mountain chains on both the east and west. The Rocky Mountain system, positioned along the east, has more than 60 peaks with elevations exceeding 4266 m (14,000 ft). The only major gap in these mountains is in central Wyoming, where there is a broad transition from shrub steppe to shortgrass prairie on high, rolling plain and basin topography (Knight 1994). In eastern Washington, eastern Oregon, and southern Idaho are various depths of loess and volcanic tephra atop largely basaltic plains of Miocene-Pliocene age. The kinds and depths of the surface materials depend mainly on distance and direction from sites where geologic catastrophes have struck. For instance, the melting of Pleistocene ice plugs on the Snake–Columbia river system and the spilling of ancient lakes over earth barriers elsewhere caused dramatic floods that scoured down to basaltic bedrock, where only scattered shrubs can now grow. Elsewhere, alluvial and loessial deposits of fines now promote the success of native bunchgrass or agronomic crops. Volcanic deposits have also affected present-day soils to a considerable degree, the 1980 eruption of Mount St. Helens being an example.

The nearly flat to rolling topography of the Columbia–Snake River Plateau is punctured by a few higher mountains, particularly in northeastern Oregon, but most of the land surface there is elevationally lower than in other intermountain areas. About 70% of that area is <1500 m (Hunt 1974).

Moving west from the Rocky Mountains in western Colorado requires one to traverse two other distinctive physiographic provinces: the Colorado Plateau and the Great Basin (Morris and Stubben 1994).

The Colorado Plateau is a vast epeirogenic upwarp of largely horizontal Mesozoic sedimentary strata that has been deeply dissected by the Colorado River and its tributaries. The elevation of the plateau varies from about 1200 m to 3000 m, with 45% being <1850 m (West 1969). Soils vary from badlands on marine shales to skeletal residual profiles on mesa tops. Small areas of colluvium next to cliffs, sand dunes, loess-covered tablelands, and alluvial fill along the rivers and washes also occur.

West of the high, generally forested, plateaus running north-south through central Utah is the Great Basin, a vast tract of broad valleys situated on a high plateau (all >1200 m). The Great Basin contains more than 200 fault-block mountain ranges, usually oriented north-south, that were created largely by expansional faulting. A few of these mountains exceed 4000 m in elevation. Mountain masses above 1850 m (the approximate average lower boundary of present pinyon-juniper woodlands) occupy about 40% of the area (Billings 1978a). The other 60% of the landscape has been largely coated with gently sloping loessial, alluvial, or lacustrine fill. The Great Basin has no drainage to the sea, and therefore many lakes existed in its valleys during the Pleistocene; some (such as Bonneville in Utah and Lahontan in Nevada) were quite large, and their present-day exposed beds explain much of the fine-textured, halomorphic soil in the valleys.

The current Great Basin has been stretched to about twice the width it had in the Miocene (Fiero 1986), and a great many geothermal springs have resulted. These specialized habitats have an inordinate proportion of threatened and endangered species. The hydrothermally altered andesitic surfaces and endemic floras of the western Great Basin have also received special attention (Billings 1950, 1992).

Soils

The temperate deserts of the intermountain region are not noted for extensive active areas of sand fields and dunes. Exceptions are the local topography-induced dunes such as Sand Mountain in the Carson Desert and the rapidly moving dune fields in Silver State Valley northwest of Winnemucca, Nevada (Morrison 1964); the Little Sahara area north of Delta, Utah; and on the Dugway

Proving Ground and the Navajo Indian Reservation. Most of the sand in the Great Basin portion of the region originated from coarse-textured deltas that formed where rivers entered pluvial lakes during the Pleistocene. These sand fields support diverse and productive plant communities compared to adjacent very fine-textured halomorphic, deep, water-deposited sediments (Pavlik 1989).

The crests of stabilized dunes in Silver State Valley are outlined with *Tetradymia tetrameres* (sand horsebrush), a creeping root-stock species obviously adapted to continuous burial. A unique genetic form, termed *gigas*, of *Atriplex canescens*, occurs in dune slacks at Little Sahara (Stutz, Melby, and Gordon 1975) and sands throughout the Great Basin.

The productivity of the sands is related to their rapid permeability to water compared to the fine sediments dominating most of the lake plain. Precipitation rapidly percolates deep into the sands where it is not evaporated but is available for uptake by plant roots. Permeability also contributes to leaching, which almost always results in low soluble salt accumulations in sand-dominated soil profiles. In contrast, much of the precipitation that falls on fine-textured lake-plain sediments is lost to evaporation from the soil surface, and salts accumulate.

Oryzopsis hymenoides (Indian ricegrass) is the dominant perennial herbaceous species that is constant to most North American temperate desert plant communities where sandy substrates prevail. This bunchgrass produces abundant polymorphic caryopses, many forms of which contribute to the building of seed banks (Young, Evans, and Roundy 1983; Blank and Young 1992). Its caryopses have an indurate lemma and palea that resist sand abrasion and multiple types and levels of dormancy. Although the caryopses are small, seedlings are capable of emerging from great depths in the sands (Young, Blank, Longland, and Palmquist 1994). Natural regeneration of Indian ricegrass stands is significantly influenced by the gathering and scatter-hoard caching activities of desert rodents (McAdoo et al. 1983).

Present Climate

The major reason semideserts prevail on most landscapes of the intermountain lowlands is the orographic "rainshadows" created by the Sierra Nevada and Cascade Mountains on the western boundary of the region that intercept moist winter air masses brought by prevailing westerly winds (Fig. 7.1).

The Rocky Mountains on the east also block some of the weather fronts developing in the Great

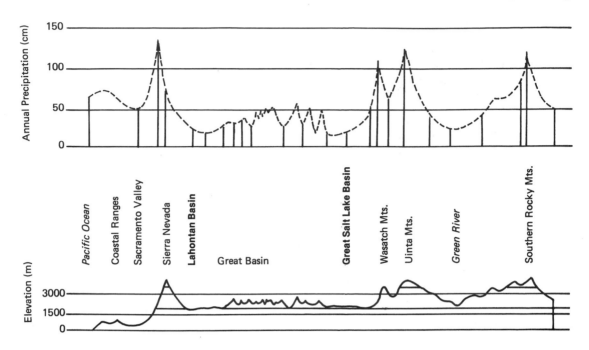

Figure 7.1. Profiles of average elevation (solid line) and mean annual precipitation (dotted line) along 40° N latitude in the western United States (after USDA Yearbook 1941). (Reprinted from West: Temperate deserts and semideserts, 1983, p. 323, with permission of Elsevier Science.)

Plains. Some "monsoonal" storms, however, develop to the south, and precipitation coming from sources other than the westerlies increases in the southeastern portion of the region, especially during the late summer and early autumn.

Whereas the snow-dominated precipitation of winter usually melts slowly, summer storms are high-intensity, short-duration rainfall events. Infiltration of water from summer storms into fine-textured soils is usually minimal and of marginal value to most vascular plants. Gradual snowmelt from winter precipitation, however, usually results in deeper infiltration. The dominance of winter precipitation, combined with either fine-textured or rocky soils, is the main reason for shrub dominance in the intermountain region (Comstock and Ehleringer 1992). Where rarer deep loams or sands prevail, perennial grasses thrive, so long as they are not excessively grazed or burned.

The relatively high elevations of the intermountain "lowlands" produce cool average temperatures. Even though generally less precipitation falls here than in the "warm" deserts to the south, it comes largely during winter and spring when evaporation and transpiration are minimal. Thus, about half of the precipitation enters the soil profile on most intermountain lowlands. "Warm" deserts to the south have much higher average temperatures, evaporation, and runoff rates and therefore

store much less moisture in the soil per unit of precipitation (see Chapter 8). Plant water-use efficiencies are, therefore, greater here than in the southern deserts (Comstock and Ehleringer 1992).

Paleoecological Influences

The present is at least partially determined by conditions of the past, especially in the intermountain lowlands. Some dramatic changes in environment have occurred here over comparatively short time spans (Petersen 1994). The evolution of the present flora is linked most strongly to conditions during the Cenozoic era. During the early Tertiary period, the western half of North America was mostly a level plain occupied largely by forests (Wolfe 1978; Axelrod 1979; Barnosky 1984). Beginning in the Oligocene, there was an uplift of the region involving especially the higher portions of the Sierras and Cascades in the Pliocene—that created drier, cooler environments favoring the evolution of semidesert plants and the extinction of many mesic precursors (Nowak, Nowak, Tausch, and Wigand 1994). There also was an immigration of genera from Eurasia (Shmida and Whittaker 1979).

During the Miocene, there was an evolutionary explosion of mammals in the intermountain region (Grayson 1994). For instance, the fossil beds of the Carson Desert have so far yielded evidence of 18 cameloid species alone. These browsers must have had a profound influence on the balance of shrub species. In the Quaternary, only the extreme northern portion of the intermountain lowlands was glaciated, but there were many glaciers on mountains exceeding 3000 m. The decrease in temperature was linked to an increase in precipitation. Lower summer temperatures meant less evaporation from the approximately 90 lake surfaces in the Great Basin. Conifer forests became established on lower mountain slopes and possibly, in some places, around lakes in valleys (Wells 1983). Diminishment of the trees and expansion of grasslands and shrublands likely took place during the warm interglacials. Holocene climatic changes caused some less pronounced movements of vegetation boundaries (LaMarche 1974; Van Devender and Spaulding 1977; Betancourt 1984; Davis 1984, Miller and Wigand 1994; Nowak et al. 1994).

Although data are ambivalent, the prevailing view is that there was a relatively drier and warmer (xerothermic or altithermal) interval from about 7000–4000 BP (Mehringer 1977). Postxerothermic cooling may then have caused elevational and latitudinal depression of vegetational zones. The most recent evidence (Davis 1984; Nowak et al. 1994) indicates that these effects differed between vegetation types, depending on the climatic variables controlling growth of their dominants.

The migration of humans into the region at least by 13,000 BP (Grayson 1993) may have influenced vegetation, mainly by reducing the numbers of browsers (Grayson 1994) and changing the fire regime (Kay 1995). A subsequent switch from hunter-gatherer to sedentary life styles in the last millennium brought locally heavy impacts, at least in the Four Corners region of the Colorado Plateau (Samuels and Betancourt 1982). Europeans, in only about 300–150 yr (depending on location) of occupation, have brought about more profound changes than all those of the previous 13,000 yr (Miller, Svejcar, and West 1994; Young 1994a, 1994b). Some of these changes are described in the following sections, organized by vegetation types.

MAJOR VEGETATION TYPES

Sagebrush-Dominated Vegetation in General

Woody species of *Artemisia* (sagebrush) are the most characteristic and widespread vegetation dominants in the intermountain lowlands. We therefore first need to know some facts about their taxonomy, autecology, and synecology. There are two groups of sagebrush – tall and low (Fig. 7.2) –

usually segregated by affinity to differing soils (Miles and Leonard 1984). The major species, *Artemisia tridentata* (big sagebrush) has four major subspecies that should be recognized if one is to fully interpret site differences (Winward 1983; Welch and McArthur 1990). These are *Artemisia tridentata* ssp. *tridentata* (basin big sagebrush), *Artemisia tridentata* ssp. *wyomingensis* (Wyoming big sagebrush), *Artemisia tridentata* ssp. *vaseyana* (mountain big sagebrush), and *Artemisia tridentata* ssp. *xericensis* (scabland big sagebrush).

The dominance of *Artemisia* is due to many factors, not the least of which is its seasonal dimorphism of leaves (McDonough, Harniss, and Campbell 1975; Miller and Shultz 1987). That is, large, ephemeral leaves develop in the spring and remain on the plant until soil moisture stress develops in the summer, and smaller, persistent, overwintering leaves develop in late spring and last through the winter, even carrying on photosynthesis then (Caldwell 1979). When water in the soil is frozen, leaf mortality may result (Hanson, Johnson, and Wight 1982).

Artemisia tridentata has both a fibrous root system that can draw water and nutrients near the surface and a taproot that can obtain nutrients and water from deep in the soil profile (Sturges 1977). Thus, if sagebrushes are removed, soil moisture usually increases at depth (Link et al. 1990). However, if flooding creates anaerobic conditions for more than 2–3 days, big sagebrush does not survive (Lunt, Letey, and Clark 1973; Caldwell 1979; Ganskopp 1986). Other *Artemisia* species, namely *A. arbuscula* and *A. cana*, may owe their dominance to being more tolerant of occasionally supersaturated soils (Passey, Hugie, Williams, and Ball 1982).

None of the major sagebrushes has the capacity to resprout after being burned. As we shall see, this is an important feature in explaining successional patterns.

Artemisia plants are long-lived once they make it past the seedling stage. West, Rea, and Harniss (1979) found that individuals of *A. tripartita* in southeastern Idaho lived an average life span of about 4 yr once they survived the first year. The maximum longevity for this species at the site studied exceeded 40 yr. Plants of *A. tridentata* can live to be 100 yr of age (Ferguson 1964). Regeneration is probably episodic, with survival greatest in wetter-than-average periods (Cawker 1980).

All tall sagebrush individuals flower in the fall, at the end of the summer drought period. Seedlings can flower and produce abundant seeds in their second year, with seed production continuing for the life of the plant (Young, Evans, and Palm-

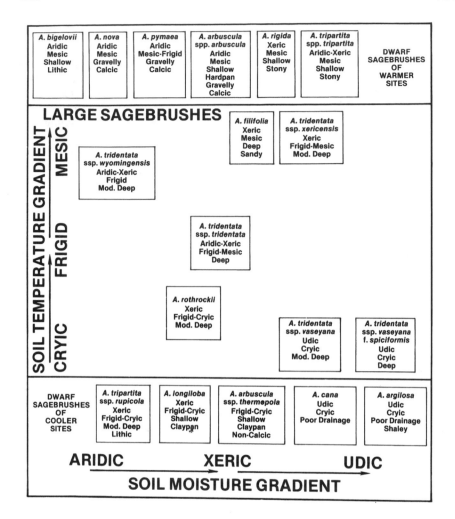

Figure 7.2. Ordination of major sagebrush taxa against gradients of soil temperature and soil moisture (adapted from Hironaka 1979, with additions by the authors; from Robertson et al. 1966; McArthur 1983; Miles and Leonard 1984; Rosentreter and Kelsey 1991). For explanation of terms relating to soil temperature and moisture regimes, see Soil Survey Staff (1994).

quist 1989). During years of severe moisture stress, *Artemisia* plants do not flower. The small (1 by 1.5 mm) achenes mature by midfall to early winter. The pappus is deciduous, and seed dispersal distances are limited except when snow covers the ground and winds can carry achenes along with drifting snow. Seed viability and germination capacity are usually quite high (Young and Evans 1989a, 1989b). Germination occurs in late winter or very early spring. The achenes germinate on the surface of the seedbed, often sheltered by slight microtopography. By 6 mo after seed production, virtually no *Artemisia tridentata* ssp. *wyomingensis* achenes are detectable in seed banks (Hassan and West 1986). Where they go is not known; but consumption and dispersal by vertebrate or inverte-

brate granivores are apparently not major factors in the short half-life of *Artemisia* achenes.

The sizes and degrees of dominance by plants of the several species and subspecies of *Artemisia* vary greatly with site and disturbance history. Sagebrush density is generally greater (Lentz and Simonson 1987a, 1987b; Jensen, Peck, and Wilson 1988; Jensen, Simonson, and Dosskey 1990; Burke 1989), but stature shorter on more xeric sites (Young and Palmquist 1992). Sagebrush also increases in abundance with excessive livestock grazing (Whisenant 1990) combined with lowered fire frequency (Daddy, Trlica, and Bonham 1988).

One can find a variety of situations, from almost pure, low sagebrush on the most xeric sites to increasingly more robust stands of tall brush with

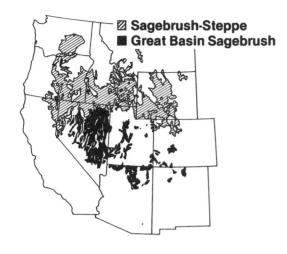

☒ Sagebrush-Steppe
■ Great Basin Sagebrush

Figure 7.3. Map of the sagebrush steppe and the Great Basin sagebrush types (adapted from Küchler 1970).

increasing amounts of perennial herbaceous species on more mesic sites. Given space limitations, we follow the lead of Küchler (1970), who recognized only two potential natural vegetation types in which sagebrush is a dominant: the sagebrush steppe and the Great Basin sagebrush types (Fig. 7.3).

Throughout both the sagebrush steppe and Great Basin sagebrush types, a series of dwarf or low species of *Artemisia* occur (Zamora and Tueller 1973). In the western Great Basin, these occupy about 5% of the total landscape, 40% being covered by big sagebrush (Young, Evans, and Tueller 1976). On the Modoc Plateau, in northeastern California, the distribution of dwarf and big sagebrush plant communities is about equal (Young, Evans, and Major 1977).

Low sagebrushes usually occur where surface soil erosion in former big sagebrush communities has exposed the clay-textured and/or calcified horizons in the subsoil. This is especially apparent in *Artemisia nova* (black sagebrush) and in some *A. arbuscula* (low sagebrush) communities. Big sagebrush seedlings can establish on coarse-textured sediments deposited during storms on top of the mature soils of black sagebrush communities (Young and Palmquist 1992). Many of the dwarf sagebrush sites are often partially flooded during the spring snowmelt period. The flooding is a product of local topography and the lack of permeability of the clay-textured soils and drainage restrictions where petrocalcic layers exist at 30–50 cm (Young et al. 1977).

Basin big sagebrush, often considered to be the progenitor of much of the modern section *Triden-*

tatae of the genus *Artemisia* (Hanks, McArthur, Stevens, and Plummer 1973), does occur on the margin of pluvial lake plains on clay-textured sediments (Young, Evans, Roundy, and Brown 1986). Thus, the most primitive form of the evolutionary group occurs, atypically, on geologically young sites most available for colonization. Basin big sagebrush very rarely occurs on upland soils with clay-textured surface horizons.

At higher elevations, sagebrushes become understory to either scattered juniper, pinyon, or ponderosa pine. However, on the moderately massive mountains of the Great Basin that lack montane forests, sagebrushes form associations with *Purshia* (bitterbrush) (Hormay 1943, Nord 1965), *Symphoricarpos* (snowberry), *Cercocarpus* (mountain mahogany), and *Prunus* (chokecherry) (Tueller and Eckert 1987).

Dwarf sagebrush communities partially parallel adjacent big sagebrush communities in associated herbaceous species composition (Franklin and Dyrness 1973; Young et al. 1976). At higher elevations *Pseudoroegneria spicata* (bluebunch wheatgrass) is the dominant understory species with both mountain big sagebrush communities and adjacent low sagebrush communities. At lower, more xeric situations, *Stipa thurberiana* (Thurber's needlegrass) is the dominant herbaceous species in both basin big sagebrush and adjacent low sagebrush communities. This parallelism does not hold true for all perennial grasses or for all minor broadleaf herbaceous (forb) species. This complex relationship between low and big sagebrush species may offer a window through which to understand the evolution of landscapes, soils, plant communities, and even species diversity in temperate desert environments.

Sagebrush Steppe

The sagebrush steppe occurs predominantly in the northern portion of the Intermountain Region (Fig. 7.3) but also at increasingly higher elevations to the south. Sagebrush steppe occurs where there is an herb-dominated phase of pristine vegetation (Community A, Fig. 7.4) as well as where there is a co-dominance of *Artemisia* with perennial bunchgrasses (Community B, Fig. 7.4). Presence of herbs depends on time since the last fire, outbreak of *Aroga* moth, extremely wet springtime conditions, or very cold winters without snow cover.

The sagebrush steppe once occupied more area than any other North American semidesert vegetation type, 44.8×10^6 ha (West 1983b). Some of this area is now farmland, and some of it has been so degraded by excessive livestock grazing and

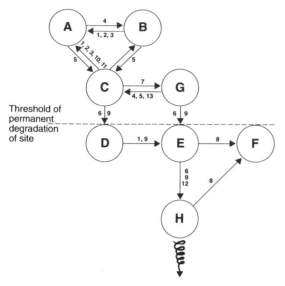

Figure 7.4. Major pathways of progressive and retrogressive succession in lower-elevation sagebrush steppe and Great Basin sagebrush vegetation. Community Types (circles) A-Herb-dominated phase of pristine (without livestock impact) sagebrush steppe; B-Shrub-dominated phase of pristine sagebrush steppe, all growth forms present; C-Denser, larger sagebrush with remnants of native, perennial herbs, after moderate livestock grazing but no episodes of recent fire. Aroga, poor soil aeration, snow mold or vole girdling damage to shrubs; D. Densest, largest sagebrush, understory dominated by introduced annuals, after lengthy, heavy grazing. E-Cheatgrass medusahead other introduced annuals (e.g. bur buttercup) dominate F. Introduced perennial grass (crested wheatgrass, Russian wildrye, etc.) monocultures or polycultures with seeded shrubs and forbs. G-Herbicide-induced native grasslands. H-Weedy, introduced biennials and perennials such as yellow starthistle, leafy spurge, knapweeds, etc. dominate. Force (arrows) 1. Wildfire. 2. Aroga moth 3. Much wetter than average years. (Promotes snow mold, poor soil aeration and vole girdling.) 4. Cessation 1, 2, and/or 3. 5. Moderates livestock grazing. 6. Lengthy, heavy livestock grazing. 7. Herbicides applied to kill broad-leaved plants. 8. Tillage followed by seeding. 9. Accelerated soil erosion. 10. Prescribed burning. 11. Sheep, deer, and/or elk use in fall or winter. 12. Further weed invasion. 13. Interseeding of native shrubs and forbs.

burning that its relationship to its origins is no longer easily recognizable. After repeated fire, sagebrush steppe has in many places been replaced by exotics, especially the European annual grasses *Bromus tectorum* (cheatgrass) and *Taeniatherum caput-medusae* (medusahead) (Whisenant 1990).

The floristic diversity of the sagebrush steppe is only moderate (West 1983b). On relict sites in central Washington, Daubenmire (1975) found an average of 20 vascular plant species on several plots of size 1000 m². Tisdale, Hironaka, and Fosberg (1965) reported 13–24 higher plant species in three

ungrazed stands in southern Idaho. Zamora and Tueller (1973) found a total of 54 vascular plant species in a set of 39 high-condition low sagebrush steppe stands in the mountains of northern Nevada.

In relatively undisturbed examples of this vegetation type, the shrub layer reaches approximately 0.5–1.0 m in height (Fig. 7.5) and has a cover of about 10–80%, depending on site and successional status (Fig. 7.4). The grass and forb stratum reaches to about 30–40 cm during the growing season (Fig. 7.6), and its cover varies widely, depending on site and successional status. On relict sites, the sum of absolute cover values, species by species, usually exceeds 80% and can approach 200% on the most mesic sites (Daubenmire 1970).

The herbaceous life form most prevalent on relict sagebrush steppe sites is the hemicryptophyte (Daubenmire 1975), although the proportion of therophytes has increased markedly even where livestock grazing has been removed (Brandt and Rickard 1994). The floristic proportion of geophytes is around 20%. A microphytic crust dominated by mosses, lichens, and algae is commonly found in the interspaces between the perennials (West 1990).

Perennial grasses associated with *Artemisia* steppe vary greatly throughout the vegetation type (West 1983b). *Festuca idahoensis* (Idaho fescue) and congeners are common in the northwest and at higher elevations and latitudes elsewhere. Bluebunch wheatgrass (*Pseudoroegneria spicata*) is probably the most widespread and important herbaceous component of this vegetation type. Sod-forming warm-season Triticeae, such as *Pascopyrum smithii* (western wheatgrass) and *Elymus lanceolatus* (thickspike wheatgrass), become more important in the eastern portion, apparently because of greater growing-season precipitation there. *Stipa* species are important codominants along the southwestern boundary of the sagebrush steppe: *S. thurberiana*, *S. arida*, and *S. speciosa* sort out along a complex gradient related to decreasing elevation (Young et al. 1977).

Perennial grasses must build most of their aboveground tissues during a narrow window of favorable temperature and adequate soil moisture in late spring and early summer (Fig. 7.7). Perennial forbs are subject to the same constraints but can apparently store more reserves in their fleshier roots.

Phenological progression is more rapid for forbs than for grasses or shrubs (Fig. 7.8). This could be related to greater forb leaf size requiring greater transpiration for cooling. That is, as moisture stress develops, forbs lose leaf turgor and vitality earlier. Shrub roots go deeper and thus draw on soil mois-

Figure 7.5. Pristine example of sage-brush steppe on the Carey kipuka, Craters of the Moon National Monument, Idaho (from Tisdale et al. 1965; photo courtesy M. Hironaka. (Reprinted from West: Temperate deserts and semi-deserts, 1983, p. 351, with permission from Elsevier Science.)

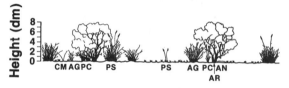

Figure 7.6. *A bisect of relict* Artemisia tridentata–Pseudo-roegneria spicata *stand, Yakima County, Washington (from Daubenmire 1970). Drawings to scale, including all vascular plants with any basal area impinging on a transect 2 cm wide by 330 cm long. Height (0–8 dm) is on the vertical axis. Key to species symbols; AG, Pseudoroegneria spicata; An, Antennaria dimorpha; AR, Artemisia tridentata; CM, Calochortus macrocarpus; PC, Poa fendleriana; PS, Poa secunda.*

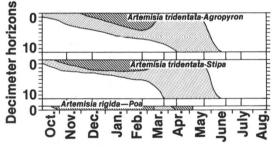

Figure 7.7. *Soil moisture status for three climax sagebrush steppe communities in eastern Washington from 2 October 1962 to 17 August 1963. Limits of horizons (in decimeters) shown on ordinate. Shaded areas indicate water content in excess of field capacity; stippled areas indicate water content between field capacity and wilting coefficient; unshaded areas lack growth water. Vertical lines between horizontal panels show actual dates of sampling (from "Annual cycles of soil moisture and temperature as related to grass development in the steppe region of eastern Washington" by R. F. Daubenmire, Ecology, 1972, 53, 419–424. Copyright 1972 Duke University Press, Durham, N.C., reprinted by permission).*

ture that is recharged over the winter. Summer precipitation usually is light and ineffective for recharging the soil moisture profile to depths where it contributes much to vascular plant growth. Instead, most of the moisture from summer precipitation is simply lost through evaporation.

The pristine sagebrush steppe evolved with large browsers, such as the Shasta ground sloth, mastodon, and camels, but most of them had disappeared by about 12,000 yr BP (Grayson 1994). Graminivore populations were also low in the Great Basin and inland Pacific Northwest (Mack and Thompson 1982) when Europeans first explored there. The small populations of aboriginal hunter-gatherers probably had little direct impact on the vegetation, except through fire (Kay 1995). The aboriginals set fires, probably to drive or attract game. Fires diminished sagebrush and favored herbaceous species.

It took European colonization to change the native vegetation more drastically. Valleys were converted to farms, and some of the upland is still being converted to intensive agriculture with sprinkler irrigation. The most widespread early influence, however, was that of livestock, whose populations built up rapidly in the late nineteenth century (Kennedy 1903; Robertson 1954). Griffiths (1902) judged that the grazing capacity of these ranges had definitely been exceeded by 1900. Hull (1976) examined historical documents and con-

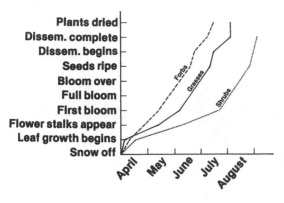

Figure 7.8. Average phenological progression for grasses, forbs, and shrubs at the U.S. Sheep Experimental Station, Dubois, Idaho, 1941–1947 (From Blaisdell 1958; reprinted from West: Temperate deserts and semi-deserts, 1983, p. 357, with permission from Elsevier Science.)

cluded that the loss of native perennial grass and expansion of shrubs took only 10–15 yr.

The primeval vegetation was only weakly stable in the face of herbivory because of the ecophysiological and reproductive disadvantages of the native perennial grasses and the great competitive advantage of the shrubs. Herbaceous species have to rebuild all their aboveground tissues each growing season. This requires upward translocation of a considerable amount of food reserves, water, and nutrients during the short growing season (late spring and early summer). To maintain their dominance, shrubs have to rebuild only enough tissues to replace the comparatively smaller amount that was shed as litter during the previous year.

Brush foliage, furthermore, has chemical defenses against herbivory, whereas grass is extremely palatable, particularly when it is green. The native bunchgrasses sustain high mortality when grazed heavily in the spring (Stoddart 1946). In addition, they rarely produce good seed crops (Young et al. 1977).

The only time that the grasses and forbs have an advantage over brush is when this vegetation is burned. Herbs resprout easily, especially if fires occur during the summer. The major species of *Artemisia* have to reestablish from seed.

Europeans introduced another force: aggressive weeds, largely from Eurasia. Foremost among these is cheatgrass, *Bromus tectorum* (Billings 1990, 1994). Although introduced to the Pacific Coast in the 1870s (Mack 1981), it was not until 1894 that M. E. Jones collected it in Utah and in 1906 that P. B. Kennedy collected it in the Reno region (Billings 1994). By 1928, it reached its present regional distribution, and during the late 1930s, 1940s, and

1950s, it became dominant. This annual grass outcompetes the native grasses, mainly by beginning growth in the fall and growing root systems throughout the winter (Harris 1977). Cheatgrass completes its life cycle and produces seeds by late June or early July. The fine, continuous fuel load it creates makes the region susceptible to earlier and more frequent fires than those that occurred in the past, and many of the native herbs and shrubs are destroyed by these fires (Wright and Bailey 1982). If soils are without cover during summer convectional storms, soil erosion can be severe. The result, over much of this area, has been a downward spiral of site degradation (Fig. 7.4).

Other groups of species that have increased because of excessive livestock grazing and an accelerated frequency of fire are members of the shrub genera *Chrysothamnus* (rabbitbrush), *Ephedra* (Mormon tea), *Tetradymia* (horsebrush), and *Gutierrezia* (snakeweed). This is because they can resprout after fire (Tisdale and Hironaka 1981). *Chrysothamus* spp. are shorter-lived shrubs than *Artemisia* and they have taken over vast reaches of the Intermountain West. Anderson, Ruppel, Glennon, Holte, and Rope (1996) conclude that *Chrysothamnus*—because it has not diminished and been taken over by sagebrush in over 40 years on long-term ungrazed plots studied on the Idaho National Engineering Laboratory—should be regarded as a climax dominant, at least on some sandy and loessial sites.

On heavier soils, the exotic annual grass *Taeniatherum caput-medusae* has even replaced cheatgrass (Young 1992). Other problem taxa that currently are localized but are expanding are *Aegilops cylindrica* (goat grass), *Salvia aethiopis* (Mediterranean sage), *Isatis tinctoria* (dyer's woad), *Ceratocephala testiculata* (bur buttercup), and several species of *Centaurea* and *Acroptilon* yellow star thistle and knapweed). There has thus been a vast replacement of native long-lived perennials by shorter-lived taxa, especially by introduced annual grasses.

Complete fire suppression is an impossibility, and reductions in livestock will not necessarily result in a return to vegetation similar to that in the pristine condition (Fig. 7.4) (Anderson and Holte 1981; Brandt and Rickard 1994). The new annual-dominated vegetation appears to have created its own new equilibrium (Hanley 1979). Our major means of obtaining greater dependability of forage production, while at the same time reducing the chance of fire and soil erosion, is to plant introduced Triticeae (wheat grasses and rye grasses) (Keller 1979). This usually results in more simplified vegetation and can be accomplished only on relatively level sites with deep soils. The remaining, rougher topography awaits more imaginative management and restoration that could benefit from basic research in plant ecology.

The rapid conversion of intermountain rangelands to dominance by adventive annual grasses has reached a level where vast areas of the sagebrush steppe may not have the potential to return to dominance by native forbs, perennial bunch grasses, and even woody sagebrush unless large amounts of capital and labor are spent on restoration practices and/or there are significant breakthroughs in revegetation technology (Monsen and Kitchen 1994).

Ben and Cindy Roche (1991) have offered the startling but logical hypothesis that dominance by annual grasses is not an ecological endpoint, merely a stage toward the eventual dominance by adventive biennials or perennials. In portions of the sagebrush steppe, such as the more mesic areas of the Columbia Basin, the northern valleys of Utah, and northeastern California, their hypothesis is supported by numerous examples of successful adventive forbs that can replace alien annual grasses that came earlier. These include a host of annual and biennial species of *Centaurea* and *Acroptilon*, with *C. solstitalis* (yellow star thistle) being the most widespread and colonizing the most diverse habitats. Several biennial adventive species that are emerging as dominants – *Isatis tinctoria*, *Salvia aethiopis*, *Onopordum acanthium*, and *Carduus nutans* – generally suppress competition during their first growing season, through either smothering rosettes or allelopathy (Young, Evans, and Kay 1971; Dewey 1991).

At the top of this array of adventive species are numerous perennials. *Centaurea biebersteinii* (spotted knapweed) is a relatively short-lived perennial that has spread over an estimated 2.8 M ha in the western United States during the twentieth century (Roche and Roche 1991). Long-lived herbaceous perennials such as *Acroptilon repens* (Russian knapweed) are serious pernicious weeds that are difficult to eradicate once established. In the same class are *Cirsium arvense* (Canada thistle), *Euphorbia esula* (leafy spurge), *Cardaria draba* (hoary cress or whitetop), and *Lepidium latifolium* (perennial pepperweed).

Most of the adventive perennials produce abundant seed crops and are vegetatively propagated by creeping rootstocks on sites usually associated with areas of more abundant moisture. Therefore, they are most widely established in more northern latitudes and at the margins or wetlands, meadows, and riparian ecosystems elsewhere within the region.

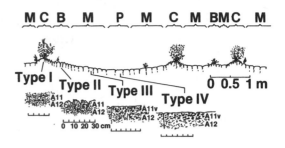

Figure 7.9. Schematic cross-sectional diagrams of the microtopographic positions and associated surface soil morphological types of gently sloping, shallowly loess-mantled Argids of the Humboldt loess belt with plant communities dominated by Wyoming big sagebrush (Artemisia tridentata ssp. wyomingensis). Microtopographic positions: C, coppice; B, coppice bench; M, intercoppice microplains; P, playette. Vertical scale somewhat exaggerated; intercoppice microplains and playettes may be much wider than shown here; several coppices may be linked together. Vertical lines under soil indicate crust polygons (AIIv) that continue downward as prisms in the compoundly weak prismatic and moderately platy A12 horizon. Type 1 is covered by litter. Circles indicate vesicles in crust (AIIv). Only types III and IV are significantly crusted (From Eckert et al. 1978; reprinted from West: Temperate deserts and semi-deserts, 1983, p. 334 with permission from Elsevier Science.)

Great Basin Sagebrush

The second vegetation type dominated by *Artemisia* is the Great Basin sagebrush found to the south of the sagebrush steppe (Fig. 7.3). This vegetation type potentially occupies 17.9×10^6 ha of the Great Basin, Colorado Plateau, and adjacent physiographic provinces. The sagebrush steppe and Great Basin sagebrush vegetation types have often been considered together in reviews (College of Natural Resources 1979; Tisdale and Hironaka 1981; Blaisdell, Murray, and McArthur 1982), but we consider them distinctive in terms of floristic diversity, production, and responses to perturbations.

Most of the Great Basin sagebrush type is much more arid and thus akin to true deserts, whereas the sagebrush steppe is similar to a semiarid grassland. In the Great Basin sagebrush type, *Artemisia* prevails, accompanied by few grasses even in pristine or late seral conditions. The A phase in Fig. 7.4 is, thus, usually lacking here.

The shrub layer seldom reaches over 1 m in height. The shrubs usually are less densely spaced than in the sagebrush steppe, and a considerable portion of the interspaces is composed mainly of vesicular "foamy" surface soils (Fig. 7.9) and/or microphytic crusts, whereas herbaceous plants are usually located near or under the shrubs. The shrubs are commonly located on hummocks of el-

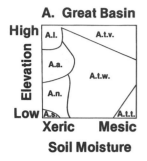

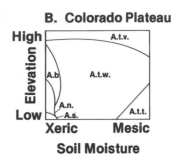

Figure 7.10. Ordinations of major sagebrush taxa found in the Great Basin (left) and Colorado Plateau (right) against gradients of elevation and effective soil moisture. A.b = Artemisia bigelovii, and so forth, as derived from the first letters of the genus, species, and subspecies names from Fig. 7.2 (Adapted from West 1979; reprinted from West: Temperate deserts and semi-deserts, 1983, p. 335 with permission of Elsevier Science.)

evated microrelief because of differential erosion and deposition. These serve as "islands of fertility," providing an ameliorated microclimate within a more hostile matrix (Pierson and Wight 1991). Anything that alters these shrub-centered patterns also diminishes productivity on a landscape basis, because these desert-like environments have too little moisture, nutrient input, and storage to carry on a high level of production across the entire landscape.

Artemisia commonly makes up more than 70% of the relative plant cover and more than 90% of the phytomass, regardless of successional status. This high degree of shrub dominance lends a characteristic grayish-green color to the landscape. An ordination of major *Artemisia* taxa in this vegetation type against elevation and soil moisture gradients is shown in Figure 7.10.

In the sparse understory there is a trend from cool-season grasses in the Great Basin to warm-season sod grasses on the Colorado Plateau (West 1979). Temperature and moisture patterns generally are more variable than those illustrated in Figure 7.7.

The Great Basin sagebrush type appears deceptively monotonous and simple at first glance. A low floristic richness and concentration of dominance in so few species leads to loose species packing (Yorks, West, and Capels 1992), apparently compensated for by great intraspecific variation in major taxa such as *Artemisia* (McArthur 1983), *Chrysothamnus* (Anderson 1975), and *Elymus elymoides* (Clary 1975). Taxa with only slight morphological variations actually have considerable genetic and ecological variants that dominate functionally different communities.

The same ecophysiological and autecological attributes of *Artemisia*, the native perennial bunchgrasses, and the introduced annuals discussed previously also apply to successional pathways here (Fig. 7.4). Original vegetation was more shrub-dominated, and stability probably was even more precarious than in the sagebrush steppe. Modern soil erosion has generally been more severe than

in sagebrush steppe because there has been less plant and litter cover.

Removal of livestock does not usually result in a decline in sagebrush and an increase of perennial grass (Potter and Krenetsky 1967; Rice and Westoby 1978; West, Provenza, Johnson, and Owens 1984; Daddy et al. 1988; Yorks et al. 1992), because of lowered site potential. The economic feasibility of vegetation type conversion is much less here than in the sagebrush steppe (West 1983c).

Undesirable plants have increased in abundance in Great Basin sagebrush communities. In addition to those already mentioned for sagebrush steppe is *Halogeton glomeratus* (halogeton), which is poisonous to livestock. *Centaurea triumfeltii* (squarrose knapweed) is spreading rapidly in big sagebrush communities in western Utah. The conversion to annual dominants seems to be a widespread and permanent trend (Rogers 1982; Sparks, West, and Allen 1990). Herbaceous species that are invasive colonizers in sagebrush are often extremely variable in genetic character. This variability may be the result of repeated introductions, evolution from isolated populations, and/or breeding systems that respond to changing environmental conditions with hybridization and recombination (e.g., Stebbins 1950, 1957). We have reviewed here only a small number of introduced perennial weeds that could potentially replace sagebrush and then become characteristic landscape plants. For example, in central Asia, the other major land area of the world with a similar temperate desert environment (West 1983a), a considerable number of perennial species of *Salsola* have evolved. Only *S. vermiculata* has been introduced to western North America, and the ecotype introduced is not winter hardy in the northern Great Basin.

Native herbivores have also been affected both directly and indirectly by the introduction of concentrations of domestic livestock. Mule deer (*Odocoileus hemionus*) populations dramatically increased on Great Basin mountain ranges following the introduction of domestic livestock (Julander and Low 1976). This increase has been largely attributed to the increase in shrubs (the main winter diet of the deer) and reduction in wildfires because of the diminishment of fine, continuous herbaceous fuel.

During the same period, pronghorn (*Antilocapra americana*) numbers first dramatically declined, then sharply rebounded. Pronghorns use significant amounts of sagebrush browse in their diets at all seasons of the year. They can, however, also exist in grassland environments without sagebrush (Ferrel and Leach 1950). This recovery in pronghorn populations has been attributed to overlap in dietary requirements of pronghorn and domestic sheep (*Ovis aries*). Once the range sheep industry declined (post–World War II), the pronghorn population increased sharply. The decline and subsequent increase in pronghorn populations may also reflect the initial increase in grasses that followed extensive removal of sagebrush via mechanical or herbicidal treatments (Ferrel and Leach 1950, 1952) as well as by increased fire frequency and spread of alien herbs.

Before contact with Euro-Americans, human populations in the Great Basin depended on grass seeds and jackrabbits (*Lepus* spp.) (Grayson 1993; Wilde 1994). Jackrabbit numbers are cyclic in nature. When high, they are prodigious consumers of herbaceous vegetation, lower-statured shrubs, and seedlings of almost all species (McAdoo and Young 1980). Artificial regeneration of shrubs is impossible on a small scale without protection from jackrabbits. The only exception is certain ecotypes of big sagebrush. Obviously, jackrabbits are a very important herbivore in sagebrush vegetation. Since the introduction of domestic livestock into the Intermountain Region, the dominant species of jackrabbit has changed from whitetailed (*Lepus townsendi*) to blacktailed (*L. californicus*). The reasons for this shift are not completely known, but they are correlated with changes in vegetation.

Saltbush-Greasewood

Vegetation dominated by perennial chenopod shrubs and half-shrubs makes up another considerable portion of the intermountain lowlands. Because such communities are usually associated with halomorphic soils, the descriptor "salt-desert shrub" (also "salt-desert scrub") has gained wide usage (West 1983d). This is synonymous with Küchler's (1970) saltbush-greasewood vegetation type, which potentially occupies a total area of 16.9 × 10⁶ ha scattered over all four of the regional deserts of North America, plus particular locations in the Central Valley of California, the upper Rio Grande drainage, and the Great Plains (Fig. 7.11).

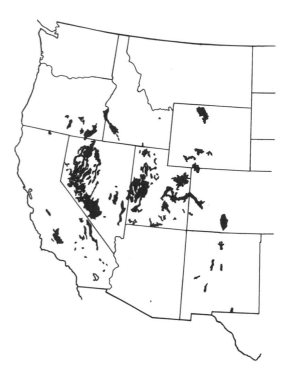

Figure 7.11. Map of the saltbush-greasewood vegetation type (adapted from Küchler 1970).

The major salt-desert shrub species and their habitat preferences are listed in Table 7.1.

Although the correlation of this vegetation with halomorphic soils is strong, it is not universal. For example, this vegetation dominates dry nonsaline soil covered by desert pavement in the extreme rainshadow of the Sierra Nevada in western Nevada and adjacent California (Billings 1949). *Artemisia* species cannot thrive there because of drought, even though halomorphic soils are lacking.

Four major types of habitats are found at or near the bottom of Great Basin valleys: uplands dominated by xerohalophytes, where the permanent water table is deeper than 1 m (Fig. 7.12); lowlands dominated by hydrohalophytes, where the water table remains within 1 m of the soil surface; marshlands and meadows dominated by grasses and grass-like plants, where the water table is above the soil surface for much of the year; and occasionally flooded barren playas. Figure 7.13 shows the positioning of plant associations in relation to water table gradients in a valley in western Nevada.

Halomorphic soils develop either from lake beds (lacustrine deposits) or from badlands on marine shales that are excessively drained as well as

Table 7.1. *Major vascular species of the salt-desert shrub type according to growth form and preference in regard to subsurface moisture*

Habitat–habit	Growth forms		
	Shrubs	Half-shrubs	Herbs
Lowland hygrohalopytes: free water table at least occasionally present at the surface and usually remaining within about 1 m	*Sarcobatus vermiculatus*	*Allenrolfea occidentalis* *Salicornia utahensis* *Suaeda moquinii*	*Distichlis stricta* *Sporobolus airoides* *Suaeda calceoliformis*
Upland xerohalopytes: water table well below 1 m	*Artemisia spinescens* *Atriplex confertifolia* *Grayia spinosa*	*Atriplex corrugata* *A. cuneata* *A. falcata* *A. gardneri* *A. tridentata* *Krascheninnikovia lanata* *Kochia americana*	*Bromus tectorum* *Leymus cinereus* *Halogeton glomeratus* *Lepidium perfoliatum* *Oryzopsis hymenoides* *Salsola kali* *Elymus elymoides*

Figure 7.12. Atriplex confertifolia–Krascheninnikovia stand in Curlew Valley, northwestern Utah, July 1964.

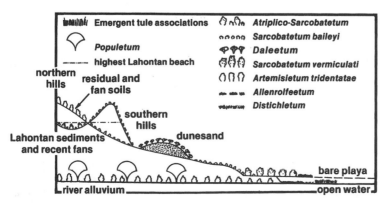

Figure 7.13. Diagrammatic representation of the topographic and geologic positions of the principal plant associations in the Carson Desert region of western Nevada (redrawn from Billings, 1945).

being halomorphic. Variations in vegetation composition and productivity are intimately related to soil characteristics in such environments (West 1983d; Knight 1994).

In the western Great Basin, above the pluvial lake plains, expansive shrub communities occur dominated by *Sarcobatus baileyi* (Bailey greasewood), sometimes considered a subspecies of *Sarcobatus vermiculatus* (black or big greasewood). Mozingo (1986), in his definitive study of Great Basin shrubs, recognized Bailey's greasewood as a separate species but Kartesz (1994) does not. Bailey's greasewood is morphologically and ecologically a very distinct plant: It is not a phreatophytic species, nor does it occupy soils extremely influenced by accumulations of soluble salts; it occupies dry sites within the rainshadow of the Sierra Nevada (Billings 1949). Black greasewood is a phreatophyte that dominates near-monocultures over vast stretches of intermountain valleys where both salts and water accumulate each year (Billings 1978b).

Most of the vascular plants found in salt-desert shrub areas are members of the Chenopodiaceae. Only occasional representatives of the Asteraceae, Brassicaceae, Fabaceae, and Poaceae are found, and none can be called a dominant. About 20% of the vegetation can be described as essentially mosaics of monocultures. This is because few plants can tolerate such dry and salty habitats, and topographic gradients are often very gentle, perhaps allowing competitive sorting to become evident (West 1983d). Because the positioning of species may follow different sequences in different valleys, at least some of the variation may be caused by ecotypic variation. Recent evolution of these Chenopodiaceae is suspected (Stutz 1978).

Total cover of vascular plants varies from zero on the most saline shale badlands or saltpans up to 25% on some upland sites. Soil texture in the uplands greatly influences water infiltration and evaporative losses, with gravelly soils having the greatest cover, richness, and production because of the inverse-texture principle (Noy-Meir 1973).

Variation in cover on sites with seasonal or permanent water tables depends on the chemical characteristics of the soil and water. Species apparently sort along seasonal patterns of matric and osmotic potentials (Detling and Klikoff 1973). The dominants of salt deserts either have means to exclude uptake of salts, or they take salts up but anatomically isolate and/or excrete them.

Shrubs of salt deserts are usually <50 cm tall (except *Sarcobatus vermiculatus*), widely spaced, and clustered (West and Goodall 1986). Regeneration of new plants occurs where old plants exist or

have existed, rather than in the interspaces (West 1983d), because of a more favorable soil moisture, soil organic matter, temperature, and nutrients in these spots. The interspaces usually are covered with soft, rugose microphytic crusts if the soil has not been compacted by animals or vehicles. These crusts reduce soil erosion (Williams, Dobrowolski, and West 1995; Williams, Dobrowolski, West, and Gillette, 1995) and may contribute nitrogen (Harper and Pendleton 1993).

Upland xerohalophytes experience a surge of growth in the spring (West and Gasto 1978), because they can draw in only freely draining (vadose) soil moisture. Several of the dominants feature leaf dimorphism: They have a set of larger, spring leaves that are lost as soil drought develops, and a second set of much smaller, overwintering leaves that can carry on some photosynthesis over the winter. The hydrohalophytes of high-water-table sites are more completely deciduous, developing new leaves in midsummer. Most (except *Sarcobatus*) are C_4 species.

The harsh environments of salt deserts slow down community dynamics. The same species or species similar in appearance and stature often succeed each other after disturbances, therefore autosuccession probably best describes succession over decades (West 1982). On a longer time scale, primary succession occurs with transgression and regression of saline lakes. The primary driving factor is either flooding, which destroys previous vegetation, or a deepening water table during drying periods. Figure 7.14 outlines the usual sequence of native plant species located along a temporal gradient. Species change largely in response to variations in salinity. For instance, Rickard (1964) described how *Sarcobatus* can cause *Artemisia* to decline as microtopography is eroded downward following salinization.

Livestock grazing over the past 140 yr has been the major instrument for change in these communities. We have little concrete evidence of change, however, because the topography of the area does not lend itself to preservation of many relict areas (Svejcar and Tausch 1991). Productivity was so low, forage so coarse, and water so scarce that overwintering sheep were the main animals Euro-Americans grazed here. This meant that there were few fences to create the unintentional "experiments" that fence-line contrasts can provide. Control over livestock numbers did not begin until after 1934 when the Taylor Grazing Act became law over the last remnants of the public domain.

Invasion of *Halogeton glomeratus*, an annual forb from Eurasia that is poisonous to livestock, began in the 1940s. This stimulated the first ecological re-

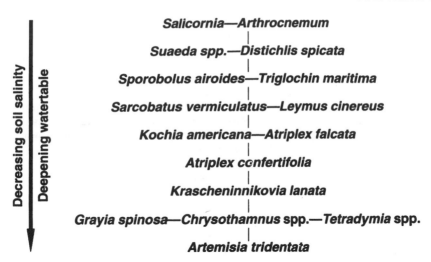

Figure 7.14. Probable halosere around retreating saline lakes of the Great Basin (adapted from Flowers and Evans 1966).

search. Thus, our perspective is short, and the data are inadequate to fully decipher retrogression. We do know that the most palatable shrubs – *Artemisia spinescens* (bud sage), *Krascheninnikovia lanata* (winterfat), and *Kochia americana* (gray molly) – declined substantially, especially when grazing use extended well into the spring (Blaisdell and Holmgren 1984). Unfortunately, these shrubs also had the least reproductive capacity. The less-palatable saltbush species (*Atriplex confertifolia, A. gardneri, A. falcata, A. tridentata, A. cuneata, A. corrugata*) have come back more rapidly after control of livestock grazing. These trends, however, are difficult to distinguish from annual fluctuations and the effects of longer-term climatic influences (Norton 1978; Yorks et al. 1992).

Population explosions of insects such as round-headed borers and cutworms, as well as mammals such as jackrabbits, can greatly change a salt desert plant community even in the absence of livestock (West 1982; Sharp, Sanders, and Rimbey 1990). Interactions between animal population irruptions and drought or excessive soil moisture are likely but cannot yet be verified. Extensive die-off of xerohalophytic (upland) shrubs was observed during the wet years of the mid-1980s (Nelson et al. 1989; Haws, Bohart, Nelson, and Nelson 1991; Price, Pyke, and Mendez 1992). A few days of water-saturated anoxic soils is the likely trigger factor leading to shrub mortality. Only partial recovery from these stand-replacing events has occurred since.

Because vascular plant cover is generally so sparse, this has historically been one of the few ex-

tensive western vegetation types in which wildfire was rare. Since 1983, however, fires have occurred in salt-desert shrub communities of the Great Basin because of the profusion of (mostly exotic) annuals after a sequence of years in which annual precipitation was twice or more the average. We shall soon learn how the various species can respond to these new kinds of impacts (West 1995). The salt deserts of western North America have their own native annual forbs, particularly in the chenopod genera *Atriplex, Chenopodium, Monolepis,* and *Suaeda* and the crucifer genera *Lepidium* and *Alyssum.* The major pioneers following disturbance are, however, exotics.

Unfortunately, the expansion of annuals does more than bring fire to these communities. *Halogeton,* for instance, may permanently change the soil surface by pumping salt to the soil surface (Eckert and Kinsinger 1960), which impedes germination of seeds of other species as well as reducing moisture infiltration and enhancing evaporation. Fortunately, *Halogeton* and the other major invaders – *B. rubens* (red brome), *Salsola kali* (Russian thistle), *Lepidium perfoliatum* (shieldcress), *Malcolmia africana* (African mustard), bur buttercup, and several annual *Atriplex* and *Chenopodium* species – are not so aggressive that they invade mid-to late-seral communities. Nevertheless, the spread of cheatgrass has been noted even under moderate livestock grazing (Yorks et al. 1992) and in one case (Svejcar and Tausch 1991) where no livestock grazing had apparently ever taken place. The prognosis for maintaining a diverse, stable, and economically valuable plant cover in some of these areas is not good.

Russian thistle (*Salsola kali*) was the first of the alien chenopods to spread across our temperate deserts (Young 1988), late in the nineteenth century. The first infestations occurred along railroads. This annual has a seed dispersal system that virtually requires tumbling of the mature plants to release the seeds. An abundance of small, poorly protected seeds that consist of little more than embryonic plants are produced by each plant. The seeds have temperature-related after-ripening requirements that disappear gradually 3–4 mo after maturity (Young and Evans 1972), protecting the tender seedlings from frost damage. Once transitory dormancy requirements are satisfied, Russian thistle seeds can germinate from 0–40° C and at diurnal fluctuations containing both extremes. Russian thistle seedlings are not highly competitive, and the dispersal system helps to assure that seeds reach disturbed habitats where competition is not a significant factor. Russian thistle plants provide significant winter forage for large herbivores, despite their spiny nature. The seeds are a major resource for granivores, especially small mammals and birds.

A second kind of Russian thistle (*Salsola paulsenii*, barbwire Russian thistle) invaded the salt desert and lower *Artemisia* portions of the temperate deserts during the 1960s (Young and Evans 1979). This annual chenopod has replaced virtually all Russian thistle infestation in this environment and has also suppressed *Halogeton* populations.

The most dramatic invasive and colonizing chenopod species to invade American temperate deserts is halogeton, a native of Kazakhstan in west-central Asia, where it occurs on solonetz and solonchak soils in deserts. It was first collected in North America near Wells, Nevada, by A. H. Holmgren in 1934 (Young 1988). In 1942, it was determined that this fleshy annual was poisonous to sheep, after the weed was already widely established in Nevada and northeastern California. It was well adapted to salt-affected soils and colonized extreme environments on the margins of playas where no native species occur. Halogeton is capable of producing high quantities of dimorphic seeds. The predominant form, known as black seed, is capable of near-instantaneous germination when soaked. The other form, brown seed, can remain dormant and viable in seed banks for nearly 10 yr. Halogeton's range currently reaches the Missouri River in Nebraska, the Canadian border in Montana, and northern New Mexico.

Early in the twentieth century, *Bassia hyssopifolia* (five-hook bassia) was accidentally introduced near Fallon, Nevada (Robbins 1940). This weed has spread along the margins of wetland areas in the temperate deserts throughout western North America.

The diminishment of native perennials and the dominance of exotic annuals in salt-desert shrublands is particularly serious because there is no successful way to reseed perennials into these rigorous environments. The only promising material is *Kochia prostrata* (forage kochia), an exotic, long-lived half-shrub (Horton, Monsen, Newhall, and Stevens 1994).

Blackbrush-dominated Vegetation

In the lower but nonsaline parts of the Colorado Plateau, and particularly where the Mojave and Great Basin deserts merge, blackbrush (*Coleogyne ramosissima*) dominates semidesert parts of the landscape. Blackbrush also occurs at higher elevations, dominating inclusions within grasslands of the Colorado Plateau or as pinyon-juniper woodland understory wherever soil depth is restricted. Blackbrush is one of the few plants in our region that seem to prefer old pediment slopes, bajadas, and ballenas with petrocalcic (caliche) horizons. Thatcher, Doughty, and Richmond (1976) and King (1981) have noted a gradient of declining brush and increasing amounts of grasses with increasing soil depth to the petrocalcic (or other types of restrictive) layers, even on areas that have never been grazed by livestock (Fig. 7.15). Not many species thrive in these difficult environments transitional between "cold" and "warm" deserts. Only in the spring is there much of a showing of annuals.

Coleogyne is a roundish shrub seldom reaching a height >0.5 m. The round shape is the result of terminal twigs dying back and forming spines; all subsequent twig growth is from subterminal buds (Bowns and West 1976). Also, intercalary cork is formed between stem segments during drought, creating multiple stems. Other kinds of shrubs are rare where *Coleogyne* occurs. Only *Prunus fasciculata*, *Grayia spinosa*, *Thamnosma montana*, *Ephedra* species, and *Gutierrezia* species are worthy of mention. Some occasional *Yucca*, *Opuntia*, and *Agave* can be found interspersed. Perennial herbs are limited to the grasses *Hilaria rigida*, *H. jamesii*, *Oryzopsis hymenoides*, *Aristida* species, *Stipa* species, *Bouteloua eriopoda*, and *Muhlenbergia porteri* (in order of declining abundance). The amounts of these are ordinarily low, and very little of the vegetation type can be described as shrub steppe.

Despite the simple flora, the total vascular plant cover can be surprising, reaching 37–51% (Beatley 1975), especially after wet conditions. The cover

Figure 7.15. Coleogyne ramosissima–dominated vegetation on pristine site never grazed by livestock; isolated mesa in the western portion of Grand Canyon National Park, Arizona, June 11, 1981.

and the resinous nature of the brush promote fire. There is typically a well-developed microphytic crust on the soil surface between shrubs, particularly where livestock grazing, human traffic, and fires have not been excessive.

Winter recharge of soil moisture from light snowstorms apparently drives development of shrubs and winter annuals. Their aboveground growth appears quite early (mid-March). Summer rainfall activates the microphytic crusts and the warm-season C$_4$ grasses *Hilaria, Bouteloua*, and *Aristida. Coleogyne* is protected against heavy ungulate herbivory by the combination of its woodiness, low nutrient quality, and secondary compounds (Provenza, Bowns, Urness, and Butcher 1983).

Although primary production is low and variable, this vegetation has been greatly influenced by livestock grazing and fire. Those interested in promoting livestock production have regarded fire as desirable (Bates and Menke 1984). *Coleogyne* does not ordinarily resprout after fire (Callison, Brotherson, and Bowns 1985), and it reseeds itself with difficulty. Various stem-sprouters and annuals (rarely perennial grasses) occur after fire, foremost among them being *Gutierrezia* species, *Bromus tectorum*, and *Bromus rubens*. Once these gain dominance, the probability of recurring fire increases. Production is highest during wet years on the sites not recently burned; it is almost nil during drought. Soil erosion has been accelerated by widespread prescribed burning and uncontrolled wildfires (West 1983e; Callison et al. 1985). Feral horses and burros have also had undesirable impacts. We currently do not have either promising replace-

ment species or the technology to consistently revegetate blackbrush.

Galleta–Three-Awn Shrub Steppe

This steppe occupies a small area (0.5 × 10^6 ha) centered in southeastern Utah (Küchler 1970). It occupies relatively deep and undeveloped sandy soils of the Canyonlands section of the Colorado Plateau (Fig. 7.16). Comparing relict areas to those that have been accessible to livestock leads us to believe that during pre-Columbian times there was a more abundant perennial grass cover and less shrubby and weedy annual cover (West 1983f). There is some affinity between this community type and grama-galleta steppes to the south, as discussed in Chapter 9.

The flora here is a mixture of about a dozen bunch- and sod-forming grasses and palatable half-shrubs. *Hilaria jamesii* (galleta) is a major long-lived sod-forming grass. *Aristida* species (three-awn) are short-lived and often weedy bunchgrasses. Both are known to increase following grazing disturbance (West 1983f). The shrubs and half-shrubs are quite scattered and also generally increase with disturbance. The major reason that grasses prevail here is that they are efficient in quickly using moisture that rapidly infiltrates the deep, sandy soils. The bulk of precipitation comes in high-intensity late-summer storms, with relatively little soil moisture recharge in winter to favor shrub growth.

The total vascular plant cover found here is

Figure 7.16. Hilaria–Aristida shrub steppe near Looking Glass Rock, San Juan County, Utah, July 1967. Shrubs are Krascheninnikovia lanata in foreground and Atriplex canescens in background.

high (25–60%) for a semidesert environment. The remaining space usually is covered by microphytic crust if disturbance has been low. In ungrazed areas, the vegetation is a mosaic of nearly monospecific patches of a few to tens of square meters (Kleiner and Harper 1972). Livestock grazing breaks down the crusts, modifies competitive interactions, and creates a more spatially mixed community. The palatable shrubs and half-shrubs decline and less desirable species invade, especially the short-lived, less palatable grasses *Aristida* spp. and *Muhlenbergia torreyana*, and the subshrub *Gutierrezia*. If the cover is broken too much, microdunes start to form; less palatable shrubs such as *Coleogyne, Ephedra, Artemisia filifolia*, and *Quercus havardii* catch the sand.

This steppe was only weakly stable prior to the introduction of livestock. Successional recovery has been slow following livestock removal (Loope 1975). Whether recovery will lead to an approximation of pre-Columbian communities or to different kinds of quasi-stable communities is unknown. Fortunately, the fuel loads rarely are great enough to carry fires, and few exotic weed species are abundant. Nevertheless, we have no proven methods of artificially restoring such vegetation.

Pinyon-Juniper Woodlands

At elevations slightly above those for the vegetation types so far considered lie woodlands dominated by scattered *Juniperus* and cembroid pines. The collective area of such woodlands is huge, at least 17×10^6 ha, and its range is far-flung (Fig. 7.17). Although all of this vegetation type is on semiarid sites with 25–50 cm total annual precipitation, the seasonality and effectiveness of this precipitation varies greatly over its expanse. Consequently, the species of *Juniperus* and *Pinus* vary geographically (Table 7.2), and the understory varies even more (West, Rea, and Tausch 1975).

To the uninitiated, all pinyon-juniper woodlands look much alike. The junipers and pines are similar in height, 10–15 m at maturity. Junipers are more widespread geographically and elevationally, going into both drier and colder habitats (Fig. 7.18). In the central parts of the vegetation type and at intermediate elevations within the woodland belt, usually one juniper species and one pinyon species form the tree guild – hence the name for these woodlands.

The understory varies so greatly over the vegetation type (West et al. 1975) that it is easier to say that it is similar to adjacent grasslands and

Figure 7.17. Map of the juniper-pinyon woodland vegetation type (adapted from Küchler 1970).

Table 7.2. *Distribution of principal tree species in pinyon-juniper woodland in United States.*

Colorado Plateau (eastern Utah, western Colorado, northern Arizona, northwestern New Mexico)
  Utah juniper (*Juniperus osteosperma*)
  Colorado pinyon (*Pinus edulis*)
  Single-seed juniper (*Juniperus monosperma*)

Great Basin (Nevada, western Utah, California east of the Sierra Nevada)
  Utah juniper (*Juniperus osteosperma*)
  Single-needle pinyon (*Pinus monophylla*)

Mojave border (southern and Baja California)
  Single-needle pinyon (*Pinus monophylla*)
  California juniper (*Juniperus californica*)
  Sierra Juarez pinyon (*Pinus quadrifolia*)

Pacific Northwest (eastern Oregon, southwestern Idaho, northeastern California)
  Western juniper (*Juniperus occidentalis*)

Northern Rockies (Wyoming, northern Colorado Front Range, Montana, eastern Idaho)
  Utah juniper (*Juniperus osteosperma*)
  Rocky Mountain juniper (*Juniperus scopulorum*)

Southern Rockies (southern Colorado, northern New Mexico)
  Single-seed juniper (*Juniperus monosperma*)
  Colorado pinyon (*Pinus edulis*)

Mogollon rim (central Arizona)
  Utah juniper (*Juniperus osteosperma*)
  Single-needle pinyon (*Juniperus monosperma*)
  Alligator-bark juniper (*Juniperus deppeana*)
  Arizona Cypress (*Cupressus arizonica*)

Sonoran Desert border (southern Arizona)
  Mexican pinyon (*Pinus cembroides* var. *bicolor*)
  Alligator-bark juniper (*Juniperus deppeana*)
  Colorado pinyon (*Pinus edulis*)
  Redberry juniper (*Juniperus erythrocarpa*)

Edwards Plateau (central Texas)
  Pinchot's juniper (*Juniperus pinchotii*)
  Ashe's juniper (*Juniperus ashei*)
  Texas pinyon pine (*Pinus cembroides* var. *remota*)

Trans-Pecos–Chihuahuan Desert border (western Texas and southeastern New Mexico)
  Texas pinyon pine (*Pinus cembroides* var. *remota*)
  Redberry juniper (*Juniperus erythrocarpa*)
  Pinchot's juniper (*Juniperus pinchotii*)

shrub steppes; that is, cool-season bunchgrasses and sagebrushes prevail in the northern and western Great Basin, whereas warm-season sod grasses and fewer shrubs prevail in the woodlands of the Colorado Plateau and upper Rio Grande basin where "monsoonal" rainfall patterns occur.

The tree crowns rarely touch in these open woodlands, but tree root systems extend out two to three times as far as crown diameters. Tierney and Foxx (1982) have reported that *Juniperus monosperma* and *Pinus edulis* have rooted to depths of 6.4 m in cracks in tuff near Los Alamos, New Mexico. Tree height and density increase with site favorableness, which is correlated strongly to elevation (West 1984). Total vascular plant cover varies from 40–80%. Most of the shrubs, herbs, and microphytes occur in the interspaces rather than under the tree crowns where litter accumulates in mounds.

Pinyon-juniper woodlands have changed enormously because of human uses, and these impacts have by no means been limited to Euro-Americans. There is mounting evidence that Native Americans brought about some localized changes because of hunting, farming, burning, and tree harvesting (Samuels and Betancourt 1982). However, the more pervasive influences have been due to livestock brought by EuroAmericans.

Examination of relict areas, tree age–class struc-ture, fire scars, and historical documents (West 1984) lead us to believe that much of the pinyon-juniper woodland was once more like savanna. Fires were frequent enough to keep the oldest trees restricted to steep, rocky, and/or dissected topography (Miller and Rose 1995). The ability of vegetation to carry fire on gentler topography was due to the former abundance of fine fuel, mainly grasses. Livestock found the grasses virtually the only edible component of this community. When grazing was excessive, fire could no longer carry and perform its natural thinning function. Shrubs

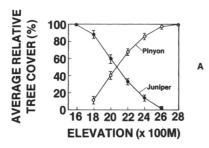

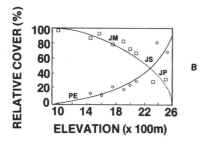

Figure 7.18. Elevational sorting of pinyon and juniper. (A) In Great Basin (from Tueller et al. 1979). The juniper is J. osteosperma; most of the pinyon is P. monophylla, except in the eastern extremities, where P. edulis occurs. Vertical lines are 95% confidence intervals. (B) in southern Rocky Mountains in New Mexico from "Juniper-pinyon east of the Continental Divide analyzed by the line strip method," by H. E. Woodin and A. A. Lindsey, Ecology, 1954, 35, 473–489. Copyright 1954 by the Ecological Society of America, reprinted by permission). The circular points represent relative cover of P. edulis; the solid squares represent relative cover of Juniperus; J. monosperma is associated with the solid line; J. scopulorum is associated with the dashed line; and J. deppeana with the dotted line.

and then trees increased in abundance, with shrubs often serving as nurse plants for tree seedlings (West et al. 1975).

Because of greater average tissue longevity, trees and shrubs are able to build more phytomass per unit of soil moisture and nutrients. Their roots are more extensive, both vertically and horizontally. They out-compete herbaceous species through the casting of shade and litter that have physical and chemical inhibiting properties. Euro-Americans exacerbated this "lignification" of the vegetation by fire suppression. This has been countered in some places by harvesting of trees for fence posts, mine props, firewood, and charcoal (Budy and Young 1979; Bahre 1991). The net effect, however, has been an increase in density of trees as well as expansion upslope and downslope into grasslands and shrub steppes that were being simultaneously degraded (Miller and Rose 1995).

In addition to the loss of forage for livestock and wild ungulates by this retrogressive succes-

sion, there may have been an increase in soil erosion rates. Much of the woodland is now composed of hardly anything but trees on mounded microtopography, with rills or gullies in the interspaces. This could mean that the system is degrading to a new level of lower potential. Because individual trees can live for hundreds of years, it will take a long time to decipher the trajectories of retrogression. Carrara and Carroll (1979) have estimated that soil erosion rates in at least one part of the pinyon-juniper woodland have increased >400% during the past century, as compared with the previous three centuries. This accelerated erosion coincides with the introduction of livestock in the vicinity of their northwestern Colorado study area. Recent reductions in livestock and increases in wood harvest have led to only slow recovery of understory vegetation (Yorks, West, and Capels 1994), apparently because seed banks and sources have been lost (Koniak and Everett 1982).

Land managers intent on promoting livestock production began to remove trees on relatively level sites by mechanical means in the 1950s and 1960s. Chaining may have been preceded by aerial seeding of grasses, or the tree debris may have been ricked into piles, and then drills later used to implant seeds. The result was an enormous increase in grasses on the most thoroughly treated sites. Most treatments were carried out before petroleum prices soared and environmental impact statements were required. Economic, environmental, archaeological, and aesthetic considerations (along with lowering demand for red meat) have recently reduced such kinds of conversions.

Successional trajectories set in place 100 yr ago are still leading to increased tree dominance. The accumulating wood is not valuable enough to manage, except in northern New Mexico and around the Navajo Indian Reservation where wood harvests are exceeding annual growth increments (Gray, Fowler, and Bray 1982). Prescribed burning usually cannot be done because fine fuels are sparse to absent. In the last several decades, *Bromus tectorum* (cheatgrass) has invaded upward from sagebrush vegetation into pinyon-juniper (*Pinus monophylla–Juniperus osteosperma*) woodlands on mountainsides in the western Great Basin. The result is midsummer crown fires and the loss of much of the pinyon-juniper ecosystem to annual grassland, leading to accelerated soil erosion and a lack of habitat for birds and mammals (Billings 1994).

If we are ready for the contingency of wildfires, we can artificially seed more desirable species on such areas (Koniak 1983; West 1984). If we do not purposely reseed perennials, annuals and fire frequency will both increase and result in accelerated soil erosion. Much more investigation is needed to develop acceptable means to restore the role of more frequent ground fires during spring or fall. Prescribed burning during the cooler seasons opens up the woodland to establishment of a more diverse and soil-protective understory.

Mountain Mahogany–Oak Scrub

The transition zone from montane coniferous forest to treeless plains and plateaus at margins of the Rocky Mountains is usually occupied by broadleaf scrub, a chaparral-like vegetation that can also be found above or intermingled with pinyon-juniper woodlands on the largest mountain ranges in both the Great Basin and Colorado Plateau. This petran chaparral (Shelford 1963) is most widespread and best developed in the southern Rockies and along the south side of the Mogollon Rim in central Arizona (Fig. 7.19). The belt narrows and becomes discontinuous farther north. This vegetation has also been discussed in Chapters 3 and 6.

The most widespread dominant is *Cercocarpus ledifolius* (curl-leaf mountain mahogany). Deciduous oaks (*Quercus gambelii, Q pauciloba*, and others) share dominance with *Cercocarpus montanus* in the Central Rockies. The evergreen oaks (*Q. turbinella, Q. emoryi, Q. dumosa, Q. chrysolepis* var. *palmeri*), along with more *Cercocarpus montanus* begin to appear and to become dominant from south of Utah (Carmichael, Knipe, Pase, and Brady 1978; Pase and Brown 1982).

Other shrubby associates include *Arctostaphylos pringlei, Acer grandidentatum, Amelanchier* species, *Arctostaphylos pungens, Artemisia tridentata* ssp. *vaseyana, Berberis* species, *Ceanothus greggii, Eriodictyon angustifolium, Fallugia paradoxa, Garrya wrightii, G. flavescens, Purshia mexicana, Purshia tridentata, Rhamnus crocea, Rhus trilobata*, and *Symphoricarpos* species, any one of which may assume local dominance. For instance, many of the upper elevations on medium-size mountain ranges in Nevada—apparently because of paleoecological and migrational influences—lack montane forests above the pinyon-juniper belts. The tops of these mountains are covered by *Artemisia–Symphoricarpos* chaparral (Tueller and Eckert 1987), and *Cercocarpus ledifolius* is restricted to small patches. Billings (1951) termed these "bald" mountains, due to their lack of trees in this zone. There are at least 100 such mountains in Nevada alone, many named "bald."

The taller shrubs rarely exhibit continuous cover but occur as dense clumps ("mottes") separated by areas of grassland or low shrub steppe.

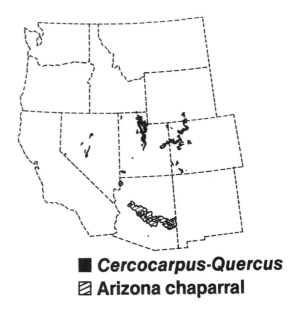

■ *Cercocarpus*-Quercus
◪ Arizona chaparral

Figure 7.19. Map of mountain mahogany–oak scrub and Arizona chaparral vegetation types (adapted from Küchler 1970).

The height of the shrub cover is 1–5 m, depending on species, site, and recent fire history. Most of the shrubs resprout readily after burning or their seed germination is stimulated by fires (e.g., *Ceanothus, Arctostaphylos*) (see Chapter 6).

There has been a marked increase in woody species as herbaceous components have been reduced by livestock grazing over the past 100–130 yr (Pase and Brown 1982). Loss of fine fuels and suppression of fire (even by road construction) have led to increased height of brush and its expansion into the interspaces between the mottes (Rogers 1982). Excessive deer and elk browsing has "high-lined" some species, especially *Cercocarpus ledifolius*, and prevented regeneration. Occasional late spring frosts can set back deciduous oak growth (Nielsen and Wullstein 1983) and provide more fuel for wildfire.

Where tall, dense oaks dominate, livestock and big-game interests have attempted to reduce the brush by burning or by chemical or mechanical means followed by seeding of herbaceous species (Plummer, Christensen, and Monsen 1968). Many introduced species can be seeded successfully (Plummer et al. 1968), because soils on relatively level sites usually are rich in nutrients and have high water-holding capacity. The resprouting characteristics of the shrubs, however, have thwarted all but the most persistent efforts to maintain herbaceous dominance.

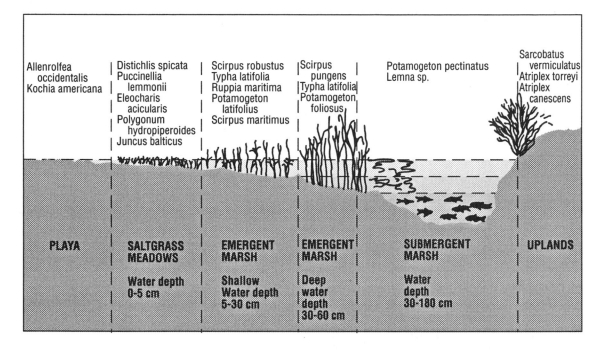

Figure 7.20. Diagram of vegetation resources of typical western Great Basin wetlands system (adapted from Hamilton and Auble 1993).

Wetlands

Obviously, wetlands are not abundant in temperate desert environments. The scarcity of wetlands, however, enhances their biological importance. Wetlands are much more restricted in the Colorado Plateau than in the Great Basin because of external drainage in the former. Humans have augmented the extent of wetlands throughout the Intermountain West by diking and spreading the water. In both the western and eastern Great Basin, rivers that predominantly originate in the highlands of the hydrologic basin terminate in evaporation surfaces in the interior of the basin. Wetlands associated with some of these rivers provide habitat for migratory songbirds, whereas the terminal lakes host waterfowl. Because the sink wetlands are terminal evaporation surfaces, they accumulate soluble salts from the inflowing river water. For many of these wetlands, the salt load exceeds the tolerance of most vascular plants; to remain productive the wetlands must periodically desiccate, because when they are dry, significant amounts of salt are removed by wind erosion (Young and Evans 1986).

In the western Great Basin, the Carson Sink has the most extensive wetlands. This area is fed with water from the Carson River, originating in the Sierra Nevada, and from the Humboldt River, the only major stream that originates in the highlands of the central Great Basin. In the early 1980s, historically high runoff occurred in the Humboldt River system. The Humboldt Sink spilled into the Carson Desert, forming a 100,000 ha lake system. This flooding was followed by 6 yr of record drought and by the mid-1990s the wetlands of the Carson Desert were nearly completely desiccated.

Great Basin wetlands are often surrounded by vast expanses of salt flats that are both atmospherically and osmotically very dry. The most salt-tolerant plant community bordering wetland is dominated by *Allenrolfea occidentalis* (iodine bush). Virtually the only herbaceous species in these communities is *Distichlis spicata* (desert saltgrass). Iodine bush is a facultative phreatophyte. It can grow where saturated brines reach the soil surface (Young et al. 1986) or exist in mounds on playa sediments without contact with groundwater (Blank, Palmquist, and Young 1992).

Depending on the topographic gradient around the wetlands, a narrow-to-broad zone of saline/alkaline meadow exists (Fig. 7.20). The dominant species are *Juncus balticus* (wiregrass), *Puccinellia lemmonii* (alkaligrass), and desert saltgrass. Iodine bush plants and black greaseweed occur on mounds. This kind of meadow may be very wet in the early spring but nearly desiccated by fall.

Below this saline/alkaline meadow zone is an area that floods in the spring to a depth of 0–5 cm. The dominant vegetation consists of *Eleocharis acicularis* (spike rush), *Polygonum* spp. (smartweed), wiregrass, and *Echinochloa crus-galli* (watergrass). An emergent marsh exists where water levels reach 5–50 cm depth until mid-summer. The dominant vegetation consists of *Scirpus acuminatus* (alkali bullrush), *Typha latifolia* (cattail), *Ruppia maritima* (widgeongrass), and *Potamogeton latifolius*. As the marsh deepens to 1.25 m, either the tall bullrushes or cattails dominate, depending on the salt content of the water (alkali bullrush is more tolerant of dissolved salts than cattail).

Open-water marshes support *Potamogeton pectinatus* (sago pondweed), floating *Elodea* (duckweed), and *Chara* (algae). Ponds that exceed 2 m in depth usually retain water during the entire year and support warm-water fish. For a more detailed treatment of vegetational and related edaphic and zoological zonation around a brackish water spring in the eastern Great Basin, see Bolen (1964).

Water from the temperate desert wetlands in the Great Basin is channeled to sinks through a gallery forest of *Populus fremontii* (Fremont cottonwood) and shrub- and tree-sized willows (*Salix* spp.). *Artemisia tridentata* subsp. *tridentata* (basin big sagebrush) occurs under the trees along with an herbaceous layer often dominated by the rhizomatous native *Leymus triticoides* (creeping wild rye).

Exotic plant species can be important components of wetlands. The wet saline/alkaline meadows contain *Lactuca serriola* (prickly lettuce) and *Taraxacum officinale* (dandelion). *Eleagnus angustifolia* (Russian olive) and *Tamarix ramosissima* (tamarisk) trees have also become established. In the Humboldt Sink, extensive areas of native vegetation were killed by flooding in the 1980s, and natives have been replaced by huge expanses of *Tamarix* and perennial pepperweed (*Lepidium latifolium*).

Perennial pepperweed has also invaded the Malheur wetlands in the northwestern Great Basin, the delta of the Susan River in the western Great Basin, wetlands along the Snake River in Idaho, the Great Salt Lake and Sevier Lake valleys, and smaller areas in the Uintah Basin of northeastern Utah and the upper Green River of Wyoming. This rhizomatous perennial dies back to the soil surface annually, but the semiwoody stems persist for several seasons, forming near-impenetrable thickets. Perennial pepperweed is adapted to a broad spectrum of wetland and riparian soils, including even saline soils. Through this range of communities, perennial pepperweed excludes virtually all other herbaceous species, including such perennial weeds as *Elytrigia repens* (quackgrass).

The most extensive wetlands in the intermountain lowlands are native hay meadows on flood plains. These meadows have usually been enhanced by flood irrigation during the spring runoff period. They receive minimum tillage and virtually no agronomic chemical input. Only a single, midsummer cutting is taken in order to conserve forage for winter maintenance of livestock. The stubble is then grazed by livestock in the fall. The bulk of the vegetation consists of a matrix of creeping native grasses (*Leymus triticoides*) and rushes (*Juncus balticus*), with some exotic grasses such as *Phleum pratense* (timothy). These meadows are important habitat for waterfowl and shorebirds and serve as sediment traps enhancing downstream water quality (Knight 1994). They are among the least-studied vegetation in the region, probably because of private ownership and their generally highly altered state, both biologically and environmentally (mainly hydrologic regime).

AREAS FOR FUTURE RESEARCH

There is great variation in the vegetation across the comparatively lower elevations of the Intermountain West. These are vast areas sparsely populated by humans between a few urban oases at the base of the higher mountains where perennial streams emerge. The generally low cover and productivity of the desert-to-woodland communities have limited research compared to the attention given to more mesic parts of the continent. In fact, those residing elsewhere often have regarded the region as an empty wasteland roamed only by a few cowboys, miners, and shepherds, and where "nuisance" activities such as military testing and training, coal-burning electricity-generating stations, and storage of toxic or nuclear wastes might occur.

The striking geology showing through the generally low, sparse vegetation of the intermountain lowlands has attracted mineral and recreational exploration. The geologic, topographic, and climatic complexities have also led to the evolution of some remarkable plants. This region has experienced dramatic vegetation change in the past 300–150 yr, mainly because of human-modified fire and livestock grazing regimes. Future research directions should include development of more economical and biologically effective techniques for revegetation of most of the types, particularly the xeric ones. Whether the objective is to reconstitute pristine-like indigenous vegetation or to replace degraded types with something more productive and protective of the soil, we need studies in basic plant autecology and vegetation dynamics. New information is especially critical for wetland vegetation, which has been so completely modified in a short span of time that its original structure and composition can only be surmised.

REFERENCES

Anderson, J. E., and K. E. Holte. 1981. Vegetation development over 25 years without grazing on sagebrush-dominated rangeland in southeastern Idaho. J. Range Mgt. 34:25–29.

Anderson, J. E., K. T. Ruppel, J. M. Glennon, K. E. Holte, and R. C. Rope. 1996. Plant Communities, ethnoecology, and Flora of the Idaho National Engineering Laboratory. Environmental Science and Research Foundation, report ESRF-005, Idaho Falls, Idaho.

Anderson, L. C. 1975. Models of adaptation to desert conditions in *Chrysothamnus*, p. 141 in H. C. Stutz (ed.), Proceedings of symposium and workshop on wildland shrubs. USDA, Forest Service, Shrub Sciences Laboratory, Provo, Utah.

Axelrod, D. I. 1979. Desert vegetation, its age and origin, pp. 1–72 in J. R. Goodin and D. K. Northington (eds.), Arid land plant resources. International Center for Arid and Semiarid Land Studies, Texas Technological University, Lubbock.

Bahre, C. J. 1991. Legacy of change: historic human impact on vegetation in the Arizona borderlands. University of Arizona Press, Tucson.

Barnosky, C. W. 1984. Late Miocene vegetational and climatic variations inferred from a pollen record in northwest Wyoming. Science 223:49–51.

Bates, P. A., and J. W. Menke. 1984. Role of fire in blackbrush succession. Abstracts, annual meeting of the Society of Range Management, Rapid City, S. Dak., No. 149.

Beatley, J. C. 1975. Climates and vegetation patterns across the Mojave/Great Basin transition of southern Nevada. Amer. Midl. Nat. 93:53–70.

Betancourt, J. L. 1984. Late Quaternary plant zonation and climate in southeastern Utah. Great Basin Nat. 44:1–35.

Billings, W. D. 1945. The plant associations of the Carson Desert region, western Nevada. Butler Univ. Bot. Studies 8:89–123.

Billings, W. D. 1949. The shadescale vegetation zone of Nevada and eastern California in relation to climate and soils. Amer. Midl. Nat. 42:87–109.

Billings, W. D. 1950. Vegetation and plant growth as affected by chemically altered rocks in the western Great Basin. Ecology 31:62–74.

Billings, W. D. 1951. Vegetational zonation in the Great Basin of western North America, pp. 101–122 in Compt. Rend. du Colloque sur les Bases Ecologiques de la Regeneration de la Vegetation des Zones Arides. Union Intern. Soc. Biol., Paris.

Billings, W. D. 1978a. Alpine phytogeography across the Great Basin. Great Basin Nat. Mem. 2:105–118.

Billings, W. D. 1978b. Plants and the Ecosystem, 3rd ed. Wadsworth, Belmont, Calif.

Billings, W. D. 1990. *Bromus tectorum*, a biotic cause of ecosystem impoverishment in the Great Basin, pp. 301–322 in G. M. Woodwell (ed.), The earth in transition: patterns and processes of biotic impoverishment. Cambridge University Press, Cambridge.

Billings, W. D. 1992. Islands of Sierran plants on the arid slopes of Peavine Mountain, Nevada. Mentzelia 6:32–39.

Billings, W. D. 1994. Ecological impacts of cheatgrass and resultant fire on ecosystems in the western Great Basin, pp. 22–30 in S. B. Monsen and S. G. Kitchen (compilers), Proceedings – Ecology and Management of Annual Rangelands. USDA Forest Service, Gen. Tech. Rep. INT-GTR-313, Intermountain Research Sta, Ogden Utah.

Blaisdell, J. P. 1958. Seasonal development and yield of native plants in the upper Snake River Plains and their relation to certain climatic factors. USDA Technical Bulletin 1190.

Blaisdell, J. P., and R. C. Holmgren. 1984. Managing intermountain rangelands–salt desert shrub ranges. USDA Forest Service, Gen. Tech. Rep. INT-163, Intermountain Forest and Range Experiment Station, Ogden, Utah.

Blaisdell, J. P., R. B. Murray, and E. D. McArthur. 1982. Managing intermountain rangeland–sagebrush-grass ranges. USDA Forest Service, General Technical Report INT-134, Intermountain Forest and Range Experiment Station, Ogden, Utah.

Blank, R. R., and J. A. Young. 1992. Influence of matric potential and substrate characteristics on germination of Nez Par Indian ricegrass. J. Range Mgt. 45: 205–209.

Blank, R. R., D. E. Palmquist, and J. A. Young. 1992. Plant-soil relationships of greasewood, Torrey saltbush, and *Allenrolfea* that occur on coarse textured mounds on playas, pp. 194–197. Proceedings, symposium on ecology and management of riparian and marsh communities. USDA, Forest Service, General Technical Report 289. Ogden, Utah.

Bolen, E. G. 1964. Plant ecology of spring-fed salt marshes in western Utah. Ecol. Mongr. 34:143–166.

Bowns, J. E., and N. E. West. 1976. Blackbrush (*Coleogyne ramosissima* Torr.) on southwestern Utah rangelands. Utah State University Agricultural Experiment Station Research Report 27. Logan, Utah.

Brandt, C. A., and W. H. Rickard. 1994. Alien taxa in the North American shrub-steppe four decades after cessation of livestock. Biol. Cons. 68:95–105.

Budy, J. D., and J. A. Young. 1979. Historical use of Nevada's pinyon-juniper. J. For. Hist. 23:112–121.

Burke, I. C. 1989. Topographic control of vegetation in a mountain big sagebrush steppe. Vegetation 84:77–86.

Caldwell, M. M. 1979. Physiology of sagebrush, pp. 74–85 in The sagebrush ecosystem: a symposium. College of Natural Resources, Utah State University, Logan, Utah.

Callison, J., J. D. Brotherson, and J. E. Bowns. 1985. The effects of fire on the blackbrush (*Coleogyne ramosissima*) community of southwest Utah. J. Range Mgt. 38:535–538.

Carmichael, R. S., O. D. Knipe, C. P. Pase, and W. W. Brady. 1978. Arizona chaparral: plant associations and ecology. USDA Forest Service Research Paper RM-202. Rocky Mountain Forest and Range Experiment Station, Ft. Collins, Colo.

Carrara, P. E., and T. R. Carroll. 1979. The determination of erosion rates from exposed tree roots in the Piceance Basin, Colorado. Earth Sur. Proc. 4:307–317.

Cawker, K. B. 1980. Evidence of climatic control from population age structure of *Artemisia tridentata*

Nutt. in southern British Columbia. J. Biogeogr. 7: 237–248.

Clary, W. P. 1975. Ecotypic variation in *Sitanion hystrix*. Ecology 56:1407–1415.

Cline, G. G. 1974. Peter Skene Ogden and the Hudson's Bay Company. University of Oklahoma Press, Norman, Okla.

College of Natural Resources (CNR). 1979. The sagebrush ecosystem: a symposium. CNR, Utah State University, Logan.

Comstock, J. P., and J. R. Ehleringer. 1992. Plant adaptation in the Great Basin and Colorado Plateau. Great Basin Nat. 52:195–215.

Daddy, F., M. J. Trlica, and C. D. Bonham. 1988. Vegetation and soil water differences among big sagebrush communities with different grazing histories. Southwestern Nat. 33:413–424.

Daubenmire, R. 1970. Steppe vegetation of Washington. Washington State University Agricultural Experiment Station Technical Bulletin 62. Pullman, Wash.

Daubenmire, R. F. 1972. Annual cycles of soil moisture and temperature as related to grass development in the steppe of eastern Washington. Ecology 53:419–424.

Daubenmire, R. 1975. An analysis of structural and functional characters along a steppe-forest catena. Northwest Sci. 49:120–140.

Davis, O. K. 1984. Multiple thermal maxima during the Holocene. Science 225:617–619.

Detling, J. K., and L. G. Klikoff. 1973. Physiological response to moisture stress as a factor in halophyte distribution. Amer. Midl. Nat. 90:307–318.

Dewey, S. A. 1991. Weedy thistles in the western United States, pp. 247–253 in L. F. James, J. O. Evans, M. H. Ralphs, and R. D. Child (eds.), Noxious range weeds. Westview Press, Boulder, Colo.

Eckert, R. E., Jr., and F. E. Kinsinger. 1960. Effects of *Halogeton glomeratus* leachate on chemical and physical characteristics of soils. Ecology 41:764–772.

Eckert, R. E., Jr., M. K. Wood, W. H. Blackburn, F. F. Peterson, J. L. Stephens, and M. S. Meurisse. 1978. Effects of surface-soil morphology on improvement and management of some arid and semi-arid rangelands, pp. 299–302 in D. N. Hyder (ed.), Proceedings of the First International Rangeland Congress. Society for Range Management, Denver.

Evans, R. A., and J. A. Young. 1970. Plant litter and establishment of alien annual species in rangeland communities. Weed Sci. 18:697–703.

Evans, R. A., and J. A. Young. 1972a. Microsite requirements for establishment of annual rangeland weeds. Weed Sci. 20:350–356.

Evans, R. A., and J. A. Young. 1972b. Germination and establishment of *Salsola* in relation to seedbed environment. II. Seed distribution, germination and seedling growth of *Salsola* and microenvironmental monitoring of the seedbed. Agron. J. 64:219–224.

Ferguson, C. W. 1964. Annual rings in big sagebrush. University of Arizona Press, Tucson.

Ferrel, C. M., and H. R. Leach. 1950. Food habits of the pronghorn antelope of California. California Division of Fish and Game 36:21–26.

Ferrel, C. M., and H. R. Leach. 1952. The prong-horn antelope of California with special reference to food habits. California Division of Fish and Game 38:285–293.

Fiero, B. 1986. Geology of the Great Basin: a natural history. University of Nevada Press, Reno.

Flowers, S., and F. R. Evans. 1966. The flora and fauna of the Great Salt Lake region, Utah, pp. 367–393 in H. Boyko (ed.), Salinity and aridity, Monographics Biologicae 16. Junk, The Hague.

Franklin, J. F., and C. T. Dyrness. 1973. Natural vegetation of Oregon and Washington. Forest Service, USDA, Pacific Northwest Forest Range Experiment Stations, General Technical Report 8, Portland, Ore.

Ganskopp, D. C. 1986. Tolerances of sagebrush, rabbitbrush, and greasewood to elevated water tables. J. Range Mgt. 39:334–337.

Gray, J. R., J. F. Fowler, and M. A. Bray. 1982. Free-use fuelwood in New Mexico: inventory, exhaustion and energy equations. J. For. 80:23–26.

Grayson, D. K. 1993. The desert's past: a natural prehistory of the Great Basin. Smithsonian Institution Press. Washington, D.C.

Grayson, D. K. 1994. The extinct Late Pleistocene mammals of the Great Basin, pp. 55–85 in K. T. Harper, L. L. St. Clair, K. H. Thorne, and W. M. Hess (eds.), Natural history of the Colorado Plateau and Great Basin. University Press of Colorado, Boulder.

Griffiths, D. 1902. Forage conditions on the northern border of the Great Basin. USDA Bureau of Plant Industry Bulletin 15.

Hamilton, D. A., and G. T. Auble. 1993. Wetland modeling and intermountain needs at Stillwater National Wildlife Refuge, U.S. Fish and Wildlife Service, Fallon, Nev.

Hanks, D. L., E. D. McArthur, R. Stevens, and A. P. Plummer. 1973. Chromatographic characteristics and phylogenetic relationships of *Artemisia* section *Tridentatae*. Res. Paper 141. USDA Forest Service, Intermountain Forest and Range Experimental Station Ogden, Utah.

Hanley, T. A. 1979. Application of an herbivore-plant model to rest-rotation grazing management on shrub-steppe ranges. J. Range Mgt. 32:115–118.

Hanson, C. L., C. W. Johnson, and J. R. Wight. 1982. Foliage mortality of mountain big sagebrush (*Artemisia tridentata* ssp. *vaseyana*) in southwestern Idaho during the winter of 1976–77. J. Range Mgt. 35:142–145.

Harper, K. T., and R. L. Pendleton. 1993. Cyanobacteria and cyanolichens: Can they enhance availability of essential minerals for higher plants? Great Basin Nat. 53:59–72.

Harris, G. A. 1977. Root phenology as a factor of competition among grass seedlings. J. Range Mgt. 30: 172–176.

Hassan, M. A., and N. E. West. 1986. Dynamics of soil seed pools in burned and unburned sagebrush semi-deserts. Ecology 67:269–272.

Haws, B. A., G. E. Bohart, C. R. Nelson, and D. L. Nelson. 1991. Insects and their relationship to widespread shrub die-off in the Intermountain West of North America, pp. 474–477 in A. Gaston et al. (eds.) Proc. Fourth International Rangeland Congress, Vol. 1, Montpellier, France.

Hironaka, M. 1979. Basic synecological relationships of the Columbia River sagebrush type, pp. 27–30 in The sagebrush ecosystem: a symposium. College of Natural Resources, Utah State University, Logan.

Hormay, A. L. 1943. Bitterbrush in California. USDA, Forest Service, Pacific Southwest Forest and Range Experiment Station, Res. Note 39. Berkeley, Calif.

Horton, H. S. Monsen, R. Newhall, and R. Stevens.

1994. Forage Kochia holds ground against cheatgrass, erosion. Utah Sci. 35:10–11.

Hull, A. C., Jr. 1976. Rangeland use and management in the Mormon West. In Symposium on agriculture, food and man – a century of progress. Brigham Young University, Provo, Utah.

Hunt, C. B. 1974. Natural regions of the United States and Canada. Freeman, San Francisco.

Jennings, J. D. 1978. Prehistory of Utah and the eastern Great Basin. University of Utah Anthropological Papers 98, Salt Lake City.

Jensen, M. E., L. S. Peck, and M. V. Wilson. 1988. Vegetation characteristics of mountainous northeastern Nevada sagebrush community types. Great Basin Nat. 48:403–421.

Jensen, M. E., G. H. Simonson, and M. Dosskey. 1990. Correlation between soils and sagebrush-dominated plant communities of northeastern Nevada. Soil Sci. Soc. Am. J. 54:902–910.

Julander, O., and J. B. Low. 1976. A historical account and present status of mule deer in the West, pp. 3–19 in G. W. Workman and J. B. Low (eds.), Mule deer decline in the West: a symposium. Utah State University, Agricultural Experiment Station. Logan, Utah.

Kartesz, J. T. 1994. A synonymized checklist of the vascular flora of the United States, Canada and Greenland. Vol. 1 Checklist, Vol. 2, Synonymy. Timber Press, Portland, Ore.

Kay, C. E. 1995. Aboriginal overkill and native burning: implications for modern ecosystem management. Proceedings 8th George Wright Society Conference on Research and Resource Management on Public Lands. Portland, Ore.

Keller, W. 1979. Species and methods for seeding in the sagebrush ecosystem, pp. 129–136 in The sagebrush ecosystem: a symposium. College of Natural Resources, Utah State University, Logan.

Kennedy, P. B. 1903. Summer ranges of eastern Nevada sheep. Nevada Agricultural Experiment Station Bulletin 55.

King, R. S. 1981. Ecological surveys of relict and long-protected range sites in the blackbrush formation in northern Arizona. Master's thesis, Northern Arizona University, Flagstaff.

Kleiner, E. F., and K. L. Harper. 1972. Environmental and community organization in grasslands of Canyonlands National Park. Ecology 53:299–309.

Knight, D. H. 1994. Mountains and plains: the ecology of Wyoming landscapes. Yale University Press, New Haven.

Koniak, S. 1983. Broadcast seeding success in eight pinyon-juniper stands after wildlife. USDA Forest Service Research Note INT-334, Intermountain Forest and Range Experiment Station, Ogden, Utah.

Koniak, S., and R. L. Everett. 1982. Seed reserves in soils of successional stages of pinyon woodlands. Amer. Midl. Nat. 108:295–303.

Küchler, A. W. 1970. Potential natural vegetation (map at scale 1:7,5000,000), pp. 90–91 in The national atlas of the U.S.A. U.S. Government Printing Office, Washington, D.C.

La Marche, V. C., Jr. 1974. Paleoclimatic inferences from long tree-ring records. Science 198:1043–1048.

LaTourrette, J. E., J. A. Young, and R. A. Evans. 1971. Seed dispersal in relation to rodent activities in seral big sagebrush communities. J. Range Mgt. 24:451–454.

Lentz, R. D., and G. H. Simonson. 1987a. Correspondence of soil properties and classification units with sagebrush communities in southeastern Oregon. I. Comparisons between mono-taxa soil-vegetation units. Soil Sci. Soc. Am. J. 51:1263–1271.

Lentz, R. D., and G. H. Simonson. 1987b. Correspondence of soil properties and classification units with sagebrush communities in southeastern Oregon. II. Comparisons within a multi-taxa soil-vegetation unit. Soil Sci. Soc. Am. J. 51:171–176.

Link, S. O., G. W. Gee, M. E. Thiede, and P. E. Beedlow. 1990. Response of a shrub-steppe ecosystem to fire: soil water and vegetational change. Arid Soil Res. Reclam. 4:163–172.

Loope, W. L. 1975. Vegetation in relation to environment of Canyonlands National Park, Utah. Ph.D. dissertation, Utah State University, Logan.

Lunt, O. R., J. Letey, and S. B. Clark. 1973. Oxygen requirements for root growth in three species of desert shrubs. Ecology 54:1356–1362.

McAdoo, J. K., and J. A. Young. 1980. Jack rabbits. Rangelands 2:135–138.

McAdoo, J. K., C. C. Evans, B. A. Roundy, J. A. Young, and R. A. Evans. 1983. Influence of heteromyid rodents on *Oryzopsis hymenoides* germination. J. Range Mgt. 36:82–86.

McArthur, E. D. 1983. Taxonomy, origin and distribution of big sagebrush (*Artemisia tridentata*) and allies (subgenus *Tridentatae*), pp. 3–11 in R. L. Johnson (ed.), Proceedings of first Utah shrub ecology workshop. College of Natural Resources, Utah State University, Logan.

McDonough, W. T., R. O. Harniss, and R. B. Campbell. 1975. Morphology of perennial and persistent leaves of three subspecies of big sagebrush grown in a uniform environment. Great Basin Na. 35:325–326.

Mack, R. N. 1981. Invasion of *Bromus tectorum* L. into western North America: an ecological chronicle. Agro-ecosystems 7:145–165.

Mack, R. N., and J. N. Thompson. 1982. Evolution in steppe with few large, hooved animals. Amer. Nat. 119:757–773.

Mehringer, P. J., Jr. 1977. Great Basin Late Quaternary environments and chronology, pp. 113–167 in D. D. Fowler (ed.), Models and Great Basin prehistory. Desert Research Institute Publications in Social Sciences 12. University of Nevada, Reno.

Miles, R. L., and S. G. Leonard. 1984. Documenting soil-plant relationships of selected sagebrush species using the soil resource information system. Soil Sur. Hori. 25:22–26.

Miller, R. F. and J. A. Rose. 1995. Historic expansion of *Juniperus occidentalis* (western juniper) in southeastern Oregon. Great Basin Naturalist 55:37–45.

Miller, R. F., and L. M. Shultz. 1987. Development and longevity of ephemeral and perennial leaves on *Artemisia tridentata* Nutt. ssp. *wyomingensis*. Great Basin Nat. 47:227–230.

Miller, R. F., T. J. Svejcar, and N. E. West. 1994. Implications of livestock grazing in the Intermountain sagebrush region: plant composition, pp. 101–148 in M. Vavra, W. A. Laycock, and R. D. Pieper (eds.), Ecological implications of livestock herbivory in the West. Soc. for Range Mgt. Denver, Colo.

Miller, R. F., and P. F. Wigand. 1994. Holocene changes in semiarid pinyon-juniper woodlands. BioScience 44:465–474.

Monsen, S. B., and S. G. Kitchen (compilers). 1994. Proc. Symposium on Ecology and Management of Annual Rangelands. Intermountain Research Sta., Forest Service, USDA, General Technical Report INT-313, Ogden, Utah.

Morris, T. H., and M. A. Stubben. 1994. Geologic contrasts of the Great Basin and Colorado Plateau, pp. 9–25 in K. T. Harper, L. L. St. Clair, K. H. Thorne, and W. M. Hess (eds.), Natural history of the Colorado Plateau and Great Basin. University Press of Colorado, Boulder.

Morrison, R. B. 1964. Lake Lahontan: geology of the southern Carson Desert. U.S. Geological Survey Prof. Paper 401.

Mozingo, H. 1986. Shrubs of the Great Basin. University of Nevada Press, Reno.

Mulligan, G. A., and J. D. Findlay. 1974. The biology of Canadian weeds. 3. *Cardaria draba, C. chalepensis*, and *C. pubescens*. Can. J. Plant Sci. 54:149–160.

Nelson, D. L., K. T. Harper, K. C. Boyer, D. J. Weber, B. A. Haws, and J. R. Marble. 1989. Wildland shrub die-offs in Utah: an approach to understanding cause, pp. 119–135 in Proceedings, Symposium on Shrub Ecophysiology and Biotechnology (A. Wallace, E. D. McArthur, and M. R. Haferkamp (compilers). 30 June–2 July 1987, Logan, Utah. USDA Forest Service, General Technical Report INT-256. Ogden, Utah.

Nielsen, R. P., and C. H. Wullstein. 1983. Biogeography of two southwest American oaks in relation to atmospheric dynamics. J. Biogeog. 10:275–298.

Nord, E. C. 1965. Autecology of bitterbrush in California. Ecol. Monogr. 38:307–334.

Norton, B. E. 1978. The impact of sheep grazing on long-term successional trends in salt desert shrub vegetation of southwestern Utah, pp. 610–613 in D. N. Hyder (ed.), Proceedings of First International Rangeland Congress. Society for Range Management, Denver.

Nowak, C. L., R. S. Nowak, R. J. Tausch, and P. E. Wigand. 1994. A 30,000 year record of vegetation dynamics at a semi-arid locale in the Great Basin. J. Veg. Sci. 5:579–590.

Noy-Meir, I. 1973. Desert ecosystems: environment and producers. Ann. Rev. Ecol. Syst. 4:25–51.

Pase, C. P., and D. E. Brown. 1982. Interior chaparral, pp. 95–99 in D. E. Brown (ed.), Biotic communities of the American Southwest–United States and Mexico. Boyce Thompson Institute of Plant Biology, Superior, Ariz.

Passey, H. B., V. K. Hugie, E. W. Williams, and D. E. Ball. 1982. Relationships between soil, plant community, and climate on rangelands of the Intermountain West. USDA Soil Conservation Service, Technical Bulletin 1669. Washington, D.C.

Pavlik, B. M. 1989. Phytogeography of sand dunes in the Great Basin and Mojave Desert. J. Biogeog. 16:227–238.

Petersen, K. L. 1994. Modern and Pleistocene climatic patterns in the West, pp. 27–53 in K. T. Harper, L. L. St. Clair, K. H. Thorne, and W. M. Hess (eds.), Natural history of the Colorado Plateau and Great Basin. University Press of Colorado, Boulder.

Pierson, F. B., and J. R. Wright. 1991. Variability of near-surface soil temperature on sagebrush rangeland. J. Range Mgt. 44:491–497.

Plummer, A. P., D. R. Christensen, and S. B. Monsen. 1968. Restoring big-game range in Utah. Utah Division of Fish and Game publication 68–3, State of Utah Department of Natural Resources, Salt Lake City.

Potter, L. D., and J. C. Krenetsky. 1967. Plant succession with released grazing on New Mexico rangelands. J. Range Mgt. 20:145–151.

Price, K. A., D. A. Pyke, and L. Mendes. 1992. Shrub dieback in a semiarid ecosystem: the integration of remote sensing and geographic information systems for detecting vegetation change. Photogram Engineer and Remote Sensing 58:455–463.

Provenza, F. D., J. E. Bowns, P. J. Urness, and J. E. Butcher. 1983. Biological manipulation of blackbrush by goat browsing. J. Range Mgt. 36:513–518.

Rice, B., and M. Westoby. 1978. Vegetation responses of some Great Basin shrub communities protected against jackrabbits or domestic stock. J. Range Mgt. 31:28–33.

Rickard, W. H. 1964. Demise of sagebrush through soil changes. BioScience 14:43–44.

Robbins, W. W. 1940. Alien plants growing without cultivation in California. California Agricultural Experiment Station Bulletin 637, Berkeley.

Robertson, D. R., J. L. Nielsen, and N. H. Bare. 1966. Vegetation and soils of alkali sagebrush and adjacent big sagebrush ranges in North Park, Colorado. J. Range Mgt. 19:17–20.

Robertson, J. H. 1954. Half-century changes on northern Nevada ranges. J. Range Mgt. 7:117–121.

Roche, B. F., Jr., and C. E. Roche. 1991. Identification, introduction, distribution, ecology, and economics of *Centaurea* species, pp. 274–291 In L. F. James, J. O. Evans, M. H. Ralphs, and R. D. Child (eds.), Noxious range weeds. Westview Special Studies in Agricultural Science and Policy, Westview Press, Boulder, Colo.

Rogers, G. G. 1982. Then and now: a photographic history of vegetation change in the central Great Basin desert. University of Utah Press, Salt Lake City.

Rosentreter, R., and R. G. Kelsey. 1991. Xeric big sagebrush, a new subspecies in the *Artemisia tridentata* complex. J. Range Mgt. 44:330–335.

Samuels, M. L., and J. L. Betancourt, 1982. Modeling the long-term effects of fuel wood harvests on pinyon-juniper woodlands. Environ. Mgt. 6:505–515.

Selleck, G. W. 1965. An ecological study of lens- and globe-podded hoary cresses in Saskatchewan. Weeds 13:1–5.

Sharp, L. A., K. Sanders, and N. Rimbey. 1990. Forty years of change in a shadscale stand in Idaho. Rangelands 12:213–228.

Shelford, V. E. 1963. The ecology of North America. University of Illinois Press, Urbana.

Shmida, A., and R. H. Whittaker. 1979. Convergent evolution of arid regions in the New and Old Worlds, pp. 437–450 in R. Tuxen (ed.), Vegetation and history. International Vereing Vegetationskunde, Rinteln, Germany.

Soil Survey Staff. 1995. Keys to soil taxonomy, 6th ed. Soil Conservation Service. USDA, Washington, D.C.

Sparks, S. R., N. E. West, and E. B. Allen. 1990. Changes in vegetation and land use at two townships in Skull Valley, western Utah, pp. 26–36 in USDA, Forest Service, Intermountain Research Station (Ogden, Utah) Gen. Tech. Rep. INT-276.

Stebbins, G. L., Jr. 1950. Variation and evolution in plants. Columbia University Press, New York.

Stebbins, G. L., Jr. 1957. Self-fertilization and population variability in the higher plants. Amer. Nat. 91:299–324.

Stoddart, L. A. 1946. Some physical and chemical responses of *Agropyron spicatum* to herbage removal at various seasons. Utah State University Agricultural Experiment Station Bulletin 324, Logan, Utah.

Sturges, D. L. 1977. Soil water withdrawal and root characteristics of big sagebrush. Amer. Midl. Nat. 98:257–274.

Stutz, H. C. 1978. Explosive evolution of perennial *Atriplex* in western North America. Great Basin Nat. Mem. 2:161–168.

Stutz, H. C., J. M. Melby, and G. K. Gordon. 1975. Evolutionary studies of *Atriplex*: A relic gigas diploid formation of *Atriplex canescens*. Amer. J. Bot. 62:236–245.

Svejcar, T., and R. Tausch. 1991. Anaho Island, Nevada: A relict area dominated by annual invader species. Rangelands 13:165–167.

Thatcher, A. P., J. W. Doughty, and D. L. Richmond. 1976. Amount of blackbrush in the pristine plant communities controlled by edaphic conditions, p. 12, vol 27. Abstracts of annual meeting of the Society of Range Management, Omaha, Nebr.

Tierney, G. D., and T. S. Foxx. 1982. Floristic composition and plant succession on near-surface radioactive waste disposal facilities in the Los Alamos National Laboratory. Los Alamos National Laboratory report LA-9212-MS.

Tisdale, E. W., and M. Hironaka. 1981. The sagebrush-grass region: a review of the ecological literature. Forest, Wildlife and Range Experiment Station Bulletin 33, University of Idaho, Moscow.

Tisdale, E. W., M. Hironaka, and F. A. Fosberg. 1965. An area of pristine vegetation in Craters of the Moon National Monument, Idaho. Ecology 46:349–352.

Trimble, S. 1989. The sagebrush ocean: a natural history of the Great Basin. University of Nevada Press, Reno.

Tueller, P. T., and R. E. Eckert Jr. 1987. Big sagebrush (*Artemisia tridentata vaseyana*) and longleaf snowberry (*Symphoricarpos oreophilus*) plant associations in northeastern Nevada. Great Basin Nat. Mem. 47: 117–131.

Tueller, P. T., C. D. Beeson, R. J. Tausch, N. E. West, and K. H. Rea. 1979. Pinyon-juniper woodlands of the Great Basin: distribution, flora, vegetal cover. USDA Forest Service Research Paper INT-229, Intermountain Forest and Range Experiment Station, Ogden, Utah.

U.S. Department of Agriculture. 1941. Climate and man. Agricultural yearbook. USDA, Government Printing Office, Washington, D.C.

Van Devender, T. R., and W. G. Spaulding. 1977. Development of vegetation and climate in the southwestern United States. Science 204:701–710.

Van Pelt, N. 1978. Woodland parks of southeastern Utah. Master's thesis, University of Utah, Salt Lake City.

Welch, B. L., and E. D. McArthur. 1990. Big sagebrush – its taxonomy, origin, distribution and utility, pp. 3–19 in H. G. Fisser (ed.), Proc. 14th Wyoming Shrub Ecology Workshop, Dept. Range Mgt., University of Wyoming, Laramie.

Wells, P. V. 1983. Paleobiogeography montane islands in the Great Basin since the last glaciopluvial. Ecol. Monogr. 53:341–382.

West, N. E. 1969. Soil-vegetation relationships in arid southeastern Utah. Abstract, International conference on arid lands in a changing world. University of Arizona, Tucson.

West, N. E. 1979. Basic synecological relationships of sagebrush-dominated lands in the Great Basin and the Colorado Plateau, pp. 33–41 in The sagebrush ecosystem: a symposium. College of Natural Resources, Utah State University, Logan.

West, N. E. 1982. Dynamics of plant communities dominated by chenopod shrubs. Int. J. Ecol. Environ. Sci. 8:73–84.

West, N. E. 1983a. Overview of North American temperate deserts and semi-deserts, pp. 321–330 in N. E. West (ed.), Temperate deserts and semi-deserts, vol. 5, Ecosystems of the world. Elsevier, Amsterdam.

West, N. E. 1983b. Western intermountain sagebrush steppe, pp. 351–374 in N. E. West (ed.), Temperate deserts and semi-deserts, vol. 5, Ecosystems of the world. Elsevier, Amsterdam.

West, N. E. 1983c. Great Basin – Colorado Plateau sagebrush semi-desert, pp. 331–349 in N. E. West (ed.), Temperate deserts and semi-deserts, vol. 5, Ecosystems of the world. Elsevier, Amsterdam.

West, N. E. 1983d. Intermountain salt-desert shrubland, pp. 375–397 in N. E. West (ed.), Temperate deserts and semi-deserts, vol. 5, Ecosystems of the world. Elsevier Amsterdam.

West, N. E. 1983e. Colorado Plateau–Mohavian blackbrush semi-desert, pp. 399–411 in N. E. West (ed.), Temperate deserts and semi-deserts, vol. 5, Ecosystems of the world. Elsevier, Amsterdam.

West, N. E. 1983f. Southeastern Utah galleta-three awn shrub steppe, pp. 413–421 in N. E. West (ed.), Temperate deserts and semi-deserts, vol. 5, Ecosystems of the world. Elsevier, Amsterdam.

West, N. E. 1984. Successional patterns and productivity potentials of pinyon-juniper ecosystems, pp. 1301–1332 in Developing strategies for range management. Westview Press, Boulder, Colo.

West, N. E. 1990. Structure and function of microphytic soil crusts in wildland ecosystems of arid to semi-arid regions. Adv. Ecol. Res. 20:180–223.

West, N. E. 1994. Effects of fire on salt-desert shrub rangelands, pp. 71–74 in S. B. Monsen and S. G. Kitchen (compilers) Proceedings – Ecology and Management of Annual Rangelands. USDA; Forest Service, Intermountain Res. Sta. Gen. Tech. Rep. INT-313.

West, N. E. 1996. Strategies for maintenance and repair of biotic community diversity on rangelands, pp. 326–346, in R. C. Szaro and D. W. Johnston (eds.), symposium on Biodiversity managed landscapes: theory and practice. Oxford University Press, New York.

West, N. E., and J. Gasto. 1978. Phenology of the aerial portions of shadscale and winterfat in Curlew Valley, Utah. J. Range Mgt. 31:43–45.

West, N. E., and D. W. Goodall. 1986. Dispersion patterns in relation to successional status of salt desert shrub vegetation. Abstracta Botanica 10:87–201.

West, N. E., F. D. Provenza, P. S. Johnson, and M. K. Owens. 1984. Vegetation change after 13 years of livestock grazing exclusion on sagebrush semidesert in west central Utah. J. Range Mgt. 37:262–264.

West, N. E., K. H. Rea, and R. O. Harniss. 1979. Plant

demographic studies in sagebrush-grass communities of southeastern Idaho. Ecology 60:376–388.

West, N. E., K. H. Rea, and R. J. Tausch. 1975. Basic synecological relationships in pinyon-juniper woodlands, pp. 41–58 in G. F. Gifford and F. E. Busby (eds.), The pinyon-juniper ecosystem: a symposium. Utah State University Agricultural Experiment Station, Logan.

West, N. E., and J. Skujiņš. 1977. The nitrogen cycle in North American cold-winter semi-desert ecosystems. Oecologia Plantarum 12:45–53.

Whisenant, S. G. 1990. Changing fire frequencies on Idaho's Snake River Plains: Ecological and management implications, pp. 4–10 in E. D. McArthur, E. M. Romney, S. D. Smith, and P. T. Tueller (compilers). Symposium on cheatgrass invasion, shrub die-off and other aspects of shrub biology and management. USDA Forest Service General Technical Report INT-276.

Wilde, J. D. 1994. Western Great Basin prehistory, pp. 87–111 in K. T. Harper, L. L. St. Clair, K. H. Thorne, and W. M. Hess (eds.), Natural history of the Colorado Plateau and Great Basin. University Press of Colorado, Boulder.

Williams, J. D., J. P. Dobrowolski, and N. E. West. 1995. Microphytic crust influence on internill erosion and infiltration capacity. Trans. Amem. Soc. Agric. Engineers 38:139–146.

Williams, J. D., J. P. Dobrowolski, N. E. West, and D. A. Gillette. 1995. Microphytic crust influence on wind erosion. Trans. Amer. Soc. Agri. Engineers 38:139–146.

Winward, A. H. 1983. Using sagebrush ecology in wildland management, pp. 15–19 in K. L. Johnson (ed.), First Utah shrub ecology workshop. College of Natural Resources, Utah State University, Logan.

Wolfe, J. A. 1978. A paleobotanical interpretation of Tertiary climates in the northern hemisphere. Amer. Sci. 66:694–703.

Woodin, H. E., and A. A. Lindsey. 1954. Juniper-pinyon east of the continental divide analyzed by the line-strip method. Ecology 35:473–489.

Wright, H. A., and A. W. Bailey. 1982. Fire ecology: United States and southern Canada. Wiley, New York.

Yorks T. P., N. E. West, and K. M. Capels. 1992. Vegetation differences in desert shrublands of western Utah's Pine Valley between 1933 and 1989. J. Range Mgt. 46:569–578.

Yorks, T. P., N. E. West, and K. M. Capels. 1994. Changes in pinyon-juniper woodlands in Western Utah's Pine Valley between 1933–1989. J. Range Mgt. 47:359–364.

Young, J. A. 1988. The public response to the catastrophic spread of Russian thistle (1880) and halogeton (1945). Agric. Hist. 62:122–130.

Young, J. A. 1992. Ecology and management of medusahead (Taeniatherum caput-meduase ssp. asperum). Great Basin Nat. 52:313–320.

Young, J. A. 1994a. Changes in plant communities in the Great Basin induced by domestic livestock grazing, pp. 113–123 in K. T. Harper, L. L. St. Clair, K. H. Thorne, and W. M. Hess (eds.), Natural history of the Colorado Plateau. University Press of Colorado, Boulder.

Young, J. A. 1994b. History and use of semiarid plant communities – changes in vegetation, pp. 5–8 in S. B. Monsen and S. G. Kitchen (compilers), Proc.

Ecology and Management of Annual Rangelands. U.S. Dept. Agric., Forest Service, Intermountain Res. Sta. Gen. Tech. Rep. INT-GTR-313. Ogden, Utah.

Young. J. A., and R. A. Evans. 1972. Germination and establishment of Salsola in relation to seedbed environment. I. Temperature, afterripening, and moisture relations of Salsola seeds as determined by laboratory studies. Agron. J. 64:214–218.

Young, J. A., and R. A. Evans. 1973. Downy brome – Intruder in the plant succession of big sagebrush plant communities of the Great Basin. J. Range Mgt. 26:410–415.

Young, J. A., and R. A. Evans. 1978. Population dynamics after wildfires in sagebrush grasslands. J. Range Mgt. 31:283–289.

Young, J. A., and R. A. Evans. 1979. Barbwire Russian thistle seed germination. J. Range Mgt. 32:390–394.

Young, J. A., and R. A. Evans. 1981. Demography and fire history of a western juniper stand. J. Range Mgt. 34:501–506.

Young, J. A., and R. A. Evans. 1986. Erosion and deposition of fine sediments from playas. J. Arid Environ. 10:103–115.

Young, J. A., and R. A. Evans. 1989a. Dispersal and germination of big sagebrush (Artemisia tridentata). Weed Sci. 37:319–325.

Young, J. A., and R. A. Evans. 1989b. Reciprocal common garden studies of the germination of seeds of big sagebrush (Artemisia tridentata). Weed Sci. 37: 319–325.

Young, J. A., R. A. Evans, and B. L. Kay. 1971. Germination of dyer's woad. Weed Sci. 19:76–78.

Young, J. A., R. A. Evans, and J. J. Major. 1977. Sagebrush steppe, pp. 763–797 in: M. G. Barbour, J. Jack Major (eds.), Terrestrial vegetation of California. Wiley, New York.

Young, J. A., R. A. Evans, and D. E. Palmquist. 1989. Big sagebrush (Artemisia tridentata) seed production. Weed Sci. 37:47–53.

Young, J. A., R. A. Evans, and B. A. Roundy. 1983. Quantity and germinability of Oryzopsis hymenoides seeds in Lahontan sands. J. Range Mgt. 36:82–86.

Young, J. A., R. A. Evans, and P. T. Tueller. 1976. Great Basin plant communities – pristine and grazed, pp. 187–215 in Robert Elston (ed.), Holocene environmental change in the Great Basin. Nev. Arch. Res. Paper 6. Reno, Nev.

Young, J. A., and D. E. Palmquist. 1992. Plant age/size distributions in black sagebrush (Artemisia nova): effects on community structure. Great Basin Nat. 52: 313–320.

Young, J. A., D. E. Palmquist, and R. A. Evans. 1991. Temperature profiles for germination of big sagebrush seeds from native stands. J. Range Mgt. 44: 385–390.

Young, J. A., R. R. Blank, W. S. Longland, and D. E. Palmquist. 1994. Seeding Indian ricegrass in arid environments in the Great Basin. J. Range Mgt. 47:2–7.

Young, J. A., R. A. Evans, R. E. Eckert, Jr., and B. L. Kay. 1987. Cheatgrass. Rangelands 9:266–272.

Young, J. A., R. A. Evans, B. A. Roundy, and J. Brown. 1986. Dynamic landforms and plant communities in a pluvial lake basin. Great Basin Nat. 46:1–21.

Zamora, B., and P. T. Tueller. 1973. Artemisia arbuscula, A. longliloba, and A. nova habitat types in northern Nevada. Great Basin Nat. 33:225–242.

Chapter

8

Warm Deserts

JAMES A. MacMAHON

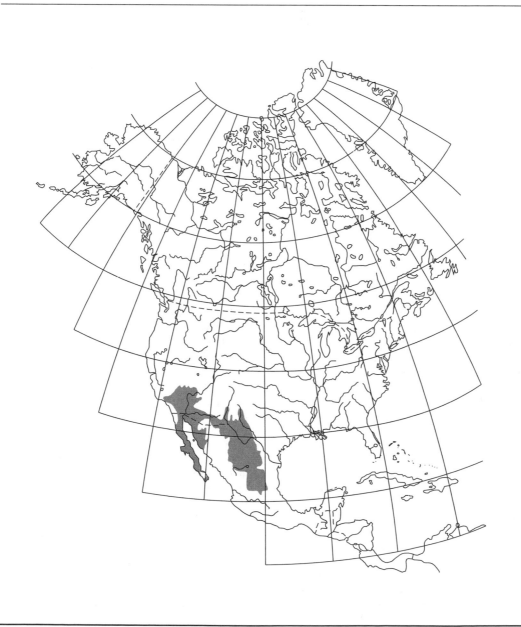

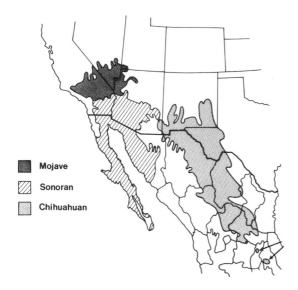

Mojave

Sonoran

Chihuahuan

Table 8.1. *Approximate areas of North American deserts based on boundaries presented by MacMahon (1979). Note that West and Young (Chapter 7 of this book) estimate the Great Basin Desert covers 448,000 km² steppe & 179,000 km² (scrub).*

Desert	Area (km²)	Percentage of North American desert area
Great Basin	409,000	32.0
Sonoran	275,000	21.5
Mojave	140,000	11.0
Chihuahuan	453,000	35.5
Total	1,277,000	100.0
Warm deserts	868,000	68.0
Cold deserts	409,000	32.0
Basin and Range Province	1,717,000	—

Figure 8.1. Distributions of Mojave, Sonoran, and Chihuahan warm deserts as adopted in this chapter. Details of this delimitation can be found in MacMahon (1979) and MacMahon and Wagner (1985).

INTRODUCTION

This chapter treats the Mojave, Sonoran, and Chihuahuan deserts, the "warm deserts" of North America (Fig. 8.1). These three quite different vegetation units occupy 68% of the area indicated as desert by MacMahon (MacMahon 1979) (Table 8.1). In general, these deserts are low-elevation, xeric sites within the southwestern portion of the United States and the northern quarter of Mexico. Based on climatic considerations, the Mojave and Chihuahuan deserts are termed warm-temperate deserts, whereas the Sonoran Desert is subtropical in nature. The three are lumped under the term "warm deserts" not so much because of their average annual temperatures but more because their precipitation is in the form of rain, even if it occurs in the winter. This contrasts with the Great Basin Desert (see Chapter 7), generally called a cold desert, that usually has over 60% of its precipitation in the form of snow.

Because of their wide elevational span (−86 m in Death Valley in the Mojave to 1525 m in southern sections of the Chihuahuan Desert in Mexico) and latitudinal span (36½–22°N latitude), the warm deserts form transitions with a wide variety of vegetation types: the Great Basin Desert in the north (see Chapter 7), grasslands in the east and at higher elevations (see Chapter 9), and subtropical thorn forests to the south (see Chapter 15). It should be noted that this chapter uses a simple vegetation-based classification for subdivisions of

the three deserts. Several other subdivision classifications exist. The most recent, detailed, and quantitatively derived is that of Peinado (Peinado, Alcaraz, Aguirre, and Dalgadillo 1995), who recognized 19 subdivisions, using the Braun-Blanquet method of phytosociological releves. Although it is useful scientifically, I deem the terminology of this system cumbersome for the current treatment.

In the past 20 yr a great deal has been written about North American deserts. This coverage has included consideration of a range of topics: all North American deserts (MacMahon 1979; Bender 1982); all or a significant part of the warm deserts (Barbour and Major 1977; Wauer and Riskind 1977; Rzedowski 1978; McGinnies 1981; Brown 1982b; MacMahon and Wagner 1985); comparisons between our deserts and other warm deserts (Orians and Solbrig 1977); water (MacMahon and Schimpf 1981) and nitrogen (West and Skujins 1978) as they relate to desert vegetation; the biology of major plants such as mesquite (Simpson 1977), creosote bush (Mabry, Hunziker, and DiFeo. 1977), ironwood (Nabhan and Carr 1994), cacti (Gibson and Nobel 1986; Nobel 1988), and agaves (Gentry 1978; Gentry 1982; Pinkava and Gentry 1985; Nobel 1988); the distribution and ecology of Sonoran Desert perennials (Turner, Bowers, and Burgess 1995); the general biology of deserts (Polis 1991); and the paleoecology of warm deserts (Axelrod 1979; Van Devender 1990; Van Devender, Thompson, and Betancourt 1987).

Additionally, identification of warm-desert plants has been made easier by the availability of a number of books that, when used with the various state floras, cover many plants in detail and provide a wealth of illustrations. Such works cover cacti (Benson 1982), woody perennials (Benson and Darrow 1981), wildflowers (Niehaus and Ripper 1976; Niehaus, Ripper, and Savage 1984), and a variety of aspects of desert biology (MacMahon 1985).

Because of the intensity of recent coverage of warm deserts, as indicated by the references cited, this chapter emphasizes work published in the past 20 yr; the earlier references cited should give the reader entrée to thousands of older studies. This chapter briefly considers the physiography, climate, and soils of our warm deserts, followed by an analysis where appropriate of the vegetation for each of the three warm deserts and their subdivisions. Finally, those vegetation types common to all three of the deserts, here termed "azonal vegetation," are considered in a separate section.

PHYSIOGRAPHY

The physiographic area that is referred to as the Basin and Range Province of North America contains essentially all of our deserts, warm and cold. The main exceptions are (1) a piece of the Great Basin desert that laps over on the Columbia Plateau Province and (2) Baja California, generally considered to be a province by itself (West 1964), the Peninsular Range Province. The Basin and Range Province extends over 30° of latitude and 12° of longitude, encompassing an area of 1,717,000 km² (Table 8.1). The province is essentially an area surrounded by the main western North American mountain masses, the Rockies and the Sierra Nevada in the United States and the Sierra Madre Occidental and Sierra Madre Oriental in Mexico (Fig. 8.2).

The name, Basin and Range Province, derives from its general appearance from the air. The entire landscape is composed of rather large basins dotted with much smaller mountain ranges that generally trend north-south. Usually, more than 50% of the land surface is covered by the basins, though the figure may exceed 80%, especially in the Sonoran Desert. For example, at Organ Pipe Cactus National Monument, the basins occupy 82.6% of the area. The mountain ranges number more than 300. One worker, seeing them on a physiographic map, said they looked like "an army of caterpillars crawling northward out of Mexico" (King 1959). Elevations of the basins may be lower than −83 m in Death Valley and up to about 1525 m in Utah. The surrounding ranges may exceed 3950 m.

The basins of the province are dotted by lakes

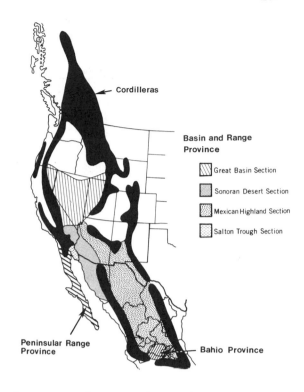

Figure 8.2. North American physiographic provinces and their sections as they apply to desert areas. Data from MacMahon (1979) and other sources.

Labels on figure:
Cordilleras
Basin and Range Province
Great Basin Section
Sonoran Desert Section
Mexican Highland Section
Salton Trough Section
Peninsular Range Province
Bahio Province

or their remnants, depending on the season and year. These lakes are dry for at least part of the year and often remain dry for years at a time. More than 200 examples of such lakes, generally referred to as "playas," occur in the province. Playas form when the runoff of precipitation from the mountains accumulates in the depressions of the surrounding basins. The standing water evaporates, leaving deposits high in various salts of calcium or sodium. Playas are usually small (less than 100 km²), and though they often look similar, the 50,000 or so that occur worldwide often are geomorphically quite distinct (Neal 1969; Neal 1975).

Another structural feature of the desert landscape is the alluvial fan. These are cone-like deposits that originate from canyons in the mountains and fan out from the mouth of a canyon into a valley. Where several alluvial fans from adjacent canyons coalesce, a bajada is formed. Bajadas (Spanish for "slope") have smooth surfaces and may cover up to 75% of a valley. They have slopes of 6–9% where they meet mountain ranges, lessening to about 1% where they flow onto the plains of the playas of valley floors. Upper bajadas have coarse-textured soils, whereas their lower ends and the associated playas may have very fine soils (e.g.,

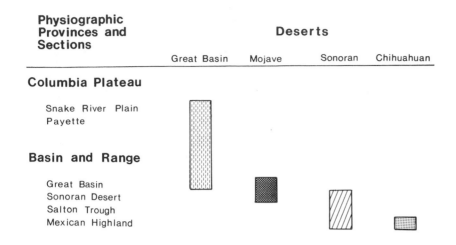

Figure 8.3. Differences between the extent of the biologically based desert subdivisions of North America and the physiographic provinces and sections of the geographers, as depicted in Fig. 8.2. The spans of sections in which the deserts occur are indicated by bars. Clearly, the biological deserts overlap some physiographic sections (i.e., one physiographic boundary may contain more than one desert type).

silts). I will emphasize the importance of bajadas often in this discussion of the distribution of plants.

The Basin and Range Province is subdivided into a number of units called sections whose boundaries are somewhat arbitrary (Fig. 8.2), and they do not coincide with the biotically defined desert boundaries (Fig. 8.3). The Great Basin Section occupies about one-third of the province, including most of Nevada, the western half of Utah, and minute portions of Idaho, Oregon, and California. It was a focus for Chapter 7. This section has few drainage outlets other than an area emptying into the Snake River in the northeast, a portion of the northwest that drains via the Pit River to the Sacramento River, and a part of the southeast corner that empties into the Colorado River via the Virgin River. Part of the northern Mojave Desert of Nevada and southwestern Utah is contained in this section. The Death Valley region is also in this section (Fiero 1986).

The Sonoran Desert Section includes the southwestern quarter of Arizona, the adjacent desert areas of California, and about one-third of Sonora, Mexico. Most of the Mojave and Sonoran deserts are in this section. Common rock substrates of the section include Precambrian granites and gneisses. Lakes were common in the Pleistocene. Because slopes often are composed of metamorphic rocks, they may have rises of up to 20%. The elevations of the valleys are low, seldom exceeding 650 m.

The Salton Sea Trough Section is an extension onto the land of the 1600 km trough occupied by the Gulf of California. In the United States, the section includes a very small area centered around Brawley and El Centro, California. It occurs just south of the mountains forming the southern boundary of Joshua Tree National Monument. A main feature of the trough is the Salton Sea. At one time, perhaps only a few hundred years ago, the trough was occupied by Lake Cahuilla.

The Mexican Highland Section occurs as a diagonal band across the middle third of Arizona, the southwest quarter of New Mexico, and southward into Mexico, where it abuts the Bahio Province near the Transverse Volcanic Belt. The Chihuahuan Desert and the Chihuahuan-Sonoran transitions are essentially within this section. The area is characterized by high valleys (up to 1525 m). Generally, the lowest elevations of the section are along the Rio Grande, rising both to the north and to the south. Unlike other portions of the Basin and Range Province, this section has well-developed drainage systems, though internal drainages and their associated playas do occur.

The Basin and Range Province is so vast and its geologic history so varied that it is difficult to generalize about it. The landscape throughout is composed of a mosaic of sedimentary rocks high in calcium carbonate patched with acidic volcanic materials. These substrates have created complex soil mosaics that influence the patterns of vegetation. Add to this the major changes in surface geomorphology caused by arroyo cutting in the past 100 yr (Cooke and Reeves 1976), and one can see that vegetation analysis must be conducted in a circumspect manner. For example, where complexes of geomorphic surfaces of different ages occur, soil chemistry and, consequently, plant species may vary (Lajtha and Schlesinger 1988). General treatments of desert geomorphic features are readily available (Cooke and Warren 1973; Goudie and Wilkinson 1977; Mabbutt 1977).

SOILS

Generally, soils in arid zones are low in organic matter, have slightly acidic to alkaline surface soils, and develop calcium carbonate accumulations in the upper 2 m of soil. They often do not show strong profile development (Dregne 1976, 1979; Hendricks 1985). Additionally, they may have long periods of low biological activity interspersed with periods of extreme activity, usually following rains during the warm season. Soils having all these characteristics are generally termed Aridisols, and they form under strong influences from wind, water, and high temperatures.

Some soil horizons in desert areas form hardpans because of cementing action by calcium carbonate, silica, or even iron compounds. Calcium carbonate cementation may form a water-impervious petrocalcic layer termed a "caliche" (Schlesinger 1985). Caliches can be quite thick (90 m), and they can be at the surface or buried to great depths (230 m) in some valleys (Shreve and Mallery 1933). It should be noted that carbonate layers, because of complex solution dynamics, can occur in desert soils that are not especially high in carbonate (Gile, Peterson and Grossman 1966).

The presence of a caliche layer greatly influences the distribution of plant species. Creosote bush (*Larrea tridentata*), a dominant warm-desert shrub, requires high-calcium, gravelly soils with a relatively deep caliche layer (Hallmark and Allen 1975).

The carbon in calcium carbonate may represent a storage pool important for the global balance of carbon. In most of the world's soils, the main carbon fraction is in the form of organic matter from decomposition of plants and animals. In deserts, the ratio of carbonate to organic carbon may exceed 10:1, and for desert areas the size of Arizona this can mean carbonate (C) stores of 396×10^7 t (Schlesinger 1982).

Many North American desert landscapes, especially those of southeastern California, adjacent Mexico, and scattered localities in the Chihuahuan Desert, are dominated by dunes (Smith 1982). The soils in these areas are composed of sand-size particles of either silica or gypsum. Both types of dune systems support characteristic floras. Dune soils (Regosols) allow for rapid percolation and subsequent storage of soil water (Bowers 1982), and thus these are among the most mesic sites in deserts (Danin 1996).

One important soil catena in deserts is represented by the particle size and salinity gradients seen on bajadas. Upper bajadas have a more diverse vegetation, a higher proportion of large soil particles, and lower salinity than the lower portions of bajadas (Solbrig et al. 1977; Phillips and MacMahon 1978). The presence of arroyos or even shallow rills on a bajada can modify vegetation and soil patterns.

Two surface phenomena of desert soils should be mentioned. In many areas, the surface of the soil is covered with stones or pebbles of relatively uniform size spaced so closely together that the soil surface is obscured. These areas, termed "desert pavement," may cover 1000 ha in some areas. Their exact origin is not clear, but it probably involves the effects of wind removing fine particles on these generally flat surfaces (Fuller 1975).

The stones forming the pavement often are covered with a dark patina composed of 60% clay and 20–30% oxides of iron and manganese that give them their dark color (Dorn 1991). It has been argued that these varnishes have completely abiotic origins (Moore and Elvidge 1982); however, Dorn and Oberlander (1981) argue for the involvement of microorganisms. There is no reason to believe that these two mechanisms are mutually exclusive. These varnishes can furnish historical evidence of geomorphic events and processes (Dorn 1988, 1991).

The second soil surface phenomenon of interest is the presence of hardened soil crusts that are inhabited and generated by cyanobacteria, algae, and lichens. These cryptogamic crusts form most often on clay or silty soils. The crusts stabilize the soil surface and may fix atmospheric nitrogen in low quantities but quantities that are significant from the perspective of ecosystem-wide nitrogen cycling (Skujins 1984).

Below the surface, desert soils exhibit a mosaic of areas of high nutrient concentration, especially nitrogen and organic matter, in a matrix of low nutrient, interplant soil. These "islands of fertility" usually form around existing plants and are thought to be enriched by the concentration and incorporation of organic matter transported by animals to the vicinity of shrub bases (Garner and Steinberger 1989). The development of this pattern may be a good indicator of grassland-to-desert transformations (Schlesinger, Raikes, Hartley, and Cross 1996).

Saline soils are common in warm deserts on the flats and valley bottoms, especially around playas.

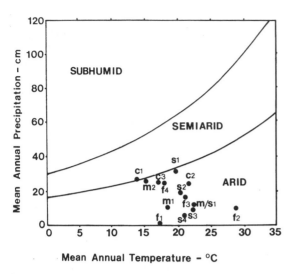

Figure 8.4. Depiction of arid and semiarid moisture-province boundaries as presented by Bailey (1979), with data from some desert localities superimposed. Mojave desert = m1 (Las Vegas, Nevada); m2 (St. George, Utah); m/s1 (Needles, California, a Mojave-Sonora transition site); Sonora desert = s1 (Tucson, Arizona): s2 (Phoenix, Arizona; s3 (Yuma, Arizona); s4 (Brawley, California); Chiuahua desert = c1 (Socorro, New Mexico); c2 (Ojinaga, Chihuaha); c3 (El Paso, Texas). Sites on other continents = f1 (Atacama Desert, Antofagasta, Chile); f2 (southern Sahara Desert, Tessalit, Mali); f3 (northern Sahara Desert, Biskra, Algeria); f4 (Karroo Desert, Beaufort West, South Africa). Climagrams are shown in Figure 8.5 and in MacMahon and Wagner (1985). Data for other sites are from Walter (1971). Note that the aridity of North American deserts is of a magnitude similar to that for many "true" deserts throughout the world.

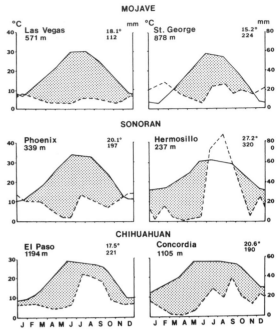

Figure 8.5. Simplified climate diagrams for some weather stations within the North American warm deserts. Adapted from Walter and Lieth (1967). The number in the upper right of each station's graph is mean annual temperature (° C); the number below that is mean annual precipitation (mm). Dashed line depicts monthly rainfall; solid line depicts mean monthly temperature; shaded area = months when potential evaporation exceeds precipitation.

Their significant interactions with vegetation are discussed in a later section.

Soils of desert areas are especially susceptible to damage by human use. Improper irrigation can cause salinity; hiking and the use of off-road vehicles can cause compaction (Eckert, Wood, Blackburn, and Peterson 1979; Webb and Wilshire 1983), as can overgrazing (Webb and Stielstra 1979). All these effects can be reversed only very slowly (Webb and Wilshire 1980), and thus they may have long-range effects on the rate of recovery or persistence of native vegetation and potential land uses.

CLIMATE

By definition, deserts are warm areas with low rainfall and high rates of evapotranspiration. There have been various schemes to classify the world into arid, semiarid, humid zones, and so forth. Often the North American deserts are thought to be semideserts, rather than true deserts, because of

their relatively lush vegetation. Figure 8.4 clearly suggests, however, that although many sites in the Chihuahuan Desert border on being semiarid, there are sites within each of our warm deserts that are equal in aridity to those elsewhere in the world. Different results might be obtained if one used a different aridity index, such as one based on water-balance parameters (Oberlander 1979).

Precipitation is low in amount and highly variable from year to year in deserts (Fig. 8.5). For example, Furnace Creek, in Death Valley, California (Mojave Desert), has a long-term mean rainfall of 4.2 cm yr^{-1}. However, between 1912 and 1962, the averages for approximately 10 yr periods showed wet episodes in which rainfall was 6.7 cm yr^{-1} and dry periods with one-third of that (2.1 cm yr^{-1}). In two of those years, 1929 and 1954, there was no rainfall over 12 mo periods (Hunt 1975).

Seasonal rainfall patterns vary significantly among our warm deserts. North American deserts lie in a zone of the earth's surface in which strong seasonal shifts occur in storm tracks because of seasonal heating of the earth's surface and global wind patterns. In the winter, storms originating in the Pacific Ocean move inland and are pushed across the mountain chains of the Coast Ranges, Sierra Nevada, and Sierra Madre Occidental, causing adiabatic cooling, condensation, and rain. Thus, the areas near the coast of western North America receive fall-to-spring rains (winter precipitation). Storm tracks responsible for winter precipitation move north of the United States in the spring, decreasing the flow of air from the Pacific Ocean. At this time, stronger storm cells originate from the Gulf of Mexico, moving westward and northwestward. This movement causes spring-to-fall rains (summer rainfall). Each of these patterns becomes less pronounced as one moves away from the places where the storms originate. The result is that the Mojave Desert, being close to the Pacific Ocean, gets winter rain, whereas the Chihuahuan Desert farther east is more directly in the path of the summer-rainfall track. The Sonoran Desert is intermediate between the two storm systems, and so it receives biseasonal rainfall the exact proportions of which depend on location along the Mojave–Chihuahuan axis (Fig. 8.6).

Summer and winter rains differ not only in seasonality but also in other attributes: Winter rains are of long duration and low intensity, and they cover large areas at a time; summer rains are cyclonic thunderstorms of short duration (minutes to hours) and high intensity, and they are limited in areal extent.

High-intensity, short-duration rainfall can cause surface runoff of water, and so the rain that falls on a given site may not be a good measure of the water that percolates to a plant's rooting zone at that site (Schlesinger and Jones 1984), with topography being an important determining factor (Osborn, Shirley, Davis, and Koehler 1980). This lack of correlation between rainfall and the amount of water available for plant growth makes it difficult to use certain indices of plant–growth–rainfall relationships, such as the rain-use efficiency metric (Le Houerou 1984).

Whereas average temperatures in our warm deserts are high (Fig. 8.5), there are differences among the deserts. The Sonoran Desert is generally the warmest. In part, this is because of its low elevation (less than 600 m) as compared with the Mojave Desert (three-quarters of the area is at 600–1200 m) or the Chihuahuan Desert (more than half above 1200 m). In contrast to this generalization, Death Valley in the Mojave Desert has the highest absolute temperatures and the lowest elevations of our warm deserts.

Of equal importance to the high temperatures are the periods in which freezing temperatures persist in excess of 36 hr. These bouts cause sig-

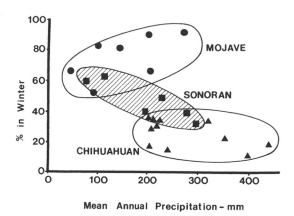

Figure 8.6. Percentage of winter rainfall plotted against mean annual precipitation for a variety of sites. Adapted from MacMahon and Wagner (1985).

nificant mortality among many desert species (Hastings and Turner 1965), the results of which are important in some areas, such as the Sonoran Desert (Lowe and Steenbergh 1981). Thus, the composition of vegetation in some areas may be due to unusual occurrences of prolonged freezes, rather than the result of tens of years of "normal" temperature and precipitation.

Evaporation and wind are important to the vegetation in desert areas. Interestingly, water loss from a plot containing *Larrea tridentata* was found to be virtually the same as from a plot of bare soil (Sammis and Gay 1979). Transpiration accounted for only 7% of the total evapotranspiration.

Fog and dew are seldom considered to be important climatic variables in North American deserts. It is clear, however, that many desert plants, including mosses, lichens, and vascular plants, can use these two forms of water for hydration. Although conditions necessary for fog or dew development are rare in inland portions of warm deserts, coastal areas, such as some of the Sonoran Desert portions of Baja California, experience both fog and dew, and these can be locally significant in determining the nature of plant communities (Nash, Niebecker, Moser, and Reeves 1979).

MOJAVE DESERT

Many scientists have suggested that the Mojave Desert does not represent a discrete vegetation type but, rather, is a zone of transition or an ecotone. This position emphasizes the Great Basin–like character of the vegetation in the north and the Sonoran Desert–like aspect in the south. However, one-fourth of the plant species of the Mojave Desert are endemics; of the annuals, nearly 80% of the 250 or so species are endemics. Because of the large number of Mojave Desert endemics, it seems to me that the vegetation is sufficiently circumscribed to be discussed as an entity in its own right.

In the north, the Mojave Desert is limited by the higher elevations of the Great Basin. If one starts in Las Vegas, Nevada (670 m), in typical Mojave Desert vegetation, and drives northward, one is in the Mojave – Great Basin transition at Beatty (1170 m), and in the Great Basin at Tonopah (1840). The transitional vegetation has been well studied in the area of the Nevada Test Site at Mercury (Beatley 1976; Rundel and Gibson 1996). Here, *Larrea*-dominated communities give way to those dominated by *Artemisia tridentata* (big sagebrush) and *Atriplex confertifolia* (shadscale). Also typical of these sites is *Coleogyne ramosissima* (blackbrush) (see Chapter 7), *Lycium* species (wolfberries), and *Atriplex (Grayia) spinosa* (hopsage).

This transitional vegetation is intermingled in complex ways with typical Mojave Desert plants. For example, two plots 90 m apart, each having 12% cover and separated in elevation by only 1.5 m, differed in that one had *Larrea* as 2.7% cover and *Lycium shockleyi* at 4.5%, whereas the other had no *Larrea* cover and 10.1% *Lycium* cover (Beatley 1974). Typical Mojave Desert sites differ from our other warm deserts in that they are dominated by low-growing, often widely spaced perennial shrubs, representing relatively few species, and although cacti are present, they are generally of low stature. Yuccas can be locally common.

An excellent summary of ecological processes and communities of the Mojave, based primarily on data from Rock Valley, Nevada, is available (Rundel and Gibson 1996). The response of native vegetation following disturbances has been documented for a variety of sites in the Mojave (Vasek 1983; Carpenter, Barbour, and Bahre 1986; Prose, Metzger, and Wilshire 1987; Webb, Steiger, and Turner 1987; Webb, Steiger, and Newman 1988). The pattern, as expected, varies with the nature and intensity of the disturbance, the nature of the substrate, and the species involved.

Subdivisions of the Mojave Desert vegetation have been proposed in several studies. Vasek and Barbour (1977) proposed and Turner (1982) accepted a classification including five general vegetation types, termed "series" – creosote bush, shadscale, saltbush, blackbrush, and Joshua tree. Rowlands and associates (1982) referred to 11 vegetation types. The difference between the two systems is the further subdivision of the vegetation of saline areas by Rowlands and colleagues. For our purposes, the five-unit system suffices. The shad-

scale series was discussed by West in Chapter 7, as was the blackbrush series. The saltbush series, and saline-soil vegetation in general, are covered in a later section of this chapter. Thus, we are left with two broad vegetation types: the creosote bush and the Joshua tree series.

Creosote Bush Series

The most common association of plants is dominated by *Larrea tridentata* and *Ambrosia dumosa* (white bursage) (Fig. 8.7). Perhaps 70% of the Mojave Desert is covered with these two species as co-dominants, especially on lower portions of bajadas and valley floors.

Creosote bush is virtually synonymous with the warm deserts, because its distribution coincides closely with their distribution. Even the approximate boundaries between our various warm deserts are roughly indicated by the distribution of chromosome numbers within creosote bush populations (Mojave, N = 39; Sonora, N = 26; Chihuahua, N = 13) (Yang 1970). Creosote bush occurs from 73 m below sea level in Death Valley to 1585 m on southern exposures (Hunt 1966). Although it occurs on soils ranging from sand dunes to quite rocky soils, creosote bush is limited to areas that are fairly well aerated, that have low salinity, that receive less than 18 cm of rainfall, and that experience freezing temperatures for no more than 6 d consecutively (Beatley 1976). Creosote bush orients its foliage clusters mainly to the southeast in the United States but shows no predominant orientation at its southern limit in Mexico (Neufeld et al. 1988). Its canopy architecture may minimize self-shading in the morning, resulting in greater water-use efficiency (Ezcurra et al. 1992). Because of its wide ecological and geographical distribution, creosote bush is associated with a variety of other species, especially on bajadas and nonsaline flats.

Typical bajadas in parts of Death Valley National Monument show significant vegetation changes from bottom to top (Hunt 1975). On the fine soils of the bajada bottom, one of two saltbush species often forms nearly pure stands. Soils with high percentages of carbonate rocks are dominated by *Atriplex hymenelytra* (desert holly), which may average densities of 120 individuals per ha but can reach 550 individuals per ha. Where there is less carbonate, *Atriplex polycarpa* (cattle spinach), a large species, predominates at about the same density. Farther up the bajada, creosote bush appears and mixes in some places with cattle spinach or desert holly.

Creosote bush densities vary considerably,

Figure 8.7. Central view of a typical Mojave Desert creosote bush flat. The taller shrub is creosote bush (Larrea tridentata) and the shorter, lighter bush is white bursage (Ambrosia dumosa). Arizona–Utah border, Washington County, Utah.

Figure 8.8. Three Mojave Desert sites: (a) Joshua tree (Yucca brevifolia) visually dominates an upper-elevation site near Wikieup, Mohave County, Arizona; (b) a creosote bush site with some Yucca schidigera north of Las Vegas, Clark County, Nevada; (c) a bajada site in Spring Mountains, Nye County, Nevada, which contains 13 shrub species in such genera as Ephedra, Thamnosma, Salazaria, and Ceratoides.

Table 8.2. *Density (D, per hectare) and cover (C, percentage) of perennials at five southern California localities dominated by creosote bush. Beneath the locality name, in parentheses, are millimeters of mean annual precipitation.*

Species	Baker (79)[a] (D/C)	Inyokern (86) (D/C)	Barstow (96) (D/C)	Lucerne Valley (106) (D/C)	Mojave (126) (D/C)
Acamptopappus sphaerocephalus		348/0.7			
Ambrosia dumosa	404/0.2	2876/8.6	2284/0.5	568/0.5	524/0.9
Atriplex polycarpa	148/0.2			44/0.6	
Atriplex spinifera			932/1.5		
Atriplex (Grayia) spinosa		40/0.2		4/<0.1	
Cassia armata				48/<0.1	
Chrysothamnus paniculatus					272/0.3
Psorothamnus fremontii		308/1.7			
Ephedra nevadensis				24/0.1	
Haplopappus linearifolius				124/0.1	8/<0.1
Larrea tridentata	196/1.3	400/7.8	144/0.9	392/3.0	400/7.8
Lepidium fremontii				4/<0.1	
Lycium andersonii	12/0.1				
Lycium pallidum			36/0.2		
Opuntia echinocarpa				4/<0.1	
Total					
Species	3	6	4	9	4
Density	748	3984	3396	1212	1204
Cover	1.7	19.1	3.1	4.3	9.0

[a] Annual ppt (mm).
Source: Based on Phillips and MacMahon (1981).

ranging from 100 to over 1000 per ha. It is difficult to count individual creosote bushes because they reproduce vegetatively; clones produced in this manner may persist for thousands of years (Vasek, Johnson, and Brum 1975; Vasek 1980).

The middle portions of bajadas are typical *Larrea–Ambrosia dumosa* communities, though variations do occur. In some sections of the Mojave Desert, including southern Death Valley, *Encelia farinosa* (brittlebush) may be abundant.

Other shrubs that are locally abundant associates of creosote bush on bajadas include *Menodora spinescens* (spiny menodora), *Lycium pallidum* or *L. andersonii* (wolfberries), *Ephedra* (Mormon tea), *Krameria parvifolia* (ratany), *Acamptopappus schlockeyi* (goldenhead), *Psorothamnus fremontii* (Fremont dalea), and *Psilostrophe cooperi* (yellow paper daisy) (Fig. 8.8c).

Three yuccas commonly occur on middle portions of bajadas: *Yucca brevifolia* (Joshua tree) (Fig. 8.8a), *Y. schidigera* (Mojave yucca) (Fig. 8.8b), and *Y. baccata* (banana yucca). A fourth species, *Y. whipplei* (desert Spanish bayonet), occurs along the western edge of the Mojave Desert.

Many cacti occur on the bajadas, and some species or varieties are Mojave Desert endemics. Benson (1982) listed 23 cacti occurring primarily in the Mojave Desert. Conspicuous species include *Opuntia basilaris* (beavertail), *O. echinocarpa*, and *O. acanthocarpa* (chollas), the many-stemmed *Echinocactus polycephalus*, and *Ferocactus acanthodes* (a barrel cactus).

Where there are rills or small channels, *Hymenoclea salsola* (cheesebush) may dominate, in association with *Cassia armata*, *Ambrosia eriocentra*, *Brickellia incana*, and *Acacia greggii* (catclaw).

On highly calcareous soils, frequently with well-developed pavements, creosote bush often associates with *Atriplex confertifolia*. Such sites often develop petrocalcic layers (caliche) that can inhibit deep penetration of plant roots. Such sites recur over major portions of the Mojave Desert. Locally, the co-dominant shrubs vary considerably. However, common associates include *Krameria, Ephedra, Lycium,* and one or another *Yucca.* Beatley (1976) pointed out that *Opuntia ramosissima* is known in the Mojave Desert only from this type of vegetation. She also listed some 40 species of herbs found in this association.

Perennial plants in the Mojave often exhibit a clumped dispersion pattern that is nonrandom in its species composition (Cody 1986a, 1986b). Reasons for this are complex (McAuliffe 1988) but include the "nurse plant" phenomenon that occurs in all our warm deserts. An interesting example is the positive association between three species of chollas (*Opuntia*) and a perennial grass (*Hilaria rigida*) (Cody 1993).

Higher-elevation sites or northern sites, with poorly developed pavements and essentially no caliche layer, support stands of creosote bush mixed with *Lycium andersonii* and *Atriplex (Grayia) spinosa.* These sites often contain Joshua tree (*Yucca brevifolia*) and thus represent a transition to the next series.

Representative stand data for some *Larrea*-dominated sites in the Mojave Desert are presented in Table 8.2.

Moderately successful models of the phenology of some shrubs in this series were developed by Turner and Randall (1987). Observations of plant–soil water relations are available (Amundson, Chadwick, and Sowers 1989; Smith, Herr, Leary, and Piorkowski 1995). Parallel studies for succulents and their rooting patterns have been conducted (Nobel, Miller, and Graham 1992).

Joshua Tree Series

As the series name implies, the Joshua tree series is usually dominated by *Yucca brevifolia,* the tallest nonriparian plant of the Mojave Desert (Fig. 8.8a). This conspicuous species is, in many people's minds, the most characteristic species of the Mojave Desert. Although its mapped distribution (Fig. 8.9) approximates the Mojave Desert boundary, the Joshua tree is actually confined to higher elevations of slightly cooler and moister sites, rather than occurring over the low-elevation, drier sites that predominate in the Mojave Desert. At the upper elevations of this vegetation type, a transition occurs with pinyon-juniper woodland.

The Joshua tree occurred at much lower elevations during the cool periods of the Pleistocene. For example, in Death Valley, where the Joshua tree currently occurs only above about 1700 m, it was present near the valley floor around 19,550 BP (Wells and Woodcock 1985).

Interestingly, despite what would seem to be a sun-drenched environment, Joshua tree orients its

Figure 8.9. *Distribution of Joshua tree, a species whose distribution essentially outlines the Mojave Desert, but it is elevationally restricted to higher sites. Adapted from Benson and Darrow (1981).*

branches to maximize interception of sunlight (Rasmuson, Anderson, and Huntly 1994).

Despite its visual dominance on some sites, Joshua tree densities may be as low as 104–125 per ha, with a cover of 0.20% in a vegetation containing 13 species of perennial shrubs with a total coverage of 10.15% and a density of 7250 plants per ha (Vasek and Barbour 1977).

The perennials associated with the Joshua tree vary considerably, depending on whether one samples sites at its lower-elevation limit, where it may mix with *Larrea* and *Ambrosia dumosa* (e.g., southern Utah), or sites at middle to upper elevations, where *Ephedra nevadensis, Ceratoides lanata, Lycium, Salazaria mexicana, Thamnosma montana,* or *Coleogyne ramosissima* may be numerical dominants. Additionally, several cacti, both opuntias and barrels (*Ferocactus*), and several perennial grasses (*Hilaria rigida, Muhlenbergia porteri,* and *Stipa* species) occur.

Where the Sonoran and Mojave deserts abut (northeastern Arizona), saguaros, junipers, Joshua trees, and paloverdes may co-occur in stands of great complexity and beauty. Also characteristic of these transitional sites is crucifixion thorn (*Canotia holacantha*).

SONORAN DESERT

The Sonoran Desert, with its biseasonal rainfall and tropical affinities, has a complex biota. Numerous plant species representing a variety of life

Figure 8.10. Sonoran Desert scenes: (a) typical middle to upper bajada site in southern Arizona with sguaro (Carnegiea gigantea), ocotillo (Fouquieria splendens), creosote bush (Larrea tridentata), bursage (Ambrosia deltoidea), and foothill paloverde (Cercidium microphyllum); (b) plains site in the Lower Colorado Valley subdivision, souther Arizona, dominated by creosote bush; (c) an upper-elevation site with a special abundance of saguaros in the Tucson Mountains, Arizona.

forms (Crosswhite and Crosswhite 1984) often co-exist, creating architecturally varied environments (Fig. 8.10a and c). On some soils, however, the mix of species and their physiognomy is virtually the same as in the Mojave Desert (Fig. 8.10b).

This complex of life forms relates to the complexity of the abiotic variables, species–species interactions, as well as to evolutionary constraints (Solbrig 1986; Shmida and Burgess 1988; Cody 1989).

This great diversity of associations was subdivided by Shreve (1951) into seven divisions. Brown and Lowe (1980) redefined subdivision boundaries,

relegated some areas to a thornscrub series, and settled on six major subdivisions that were accepted by Turner and Brown (1982) and that I follow here (Fig. 8.11). Turner and Brown used data from 162 weather stations to show that major changes in mean annual temperature and precipitation, as well as continentality (the difference between average temperatures of summer and winter), coincide generally with vegetation boundaries determined by physiognomy and species composition.

The conceptualization of the boundaries, as well as the subdivisions, of the Sonoran Desert rests on the data and assumptions that one uses. Schmidt

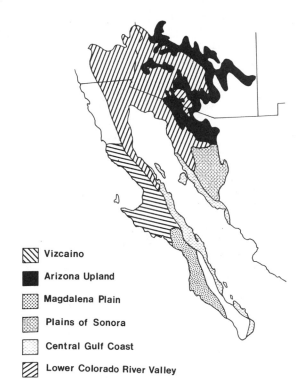

Vizcaino

Arizona Upland

Magdalena Plain

Plains of Sonora

Central Gulf Coast

Lower Colorado River Valley

Figure 8.11. Subdivisions of the Sonora Desert, based originally on ideas of Forrest Shreve but modified most recently by Turner and Brown (1982).

(1989) nicely summarized 12 conceptualizations of the 17 he recognized, and presented his own based on calculation of the Martonne Aridity Index from 250 weather stations. For Baja California see Peinado, Alcaraz, Delgadillo, and Aguado (1993), whose results, based on abiotic variables, parallel but differ from my delineations based on distribution patterns of plants and animals.

Many Sonoran Desert genera also occur in the Monte Desert of Argentina (Solbrig 1972), and these two deserts are, therefore, structurally and functionally similar (Orians and Solbrig 1977).

Lower Colorado Valley Subdivision

Because of its relatively high winter and summer temperatures and low annual precipitation, this largest Sonoran Desert subdivision is generally the most xeric. Adding to the stress produced by its low rainfall amounts is the fact that the rainfall frequency is low and its pattern highly irregular, a situation that contrasts with that in the adjacent Arizona Upland subdivision (Ezcurra and Rodrigues 1986). The vegetation in many areas reflects these environmental extremes by its simple

species composition and lack of structural complexity. Large areas are dominated by *Ambrosia dumosa* and *Larrea*, with few other perennial shrubs present (Fig. 8.10b, Table 8.3). Throughout the Sonoran Desert, and others also, *Larrea* is strongly associated with calcareous soils that have a caliche layer. Some areas, especially those covered with pavements, support virtually no shrubs unless the landscape is traversed by rills or runnels (Turner and Brown 1982).

Many of North America's most extensive sand dune systems occur in this subdivision. Adjacent flats with conspicuously sandy soils support stands of perennial grasses (e.g., *Hilaria rigida*) and shrubs such as *Ephedra trifurca*, *Psorothamnus schottii*, *P. emoryi*, *Eriogonum deserticola*, and *Petalonyx thurberi* (Felger 1980).

In areas with significant development of washes, *Cercidium floridum*, *Psorothamnus spinosus*, *Sapium biloculare*, *Condalia globosa*, and *Prosopis glandulosa*, or any one of some 20 other species, may dominate locally (Crosswhite and Crosswhite 1982).

In some areas of the subdivision, brittlebush (*Encelia farinosa*) is abundant and especially conspicuous because its light gray leaves and showy yellow flowers stand out against the dark volcanic soils it frequently inhabits. Also on volcanic substrates one encounters *Agave deserti*, *Trixis californica*, *Hyptis emoryi*, and *Peucephyllum schottii* (Zabriskie 1979).

On bajadas, the simple flora of the plains gives way to a complex that includes *Fouquieria splendens*, *Cercidium microphyllum*, *Acacia greggii*, and saguaro *Carnegiea gigantea*), a mixture also encountered in the Arizona Upland.

The vegetation of bajadas throughout the Sonoran Desert increases in physiognomic and species complexity as one ascends them (Fig. 8.12). These plant associations are composed of nonrandom sets of species with regard to form (Bowers and Lowe 1986). Patterns of composition change correlate well with soil physical properties and soil–water relations (Phillips and MacMahon 1978), with some caveats mentioned earlier.

In flat areas associated with major rivers, where periodic floods historically have taken place, soils are often slightly to moderately saline. Here, *Atriplex polycarpa* may be an obvious dominant, often forming nearly pure stands. Other species that do well in such sites include species of *Isocoma*, a *Lycium* (often *L. fremontii*), and another *Atriplex* (usually either *A. canescens* or *A. lentiformis*). In some areas, especially on sandy soils, screwbean mesquite (*Prosopis pubescens*) is abundant, and it may occur with *Tessaria sericea*, as

Table 8.3. *Density (D, per hectare) and cover (C, percentage) of perennials at four Sonoran Desert sites. Dash indicates cover <0.1%.*

Species	Lower Colorado Valley (D/C)	Arizona Upland (D/C)	Central Gulf Coast (D/C)	Plains of Sonora (D/C)
Shrubs				
Ambrosia ambrosioides		1.01/—		
Ambrosia deltoidea		87.8/0.5		
Ambrosia dumosa	84/0.1	549.7/2.7		
Caesalpinia pumila			104.0/2.5	
Calliandra eriophylla			45.6/0.2	
Coursetia glandulosa			5.3/0.3	
Croton sonorae			104.0/0.4	
Encelia farinosa		316.7/2.0	1067.4/4.9	2882/1.6
Hymenoclea monogyra		151.3/1.1		
Jatropha cuneata			40.3/2.9	
Krameria parvifolia			58.4/0.3	
Larrea tridentata	448/5.5	437.7/6.3	1.1/—	
Lycium andersonii		88.8/1.1	1.1/—	
L. berlandieri			9.5/0.6	
Mimosa laxiflora			37.1/0.4	
Trixis californica		3.0/—		
Other shrubs			54.1/0.7	
Subtrees				
Acacia constricta			82.8/4.6	
Bursera laxiflora			33.0/2.3	1/0.3
B. microphylla			17.0/2.2	
Cericidium microphyllum		26.2/6.7	19.2/3.7	24/7
Eysenhardtia orthocarpa			4.2/0.1	
Fouquieria macdougalii			4.2/0.3	17/0.7
Forchammeria watsoni			1.1/—	
Guaiacum coulteri				5/0.1
Olneya tesota		1.0/0.1	17.0/2.4	65/9
Pithecellobium mexicanum			6.8/0.2	
Prosopis glandulosa				
Large cacti				
Carnegiea gigantea		11.0/—		
Stenocereus thurberi			0.1/—	
Lophocereus schottii				8/0.5
Pachycereus pringlei				1/0.1
Small cacti				
Opuntia acanthocarpa		27.2/0.2		
Opuntia fulgida			20.2/—	
Opuntia ramosissima	144/0.5			
Totals				
Species	3	12	23	8
Density	676	1701	1734	3003
Cover	6.1	20.7	29.0	19.3

along the Gila River of Arizona (Rea 1983). Many of these areas, where salinity is not extreme, currently are used for agricultural purposes and are quite productive. One such is California's Coachella Valley.

Arizona Upland Subdivision

The Arizona Upland subdivision is remarkable for its diversity of species and life forms. Many areas are visually dominated by subtrees and tall cacti

Figure 8.12. A stylized cross section of vegetation change along a Sonora Desert bajada in the Arizona Upland subdivision. From right to left (ascending the bajada), species from Larrea-Ambrosia flats become mixed with subtrees, usually Cercidium, Olneya, and Prosopis, with ocotillo (Fouquieria) and a variety of prickly pear and cholla cacti (Opuntia) and columnar cacti such as saguaro (Carnegiea gigantea). Middle to upper bajada sites typify the Sonora Desert; lower bajadas physiognomically resemble vast areas of the Mojave and Chihuahua deserts.

such as saguaros. In fact, the saguaro-dominated landscape is synonymous in the minds of most North Americans with the word "desert." The biology of saguaro has been extensively documented (Steenbergh and Lowe 1976, 1977, 1983; Lajtha, Kolberg, and Getz 1997).

The annual rainfall is 10–300 mm and has a biseasonal distribution of about equal amounts. This pattern supports an annual flora in each period of mesic conditions, adding to the species richness of sites in the subdivision.

As might be expected, several successive years of wet winters cause fuel buildup. Consequently, fires are more numerous and of greater areal extent than under "usual" rainfall patterns (Rogers and Vint 1987; Schmid and Rogers 1988). Such unusual fire regimes could cause the local extinction of many desert perennials (e.g., saguaro) if the pattern persisted (Rogers 1985).

Valleys and lower portions of bajadas are dominated by creosote bush; the density in pure stands may reach 3700 per ha and represent 29% cover. Although white bursage co-dominates in many areas, *Ambrosia deltoidea* is a frequent dominant on coarse soils (Niering and Lowe 1984). Acacias and several chollas, such as *Opuntia fulgida* (Fig. 8.13a), are locally common.

On the slopes, saguaros are common (Fig. 8.10c), especially where larger shrubs and subtrees occur. Young saguaros (and other cacti, large and small) frequently establish beneath the canopy of a "nurse" plant that perhaps affords some advantage in avoiding heat stress and the effects of frost, predators, and so forth (Fig. 8.13d) (Nobel 1980; Vandermeer 1980). As the saguaro grows, it may impede its nurse and cause an increase in stem dieback, at least for one common nurse, *Cercidium microphyllum* (McAuliffe 1984). This case of facilitation for the cactus, followed by competition, is but one example of a complex of biotic interactions that

have been documented within this subdivision. Yeaton and associates (1977) detailed a complex competitive milieu involving five species in Organ Pipe Cactus National Monument. In their study, saguaro did not compete with any of the other species (*Larrea, Ambrosia deltoidea, Opuntia fulgida, Fouquieria splendens*), whereas *Larrea* competed with all species except saguaro.

As one travels up the slopes of this subdivision, subtrees increase in abundance. Species include *Cercidium microphyllum, Fouquieria splendens, Olneya tesota, Celtis pallida,* and *Condalia warnockii*. In the northern localities of this subdivision, a subtree characteristic of transitional sites, *Canotia holacantha,* is abundant. In southern sites, *Bursera microphylla* and two large cacti, *Lophocereus schottii* and *Stenocereus thurberi* (Fig. 8.13c), are locally conspicuous. The pattern on bajadas is modified by a variety of site-specific lithological factors (Parker 1991) and is influenced by the historic pools of species' availability (Bowers 1988).

Shrubs of the middle to upper bajadas include *Calliandra eriophylla, Zinnia grandiflora, Psilostrophe cooperi, Jatropha cardiophylla, Simmondsia chinensis, Encelia farinosa, Krameria parvifolia,* and a host of others. *Simmondsia* (jojoba) is of special interest because of its economic importance and so its biology is well summarized (Benzioni and Dunstone 1986).

A variety of small- to medium-size cacti also respond to the soil gradient of bajadas (Yeaton and Cody 1979).

It should be noted that three subtrees (*Cercidium floridum, C. microphyllum,* and *Olneya tesota*) and the saguaro (*Carnegiea gigantea*) that so typify this subdivision also occur prominently in several other subdivisions of the Sonoran Desert (Fig. 8.14). Especially widespread are *C. microphyllum* and *O. tesota* (Fig. 8.14c and 8.14d), two species that occur in nearly every subdivision except on the Pacific Ocean side of Baja California.

*Figure 8.13. Some conspicuous plant forms found in the Sonora Desert: (a) cholla (*Opuntia fulgida*); (b) ocotillo (*Fouquieria splendens*): (c) columnar organ pipe cactus* (Lamaireocereus thurberi*); (d) foothill paloverde subtree (*Cercidium microphyllum; *note young saguaros using the paloverde as a nurse plant).*

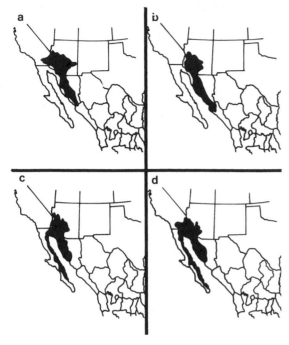

Figure 8.14. Distributions of four widely ranging Sonora Desert dominants that impart this desert's unique physiognomy: (a) blue paloverde (Cercidium floridum); (b) saguaro (Carnegiea gigantea); foothill paloverde (Cercidium microphyllum); (d) ironwood (Olneya tesota). Maps are based on data from Benson and Darrow (1981) and Hastings et al. (1972).

Finally, a long-term study (72 yr) within permanent plots in this subdivision (Tumamock Hill, near Tucson, Arizona) showed no consistent directional changes in vegetation composition despite large fluctuations in cover and density of species. The relative cover by most dominants remained about the same, albeit absolutely varying with rainfall regimes (Goldberg and Turner 1986). Other sites, however, especially disturbed sites, may change directionally (Hastings and Turner 1965).

Central Gulf Coast Subdivision

This subdivision is split into two portions, one on each side of the Gulf of California. The mainland portion in Sonora consists of a narrow band of vegetation ranging from 27–30° N latitude. A similar band exists in Baja California, extending nearly as far north but reaching the tip of Baja at 23° N.

The climate patterns, especially the seasonal distribution and reliability of rainfall, vary greatly along this rather large north-south gradient (Humphrey 1974; Turner and Brown 1982).

The vegetation of the area is dominated by subtrees and tall cacti, with some medium-size shrubs. Conspicuously absent is the layer of small shrubs (e.g., one of the *Ambrosia* species) that is so obvious in the Sonoran Desert in the United States. The result is that the vegetation has a sparse, coarse appearance. Soils are often rocky.

The composition of the vegetation on the two sides of the Gulf differs in that the mainland portion does not contain all the conspicuous dominants found on the peninsula side. Some species that do occur on the mainland have very limited distribution. The highly photogenic boojum (*Fouquieria columnaris*) is such a species. In Baja, boojum spans a north-south distance of 400 km, whereas in Sonora it is confined to a 50 km section of the Sierra Bacha, and there it is confined to within a few kilometers of the coast (Humphrey 1974), perhaps because of humidity (Humphrey and Marx 1980).

Shreve and Wiggins (1964) referred to this subdivision as the *Bursera–Jatropha* region. This designation indicates the dominance by *Bursera microphylla*, *B. hindsiana*, *Jatropha cuneata*, and *J. cinerea*, in association with the more widely distributed *Olneya tesota*, *Cercidium floridum*, *Fouquieria splendens*, and teddy bear cholla (*Opuntia bigelovii*). Creosote bush is much less prominent than in the Arizona Upland subdivision. Close to shores, *Frankenia palmeri* mixes with *Atriplex*, *Lycium*, *Suaeda*, *Encelia*, and *Ambrosia* to form an association with a physiognomy of very low form that extends north into the Lower Colorado Valley subdivision and into the Vizcaino subdivision in Baja.

On bajadas and inland, the *Bursera–Jatropha* admixture is punctuated by a variety of species, including *Opuntia cholla*, *Lysiloma candida*, *Fouquieria columnaris*, *Ficus palmeri* (one of several *Ambrosia* species), *Viscainoa geniculata*, *Solanum hindsianum*, *Hyptis emoryi*, and *Justicia californica* (Humphrey 1974; Bratz 1976; Felger and Lowe 1976; Stromberg and Krischan 1983). Sampling at Punta Cirio, Sonora, yielded 27 species of perennials with densities of 2080 per ha in arroyos, 3295 on south-facing pediments, and 4152 on north-facing pediments (Leitner 1987).

Turner and Brown (1982) recognized three series within this subdivision: two are a *Bursera–Pachycereus pringlei* grouping on deep granite soils and an ocotillo–*Jatropha*–creosote bush series on bjadas. Both of these occur on both sides of the Gulf of California. A third series, cactus–mesquite–saltbush, occurs on the coastal plain between Empalme and Potam, Sonora. This grouping contains five columnar cacti (saguaro, organ pipe, senita,

Pachycereus pecten–aboriginum, and *Stenocereus alamosensis*) in association with burseras, jatrophas, mesquites, and paloverdes.

Plains of Sonora Subdivision

The Plains of Sonora subdivision is the smallest and least diversified region of the Sonoran Desert. "Over most of the area the impression given by the vegetation is that of a very open forest of small, low-branching trees, with irregularly placed colonies of shrubs which are not tall enough to impair the view, and with large but very widely spaced columnar cacti" (Shreve and Wiggins 1964, p. 80).

Ironwood (*Olneya tesota*) is a very prominent member of this association, as are paloverdes (*Cercidium floridum*, *C. microphyllum*, *C. praecox*), *Parkinsonia aculeata*, *Atamisquea emarginata*, *Fouquieria splendens*, and mesquite (*Prosopis*). Other subtrees reach their northern terminus in this association, especially along washes. Included in this mixture are *Forchammeria watsoni*, *Jatropha cordata*, *Bursera laxiflora*, *Guaiacum coulteri*, *Fouquieria macdougalii*, *Piscidia mollus*, *Ipomoea arborescens*, and *Ceiba acuminata* (Table 8.3).

Brittlebush (*Encelia*) is widespread and often common in some places, as are several columnar cacti. Several chollas (*Opuntia fulgida*, *O. arbuscula*, *O. leptocaulis*) are locally abundant.

Common shrubs include *Caesalpinia pumila*, *Coursetia glandulosa*, *Calliandra*, *Eysenhardtia*, and *Mimosa laxiflora*, among others.

Locally, *Cordia parvifolia*, *Croton sonorae*, *Jacobinia ovata*, *Tecoma stans*, and *Zizyphus obtusifolia* are conspicuous.

Magdalena Plain Subdivision

The Magdalena Plain of southern Baja California is the southernmost section of the Sonoran Desert, reaching north to only 27°N latitude. Rainfall is low (generally less than 200 mm). Turner and Brown (1982) emphasized the affinity between some Magdalenan sites and thornscrub or San Lucan deciduous shrubs. The presence of *Larrea* as the dominant in many sites causes me to include it as a Sonoran Desert complex, albeit somewhat transient in its nature at some sites.

Several conspicuous cacti, including *Pachycereus pringlei*, *Lophocereus schottii*, *Stenocereus thurberi*, and *Stenocereus gummosus*, occur commonly.

Paloverdes (*Cercidium praecox*, *C. microphyllum*, or hybrids) occur in combination with *Jatropha cuneata*, *J. cinerea*, *Bursera microphylla*, *Fourquieria peninsularis*, *Lycium brevipes*, *Fagonia californica*, *Encelia farinosa*, and *Krameria parvifolia*.

All these and *Lysiloma candida* give this area its characteristic appearance of scattered subtrees within a sparse shrub matrix. This rather stark visage contrasts with that of the adjacent Vizcaino subdivision, with its architecture accented by *Fouquieria columnaris* and *Pachycormus discolor*. Also absent are most yuccas and agaves, with the stark exception of *Yucca valida*, especially in the south.

A cholla, *Opuntia cholla*, occurs throughout the subdivision, and it is sometimes associated with only one or two other perennial species on fine soils of volcanic origin.

Vizcaino Subdivision

The Vizcaino subdivision lies to the west of the Lower Colorado Valley subdivision in the north and the Central Gulf Coast subdivision to the south. Although it extends inland, it is generally open toward the Pacific Ocean side and thus receives the cooling effect of westerly breezes and the low-vapor-pressure deficits of moist air and fog, despite a generally low mean annual rainfall of about 100 mm (Turner and Brown 1982).

These coastal-inland gradients for temperature and humidity influence one of the visually conspicuous aspects of the Vizcaino: the lichen flora. Going from the coast to 70 km inland (at four sites), lichen species density dropped (25, 21, 9, and 2 species per site), as did cover (15.2%, 6.3%, 0.9%, and 0.1%). This was the reverse of the pattern for the richness of the higher-plant species (12, 9, 17, and 16 species per site) and cover (38.9%, 17.9%, 45.4%, and 45.1%), which increased inland. In general, fruticose species dominated the coasts, whereas foliose and crustose species dominated inland areas (Nash et al. 1979).

Similar patterns for richness and abundance are observed for all the warm deserts of North America as one moves from the ocean, inland (Nash and Moser 1982). Differences in lichen biomass between ocean and inland sites may reach 5000-fold.

The higher plants of the Vizcaino subdivision present a very complex pattern of species richness and physiognomy. Boojum dominates many areas visually because of its height. Locally it may reach coverages of up to 2% and densities close to 200 per ha. Its relative role and the general pattern of Vizcaino vegetation can be seen by summarizing data from two series presented by Turner and Brown (1982) (Table 8.4).

Shreve and Wiggins (1964) termed this subdivision the *Agave–Franseria* (*Ambrosia*) region, in deference to the dominance by these two genera. On gentle slopes and loams, *Ambrosia chenopodifolia* predominates, but it is replaced on clays by *Am-*

Table 8.4. *Density (D, per hectare) and cover (C, percentage) of perennials at two Vizcaino subdivision sites. Dash indicates cover <0.1%.*

Species	Catavina (D/C)	Punta Prieta (D/C)
Agave cerulata	156/0.3	1580/5.7
Ambrosia chenopodifolia	972/3.0	182/0.7
Ambrosia magdalenae	—	382/3.6
Atriplex polycarpa	6/0.1	233/1.2
Encelia californica	204/0.7	—
Encelia farinosa	—	449/0.5
Eriogonum fasciculatum	2864/0.9	—
Euphorbia misera	—	151/1.4
Fagonia californica	228/0.1	1806/0.7
Fouquieria columnaris	154/1.3	63/1.1
Fouquieria splendens	215/—	13/0.5
Krameria grayi	196/0.8	—
Larrea tridentata	108/1.1	45/1.0
Pachycormus discolor	22/0.7	—
Pedilanthus macrocarpus	—	158/0.6
Prosopis juliflora	7/0.4	383/1.8
Simmondsia chinensis	107/0.5	7/0.1
Stenocereus gummosus	—	366/3.3
Viguiera laciniata	2447/9.0	184/1.8
Totals		
Species	36	37
Density	9594	7268
Cover	21.6	27.8

Source: Data from Turner and Brown (1982).

brosia camphorata, though the two species co-occur. Maguey (*Agave shawii*) can be quite abundant, especially in the north and along the coasts. Also, in inland areas, one finds *Viguiera laciniata, Simmondsia, Eriogonum fasciculatum, Opuntia prolifera, Yucca valida, Larrea, Agave deserti*, and several large cacti.

In areas of the north, a stark vegetation of *Ambrosia* and *Opuntia* persists, whereas nearby areas that are on granite support complex groups such as the ragged-leaf goldeneye–boojum admixture described by Table 8.4.

South of Catavina, valleys contain creosote bush and *Atriplex polycarpa*, and the slopes include *Lycium californicum, Fouquieria splendens, Yucca whipplei*, and *Ephedra*. Chollas are noticeably uncommon, as are columnar cacti, but barrels (*Ferocactus*) are prominent.

Boojum and *Pachycormus* increase in importance southward in the subdivision. On some sites, 75% of the individuals are *Pachycormus*, associated with ocotillo and elephant trees (*Bursera*).

The Vizcaino plain, below 100 m, contains many of the species in Table 8.3, but one also encounters *Fouquieria peninsularis, Opuntia calmalliana*, and *Triteliopsis palmeri*. The transition of this region with mediterranean-type shrublands is summarized by Peinado, Alcaraz, Aguirre, Dalgadillo, and Aguado (1995).

The vegetation of the Sonoran Desert, even in protected areas, has varied in response to changing climate. Turner (1990) found that, in the Sierra del Pinacate Reserve, Sonora (Ezcurra, Equihue, and Lopez-Porfillo 1987) during the first half of this century, *Larrea* populations decreased 50–90%, *Cercidium* decreased 60%, and *Carnegiea* increased four-fold. Declines seem to relate to drought, whereas establishment to unusually high rainfall. Over longer periods, community composition has been shown to change in response to soil development and landscape evolution (McAuliffe 1994), modified by successional processes (McAuliffe 1991).

CHIHUAHUAN DESERT

The general vegetation of the Chihuahuan Desert (Fig. 8.15) has been studied less well than that of any other North American desert. In stark contrast to this relative dearth of published broad-scale information is a series of detailed studies of the functioning of desert communities and their plant and animal components that have been conducted by Robert Chew and James Brown on the Arizona–New Mexico border near Portal, Arizona, by a large group of workers on the Jornada Experimental Range, near Las Cruces, New Mexico, and a group working on the Sevilleta Long Term Ecological Research (LTER) site near Albuquerque, New Mexico.

The Chihuahuan Desert is elevationally high, with many sites in the Mexican basin being above 1000 m, and up to 2000 m in the south. Most sites are between 1100 and 1500 m. Its lowest elevations, those along the Rio Grande, are near 400 m. Much of the landscape is dominated by limestone, though gypsum and igneous rocks are the parent materials in some areas. Gypsic "sand" dunes, a characteristic feature of the Chihuahuan Desert, contain a host of endemic plant species.

The Chihuahuan Desert is cooler, partly because of high elevations, and has more rainfall than our other warm deserts: The yearly mean temperature for a suite of stations across the whole desert is 18.6° C (14–23° C), and the average precipitation is 235 mm (150–400 mm) (Schmidt 1986). The cool, moist aspect of the desert, coupled with the limestone substrates, probably are responsible for a significant grass component in this desert as compared with other warm deserts.

Interestingly, though the area is relatively unstudied, generally similar boundaries for the desert have been recognized, whether based on climate

Figure 8.15. Chihuahua Desert in Big Bend National Park, Texas. Leaf succulents, especially lechuguilla (Agave lechuguilla), and a yucca (Yucca torreyi) are apparent. Creosote bush is the dominant shrub. Dasylirion leiophyllum is in the background. Grasses include fluff grass (Erioneuron pulchellum) and Muchelenbergia porteri.

(Schmidt 1979), reptile and amphibian distributions (Morafka 1977), or vegetation (Henrickson and Straw 1976; Brown 1982a). Morafka (1977) subdivided the Chihuahuan Desert into three regions. The northernmost region, the Trans-Pecos, encompasses about 40% of the desert and includes all the sections in the United States and more than half of the desert areas of Chihuahua. The middle region, the Mapimian, includes parts of Chihuahua, Coahuila, and Durango. This area is dominated by basin and limestone range topgraphy and contains many playas. The third and most southern region, the Saladan, includes Zacatecas and San Luis Potosí and small parts of other states. Elevational ranges here are extreme, with valley floors near 500 m and mountain peaks, both limestone and igneous, exceeding 3000 m.

Henrickson and Johnston (1986) accepted these three divisions based on the flora and outlined a system of plant community delineation. Their system includes eight primary subdivisions of desert scrub and woodlands: Chihuahuan Desert Scrub, Lechuguilla Scrub, Yucca Woodland, Prosopis–Atriplex Scrub, Alkali Scrub, Gypsophilous Scrub, Cactus Scrub, and Riparian Woodland. Each of these subdivisions is a conspicuous floristic unit. On a more local scale of classification, the Chi-

huahuan Desert Scrub is subdivided into five phases: Larrea Scrub, Mixed Desert Scrub, Sandy Arroyo Scrub, Canyon Scrub, and Sand Dune Scrub. Obviously, some of these community types and phases are part of what I have termed azonal vegetation (e.g., the Riparian Woodland community type and the Sand Dune Scrub phase of the Chihuahuan Desert Scrub). However, the general classification is useful and is followed here to some extent. Valverde, Zavala-Hurtado, and Montaña (1996) have additionally described, on the basis of a numerical analysis, four physiognomic associations in the southern Chihuahuan.

The five phases of the Chihuahuan Desert Scrub cover about 70% of the entire area of the Chihuahuan Desert. The Larrea Scrub phase covers 40% of the area. This phase varies in its composition, but over vast areas 40–80% of the plant cover can be composed of Larrea or Flourensia cernua or a mixture of the two (Figs. 8.15 and 8.16). In this community type, Flourensia and Larrea shrubs both increase the availability of soil water under their canopies by significant stemflow (Mauchamp and Janeau 1993; Martinez-Meza and Whitford 1996). Often mixed in with these two dominants are Parthenium incanum, Jatropha dioica, Koeberlinia spinosa, a Lycium species, perhaps an acacia (often

Figure 8.16. Rolling limestone hills of the Big Bend area, Texas: (a) flats are dominated by Larrea and Flourensia, whereas slopes are more diverse; (b) monotonous creosote bush flat with some lechuguilla.

Figure 8.17. Three conspicuous plants of the Chihuahua Desert: (a) soaptree (Yucca elata), a species of grassland-desert transitions or on mesic desert sites; (b) Opuntia phaeacantha, a widespread prickly pear cactus; (c) lechuguilla (Agave lechughuilla).

A. neovernicosa), Mortonia scabrella, a Psorothamnus (e.g., P. formosa), a Krameria, and an Ephedra.

In the Saladan region, a Yucca filifera woodland may develop. Some individuals of this species may reach nearly 15 m in height. In the north, some sites contain smaller species of yucca, such as Y. torreyi or Y. elata (soaptree, Fig. 8.17a). The latter is most abundant in desert grasslands but occurs commonly and conspicuously in "true" desert on mesic sites. It tends to be clumped in its dispersion because of vegetative reproduction, cattle, and fire (Smith and Ludwig 1978).

Ascending the slopes of bajadas, one encounters an enrichment of shrub species and a greater sharing of dominance. This is equally true for the Trans-Pecos (Wierenga, Hendrickx, Nash, Ludwig, and Daugherty 1987; Cornelius, Kemp, Ludwig, and Cunningham 1991), the Mapimian (Montana 1990), and the Saladan. On those sites that become the Mixed Desert Scrub phase, cacti (Fig. 8.17b), ocotillo (Fouquieria splendens), and lechuguilla (Agave lechuguilla) (Fig. 8.17c) become common (Plumb 1991). Ocotillo and lechuguilla occur over most of the Chihuahuan Desert and are among the best Chihuahuan Desert "indicators" (Fig. 8.18). Also locally common are allthorn (Koeberlinia spinosa) and species of Zinnia. Sites where the Mixed Desert Scrub phase might be expected are often dominated by dense stands of lechuguilla and are differentiated by Henrickson and Johnston as the Lechuguilla Scrub. This community type contains species from surrounding communities and is recognized only by the relative importance of its members.

The patchwork of limestone and igneous rocks obviously influences local vegetation. On slopes, desert scrub communities prevail on limestone at elevations that support grassland on igneous rocks (Wentworth 1981; Aide and Van Auken 1985). Similar relationships have been described for Sonoran Desert mountains (Whittaker and Niering 1968).

On limestone, Agave lechuguilla and Fouquieria are common, as are Hechtia scariosa and Leucophyllum species (e.g., L. frutescens). Euphorbia antisyphylitica, a species that is economically important for its wax, is common on, but not confined to, limestone. Also occurring with these species are dogweeds (Dyssodia species), several species of Condalia and Viguiera stenoloba. At ground level, the holly-like leaves of Perezia nana are often abundant (Rzedowski 1966, 1978). One species of particular

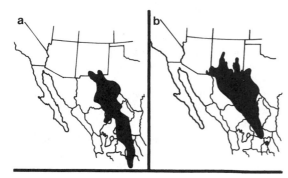

Figure 8.18. Distributions of two Chihuahua Desert indicator species: (a) Agave lechuguilla (map adapted from Gentry 1982); (b) tarbush (Flourensia cernua, map adapted from Dillon 1984).

interest on the limestone slopes, and less commonly on the plains, is guayule (*Parthenium argentatum*), a locally common, widespread species whose rubber is nearly equal in quality to that of *Hevea brasiliensis* (Miller 1986).

Sites with igneous substrates in Mexico contain many *Opuntia* species, large cacti such as *Myrtillocactus geometrizans*, and *Yucca carnerosana*. In their best development, such sites form the arborescent Cactus Scrub subdivision of Henrickson and Johnston. Two areas of this vegetation type occur, both in the Saladan region. In the southern Saladan, forests of these species occur. Also present are *Acacia schaffneri*, *Agave filifera*, *Opuntia robusta*, *O. leucotricha*, and *O. streptacantha*), and an outer zone of decumbent platyopuntias (*O. rastrera* or *O. robusta*) near the boundary. Also associated with these areas, but not in such a patterned manner, are *Agave salmiana*, *Jatropha dioica*, and *Mimosa biuncifera* (Yeaton and Manzanares 1986).

Grasses are obvious in most areas of the Chihuahuan Desert. Large swale areas can be covered by tobosa (*Hilaria mutica*). Genera such as *Sporobolus*, *Muhlenbergia*, and *Bouteloua* are often common. Bush muhly (*Muhlenbergia porteri*) frequently occurs growing among the branches of creosote bush (Welsh and Beck 1976). In sandy areas, *Oryzopsis hymenoides* is common and conspicuous. Throughout many areas, even in creosote bush flats, *Erioneuron pulchellum* occurs, sometimes with species such as *Bouteloua eiopoda* or *Sporobolus airoides*.

One community type that I think is a transition to grasslands is the *Yucca* woodland. In Mexico, some areas just above the Chihuahuan Desert Scrub are dominated by large yuccas and sotols (*Dasylirion*). Especially prominent are *Yucca carnerosana*, *Y. faxoniana*, *Y. torreyi*, *Y. filifera*, and either

Dasylirion leiophyllum or *D. texanum*. In the northern portions of the desert, the same genera, or even species (e.g., *D. leiophyllum*), form a transition; however, it usually does not take on a dense, woodland appearance.

Plains areas with fine soils are dominated by one of three community types. On gypsum, a unique Gypsophilous Scrub develops. This type is treated in the later section on azonal vegetation. On alluvium with low salinity, vast stands of *Prosopis glandulosa* or, in the Saladan region, *P. laevigata* dominate. These are joined by *Atriplex canescens*, *Lycium* species, and *Ziziphus*, often *Z. obtusifolia*. In areas of southeastern New Mexico, this habitat type also includes *Microrhamnus ericoides* and *Xanthocephalum sarothrae* (Secor, Shamash, Smeal, and Gennaro 1983). As salinity increases, an Alkali Scrub community develops. *Atriplex* species (*A. acanthocarpa*, *A. canescens*, *A. obovata*) mix with *Allenrolfea occidentalis*, *Suaeda* species, *Sesuvium verrucosum*, and salt grass (*Distichlis spicata*).

In the Chihuahuan Desert, many species of cacti prevail. *Opuntia phaeacantha* can occur in dense stands over large areas, as can *Echinocactus horizonthalonius*. Many small- to medium-size species occur in complex mixes and represent such a great variety of forms that the Chihuahuan Desert is a cactophile's paradise. Some species (e.g., *Opuntia leptocaulis*) have cyclical replacement relationships with shrubs such as creosote bush (Yeaton 1978). *Larrea* populations show north-south geographic variation in their response to water, suggesting at least some systematic genetic differentiation within the region (McGee and Marshall 1993).

Despite the profusion of cacti in the Chihuahuan Desert, the leaf succulents dominate the visual landscape and give the Chihuahuan Desert its special appearance, which contrasts with the Sonoran Desert's subtree–columnar cactus profile or the low, scattered aspect of the Mojave Desert.

Plants in the Chihuahuan Desert show the same clumped species patterns as the two deserts discussed previously (Silvertown and Wilson 1994), as well as the nurse plant phenomenon. In parts of the Mapimian region, vegetation stripes occur. Similar bands of alternating dense vegetation and barren spaces, oriented parallel to contour lines, are known from deserts worldwide (Montana, Lopez-Portillo, and Mauchamp 1990).

Researchers in the Chihuahuan Desert have elucidated important interactions among plants, animals, soil, and microbes. The same processes occur in all deserts but are extremely well documented in the Chihuahuan. A few examples suffice to suggest the richness of this literature.

A complex set of experimental enclosures designed by Jim Brown has elucidated the influence of animals on vegetation. Exclusion of kangaroo rats (*Dipodomys*) increased the abundance of grasses, but decreased the diversity of summer annuals (Heske, Brown, and Guo 1993). Exclusion of birds had effects on summer plant communities (Guo, Thompson, Valone, and Brown 1995). In other studies, the mounds of kangaroo rats positively affected the growth and flower production of *Larrea* (Chew and Whitford 1992) and the cover of annuals (Moorhead, Fisher, and Whitford 1988; Guo 1996), and their burrows were hotspots for desert soil fungi (Hawkins 1996). Even termites, usually associated with decomposition and nutrient cycling, influence hydrological characteristics of soils and, in turn, plant communities (Elkins et al. 1986).

Finally, the influence of grazing on plant communities has been extensively studied, partly because of the high proportion of grasses in the Chihuahuan Desert. In some cases, grazing is thought to change tension zones from grassland to shrubland (Bahre and Shelton 1993), but for annuals there was little response (Kelt and Valone 1995) to grazing. For grasses and shrubs, desirable species do not always recover rapidly following grazing suppression (Wester and Wright 1987). This simplistic story is greatly modified by soil type that seems to alter species composition, productivity, and leaf area index (Dugas, Hicks, and Gibbens 1996; Gibbens, Hicks, and Dugas 1996).

AZONAL VEGETATION

Some habitats occur on similar sites across all our deserts; that is, they are not confined to only one geographical zone. The plant species occupying these sites, regardless of the desert involved, are often similar or even the same species. Because of this similarity, I treat some of these vegetation types separately here.

Sand Dunes

Sand dunes occur in all our deserts but are especially apparent in southern California and southwestern Arizona (Bowers 1984; Lancaster 1993). Despite their commonness, the conditions under which dunes form and the factors that control their size and spacing are not well understood (Lancaster 1984). Sand movement characterizes most dunes. The degree of this movement can alter the vegetation significantly (compare Fig. 8.19a and c). When sand moves, plants can be covered in zones of accumulation or uncovered in zones of deflation. Many plants that grow in dunes show rapid

growth as an adaptation to sand movement (Bowers 1982), as well as a suite of morphological adaptations (Danin 1996).

Dune vegetation is often formed of a mix of widely distributed species, species occurring in a limited geographical area and those endemic to a particular dune system. Widely distributed species include sandy-soil specialists such as sand sage (*Artemisia filifolia*), a species occurring from the Mojave Desert through the Sonoran Desert and Chihuahuan Desert as far south as Coahuila, Mexico, and species of more catholic substrate affinities, such as *Larrea* and *Prosopis glandulosa* (Fig. 8.19b), both species of wide occurrence. Species of more constrained geographical occurrence, such as those limited to one desert type, are exemplified by species that are essentially restricted to parts of the Sonoran Desert, such as *Ammobroma sonorae, Tiquilia palmeri, Eriogonum deserticola*, and *Hesperocallis undulata*. At a finer level of geographical differentiation, various dunes have their share of endemics, usually ranging from 5–15% of the species in that system (Bowers 1982). Good examples are the grass *Swallenia alexandra* and forms of *Oenothera avita* and *Astragalus lentiginosus* that occur only in the Eureka Valley dunes in California (Pavlik and Barbour 1988).

Dune floras seem to break into eastern and western groups, with little overlap of species. These two floras correlate, roughly, with winter versus summer rainfall regimes (Bowers 1984).

One specialized "dune" system is formed where *Prosopis glandulosa* occurs in sandy areas. The mesquite takes on a prostrate, coppice form that traps sand, whereas deflation occurs in the poorly vegetated interdunal areas (Fig. 8.19b). Soils of the dunes and the interdunes are virtually identical (Hennessy, Gibbens, Tromble, and Cardenas 1985). Microbial populations are higher under mesquite than in the interdune soils (Baker and Wright 1988). These areas probably developed in the past 100 yr in response to overgrazing (Bahre and Shelton 1993), though the cause-effect relationships for the dune origins are complex (Wright 1982).

All dune systems support large numbers of showy, annual herbs during rainy periods. Especially conspicuous are species of *Euphorbia, Pectis, Boerhavia, Allionia*, and *Oenothera*.

Gypsum Soils

Some dune areas of the warm deserts, especially in the Chihuahuan Desert, are composed of gypsum (hydrous calcium sulfate) (Fig. 8.20a) rather than silica sands. These areas contain several

Figure 8.19. Azonal sandy desert sites: (a) sand dunes dominated by sand sage (Artemisia filifolia, near St. George, Utah), a light-colored shrub that occurs on loose sands in all North American warm deserts, and some creosote bush; (b) the prostrate shrub form of mesquite (Prosopis glandulosa, east of El Paso, Texas) that dominates vast areas of loose sands throughout North American warm deserts; (c) stabilized sand area only 100 m distant from site (a) that supports increased importance of creosote bush and the shrubs Krameria, Ambrosia dumosa, and Hymenoclea.

widely distributed species such as *Yucca elata*, Indian rice grass (*Oryzopsis hymenoides*), *Atriplex canescens*, *Hilaria mutica*, *Rhus trilobata*, and even an occasional cottonwood (*Populus*) (Shields 1953). There is a significant degree of endemism among members of plant communities in gypsum soil. Some 70 species are confined to Chihuahuan Desert gypsums (Henrickson and Johnston 1986). These include some endemic genera (*Dicranocarpus, Marshalljohnstonia, Strotheria*) and others mainly associated with gypsum (*Selinocarpus, Nerisyrenia, Sartwellia,* and *Pseudoclappia*). Additionally, many widespread genera contain species more or less endemic to gypsum soils (Fig. 8.20a and b).

Most gypsum communities are depauperate compared with adjacent areas. For example, at White Sands National Monument, New Mexico (Fig. 8.20c), 28 species common to the surrounding areas are absent from the gypsum dunes. Most gypsum species seem to be herbs or dwarf shrubs; annuals are uncommon (Parsons 1976; Powell and Turner 1977).

Some genera of plants in gypsum soil show significant regional differentiation of species within the Chihuahuan Desert subregions. In the Trans-

Figure 8.20. (a) Gypsum "sand" area, east side of Guadalupe Mountains National Park, Texas. Prominent species
include Yucca elata, Ephedra torreyana, and several
grasses, including Bouteloua breviseta; (b) Indian blanket

(Gaillardia multiceps), confined mainly to gypsum soils
and present at site (a); (c) White Sands National Monument, New Mexico, a shifting gypsum dune system.

Pecos region, *Nerisyrenia linearifolia* occurs, but it is
replaced by *N. castillonii* in the Mapimian region
and *N. gracilis* in the Saladan region. Similar species replacements occur for *Sartwellia* and *Nama*.
Gypsum soils in our other deserts contain species
in many of the genera mentioned earlier. The cause
of gypsophile endemism is unknown; Meyer (1986)
suggests that gypsum chemistry per se may be less
important than water relations.

Riparian and Wash Vegetation

Water channels ranging from small rills to major
river courses influence desert vegetation. In some

creosote bush stands, on what appear to be flat
plains, one can find cheesebush (*Hymenoclea salsola*) arrayed in lines along almost imperceptible
rills.

In contrast to this situation, large river courses
can be surrounded by extensive, dense forests
(bosques) of willows (*Salix*, e.g., *S. gooddingii*),
mesquites (*Prosopis glandulosa* or *P. pubescens*), and
the aggressive introduced tamarisks or salt cedars
(*Tamarix*, e.g., *T. chinensis* or *T. ramosissima*). Often
the banks of the river will be dominated by large
cottonwoods (*Populus fremontii*). Various ashes
(*Fraxinus*) and desert willow (*Chilopsis linearis*) are
also common. In some areas, thickets of the native

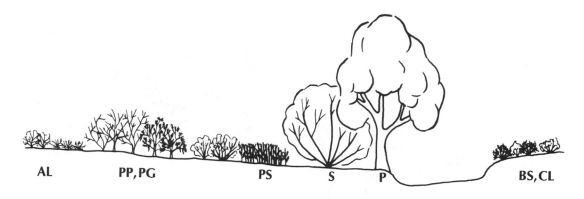

AL   PP,PG   PS   S   P   BS,CL

*Figure 8.21. Stylized transect from a Sonora Desert river channel upslope (right to left), showing riparian vegetation. Plants include cottonwood (*Populus*) and willow (*Salix*), often followed by a band of arrow weed (*Pluchea* = *Tessaria sericea*) and in turn by mesquite (*Prosopis pubescens*, P.* glandulosa*) and perhaps quail bush (*Atriplex lentiformis*). Minor channel species (right of riverbed) include seep willow (*Baccharis sarothroides*) and desert willow (*Chilopsis linearis*).*

arrow weed (*Pluchea sericea*) or the introduced *Arundo donax* border rivers.

Along washes, several of these species are common, as are hackberries (*Celtis* species), species of *Rhus*, *Dalea*, and *Lycium*, and Apache plume (*Fallugia paradoxa*). Several acacias occur in similar sites. In the Mojave and Sonoran deserts, catclaw (*Acacia greggii*) is abundant, whereas *A. constricta* is often the dominant in the Chihuahuan Desert.

Gardner (1951) listed the plants occurring in 93 New Mexico washes. Nine species were common, and an additional 22 species were recorded. The number of washes in which particular species occurred were: *Prosopis glandulosa* 54, *Hymenoclea monogyra* 45, *Fallugia paradoxa* 44, *Rhus microphylla* 40, *Larrea* 36, *Brickellia laciniata* 33, and *Chilopsis linearis* 12. In the Mojave Desert, Beatley (1976) listed 22 arroyo shrub species and 30 herbs. A "typical" riparian zone is pictured in Figure 8.21.

Throughout North American deserts native riparian vegetation is declining and being replaced by naturalized species such as *Tamarix ramosissima* (Busch and Smith 1995). Although the physiognomy of the aliens is similar to that of the natives, stand diversity for plants and animals decreases. In part this can be ascribed to the rapid increase in tamarisk density and a lowering of the water table (Cleverly, Smith, Sala, and Devitt 1997).

Saline Areas

In saline sites (more than 2% salt), the number of species of plants is rather limited. In the entire Chihuahuan Desert, Henrickson (1977) found only 40 halophytic vascular plant taxa. Unexpectedly, 25 of these were Chihuahuan endemics, including three genera: *Meiomeria*, *Reederochloa*, and *Pseudoclappia*. More typically, halophytic genera and species occur in saline sites in all our deserts, as well as in nondesert habitats. These include *Allenrolfea occidentalis*, *Atriplex canescens*, *Distichlis spicata*, and *Sporobolus airoides*. The genus *Atriplex* has many salt-tolerant species, though the presence of the genus cannot be used as a sign of high soil salinity. Similarly, *Tidestromia*, *Suaeda*, *Juncus*, *Salicornia*, and *Nitrophila* all contain common salt-tolerant species. In the Mojave Desert, and somewhat into the Sonoran Desert, greasewood (*Sarcobatus vermiculatus*) occupies saline sites that have standing water or are periodically inundated.

Mesquite often occurs in quite saline areas and thus is frequently a conspicuous plant around the edges of playas, a zone of moderate salinity.

Transitions

At the upper elevations or eastern limits of our deserts, grasslands usually prevail. Elevation transitions often include xeric-adapted trees such as oaks (*Quercus*) or junipers (*Juniperus*) (Fig. 8.22a), whereas the eastern boundaries are more dominantly grasslands (Fig. 8.22b) (Burgess and Northington 1977; Marroquin 1977). Desert grasslands are ecotones whose location, site, and composition have been affected by human activities in the past 200 yr. These activities include fire management, grazing, and wood gathering, as recently reviewed by McClaran and Van Devender (1995).

To the south, both the Chihuahuan and Sonoran deserts blend into subtropical thorn scrub (Fig. 8.22c), a vegetation type that shares many genera

Figure 8.22. Examples of three North American desert transitions: (a) Chihuahua Desert upper-elevation transition to grassland–oak parkland, Big Bend area, Texas; (b) grass- land-desert transition in the Sonora Desert on the Santa Rita Experimental Range near Tucson, Arizona; (c) Sonora Desert–thorn scrub transition, southern Sonora, Mexico.

with the adjacent deserts. To the north, the Sonoran Desert blends with the Mojave Desert, and the Mojave with the Great Basin Desert. Much of the northern Chihuahuan Desert grades to grasslands. All these vegetation types are covered elsewhere in this volume (Chapters 7, 9, 15).

POTPOURRI OF VEGETATION TOPICS

This chapter has emphasized the floristic and vegetational aspects of warm deserts. In using this approach, several important topics have been omitted, including consideration of dispersion patterns, primary productivity, patterns of photosynthetic pathways, and annuals. Each of these topics is considered briefly in the following sections in order to provide an introduction to an extensive literature related to these topics, a literature beyond the scope of this book.

Finally, because deserts are often considered to be wastelands, I want at least to mention some potentially important economic factors related to deserts and their use for and by humans.

Productivity

Primary productivity is generally thought to be low in deserts (Ludwig 1987). A range of 30–300 g dry weight per square meter per year has generally been accepted as encompassing North American deserts, although low values may fall to 3 (a sand dune) and high values may reach 800 (Chihuahuan Desert) (Szarek 1979; Hadley and Szarek 1981). An especially productive site was studied by Sharifi and associates (Sharifi, Nilsen, and Rundel 1982): a *Prosopis glandulosa* stand (90% of the plant cover was mesquite) in the Sonoran Desert of southern California. Stand biomass ranged from 23,000 kg ha^{-1} nearest a wash to 3500 kg ha^{-1} at the fringe of the stand. Mean dry-weight production in 1980 was 3650 kg ha^{-1}; 51% of that was allocated to wood and 33.6% to leaves.

On a Mojave Desert site net annual primary productivity (NPP) over a 10 yr period ranged from <0.1 to 64/m^2 for winter annuals and 20 to 68 g/m^2 (Turner and Randall 1989). For both forms combined values ranged from 0.02–68.8 g (Turner and Randall 1989; Rundel and Gibson 1996).

Although desert primary production is, in general, related to abiotic variables (Lane, Romney, and Hakonson 1984), the specific stand composition is also very important (Webb and Wilshire 1983).

Dispersion

The pattern of distribution of plants with respect to one another is termed "dispersion." Three general dispersion patterns are usually recognized: random, regular, and clumped. Desert shrubs were at one time alleged to show almost classic regularity in their distribution. This was supposed to have been the result of competition for water or nutrients and/or a response to water-soluble toxins exuded through the plant's roots. Barbour (1973) questioned the generality of a regular dispersion pattern. Subsequently it has become clear that not all desert shrubs show regular patterns (Fonteyn and Mahall 1981; Phillips and MacMahon 1981), even though competition for water may occur under stress conditions (Fonteyn and Mahall 1978; Fonteyn and Mahall 1981). Probably, as is the case for most natural phenomena, the answer to the question of whether or not desert shrubs are regularly dispersed is, "It depends." For example, it depends on the species involved, their size, the values for the abiotic variables of the plant's environment, and a host of other factors. Clearly, many plants are clumped at one time in their life cycles (e.g., the nurse–tree–cactus interaction). Just as clearly, this pattern may change.

Pattern regularity has been inferred to be due in part to the methodological problems of discerning when a clone-forming plant like creosote bush is an individual as opposed to being part of a clonal clump (Ebert and McMaster 1981; King and Woodell 1984). McAuliffe (1988) has shed considerable light on these patterns. He found that recruitment of plants can be modeled as a Markov chain. Furthermore, he found that some plants, especially *Ambrosia* species, can colonize open space whereas many others require the cover of another plant, causing clumped patterns overall. Importantly, he shows that the mechanism for pattern development is not related to competition alone but, as predicted by some workers, relies on a diverse set of mechanisms that influence seed distribution, germination success, and postgermination mortality.

More experimentation is needed to elucidate the factors involved in an interaction as complex as the development of a community pattern, and it is highly unlikely that one factor will explain dispersion or that one dispersion pattern will exist to the exclusion of all others.

Metabolic Pathways in Relation to Vegetation

Plants vary with regard to the first chemical products associated with the process of photosynthesis. In many plants, a 3-carbon acid (phosphoglyceric acid) is the first, whereas in others there is a 4-carbon acid. Additionally, there are two different pathways that use a 4-carbon acid. In one case, the assimilation proceeds in light; in the other, CO_2 is changed to malate at night and stored until day. These three pathways are termed C_3, C_4, and CAM (crassulacean acid metabolism), respectively. C_3 plants are represented by the vast majority of common plants in a host of families. C_4 plants have a special sheath surrounding the vascular traces in their leaves and include families such as Asteraceae, Chenopodiaceae, Euphorbiaceae, Nyctaginaceae, Poaceae, and Zygophyllaceae, to mention a few. CAM plants are usually succulents, either leaf or stem forms, and include the Agavaceae, Cactaceae, and Crassulaceae. C_4 plants demonstrate more favorable high-temperature physiological performance and efficiency of water use than do C_3 forms. CAM plants also have high efficiencies of water use. Clearly, we would anticipate that C_4 and CAM plants should be better suited to arid conditions, and a number of workers have used transect data to demonstrate this point. It is possible that at any site, various communities can be formed of any combination of these three types, ranging from pure stands of only one type to mixtures of all pair-wise combinations and even to combinations of all three.

In Johnson's analysis (1976) of about 1000 desert species in California, 85% were C_3, 11% were C_4, and 4% were CAM. More than half of the C_4 species were grasses. Johnson found that communities on alkali soils contained mostly C_4 plants, that many desert sites contained high proportions of C_3 species, and that mixes of all three strategies occurred throughout the Mojave Desert. Sites dominated by only communities of CAM species were rare.

A study of plants along an aridity gradient that included the Chihuahuan Desert in the Big Bend area of Texas showed that – for 88 nonherbaceous species – as aridity decreased, the dominant plants changed in order from CAM to C_4 to C_3 species, as we would infer (Eickmeier 1978) from what was said previously. CAM species had somewhat of a bimodal distribution (Eickmeier 1979).

Looking only at evergreen rosette plants of the Liliaceae in the Chihuahuan Desert, a group one would expect to be C_3 species, Kemp and Gardetto (1982) found that *Yucca baccata* and *Y. torreyi* were CAM species, whereas *Yucca elata, Y. campestris, Nolina microcarpa*, and *Dasylirion wheeleri* were C_3. Despite the fact that some species are known to be able to shift between C^3 and CAM, none of these species did. The important point is that although C_4 and CAM appear to be the species best adapted to deserts, many desert species are C_3, and one needs to look carefully at the specific microhabitat when interpreting relationships between patterns of distribution and photosynthetic pathways. Mesic microsites or times of the year permit C_3 species to flourish in areas that appear in general to be very arid.

The importance of seasonally mesic conditions was well demonstrated by Kemp (1983), who studied the phenological patterns of Chihuahuan Desert plants in relation to the timing of water availability. The perennials on his study site consisted of C_3 and C_4 forbs, C_4 grasses, C_3 shrubs, and CAM shrubs. Remembering that rainfall occurs predominantly in the summer in the Chihuahuan Desert, it is interesting that C_3 forbs showed greatest activity in spring or fall, whereas C_4 grasses were active in the summer and fall. The C_3 and CAM shrubs were active at various times and thus were not as dependent on soil moisture. These results suggest that the differences in responses between C_3 and C_4 species were greater for variations in temperature regime than for variations in available moisture. This last point is important. Many general studies of aridity gradients have suggested that the proportions of the metabolic pathway that involve the various photosynthetic types should segregate along the gradient. And whereas in some cases this generally occurs, often there is considerable variability in the relationships. "Aridity" gradients often involve parallel changes in moisture availability and temperature; these two factors are not independent of each other. A plant establishes and lives at a point in space and time, not on an "average" landscape. Thus, our crude correlations of C_3, C_4, or CAM strategies to yearly or monthly measurements of average values of abiotic variables across a whole landscape may obscure the biologically most important variables. In a sense, we measure the abiotic factors on a broad integrated scale, but the plant is responding to a highly localized series of variables acting where the plant happens to occur.

Burgess and Shmida (1989) present a hypothesis to explain the worldwide distribution of CAM plants. They posit that CAM is best suited to coastal deserts cooled by polar currents and to semiarid subtropical montane areas. Within such areas, rocky slopes are especially suitable sites. Their hypothesis fits worldwide patterns well.

Annuals

I have ignored annual plants in this treatment because in many years they are virtually nonexistent as adults, although in other years they dominate the landscape. This variability makes characterization of deserts based on annuals difficult. This omission is not meant to demean their importance. In terms of number of species on a site, annuals often dominate. For a Mojave Desert site at Rock Valley, Nevada, there are 10 common perennials, but 41 annuals; at a Sonoran Desert site near Tucson, the ratio is 8 to 48. The productivity of annuals in favorable rainfall years can exceed that of perennials. For example, at Rock Valley, Nevada, annuals produced a dry-weight biomass of 616 kg ha^{-1} yr^{-1}, compared with less than 600 for perennials (MacMahon and Wagner 1985). One specific site in the Mojave Desert makes the importance of annuals quite clear: Beatley (1976) studied sites that, in 1000 m^2, contained 30 species, 30% cover, 975 plants pr m^2, and a standing crop of 616 kg ha^{-1}. Soil seed banks may contain up to 187,000 seeds/m^2 (Inouye 1991).

Annuals are generally C_3 or C_4 species. In the Mojave Desert, some sites contain virtually all winter annuals with C_3 metabolism, whereas other sites contain some summer annuals that are either C_3 or C_4 (Mulroy and Rundel 1977). This is generally the case across all our deserts: The majority of winter annuals are C_3; summer annuals are C_4 or a mix of C_3 and C_4.

Winter annuals tend to have long lives and high mortality rates and occur both under shrubs and in the interspaces. Summer annuals are short-lived and occur mainly in the interspaces. All deserts have mixtures of winter and summer annuals, but usually one group predominates. For example, because the west coast of Baja California receives abundant winter precipitation, it supports 114 winter-annual species out of a total of 400–500 annuals. The Plains of Sonora subdivision shares about 100 annual species with southern Arizona; of the 84 species listed by Shreve and Wiggins (1964) as common, 56 species (66.7%) also occur in Texas; they are predominantly summer annuals.

Numerous genera contain annual species that can be seen in all three deserts. Some conspicuous examples are *Astragalus, Dyssodia, Eriogonum, Plantago, Eriophyllum, Phacelia*, and *Cryptantha*.

Some species of annuals occur in all our warm

deserts. Winter forms include *Monolepis nuttalliana, Lepidium lasiocarpum, Descurainia pinnata,* and *Calycoseris wrightii.* Summer forms include *Bouteloua barbata, Cenchrus echinatus, Tidestromia lanuginosa,* several *Boerhavia* species, *Euphorbia micromera, Baileya multiradiata* (one of the most conspicuous plants of desert roadsides), *Bahia absinthifolia,* and *Pectis papposa.*

Early workers suggested that annuals did not grow under the canopies of many shrub and subtree species. Although this may be true in some cases, it is more generally true that annuals do occur under a wide variety of shrubs (Ludwig, Cunningham, and Whitson 1988). In fact, in relatively dry years, annuals may be more abundant under shrubs than in adjacent interspaces. The density of annuals even varies under the shrub, depending on compass position (Patten 1978).

Richard Inouye (1991) presents a masterful review of desert annuals that should be consulted for elaboration of the points I have made, and Pake and Venable (1996) relate the annual habit to theories of life history evolution.

Human Uses of Deserts and Their Plants

There are myriad potential uses of deserts and their plants. As human populations expand, old cities will expand, and new ones will form in the Sun Belt of the Americas. Expansion of human populations requires construction of power-transmission lines (Vasek, Johnson, and Brum 1975), pipelines (Vasek, Johnson, and Eslinger 1975), roadways, and reservoirs (Rea 1983). All these activities influence desert vegetation, often in ways that have long-term deleterious effects. The presence of humans and their projects can alter climatological factors such as albedo, vapor pressure deficit, and the like, with the consequent alteration of natural climatic regimes. Bahre (1991, 1995) and Bahre and Shelton (1993) have clearly shown that major vegetational changes in the warm deserts have been due to human activity, not to any recent climate change.

We cannot afford to alter desert vegetation in an unthinking manner. For one thing, many species of desert plants contain highly desirable chemicals; for example, an oil in jojoba (*Simmondsia chinensis*) can replace whale oil, and a rubber substitute can be produced from guayule (*Parthenium argentatum*). Many other species can be used as food, forage, and sources of medicines and industrial goods (McKell 1985). Clearly, desert species represent a vast, relatively untapped source of products for human use.

Currently, desert landscapes are used as sites for recreation, especially by off-road-vehicle enthusiasts, and as places to raise cattle. Both activities can have damaging effects. For example, on some sites grazing can reduce the diversity of annuals (Waser and Price 1981) or alter the composition of the community so that undesirable species predominate (Chew 1982). Uncontrolled use of off-road vehicles can alter desert soil in a way that accelerates erosion, causes compaction, produces dust, and ultimately destroys native vegetation (Luckenbach and Bury 1983; Sharifi, Gibson, and Rundel 1997; Webb and Wilshire 1983).

The examples cited mandate that we use deserts in a prudent fashion. There are many positive benefits that we can derive from deserts if we temper our activities with knowledge of the potential influences of our actions. For many of us, the best use of the desert is as an environment in which to find solitude and contemplate life. I have never been to a place that permits clear thinking and introspection better than does a desert.

AREAS FOR FUTURE RESEARCH

Clearly, from a vegetation perspective, warm deserts are both well known and unknown. They are well known in a floristic sense; that is, the distributions and co-occurrences of species are reasonably well documented. They are better known now than 17 years ago in the sense that quantitative data on the composition of vegetation at the level of species, life forms, or functional groups have accumulated between 1980 and now. The definitive knowledge of the causes of these documented patterns still eludes us.

Correlation, in a cause–effect sense, between vegetation attributes and abiotic factors has been attempted for a few species, as well as for numerous species in a very local area or over a broad area, but it has been based on very superficial measurements. To truly understand the causes of the structure and functioning of our deserts, we must accumulate more data to permit correlations between vegetation characteristics and physical factors over a greater variety of habitat types across the full span of our deserts. The work of Joe McAuliffe has taken us in this direction (1991, 1994).

Additionally, we know that the biotic environment is extremely important to species and consequently to communities. Additional studies of species–species interactions are needed to further our understanding of community origins and organization.

Finally, because deserts show extreme variation, both biotically and abiotically, over time spans of

1 yr or less, more long-term ecological research of even a simple monitoring nature must be conducted. The responses of vegetation to episodic, extreme conditions may give us better clues as to what drives these communities than do studies conducted in "average" years. This alone justifies the establishment of the Long Term Ecological Research sites by the National Science Foundation. Unfortunately, both of the desert sites are in the Chihuahuan Desert.

The foregoing comments are directed only to consideration of studies that might explain community composition. Community functioning is beyond the scope of this treatment, although it is of great importance to understanding these ecosystems. Syntheses such as those of Rundel and Gibson (1996) have helped, as have a host of individual research papers. Despite this burgeoning set of studies, we know too little about deserts. They are too important to ignore.

REFERENCES

Aide, M., and O. W. Van Auken. 1985. Chihuahuan desert vegetation of limestone and basalt slopes in west Texas. Southw. Nat. 30:533–542.

Amundson, R. G., O. A. Chadwick, and J. M. Sowers. 1989. A comparison of soil climate and biological activity along an elevation gradient in the eastern Mojave Desert. Oecologia 80:395–400.

Axelrod, D. I. 1979. Age and origin of Sonoran Desert vegetation. California Academy of Sciences. Occasional papers 132:1–74.

Bahre, C. J. 1991. Legacy of change: historic human impact on vegetation in the Arizona borderlands. University of Arizona Press, Tucson.

Bahre, C. J. 1995. Human impacts on the grasslands of southeastern Arizona, pp. 230–264 in M. P. McClaran and T. R. van Devender (eds.), The desert grassland. University of Arizona Press, Tucson.

Bahre, C. J., and M. L. Shelton. 1993. Historic vegetation change, mesquite increases, and climate in southeastern Arizona. J. Biogeog. 20:489–504.

Baker, E. H., and R. A. Wright. 1988. Microbiology of a duneland ecosystem in southern New Mexico, U.S.A. J. Arid Environ. 15:253–259.

Barbour, M. G. 1973. Desert dogma reexamined: root/shoot productivity and plant spacing. Amer. Midl. Nat. 89:41–57.

Barbour, M. G., and J. E. Major. 1977. Terrestrial vegetation of California. Wiley, New York.

Beatley, J. C. 1974. Effects of rainfall and temperature on the distribution and behavior of *Larrea tridentata* (creosote-bush) in the Mojave Desert of Nevada. Ecology 55:245–261.

Beatley, J. C. 1976. Vascular plants of the Nevada test site and central-southern Nevada: ecologic and geographic distributions, TID-26881. Energy Research and Development Administration, NTIS, Springfield, Va.

Bender, G. L. 1982. Reference handbook on the deserts of North America. Greenwood Press, Westport, Conn.

Benson, L. 1982. The cacti of the United States and Canada. Stanford University Press, Stanford, Calif.

Benson, L., and R. A. Darrow. 1981. Trees and shrubs of the southwestern deserts, 3rd ed. University of Arizona Press, Tucson.

Benzioni, A., and R. L. Dunstone. 1986. Jojoba: adaptation to environmental stress and the implications for domestication. Quart. Rev. Biol. 61:177–199.

Bowers, J. E. 1982. The plant ecology of inland dunes in western North America. J. Arid Environ. 5:199–220.

Bowers, J. E. 1984. Plant geography of southwestern sand dunes. Desert Plants 6:31–42, 51–54.

Bowers, M. A. 1988. Plant associations on a Sonoran Desert bajada: geographical correlates and evolutionary source pools. Vegetatio 74:107–112.

Bowers, M. A., and C. H. Lowe. 1986. Plant-form gradients on Sonoran Desert bajadas. Oikos 46:284–291.

Bratz, R. D. 1976. The central desert of Baja California. J. Idaho Acad. Sci. 12:58–72.

Brown, D. E. 1982a. Chihuahuan desertscrub. Desert Plants 4:169–179.

Brown, D. E. 1982b. Biotic communities of the American Southwest—United States and Mexico. Desert Plants 4:1–342.

Brown, D. E., and C. H. Lowe. 1980. Biotic communities of the Southwest. USDA Forest Service General Technical Report RM-78. Rocky Mountain Forest and Range Experiment Station, Ft. Collins, Colo.

Burgess, T. L., and D. K. Northington, 1977. Desert vegetation in the Guadalupe Mountains region, pp. 229–242 in R. H. Wauer and D. H. Riskind (eds.), Transactions of the symposium on the biological resources of the Chihuahuan Desert region, United States and Mexico USDI, National Park Service, Transactions and Proceedings Series, No. 3. Government Printing Office, Washington, D.C.

Burgess, T. L., and A. Shmida. 1989. Succulent growth forms in arid environments, pp. 383–395 in E. E. Whitehead, C. F. Hutchinson, B. N. Timmermann, and R. G. Verity (eds.), Arid lands: today and tomorrow. University of Arizona, Tucson.

Busch, D. E., and S. D. Smith. 1995. Mechanisms associated with decline of woody species in riparian ecosystems of the southwestern U.S. Ecol. Monogr. 65: 347–370.

Carpenter, D. E., M. G. Barbour, and C. J. Bahre. 1986. Old field succession in Mojave Desert scrub. Madroño 33:111–122.

Chew, R. M. 1982. Changes in herbaceous and suffrutescent perennials in grazed and ungrazed desertified grassland in southeastern Arizona, 1958–1978. Amer. Midl. Nat. 108:159–169.

Chew, R. M., and W. G. Whitford. 1992. A long-term positive effect of kangaroo rats (*Dipodomys spectabilis*) on creosote bushes (*Larrea tridentata*). J. Arid Environ. 22:375–386.

Cleverly, J. R., S. D. Smith, A. Sala, and D. A. Devitt. 1997. Invasive capacity of *Tamarix ramosissima* in a Mojave Desert floodplain: the role of drought. Oecologia 111: 12–18.

Cody, M. L. 1986a. Spacing in Mojave Desert plant communities. II. Plant size and distance relationships. Isr. J. Bot. 35:109–120.

Cody, M. L. 1986b. Spacing patterns in Mojave Desert plant communities: near-neighbor analyses. J. Arid Environ. 11:199–217.

Cody, M. L. 1989. Growth-form diversity and community structure in desert plants. J. Arid Environ. 17: 199–209.

Cody, M. L. 1993. Do cholla cacti (*Opuntia* spp., subgenus Cylindropuntia) use or need nurse plants in the Mojave Desert? J. Arid Environ. 24:139–154.

Cooke, R. U., and R. W. Reeves. 1976. Arroyos and environmental change in the American south-west. Clarendon Press, Oxford.

Cooke, R. U., and A. Warren. 1973. Geomorphology in deserts. University of California Press, Berkeley.

Cornelius, J. M., P. R. Kemp. J. A. Ludwig, and G. L. Cunningham. 1991. The distribution of vascular plant species and guilds in space and time along a desert gradient. J. Vege. Sci. 2:59–72.

Crosswhite, F. S., and C. D. Crosswhite. 1982. The Sonoran Desert, pp. 163–319 in G. L. Bender (ed.), Reference handbook on the deserts of North America. Greenwood Press, Westport, Conn.

Crosswhite, F. S., and C. D. Crosswhite. 1984. A classification of life forms of the Sonoran Desert, with emphasis on the seed plants and their survival strategies. Desert Plants 5:131–161.

Danin, A. 1996. Plants of desert dunes: adaptations of desert organisms. Springer-Verlag, Berlin.

Dillon 1984. A systematic study of *Flourensia* (Asteraclae, Helianthal). Fieldiana Publications 1357, botany, new series, Mo. 16. Field Museum of Natural History, Chicago.

Dorn, R. I. 1988. A rock varnish interpretation of alluvial-fan development in Death Valley, California. Natl. Geogr. Res. 4:56–73.

Dorn, R. I. 1991. Rock varnish. Amer. Sci. 79:542–553.

Dorn, R. I., and T. M. Oberlander. 1981. Microbial origin of desert varnish. Science 213:1245–1247.

Dregne, H. E. 1976. Soils of arid regions. Elsevier, Amsterdam.

Dregne, H. E. 1979. Desert soils, pp. 73–81 in J. R. Goodin and D. K. Worthington (eds.), Arid land plant resources. International Center for Arid and Semi-arid Land Studies. Texas Technological University, Lubbock.

Dugas, W. A., R. A. Hicks, and R. P. Gibbens. 1996. Structure and function of C_3 and C_4 Chihuahuan Desert plant communities. Energy balance components. J. Arid Environ. 34: 63–79.

Ebert, T. A., and G. S. McMaster. 1981. Regular pattern of desert shrubs: a sampling artefact? J. Ecol. 69:559–564.

Eckert, R. E., Jr., M. K. Wood, W. H. Blackburn, and F. F. Peterson. 1979. Impacts of off-road vehicles on infiltration and sediment production of two desert soils. J. Range Mgt. 32:394–397.

Eickmeier, W. G. 1978. Photosynthetic pathway distributions along an aridity gradient in Big Bend National Park, and implications for enhanced resource partitioning. Photosynthetica 12:290–297.

Eickmeier, W. G. 1979. Eco-physiology differences between high and low elevation CAM species in Big Bend National Park, Texas. Amer. Midl. Nat. 101: 118–126.

Elkins, N. Z., G. V. Sabol, T. J. Ward, and W. G. Whitford. 1986. The influence of subterranean termites on the hydrological characteristics of a Chihuahuan desert ecosystem. Oecologia 68:521–528.

Ezcurra, E., S. Arizaga, P. L. Valverde, C. Mourelle, and A. Flores-Martinez. 1992. Foliole movement and canopy architecture of *Larrea tridentata* (DC.) Cov. in Mexican deserts. Oecologia 92:83–89.

Ezcurra, E., M. Equihua, and J. Lopez-Portillo. 1987. The desert vegetation of El Pinacate, Sonora, Mexico. Vegetatio 71:49–60.

Ezcurra, E., and V. Rodrigues. 1986. Rainfall patterns in the Gran Desierto, Sonora, Mexico. J. Arid Environ. 10:13–28.

Felger, R. S. 1980. Vegetation and flora of the Gran Desierto, Sonora, Mexico. Desert Plants 2:87–114.

Felger, R. S., and C. H. Lowe. 1976. The island and coastal vegetation and flora of the northern part of the Gulf of California. Contributions in Science 285, Natural History Museum of Los Angeles County, Calif.

Fiero, B. 1986. Geology of the Great Basin. University of Nevada Press, Reno.

Fonteyn, P. J., and B. E. Mahall. 1978. Competition among desert perennials. Nature 275:544–545.

Fonteyn, P. J., and B. E. Mahall. 1981. An experimental analysis of structure in a desert plant community. J. Ecol. 69:883–896.

Fuller, W. H. 1975. Soils of the desert Southwest. University of Arizona Press, Tucson.

Gardner, J. L. 1951. Vegetation of the creosotebush area of the Rio Grand valley in New Mexico. Ecol. Monogr. 21:379–403.

Garner, W., and V. Steinberger. 1989. A proposed mechanism for the formation of "Fertile Islands" in the desert ecosystem. J. Arid Environ. 16:257–262.

Gentry, H. S. 1978. The agaves of Baja California. Occasional Papers of the California Academy of Science, No. 130, San Francisco.

Gentry, H. S. 1982. Agaves of continental North America. University of Arizona Press, Tucson.

Gibbens, R. P., R. A. Hicks, and W. A. Dugas. 1996. Structure and function of C_3 and C_4 Chihuahuan Desert plant communities. Standing crop and leaf area index. J. Arid Environ. 34: 47–62.

Gibson, A. C., and P. S. Nobel. 1986. The cactus primer. Harvard University Press, Cambridge, Mass.

Gile, L. H., F. F. Peterson, and R. B. Grossman. 1966. Morphological and genetic sequences of carbonate accumulation in desert soils. Soil Sci. 101:347–360.

Goldberg, D. E., and R. M. Turner. 1986. Vegetation change and plant demography in permanent plots in the Sonoran Desert. Ecology 67:695–712.

Goudie, A., and J. Wilkinson. 1977. The warm desert environment. Cambridge University Press, Cambridge.

Guo, Q. 1996. Effects of bannertail kangaroo rat mounds on small-scale plant community structure. Oecologia 106: 247–256.

Guo, Q., D. B. Thompson, T. J. Valone, and J. H. Brown. 1995. The effects of vertebrate granivores and folivores on plant community structure in the Chihuahuan Desert. Oikos 73:251–259.

Hadley, N. F., and S. R. Szarek. 1981. Productivity of desert ecosystems. BioScience 31:747–753.

Hallmark, C. T., and B. L. Allen. 1975. The distribution of creosotebush in west Texas and eastern New Mexico as affected by selected soil properties. Soil Sci. Soc. Amer. Proc. 39:120–124.

Hastings, J. R., and R. M. Turner. 1965. The changing mile: an ecological study of vegetation change with time in the lower mile of an arid and semi-arid region. University of Arizona Press, Tucson.

Hastings, J. R., R. M. Turner, and D. K. Warren. 1972. An atlas of some plant distributions in the Sonoran Desert. Technical Reports on the Meteorology and

Climatology of Arid Regions, no. 21. University of Arizona Institute of Atmospheric Physics, Tucson.

Hawkins, L. K. 1996. Burrows of kangaroo rats are hotspots for desert soil fungi. J. Arid Environ. 32:239–249.

Hendricks, D. M. 1985. Arizona soils. College of Agriculture, University of Arizona, Tucson.

Hennessy, J. T., R. P. Gibbens, J. M. Tromble, and M. Cardenas. 1985. Mesquite (*Prosopis glandulosa* Torr.) dunes and interdunes in southern New Mexico: a study of soil properties and soil water relations. J. Arid Environ. 9:27–38.

Henrickson, J. 1977. Saline habitats and halophytic vegetation of the Chihuahuan Desert region, pp. 289–314 in R. H. Wauer and D. H. Riskind (eds.), Transactions of the symposium on the biological resources of the Chihuahuan Desert Region, United States and Mexico USDI, National Park Service, Transactions and Proceedings Series, No. 3. Government Printing Office, Washington, D.C.

Henrickson, J., and M. C. Johnston. 1986. Vegetation and community types of the Chihuahuan Desert, pp. 20–39 in J. C. Barlow, A. M. Powell, and B. N. Timmermann (eds.), Chihuahuan Desert – U.S. and Mexico, Vol. 11 Chihuahuan Desert Research Institute, Sul Ross State University, Alpine, Texas.

Henrickson, J., and R. M. Straw. 1976. A gazetteer of the Chihuahuan Desert region. A supplement to the Chihuahuan Desert flora. California State University, Los Angeles.

Heske, E. J., J. H. Brown, and Q. Guo. 1993. Effects of kangaroo rat exclusion on vegetation structure and plant species diversity in the Chihuahuan Desert. Oecologia 95:520–524.

Humphrey, R. R. 1974. The boojum and its home. University of Arizona Press, Tucson.

Humphrey, R. R., and D. B. Marx. 1980. Distribution of the boojum tree (*Idria columnaris*) on the coast of Sonora, Mexico as influenced by climate. Desert Plants 2:183–196.

Hunt, C. B. 1966. Plant ecology of Death Valley, California. USDI, Geological Survey Professional Publication 509, Government Printing office, Washington, D.C.

Hunt, C. B. 1975. Death Valley. Geology, ecology, archaeology. University of California Press, Berkeley.

Inouye, R. S. 1991. Population biology of desert annual plants, pp. 27–54 in Gary A. Polis (ed.), The ecology of desert communities (Desert Ecology Series). University of Arizona Press, Tucson.

Johnson, H. B. 1976. Vegetation and plant communities of southern California deserts – a functional view, pp. 125–164 in J. Latting (ed.), Vegetation and plant communities of southern California deserts – a functional view. California Native Plant Society, Berkeley, Calif.

Kelt, D. A., and T. J. Valone. 1995. Effects of grazing on the abundance and diversity of annual plants in Chihuahuan desert scrub habitat. Oecologia 103:191–195.

Kemp, P. R. 1983. Phenological patterns of Chihuahuan Desert plants in relation to the timing of water availability. J. Ecol. 71:427–436.

Kemp, P. R., and P. E. Gardetto. 1982. Photosynthetic pathway types of evergreen rosette plants (Liliaceae) of the Chihuahuan Desert. Oecologia 55:149–156.

King, P. B. 1959. The evolution of North America. Princeton University Press, Princeton, N.J.

King, T. J., and S. R. J. Woodell. 1984. Are regular patterns in desert shrubs artefacts of sampling? J. Ecol. 72:295–298.

Lajtha, K., K. Kolberg, and J. Gety. 1997. Ecophysiology of the saguaro cactus (*Carenegiea gigantea*) in Saguaro National Monument: relationship to symptoms of decline. J. Arid Environ. 36:579–590.

Lajtha, K., and W. H. Schlesinger. 1988. The biogeochemistry of phosphorus cycling and phosphorus availability along a desert soil chronosequence. Ecology 69:24–39.

Lancaster, N. 1984. Aeolian sediments, processes and landforms. J. Arid Environ. 7:249–254.

Lancaster, N. 1993. Kelso dunes. Natl. Geogr. Res. 9:444–459.

Lane, L. J., E. M. Romney, and T. E. Hakonson. 1984. Water balance calculations and net production of perennial vegetation in the northern Mojave Desert. J. Range Mgt. 37:12–18.

Le Houerou, H. N. 1984. Rain use efficiency: a unifying concept in arid-land ecology. J. Arid Environ. 7:1–35.

Leitner, L. A. 1987. Plant communities of a large arroyo at Punta Cirio, Sonora. Southw. Nat. 32:21–28.

Lowe, C. H., and W. F. Steenbergh. 1981. On the Cenozoic ecology and evolution of the Sahuaro. Desert Plants 3:83–86.

Luckenbach, R. A., and R. B. Bury. 1983. Effects of off-road vehicles on the biota of the Algodones Dunes, Imperial County, California. J. Appl. Ecol. 20:265–286.

Ludwig, J. A. 1987. Primary productivity in arid lands: myths and realities. J. Arid Environ, 13:1–7.

Ludwig, J. A., G. L. Cunningham, and P. D. Whitson. 1988. Distribution of annual plants in North American deserts. J. Arid Environ. 15:221–227.

Mabbutt, J. A. 1977. Desert landforms, Vol. 2. MIT Press, Cambridge, Mass.

Mabry, T. J., J. H. Hunziker, and D. R. DiFeo. 1977. Creosote bush. Biology and chemistry of *Larrea* in New World deserts. United States/International Biological Program (US/IBP) 6, Dowden, Hutchinson & Ross, Stroudsburg, Pa.

MacMahon, J. A. 1979. North American deserts: their floral and faunal components, pp. 21–82 in D. W. Goodall and R. A. Perry (eds.), Arid-land ecosystems: structure, functioning and management, Vol. 1, US/IBP 16 Cambridge University Press, Cambridge.

MacMahon, J. A. 1985. The Audubon Society field guide to North American desert habitats. Chanticleer Press, New York.

MacMahon, J. A., and D. J. Schimpf. 1981. Water as a factor in the biology of North American desert plants, pp. 119–171 in D. D. Evans and J. L. Thames (eds.), Water in desert ecosystems, US/IBP 12, Dowden, Hutchinson and Ross, Stroudsburg, Pa.

MacMahon, J. A., and F. H. Wagner. 1985. The Mojave, Sonoran and Chihuahuan deserts of North America, pp. 105–202 in M. Evenari et al. (eds.), Hot deserts and arid shrublands. Elsevier, Amsterdam.

Marroquin, J. S. 1977. A physiognomic analysis of the types of transitional vegetation in the eastern parts of the Chihuahuan Desert in Coahuila, Mexico, pp. 249–272 in R. H. Wauer and D. H. Riskind (eds.), Transactions of the symposium on the biological resources of the Chihuahuan Desert region, United States and Mexico USDI, National Park Service, Transactions and Proceedings Series, No. 3. Government Printing Office, Washington, D.C.

Martinez-Meza, E., and W. G. Whitford. 1996. Stemflow, throughfall and channelization of stemflow by roots in three Chihuahuan desert shrubs. J. Arid Environ. 32:271–287.

Mauchamp, A., and J. L. Janeau. 1993. Water funnelling by the crown of *Flourensia cernua*, a Chihuahuan Desert shrub. J. Arid Environ. 25:299–306.

McAuliffe, J. R. 1984. Sahuaro – nurse tree associations in the Sonoran Desert: competitive effects of sahuaros. Oecologia 64:319–321.

McAuliffe, J. R. 1988. Markovian dynamics of simple and complex desert plant communities. Amer. Midl. Nat. 131:459–490.

McAuliffe, J. R. 1991. Demographic shifts and plant succession along a late Holocene soil chronosequence in the Sonoran Desert of Baja California. J. Arid Environ. 20:165–178.

McAuliffe, J. R. 1994. Landscape evolution, soil formation, and ecological patterns and processes in Sonoran Desert bajadas. Ecol. Monogr. 64:111–148.

McClaran, M. P., and T. R. Van Devender. 1995. The desert grassland. University of Arizona Press, Tucson.

McGee, K. P., and D. L. Marshall. 1993. Effects of variable moisture availability on seed germination in three populations of Larrea tridentata. Amer. Midl. Nat. 130:75–82.

McGinnies, W. G. 1981. Discovering the desert. University of Arizona Press, Tucson.

McKell, C. M. 1985. North America, pp. 187–232 in J. R. Goodin and D. K. Northington (eds.), Plant resources of arid and semiarid lands, a global perspective. Academic Press, Orlando, Fla.

Meyer, S. E. 1986. The ecology of gypsophile endemism in the eastern Mojave Desert. Ecology 67: 1303–1313.

Miller, J. M. 1986. Phytogeography and potential economic use of the guayule rubber plant on Chihuahuan Desert limestone geologic formations. J. Arid Environ. 10:153–162.

Montana, C. 1990. A floristic-structural gradient related to land forms in the southern Chihuahuan Desert. J. Vege. Sci. 1:669–674.

Montana, C., J. Lopez-Portillo, and A. Mauchamp. 1990. The response of two woody species to the conditions created by a shifting ecotone in an arid ecosystem. J. Ecol. 78:789–798.

Moore, C. B., and C. Elvidge. 1982. Desert varnish, pp. 527–536 in G. L. Bender (ed.), Reference handbook on the deserts of North America. Greenwood Press, Westport, Conn.

Moorhead, D. L., F. M. Fisher, and W. G. Whitford. 1988. Cover of spring annuals on nitrogen-rich kangaroo rat mounds in a Chihuahuan Desert grassland. Amer. Midl. Nat. 120:443–447.

Morafka, D. J. 1977. A biogeographical analysis of the Chihuahuan Desert through its herpetofauna. Biogeographic 9. Junk, The Hague.

Mulroy, T. W., and P. W. Rundel. 1977. Annual plants: adaptations to desert environments. BioScience 27: 109–114.

Nabhan, G. P., and J. L. Carr. (eds.). 1994. Ironwood: an ecological and cultural keystone of the Sonoran Desert. Occasional Papers in Conservation Biology No. 1. Conservation International, Washington, D.C.

Nash, T. H., III, and T. J. Moser. 1982. Vegetational and physiological patterns of lichens in North American deserts. J. Hattori Bot. Lab. 53:331–336.

Nash, T. H., III, G. T. Nebeker, T. J. Moser, and T. Reeves. 1979. Lichen vegetational gradients in relation to the Pacific coast of Baja California: the maritime influence. Madroño 26:149–163.

Neal, J. T. 1969. Playa variation, pp. 15–44 in W. G. McGinnies and B. J. Goldman (eds.), Arid lands in perspective. University of Arizona Press, Tucson.

Neal, J. T. 1975. Playas and dried lakes: Occurrence and development. Dowden, Hutchinson & Ross, Stroudsburg, Pa.

Neufeld, H. S., F. C. Meinzer, C. S. Wisdom, M. R. Sharifi, P. W. Rundel, M. S. Neufeld, Y. Goldring, and G. L. Cunningham. 1988. Canopy architecture of *Larrea tridentata* (DC.) Cov., a desert shrub: foliage orientation and direct beam radiation interception. Oecologia 75:54–60.

Niehaus, T. F., and C. L. Ripper. 1976. A field guide to Pacific states wildflowers. Houghton Mifflin, Boston.

Niehaus, T. F., C. L. Ripper. and V. Savage. 1984. A field guide to southwestern and Texas wildflowers. Houghton Mifflin, Boston.

Niering, W. A., and C. H. Lowe. 1984. Vegetation of the Santa Catalina Mountains: community types and dynamics. Vegetatio 58:3–28.

Nobel, P. S. 1980. Morphology, nurse plants, and minimum apical temperatures for young *Carnegiea gigantea*. Bot. Gaz. 141:188–191.

Nobel, P. S. 1988. Environmental biology of agaves and cacti. Cambridge University Press, Cambridge.

Nobel, P. S., P. M. Miller, and E. A. Graham. 1992. Influence of rocks on soil temperature, soil water potential, and rooting patterns for desert succulents. Oecologia 92:90–96.

Oberlander, T. M. 1979. Characterization of arid climates according to combined water balance parameters. J. Arid Environ. 2:219–241.

Orians, G. H., and O. T. Solbrig. 1977. Convergent evolution in warm deserts. US/IBP 3 Dowden, Hutchinson & Ross, Stroudsburg, Pa.

Osborn, H. G., E. D. Shirley, D. R. Davis, and R. B. Koehler. 1980. Model of time and space distribution of rainfall in Arizona and New Mexico. Agricultural Reviews and Manuals, ARM-W-14. USDA, Science and Education Administration, Oakland, Calif.

Pake, C. E., and D. L. Venable. 1996. Seed banks in desert annuals: implications for persistence and coexistence in variable environments. Ecology 77:1427–1435.

Parker, K. C. 1991. Topography, substrate, and vegetation patterns in the northern Sonoran Desert. J. Biogeogr. 18:151–163.

Parsons, R. F. 1976. Gypsophily in plants – a review. Amer. Midl. Nat. 96:1–20.

Patten, D. T. 1978. Productivity and production efficiency of an upper Sonoran Desert ephemeral community. Amer. J. Bot. 65:891–895.

Pavlik, B. M., and M. G. Barbour. 1988. Demographic monitoring of endemic sand dune plants, Eureka Valley, California. Biol. Conserv. 46:217–242.

Peinado, M., F. Alcaraz, J. L. Aguirre, and J. Dalgadillo.

1995. Major plant communities of warm North American deserts. J. Vege. Sci. 6:79–94.

Peinado, M., F. Alcaraz, J. Delgadillo, and I. Aguado. 1993. Fitogeografia de la peninsula de Baja California, México. Anales del Jardin Botanico de Madrid 51:255–277.

Peinado, M., F. Alcaraz, J. L. Aguirre, J. Dalgadillo, and I. Aguado. 1995. Shrubland formations and associations in mediterranean-desert transitional zones of northwestern Baja California. Vegetatio 117:165–179.

Phillips, D. L., and J. A. MacMahon. 1978. Gradient analysis of a Sonoran Desert bajada. Southw. Nat. 23:669–680.

Phillips, D. L., and J. A. MacMahon. 1981. Competition and spacing patterns in desert shrubs. J. Ecol. 69:97–115.

Pinkava, D. J., and H. S. Gentry. 1985. Symposium on the genus *Agave*, Desert Botanical Garden, Phoenix, March 7–9, 1985. Desert Plants 7:1.

Plumb, G. A. 1991. Assessing vegetation types of Big Bend National Park, Texas, for image-based mapping. Vegetatio 94:115–124.

Polis, G. A. 1991. The ecology of desert communities. University of Arizona Press, Tucson.

Powell, A. M., and B. L. Turner. 1977. Aspects of the plant biology of the gypsum outcrops of the Chihuahuan Desert, pp. 315–325 in R. H. Wauer and D. H. Riskind (eds.), Transactions of the symposium on the biological resources of the Chihuahuan Desert region, United States and Mexico. USDI, National Park Service, Transactions and Proceedings Series, No. 3. Government Printing Office, Washington, D.C.

Prose, D. V., S. K. Metzger, and H. G. Wilshire. 1987. Effects of substrate disturbance on secondary plant succession; Mojave Desert, California. J. Appl. Ecol. 24:305–313.

Rasmuson, K. E., J. E. Anderson, and N. Huntly. 1994. Coordination of branch orientation and photosynthetic physiology in the Joshua tree (*Yucca brevifolia*). Great Basin Nat. 54:204–211.

Rea, A. M. 1983. Once a river. Bird life and habitat changes on the middle Gila. University of Arizona Press, Tucson.

Rogers, G. F. 1985. Mortality of burned *Cereus giganteus*. Ecology 66:630–632.

Rogers, G. F., and M. K. Vint. 1987. Winter precipitation and fire in the Sonoran Desert. J. Arid Environ. 13:47–52.

Rowlands, P., H. Johnson, E. Riter, and A. Endo. 1982. The Mojave Desert, pp. 103–145 in G. L. Bender (ed.), Reference handbook on the deserts of North America. Greenwood Press, Westport, Conn.

Rundel, P. W., and A. C. Gibson. 1996. Ecological communities and processes in a Mojave Desert ecosystem: Rock Valley, Nevada. Cambridge University Press, Cambridge.

Rzedowski, J. 1966. Vegetacion del Estado de San Luis Potosí. Acta Cientifica Potosina 5:5–291.

Rzedowski, J. 1978. Vegetacion de México. Editorial Limusa, México City.

Sammis, T. W., and L. W. Gay. 1979. Evapotranspiration from an arid zone plant community. J. Arid Environ. 2:313–321.

Schlesinger, W. H. 1982. Carbon storage in the caliche of arid soils: a case study from Arizona. Soil Sci. 133:247–255.

Schlesinger, W. H. 1985. The formation of caliche in soils of the Mojave Desert, California. Geochimica et Cosmochimica Acta 49:57–66.

Schlesinger, W. H., and C. S. Jones. 1984. The comparative importance of overland runoff and mean annual rainfall to shrub communities of the Mojave Desert. Bot. Gaz. 145:116–124.

Schlesinger, W. H., J. A. Raikes, A. E. Hartley, and A. F. Cross. 1996. On the spatial pattern of soil nutrients in desert ecosystems. Ecology 77:364–374.

Schmid, M. K., and G. F. Rogers. 1988. Trends in fire occurrence in the Arizona upland subdivision of the Sonoran Desert, 1955 to 1983. Southw. Nat. 33:437–444.

Schmidt, R. H., Jr. 1979. A climatic delineation of the "real" Chihuahuan Desert. J. Arid Environ. 2:243–250.

Schmidt, R. H., Jr. 1986. Chihuahuan climate, pp. 40–63 in J. C. Barlow, A. M. Powell, and B. N. Timmermann (eds.), Chihuahuan Desert – U.S. and Mexico, Vol. II Chihuahuan Desert Research Institute, Sul Ross State University, Alpine, Texas.

Schmidt, R. H. 1989. The arid zones of Mexico: climatic extremes and conceptualization of the Sonoran Desert. J. Arid Environ. 16:241–256.

Secor, J. B., S. Shamash, D. Smeal, and A. L. Gennaro. 1983. Soil characteristics of two desert plant community types that occur in the Los Medanos area of southeastern New Mexico. Soil Sci. 136:133–144.

Sharifi, M. R., A. C. Gibson, and P. W. Rundel. 1997. Surface dust impacts on gas exchange in Mojave Desert shrubs. J. Appl. Ecol. 34:837–846.

Sharifi, M. R., E. T. Nilsen, and P. W. Rundel. 1982. Biomass and net primary production of *Prosopis glandulosa* (Fabaceae) in the Sonoran Desert of California. Amer. J. Bot. 69:760–767.

Shields, L. M. 1953. Gross modifications in certain plant species tolerant of calcium sulfate dunes. Amer. Midl. Nat. 50:224–237.

Shmida, A., and T. L. Burgess. 1988. Plant growth-form strategies and vegetation types in arid environments, pp. 211–241 in M. J. A. Werger, P. J. M. van der Aart, H. J. During, and J. T. A. Verboeven (eds.), Plant growth-form strategies and vegetation types in arid environments. SPB Academic Publishing, The Hague.

Shreve, F. 1951. Vegetation of the Sonoran Desert. Carnegie Institution of Washington pub. 591.

Shreve, F., and T. D. Mallery. 1933. The relation of caliche to desert plants. Soil Sci. 35:99–113.

Shreve, F., and I. L. Wiggins. 1964. Vegetation and flora of the Sonoran Desert, Vol. 1. Stanford University Press, Stanford, Calif.

Silvertown, J., and J. B. Wilson. 1994. Community structure in a desert perennial community. Ecology 75:409–417.

Simpson, B. B. 1977. Mesquite. Its biology in two desert scrub ecosystems. US/IBP 4, Dowden, Hutchinson & Ross, Stroudsburg, Pa.

Skujins, J. 1984. Microbial ecology of desert soils, pp. 49–91 in C. C. Marshall (ed.), Microbial ecology of desert soils. Plenum, New York.

Smith, R. S. U. 1982. Sand dunes in the North American deserts, pp. 481–524 in G. L. Bender (ed.), Reference handbook on the deserts of North America. Greenwood Press, Westport, Conn.

Smith, S. D., C. A. Herr, K. L. Leary, and J. M. Piorkowski. 1995. Soil–plant water relations in a Mojave Desert mixed shrub community: a comparison of three geomorphic surfaces. J. Arid Environ. 29: 339–351.

Smith, S. D., and J. A. Ludwig. 1978. The distribution and phytosociology of *Yucca elata* in southern New Mexico. Amer. Midl. Nat. 100:202–212.

Solbrig, O. T. 1972. The floristic disjunctions between the "Monte" in Argentina and the "Sonoran Desert" in Mexico and the United States. Ann. Mo. Bot. Gard. 59:218–223.

Solbrig, O. T. 1986. Evolution of life-forms in desert plants, pp. 89–105 in N. Polunin (ed.), Evolution of life-forms in desert plants. Wiley, New York.

Solbrig, O. T., M. G. Barbour, J. Cross, G. Goldstein, C. H. Lowe, J. Morello, and T. W. Yang. 1977. The strategies and community patterns of desert plants, pp. 67–106 in G. H. Orians and O. T. Solbrig (eds.), Convergent evolution in warm deserts. US/IBP 3. Dowden, Hutchinson & Ross, Stroudsburg, Pa.

Steenbergh, W. F., and C. H. Lowe. 1976. Ecology of the saguaro. I. The role of freezing weather on a warm-desert plant population. Research in the Parks, National Park Service Symposium Series 1:49–92.

Steenbergh, W. F., and C. H. Lowe. 1977. Ecology of the saguaro. II. Natural Part Service, Washington, D.C. Reproduction, germination, establishment, growth and survival of the young plant. National Park Service, Scientific Monograph Series, No. 8. Government Printing Office, Washington, D.C.

Steenbergh, W. F., and C. H. Lowe. 1983. Ecology of the saguaro. III. Growth and demography. National Park Service, Scientific Monograph Series, No. 17. Government Printing Office, Washington, D.C.

Stromberg, J., and T. M. Krischan. 1983. Vegetation structure at Punta Cirio, Sonora, Mexico. Southw. Nat. 28:211–214.

Szarek, S. R. 1979. Primary production in four North American deserts: indices of efficiency. J. Arid Environ. 2:187–209.

Turner, F. B., and D. C. Randall. 1987. The phenology of desert shrubs in southern Nevada. J. Arid Environ. 13:119–128.

Turner, F. B., and D. C. Randall. 1989. Net production by shrubs and winter annuals in southern Nevada. J. Arid Environ. 17:23–36.

Turner, R. M. 1982. Mohave desert scrub. Desert Plants 4:157–168.

Turner, R. M. 1990. Long-term vegetation change at a fully protected Sonoran Desert site. Ecology 71: 464–477.

Turner, R. M., and D. E. Brown. 1982. Sonoran desert scrub. Desert Plants 4:181–222.

Turner, R. M., J. E. Bowers, and T. L. Burgess. 1995. Sonoran desert plants: an ecological atlas. University of Arizona Press, Tucson.

Valverde, P. L., J. Zavala-Hurtado, C. Montaña, and E. E. zeurra. 1996. Numerical analyses of vegetation based on environmental relationships in the southern Chihuahuan desert. Southwestern Natur. 41:424–433.

Vandermeer, J. 1980. Saguaros and nurse trees: A new hypothesis to account for population fluctuations. Southw. Nat. 25:357–360.

Van Devender, T. R. 1990. Late Quarternary vegetation and climate of the Chihuahuan and Sonoran Deserts, United States and Mexico, pp. 104–164 in J. L. Van Devender, and P. S. Martin (eds.), Packrat Middens, the last 40,000 years of biotic change. Betancourt, T. R. University of Arizona Press, Tucson.

Van Devender, T. R., R. S. Thompson, and J. L. Betancourt. 1987. Vegetation history of the deserts of southwestern North America, pp. 323–352 in W. F. Ruddiman and H. E. Wright (eds.), The Geology of North America, Vol. K-3. Geological Society of America, Boulder, Colo.

Vasek, F. C. 1980. Creosote bush: Long-lived clones in the Mojave Desert. Amer. J. Bot. 67:246–255.

Vasek, F. C. 1983. Plant succession in the Mojave Desert. Crossosoma 9:1–23.

Vasek, F. C., H. B. Johnson, and G. D. Brum. 1975. Effects of power transmission lines on vegetation of the Mojave Desert. Madroño 23:114–130.

Vasek, F. C., H. B. Johnson, and D. H. Eslinger. 1975. Effects of pipeline construction on creosote bush scrub vegetation of the Mojave Desert. Walter 1971 Madroño 23:1–13.

Walter, H. 1971. Ecology of tropical and subtropical vegetation. Oliver & Boyd, Edinburgh.

Walter, H., and H. Lieth. 1967. Klimadiagramm-Weltatlas. Gustav Fischer-Verlag, Jena.

Waser, N. M., and M. V. Price. 1981. Effects of grazing on diversity of annual plants in the Sonoran Desert. Oecologia 50:407–411.

Wauer, R. H., and D. H. Riskind. 1977. Transactions of the symposium on the biological resources of the Chihuahuan Desert Region, United States and Mexico. USDI, National Park Service Transactions and Proceedings Series, No. 3. Government Printing Office, Washington, D.C.

Webb, R. H., J. W. Steiger, and E. B. Newman. 1988. The response of vegetation to disturbance in Death Valley National Monument, California. U.S. Geological Survey Bulletin 1793, U.S. Geological Survey, Denver.

Webb, R. H., J. W. Steiger, and R. M. Turner. 1987. Dynamics of Mojave Desert shrub assemblages in the Panamint Mountains, California. Ecology 68:478–490.

Webb, R. H., and S. S. Stielstra. 1979. Sheep grazing effects on Mojave Desert vegetation and soils. Environ. Mgt. 3:517–529.

Webb, R. H., and H. G. Wilshire. 1980. Recovery of soils and vegetation in a Mojave desert ghost town, Nevada, U.S.A. J. Arid Environ. 3:291–303.

Webb, R. H., and H. G. Wilshire. 1983. Environmental effects of off-road vehicles. Springer-Verlag, New York.

Wells, P. V., and D. Woodcock. 1985. Full-glacial vegetation of Death Valley, California: juniper woodland opening to *Yucca* semidesert. Madroño 32:11–23.

Welsh, R. G., and R. F. Beck. 1976. Some ecological relationships between creosotebush and bush muhly. J. Range Mgt. 29:472–475.

Wentworth, T. R. 1981. Vegetation on limestone and granite in the Mule Mountains, Arizona. Ecology 62:469–482.

West, N. E., and J. J. Skujins. 1978. Nitrogen in desert ecosystems. US/IBP 9. Dowden, Hutchinson & Ross, Stroudsburg, Pa.

West, R. C. 1964. Surface configuration and associated geology of Middle America, pp. 33–83 in R. C. West (ed.), Natural environment and early cultures. University of Texas Press, Austin.

Wester, D. B., and H. A. Wright. 1987. Ordination of

vegetation change in Guadalupe Mountains, New Mexico, USA. Vegetatio 72:27–33.

Whittaker, R. H., and W. A. Niering. 1968. Vegetation of the Santa Catalina Mountains, Arizona. IV. Limestone and acid soils. J. Ecol. 56:523–544.

Wierenga, P. J., J. M. H. Hendrickx, M. H. Nash, J. Ludwig, and L. A. Daugherty. 1987. Variation of soil and vegetation with distance along a transect in the Chihuahuan Desert. J. Arid Environ. 13:53–63.

Wright, R. A. 1982. Aspects of desertification in *Prosopis* dunelands of southern New Mexico, U.S.A. J. Arid Environ. 5:277–284.

Yang, T. W. 1970. Major chromosome races of *Larrea divaricata* in North America. J. Ariz. Acad. Sci. 6:41–45.

Yeaton, R. I. 1978. A cyclical relationship between *Larrea*

tridentata and *Opuntia leptocaulis* in the northern Chihuahuan desert. J. Ecol. 66:651–656.

Yeaton, R. I., and M. L. Cody. 1979. The distribution of cacti along environmental gradients in the Sonora and Mohave deserts. J. Ecol. 67:529–541.

Yeaton, R. I., and A. R. Manzanares. 1986. Organization of vegetation mosiacs in the *Acacia schaffneri–Opuntia streptacantha* association, southern Chihuahuan desert, Mexico. J. Ecol. 74:211–217.

Yeaton, R. I., J. Travis, and E. Gilinsky. 1977. Competition and spacing in plant communities: the Arizona upland association. J. Ecol. 65:587–595.

Zabriskie, J. G. 1979. Plants of Deep Canyon and the Central Coachella Valley, California. University of California, Riverside.

Chapter
9

Grasslands

PHILLIP L. SIMS   PAUL G. RISSER

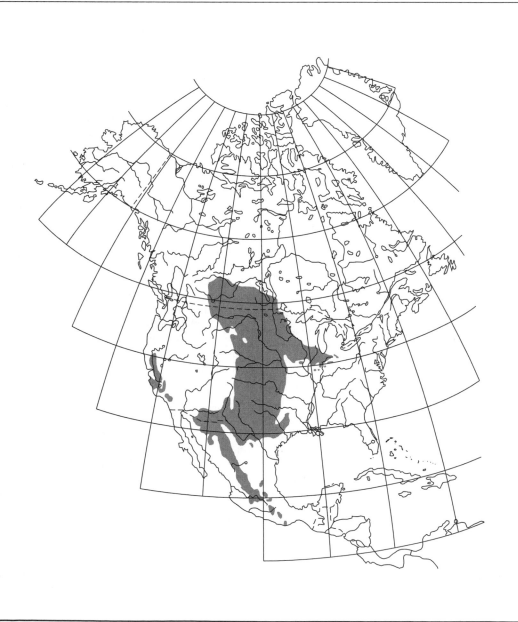

INTRODUCTION

Grassland is the largest of the four major natural vegetation formations that cover the earth's land surface (Gould 1968; Gould and Shaw 1983). Grasslands occur on most continents and account for 24% of the earth's vegetation and cover more than 4.6 billion ha (Shantz 1954). Grassland is accompanied by certain consistencies in climate, flora, fauna, plant growth form, and vegetational physiognomy (Carpenter 1940; Ripley 1992). This formation, including steppes (Russia), velds (South Africa), pampas (South America), puszta (Hungary), and prairies (North America), has immense aesthetic, economic, and watershed values, and provides habitat for large numbers of domestic and wild animals. Grasses are the most widely adapted of all plants (Archer and Bunch 1953), the primary source of germplasm for food crops, and an incredibly important germplasm reservoir for the future.

Originally, grasslands dominated central North America and occurred as important and sometimes extensive islands of vegetation throughout the western United States, Canada, and Mexico (Küchler 1975; Bailey 1976; Rzedowski 1978). They stretched from southeastern Alberta, central Saskatchewan, and southeastern Manitoba to the highlands of central Mexico and from eastern Indiana to California. They were the largest of the North American vegetation formations, originally covering almost 300 million (m) ha in the United States (Küchler 1964; Samson and Knopf 1994) and about 50 and 20 M ha in Canada and Mexico, respectively, as estimated from the data of Rowe (1972) and Rzedowski 1978). Even today, grasslands remain the largest natural biome in the United States, covering more than 125 M ha (U.S. Forest Service 1980). Most of the productive, arable lands in North America were once grasslands. The grasslands that remain, along with desert and mountain shrublands, savannas, and alpine areas, make up the nation's rangeland resources. More than 3000 species of mammals, birds, reptiles, fish, and amphibians live on these resources (Krausman 1996).

The central North American grasslands arose some 20 M yr ago as aridity increased, leaving the drought-tolerant grasses and forbs to dominate as the forest retreated eastward (Dix 1964; Axelrod 1985). This grassland formation includes the tallgrass, mixed-grass, and shortgrass prairies of the

central plains, desert grasslands of the southwestern United States and Mexico, California grasslands, Palouse prairie in the Intermountain Region of the northwestern United States and British Columbia, and the fescue prairie of northern Montana, southern Alberta, and central Saskatchewan (Fig. 9.1). The coastal, California, fescue, and desert grasslands account for about 10, 2, 9, and 8%, respectively, of all grasslands, with the remaining 80% rather uniformly distributed as shortgrass, mixed-grass, tallgrass, and Palouse prairies (Table 9.1). Small pockets of mountain grasslands occur within the western coniferous forest, with about 27 M ha of grassland associating with *Pinus ponderosa* in the Rocky Mountains (Risser et al. 1981).* These mountain grasslands are valuable scenic resources and important watersheds in western North America. Extensive grassland–forest combinations are also found in the deciduous forest–grassland transition zone east of the Great Plains.

Three or four grass species usually produce most of the biomass (Coupland 1974; Sims, Singh, and Lauenroth 1978; Barbour, Burk, and Pitts 1980), although grasses often compose no more than 20% of the total number of species in a typical grassland community. Forbs (herbaceous plants other than grass) may be seasonally important and, along with some dwarf shrubs, often dominate the grassland aspect. *Asteraceae* species are the most numerous plants, followed by *Fabaceae* (Coupland 1979).

North American grasslands contain approximately 7500 plant species from about 600 genera of grasses, plus numerous grass-like plants, forbs, and woody plants (Hartley 1950, 1964; Risser 1985). The graminoids rank third in number of genera, fifth in number of species, and first in geographical distribution, and they compose the greatest percentage of the total world vegetation biomass (Gould 1968).

Adaptive Strategies

Grasses adapt to a myriad of environments because of their tremendous genetic amplitude. Genetic diversity is expressed through the morphological and physiological response mechanisms that operate within a complex of ecological processes. Although no generalized morphological-physiological model fully characterizes species re-

The authors acknowledge the assistance of Sherry Dewald for word and manuscript processing and James Bradford for assistance with the figures and in updating of scientific nomenclature.

*Nomenclature of scientific plant names generally follow Kartesz (1994). Authorities are cited only on first mention of a species.

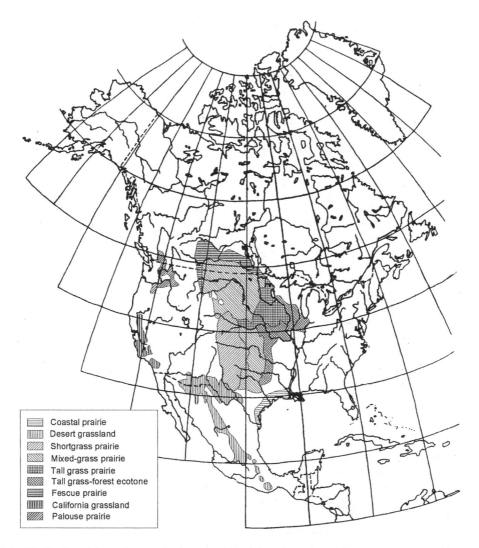

Figure 9.1. Distribution of the major grasslands types of North America (adapted from Rowe 1972, Küchler 1975, Wright and Bailey 1982, and Rzedowski 1978).

Table 9.1. *Extent of the major grassland types of the contiguous United States.*

Grassland	Area (ha)	Percentage
Tallgrass prairie	57,351,100	19
Mixed-grass prairie	56,617,400	19
Shortgrass prairie	61,522,300	21
Coastal prairie	3,800,000	1
California grassland	9,200,000	3
Palouse prairie	64,471,600	22
Fescue prairie	25,500,000	8
Desert grassland	20,756,500	7
Total	299,222,900	100

Source: Data from Küchler (1964) as planimetered by Risser et al. (1981) plus additions by the present authors.

sponses in grassland communities, there are, several recurring adaptive strategies (Risser 1985), as follows:

1. Grassland plants heavily invest carbohydrates in structural development early in the growing season when moisture conditions are generally adequate. During drought or following heavy grazing, grassland plants use stored labile carbohydrates, decrease dark respiration to conserve substrate, and can maintain gas-exchange processes under water potentials as low as −4 MPa.

2. Grassland plants physically respond to drought stress by closing stomata and curling leaves to reduce water losses. Pubescence and paraheli-

otropism also contribute to the efficiency of water use. Warm-season C_4 plants, generally favored in drier climates, tend to be more efficient in use of water than cool-season C_3 plants.

3. Many grassland plants produce seeds or other propagules that become dormant during adverse conditions or are able to germinate in relatively dry soil conditions. The primary and adventitious roots of grasses undergo rapid expansion when moisture and temperature conditions permit, increasing their ability to cope with ensuing stress. Physically, the roots of grassland plants have the strength to withstand the shrink-swell characteristics of clay soils.

4. Nutrient uptake is rapid when moisture is available and plants are growing. At that time, forage quality is highest, and the plants are palatable and most tolerant of grazing. As plants mature, the relative proportion of coarse stems increases, and they become less palatable. Removal of a portion of the growing tissue may increase the rate of photosynthesis in the remaining tissue. Basal intercalary meristems of grasses permit growth following grazing or other forms of foliage removal.

Climate

The North American grassland climates range from continental in the central grasslands, to mediterranean in the California and Palouse grasslands, and to dry subtropical in the desert grassland. All have distinct wet and dry seasons and are noted for temperature and precipitation extremes. The Great Plains climate has severe, windy, dry winters, with little snow accumulation, relatively moist springs, and summers that are often droughty and punctuated by thunderstorms (Borchert 1950). Except in the California grasslands and the Palouse prairie, approximately two-thirds of the precipitation in the North American grasslands falls during the growing season (Sims et al. 1978). Although fire, soils, and other factors are involved, a deficiency of rainfall late in the growing season tends to support a grassland vegetation in the central plains over the deciduous forests to the east.

The seasonal dynamics and movements of air masses (maritime polar, maritime tropical, continental polar, and continental tropical) over long distances and physiographic features (Thornthwaite 1933; Willet 1949; Brunnschweiler 1952; Harlan 1956) create a greater range of climates than for any other North American biome (Collins 1969).

The western cordillera constrains the flow of moist Pacific air masses across the continent. Consequently, the grasslands of the central region are primarily impacted by cold arctic air in winter and frequent incursions of tropical air in summer (Ripley 1992), creating wide temperature extremes and relatively low humidity.

The climate of the Great Plains generally follows a north-south and east-west pattern (Singh, Lauenroth, Heitschmidt, and Dodd 1983). Precipitation and relative humidity decrease, and solar radiation, rainfall variability, water stress, and potential evaporation increase from east to west. Air temperature, number of frost-free days, potential evapotranspiration, fraction of precipitation occurring in summer, and solar radiation all increase from north to south. The distribution of the grasslands roughly parallels a prevailing northeast-southwest precipitation gradient (Collins 1969).

Solar radiation across the North American grasslands ranges from 5×10^9 J m^{-2} yr^{-1} in the northern mixed-grass prairie to 8×10^9 J m^{-2} yr^{-1} in the desert grassland (Sims et al. 1978). Mean annual air temperatures vary from about 3° C in montane grassland, to 4–8° C in northern mixed-grass prairie, and to 15° C in tallgrass, southern shortgrass, and desert grasslands.

Average annual precipitation is greatest in the tallgrass prairie (100 cm) and decreases westward across the mixed-grass prairie (50 cm) and shortgrass prairies (30 cm) to a low in the southwestern desert grassland (20 cm). Precipitation equals or exceeds potential evaporation only in the eastern tallgrass prairie and in the montane grassland; elsewhere, evaporation exceeds precipitation. In the desert grassland, potential evaporation is four times greater than precipitation. The relatively high wind velocities across the plains major contributors to evaporative stress.

The length of the potential growing season (the number of consecutive days with a 15-day-running mean annual temperature $\geqslant 4.4°$ C) ranges from a minimum of 100 days in montane grassland to a maximum of 335 days in desert grassland (Sims et al. 1978). The growing season begins as early as mid-January in desert grassland and as late as June in montane grassland. In the mixed-grass prairie of the Great Plains, the growing season varies from 168 days in North Dakota to 226 days in central Kansas. The growing season in the central Great Plains shortgrass prairie averages 193 days, compared to about 270 days in the southern plains shortgrass prairie and the tallgrass prairie in northeastern Oklahoma.

Natural and aborigine-caused fires moved uninterrupted across the relatively level plains at sufficient frequency to restrict the occurrence of trees and shrubs. These grasslands had ample flammable plant material during the dry periods following years of adequate rainfall. Thus, the grasslands as the European settlers first saw them and as we know them today may be as much the result of recurrent fires (Gleason 1913, 1923; Sauer 1950; Curtis 1962; Axelrod 1985) as of climate (Clements 1916; Thornthwaite 1933; Borchert 1950; Coupland 1979).

The responses of grasslands to fire varies with climate and influences their distribution (Anderson 1982; Sauer 1950; Wright and Bailey 1982; Axelrod 1985). In addition, grassland distribution is also a function of the interaction of soils, topography, and herbivory on energy flow and nutrient cycling processes, precipitation-evaporation ratios, seasonal precipitation-temperature regimes, precipitation-soil interactions (McMillan 1959a; Risser et al. 1981; McNaughton, Coughenhour, and Wallace 1982), and the impact of both large prehistoric browsers (Axelrod 1985) and more recent domesticated livestock (Heady 1975; Anderson 1982).

Physiography

The central grasslands of the Great Plains are located in the Coastal Plains and Central Lowlands physiographic regions. These compose a vast plain, from the Mackenzie River delta in Canada to the Balcones Fault in south Texas which slopes gently downward to the east at a rate of about 1.2 m per kilometer from the base of the Rocky Mountains to the Central Lowlands on the banks of the Mississippi River in the upper Midwest (Hunt 1972; Lewis and Engle 1980).

The pristine grasslands of California occupied the central valleys of the Pacific Border physiographic province between the Coast Range to the west and the Sierra Nevada on the east (Austin 1965). These original grasslands also occupied a narrow coastal strip in central to northern California and a broader coastal zone in southern California now covered by the Los Angeles–San Diego urban area (Barry 1972).

The Palouse prairie of eastern Washington and Oregon and western Idaho lie, principally in the lava-rich Walla Walla section of the Columbia Plateau between the Rocky and Cascade mountains. Loess and ash deposits over the basalt plain have been moderately dissected by deep canyons along the major streams, with resulting hills and steep slopes (Austin 1965).

The desert grasslands are found on the Colorado Plateau and the southern and eastern extensions of the Basin and Range physiographic regions. They lie between the Colorado River in

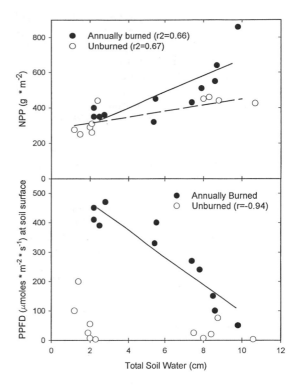

Figure 9.2. Relationships between net primary production (NPP) and sunlight reaching the soil surface (PPFD) and the total soil water content in annually burned and unburned tallgrass prairie watersheds (adapted from Knapp et al. 1993).

Arizona and the Pecos River in New Mexico and extend from the Rocky Mountains southward into Mexico (Hunt 1972; Western Land Grant Universities and Colleges and Soil Conservation Service, 1964). In Mexico, the desert grasslands are found in northern Sonora and along the east slope of the Sierra Madre Occidental in Chihuahua, Durango, and Zacatecas (Rzedowski 1978).

Landscape features greatly affect grassland function (Knapp 1984; Cale, Henebry, and Yeakley 1989; Schimel et al. 1991; Knapp et al. 1993). Understanding relationships between soil water, light interception, and net primary production (NPP) is key to predicting production across the topographically variable grasslands (Fig. 9.2). "Scaling-up" process-level measures, such as leaf gas exchange data, from a point measure to the landscape level is a formidable task, as the International Biological Program (IBP) discovered in the 1970s (Golley, 1993).

Soils

The diversity in grassland soils illustrates the influences of temperature and moisture on soil morphogenesis from diverse parent materials (Aandahl 1982; Looman 1980, 1981). Nearly all the wetter and cooler grasslands are associated with Mollisols, soils developed through processes unique to grasslands. Aridisols are the common soils of the arid grasslands (see Chapters 7 and 8). Entisol and Inceptisol soils have little or limited profile development and are often associated with Mollisols and Aridisols.

The dominant soil-forming process for Mollisols is melanization: darkening of the soil profile by addition of organic matter (Buol, Hole, and Mc-Cracken 1980), thus forming a mollic surface horizon that extends to varying depths depending on the amount of rainfall and temperature. Melanization results from root penetration into the developing soil and a partial decaying of this root material, leaving dark, relatively stable organic compounds (Hole and Nielsen 1968). Rodents, ants, earthworms, moles, and cicada nymphs rework the soil and organic matter and form characteristic dark soil organic matter complexes, krotovinas, and mounds. Through further eluviation and illuviation of organic and mineral colloids, the surfaces become coated with dark cutans (decay-resistant lignoprotein residues) that impart the dark colors that remain in many grassland soils even after long periods of disturbance.

Aridisol development is similar to that for Mollisols, except the reactions are less intense. Aridisols are dry most of the year, even when temperatures are adequate for plant growth (Buol et al. 1980). These soils form where potential evapotranspiration greatly exceeds precipitation during most of the year and very little if any water percolates through the soil (Buol et al. 1980). Leaching that may be evident in Aridisols was probably caused in periods of humid paleoclimates (Smith 1965).

Glacial deposits and associated outwash sands and gravels are the principal parent materials north of the Missouri River in the northern Great Plains. Soils derived from sandstone and shale are found south and west of the Missouri River in Montana and the Dakotas and in northeastern Wyoming and northwestern Nebraska. Extensive loess and eolian sand deposits are found in Nebraska, Kansas, eastern Colorado, southeastern Wyoming, the southern High Plains, and along major streams. Areas just east of the Rocky Mountains have soils that are deep, loamy sediments of loess, eolian sand, alluvium, and mountain-outwash origin. Large areas of fine-textured soils are found in the Texas panhandle and adjacent areas.

Soils of the southern plains vary from fine sands to clay soils. A strip of relatively fertile, medium-brown Alfisols stretches from southeastern Kansas

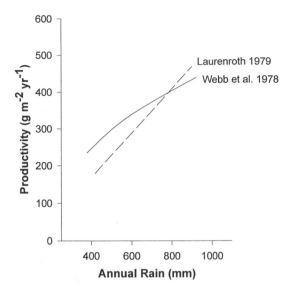

Figure 9.3. Relationship of net primary productivity and annual rainfall across grassland ecosystems (adapted from Lauenroth 1979 and Webb et al. 1978).

southward across east-central Oklahoma into central Texas, generally following the cross timbers vegetation type, as mapped by Küchler (1975). Alfisols are also found in much of the southern High Plains shortgrass prairie in west Texas and eastern New Mexico. Mollisols dominate the remainder of the southern plains. Mollisols are also the primary soils of the Palouse prairie, with Aridisols being an important secondary soil group. California grassland soils are principally Entisols, whereas the desert grassland soils are Aridisols.

Ecosystem Processes

The structural and functional organization of natural grassland ecosystems is primarily dominated by the interactive processes of carbon and nitrogen assimilation and allocation in relation to precipitation-evapotranspiration fluxes (McNaughton et al. 1982). Interactions among these processes are not always direct and may even be bidirectional. For example, grazers influence grassland plant responses to climate, and herbivores are impacted by changes in the vegetation. These interactions influence such fundamental mechanisms as carbon and nitrogen assimilation and efficiency of water use.

The relationship between rainfall and grassland production is generally linear (Fig. 9.3), and there is strong correlative evidence that water availability and use are the fundamental regulators of energy flow in grassland ecosystems (Lauenroth et al.

1979). The relationship tends to be asymptotic for forest ecosystems (Webb, Szarek, Lauenroth, Kinerson, and Smith 1978), probably because the more fertile grassland soils support plant productivity at higher levels of rainfall than do soils under forest ecosystems. In addition, herbivory has a much greater influence on energy and nutrient flow pathways in grasslands than in forests. The detritus pathway dominates material and energy flows in most terrestrial ecosystems where less than 5% of the flow goes through the grazing food web (Golley 1971; Weigert and Owen 1971), Grasslands, by contrast, have up to 60% of materials and energy flowing through the grazing pathway as in some African grassland ecosystems.

Grasslands occur in climates with high light intensity, warm temperatures, and at least one annual dry season. They have high leaf area indices (LAI) and are generally dominated by warm season C_4 grasses in the southern latitudes, a pattern attributable in part to their higher efficiency of water use and their relative tolerance of higher temperatures (Slack, Roughan, and Basset 1974; Teeri and Stowe 1976; Ehleringer and Bjorkman 1977; Ehleringer 1978; Risser 1985) (Fig. 9.4a, b). Grazing can also influence the relative proportion of C_3 and C_4 species. For example, grazing by domestic livestock in historic times has shifted the species composition from cool-season to warm-season species in the northern shortgrass and mixed-grass prairies, presumably because grazing resulted in a drier, warmer habitat suitable for the plants of more southerly origin (Sims et al. 1978; Smolik and Lewis 1982).

Knapp (1993) evaluated the energetic costs in different stomatal conductance responses between C_3 and C_4 grasses during periods of sunlight variability. The C_4 species responded more rapidly to short-term fluctuation (4–10 min intervals) in sunlight, thus reducing the loss of water via transpiration. In water-limited conditions, this benefit of the C_4 photosynthetic pathway may contribute to the relative success of C_4 grasses.

Grassland processes operate differently between local and regional scales, and this variability may significantly affect community heterogeneity. For example, grazing tallgrass prairie reduces the influence of the dominant plants and encourages colonization by other species (Glenn, Collins, and Gibson 1992). These plants migrate both from within the same community and from surrounding communities. Local origins are largely a function of distance to the parent plant; however, the processes that bring propagules from more distant locations include animal movements, wind speed and direction, and other factors such as the size of

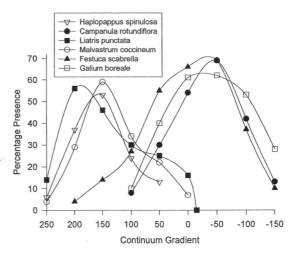

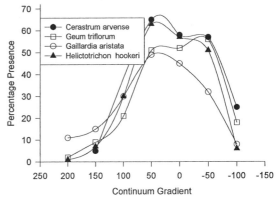

Figure 9.4. Distribution of plants across a synthetic moisture gradient for species with (a) different tolerances (partial and overlapping) and (b) similar tolerances to soil moisture (adapted from Looman 1980).

propagules and their propensity to attach to and detach from moving animals.

In established grasslands, competition between herbaceous and woody components may have different outcomes, depending on the climate dynamics and management treatments. For example, in relatively dry savannas competition between the woody overstory and the herbaceous understory reduces the productivity of the grassland component. However, understory production can increase along with the overstory when a more favorable microclimate occurs under the woody canopy or when increased nutrients are available from decomposed nutrient-rich tree leaves (Belsky 1994).

Temperature and precipitation patterns explain some grassland ecosystem processes (Sims and Singh 1978a, 1978b, 1978c). Grasslands with a mean annual temperature of 10° C or less are naturally dominated by cool-season grasses (C_3); those

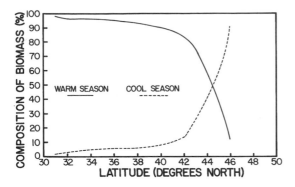

Figure 9.5. Changes in the relative compositions of warm-season and cool-season plants across the North American grasslands from the southern to the northern latitudes (based on data from Sims, Singh, and Lauenroth 1978).

Table 9.2. *Genera of cool-temperate and warm-temperate origins that are important in the grasslands of North America.*

Cool-temperate	Warm-temperate
Agrostis	Andropogon
Bromus	Bouteloua
Dactylis	Buchloë
Danthonia	Chloris
Deschampsia	Eragrostis
Elymus	Muhlenbergia
Festuca	Panicum
Hordeum	Paspalum
Nassella	Schizachyrium
Pascopyrum	Setaria
Poa	Sorghastrum
Pseudoroegneria	Sporobolus
Secale	
Stipa	
Triticum	

at southern latitudes, with mean annual temperatures higher than $10°$ C, are dominated by warm-season grasses (C_4) (Fig. 9.5 and Table 9.2). Cool-season forbs and shrubs compose 15–40% of the biomass in the cooler grasslands and warm-season forbs are similarly important in the southern grasslands. The factors controlling cool-season and warm-season plant production seem to be long-term mean annual temperature, growing season precipitation, annual or growing season usable solar radiation, and annual actual evapotranspiration or growing season actual evapotranspiration (Sims et al. 1978). The accuracy of these simple or multiple regression relationships under changing climate or with the invasion of alien species will require additional investigation (Seastedt, Coxwell, Ojima, and Parton 1994).

Ecosystem processes are a function of the interactions between abiotic and biotic components. For example, soil organisms play a significant role in carbon and nutrient dynamics (Anderson, Hetrick, and Wilson 1994). On a global basis, soil organisms are thought to store as much as 1.5% of the carbon and 3.0% of the nitrogen within terrestrial ecosystems (Wardle 1992). The metabolism of soil organisms is constrained by carbon inputs from plant litter production. On an annual basis, we would expect that all labile carbon from plant litter and in the mineral soil should be metabolized and that there would be little or no net annual growth of soil microorganisms. Zak et al. (1994) found, across broad spatial scales and several types of ecosystems, that annual aboveground net primary productivity was significantly related to labile soil organic carbon pools and microbial biomass. Soil carbon storage is generally at or near equilibrium, and net carbon assimilation in primary production approximately equals the amount of carbon enter-

ing the soil in late successional communities. In the shortgrass prairie, where 21% of the soil carbon was labile, carbon metabolism is probably limited by soil water potential. A more mesic climate would stimulate microbial activity to process soil carbon more rapidly.

The nitrogen cycle in grasslands is relatively well known, especially how grazing intensity affects nitrogen availability for plant growth (Shariff, Biondini, and Grydiel 1994). Denitrification rates in grasslands are affected both by site conditions and by management techniques such as fertilization and burning. On fertile sites, denitrification rates can reach 1 gm m^{-2} yr^{-1}, a rate nearly equal to precipitation inputs minus losses associated with volatilization during the burning of grasslands (Ojima, Parton, Schimel, and Owensby 1989). Groffman, Rice, and Tiedje (1993) found that unburned tallgrass prairie was wetter and had higher concentrations of nitrate in soil solution than did burned sites. Denitrification rates were higher in unburned sites than in burned, burned and grazed, or cultivated sites. There is, however, great temporal and spatial variability in denitrification rates, and future research is needed to confirm the expected relationships between denitrification and site characteristics and management. This is especially true if denitrification rates are to be extrapolated to regional and higher scales.

Landscape and Global-Scale Processes

Understanding the ecological implications of spatial variation in grassland ecosystems is critical to effective management. Knapp et al. (1993) and

Schimel et al. (1991) examined variation of above-ground biomass, leaf-level net photosynthesis, light interception, leaf nitrogen, soil water availability, and plant water stress across landscapes. Knapp et al. (1993) found less variability in net primary production (NPP) across topographic gradients in the unburned tallgrass prairie compared to the burned prairie. The greater homogeneity in NPP was correlated with increased litter accumulation on unburned sites that both absorbed and reflected sunlight.

At the global scale, differential species response to increased atmospheric CO_2 levels might be expected to cause changes in the composition of grasslands. These interpretations are complicated by species interactions and competitive pressures. When honey mesquite (*Prosopis glandulosa*) was grown alone, higher levels of CO_2 stimulated nitrogen fixation, belowground biomass, and efficiencies in whole-plant water and nitrogen use (Polley, Johnson, and Mayeux 1994). When grown with little bluestem (*Schizachyrium scoparium*, honey mesquite belowground biomass, nitrogen fixation, and nitrogen-use efficiency were not affected by increased levels of ambient CO_2. Thus, the response of one species to changes in CO_2 may be strongly affected by competitors. In other ecosystems, the effects of enriched CO_2 environments has been transient, with ecosystem processes returning to near original conditions after one or more seasons. Nevertheless, grassland soils could operate as large sinks for carbon, a process of increasing significance with increasing global atmospheric CO_2 levels (Seastedt and Knapp 1993).

Future study of North American grasslands will focus on the variability in ecosystem processes across broader spatial scales and longer temporal periods. Further work is needed to address the interactions between grasslands and atmospheric processes, especially under various scenarios of climate change.

Classification

The North American grassland formation has been classified as the *Stipa-Bouteloua* formation (Clements 1920; Weaver and Clements 1929, 1938) and as the *Stipa-Antelocapra* biome (Clements and Shelford 1939). The grassland formation has been further subdivided into the following:

1. *Stipa-Bouteloua* mixed prairie association that includes the *Buchloë-Bouteloua* short grass plains.
2. *Stipa-Sporobolus* true prairie association that includes the *Andropogon* subclimax prairie.
3. *Stipa-Andropogon* coastal prairie association;

Elymus-Pseudoroegneria-Pascopyrum-Festuca, bluestem prairie association.
4. *Stipa-Poa*, California prairie association.
5. The *Aristida-Bouteloua* desert grassland association.

Clements and his colleagues developed descriptions that encompassed all the grasslands of the western United States. For example, Carpenter (1940), who limited the grassland biome in North America to the plains east of the Rocky Mountains, called the grasslands the *"Andropogon-Bouteloua-Bison-Cannus"* biome and divided it into three associations: the *Andropogon-Bison-Cannus* tallgrass prairie; the *Andropogon-Bouteloua-Bison-Antelocapra* mixed-grass prairie association; and the *Bouteloua-Buchloë-Bison-Antelocapra* shortgrass association.

These two classifications differ primarily on whether the shortgrass plains were an overgrazed mixed prairie or an association of its own. Clements included the shortgrass prairie in the mixed-grass prairie and believed that the shortgrass dominance came with overgrazing. Carpenter, on the other hand, believed that the shortgrass prairie was a component of the presettlement vegetation under bison grazing. Most probably, there are areas of shortgrass prairie because of soils and climate and other sites that are indeed overgrazed mixed-prairie grasslands.

Nomenclature in this chapter follows Kartesz (1994).

The North American grasslands are bound together by a small array of perennial grass species that occur in at least three or more associations. Clements and Shelford (1939) listed these widespread species in order of decreasing range:

Sporobolus cryptandrus
Koeleria macrantha
Stipa comata
Nassella viridula
Pascopyrum smithii
Bouteloua gracilis
Bouteloua curtipendula
Bouteloua hirsuta
Elymus elymoides
Poa secunda
Festuca ovina, Schizachyrium scoparium, and *Buchloë dactyloides*

The numbers of plant species that occur in grasslands increase as the growing-season environment becomes more mesic, where topographic variations increase, and in some cases, where humans have had the least impact (Coupland 1979). Coupland, Ripley, and Robbins (1973) recorded 50 vascular species in a temperate, semiarid, nearly level grassland. Steiger

(1930) found more than 200 species in a subhumid prairie. Over 330 plant species contributed significantly to the biomass structure of the 10 western North American grasslands (Sims et al. 1978) studied during the IBP's Grassland Biome Project (Van Dyne 1971; Breymeyer and Van Dyne 1980). Generally, forbs outnumbered grasses by three- to four-fold, and shrubs, half-shrubs, succulents, and trees were minor components that provided important differentiating features to the grasslands.

The central grasslands of North America have few unique taxa. Wells (1970) reported no plant families and few genera and species (and these were primarily forbs) endemic to the central plains. The same is true for insects and birds (Axelrod 1985; Bolen and Crawford 1996). In contrast, numerous endemic plant and animal species occur in the forests bordering grasslands and the deserts to the south. The lack of endemism is attributed primarily to the youthfulness of these grasslands, that only attained their present distribution in the postglacial period (Sears 1935, 1948; Schmidt 1938; Hartley 1950; Axelrod 1985), developing primarily during the late Miocene or Pliocene (Dix 1964). The dominant grasses had Arcto-Tertiary (*Elymus, Pseudoroegneria, Pascopyrum, Koeleria,* and *Poa*), Neotropical (*Bouteloua, Sporobolus, Stipa,* and *Hilaria*), and Madro-Tertiary (*Stipa*) origins. *Andropogon, Panicum, Sorghastrum,* and *Liatris* genera had their origin in Arcto-Tertiary forests but appeared as remnants of the eastern deciduous forest somewhat later.

Origin and history of the North American grasslands vegetation is complicated because of the dynamics of its climate (Dix 1964). The widely fluctuating Pleistocene climates resulted in the migration of vegetational units throughout the grasslands, particularly in the eastern regions of the Great Plains. Even at an earlier time, after the "early Rocky Mountains had been periplained" (Dix 1964), before the Cascadian uplift, a more uniformly warm and humid climate from the Pacific may have encouraged east-west vegetation migrations. During the Ice Age, flora of the grasslands migrated, mingled, competed, and evolved, and some became extinct. Consequently, the vegetation history of the grasslands is complex and obscure.

Paleobotanical evidence indicates a gradual shift in vegetation in the central plains between the middle Miocene and the early Holocene, from a largely semiopen forest and woodland with scattered grassy areas, to an open grassland with trees and woodlands limited to breaks and escarpments (Axelrod 1985). The shift to grasslands occurred as aridity increased and drought became more frequent west of the hundredth meridian.

Although the extent and nature of prehistoric animal populations are largely unknown, large numbers of herbivores were present in the North American grasslands at the time of settlement by Europeans (Maller-Beck 1966; Sims, Soseby, and Engle 1982). Large herds of migratory bison, pronghorn, elk, and deer and vast colonies of nonmigratory prairie dogs were present, and they had a significant impact on local areas at times (Nelson 1925; Seton 1927; Larson 1940; England and DeVos 1969; Vankat 1979; McDonald 1981). The large herbivores migrated in search of green forage and in response to patterns of precipitation, drought, and fire. The impact of the migrating pristine herbivores was intermittent, allowing rest between grazing periods (Curtis 1962; Heady 1975). For example, bison used a particular area for a short period, then moved to a new area – a pattern that became more or less repetitive from year to year (Heady and Child 1994). This relatively fixed pattern of migration exerted seasonal grazing pressures on the vegetation, and the plant species evolved with these pressures through natural selection.

MAJOR GRASSLAND TYPES

For convenience, grassland vegetation has been abstracted into discrete communities such as tallgrass, mixed-grass, and shortgrass prairies, and others. The east-west precipitation gradient of central North America, overlain by a north-south temperature gradient, gives rise to diversity in soils (Jenny 1930) that supports floristically and functionally complex plant communities. The abiotic gradients that exist across the landscape are inhabited by genetically specialized plants that match the constraints of the environment (Beetle 1947; McMillan 1959b; Transeau 1935; Harlan 1956; Axelrod 1985).

Plains Grasslands

Tallgrass prairie. The tallgrass prairie is the most mesic of the central plains grasslands. It has the greatest north-south species diversity and has more dominant species than any other grassland formation (Risser et al. 1981). Although the tallgrass prairie is relatively homogeneous (Fig. 9.6), its landscape varies with changing climate and soils (Weaver 1954). Three grassland associations are commonly differentiated within the tallgrass prairie (Küchler 1975). The bluestem, or "true," prairie extends from the southern tip of Manitoba through eastern North Dakota and western Minnesota southward to eastern Oklahoma. It is dom-

Figure 9.6. A tallgrass prairie site in the Flint Hills of Kansas showing the grassland vegetation interspersed with trees and shrubs along the breaks.

inated by *Andropogon gerardii, Panicum virgatum,* and *Sorghastrum nutans.* A second community, dominated by *Elymus, Pseudoroegneria, Pascopyrum, Andropogon,* and *Nessella* (formerly *Stipa*) species, originally occurred from south-central Canada down through east-central North and South Dakota and Nebraska to north-central Kansas. A third community occurs in the Nebraska Sandhills, dominated by *Andropogon, Calamovilfa,* and *Nessella* species. Bailey (1976, 1980) combined the two latter communities to form a wheatgrass, bluestem, and needlegrass prairie.

Andropogon gerardii dominates the tallgrass prairie, particularly on the lowlands and other wetter sites (Weaver 1954). *Schizachyrium scoparium* dominates the uplands, especially on the more shallow slopes, whereas *Stipa spartea* Trin. and *Sporobolus heterolepis* (Gray) Gray are important on upland associations where soils are more shallow, rocky, or sandy. *Sporobolus compositus* is an important intermediate-height grass, especially in grazed areas.

Küchler (1975) identified two grassland-forest ecotone communities east of the tallgrass prairie; (1) the upper Midwest *Quercus–Andropogon* type in North Dakota and around the Great Lakes and (2) the *Juniperus–Quercus–Sporobolus–Andropogon* type of west Tennessee, Alabama, Missouri, and Arkansas. Grimm (1983) described climate changes and fire frequencies that trigger transitions between prairie and woodland (Fig. 9.7a). A reduction in fire frequency lowers the degree of climatic change needed for established woodland to persist. Tree seedlings are more susceptible to fire than mature trees, and the oak woodland is less flammable than prairie; therefore, it is less likely to burn under similar climatic conditions. The thresholds for the changes from oak woodland to prairie and from a

tallgrass prairie to oak woodland change over time (Fig. 9.7b). Oak woodland persists above a certain precipitation level, whereas only the prairie vegetation persists below a certain level of precipitation (Grimm 1983).

Historically, fire suppressed the encroachment of trees and shrubs, reduced competition from cool-season invaders such as *Poa pratensis* and *Bromus inermis,* and improved the palatability and nutritional value of the grazable forage (Wright and Bailey 1980). Late spring burning generally benefits the major tallgrasses; forbs may be reduced in growth, but their diversity is unaffected (Anderson 1965; McMurphy and Anderson 1965; Kucera 1970). Long-term protection of the tallgrass prairie from fire has led to an increase in woody vegetation (Kucera 1960; Penfound 1964; Ojima et al. 1989).

Most of the tallgrass prairie canopy is less than 1 m tall. The leaf area for live tissue peaks at about 1.3 cm^2 cm^{-2} in the middle of the growing season (Fig. 9.8). There is a large amount of standing dead material in the canopy at all times – as much as 1.8 cm^2 cm^{-2} in the ungrazed tallgrass prairie of northeastern Oklahoma (Conant and Risser 1974).

The tallgrass prairie remains as important grasslands in the Osage and Flint Hills of Oklahoma and Kansas, in the Nebraska Sandhills, and in isolated locations throughout the Central Lowlands geographical region. Isolated tracts of tallgrass prairie are preserved in the Great Plains because topography or rockiness prevents farming. Watershed, wildlife, and domestic livestock grazing are the primary uses. Most users and ranchers take pride in sustaining this grassland vegetation similar to what it was prior to settlement of the Great Plains.

Most of the original tallgrass prairie is now un-

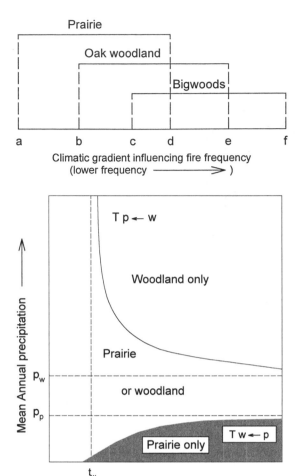

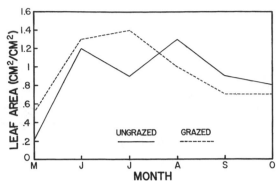

Figure 9.8. Seasonal progression of the actual leaf area during the growing season for ungrazed and grazed tallgrass prairie (adapted from Conant and Risser 1974).

Figure 9.7. (a) Diagram of the ranges over which prairie, oak woodland, and bigwoods communities persist on a one-dimensional climate gradient that influences fire frequency, and (b) thresholds of fire frequency and its influence on changes from oak woodland to prairie and from tallgrass prairie to oak woodland across a two-dimensional gradients of mean annual precipitation and time (adapted form Grimm 1983).

der cultivation. Earlier studies of the tallgrass prairie following cultivation generally found secondary succession to go through a pioneer weed, annual grass, bunchgrass, to a mature prairie stage (Booth 1941; Perino and Risser 1972). Collins and Adams (1983) followed 32 yr of secondary succession on a central Oklahoma tallgrass prairie that was either protected from grazing, plowed once in the fall of 1949, or plowed for 5 consecutive years beginning in 1949. In these conditions, succession was less predictable and species composition on the three treatments was heterogeneous. In the absence of fire, instead of a mature prairie, the site

became dominated by shrubs and trees. As more successional sequences are examined, it appears that the process of succession is complex and more difficult to predict (Glenn-Lewin 1980) than earlier studies had indicated.

Livestock preferentially select the most palatable species when grazing and are set in motion by competitive processes (Dyksterhuis 1949; Voight and Weaver 1951; Sims and Dwyer 1965). In general, the more palatable plants may decrease from 60–90% cover to < 2% with excessive livestock grazing. Schizachyrium scoparium, Andropogon gerardii, Sorghastrum nutans, and Panicum virgatum were the principal grasses in tallgrass prairie that was in excellent condition (Fig. 9.9). With overgrazing, Panicum virgatum and Sorghastrum nutans were the first to decline, followed by Schizachyrium scoparium. Andropogon gerardii, the second most abundant species in well-managed grasslands, persisted even on some of the most overgrazed sites, probably because of its short, strong rhizomes. A. gerardii declined more with mowing for hay production than from heavy grazing. In contrast to Andropogon gerardii, Sorghastrum nutans seems to be favored by mowing and reduced by grazing.

The less palatable species increased from about 15% by three-to sixfold with heavy grazing. Species of intermediate palatability then declined to 10% cover as grazing pressure increased. Invader plants, those that were not present originally, increased from 2–47% on these overgrazed prairies. Responses of annual forb populations to grazing pressure are somewhat more erratic and fluctuate more than the perennial grasses as current moisture conditions change. Mulch cover declines with grazing intensity, resulting in less rainfall infiltration and more runoff. Consequently, with equal amounts of rainfall, soil moisture conditions are

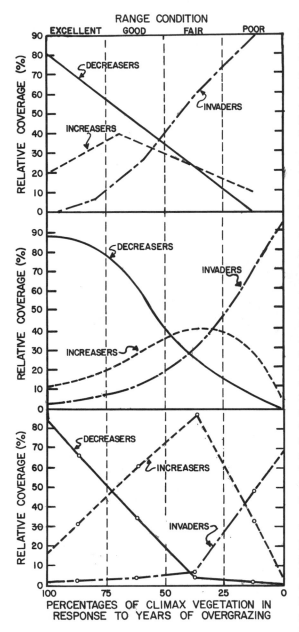

Figure 9.9. Relative changes in species compositions for (top to bottom) southern (Fort Worth Prairie) (Dyksterhuis 1949), central (north central Oklahoma) Sims and Dwyer 1965), and northern (eastern Nebraska) (Voight and Weaver 1951) tallgrass prairie sites.

lower on heavily grazed compared to properly grazed tallgrass prairie.

Generally, *Bouteloua curtipendula* and *B. gracilis* became more abundant as the tallgrasses declined. *Buchloë dactyloides*, which is rare on most pastures in excellent condition, increases to as much as 70% relative cover on the overgrazed tallgrass prairie.

Changes in the Nebraska sandhills tallgrass prairie are generally less marked than in other areas. The fragile, sandy soils that are susceptible to wind erosion may have cautioned ranchers to avoid overgrazing. Changes that have occurred in the sandhills under grazing include decreases in *Schizachyrium scoparium*, *Stipa comata*, and *Andropogon gerardii* and increases in *Bouteloua hirsuta*, *Calamovilfa longifolia*, and *Sporobolos cryptandrus*.

The Cross Timbers (Küchler 1975), often associated with the tallgrass prairie, lies in a band 10–180 km wide from the southern edge of the bluestem prairie in Kansas southward across east-central Oklahoma to the Trinity River in east Texas. The overstory vegetation consists primarily of *Quercus stellata* and *Q. marilandica* with the former frequently more common than the latter. Grazing, erosion, and reduction of fire have drastically changed the vegetation of Cross Timbers. In the undisturbed Cross Timbers, Dyksterhuis (1948) found an understory of *Schizachyrium scoparium* (65%), *Sorghastrum nutans* (6%) with 2–3% each of *Andropogon gerardii*, *Bouteloua hirsuta*, *B. curtipendula*, and *Sporobolus asper*. He reported that much of the Cross Timbers had been burned. Basal plant cover increased 4–7% following grazing and burning, mostly from the invasion of annual grasses and forbs and warm-season perennial grasses such as *Buchloë dactyloides*, *Aristida* species, *Paspalum setaceum*, *Nasella leucotricha*, *Bothriochloa saccharoides*, and *Cynodon dactylon*. The original tallgrasses were virtually eliminated, *Bouteloua hirsuta* increased, and *B. curtipendula* remained as in the original undisturbed stands.

The Blackland Prairie (Fig. 9.10) intermingles with the Cross Timbers at the northern end of this prairie. Although often associated with the tallgrass prairie (Gould 1962), its soils are Vertisols rather than Mollisols. These immature soil profiles overlie soft limestone parent material. This prairie has been largely converted to farmland, although a few relicts remain in wildlife habitat and on ranches. *Schizachyium scoparium* is a community dominant along with *Andropogon gerardii*, *Sorghastrum nutans*, *Panicum virgatum*, *Bouteloua curtipendula*, *B. hirsuta*, *Sporobolus asper*, *Bothriochloa saccharoides*, and *Nassella leucotricha*. Heavy grazing causes *Nassella leucotricha* to increase and *Buchloë dactyloides* and *Bouteloua rigidiseta* to invade. Annual grasses such as *Hordeum pusillum*, *Bromus japonicus*, *B. catharticus*, and *Vulpia octoflora* invade much the same as in the Cross Timbers.

Coastal prairie. A crescent-shaped coastal prairie occurs along the Gulf of Mexico from southwestern Louisiana to the Mexican border near Brownsville,

Figure 9.10. The level Blackland Prairie of south Texas.

Texas, varying from 40 to 160 km in width in the Coastal Plain physiographic province (Hunt 1967; Lytle 1968; Godfrey McKee, and Oakes 1973). The primary soils are Vertisols along the Texas coast and Alfisols in Louisiana and inland from the coast. Coastal prairie soils are poorly drained because of limited relief and dense clay subsoils. Consequently, they are waterlogged much of the winter and even may be supersaturated following heavy rains.

The warm, moist, tropical climate, dominated by southeasterly winds off the Gulf of Mexico, has precipitation that ranges from over 1400 mm in Louisiana to half that in southeastern Texas. The precipitation is seasonal and ranges from summer rains in the upper coast to a late spring, early autumn mix in the southern Coastal Prairie. Relative humidity is 55–75% (Thornthwaite 1948; Carr 1967).

The principal vegetation of the coastal prairie is medium to tall grasses, with a mixture of woody vegetation along the rivers and streams (Smeins, Diamond, and Hanselka 1992). Trees and shrubs may have covered much of this area since the Pleistocene (Bogush 1952). Others argue that trees and shrubs have increased significantly since settlement (Bray 1906; Tharp 1926; Lehmann 1965; Scifres 1980). Because little of the coastal prairie remains undisturbed, it is difficult to determine

the nature of the original vegetation. Smeins et al. (1992) concluded that the species richness in this prairie was originally very high, with more than 70 vascular plant species per ha. *Schizachyrium scoparium* is the dominant grass, followed by *Sorghastrum nutans* and *Paspalum plicatulum* Michx.

In the lowland coastal prairie, with only a slight drop in elevation, *Panicum virgatum* and *Tripsacum dactyloides* (L.) L. make up to 60% of the biomass. *Paspalum floridanum* Michx. and *S. nutans*, along with *Andropogon gerardii*, are also present in the better drained sandy areas. *Acacia farnesiana* (L.) Willd., *Prosopis glandulosa*, and *Quercus virginiana* P. Mill. were the original woody dominants that occurred in a narrow band along the many streams (Inglis 1964).

Mixed-grass prairie.   The mixed-grass prairie, recognized as a distinct plant association by Clements (1920), is a blend of tallgrass and shortgrass prairie vegetation. The boundaries of this grassland are less well defined, because it is a broad and historically shifting ecotone. The northern Great Plains mixed-grass prairie lies to the west of the tallgrass prairie in western North and South Dakota, northeastern Wyoming, eastern Montana, and into the south-central Canadian provinces (Fig. 9.11). A broad belt of mixed-grass prairie extends south-

Figure 9.11. A mixed-grass prairie site near Antelope, South Dakota, where the overstory of cool-season grasses hides the shorter warm-season grasses.

Figure 9.12. Gently rolling sandhills mixed-grass prairie in northwestern Oklahoma that has been grazed at a moderate grazing intensity for over 50 yr.

west from the Nebraska sand-hills through western and central Kansas and Oklahoma (Fig. 9.12) to central Texas.

The mixed-grass prairie has the richest floristic complexity of all the grasslands (Barbour et al. 1980) with an array of tall, intermediate, and short grasses, a large number of forbs, a few suffrutescents, and a scattering of low-growing shrubs. Typical dominant species are *Schyzachrium, Stipa, Elymus, Pseudoroegneria, Pascopyrum, Calamovilfa, Bouteloua,* and *Sporobolus.* Primary shortgrass species are *Bouteloua, Buchloë, Muhlenbergia,* and *Aristida.* Sedges (*Carex* spp.) are also important in the mixed-grass prairie.

In North Dakota, the mixed-grass prairie is dominated by *Bouteloua gracilis, Schizachyrium scoparium, Stipa comata, Pascopyrum smithii, Carex filifolia, Koelaria macrantha,* and *Poa secunda* (Whitman 1941).

Redmann (1975) grouped mixed-prairie stands in excellent condition into three communities corresponding to soil type, topography, and position along a moisture gradient. The rolling upland with fine-textured soils was dominated by *Pascopyrum smithii, Carex pensylvania,* and *Stipa comata.* Stands at lower topographic positions on medium-textured soils were dominated by *Sporobolus heterolepis,* whereas on sites at higher topographic positions on coarse-textured soils *Schizachyrium scoparium* dominated. In west-central Kansas, Albertson (1937) identified three distinct mixed-prairie communities, with *Bouteloua gracilis–Buchloë dactyloides, Schizachyrium scoparium,* or *Andropogon gerardii* dominating across a habitat gradient from dry upland, to more mesic slopes, down to a well-watered, lowland site.

The seasonal distribution of live biomass (Table 9.3) depends on the mix of cool-season (C_3) and

Table 9.3. *Comparison of the relative contributions (%) of warm- and cool-season grasses to season-long biomass production in ungrazed and grazed areas for three mixed-grass prairies in the central grasslands of North America.*

| Prairie site | Grasses | | | | Forbs and other plants | |
| | Warm-season | | Cool-season | | | |
	Ungrazed	Grazed	Ungrazed	Grazed	Ungrazed	Grazed
Dickinson, North Dakota	17	27	58	45	25	28
Cottonwood, South Dakota	19	74	78	18	3	8
Hays, Kansas	85	80	<1	4	15	16

Note: Primarily warm-season forbs.

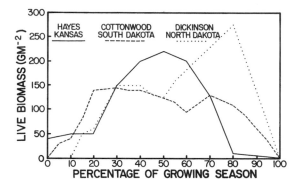

Figure 9.13. *Distribution of seasonal live biomass at three locations of the mixed-grass prairie in the central and northern Great Plains (based on data from Sims and Singh (1978a).*

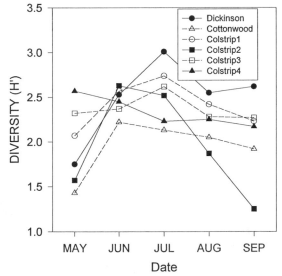

Figure 9.14. *Seasonal diversity indices, based on live and current dead biomass, throughout the growing season for the northern mixed prairie (adapted from Singh et al. 1983).*

warm-season (C_4) plants (Sims et al. 1978; Singh et al. 1983). A single peak in live biomass dominated by warm-season plants occurred in central Kansas (Fig. 9.13), whereas in North and South Dakota, a bimodal pattern emerged from first the cool-season and then the warm-season plants (Sims and Singh 1978b). Maximum plant diversity occurs at midseason when the biomass is more equitably distributed among both warm-season and cool-season species (Fig. 9.14).

Ode, Tieszen, and Lerman (1980) found 27 C_4 and 278 C_3 species in the mixed Ordway prairie in north-central South Dakota. The productivity of C_3:C_4 plants—as measured by delta ^{13}C values (Tieszen, Hein, Qvortrup, Troughton, and Imbamba 1979; Tieszen, Senyimba, Imbamba, and Troughton 1979) early, midway, and late in the growing season for an upland site—were about 90:10, 55:45, and 85:15, respectively. Corresponding figures for a lowland site were 100:0, 85:15, and 95:5.

The mixed-grass prairie is the most dynamic of the central grasslands because of the climatic extremes across this layered vegetation of diverse origins. The mixed-grass prairie is a tension zone between the tallgrass and shortgrass prairies, with the vegetation fluctuating with prevailing climate, suppression of fire, and the degree and frequency of grazing by domestic livestock and wildlife. As the climate becomes drier and grazing further increases the aridity of the environment, the shorter, more drought-tolerant species increase in the plant community (Albertson 1937). The uplands are co-dominated by the sod-forming *Bouteloua gracilis* and *Buchloë dactyloides* during more normal climates (Table 9.4). Associated secondary taller

Table 9.4. *Dominant, principal, and secondary species for short grass, little bluestem, and big bluestem habitat types in the mixed-grass prairie in west-central Kansas.*

Species type	Short grass	Little bluestem	Big bluestem
Dominants	*Bouteloua gracilis* *Buchloë dactyloides*	*Schizachyrium scoparium*	*Andropogon gerardii* *Pascopyrum smithii* *Bouteloua curtipendula* *Sporobolus compositus*
Principal grass and sedge	*Pascopyrum smithii* *Schizachyrium scoparium* *Aristida purpurea* *Bouteloua curtipendula* *Carex praegracilis* *Elymus elymoides*	*Andropogon gerardii* *Bouteloua curtipendula* *Bouteloua gracilis* *Bouteloua hirsuta* *Panicum virgatum* *Sorghastrum nutans*	*Bothriochloa saccharoides* *Carex gravida* *Elymus canadensis* *Elymus virginicus* *Panicum virgatum* *Poa arida* *Sorghastrum nutans*
Grasses of secondary importance	*Alopecurus carolinianus* *Andropogon gerardii* *Distichlis spicata* *Vulpia octoflora* *Hordeum pusillum* *Munroa squarrosa* *Sporobolus compositus* *Sporobolus cryptandrus*	*Buchloë dactyloides* *Sphenopholis obtusata* *Vulpia octoflora* *Koeleria macrantha* *Sporobolus compositus* *Sporobolus cryptandrus* *Erioneuron pilosum*	*Bouteloua gracilis* *Buchloë dactyloides* *Sporobolus cryptandrus*
Principal forbs	*Ambrosia psilostachya* *Anemone caroliniana* *Antennaria neglecta* *Aster ericoides* *Astragalus spp.* *Cirisium undulatum* *Gaura coccinea*	*Ambrosia psilostachya* *Amorpha canescens* *Echinacea pallida* *Liatris punctata* *Calylophus serrulatus* *Psoralidium tenuiflorum* *Tetraneuris scaposa*	*Amorpha canescens* *Aster ericoides* *Erigeron strigosus* *Psoralidium tenuiflorum* *Salvia azurea* *Verbena stricta* *Vernonia baldwini*

grasses, such as *Pascopyrum smithii*, *Schizachyrium scoparium*, and *Bouteloua curtipendula*, form a distinct middle layer but are rarely abundant.

Drought reduces the density and cover of the shortgrasses by as much as 70–80% and of the bunchgrasses and many of the native forbs to almost zero (Albertson 1937). When normal precipitation returns, the more aggressive stoloniferous *Buchloë dactyloides* assumes a more dominant position than the slower recovering *Bouteloua gracilis*. *Sphaeralcea coccinea*, an important forb of the shortgrass habitat in the mixed-grass community, is tolerant of drought. During dry cycles, however, most native forbs are readily replaced by an array of weedy and invasive forbs.

The *A. gerardii* community in the mixed-grass prairie is closely allied to the tallgrass prairie (Weaver and Albertson 1956). It occurs at the base of the slopes and in ravines and valleys where soil moisture is enhanced by runoff from the uplands and slopes, where snow accumulates, and where mois-

ture losses are reduced because of decreased wind (Weaver and Albertson 1956). *A. gerardii* composes 50–90% of the vegetation, with *S. nutans* being a common but not abundant associate. *P. virgatum*, *Elymus canadensis* L., and *E. virginicus*, along with the tall sedge *Carex gravida*, were often intermixed but were not abundant except on the wetter sites. Many large forbs are also present (Table 9.4). Insufficient light penetrates the dense overstory of the tall species to allow the shortgrasses to exist, even though the basal cover of the tallgrass prairie is only 8–13%.

The major vegetation changes in the mixed-grass and shortgrass prairies are associated with normal climate dynamics, especially drought, grazing, with fire of secondary importance. During the severe droughts of the 1930s and the 1950s, the basal cover of grasses on even moderately grazed mixed prairie declined from 80% to less than 10% in 3–5 yr. The perennial grasses were replaced by annuals such as *Salsola kali*, which dominated cer-

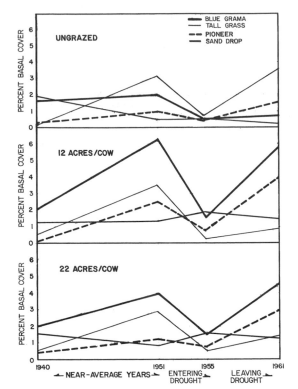

Figure 9.15. Changes in an Oklahoma mixed-grass prairie in relation to changing climatic conditions over time and grazing intensity (McIlvain and Shoop, unpublished data).

tain sites (E. H. McIlvain and M. C. Shoop, unpublished data). Following drought, the basal cover of the tallgrasses *Bouteloua gracilis* and *Paspalum setaceum* declined as *Sporobolus cryptandrus*, an aggressive drought-tolerant plant, increased (Fig. 9.15). Thus, the dynamics of species changes in the mixed-grass prairie are functions of climate, but the magnitudes of these changes are greatly influenced by the intensity of grazing.

McMillan (1956a, 1956b, 1957, 1959a), Olmsted (1944), Larsen (1947), Riegel (1940), and Rice (1950) identified some relationships among vegetation genetics, habitat gradients, and climate. The photoperiodic responses of a wide array of clones of warm- and cool-season grasses collected from Canada to Mexico were evaluated, and the general flowering and growth characteristics were defined.

The most prevalent pattern was the earlier flowering of northern clones, followed by progressively later flowering of clones collected farther south and east. Major species with this flowering pattern were *Panicum virgatum, Schizachyrium scoparium,* the *Andropogon gerardii–hallii* complex, *Sorghastrum nutans, Sporobolus heterolepis, Koeleria macrantha, Bouteloua gracilis,* and *B. curtipendula. Elymus canadensis* essen-

tially had a reverse flowering pattern. *Stipa spartea, S. comata,* and *Oryzopsis hymenoides* displayed no geographically oriented flowering trends.

Fire has been used to increase forage production and the palatability of coarse grasses in the mixed-grass prairie of the southern plains (Wright and Bailey 1980). Planned fires are also used to reduce undesirable annual grasses and forbs and to suppress *Prosopis glandulosa* var. *glandulosa, Juniperus virginiana,* and *Opuntia* species. Most grasses, except *Bouteloua curtipendula* and the cool-season *Poa* species, appear to be relatively tolerant of prescribed fires (Wright 1974). In the northern mixed-grass prairie, early spring burns may enhance stands of *Andropogon gerardii* and *Schizachyrium scoparium* and certain forbs (Gartner and Thompson 1972). Fires later in the growing season reduced both warm- and cool-season grasses (Schripsema 1977). The *Andropogon*-forest ecotone in the South Dakota Black Hills may be more tolerant of fire than other northern plains communities. Winter burns had the opposite effect, increasing the cool-season *Pascopyrum smithii* and *Stipa* species at the expense of the warm-season grasses.

Shortgrass prairie. The shortgrass prairie is a large contiguous expanse of vegetation that lies to the east of the Rocky Mountains from the Nebraska panhandle and southeastern Wyoming through eastern Colorado and western Kansas southward through the High Plains of Oklahoma, Texas, and New Mexico. A remnant of the shortgrass prairie is also found in northern Arizona (Stoddart et al. 1975). The intermediate-height grasses of the mixed-grass prairie are secondary in importance, and only a sprinkling of tallgrasses can be found on wetter sites.

The shortgrass prairie has been called a grazing disclimax of the mixed-grass prairie (Weaver and Clements 1938; McComb and Loomis 1944) or a true climax shortgrass prairie (Larson 1940). Sampson (1950) indicated that, although grazing had an impact on the distribution of the shortgrass prairie in the more mesic areas, the drought-enduring shortgrasses were climax over vast areas of the Great Plains. As much as 7 M ha of shortgrass prairie were originally distributed across the eastern plains and mesas of New Mexico (Dick-Peddie 1993). *Bouteloua gracilis* and *Buchloë dactyloides* dominate the shortgrass prairie (Fig. 9.16), and *Pascopyrum smithii, Sporobolus cryptandus, Muhlenbergia torreyana, Stipa comata, Koeleria macrantha,* and *Hilaria jamesii* are also important components. In northern New Mexico, *Pascopyrum smithii* and *Hilaria jamesii* or *Oryzopsis hymenoides* share dominance with blue grama on the fine-textured soils.

Figure 9.16. A shortgrass prairie in eastern Colorado with Opuntia species interspersed among the Bouteloua gracilis and Buchloë dactyloides grass cover.

Tobosa grass (*Hilaria mutica* (Buckl.) Benth.) communities that occur on fine-textured soils in swales surrounded by the plains grasslands are unique to the southwestern shortgrass prairie that extends into the Chihuahua region of Mexico. *Stipa neomexicana* and *S. comata* can also be important in some *B. gracilis* communities. The vegetation patterns of the New Mexico shortgrass prairies have warm-season, cool-season gradients similar to those found on the larger scale of the Great Plains.

In eastern Colorado and in Montana and Wyoming, *Artemisia frigida*, *Carex filifolia*, and *Koeleria macrantha* are more important to the shortgrass community (Sampson 1950). The shortgrass prairie extends into Canada where it takes on the aspect of a mixed-grass prairie (Stoddart, Smith, and Box 1975). Across the widely distributed shortgrass prairie, the effects of decreasing temperature from south to north and increasing precipitation from west to east is quite apparent. Consequently, cool-season species are more important in the north, shortgrasses dominate in the western extremes, and midheight grasses and tallgrasses play larger roles at the eastward edge of the shortgrass prairie. The original grasses persisted when the shortgrass prairie was first grazed by domestic livestock, probably because of their low stature and natural resistance to grazing. As abuse continued and increased, the grasses declined and the weedy perennial species *Opuntia*, *Gutierrezia*, and *Yucca* increased, along with annual invaders such as

Bromus, *Salsola*, *Hordeum*, and *Festuca* species (see Chapters 7 and 8).

Dryland cultivation was unsuccessful on large areas of the shortgrass prairie (Stoddart et al. 1975), but with irrigation, high levels of crop production were attained. The Dust Bowl of the 1930s was centered in southeastern Colorado, southwestern Kansas, and the panhandles of Texas and Oklahoma, where the shortgrass prairie was plowed for dryland farming. Many "old fields" remain today, more than 50 yr after cultivation attempts failed and the fields were left to revegetate naturally. *Aristida* species persist in many of these old fields because of changes in plowed soils that require long periods for natural restoration. Reductions in soil phosphorus and other nutrients may perhaps leave such sites more suited to this species than to the original plants.

Much of the shortgrass prairie in the Texas High Plains either has been converted to irrigated farming or has been invaded by *Prosopis glandulosa* var. *glandulosa* to form a shrubland or savanna with an understory of the shortgrasses. *Prosopis glandulosa* stands on upland sites, with finer-textured soils often associated with other brush and suffrutescent species: *Opuntia imbricata* var. *imbricata*, *O. polycantha*, *O. leptocaulis*, *Ziziphus obtusifolia*, and *Mahonia trifoliata* (Scifres 1980). *Quercus harvardii* and *Artemisia filifolia* are the dominant brush species in localized areas on sandy soils of the High Plains.

Fire generally is detrimental to shortgrass prairie plants (Wright and Bailey 1980, 1982). Reduced forage yields (Launchbaugh 1972) were attributed to reduced numbers of tillers and shorter plant growth caused by increased runoff and less infiltration of soil moisture after burning (Launchbaugh 1964). Much of our understanding of the effects of fire on the shortgrass prairie came from studies of wildfires; knowledge of prescribed fires, following precise conditions, is limited.

California Grasslands

Originally, grasslands covered some 9.2 M ha (Table 9.1) of the Central Valley and low elevations along the coast of California (Barbour, Pavlik, Drysdale, and Lindstrom 1993). The California grasslands extended upslope into the oak woodland, forming an herbaceous understory beneath the trees. These transitional grasslands provided natural migration routes for the native herbivores to move to higher elevations in the dry seasons (Barbour and Whitworth 1992).

The Central Valley has a mediterranean climate of hot, dry summers and cool, wet winters; annual precipitation is 25–75 cm, with 90% falling between October and April. Climate in the coastal prairie has similar rainfall, but more moderate temperatures.

The soils under both the Central Valley grasslands and those along the coast are generally deep, brown, fertile soils that are highly suited to grasses (Barbour et al. 1993). The Central Valley soils were deposited as sediments eroded from the coastal and Sierra Nevada mountains and volcanic material ejected from Mt. Shasta, Mt. Lassen, and Sutter Buttes along the floor of an ancient sea. The coastal prairie soils were derived from uplifted marine terraces.

Vegetation on the coastal terraces is an assortment of perennial grasses and flowering broadleaf herbs. Originally, the Pacific prairie was dominated by cool-season, perennial bunchgrasses such as *Nassella pulchra*, *N. cernua*, *Elymus* spp. *Poa secunda*, *Aristida* spp., *Koeleria macrantha*, *Melica imperfecta*, and *Muhlenbergia rigens* (Weaver and Clements 1938; Burcham 1957; Barry 1972; Stoddart et al. 1975; Heady 1977; Heady et al. 1977). Some California grassland dominants may have extended well north of California. Oregon hairgrass (*Deschampsia caespitosa* ssp. *holciformis*, Idaho fescue (*Festuca idahoensis*), California oatgrass (*Danthonia californica*) were important native perennial bunchgrasses.

The perennial bunchgrasses in the California grasslands were complemented by an array of annual and perennial grasses and late spring-flowering forbs (Barbour et al. 1993). Important forbs included *Iris douglasiana*, *Lasthenia* spp., *Ranunculus californicus*, *Nemophila menziesii*, and *Layia platyglossa* ssp. *platyglossa*. On more shallow, drier soils, the vegetation of the coastal prairie resembles that of the Central Valley where *Nassella pulchra*, *Poa secunda*, and other annual grasses and herbs dominate. The seaward edge of the Coastal Prairie is influenced heavily by the ocean's wind and salt. This narrow strip of grassland was dominated by *Armeria maritima* ssp. *californica* and *Erigeron glaucus* (Barbour et al. 1973).

The Pacific prairie shares an ecotone with the Palouse prairie in northern California and exhibits some similarity in species relationships (Beetle 1947). Weaver and Clements (1938) contended that this association also had an affinity with the central mixed-grass prairie, because relicts of *Bouteloua*, *Andropogon*, *Hilaria*, and *Aristida* species could be found in southern California and because *Nassella lepida* and *Stipa speciosa* from the Pacific grassland were also found in Arizona and New Mexico.

Although descriptions of the California grasslands are vague, scattered reports indicate that these grasslands were luxuriant with deer, pronghorn antelope, tule elk, and bear, along with other big game species (Edwards 1992). The original grasslands survived under seasonal grazing coupled with fires set naturally and by the native people to increase the productivity of these grasslands and to enhance the wildlife on which they depended (Barry 1972; Barbour et al. 1993). Seasonal grazing allowed these grasslands species to recover and to sustain grazing of the vegetation over long periods of geological time. Fires prevented invasion of the woody plants that reproduced by aboveground exposed buds.

Fire suppression by European settlers and the year-long, heavy grazing by cattle and sheep transformed these coastal grasslands into almost pure stands of nonnative *Holcus lanatus*, *Anthoxanthum odoratun*, *Bromus hordeaceus*, *Lolium perenne*, *Avena fatua*, and *Hordeum* spp. Forbs, largely unpalatable to cattle, also spread rapidly in the absence of fire: Milk thistle (*Silybum marianum*), wild artichoke (*Cynara scolymus*), and Klamath weed (*Hypericum perforatum*) became dominant on much of the former coastal grasslands. Introduced weeds dramatically changed this grassland in a short time from a palatable, nutritious, native perennial canopy to that of an annual grassland of primarily introduced species (Barbour and Whitworth 1992). About 90% of the biomass was introduced from other continents.

Species from similar environments (southern

Europe, Chile, South Australia, and South Africa) were preadapted to this grassland environment (Jackson 1985). The native grasses lacked the ability to withstand heavy, continuous grazing and to compete with the exotic introductions. These exotics quickly spread and formed stable communities. Heady (1977) called them "new natives" to emphasize their permanent nature.

Vegetation changes in the California grasslands have been more drastic and perhaps more rapid than in any other North American grassland (Biswell 1956). By 1918, *Avena fatua* had invaded most of the Central Valley and was dominant throughout (Beetle 1947); less than 5% of this area remained as original perennial vegetation (Sampson 1950).

The impact of fire on the original perennials is largely unknown, although *Nasella pulchra* probably was fire-tolerant (Wright and Bailey 1982). This perennial grass species persisted along railroads, where it was protected from grazing but frequently burned. Fire shifted the annual grassland species from grasses to legumes and forbs. Yields of annual grasses may be reduced by about 25% the first year following fire, whereas yields of *Medicago polymorpha*, *Erodium cicutarium* (L.), and *E. botrys* increase more than five-fold (Hervey 1949).

Ranching began in California with the Spanish colonists in 1769 (Burcham 1951 and 1981). Prior to that, the grasslands were transitionally grazed by pronghorn (*Antilocapra americana*), elk (*Cervus elaphus nannodes*), and deer (*Odocoileus hemionus*) (Barry 1972). Although pronghorn may have been abundant in the Central Valley, evidently the grazing pressure was either in balance with the vegetation or the pronghorns followed a natural rotational grazing system and moved upslope in the summer. Extensive heavy grazing began with the large Spanish land grants of 1824 (Barry 1972). Livestock numbers grew gradually until the gold rush of 1849, when demand for meat brought rapid expansion. Cattle numbers in California peaked at about 3.5 M head in about 1860, as farming began to expand (Branson 1985). More recently, large areas of the Central Valley have been used for cultivation, large urban centers, and industrial expansion.

Palouse Prairie

Palouse, an intermountain bunchgrass vegetation type, originally extended throughout southwest Canada, eastern Washington and Oregon, southwestern Idaho, northern Utah, and into western Montana (Stoddart et al. 1975). The Palouse prairie grassland was dominated originally by *Pseudoroegneria spicata*, *Festuca idahoensis*, *Poa secunda*, and *Leymus condensatus*, and the associated species *Koeleria macrantha*, *Elymus elymoides*, *Stipa comata*, and *Pascopyrum smithii* from the annual grasslands of California and the mixed-grass prairie of the central grasslands (Weaver and Clements 1938). Although Palouse flourishes in a climate similar to that of the California grassland, its aspect more nearly resembles mixed-grass vegetation, except that there are no shortgrasses present. Cool-season grasses dominate and compose more than 80% of the flora, with the remainder being predominantly C_4 grasses (Barbour et al. 1987; Barbour and Christiansen 1993).

Excessive grazing in the past resulted in the demise of many of the perennial grasses and led to an abundance of *Artemisia* and *Bromus* species. Intense fires that occurred during the summer caused considerable damage to the perennial grasses (Daubenmire 1970). Annual forbs (*Amsinckia*, *Helianthus*, *Lactuca*, *Conyza*, and *Sisymbrium*) rapidly invaded the community following such fires. Even cool-burning fires, though not so damaging, favored annual *Bromus* species (Wright and Bailey 1982) that became widespread in the Palouse grassland. Fire also reduced *Descurania pinnata*, *Sisymbrium altissimum*, and *Opuntia* species, while favoring *Lithophragma glabrum* and *Plantago patagonia*.

Today the Palouse prairie is a shrub-steppe grassland, with dominance shared by remnants of perennial grass and *Artemisia* and *Bromus* species. The invading *Bromus* species changed these grasslands from a year-long forage resource to seasonal only, because the introduced ephemerals have a shorter growth period during which nutritious forage is available.

Massive changes in the Palouse prairie vegetation have been caused by cultivation, grazing, and plant introduction (Franklin and Dyrness 1969; Young, Evans, and Major 1977; Mack 1981). The major dominants, *Festuca idahoensis*, *F. campestris*, and *Poa secunda*, were replaced by *Bromus tectorum* and *Poa pratensis*. The original perennials were not resistant to grazing, and soil moisture patterns favored the cool-season annuals over the perennials (Branson 1985). Fire and grazing favored annuals over perennials, and there is evidence that annuals are better competitors for soil moisture (Harris 1967; Stoddart et al. 1975).

In the more moist and cooler northern areas, overgrazing decreased *Festuca* species and *Pseudoroegneria spicata* but increased *Bromus tectorum*, *Stipa nelsonii*, *Poa pratensis*, *Antennaria parvifolia*, and *Poa secunda* (Tisdale 1947; Branson 1985). Farther south, in the warmer and drier zones, *B. tectorum*, primarily, along with *Poa secunda*, replaced *Festuca idahoensis*. *Artemisia* species and *B. tectorum* have increased at the expense of the perennial grasses.

Fescue Prairie

The fescue prairie is a transition grassland between the mixed prairie and the western forest regions (Coupland 1992). It is a grassland component of aspen parkland (*Populus tremuloides*) that occurs from central Saskatchewan west and south into the Rocky Mountains of southern Alberta and northern Montana. The fescue prairie grades into the tallgrass prairie to the east and the Palouse prairie to the west. Along the northern and eastern edges, the fescue prairie intermingles with elements of both the mixed prairie and the tallgrass prairie. Although Clements and Clements (1939) included the fescue prairie as a submontane prairie, others who studied this region more closely recognized a distinct grassland dominated by *Festuca campestris* (Moss 1944; Moss and Campbell 1947; Coupland and Brayshaw 1953).

Although the origin of this upland grassland is speculative, it has relationships with the Palouse prairie of British Columbia (Tisdale 1947; Moss 1955). *Festuca scabrella*, its principal species, is found as far south and east as western North Dakota (Dix 1964). The annual precipitation for the fescue prairie ranges from 400–600 mm. Both summer temperatures and annual precipitation increase with elevation that ranges from 500 up to 1400 m. Temperatures tend to increase in the eastern portion of the fescue prairie.

In this glaciated landscape where the fescue prairie occurs, numerous shallow depressions are common with wetland grasses, sedges, and rushes (Walker and Coupland 1970). These depressions are often characterized by a perimeter of *Salix* species or *Populus tremuloides* where the site is somewhat drier.

Livestock grazing on the fescue prairie has decreased *Festuca scabrella* (Moss and Campbell 1947; Johnston and MacDonald 1967). These studies indicated that *Festuca scabrella* could decline significantly in short periods with close grazing in the summer. It is more tolerant of close grazing in the winter. This grass was the principal winter grass for bison on the northern Great Plains. Willms, Smoliak, and Dormaar (1985) found that *Fescuca campestris* declined in abundance after 5 yr of very heavy grazing in northern Montana; *Danthonia parryi* Scribn. tended to replace fescue, and there was an increase in less palatable forbs and grasses. Grazing also changed the color of the soils from black to dark brown (Johnston, Dormaar, and Smoliak 1971). This coincided with a reduction in organic matter and an increase in nitrate nitrogen, ammonium nitrogen, and available phosphorus (Willms, Dormaar, and Schaalje 1988). These grazing studies, however, indicated that moderate use is possible and that the fescue grassland can be sustained with proper and appropriate grazing on the 25.5 m ha original fescue prairie (of which an estimated 1.28 (Trottier 1992) to 2 m ha (Looman 1969) remain).

Desert Grassland

The desert grasslands are transitional plant communities (Dick-Peddie 1993) between the plains grassland and the desert regions that extend from the shortgrass area in Texas, New Mexico, and through southwestern Arizona into northern and central Mexico (Fig. 9.17). This grassland forms the southern periphery of the North American grassland and is dominated by *Bouteloua*, *Hilaria*, and *Aristida* species (Dix 1964). Ice Age climatic functions least influenced this grassland, which remained comparatively stable until the arrival of Europeans.

Originally, *Bouteloua eriopoda*, *Hilaria belangeri*, *Oryzopsis hymenoides*, and *Muhlenbergia porteri* were the important species because of their forage value and their coverage of vast areas of land (Stoddart et al. 1975). Those plateau grasslands that occur at the edge of the Sonora and Chihuahua deserts above 1000 m elevation originally had a short-bunchgrass appearance, with desert scrub restricted to ravines, knolls, and sites where soils were particularly poor, thus limiting the more productive grasses (Barbour et al. 1987).

Changes in the vegetation of the desert grassland have been documented in several places across the southwestern United States (Humphrey 1953, 1958, 1962; Humphrey and Mehrhoff 1958; Buffington and Herbel 1965; Branson 1985). Grasses have gradually been replaced by desert shrubs such as *Larrea tridentata*, *Flourensia cernua*, and varieties of *Prosopis*. Three varieties of *Prosopis* (mesquite) are distributed across the Southwest (Parker and Martin 1952): *P. glandulosa* var. *glandulosa* is an abundant tree or shrub in Texas; *P. glandulosa* var. *torreyana* is a common shrub in west Texas, southern New Mexico, and northern Mexico; and *P. velutina* is common in Arizona. *P. glandulosa* var. *glandulosa* has spread into *Bouteloua eriopoda* grasslands of New Mexico and displaced

Figure 9.17. Excellent cover on a desert grassland site in the Big Bend National Park in southwestern Texas.

numerous forbs and shrubs (Hennessy, Gibbens, Tromble, and Cardenas 1983).

Overgrazing by domestic livestock and subsequent erosion of the thin topsoil, fire suppression, and perhaps a gradual warming of the climate of the Southwest are correlated with these marked vegetation changes. The grasses have disappeared on both grazed and ungrazed areas, and *Sporobolus flexuosus*, *Erioneuron pulchellum*, and *Gutierrezia sarothrae* have increased with the mesquite. Mesquite has increased on all soil types in the desert grassland but more so on sandy soils. In certain areas the increase in this shrub was associated with the development of a sand dune aspect (Gibbens, Tromble, Hennessy, and Cardenas 1983). Buffington and Herbel (1965) suggested that a generally drier climate and increased seed dispersal achieved by grazing animals, particularly during droughts, hastened the invasion and spread of *P. glandulosa* var. *glandulosa* across the Jornada Experimental Range between 1858 and 1963 (see also Chapter 8).

Many areas of the desert grassland have been modified for so long that no standard for comparison exists (Martin 1975). Generally, shifts from grassland to shrubland have occurred since domestic livestock were first introduced, probably earlier in the Southwest than in other areas of North America. Although overgrazing is generally implicated as the cause of increases in *Prosopis*, Malin (1953) believes that *P. glandulosa* has been widely distributed across Texas as a savanna element since 1800. Shrubs have invaded grazed Arizona grasslands as well as those where both large and small herbivores have been excluded. As a result, Brown (1950) concluded that the desert grass-

land is a subclimax to the desert shrub climax in southern Arizona.

HABITAT LOSS, CONSERVATION, AND RESTORATION ECOLOGY

Habitat Loss

Grasslands of the Great Plains once covered at least 300 (Table 9.1) and perhaps as much as 370 M ha. Even now the grasslands cover more than 125 M ha. Homesteading began in the 1830s, with widespread conversion to croplands in subsequent decades. Estimates of the loss of Great Plains grasslands range from 82–99% for the tallgrass prairie, 30–99% for the mixed-grass prairie, and 20–86% for the shortgrass prairie (Table 9.5). Currently, 55 grassland species in the United States are listed as threatened or endangered, 728 are candidates for listing, and one-third of the species considered as endangered in Canada are found in the grasslands (Samson and Knopf 1994). There are also numerous invertebrate species in grassland ecosystems, reaching 4000 per m^2 aboveground and many times that number belowground (Risser et al. 1981).

The plains bison and the prairie dog, two animals that characterize North American grasslands, have been drastically reduced. Although difficult to determine, bison numbers before the arrival of the Europeans in North America have been estimated as 12–60 M. Shaw (1996) says "there were tens of millions of bison between the Mississippi River and the Rocky Mountains." Roe (1970) reports that one herd alone near Dodge City, Kansas, was estimated at 4 M bison. Relatively few remain

Table 9.5. *Estimated current and historic area[a] and percentage of area that has declined and is under protection for the tallgrass, mixed-grass, and shortgrass prairies.*

Prairie	Province or state	Historical	Current	Decline	Protected
Tallgrass	Manitoba	600,000	300	99.9	N/A
	Illinois	8,900,000	930	99.9	<.01
	Indiana	2,800,000	404	99.9	<.01
	Iowa	12,500,000	12,140	99.9	<.01
	Kansas	6,900,000	1,200,000	82.6	N/A
	Minnesota	7,300,000	30,350	99.6	<.01
	Missouri	5,700,000	30,350	99.5	<.01
	Nebraska	6,100,000	123,000	98.0	<.01
	North Dakota	1,200,000	1,200	99.9	N/A
	Oklahoma	5,200,000	N/A	N/A	N/A
	South Dakota	3,000,000	449,000	85.0	N/A
	Texas	7,200,000	720,000	90.0	N/A
	Wisconsin	971,000	4,000	99.9	N/A
Mixed grass	Alberta	8,700,000	3,400,000	61.0	<.01
	Manitoba	600,000	300	99.9	<.01
	Saskatchewan	13,400,000	2,500,000	81.3	<.01
	Nebraska	7,700,000	1,900,000	77.1	N/A
	North Dakota	13,900,000	3,900,000	71.9	N/A
	Oklahoma	2,500,000	N/A	N/A	N/A
	South Dakota	1,600,000	N/A	N/A	N/A
	Texas	14,100,000	9,800,000	30.0	N/A
Shortgrass	Saskatchewan	5,900,000	840,000	85.8	N/A
	Oklahoma	1,300,000	N/A	N/A	N/A
	South Dakota	179,000	N/A	N/A	N/A
	Texas	7,800,000	1,600,000	80.0	N/A
	Wyoming	3,000,000	2,400,000	20.0	N/A

Source: [a]Estimates of current and historical prairie areas are based on information from The Nature Conservancy's Heritage Program; U.S. Department of Interior Fish and Wildlife Service; U.S. Department of Agriculture Forest Service; Canadian Wildlife Service; Provinces of Alberta, Manitoba, and Saskatchewan; and state conservation agencies.

(Risser et al. 1981), and many of these are in national parks. Bison population continues to increase, especially in private herds. Estimates exceeded 75,000 in 1989 (Hawley 1989), and they continue to increase. Bison losses came from disease brought in by domestic cattle and by severe over hunting, especially between 1870 and 1883 (Koucky 1983).

Prairie dog populations have declined by 98% since European settlement. This decline has been caused by habitat conversion and by poisoning by stockmen who fear competition between prairie dogs and cattle for grass forage (Krueger 1986), although there is mixed evidence of such competition (Sampson and Knopf 1994). The problem, however, is controlling the continual spread of prairie dog populations across the grasslands. In the southern plains mixed-prairie, O'Meilia, Knopf, and Lewis (1982) found that burrowing owls nested in prairie dog holes and preyed on grasshopper mice, resulting in increases in insect populations. In this study, prairie dogs tended to shift the grassland to a shortgrass aspect by their foraging habits.

The biological diversity of grassland birds has significantly declined. Of the 435 breeding bird species in the United States, as many as 330 may reproduce within the Great Plains. Declines over the past two or three decades range from 17–91%, with declines of 25–65% from 1980 to 1989 (Knopf 1992, 1994). As an example, half of the 26 grassland birds known to breed in Illinois are declining nationally and/or regionally (Herkert 1994). Declines in bird populations are caused by the reduction in grasslands and by continual fragmentation of the remaining prairie, which enhances hybridization among populations and competition and nest predation from forest-edge bird species (Knopf 1986).

Herkert (1994) studied 24 native prairie, restored prairie, and cool-season grasslands in northern Illinois that ranged in size from 0.5 to 650 ha. The number of grassland bird species was proportional to the size of the remaining grassland fragment (Fig. 9.18). This relationship was consistent among the different types of grasslands and across the three-year sampling period. Eighty-four percent of the variability in the number of bird species were accounted for by the size of the prairie fragment. For some bird species, the incidence was also correlated with vegetation structure. Transects located in a small grassland fragment that was an appropriate habitat for a particular species were frequently unoccupied. Both Henslow's sparrow (*Ammódramus hénslowii*) and the bobolink (*Dolichónyx oryzívorus*) are examples of area-sensitive species. In large grasslands, the Henslow's sparrow occupied 73% of the suitable transects; it was not found in any of the suitable small grassland fragments. The bobolink was found in 88% of the suitable transects in large grasslands but in only 7% of suitable small grassland fragments.

The substantial conversion of the North American grasslands has reduced and in some cases jeopardized a number of species. Our understanding of the consequences of these habitat losses depends on our knowledge of the taxonomy of grassland vertebrates and vascular plants. Basic knowledge of their abundance, rarity, and distribution is generally adequate in the more developed nations (West 1993). The same cannot be said, however, for even these organisms in grasslands in the developing world. Knowledge of invertebrates, nonvascular plants, and microorganisms is deficient everywhere in the world.

Conservation and Restoration

Because of significant conversion of grassland to other uses, it is important to consider preservation of the remaining North American grasslands (West 1993). The need for sustained biological diversity has further increased the demand for the conservation of grasslands (Samson and Knopf 1994) and for conservation in general (Turner Romme, and Gardner 1994). There are many examples of grassland conservation efforts, and numerous small to large protected prairies exist.

Because not all can be saved, the most ecologically important should be identified and conservation efforts concentrated on these sites and species (West 1993). Those species that play key roles in ecosystem processes and in maintaining the integrity of grasslands are critical. For example, prairie vegetation holds the soils against wind and water erosion; without this cover, soil losses such as those that occurred in the droughts of the 1930s can be substantial (Weaver 1954, 1968). Similarly, certain species are important for nitrogen fixation, for providing food for herbivores, and for sequestering

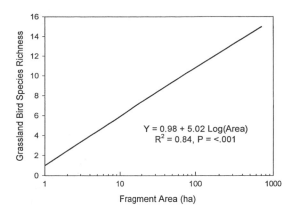

$$Y = 0.98 + 5.02 \, \mathrm{Log(Area)}$$
$$R^2 = 0.84, \, P = <.001$$

Figure 9.18. Number of species breeding bird on grassland fragments of different size in a 1987–1989 census (adapted from Heckert 1994).

and cycling nutrients. Understanding the role of individual species helps to characterize the guild structure of populations, thus guiding conservation efforts to ensure that key ecological processes are preserved.

Care must be taken in managing preserved grasslands. For example, from 1938 to 1941, the Civilian Conservation Corps planted numerous windbreaks, but in some instances aggressive alien species were inadvertently introduced. Federal land management agencies and private landowners planted hundreds of hectares with crested wheatgrass, imported from Siberia, on marginal farmland and abused rangeland. This species replaced large areas of native grasslands (Samson and Knopf 1994). Although these introduced grasses conserved the soil, met some economic goals of producers, and helped to preserve rural communities, they did change the landscape.

Retaining the natural processes usually enhances the quality of the grassland. For example, livestock grazing alone does not necessarily reduce biological diversity (West 1993; Sims, Berg, and Bradford 1995), and in many grasslands, species diversity is maximized with light to moderate grazing and by periodic burning (Anderson 1990; Risser 1988, 1990).

Restoration of native communities (Falk and Olwell 1992) has received much attention at the annual North American Prairie Conference (e.g., Schramm 1990). There have been substantial efforts to restore native grasslands in North America, especially the tallgrass and to a lesser extent the mixed-grass and shortgrass prairies. There are also many restoration projects in oak savannas along the eastern portion of the central prairies (Anderson 1990; Tester 1989). The most common techniques for prairie restoration include planting native species and removing or reducing competing species by the use of fire, herbicides, or mechanical means (Fig. 9.19).

Sustainable Grasslands

Understanding and designing sustainable grassland systems must include recognition of the history of the vegetation and the animals (Barnard and Frankel 1966; Tausch, Wigand, and Burkhardt 1993). Grazing, originally by native and later by domestic animals, has always been a component of grasslands and it influenced evolutionary processes as well as the very nature of grasslands themselves (Fig. 9.20). Platou and Tueller (1985) concluded that a better understanding of pristine grazing systems could improve the design and implementation of contemporary grazing management strategies, help to design innovative grazing systems, and lead to more efficient use of watersheds, grazing lands, and recreational resources. In addition to being grazed, grasslands have always burned (Wright and Bailey 1982; Axelrod 1985; Anderson 1990; Collins and Wallace 1990). Currently, fires are controlled by human efforts and by natural and human-produced firebreaks. Prior to human intervention, and depending in part on fuel load and weather conditions, these fires were sometimes very large and intense. Under natural conditions, a given hectare of the North American prairie may have burned as often as every 5 or as infrequently as every 30 yr (Wright and Bailey 1982). Thus, periodic drought, grazing, and fires have always been a part of North American grasslands.

In recent years, more has been learned about how grasslands respond to drought, grazing, and fire (Risser et al. 1981; Risser 1985; Collins and Wallace 1990; Seastedt and Ramundo 1990; Bidwell and Engle 1992). It has become increasingly clear that the biological diversity of grasslands not only is a characteristic of these communities but also plays a role in the sustainability of grasslands. For example, Tilman and Downing (1994) found different responses to drought in a sandy Minnesota prairie that had varying levels of plant biodiversity resulting from different fertilizer treatments. Biomass production on species-rich plots was greater and more resilient to drought than on the species-poor plots. Similarly, results from a Yellowstone National Park study showed that drought had less effect on the more diverse plant communities (Frank and McNaughton 1991). From very long-term experiments at Rothamsted in England, grassland biomass yields on acid soils showed more variation among years than yields on other soils, and species-rich plots

tended to be less variable than those with a smaller number of species (Silvertown, Dodd, McConway, Potts, and Crawley 1994).

Classification of rangeland conditions and trends has long been a central concept in the management of grasslands. In its original and simplest concept, rangeland condition classes are based on vegetation similarity to climax vegetation (i.e., vegetation communities that perpetuate themselves and would be considered sustainable). The system is predicated on the assumption that if the species composition is similar to the original undisturbed vegetation, then the functional and structural quality of the rangeland is excellent and presumably the grassland would persist indefinitely. A number of issues have been raised about the classification of rangeland condition (Risser 1989). To be most useful, classification of rangeland condition should take into account soil conditions and other components of the biota. Grasslands are managed for many different purposes, and integrated methods are needed to quantify multiple options and values (Bernardo, Engle, Lochmiller, and McCollum 1992; Bosch and Booysen 1992; Risser 1989; Scarnecchia 1994). Risser (1989) has argued that these improvements in the rangeland classification systems will occur as our understanding of the ecology of the grasslands improves. Joyce (1993) suggests that improved classification and condition measures will come through the development of new concepts that lead to improved field investigation techniques (Pickup, Bastin, and Chewings 1994).

AREAS FOR FUTURE RESEARCH

Although our current knowledge of North American grasslands is substantial, additional information is needed to manage these grasslands more wisely in the future. Knowledge gaps have appeared because grasslands did not respond to changing climate or various management techniques as we had predicted. In other instances, we have not understood key processes at the organism, population, community, ecosystem, landscape, or global levels. It is frequently impossible to predict how the processes at one organizational level will affect those at another level.

As we characterize grassland processes in the human context, it becomes obvious that understanding grassland ecosystem functions at all organizational levels is critical to maintaining a productive biosphere. The ultimate test of our knowledge of North American grasslands is whether we can preserve, restore, and maintain them for future generations.

Basic knowledge of how grasslands were orig-

Figure 9.19. Sequences of restoration of tallgrass prairie using fire to control Eastern red cedar.

Figure 9.20. Tripsacum dactyloides is a grass native to the southern plains and the south and eastern United States that can be a perennial replacement for annuals now being grown on erodible soils.

inally distributed across the North American continent and how the plants, animals, and microorganisms adapted to the various site conditions (Anderson et al. 1994; Tausch et al. 1993) is still needed. Although much is known (McMillan 1959a; Risser et al. 1981; Weaver 1954, 1968), additional research is required to further define the mechanisms that support the geographical repetition of an apparently uniform vegetation array across a number of diverse, widespread habitats. Although the general distribution of carbohydrates within plants is relatively well known, we still lack reliable models to describe the distribution of various organic compounds in plants subjected to different climates or management conditions. Extension of carbon-assimilation and carbon-allocation studies (Caldwell, Osmond, and Nott 1977; Caldwell, White, Moore, and Camp 1977) should provide a data base for generalizing and then modeling these processes in relation to changes in climate (Zak et al. 1994), herbivory (McCollum, Gillen, and Brummer 1994; Caswell, Reed, Stephenson, and Werner 1973; Sims and Singh 1978b), and fire (Ode et al. 1980; Ojima, Parton, Schimel, and Owensby 1989; Ojima, Schimel, Parton, and Owensby 1994).

Although the general advantages of the C_3 and C_4 carbon fixation patterns are presumably understood, more attention must be devoted to the detailed implications of these processes, such as their behavior in fluctuating light conditions and the ways in which these processes respond to competitive relationships, changing climate, and increasing atmospheric CO_2 levels (Tilman 1982; Schimel et al. 1991; Polley et al. 1994; Silvertown et al. 1994; Winner 1994). If we are to understand the origin of variability in grassland composition, it will be

necessary to evaluate the dynamics of both the local and more regional processes affecting potential seed sources.

Much is known about grassland species and their interactions at the population and community levels. However, there are significant unanswered questions about the processes that determine photosynthetic pathways and nutrient flows in grassland and savanna ecosystems. Similarly, the fundamental ecosystem processes of carbon dynamics and nutrient cycles are reasonably well understood for a few grassland sites, but large voids, especially for belowground processes, require additional study. In virtually all instances, the spatial and temporal variability in these ecosystem processes is not sufficiently quantified and described. Future research on the sustainability of grasslands will incorporate studies on how ecological conditions affect, and are affected by, biodiversity. Further work is needed on the role biodiversity plays in specific ecological processes. Grassland research will continue to include studies on ecological characteristics and processes that operate at all levels of organization and at all scales. These more traditional studies should be augmented by integrated studies that incorporate economic and cultural issues and values (Bernardo et al. 1992).

Understanding and managing sustainable grasslands in the future will require a fundamental change in methods used for determining the value of natural resources and ecological processes (Costanza and Daley 1992). Methodology must more explicitly include the values for ecological goods and services produced by grassland ecosystems. Future research will also involve interaction with various stakeholders and with individuals and organizations responsible for preserving, restoring, and managing North American grasslands.

REFERENCES

Aandahl, A. R. 1982. Soils of the Great Plains – land use, crops and grasses. University of Nebraska Press, Lincoln.

Albertson, F. W. 1937. Ecology of mixed prairie in west central Kansas. Ecol. Monogr. 7:481–547.

Anderson, K. L. 1965. Fire ecology – some Kansas prairie forbs. Proc. Tall Timbers Fire Ecol. Conf. 4:153–160.

Anderson, R. C. 1982. An evolutionary model summarizing the roles of fire, climate, and grazing animals in the origin and maintenance of grasslands, pp. 297–308 in J. R. Estes, R. J. Tyrl, and J. N. Brunken (eds.), Grasses and grasslands, systematics and ecology. University of Oklahoma Press, Norman.

Anderson, R. C. 1990. The historic role of fire in the North American grasslands, pp. 8–18 in S. L. Collins, and L. Wallace (eds.). 1990. Fire in North American tallgrass prairies. Oklahoma University Press, Norman.

Anderson, R. C., B. A. D. Hetrick, and G. W. T. Wilson. 1994. Mycorrhizal dependence of *Andropogon gerardii* and *Schizachyrium scoparium* in two prairie soils. Amer. Mid. Nat. 132: 366–376.

Archer, S. G., and C. E. Bunch. 1953. The American grass book. A manual of pasture and range practices. University of Oklahoma Press, Norman.

Austin, M. E. 1965. Land resource regions and major land resources areas of the United States (exclusive of Alaska and Hawaii). USDA, Soil Conservation Service, agric. hdbk 296.

Axelrod, D. I. 1985. Rise of the grassland biome, central North America. Bio. Rev. 51: 163–201.

Bailey, R. G. 1976. Ecoregions of the United States [Map]. USDA Forest Service, Ogden, Utah.

Bailey, R. G. 1980. Description of the ecoregions of the United States. USDA Forest Service, Ogden, Utah.

Barbour, M. G., R. B. Craig, F. R. Drysdale, and M. T. Ghiselin. 1973. Coastal ecology: Bodega Head. University of California Press, Berkeley.

Barbour, M. G., J. H. Burk, and W. D. Pitts. 1987. Terrestrial plant ecology 2nd ed. Benjamin/Cummings Publishing, Menlo Park, Calif.

Barbour, M., B. Pavlik, F. Drysdale, and S. Lindstrom. 1993. California's changing landscapes. California Native Plant Society, Sacramento.

Barbour, M., and V. Whitworth. 1992. California's grassroots: native or European? Pacific Discovery 45:8–15.

Barbour, M. G., and N. L. Christensen. 1993. Vegetation, pp. 97–131 in Flora of North America Editorial Committee, Flora of North America North of Mexico. Vol. 1, Oxford University Press, New York.

Barnard, C., and O. H. Frankel, 1966. Grass, grazing animals and man, historic perspective of grasses and grasslands, pp. 1–12 in C. Barnard (ed.), Grasses and grasslands. Macmillan, New York.

Barry, W. J. 1972. The Central Valley prairie, vol. 1, California Prairie Ecosystem. California Department of Parks and Recreation, Sacramento.

Beetle, A. A. 1947. Distribution of the native grasses of California. Hilgardia 17:309–357.

Belsky, A. J. 1994. Influences of trees on savanna productivity: tests of shade, nutrients, and tree grass competition. Ecology 75:922–932.

Bernardo, D. J., D. M. Engle, R. L. Lochmiller, and F. T. McCollum. 1992. Optimal vegetation management under multiple-use objectives in the Cross Timbers. J. Range Mgt. 45:462–465.

Bidwell, T. G., and D. M. Engle. 1992. Relationship of fire behavior to tallgrass prairie herbage production. J. Range Mgt. 45:579–584.

Biswell, H. H. 1956. Ecology of California grasslands. J. Range Mgt. 9:19–24.

Bogush, E. R. 1952. Brush invasion in the Rio Grande Plain of Texas. Texas J. Sci., 4:85–91.

Bolen, E. G., and J. A. Crawford. The birds of rangelands, pp. 15–27 in P. R. Krausman (ed.), Rangeland wildlife. The Society for Range Management, Denver, Colo.

Booth, W. E. 1941. Revegetation of abandoned fields in Kansas and Oklahoma. Am. J. Bot. 28:415–422.

Borchert, J. R. 1950. The climate of the central North American grassland. Ann. Assoc. Amer. Geogr. 40:1–39.

Bosch, O. J. H. and J. Booysen. 1992. An integrative approach to rangeland condition and capability assessment. J. Range Mgt. 45:116–122.

Branson, F. A. 1985. Vegetation changes on western rangelands. Society for Range Management, Denver, Colo.

Bray, W. L. 1906. Distribution and adaptation of the vegetation of Texas. University of Texas Bull. 82.

Breymeyer, A. I., and G. M. Van Dyne (eds.) 1980. Grasslands, systems analysis and man. Cambridge University Press, Cambridge.

Brown, A. L. 1950. Shrub invasion of southern Arizona desert grassland. J. Range Mgt. 3:172–177.

Brunnschweiler, D. H. 1952. The geographic distribution of air masses in North America. Vierteljahrsschr. Naturforsch. Ges. Zür. 97:42–49.

Buffington, L. C., and C. H. Herbel. 1965. Vegetational changes of a semidesert grassland range from 1858 to 1963. Ecol. Monogr. 35:139–164.

Buol, S. W., F. D. Hole, and R. J. McCracken. 1980. Soil genesis and classification, 2nd ed. Iowa State University Press, Ames.

Burcham, L. T. 1951. Cattle and range forage in California: 1770–1880. Agric. Hist. 35: 140–149.

Burcham, L. T. 1957. California rangeland. California Division of Forestry, Sacramento.

Burcham, L. T. 1981. California rangelands in historical perspective. Rangelands 3:95–104.

Caldwell, M. M., C. B. Osmond, and D. L. Nott. 1977. C_4 pathway photosynthesis at low temperature in cold tolerant Atriplex species. Plant Physiol. 60:157–164.

Caldwell, M. M., R. S. White, R. T. Moore, and L. B. Camp. 1977. Carbon balance, productivity, and water use of cold-winder desert shrub communities dominated by C_3 and C_4 species. Oecologia 29:275–300.

Cale, W. G., G. M. Henebry, and J. A. Yeakley. 1989. Inferring process from patterns in natural communities. BioScience 34:363–367.

Carpenter, J. R. 1940. The grassland biome. Ecol. Monogr. 10:617–684.

Carr, J. T. 1967. The climate and physiography of Texas. Report 53, Texas Water Development Board, Austin, Texas.

Caswell, H., F. Reed, S. N. Stephenson, and P. A. Werner. 1973. Photosynthetic pathways and selective herbivory: a hypothesis. Amer. Nat. 107:465–481.

Clements, F. E. 1916. Plant succession: an analysis of the development of vegetation. Carnegie Institution of Washington, publication 242, Washington, D.C.

Clements, F. E. 1920. Plant indicators. Carnegie Institution of Washington, publication 290, Washington, D.C.

Clements, F. E., and Clements, E. S. 1939. Climate, climax, and conservation. Carnegie Instit. Wash. Yearb. 38:137–140.

Clements, F. E., and V. E. Shelford. 1939. Bio-ecology Wiley, London.

Collins, D. D. 1969. Macroclimate and the grassland ecosystem, pp. 29–39 in R. L. Dix (ed.), The grassland ecosystem, a preliminary synthesis. Range Science Department, Science Series no. 2, Colorado State University, Fort Collins.

Collins, S. L., and D. E. Adams. 1983. Succession in grasslands: Thirty-two years of change in a central Oklahoma tallgrass prairie. Vegetatio 51:181–190.

Collins, S. L., and L. Wallace, editors. 1990. Fire in North American tallgrass prairies. Oklahoma University Press, Norman.

Conant, S., and P. G. Risser. 1974. Canopy structure of a tall-grass prairie. J. Range Mgt. 27:313–318.

Costanza, R., and H. E. Daley. 1992. Natural capital and sustainable development. Conser. Biol. 6:1–10.

Coupland, R. T. 1974. Grasslands, pp. 280–294 in The New Encyclopedia Britannica, vol. 8. Encyclopedia Britannica, Inc., Chicago.

Coupland, R. T. 1979. Grassland ecosystems of the world: analysis of grassland and their uses. Cambridge University Press, Cambridge.

Coupland, R. T. 1992. Fescue prairie, pp. 291–295 in R. T. Coupland (ed.), Natural grasslands, introduction and western hemisphere (Ecosystems of the World 8A). Cambridge University Press, Cambridge.

Coupland, R. T., and Brayshaw, T. C. 1953. The fescue grassland in Saskatchewan. Ecology 34:386–405.

Coupland, R. T., E. A, Ripley, and P. C. Robbins. 1973. Description of site, I, Floristic composition and canopy architecture of the vegetative cover. Canadian Committee for IBP, Matador Project report 11, University of Saskatchewan, Saskatoon.

Curtis, J. T. 1962. The modification of mid-latitude grasslands and forests by man, pp. 721–736 in W. L. Thomas, Jr. (ed.), Man's role in changing the face of the earth. University of Chicago Press, Chicago.

Daubenmire, R. 1970. Steppe vegetation of Washington. Technical Bulletin no. 62, Washington Agricultural Experiment Station, Pullman.

Dick-Peddie, W. A. 1993. New Mexico vegetation, past, present, and future. University of New Mexico Press, Albuquerque.

Dix, R. L. 1964. A history of biotic and climatic changes within the North American grassland, pp. 71–89 in D. J. Crisp (ed.), Grazing in terrestrial and marine environments, a symposium of the British Ecological Society. Blackall Scientific Publ., Oxford.

Dyksterhuis, E. J. 1948. The vegetation of the western cross timbers. Ecol. Monogr. 18:27–376.

Dyksterhuis, E. J. 1949. Condition and management of rangeland based on quantitative ecology. J. Range Mgt. 2:104–115.

Edwards, S. W. 1992. Observations on the prehistory

and ecology of grazing in California. Fremontia 20(1): 3-11.

Ehleringer, J. R. 1978. Implications of quantum yield differences on the distributions of C_3 and C_4 grasses. Oecologia 31:255–267.

Ehleringer, J. R., and O. Bjorkman, 1977. Quantum yields for CO_2 uptake in C_3 and C_4 plants: dependence on temperature, CO_2 and O_2 concentration. Plant Physiol. 59:86–90.

England, R. C., and A. DeVos. 1969. Influence of animals on the pristine conditions on the Canadian grasslands. J. Range Mgt. 22:87–94.

Falk, D. A., and P. Olwell. 1992. Scientific and policy considerations in restoration and reintroduction of endangered species. Rhodora 94:287–315.

Frank, D. A., and S. J. McNaughton. 1991. Stability increases with diversity in plant communities: empirical evidence from the 1988 Yellowstone drought. Oikos 62:360–362.

Franklin, J. F., and C. T. Dyrness. 1969. Vegetation of Oregon and Washington. USDA Forest Service Research Paper PNW-80.

Gartner, F. G., and W. W. Thompson. 1972. Fire in the Black Hills forest-grass ecotone. Proc. Tall Timbers Fire Ecol. Conf. 12:37–68.

Gibbens, R. P., J. M. Tremble, J. T. Hennessy, and M. Cardenas. 1983. Soil movement in mesquite dunelands and former grasslands of southern New Mexico from 1933 to 1980. J. Range Mgt. 36:145–148.

Gleason, H. A. 1913. The relation of forest distribution and prairie fires in the middle west. Torreya 13:173–181.

Gleason, H. A. 1923. The vegetational history of the Middle West. Ann. Assoc. Amer. Geogr. 12:39–85.

Glenn, S. M., S. L. Collins, and D. J. Gibson. 1992. Disturbances in tallgrass prairie: local and regional effects of community heterogeneity. Landscape Ecol. 7:243–251.

Glenn-Lewin, D. C., 1980. The individualistic nature of plant community development. Vegetatio 43:141–146.

Godfrey, C. L., G. S. McKee, and H. Oakes. 1973. General Soils Map of Texas, 1/1 500 000. Texas Agric. Expt. Station, College Station.

Golley, F. B. 1971. Energy flux in ecosystems, pp. 699–788 in J. A. Weins (ed.), Ecosystem structure and function. Proceedings of the 31st Annual Biology Colloquium, Oregon State University, 1970. Oregon State University Press, Corvallis.

Golley, F. B. 1993. A history of the ecosystem concept. Yale University Press, New Haven.

Gould, F. W. 1962. Texas plants – a checklist and ecological summary. Texas Agricultural Experiment Station bulletin MP-585.

Gould, F. W. 1968. Grass systematics. McGraw-Hill, New York.

Gould, F. W., and R. B. Shaw. 1983. Grass systematics, 2nd ed. Texas A&M University Press, College Station.

Grimm, E. C. 1983. Chronology and dynamics of vegetation change in the prairie-woodland region of southern Minnesota, U.S.A. New Phytol. 93:311–350.

Groffman, P. M., C. W. Rice, and J. M. Tiedje. 1993. Denitrification in a tallgrass prairie landscape. Ecology 74:855–862.

Harlan, J. R. 1956. Theory and dynamics of grassland agriculture. D. Van Nostrand, New York.

Harris, G. A. 1967. Some competitive relationships between *Agropyron spicatum* and *Bromus tectorum*. Ecol. Monogr. 37:89–111.

Hartley, W. 1950. The global distribution of tribes of Gramineae in relation to historical and environmental factors. Austr. J. Agric. Res. 1:355–373.

Hartley, W. 1964. The distribution of grasses, pp. 29–64 in C. Barnard (ed.), Grasses and grasslands. Macmillan, New York.

Hawley, A. W. L. 1989. Bison farming in North America, pp. 346–361 in R. J. Hudson, K. R. Drew, and L. M. Baskin (eds.), Wildlife production systems. Cambridge University Press, Cambridge.

Heady, H. F. 1975. Rangeland management. McGraw-Hill, New York.

Heady, H. F. 1977. Valley grassland, pp. 491–514 in M. G. Barbour and J. Major (eds.), Terrestrial vegetation of California. Wiley, New York.

Heady, H. F., and R. D. Child. 1994. Rangeland ecology and management. Westview Press, Boulder.

Heady, H. F., T. C. Foin, J. J. Kektner, D. W. Taylor, M. G. Barbour, and W. J. Berry. 1977. Coastal prairie and northern coastal scrub, pp. 733–760 in M. G. Barbour and J. Major (eds.), Terrestrial vegetation of California. Wiley, New York.

Hennessy, J. T., R. P. Gibbens, J. M. Tromble, and M. Cardenas. 1983. Vegetation changes from 1935 to 1980 in mesquite dunelands and former grasslands of southern New Mexico. J. Range Mgt. 36:370–374.

Herkert, J. R. 1994. The effects of habitat fragmentation on midwestern grassland bird communities. Ecolo. Applic. 4:461–471.

Hervey, D. F. 1949. Reaction of a California annual-plant community to fire. J. Range Mgt. 2:116–121.

Hole, F. D., and G. A. Nielsen. 1968. Some processes of soil genesis under prairie, pp. 28–34 in P. Schramm (ed.), Proceedings of a symposium on prairie and prairie restoration. Knox College, Galesburg, Ill.

Humphrey, R. R. 1953. The desert grassland, past and present. J. Range Mgt. 6:159–164.

Humphrey, R. R. 1958. The desert grassland. Bot. Rev. 24:193–252.

Humphrey, R. R. 1962. Range ecology. Ronald Press, New York.

Humphrey, R. R., and L. A. Mehrhoff. 1958. Vegetation changes on a southern Arizona grassland range. Ecology 39:720–726.

Hunt, C. B. 1967. Physiography of the United States. Freeman, San Francisco.

Hunt, C. B. 1972. Physiography of the United States, 2d ed. Freeman, San Francisco.

Inglis, J. 1964. A history of vegetation on the Rio Grande Plains. Texas Parks Wildl. Dept. Bull. 45. Austin.

Jackson, L. E. 1985. Ecological origins of California's Mediterranean grasses. J. Biogeog. 12:349–361.

Jenny, H. 1930. A study of the influence of climate upon the nitrogen and organic matter content of soil. University of Missouri Agricultural Experiment Station bulletin 152, Columbia.

Johnston, A., and M. D. MacDonald. 1967. Floral initiation and seed production in *Festuca scabrella* Torr. Can. J. Plant Sci. 47:577–583.

Johnston, A., J. F. Dormaar, and S. Smoliak. 1971. Long-term grazing effects on fescue grassland soil. J. Range Mgt. 24:185–188.

Joyce, L. A. 1993. The life cycle of the range condition concept. J. Range Mgt. 46:132–138.

Kartesz, J. T. 1994. A synonymized checklist of vascular flora of the United States, Canada, and Greenland. Vol. 1 – Checklist. 2nd ed. Timber Press, Portland, Ore.

Knapp, A. K. 1984. Post-burn differences in solar radiation, leaf temperature and water stress in influencing production in a lowland tallgrass prairie. Amer. J. of Bot. 71:220–227.

Knapp, A. K. 1993. Gas exchange dynamics in C_3 and C_4 grasses: consequences of differences in stomata conductances. Ecology 74:113–123.

Knapp, A. K., J. T. Fahnestock, S. P. Hambur, L. B. Statland, T. R. Seastedt, and D. S. Schimel. 1993. Landscape patterns in soil-plant water relations and primary production in tallgrass prairie. Ecology 74:549–560.

Knopf, F. L. 1986. Changing landscapes and the cosmopolitanism of eastern Colorado avifauna. Wildl. Soc. Bull. 14:132–142.

Knopf, F. L. 1992. Faunal mixing, faunal integrity, and the biopolitical template for diversity conservation. Transactions of the North American Wildland Natural Resource Conference 57:330–342.

Knopf, F. L. 1994. Avian assemblages on altered grasslands. Stud. in Avian Biol. 15:247–257.

Koucky, R. W. 1983. The buffalo disaster of 1882. N. Dak. Hist. 50:23–37.

Krausman, P. R., (ed.). 1996. Rangeland wildlife. Society for Range Management, Denver, Colo.

Krueger, K. 1986. Feeding relationships among bison, pronghorn, and prairie dogs: an experimental analysis. Ecology 67:760–770.

Kucera, C. L. 1960. Forest encroachment in native prairie. Iowa State J. Sci. 34:635–639.

Kucera, C. L. 1970. Ecological effects of fire on tallgrass prairie. Ecology 43:334–336.

Küchler, A. W. 1964. Potential natural vegetation of the conterminous United States. American Geographical Society special publication 36, New York.

Küchler, A. W. 1975. Potential natural vegetation of the conterminous United States [Map]. American Geographical Society, New York.

Larsen, E. C. 1947. Photo-periodic responses of geographical strains of *Andropogon scoparius*. Bot. Gaz. 109:132–149.

Larson, F. 1940. The role of bison in maintaining the short grass plains. Ecology 21:113–121.

Lauenroth, W. K. 1979. Grassland primary production: North American grasslands in perspective, pp. 3–24 in N. R. French, ed., Perspectives in grassland ecology. Springer-Verlag, New York.

Launchbaugh, J. L. 1964. Effects of early spring burning on yields of native vegetation. J. Range Mgt. 17:5–6.

Launchbaugh, J. L. 1972. Effects of fire on shortgrass and mixed prairie species. Proc. Tall Timbers Fire Ecol. Conf. 12:129–151.

Lehmann, V. W. 1965. Fire in the range of Attwater's prairie chicken. Tall Timbers Fire Ecol. Conf., 4:127–143.

Lewis, J. K., and D. M. Engle. 1980. Impacts of technologies of productivity and quality of rangelands in the Great Plains region: Background paper no. 19, in Impacts of technology on U.S. cropland and rangeland productivity. Vol. 2. Background papers, Part D. United States Congress, Office of Technology Assessment. Government Printing Office, Washington, D.C.

Looman, J. 1969. The fescue grasslands of western Canada. Vegetatio 19:128–145.

Looman, J. 1980. The vegetation of the Canadian Prairie Provinces II. The grasslands, Part 1. Phytocoenologia 8:153–190.

Looman, J. 1981. The vegetation of the Canadian Prairie Provinces II. The grasslands, Part 2. Mesic grasslands and meadows. Phytocoenologia 9:1–26.

Lytle, S. A. 1968. The morphological characteristics and relief relationships of representative soils in Louisiana. La. Agric. Exp. Stn. Bull. 631.

Mack, R. N. 1981. The invasion of *Bromus tectorum* L. into western North America: an ecological chronicle. Agro-ecosystems 7:145–165.

Malin, J. C. 1953. Soil, animal, and plant relations of the grassland, historically reconsidered. Sci. Monthly 76: 207–220.

Martin, S. C. 1975. Ecology and management of southwestern semi-desert grass-shrub ranges: the status of our knowledge. USDA Forest Service research paper RM-156. Rocky Mountain Forest and Range Experiment Sta., Fort Collins.

Maller-Beck, H. 1966. Paleohunters in America: origins and diffusion. Science 152:1191–1210.

McCollum F. T. III, R. L Gillen, and J. E. Brummer. 1994. Cattle diet quality under short duration grazing on tallgrass prairie. J. Range Mgt. 47:489–493.

McComb, A. L., and W. E. Loomis. 1944. Subclimax prairie. Bull. Torrey Bot. Club 71:46–76.

McDonald, J. N. 1981. North American bison; their classification and evolution. University of California Press, Berkeley.

McMillan, C. 1956a. Nature of the plant community, I, Uniform garden and light period studies of five grass taxa in Nebraska. Ecology 37:330–340.

McMillan, C. 1956b. Nature of the plant community, II, Variation in flowering behavior within populations of *Andropogon scoparius*. Amer. J. Bot. 43:429–436.

McMillan, C. 1957. Nature of the plant community, III, Flowering behavior within two grassland communities under reciprocal transplanting. Amer. J. Bot. 44:144–153.

McMillan, C. 1959a. Nature of the plant community, V, Variation within the true prairie community type. Amer. J. Bot. 46:418–424.

McMillan, C. 1959b. The role of ecotypic variation in the distribution of the central grassland of North America. Ecol. Monogr. 29:287–308.

McMurphy, W. E., and K. L. Anderson. 1965. Burning Flint Hills range. J. Range Mgt. 18:265–269.

McNaughton, S. J., M. B. Coughenhour, and L. L. Wallace. 1982. Interactive processes in grassland ecosystems, pp. 167–193 in J. R. Estes, R. J. Tyrl, and J. N. Brunken (eds.), Grasses and grasslands, systematics and ecology. University of Oklahoma Press, Norman.

Moss, E. H. 1944. The prairie and associated vegetation of southwestern Alberta. Can. J. Res. Sect. C. 22:11–31.

Moss, E. H. 1955. The vegetation of Alberta. Bot. Rev. 21:493–567.

Moss, E. H., and J. A. Campbell. 1947. The fescue grassland of Alberta. Can. J. Res. Sect. C. 25:200–227.

Nelson, E. W. 1925. Status of the pronghorned antelope, 1922–1924. USDA bulletin 1346. Government Printing office, Washington, D.C.

Ode, D. J., L. L. Tieszen, and J. C. Lerman. 1980. The seasonal contribution of C_3 and C_4 plant species to primary production in a mixed prairie. Ecology 61: 1304–1311.

Ojima, D. S., W. J. Parton, D. S. Schimel, and C. E. Owensby. 1989. Simulating impacts of annual burning on prairie ecosystems, pp. 118–132 in S. L. Collins and L. Wallace, ed, Fire in North American tallgrass prairies. Oklahoma University Press, Norman.

Ojima, D. S., D. S. Schimel, W. J. Parton, and C. E. Owensby. 1994. Long-and short-term effects of fire on nitrogen cycling in tallgrass prairie. Biogeochem. 24: 67–84.

Olmsted, C. E. 1944. Growth and development in range grasses, IV, Photoperiodic responses in twelve geographic strains of side-oats grama. Bot. Gaz. 106:46–74.

O'Meilia, M. E., F. L. Knopf, and J. C. Lewis. 1982. Some consequences of competition between prairie dogs and beef cattle. J. Range Mgt. 35:580–585.

Parker, K. W., and S. C. Martin. 1952. The mesquite problem on southern Arizona ranges. USDA circular 908. Rocky Mountain Forest and Range Experiment Sta., Fort Collins.

Penfound, W. T. 1964. The relation to grazing to plant succession in the tallgrass prairie. J. Range Mgt. 17: 256–260.

Perino, J. V., and P. G. Risser. 1972. Some aspects of structure and function in Oklahoma old-field succession. Bull. Torrey Bot. Club 99:233–239.

Pickup, G., G. N. Bastin, and V. H. Chewings. 1994. Remote-sensing-based condition assessment for nonequilibrium rangelands under large-scale commercial grazing. Ecolo. Applica. 4:497–517.

Platou, K. A., and P. T. Tueller. 1985. Evolutionary implications for grazing management systems. Rangelands 7:57–61.

Polley, H. W., H. B. Johnson, and H. S. Mayeux. 1994. Increasing CO_2: comparative response of the C_4 grass Schizchyrium and grassland invader Prosopis. Ecology 75:976–988.

Redmann, R. E. 1975. Production ecology of grassland plant communities in western North Dakota. Ecol. Monogr. 45:83–106.

Rice, E. L. 1950. Growth and floral development of five species of range grasses in central Oklahoma. Bot. Gaz. 3:361–377.

Riegel, A. 1940, A study of the variations in the growth of blue grama grass from seed produced in various sections of the Great Plains Region. Kan. Acad. Sci. Trans. 43:155–171.

Ripley, E. A. 1992. Grassland climate, pp. 7–24 in R. T. Coupland (ed.), Natural grasslands, introduction and western hemisphere, Vol. 81, Ecosystems of the World. Elsevier, New York.

Risser, P. G. 1985. Grasslands, pp. 232–256 in B. F. Chabot and H. A. Mooney (eds.), Physiological ecology of North American plant communities. Chapman and Hall, New York.

Risser, P. G. 1988. Diversity in and among grasslands, pp. 176–180 in E. O. Wilson (ed.), Biodiversity. National Academy Press, Washington, D.C.

Risser, P. G. 1989. Range condition analysis: past, present, and future, pp. 143–155 in W. K. Lauenroth and W. A. Laycock (eds.), Secondary succession and the evaluation of rangeland condition. Westview Press, Boulder, Colo.

Risser, P. G. 1990. Landscape processes and the vegetation of the North American grassland, pp. 133–146 in S. L. Collins, and L. Wallace (eds.), Fire in North

American tallgrass prairies. Oklahoma University Press, Norman.

Risser, P. G., E. C. Birney, H. D. Blocker, S. W. May, W. J. Parton, and J. A. Wiens. 1981. The true prairie ecosystem. Hutchinson Ross, Stroudsburg, Pa.

Roe, F. 1970. The North America buffalo, 2nd ed. University of Toronto Press, Toronto.

Rowe, J. S. 1972. Forest regions of Canada. Department of Environment, Canadian Forest Service, publication 1300.

Rzedowski, J. 1978. Vegetacion de México. Editorial Limusa, México City.

Sampson, A. W. 1950. Application of ecologic principles in determining condition of range lands, pp. 509–514 in United Nations conference on conservation and utilization of resources, Lake Success, N.Y., vol. 6, Land resources. Department of Economic Affairs, United Nations, New York.

Sampson, F., and F. Knopf. 1994. Prairie conservation in North America. BioScience 44:418–421.

Sauer, C. O. 1950. Grassland, climax, fire, and man. J. Range Mgt. 3:16–22.

Scarnecchia, D. L. 1994. A viewpoint: using multiple variables as indicators in grazing research and management. J. Range Mgt. 47:107–111.

Schmidt, K. P. 1938. Post glacial steppes in North America. Ecology 19:396–407.

Schimel, D. S., T. G. F. Kittel, A. K. Knapp, T. R. Seastedt, W. J. Parton, and V. B. Brown. 1991. Physiological interactions along resource gradients in a tallgrass prairie. Ecology 72:672–684.

Schramm, P. 1990. Prairie restoration: a 25-year perspective on establishment and management, pp. 169–177, in Proceedings of the Twelfth North American Prairie Conference, Cedar Falls, Iowa.

Schripsema, J. R. 1977. Ecological changes on pine-grassland burned in the spring, late spring, and winter. Master's thesis, Biology Department, South Dakota State University, Brookings.

Scifres, C. J. 1980. Brush management, principles and practices for Texas and the Southwest. Texas A&M University Press, College Station.

Sears, P. B. 1935. Glacial and postglacial vegetation. Bot. Rev. 1:37–51.

Sears, P. B. 1948. Forest sequence and climatic change in northeastern North America since early Wisconsin time. Ecology 29:326–333.

Seastedt, T. R., and A. K. Knapp. 1993. Consequences of nonequilibrium resource availability across multiple time scales: the transient maxima hypothesis. Amer. Nat. 141:621–633.

Seastedt, T. R., and R. A. Ramundo. 1990. The influence of fire on belowground processes of tallgrass prairie, pp. 99–117 in S. L. Collins and L. Wallace, (eds.), Fire in North American tallgrass prairies. University of Oklahoma Press, Norman.

Seastedt, T. R., C. C. Coxwell, D. S. Ojima, and W. J. Parton. 1994. Controls of plant and soil carbon in a semihumid temperate grassland. Ecolog. Applica. 4:344–353.

Seton, E. T. 1927. Lives of game animals, Vol. III, hoofed animals. Doubleday, New York.

Shantz, H. L. 1954. The place of grasslands in the earth's cover of vegetation. Ecology 35:143–145.

Shariff, A. R., M. E. Biondini, and C. E. Grygiel. 1994. Grazing intensity effects on litter decomposition and soil nitrogen mineralization. J. Range Mgt. 47: 444–449.

Shaw, J. H. 1996. Bison, pp. 227–236 in P. R. Krausman (ed.), Rangeland wildlife. Society for Range Management, Denver, Colo.

Silvertown, J., M. E. Dodd, K. McConway, J. Potts, and M. Crawley. 1994. Rainfall, biomass variation, and community composition in the Park Grass Experiment. Ecology 75:2430–2437.

Sims, P. L., and D. D. Dwyer. 1965. Pattern of retrogression of native vegetation in north-central Oklahoma. J. Range Mgt. 18:20–25.

Sims, P. L., and J. S. Singh. 1978a. The structure and function of ten western North American grasslands, II, Intraseasonal dynamics in primary producer compartments. J. Ecol. 66:547–572.

Sims, P. L., and J. S. Singh. 1978b. The structure and function of ten western North American grasslands, III, Net primary production, turnover, and efficiencies of energy capture and water use. J. Ecol. 66:573–597.

Sims, P. L., and J. S. Singh. 1978c. The structure and function of ten western North American grasslands, IV, Compartmental transfers and energy flow within the ecosystem. J. Ecol. 66:983–1009.

Sims, P. L., J. S. Singh, and W. K. Lauenroth. 1978. The structure and function of ten western North American grasslands, I, Abiotic and vegetational characteristics. J. Ecol. 66:251–285.

Sims, P. L., R. E. Sosebee, and D. M. Engle. 1982. Plant responses to grazing management, pp. 4–31 in D. D. Briske and M. M. Kothmann (eds.), Proceedings of a national conference on grazing management technology. Texas A&M University, College Station.

Sims, P. L., W. A. Berg, and J. A. Bradford. 1995. Vegetation of sandhills under grazed and ungrazed conditions, pp. 129–135 in D. C. Hartnett (ed.), Proceedings of the Fourteenth North American Prairie Conference: Prairie Biodiversity, Kansas State University, Manhattan.

Singh, J. S., W. K. Lauenroth, R. K. Heitschmidt, and J. L. Dodd. 1983. Structural and functional attributes of the vegetation of the mixed prairie of N. Amer. Bot. Rev. 489:117–149.

Slack, C. R., P. G. Roughan, and H. C. M. Basset. 1974. Selective inhibition of mesophyll chloroplast development in some C_4 pathway species by low night temperature. Planta 118:57–73.

Smeins, F. E., D. D. Diamond, and C. W. Hanselka. 1992. Coastal prairie, pp. 269–290, in R. T. Coupland (ed.), Natural grasslands. introduction and Western Hemisphere. Vol. 81, Ecosystems of the World. Elsevier, New York.

Smith, G. D. 1965. Lectures on soil classification. Pedologie special issue 4. Belgium Soil Science Society Rozier 6, Ghent, Belgium.

Smolik, J. D., and J. K. Lewis. 1982. Effect of range condition on density and biomass of nematodes in a mixed prairie ecosystem. J. Range Mgt. 35:657–663.

Steiger, T. L. 1930. Structure of prairie vegetation. Ecology 11:170–217.

Stoddart, L. A., A. D. Smith, and T. W. Box. 1975. Range management. McGraw-Hill, New York.

Tausch, R. J., P. E. Wigand, and J. W. Burkhardt. 1993. Viewpoint: Plant community thresholds, multiple steady states, and multiple successional pathways: legacy of the Quaternary. J. Range Mgt. 46: 439–447.

Teeri, J. A., and L. G. Stowe. 1976. Climatic patterns

and the distribution of C_4 grasses in North America. Oecologia 23:1–2.

Tester, J. R. 1989. Effects of fire frequency on oak savanna in east-central Minnesota. Bull. Torrey Bot. Club 116:134–144.

Tharp, B. C. 1926. Structure of the Texas vegetation east of the 98th meridian. University of Texas Bull. 2606. University of Texas, Austin.

Thornthwaite, C. W. 1933. The climates of the earth. Georgr. Rev. 23:433–440.

Thornthwaite, C. W. 1948. An approach towards a rational classification of climate. Georgr. Rev. 38:55–94.

Tieszen, L. L., D. Hein, S. A. Qvortrup, J. H. Troughton, and S. K. Imbamba. 1979. Use of delta 13 C values to determine vegetation selectivity in East African herbivores. Oecologia 37:351–359.

Tieszen, L. L., M. M. Senyimba, S. K. Imbamba, and J. H. Troughton. 1979. The distribution of C_3 and C_4 grasses and carbon isotope discrimination along an altitudinal and moisture gradient in Kenya. Oecologia 37:337–350.

Tilman, D. 1982. Plant strategies and the dynamics and structure of plant communities. Princeton University Press, Princeton, N. J.

Tilman, D., and J. A. Downing. 1994. Biodiversity and stability in grasslands. Nature 367:363–365.

Tisdale, E. W. 1947. The grasslands of southern British Columbia. Ecology 28:346–382.

Transeau, E. N. 1935. The prairie peninsula. Ecology 16:423–437.

Trottier, G. C. 1992. Conservation of Canadian prairie grasslands – a landowner's guide. Environment Canada. Canadian Wildlife Service, Edmonton, Alberta.

Turner, M. G., W. H. Romme, and R. H. Gardner. 1994. Landscape disturbance models and the long-term dynamics of natural areas. Nat. Areas J. 14:3–11.

U.S. Forest Service. 1980. An assessment of the forest and range land situation in the United States. USDA publication FS-345. Forest Service, Washington, D.C.

Van Dyne, G. M. 1971. The U.S. IBP grassland biome study – an overview, pp. 1-9, in N.R. French (ed.), Preliminary analysis of structure and function in grasslands. Range Science Department, Science Series 10, Colorado State University, Fort Collins.

Vankat, J. L. 1979. The natural vegetation of North America, an introduction. Wiley, New York.

Voight, J. W., and J. E. Weaver. 1951. Range condition classes of native midwestern pasture; an ecological analysis. Ecol. Monogr. 21:39–60.

Walker, B. H., and R. T. Coupland. 1970. Herbaceous wetland vegetation in the aspen grove and grassland regions of Saskatchewan. Can. J. Bot. 48:1861–1878.

Wardle, D. A. 1992. A comparative assessment of the factors which influence microbial biomass carbon and nitrogen in soil. Biolog. Rev. 67:321–358.

Weaver, J. E. 1954. North American prairie. Johnsen Publishing, Lincoln, Nebr.

Weaver, J. E. 1968. Prairie plants and their environment. A fifty-year study in the Midwest. University of Nebraska Press, Lincoln.

Weaver, J. E., and F. W. Albertson. 1956. Grasslands of the Great Plains: their nature and use. Johnsen Publishing, Lincoln, Nebr.

Weaver, J. E., and F. E. Clements. 1929. Plant ecology. McGraw-Hill, New York.

Weaver, J. E., and F. E. Clements. 1938. Plant ecology, 2nd ed. McGraw-Hill, New York.

Webb, W. S., S. Szarek, W. Lauenroth, R. Kinerson, and M. Smith. 1978. Primary productivity and water use in native forest, grassland, and desert ecosystems. Ecology 59:1239–1247.

Wells, P. V. 1970. Historical factors controlling vegetation patterns and floristic distributions in the Central Plains region of North America, pp. 211–221 in W. Dort, Jr., and J. K. Jones (eds.), Pleistocene and recent environments of the central Great Plains. University of Kansas, Department of Geology special publication 3. University of Kansas Press, Lawrence.

West, N. E. 1993. Biodiversity of rangelands. J. Range Mgt. 46:2–13.

Western Land Grant Universities and Colleges and Soil Conservation Service. 1964. Soils of the western United States. Regional publications, Washington State University, Pullman.

Whitman, W. C. 1941. The native grassland, pp. 5–7 in Grass. North Dakota Agricultural Experiment Station research bulletin 300. North Dakota State University, Fargo.

Wiegert, R. G., and D. F. Owen. 1971. Trophic structure, available resources and population density in terrestrial versus aquatic systems. J. Theor. Biol. 30:69–81.

Willet, H. C. 1949. Long period fluctuations in the general circulation of the atmosphere. Meteorol. 6:1.

Willms, W. D., S. Smoliak, and J. F. Dormaar. 1985. Effects of stocking rate on rough fescue grassland vegetation. J. Range Mgt. 38:220–225.

Willms, W. D., J. F. Dormaar, and G. B. Schaalje. 1988. Stability of grazed patches on rough fescue grasslands. J. Range Mgt. 41:503–508.

Winner, W. E. 1994. Mechanistic analysis of plant responses to air pollution. Ecolog. Applica. 4:651–661.

Wright, H. A. 1974. Effect of fire on southern mixed prairie grasses. J. Range Mgt. 27:417–419.

Wright, H. A., and A. W. Bailey. 1980. Fire ecology and prescribed burning in the Great Plains – a research review. USDA Forest Service general technical report INT-77. Intermountain Forest and Range Experiment Station, Ogden, Utah.

Wright, H. A., and A. W. Bailey. 1982. Fire ecology, United States and southern Canada. Wiley, New York.

Young, J. A., R. A. Evans, and J. Major. 1977. Alien plants in the Great Basin. J. Range Mgt. 25:194–201.

Zak, D. R., D. Tilman, R. R. Parmenter, C. W. Rice, F. M. Fisher, J. Vose, D. Milchunas, and C. W. Martin. 1994. Plant production and soil microorganisms in late successional ecosystems: a continental-scale study. Ecology 75:2333–2347.

Chapter
10

Eastern Deciduous Forests

HAZEL R. DELCOURT   PAUL A. DELCOURT

INTRODUCTION

Deciduous forest once occupied some 2,560,000 km^2 across the central portion of eastern North America, extending from 32–48° N latitude and from 70–98° W longitude (Shantz and Zon 1924). The presettlement expanse of hardwood forest formed a largely unbroken canopy cover extending from the central Atlantic Coastal Plain to the eastern edge of the Great Plains (Williams 1989; Whitney 1995; Figs. 10.1–10.3). Exceptions included grassy barrens and prairie outliers in the Prairie Peninsula region of the lower Midwest (Transeau 1905; Stuckey 1981; DeSelm and Murdock 1993), localized herbaceous glades on outcrops of limestone, sandstone, and granite scattered across the Southeast (Baskin and Baskin 1986; Quarterman, Burkbanck, and Shure 1993), and old fields once cultivated by Native Americans, including the Cherokee and other eastern tribes along the bottomlands of major river systems (Smith 1978; Morse and Morse 1983; Delcourt, Cridlebaugh, and Chapman 1986; Delcourt 1987). In this chapter, we describe the deciduous forests of eastern North America in relation to the physiological, environmental, and historical factors that have resulted in the present-day distribution, structure, and composition of the vegetation.

The eastern deciduous forest is readily recognized by its physiognomy as well as the physiological adaptations of its species (Braun 1950; Barnes 1991). Deciduous forest stands are composed primarily of winter-deciduous woody angiosperms but include some evergreen gymnosperms such as *Pinus* (pines) and *Tsuga* (hemlock). The forest is typically multistoried, with canopy trees commonly reaching heights of 25–40 m and with maximum longevity of 200–600 yr for dominant tree species (Burns and Honkala 1990a, 1990b). The understory is diverse and characterized by a rich assemblage of spring ephemeral herbs that are largely perennials (Braun 1950; Rogers 1981). Physiological adaptations of deciduous forest species to seasonal environments include both strategies of avoidance and of tolerance of stresses that exist under a closed canopy, and community structure is a result of either complementary or out-of-phase interactions among populations of a large number of species (Hicks and

We thank Michael Barbour, Burton Barnes, and the late William Dwight Billings for their critical reviews and constructive suggestions for improving this chapter. We also thank Carol and Jerry Baskin, Alton Lindsey, Ronald Stuckey, and Peter White for their suggestions for studies to include within this chapter. Contribution Number 62, Quaternary Ecology, University of Tennessee, Knoxville.

Chabot 1985; Peet 1992). For example, spring ephemeral herbs complete much of their annual growth cycle within the few weeks of the early growing season before canopy leafout, taking advantage of high light intensities, as well as available water and nutrients. Unable to modulate photosynthetic and respiratory physiology in response to the decreased light levels that accompany canopy closure, spring ephemerals grow dormant soon after canopy species leaf out, thus exhibiting a seasonal avoidance strategy (Hicks and Chabot 1985). Tree species differ in their responses to temperature; bud break and development of forest canopy occur over 4 + wk. Duration and timing of the growth flush are correlated with successional status and hydraulic architecture of individual hardwood species, with early successional species often having ring-diffuse wood with continuously active xylem, a long period of shoot growth, and an indeterminate, opportunistic growth pattern. In contrast, mid- to late-successional species tend to be more conservative. Species of *Quercus* (oaks), for example, are determinate in growth, with a shorter growth period and ring-porous wood for which a new increment of annual conducting tissue must be differentiated before bud break can occur (Hicks and Chabot 1985).

The boundaries of forests dominated by winter-deciduous hardwoods can be expressed as: (1) the overall ranges and dominance distributions of canopy and understory species (Transeau 1905 [Fig. 10.1]; Shantz and Zon 1924 [Fig. 10.2]; Braun 1950 [Fig. 10.3]; Little 1971; Delcourt, Delcourt, and Webb 1984); (2) the ranges of indicator species of high fidelity (restriction) to the forest communities (Braun 1950; Greller 1988); or (3) the distribution of endemics (Harshberger 1911). Deciduous forests can also be characterized by what they are *not*. That is, they are bounded regionally by adjacent, contrasting vegetation types forming mutual ecotones (Smith 1980; Barnes 1991). For example, eastern deciduous forest is bounded to the north by boreal coniferous forest in which predominant gymnosperm species are tolerant of short growing seasons and are cold-hardy beyond a winter minimum temperature of −40° C (Sakai and Weiser 1973; Larsen 1980; Delcourt and Delcourt 1987; Woodward 1987; Arris and Eagleson 1989). Certain needle-leaf evergreen trees characteristic of the boreal forest are more tolerant of nutrient-deficient, water-logged soils than are broadleaf deciduous trees. Other boreal conifers are more tolerant of droughty soils and frequent disturbances by fire (Larsen 1980; Engstrom and Hansen 1985; Burns and Honkala 1990a, 1990b). The southern boundary of deciduous forest is a transition zone located

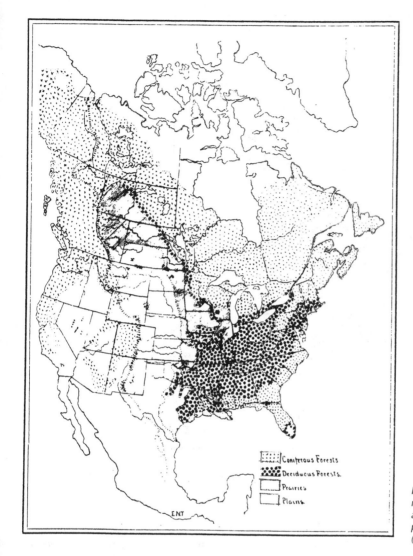

Figure 10.1. Edgar Transeau's 1903 map of North American deciduous and coniferous forests, prairies, and plains, as modified from Sargent (1884) (from Stuckey 1981).

approximately across the physiographic boundary between the Piedmont and the Inner Coastal Plain, where the change in dominant physiognomy from broadleaf deciduous to broadleaf evergreen forest may be related in part to the frost-hardiness of coastal plant species (see also Chapter 11).

Eastern deciduous forest communities can also be characterized by a long recurrence interval of disturbance (relative to successional recovery time), as well as a small disturbance area relative to spatial extent of the forest type (Turner, Romme, Gardner, O'Neill, and Kratz 1993). For instance, in the southern Appalachian Mountains, windthrown of individual canopy trees is the major former of light gaps in the forest canopy (Runkle 1985); mixed deciduous forests of that region are in dynamic successional equilibrium or stable steady state, with minimal probability of catastrophic and

spatially extensive natural disturbance (Turner et al. 1993).

Deciduous forests are not only defined by their spatial extent and responses to present-day environmental variables, but they are also a direct product of a long-term history of development on time scales of thousands to millions of years. Floristically, eastern deciduous forests are closely related to now widely disjunct temperate forests in Europe, Japan, and eastern China (Graham 1972, 1993; Wolfe 1979; Davis 1983; White 1983; Miyawaki, Iwatsuki, and Grandtner 1994), with a common history extending back to the Tertiary Period. In the Eocene Epoch, broadleaf and needle-leaf evergreen coniferous trees were intermixed with broadleaf deciduous angiosperm trees in forests that were contiguous across much of the Northern Hemisphere (Graham 1993). Deciduous forests per-

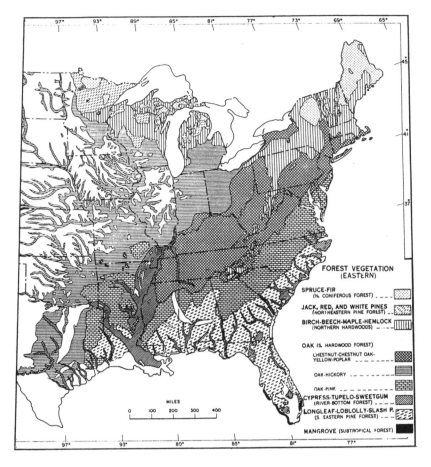

FOREST VEGETATION
(EASTERN)

SPRUCE-FIR
(N. CONIFEROUS FOREST)

JACK, RED, AND WHITE PINES
(NORTHEASTERN PINE FOREST)

BIRCH-BEECH-MAPLE-HEMLOCK
(NORTHERN HARDWOODS)

OAK (S. HARDWOOD FOREST)

CHESTNUT-CHESTNUT OAK-
YELLOW-POPLAR

OAK-HICKORY

OAK-PINE

CYPRESS-TUPELO-SWEETGUM
(RIVER-BOTTOM FOREST)

LONGLEAF-LOBLOLLY-SLASH P.
(S. EASTERN PINE FOREST)

MANGROVE (SUBTROPICAL FOREST)

MILES
0   100   200   300   400

Figure 10.2. Homer Shantz and Raphael Zon's 1924 map of the natural forests and grasslands of the eastern United States.

sisted in western North America until Miocene times (15–10 M yr ago) (Axelrod 1992; Axelrod and Schorn 1994). In the past 2 M yr of the Quaternary Period, a predominant glacial-interglacial climatic cycle, with a periodicity of 100,000 yr, has served as the impetus for repeated dissassembly and reassembly of temperate deciduous forest communities, as species have responded individualistically to changes in temperature and seasonality of climates (Davis 1983; Watts 1988; Huntley and Webb 1989; Prentice 1992; Wright et al. 1993; Delcourt and Delcourt 1994). The total area of mesic deciduous forests in eastern North America has fluctuated from <600,000 km^2 during glacial ages, when many of the now widespread species were restricted to small "pocket" refuges in the southeastern United States, to 1,700,000 km^2 in early interglacial times, when deciduous trees were constituents of ephemeral, rapidly changing communities, to <2,500,000 km^2 in late-interglacial

times, when temperate deciduous forest became dominant over a broad region (Delcourt and Delcourt 1987, 1994).

Today the eastern deciduous forest region is defined by land use and forest management practices (Barrett 1980; Eyre 1980). For example, much of the fertile forest soils of the lower Midwest are in cultivation. The now-eroded, rocky slopes of the Appalachian Mountains are in national forest or commercial timber holdings (Martin and Boyce 1993), as is much of the land in the Upper Great Lakes region and northern New England, and are marginal for cash crops (Stearns 1987; Williams 1989; Whitney 1995).

The various viewpoints on what constitute the essential qualities that make eastern deciduous forests distinct from other vegetation regions reflect different perspectives on the relative influences of process on pattern, each of which changes in importance as different spatial and temporal scales

Figure 10.3. Emma Lucy Braun's 1950 map of principal forest regions in the deciduous forest formation across eastern North America (from Stuckey 1994).

are emphasized (Delcourt, Delcourt, and Webb 1983; Urban, O'Neill, and Shugart 1987; Delcourt and Delcourt 1988). Physiognomy, floristics, and evolution of physiological tolerances of terrestrial vascular plants to climate are a reflection of long-term, megascale processes that include progressive global climate change, shifting distributions on continents as driven by plate tectonics, and episodes of mountain building (Barnosky 1987; Graham 1993). Present-day community composition and geographic distribution of forest types as well as of individual species are a product of species responses to macroscale environmental changes, including cyclic changes in seasonality of climates and soil development (Birks 1986). The expression of deciduous forest vegetation on the modern landscape is a mesoscale to microscale phenomenon (Delcourt and Delcourt 1988) that is reflected as a series of successional stages arrayed along gradients of soil moisture, aspect, and elevation, and responding to prevailing disturbance regimes, including windstorms, fire, and human land use (Whittaker 1956, 1975; Oliver 1981; Barnes 1991).

CLASSIFICATION AND DESCRIPTION OF MAJOR VEGETATION TYPES

Historical Vegetation Mapping

The first map depicting the distribution of forest and nonforest vegetation of the conterminous United States delineated the general distribution of three types of eastern forest, designated as "foliaceous, coniferous, and deciduous" (Henry 1858; Williams 1989). By the mid- to late-1800s, the American General Land Office Surveys (White 1984), combined with numerous plant collections made during botanical explorations (e.g., Michaux 1803; Bartram 1791; Reveal and Pringle 1993) and investigations of the timber and agricultural potential of the land (e.g., Hilgard 1860; Sargent 1884), provided a basis for delineating the distributions of important forest trees (Cooper 1860) and the bounds between eastern deciduous and coniferous forests and midwestern prairies (Henry 1858; Sargent 1884).

Plant ecologists of the early twentieth century

recognized that even the earliest vegetation maps of eastern North America indicated a fundamental relationship between climate and the distribution of vegetation on a subcontinental scale (Merriam 1898; Transeau 1903; Livingston and Shreve 1921; Stuckey 1981; Greller 1989). Transeau (1903, 1905), studying plant distributions in relation to geography and climate, determined and mapped four great "centers of distribution" for forest assemblages in eastern North America (Transeau 1905). Transeau's use of the term "center of distribution" carried no implication that the plants necessarily evolved and spread from these centers but that the centers of plant distribution correspond with climatic centers defined by present-day gradients in temperature and in the ratio of precipitation to evaporation (Transeau 1905; Stuckey 1981). Transeau (1903; Fig. 10.1) used Sargent's (1884) mapping of forest conditions to draw a detailed map depicting the boundaries of the eastern deciduous forest region.

The first map to illustrate the boundaries of natural forest types of North America designated by important or characteristic tree taxa was compiled by H. L. Shantz and R. Zon (1924; Fig. 10.2), who drew information from published vegetation studies and maps, local floras, Forest Service maps, and other reports (e.g. Sargent 1884; Merriam 1898; Harshberger 1911). Shantz and Zon (1924) used vegetation as the biological unit that formed the basis of classification of natural regions. Boundaries between vegetation types were largely based on the distributions of areally dominant canopy trees in addition to indicator species. For example, the chestnut–chestnut oak–yellow poplar (tuliptree) forest of the central and southern Appalachian Mountains was distinguished from the oak–hickory forest of the lower Midwest region by floristic differences that included differences in species richness (Shantz and Zon 1924). Overall, Shantz and Zon (1924) characterized the vegetation of the eastern deciduous forest as being essentially fagaceous in nature, with species of *Quercus* making up the majority of the composition. They delineated the oak-hardwood forest as central in distribution in eastern North America, with hardwood forest becoming mixed with pines and other conifers both to the north (especially with *Pinus strobus* [eastern white pine] and *Tsuga canadensis* [eastern hemlock] becoming important in the Great Lakes region) and to the south (with *Pinus echinata* [shortleaf pine] becoming mixed with oaks in the Piedmont region).

The most widely recognized map of vegetation regions of temperate eastern North America is that of E. Lucy Braun (1950; Fig. 10.3). Braun was a bot-

anist who had both plant ecological and geological training (Wright 1974; Stuckey 1994) and who brought two fundamentally new concepts to vegetation mapping: (1) the climatic climax of Clements (1916); and (2) the peneplain erosion cycle of Davis (Wright 1974). Both schemes emphasized progression in development of natural systems by analogy to growth of an organism, progressing in stages from youth through maturity to old age. Through plant succession, vegetation eventually became characterized by diverse, old-growth communities reaching ultimate development on mesic, nutrient-rich sites. Through the geomorphic erosion cycle, over long periods, mountains were worn down to become flat-lying, coastal landforms (peneplains) that persisted until uplift produced new highlands and once again rejuvenated the geomorphic system, causing the next cycle of erosion to begin. Braun classified forest types and broader forest regions largely on the composition of old-growth stands on mesic sites, using indicator species of vascular plants occurring in both the overstory and understory. Boundaries of forest types were drawn primarily on physiographic and geologic transitions, with certain exceptions such as the northern boundary of northern hardwoods forest, which was based on the distributional limit of *Betula alleghaniensis* (yellow birch). Braun equated centers of diversity in vegetation with floristic centers of origin (evolution of new species). She maintained that the species-rich "mixed mesophytic forests" located on ancient Appalachian landscapes of the Cumberland and Allegheny plateaus were relict communities originating in and persisting unchanged for millions of years since the Tertiary Period (Braun 1950). Her 1950 classification of vegetation of the eastern deciduous forest thus incorporated an idealized view of long-term vegetation development that has been largely discounted in recent decades on the basis of fossil evidence (Delcourt 1979; Watts 1970, 1980). Nevertheless, E. Lucy Braun's descriptions of forest communities (Braun 1950), which were based mainly on her own vegetation sampling of old-growth stands distributed across the region, remain the most comprehensive compilation of compositional data for virgin forest communities of the eastern deciduous forest region.

Maps of late twentieth-century forest vegetation of eastern North America (e.g., Eyre 1980) clearly show the effects of centuries of deforestation and conversion of original forests to agriculture and commercial lumber and paper production. By 1924, it was estimated that half of the original area of forest east of the hundredth meridian had been cleared for farmland and settlements, with only a

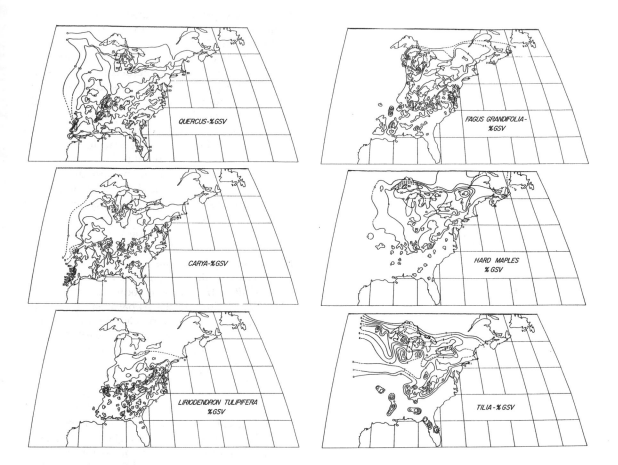

Figure 10.4. Contemporary population maps of forest canopy dominants of eastern deciduous forests (contoured isophyte lines represent values of forest composition, expressed as percentage of growing-stock volume [%GSV] of merchantable wood that can be harvested from commercial forests): Quercus species (oak); Carya species (hickory); Liriodendron tulipifera (tuliptree) (from Delcourt et al. 1984).

Figure 10.5. Contemporary population maps of forest canopy dominants of eastern deciduous forests (contoured isophyte lines for percentage of growing-stock volume [%GSV]): Fagus grandifolia (American beech); "hard maples," Acer saccharum and its subspecies (sugar maple); Tilia species (basswood) (from Delcourt et al. 1984).

fourth of the original forest area remaining in merchantable forest lands (Shantz and Zon 1924). Originally unbroken forests became progressively fragmented into forest islands across much of the eastern deciduous forest (Curtis 1959; Burgess and Sharpe 1981; Williams 1989). Virgin forests became progressively restricted to relatively small preserves such as the Boundary Waters Canoe Area (BWCA) of northern Minnesota (Heinselman 1973; Baker 1989) and the Great Smoky Mountains National Park of Tennessee–North Carolina (Whittaker 1956; Martin and Boyce 1993). The decline in forest area continued until well into this century, with a slight increase in biomass because of deliberate afforestation taking place only since the 1960s (Delcourt and Harris 1980). Resultant patterns of

forest distribution (Eyre 1980) still reflect to some extent original physiographic and climatic bounding conditions. Forest composition today is, however, in large part a product of succession after logging and afforestation in commercially viable tracts (Eyre 1980; Delcourt et al. 1981; Williams 1989; Whitney 1995).

Distributions of Dominant Tree Taxa

The deciduous forest formation is broadly characterized by canopy dominance of six broadleaf deciduous tree taxa (Braun 1950; Daubenmire 1978; Vankat 1979; Delcourt et al. 1984): Quercus, Carya (hickory); Liriodendron tulipifera (yellow poplar, or tuliptree, all in Fig. 10.4); Fagus grandifolia (American beech); Acer saccharum (sugar maple, or "hard maples") and Tilia (basswood, all in Fig. 10.5).

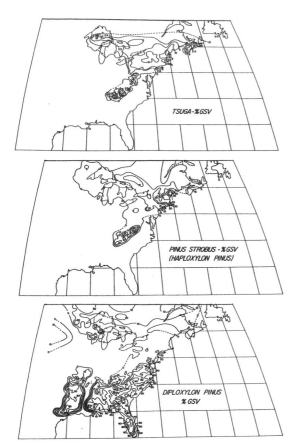

Figure 10.6. Contemporary population maps of forest canopy dominants of hemlock–white pine–northern hardwoods forests and eastern coniferous forests (contoured isophyte lines for percentage of growing-stock volume [%GSV]): Tsuga canadensis (eastern hemlock); Pinus strobus (eastern white pine); Subgenus Diploxylon Pinus, with both a northern group of boreal pines (north of 42° N latitude) and a southeastern group of warm-temperate pines (south of 42° N) (from Delcourt et al. 1984).

Conifers important in mixed stands within the deciduous forest region include *Tsuga canadensis* (Fig. 10.6) and *Pinus strobus* (Haploxylon *Pinus* in Fig. 10.6). The eastern deciduous forest region is bounded by the distribution of other species of *Pinus* (Diploxylon *Pinus* in Fig. 10.6).

The patterns of dominance distribution of trees in commercial forest stands are represented in Figures 10.4–10.6 as percentages of growing-stock volume, the harvestable wood volume in tree boles (data summarized from Continuous Forest Inventories of the United States and Canadian Forest Services; Delcourt et al. 1984). The contoured tree-population maps of percentages of dominance show that overall population distributions of major

tree taxa are primarily related to broad environmental constraints, despite the overprint of recent land use (Eyre 1980).

For example, *Quercus* (Fig. 10.4) is prevalent in the geographic center of the deciduous forest region, reaching ⩾80% of the forest composition in a broad latitudinal belt extending from the southern Appalachian Mountains to the Ozark Highlands. *Quercus* species diminish in importance toward the northern distributional limits of the genus in the central Great Lakes region and northern New England. This dominance-distribution pattern reflects in part the decrease in number of *Quercus* species from the central deciduous forest to its northern limit. *Carya* (Fig. 10.4) is most important to the south and southwest, with dominance values of 30–40% from the southern Appalachian Mountains to the Ozark Highlands to the Gulf Coastal Plain of southeastern Texas. *Liriodendron tulipifera* (Fig. 10.4) is locally 20–30% of the forest composition in the Appalachian Mountains and in the Ozark Highlands; elsewhere throughout its broad range, it is often <5%. Nevertheless, it is a characteristic large tree of much of the deciduous forest even at the lower elevations away from the mountains.

Fagus grandifolia (Fig. 10.5) comprises as much as 10% of the forests on a regional basis in the central Appalachian Mountains and in southern Michigan. It is widespread throughout the eastern deciduous forest at about 5% of the forest composition. *Acer saccharum* ("hard maples," Fig. 10.5) is most important in the northern half of the eastern deciduous forest region, reaching 30–40% of the forest composition in the Great Lakes and New England regions. *Tilia* (Fig. 10.5) has two population centers where it reaches 6–10% of the forest: in the southern Appalachian Mountains and in the western Great Lakes region.

Castanea dentata (American chestnut) was formerly a dominant deciduous tree throughout the Appalachian Mountains (Shantz and Zon 1924; Braun 1950). It was virtually eliminated during the early decades of the twentieth century following introduction from Europe of the fungal blight *Endothia parasitica* (Anderson 1974). For trees such as *Liriodendron tulipifera*, which in presettlement times were restricted to infrequent occurrence as successional trees within large canopy gaps in Appalachian forests (Schenck 1974; Buckner and McCracken 1978; Barden 1980, 1981; Runkle 1982), current high relative abundances within forest communities (e.g., Fig. 10.4) might reflect both adjustments to widespread loss of *Castanea dentata* (McCormick and Platt 1980) and establishment of extensive successional stands following land clearance that opened large gaps in the once-closed forest canopy (Clebsch and Busing 1989; Busing 1995).

Two needleleaf evergreen conifers are locally important within the eastern deciduous forest region. *Tsuga canadensis* (Fig. 10.6) and *Pinus strobus* (Fig. 10.6) range generally northward from the Appalachian Mountains to New England and the Great Lakes.

The northern ecotone of eastern deciduous forest with boreal forest, which extends through the northern Great Lakes region to New England, is marked by the 10% contour in boreal Diploxylon *Pinus* (primarily the southern distribution of *Pinus banksiana*, jack pine, Fig. 10.6). The southern ecotone of deciduous forests with the southeastern evergreen forest region of Braun (1950) generally coincides with the northern limits of distribution of a number of species of *Pinus* (southern Diploxylon *Pinus*, Fig. 10.6). Southern species of *Pinus* today generally comprise 40–80 + % of the commercial forests across the sandy, rolling uplands of the Gulf and southern Atlantic coastal plains. Northward, southern *Pinus* species diminish in importance to 10% or less in the Ozark Highlands of Arkansas and Missouri, the Interior Low Plateaus of northern Alabama and Tennessee, and the southern Appalachian Mountains of northern Georgia, western North Carolina, and East Tennessee.

Description of Braun's Forest Regions

Braun mapped nine major forest regions within the deciduous forest formation (Braun 1950; Fig. 10.3). Her evaluation of compositional variability within each of the forest regions was based on data from a large array of sampled virgin and mature, old-growth stands (representative examples are summarized in Table 10.1; taxonomic nomenclature follows Kartesz 1994). This summary is based on Braun's original descriptions of the forest regions, followed by examples from contemporary literature to illustrate successional trends, vegetation distributions along environmental gradients, and disturbance regimes representative of each region. For more detailed site-specific descriptions of representative communities and their successional relationships within each region, as well as additional literature references that space limitations do not permit listing here, we refer the reader to the following: Rowe (1972); Eyre (1980); Hicks and Chabot (1985); Greller (1988); Barnes (1991); and chapters in Martin, Boyce, and Echternacht (1993a, 1993b), and by Barbour and Christensen (1993).

Mixed mesophytic forests. Nearly coextensive with the unglaciated Cumberland and Allegheny

Figure 10.7. Mixed mesophytic forest stand composed primarily of Fagus grandifolia (beech), Acer saccharum (sugar maple), Aesculus flava (yellow buckeye), Tilia americana var. americana (white basswood), species of Quercus (oaks), and Carya (hickories), with Tsuga canadensis (eastern hemlock) on north-facing mid- and lower slopes. Caney Fork in Fall Creek Falls State Park, Cumberland Plateau, Tennessee (photograph by Hazel and Paul Delcourt).

plateaus and characterized by the most complex plant association of the eastern deciduous forest region, the mixed mesophytic forest region is geographically central (Fig. 10.3). Dominant trees within the mixed mesophytic forest climax community (Fig. 10.7) are *Fagus grandifolia, Liriodendron tulipifera, Tilia americana* var. *heterophylla* (white basswood), *Acer saccharum*, (formerly) *Castanea dentata, Aesculus flava* (yellow buckeye), *Quercus rubra* var. *rubra* (northern red oak), *Q. alba* (white oak), and *Tsuga canadensis* (Table 10.1, Stands 1 and 2). In outliers within the southern Appalachian Mountains, Braun (1950) considered *Halesia tetraptera* (silverbell) an indicative species in communities of mixed mesophytic cove hardwoods. Other locally abundant species within the mixed mesophytic forest include *Betula alleghaniensis, Prunus serotina* (black cherry), *Magnolia acuminata* (cucumber magnolia), *Fraxinus americana* (white ash), and *Acer rubrum* (red maple) (Table 10.1, Stands 1 and 2).

Overall, mixed mesophytic forest is character-

Table 10.1. *Canopy Composition (%) of Old-Growth, Mesic Forest Stands (trees ≥ 10 inches d.b.h.) from the Deciduous Forest Formation (data from Braun 1950). The asterisk (*) denotes an evergreen habit. Scientific nomenclature follows Kartesz (1994).*

	Tree Species Letter Code (Fig. 10.12)	Mixed Mesophytic Forest Region: ○ Cumberland Mts., Letcher Co., Ky.	○ Allegheny Mts., Webster Co., W.Va.	Western Mesophytic Forest Region: ○ Bluegrass Section, Grant Co., Ky.	○ Mississippian Plateau Section, Wayne Co., Ky.	○ Loess Hills Mississippi Embayment Section, Obion Co., Tenn.	Oak-Hickory Forest Region: *Southern Division, Interior Highlands–* ○ Ozark Plateau, Carter Co., Mo.	○ Western Border, eastern Okla.	*Northern Division–* ○ Prairie Peninsula Section, DuPage Co., Ill.	Oak-Chestnut Forest Region: ○ Southern Appalachians, Mesic Cove Hardwoods, Joyce Kilmer, Graham Co., N.C.	○ Montane Spruce-Fir, Roan Mt., Avery Co., N.C.
Forest Stand Number		1	2	3	4	5	6	7	8	9	10
(Braun Table, Site)		(T1–1)	(T9–C)	(T19)	(T29–A)	(T31)	(T32–5)	(T34)	(T38–4)	(T43–3)	(T41–2)
Braun Designation for Forest Region – Type		1A	1B	2A	2E	2F	3A(1)	3A(2)	3D	4A	4A(SF)
Canopy Species Richness		19	9	16	14	13	12	6	8	12	6
*Abies balsamea**	AB	—	—	—	—	—	—	—	—	—	—
*Abies fraseri**	AFr	—	—	—	—	—	—	—	—	—	19.4
Acer rubrum	AR	0.7	—	—	—	—	—	—	—	—	—
Acer saccharum	ASa	31.1	22.1	15.0	7.8	1.0	3.8	—	26.1	10.6	—
Acer spicatum	ASp	—	—	—	—	—	—	—	—	—	0.8
Aesculus flava (SYN = A. octandra)	AFl	16.7	—	—	6.6	—	—	—	—	—	0.8
Betula alleghaniensis (B. lutea)	BA	0.7	1.3	—	—	—	—	—	—	—	21.0
Betula lenta	BL	—	—	—	—	—	—	—	—	1.4	—
Betula papyrifera	BP	—	—	—	—	—	—	—	—	—	—
Carya cordiformis	CC	—	—	—	—	—	5.8	—	—	—	—
Carya glabra	CG	—	—	1.2	—	—	—	—	—	—	—
Carya ovata	COv	1.0	—	4.2	2.6	—	—	—	—	—	—
Carya undiff. spp.	CU	0.3	—	0.6	—	7.7	—	15.1	—	0.7	—
Castanea dentata	CD	9.4	37.6	—	—	—	—	—	—	16.9	—
Celtis occidentalis	COc	—	—	0.6	—	—	—	—	—	—	—
Diospyros virginiana	DV	—	—	—	—	—	—	—	—	—	—
Fagus grandifolia	FG	5.4	13.0	22.2	24.7	46.2	—	—	—	12.7	0.8
Fraxinus americana	FA	1.3	1.3	6.6	6.6	3.8	5.8	—	20.4	1.4	—
Fraxinus nigra	FN	—	—	—	—	—	—	—	—	—	—
Fraxinus quadrangulata	FQ	—	—	0.6	—	—	—	—	—	—	—
Fraxinus undiff. spp.	FU	—	—	—	—	—	—	—	—	—	—
Gleditsia triacanthos	GT	—	—	—	—	—	—	—	—	—	—
Halesia tetraptera (H. monticola)	HT	—	—	—	—	—	—	—	—	12.7	—
Juglans cinerea	JC	0.3	—	—	2.6	—	3.8	—	—	—	—
Juglans nigra	JN	0.3	—	9.6	1.3	2.9	7.7	—	3.4	—	—
Liquidambar styraciflua	LS	—	—	—	—	1.0	—	—	—	—	—
Liriodendron tulipifera	LT	4.0	—	—	11.7	9.6	—	—	—	16.2	—
Magnolia acuminata	MA	1.0	2.6	—	2.6	6.7	—	—	—	0.7	—
*Magnolia grandiflora**	MG	—	—	—	—	—	—	—	—	—	—
Magnolia fraseri	MF	0.3	1.3	—	—	—	—	—	—	1.4	—
Morus rubra	MR	—	—	—	—	—	3.8	—	—	—	—
Nyssa sylvatica	NS	0.3	—	1.8	—	4.8	1.9	—	—	—	—
Nyssa undiff. spp.	NU	—	—	—	—	—	—	—	—	—	—
Ostrya virginiana	OV	—	—	—	—	—	—	—	—	—	—
*Picea rubens**	PR	—	—	—	—	—	—	—	—	—	57.3
*Pinus echinata**	PE	—	—	—	—	—	—	2.7	—	—	—
*Pinus strobus**	PS	—	—	—	—	—	—	—	—	—	—
*Pinus taeda**	PT	—	—	—	—	—	—	—	—	—	—

Northern Blue Ridge Mts., Oak-Chestnut, Rappahannock Co., Va.	Ridge and Valley Section, Oak-Tuliptree, Bath Co., Va.	Piedmont Section, Beech-Tuliptree, Northern Md.	Oak-Pine Forest Region: Atlantic Slope Section, Duke Forest, Durham, N.C.	Southeastern Evergreen Forest Region: Bottomland Forest, Mississippi Alluvial Plain, Southern Ill.	Upland Hardwood Forests, Loess Hills, Tunica Hills, West Feliciana Parish, La.	Beech-Maple Forest Region: Berrien Co., Mich.	Maple-Basswood Forest Region: Big Woods Section, SW Minn.	Hemlock-White Pine-Northern Hardwoods Forest Region: Great Lakes-St. Lawrence Division– Great Lakes Section, Chippewa-Luce Co., Mich.	Minnesota Section, Beltrami Co., Minn.	Northern Appalachian Highland Division– Allegheny Section, Warren Co., Pa.	Adirondack Section, Western Adirondack Mts., N.Y.	New England Section, Pittsburg Township, N.H.	Taxon Frequency (No. stands/23 stands)
11 (T45–3)	12 (T46–1)	13 (T51–2)	14 (T52–A)	15 (T54–3)	16 (T57–2)	17 (T63–1)	18 (T67–M)	19 (T71–7C)	20 (T79–5)	21 (T81–1a)	22 (T87–5)	23 (T90–B)	
4B	4C	4D	5A	6A	6B	7	8B	9A	9C	9E	9F	9G	
8	8	11	10	10	9	8	5	9	3	8	11	6	
—	—	—	—	—	—	—	—	—	—	—	5.7	5.2	2/23
—	—	—	—	—	—	—	—	—	—	—	—	—	1/23
2.6	—	1.3	—	—	—	—	—	5.3	—	0.7	3.5	—	6/23
—	—	—	—	—	—	31.0	47.2	42.1	27.8	15.9	8.3	51.3	15/23
—	—	—	—	—	—	—	—	—	—	—	—	—	1/23
—	—	—	—	—	—	—	—	—	—	—	—	—	3/23
—	—	—	—	—	—	—	—	8.8	—	0.2	19.1	27.6	7/23
1.9	—	—	—	—	—	—	—	—	—	1.5	—	—	3/23
—	—	—	—	—	—	—	—	—	—	—	—	0.2	1/23
—	—	—	—	—	—	—	—	—	—	—	—	—	1/23
—	—	—	—	—	—	—	—	—	—	—	—	—	1/23
—	—	—	—	—	—	0.4	—	—	—	—	—	—	4/23
—	3.8	2.5	31.7	21.1	2.1	—	—	—	—	—	—	—	10/23
34.8	1.1	43.0	—	—	—	—	—	—	—	—	—	—	6/23
—	—	—	—	—	—	—	—	—	—	—	—	—	1/23
—	—	—	1.2	—	—	—	—	—	—	—	—	—	1/23
—	—	16.3	—	—	51.1	62.1	—	8.8	—	36.6	13.6	1.5	14/23
0.6	—	—	—	—	—	—	—	2.6	—	0.7	—	—	11/23
—	—	—	—	—	—	—	—	—	—	—	0.2	—	1/23
—	—	—	—	—	—	—	—	—	—	—	—	—	1/23
—	—	—	—	4.3	—	—	—	—	—	—	—	—	1/23
—	—	—	—	0.6	—	—	—	—	—	—	—	—	1/23
—	—	—	—	—	—	—	—	—	—	—	—	—	1/23
—	—	—	—	—	—	—	—	—	—	—	—	—	3/23
—	—	2.5	—	1.9	—	—	—	—	—	—	—	—	8/23
—	—	1.3	—	19.2	—	—	—	—	—	—	—	—	3/23
—	18.8	16.3	—	—	6.4	0.9	—	—	—	—	—	—	8/23
—	—	—	—	—	—	—	—	—	—	—	—	—	5/23
—	—	—	—	—	14.9	—	—	—	—	—	—	—	1/23
—	—	—	—	—	—	—	—	—	—	—	—	—	3/23
—	—	—	—	—	—	—	—	—	—	—	—	—	1/23
—	1.6	5.1	1.2	—	—	—	—	—	—	—	—	—	7/23
—	—	—	—	6.2	—	—	—	—	—	—	—	—	1/23
—	—	—	—	2.1	—	—	—	—	—	—	—	—	1/23
—	—	—	—	—	—	—	—	—	—	—	42.8	12.2	3/23
—	—	—	—	—	—	—	—	—	—	—	—	—	1/23
—	—	—	—	—	—	—	—	0.9	—	—	0.2	—	2/23
—	—	—	1.2	—	—	—	—	—	—	—	—	—	1/23

Table 10.1. *Cont.*

	Tree Species Letter Code (Fig. 10.12)	Mixed Mesophytic Forest Region: Cumberland Mts., Letcher Co., Ky.	Allegheny Mts., Webster Co., W.Va.	Western Mesophytic Forest Region: Bluegrass Section, Grant Co., Ky.	Mississippian Plateau Section, Wayne Co., Ky.	Loess Hills Mississippi Embayment Section, Obion Co., Tenn.	Oak-Hickory Forest Region: Southern Division, Interior Highlands— Ozark Plateau, Carter Co., Mo.	Western Border, eastern Okla.	Northern Division— Prairie Peninsula Section, DuPage Co., Ill.	Oak-Chestnut Forest Region: Southern Appalachians, Mesic Cove Hardwoods, Joyce Kilmer, Graham Co., N.C.	Montane Spruce-Fir, Roan Mt., Avery Co., N.C.
Forest Stand Number		1	2	3	4	5	6	7	8	9	10
(Braun Table, Site)		(T1–1)	(T9–C)	(T19)	(T29–A)	(T31)	(T32–5)	(T34)	(T38–4)	(T43–3)	(T41–2)
Braun Designation for Forest Region – Type		1A	1B	2A	2E	2F	3A(1)	3A(2)	3D	4A	4A(SF)
Canopy Species Richness		19	9	16	14	13	12	6	8	12	6
Platanus occidentalis	PO	—	—	—	—	—	—	—	—	—	—
Prunus serotina	PSr	0.3	16.9	—	—	—	—	—	—	—	—
Quercus alba	QA	—	—	16.2	2.6	3.8	—	17.8	3.0	—	—
Quercus rubra var. rubra (*Quercus borealis var. maxima*)	QR	6.3	3.9	—	3.9	8.6	38.5	—	22.7	—	—
Quercus coccinea	QC	—	—	—	—	—	—	—	—	—	—
Quercus falcata	QF	—	—	—	—	—	—	—	—	—	—
Quercus macrocarpa	QMc	—	—	—	—	—	—	—	5.7	—	—
Quercus marilandica	QMr	—	—	—	—	—	—	39.7	—	—	—
Quercus prinus (*Quercus montana*)	QP	0.3	—	—	—	—	—	—	—	—	—
Quercus muhlenbergii	QMu	—	—	2.4	—	—	15.4	—	—	—	—
Quercus nigra	QN	—	—	—	—	—	—	—	—	—	—
Quercus shumardii	QSh	—	—	1.8	—	—	—	—	—	—	—
Quercus stellata	QSt	—	—	—	—	—	—	15.1	—	—	—
Quercus velutina	QV	—	—	9.0	—	2.9	—	19.6	—	—	—
SUMMARY VALUE FOR *QUERCUS* (*TOTAL* QUERCUS *Spp.*)		(6.6)	(3.9)	(29.4)	(6.5)	(15.3)	(53.9)	(92.2)	(36.4)	(0.0)	(0.0)
Sassafras albidum	SA	—	—	—	—	1.0	1.9	—	—	—	—
*Thuja occidentalis**	TO	—	—	—	—	—	—	—	—	—	—
Tilia americana var. americana (*T. neglecta*)	TAa	—	—	5.4	—	—	—	—	12.5	—	—
Tilia americana var. caroliniana (*Tilia floridana*)	TAc	—	—	—	—	—	9.6	—	—	—	—
Tilia americana var. heterophylla (*Tilia heterophylla*)	TAh	20.1	—	—	7.8	—	—	—	—	—	—
Tilia undiff. spp.	TU	—	—	—	—	—	—	—	—	21.8	—
*Tsuga canadensis**	TC	—	—	—	18.2	—	—	—	—	3.5	—
Ulmus americana	UA	—	—	—	—	—	—	—	—	—	—
Ulmus rubra (*U. fulva*)	UR	—	—	3.0	1.3	—	1.9	—	1.1	—	—

ized by a large number of canopy co-dominants, although the composition and relative abundances of the species may vary greatly from place to place (termed "association-segregates"). In the geographic area included within the mixed mesophytic forest region, late-successional (climax) communities of mixed mesophytic forest were described by Braun (1950) as originally covering most of the land surface except for dry ridge tops and upper south-facing slopes, floodplains, and edaphically unsuitable limestone or sandstone outcrops. Beyond the limits mapped for this forest region (Fig. 10.3), mixed mesophytic forest communities were less continuous and occupied mesic, nutrient-rich sites. Toward the limits of mixed mesophytic forest distribution, only a small part of the landscape was originally occupied by this forest type.

	Northern Blue Ridge Mts., Oak-Chestnut, Rappahannock Co., Va.	Ridge and Valley Section, Oak-Tuliptree, Bath Co., Va.	Piedmont Section, Beech-Tuliptree, Northern Md.	Oak-Pine Forest Region: Atlantic Slope Section, Duke Forest, Durham, N.C.	Southeastern Evergreen Forest Region: Bottomland Forest, Mississippi Alluvial Plain, Southern Ill.	Upland Hardwood Forests, Loess Hills, Tunica Hills, West Feliciana Parish, La.	Beech-Maple Forest Region: Berrien Co., Mich.	Maple-Basswood Forest Region: Big Woods Section, SW Minn.	Hemlock-White Pine-Northern Hardwoods Forest Region: Great Lakes-St. Lawrence Division– Great Lakes Section, Chippewa-Luce Co., Mich.	Minnesota Section, Beltrami Co., Minn.	Northern Appalachian Highland Division– Allegheny Section, Warren Co., Pa.	Adirondack Section, Western Adirondack Mts., N.Y.	New England Section, Pittsburg Township, N.H.	Taxon Frequency (No. stands/23 stands)
	11 (T45–3)	12 (T46–1)	13 (T51–2)	14 (T52–A)	15 (T54–3)	16 (T57–2)	17 (T63–1)	18 (T67–M)	19 (T71–7C)	20 (T79–5)	21 (T81–1a)	22 (T87–5)	23 (T90–B)	
	4B	4C	4D	5A	6A	6B	7	8B	9A	9C	9E	9F	9G	
	8	8	11	10	10	9	8	5	9	3	8	11	6	
	—	—	1.3	—	—	—	0.4	—	—	—	—	—	—	2/23
	—	—	—	—	—	—	—	—	—	—	2.0	0.1	—	4/23
	20.0	53.0	1.3	43.9	29.2	—	—	—	—	—	—	—	—	10/23
	14.8	1.1	8.9	1.2	—	—	0.4	5.7	—	—	—	—	—	12/23
	—	—	3.7	—	—	—	—	—	—	—	—	—	—	1/23
	—	—	2.4	—	—	—	—	—	—	—	—	—	—	1/23
	—	—	—	—	—	—	—	—	—	—	—	—	—	1/23
	—	—	—	—	—	—	—	—	—	—	—	—	—	1/23
	18.1	8.3	—	—	—	4.2	—	—	—	—	—	—	—	4/23
	—	—	—	—	—	—	—	—	—	—	—	—	—	2/23
	—	—	—	—	—	10.6	—	—	—	—	—	—	—	1/23
	—	—	—	—	—	2.1	—	—	—	—	—	—	—	2/23
	—	—	—	8.5	—	—	—	—	—	—	—	—	—	2/23
	—	12.1	—	4.9	8.1	—	—	—	—	—	—	—	—	6/23
	(52.9)	(74.5)	(10.2)	(64.6)	(37.3)	(16.9)	(0.4)	(5.7)	(0.0)	(0.0)	(0.0)	(0.0)	(0.0)	(16/23)
	—	—	—	—	2.5	—	—	—	—	—	—	—	—	3/23
	—	—	—	—	—	—	—	—	—	—	—	0.2	—	1/23
	—	—	—	—	—	—	1.3	30.2	14.0	55.5	—	—	—	6/23
	—	—	—	—	—	—	—	—	—	—	—	—	—	1/23
	—	—	—	—	—	—	—	—	—	—	—	—	—	2/23
	—	—	—	—	—	6.4	—	—	—	—	—	—	—	2/23
	7.1	—	—	—	—	—	—	—	5.3	—	42.3	6.3	—	6/23
	—	—	—	—	6.8	—	3.4	9.4	13.1	16.7	—	—	—	5/23
	—	—	—	—	—	—	—	7.5	—	—	—	—	—	5/23

A summary of contemporary plant ecological studies within the mixed mesophytic forest region is given by Hinkle, McComb, Safley, and Schmalzer (1993). In the most areally extensive study to date of vegetation of the southern portion of the mixed mesophytic forest region, Hinkle (1989) identified 24 forest community types distributed along topographic and edaphic gradients across the Cumberland Plateau of Tennessee. Landscape (slope) position reflecting a soil-moisture gradient from xeric, to mesic, to hydric sites was the most important factor in plant community distribution. Mixed oak vegetation was originally areally dominant across the upland, with rich mixed oak forests containing mixed mesophytic forest species as minor components restricted to sheltered slopes with V-shaped ravines or gorges deeply eroded into the plateau surface. Forests similar to those of the mixed mesophytic included communities dominated by only *Fagus grandifolia*, by both *Fagus grandifolia* and *Liriodendron tulipifera*, by *Liriodendron tulipifera, Carya laciniosa* (shellbark hickory), and *Quercus rubra* var. *rubra*, or by the assemblage of *Acer saccharum, Tilia americana* var. *heterophylla, Fraxinus*, and *Aesculus flava*. These communities were confined to middle and lower slope characterized by deep, moist, nutrient-rich soils. Upland species were segregated along a moisture gradient related to soil depth, total available water in the solum, and slope position. Ravine canopy species were segregated along gradients of pH and nutrients (particularly soil P and K) as well as slope position. Hinkle's (1989) study of the natural vegetation of the southern portion of the mixed mesophytic forest region illustrates that Braun's (1950) conceptual climatic climax community was not areally dominant throughout the mapped region, but, rather, it was highly restricted to favorable sites. The southern Cumberland Plateau is thus an area in which presettlement mixed-oak forests prevailed on flat to rolling uplands, with mixed mesophytic forest communities and rich oak vegetation restricted to escarpment slopes, sheltered coves, and deeper ravines (Hinkle 1989; Hinkle et al. 1993).

Western mesophytic forests. The eastern boundary of this forest region is located at the western escarpment of the Cumberland and Allegheny plateaus. Its western boundary coincides with the loess-mantled upland bluffs ("blufflands") bordering the eastern valley wall of the Mississippi River (Fig. 10.3). The region extends from northern Alabama and Mississippi to southern Ohio and eastern Indiana, generally coinciding with the Interior Low Plateau physiographic region (Fig. 10.3).

Braun (1950) considered the western mesophytic forest region to be a transitional region that is not characterized by a single climax forest type. On the whole, forests are less luxuriant than in the mixed mesophytic forest region to the east. In the eastern part of the Interior Low Plateaus, mixed mesophytic forest communities are frequent, but to the west they become more limited in areal extent and contiguity. These western mesic communities are restricted to favorable habitats and display a greater tendency for only a few species to dominate the forest (Table 10.1, Stands 3–5). Oak and oak-hickory communities become more spatially prevalent toward the west, and forest openings including *Juniperus virginiana* (red cedar)– dominated cedar glades and grassy barrens are characteristic of Middle Tennessee and central Kentucky. Several characteristic mixed mesophytic forest species

drop out to the west, including *Tilia americana* var. *heterophylla* and *Aesculus flava*.

Hinkle et al. (1993) and Bryant, McComb, and Fralish (1993) summarized compositional data from forest stands within the western mesophytic forest region. Studies of vegetation in the Shawnee Hills of southern Illinois (Fralish, Crooks, Chambers, and Harty 1991) and the Land Between the Lakes, in Kentucky and Tennessee (Fralish and Crooks 1989) identified a number of forest communities arrayed along moisture gradients. In the Shawnee Hills, the xeric to mesic edaphic gradient supported communities dominated by *Juniperus virginiana, Quercus stellata* (post oak), *Q. alba, Q. rubra* var. *rubra*, mixed hardwoods, and *Acer saccharum* (Fralish et al. 1991). Seven community types in the Land Between the Lakes are dominated by (from xeric to mesic sites, respectively) *Pinus echinata, Quercus stellata, Q. prinus* (chestnut oak), *Q. alba, Fagus grandifolia – Acer saccharum*, and *Acer saccharum* – mesophytic hardwoods (Fralish and Crooks 1989). The General Land Office Survey of 1820 of land west of the Tennessee River included descriptions of a vegetation mosaic of barrens, post oak – blackjack oak (*Quercus marilandica*) savannas, white oak uplands, elm – ash – maple forests along streams, and swamps of bald cypress (*Taxodium distichum*) (Bryant and Martin 1988).

Literature documenting the floristics and vegetation of cedar glades, barrens, and nonforest vegetation of sandstone and granitic outcrops is summarized in papers by Baskin and Baskin (1986), DeSelm and Murdock (1993), and Quarterman et al. (1993). Shallow soils developed over limestone, sandstone, and granite outcrops support nonforest plant communities that are scattered across the eastern deciduous forest region from the Ozark Highlands to the Piedmont of the Carolinas (Quarterman et al. 1993). The most areally extensive of herbaceous outcrop communities are the cedar glades (Baskin and Baskin 1986), which originally covered 5–6% of the central basin of Middle Tennessee on the thin-bedded, dolomitic Lebanon Limestone (Fig. 10.8). The cedar glade flora is dominated by winter annuals, particularly endemic species of *Leavenworthia, Sedum pulchellum* (Texas stonecrop), and *Minuartia patula* (sandwort) and also includes a number of prairie grasses and forbs. Deeper soils support open stands of *Juniperus virginiana* (Baskin and Baskin 1986; Quarterman et al. 1993). Grassland vegetation, locally termed "barrens" to refer to brushy-grassy forest openings, was extensive in presettlement times through the Interior Low Plateau region of western and central Kentucky (Baskin and Baskin 1981). The Kentucky Barrens are dominated by grasses including *Schi-*

Figure 10.8. Cedar glade with open stands of Juniperus virginiana *(red cedar)* on shallow soils over limestone bedrock, Interior Low Plateau physiographic province, Cedars of Lebanon State Park, Middle Tennessee (photograph by Hazel and Paul Delcourt).

zachyrium scoparium (little bluestem) and *Andropogon gerardii* (big bluestem), along with *Bouteloua curtipendula* (side-oats grama), and *Sorghastrum nutans* (Indian grass) (Baskin and Baskin 1981; DeSelm and Murdock 1993).

Oak-hickory forests. This is the most westerly of Braun's (1950) deciduous forest regions, extending from Texas to Canada and varying in width along its irregular border with prairie (Fig. 10.3). Forests dominated by oaks and containing numerous species of hickory characterize the Ozark Highlands of Missouri and Arkansas. To the south, characteristic species include *Quercus stellata, Q. marilandica, Q. shumardii* (shumard oak), *Carya texana,* and in the bottomlands *Quercus nigra* (water oak), *Quercus lyrata* (overcup oak), *Carya illinoinensis* (pecan), *Carya myristiciformis* (nutmeg hickory) and *Carya aquatica* (water hickory) (Table 10.1, Stands 6 and 7). In the north, *Quercus macrocarpa* (bur oak) is prominent along with *Q. ellipsoidalis* (jack oak) (Table 10.1, Stand 8). Species of oaks and hickories that are found throughout the region include *Quercus alba, Q. rubra* var. *rubra, Q. velutina* (black oak), *Carya cordiformis* (butternut hickory), and *C. ovata* (shagbark hickory) (Table 10.1, Stands 6–8).

Summary papers by Skeen, Doerr, and Van Lea (1983) and Bryant et al. (1993) contain literature citations of contemporary studies of vegetation from this forest region. In the center of development of Braun's oak-hickory forest region in the Ozark Highlands of Missouri and Arkansas, Pell (1984) found the white oak–black oak–southern red oak forest type and its variants to be the most common. Mesic forests containing *Acer saccharum, Quercus muhlenbergii* (chinquapin oak), *Q. alba, Carya cordiformis, Fagus grandifolia,* and other mixed meso-

phytic forest species are local in occurrence and confined to moist habitats, north-facing slopes, and coves (Dale 1986).

The ecotone between these western forests and the central prairies is discussed in Chapter 9.

Oak-chestnut forests. This forest region occupies the Blue Ridge and Ridge and Valley physiographic provinces from southern New England and the Hudson River Valley to northern Georgia (Fig. 10.3). To the west it borders the mixed mesophytic forest region, which was included by Shantz and Zon (1924) in their more extensive chestnut–chestnut oak–yellow poplar forest region (Fig. 10.2). To the south, the oak-chestnut forest region is bounded by the oak-pine forest region across the southern Ridge and Valley and the Blue Ridge physiographic provinces and along the inner border of the Piedmont. This region has a steeply dissected, mountainous landscape except in its northeasternmost part, where Quaternary continental glaciers scoured the terrain, locally reducing topographic relief. Originally, oak-chestnut communities occupied slopes of ridges rather than valleys, where they were replaced by forests dominated by *Quercus alba* (Table 10.1, Stands 11 and 12). Mesic lower slopes along streams and cool, northeast-facing coves within the southern Appalachian Mountains were occupied by mixed mesophytic forest communities (Table 10.1, Stand 9).

In the southern Appalachian Mountains, the oak-chestnut forest region includes altitudinal gradients along which oak-chestnut communities (in early settlement times) dominated mountain slopes at 400–1460 m, as in the Great Smoky Mountains of Tennessee and North Carolina (Ayres and Ashe 1905; Whittaker 1956). In many areas, *Castanea den-*

tata was the dominant tree, often occurring in almost pure stands; at lower elevations the oak-chestnut forest on northerly, moist sites was a dense forest of very large trees, mostly *Castanea dentata* and *Liriodendron tulipifera* (Table 10.1, Stand 13). Ericaceous shrubs were characteristic understory species, including evergreen and deciduous species of *Rhododendron* as well as *Kalmia latifolia* (mountain laurel). *Liriodendron tulipifera* decreased in importance with elevation, replaced by oaks including *Quercus prinus*, *Q. rubra* var. *rubra*, *Q. alba*, *Q. velutina*, and *Q. coccinea* (scarlet oak).

Vegetation studies from throughout the former oak-chestnut forest region, now known as the Appalachian Oak Forest Region, are summarized in Stephenson, Ash, and Stauffer (1993). The following six community types are now recognized as most important, based on areal extent and compositional distinctiveness (Stephenson et al. 1993):

1. Oak-pine forest, dominated by *Quercus prinus*, *Q. coccinea*, *Pinus rigida* (pitch pine), *P. virginiana* (Virginia pine), and *P. pungens* (table mountain pine) on xeric sites (Zobel 1969), particularly upper south- to southwest-facing slopes and ridgetops, usually associated with shallow, rocky soils.
2. Chestnut oak forest, dominated by *Quercus prinus*, *Q. velutina*, *Q. coccinea*, *Q. alba*, and *Carya glabra* on subxeric sites, including middle and upper slopes at lower and moderate elevations.
3. Red oak forest, dominated by *Quercus rubra* var. *rubra*, *Q. alba*, *Acer rubrum*, *Liriodendron tulipifera*, and *Betula lenta* (cherry birch) on submesic sites, particularly north-facing slopes, and becoming more common toward higher elevations.
4. White oak forest, dominated by *Quercus alba*, *Q. prinus*, *Carya glabra*, *C. ovalis* (sweet pignut), *C. alba* (mockernut hickory), and *Quercus rubra* var. *rubra* on subxeric to submesic sites, usually valley floors and gentle slopes.
5. Oak-hickory forest, dominated by *Carya glabra*, *C. ovalis*, *C. ovata*, and *C. alba*, *Quercus prinus*, *Q. rubra* var. *rubra*, and *Q. alba* on subxeric to submesic sites, usually south- and west-facing slopes at lower to moderate elevations, often on less acidic soils.
6. Mixed mesophytic forest, dominated by *Acer saccharum*, *Fagus grandifolia*, *Tilia*, *Carya* spp., *Quercus rubra* var. *rubra*, *Liriodendron tulipifera*, *Tsuga canadensis*, and *Aesculus flava* on mesic sites, including north-facing slopes, coves, ravines, and draws.

Oak-pine forests. The oak-pine forest region extends from southern New Jersey southward across the Piedmont of Virginia and the Carolinas to Georgia and westward to the Lower Mississippi Alluvial Valley. West of the Mississippi Valley, it extends through southern Arkansas, northwestern Louisiana, and eastern Texas (Fig. 10.3). Much of the region has a rolling topography with gentle slopes. Including southern pines such as *Pinus taeda* (loblolly pine) and *Pinus echinata* as conspicuous dominants in its communities, Braun (1950) considered this forest region to be transitional between the central deciduous forest and the predominantly evergreen forests of the southeastern coastal plains (Table 10.1, Stand 14). Skeen et al. (1993) summarized studies of plant succession in the oak-pine forest region.

Widespread destruction of original forests throughout this region since the time of Euro-American settlement left a patchwork of fields and successional forests of *Pinus taeda* (loblolly pine) and *P. echinata*, with the most mature hardwood stands occupying marginal sites. In the Piedmont of North Carolina before 1950, the most widespread type of upland deciduous forest was dominated by *Quercus alba* and *Carya* spp. as subdominants. *Quercus velutina*, *Q. stellata*, *Q. rubra* var. *rubra*, *Q. falcata* (southern red oak), and *Q. coccinea* were well represented (Table 10.1, Stand 14). On drier sites with poor soils, hardwood forests were dominated by *Quercus stellata* with *Quercus alba* and *Carya* spp. as subdominants (Oosting 1942). Successional hardwood stands dominated in the 1930s by *Pinus echinata* (Billings 1938) are, however, being invaded by *Fagus grandifolia*, *Acer saccharum*, and other species that were rare or absent previously (Billings, unpublished data).

In the Duke University Forest, Peet and Christensen (1980) found that wet alluvial and swamp forests were characterized by abundant moisture and nutrients and high species richness, dominated by *Liquidambar styraciflua* (sweetgum), *Liriodendron tulipifera*, *Platanus occidentalis* (sycamore), *Carpinus caroliniana* (American hornbeam), *Fagus grandifolia*, *Ulmus rubra* (slippery elm), *Acer rubrum*, and *Fraxinus pennsylvanica* (green ash). Low, moist ravines with abundant nutrients and moisture were dominated by *Fagus grandifolia*, *Liriodendron tulipifera*, *Quercus alba*, *Quercus rubra*, *Fraxinus americana* (white ash), and *Acer saccharum*. Dry sites on upper slopes, hilltops, and ridges included fertile sites supporting species-rich forests dominated by *Quercus stellata*, *Q. alba*. *Carya* spp., *Cercis canadensis* (eastern redbud), and *Fraxinus americana* as well as nutrient-poor sites with shallow soils (developed over acidic, crystalline rocks) characterized by very low species richness and including *Quercus prinus*, *Oxydendrum arboreum* (sourwood), *Quercus alba*, *Q.*

coccinea, Acer rubrum, and *Carya tomentosa* as important trees. Upland sites over shallow montmorillonitic subsoil with impeded drainage and alternating wet/dry conditions supported a forest with intermediate species richness with a mixed composition including swamp and upland species of *Quercus phellos* (willow oak), *Ulmus alata* (winged elm), *Fraxinus* spp., *Carya ovata, Quercus stellata, Q. marilandica, Diospyros virginiana* (common persimmon), *Juniperus virginiana, Pinus taeda,* and *Pinus echinata* (Peet and Christensen 1980; Barnes 1991).

The vegetation of this southeastern region is treated in more detail in Chapter 11.

Southeastern evergreen forests. Temperate deciduous forest communities extending southward onto the Gulf and southern Atlantic coastal plains include bottomland forests of the Lower Mississippi Alluvial Valley and mesic hardwood forests of uplands and slopes such as the loess hills, or "blufflands," of West Tennessee, western Mississippi, and southeastern Louisiana (Fig. 10.3; Table 10.1, Stands 15 and 16). A comprehensive treatment of the vegetation of the Southeastern Evergreen Forest Region is provided in Chapter 11.

In the Mississippi Alluvial Valley, ridge bottoms and well-drained topographic benches provide mesic habitats for forests dominated by *Quercus alba, Carya* spp., and *Liquidambar styraciflua* (Table 10.1, Stand 15). In presettlement times, at the southernmost limit of the eastern deciduous forest formation, mesic silt-loam loessial soils supported a diverse community dominated by *Fagus grandifolia* and evergreen *Magnolia grandiflora* (southern magnolia) in the Tunica Hills of southeastern Louisiana (Braun 1950; Quarterman and Keever 1962; Delcourt and Delcourt 1974, 1977; Table 10.1, Stand 16).

In the loess hills of west-central Mississippi, Caplenor et al. (1968) documented the relationship of forest community composition to thickness of calcareous silt-loam soils formed from loess that was deposited by wind during the late Pleistocene. Forests on thick loess (>32 ft [9.8 m] thick) near Vicksburg at the west end of the transect were dominated by *Liquidambar styraciflua, Tilia, Quercus nigra, Liriodendron tulipifera, Quercus pagoda* (cherrybark oak), and *Carya cordiformis,* with an understory of *Ostrya virginiana* (eastern hophornbeam), and *Carpinus caroliniana.* Forests developed to the east on thin loess (4–32 [1.2–9.8 m] thick) were dominated by mixed hardwoods including *Fagus grandifolia, Nyssa sylvatica* (black gum), *Quercus velutina, Carya tomentosa, Quercus alba, Oxydendrum arboreum,* and *Liquidambar styraciflua.* Farther east, forests near Jackson, Mississippi, in nonloessial, acidic, sandy uplands were mixed southern pine–

oak forest with *Pinus taeda, P. echinata, Quercus falcata, Q. alba, Carya tomentosa, Quercus stellata, Q. phellos* (willow oak), and *Fraxinus americana.* Overall, Caplenor et al. (1968) found a west to east diminishment in tree species richness that was related to pH of the soil.

Beech-maple forests. The beech-maple forest region lies immediately north of the mixed mesophytic and western mesophytic forest regions and extends northward to the central Lower Peninsula of Michigan (Fig. 10.3), thus occupying formerly glaciated portions of Ohio, Indiana, southern Ontario, and southern Michigan. The northern boundary is transitional to the hemlock–white pine–northern hardwoods forest region, and the western boundary interfingers with oak-hickory forests through Indiana. This forest region is characterized by mesic, late-successional forest communities dominated by *Fagus grandifolia* and *Acer saccharum.*

With the exception of wetlands, much of the flat to rolling terrain of the glacial till and lake plains within this region was originally covered by forest dominated by *Fagus grandifolia* or *Acer saccharum* or both. Average dominance of *Fagus grandifolia* plus *Acer saccharum* in the canopy was 80% (Table 10.1, Stand 17). In contrast, within mixed mesophytic forest communities to the south, the two species together were typically no more than 50%, sharing dominance with a large number of other deciduous trees (Table 10.1, Stands 1–5). The number of canopy species in beech-maple forests ranges from 3 to 14, compared with 12 to 23 in mixed mesophytic forests. Differences in dominance and number of species are primarily responsible for differences in appearance of mesic forests in the two regions (Braun 1950).

Lindsey and Escobar (1976) provided a summary of ecological descriptions of forest stands from throughout the beech-maple forest region. They found that within the broad limits of the beech-maple type, differences of composition are due to the limits of distribution of some of the component species. Although the dominants on mesic sites tend to be *Fagus grandifolia* and *Acer saccharum* (Fig. 10.9), the proportions of those two species vary according to local factors. In Indiana, for example, *Fagus grandifolia* is reported to be the most important tree in mesic hardwood stands (Petty and Jackson 1966), in part because of selective cutting that has removed other species. A study of 58 old-growth stands in Indiana (Lindsey and Schmelz 1970) documented that approximately one-third of all the stands contained more than 80% *Fagus grandifolia* and *Acer saccharum;* the re-

Figure 10.9. Multistoried, secondary forest of Fagus grandifolia (American beech) and Acer saccharum (sugar maple), occupying protected sites in the lee of sand dunes along the shore of Lake Michigan, Hoffmeister State Park, southwestern Lower Michigan (photograph by Hazel Delcourt).

mainder of the stands represented examples of (1) upland oak-hickory forest in which Carya ovata, C. glabra, Quercus alba, Q. velutina, and Q. prinus were important, and (2) lowland forests dominated by Acer rubrum, A. saccharinum (silver maple), Carya laciniosa, Cercis canadensis, Fraxinus pennsylvanica, Liquidambar styraciflua, Nyssa sylvatica, Platanus occidentalis, Quercus macrocarpa, Q. palustris (pin oak), Q. shumardii, and Q. bicolor (swamp white oak) (Lindsey and Escobar 1976; Lindsey and Schmelz 1970). The composition of beech-maple forests in southern Michigan (Cain 1935) (Fig. 10.9) is transitional between that of the southern beech-maple forest and the hemlock–white pine–northern hardwoods forest, with Quercus alba important only in the most southern of Michigan stands and with Betula alleghaniensis and Tsuga canadensis included in the northernmost of the beech-maple stands (Quick 1923; Lindsey and Escobar 1976).

Maple-basswood forests. Late-successional forest dominants in the northwestern part of the deciduous forest are Acer saccharum and Tilia americana var. americana (American basswood). The extent of the maple-basswood forest region in southwestern Wisconsin and adjacent portions of southeastern Minnesota (Fig. 10.3) is primarily determined by both local topography and the locations of peatlands, streams, and lakes that constitute fire breaks. In the Big Woods of southeastern Minnesota (Table 10.1, Stand 18), important tree species in addition to Acer saccharum and Tilia americana var. americana include Quercus rubra var. rubra and Ulmus americana. Other trees important locally in the region include Ostrya virginiana and Carya cordiformis (Table 10.1, Stand 18).

Papers by Daubenmire (1936) and Grimm (1983,

1984) documented the relationships of deciduous forest communities to fire and topography within this region. Daubenmire (1936) found that the climax forest community on mesic till in the Big Woods region of southeastern Minnesota was dominated by Acer saccharum and Tilia americana var. americana. He hypothesized that mesic maple-basswood forests were restricted in distribution by depth in the soil to water table, with soils situated south of the Minnesota River more droughty, more fire-prone, and therefore less capable of supporting a rich forest. Grimm (1984) found that the most common tree in the southeastern Minnesota region that included the Big Woods in presettlement times was Ulmus americana (27% of 18,512 bearing trees recorded in General Land Office Survey records for 1847–1856). Grimm's (1984) study area extended across low-lying stream drainages and hummocky till plains in addition to the well-drained, mesic till studied by Daubenmire (1936). Grimm (1984) found that, regionally, Tilia americana var. americana was 14% of the overall forest composition, and Acer saccharum was 12%. Tilia americana var. americana was the most evenly distributed of the tree taxa, and Acer saccharum was more restricted, nearly absent from the westernmost part of the study area. Acer saccharum dominated in some areas, and Tilia americana var. americana was a common associate of Acer saccharum where the latter occurred. Oaks dominated the forests east and south of the Big Woods region, and Populus spp. (aspens) dominated the woodlands to the west. Grimm (1983, 1984) concluded that, on the basis of bearing tree records, maple-basswood forests were not the areally dominant vegetation type across southeastern Minnesota. Rather, the landscape of the prairie-woodland border of Minnesota was

Figure 10.10. Mosaic of coniferous and deciduous forest in the western-most hemlock–white pine–northern hardwoods forest region, occupying thin glacial deposits mantling ledges of exposed sandstone bedrock. Head-waters of the St. Croix River, marking the boundary between eastern Minnesota and northwestern Wisconsin (photograph by Hazel Delcourt).

originally occupied by a heterogeneous vegetation mosaic (Marschner 1974), exhibiting multiple stable states that were related to historical factors as well as the pattern of occurrence of wildfire, locations of streams that served as firebreaks, and soils derived from glacially deposited materials and bedrock (Grimm 1983, 1984).

Hemlock–white pine–northern hardwoods forests. This forest region extends across a broad geographic region from northern Minnesota through the Upper Great Lakes region and eastward across southern Canada and New England to the Appalachian Plateau of New York and northern Pennsylvania (Fig. 10.3). The physiographic boundaries of this largely deglaciated region are not well defined and are gradational, but they generally correspond with the geographic region of Spodosols.

Late-successional deciduous forest communities are dominated by combinations of *Acer saccharum, Fagus grandifolia,* and *Tilia americana* var. *americana. Betula alleghaniensis, Ulmus americana,* and *Acer rubrum* are frequent (Stands 19 and 20). Early successional deciduous communities are dominated by *Populus tremuloides* (quaking aspen), *P. balsamifera* (balsam poplar), *Betula papyrifera* var. *papyrifera* (paper birch) and, in addition in New England, *B. populifolia* (gray birch). Fire- and drought-adapted evergreen conifers including *Pinus strobus, Pinus resinosa* (red pine), and *P. banksiana* occupy dry, sandy glacial outwash plains and relict beach ridges. A second set of conifers characteristic of poorly drained bogs and muskegs includes *Picea mariana, Thuja occidentalis* (northern white cedar), and *Larix laricina* (tamarack). *Picea rubens* occupies mesic flats and slopes in the part of the region generally east of

80° W and south of 47° N (White and Cogbill 1992). Mixed conifer-hardwoods communities (Fig. 10.10) include those dominated by *Tsuga canadensis* and mesic deciduous trees (Table 10.1, Stand 21) and those characterized by *Picea rubens* and a variety of hardwoods, including *Acer saccharum, Fagus grandifolia, Tilia americana* var. *americana,* and *Betula alleghaniensis* (Table 10.1, Stands 22 and 23).

Classic and contemporary syntheses of vegetation of the hemlock–white pine–northern hardwoods region have been provided by McIntosh (1962), Siccama (1971), Rowe (1972), Bormann and Likens (1979), Stearns (1987), Whitney (1986, 1987, 1990, 1995), and Frelich and Lorimer (1991). Whitney (1986, 1987) documented the original distribution of upland conifer, lowland conifer, and upland mesic hardwood communities in northern Lower Michigan based on witness tree records from the General Land Office Survey of 278,000 ha within 32 townships surveyed from 1836 to 1859. He found that the abundance of pines (*Pinus banksiana, P. resinosa, P. strobus*) in that region was correlated with coarse-textured soils derived from outwash and ice-contact deposits. Pine forests were very susceptible to fire; average return time for severe crown fires ranged from 80 yr for *Pinus banksiana* forest, to 120–240 yr for mixed pine forest, and to 1200 yr for hemlock–white pine–northern hardwoods forest. Windfalls were the primary form of disturbance in swamp conifer and hemlock–white pine–northern hardwoods types (Whitney 1986). Whitney's 1986 study showed that the distribution of forest communities in this portion of the hemlock–white pine–northern hardwoods forest region of Braun (1950) is fundamentally related not only to substrate and topography but also to incidence of natural disturbance.

Frelich and Lorimer (1991) documented natural disturbance regimes in hemlock-hardwood forests of Upper Michigan. In stands dominated by *Acer saccharum* and *Tsuga canadensis* located in the Huron Mountains, Porcupine Mountains, and the Sylvania tract in the western Upper Peninsula, mesic hemlock-hardwoods forests included *Betula alleghaniensis, Acer rubrum*, and *Tilia americana* var. *americana* on substrates ranging from former proglacial lake plains to glacial end moraines and bedrock of shale, sandstone, and granitic gneiss. The forests of Upper Michigan appear to be in quasi-equilibrium with a modern disturbance regime that includes periodic windfalls from downburst storms but in which natural fires are rare (Frelich and Lorimer 1991).

Spruce-fir forests. Montane forests of the Appalachians include high-elevation spruce-fir forests and intermediate-elevation northern hardwoods (White et al. 1993). Although included by Braun (1950) within the oak-chestnut forest region, Appalachian spruce-fir forests constitute a distinctive ecosystem that we treat separately here. Spruce-fir forest occurs in the southern Appalachian Mountains generally 1350 m, in central West Virginia and Virginia extending down to 975 m. It originally covered between 69 and 300 km² in the central and southern Appalachians, an areal extent of about 1–5% of the southeastern United States. Southern Appalachian spruce-fir forests, dominated by *Picea rubens* (red spruce) and the endemic species *Abies fraseri* (Fraser fir), are related to, but a distinct variant of the boreal forest biome (Oosting and Billings 1951; Cogbill and White 1991; White and Cogbill 1992).

The spruce-fir/deciduous forest ecotone (Fig. 10.11) is often very narrow, with the mean contour rising upward from north to south along the Appalachian Mountains by 100 m/degree latitude (Cogbill and White 1991; Delcourt and Delcourt 1998b). Locally in the southern Appalachians, this ecotone can vary from about 1400 to 1680 m depending on aspect (up to 200 m lower on north-facing than on south-facing slopes), steepness, slope shape, and disturbance history. The central Appalachians represent a topographic gap where elevations of summits are typically less than 1000 m and spruce-fir forests are not well developed. At the ecotone with high-elevation spruce-fir forest (1460 m; Fig. 10.11), *Quercus rubra* var. *rubra* was a dominant, and *Betula alleghaniensis, Acer rubrum*, and *Picea rubens* occurred in mixed stands along with *Castanea dentata*. Summits of ten high mountain peaks of the southern Appalachian Mountains are occupied by high-elevation spruce-fir forest communities dominated by *Picea rubens* and *Abies fraseri* (Oosting and Billings 1951) and including indicator species such as *Acer spicatum*

Figure 10.11. Southern Appalachian ecotone (1460 m elevation) between the intermediate elevation northern hardwoods forest and the high-elevation montane spruce-fir forest dominated by Picea rubens *(red spruce)* and Abies fraseri *(Fraser fir), western Great Smoky Mountains National Park, Tennessee (photograph by Hazel and Paul Delcourt).*

(mountain maple) and *Betula alleghaniensis* (Table 10.1, Stand 10).

Northern and southern high-elevation spruce-fir forests are similar in physiognomy and vascular plant richness, but the forests of the southern Appalachians have greater average height and growth rate of species, greater herb and bryophyte cover, and commonly a broadleaf, evergreen understory of *Rhododendron* (Oosting and Billings 1951; Whittaker 1956). *Picea glauca* (white spruce) and *Picea mariana* (black spruce) dominate in northern lowland boreal forest (see Chapter 2), but *Picea rubens* is characteristic of the Appalachian spruce-fir forest. *Abies balsamea* (balsam fir) is widespread from the central Appalachians northward into the Maritime Provinces of Canada (see Chapter 2), but it is replaced by the endemic *Abies fraseri* in the southern Appalachians. *Betula papyrifera* var. *cordifolia* (paper birch) is an important disturbance-dependent tree of the northern Appalachians, but it is rare in the southern high peaks. A high percentage of endemic plants, many of which are pre-

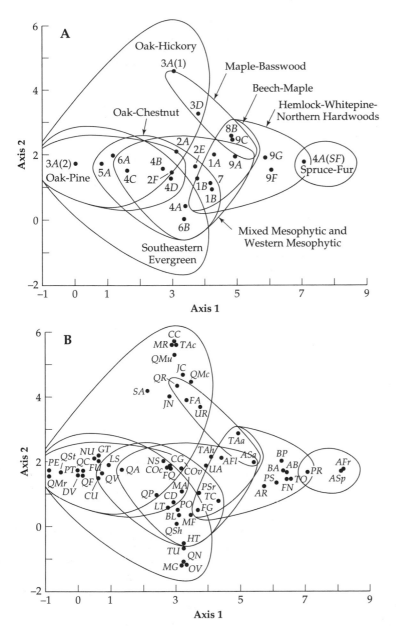

Figure 10.12. Detrended Correspondence Analysis (DCA, program DE-CORANA) ordination of forest canopy composition from 23 undisturbed, old-growth stands, considered by Braun (1950) to represent premier examples of "climatic climax" in all major forest regions of the deciduous forest formation (see Table 10.1 for supporting documentation of stand location and species composition). (A) Sample scores; (B) unweighted species scores.

sumed to be Ice Age relics of alpine tundra communities, is another distinguishing feature of the high-elevation vegetation of the southern Appalachian Mountains (White, Buckner, Pittillo, and Cogbill 1993; Cogbill White, and Wiser 1997).

Ordination of Braun's data. Braun's work was completed in 1950 before multivariate statistical methods came into common use in plant ecology. Relationships of both sites and species as distributed across environmental gradients are clarified by ordination of such large data sets (Whittaker

1956; Ware 1982). Because we base much of the vegetation description in this chapter on Braun's original data (e.g., Table 10.1), we used Detrended Correspondence Analysis (DCA, computer program DECORANA) (Gauch 1982; Clampitt 1985) to array both sample scores (Fig. 10.12a) and (unweighted) species scores (Fig. 10.12b) in ordination space. Our data set included 23 stands and 61 overstory tree species (Table 10.1) chosen to represent undisturbed, mature old-growth forests developed on mesic, fertile sites and distributed across each of Braun's forest regions. These representative stands thus exemplify her "type concept" of climatic climax for each region (Braun 1950). The results of DCA analysis lend support to Braun's original interpretations of vegetation distribution.

Ordination axes 1 and 2 account for most of the variation in composition represented in the data set (eigenvalue for DCA axis 1 = 0.814; eigenvalue for DCA axis 2 = 0.516). Variation along axis 1 reflects the north-to-south macroclimatic gradient in average mean and extreme seasonal values for temperature across the latitudinal bands of cold-hardiness zones (USDA 1990) across eastern North America, with samples of forest stands from the mixed mesophytic forest region plotting in a central location in ordination space, samples of hemlock–white pine–northern hardwoods forest clustering to the right of DCA axis 1, and oak–hickory–pine forest samples clustering along with southeastern evergreen forest samples to the left (Fig. 10.12a). Sample scores were spread out along DCA axis 2 along what we infer is a gradient of increasing east-to-west continentality (decreasing absolute values of total precipitation and decreasing ratios of precipitation to evapotranspiration), with mixed mesophytic forest and southeastern evergreen forest samples clustered toward the lower part of axis 2 and with oak-hickory forest and maple-basswood forest samples arrayed toward the top of the diagram (Fig. 10.12a).

Individual species scores exhibit a complex pattern (Fig. 10.12b), with clusters that represent species listed as characteristic of particular forest regions (Braun 1950). For instance, Quercus alba (QA), Fagus grandifolia (FG), and Acer saccharum (ASa) each plot within the bounds of the mixed mesophytic forest but at locations in ordination space that overlap broadly the bounds of several other forest types. Certain other species occur within only one forest type and represent endpoints on the ecological gradient. Examples include Abies fraseri (AFr), characteristic of high-elevation spruce-fir forest in the southern Appalachian high peaks, and Magnolia grandiflora (MG), restricted in distribution to the Gulf and southern Atlantic coastal plains at the southern periphery of the eastern deciduous forest region.

Ware (1982) ordinated Braun's original data from 34 mixed mesophytic forest stands in the Cumberland Mountains. On this more fine-grained spatial scale in the geographic center of the eastern deciduous forest, Fagus grandifolia and Acer saccharum were the most important species; however, they occurred within a large admixture of mesic species. Tilia americana var. heterophylla and Aesculus flava were strongly associated with

one another and with Acer saccharum, but they were strongly dissociated from the suite of Tsuga canadensis, Quercus alba, and Fagus grandifolia, showing strong overlap only with Liriodendron tulipifera and, formerly, Castanea dentata. Ware's (1982) results confirmed Braun's interpretation that all the Cumberland Mountain stands were part of one highly variable mixed mesophytic forest complex.

VEGETATION, CLIMATE, AND DISTURBANCE

Bryson (1966) showed that the distributional limits of broad vegetation regions across North America are related to both seasonal mean and extreme positions of frontal zones between contrasting air masses. Three air masses predominate through part or all of the year across eastern North America: (1) the dry, frigid polar or arctic air mass, originating in the north polar region of Alaska and northern Canada; (2) the relatively dry Pacific air mass, originating in the North Pacific Ocean and losing moisture as it sweeps eastward across the mountain barriers of western North America, generating a dry rain shadow across the continental interior; and (3) the moist, warm maritime tropical air mass originating in the Gulf of Mexico and the subtropical Caribbean and western Atlantic Ocean (Bryson 1966; Bryson and Hare 1974; Brouillet and Whetstone 1993). Precipitation is concentrated along the maritime coastal zone and the southern Appalachians, where it is particularly heavy (Billings and Anderson 1966). Precipitation is high also where contrasting air masses converge over the continental land mass – for example, along the Polar Frontal Zone between arctic and maritime tropical air masses. Because of seasonal shifts in positions of air mass boundaries across eastern North America, average values for mean annual temperature, "temperateness" as a measure of the annual range of seasonal temperatures, and length (and especially summer warmth) of frost-free growing season all diminish northward (Greller 1989). Effective precipitation decreases northwestward from the southeastern seaboard to the continental interior (Greller 1989). These subcontinental-scale atmospheric circulation patterns result in strong latitudinal gradients in temperature and longitudinal gradients in precipitation. The eastern deciduous forest region is characterized by a mild to warm, temperate, and humid climate that is shaped by the dominance of the Pacific air mass in autumn and winter (the dormant interval for winter-deciduous trees), and by dominance of the tropical air mass in spring through late summer (the growing season) (Bryson and Hare 1974; Greller 1989).

The correspondence of modal positions of air mass boundaries with major ecotones is related to physiological tolerance limits to winter cold (Greller 1989). For example, the northern limit of the eastern deciduous forest formation, defined by the range limits of numerous characteristic species, occurs today at about 48° N latitude, coincident with cold-hardiness Zone 2, wherein average annual minimum temperatures are lower than −40° C (Woodward 1987; Arris and Eagleson 1989; USDA 1990). Many woody vascular plants survive such cold temperatures by deep supercooling, wherein water in the protoplasm of the plant cells remains in liquid form because of a lack of nucleation sites for formation of ice crystals (Burke et al. 1975; Sakai and Weiser 1973; George et al. 1974; Arris and Eagleson 1989). Many species of trees native to the eastern deciduous forest thus exhibit low exotherms (the temperature limit at which ice formation occurs in plant tissue) from −41 to −53° C (Burke, George, and Bryant 1975; Sakai and Weiser 1973; George, Burke, Pellet, and Johnson 1974; Arris and Eagleson 1989). The northern distributional limits of deciduous trees are determined in part by their least hardy overwintering tissue, in most cases xylem (Arris and Eagleson 1989). Boreal conifers and some boreal deciduous trees, in contrast, can withstand temperatures below −80° C (Sakai and Weiser 1973).

Barnes (1991) suggests, however, that the northern limits of distribution for many eastern deciduous forest trees may not be directly controlled by their ability to withstand the absolute minimum winter temperature to which they are exposed but, rather, by their ability to avoid frost injury by ceasing growth and initiating cold hardiness in late summer and autumn when temperatures are decreasing. The shortening of the length of the day in late summer and early autumn induces dormancy in many north-temperate deciduous tree species regardless of temperature (Kramer 1936). An additional factor to consider (Barnes 1991) is bud hardiness, especially reproductive buds, which are less resistant to low temperatures than vegetative buds or stems. Vascular plant species may be prevented from occupying their full potential ranges by their failure to reproduce sexually at their northern range limits; germinating shoots and roots of seedlings have less freezing resistance than do bud, leaf, or stem tissues of mature trees (Sakai and Larcher 1987).

The southern boundary of the eastern deciduous forest may also be partly determined by cold hardiness, in this case lack of extreme chilling essential to the seasonal hardening-off process for cool-temperate deciduous trees (Barnes, personal communication). This biotic limit also corresponds with discrete terrain boundaries in soils and landforms, from nutrient-poor sands and gravels distributed across the Gulf and southern Atlantic coastal plains to more fertile, loamy soils formed in weathered regolith of metamorphic and igneous bedrock of the Piedmont and Blue Ridge physiographic provinces (see also Chapter 11). The change in physiognomy from broadleaf deciduous to broadleaf evergreen forest vegetation that occurs across the boundary between the Piedmont and the Inner Coastal Plain of the southeastern United States may be related to average minimum temperatures averaging zero to −15° C. Woodward (1987) suggested that frost-resistant vegetation should show a sharply defined geographical limit where minimum temperatures reach −15° C but a less well-defined limit in areas without freezing temperatures. In eastern North America the −15° C minimum winter isotherm corresponds with cold-hardiness Zone 7, which has its southern boundary at the "fall line" between the Piedmont and the Inner Coastal Plain (USDA 1990). In addition, the general northern limits of southern *Pinus* (Fig. 10.6) coincide with the distribution of frequent winter-glaze ice storms (USDA 1969) that result in the breakage of softwoods whose wood architecture limits their strength and elasticity (Lemon 1961).

The western limit of deciduous forest (and the eastern limit of prairie) corresponds with the geographic limit of dominance of the dry Pacific air mass (important for no more than 4 mo per yr) and its replacement, especially in spring, by incursions of moist tropical air (Bryson and Hare 1974). The correspondence between increasing climatic continentality (decreasing ratio of precipitation to evapotranspiration) and the distribution of North American grasslands has long been recognized (Fig. 10.1; Transeau 1905; Borchert 1950; Risser 1985; see also Chapter 9). The western limits of distribution of deciduous trees differ with latitude. To the north, evapotranspiration rates are lower, and 60 cm of annual precipitation is sufficient for maintaining deciduous forest; to the south, deciduous trees require 90–100 cm of annual precipitation because of higher overall temperatures, greater evapotranspiration, and longer growing seasons (Barbour and Christensen 1993). Incidence of fire, grazing, and local soil and hydrology play important roles in determining the distribution of com-

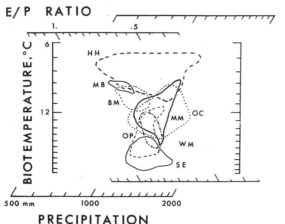

E/P RATIO

PRECIPITATION

Figure 10.13. Regional climatic characterization of Braun's (1950) forest regions of the deciduous forest formation, using the Holdridge system of life zones. Climatic summaries for more than 1400 weather stations across eastern North America were used to characterize each forest region with respect to average annual biotemperature in ° C, average annual precipitation in mm, and the ratio of potential evapotranspiration to precipitation. On this triangular graph, the scale of each axis is logarithmic on the base 2. Letter code for eight of Braun's forest regions: HH, hemlock–white pine–northern hardwoods; MB, maple-basswood; BM, beech-maple; MM, mixed mesophytic; WM, western mesophytic; OC, oak-chestnut; OP, oak-pine; and SE, southeastern evergreen. Oak-hickory forest type was omitted because of difficulty in determining its western boundary (from Lindsey and Escobar 1976).

munities along the prairie-forest ecotone (Grimm 1984; Barbour and Christensen 1993; see also Chapter 9).

Within the deciduous forest formation, Lindsey and Sawyer (1971) used the Holdridge system of climate classification for life zones (Holdridge and Tosi 1964; Sawyer and Lindsey 1964) to characterize the bioclimate of eight of Braun's (1950) forest regions. Results of this analysis (Fig. 10.13) illustrated the climatically central position of Braun's mixed mesophytic forests (also reflected in DCA ordination results in Fig. 10.12). Oak-chestnut forests were very similar climatically to mixed mesophytic forests, indicating that oak-chestnut forests were segregated from mixed mesophytic forests by edaphic or disturbance-related factors rather than by macroclimate. Oak-pine forests showed broad overlap with other types in climate space and was interpreted as a transitional or seral type between southeastern evergreen forests and one or more deciduous forest types to the north. Hemlock–white pine–northern hardwoods forests covered a broad climatic range, with relatively little overlap with other deciduous forest types (Fig. 10.13). Beech-

maple forests showed a compact, narrow envelope in Holdridge climate space that was overlapped at its cooler, drier end by climatic conditions associated with maple-basswood forests. Overall, the north-south orientation of Braun's forest types in Holdridge climate space (Fig. 10.13) corresponded latitudinally with the graphed axis of increasing biotemperature. These forest types also sorted out east to west along a trend of increasing ratios of evapotranspiration to precipitation and a trend toward reduced levels of absolute annual precipitation (Lindsey and Sawyer 1971; Lindsey and Escobar 1976).

Landscape Disturbance Profiles

Natural disturbances, including windstorms, fires, and pathogen outbreaks, are important agents in shaping vegetation patterns and successional processes of forested landscapes across eastern North America (Runkle 1985).

In presettlement times, wildfire was the dominant form of disturbance along the border between prairie and deciduous forest (Daubenmire 1936; Grimm 1984), as well as along the northwestern limit of hemlock–white pine–northern hardwoods forest at its ecotone with southern boreal forest (Heinselman 1973; Baker 1989). Elsewhere in the Great Lakes region, catastrophic downburst windstorms (Frelich and Lorimer 1991) and tornadoes created large windfalls up to 3784 ha in extent (Canham and Loucks 1984), resulting in a patchwork of forest regeneration in poorly drained lakeplain wetlands and on mesic ground moraines of glacial till. Fires set by lightning and Native Americans locally enhanced the diversity of successional forest patches on xeric sandy, glacial outwash plains and relict beach ridges (Whitney 1986, 1987; Delcourt and Delcourt 1996a). Early accounts by land surveyors in both Maine (ca. 1793–1827) and Michigan (ca. 1840–1850) include descriptions of treefalls and standing dead trees of conifers including *Picea* that can be attributed to outbreaks of pathogens such as spruce budworm (Lorimer 1977).

In the more humid climates of the Appalachian Mountains, the predominant form of natural disturbance in presettlement times was windfall rather than wildfire (Runkle 1985; Whitney 1990). Cove forests of the southern Appalachian Mountains were affected almost entirely by frequent, local-scale, single-tree or multiple-tree windfalls that created relatively small gaps in the forest canopy. In contrast, forests of the Allegheny Plateau in Pennsylvania were affected by both frequent, lo-

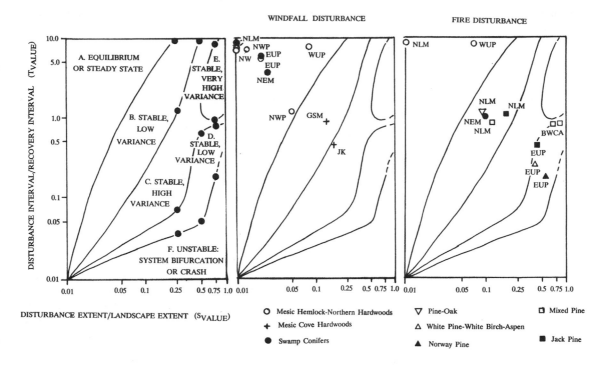

Figure 10.14. Space-time diagram of landscape state (Turner et al. 1993), depicting scaled values for temporal extent (T) and spatial extent (S) of wind and wildfire disturbance and of recovery in old-growth, virgin eastern deciduous forests and hemlock–white pine–northern hardwoods forests. T represents the ratio of the disturbance recurrence interval to the recovery time required for disturbed sites to return to mature, late-successional forest. S is the ratio of size of disturbance to the total areal extent of the landscape. Landscape disturbance profiles illustrate forest recovery from natural disturbance regimes in the Great Smoky Mountains of East Tennessee–western North Carolina (GSM) (Runkle 1982, 1985); Joyce Kilmer Wilderness Area, western North Carolina (JK) (Lorimer 1980; Runkle 1982); northwestern Pennsylvania (NWP) (Runkle 1982, 1985; Whitney 1990); northeastern Maine (NEM) (Lorimer 1977); northern Lower Michigan (NLM) (Whitney 1986); eastern Upper Michigan (EUM) (Delcourt and Delcourt 1996a); western Upper Michigan (WUM) (Frelich and Lorimer 1991); northern Wisconsin (NW) (Canham and Loucks 1984); and the Boundary Waters Canoe Area of northeastern Minnesota (BWCA) (Heinselman 1973; Baker 1989). The dashed lines on the left panel are landscape–state boundaries as extrapolated from Turner et al. (1993).

cal-scale windthrows and less frequent but larger storms that uprooted trees on tracts of many hectares (Runkle 1982, 1985; Whitney 1990).

Regional differences in natural disturbance regimes and their relationship to vegetation stability can be compared and contrasted directly by scaling both spatial and temporal scales of disturbance and recovery of vegetation (Turner et al. 1993; Fig. 10.14). In mixed mesophytic or cove hardwoods forests of the Great Smoky Mountains, for example, with only fine-scale disturbance and a rotation period of 83 yr, the vegetation is stable, with low variance (T = 0.90, S = 0.14, Fig. 10.14; data from Runkle 1982, 1985). In the Joyce Kilmer Forest of the nearby Nantahala Mountains, mesic cove forests are stable but with higher variance (T = 0.46, S = 0.17). Runkle (1982, 1985) suggested that the larger average gap size caused by more severe storms occurring in and near the Piedmont has im-

portant implications for differences in successional trends observed in the two locations. At Joyce Kilmer, *Liriodendron tulipifera* forms even-aged stands in windfall gaps >400 m². This species is virtually absent in single-tree gaps that characterize virgin coves of the Great Smoky Mountains, where many small gaps result in a large ratio of disturbed edge relative to total area of forest. High rates of repeat disturbances, which are a function of both size and age distribution of gaps, favor survival of tree species whose saplings are able to alternate between periods of moderate to rapid growth while in light gaps and periods of slow growth during times between gap formation. Examples of species tolerant to alternating suppression and release include *Acer saccharum*, *Aesculus flava*, *Fagus grandifolia*, and *Tsuga canadensis* (Runkle 1982, 1985). *Liriodendron tulipifera* can grow fast within larger gaps, but it cannot withstand the periodic growth suppression characteristic of smaller light gaps (Buckner and McCracken 1978; Runkle 1985; Busing 1995). Within the heartland of Braun's mixed mesophytic forest region, small, frequently generated windthrow gaps represent a form of intermediate disturbance (Pickett and White 1985) that provides a mechanism for enhanced diversity and codominance of those late-successional forest trees characteristic of mixed mesophytic forest communities (Braun 1950; Hinkle et al. 1993).

On the Allegheny Plateau of northwestern Pennsylvania (Whitney 1990), fine-scale, frequent gap formation (natural rotation period of 167 yr) occurs within late-successional, equilibrium beech-hemlock forests (T = 1.18, S = 0.05). Catastrophic windthrow occurs much less often (natural rotation period of 1060 yr) and extends over a relatively small proportion (1.4%) of the total forest area (T = 7.52, S = 0.01) (Fig. 10.14). This equilibrium/steady-state pattern is also characteristic of mesic northern hardwoods forests of the Upper Great Lakes region (Fig. 10.14; Whitney 1986; Frelich and Lorimer 1991) and northeastern Maine (Lorimer 1977; Fig. 10.14).

In regions where wildfire is an important natural disturbance agent, windthrow can propagate fire disturbance by adding fallen timber and dead snags to the natural fuel load (Turner et al. 1993). In the Upper Great Lakes region, fire was particularly important in determining the successional status of presettlement conifer forests on dry, sandy outwash plains (Whitney 1986, 1987; Delcourt and Delcourt 1996a; Fig. 10.14). There, the combination of recurring fire, nutrient-deficient substrate, and coarse soil texture led to droughty edaphic conditions that have perpetuated xeric, species-poor forests dominated by *Pinus banksiana* for many thousands of years (Brubaker 1975). In the Boundary Waters Canoe Area (BWCA) of northern Minnesota, recurrent fires have been the predominant disturbance regime over the past 9500 yr (Craig 1972; Swain 1973). Relatively frequent fires of broad spatial extent have resulted in early-successional stands of conifers and fire-tolerant deciduous trees such as *Betula papyrifera* var. *papyrifera* and *Populus tremuloides* (Heinselman 1973; Baker 1989). The landscape disturbance profile of wildfire for the BWCA is characterized as stable, with low variance (T = 0.79, S = 0.84), but very close to unstable, verging on system bifurcation or crash, with potential for abrupt change to an alternate landscape state given a change in disturbance regime caused by human interference. (Fig. 10.14). Baker (1989) characterized this landscape as a mosaic of non-steady-state ecosystems.

He found that vegetation in the BWCA did not attain equilibrium even at a scale 87 times the size of the mean disturbance patch. He concluded that a temporally stable patch-mosaic was not evident at any scale studied because of: (1) spatial heterogeneity in the fire regime and/or environment, and (2) a mismatch between the grain of fire patches and the grain of the environment (individual fires burn relatively little area and are spatially heterogeneous).

VEGETATION HISTORY

Although broadleaf deciduous angiosperm trees are known from high latitudes of North America as early as the late Cretaceous Period (82–65 M yr ago), modernization of the flora of the present eastern deciduous forest region has taken place only within the past few million years. The deciduous habit is thought to have evolved in flowering plants as a response to seasonally dry environments in the tropics; hence, deciduous angiosperms were "pre-adapted" to temperate climates that began to develop at middle and high latitudes during the Tertiary Period (Whitehead 1969). The following discussion of plant evolution is drawn largely from the comprehensive review by Graham (1993).

In late Cretaceous times, vegetation north of 55° N in Alaska was a mosaic of deciduous angiosperm and gymnosperm forests in a climate with mean annual temperatures of about 13° C, with a 1° C isotherm for the coldest month. The flora included *Fagopsis*, a genus of Fagaceae that became extinct at the end of the Eocene epoch, and also included members of the Aceraceae, as well as *Alnus* and *Betula*, with fossil wood showing well-defined growth rings indicating seasonally dry climate. During the early to middle Tertiary Period (65–35 M yr ago), North American climates remained warm and winter-dry. Tropical arboreal members of the Betulaceae, Fagaceae, Juglandaceae, and Ulmoidae diversified in middle and high latitudes. Broadleaf deciduous vegetation extended across the North American midcontinent. By Eocene times, forests of the Appalachian and Ozark Highlands included modern temperate genera such as *Pinus*, *Alnus*, *Betula*, *Castanea*, *Corylus*, *Nyssa*, and *Tilia*, as well as such tropical genera as *Annona* and *Engelhardia*.

North America was connected to Asia for much of the early Tertiary by the Bering land bridge and to Europe by a second land bridge through Greenland (Tiffney 1985), facilitating migrations and interchange of floristic elements across continents, particularly during the warm interval of the early Eocene Epoch. Global temperature cooling during the mid–Tertiary Period was marked in southeastern North America by increases in fagaceous trees such as *Quercoidites* and a decline in overall diversity. Middle Miocene forests of eastern North America included temperate genera such as *Abies, Picea, Pinus, Podocarpus, Tsuga, Alnus, Betula, Carya, Castanea, Fagus, Ilex, Liquidambar, Nyssa, Quercus,* and *Ulmus-Zelkova.*

By the beginning of the late Miocene Epoch (10 M yr ago), tectonic events disrupted the North Atlantic land bridge, and uplift of the Rockies resulted in aridity in the emerging rain shadow of the continental interior that favored evolution and expansion of grassland species at the expense of broadleaf deciduous trees. Many Asian, neotropical, and paleotropical floral elements disappeared from eastern North America. With continued cooling and increasingly seasonal rainfall, coniferous evergreen forests expanded at high elevations in the Appalachians and at high latitudes. Continued cooling of climates through the Pliocene Epoch (until 2 M yr ago) resulted in evolution of tundra at high northern latitudes. During Pliocene times, broadleaf deciduous forests were extirpated from many areas of western North America. Temperate deciduous forest species that had spread into Mexico during the Miocene were isolated along the eastern escarpment of the Mexican Plateau, resulting in a number of disjunctions between the southeastern United States and eastern Mexico (Miranda and Sharp 1950).

The so-called Arcto-Tertiary Geoflora, a broadleaf deciduous forest vegetation that had been extensive across middle latitudes and high elevations of continents across the northern hemisphere since the Eocene, became widely disjunct in eastern North America, eastern Asia, and Europe. The modern floristic relationships of the broadleaf deciduous forests of eastern North America and eastern Asia thus resulted from maximum extension of temperate deciduous forest in the mid–Tertiary and its disruption in western North America during the late Miocene and Pliocene (Axelrod 1992; Axelrod and Schorn 1994) and in western Europe during the 2 m yr of the Quaternary Period (Wolfe 1979; White 1983; Graham 1972, 1993).

Emergence of Modern Communities

The Quaternary Period is characterized by alternating glacial-interglacial intervals. Each climatic cycle has a period of approximately 100,000 yr, with 90,000 yr of progressive climate cooling followed by 10,000 yr of warmer interglacial climate. The cycles had profound effects on the distribution and abundance of terrestrial plants (Kutzbach and Guetter 1986; Kutzbach and Webb 1991; Wright et al. 1993). Species populations migrate in response to changes in mean annual temperatures, length of growing season, and seasonal contrasts (Davis 1983; Delcourt and Delcourt 1987, 1994; Jacobson, Webb, and Grimm 1987; Prentice 1992). Species responded individualistically to environmental change according to their physiological tolerances and dispersal capabilities (Delcourt and Delcourt 1993).

Vegetation distributions during glacial times of the Pleistocene Epoch (2 M–10,000 yr ago) were relatively stable (Birks 1986) but very different from those of interglacials – for example, the last 10,000 yr (Delcourt and Delcourt 1987). The effect of cool, equable glacial climates on the vegetation of eastern North America was to favor dominance of boreal-like coniferous forests of *Pinus banksiana* and *Picea* spp. over broad areas south of the glacial margin of the Laurentide Ice Sheet (Fig. 10.15). During glacial intervals, mesic deciduous forests were distributed south of 33° N and were greatly restricted in extent to favorable pocket refuges within a vegetation mosaic of southeastern evergreen forest – for example, in the loess hills adjacent to the Mississippi Alluvial Valley (Fig. 10.15; Delcourt and Delcourt, 1975, 1979, 1981, 1991, 1993, 1996b). Species of today's mixed mesophytic forests probably survived the ice ages in numerous localized sites – for example, around sinkholes in northern Florida (Watts et al. 1992), with "founder" populations scattered across the southeastern coastal plains. There is no fossil evidence of a "Texas corridor" connection (Martin and Harrell 1957) for plant distributions between the southeastern United States and eastern Mexico since the Pliocene. During glacial advances, southern Florida was covered by dry sand-dune scrub vegetation and so was not available as a refuge for mesic deciduous forest species (Fig. 10.15; Watts 1980).

Broad-scale vegetational changes in response to postglacial climate warming include biome-level shifts of ecotones in response to changes in positions of dominant air masses (Fig. 10.15). Each glacial age was terminated by both heightened seasonal contrast and an increase in average temperatures. Species responded by spreading outward at different rates and in different directions from refuge areas, forming temporary communities that continually changed in composition as constituent species continued to migrate northward (Davis 1983; Delcourt and Delcourt 1987; Ja-

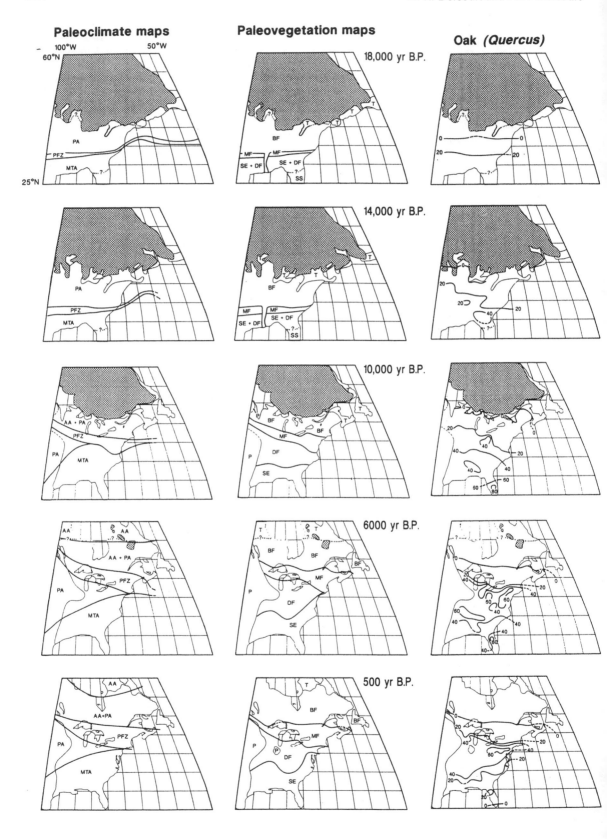

Paleoclimate maps

Paleovegetation maps

Oak *(Quercus)*

18,000 yr B.P.

14,000 yr B.P.

10,000 yr B.P.

6000 yr B.P.

500 yr B.P.

cobson et al. 1987). Temperate, broadleaf deciduous forest thus reemerged as an extensive plant formation during each interglacial interval (Fig. 10.15) (Jacobson et al. 1987; Delcourt and Delcourt 1993; Wright et al. 1993).

Different deciduous forest species have been favored during different parts of interglacial cycles (Watts 1988; Huntley and Webb 1989; Wright 1992; Delcourt and Delcourt 1994). In the Holocene, seasonality of climates peaked between 12,000 and 6000 yr ago (Kutzbach and Guetter 1986). During the early Holocene, *Ostrya virginiana* and *Carpinus caroliniana*, which are today minor understory species, composed 30% or more of the forests across the region from the southern Appalachian Mountains to the Ozark Highlands (Delcourt and Delcourt 1994). In the present-day maple-basswood forest region, early Holocene forests were dominated by *Ulmus* (Wright 1992). Oaks, hickories, beech, and other trees characteristic of the central deciduous forest region today came into prominence only after 6000 yr ago (Fig. 10.15). Certain tree species that were important in the early Holocene – for example, *Ostrya virginiana* and *Carpinus caroliniana* – have ring-diffuse wood, are indeterminate in growth, and have continuously active xylem. This hydraulic wood architecture may have enabled them to take advantage of early spring warm spells (Delcourt and Delcourt 1994). In contrast, ring-porous tree species, for example *Quercus*, begin growth later in the growing season and therefore may be better adapted to climates of the latter part of the interglacial cycle when climates were less seasonal (analogous to European "terminocratic" species of Watts 1988). Additional

Figure 10.15. Glacial-interglacial changes in past climate, areal extent of glacial ice (shaded areas), vegetation, and Quercus (oak) tree populations (with contoured values expressed as % forest composition) mapped across eastern North America for the past 18,000 yr. The maps reflect full-glacial times (18,000 yr BP), the late-glacial interval (14,000 yr BP), and the present Holocene interglacial interval (10,000 yr BP, 6000 yr BP, and 500 yr BP). Paleoclimatic maps identify climatic regions dominated by arctic (AA), Pacific (PA), and maritime tropical (MTA) air masses. The typical path of the Polar Jet Stream corresponds with the Polar Frontal Zone (PFZ). The paleovegetation and paleopopulation maps are based on paleoecological reconstructions from 162 radiocarbon-dated plant and fossil sites (Delcourt and Delcourt 1987). Forest types include boreal (BF); mixed conifer–northern hardwoods (MF); deciduous (DF); and southeastern evergreen (SE). Tree populations are not mapped across the nonforested regions of tundra (T), prairie (P), or sand dune scrub vegetation (SS) (modified from Delcourt and Delcourt 1993; From Flora of North America: North of Mexico, Vol. 1: Introduction, edited by Flora of North America. Copyright © 1993 by Flora of North America Association. Used by permission of Oxford University Press, Inc.).

physiological responses to highly seasonal climates that may affect the competitive interactions and rate of northward spread of temperate deciduous tree species include timing of bud break after last frost and hardening-off of flower buds before first frost (Barnes 1991).

Overall, the postglacial rates of spread of temperate broadleaf deciduous trees seem to have been controlled more by individual responses to climate change than to differences in dispersal capabilities (Davis 1983; Delcourt and Delcourt 1987; Jacobson et al. 1987). Some species reached their modern distributional ranges only in the last few thousand years, e.g. *Castanea dentata* in the northern Appalachian Mountains (Davis 1983) and *Pinus echinata* in the Ozark Highlands (Delcourt and Delcourt 1991). Some tree species were still adjusting their range limits at the time of EuroAmerican settlement (e.g., *Fagus grandifolia* in the western Great Lakes region) (Davis, Woods, Webb, and Futyma 1986).

In addition to climatic change and species immigrations, late-Quaternary trends of regional forest development have been shaped by postglacial changes in disturbance regimes, microbial assemblages colonizing the organic litter (Perry, Borchers, Borchers, and Ameranthus 1990), soil pH, leaching, and nutrient availability. For example, in the White Mountains of New Hampshire, 14,000 yr ago tundra plants colonized unweathered glacial debris (Spear, Davis, and Shane 1994). Rigorous permafrost environments persisted until 11,500 yr ago and were characterized by periglacial freeze–thaw processes of continuous soil creep and episodic avalanches that would have favored shade-intolerant pioneer species of annual and perennial herbs. *Picea*-dominated woodlands expanded up to intermediate elevations between 11,500 and 9000 yr ago in response to longer growing seasons and stabilization of hillslopes. Fire-tolerant *Pinus strobus*, *Quercus* spp., and *Betula papyrifera* established during the warm, dry interval between 9000 and 7000 yr ago. Alpine tundra persisted thereafter only above 1700 m elevation. A disturbance regime including infrequent but catastrophic windfall resulting from storms and hurricanes became more important after 7000 yr ago and favored mesic northern hardwoods such as *Fagus grandifolia, Acer saccharum*, and *Betula alleghaniensis* (Spear et al. 1994; Delcourt and Delcourt, 1996b).

Impact of Native Americans

Since the arrival of nomadic PaleoIndian populations >12,000 yr ago across eastern North America (Morse and Morse 1983), Native American impacts have been cumulative within deciduous and mixed conifer–hardwood forest regions. Initially, hunting activities of PaleoIndians focused on and contributed to the extinction of late-Pleistocene megafauna (Martin and Klein 1984). By early Holocene times, selective extermination of many genera of large mammals disrupted co-evolutionary plant–animal linkages, favoring generalist herbivores over specialists (Graham and Lundelius 1984). Thereafter, dispersal of plant propagules may have been limited for some large-seeded temperate tree species, leading to their progressive geographic isolation (Janzen and Martin 1982; Delcourt and Delcourt 1991).

Over the Holocene interglacial period, aboriginal human populations shaped vegetation patterns directly by augmenting the natural fire regime (McAndrews 1988; Wilkins et al. 1991; Abrams 1992; Clark and Royall 1994; Delcourt and Delcourt 1997, Delcourt and Delcourt, 1998a; Delcourt, Delcourt, Ison, Sharp, and Gremillion, in press). Native Americans also affected the native vegetation because of cultural preferences (and taboos) for exploiting certain plant resources for habitation, fuel, and food (Bareis and Porter 1984; King 1985; Delcourt et al. 1986). During the past 5000 yr, domestication of native plant species and introduction of exotic cultigens have resulted in the expansion of agricultural fields at the expense of forest lands, particularly along major river systems and coastal areas of the southeastern United States (Chomko and Crawford 1978; Smith 1978; Ford 1985; Fearn and Liu 1995).

Prehistoric Native Americans influenced the vegetation of the eastern deciduous forest region in a number of ways (Delcourt 1987). First, Native Americans changed the dominance structure of forest communities through exploitation of wood for fuel for hearth fires and timbers for construction of individual habitations, palisaded villages, and fortified cities. During the Mississippian period (A.D. 800–1500), sedentary human populations were relatively large along riparian corridors. For example, about 600 yr ago human population densities are estimated at 2000 individuals per km² in the Mississippian heartland of Cahokia, a vast prehistoric city with hundreds of earthen temple mounds and plazas, covering >13 km² of alluvial bottomlands bordering the Mississippi River in present-day East St. Louis, Illinois (Bareis and Porter 1984). For unknown reasons that possibly included warfare among aboriginal tribes, Cahokia was abandoned about A.D. 1450 (Bareis and Porter 1974; Smith 1978). Extensive areas of wet meadows and bottomland prairies mapped by General Land Office surveyors along the Mississippi and other rivers in southern Illinois (Anderson 1970; Iverson, Tucker,

Risser, Burnett, and Rayburn 1989) may have represented Indian old fields that had not reverted to forest by the time of EuroAmerican settlement.

With slash and burn clearance of forests, Native Americans changed portions of the landscape mosaic from primarily mid- to late-successional deciduous forest to a more open, early successional brushy thicket and immature forest. In some cases, these early seral patches were maintained for millennia. Even localized Indian-set fires altered forest succession. For example, near Crawford Lake in south-central Ontario, the Iroquois used fire to clear gaps in mature beech-maple forest for cultivating garden plots of *Zea mays* (maize), *Phaseolis vulgaris* (beans), and *Cucurbita pepo* (squash) (McAndrews 1988). After A.D. 1650, the Iroquois moved away and abandoned the garden plots, which became sites for colonization of disturbance-favored trees such as *Pinus strobus* and *Quercus rubra* var. *rubra*, which persisted for centuries thereafter. The combination of a prehistoric human-augmented fire regime (Clark and Royall 1995) and the onset of cool and moist climatic conditions of as much as 2° C during the Little Ice Age (Campbell and McAndrews 1993) favored local establishment of *Pinus strobus* and also promoted a southward shift of the hemlock–white pine–northern hardwoods forest type in southern Ontario (McAndrews 1988).

In north-central Kentucky, oak savanna and xeric oak-hickory forests alternated with upland grasslands known as the "Big Barrens" (Baskin and Baskin 1981). Once thought to represent a mid-Holocene extension of the Prairie Peninsula (Stuckey 1981; DeSelm and Murdock 1993), grasslands in the Big Barrens of Kentucky have been reinterpreted as the product of Indian-set fires during the past 3000 years, based on fossil pollen and charcoal preserved in lake sediments (Wilkins et al. 1991). In an Indian-augmented fire regime, *Quercus* regeneration could well have been favored on fire-prone, xeric or nutrient-poor sites across much of the eastern deciduous forest region (Abrams 1992). In both the Great Lakes and New England regions however, the extent of Native American impact on forest succession through deliberate use of fire is a continuing subject of debate (Russell 1983; Whitney 1995; Clark and Royall 1995, 1996; Abrams and Seischab 1997).

In the heartland of the eastern deciduous forest, Native Americans have continuously occupied the rich floodplains and low-level terraces along major river systems for the past 10,000 yr. Along the Little Tennessee River Valley, ever-expanding prehistoric human populations gradually deforested the floodplain, then the low and intermediate-level terraces, spreading to uplands near natural ponds in late prehistoric times (Delcourt et al. 1986; Delcourt and Delcourt 1988). At the transition from the Woodland to the Mississippian cultural period times in A.D. 800 and again at the transition from the Mississippian to the Historic cultural period in A.D. 1600, use of fire increased by an order of magnitude, and landscapes along the Little Tennessee River were destabilized by sheetwash erosion of soils. Deforestation of low river terraces and cultivation of garden plots favored the spread of ruderal and forest-edge plant species as weeds in Indian old fields (Delcourt et al. 1986; Delcourt 1987; Delcourt and Delcourt 1988).

Prehistoric human impact on deciduous forests extended beyond their agricultural transformation of riparian corridors along major river valleys (Smith 1978). For example, in eastern Kentucky, within the steep gorges cut through the western escarpment of the Cumberland Plateau, prehistoric humans occupied cliff-edge rock shelters and seasonally collected plant foods such as acorns, hickory nuts, chestnuts, walnuts, and butternuts. Through ethnobotanical investigations of rock shelter sites and fossil pollen and charcoal assemblages from Cliff Palace Pond, Jackson County, Kentucky, Delcourt et al. (1998) demonstrate that human impact on the upland vegetation began after about 3000 BP. Prehistoric Native Americans used fire to clear gaps within the forest, and they cultivated both introduced and native, domesticated, weedy herb species within local garden plots. Cultigens within the Eastern Agricultural Complex included *Helianthus annuus* (sunflower), *Iva annua* var. *macrocarpa* (sumpweed), *Chenopodium berlandieri* (goosefoot), *Phalaris caroliniana* (maygrass), and *Polygonum erectum* (knotweed). Additional weedy plants such as *Ambrosia* (ragweed) spread into garden plots from open habitats such as point bars located in the Kentucky River. By 3000 yr BP, increased cultural use of fire converted the regional forest composition from fire-intolerant canopy dominants such as red cedar to fire-promoted species of oak and chestnut, and favored development of ridge-top stands of endemic, fire-adapted pines including *Pinus rigida*. The ethnobotanical and pollen evidence of tuliptree corresponds with the cultivation of garden plots in Late Archaic and Woodland times. On the basis of experimental data, Cowan (1985) proposed that cleared areas of about 400 m² located on midslopes near rock shelters would have provided sufficient plant food for winter survival of individual families. Successive clearing of new forest gaps and abandonment of old garden plots would have created light gaps of sufficient extent for colonization

of landscape patches by tuliptree and white pine. We speculate that some of the large, remnant old-growth stands of tuliptree, such as those at Joyce Kilmer Forest, North Carolina (see Stand #9, Table 10.1; Braun 1950; Lorimer 1980), may represent secondary forests established on abandoned Indian old fields.

In the southernmost Appalachian Mountains near Highlands, North Carolina, the late-Holocene sequence of fossil pollen and charcoal from Horse Cove Bog demonstrates the importance of prehistoric human-set fires in forest dynamics of the oak-chestnut forest region (Delcourt and Delcourt 1997). With a mean annual precipitation up to 254 cm, natural wildfire is extremely rare. Ignitions caused by lightning strikes are limited to ridge crests and south-facing, upper hill slopes. The record of charcoal influx from Horse Cove Bog shows that prehistoric human-set wildfires were concentrated on two portions of the landscape: (1) alluvial bottoms in intermontane valleys where forests were cleared for habitations and garden plots; and (2) relatively xeric, exposed upper slopes dominated in presettlement times by fire-tolerant oak, chestnut, and pine. Human use of fire thus appears to have contributed to overall landscape heterogeneity and to have been a major factor favoring the regional rise to dominance of Appalachian oak-chestnut forests during the past 3000 years (Delcourt and Delcourt 1997; 1998b).

Native Americans also altered the distributional ranges of species, both by introducing domesticated plants from northern Mexico and the American Southwest (Ford 1985; King 1985) and by causing local extinctions from overexploitation of native plants. For example, the northern range limit of *Taxodium distichum* in the lower Illinois River Valley receded southward in Mississippian times because of extensive use of the trees for canoes (Morse and Morse 1983). By A.D. 1500, landscape-level patterns of eastern deciduous forests were the collective "result of continued individualistic responses of species' populations to long-term changes in climate, prevailing disturbance regimes, and prehistoric human activities that included the use of fire and the development of agriculture" (Delcourt et al. 1993, p. 72).

Influences of EuroAmerican Settlement

Post-Columbian settlement was concentrated initially along the Atlantic Seaboard, the inland route of the St. Lawrence River to the Great Lakes, and the Mississippi–Missouri–Ohio rivers system. The western frontier of deciduous forest was opened for settlement after the Congressional Act of 1796 that formally established the administrative structure (for what would become the General Land Office) and methodology for federal survey of these public lands (White 1984; Cronon 1983; Williams 1989; Whitney 1995). The subsequent population increases on expanding homesteads provided both more extensive and intensive use of forests. This land use led to progressive forest fragmentation and transformation of a natural to a cultural landscape (Marschner 1959; Curtis 1959; Burgess and Sharpe 1981; Whitney 1995). From Maine in the 1770s to Minnesota in the 1890s, loggers first selectively harvested *Pinus strobus* and *P. resinosa*. After Great Lakes pineries were exhausted, most of the *Pinus echinata* was removed from the southeastern Ozarks between 1890 and 1910 (Rafferty 1980). Hardwoods were harvested in a second cutting wave from the 1880s to the 1920s for kiln-manufacture of commercial charcoal for smelting metal and fuel for factories, steam-driven locomotives, and steamships, for the production of structural timbers to build cities, for furniture, and for wood distillation of organic chemicals (Whitney 1987, 1990, 1995; Williams 1989). Acidic tanbark, stripped from trunks of *Castanea dentata, Quercus prinus,* and *Tsuga canadensis,* was harvested for leather tanning (Burns and Honkala 1990a, 1990b). Leather belts ran the factory machines and leather straps harnessed the ox- and horse-power on farms (Whitney 1990).

Widespread fires ignited in the logging slash debris resulted in broad burnt areas of bare mineral soil, large tracts of brushlands, and even-aged stands of immature forest throughout the northern deciduous forest region. Postfire communities were often dominated by fire-resistant root sprouts of *Quercus* ("oak grubs"), *Betula, Populus,* and *Pinus banksiana. Liriodendron tulipifera* became widespread after fire in successional forests of the Appalachian Mountains (Ayres and Ashe 1905).

The expanding area of cultivated land, pasture, and immature forest throughout the eastern deciduous forest region (Delcourt and Harris 1980) produced more available browse in forest "edge" habitats and triggered near-exponential population growth of white-tailed deer. Increased herbivory is one possible factor prohibiting regeneration of populations of many former forest dominants – for example, limiting the reestablishment of *Tsuga canadensis* within mesic mixed conifer–northern hardwoods stands (Alverson, Waller, and Solheim 1988; Whitney 1990; Mladenoff and Stearns 1993). Successive historic introductions of exotic pathogens have eliminated other forest canopy species such as *Castanea dentata* and *Ulmus americana* and now threaten populations of *Fagus grandifolia,*

Quercus rubra, Ulmus, and *Abies fraseri* (Castello, Leopold, and Smallidge 1995).

Changing technology for access to forest stands (the invention of the chain saw and logging trucks) and new uses of wood products has favored both the traditional need for fast-growing trees for timber, as well as the post–World War II need for pulpwood. Commercial incentives for efficient timber harvest have reinforced the forestry practice of clear-cutting stands in rotation cycles of 30–60 yr. Management includes prescribed fire and chemical application to minimize competition by undesirable woody species (Whitney 1987). One result of current forest management practices is a change to a more coarse-grained landscape, with frequent, spatially extensive episodes of clear-cut harvesting and restocking in extensive plantations (Stearns 1987; Turner 1990; Hansen, Spies, Swanson, and Ohmann 1991).

The post-1950 harvesting regime has emphasized establishment of monocultures of homogeneous genetic stock. These forests may become vulnerable to infestations of pesticide-resistant pathogens and to increased climatic variability in the next century (Fowells and Means 1990; Root and Schneider 1993). Managed forests may plot on Figure 10.14 in the landscape states of either (D)— stable system, low variance (hypothetical example: 60 yr rotation period, 91 yr recovery to mature state, T = 0.7; total of 90% area harvested in commercial forest, S = 0.9)—or (F), an unstable system, susceptible to crash or bifurcation to new states (hypothetical example: 30 yr rotation period, 91 yr recovery period, T = 0.3; total of 90% forested area harvested, S = 0.9).

Achievement of long-term sustainability of forest resources (Lubchenco et al. 1991) is questionable given the inability of understory herbs to recover following the harvesting of Appalachian deciduous forests (Duffy and Meier 1992; Duffy 1993a, 1993b). Overall, since the 1950s, logging practices have had five major impacts on natural forests in eastern North America:

1. Changing the scale of disturbance and increasing the size of forest openings.
2. Increasing the frequency of large openings in the forest canopy.
3. Removing woody debris (germination sites for seedlings) from the ecosystems.
4. Compacting and eroding soil by modern logging equipment;
5. Impacting understory species by trampling (Bratton 1994).

Preservation of biodiversity in eastern deciduous forests may require ecologists to ask increasingly sophisticated questions about ecological processes scaled across space and time, as well as the long-term consequences of human use and management including prescribed burning, of remaining forest lands (Bratton 1994).

AREAS FOR FUTURE RESEARCH

Although the eastern deciduous forest is one of the most thoroughly studied vegetation regions of North America, significant gaps remain in understanding pattern and process on several spatial-temporal scales. The effects of geographic isolation during the Pleistocene on the genetics and distributional history of deciduous forest species, as well as the extent to which modern plant communities are typical of those that have existed during previous interglacials, await additional paleoecological studies in the southeastern United States (Watts 1980, 1988; Watts et al. 1992).

The role of natural and anthropogenic fire regimes in determining the composition and distribution of presettlement vegetation remains to be clarified by additional paleoecological studies, particularly as relevant to forests dominated by *Quercus* and *Castanea* in the central deciduous forest region (Russell 1983; Delcourt et al. 1986; Abrams 1992; Clark and Royall 1995, 1996; Abrams and Seischab 1997; Delcourt and Delcourt 1998a).

Because of continuing land use, pollution stress, and possible near-future climatic changes, a current imperative is to conserve biodiversity at all levels of biological organization (Lubchenco et al. 1991; Delcourt and Delcourt, 1998b). Studies of genetics and demography of rare and endangered plants, of successional trends within virgin and second-growth forests, and of the responses of plants and animals to changes in landscape heterogeneity are avenues for continuing research that are all essential for conserving eastern deciduous forests into the future.

REFERENCES

Abrams, M. D. 1992. Fire and the development of oak forests. BioScience 42:346–353.

Abrams, M. D., and F. K. Seischab. 1997. Does the absence of sediment charcoal provide substantial evidence against the fire and oak hypothesis? J. Ecology 85:373–375.

Alverson, W. S., D. M. Waller, and S. L. Solheim. 1988. Forests too deer: edge effects in northern Wisconsin. Cons. Biol. 2:348–358.

Anderson, R. C. 1970. Prairies in the prairie state. Trans. Ill. State Acad. Sci 63:214–221.

Anderson, T. W. 1974. The chestnut pollen decline as a time horizon in lake sediments in eastern North America. Can. J. Earth Sci. 11:678–685.

Arris, L. L., and P. S. Eagleson. 1989. Evidence of a physiological basis for the boreal-deciduous forest ecotone in North America. Vegetatio 82:55–58.

Axelrod, D. I. 1992. The Middle Miocene Pyramid Flora of western Nevada. University of California Publ. Geol. Sci. 137:1–50.

Axelrod, D. I., and H. E. Schorn. 1994. The 15 Ma floristic crisis at Gillam Spring, Washoe County, northwestern Nevada. PaleoBioscience 16:1–10.

Ayres, H. B., and W. W. Ashe. 1905. The southern Appalachian forests. U.S. Geol. Surv. Prof. Paper No. 37 (Series H, Forestry, 12):1–291.

Baker, W. L. 1989. Landscape ecology and nature reserve design in the Boundary Waters Canoe Area, Minnesota. Ecology 70:23–35.

Barbour, M. G., and N. L. Christensen. 1993. Vegetation, pp. 97–131 in N. R. Morin (conv. ed.), Flora of North America north of Mexico, Vol. 1. Oxford University Press, New York.

Barden, L. S. 1980. Tree replacement in a cove hardwood forest of the southern Appalachians. Oikos 35:16–19.

Barden, L. S. 1981. Forest development in canopy gaps of a diverse hardwood forest of the southern Appalachian Mountains. Oikos 37:205–209.

Bareis, C. J., and J.W. Porter. 1984. American Bottom archaeology. University of Illinois Press, Urbana.

Barnes, B. V. 1991. Deciduous forests of North America, pp. 219–344 in E. Rohrig and B. Ulrich (eds.), Ecosystems of the world 7: temperate deciduous forests. Elsevier, Amsterdam.

Barnosky, C. W. 1987. Response of vegetation to climate changes of different duration in the late Neogene. Trends Ecol. Evol. 2:247–250.

Barrett, J. W. 1980. Regional silviculture of the United States. Wiley, New York.

Bartram, W. 1791. Travels through North and South Carolina, Georgia, East and West Florida, the Cherokee Country, the Extensive Territories of the Muscogulges or Creek Confederacy, and the Country of the Choctaws: containing an account of the soil and natural productions of those regions, together with observations on the manners of the Indians. James and Johnson, Philadelphia.

Baskin, J. M., and C. C. Baskin. 1981. The Big Barrens of Kentucky not a part of Transeau's Prairie Peninsula, pp. 43–48 in R. L. Stuckey and K. J. Reese (eds.), The Prairie Peninsula – in the "Shadow" of Transeau. Ohio Biol. Surv., Biol. Notes 15, Ohio State University, Columbus.

Baskin, J. M., and C. C. Baskin. 1986. Distribution and geographical/evolutionary relationships of cedar glade endemics in southeastern United States. Assoc. Southeast. Biol. Bull. 33:138–154.

Billings, W. D. 1938. The structure and development of old field shortleaf pine stands and certain associated physical properties of the soil. Ecol. Monogr. 8:437–499.

Billings, W. D., and L. E. Anderson. 1966. Some microclimatic characteristics of habitats of endemic and disjunct bryophytes in the southern Blue Ridge. Bryologist 68:76–95.

Birks, H. J. B. 1986. Late-Quaternary biotic changes in terrestrial and lacustrine environments, with particular reference to north-west Europe, pp. 3–65 in B. E. Berglund (ed.), Handbook of Holocene palaeoecology and palaeohydrology. Wiley, New York.

Borchert, J. R. 1950. The climate of the central North

American grassland. Ann. Assoc. Amer. Geogr. 40:1–39.

Bormann, F. H., and G. E. Likens. 1979. Pattern and process in a forested ecosystem. Springer-Verlag, New York.

Bratton, S. P. 1994. Logging and fragmentation of broadleaved deciduous forests: are we asking the right ecological questions? Cons. Biol. 8:295–297.

Braun, E. L. 1950. Deciduous forests of eastern North America. Blakiston, Philadelphia.

Brouillet, L., and R. D. Whetstone. 1993. Climate and physiography, pp. 15–46 in N. R. Morin (conv. ed.), Flora of North America north of Mexico, Vol. 1. Oxford University Press, New York.

Brubaker, L. B. 1975. Postglacial forest patterns associated with till and outwash in northcentral Upper Michigan. Quat. Res. 9:349–362.

Bryant, W. S., and W. H. Martin. 1988. Vegetation of the Jackson Purchase of Kentucky based on the 1820 General Land Office Survey, pp. 264–276 in D. H. Synder (ed.), Proceedings of the first annual symposium on the natural history of lower Tennessee and Cumberland River valleys. Center for Field Biology of Land Between the Lakes, Austin Peay State University, Clarksville, Tenn.

Bryant, W. S., W. C. McComb, and J. S. Fralish. 1993. Oak-hickory forests (western mesophytic/oak-hickory forests), pp. 143–201 in W. H. Martin, S. G. Boyce, and A. C. Echternacht (eds.), Biodiversity of the southeastern United States: upland terrestrial communities. Wiley, New York.

Bryson, R. A. 1966. Air masses, streamlines, and the boreal forest. Geogr. Bull. 8:228–269.

Bryson, R. A., and F. K. Hare. 1974. The climates of North America, pp. 1–47 in R. A. Bryson and F. K. Hare (eds.), Climates of North America, world survey of climatology, Vol. 11. Elsevier, Amsterdam.

Buckner, E., and W. McCracken. 1978. Yellow poplar: a component of climax forests? J. Forestry 76:421–423.

Burgess, R. L., and D. M. Sharpe. 1981. Forest island dynamics in man-dominated landscapes. Ecological Studies, Vol. 41. Springer-Verlag, New York.

Burke, M. J., M. F. George, and R. G. Bryant. 1975. Water in plant tissues and frost hardiness, pp. 111–135 in R. B. Duckworth (ed.), Water relations of foods. Academic Press, New York.

Burns, R. M., and B. H. Honkala (eds.). 1990a. Silvics of North America, Vol. 1, Conifers. U.S. Dept. Agric. For. Serv. Handbook 654. Govt. Printing Office, Washington, D.C.

Burns, R. M., and B. H. Honkala (eds.). 1990b. Silvics of North America, Vol. 2, Hardwoods. U.S. Dept. Agric. For. Serv. Handbook 654. Govt. Printing Office, Washington, D.C.

Busing, R. T. 1995. Disturbance and the population dynamics of Liriodendron tulipifera: simulations with a spatial model of forest succession. J. Ecology 83:45–53.

Cain, S. A. 1935. Studies on virgin hardwood forest: III. Warren's Woods, a beach-maple climax forest in Berrien County, Michigan. Ecology 16:500–513.

Campbell, I. D., and J. H. McAndrews. 1993. Forest disequilibrium caused by rapid Little Ice Age cooling. Nature 366:336–338.

Canham, C. D., and O. L. Loucks. 1984. Catastrophic windthrow in the presettlement forests of Wisconsin. Ecology 65:803–809.

Caplenor, C. D., R. E. Bell, J. Brook, D. Caldwell, C. Hughes, A. Regan. A. Scott, S. Ware, and M. Wells. 1968. Forests of west central Mississippi as affected by loess. Miss. Geol., Econ., and Topogr. Surv. Bull. 111:205–267.

Castello, J. D., D. J. Leopold, and P. J. Smallidge. 1995. Pathogens, patterns, and processes in forest ecosystems. BioScience 45:16–24.

Chomko, S., and G. Crawford. 1978. Plant husbandry in prehistoric North America. Amer. Antiquity 43:405–408.

Clampitt, C. 1985. DECORANA for IBM-PCs. BioScience 35:738.

Clark, J. S., and P. D. Royall. 1995. Transformation of a northern hardwood forest by aboriginal (Iroquois) fire: charcoal evidence from Crawford Lake, Ontario, Canada. The Holocene 5:1–9.

Clark, J. S., and P. D. Royall. 1996. Local and regional sediment charcoal evidence for fire regimes in presettlement north-eastern North Amer. J. Ecology 84:365–382.

Clebsch, E. E. C., and R. T. Busing. 1989. Secondary succession, gap dynamics, and community structure in an Appalachian cove forest. Ecology 70:728–735.

Clements, F. E. 1916. Plant succession. Carnegie Institute of Washington, Publ. 242, Washington, D.C.

Cogbill, C. V., and P. S. White. 1991. The latitude-elevation relationship for spruce-fir forest and treeline along the Appalachian mountain chain. Vegetatio 94:153–175.

Cogbill, C. V., White, P. S., and S. K. Wiser. 1997. Predicting treeline elevation in the Southern Appalachians. Castanea 62:137–146.

Cooper, James G. 1860. The forests and trees of North America, as connected with climate and agriculture. Report of the Commissioner of Patents for 1860. U.S. Congress, House, 36th Cong., 2nd Sess., 1861 H. Ex. Doc. 48 (serial no. 1099), 416–445. Government Printing Office, Washington, D.C.

Cowan, C. W. 1985. From foraging to incipient food production: Subsistence change and continuity on the Cumberland Plateau of eastern Kentucky. Ph.D. dissertation, University of Michigan, University Microfilms # 8600429–03600, Ann Arbor.

Graig, A. J. 1972. Pollen influx to laminated sediments: a pollen diagram from northeastern Minnesota. Ecology 53:46–57.

Cronon, W. 1983. Changes in the land, Indians, colonists, and the ecology of New England. Hill & Wang, Div. of Farrar, Straus, and Giroux, New York.

Curtis, J. T. 1959. The vegetation of Wisconsin: an ordination of plant communities. University of Wisconsin Press, Madison.

Dale, E. E. 1986. The vegetation of Arkansas (including an inserted vegetation map). Ark. Nat. 4:7–27.

Daubenmire, R. F. 1936. The "Big Woods" of Minnesota: its structure, and relation to climate, fire, and soils. Ecol. Monogr. 6:233–268.

Daubenmire, R. F. 1978. Plant geography, with special reference to North America. Academic Press, New York.

Davis, M. B. 1983. Quaternary history of deciduous forests of eastern North America and Europe. Ann. Mo. Bot. Gard. 70:550–563.

Davis, M. B., K. Woods, S. Webb, and R. P. Futyma. 1986. Dispersal versus climate: expansion of Fagus and *Tsuga* into the Upper Great Lakes region. Vegetatio 67:93–103.

Delcourt, H. R. 1979. Late Quaternary vegetation history of the Eastern Highland Rim and adjacent Cumberland Plateau of Tennessee. Ecol. Monogr. 49: 255–280.

Delcourt, H. R. 1987. The impact of prehistoric agriculture and land occupation on natural vegetation. Trends Ecol. Evol. 2:39–44.

Delcourt, H. R., and P. A. Delcourt. 1974. Primeval magnolia-holly-beech climax in Louisiana. Ecology 55:638–644.

Delcourt, H. R., and P. A. Delcourt. 1975. The Blufflands: Pleistocene pathway into the Tunica Hills. Amer. Midl. Nat. 94:385–400.

Delcourt, H. R., and P. A. Delcourt. 1977. Presettlement magnolia-beech climax of the Gulf Coastal Plain: quantitative evidence from the Apalachicola River Bluffs, north-central Florida. Ecology 58:1085–1093.

Delcourt, H. R., and P. A. Delcourt. 1988. Quaternary landscape ecology: relevant scales in space and time. Landscape Ecol. 2:23–44.

Delcourt, H. R., and P. A. Delcourt. 1991. Late-Quaternary vegetation history of the Interior Highland region of Missouri, Arkansas, and Oklahoma, pp. 15–30 in D. Henderson and L. D. Hedrick (eds.), Restoration of old growth forests in the Interior Highlands of Arkansas and Oklahoma. Quachita National Forest and Winrock Institute for Agricultural Development. Winrock Institute. Morrilton, Ark.

Delcourt, H. R., and P. A. Delcourt. 1994. Postglacial rise and decline of *Ostraya virginiana* (Mill.) K. Koch and *Carpinus caroliniana* Walt. in eastern North America: predictable responses of forest species to cyclic changes in seasonality of climates. J. Biogeogr. 21:137–150.

Delcourt, H. R., and P. A. Delcourt. 1996a. Presettlement landscape heterogeneity: evaluating grain of resolution using General Land Office Survey data. Landscape Ecol. 11:363–381.

Delcourt, H. R., and P. A. Delcourt. 1997. Pre-Columbian native American use of fire on Southern Appalachian landscapes. Cons. Biol. 11:1010–1014.

Delcourt, P. A., and H. R. Delcourt. 1996b. Quaternary vegetation history of the Lower Mississippi Valley. Engineering Geol. 45:219–242.

Delcourt, H. R., P. A. Delcourt, and T. Webb III. 1983. Dynamic plant ecology: the spectrum of vegetational change in space and time. Quat. Sci. Rev. 1: 153–175.

Delcourt, H. R., and W. F. Harris. 1980. Carbon budget of the southeastern U.S. biota: analysis of historical change in trend from source to sink. Science 210: 321–323.

Delcourt, H. R., D. C. West, and P. A. Delcourt. 1981. Forests of the southeastern United States: quantitative maps for aboveground woody biomass, carbon, and dominance of major tree taxa. Ecology 62: 879–887.

Delcourt, P. A., and H. R. Delcourt. 1979. Late Pleistocene and Holocene distributional history of the deciduous forest in the southeastern United States. Veroff. des Geobot. Inst. ETH, Stift. Rubel (Zurich) 68:79–107.

Delcourt, P. A., and H. R. Delcourt. 1981. Vegetation maps for eastern North America: 40,000 yr BP to the present, pp. 123–165 in R. C. Romans (ed.), Geobotany II. Plenum, New York.

Delcourt, P. A., and H. R. Delcourt. 1987. Long-term forest dynamics of the Temperate Zone. Ecological Studies 63. Springer-Verlag, New York.

Delcourt, P. A., and H. R. Delcourt. 1993. Paleoclimates, paleovegetation, and paleofloras during the Late Quaternary, pp. 71–96 in N. R. Morin (conv. ed.), Flora of North America north of Mexico, Vol. 1. Oxford University Press, New York.

Delcourt, P. A., and H. R. Delcourt. 1998a. The influence of prehistoric human-set fires on oak-chestnut forests in the southern Appalachians. Castanea.

Delcourt, P. A., and H. R. Delcourt. 1998b. Conservation of biodiversity in light of the Quaternary paleoecological record: should the focus be on species, ecosystems, or landscapes? Ecol. Appl.

Delcourt, P. A., H. R. Delcourt, P. A. Cridlebaugh, and J. Chapman. 1986. Holocene ethnobotanical and paleoecological record of human impact on vegetation in the Little Tennessee River Valley, Tennessee. Quat. Res. 25:330–349.

Delcourt, P. A., H. R. Delcourt, C. R. Ison, W. E. Sharp, and K. J. Gremillion. 1998. Prehistoric human use of fire, the Eastern Agricultural Complex, and Appalachian oak-chestnut forests: Paleoecology of Cliff Palace Pond, Kentucky. Amer. Antiquities.

Delcourt, P. A., H. R. Delcourt, D. F. Morse, and P. A. Morse. 1993. History, evolution, and organization of vegetation and human culture, pp. 47–79 in W. H. Martin, S. G. Boyce, and A. C. Echternacht (ed.), Biodiversity of the southeastern United States: lowland terrestrial communities. Wiley, New York.

Delcourt, P. A., H. R. Delcourt, and T. Webb III. 1984. Atlas of mapped distributions of dominance and modern pollen percentages for important tree taxa of eastern North America. Am. Assoc. Strat. Palynol. Contrib. Ser. No. 14:1–131.

DeSelm, H. R., and N. Murdock. 1993. Grass-dominated communities, pp. 87–142 in W. H. Martin, S. G. Boyce, and A. C. Echternacht (eds), Biodiversity of the southeastern United States: upland terrestrial communities. Wiley, New York.

Duffy, D. C. 1993a. Herbs and clearcutting: reply to Elliot and Loftis and Steinbeck. Cons. Biol. 7:221–223.

Duffy, D. C. 1993b. Seeing the forest for the trees: response to Johnson et al. Cons. Biol. 7:436–439.

Duffy, D. C., and A. J. Meier. 1992. Do Appalachian herbaceous understories ever recover from clearcutting? Cons. Biol. 6:196–201.

Engstrom, D. R., and B. C. S. Hansen. 1985. Postglacial vegetational change and soil development in southeastern Labrador as inferred from pollen and chemical stratigraphy. Can. J. Bot. 63:543–561.

Eyre, F. H. 1980. Forest cover types of the United States and Canada. Society of American Foresters, Washington, D.C.

Fearn, M. L., and K.-B. Liu. 1995. Maize pollen of 3500 B. P. from southern Alabama. Amer. Antiquity 60: 109–117.

Ford, R. I. 1985. Patterns of prehistoric food production in North America, pp. 341–364 in R. I. Ford (ed.), Prehistoric food production in North America. Anthropological Papers No. 75, Museum of Anthropology, University of Michigan, Ann Arbor.

Fowells, H. A., and J. E. Means. 1990. The tree and its environment, pp. 1–16 in R. M. Burns and B. H. Honkala (eds.). Silvics of North America, Vol. 1,

Conifers. U.S. Dept. Agric. For. Serv. Handbook 654. Government Printing Office, Washington, D.C.

Fralish, J. S., and F. B. Crooks. 1989. Forest composition, environment and dynamics at Land Between the Lakes in northwest Middle Tennessee. J. Tenn. Acad. Sci. 64:107–111.

Fralish, J. S., F. B. Crooks, J. L. Chambers, and F. M. Harty. 1991. Comparison of presettlement, second-growth and old-growth forest on six site types in the Illinois Shawnee Hills. Amer. Midl. Nat. 125:294–309.

Frelich, L. E., and C. G. Lorimer. 1991. Natural disturbance regimes in hemlock-hardwood forests of the Upper Great Lakes region. Ecol. Monogr. 61:145–164.

Gauch, H. 1982. Multivariate analysis in community ecology. Cambridge University Press, Cambridge.

George, M. F., M. J. Burke, H. M. Pellet, and A. G. Johnson. 1974. Low temperature exotherms and woody plant distribution. Hort. Science 9:519–522.

Graham, A. (ed.). 1972. Floristics and paleofloristics of Asia and eastern North America. Elsevier, Amsterdam.

Graham, A. 1993. History of the vegetation: Cretaceous (Maastrichtian) – Tertiary, pp. 57–70 in N. R. Morin (conv. ed.). Flora of North America north of Mexico, Vol. 1. Oxford University Press, New York.

Graham, R. W., and E. L. Lundelius, Jr. 1984. Coevolutionary disequilibrium and Pleistocene extinctions, pp. 223–249 in P. S. Martin and R. G. Klein (ed.), Quaternary extinctions – a prehistoric revolution. University of Arizona Press, Tucson.

Greller, A. M. 1988. Deciduous forest, pp. 287–316 in M. G. Barbour and W. D. Billings (eds.), North American terrestrial vegetation, 1st ed. Cambridge University Press, Cambridge.

Greller, A. M. 1989. Correlation of warmth and temperateness with the distributional limits of zonal forests in eastern North America. Bull. Torrey Bot. Club 116:145–163.

Grimm, E. C. 1983. Chronology and dynamics of vegetation change in the prairie-woodland region of southern Minnesota, U.S.A. New Phytol. 93:311–350.

Grimm, E. C. 1984. Fire and other factors controlling the Big Woods vegetation of Minnesota in the mid-nineteenth century. Ecol. Monogr. 54: 291–311.

Hansen, A. J., T. A. Spies, F. J. Swanson, and J. L. Ohmann. 1991. Conserving biodiversity in managed forests. BioScience 41:382–392.

Harshberger, J. W. 1911. Phytogeographic survey of North America, pp. 1–790 in A. Engler and O. Drude (comp.), Die vegetation der erde. Wilhelm Engelmann, Leipzig, and G. E. Strickland, New York.

Heinselman, M. L. 1973. Fire in the virgin forests of the Boundary Waters Canoe Area, Minnesota. Quat. Res. 3:329–382.

Henry, J. 1858. Meteorology and its connection with agriculture. U.S. Congress, House, Report of the Commissioner of Patents for 1858, 35th Cong., 1st sess., 1858, H. Ex. Doc. 32 (serial no. 954), 429–93 and frontispiece. Government Printing Office, Washington, D.C.

Hicks, D. J., and B. F. Chabot. 1985. Deciduous forest, pp. 257–277 in B. F. Chabot and H. A. Mooney (eds.), Physiological ecology of North American plant communities. Chapman and Hall, New York.

Hilgard, E. W. 1860. Report on the geology and agriculture of the state of Mississippi. E. Barksdale, State Printer, Jackson, Miss.

Hinkle, C. R. 1989. Forest communities of the Cumberland Plateau of Tennessee. J. Tenn. Acad. Sci. 64: 123–129.

Hinkle, C. R., W. C. McComb, J. M. Safley, Jr., and P. A. Schmalzer. 1993. Mixed mesophtyic forests, pp. 203–254 in W. H. Martin, S. G. Boyce, and A. C. Echternacht (eds.), Biodiversity of the southeastern United States: upland terrestrial communities. Wiley, New York.

Holdridge, L. R., and J. A. Tosi, Jr. 1964. Life zone ecology. Tropical Science Center, San José, Costa Rica.

Huntley, B., and T. Webb III. 1989. Migration: species' response to climatic variations caused by changes in the earth's orbit. J. Biogeogr. 16:5–19.

Iverson, L. R., D. P. Tucker, R. G. Risser, C. D. Burnett, and R. G. Rayburn. 1989. The forest resources of Illinois: an atlas and analysis of spatial and temporal trends. Ill. Nat. Hist. Surv. Special Publ. 11:1–181.

Jacobson, G. L., T. Webb III, and E. C. Grimm. 1987. Patterns and rates of vegetation change during the deglaciation of eastern North America, pp. 277–288 in W. F. Ruddiman and H. E. Wright, Jr. (eds.), North America and adjacent oceans during the last deglaciation, Vol. K–3, The geology of North America. Geological Society of America, Boulder, Colo.

Janzen, D. H., and P. S. Martin. 1982. Neotropical anachronisms: the fruits the Gomphotheres ate. Science 215:19–27.

Kartesz, J. T. 1994. A synonymized checklist of the vascular flora of the United States, Canada, and Greenland, 2nd ed., Vol. 1 – Checklist. Timber Press, Portland, Ore.

King, F. B. 1985. Early cultivated cucurbits in eastern North America, pp. 73–97 in R. I. Ford (ed.), Prehistoric food production in North America. Anthropological Papers No. 75. Museum of Anthropology, University of Michigan, Ann Arbor.

Kramer, P. J. 1936. Effect of variation in length of day on growth and dormancy of trees. Plant Physiol. 11: 127–137.

Kutzbach, J. E., and P. J. Guetter. 1986. The influence of changing orbital parameters and surface boundary conditions on climate simulations for the past 18,000 years. J. Atmosph. Sci. 43:1726–1759.

Kutzbach, J. E., and T. Webb III. 1991. Late Quaternary climatic and vegetational change in eastern North America: concepts, models, and data, pp. 175–217 in L. C. K. Shane and E. J. Cushing (eds.), Quaternary landscapes. University of Minnesota Press, Minneapolis.

Larsen, J. A. 1980. The boreal ecosystem. Academic Press, New York.

Lemon, P. C. 1961. Forest ecology of ice storms. Bull. Torrey Bot. Club 88:21–29.

Lindsey, A. A., and L. K. Escobar. 1976. Eastern deciduous forest, Vol. 2, Beech-maple region. Inventory of natural areas and sites recommended as potential natural landmarks, Natural History Theme Studies 3. U.S. Dept. Interior, National Park Service. Government Printing Office, Washington, D.C.

Lindsey, A. A., and J. O. Sawyer, Jr. 1971. Vegetation-climate relationships in the eastern United States. Proc. Indiana Acad. Sci. 80:210–214.

Lindsey, A. A., and D. V. Schmelz. 1970. The forest types of Indiana and a new method of classifying midwestern hardwood forests. Proc. Indiana Acad. Sci. 79:198–204.

Little. E. L., Jr. 1971. Atlas of United States trees, Vol. 1, Conifers and important hardwoods. U.S. Dept. Agric. For. Serv. Misc. Publ. 1146. Government Printing Office, Washington, D.C.

Livingston, B. E., and F. Shreve. 1921. The distribution of vegetation in the United States, as related to climatic conditions. Carnegie Institution of Washington Publ. No. 284. Gibson Brothers, Washington, D.C.

Lorimer, C. G. 1977. The presettlement forest and natural disturbance cycle of northeastern Maine. Ecology 58:141–148.

Lorimer, C. G. 1980. Age Structure and disturbance history of a southern Appalachian virgin forest. Ecology 61:1169–1184.

Lubchenco, J., A. M. Olson, L. B. Brubaker, S. R. Carpenter, M. M. Holland, S. P. Hubbell, S. A. Levin, J. A. MacMahon, P. A. Matson, J. R. Melillo, H. A, Mooney, C. H. Peterson, H. R. Pulliam, L. A. Real, P. J. Regal, and P. G. Risser. 1991. The sustainable biosphere initiative: an ecological research agenda. Ecology 72:371–412.

McAndrews, J. H. 1988. Human disturbance of North American forests and grasslands, the fossil pollen record, pp. 673–697 in B. Huntley and T. Webb III (eds.), Vegetation history. Kluwer Academic, Dordrecht.

McCormick, J. F., and R. B. Platt. 1980. Recovery of an Appalachian forest following the chestnut blight, or Catherine Keever – you were right! Amer. Midl. Nat. 104:264–273.

McIntosh, R. P. 1962. The forest cover of the Catskill Mountain region, New York, as indicated by land survey records. Amer. Midl. Nat. 68:409–423.

Marschner, F. J. 1959. Land use and its patterns in the United States. U.S. Department of Agriculture Handbook No. 153. Government Printing Office, Washington, D.C.

Marschner, F. J. 1974. Map of the original vegetation of Minnesota, Scale 1:500,000 (redrafted by P. J. Burwell and S. J. Haas), with accompanying text by M. L. Heinselman. North Central Forest Expt. Sta. St. Paul, Minn.

Martin, P. S., and R. G. Klein (eds.). 1984. Quaternary extinctions – a prehistoric revolution. University of Arizona Press, Tucson.

Martin, P. S., and B. E. Harrell. 1957. The Pleistocene history of temperate biotas in Mexico and eastern United States. Ecology 38:468–480.

Martin, W. H., and S. G. Boyce. 1993. Introduction: the southeastern setting, pp. 1–46 in W. H. Martin, S. G. Boyce, and A. C. Echternacht (eds.), Biodiversity of the southeastern United States: lowland terrestrial communities. Wiley, New York.

Martin, W. H., S. G. Boyce, and A. C. Echternacht (eds.). 1993a. Biodiversity of the southeastern United States: lowland terrestrial communities. Wiley, New York.

Martin, W. H., S. G. Boyce, and A. C. Echternacht (eds.), 1993b. Biodiversity of the southeastern United States: upland terrestrial communities. Wiley, New York.

Merriam, C. H. 1898. Life zones and crop zones of the United States. United States Department of Agriculture Div. Biol. Surv. Bull. 10. Government Printing Office, Washington, D.C.

Michaux, A. 1803. Flora Boreali-Americana, 2 vols. Frates Levrault, Paris, France.

Miranda, F., and A. J. Sharp. 1950. Characteristics of the vegetation in certain temperate regions of eastern Mexico. Ecology 31:313–333.

Miyawaki, A., K. Iwatsuki, and M. M. Grandtner. 1994. Vegetation in eastern North America. University of Tokyo Press, Tokyo.

Mladenoff, D. J., and F. Stearns. 1993. Eastern hemlock regeneration and deer browsing in the northern Great Lakes region: a re-examination and model simulation. Cons. Biol. 7:889–900.

Morse, D. F., and P. A. Morse. 1983. Archaeology of the Central Mississippi Valley. Academic Press, New York.

Oliver, C. D. 1981. Forest development in North America following major disturbances. For. Ecol. Mgt. 3: 153–168.

Oosting, H. J. 1942. An ecological analysis of the plant communities of Piedmont, North Carolina. Amer. Midl. Nat. 28:1–126.

Oosting, H. J., and W. D. Billings. 1951. A comparison of virgin spruce-fir forest in the northern and southern Appalachian system. Ecology 32:84–103.

Peet, R. K. 1992. Community structure and ecosystem function, pp. 103–151 in D. C. Glenn-Lewin, R. K. Peet, and T. T. Veblen (eds.), Plant succession, theory and prediction. Population and community biology series 11. Chapman and Hall, New York.

Peet, R. K., and N. L. Christensen. 1980. Hardwood forest vegetation of the North Carolina Piedmont. Veroff. Geobot. Inst. ETH, Stift. Rubel (Zurich) 69:14–39.

Pell, W. F. 1984. Plant communities, in B. Shepherd (ed.), Arkansas's natural heritage. August House, Little Rock, Ark.

Perry, D. A., J. G. Borchers, S. L. Borchers, and M. P. Amaranthus. 1990. Species migrations and ecosystem stability during climate change: the belowground connection. Cons. Biol. 4:266–274.

Petty, R. O., and M. T. Jackson. 1966. Plant communities, pp. 264–296 in A. A. Lindsey (ed.). Indiana Sesquicentennial Volume. Natural features of Indiana, Indiana Academy of Sciences, Indianapolis, Ind.

Pickett, S. T. A., and P. S. White (eds.). 1985. The ecology of natural disturbance and patch dynamics. Academic Press, Orlando, Fla.

Prentice, I. C. 1992. Climate change and long-term vegetation dynamics, pp. 293–339 in D. C. Glenn-Lewin, R. K. Peet, and T. T. Veblen (eds). Plant succession, theory and prediction. Population and community biology series 11. Chapman and Hall, New York.

Quarterman, E., and C. Keever. 1962. Southern mixed hardwood forest: climax in the southeastern coastal plain: U.S.A. Ecol. Monogr. 32:167–185.

Quarterman, E., M. P. Burbanck, and D. J. Shure. 1993. Rock outcrop communities: limestone, sandstone, and granite, pp. 35–86 in W. H. Martin, S. G. Boyce, and A. C. Echternacht (eds.), Biodiversity of the southeastern United States: upland terrestrial communities. Wiley, New York.

Quick, B. E. 1923. A comparative study of the distribution of the climax association in southern Michigan. Pap. Mich. Acad. Sci. Arts Lett. 3:211–243.

Rafferty, M. D. 1980. The Ozarks, land and life. University of Oklahoma Press, Norman.

Reveal, J. L., and J. S. Pringle. 1993. Taxonomic botany

and floristics, pp. 157–198 in N. R. Morin (con. ed.), Flora of North America north of Mexico, Vol. 1. Oxford University Press, New York.

Risser, P. G. 1985. Grasslands, pp. 232–256 in B. F. Chabot and H. A. Mooney (eds), Physiological ecology of North American plant communities. Chapman and Hall, New York.

Rogers, R. S. 1981. Mature mesophytic hardwood forest: community transitions, by layer, from east-central Minnesota to southeastern Michigan. Ecology 62: 1634–1647.

Root, T. L., and S. H. Schneider. 1993. Can large-scale climatic models be linked with multiscale ecological studies? Cons. Biol. 7:256–270.

Rowe, J. S. 1972. Forest regions of Canada. Canadian Forest Service, Department of Environment Publ. 1300. Ottawa.

Runkle, J. R. 1982. Patterns of disturbance in some old-growth mesic forests of eastern North America. Ecology 63:1533–1546.

Runkle, J. R. 1985. Disturbance regimes in temperate forests, pp. 17–33 in S. T. A. Pickett and P. S. White (eds.), The ecology of natural disturbance and patch dynamics. Academic Press, Orlando, Fla.

Russell, E. W. B. 1983. Indian-set fires in the forests of the northeastern United States. Ecology 64:78–88.

Sakai, A., and W. Larcher. 1987. Frost survival of plants. Springer-Verlag, New York.

Sakai, A., and C. J. Weiser. 1973. Freezing resistance of trees in North America with reference to tree regions. Ecology 54:118–126.

Sargent, C. S. 1884. Report on the forests of North America (exclusive of Mexico). United States Census Office, 10th Census, 1880, vol. 9. Government Printing Office, Washington, D.C.

Sawyer, J. O., Jr., and A. A. Lindsey. 1964. The Holdridge bioclimatic formations of the eastern and central United States. Proc. Ind. Acad. Sci. 72:105–112.

Schenck, C. A. 1974. The birth of forestry in America, Biltmore Forest School 1898–1913. Forest History Society and the Appalachian Consortium, Santa Cruz, Calif.

Shantz, H. L., and R. Zon. 1924. The natural vegetation of the United States, pp. 1–29 in O. E. Baker (comp.), Atlas of American Agriculture, 1936. United States Department of Agriculture. Government Printing Office, Washington, D.C.

Siccama, T. G. 1971. Presettlement and present forest vegetation in northern Vermont with special reference to Chittenden County. Amer. Midl. Nat. 85: 153–172.

Skeen, J. N., P. D. Doerr, and D. H. Van Lear. 1993. Oak-hickory-pine forests, pp. 1–33 in W. H. Martin, S. G. Boyce, and A. C. Echternacht (eds.), Biodiversity of the southeastern United States: upland terrestrial communities. Wiley, New York.

Smith, B. D. 1978. Mississippian settlement patterns. Academic Press, New York.

Smith, D. M. 1980. The forests of the United States, pp. 1–23 in J. W. Barrett (ed.), Regional silviculture of the United States. Wiley, New York.

Spear, R. W., M. B. Davis, and L. C. K. Shane. 1994. Late Quaternary history of low- and mid-elevation vegetation in the White Mountains of New Hampshire. Ecol. Monogr. 64:85–109.

Stearns, F. 1987. The changing forests of the Lake States, pp. 24–35 in Conversation Foundation, Lake States Governor's Conference on Forestry (including Lake States maps for presettlement forests and for modern major forest types, prepared by F. Stearns and G. Guntenspergen, UWM Cartographic Laboratory). Milwaukee, Wisc.

Stephenson, S. L., A. N. Ash, and D. F. Stauffer. 1993. Appalachian oak forests, pp. 255–303 in W. H. Martin, S. G. Boyce, and A. C. Echternacht (eds.), Biodiversity of the southeastern United States: upland terrestrial communities. Wiley, New York.

Stuckey, R. L. 1981. Origin and development of the concept of the Prairie Peninsula, pp. 4–23 in R. L. Stuckey and K. J. Reese (eds.), The Prairie Peninsula – in the "shadow" of Transeau. Ohio Biol. Surv. Biol. Notes 15. Ohio State University, Columbus.

Stuckey, R. L. 1994. E. Lucy Braun: Ohio's foremost woman botanist. Compiled for Symposium: the eastern deciduous forest since E. Lucy Braun 1950, Botanical Society of America, AIBS Meeting, Knoxville, Tenn.

Swain, A. M. 1973. A history of fire and vegetation in northeast Minnesota as recorded in lake sediments. Quat. Res. 3:383–396.

Tiffney, B. H. 1985. Perspectives on the origin of the floristic similarity between eastern Asia and eastern North America. J. Arnold Arbor. 66:73–94.

Transeau, E. N. 1903. On the geographic distribution and ecological relations of the bog plant societies of northern North America. Bot. Gaz. 36:401–420.

Transeau, E. N. 1905. Forest centers of eastern America. Amer. Nat. 39:875–889.

Turner, M. G. 1990. Landscape changes in nine rural counties in Georgia. Photogramm. Eng. Remote Sens. 56:379–386.

Turner, M. G., W. H. Romme, R. H. Gardner, R. V. O'Neill, and T. K. Kratz. 1993. A revised concept of landscape equilibrium: disturbance and stability on scaled landscapes. Landscape Ecol. 8:213–227.

Urban, D. L., O'Neill, R. V., and H. H. Shugart. 1987. Landscape ecology: a hierarchical perspective can help scientists understand spatial patterns. BioScience 37:119–127.

USDA 1969. A forest atlas of the south. Southern and Southeastern Forest Experiment Stations, U.S. Dept. Agriculture Forest Service. Government Printing Office, Washington, D.C.

USDA 1990. USDA Plant hardiness zone map. U.S. Dept. Agriculture, Agricultural Res. Serv. Misc. Publ. 1475. Government Printing Office, Washington, D.C.

Vankat, J. L. 1979. The natural vegetation of North America. Wiley, New York.

Ware, S. 1982. Polar ordination of Braun's mixed mesophytic forest. Castanea 47:403–407.

Watts, W. A. 1970. The full-glacial vegetation of northern Georgia. Ecology 51:17–33.

Watts, W. A. 1980. The late Quaternary vegetation history of southeastern United States. Ann. Rev. Ecol. Syst. 11:387–409.

Watts, W. A. 1988. Late-Tertiary and Pleistocene vegetation history – 20 My to 20 ky: Europe, pp. 155–192 in B. Huntley and T. Webb III (eds.), Vegetation history. Kluwer Academic Publishers, Dordrecht, The Netherlands.

Watts, W. A., B. C. S. Hansen, and E. C. Grimm. 1992. Camel Lake: a 40 000-yr record of vegetational and forest history from northwest Florida. Ecology 73: 1056–1066.

White, C. A. 1984. A history of the rectangular survey system. Bureau of Land Management, U.S. Dept. Interior. Government Printing Office, Washington, D.C.

White, P. S. 1983. Eastern Asian – eastern North American floristic relations: the plant community level. Ann. Missouri Bot. Gard. 70:734–747.

White, P. S., and C. V. Cogbill. 1992. Spruce-fir forests of eastern North America, pp. 3–39 in C. Eagar and M. B. Adams (eds.), Ecology and decline of red spruce in the eastern United States. Ecological Studies 96. Springer-Verlag, New York.

White, P. S., E. Buckner, J. D. Pittillo, and C. V. Cogbill. 1993. High-elevation forests: spruce-fir forests, northern hardwoods, and associated communities, pp. 339–366 in W. H. Martin, S. G. Boyce, and A. C. Echternacht (eds.), Biodiversity of the southeastern United States: upland terrestrial communities. Wiley, New York.

Whitehead, D. R. 1969. Wind pollination in the angiosperms: evolutionary and environmental considerations. Evolution 23:28–35.

Whitney, G. G. 1986. Relation of Michigan's presettlement pine forests to substrate and disturbance history. Ecology 67:1548–1559.

Whitney, G. G. 1987. An ecological history of the Great Lakes Forest of Michigan. J. Ecol. 75:667–684.

Whitney, G. G. 1990. The history and status of the hemlock-hardwood forests of the Allegheny Plateau. J. Ecol. 78:443–458.

Whitney, G. G. 1995. From coastal wilderness to fruited plain: a history of environmental change in temperate North America from 1500 to the present. Cambridge University Press, Cambridge.

Whittaker, R. H. 1956. Vegetation of the Great Smoky Mountains. Ecol. Monogr. 26:1–80.

Whittaker, R. H. 1975. Communities and ecosystems. Macmillan, New York.

Wilkins, G. R., P. A. Delcourt, H. R. Delcourt, F. W. Harrison, and M. R. Turner. 1991. Paleoecology of central Kentucky since the last glacial maximum. Quat. Res. 36:224–239.

Williams, M. 1989. Americans and their forests, a historical geography. Cambridge University Press, Cambridge.

Wolfe, J. A. 1979. Temperature parameters of humid to mesic forests of eastern Asia and relation to forests of other regions of the northern hemisphere and Australasia. U.S. Geol. Surv. Prof. Paper 1106:1–37.

Woodward, F. I. 1987. Climate and plant distribution. Cambridge studies in ecology, Cambridge University Press, Cambridge.

Wright, H. E., Jr. 1974. Landscape development, forest fires, and wilderness management. Science 186:487–495.

Wright, H. E., Jr. 1992. Patterns of Holocene climatic change in the midwestern United States. Quat. Res. 38:129–134.

Wright, H. E., Jr., J. E. Kutzbach, T. Webb III, W. F. Ruddiman, F. A. Street-Perrott, and P. J. Bartlein (eds.). 1993. Global climates since the last glacial maximum. University of Minnesota Press, Minneapolis.

Zobel, D. B. 1969. Factors affecting the distribution of *Pinus pungens*, an Appalachian endemic. Ecol. Monogr. 39:271–301.

Chapter
11

Vegetation of the Southeastern Coastal Plain

NORMAN L. CHRISTENSEN

INTRODUCTION

The climax vegetation over most of the eastern United States is generally considered to be broad-leaved deciduous forest (Kuchler 1964). The Coastal Plain of the Atlantic and Gulf states is the most striking and extensive exception to this statement. Community physiognomy varies across this landscape from grasslands and savannas to shrublands, to needle- and broadleaf sclerophyllous woodlands and to rich mesophytic forest. These differences can be observed over a distance of only a few hundred meters and an elevational gradient of only 10 m. In addition, the southeastern Coastal Plain has the most diverse assemblage of freshwater wetland communities in North America, and its lengthy and complex shoreline is vegetated by a rich array of maritime ecosystems. Much of this variation is a consequence of dramatic gradients in physical and chemical characteristics of soils and hydrology. However, as Wells (1946) suggested, "succession would simplify the mosaic just outlined considerably were it not for fire," and where fire is not important, as in maritime and alluvial communities, other forms of chronic disturbance complicate community structure and function. Thus, succession and patch dynamics are an integral part of any discussion of the vegetation of this province.

The southeastern Coastal Plain was an important laboratory for many prominent ecologists of the first half of this century, including J. W. Harshberger, R. M. Harper, B. W. Wells, and H. J. Oosting. In the intervening years, North American community ecologists have come to accept the gradient nature of community variation, to deal with the complexity of successional pathways and mechanisms, and to recognize the role of natural disturbance within communities and across landscapes. All these lessons were obvious to those early Coastal Plain ecologists.

My interest in Coastal Plain ecology was first kindled and has subsequently been encouraged by Lewis E. Anderson. Although he may disagree with some of the notions presented here, he has shaped my understanding of this region more than any other person. However, I suspect that many of the ideas that I thought were mine were stimulated by discussions with Dwight Billings, Paul Godfrey, Robert Peet, Bill Ralston, and Curtis Richardson. Warren Abrahamson, Ralph Good, Catherine Keever, Suzanne McAlister, Pat Peroni, Elsie Quarterman, Richard Schneider, Joan Walker, Stewart Ware, and Rebecca Wilbur provided original data and suggestions for which I am grateful. Various iterations of this manuscript benefitted from reviews and criticism by Michael Barbour, Dwight Billings, Ralph Good, and Carl Monk. I especially thank Joan Walker for her comments on the structure and content of this paper.

The past decade has witnessed increased interest in Coastal Plain ecosystems. As the "piney woods" of the Southeast have become North America's number one source of wood fiber, concerns about the status and sustainability of these ecosystems have also increased. The unique flora and high degree of endemism displayed in some communities have led to laws protecting rare and endangered species. Agricultural and urban development now dissect this complex landscape, altering movement of animals, flows of water, and patterns of natural disturbance. Much recent research has been directed toward conservation of remaining habitat and restoration of altered ecosystems.

VEGETATION HISTORY

The Coastal Plain of the southeastern United States had its beginnings during the Triassic period (~200 M yr BP) as North America separated from North Africa, creating the Atlantic Ocean. During the previous 400 M yr, the southern Appalachians had grown into a mountain range rivaling the present-day western Cordillera in majesty. Mountain building has virtually ceased in eastern North America since the Triassic, and the Coastal Plain Geological Province has been built from the sediments eroded from that great massif. It is the distribution of these sediments that defines the extent of this province.

The oldest fossils of terrestrial vegetation on the Coastal Plain are from the Paleocene and Eocene. Macrofossils from the Mississippi embayment reveal a distinctly tropical flora in this region during that time (Dilcher 1973a, 1973b), and they suggest a mean annual temperature of about 27° C (Wolfe 1978, 1985). The general cooling trend that culminated in the Pleistocene Ice Age appears to have begun during the Oligocene. Little is known of middle Tertiary vegetation of the southeast; however, shallow marine deposition of phosphatic carbonates along the southeastern coast during the Miocene suggests considerable upwelling of cold, nutrient-rich water. Data from Miocene floras elsewhere in North America support the hypothesis that southeastern climate was tropical to subtropical during this period (Axelrod and Bailey 1969).

Palynological analysis of Coastal Plain lake sediments has given us a detailed picture of the late Pleistocene history of this region (see Chapter 10 for a discussion of the relevance of these data to the entire Eastern Deciduous Forest Province). During the Altonian Subage of the Wisconsinan (40,000 BP), most of the Coastal Plain north of central South Carolina was dominated by a mixture of jack pine (*Pinus banksiana*) and spruce (*Picea* spp.)

with, perhaps, a zone of mixed conifer-northern hardwoods extending across southern South Carolina (Watts 1980a, 1980b; Delcourt and Delcourt 1981). Communities similar to the Coastal Plain communities of today occupied a broad band extending from the coast of Georgia across the Mississippi embayment and into east Texas (Delcourt, Delcourt, Brister, Lackey 1980; Delcourt and Delcourt 1981). Peninsular Florida, greatly enlarged by the lowering of sea level, was dominated by xeric sand dune communities that appear to have no modern analogue (Watts 1975). Through the remainder of the Wisconsinan Ice Age, these vegetation zones waxed and waned in relation to shifts in the ice mass to the north. Although there was clearly no single refuge for the mixed mesophytic forest during full glacial periods (cf. Braun 1955; Delcourt and Delcourt 1979, Davis 1983; Delcourt, Delcourt, Morse, and Morse 1993), there is little doubt that many of the vascular plant constituents of this forest association could be found in river bluff habitats throughout the Southeast (Delcourt and Delcourt 1977a, 1980).

The northern extent of mixed mesophytic forest species along the Atlantic limb of the Coastal Plain is difficult to determine. During full glacial periods, sea level was >150 m lower, and the coastline in the middle Atlantic region was 50–150 km east of its present location. Perhaps the southern forest elements extended farther north near the coast, much as they do today, but evidence to test this hypothesis is now submerged and perhaps obliterated by coastal processes.

During retreat of the continental glaciers 12,000–14,000 yr BP, the Southeast saw a marked warming trend (the Hypsithermal or Xerothermic Period; Wright 1976) and regional extinction of boreal elements. Mixed deciduous forest dominated most of the northern Coastal Plain (Watts 1980a, 1980b; Delcourt and Delcourt 1981; Delcourt et al. 1993). Increased rainfall over the Florida peninsula favored widespread oak savanna in this area.

Sea level returned to near its modern position between 5000 and 3500 yr BP (Field, Meisburger, Stanley, and Williams 1979), resulting in decreased drainage over many areas of the lower Coastal Plain and coincident with the initiation of a general cooling trend (Watts 1971, 1980b), which probably lowered regional evapotranspiration rates. These two factors are likely responsible for the initiation of a period of paludification and bog formation in several localities along the Atlantic Coastal Plain (Daniels, Gamble, Wheeler, and Holzher 1977; Whitehead 1972, 1981; Cohen 1973, 1974; Cohen, Andrejko, Spackman, and Corrinus 1984). Abundant charcoal in sediments indicates that fires were common across the Coastal Plain (Buell 1939, 1946; Delcourt 1980; Cohen et al. 1984).

Human habitation of the Coastal Plain began no less than 12,000 BP. At this time, hunters of the PaleoIndian culture moved into the Southeast in pursuit of large mammals driven eastward by the Hypsithermal drying of the Great Plains (Hudson 1976). The Archaic Indian culture was well established by 8000–10,000 BP coincident with the extinction of the megafauna. These people were hunter-gatherers whose food sources included deer, small mammals, fish, shellfish, and wild vegetables (Hudson 1976; Cowdrey 1983). Archaic Indian populations were largest near the coast and, as paludification proceeded, near swamp complexes such as the Great Dismal Swamp, the Green Swamp, and the Okefenokee Swamp (Hudson 1976; Wright 1984). These swamp complexes offered a wide variety of vegetable and wildlife resources (Wright 1984). Although they were seminomadic, Archaic Indians intensively managed the Coastal Plain landscape, particularly with the use of fire (Pyne 1982). Buell (1946) suggested that an abrupt increase in the charcoal content of strata in Jerome Bog, a peat-filled bay lake in North Carolina, might coincide with either the advent of the Indian or climatic change. Undoubtedly fire frequency increased as a consequence of both anthropogenic and climatic influences.

The Woodland culture, an Indian tradition incorporating a mixture of hunting, gathering, and primitive agriculture (including corn and squash), took shape along the Mississippi river about 3000 BP. Woodland Indians apparently relied most heavily on natural resources, using agriculture to subsidize their needs and perpetuating the land-use practices of their Archaic predecessors (Hudson 1976; Cowdrey 1983).

The Mississippian tradition, a culture that was primarily sedentary and agricultural, arose in the floodplains of the Mississippi embayment approximately 1300 BP. This was the most highly developed culture on the Coastal Plain at the time of EuroAmerican contact. It was noted for its large walled towns and elegant tools. These people apparently never discovered the use of fertilizer, and the sterile Coastal Plain soils confined their agricultural activities to nutrient-subsidized floodplains of large rivers (Cowdrey 1983). Up to European colonization, the Archaic, Woodland, and Mississippian cultures coexisted in different habitats and regions of the southeastern Coastal Plain.

It was the Coastal Plain where the first European colonists to temperate North America settled. This landscape must have been a magnificent sight to a people coming from a land where two millenia of intensive land use had removed every vestige of pristine forest and where most major industries were limited by the availability of wood, especially tall spars for shipbuilding (Williams). Virtually every account by early explorers of the province emphasizes the extensive and majestic forests and abundant natural resources. Just as frequently, early travelers in the region commented on the frequency of natural and man-caused fire (Catesby 1654; Lawson 1714; Byrd 1728; Bartram 1791). The impact of the subsequent exploitation of Coastal Plain natural resources on specific plant communities will be discussed later and are outlined in detail by Cowdrey (1983).

The effects of colonization on the various fire regimes of the Coastal Plain varied with location and the cultural tradition of the colonists. For example, settlers of Scotch and Irish descent were accustomed to using fire in the management of heaths and agricultural fields, so in many locations they perpetuated Indian burning practices. However, immigrants from central Europe were more apt to exclude fire from lands they managed (Pyne 1982). In general, however, most Coastal Plain colonists quickly learned the value of fire as a tool to manipulate the landscape and from this region the accepted twentieth-century fire management practices later developed (Pyne 1982).

THE COASTAL PLAIN ENVIRONMENT

Physiography and Geology

The boundaries of the Coastal Plain are shown in Fig. 11.1. On the Atlantic seaboard the Coastal Plain merges with the Piedmont plateau along the fall line, or zone, a ragged line running roughly parallel to the trends of the mobile belts comprising the Piedmont and Appalachian Highlands. In addition to rather abrupt changes in soils and their parent rocks, this zone is often marked by a marked change in topography from rolling hills to a rather flat plain. The Coastal Plain border turns westward across Georgia and Alabama, then northward to the head of the Mississippi embayment in southern Illinois. The western margin of the Coastal Plain is not as clearly defined, either geologically or botanically, and the border indicated here is that suggested by Murray (1961).

The Atlantic Coastal Geologic Province extends from the fall line to the edge of the continental shelf as a geosynclinal wedge of alluvial and marine sediments. These strata rest on a basement of Paleozoic and Precambrian rocks that is actively subsiding in many places under the weight of this growing wedge. Much of the sediment, particu-larly at the surface, is siliceous alluvium. In addition, considerable carbonaceous sediment, often rich in phosphates, were deposited during a Miocene episode of coastal upwelling. These carbonates outcrop in many localities but are most prominent throughout peninsular Florida and across the Gulf states, where they give rise to karst topography and produce comparatively fertile soils.

Several physiographic sections are recognized (Fig. 11.1), although they are not necessarily separated by distinct boundaries. The Chesapeake–Delaware downwarp lies north of central North Carolina. In this region, basement crystalline rocks are subsiding and major river valleys have been "drowned" or embayed, forming extensive bays and sounds (e.g., the Chesapeake Bay and Albermarle and Pamilico sounds). Toward the south, these basement rocks warp upward to form the Cape Fear arch, which reaches its peak along the border of North and South Carolina. Uplift in this region has resulted in erosion of Cenozoic sediments, exposing sandy Cretaceous sediments beneath. Crystalline rocks appear to be subsiding beneath the Sea Island's downwarp section, and rising sea level in this region has created extensive tidal marshes and swamps. The peninsular Florida section is a broad sandy plain resting on the Ocala arch. The Mississippi and East Texas embayments are characterized by broad deltaic plains along the coast. Inland is a "staircase" of Pleistocene depositional terrace surfaces and an inner region of belted topography developed on differentially eroded sediments (Murray 1961).

The transition from terrestrial to marine ecosystems along the thousands of kilometers of southeastern coastline is gradual and complex. In addition to abundant river deltas, the coast is often bordered by long ribbons of barrier or sea islands, which protect extensive estuaries, lagoons, and sounds. These coastal features are in a constant state of change owing to erosion by wind and water and changes in the influx of sediment (Pilkey et al. 1980).

The surface of the Coastal Plain has been reworked considerably by coastal and fluvial processes during the last 2–3 M yr. The areal extent of the Coastal Plain has varied with changes in global climate and sea level. When sea level was highest, Florida was nearly submerged, and when sea level was lowest, the peninsula was nearly 600 km across. Seven terraces and scarps have been recognized on the Atlantic segments of the Coastal Plain (Cooke 1931). These were considered to be high-water marks formed by shifts in sea level during the Pleistocene. The elevation of the highest of these terraces is approximately 85 m above sea

Figure 11.1. Boundaries and major physiographic features of the southeastern Coastal Plain.

level. The origin and location of these terraces, particularly those above 25 m, are still debated (Doering 1958; Murray 1961; Cronin et al. 1981). The fluvatile terraces along major river systems (especially the Mississippi) have been correlated with the coastal terraces and with alleged interglacial periods (Murray 1961). Coastal features such as dune fields, swales, and remains of mud flats are obvious on lower terraces and greatly influence vegetation patterns (Colquhoun 1969).

Hundreds of thousands of elliptical depressions, the so-called Carolina Bays, occur on the Coastal Plain from southern Virginia to northern Florida. The axes of the ellipses are all oriented

within a few degrees of a line trending northwest to southeast, and rims of coarse sterile sand often border them. The origin of these bays has been a matter of considerable debate (Cooke 1933; Johnson 1942; Prouty 1952; Wells and Boyce 1953; Sharitz and Gibbons 1982; Savage 1982) and the most widely accepted theories involve combinations of wind orientation and deflation, karst subsidence, and steamlining by groundwater, although some have argued they were created by a meteor shower or comet impact (Murray 1961; Savage 1982). These features have a prominent effect on vegetation where they occur: The depressions vary from peat-filled bogs to open lakes or, where sediment is clayey, to savannas and flatwoods, and the wind-worked sands of the bay rims are among the most sterile and dry Coastal Plain habitats.

Soils

The large-scale distribution of major soil orders across the Coastal Plain is described in detail in Buol (1973). Entisols (soils with virtually no profile development) are common on the very well-drained sands throughout the region; they represent the low extreme in water retention and they have very low mineral adsorptive capacity, hence are quite infertile. Inceptisols (soils with weakly developed horizons) are most common on alluvial plains; they are highly variable in texture and drainage and are generally infertile (although nutrient subsidization from river floods may increase fertility). Typical cool-climate Spodosols are found only in the northern Coastal Plain (Tedrow 1979), but Aquods (groundwater podosols), are common throughout the region. Aquod soils form when soluble organic compounds, iron, and aluminum are leached to shallow water tables. Alfisols (soils with light-colored surface layers and a definite clay-enriched B horizon) border the alluvial plain of the Mississippi River and occur at other locations in Alabama, Florida, and South Carolina. Ultisols (highly weathered soils with a B horizons containing appreciable amounts of translocated clay), occur widely over the Coastal Plain. These soils vary in fertility in relation to parent material. Soils from siliceous lagoonal sediments are nutrient poor, whereas soils from carbonaceous sediment are fertile. Because Alfisols are less weathered than Ultisols, they retain considerable quantities of calcium and magnesium and are comparatively fertile. Drainage on Alfisols and Ultisols is highly variable; clay horizons may impede percolation and result in poor soil aeration. Histosols (organic soils) are associated with paludal wetlands throughout the Coastal Plain. Saprist or mucky peats form from herb wetlands such as marshes, whereas hemist and fibrist peats are characteristic of woody vegetation. Many of these soil orders may be found within a distance of a few hundred meters along catenas related to topography or hydrology (Daniels et al. 1984).

Climate

According to Köppen's classification (Trewartha 1968), the climate of the Coastal Plain is humid subtropical. Mean daily temperatures are between 0 and 18° C in the coldest month and >22° C in the warmest month; rainfall is distributed evenly throughout the year. Winter rainfall in this region is primarily a consequence of frontal cyclonic storms, whereas summer rain is usually associated with convectional thunderstorms. Although there is considerable uniformity over this rather extensive area (Fig. 11.2), several trends should be noted. Seasonal variation in temperature increases away from the coast. Length of the frost-free season tends to increase toward the coast and to the south. Rainfall is highest in the southeastern section where the seasonal pattern is decidedly tropical (i.e., winter drought and summer peak). Rainfall tends to decrease away from the coast. Potential evapotranspiration (based on temperature; Thornthwaite, Mather, and Carter 1958) varies from 700–1300 mm along a north-south gradient. Although annual precipitation exceeds potential evapotranspiration by 50–400 mm throughout most of the Coastal Plain, during much of the summer actual evapotranspiration is often less than potential, owing to depleted soil reserves (Thornthwaite et al. 1958). In general, water deficits (i.e., potential minus actual evapotranspiration) increase to the south and west; topographic and soil factors make variability in water deficit within any local area as great as that over the entire region.

Violent weather associated with hurricanes and convectional storms contributes to the disturbance mosaic in many Coastal Plain forests. Much tree windthrow is associated with such storms. Furthermore, this region has the highest frequency of lightning strikes of any region in North America (Komarek 1968), providing an ignition source for frequent fires.

UPLAND PINE FOREST VEGETATION

Northern Pine Barrens

The pine forests of the Atlantic Coastal Plain north of Delaware Bay are quite distinct from their counterparts to the south. The dominant pine species is pitch pine (*Pinus rigida*) although shortleaf pine (*Pinus echinata*) may co-occur in some localities. These forests have strong floristic affinities to the ridge-top pitch pine forests of the southern Appalachians (Whittaker 1979; see also Chapter 10). For example, *Comptonia peregrina*, *Kalmia latifolia*, *Gaylussacia baccata*, *Quercus ilicifolia*, and *Q. prinoides*, species characteristic of the shallow soils in the Appalachians, are common here as well; and so are southern Atlantic Coastal Plain species such as *Ilex glabra*, *Gaylussacia frondosa*, *G. dumosa*, and *Clethra alnifolia* are also abundant. In fact, 109 species with centers of abundance on the southeastern Coastal Plain reach their northernmost limit in this region (Little 1979).

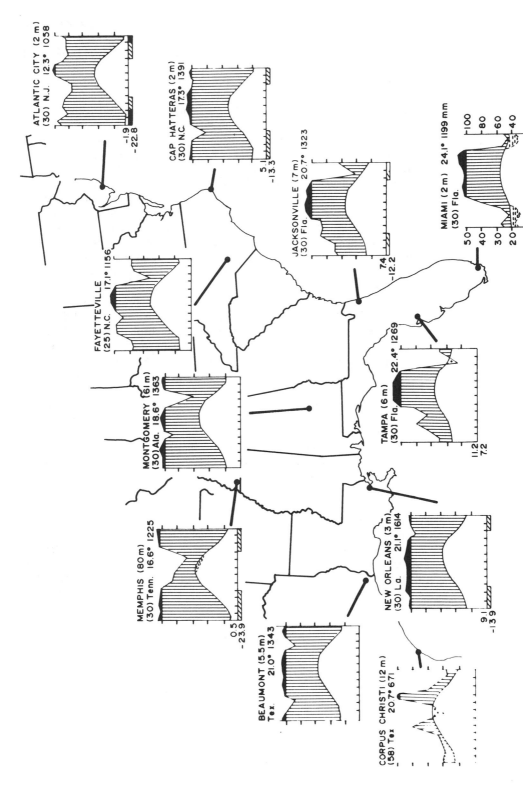

Figure 11.2. Climate diagrams for representative Coastal Plain stations. Graphs were prepared from Walter and Lieth (1967) using conventions outlined by Walter (1979). Data following place name include elevation (m), years of record, mean annual temperature (°C), and annual precipitation (mm). See the graph of Miami (lower right) for axes labels. Left axis, mean monthly temperature; right axis, monthly precipitation. Vertical lines show months with precipitation >evaporation; black area shows months with precipitation >100 mm.

The unique features of the pine barrens flora were recognized by several early authors (Stone 1911; Harshberger 1916). Braun (1950) considered that these features were a consequence of colder climes, shorter growing season, and the podsolic nature of the soils in this region. The vegetation and ecology of pine barren ecosystems have been reviewed in a set of excellent papers edited by Forman (1979). Buchholz and Good (1982a) have compiled a complete annotated bibliography of Pine Barrens literature.

Vegetation composition and structure. Lutz (1934) divided the pine-dominated ecosystems of this region into three types differentiated by stature and composition: the pine-oak or pine barrens forests; the pine–shrub oak forests (transitional forests); and the dwarf pine plains. This classification has been adopted by most later ecologists (Olsvig, Cryan, and Whittaker 1979; Whittaker 1979; Forman and Boerner 1981).

Pine-oak forests, which correspond to the pine barrens community of Lutz (1934), have a well-developed tree stratum with scattered individuals of *Quercus stellata* and *Q. marilandica* (Table 11.1). There is normally an extensive shrub layer dominated by *Quercus ilicifolia*, *Gaylussacia baccata*, and *Vaccinium* spp. (Buell and Cantlon 1950; Lutz 1934; Olsvig et al. 1979). On finer-textured soils (Olsvig et al. 1979) or with decreased fire frequency (Forman and Boerner 1981), these forests grade into oak-pine forests. In more mesic sites *Acer rubrum* and *Nyssa sylvatica* may occur (Whittaker 1979).

Pine–shrub oak forests are transitional between the pine plains and pine-oak forests. Tall-stature pitch pines (<10 m) are dominant, whereas tree oaks (*Quercus stellata* and *Q. marilandica*) are stunted or absent (Olsvig et al. 1979). The domi-nant shrubs are the same as those found in the plains.

Dwarf pine plains are dominated by a scrub ecotype of *Pinus rigida* and a dwarfed form of *Quercus marilandica*, both multistemmed from large irregularly shaped stools that are considerably older than the current stems (McCormick and Buell 1968; Little 1979). *Quercus ilicifolia* and other shrub species are also common. The shrub canopy is usually 0.5–1.5 m high and only rarely exceeds 3 m (Fig. 11.3). Although somewhat less conspicuous, the pyxie moss (*Pyxidanthera barbulata*, Diapensiaceae) is characteristic of these ecosystems (Good, Good, and Andresen 1979). Olsson (1979) distinguished two communities. In open denuded sites created by severe fire or other extreme disturbance, *Corema conradii* and *Arctostaphylos uva-ursi* are common; such areas are similar in composition to the Long Island heaths described by Olsvig et al. (1979). On more typical sites, the shrub canopy is closed, and the understory is characterized by the lichens *Cladonia caroliniana* and *C. strepsilis*.

Lutz (1934) listed 51 species of vascular plants from dwarf pine plains, 47 from the transition forests, and 39 from the pine-oak forests. Lichens, especially species of *Cladonia*, and mosses were also more abundant in the plains communities. Archard and Buell (1954), in a study of life-form spectra along this vegetational gradient, found that cryptophytes and hemicryptophytes were a larger component of the flora in plains shrublands than in pine-oak forests.

Community dynamics and succession. Considerable effort has been directed toward understanding the factors responsible for the gradient from pine plains to pine-oak forests (Lutz 1934; Olsvig et al. 1979; Good et al. 1979). Factors proposed in-

Table 11.1. *Density, basal area (BA), and percentage cover for trees > 1.2 cm dbh in pine-oak and oak-pine communities in the New Jersey pine barrens*

Species	Pine-oak			Oak-pine		
	Density (ha^{-1})	BA $(m^2\ ha^{-1})$	Cover (%)	Density (ha^{-1})	BA $(m^{2\ -1})$	Cover (%)
Pinus rigida	388	11.3	29.8	204	3.79	16.6
Pinus echinata	24	0.89	2.9	—	—	2.0
Quercus marilandica	44	0.04	0.8	12	0.03	—
Quercus stellata	44	0.74	1.6	28	0.11	1.4
Quercus velutina	8	0.78	6.1	212	5.16	39.9
Quercus prinus	—	—	—	476	6.31	43.0
Quercus alba	—	—	0.2	—	—	0.9
Sassafras albidum	4	0.02	0.4	4	0.001	—
Totals	513	13.8	45.5	936	15.4	103.6

Source: Data from Buell and Cantlon (1950).

Figure 11.3. View of pine plains vegetation near Lebanon, New Jersey. Shrubs and pines in the foreground are approximately 2 m tall (photo by W. D. Billings).

clude nutrient limitations, impervious subsoil, aluminum toxicity, and water deficits, but these factors are neither unique to the pine plains nor sufficient to explain observed patterns. Good et al. (1979) instead concluded that the pattern is a consequence of variation in fire history. Lutz (1934) noted that fire return intervals in pine plains communities averaged 8 yr compared to 16–26 yr in other pine barrens communities, and that the pine plains fires are intense and crown-killing, whereas pine-oak fires are lighter and more variable (Forman and Boerner 1981).

Good et al. (1979) proposed that severe, frequent fires have selected a pine plains ecotype of *Pinus rigida* whose shrubby growth form has a genetic basis (see also Andresen 1959). Common garden experiments have shown that pine plains genotypes have reduced apical dominance compared to those of larger statured types (Good and Good 1975) and that plains genotypes produce more intensely serotinous cones compared to those of other barrens genotypes (Frasco and Good 1976). Little (1979) noted that plains oaks begin producing acorns only 3–4 yr following fire, whereas tree oaks require 20 yr to begin seed production. It is

not known whether other shrubby forms of tree species, such as *Quercus marilandica*, are ecotypes, but there is little doubt that frequent fire is responsible for their coppiced physiognomy in the plains.

Woodwell (1979) suggested that frequent fires cause loss of nutrient capital from plains ecosystems. The nitrogen content of foliage and litter is lower in the plains than in other barrens communities (Lutz 1934). Lower litter nutrient content may reduce rates of decomposition (Vitousek 1982), resulting in the accumulation of flammable fuel and increased frequency of fire (see also Christensen 1985). Given that at least some pine plains plant populations are genetically distinct from those of other barrens communities, it is unlikely that plains communities would rapidly succeed to pine-oak or oak-pine forests in the absence of fire (Good et al. 1979).

The environmental effects and vegetational responses to fire in pine barrens have been reviewed by Little (1979) and Boerner (1981, 1983; Boerner and Forman 1982). These responses are correlated with fire intensity, which in turn varies with climatic and fuel conditions. Following intense fires, oak reproduction is primarily from sprouts since acorns are usually killed by heat (Little 1979; Gallagher and Good 1985). In the plains, abundant pitch pine seed released from serotinous cones may result in large postfire seedling populations. Pitch pine seeds germinate best on a mineral seed bed (Ledig and Little 1979), characteristic of intensely burned sites, whereas the oaks, with their large nutrient reserves, are quite successful in relatively thick mats of leaves (Little 1979). Shrub cover is reduced in the first few growing seasons following fire, but recovery is most rapid following fires of low intensity (Boerner 1981).

In general, herb diversity is increased by fire, but the success of particular species appears to depend on fire intensity. For example, Boerner (1981) found that growth of *Pteridium aquilinum* was greatest in pine-oak stands burned by intense wildfire compared to stands that were burned deliberately by prescription. Olsson (1979) found that floristic composition had shifted in the nearly 50 yr since Lutz's (1934) study and that herb diversity had declined in plains communities; he attributed this shift to decreased frequency of fire.

Although surface fires in pine-oak stands may increase nutrient availability, losses of minerals via volatilization and leaching may be high (Wang 1984). A minimum of 8 yr is required for nutrient inputs to compensate for these losses. Intense wildfires result in greater nutrient loss than light, prescribed fires (Boerner 1983; Boerner and Forman 1982), but in either case rapid vegetation regrowth minimizes these losses and, as suggested by Woodwell (1979), these systems become less leaky with successional age (Boerner and Forman 1982).

Urbanization of the pine barrens region has resulted in a general lengthening of fire return intervals (Forman and Boerner 1981). Regional fire frequency (the number of fires per year) actually increased from 1900 to 1940 and has since remained about the same; however, average size of fires decreased considerably. As a consequence, average fire return intervals for pine-oak forests have increased from 20 to about 65 yr in the period from 1900 to the present. Forman and Boerner suggest that longer return intervals will result in changes to community structure in plains and pine-oak communities toward greater dominance of oaks and other less fire-tolerant or dependent species.

Xeric Sand Communities

Coarse sandy soils are abundant across the Coastal Plain, and they exhibit frequent water deficits owing to poor water retention and nutrient limitations resulting from low mineral adsorptive capacity. Extensive areas of sand are found in the fall line sandhills, a more or less continuous formation

of rolling hills extending from southern North Carolina through Georgia and parts of Alabama. These highly weathered sands are the residual product of underlying Cretaceous sediment. In addition, aeolian and alluvial processes deposited sands in various locations across the middle and lower Coastal Plain during the Pleistocene; on the lower, more recently exposed terraces, relict dune fields and other coastal features provide abundant sandy habitats (Dubar, Johnson, Thom, and Hatchell 1974; Stout and Marion 1993). These areas often form a so-called ridge and swale topography, with xeric conditions on the ridges and poor fens in the swales (Woodwell 1956; Christensen 1981; Ash, McDonald, Kane, and Pories 1983).

Vegetation composition and structure. Three physiognomically and compositionally distinct ecosystems occur on these sands: xeric longleaf pine woodlands, subxeric longleaf pine woodlands (Peet and Allard 1993), and sand pine scrub (Laessle 1942).

Xeric Longleaf Pine Woodlands. Longleaf pine (*Pinus palustris*) trees are widely scattered in this ecosystem, with subcanopy trees such as turkey oak (*Quercus laevis*), persimmon (*Diospyros virginiana*), and bluejack oak (*Q. incana*) in the understory. The grass layer is typically dominated by wiregrasses, *Aristida stricta* north of the Congaree/Cooper watershed (South Carolina) and *A. beyrichiana* to the south (Peet 1993). Peet and Allard (1993) identified five series of this ecosystem type based on geographic location and variations in the composition of herb and shrub species.

1. Fall line xeric longleaf woodland. These woodlands occur on the driest ridges of the fall line region. Common associated species include *Gaylussacia dumosa, Stipulicida setacea, Cnidoscolus stimulosus, Minuartia caroliniana, Euphorbia ipecacuanhae, Carphephorus bellidifolius,* and *Pityopsis graminifolia.*

2. Atlantic xeric longleaf woodland. On coarse sands of the lower and middle Atlantic Coastal Plain *Quercus incana* is more common. In the driest sites, foliose lichens form a brittle carpet along with the sand-binding lichen *Lecidea uliginosa* in open areas. A number of lichens in the genera *Cladonia* and *Cladina* are endemic to these habitats. Mosses such as *Dicranum spurium* and very drought tolerant herbs such as *Selaginella arenicola, Minuartia caroliniana,* and *Stipulicida setacea* are common (Figure 11.4)

3. Southern xeric longleaf woodland. Across the Gulf Coastal Plain, *Sporobolus junceus* and *Licania michauxii* are common associates. Many

Figure 11.4. Pine–turkey oak sandridge near Jones Lake, North Carolina. Note the large areas of bare sand and abundant lichens and mosses. A turpentine scar is obvious on the longleaf pine at the center of the photo.

tures were selected as a consequence both of soil water deficits and of high irradiance owing to high albedo of the white sand. Indeed, the importance of vertical leaf orientation in seedlings of turkey oak has been clearly demonstrated to be related to high irradiance (Wells and Shunk 1931; Raff 1954).

Monk (1968) recognized three distinguishable phases of what he termed the Florida sandhill association dominated by turkey oak, bluejack oak, and southern red oak, respectively. Longleaf pine forms a broken canopy in each of these communities but may share dominance with slash pine (*Pinus elliottii*). The turkey oak phase occurs on driest sites and is structurally similar to the pine–turkey oak sandridge vegetation already described. Scattered individuals of *Quercus incana* and *Diospyros virginiana* may also occur in this phase. *Diospyros* is most common on sites that have not been recently burned (Laessle 1942). The herb layer is dominated by two wiregrasses: *Aristida beyrichiana* and *Sporobolus gracilis*. Shrubs are scarce in this community except for the gopher apple, *Licania michauxii*. The bluejack oak phase is characteristic of finer-texture, somewhat more fertile soils (Laessle 1942). Live oak (*Quercus virginiana*) is common in the overstory, and *Aristida beyrichiana* forms a dense ground cover (Veno 1976). The southern red oak phase is more typical of calcareous soils and grades into southern mixed hardwood forest (Monk 1960, 1968).

Marks and Harcombe (1981) and Harcombe, Glitzenstein, Knox, Orzell, and Bridges (1993) describe "sandhill pine forests" in eastern Texas as having an overstory of longleaf and loblolly (*Pinus taeda*) pines and an understory dominated by bluejack oak (*Q. incana*) sand post oak (*Q. margarettiae*), and hickories (*Carya texana* and *C. tomentosa*) but lacking wiregrass and turkey oak.

species such as *Ceanothus microphyllus, Asimina* spp., *Babtisia lecontei, Chapmannia floridana, Matalea pubiflora, Liatris chapmanii,* and *Andropogon floridanus* are unique to this series.

4. Atlantic maritime longleaf woodland. This community occurs in coastal areas along the Atlantic where climatic conditions are less extreme. The canopy may be nearly closed and the subcanopy includes such species as *Osmanthus americanus, Quercus geminata, Q. hemisphaerica, Myrica cerifera, Persea borbonia, Ilex opaca,* and *Sassafras albidum*. These woodlands grade into the maritime forests described later in this chapter.

5. Gulf maritime longleaf woodland. On the Gulf Coast these forests include a much more diverse assemblage of shrubs and subcanopy trees, including *Serenoa repens, Ilex glabra, I. vomitoria,* and *Conradina canescens*.

Many xeric longleaf pine woodland species display obvious xeromorphic features such as microphylly, glaucous pubescence, and succulence. Wells and Shunk (1931) suggested that these fea-

Subxeric longleaf pine woodlands. This ecosystem type dominated much of the upland presettlement Coastal Plain landscape (Ware, Frost, and Doerr 1993). Dominant trees include those found in xeric woodland, plus a diverse assemblage of species adapted to moister conditions. Soils on these sites are generally sandy and infertile but contain larger amounts of silt and clay than xeric woodlands. Clay horizons near the soil surface result in seasonal "perched" water tables in some areas (Wells and Shunk 1931). Species abundances for a subxeric and nearby xeric longleaf pine woodland in North Carolina are shown in Table 11.2.

Peet and Allard (1993) noted that, although variations in soil texture account for considerable variation in vegetation within localities, variations in latitude and moisture availability are more important on a regional scale. They differentiated three

Table 11.2. *Percentage cover and basal area (BA, trees) for plants in xeric and subxeric fall-line sandhill forests*

	Xeric		Subxeric	
Species	Cover (%)	BA (m^2 ha^{-1})	Cover (%)	BA (m^2 $^{-1}$)
Trees				
Quercus laevis	28.8	4.8	2.2	0.9
Quercus marilandica	1.5	—	31.3	4.3
Quercus margaretta	—	—	11.4	1.5
Quercus incana	1.8	—	3.1	0.5
Carya species	—	—	3.3	0.4
Pinus palustris	8.6	3.0	8.3	2.0
Pinus taeda	—	—	3.5	1.1
Shrubs and herbs				
Aristida stricta	15.0		13.3	
Cryptogams	3.0		1.8	
Gaylussacia dumosa	3.3		2.4	
Tephrosia virginiana	1.0		2.7	
Andropogon species	1.0		2.5	
Clethra alnifolia	—		2.1	
Lyonia mariana	—		0.4	
Rhus radicans	—		0.8	
Solidago species	—		0.4	
Carphephorus bellidifolius	0.8		—	
Totals	64.8	7.8	89.3	10.7

Source: Data from Weaver (1969) for North Carolina.

geographically segregated series: fall line, Atlantic, and Gulf, corresponding to similarly named series for the xeric longleaf pine woodlands. Longleaf pine and turkey oak are common to all three series, along with *Quercus margarettiae, Q. incana, Q. marilandica, Dispyros virgiana,* and *Carya* spp. The understory of these subxeric woodlands is usually dominated by wiregrass (*Aristida stricta* or *A. beyrichiana*) and a diverse assemblage of graminoids and herbs. It is primarily variations in this understory community that differentiate the regional series (Peet and Allard 1993).

The southern ridge sandhill community described by Abrahamson, Johnson, Layne, and Peroni (1984) is characteristic of sandy habitats in south-central Florida. It is intermediate in composition and structure between the sand pine scrub and the xeric pine woodlands, with a distinctly three-layered open canopy. The overstory is 5–10 m in height (Fig. 11.5). The dominant canopy tree is the southern Florida slash pine (*Pinus elliottii* var. *densa*), which may co-occur with scattered individuals of longleaf pine and sand pine. Turkey oak (*Quercus laevis*) and scrub hickory (*Carya floridana*) are also important canopy dominants. Abrahamson et al. (1984) divide this community into two phases based on the relative prevalence of these two species.

Sandhill pine forests have also been described for the Big Thicket region of eastern Texas (Marks and Harcombe 1981). As in other sandhill forests, pine density is low, herb cover is relatively sparse, and there is considerable exposed sand. This area is several hundred miles west of the range of turkey oak, and the dominant oaks are *Quercus incana* and post oak (*Quercus stellata*). Here longleaf pine shares dominance with shortleaf pine (*Pinus echinata*) and loblolly pine (*Pinus taeda*). Yaupon (*Ilex vomitoria*) and flowering dogwood (*Cornus florida*) are common understory trees.

The Sand Pine Scrub. The dominant overstory tree in this community is sand pine (*Pinus clausa*). The understory is dominated by a dense and rather diverse assemblage of evergreen sclerophyllous shrubs (Fig. 11.6). Abrahamson et al. (1984) differentiated two phases of the sand pine scrub community. The oak understory phase is a three-layered community, with a lower shrub layer of saw palmetto (*Serenoa repens*) and scrub palmetto (*Sabal etonia*), an upper shrub layer dominated by scrub live oaks (*Quercus geminata, Q. myrtifolia, and Q. virgininia), Carya floridana,* and rusty lyonia (*Lyonia ferruginea*), and an overstory of sand pine. Herbs are particularly scarce in this phase. The rosemary phase of this community is a somewhat

Figure 11.5. Southern ridge sandhill community near the Archbold Biological Station, central Florida. The longleaf pine in the center is approximately 10 m tall (photo by P. Peroni).

nities of peninsular Florida. This variety has serotinous cones, successfully establishes from seed primarily following fire and, as would be expected, forms very nearly even-aged stands (Little and Dorman 1952; Burns 1973).

Mulvania (1931) observed that water availability was one of the major selective factors on

Figure 11.6. Sand pine scrub communities near the Archbold Biological Station, central Florida. (A) Rosemary phase; (B) oak understory phase. Photos by P. Peroni.

more open community dominated by even-aged stands of rosemary (*Ceratiola ericoides*). Another scrub oak, *Quercus inopina*, may share dominance with the rosemary. The pine canopy is much more broken in this phase, and herbs are more abundant and diverse. The rosemary phase appears to be characteristic of drier ridges and knolls. Although not usually abundant in either phase, *Rhynchospora dodecandra* and *Andropogon floridanus* are indicative of scrub communities (Laessle 1958). These communities are floristically similar to pine flatwoods and maritime scrub forests of Florida.

Two genetic races of sand pine are recognized. Choctawatchee sand pine (*Pinus clausa* var. *immuginata*) grows naturally in a rather restricted region of northwest Florida (usually on soils derived from recent littoral deposits). This variety has nonserotinous cones, often occurs in mixed-age stands, and may be found in several different community types. Ocala sand pine (*Pinus clausa* var. *clausa*) grows exclusively in the sand pine scrub commu-

shrub morphology in sand pine scrub. For example, several shrub species (most especially *Ceratiola ericoides*) have short sclerophyllous needle-like leaves. Leaves of other species are often heavily cutinized, revolute, and sometimes tomentose underneath. Mulvania noted that shrub leaves are often not deployed horizontally in these species and that stomates are smaller and less dense than in nonscrub species. Given the low nutrient concentrations in these sandy soils, it is possible that sclerophylly may also be a response to nutrient limitations (see Loveless 1961, 1962). Shrubs and palms reproduce vigorously from underground rhizomes and burls; most shrub reproduction following fire is vegetative (Wade, Ewel, and Hofstetter 1980).

Community dynamics and succession.   On the driest sand ridges the rate of fuel accumulation is very slow, and the spatial patterning of vegetation and detritus is discontinuous. Fires in this vegetation type are infrequent and often localized near the source of ignition. Historical accounts, such as that of Bartram (1791), suggest that during precolonial times these ridges were dominated by longleaf pines with a relatively open understory. Pine litter would have provided sufficient fuel for occasional fires, which in turn would have favored pine reproduction and suppressed hardwood growth (Christensen 1979b, 1981). These generally misshapen pines were not suitable for saw timber but were an important source of turpentine. Many of the relict trees on these sites still bear the distinctive scars of this enterprise. Most pines were eventually harvested for tar extraction, allowing such hardwoods as turkey oak and persimmon to invade and initiating a vastly different fire regime. Owing to the sterility of these soils and an altered fire regime, longleaf pine has been very slow to reinvade.

Low-intensity surface fires occur at a very high frequency in wiregrass-dominated pine woods. Fire return intervals may be as brief as 1 yr in areas that are prescribe-burned for fuel reduction, but the natural interval probably was 3–10 yr (Heyward 1939; Wells 1942; Garren 1943; Parrott 1967; Christensen 1981). Three to four years are required for sufficient accumulation of dry fuel to carry a surface fire after which time the probability of fire is determined by the availability of ignition sources (Parrott 1967; Christensen 1981).

Fire has the effect of briefly increasing nutrient availability and, thereby, postfire herb production (Christensen 1977). However, reproduction during the first postfire growing season is largely vegetative (Hodgkins 1958; Arata 1959; Parrott 1967; Lewis and Harshbarger 1976). Flowering is stimu-lated by burning in many sandhill species. For example, flowering in *Aristida stricta* is confined to the first postfire growing season (Parrott 1967). Christensen (1977) found that simply clipping leaves would initiate some flowering. However, flower production was always greater in burned compared to clipped areas, presumably owing to fire-caused nutrient enrichment. Recent research has demonstrated considerable variability in herbaceous response to fire, depending on season of burning (Platt, Evans, and Rathbun, 1988; Schneider 1988; Streng, Glitzenstein, and Platt 1993) and small-scale variation in fuel distribution and fire behavior (Hermann 1993).

The life history of *Pinus palustris* is ideally suited to this high-frequency, low-intensity fire regime. Longleaf pines establish most successfully on bare mineral soil following fire (Chapman 1932; Wahlenberg 1946). During the "grass" stage (Fig. 11.7), seedlings are fire resistant and allocate most of their photosynthate to the production of an extensive root system, including a massive taproot. After 3–5 yr, apical growth is initiated, carrying the apical bud above the zone of most surface fires. Unlike pines adapted to less frequent crown fires, longleaf pine does not sprout or produce serotinous cones. If fire frequency is reduced, other pine species, such as *Pinus taeda* and *P. elliottii* may replace *P. palustris* (Chapman 1926; Little and Dorman 1954; Christensen 1981).

Mature stands of longleaf pine are generally uneven-aged, and reproduction typically occurs in 30–50 m radius patches associated with tree blowdowns or other mortality events (Platt et al. 1988, Platt and Rathbun 1993). Thus, frequent low-intensity fires facilitate the early growth and survival of longleaf pine while discouraging competitors, but successful recruitment also depends on small-scale disturbances that are often not associated with fire.

The natural return interval for fires in sand pine scrub is 30–60 yr (Webber 1935; Harper 1940; Laessle 1965; Christensen 1981). These crown fires ordinarily kill the pines and aboveground portions of the shrubs. Seed rain from the sand pine may exceed 250 seeds m^{-2} (Cooper, Schopmeyer, and McGregor 1959), and young stands often have sand pine densities exceeding 10,000 trees ha^{-1} (Price 1973; Coile unpublished data). As their crowns close, pines begin to thin, converging to densities of 400–800 trees ha^{-1} >40 yr. In the absence of fire, the pine canopy becomes sufficiently broken to allow establishment of seedlings in the understory, and the stand may become uneven-aged (Coile unpublished data; Peroni 1983). However, stands >70 yr of age are rare. Dry conditions,

Figure 11.7. Stages in the life history of Pinus palustris, longleaf pine. (A) "Grass" or seedling stage. (B) After 4–6 yr, often following fire, rapid apical growth is initiated and the apical bud, which is vulnerable to fire, is carried above the reach of flames. (C) A pole-size tree, approximately 10 yr old. Survivorship among trees this size is very high.

frequent lightning storms, and accumulation of highly flammable fuels make fire an inevitable event in these communities (Hough 1973).

Laessle (1958, 1967) argued that the Florida sandhill and sand pine scrub communities were segregated with respect to variations in soil characteristics associated with the distribution of Pleistocene shorelines. Recent work suggests that these communities may succeed one another as a consequence of variations in fire regime. Kalisz and Stone (1984) found that sites currently occupied by scrub were, in the recent past, dominated by sandhill vegetation and vice versa. Myers (1985; see also Myers and Deyrup 1983) showed that exclusion of fire from sandhill vegetation results in invasion of scrub, which tends to favor intense crown fires at 30–60 yr intervals, which in turn maintains sandhill vegetation. In the complete absence of fire, these sites are presumed to be seral to xeric hardwoods (Fig. 11.8)

Mesic Pine Communities

With increased moisture availability, xeric sandhill communities grade into pine-dominated flatwoods and savannas. Flatwoods and savannas are distinguished primarily on structural grounds (Fig. 11.9). Savannas are typified by an open canopy (pine density usually <150 ha^{-1}) with a graminoid-dominated understory (Penfound and Watkins 1937). Overstory density is greater in flatwoods, and the understory is composed of a diverse array of shrubs and subcanopy trees. Intermediate situations are, however, common, and some workers (Laessle 1942; Abrahamson et al. 1984; Abrahamson and Hartnett 1990) refer to all these communities as flatwoods.

Vegetation composition and structure. There is no generally accepted regional classification of these mesic pine woods. As with more xeric pine woods, research has been concentrated in the Carolinas,

central Florida, and east Texas. Although species composition along the flatwood-savanna gradient varies among these areas, the causes of compositional change along the gradient seem to be due to the interaction of fire frequency and soil conditions affecting moisture availability in each area (Peet and Allard 1993).

Flatwoods. The term "flatwood" is used for just about any Coastal Plain pine forest with a well-developed woody understory (Christensen 1979b). Many such forests, particularly on relatively fertile soils, are successional from cropland abandonment and are similar in composition to the successional pine stands of the Piedmont (Oosting 1942; Peet and Allard 1993). Given the multiplicity of disturbance histories and site conditions, these successional forests virtually defy classification.

Throughout the Gulf Coastal Plain, forests dominated by longleaf and/or slash pine and with an understory of shrubs and saw palmetto occur on poorly drained Ultisols. Pessin (1933) divided Gulf flatwood communities into xerophytic forests – dominated by longleaf pine, scrub oaks, and *Serenoa repens*, and mesohydrophytic forests – in which *Pinus elliottii* co-dominates in the canopy and the scrub oaks are replaced with more hydric shrubs such as *Ilex glabra*, and *Myrica cerifera*. Edmisten (1963; see also Monk 1968) recognized three flatwoods types based on the dominant canopy tree species. His longleaf pine and slash pine flatwoods categories correspond to Pessin's xerophytic and mesohydrophytic forests, respectively. He also described forests dominated by pond pine (*Pinus serotina*) that occur in very wet sites and appear to be successional to swamp forest or wet hammock (Monk 1968).

In central Florida, Abrahamson et al. (1984) differentiated two flatwoods types, scrubby flatwoods and flatwoods proper (Table 11.3). Scrubby flatwoods are transitional between scrub communities and more mesic ecosystems. Unlike other flat-

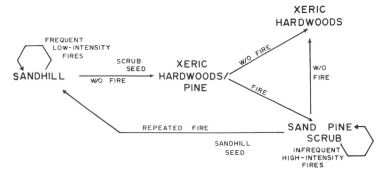

Figure 11.8. A model for successional relations among sandhill, sandpine, and xeric hardwood communities. Given the same environmental conditions, the most stable community will depend on the fire regime. Once established, fuel and environmental conditions tend to perpetuate a given community type (after Myers 1985).

Figure 11.9. Pinus elliottii *flatwoods near the Okefenokee Swamp, Georgia with a palmetto-gallberry understory (A) 3 mo following fire and (B) unburned for >5 yr (photograph* taken near Fargo, Georgia). Pinus palustris *mesic savanna in the Green Swamp, North Carolina (C) burned annually for the past 20 yr and (D) nearby unburned for >10 yr.*

woods communities, these ecosystems occur on well-drained soils (but with a higher water table than in pine scrub communities). The canopy is dominated by scattered individuals of slash and sand pine. On drier sites, *Quercus inopina* dominates the understory, whereas *Q. virginiana* and *Q. chapmanii* are dominant on moist sites. The flatwoods association occurs on poorly drained soils that vary between very dry during drought periods to flooded during rainy periods.

The flatwoods association itself has been sub-

divided into several phases (Abrahamson et al. 1984; Abrahamson and Hartnett 1990). The wiregrass phase is dominated by *Aristida stricta* in the understory. It differs from sandhill communities by the abundance of *Serenoa repens, Quercus minima,* and other shrubby species. On somewhat more favorable sites, cutthroat grass (*Panicum abscissum*) replaces wiregrass. The palmetto flatwoods phase is characterized by an overstory of *Pinus elliottii* var. *densa* and a dense 1–2 m tall sward of saw palmetto. In the gallberry-fetterbush flatwoods

Table 11.3. *Cover of dominant species in Florida flatwoods communities*

Species	Cover (%)			
	SF[a]	FLW	FLP	FLG
Pinus elliottii	—	1	83	24
Pinus clausa	1	—	—	—
Quercus geminata	4	<1	3	—
Quercus chapmanii	16	—	<1	—
Quercus inopina	37	—	—	—
Sabal etonia	8	<1	—	—
Aristida stricta	<1	54	—	1
Quercus minima	—	15	—	—
Hypericum reductum	—	5	—	<1
Serenoa repens	22	36	77	46
Ilex glabra		<1	27	17
Persea borbonia	—	—	12	—
Lyonia lucida	3	3	7	20
Panicum abscissum	—	—	—	86

[a]SF = scrubby flatwoods, FLW = flatwoods/wiregrass, FLP = flatwoods/palmetto, FLG = flatwoods/galberry.
Source: Data from Abrahamson et al. (1984).

phase, *Ilex glabra* and *Lyonia lucida* share dominance in the understory with saw palmetto. The fern phase understory is dominated by chain fern (*Woodwardia virginica*), as well as *Osmunda regalis*, *Andropogon* spp., and *Panicum hemitomon*. In addition, woody shrubs typical of hardwood swamps occur in the understory, suggesting that fire exclusion from this phase might result in succession to swamp forest or bayhead (Abrahamson et al. 1984; Abrahamson and Hartnett 1990).

The shrub oaks *Quercus chapmanii* and *Q. minima*, slash pine, and saw palmetto reach their northernmost limits in southeastern South Carolina. Nevertheless, pine-dominated communities with a shrubby hardwood understory are found on heavier soils north into Virginia. *Pinus taeda* is often the dominant pine, although longleaf pine may be found on sandier soils and pond pine occurs in moist habitats. The prevalence of loblolly pine in such forests today may be a consequence of historical factors (Snyder 1980; Christensen 1981). Ashe (1915) suggested that forests dominated by loblolly pine were common only in extreme northeastern North Carolina. Extensive use of this species as a plantation tree has increased its seed rain throughout the Southeast, and fire exclusion from many areas has given it a competitive edge against longleaf pine.

Wells (1928) classified mid-Atlantic flatwoods as part of a *Quercus–Carya–Pinus* associes, which he said "could be found on virtually any upland site." The subcanopy of these forests is well developed, particularly where fire has been excluded.

Trees, such as several gum and oak species, may reach 5–10 m in height. Shrubs such as *Ilex glabra*, *I. coriacea*, *Myrica cerifera*, *Lyonia lucida*, *L. mariana*, and *Clethra alnifolia* indicate floristic affinity to more southern flatwoods. Herb cover is generally <5% in unburned woods, but when burned frequently, herb cover may exceed 20% and the vegetation may become savanna-like (Lewis and Harshbarger 1976). In moist locations, switch cane (*Arundinaria gigantea*) may form dense canebrakes (Hughes 1966).

The upland pine forests and the upper slope pine-oak forests described by Marks and Harcombe (1981), Harcombe et al., 1993), and Matos and Rudolph (1985) for east Texas roughly correspond to the flatwoods of the mid-Atlantic states. Longleaf or shortleaf pine dominates the overstory, with scattered individuals of loblolly pine, several oaks, and *Liquidambar styraciflua*. Understory species include *Callicarpa americana*, *Ilex vomitoria*, and *Cornus florida*.

Savannas. Savanna vegetation of the Coastal Plain is often transitional between xeric pine communities and wetland pocosins and bayheads (Harper 1906, 1922; Wells 1928, 1946; Wells and Shunk 1928; Woodwell 1956; Eleuterius and Jones 1969; Kologiski 1977; Wilson 1978; Christensen 1979b; Snyder 1980; Walker and Peet 1983; Jones and Gresham 1985; Peet and Allard 1993). References to savannas as "southern upland grass-sedge bogs" (Wells and Shunk 1928), "pine-barren bogs" (Harper 1922), "pitcher plant bogs" (Eleuterius 1968; Eleuterius and Jones 1969), and "wetland pine savannas" (Marks and Harcombe 1981; Harcombe et al. 1993) indicate their seasonally wet condition.

Woodwell (1956) divided savannas in the Carolinas into those dominated by longleaf pine and those dominated by pond pine (*Pinus serotina*). Walker and Peet (1983) recognized a gradient from xeric savannas into mesic savannas and wet savannas. In xeric savannas, the tree canopy contains only longleaf pine and *Aristida stricta* dominates the herb layer. Legumes, including *Cassia fasciculata, Lespedeza capitata, Clitoria mariana*, and *Amorpha herbacea*, are common in these areas. Characteristic woody species include *Myrica cerifera* and *Smilax glauca*.

In mesic savannas (Fig. 11.9) graminoid diversity, indeed herb diversity in general, is very high. Common graminoids include *Sporobolus teretifolius, Muhlenbergia expansa, Ctenium aromaticum, Andropogon* spp., and *Rhynchospora plumosa*. Other indicative species include *Lycopodium carolinianum, Lachnocaulon anceps*, and *Xyris smalliana*.

Perhaps the most distinctive occupants of mesic savannas are insectivorous plants, including *Drosera capillaris, Pinguicula* spp., *Sarracenia* spp., and the celebrated Venus flytrap (*Dionaea muscipula*). The Venus flytrap is found only in savannas of the outer coastal plain of southeastern North Carolina and at a few localities in South Carolina (Roberts and Oosting 1958).

Wet savannas occupy depressions and the ecotones between mesic savannas and shrub bogs. Walker and Peet (1983) point out that they are most likely to be found where frequent fires on mesic savannas have spread into adjacent pocosins, eliminating the shrubs. *Drosera intermedia, Coreopsis falcata, Rhynchospora chalarocephala, Oxypolis filiformis, Iris tridentata, Aristida affinis,* and *Anthaenantia rufa* are indicative of this portion of the gradient. Sprouts of shrub bog species such as *Cyrilla racemiflora* and *Vaccinium corymbosum* are common. Longleaf pine may be missing from this zone, and one finds only scattered individuals of pond pine, pond cypress (*Taxodium ascendens*), and tupelo (*Nyssa sylvatica* var. *biflora*).

Peet and Allard (1993) differentiated among savanna types across the entire Coastal Plain region. Atlantic longleaf savannas in the Carolinas are typified by a number of endemic species such as *Dionaea muscipula, Gentiana autumnalis,* and species of *Lysimachia, Solidago,* and *Tofieldia*. Southern longleaf savannas of the Gulf Coast are typified by endemic species of *Aster, Pinguicula, Sarracenia,* and *Xyris* (Peet and Allard 1993; Walker 1993).

Streng and Harcombe (1982) identified two savanna types in the Big Thicket region of east Texas. Wet meadow savannas occur on poorly drained soils and contain sparsely scattered individuals of longleaf and loblolly pine, as well as *Nyssa sylvatica, Liquidambar styraciflua,* and *Magnolia virginiana*. This savanna type floristically resembles other southeastern savannas, save that *Andropogon scoparius* is the dominant grass. The second type, pine-bluestem savanna, is also dominated by *A. scoparius* and is floristically most similar to nearby prairies (Vogl 1973).

Community dynamics and succession. The successional status of flatwoods communities appears to depend largely on site conditions. On comparatively fertile and well-drained sites, where broad-leaved trees invade the understory, stands will almost certainly be seral to southern mixed hardwood forest in the absence of fire (Quarterman and Keever 1962; Monk 1968; Blaisdell, Wooten, and Godfrey 1974; Hebb and Clewell 1976; Delcourt and Delcourt 1977b). On sandy soils and in poorly drained areas these communities display little evidence of compositional change, even when fire has been excluded for long periods (Abrahamson 1984a, 1984b). Givens, Layner, Abrahamson, and White-Schuler (1984), studying permanent plots in flatwoods from which fire had been excluded for nearly 50 yr, observed substantial thinning of herb and shrub layers with the loss of a few shade-intolerant species, an increase in the richness of tree and shrub species, an increase in canopy coverage, and an increase in litter coverage and depth. However, they noted no appreciable succession toward hardwood forest (see also Veno 1976). Streng and Harcombe (1982) found that Texas pine-bluestem savannas that occur on well-drained sites would succeed to forest in the absence of fire but that poor drainage in wet meadow savannas prevents tree and shrub invasion, regardless of fire regime.

Savannas in North Carolina from which fires have been excluded are consistently less rich than their frequently burned counterparts (Walker and Peet 1983). Roberts and Oosting (1958) found that fire exclusion resulted in the decline of shade-intolerant savanna species, including the Venus flytrap. But there was no evidence that these savannas would succeed to some other forest type. Normally the ecotone between savanna and adjacent pocosin is very abrupt because frequent surface fires in the savanna burn to the edge of the pocosin, thus sharpening the line (Christensen 1981; Christensen, Burchell, Liggett, and Sims 1981).

The consequences of fire exclusion vary along the xeric to wet savanna gradient. On dry sites, sandhill species, such as *Quercus laevis and Diospyros virginiana,* may invade (Christensen 1979b), whereas mesic savannas may succeed to flatwood in the absence of fire (Kologiski 1977). Wet savannas are quickly invaded by pocosin shrubs if not frequently burned (Kologiski 1977; Walker and Peet 1983).

Flatwoods and savannas demonstrate the nearly circular relationship between vegetation structure and composition and fire regime on the Coastal Plain. Moist fuel conditions on heavy soils in flatwoods diminish the probability of ignition, but this is offset by the flammability of plants such as *Myrica cerifera, Ilex* spp., and scrub oaks (Hough and Albini 1978; Shafizadeh Chin, and De Groot 1977). During interfire years, understory growth produces an increasingly continuous vertical distribution of flammable fuel. Thus, the longer the interfire interval, the more likely that ignition will result in a severe crown-killing fire that initiates a prolonged successional sere. Frequent fires create a distinctly discontinuous vertical fuel distribution, and low-intensity surface fires are the rule (Heyward 1939; Wahlenberg, Greene, and Reed 1939; Lemon 1949; Komarek 1974; Christensen 1981). Lewis and Harshbarger (1976) examined vegetation response to different fire regimes in South Carolina pine woods and found that frequent (1–3 yr) summer fires favored an herb-dominated savanna-like community, whereas less frequent (5–10 yr) fires resulted in a dense shrubby understory. Komarek (1977) documented similar relationships between community structure and fire frequency in north Florida.

UPLAND HARDWOOD FORESTS

The broadleaved deciduous forest is described in Chapter 10. However, because the structure and successional status of deciduous forests on the Coastal Plain have been a matter of considerable debate, I include a somewhat abbreviated discussion of them here.

Vegetation Composition and Structure

Harper (1906, 1911) considered that the climax for most of northern Florida and southern Georgia was mixed deciduous and evergreen forest. Pessin (1933) concluded that oak and hickory forests would replace the longleaf pine forests of the Gulf States but would themselves be replaced by forests dominated by beech (*Fagus grandifolia*) and magnolia (*Magnolia grandiflora*). This view was parallel to Wells's (1928) view that beech-maple forests would, in the absence of disturbance, dominate the Coastal Plain of North Carolina where *Magnolia grandiflora* is absent. As data accumulated on the extremes of soil and hydrology, most ecologists later conceded that a variety of edaphic hardwood climaxes might exist (Wells 1942; Harper 1943).

Data on stand structure in mixed hardwood stands can be found in Pessin (1933), Laessle (1942), Kurz (1944), Braun (1950), Quarterman and Keever (1962), Monk (1965), Beckwith (1967), Ware (1970, 1978), Nesom and Treiber (1977), Marks and Harcombe (1981), and Matos and Rudolph (1985). Representative data for such forests in Louisiana appear in Table 11.4. The diversity of tree species is highest in northern Florida and the central Gulf States and diminishes to the north and west (Monk 1967). Although this forest type is often designated as beech-magnolia (Marks and Harcombe 1981; Harcombe et al. 1993), beech is absent from southeast Georgia and eastern Florida (Monk 1965), and *Magnolia grandiflora* is not abundant in the Carolinas and Virginia (Ware 1978; Ware et al. 1993).

Table 11.4. *Relative densities of canopy (C) and understory (U) trees in two Louisiana coastal plain mesic hardwood stands*

Species	Stand 1 C (%)	Stand 1 U (%)	Stand 2 C (%)	Stand 2 U (%)
Magnolia grandiflora	33.8	8.1	14.9	17.6
Fagus grandifolia	6.5	1.6	51.1	17.6
Quercus nigra	2.6	3.2	10.6	11.7
Quercus michauxii	18.2	8.1	4.2	5.9
Pinus taeda	13.0	32.3	—	—
Liquidambar styraciflua	10.4	11.3	—	—
Liriodendron tulipifera	—	—	6.4	5.9
Tilia species	—	—	6.4	—
Quercus falcata var. pagodaefolia	6.5	1.6	—	—
Quercus shumardii	—	—	2.1	11.7
Quercus alba	2.6	—	—	—
Quercus virginiana	1.3	—	—	—
Carya species	1.3	3.2	2.1	—
Morus rubra	1.3	6.4	—	—
Ulmus alata	1.3	—	—	—
Ilex opaca	1.3	1.6	—	—
Ostrya virginiana	—	—	2.1	11.7
Carpinus caroliniana	—	17.7	—	11.7
Cornus florida	—	1.6	—	—
Persea borbonia	—	1.6	—	—
Prunus caroliniana	—	—	—	5.9
Nyssa sylvatica	—	1.6	—	—

Source: Data from Braun (1950).

There is considerable variation in community structure and composition within particular regions related to gradients of moisture and nutrient availability (Monk 1965; Ware 1978; Ware et al. 1993). Monk noted that species diversity was highest on mesic calcareous soils and decreased on wet, dry, and less fertile sites. The relative importance of evergreen species varied as a function of nutrient and water availability. On the driest, most sterile sites evergreen plants accounted for >80% of total importance, whereas on the most fertile sites they accounted for only 10–30% (Monk 1965). It was this pattern that led Monk (1966a) to suggest that evergreen plants were more fit on nutrient-poor sites because year-round leaf fall in evergreens, coupled with sclerophylly, resulted in constant rates of nutrient turnover and reduced nutrient loss. In the Carolinas, Ware (1978) put beech and white oak (*Quercus alba*) at one end of a vegetational gradient, with *Quercus laurifolia, Liquidambar styraciflua*, and *Carya glabra* at the other, but he did not discover what environmental factors corresponded with the gradient.

Community Dynamics and Succession

Upland mixed hardwoods are best developed on fertile soils, often derived from limestone, phosphatic deposits, or finer-texture sediments (Monk 1965; Nesom and Treiber 1977; Ware 1978; Ware et al. 1993). They are also found on somewhat more sterile soils where fire has been excluded. Nesom and Treiber (1977) considered such communities in North Carolina to be unique "topo-edaphic" climaxes confined primarily to river bluffs, but Ware (1970, 1978; Ware et al. 1993) suggested that the present restricted distribution of these forests is an artifact of 350 yr of agriculture. DeWitt and Ware (1979) and Monette (1975) observed considerable invasion of mixed hardwood forest species into sites that had been previously farmed.

Although the mixed hardwood forest is considered by many to be the climatic climax for this region, little work has been done on the successional dynamics within these forests. *Magnolia grandiflora* seedlings do not survive beneath adult magnolias (Kurz 1944; Quarterman and Keever 1962; Blaisdell et al. 1974), and most of the associated oaks and hickories are intolerant of shade. Consequently, these tree populations become depleted in sapling and subcanopy size classes (Blaisdell et al. 1974; Marks and Harcombe 1975, 1981), and it is presumed that gap-phase replacement plays a significant role in continued stand maintenance.

Fires are infrequent in these forests (Christensen 1981), and it is generally assumed that such a disturbance would initiate a prolonged successional sere with an early herbaceous stage, followed by an even-aged pine stage, and eventually reestablishing mixed hardwood forest (Laessle 1942; Kurz 1944; Monk 1968). However, in the only study of fire response in such a community, Blaisdell et al. (1974) found that most of the larger trees survived a canopy fire, although beech suffered considerable trunk scarring. There was a large increase in beech seedling and sprout density following fire, suggesting that these forests are self-replicating, even in the face of major disturbance.

WETLANDS

As a consequence of subdued topography and complex drainage patterns, wetlands cover over 10^7 ha of the southeastern Coastal Plain, or 15% of the total land surface in the region (Turner, Forsythe, and Craig 1981; Wharton, Kitchens, Pendleton, and Sipe 1982). Because they are such a prominent part of the Coastal Plain landscape, wetlands are discussed here as well as in Chapters 12 and 13, which are devoted exclusively to North American wetlands.

Wetlands of the southeastern Coastal Plain vary widely with respect to virtually every community characteristic, and a variety of classification schemes have been devised to account for that variation (e.g., Shaler 1885; Wells 1928; Penfound 1952; Cowardin, Carter, Golet, and LaRoe 1979; Kologiski 1977; Wharton 1978; Huffman and Forsythe 1981; Wharton et al. 1982; Mitsch and Goselink 1986; Christensen Wilbur, and McLean 1988). Wells (1928, 1942) considered that the length of time each year that soil is saturated is the primary environmental factor underlying wetland vegetation gradients. However, it is now recognized that other characteristics of the inundation regime, such as water quality, velocity, and periodicity, affect the relative success of southeastern wetland species (Whitlow and Harris 1979; Wharton et al. 1982; Sharitz and Mitsch 1993).

Following this discussion, we describe these ecosystems as either alluvial wetlands (associated with a stream or river) and paludal wetlands (without significant surface water flow).

Alluvial Wetlands

Alluvial wetlands have received intensive study, and vegetation gradients along river courses have been described in detail for many southeastern river systems; consequently, there is considerable agreement on the classification of riverine communities (Huffman and Forsythe 1981; Wharton et al. 1982; Mitsch and Goselink 1986; Sharitz and Mitsch 1993).

Coastal Plain rivers are divisible into three types. (1) "Brownwater" rivers have their headwaters in the mountains and the Piedmont and comprise the major river systems of this region. Winter rainfall on these watersheds, coupled with low evapotranspiration rates during these months, result in peak flows and flooding of backswamp areas during the winter and spring months. During the summer months, evapotranspiration limits flows, even during periods of heavy rain (Wharton et al. 1982). (2) Blackwater rivers arise on the Coastal Plain and may discharge into the sea or into other rivers. These streams often drain shrub bogs and bay swamps and generally have less discharge than other river types. Flow in these rivers is highly variable, being highest when rainfall exceeds the water storage capacity of the bogs they drain. Flooding may occur during any season due to heavy storms. (3) Spring-fed streams originate where Tertiary limestone outcrops occur on the Coastal Plain.

Figure 11.10. Old-growth zone II alluvial forest near Merchant's Mill Pond, North Carolina. The large tree at the center of the photo is an ancient specimen of Taxodium distichum. Today, trees of this stature are very rare anywhere on the Coastal Plain.

Such outcrops are especially common in northwest Florida and at scattered locations in Georgia and the Carolinas. Water in these streams is quite clear and usually alkaline. Because they are fed by groundwater, flows are quite constant.

Many southeastern Coastal Plain rivers are "underfitted." That is, the modern floodplain is considerably smaller than the historical floodplain. The result is a terraced landscape where higher terraces represent "paleofloodplains" (Dury 1977; Wharton 1978). For example, the second terrace of many large rivers on the lower Coastal Plain of the Carolinas and Georgia corresponds to a floodplain formed during a fluvial period at the close of the Wisconsinan Ice Age (14,000 BP) when river discharge was as much as 18 times that of today (Dury 1977). Alluvial features and soils thus appear well above the present dominion of the river.

As rivers flow onto the Coastal Plain, their floodplains become more sinuous and complex. The concave bank of meander loops often erodes into relatively undisturbed floodplain forest, and sandbars form on the convex bank, creating new habitat for colonization. Numerous oxbow lakes along many southeastern rivers are the remains of abandoned meanders (see Leopold and Wolman 1957; Leopold, Wolman, and Miller 1964; Dury 1977 for discussion of floodplain geomorphology).

When a river breaches its banks, sediment is deposited across the floodplain. Such deposition is greatest near the low water channel, resulting in the formation of levees. Levee formation varies with sediment load and river velocity. Blackwater rivers, for example, carry very little sediment and have poorly formed levees. The areas between the valley walls and natural levees are called flats or backswamps (Wharton et al. 1982). Sediments in flats are generally composed of fine silts and clays or alternating layers of coarse or fine sediment deposited during periods of high or low flood states, respectively. Topographic relief is barely perceptible in many flat areas, but small variations in surface topography may greatly influence plant distribution.

Vegetation classification and structure. Nearly all of the classification schemes devised for floodplain plant communities are based on changes in inundation regime or the "anaerobic gradient" (Wharton et al. 1982; Mitsch and Goselink 1986). I use the zonal classification proposed by Huffman and Forsythe (1981; see also Mitsch and Goselink 1986) and adopted by the National Wetlands Technical Council. However, any such classification is rather arbitrary, and most species occur in several zones.

Zone I – Permanent Water Courses. Herbs dominate this zone, which includes river channels, oxbow lakes, and other permanently inundated areas. Species composition is greatly affected by water flow. In areas of high water velocity, submerged aquatics, often with streamlined leaves, predominate. Where flow is more sluggish, plants with floating or emergent leaves are more common. If there is no current, floating mats of duckweed (*Lemna* and *Spirodela* spp.) and water fern (*Azolla caroliniana*) may occur.

Several nonnative species have become important "weeds" in this zone. Alligator weed (*Alternanthera philoxeroides*) is a submerged aquatic that dominates and, in some cases, chokes many river channels throughout the Southeast. In southern Georgia and Florida, water hyacinth (*Eichornia crassipes*) forms dense floating mats that clog drainage in many slow-flowing streams.

Zone II – River Swamp Forest.   This zone, which includes the wettest flats and sloughs, is generally dominated by cypress-gum swamp forest. Although the silty to sandy soils in this zone may be occasionally exposed, they are saturated with water and typically anoxic.

Bald cypress (*Taxodium distichum*) is the most typical tree species of this habitat. Its buttressed boles and "knees" are a central feature of everyone's concept of a swamp forest (Fig. 11.10). Pond cypress (*Taxodium ascendens*) may replace bald cypress on sandy substrates or in impounded areas. The cypresses often grow in association with one of three species of gum or tupelo. Water tupelo (*Nyssa aquatica*) is most common where water is relatively deep and inundation periods are long. In shallower, less frequently inundated areas, swamp tupelo (*N. sylvatica* var. *biflora*) is abundant. Ogeechee tupelo (*N. ogeche*) is found along blackwater rivers and sloughs of Georgia and Florida. Eastern white cedar (*Chamaecyparis thyoides*) grows along some blackwater rivers, particularly on organic substrates underlain by sand (Wharton et al. 1982). Several other tree species that typically occur in less frequently inundated zones, such as laurel oak (*Quercus laurifolia*), red maple (*Acer rubrum*), American elm (*Ulmus americana*), and sweetgum (*Liquidambar styraciflua*), may become established here on the stumps of cypress or tupelo or on slightly elevated hummocks.

Trees of smaller stature are common in the understory of these forests. On mineral soils, water elm (*Planera aquatica*), pop ash (*Fraxinus caroliniana*), and pumpkin ash (*F. profunda*) are particularly common. On peaty soils, sweet bay *Magnolia virginiana*, red bay (*Persea borbonia*), and titi (*Cyrilla racemiflora*) predominate. Shrubs around the bases of trees and on hummocks include swamp leucothoe (*Leucothoe racemosa*), fetterbush (*Lyonia lucida*), sweet pepperbush (*Clethra alnifolia*), and several *Ilex* species.

Zone III – Lower Hardwood Swamp Forest.   This is a relatively restricted zone on southeastern floodplains, transitional to less frequently inundated backwater swamp forests of zone IV. Soils in this zone are saturated for 40–50% of the year but may become quite dry during the late summer (Leitman, Sohm, and Franklin 1981). Here, plants must be able to tolerate inundation in the early part of the growing season and drydown in the late summer. Zone III habitats include wet flats, low levees, and depressions in higher zones (Wharton et al. 1982).

Overcup oak (*Quercus lyrata*) and water hickory (*Carya aquatica*) are the most typical tree species in this zone. These trees remain dormant well into the spring, which may account in part for their tolerance of flooding in this part of the growing season. Common understory trees and shrubs include winterberry (*Ilex verticillata*), water locust (*Gleditsia aquatica*), Virginia "willow" (*Itea virginica*), American snowbell (*Styrax americanum*), and stiff dogwood (*Cornus foemina*). In disturbed areas, black willow (*Salix nigra*), may haw (*Crataegus aestivalis*), and water elm are often common.

Zone IV – Forests of Backwaters and Flats.   This zone comprises the greatest area of most southeastern floodplains. These areas are inundated for most of the winter and spring but only briefly during the growing season; thus soils are saturated only 20–30% of the year (Leitman et al. 1981). There is little variation in surface topography over large areas aside from minibasins, hummocks, and scour channels, but this microtopographic relief contributes significantly to patterning of species. Inpenetrable soil layers are common and may cause ponding of water.

Quercus laurifolia is the dominant tree species over much of this zone, with willow oak (*Q. phellos*), *Fraxinus pennsylvanica*, *Ulmus americana*, and *Liquidambar styraciflua* as common associates. Several shrubs and small trees are indicative of this inundation zone, including possum haw (*Ilex decidua*), *Crataegus* spp., *Viburnum obovatum*, ironwood (*Carpinus caroliniana*), and several *Rhododendron* species. Dwarf palm (*Sabal minor*) often forms dense thickets here. Vine diversity is quite high in this community and includes such taxa as poison ivy (*Rhus radicans*), greenbriar (*Smilax* spp.), supplejack (*Berchemia scandens*), and cross vine (*Anisostichus capriolata*).

Zone V – Transition to Upland.   Zone V comprises the highest locations of the active floodplain and includes natural levees, higher terraces and flats, and Pleistocene ridges and dunes (Wharton et al. 1982). Inundation is infrequent here and soils are saturated 15% of the year and not at all during the growing season. Zone V soils are often more sandy and less fertile than those of lower zones.

On flats and old levee ridges, basket oak (*Quercus michauxii*) and cherry bark oak (*Quercus falcata* var. *pagodaefolia*) are usually dominant. Occasional individuals of water oak (*Quercus nigra*) and live oak (*Quercus virginiana*) become established here on local high spots, such as logs, stumps, or mounds of soil created by tree falls. Several hickory species grow into the canopy in this area. Spruce pine (*Pinus glabra*), one of the few truly shade-tolerant pines, is typically found in wetter portions of this zone and sometimes into zone IV. Loblolly pine is common in drier areas. Understory trees and shrubs include American holly (*Ilex opaca*), pawpaw (*Asimina triloba*), and spicebush (*Lindera benzoin*). In Florida and across the Gulf States, *Sabal palmetto* and *Serenoa repens* are common in this zone. Zone V is bordered by a variety of upland forest types; however, beech and magnolia forests frequently mark the upper edge of potential inundation. On sandy soils xeric hammock vegetation – dominated by live oak – may form this border.

Even where elevation increases uniformly away from a single channel, the distribution of vegetation is often more complex than the simple classifications of the zone system. Species such as bald cypress and pop ash that are characteristic of frequently inundated zones often grow at the same elevation as inundation-intolerant live oak and pignut hickory. On the scale of just a few meters, seeps or localized clay lenses may create wet soil environments at elevations that are not frequently inundated, and small hummocks, tip-up mounds, and tree stumps provide very localized "dry" habitats in zones that are frequently inundated.

Community dynamics and succession. The most important disturbances in alluvial ecosystems are associated with flooding and the geomorphological processes that have been described. On sandbars and newly formed levees, *Salix nigra* is an important pioneer. Once stabilized, these habitats may support *Populus deltoides* and *Acer saccharinum*. Eventually these comparatively short-lived trees are replaced by Zone IV species, such as *Quercus laurifolia* and *Carya aquatica* (Wharton et al. 1982).

Succession following the formation of oxbow lakes fits the classic model for a hydrarch sere (Clements 1916; Weaver and Clements 1938). Vegetation invades the oxbow in a series of zones that gradually converge as the lake fills in with sediment and organic debris. In the deepest water, submerged aquatics such as *Ceratophyllum, Myriophyllum, Cabomba, Najas,* and *Potamogeton* dominate. Toward shore, floating leaved aquatics, such as *Brasenia, Castalia, Nelumbo,* and *Nymphaea,* are important. In some areas floating aquatics such as *Alternanthera philoxeroides, Jussiaea grandiflora,* and *Eichornia crassipes* may form a floating mat. Penfound

and Earle (1948) observed that mats of *Eichornia* could form a floating prairie with a diverse assemblage of aquatic herbs in less than 25 years. Along the shore, cattail (*Typha latifolia*), *Salix* spp., and *Cephalanthus occidentalis* form a shrubby zone, often with emergent *Taxodium ascendens, Planera aquatica,* and *Persea* spp. Occasional alluvial flooding may subsidize the sedimentation process, and oxbow lakes may fill in a few hundred years (Penfound 1952). Nonetheless, the organic soils and poor drainage associated with these old oxbows often result in an assemblage of species distinct from the adjacent backswamp.

Beaver (*Castor canadensis*) activity has become an important natural disturbance factor along southeastern rivers. Nearly extirpated from southeastern rivers by 1900, owing to habitat destruction and trapping, beaver populations have rebounded as a result of protection and restocking programs. In the ten southeastern states, nearly 200,000 ha of floodplain are impounded by beavers (Hill 1976). By damming sluggish streams, beavers inundate portions of a floodplain, killing many plants in mesic zones not adapted to prolonged anoxia. They then girdle and cut trees along the water, creating open habitat that favors the invasion of early successional shrubs and small trees, such as *Alnus serrulata, Cephalanthus occidentalis,* and willows (Hair, Hepp, Luckett, Reese, and Woodward 1979). It is these shrubs that are the mainstay of the beaver diet.

The history of human disturbance in wetlands is long and extensive. Floodplains were among the first habitats to be farmed by the American Indian and, because of the navigability of the rivers, alluvial forests were among the first southeastern forests to be heavily cut (Pinchot and Ashe 1897; Cowdrey 1983). Thus, most forested wetlands in the Southeast are in some stage of succession from human disturbance. Perhaps the most far-reaching and least understood of human perturbations to alluvial wetlands have been alterations to hydrology. Flows along most of the major river systems of the Southeast have been altered by dams, control structures, and artificial levees. These have greatly altered inundation regimes, generally decreasing average flows and, more important, diminishing seasonal variation in flow. Thus, the primary determinant of the vegetation composition of many successional bottomlands has changed and so has the course of succession.

Paludal Wetlands

Paludal wetlands occur across the southeastern Coastal Plain in association with topographic depressions, groundwater seeps, local areas with poorly drained soils, and extensive accumulations of peat. Five paludal ecosystem types – graminoid-dominated wetlands; pocosins; Atlantic white cedar swamp forest; bay forest; and cypress domes, heads, and islands – all occur across the Coastal Plain. The vegetation composition and structure, as well as the successional dynamics of these wetlands, are described in the following sections.

Graminoid-dominated wetlands. Shallow marshes dominated by a variety of grass, sedge, and rush species are common across the lower Coastal Plain on a diverse assemblage of sites. Dominant genera include *Panicum, Muhlenbergia, Carex, Rhynchospora, Cladium, Scirpus,* and *Juncus* (Penfound 1952). *Sphagnum* is represented by a diverse assemblage of species but is rarely dominant in these habitats.

Graminoid-dominated wetlands are most often early seral stages following disturbance of forested wetlands. For example, such communities are common following cutting or burning of bay forests (Wells 1942) or wet flatwoods of slash pine (Penfound and Watkins 1937). Severe fires in peatlands also create communities dominated by grass sedge (Kologiski 1977; Hamilton 1984) and are clearly responsible for the maintenance of the "prairies" of the Okefenokee Swamp (Wright and Wright 1932; Cypert 1961; Duever and Riopelle 1983).

The Everglades of southern Florida is certainly the best known and most extensive graminoid-dominated wetland in the Southeast. A massive water track flowing south from Lake Okeechobee creates an extensive wetland dominated over large areas by a single sedge species, *Cladium jamaicensis* (see Davis 1943; Gunderson and Loftus 1993 and Chapter 12, this volume, for a detailed description of Everglades vegetation patterns). Succession to forested wetland is prevented by frequent fire, and spatial variability in fire behavior has long been recognized as a major determinant of landscape patterns (Harper 1911; Wade et al. 1980).

Pocosins. Shrub-dominated wetlands, or pocosins, are a common feature of the Atlantic limb of the Coastal Plain (Richardson, Evans, and Carr 1981; Sharitz and Gibbons 1982; Richardson and Gibbons 1993). These peatlands are dominated by a dense, nearly inpenetrable cover of evergreen and deciduous shrubs with scattered emergent trees (Fig. 11.11). Based on physiographic location, four types of pocosin are recognized – interstream, Carolina bay, ridge and swale, and seep or streamhead pocosins – based on physiographic location (Woodwell 1956; Christensen et al. 1981; Ash et al. 1983; Richardson and Gibbons 1993).

Interstream pocosin complexes may extend over thousands of hectares in flat interstream "uplands." These bogs began forming in clogged stream channels approximately 6000–8000 yr ago, coincident with a shift toward cooler climates and the rise of sea level to near its present elevation. Beginning as primary mires, these bog complexes "grew," owing to slow decomposition rates and paludification, into tertiary mires or domed bogs (Whitehead 1972; Daniel 1981; Ash et al. 1983). The plant communities of these peatlands are biogeochemically separated from the mineral soil substrate and receive all nutrient inputs from rain and dryfall (i.e., they are ombrotrophic) (Wilbur and Christensen 1983). It is these raised bogs that gave rise to the term "pocosin," an Algonquin Indian word meaning "swamp-on-a-hill" (Tooker 1899).

Pocosins also occur in many of the Carolina Bays (Sharitz and Gibbons 1982). These bays are quite varied vegetationally and appear in many cases to be undergoing a typical hydrarch succession from lake to bog to forest. Ridge and bay pocosins are found in swales of relict dune fields on the lower Coastal Plain. Pocosins may occur near seeps, springs, and the margins of slow-flowing streams, particularly in sandy areas. Such seeps are often associated with clay horizons. Although bay, ridge and bay, and seep types are not ombrotrophic, the water they receive is nonetheless nutrient-poor, and all pocosin types are quite nutrient-limited, especially with respect to phosphorus (Woodwell 1958; Simms 1983, 1985; Wilbur 1985; Bridgham and Richardson 1993).

Kologiski (1977) divided pocosin vegetation into two classes: conifer-hardwood and pine-ericalean. Conifer-hardwood communities generally occur on mineral soils or shallow peats and are successional to other forest types such as flatwoods or swamp forest (Kologiski 1977; Christensen et al. 1981). Within the pine-ericalean class, Kologiski delimited three types, each briefly described in the following separate paragraphs. There is virtually complete overlap in the list of species found in each of the community types, but they differ considerably in relative abundance. Species evenness is highest in the low pocosin communities and decreases steadily as productivity increases (Woodwell 1956; Christensen 1979a).

The *Pinus serotina–Cyrilla racemiflora–Zenobia pulverulenta* type corresponds to Wells (1946) "low pocosin" (Table 11.5). These communities are typical of the most nutrient-limited sites with the deepest peats, such as occur at the very centers of bog complexes. *Sphagnum* is not the primary contributor to peat formation in most of these bogs but may account for as much as 50% of total cover in some

Figure 11.11. Top to bottom: low-, medium-, and high-stature pocosin communities of the Green Swamp, North Carolina. Left unburned, high pocosin forest will succeed to bay forest, but low pociosin communities show little evidence of successional change.

places. In bog centers, *Sphagnum magellanicum* and *S. bartlettianum* form extensive hummocks, and *S. cuspidatum* is common in depressions. As shrub cover increases in more productive sites, *Sphagnum* abundance decreases to scattered clumps.

The *Pinus serotina–Gordonia lasianthus–Lyonia lucida* community type is characteristic of elevated areas or "islands" within the type just discussed.

These somewhat more productive areas are about 5–20 m across and have a very regular distribution through the low pocosin matrix. Their history is uncertain, but they appear to be formed by the accumulation of litter and *Sphagnum* around tree boles and stumps.

The *Pinus serotina–Cyrilla racemiflora–Lyonia lucida* community type is frequently referred to as

Table 11.5. *Percentage cover of shrub species from pocosins in Dare County, North Carolina*

Species	Cover (%)
Pinus serotina	19
Ilex glabra	56
Lyonia lucida	26
Chamaedaphne calyculata	17
Zenobia pulverulenta	3
Persea borbonia	2
Gordonia lasianthus	2
Kalmia angustifolia var. *carolina*	<1
Acer rubrum	<1
Clethra alnifolia	<1
Myrica cerifera	<1
Cyrilla racemiflora	<1
Woodwardia virginica	2
Smilax laurifolia	3
Sphagnum species	6
Carex walteriana	<1
Andropogon species	<1

Source: Data from Laney and Noffsinger (1985).

"high pocosin." This community type occurs on shallower peats (generally <0.5 m deep). Trees are 10–15 m tall and shrubs reach 5 m tall.

The successional status of pocosins varies with site conditions. On shallow peat soils and in some of the Carolina Bays, white cedar and pond cypress are invading. If left unburned, these communities will probably succeed to a variety of swamp forest types (Kologiski 1977; Christensen et al. 1981). Most of these sites were heavily cut over during the nineteenth century and have been frequently burned since, thus maintaining pocosin status (Lilly 1981). On deeper peats and truly ombrotrophic sites, pocosins show no sign of invasion by swamp forest species (Christensen et al. 1981). This is consonant with Otte's (1981) proposal that these ecosystems are the endpoint of a long-term paludification process beginning with a grass-sedge bog that is subsequently invaded by swamp forest. Peat accumulation separates the ecosystem from the mineral substrate and drives succession to pocosin.

The effects of fire are highly variable in these peatland ecosystems. Wilbur and Christensen (1983) found that availability of most nutrients, including phosphorus, was increased in the first postfire year. However, the increase lasted in most cases only a single growing season. Furthermore, they found that there was considerable spatial variance in enrichment, probably associated with small-scale variations in fire behavior and ash deposition. Thus, phosphorus might be enriched 200-fold at one location and unchanged 10 m away.

Christensen et al. (1981) noted that analogous variations occur among fires. Hot summer fires may burn deep into the peat, whereas spring or autumn burns may burn only the surface litter. Burning causes severe drying of surface peat layers, rendering the surface nonwettable for several years (Wilbur 1985). Thus, growing conditions in burned pocosins after the first year may actually be poorer than in unburned areas.

Vegetation response to spring fires of moderate intensity is quite rapid and almost entirely due to vegetative reproduction. All pocosin shrubs sprout vigorously; by the end of the first growing season, the community recovers nearly half of its prefire biomass and has an LAI of 2.2, compared to a prefire index of 3.9. Production is even greater in the second growing season, but in years 3 and 4 it drops below that of unburned pocosins. This decline appears to be a consequence of decreased nutrient capital coupled with unfavorable moisture conditions. Even after a decade, production may not recover to prefire levels.

Species responses to burning are not uniform. Wells (1946) and Woodwell (1956) noted that *Zenobia pulverulenta* tends to be more common on recently burned sites and is subsequently overtopped by *Lyonia lucida* and *Cyrilla racemiflora*. I have observed (unpublished data) that *Zenobia*, *Sorbus arbutifolia*, and *Ilex glabra* are enhanced by burning and that these species flowered in pocosins only after fire. These species account for a large portion of the increased first and second years' increase in production. Production of *Cyrilla racemiflora*, the prefire dominant, was actually depressed by burning.

Christensen et al. (1981) proposed that species richness in these shrub bogs is as much dependent on the variability created by the fire as on the release from competition. They noted that, although reproduction is primarily from vegetative sprouts, prefire composition in permanent plots was frequently not related to postfire composition. They proposed that a high postfire variance in limiting nutrients exists and that species are differentially successful along this nutrient gradient.

The peat soils of these ecosystems create the possibility for very intense ground fires. During summer months, evapotranspiration dries out the surface peat layers, creating very flammable conditions. If fires start, they may burn through several decimeters of peat, killing all subterranean vegetative structures. Succession then is by invasion of herbaceous plants, forming a grass-sedge bog, often with abundant *Woodwardia virginica* (Kologiski 1977). Where peat has burned all the way

Figure 11.12. White cedar (Chamae-cyparis thyoides) *swamp forest on deep peat soils in the Dismal Swamp, North Carolina. This photo was taken at the forest edge; shrub cover beneath the dense canopy is much more sparse.*

Table 11.6. *Cover of overstory and understory species in white cedar swamp forests of Dare County, North Carolina*

Species	Cover (%)
Overstory (>5 m high)	
Chamaecyparis thyoides	64
Nyssa sylvatica var. biflora	21
Persea borbonia	7
Acer rubrum	7
Magnolia virginiana	4
Pinus taeda	4
Pinus serotina	2
Gordonia lasianthus	2
Understory (<5 m high)	
Lyonia lucida	43
Ilex coriacea	42
Persea borbonia	24
Ilex glabra	17
Smilax laurifolia	15
Sphagnum species	11
Woodwardia virginica	12
Nyssa sylvatica var. biflora	7
Clethra alnifolia	6
Amelanchier arborea	2
Myrica species	1
Magnolia virginiana	<1
Gordonia lasianthus	<1

Source: Data from Laney and Noffsinger (1985).

down to mineral soil, shallow lakes may be formed when the water table rises. Such peat-burn lakes are a common feature of the lower terraces of the North Carolina coastal plain (Whitehead 1972).

Atlantic white cedar swamp forest. Atlantic white cedar (*Chamaecyparis thyoides*) ranges from New England to northern Florida and Alabama (Clewell and Ward 1985; Belling *personal commun.* in a variety of wetland habitats (Moore and Carter 1985). However, it forms extensive stands in only a few areas, including the New Jersey Pine Barrens, lower terraces of the North Carolina and Virginia coastal plains, and northern Florida (Korstian and Brush 1931; Little 1950; Frost 1986). Swamp forests dominated by white cedar are frequently associated with deep peats over sandy substrates. Very little is known about the factors restricting the distribution of this community type. Frost (1986, 1995) cites historical evidence that such forests may have been more widespread prior to human development of the Coastal Plain.

In the Carolinas and Virginia, white cedar forms dense, even-aged populations, often with a very closed canopy (Fig. 11.12). Compositional data for a typical white cedar swamp forest in North Carolina are in Table 11.6. In the Middle Atlantic

Table 11.7. *Density (for overstory and shrub-layer species) and basal area (BA, for overstory species) of bay forest stands in South Carolina*

Species	Density (ha⁻¹)	BA (m² ha⁻¹)
Overstory (>3 cm dbh)		
Gordonia lasianthus	322	4.64
Persea borbonia	343	2.15
Nyssa sylvatica var. biflora	211	3.91
Pinus taeda	134	5.32
Pinus serotina	134	4.93
Liquidambar styraciflua	102	0.77
Vacinium species	130	0.22
Myrica cerifera	156	0.46
Cyrilla racemiflora	72	0.21
Magnolia virginiana	75	0.26
Totals	1912	24.81
Shrub layer (>1 m high, <3 cm dbh)		
Lyonia lucida	26,915	
Clethra alnifolia	6,789	
Vaccinium species	2,709	
Ilex glabra	2,409	
Persea borbonia	1,324	
Smilax species	1,118	
Sorbus arbutifolia	684	
Myrica cerifera	298	
Ilex cassine	675	
Gordonia lasianthus	553	
Total	46,278	

Source: Data from Gresham and Lipscomb (1985).

states, white cedar may comprise >95% of canopy cover (Korstian and Brush 1931; Buell and Cain 1943). In the Gulf states, however, white cedar shares dominance with a wider diversity of tree species and rarely accounts for >50% of cover or relative stem density (Harper 1926; Korstian and Brush 1931; Dunn et al. 1985). Shrub cover in the understory may reach 80% and includes the species *Lyonia lucida, Ilex coriacea, I. glabra, Clethra alnifolia,* and *Persea borbonia. Sphagnum* and *Woodwardia virginica* are the two most important constituents of the herb layer (Laney and Noffsinger 1985). Composition of the white cedar swamps of New Jersey and the Delmarva Peninsula are quite similar (Little 1950; McCormick 1979; Hull and Whigham 1985).

Although white cedar is comparatively shade-tolerant, it does not establish in the dark shade of mature stands (Korstian and Brush 1931). Successful seedling establishment occurs when wet conditions follow intense crown-killing fires (Korstian 1924; Buell and Cain 1943). Korstian (1924) found that the top layer of peat in mature white cedar stands contained enormous quantities of viable seed that could yield very high seedling densities

(>4 × 10⁶ ha⁻¹). Eight-yr-old stands with >60,000 trees ha⁻¹ are not uncommon. Buell and Cain 1943) documented later thinning in such stands, resulting in white cedar stem densities of 2500 and 1700 ha⁻¹ in 35- and 85-yr-old stands, respectively. They noted that nearly 70% of the standing stems in stands older than 80 yr were dead, and that as stands matured a considerable quantity of dead fuel accumulated. The broken canopy of these older stands permits invasion of many bay forest species, including *Ilex cassine, Persea borbonia, Gordonia lasianthus,* and *Magnolia virginiana;* in the absence of fire, succession would lead to dominance by these species. However, the accumulated dead fuel of the white cedar increases flammability of the older stands, increasing the likelihood of crown fire that returns the system to its initial condition. Frost (1986) noted that the present range of white cedar swamp forest may be restricted owing to alteration of fire regimes. Short intervals of fire return tend to favor pocosin vegetation on these sites, whereas fire exclusion may result in the replacement of white cedar by bay forest species.

Bay forests, bayheads, and baygalls. The term "bay" is widely used in the Southeast to refer to a number of evergreen trees, including sweet bay (*Magnolia virginiana*), red bay (*Persea borbonia*), and loblolly bay (*Gordonia lasianthus*). Communities dominated by these species are referred to as bay forests in the Carolinas and Georgia (Kologiski 1977; McCaffrey and Hamilton 1984; Gresham and Lipscomb 1985) and as bayheads in Florida (Gano 1917; Laessle 1942, Monk 1966b; and Abrahamson et al. 1984). Related wetlands in east Texas, in which *Gordonia* is absent, are referred to as baygalls (Marks and Harcombe 1981), a term used widely throughout the Southeast during the nineteenth century (Wright and Wright 1932). These ecosystems are often associated with shallow depressions or poorly drained interstream areas and in Carolina bays.

The species found in association with the bay trees listed earlier vary regionally. In the Carolinas they include *Cyrilla racemiflora, Ilex cassine* var. *myrtifolia, Chamaecyparis thyoides, Pinus serotina, Nyssa sylvatica* var. *biflora, Acer rubrum,* and occasional individuals of *Taxodium distichum* (Table 11.7). Kologiski (1977) differentiated between evergreen and deciduous bay forests in the Green Swamp. The latter community occurs on shallow peats and mineral soils and is probably successional to wet deciduous forest. The shrub stratum of bay forests may be quite diverse, including such species as *Cyrilla racemiflora, Lyonia lucida, Ilex coriacea, I. glabra, Myrica heterophylla, Vaccinium atrococcum, V. cor-*

Table 11.8. *Density and basal area (BA) for species in the tree and shrub strata in 18 cypress forest stands in the Okefenokee Swamp*

Species	Density (stems ha⁻¹)	BA (m² ha⁻¹)
Trees (>4 cm dbh)		
Taxodium ascendens	1467.0	69.4
Ilex cassine	189.3	1.0
Nyssa sylvatica var. biflora	134.1	0.85
Cyrilla racemiflora	59.6	0.16
Magnolia virginiana	43.0	0.31
Lyonia lucida	9.0	0.01
Persea borbonia	11.7	0.02
Clethra alnifolia	1.5	0.004
Vaccinium species	0.8	0.001
Totals	1916.0	71.90
Shrub stratum (>1 high, <4 cm dbh)		
Itea virginica	6840	56.3
Lyonia lucida	7840	38.5
Leucothoe racemosa	3340	35.8
Clethra alnifolia	2080	29.3
Ilex cassine	1180	17.8
Smilax walteri	1620	14.3
Cyrilla racemiflora	1290	13.0
Pieris phillyreifolia	760	15.8
Taxodium ascendens	380	10.3
Decodon verticillatus	260	5.8
Nyssa sylvatica var. biflora	190	3.0
Rhus radicans	270	1.3
Vaccinium species	80	1.0
Magnolia virginiana	40	0.8
Ilex glabra	70	0.5
Cephalanthus occidentalis	20	0.8
Ilex coriacea	10	0.5
Smilax laurifolia	10	0.3
Gordonia lasianthus	10	0.3
Acer rubrum	10	0.3

Source: Data from Schlesinger (1978a).

ymbosum, and *Zenobia pulverulenta. Woodwardia virginica* is the most abundant herb (Kologiski 1977; Gresham and Lipscomb 1985). In Florida bayheads, *Pinus elliottii* is found with *P. serotina,* and cinnamon fern (*Osmunda cinnamomea*) is an important understory herb (Laessle 1942; Monk 1966b; Abrahamson et al. 1984). Vines, including *Smilax* spp., *Gelsemium sempervirens,* and *Parthenocissus quinquifolia,* are an important component of bay forest communities across the Coastal Plain. The overstory dominants of the Texas baygalls include *Quercus laurifolia, Nyssa sylvatica, Magnolia virginiana, Ilex vomitoria,* and *Acer rubrum* with a shrub understory that includes *Cyrilla racemiflora* and *Ilex coriacea.*

Buell and Cain (1943) and Penfound (1952) considered bay forests to be the climax of a succession from pocosin and white cedar swamp. Monk (1966b) suggested that these communities might also arise by means of paludification in certain flatwoods types, particularly those dominated by *Pinus serotina,* or by means of hydrarch succession from open-water ponds to bogs to bayheads (see also Laessle 1942, Davis 1946, and Hamilton 1984). Kologiski (1977) and Hamilton (1984) noted that selective cutting of cypress, white cedar, and swamp hardwoods has favored bay species and increased the aerial extent of bay forest.

The effects of fire in bay forest are dependent on fire behavior and postfire hydrology (Wells 1946; Kologiski 1977; Hamilton 1984). Shallow peat burns give rise to pocosin or white cedar stands that – in the absence of further disturbance – presumably succeed back to bay forest. Deep peat burns with a high postburn water table may produce sedge bogs or "prairies" dominated by *Carex walteriana* and *Woodwardia virginica.* These areas are subject to frequent, low-intensity fires; thus tree invasion may require decades (Cypert 1961, 1972; Kologiski 1977). A deep peat fire when the water table is low can favor the establishment of deciduous species, such as *Nyssa sylvatica* and *Acer rubrum* (Kologiski 1977).

Cypress domes, heads, and islands. Cypress domes (also called cypress heads or wet hammocks) are small forested wetlands that occur in poorly drained depressions throughout the southeastern Coastal Plain (Kurz and Wagner 1953; Monk and Brown 1965; Marois and Ewel 1983). Similar communities also develop on higher points of land or accumulations of peat in the midst of the grass sedge wetlands or "prairies" of the large southern swamp complexes, such as the Everglades, Big Cypress Swamp, and Okefenokee Swamp (Davis 1943; Wright and Wright 1932; McCaffrey and Hamilton 1984). In the Okefenokee Swamp such "islands" are referred to as "houses."

Compared to bay heads, cypress domes usually have deeper water and longer periods of inundation (Monk 1966b). Monk and Brown (1965) also found that they had somewhat higher soil pH (4.1–4.6) and cation concentrations. However, compared to floodplain swamps, these forests are profoundly nutrient-limited. Brown (1981), for example, found that phosphorus inputs into cypress domes amounted to only 0.11 g m⁻² yr⁻¹ compared to 1625 g m⁻² yr⁻¹ in a floodplain cypress swamp (see also Schlesinger 1978b).

Compositional data for 17 cypress domes in the Okefenokee Swamp appear in Table 11.8. These data are similar to those for domes in the Carolinas

(Kologiski 1977), Florida (Monk and Brown 1965; Marois and Ewel 1983), and the Gulf states (Penfound 1952). Although *Taxodium distichum* dominates alluvial floodplain forests, *T. ascendens* is the canopy dominant in the domes (Neufeld 1983). Monk (1968) ascribed their distribution to differences in tolerance to low pH, but Brown (1981) suggested that *T. ascendens* is considerably more drought-tolerant owing to its reduced leaf area and sunken stomates. Neufeld (1984) found that midday xylem pressure potentials were consistently lower in *T. distichum* than in *T. ascendens* grown in equivalent moisture regimes. Because cypress domes dry down periodically, drought-tolerance may be important. The reduced leaf area of *T. ascendens* could also be an adaptation to nutrient stress (Brown 1981).

The understory of cypress domes is dominated by a diverse mixture of shrubs, many of which are evergreen sclerophyllous and found in other nutrient-limited wetlands such as pocosins. Schlesinger (1978a) found that the diversity (richness and evenness) of these understory shrub communities was enhanced by relatively frequent light surface fires.

The Spanish moss (*Tillandsia usneoides*) is a ubiquitous epiphyte in this community, often having a biomass in excess of that for total herbs in many upland forests (Schlesinger 1978b). Schlesinger noted that this species extracts considerable nutrients from incident rainfall and may play an important role in nutrient cycling in these communities. Other common epiphytes include the resurrection fern (*Polypodium polypodioides*) and – in central and southern Florida – other species of *Tillandsia*.

The successional fate of cypress domes appears to be dependent on hydrology. In areas that remain permanently inundated, cypress may form stable, dominant populations. If, however, the area receives any nutrient subsidy, such as sediment from a river, cypress may be replaced by broadleaved deciduous trees such as *Acer rubrum, Quercus* spp., and *Fraxinus caroliniana*. On drier sites, relatively shade-tolerant broadleaved evergreen trees, such as *Persea* spp., *Magnolia virginiana*, and *Gordonia lasianthus*, replace the less shade-tolerant cypress, producing a bayhead forest (Putnam, Furnival, and McKnight 1960; Monk 1968). This is consistent with the observations of Marois and Ewel (1983) that artificially drained cypress domes had increased importance of bayhead species.

Cypress seeds do not germinate in flooded soils (Mattoon 1916; Demaree 1932; Dickson and Broyer 1972), therefore cypress stands must become established on unoccupied sites during periods of drought (Kologiski 1977; Schlesinger 1978a). Such conditions exist following crown-killing fires that do not consume the surface peat layer (Hamilton 1984). At the other extreme, Ewel and Mitsch (1978) found that cypress trees survived surface fires better than competing hardwoods, therefore surface fires may favor long-term cypress dominance on sites that might otherwise succeed to bayhead forest. Gunderson (1984) found that *Salix caroliniana* had invaded recently burned cypress stands and that cypress was regenerating successfully from seed and sprouts.

Recovery from severe fires that burn into the peat may be very slow. Seed dispersal is very limited (particularly in stagnant water; Hamilton 1984). Younger cypress trees (<200 yr) may sprout if the fire is not too severe, but older trees produce few sprouts. Consequently, where severe fires have burned old-growth cypress stands, there is little sign that the communities are returning to their predisturbance composition (Cypert 1961, 1973; Hamilton 1984).

Ewel (1995) suggested that differences in fire regimes may have been an important selective factor in the evolutionary divergence of bald and pond cypress; the former is fire intolerant, whereas the latter species is able to withstand fires of low to moderate intensity.

Succession leading to the formation of cypress islands in the midst of extensive grass-sedge wetlands (such as occur in the Okefenokee Swamp) is illustrated in Fig. 11.13. Peat masses, or "batteries," break loose from the substrate and float to the water surface, forming a relatively dry habitat for colonization (Cypert 1972; Spackman, Cohen, Given, and Casagrande 1976; Duever and Riopelle 1983). Batteries may then be invaded by shrub species *Cephalanthus occidentalis, Lyonia lucida*, and *Ilex cassine*. The shrubs stabilize this habitat and increase the rate of peat accumulation, thus favoring subsequent invasion of bay species and cypress. The resulting community is similar in structure and composition to cypress domes found in closed depressions. The resulting islands ("houses") are often elliptical, perhaps streamlined by water flow (Wright and Wright 1932). Because they are raised above the surrounding minerotrophic water, they may be ombrotrophic and nutrient-limited.

Much of the cypress in the Southeast has been lost to logging (Pinchot and Ashe 1897; Izlar 1984). Stands that have been high-graded for cypress may succeed to bay forest (Kologiski 1977; Hamilton 1984). However, where cypress was the primary canopy constituent, logging opens the canopy sufficiently to allow successful seedling and sprout establishment (Gunderson 1984; Hamilton 1984). In many locations the combination of drainage and logging has created ideal conditions for intense fires. The combination of fire and logging has a devastating effect on community composition, and most sites so disturbed in the past 50 yr show no sign of returning to their predisturbance composition (Cypert 1961, 1973; Hamilton 1984; Kologiski 1977; Gunderson 1984). The varied effects of fire and logging on community succession in these wetlands is illustrated in Figure 11.14.

Because these ecosystems are often hydrologically isolated from open surface water and hydrologic residence times are relatively long, they have received considerable study as potential depositories for wastewater (Dierberg and Brezonik 1984). Such nutrient enrichment results in an immediate response among the herbs and floating aquatics. Ewel (1984) found that growth of duckweed (*Lemna* spp.) was greatly enhanced, as was dry matter production of other herbs. However, establishment and growth of *Nyssa sylvatica* var. *biflora* and *Taxodium ascendens* seedlings are inhibited by wastewater (Deghi 1984). Although net photosynthesis was increased in cypress growing in sewage-

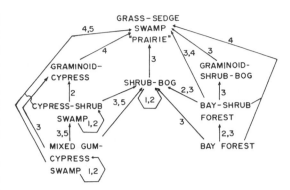

Figure 11.14. Patterns of succession among various wetland communities in Okefenokee Swamp under various intensities of fire and logging: 1, light fire; 2, moderate fire; 3, severe fire; 4, very sever or frequent fire; 5, logging followed by fire (after Hamilton 1984).

treated domes (Brown et al. 1984), there was no sign of increased tree growth even after 6 yr (Straub 1984). The long-term successional effects of such treatments have yet to be determined.

Marois and Ewel (1983) investigated the impact

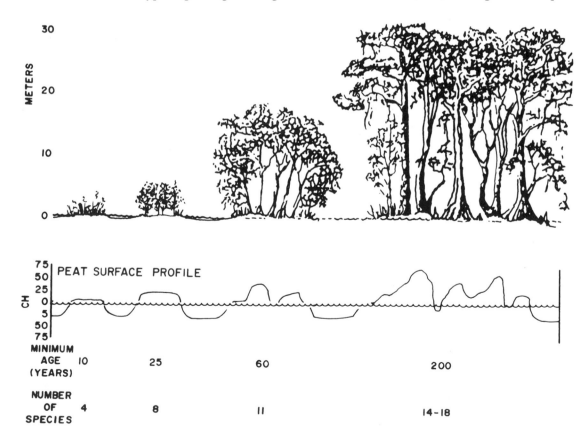

Figure 11.13. Patterns of succession on peat batteries in the Okefenokee Swamp, Georgia (from Deuver and Riopelle 1983).

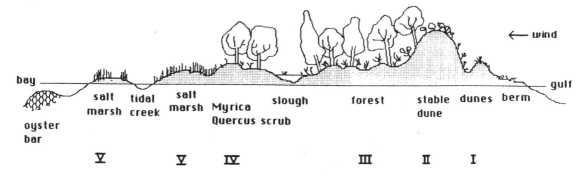

Figure 11.15. An idealized barrier island profile, Gulf Coast of the southeastern United States.

Figure 11.16. Foredune community dominated by Uniola paniculata on Shackleford Bank, North Carolina.

of artificial drainage on cypress domes and found that cypress tree growth was enhanced by drainage in some but not all cases. However, they warn that changes in soil factors and vegetation composition in drained domes inhibit cypress regeneration.

MARITIME VEGETATION GRADIENTS

In addition to river deltas, the southeastern coast is bordered by long ribbons of barrier, or sea, islands that shelter estuaries, lagoons, and sounds. Over very short distances, there are striking gradients of salinity, soil, and climate, and therefore steep gradients in community composition, struc-

ture, and physiognomy. The spatial arrangement of geomorphological features and plant communities along the strand varies considerably from location to location. However, the relationship of communities to underlying environmental factors and the composition of those communities are remarkably constant throughout the Southeast (cf. Shaler 1885; Kearney 1900; Coker 1905; Lewis 1917; Penfound and O'Neill 1934; Kurz 1942; Penfound 1952; Brown 1959; Adams 1963; Au 1974; Godfrey and Godfrey 1976; Stalter and Odum 1993). I therefore organize this discussion around a hypothetical transect from the sea strand on a typical barrier island (see Fig. 11.15), across that island, into the sound, and onto the adjacent mainland.

Vegetation Composition and Structure

Strand vegetation consists of an assemblage of short-lived plants whose spatial distribution shifts from season to season and year to year. Many of these species are salt-tolerant and have life history characteristics that allow them to invade suitable habitat when it becomes available.

The principal grass that stabilizes dunes is sea oats, Uniola paniculata (Fig. 11.16). Once established, this grass produces numerous adventitious roots from the culm. Continued apical growth and lateral rhizome and root production allow the plant to stabilize the sand that accumulates at its base. In the absence of inundation or overwash, a dune is formed. Iva imbricata, Physalis maritima, Croton punctatus, and Euphorbia polygonifolia grow among the sea oats culms

In the lee of the foredune, species richness increases. Several grass species dominate this area, including Spartina patens, Andropogon scoparius var. littoralis, Distchlis spicata, Panicum spp., and Cenchrus tribuloides. Among the most common herbs are Hydrocotyle bonariensis, Oenothera humifusa, and Opuntia compressa. Boyce (1954) noted that many

Figure 11.17. Maritime scrub vegetation in dune slack region of Shackleford Bank. Note the salt-spray pruning of vegetation. Baccharis halimifolia, Ilex vomitoria, Juniperus virginiana, and Pinus taeda are present. The relatively salt-tolerant Baccharis probably established first, creating a refuge in its lee for other species. The dominant grass in the surrounding area is Andropogon scoparius var. littoralis.

Table 11.9. Basal area (BA) of woody species in an old-growth maritime forest on Fort Smith Island, North Carolina

Species	BA (m² ha⁻¹)
Quercus virginiana	25.19
Persea borbonia	3.45
Carpinus caroliniana	2.23
Juniperus virginiana	1.78
Pinus taeda	1.50
Ilex vomitoria	1.30
Osmanthus americana	1.09
Ilex opaca	0.85
Sabal palmetto	0.69
Cornus florida	0.65
Prunus caroliniana	0.65
Morus rubra	0.56
Quercus nigra	<0.50
Callicarpa americana	<0.50
Nyssa sylvatica	<0.50
Oxydendrum arboreum	<0.50
Myrica cerifera	<0.50
Aralia spinosa	<0.50
Acer rubrum	<0.50
Total	40.64

Source: Data from Bourdeau and Oosting (1959).

species commonly found in old fields and ruderal habitats in nonmaritime locations also occur here; these include *Andropogon virginicus, Erigeron canadensis, Heterotheca subaxillaris, Solidago* spp., and *Aster* spp.

Oosting and Billings (1942), Oosting (1945), and Boyce (1954) demonstrated experimentally that dune ridge species, such as *Uniola paniculata*, were much more tolerant of salt aerosol impaction than species typically found in dune slack areas.

Freshwater habitats occur at the heads of tidal creeks and in areas where peat or clay lenses result in perched water tables. Taxa such as *Fimbristylis* spp., *Cladium jamaicense, Juncus* spp., and *Typha latifolia* are dominant. However, these communities often include a diverse assemblage of herbs (Odum, Smith, Hoover, and McIvor 1984). Such habitats are important resources for barrier island animals (Engels 1952; Rubenstein 1981).

Near the foredunes, stunted salt-pruned shrubs may be found in protected swales. The pattern of establishment is closely tied to salt-spray tolerance. *Iva imbricata* or *Baccharis halimifolia* are often early invaders. Other species, such as *Ilex vomitoria, Juniperus virginiana, Myrica* spp., and *Quercus virginiana* may then become established in the lee of early invaders (Fig. 11.17, Boyce 1954; Oosting 1954).

With reduced salt impaction and increasingly stable environments toward the mainland side of the island, these maritime heaths expand first into dwarf woodland and then into closed canopy forests (Table 11.9). *Pinus taeda, Persea borbonia,* and *Magnolia virginiana* are particularly common.

For regional differences compare Wells (1939), and Penfound and O'Neill (1934), Kurz (1942), Bourdeau and Oosting (1959), Johnson and Barbour (1990) and Barbour et al. (1987). A number of taxa with subtropical affinities that are widespread in Florida and the Gulf states extend north into the Carolinas and Virginia only in this community. These include *Sabal palmetto, Sabal minor,* and *Osmanthus americana.*

Wells (1939) referred to the maritime forest as the "salt spray climax," suggesting that salt aerosol prevented the usual inland sere from taking place. Wells and Shunk (1938) found that sandhill species such as *Aristida stricta* could thrive in these habitats when protected from such aerosol. Bourdeau and Oosting (1959) found that – aside from a modest increase in organic matter and water retention with successional age – soils from young dunes, shrub zones, and mature maritime forest were physically and chemically similar.

Considerable attention has been paid to the ecology of lagoons and marshes (Teal 1962; Peterson and Peterson 1979; Bahr and Lanier 1981; Gosselink, Bailey, Conner, and Turner 1979). Space limitations permit only a brief overview of vegetation patterns abstracted from Wells (1932), Penfound and Hathaway (1938), Penfound (1952), Hinde (1954), Adams (1963), and Johnson, Hillsted, Shanholtzer, and Shanholtzer (1974).

Vascular plants are generally absent from those areas of the lagoon that are constantly inundated. However, in shallow, low-energy depositional environments, eel grass (*Zostera marina*) may form dense stands. Saltmarsh covers vast stretches of the intertidal portions of these sounds, often dominated by dense swards of a single species, *Spartina alterniflora* (Fig. 11.18). This plant has remarkable tolerance to the widely varying salinity environment of the marsh (Longstreth and Strain 1977). On less frequently inundated flats or pannes, where evaporation may concentrate salts, strict halophytes such as *Salicornia virginica, Sueda linearis*, and *Distichlis spicata* dominate. Bordering the edge of the salt marsh, in an area inundated at only the highest tides, is a zone of black needle rush (*Juncus roemerianus*). Plants in this zone must deal with considerable soil salinity but benefit from better soil aeration (Hinde 1954; Adams 1963). Between the *Juncus* and the maritime forest are a diverse assemblage of herbs, many of which are also found in dune slack habitats.

Community Dynamics and Succession

Natural disturbance and succession in these ecosystems are closely tied to coastal geomorphic processes. Rising sea level (or coastal subsidence) has resulted in a general erosion of island foredunes and sediment deposition across barrier islands. Thus, in many localities active wind-driven dunes are devouring maritime scrub, and islands are advancing into lagoons (see Godfrey and Godfrey 1976; Dolan, Hayden, and Lins 1980; and Pilkey et al. 1980 for reviews of the dynamics of these processes). As evidence of this island movement, it is not uncommon to find dead snags of *Juniperus virginiana* emerging from the foredunes or strand, or strata of lagoonal peat being eroded along the beach strand. The rate of island turnover may be as brief as 200–400 yr in many places (Au 1974; Godfrey and Godfrey 1976).

Among the lasting reminders of the human habitation of barrier islands during the seventeenth and eighteenth centuries are large populations of feral animals, including pigs, goats, cattle, and horses. There is little doubt that these animals have altered the structure of the herb communities, particularly in the dune slack and near freshwater marshes (Engels 1952; Au 1977; Rubenstein 1981); but Baron (1982) suggested that their impact is small compared to natural disturbance processes.

Human development of these coastal habitats during the past century and especially in the past three decades has had an important impact on barrier island ecosystems (see Pilkey et al. 1980). Although much of this development is concentrated on the strand side of islands, it has directly resulted in dune erosion allowing salt spray to carry farther inland, causing deterioration of otherwise undisturbed maritime forest (e.g. Boyce 1954).

Attention has also been focused on the colonization and subsequent succession of river delta sediments (Johnson, Sasser, and Gosselink 1985;

Figure 11.18. Salt marsh near Beaufort, North Carolina. Spartina-dominated zone in the lower left is bordered by a narrow Juncus zone and then an infrequently indundated zone dominated by salt-tolerant grasses and shrubs.

Table 11.10. *References, locations, and community types for studies included in Fig. 11.19; community type refers to the nomenclature used in each study*

Sample number	Reference	Location	Community type
1	Blaisdell et al. (1974)	North Florida	Mesic Hammock
2	"	"	Mesic Hammock
3	Ehrenfeld and Gulick (1981)	New Jersey	Swamp Hardwoods
4	Gemborys and Hodgkins (1971)	Alabama	Stream-bottom Hardwoods
5	"	"	Upland-margin Hardwoods
6	Christensen (unpublished)	North Carolina	Pocosin
7	Quarterman and Keever (1962)	Southeast	Group I Hardwoods
8	Snyder (1980)	North Carolina	Xeric Flatwoods
9	"	"	Pine Savanna
10	"	"	Pocosin
11	"	"	Bottomland Hardwoods
12	Laessle (1942)	North Florida	River Swamp
13	"	"	Sand Pine Scrub
14	"	"	Sandhill
15	"	"	Xeric Hammock
16	"	"	Mesic Hammock
17	"	"	Hydric Hammock
18	"	"	Bayhead
19	Edmisten (1963)	Central Florida	Longleaf Pine Flatwoods
20	"	"	Pond Pine Flatwoods
21	"	"	Slash Pine Flatwoods
22	White (1983)	South Louisiana	Cypress-tupelo Swamp
23	"	"	Bottomland Forest
24	Bourdeau and Oosting (1959)	North Carolina	Maritime Forest
25	Kurz (1942)	Central Florida	Scrub
26	Porcher (1981)	South Carolina	Swamp Forest
27	"	"	Hardwood Bottom
28	"	"	Ridge Bottom
29	"	"	Mixed Mesophytic
30	Monk (1965)	North Florida	Cypress Dome
31	"		Pocosin
32	Monk (1966b)	North Florida	Bayhead
33	"	"	Mixed Hardwood Swamp
34	Marks and Harcombe (1981)	East Texas	Sandhill Pine
35	"	"	Upland Pine-Oak
36	"	"	Wetland Pine Savanna
37	"	"	Upper Slope Oak-pine
38	"	"	Mid-Slope Oak-pine
39	"	"	Lower Slope HW-pine
40	"	"	Floodplain Hardwoods
41	"	"	Flatwood Hardwoods
42	"	"	Baygall Thicket
43	"	"	Cypress-tupelo Swamp
44	Schlesinger (1978a)	Georgia	Cypress Swamp Forest
45	Veno (1976)	North Florida	Pine Oak Forest
46	"	"	Xeric Hammock
47	"	"	Mesic Hammock
48	"	"	Sandhill
49	Hall and Penfound (1943)	Alabama	Cypress-gum Swamp
50	Parsons and Ware (1982)	Virginia	"Dry" Alluvial Swamp
51	"	"	"Wet" Alluvial Swamp
52	Schlesinger (1976)	Georgia	Cypress Swamp
53	Kologiski (1977)	North Carolina	Pocosin
54	"	"	Pine Savanna
55	"	"	White Cedar Swamp
56	"	"	Evergreen Bay Forest
57	"	"	Deciduous Bay Forest

Table 11.10. *(cont.)*

Sample number	Reference	Location	Community type
58	Abrahamson et al. (1984)	"	Southern Ridge Sandhill
59	Coile (unpublished)	North Florida	Sand Pine Scrub
60	Abrahamson et el. (1984)	Central Florida	Southern Ridge Sandhill
61	"	"	Sand Pine Scrub
62	"	"	Sand Pine Scrub
63	"	"	Scrubby Flatwoods
64	"	"	Wiregrass Flatwoods
65	"	"	Palmetto Flatwoods
66	"	"	Gallberry Flatwoods
67	"	"	Bayhead
68	Wilbur (1985)	"	Pocosin
69	Whipple et al. (1981)	South Carolina	Oak Hickory Forest
70	"	"	Gum-red Bay Forest
71	"	"	Gum-red Maple Forest
72	"	"	Black Oak Forest
73	"	"	Laurel Oak Forest
74	"	"	Gum-ash Forest
75	"	"	Cypress-gum Swamp
76	Hilmon (1968)	Florida	Palmetto Flatwoods
77	Weaver (1969)	North Carolina	Mesic Sandhill
78	"	"	Ridge Sandhill
79	Allen (1958)	North Carolina	Swamp Tupelo-Cypress
80	"	"	Water Tupelo-Cypress
81	"	"	Bar Forest
82	Applequist (1959)	North Carolina	Cypress-gum Swamp
83	"	"	Tupelo-gum Swamp
84	Cypert (1972)	Georgia	Cypress Swamp
85	Hebb and Clewell (1976)	North Florida	Slash Pine Forest
86	McAlister (unpublished)	North Carolina	Sand Ridge
87	Caplenor (1968)	Mississippi	Hardwoods, Thick Loess
88	"	"	Hardwoods, Creek Bottom
89	"	"	Hardwoods, Non Loess
90	"	"	Hardwoods, Thin Loess
91	Ware (1970)	Virginia	Mesophytic Hardwoods
92	Christensen (unpublished)	North Carolina	Sand Ridge
93	"	South Carolina	Sand Ridge
94	Quarterman and Keever (1962)	Georgia	Southern Mixed Hardwoods
95	"	Georgia	"
96	"	South Carolina	"
97	"	Georgia	"
98	"	Georgia	"
99	"	Mississippi	"
100	"	Alabama	"
101	"	"	"
102	"	Louisiana	"
103	"	South Carolina	Southern Mixed Hardwoods
104	"	"	"
105–24	DeWitt and Ware (1979)	Virginia	Southern Mixed Hardwoods
125	Laney and Noffsinger (1985)	North Carolina	Pocosin
126	"	"	"
127	"	"	Bay
128	"	"	"
129	"	"	White Cedar Swamp
130	"	"	"
131	"	"	"
132	Conner and Day (1976)	Louisiana	Alluvial Swamp
133	Muzika et al. (1987)	South Carolina	Alluvial Swamp

(Cont.)

Table 11.10. *(Cont.)*

Sample number	Reference	Location	Community type
134	Jones and Gresham (1985)	South Carolina	Savanna
135	"	"	Pocosin
136	"	"	Bay
137	"	"	Sandy Alluvial Swamp
138	"	"	Red Water Swamp
139	"	"	Alluvial Swamp
140	Matos and Rudolph (1985)	Texas	Floodplain Forest
141	"	"	Alluvial Swamp
142	"	"	Baygall
143	"	"	"
144	"	"	Wet Transition Forest
145	"	"	Dry Transition Forest
146			Southern Mixed
	Golley et al. (1965)	South Carolina	Hardwoods

Neill and Deegan 1985). Mudflats are initially invaded by salt or freshwater marsh, depending on tidal flows and location relative to a river. Such areas accumulate additional sediments and may be invaded by woody vegetation. These areas subsequently undergo erosion and subsidence and, if abandoned by the river, may be invaded by brackish marsh vegetation. With continued erosion and subsidence, the delta lobe may deteriorate completely. This may be completed in about 4000 yr (Neill and Deegan 1985).

SYNTHESIS AND FUTURE RESEARCH NEEDS

With the single exception of Quarterman and Keever (1962), studies in Coastal Plain vegetation have been confined to communities and gradients within specific regions, such as the fall line sandhills, the central Florida highlands, and the Texas Big Thicket. Considerable amounts of unstudied territory separate localities that have received intensive study. Comparisons are made more difficult because of variations in sampling methods, measures of abundance, and community taxonomy.

With these difficulties in mind, I analyzed community data taken from studies throughout the Coastal Plain (Table 11.10). I included in this analysis only studies that reported data on species abundances. I excluded studies from strictly herb-dominated communities, such as grass-sedge marshes and salt marshes. Because sampling procedures and abundance measures varied among studies, I included only the 12 most abundant species from each study and assigned each an impor-

tance value (IV) based on the percentage of total community abundance accounted for by that species. For species accounting for >10%, 1–10%, and <1% of total community abundance, IV was set equal to 3, 2, and 1, respectively. These data were then ordinated using the DECORANA program for detrended correspondence analysis (e.g., see Hill and Gauch 1980; Gauch 1982). The ordination of sites appears in Fig. 11.19. The first axis corresponds in a general fashion to successional schemes proposed by Wells (1928), Laessle (1942), and Kurz (1944). Mixed hardwoods have low first-axis scores, whereas chronically disturbed (burned) communities, such as sandhills, sand pine scrubs, flatwoods, and pocosins, had high scores. However, this axis also corresponds to a gradient in soil fertility. Communities on coarse sterile sands or on nutrient-deficient wetlands have high scores, whereas those on fertile soils or in floodplains that receive nutrient subsidy have low scores. As would be expected from Monk's (1968) demonstration of the correlation between soil infertility and the importance of evergreen plants, the proportion of evergreens relative to deciduous plants increases with increasing first-axis score.

The second axis separated wetlands from uplands. Thus, beginning in the lower right corner of the ordination diagram, flatwoods and pocosins grade into bayheads, cypress heads, and white cedar swamps, and these in turn grade into Zone II, III, and IV alluvial wetlands. Zone IV alluvial wetlands are intermixed with communities classified as southern mixed hardwoods, (these two were clearly separated on the third ordination axis not shown here). Starting in the upper right-hand corner of the ordination, xeric sand ridges grade into sandhills and savannas, then into xeric oak woods, mixed pine-oak stands, and finally into southern mixed hardwoods.

What conclusions can we draw from this ordination? First, community types identified with specific habitats are quite similar, in composition, even though some sites were over 2000 km separate from each other. This provides hope for the development of a more uniform system of classification for the region. Second, this ordination also verifies the gradient nature of compositional change on this landscape. However, the ordination masks our ignorance of the details of variation within particular community types, either within a particular locality or across the region. We can begin to appreciate the nature and significance of these variations only with a more systematic effort to connect the intensively studied areas with extensive sampling. Given the rate of economic devel-

opment of this region, time is running out for this effort.

I have discussed the foregoing relationships between vegetation and environmental gradients as though we really understood the mechanisms underlying these relations. Nothing could be further from the truth. For example, we assume that vegetation varies with respect to gradients of soil fertility, but few data are available to indicate exactly which nutrients are limiting. Correlations between soil calcium or pH and evergreenness convince us that nutrients are regulating the composition of communities, but calcium and pH are not the proximal causes of variation in plant performance. Similarly, we can identify obvious moisture gradients, but few studies have either characterized water availability as it might be perceived by plants or compared plants with regard to efficiency of water use.

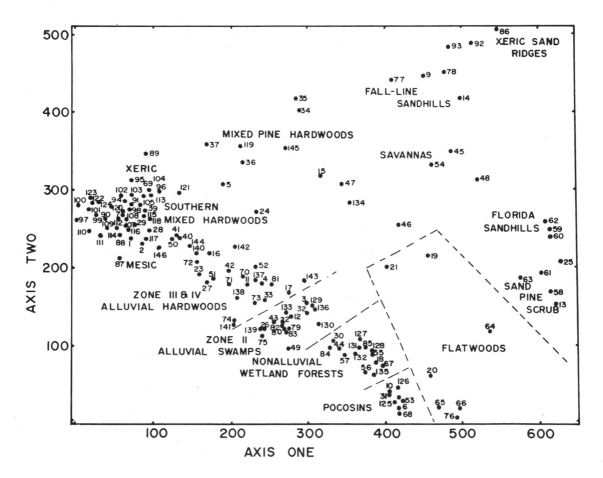

Figure 11.19. Ordination of vegetation samples from throughout the Southeast. The numbers refer to studies described in Table 11.10.

The southeastern Coastal Plain is an underused laboratory for testing theories of disturbance, succession, and landscape development. Coastal Plain ecologists were among the first to forsake the notion of the climatic monoclimax (Harper 1914; Wells 1942; Laessle 1942), realizing that it was useless to consider succession in the absence of natural disturbance cycles, especially fire. Furthermore, they were among the first ecologists to realize that such disturbance cycles might be regulated by features of the community itself. This is not to say that the seral nature of various communities has been settled beyond debate. The Coastal Plain offers community ecologists a wide range of fire regimes, from short-return, low-intensity fires of savannas to unpredictable conflagrations of swamp forests. Future studies might well focus on the factors that regulate such fire cycles, or on comparative studies of the consequences of differences in fire regimes, or on the role of disturbance on the evolution of plant life-history characteristics.

Delcourt and Delcourt (1991) point out that the Coastal Plain flora has been relatively constant for the past 9000 yr. We are beginning to appreciate that this regional stability is the consequence of a dynamic equilibrium between local successional trends, the unique variety of Coastal Plain environments, and chronic disturbance.

CONSERVATION CHALLENGES

Perhaps because of the subtlety and complexity of its vegetation gradients, or because of the long tenure and significance of human impact, little attention had been given to the conservation of southeastern Coastal Plain ecosystems until the past two decades. Much of the presettlement biological pattern and diversity of these landscapes has been greatly altered, and once-widespread ecosystem types are now greatly restricted in distribution. This is particularly true of some of the most unique ecosystem types. For example, less than half of the nearly 2M ha of paludal wetlands that dominated lower Coastal Plain landscapes remain intact. Even more remarkable, <3% of the Coastal Plain upland landscape remains in anything near its presettlement state (Croker 1987; Frost 1993, 1995). Many Coastal Plain species have been listed as threatened or endangered under the aegis of the Endangered Species Act. Important conservation challenges on Coastal Plain landscapes include habitat alteration and restoration, altered disturbance regimes, landscape fragmentation, humans as ecosystem components, and development of sustainable resource use strategies.

Habitat Alteration and Restoration

Recent legislation and judicial decisions have focused attention on the delineation and alteration of wetlands (e.g., protection under section 404 of the Clean Water Act). Indeed, in 1990 the Environmental Protection Agency and the Army Corps of Engineers agreed that there should be "no overall loss of [wetland] values and functions." In 1985, the President's Council on Environmental Quality stated that "the bottomland wetlands of the Southeast are of such importance as wildlife habitats, and are becoming so scarce, that the principle of full, in-kind replacement should override other considerations." Interest in wetland alteration and restoration has increased owing to mitigation banking programs in some states that allow some wetlands to be developed in exchange for restoration of degraded wetlands elsewhere (Mitsch and Goselink 1986).

This is an example of policy being considerably ahead of science and management know-how. We are just beginning to understand the specific relationships between hydrologic regimes and ecosystem structure, and our knowledge of the functional variations among wetland types (i.e., the ability of wetlands to retain nutrients and pollutants or buffer hydrologic flows) is rudimentary at best (Mitsch and Goselink 1986; Richardson and Gibbons 1993; Sharitz and Mitsch 1993). Very little is known regarding actual restoration strategies for most wetland types.

Because it is the primary habitat for a number of endangered species (most especially the red cockaded woodpecker), alteration and restoration of longleaf pine forests have been of particular interest to conservation biologists. Much is known about the regeneration of longleaf pine (e.g., Wahlenburg 1946; Boyer 1993; Farrar 1993), but attempts to restore the often diverse herb community have met only limited success (Myers 1993).

Altered Disturbance Regimes

During the first half of the twentieth century, natural area conservation was implicitly treated as a form of museum curation on a large scale (Christensen 1997). The goal was preservation of "vignettes of primitive America" (Leopold et al. 1964). Although never explicitly stated as such, the biological diversity generally thought worthy of preservation was that associated with so-called climax communities. The putative climatic climaxes of the southeastern Coastal Plain comprise a rather small fraction of the region's biodiversity; most of it is associated with ecosystems undergoing constant change in response to myriad disturbance regimes.

The importance of variations in the type, frequency, and intensity of natural disturbances to the biological diversity of Coastal Plain vegetation has been a recurring theme in this chapter. Although growing, our understanding of the specific disturbance regimes necessary to maintain ecosystem gradients remains limited. For example, it is certainly understood that variations in structure and composition among certain paludal wetland ecosystems or xeric pine communities are a consequence of variations in the season, frequency, and severity of fires. However, we are only now beginning to use this knowledge to develop specific prescribed fire strategies.

Human activities and patterns of land use often present difficult conflicts with the maintenance of disturbance regimes. For example, considerable energy and money are devoted to strategies to stabilize coastal ecosystems, even though we understand the inevitability and importance of the dynamics of barrier islands driven by hurricanes and rising sea level (Pilkey and Dixon 1996).

It is ironic that human interventions aimed at diminishing risks of disturbance in the short term often increase risks of catastrophe in the long term. Thus, seawalls and groins designed to minimize erosion and sediment transport in coastal areas may increase the vulnerability of these areas to catastrophic storm events (Pilkey and Dixon 1996). Similarly, exclusion of fire from many forest ecosystems results in the accumulation of flammable fuels that increase the likelihood of intense fire events (Christensen 1992).

Landscape Fragmentation

One of the most profound human impacts on Coastal Plain ecosystems is the fragmentation of the landscape mosaic into smaller and more clearly delineated patches. This trend is, of course, not unique to this region, but its impacts here are magnified by the importance of natural disturbance and successional change.

Island biogeographic theory argues that the species richness of a locality is a consequence of the balance between the immigration of new species and the local extinction of resident species. Populations are generally larger in larger preserves and less apt to go extinct. Thus, an important strategy for preservation of biological diversity is to make preserves as large as possible.

Where natural disturbance and successional change are the rule, these processes of immigration

and extinction in a given patch occur at a relatively rapid rate and depend heavily on the character of the landscape surrounding each patch. For example, recall that the development of sand pine scrub or sandhill vegetation depends on the relative proximity of established communities of these types to provide seed (Myers 1985). Thus, it is impossible to preserve such communities by simply delineating "type specimens" scattered widely across the landscape.

On the Coastal Plain, a landscape approach – that is, an approach that acknowledges the importance of spatial relationships and temporal dynamics to preserve design and management – is essential. However, the complexity of land ownership and often arbitrary boundaries of management jurisdiction in this region present formidable challenges to this approach. Less than 5% of the land base is in public ownership, and the boundaries of land ownership and management jurisdictions often have little to do with the spatial extent or behavior of ecosystem processes such as fire or hydrologic flows, or the distribution and migratory patterns of organisms (Christensen et al. 1996). Successful conservation of this region's biological diversity will demand collaborative approaches to reconcile management goals across ownership boundaries. Experiments in such collaboration among nongovernmental organizations, private industry, and public agencies are currently under way throughout the region.

Humans as Ecosystem Components

The temporal scale of human activity on landscapes is nearly the same as the time during which the southeastern vegetation has readjusted from the last Ice Age. For millennia, human activities have been a major factor shaping the structure and driving the dynamics of Coastal Plain landscapes. More than ever, humans are an inevitable and, some would argue, essential part of those landscapes. Having said this, we must also acknowledge that the role of historical and current human activities in the maintenance of Coastal Plain ecosystems is at best poorly understood and rarely explicitly acknowledged in the formulation of conservation plans or strategies.

Although the potential importance of historical human activities on southeastern landscapes has been acknowledged, such effects have received comparatively little study. Indeed, ecologists often explicitly avoid study of areas known to have had significant past human use (Christensen 1989). Historically, some notion of "naturalness" forms the core of management goals for preserves on most landscapes; the absence of human impacts is implicit in such notions (Christensen et al. 1996).

The inescapable reality of virtually all Coastal Plain landscapes is that human impacts are not only ubiquitous and increasing in scope and complexity, but also they have historical precedents over millennial time scales. It is essential that the historic role of humans, as well as the inevitable and increasing impacts of current human populations, be understood and explicitly incorporated in preserving designs and management goals and protocols.

Development of Sustainable Resource Management Strategies and Future Research Needs

Human demands on the natural resources of the southeastern Coastal Plain are growing rapidly owing to increased human population and changes in land use elsewhere. As metropolitan areas grow, once-rural landscapes are becoming increasingly suburbanized and fragmented. This has altered not only the behavior and extent of natural disturbance and successional processes but also has constrained management options with respect to those disturbances. For example, the use of prescribed fire as a surrogate for natural fire regimes is severely limited in many areas because of risks to human life and property and impact on air quality (Christensen et al. 1996). We understand very little about the ecological consequences of substituting human disturbances such as timber activities or grazing for natural disturbance processes.

During the past decade, the southeastern Coastal Plain region has become the most important source of wood fiber for the entire United States. This has happened in part as timber supply from other regions such as the Pacific Northwest has diminished owing to increased forest conservation (Christensen et al. 1996). The dynamic character of these landscapes has in many ways facilitated this transition in that early successional, even-aged pine forests represent the most important source of Coastal Plain timber. Old-growth forest types are far less important than in other regions of North America.

Nevertheless, the general trend in the management of pine forests has been toward shorter rotations and increased emphasis on plantation-style pine monoculture. The impact of this trend on the productivity of individual sites, as well as key ecosystem processes and biological diversity over large landscapes, remains poorly understood. Some argue that plantation forestry increases production on one portion of the land base, releasing other portions of the landscape for other management goals. Others argue that this can be true only if there is planning for land use at the landscape level and coordination and cooperation across ownerships.

In truth, few data are available on the impact of human management practices across scales of space and time. Most managers would acknowledge that practices viewed as sustainable at the level of a single stand might be unsustainable when their accumulated effects are taken over a watershed. Very little quantitative information is available on the trade-offs between fiber production and other ecosystem values across the range of management options. Research in this area must be a high priority.

REFERENCES

Abrahamson, W. G. 1984a. Post-fire recovery of Florida Lake Wales Ridge vegetation. Amer. J. Bot. 71:9–21.

Abrahamson, W. G. 1984b. Species responses to fire on the Florida Lake Wales Ridge. Amer. J. Bot. 71:35–43.

Abrahamson, W. G., and D. C. Hartnett. 1990. Pine flatwoods and dry prairies, pp. 103–149 in R. L. Myers and J. J. Ewel (eds.), Ecosystems of Florida. The University of Central Florida Press, Orlando.

Abrahamson, W. G., J. N. Layne, A. F. Johnson, and P. A. Peroni. 1984. Vegetation of the Archbold Biological Station, Florida: an example of the southern Lake Wales Ridge. Fla. Sci. 47:209–250.

Adams, D. A. 1963. Factors affecting vascular plant zonation in North Carolina salt marshes. Ecology 44:445–455.

Allen, P. H. 1958. A Tidewater swamp forest and succession after clearcutting. Master's thesis, Duke University, Durham.

Andresen, J. W. 1959. A study of pseudo-nanism in *Pinus rigida* Mill. Ecol. Monogr. 29:309–322.

Applequist, M. B. 1959. A study of soil and site factors affecting growth and development of swamp blackgum and tupelogum stands in southeastern Georgia. Ph.D. dissertation, Duke University, Durham.

Arata, A. A. 1959. Effects of burning of vegetation and rodent populations in a longleaf pine turkey oak association in north central Florida. Quar. J. Fla. Acad. of Sci. 22:94–104.

Archard, H. O., and M. F. Buell. 1954. Life form spectra of four New Jersey pitch pine communities. Bull. Torr. Bot. Club 81:169–175.

Art, H. W. 1976. Ecological studies of the sunken forest, Fire Island National Seashore, New York. U.S. National Park Service Scientific Monograph No. 7.

Ash, A. N., C. B. McDonald, E. S. Kane, and C. A. Pories. 1983. Natural and modified pocosins: literature synthesis and management options. U.S. Fish and Wildlife Service Report FWS/OBS-83/04.

Ashe, W. W. 1915. Loblolly or North Carolina pine. North Carolina Geological and Economic Survey Bulletin No. 24.

Au, S. 1974. Vegetation and ecological processes on

Shackleford Bank, North Carolina. U.S. National Park Service Scientific Monograph No. 6, 86 pp.

Axelrod, D. I., and H. P. Bailey. 1969. Paleotemperature analysis of Tertiary floras. Paleogeog. Paleoclima. and Paleoecol. 6: 163–195.

Bahr, L. M., and W. P. Lanier. 1981. The ecology of intertidal oyster reefs of the south Atlantic coast: a community profile. U.S. Fish and Wildlife Service Report FWS/OBS-81/15.

Barbour, M. G., M. Rejmanek, A. F. Johnson, and B. M. Parlik. 1987. Beach vegetation and plant distribution patterns along the northern Gulf of Mexico. Phytocoenologia 15:201–233.

Baron, J. 1982. Effects of feral hogs (Sus scrofa) on the vegetation of Horn Island, Mississippi. Ameri. Midl. Natur. 107:202–205.

Bartram, W. 1791. Travels Through North and South Carolina, Georgia, East and West Florida, the Cherokee country, the Extensive Territories of the Muscogulges or Creek Confederacy, and the Country of the Choctaws: Containing an Account of the Soil and Natural Productions of Those Regions, Together with Observations on the Manners of the Indians. James and Johnson, Philadelphia.

Beaven, G. F., and H. J. Oosting. 1939. Pocomoke Swamp: a study of a cypress swamp on the Eastern Shore of Maryland. Bull. Torr. Bo. Club 66:367–389.

Beckwith, S. L. 1967. Chinsegut Hill–McCarty Woods, Hernando County, Florida Quar. J. Fla. Acad. of Sci. 30:250–268.

Blaisdell, R. S., J. Wooten, and R. K. Godfrey. 1974. The role of magnolia and beech in forest processes in the Tallahassee, Florida, Thomasville, Georgia area. Proceedings of the Annual Tall Timbers Fire Ecology Conference 13:363–397.

Boerner, R. E. J. 1981. Forest structure dynamics following wildfire and prescribed burning in the New Jersey pine barrens. Ameri. Midl. Nat. 105: 321–333.

Boerner, R. E. J. 1983. Nutrient dynamics of vegetation and detritus following two intensities of fire in the New Jersey pine barrens. Oecologia 59: 129–134.

Boerner, R. E. J., and R. T. T. Forman. 1982. Hydrologic and mineral budgets of New Jersey Pine Barrens upland forests following two intensities of fire. Can. J. For. Res. 12:503–510.

Bourdeau, P. F., and H. J. Oosting. 1959. The maritime live oak forest in North Carolina. Ecology 40: 148–152.

Boyce, S. G. 1954. The salt spray community. Ecol. Monogr. 24:29–67.

Boyer, W. D. 1993. Regenerating longleaf pine with natural seeding. Proceedings of the Tall Timbers Fire Ecology Conference 18:299–310.

Braun, E. L. 1950. Deciduous forests of eastern North America. Blakiston, C Philadelphia.

Braun, E. L. 1955. The phytogeography of unglaciated eastern United States and its interpretation. Bot. Rev. 21:297–375.

Bridgham, S. D., and Richardson, C. J. 1993. Hydrology and nutrient gradients in North Carolina Peatlands. Wetlands 13:207–218.

Brown, C. A. 1959. Vegetation of the Outer Banks of North Carolina. Louisiana State University Coastal Studies Series No. 4. Baton Rouge.

Brown, S. 1981. A comparison of the structure, primary productivity, and transpiration of cypress ecosystems in Florida. Ecolog. Monogr. 51: 403–427.

Brown, S. L., E. W. Flohrschutz, and H. T. Odum. 1984. Structure, productivity, and phosphorus cycling of the scrub cypress ecosystem, pp. 304–317 in K. C. Ewel and H. T. Odum (eds.), Cypress swamps. University of Florida Press, Gainesville.

Buchholz, K., and R. E. Good. 1982a. Compendium of New Jersey Pine Barrens literature. Center for Coastal and Environmental Studies, Rutgers University, New Brunswick, N.J.

Buchholz, K., and Good, R. E. 1982b. Density, age structure, biomass, and net annual aboveground productivity of dwarfed Pinus rigida Mill. from the New Jersey barren plains. Bull. Torr. Bot. Club 109: 24–34.

Buell, M. F. 1939. Peat formation in the Carolina Bays. Bull. Torr. Bot. Club 66:483–487.

Buell, M. F. 1946. Jerome Bog, a peat-filled "Carolina Bay." Bull. Torr. Bot. Club 73: 24–33.

Buell, M. F., and R. L. Cain. 1943. The successional role of southern white cedar, Chamaecypris thyoides, in southeastern North Carolina. Ecology 24:85–93.

Buell, M. F., and Cantlon. J. E. 1950. A study of two communities of the New Jersey pine barrens and a comparison of methods. Ecology 31:567–586.

Buol, S. W. 1973. Soils of the southern states and Puerto Rico. Agricultural Experiment Stations of the Southern States and Puerto Rico Land-Grant Universities, Southern Cooperative Series Bulletin No. 174.

Burns, R. M. 1973. Sand pine: distinguishing characteristics and distribution. In Sand Pine Symposium Proceedings, pp. 13–22. USDA Forest Service General Technical Report SE-2.

Byrd, W. 1728. Histories of the dividing line betwixt Virginia and North Carolina. (Facsimile ed.). Dover, New York.

Caplenor, D. 1968. Forest composition on loessial and non-loessial soils in west-central Mississippi. Ecology 49: 322–331.

Catesby, M. 1654. The natural history of Carolina, Florida and the Bahama Islands. C. March, London.

Chapman, H. H. 1926. Factors determining natural reproduction of longleaf pine on cut-over lands in LaSalle Parish, La. Yale University School of Forestry Bulletin No. 16.

Chapman, H. H. 1932. Is the longleaf type a climax? Ecology 13:328–334.

Christensen, N. L. 1976. The role of carnivory in Sarracenia flava L. with regard to specific nutrient deficiencies. J. Elisha Mitchell Sci. Soc. 92: 144–147.

Christensen, N. L. 1977. Fire and soil-plant nutrient relations in a pine wiregrass savanna on the Coastal Plain of North Carolina. Oecologia 31: 27–44.

Christensen, N. L. 1979a. Shrublands of the southeastern United States. In Heathlands and related shrublands of the world, a. Descriptive studies, pp. 441–449 in R. L. Specht (ed.), Elsevier, Amsterdam.

Christensen, N. L. 1979b. The xeric sandhill and savanna ecosystems of the southeastern Atlantic Coastal Plain, U.S.A. Veroffentlichungen des Geobotanischen Institutes der Eidgenoessische Technische Hochschule Stiftung Rubel, in Zurich 68: 246–262.

Christensen, N. L. 1981. Fire regimes in southeastern ecosystems, pp. 112–136, in H. A. Mooney, T. M. Bonnicksen, N. L. Christensen, J. E. Lotan, and

W. A. Reiners (eds.), Fire regimes and ecosystem properties. USDA Forest Service General Technical Report WO-26.

Christensen, N. L. 1985. Shrubland fire regimes and their evolutionary consequences, pp. 85–100 in S. T. A. Pickett and P. S. White (eds.), The ecology of natural disturbance and patch dynamics. Academic Press, New York.

Christensen, N. L. 1989. Landscape history and ecological succession on the Piedmont of North Carolina. J. For. Hist. 33:116–124.

Christensen, N. L. 1992. Variable fire regimes on complex landscapes: ecological consequences, policy implications, and management strategies, pp. ix–xiii in T. Waldrop, Fire in the environment. USDA Forest Service General Technical Report SE-69

Christensen, N. L. 1993. The effects of fire on nutrient cycles in longleaf pine ecosystems. Proceedings of the Tall Timbers Fire Ecology Conference 18:205–214.

Christensen, N. L. 1997. Managing dynamic landscapes for heterogeneity and complexity. In S. T. A. Pickett and R. Ostveld (eds.), Chapman-Hall, Inc., New York.

Christensen, N. L., A. Bartuska, J. H. Brown, S. Carpenter, C. D'Antonio, R. Francis, J. F. Franklin, J. A. MacMahon, R. F. Noss, D. J. Parsons, C. H. Peterson, M. G. Turner, and R. G. Woodmansee. 1996. The scientific basis for ecosystem management. Ecolog. Applica. 6:665–691.

Christensen, N. L., R. B. Burchell, A. Liggett, and E. L. Simms. 1981. The structure and development of pocosin vegetation, pp. 43–61 in C. J. Richardson (ed.), Pocosin wetlands. Hutchinson Ross, Stroudsburg Pa.

Christensen, N. L., R. B. Wilbur, and J. S. McLean. 1988. Soil-vegetation correlations in pocosins of Croatan National Forest, North Carolina. U.S. Fish and Wildlife Service Biological Report 88 (28).

Clements, F. E. 1916. Plant succession: an analysis of the development of vegetation. Carnegie Institute of Washington, publ. No. 242, Washington, D.C.

Clewell, A. F., and D. B. Ward. 1985. White cedar forests in Florida and Alabama. In Proceedings of the Atlantic White Cedar Wetlands Symposium, ed. A. D. Laderman (in press).

Cohen, A. D. 1973. Petrology of some Holocene peat sediments from the Okefenokee Swamp–Marsh Complex of southern Georgia. Geolog. Soc. Amer. Bull. 84:3867–3878.

Cohen, A. D. 1974. Petrography and paleoecology of Holocene peats from the Okefenokee Swamp–Marsh Complex at Georgia. J. Sedimentary Petrology 44:716–720.

Cohen, A. D., M. J. Andrejko, W. Spackman, and D. Corrinus. 1984. Peat deposits of the Okefenokee Swamp, pp. 493–553 in A. D. Cohen, D. J. Casagrande, M. J. Andrejko, and G. R. Best (eds.), The Okefenokee swamp. Wetland Surveys, Los Alamos, N.M.

Coker, W. C. 1905. Observations on the flora of the Isle of Palms, Charleston, S. C. Torreya 5:135–145.

Colquhoun, D. J. 1969. Geomorphology of the lower coastal plain of South Carolina. State of South Carolina Division of Geology, State Development Board, Ms-15.

Conner, W. H., and J. W. Day, Jr. 1976. Productivity and composition of a bald cypress – water tupelo

site and a bottomland hardwood site in a Louisiana swamp. Amer. J. Bot. 63: 1354–1364.

Cooke, C. W. 1931. Seven coastal terraces in the southeastern states. Washington Acad. of Sci. J. 21:503–513.

Cook, C. W. 1933. Origin of the so-called meteorite scars of South Carolina. Washington Acad. Sci. J. 23: 569–570.

Cooper, R. W., C. S. Schopmeyer, and W. H. D. McGregor. 1959. Sand pine regeneration on the Ocala National Forest. USDA Prod. Res. Rep. 30.

Cowardin, L. M., V. Carter, F. C. Golet, and E. T. LaRoe. 1979. Classification of wetlands and deepwater habitats of the United States. U.S. Fish and Wildlife Service Report FWS/OBS-79/31.

Cowdrey, A. E. 1983. This land, this south: an environmental history. University Press of Kentucky, Lexington.

Croker, T. C., Jr. 1987. Longleaf pine, a history of man and a forest. USDA Forest Service Southern Forest Experiment Station Forestry Report R8-FR7.

Cronin, T. M., B. J. Szabo, T. A. Ager, J. E. Hazel, and J. P. Owens. 1981. Quaternary climates and sea levels of the U.S. Atlantic Coastal Plain. Science 211: 233–240.

Croom, J. M. 1978. Sandhills–Turkey Oak (*Quercus laevis*) Ecosystem: Community Analysis and a Model of Radiocesium Cycling. Ph.D. dissertation, Emory University, Atlanta, Ga.

Crum, H. A. and L. E. Anderson, 1981. Mosses of eastern North America. Columbia University Press, New York.

Cypert, E. 1961. The effects of fires in the Okefenokee Swamp in 1954 and 1955. Ameri. Midl. Nat. 66:485–503.

Cypert, E. 1972. The origin of houses in the Okefenokee prairies. Ameri. Midl. Nat. 87: 448–458.

Cypert, E. 1973. Plant succession on burned areas in Okefenokee Swamp following fires of 1954 and 1955. Proceedings of the Annual Tall Timbers Fire Ecology Conference 12: 199–217.

Dabel, C. V., and F. P. Day, Jr. 1977. Structural comparisons of four plant communities in the Great Dismal Swamp, Virginia. Bull. of the Torr. Bot. Club 104: 352–360.

Dachnowski-Stokes, A. P., and B. W. Wells. 1929. The vegetation, stratigraphy, and age of the "Open Land" peat area in Carteret County, North Carolina. Washington Acad. Sci. J. 19: 1–11.

Daniel, C. C., III. 1981. Hydrology, geology and soils of pocosins: a comparison of natural and altered systems, pp. 69–108 in C. J. Richardson, (ed.), Pocosin wetlands. Stroudsburg, Pa., Hutchinson Ross.

Daniels, R. B., E. E. Gamble, W. H. Wheeler, and C. S. Holzhey. 1977. The stratigraphy and geomorphology of the Hofmann Forest Pocosin. Soil Sci. Soc. Ameri. J. 41: 1175–1180.

Daniels, R. B., H. J. Kleiss, S. W. Buol, H. J. Byrd, and J. A. Phillips. 1984. Soil systems in North Carolina. North Carolina Agricultural Research Service, Bulletin 467, Raleigh.

Davis, J. H. 1943. The natural features of southern Florida, especially the vegetation, and the Everglades. Fla. Geolog. Surv. Bull. No. 25.

Davis, J. H. 1946. The peat deposits of Florida: their occurrence, development, and uses. Fla. Geolog. Sur. Bull. No. 30.

Davis, M. B. 1981. Quaternary history and the stability

of deciduous forests, pp. 132–177 in D. C. West, H. H. Shugart, and D. B. Botkin (eds.), Forest succession. Springer-Verlag, New York.

Davis, M. B. 1983. Holocene vegetational history of the eastern United States, pp. 166–181 in H. E. Wright, Jr. (ed.), Quaternary environments of the United States, Volume 2. The Holocene. University of Minnesota Press, Minneapolis.

Deghi, G. S. 1984. Seedling survival and growth rates in experimental cypress domes, in pp. 141–144 K. C. Ewel and H. T. Odum (eds.), Cypress swamps. University of Florida Press, Gainesville.

Delcourt, H. R. 1976. Presettlement vegetation of the North of Red River Land District, Louisiana. Castanea 41:122–139.

Delcourt, H. R., and P. A. Delcourt. 1974. Primeval magnolia – holly – beech climax in Louisiana. Ecology 55:638–644.

Delcourt, H. R., and P. A. Delcourt. 1977a. The Tunica Hills, Louisiana-Mississippi: late glacial locality for spruce and deciduous forest species. Quat. Res. 7: 218–237.

Delcourt, H. R., and P. A. Delcourt. 1977b. Presettlement magnolia-beech climax of the Gulf Coastal Plain: quantitative evidence from the Apalachicola River Bluffs, north-central Florida. Ecology 58:1085–1093.

Delcourt, H. R., and P. A. Delcourt. 1991. Quaternary ecology: a paleoecological perspective. Chapman Hall, New York.

Delcourt, P. A. 1980. Goshen Springs: late Quaternary vegetation record for southern Alabama. Ecology 61: 371–386.

Delcourt, P. A., and H. R. Delcourt. 1979. Late Pleistocene and Holocene distributional history of the deciduous forest in the southeastern United States. Veroffentlichungen des Geobotanischen Institute der Eidgenoessische Technische Hochschule Stiftung Rubel, in Zurich 68:79–107.

Delcourt, P. A., and H. R. Delcourt. 1980. Pollen preservation and Quaternary environmental history in the southeastern United States. Palynology 4:215–231.

Delcourt, P. A., and H. R. Delcourt. 1981. Vegetation maps for eastern North America: 40,000 YR B.P. to the present, pp. 123–165 in R. C. Romans (ed.), Geobotany II. Plenum Press, New York.

Delcourt, P. A., H. R. Delcourt, R. C. Brister, and L. E. Lackey. 1980. Quaternary vegetation history of the Mississippi Embayment. Quat. Res. 13:111–132.

Delcourt, P. A., H. R. Delcourt, D. F. Morse and P. A. Morse. 1993. History, evolution, and organization of vegetation and human culture, pp. 47–79 in W. H. Martin, S. G. Boyce, and A. C. Echternacht (eds.), Biodiversity of the Southeastern United States/Lowland Terrestrial Communities. Wiley, New York.

Demaree, D. 1932. Submerging experiments with Taxodium. Ecology 13:258–262.

DeWitt, R., and S. Ware. 1979. Upland hardwood forests of the central Coastal Plain of Virginia. Castanea 44:163–174.

Dickson, R. E., and T. C. Broyer. 1972. Effects of aeration, water supply, and nitrogen source on growth and development of tupelo gum and bald cypress. Ecology 53:626–634.

Dierberg, F. E., and P. L. Brezonik. 1984. Nitrogen and phosphorus mass balances in a cypess clome re-

ceiving waste water, pp. 112–118, in Cypress swamps. K. C. Ewel and H. T. Odum (eds.), University of Florida Press, Gainesville.

Dilcher, D. L. 1973a. Revision of the Eocene flora of southeastern North America. Paleobotanist 20:7–18.

Dilcher, D. L. 1973b. A paleoclimatic interpretation of the Eocene floras of southeastern North America, pp. 39–59, in A. Graham (ed.), Vegetation and vegetational history of north Latin America. Elsevier, Amsterdam.

Doering, J. 1958. Citronelle age problem. Ameri. Assn. Petrol. Geolog. Bull. 42:764–786.

Dolan, R., B., Hayden, and H. Lins. 1980. Barrier Islands. Ameri. Sci. 68:16–25.

DuBar, J. R., H. S. Johnson, Jr., B. G. Thom, and W. O. Hatchell. 1974. Neogene stratigraphy and morphology, south flank of the Cape Fear Arch, North and South Carolina, pp. 139–173 in R. Q. Oaks, Jr. and J. R. DuBar (eds.), Post-Miocene stratigraphy, Central and Southern Atlantic Coastal Plain. Utah State University Press, Logan.

Duever, M. J., and L. A. Riopelle. 1983. Successional sequences and rates on tree islands in the Okefenokee Swamp. Ameri. Midl. Nat. 110:186–193.

Duke, J. A. 1961. The psammophytes of the Carolina Fall-line Sandhills. J. Elisha Mitchell Sci. Soc. 77:3–25.

Dunn, W. J., L. N., Schwartz, and G. R. Best. 1985. Structure and water relations of the white cedar forests of north central Florida. In Proceedings of the Atlantic White Cedar Wetlands Symposium, ed. A. D. Laderman (in press).

Dury, G. H. 1977. Underfit streams: retrospect, prospect, and prospect, pp. 281–293, in K. J. Gregory (ed.), River channel changes. Wiley, New York.

Edmisten, J. A. 1963. The ecology of the Florida pine flatwoods. Ph.D. dissertation, University of Florida, Gainesville.

Ehrenfeld, J. G., and M. Gulick. 1981. Structure and dynamics of hardwood swamps in the New Jersey Pine Barrens: contrasting patterns in trees and shrubs. Ameri. Jo. of Bot. 68:471–481.

Eleuterius, L. N. 1968. Floristics and ecology of coastal bogs in Mississippi. Master's thesis, University of Southern Mississippi, Hattiesburg.

Eleuterius, L. N., and S. B. Jones, Jr. 1969. A floristic and ecological study of pitcher plant bogs in south Mississippi. Rhodora 71:29–34.

Engels, W. L. 1952. Vertebrate fauna of North Carolina coastal islands. II. Shackleford Banks. Ameri. Midl. Nat. 47:702–742.

Ewel, K. C. 1984. Effects of fire and wastewater on understory vegetation in cypress domes, pp. 119–126 in. K. C. Ewel and H. T. Odum, (eds.), Cypress swamps. University of Florida Press, Gainesville.

Ewel, K. C. 1995. Fire in cypress swamps in the southeastern United States. Proceedings of the Tall Timbers Fire Ecology Conference 19:111–116.

Ewel, K. C., and W. J. Mitsch. 1978. The effects of fire on species composition in cypress dome ecosystems. Fla. Sci. 41:25–31.

Farrar, R. M. 1993. Growth and yield in naturally regenerated longleaf pine stands. Proceedings of the Tall Timbers Fire Ecology Conference 18:311–336.

Field, M. E., E. P. Meisburger, E. A. Stanley, and S. J. Williams. 1979. Upper Quaternary peat deposits on the Atlantic inner shelf of the United States. Bull. Geolog. Soc. of Ameri. 90:618–628.

Fletcher, S. W. 1975. Adaptations of two seasonally dissimi-

lar annual plant species to the environment of the Carolina Outer Banks. Ph.D. dissertation, Duke University, Durham.

Forman, R. T. T. (ed.). 1979. Pine barrens: ecosystem
and landscape. Academic Press, New York.

Forman, R. T. T., and R. E. Boerner. 1981. Fire frequency and the pine barrens of New Jersey. Bull.
Torr. Bot. Club 108:34–50.

Frasco, B., and R. E. Good. 1976. Cone, seed and germination characteristics of pitch pine (Pinus rigida
Mill.). Bartonia 44:50–57.

Frost, C. C. 1986. Historical overview of Atlantic white
cedar in the Carolinas. In Proceedings of the Atlantic White Cedar Wetlands Symposium, ed. A. D.
Laderman.

Frost, C. C. 1993. Four centuries of changing landscape
patterns in the longleaf pine ecosystem. Proceedings of the Tall Timbers Fire Ecology Conference 18:
17–44.

Frost, C. C. 1995. Presettlement fire regimes in southeastern marshes, peatlands and swamps. Proceedings of the Tall Timbers Fire Ecology Conference 19:
39–60.

Gallagher, M. G., and R. E. Good. 1985. Two decades of
vegetation change in the New Jersey USA pine barrens. Ameri. J. Bot. 72:844.

Gano, L. 1917. A study in physiographic ecology in
northern Florida. Bot. Gaz. 63:337–372.

Garren, K. H. 1943. Effects of fire on vegetation of the
southeastern United States. Bot. Rev. 9:617–654.

Gauch, H. G. 1982. Multivariate analysis in community
ecology. Cambridge University Press, Cambridge.

Gemborys, S. R., and E. J. Hodgkins. 1971. Forests of
small stream bottoms in the coastal plain of southwestern Alabama. Ecology 52:70–84.

Givens, K. T., J. N., Layne, W. G., Abrahamson, and
S. C. White-Schuler. 1984. Structural changes and
successional relationships of five Florida Lake
Wales ridge plant communities. Bull. Torr. Bot. Club
111.

Givnish, T. 1981. Serotiny, geography and fire in the
Pine Barrens of New Jersey. Evolution 35:101–123.

Godfrey, P. J., and M. M. Godfrey. 1976. Barrier island
ecology of Cape Lookout National Seashore and vicinity, North Carolina. U.S. National Park Service
Scientific Monograph No. 9.

Gohlz, H. L., and R. F. Fisher. 1982. Organic matter
production and distribution in slash pine (Pinus ellisttii) plantations. Ecology 63:1827–1839.

Golley, F. B., G. A. Petrides, and J. F. McCormick. 1965.
A survey of the vegetation of the Boiling Spring
Natural Area, South Carolina. Bull. Torr. Bot. Club
92:355–363.

Good, R. E., and N. F. Good. 1975. Growth characteristics of two populations of Pinus rigida Mill. from
the Pine Barrens of New Jersey. Ecology 56:1215–
1220.

Good, R. E., N. F. Good, and J. W. Andresen. 1979. The
Pine Barren plains, pp. 283–295 in R. T. T. Forman
(ed.), Pine Barrens: ecosystem and landscape, Academic Press, New York.

Gosselink, J. G., C. L., Cordes, and J. W. Parsons. 1979.
An ecological characterization of the Chenier Plain
coastal ecosystem of Louisiana and Texas. U.S. Fish
and Wildlife Service Reports FWS/OBS 79/9, 79/10,
and 79/11 (three volumes).

Gosselink, J. G., S. E., Bailey, W. H., Conner, and R. E.
Turner. 1981. Ecological factors in the determina-
tion of reparian wetland boundaries, pp. 197–219 in
J. R. Clark and J. Benforado (eds.), Wetlands of bottomland hardwood forest. Elsevier, Amsterdam.

Gresham, C. A., and D. J. Lipscomb. 1985. Selected ecological characteristics of Gordonia lasianthus in
coastal South Carolina. Bull. Torr. Bot. Club 112:53–
58.

Grubb, P. J. 1977. The maintenance of species richness
in plant communities: the importance of the regeneration niche. Bio. Rev. Cambridge Philosoph. Soc. 52:
107–145.

Gunderson, L. H. 1984. Regeneration of cypress in
logged and burned stands at Corkscrew Swamp
Sanctuary, Florida, pp. 349–357 in K. C. Ewel and
H. T. Odum (eds.), Cypress swamps. University of
Florida Press, Gainesville.

Gunderson, L. H., and W. F. Loftus. 1993. The Everglades, pp. 199–256 in W. H. Martin, S. G. Boyce,
and A. C. Echternacht (eds.). Biodiversity of the
Southeastern United States/Lowland Terrestrial
Communities. Wiley, New York.

Hair, J. D., G. T., Hepp, L. M., Luckett, K. P., Reese,
and D. K. Woodward. 1979. Beaver pond ecosystems and their multiuse natural resource management, pp. 80–92 in R. R. Johnson and J. F. McCormick (eds.)., Strategies for protection and
management of floodplain wetlands and other riparian ecosystems. U.S.D.A. Forest Service General
Technical Report WO-12.

Hall, T. F., and W. T. Penfound. 1939. A phytosociological study of a cypress-gum swamp in southeastern
Louisiana. Ameri. Midl. Nat. 21:378–395.

Hall, T. F., and W. T. Penfound. 1943. Cypress-gum
communities in the Blue Girth Swamp near Selma,
Alabama. Ecology 24:208–217.

Hamilton, D. B. 1984. Plant succession and the influence
of disturbance in the Okefenokee Swamp, pp. 86–
111 in A. D. Cohen, D. J. Casagrande, M. J. Andrejko, and G. R. Best (eds.), The Okefenokee Swamp.
Wetland Surveys, Los Alamos, N. Mex.

Harcombe, P. A., and P. L. Marks. 1978. Tree diameter
distributions and replacement processes in southeast Texas forests. For. Sci. 24:153–166.

Harcombe, P. A., and P. L. Marks. 1983. Five years of
tree death in a Fagus Magnolia forest, southeast
Texas (USA). Oecologia 57:49–54.

Harcombe, P. A., J. S. Glitzenstein, R. G. Knox, S. L. Orzell, and E. L. Bridges. 1993. Vegetation of the longleaf pine region of the West Gulf Coastal Plain.
Proceedings of the Tall Timbers Fire Ecology Conference 18:83–104.

Harper, R. M. 1906. A phytogeographical sketch of the
Altamaka Grit Region of the Coastal Plain of Georgia. Ann. New York Acad. of Sci. 7:1–415.

Harper, R. M. 1911. The relation of climax vegetation to
islands and peninsulas. Bull. Torr. Bot. Club 38:515–
525.

Harper, R. M. 1914. The "pocosin" of Pike Co., Ala.,
and its bearing on certain problems of succession.
Bull. Torr. Bot. Club 41:209–220.

Harper, R. M. 1914. The geography and vegetation of
northern Florida. Fla. Geologi. Surv. Ann. Rep.
No. 6.

Harper, R. M. 1922. Some pine-barren bogs in central
Alabama. Torreya 22:57–60.

Harper, R. M. 1926. A middle Florida white cedar
swamp. Torreya 26:81–84.

Harper, R. M. 1940. Fire and forests. Ameri. Bot. 46: 5–7.

Harper, R. M. 1943. Forests of Alabama. *Geolog. Surv. of Alabama Monograph* No. 10.

Harshberger, J. W. 1916. *The vegetation of the New Jersey Pine Barrens: an ecological investigation.* Christopher Sower, Philadelphia.

Hebb, E. A., and A. F. Clewell. 1976. A remnant stand of old–growth slash pine in the Florida panhandle. *Bull. Torr. Bot. Club* 103:1–9.

Herman, S. M. 1993. Small-scale disturbances in longleaf pine forests. Proceedings of the Tall Timbers Fire Ecology Conference 18:265–274.

Heyward, F. 1939. The relation of fire to stand composition of longleaf pine forests. *Ecology* 20:287–304.

Hill, E. P. 1976. Control methods for nuisance beaver in the southeastern United States. Vertebrate Pest Control Conference 7:85–98.

Hill, M. O., and H. G. Gauch. 1980. Detrended correspondence analysis, an improved ordination technique. *Vegetatio* 42:47–58.

Hilmon, J. B. 1968. Autecology of Palmetto (*Serenea repens* (Bartr.) Small). Ph.D. dissertation, Duke University, Durham.

Hinde, H. P. 1954. The vertical distribution of phanerogams in relation to tide level. *Ecolog. Monogr.* 24: 209–225.

Hodgkins, E. J. 1958. Effects of fire on undergrowth vegetation in upland southern pine forests. *Ecology* 58:36–46.

Hough, W. A. 1973. Fuel and weather influence wildfires in sand pine forests. USDA Forest Service Research Paper SE-106.

Hough, W. A., and F. A. Albini. 1978. Predicting fire behavior in palmetto gallberry fuel complexes. USDA Forest Service Research Paper SE-174.

Hudson, C. 1976. *The Southeastern Indians.* University of Tennessee Press, Knoxville.

Huffman, R. T., and S. W. Forsythe. 1981. Bottomland hardwood forest communities and their relation to anaerobic soil conditions, pp. 187–196 in Wetlands of bottomland hardwood forests. J. R. Clark and J. Benforads (eds.), Elsevier, Amsterdam.

Hughes, R. H. 1966. Fire ecology of canebrakes. Proceedings of the Annual Fall Timbers Fire Ecology Conference 5:149–158.

Hull, J. C., and D. F. Whigham. 1985. Atlantic white cedar in the Maryland inner Coastal Plain and the Delmarva Peninsula. In Proceedings of the Atlantic White Cedar Wetlands Symposium, ed. A. D. Laderman.

Izlar, R. L. 1984. Some comments on fire and climate in the Okefenokee Swamp – Marsh complex, pp. 70–85 in A. D. Cohen, D. J. Casagrande, M. J. Andrejko, G. R. Best (eds.), *The Okefenokee Swamp.* Wetland Surveys. Los Alamos, N. Mex.

Johnson, A. F. and M. G. Barbour. 1990. Dunes and maritime forests, pp 429–480 in: R. L. Myers and J. J. Ewel (eds.), Ecosystems of Florida. Univ. Central Florida Press, Orlando.

Johnson, A. S., H. O. Hillsted, S. Shanholtzer, and G. F. Shanholtzer. 1974. Ecological survey of the coastal region of Georgia. National Park Service Scientific Monograph No. 3.

Johnson, D. 1942. The origin of the Carolina bays. Columbia University Press, New York.

Johnson, W. B., C. E. Sasser, and J. G. Gosselink. 1985. Succession of vegetation in an evolving river delta, Achafalaya Bay, Louisana. *J. Ecol.* 73:973–986.

Jones, R. H., and C. A. Gresham. 1985. Analysis of com-

position, environmental gradients, and structure in the Coastal Plain lowland forests of South Carolina. *Castanea* 50:207–227.

Kalisz, P. J., and E. L. Stone. 1984. The longleaf pine islands of Ocala National Forest, Florida: a soil study. *Ecology* 65: 1743–1754.

Kartesz, J. T. 1994. A synonymized checklist of the vascular flora of the United States, Canada and Greenland. Timber Press, Portland, Ore.

Kearney, T. H. 1900. The plant covering of Ocracoke Island; a study of the ecology of the North Carolina strand vegetation. *U.S. Natl. Herbar., Contribu.* 5:261–319.

Kearney, T. H. 1901. Report on a botanical survey of the Dismal Swamp Region. U.S. *Natl. Herbar., Contribu.* 5:321–585.

Kologiski, R. L. 1977. The phytosociology of the Green Swamp, North Carolina. North Carolina Agricultural Experiment Station Technical Bulletin No. 250.

Komarek, E. V., Sr. 1968. Lightning and lighning fires as ecological forces. Proceedings of the Annual Tall Timber Fire Ecology Conference 9:169–197.

Komarek, E. V., Sr. 1974. Effects of fire on temperate forests and related ecosystems: southeastern United States, pp. 251–277 in T. T. Kozlowski and C. E. Ahlgren (eds.), *Fire and ecosystems.* Academic Press, New York

Komarek, E. V., Sr. 1977. A quest for ecological understanding. Tall Timbers Research Station Miscellaneous Publication No. 5.

Korstian, C. F. 1924. Natural regeneration of southern white cedar. *Ecology* 5:188–191.

Korstian, C. F., and W. D. Brush. 1931. Southern white cedar. USDA Technical Bulletin No. 251.

Kuchler, A. W. 1964. Potential natural vegetation of the coterminous United States. American Geographical Society, Special Publication No. 36, Map.

Kurz, H. 1938. A physiographic study of the tree associations of the Apalachicola River. Proce. Florida Acad. of Sci. 3:78–90.

Kurz, H. 1942. Florida dunes and scrub, vegetation and geology. State of Florida Department of Conservation, Geological Bulletin No. 23.

Kurz, H. 1944. Secondary forest succession in the Tallahassee Red Hills. *Proc. Florida Acad. of Sci.* 7:59–100.

Kurz, H., and Wagner, K. A. 1953. Factors in cypress dome development. *Ecology* 34:157–164.

Laessle, A. M. 1942. The plant communities of the Welaka area. University of Florida Publication, Biological Science Series 4:5–141.

Laessle, A. M. 1958. The origin and successional relationship of sandhill vegetation and sand pine scrub. *Ecolog. Monogr.* 28:361–387.

Laessle, A. M. 1965. Spacing and competition in natural stands of sand pine. *Ecology* 46:65–72.

Laessle, A. M. 1967. Relationship of sand pine scrub to former shore lines. *Quar. J. Florida Acad. of Sci.* 30: 269–286.

Laney, R. W., and R. E. Noffsinger. 1985. Vegetative composition of Atlantic white cedar (*Chandecyparis thyoides* (L.) B.S.P.) swamps in Dare County, North Carolina. In Proceedings of the Atlantic White Cedar Wetlands Symposium, ed. A. D. Laderman.

Lawson, J. 1714. Lawson's history of North Carolina. Garrett and Massil, Richmond Va. (reprinted 1952).

Ledig, F. T., and S. Little. 1979. Pitch pine (*Pinus rigida* Mill.): ecology, physiology, and genetics, pp. 347–

372 in R. T. T. Forman (ed.), *Pine Barrens: Ecosystem and Landscape*. Academic Press, New York.

Leitman, H. M., J. E. Sohm, and M. A. Franklin. 1981. Wetland hydrology and tree distribution of the Apalachicola River flood-plain, Florida. U.S. Geological Survey Water Supply Paper No. 2196-A.

Lemon, P. C. 1949. Successional responses of herbs in the longleaf–slash pine forest after fire. *Ecology* 30: 135–145.

Leopold, L. B., and M. G. Wolman. 1957. River channel patterns: braided meandering and straight. U.S. Geological Survey Professional Paper 282-B.

Leopold, L., M. G. Wolman, and J. Miller. 1964. *Alluvial processes in geomorphology*. Freeman, San Francisco.

Lewis, C. E., and T. J. Harshbarger. 1976. Shrub and herbaceous vegetation after 20 years of prescribed burning in the South Carolina Coastal Plain. *J. Range Mgt.* 29:13–18.

Lewis, I. F. 1917. The vegetation of Shackleford Bank. North Carolina Geological and Economic Survey Paper No. 46.

Lilly, J. P. 1981. A history of swamp development in North Carolina, pp. 20–39. C. J. Richardson, (ed.), *Pocosin wetlands*. Hutchinson Ross, Stroudsberg, Pa.

Little, E. L., Jr., and K. W. Dorman. 1952. Geographic differences in cone-opening in sand pine. *J. of For.* 50:204–205.

Little, E. L., Jr., and K. W. Dorman. 1954. Slash pine (*Pinus elliotii*), including South Florida slash pine. USDA Forest Service Paper SE-36.

Little, S. 1950. Ecology and silviculture in the white cedar and associated hardwoods in southern New Jersey. Yale University School of Forestry Bulletin 56:1–103.

Little, S. 1979. Fire and plant succession in the New Jersey Pine Barrens, pp. 297–314. in R. T. T. Forman (ed.,) Pine Barrens: ecosystem and landscape. Academic Press, New York.

Longstreth, D. J., and B. R. Strain. 1977. Effects of salinity and illumination on photosynthesis and water balance of *Spartina alterniflora* Loisel. *Oecologia* 31: 191–199.

Loveless, A. R. 1961. A nutritional interpretation of sclerophylly based on differences in the chemical composition of sclerophyllous and meophytis leaves. *Ann. of Bot.* 25:168–184.

Loveless, A. R. 1962. Further evidence to support a nutritional interpretation of sclerophylly. *Ann. of Bot.* 26:551–561.

Lutz, H. J. 1934. Ecological relations in the pitch pine plains of southern New Jersey. Yale University School of Forestry, Bulletin No. 38.

McCaffrey, C. A., and D. B. Hamilton. 1984. Vegetation mapping of the Okefenokee ecosystem, pp. 201–211 in A. D. Cohen, D. J. Casagrande, M. J. Andrejko, and G. R. Best (eds.), *The Okefenokee Swamp*. Wetland Surveys, Los Alamos, N. Mex.

McCormick, J. 1979. The vegetation of the New Jersey Pine Barrens, pp. 229–244 in R. T. T. Forman (ed.), Pine Barrens: ecosystem and landscape. Academic Press, New York.

McCormick, J., and M. F. Buell. 1968. The Plains: pigmy forests of the New Jersey Pine Barrens, a review and annotated bibliography. Bull. New Jersey Acad. Sci. 13:20–34.

Marks, P. L., and P. A. Harcombe. 1975. Community diversity of Coastal Plain forests in southern East Texas. *Ecology* 56:1004–1008.

Marks, P. L., and P. A. Harcombe. 1981. Forest vegetation of the Big Thicket, southeast Texas. *Ecolog. Monogr.* 51:287–305.

Marois, K. C., and K. C. Ewel. 1983. Natural and management-related variation in Cypress domes. *For. Sci.* 29:627–640.

Matoon, W. R. 1916. Water requirements and growth of young cypress. Proceedings of the Society of American Foresters 11:192–197.

Matos, J. A., and D. C. Rudolph. 1985. The vegetation of the Roy E. Larsen Sandylands Sanctuary in the Big Thicket of Texas. *Castanea* 50:228–249.

Mitsch, W. J., and K. C. Ewel. 1979. Comparative biomass and growth of cypress in Florida wetlands. *Ameri. Midl. Nat.* 101:417–426.

Mitsch, W. J., and J. G. Gosselink. 1986. Wetlands. Van Nostrand Reinhold, New York.

Monette, R. 1975. Early forest succession in the southeastern Virginia Coastal Plain. *Va. J. Sci.* 26:65.

Monk, C. D. 1960. A preliminary study on the relationships between the vegetation of a mesic hammock community and a sandhill community. *Quart. J. of the Fla. Acad. of Sci.* 23:1–12.

Monk, C. D. 1965. Southern mixed hardwood forest of northcentral Florida. *Ecolog. Monogr.* 35:335–354.

Monk, C. D. 1966a. An ecological significance of evergreenness. *Ecology* 47:504–505.

Monk, C. D. 1966b. An ecological study of hardwood swamps in northcentral Florida. *Ecology* 47:649–654.

Monk, C. D. 1967. Tree species diversity in the eastern deciduous forest with particular reference to north central Florida. *Ameri. Midl. Nat.* 101:173–187.

Monk, C. D. 1968. Successional and environmental relationships of the forest vegetation of north central Florida. *Ameri. Midl. Nat.* 79:441–457.

Monk, C. D., and T. W. Brown. 1965. Ecological consideration of cypress heads in northcentral Florida. *Ameri. Midl. Nat.* 74:127–140.

Moore, J. H., and J. H. Carter, III. 1985. The range and habitats of Atlantic white cedar in North Carolina. In *Proceedings of the Atlantic White Cedar Wetlands Symposium*, ed. A. D. Laderman.

Moore, W. H., B. F. Swindel, and W. S. Terry. 1982. Vegetative response to prescribed fire in a north Florida flatwoods forest. *J. of Range Mgt.* 35:386–389.

Mulvania, M. 1931. Ecological survey of a Florida scrub. *Ecology* 12:528–540.

Murray, G. E. 1961. *Geology of the Atlantic and Gulf Coastal Province of North America*. Harper, New York.

Muzika, R. M., J. B. Gladden, and J. D. Haddock. 1986. Structural and functional aspects of recovery in southeastern floodplain forests following a major disturbance. *Ameri. Midl. Nat.*

Myers, R. L. 1985. Fire and the dynamic relationship between Florida sandhill and sand pine scrub vegetation. *Bull. Torr. Bot. Club* 112:241–252.

Myers, R. L. 1990. Scrub and high pine, pp. 150–193 in R. L. Myers and J. J. Ewel (eds.), Ecosystems of Florida. The University of Central Florida Press, Orlando.

Myers, R. L. 1993. Restoring longleaf pine community integrity. Proceedings of the Tall Timbers Fire Ecology Conference 18:349–350

Myers, R. L., and N. D. Deyrup. 1983. The dynamic relationship between Florida sandhill and sand pine scrub vegetation. *Bull. Ecolog. Soc. of Ameri.* 64:62.

Neill, C., and L. A. Deegan. 1985. The effect of Mississippi River delta lobe development on the habitat composition and diversity of Louisana coastal wetlands. *Ameri. Midl. Nat.* 116:296–303.

Nesom, G. L., and M. Treiber. 1977. Beech–mixed hardwoods communities: a topo-edaphic climax on the North Carolina Coastal Plain. *Castanea* 42:119–140.

Neufeld, H. S. 1983. Effects of light on growth, morphology, and photosynthesis in bald cypress (*Taxodium distichum* (L.) Rich.) and pondcypress (T. ascendens Brongn.) seedlings. *Bull. Torr. Bot. Club* 110: 43–54.

Neufeld, H. S. 1984. *Comparative ecophysiology of baldcypress* (Taxodium distichum *(L.) Rich.) and pondcypress* (Taxodium ascendens Brongn.). Ph.D. dissertation, University of Georgia, Athens.

Odum, W. E., T. J., Smith, III, J. K. Hoover, and C. C. McIvor. 1984. The ecology of tidal freshwater marshes of the United States east coast: a community profile. U.S. Fish and Wildlife Service FWS/ OBS 83/17.

Olsson, H. 1979. Vegetation of the New Jersey Pine Barrens: a phytosociological classification, in R. T. T. Forman (ed.), *Pine Barrens: ecosystem and landscape.* Academic Press, New York.

Olsvig, L. S., J. F. Cryan, and R. H. Whittaker. 1979. Vegetational gradients of the pine plains and barrens of Long Island, New York, pp. 265–282 in R. T. T. Forman (ed.), Pine Barrens: ecosystem and landscape. Academic Press, New York.

Oosting, H. J. 1942. An ecological analysis of the plant communities of piedmont, North Carolina. *Ameri. Midl. Nat.* 28:1–126.

Oosting, H. J. 1945. Tolerance to salt spray of plants of coastal dunes. *Ecology* 26:85–89.

Oosting, H. J. 1954. Ecological processes and vegetation of the maritime strand in the southeastern United States. *Bot. Rev.* 20:226–262.

Oosting, H. J., and W. D. Billings. 1942. Factors affecting vegetational zonation on coastal dunes. *Ecology* 23:131–142.

Otte, L. J. 1981. Origin, development, and maintenance of the pocosin wetlands of North Carolina. Unpublished report of North Carolina Department of Natural Resources and Community Development Natural Heritage Program, Raleigh, N. C.

Parrish, F. K., and E. J. Rykiel, Jr. 1979. Okefenokee Swamp origin: review and reconsideration. *J. Elisha Mitchell Sci. Soc.* 95:17–31.

Parrott, R. T. 1967. A study of wiregran (*Aristida stricta*) with particular reference to fire. Master's thesis, Duke University, Durham.

Parsons, S. E., and S. Ware. 1982. Edaphic factors and vegetation in Virginia Coastal Plain swamps. *Bull. Torr. Bot. Club* 109:365–370.

Peet, R. K. 1993. A taxonomic study of *Aristida stricta* and *A. Beyrichiana.* Rhodora 95:25–37.

Peet, R. K., and D. J. Allard. 1993. Longleaf pine vegetation of the southern Atlantic and eastern Gulf Coast regions: a preliminary clasification. Proceedings of the Tall Timbers Fire Ecology Conference 18: 45–82.

Penfound, W. T. 1952. Southern swamps and marshes. Bot. Rev. 18:413–446.

Penfound, W. T., and T. T. Earle. 1948. The biology of the water hyacinth. *Ecolog. Monogr.* 18:447–472.

Penfound, W. T., and E. S. Hathaway. 1938. Plant communities in the marshlands of southeastern Louisiana. *Ecolog. Monogr.* 8:1–56.

Penfound, W. T., and Howard, J. A. 1940. A phytosociological study of an evergreen oak forest in the vicinity of New Orleans, Louisiana. *Ameri. Midl. Nat.* 23:165–174.

Penfound, W. T., and M. E. O'Neill. 1934. The vegetation of Cat Island, Mississippi. *Ecology* 15:1–16.

Penfound, W. T., and A. G. Watkins. 1937. Phytosociological studies in the pinelands of southeastern Louisiana. *Ameri. Midl. Nat.* 18:661–682.

Peroni, P. A. 1983. Vegetation history of the southern Lake Wales Ridge, Highlands County, Florida. Master's thesis, Bucknell University, Lewisburg, Pa.

Pessin, L. J. 1933. Forest associations in the uplands of the lower Gulf Coastal Plain. *Ecology* 14:1–14.

Peterson, C. H., and Peterson, N. M. 1979. The ecology of intertidal flats of North Carolina: a community profile. U.S. Fish and Wildlife Service FWS/OBS-79-39.

Pilkey, O. H., Jr., and K. Dixon. 1996. The Corps and the Shore. Island Press, Washington D.C.

Pilkey, O. H., Jr., W. J. Neal, and O. H. Pilkey, Sr. 1980. From Currituck to Calabash. North Carolina Science and Technology, Center. Research Triangle Park.

Pinchot, G., and Ashe, W. W. 1897. Timber trees and forests of North Carolina. *North Carolina Geolog. Sur. Bull.* No. 6.

Platt, W. J., G. W. Evans, and S. L. Rathbun. 1988. The population dynamics of a long-lived conifer (*Pinus palustris*). Amer. Midl. Nat. 131:491–525.

Platt, W. J., J. S. Glitzenstein, and D. R. Streng. 1991. Evaluating pyrogenicity and its effects on vegetation in longleaf pine savannas. Proceedings of the Tall Timbers Fire Ecology Conference 17:143–161.

Platt, W. J., and S. J. Rathbum. 1993. *Proceedings of the Tall Timbers Fire Ecology Conference* 18:275–298.

Porcher, R. D. 1981. The vascular flora of the Francis Beilder Forest in Four Holes Swamp, Berkeley and Dorchester Counties, South Carolina. Castanea 46: 248–280.

Price, M. B. 1973. Management of natural stands of Ocala sand pine. In *Sand Pine Symposium Proceedings*, pp. 153–163. USDA. Forest Service General Technical Report SE-2.

Prouty, W. F. 1952. Carolina Bays and their origin. *Bull. Geolog. Soci. Ameri.* 63: 187–224.

Putnam, J. A., G. M. Furnival, and J. S. McKnight. 1960. Management and inventory of southern hardwoods. USDA Forest Service Agricultural Handbook No. 181.

Pyne, S. J. 1982. *Fire in America: a cultural history of wildland and rural fire.* Princeton University Press, Princeton, N.J.

Quarterman, E., and Keever, C. 1962. Southern mixed hardwood forest: climax in the southeastern coastal plain, U.S.A. Ecolog. Monogr. 32: 167–185.

Raff, P. J. 1954. Aspects of the Ecological Life-History of Turkey Oak (*Quercus laeris* Walter). Master's thesis, Duke University, Durham.

Richardson, C. J. 1991. Pocosins: an ecological perspective. *Wetlands* 11:335–354.

Richardson, C. J., R. Evans, and D. Carr. 1981. Pocosins: an ecosystem in transition, pp. 3–19 in C. J. Richardson, (ed.), *Pocosin wetlands.* Hutchinson Ross, Stroudsberg, Pa.

Richardson, C. J., and J. W. Gibbons. 1993. Pocosins, Carolina bays, and mountain bogs, pp. 257–310 in W. H. Martin, S. G. Boyce and A. C. Echternacht (eds.), Biodiversity of the Southeastern United States/Lowland Terrestrial Communities. Wiley, New York.

Richardson, C. J., and E. J. McCarthy. 1994. Effect of land development and forest management on hydrologic response in southeastern coastal wetlands: a review. *Wetlands* 14:56–71.

Roberts, P. R., and Oosting, H. J. 1958. Responses of venus fly trap (*Dionaea muscipula*) to factors involved in its endemism. *Ecolog. Monogr.* 28:193–218.

Rubenstein, D. I. 1981. Behavioral ecology of island feral horses. *Equine Veterinary.* 13:27–34.

Savage, H., Jr. 1982. *The mysterious Carolina bays.* University of South Carolina Press, Columbia.

Schlesinger, W. H. 1976. Biogeochemical limits on two levels of plant community organization in the cypress forest of Okefenokee Swamp. Ph.D. dissertation, Cornell University, Ithaca, New York.

Schlesinger, W. H. 1978a. Community structure, dynamics and nutrient cycling in the Okefenokee Cypress swamp forest. *Ecolog. Monogr.* 48:43–65.

Schlesinger, W. H. 1978b. On the relative dominance of shrubs in Okefenokee Swamp. *Ameri. Nat.* 112:949–954.

Schneider, R. E. 1988. The effect of variation in season of burning on a pine-wiregrass savanna in the Green Swamp, North Carolina. Ph.D. dissertation, Duke University, Durham.

Shafizadeh, F., P. P. S. Chin, and W. F. De Groot. 1977. Effective heat content of forest fuels. *For. Sci.* 23:81–89.

Shaler, N. S. 1885. Seacoast swamps of the eastern United States. U.S. Geolog. Surv. Ann. Rep. 6:353–398.

Sharitz, R. R., and J. W. Gibbons. 1982. The ecology of southeastern shrub bogs (pocosins) and Carolina Bays: a community profile. U.S. Fish and Wildlife Service Report FWS/OBS-82/04.

Sharitz, R. R., and W. J. Mitsch. 1993. Southern floodplain forests, pp. 311–372 in W. H. Martin, S. G. Boyce, and A. C. Echternacht (eds.), Biodiversity of the Southeastern United States/Lowland Terrestrial Communities. Wiley, New York.

Simms, E. L. 1983. The growth, reproduction, and nutrient dynamics of two pocosin shrubs, the evergreen *Lyonia lucida* and the deciduous *Zenobia pulverulenta*. Ph.D. dissertation, Duke University, Durham.

Simms, E. L. 1985. Growth response to clipping and nutrient addition in *Lyonia lucida* and *Zenobia pulverulenta*. *Ameri. Midl. Nat.* 114:44–50.

Snyder, J. R. 1980. Analysis of coastal plain vegetation, Croatan National Forest, North Carolina. *Veroffentlichongen des Geobotanischen Institutes der Eidgenoessiche Technische Hochschule Stiftung Rubel, in Zurich* 69:40–113.

Spackman, W., A. D. Cohen, P. H. Given, and D. J. Casagrande. 1976. Comparative study of the Okefenokee Swamp and the Everglades–Mangrove Complex of southern Florida. Coal Research Section, Pennsylvania State University, State College, Pa.

Stalter, R., and W. E. Odum. 1993. Maritime communities, pp. 117–164 in W. H. Martin, S. G. Boyce, and A. C. Echternacht (eds.), Biodiversity of the Southeastern United States/Lowland Terrestrial Communities. Wiley, New York.

Stephenson, S. N. 1965. Vegetation change in the Pine Barrens of New Jersey. *Bull. Torr. Bot. Club* 92:102–114.

Stone, W. 1911. The plants of southern New Jersey, with especial reference to the flora of the Pine Barrens and the geographical distribution of the species. New Jersey State Museum *Ann. Rep.* 1910:23–828.

Stout, I. J., and W. R. Marion. 1993. Pine flatwoods and xeric pine forests of the southern (lower) Coastal Plain, pp. 373–446 in W. H. Martin, S. G. Boyce, and A. C. Echternacht (eds.), Biodiversity of the Southeastern United States/Lowland Terrestrial Communities. Wiley, New York.

Straub, P. A. 1984. Effects of wastewater and inorganic fertilizer on growth rates and nutrient concentrations in dominant tree species in cypress domes, pp. 127–140 in K. C. Ewel and H. T. Odum, (eds.), *Cypress swamps.* University of Florida Press, Gainesville.

Streng, D. R., and Harcombe, P. A. 1982. Why don't east Texas savannas grow up to forest. *Ameri. Midl. Nat.* 108:278–294.

Streng, D. R., J. S. Glitzenstein, and W. J. Platt. 1993. Evaluating effects of season of burn in longleaf pine forests: a critical literature review and some results from an ongoing long-term study. Proceedings of the Tall Timbers Fire Ecology Conference 18: 227–264.

Teal, J. M. 1962. Energy flow in the salt marsh ecosystem of Georgia. *Ecology* 43:614–624.

Tedrow, J. D. F. 1979. Development of pine barrens soils, pp. 61–80 in R. T. T. Forman (ed.), *Pine Barrens: ecosystem and landscape.* Academic Press, New York.

Thornthwaite, C. W., Mather, J. R., and Carter, D. B. 1958. *Three water balance maps of eastern North America.* Resources for the Future, Washington, D.C.

Tooker, W. W. 1899. The adapted Algonquin term "Poquosin." *Ameri. Anthro.* Jan.: 162–170.

Trewartha, G. T. 1968. *An introduction to climate*, 4th ed. McGraw-Hill, New York.

Turner, R. E., S. W. Forsythe, and N. J. Craig. 1981. Bottomland hardwood forest land resources of the southeastern United States, pp. 13–28 in J. R. Clark and J. Benforado, (eds.), *Wetlands of bottomland hardwood forests.* Elsevier, Amsterdam.

Veno, P. A. 1976. Successional relationships of five Florida plant communities. *Ecology* 57:498–508.

Vitousek, P. M. 1982. Nutrient cycling and nutrient use efficiency. *Ameri. Nat.* 119:553–572.

Vogl, R. J. 1973. Fire in the southeastern grasslands. *Proceedings of the Annual Tall Timbers Fire Ecology Conference* 12:175–198.

Wade, D., J., Ewel, and R. Hofstetter. 1980. *Fire in south Florida ecosystems* USDA Forest Service General Technical Report SE – 17.

Wagner, R. H. 1964. The ecology of *Uniola paniculata* L. in the dune-strand habitat of North Carolina. *Ecolog. Monogr.* 34: 79–96.

Wahlenberg, W. G. 1946. *Longleaf pine.* Charles Lathrop Pack Forest Foundation, Washington, D.C.

Wahlenberg, W. G., S. W. Greene, and H. R. Reed. 1939. Effect of fire and cattle grazing on longleaf pine lands studied at McNeill, Mississippi. USDA Agricultural Technical Bulletin No. 683.

Walker, J. 1984. *Species diversity and production in pine-*

wiregrass savannas of the Green Swamp, North Carolina. Ph.D. dissertation, University of North Carolina, Chapel Hill.

Walker, J. 1993. Rare vascular plant taxa associated with the longleaf pine ecosystems: patterns in taxonomy and ecology. Proceedings of the Tall Timbers Fire Ecology Conference 18:105–126.

Walker, J., and R. K. Peet. 1983. Composition and species diversity of pine-wire grass savannas of the Green Swamp, North Carolina. *Vegetatio* 55:163–179.

Walter, H. 1985. Vegetation of the earth, 3rd ed. Springer-Verlag, New York.

Walter, H. and H. Leith. 1967. Klimadiagramm-Weltatlas. Gustav Fischer, Jena.

Wang, D. 1984. Fire and nutrient dynamics in a pine-oak forest ecosystem in the New Jersey Pine Barrens. Ph.D. dissertation, Yale University, New Haven.

Ware, S. 1970. Southern mixed hardwood forest in the Virginia Coastal Plain. *Ecology* 51:921–924.

Ware, S. 1978. Vegetational role of beech in the southern mixed hardwood forest and the Virginia Coastal Plain. *Va. J. Sci.* 29:231–235.

Ware, S., C. Frost, and P. D. Doerr. 1993. Southern mixed hardwood forest: the former longleaf pine forest, pp. 447–493 in W. H. Martin, S. G. Boyce, and A. C. Echternacht (eds.), Biodiversity of the Southeastern United States/Lowland Terrestrial Communities. Wiley, New York.

Watts, W. A. 1971. Postglacial and interglacial vegetation history of southern Georgia and central Florida. *Ecology* 52: 676–690.

Watts, W. A. 1975. A late Quaternary record of vegetation from Lake Annie, south-central Florida. *Geology* 3:344–346.

Watts, W. A. 1980a. The late Quaternary vegetation history of the southeastern United States. *Ann. Rev. Ecol. and System.* 11:387–409.

Watts, W. A. 1980b. Late Quaternary vegetation history of White Pond on the inner Coastal Plain of South Carolina. *Quat. Res.* 13:187–199.

Weaver, J. E., and Clements, F. E. 1938. *Plant ecology.* McGraw-Hill, New York.

Weaver, T. W., III. 1969. *Gradients in the Carolina fall-line sandhills: environment, vegetation, and comparative ecology of the oaks.* Ph.D. dissertation, Duke University, Durham.

Webber, J. 1935. Florida scrub, a fire fighting association. *Ameri. J. Bo.* 22:344–361.

Wells, B. W. 1928. Plant communities of the Coastal Plain of North Carolina and their successional relations. *Ecology* 9:230–242.

Wells, B. W. 1932. *The natural gardens of North Carolina.* University of North Carolina Press, Chapel Hill.

Wells, B. W. 1939. A new forest climax: the salt spray climax of Smith Island, N.C. *Bull. Torr. Bot. Club* 66: 629–634.

Wells, B. W. 1942. Ecological problems of the southeastern United States Coastal Plain. *Bot. Rev.* 8:533–561.

Wells, B. W. 1946. Vegetation of Holly Shelter Wildlife Management area. North Carolina Department of Conservation and Development, Division of Game and Inland Fisheries Bulletin No. 2.

Wells, B. W., and S. G. Boyce. 1953. Carolina bays: additional data on their origin, age and history. *J. Elisha Mitchell Sci. Soc.* 69:119–141.

Wells, B. W., and I. V. Shunk 1928. A southern upland grass-sedge bog. *North Carolina State College Agricultural Experimental Station Technical Bulletin* No. 32.

Wells, B. W., and I. V. Shunk. 1931. The vegetation and habitat factors of coarser sands of the North Carolina Coastal Plain: an ecological study. *Ecolog. Monogr.* 1:465–520.

Wells, B. W., and I. V. Shunk. 1938. Salt spray: an important factor in coastal ecology. *Bull. Torr. Bot. Club* 65:485–492.

Wells, B. W., and L. A. Whitford. 1976. History of stream-head swamp forests, pocosins, and savannahs in the Southeast. *J. Elisha Mitchell Sci. Soc.* 92: 148–150.

Wharton, C. H. 1978. *The natural environments of Georgia.* Georgia Department of Natural Resources, Atlanta, Ga.

Wharton, C. H., W. M. Kitchens, E. C. Pendleton, and T. W. Sipe. 1982. The ecology of bottomland hardwood swamps of the Southeast: a community profile. U.S. Fish and Wildlife Service FWS/OBS 81/37.

Whipple, S. A., L. H. Wellman, and B. J. Good. 1981. A classification of hardwood and swamp forests on the Savannah River Plant, South Carolina. U.S. Department of Energy, Savannah River Plant publication SRO/NERP-6. 36 pp.

White, D. A. 1983. Plant communities of the lower Pearl River basin, Louisiana. *Amer. Midl. Nat.* 110:381–396.

Whitehead, D. R. 1972. Development and environmental history of the Dismal Swamp. *Ecolog. Monogr.* 42: 301–315.

Whitehead, D. R. 1981. Late-Pleistocene vegetational changes in northeastern North Carolina. *Ecolog. Monogr.* 51:451–471.

Whitlow, T. H., and R. W. Harris. 1979. Flood tolerance in plants: a state of the art review. U.S. Army Corps of Engineers, Environmental and Water Quality Operational Studies Technical Report E-79-2.

Whittaker, R. H. 1979. Vegetational relationships of the Pine Barrens, pp. 315–332 in R. T. T. Forman *Pine Barrens: ecosystem and landscape.* Academic Press, New York.

Wilbur, R. B. 1985. Effects of fire on nitrogen and phosphorus availability in a North Carolina Coastal plain pocosin. Ph.D. dissertation, Duke University, Durham.

Wilbur, R. B., and N. L. Christensen. 1983. Effects of fire on nutrient availability in a North Carolina coastal plain pocosin. *Ameri. Midl. Nat.* 110:54–61.

Wilson, J. E. 1978. *A floristic study of the "savannahs" on pine plantations in the Croatan National Forest.* Master's thesis, University of North Carolina, Chapel Hill.

Wolfe, J. A. 1978. A paleobotanical interpretation of Tertiary climates in the Northern Hemisphere. *Ameri. Sci.* 66:694–704.

Wolfe, J. A. 1985. Distribution of major vegetational types during the Tertiary, pp. 357–375 in E. T. Sundquist and W. S. Broecker (eds.), The carbon cycle and atmospheric CO_2: natural variations archean to present. American Geophysical Union, Monograph 32. Washington, D.C.

Woodwell, G. M. 1956. *Phytosociology of Coastal Plain*

wetlands in the Carolinas. Master's thesis, Duke University, Durham.

Woodwell, G. M. 1958. Factors controlling growth of pond pine seedlings in organic soils of the Carolina. *Ecolog. Monogr.* 28:219–236.

Woodwell, G. M. 1979. Leaky ecosystems: nutrient fluxes and succession in the Pine Barrens vegetation, pp. 333–343 (ed.), in R. T. T. Forman, (ed.), Pine Barrens: ecosystem and Landscape. Academic Press, New York.

Wright, A. H., and A. A. Wright. 1932. The habitats and composition of the vegetation of Okefenokee Swamp, Georgia. *Ecolog. Monogr.* 2:110–232.

Wright, H. E., Jr. 1976. The dynamic nature of Holocene vegetation, a problem in paleoclimatology, biogeography and stratigraphic nomenclature. *Quat. Res.* 6: 581–596.

Wright, N. O. 1984. A cultural history of the Okefenokee, pp. 58–69 in A. D. Cohen, D. J. Casagrande, M. J. Andrejko, and G. R. Best (eds.), *The Okefenokee Swamp*. Wetland Surveys, Los Alamos, N. Mex.

Chapter
12

Freshwater Wetlands

CURTIS J. RICHARDSON

INTRODUCTION

The generic term "wetland" is now used worldwide and includes specific ecosystems known regionally in North America as bayous, bogs, bottomlands, flats, fens, floodplains, mangroves, marshes, mires, moors, muskegs, playas, peatlands, pocosins, potholes, reedswamps, sloughs, swamps, wet meadows, and wet prairies. They are found in every biome, including the desert. These ecosystems are transitional between terrestrial and aquatic systems and have a similarity of excessive water supply such that these lands have water at or near the surface of the ground for much of the year. The native plants living in wetlands are uniquely adapted to live under conditions of intermittent flooding, lack of oxygen (anoxia), often a lack of essential nutrients (such as N, P, or K), harsh (even toxic) conditions of reduced chemical species (e.g., H_2S rather than SO_4), and extremely low or high pH. The complexity of these diverse wetland systems has only recently been explored by ecologists, hydrologists, and soil scientists, and much remains to be understood about the environmental factors that control their plant and animal communities.

Because of the great diversity of wetland types and the vast area they represent, it is not possible to cover in detail even all of freshwater wetland ecosystems in North America in this chapter. The reader is referred to regional books on such topics as cypress swamps (Ewel and Odum 1984), the Everglades (Davis and Ogden 1994), freshwater wetlands (Good, Whigham, and Simpson 1978), freshwater marshes (Weller 1994), prairie potholes (van der Valk 1989), pocosins (Richardson 1981, 1993), as well as an excellent series of wetland community profiles by the U.S. Fish and Wildlife Service on peatlands (Glaser 1987), and peat bogs (Damman and French 1987), for more details. Other important freshwater areas with extremely interesting and often rare wetland plant species not covered in this chapter include the Atlantic white cedar wetlands (Laderman 1987), the northern swamp forests, and ephemeral wetlands (playas) of the western United States (Bolen 1982). In addition, other chapters in this book have provided some key information on the wetland areas found within their region. For example, Chapters 1 and 2 cover certain components of the bog areas of the Arctic, Chapter 11 provides extensive analysis of southeastern coastal wetlands, alluvial bottomland hardwood systems, Carolina bays, and pocosins, and Chapter 13 describes tidal wetlands.

In this chapter I present the current status of the major freshwater wetlands in North America, provide a hydrogeomorphic framework to help the reader organize the myriad wetland types into functional ecosystems on the landscape, present background information on the ecological processes controlling the vegetation in each wetland type, and give sufficient data to enable the reader to compare plant community dynamics, community structure, and successional patterns. Areas of future research are briefly explored.

Wetland Areas

Wetlands cover approximately 6% ($8.5 \ km^2 \times 1000$) of the world's land surface and are found in every climate from the tropics to the tundra (Maltby and Turner 1983). Freshwater wetlands comprise more than 95% of the total area of wetlands in North America, with the peatlands (mires) of Canada, Alaska, and the lower 48 states making up the vast majority of area at 150 M ha, 49.2 M ha, and 10.2 M ha, respectively (Bord na Mona 1984). The intensive conversion of wetlands to agriculture, forestry, and urban areas has resulted in the loss of 53% of the wetland habitats in the conterminous United States during the period 1780-1980 (Dahl 1990). By the mid-1980s, an estimated 55 M ha of freshwater (inland) wetlands remained in the conterminous United States out of an original 89 M ha (Dahl and Johnson 1991). Alaska, however, has lost only 0.1% of its total wetland area to development. Of the remaining freshwater wetlands in the 48 states 52.9% are forested, 25.1% are emergent herblands, and 15.7% are dominated by shrubs.

The major wetland types covered in this chapter are: (1) the boreal peatlands (bogs and fens) of Canada, Alaska, and the northern United States; (2) northern marsh systems, the prairie pothole region of the Midwest; (3) southern peatlands (fens), the Everglades mires of southern Florida; and (4) the riparian systems (southern bottomland hardwood swamps) of the Mississippi and Coastal Plains of the southeast, northeastern swamps, and western riparian systems. A number of wetlands, such as the ephemeral playas of the West, the riverine cattail marshes of the central United States, the tule marshes of California and Oregon, and the marshes of the Mississippi delta, are only briefly compared; for more imformation see excellent reviews by Klopatek (1978), Herdendorf (1987), and Gosselink (1984). The marshes of the western United States and the tidal freshwater marshes of the East Coast are not discussed in detail, and the reader is referred to volumes by Hofstetter (1983), Faber, Keller, Sands, and Massey (1989), and Good et al (1978).

What Defines a Wetland?

Wetlands are not easily defined. It is a collective term used to describe a great diversity of ecosystems worldwide whose formation and existence are dominated by water. Permanently flooded deepwater areas (generally to > 2m water depth) are not considered wetlands. Wetlands are part of a continuous water gradient from upland to continuously wet open-water ecosystems. Thus, the upper and lower limits of a wetland are somewhat arbitrary. Historically, wetlands were defined by specialists like botanists and foresters who focused on plants adapted to flooding and/or saturated soil conditions. An hydrologist's definition emphasized the position of the water table relative to the ground surface over time. Wildlife biologists focused on the habitats of waterfowl, wading birds, and game and fish species. No formal or universally recognized definition of wetlands existed until the U.S. Fish and Wildlife Service (FWS) in 1979, after years of review, proposed a comprehensive definition in a report by Cowardin, Carter, Golet, and LaRoe (1979) entitled "Classification of Wetlands and Deepwater Habitats of the United States." The wetland definition in this report was used as the criterion for a new U.S. Wetlands Inventory and stated:

Wetlands are lands transitional between terrestrial and aquatic systems where the water table is usually at or near the surface or the land is covered by shallow water. Wetlands must have one or more of the following three attributes: (1) at least periodically, the land supports predominantly hydrophytes (water-loving plants), (2) the substrate predominantly undrained hydric soil (wet soils), and (3) the substrate is nonsoil and is saturated with water or covered by shallow water at some time during the growing season of each year. [p. 3]

This definition is widely accepted today by both scientists and land managers and has been adopted internationally in many countries. However, Canada has not adopted this definition and continues to use the more traditional terms – bog, fen, swamp, and marsh – to define wetlands as:

... areas where wet soils are prevalent, having a water table near or above the mineral soil for the most part of the thawed season, supporting a hydrophylic vegetation. [Zoltai 1979, p. 1]

Many wetland classification systems based on vegetation have been used worldwide, but the most useful are based on hydrology, geologic origin, and ecological characteristics.

The modern FWS classification of wetlands divides wetland and deepwater habitats into five ecological systems: (1) marine, (2) estuarine, (3) riverine, (4) lacustrine, and (5) palustrine, with a number of subsystems and classes (Fig. 12.1). Freshwater wetlands, the focus of this chapter, comprise the last three systems and make up over 95% of the world's wetlands. The marine system consists of open ocean and its associated coastline. Mostly deepwater habitat, marine wetlands are limited to intertidal areas, rocky shores, beaches, and some coral reefs with salinities >30 parts per 1000. The estuarine system is more closely associated with land and is comprised of salt and brackish tidal marshes, mangroves, swamps, intertidal mud flats, as well as bays, sounds, and coastal rivers to where ocean-derived salts are <0.5 parts per 1000.

The riverine system is limited to lotic (flowing) freshwater river and stream channels and is mainly a deepwater habitat. The lacustrine (lake) system includes wetlands situated in lentic (nonflowing) water bodies such as lakes, reservoirs, and deep pond habitats where trees, shrubs, and emergent plants do not make up more than 30% areal coverage in areas <2 m in depth. The palustrine system comprises the vast majority (>90%) of the world's inland marshes, bogs, mires, and swamps and does not include any deepwater habitats. Unfortunately, the palustrine system is so broad in scope that it does not allow for an easy separation of the many types of forested or peatland systems. The major wetland areas covered in this chapter are palustrine or lacustrine.

Why Do So Many Types of Freshwater Wetlands Exist?

Part of the confusion is owing to the fact that ecologically similar wetland types simply are referred to by many different names throughout the world. For example, a swamp in North America refers to a wetland with trees or shrubs, but in Europe it is called a "carr." The pocosins (Algonquin Indian word for swamp-on-a-hill) of the southeastern coastal plains of the United States are classified as evergreen shrub bogs (see Christensen, Chapter 11 this volume). Likewise, a bottomland refers to a floodplain wetland generally along a stream, and both are often called riparian wetlands in the western United States, even though this term represents only the area of streamside zone influence. In Canada, peatlands are often called muskegs, another Algonquin Indian term. Thus, the use of local and regional names greatly increases the confusion about the true number of wetland types.

A classification of distinct wetland groups characterized according to their sources of water, nutrients, ecological similarities, and topographic position can be used to distinguish wetland types (Brinson 1993). The key to ordering these systems

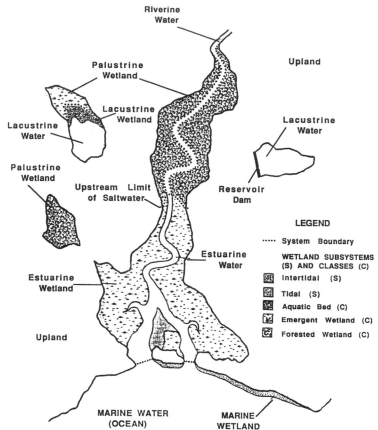

Figure 12.1. Major wetland systems, according to the U.S. Fish and Wildlife Service classification system (from Tiner 1984).

is based on their dominant sources of water. For classification purposes the water inflows can be simplified to inputs from (1) precipitation, (2) groundwater discharge, and (3) surface and near-surface inflow (e.g., overbank flow from stream channels or overland flow) (Fig. 12.2). A particular wetland may have only one dominant source of water input or a combination of sources. The relative positions of major wetland types, when scaled according to the relative importance of water sources, shows that bogs and pocosins receive their water almost exclusively from rainfall (ombrotrophic) and for this reason are nutrient poor (oligotrophic). By contrast fens, seeps, and some marsh types are controlled by groundwater inputs. These wetlands are nourished by minerals from the ground (minerotrophic) and are often more nutrient-rich (eutrophic). The wetland types dominated by surface flow are very diverse and include swamps, salt marshes, and wetlands along the fringes of lakes and streams.

The relative position of these wetland types to each other on the landscape is best displayed when scaled out along axes of both water regime and nutrients (Fig. 12.3). The nutrient content of the vegetation and soils is lowest in bogs and increases along the fen, marsh, and swamp gradient. By contrast, the amount of peat formation decreases along this gradient because of increased decomposition of organic matter in enriched sites with higher pH. The biomass of trees in proportion to shrubs in bogs is generally low, but forest species increase to become dominant in the swamps. The duration of hydroperiod (seasonal pattern of water level at or near the surface) ranges from permanently flooded in the lacustrine (shallow lake) wetland, to periods of 6–9 mo in the bogs and to as little as 2 wk in some swamp systems. The seasonal magnitude of changes in water level ranges from as small as 0.5 m in the bogs to over 7 m in the tropical swamp forest systems. Thus, the amount of rainfall versus surface or groundwater can be used to separate the general types of wetlands (bogs, fens, marshes, and swamps) and place them into functional groupings on the landscape. The amount of peat formation in each wetland grouping is related to the net difference in plant productivity minus decomposition, and these processes are related directly to the

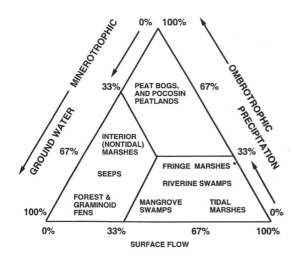

Figure 12.2. The percentage of water contribution to a wetland from precipitation, groundwater, or surface flow creates different types of wetlands. Fringe marshes (*) include both lacustrine and estuarine marshes (modified from Brinson 1993).

amount of oxygen in the soil and water, acidity, and nutrient content (Mitsch and Gosselink 1993).

WETLAND FORMATION AND PROCESSES

Compared with mountains and most rivers, our present wetlands are a young and dynamic ecosystem on the landscape. Most of the current peat-based wetlands date from the past 12,000 yr (postglacial), although wetland habitats, especially riverine, may exist from the Pleistocene ice ages (2 M to 10,000 yr ago) near lakes and former glaciers. Any process that produces a hollow or depression in the landscape and holds sufficient water may result in wetland formation. Wetlands are found in deserts near springs and in high rainfall or runoff areas of the mountains. Many of the northern bogs and prairie potholes of the midwestern United States and Canada were formed in depressions left by buried ice blocks (kettle holes) from retreating glaciers 10,000 yr ago. Along rivers and streams, periodic flooding lays down layers of silt and mud (alluvial deposits) along the banks and floodplains, creating bottomlands or swamp forests. Beavers also play a vital role in creating thousands of acres of new wetlands each year.

Wetland formation can happen as suddenly as when a major flood on the Mississippi in 1973 created new wetlands along the Atchafalaya delta or when debris dams up local rivers or streams. Much slower formation processes are at work in the Arc-

tic, where only the upper frozen layer of ice melts in the summer (see Chapter 1, this volume). Because of permafrost, cool conditions, and low evaporation, the small amount of annual rainfall and ice melt creates waterlogged soils and, in turn, the world's most extensive peatlands, primarily in Russia, Canada, and Alaska. Humans have also contributed greatly to the formation of wetlands in Canada and Alaska by cutting the forests, which causes the water table to rise and form new peatlands.

Wetlands are usually found in depressions or along rivers, lakes, and coastal waters, where they are subjected to periodic flooding. They can also occur on slopes adjacent to groundwater seeps or can cover an entire landscape in areas like large portions of northern Minnesota or coastal North Carolina, where precipitation levels exceed evapotranspiration (ET) and blocked drainage creates peat soils, which can be over 90% water by volume. Surface water depressions receive precipitation and overland flow. Groundwater depressions are in contact with the water table, so they receive groundwater plus rainfall and overland flow. Seepage wetlands occur on slopes where ground water flows near the surface. They differ from ground water depressions in that they have an outlet. The size of these wetlands depends directly on the amount of groundwater and overload discharge into these wetlands. Finally, overflow wetlands receive water from river flooding, lake water, or even tidal influences. The water level in these wetlands closely follows the water levels or flooding frequency of the water source.

A Comparison of Wetland Ecosystem Processes

Wetland functions or processes can be placed in five ecosystem-level categories: hydrologic flux, productivity, biogeochemical cycling, decomposition, and community/wildlife habitat (Richardson 1994, 1995). These processes directly influence the vegetation patterns found in each wetland type, and in turn the rates for these functions are altered by plant processes (e.g., photosynthesis, transpiration, nutrient absorption). The importance of each process depends on the wetland type because not all processes function at the same rate in each wetland. Thus, a brief comparison of ecosystem processes among the common wetland types found in North America provides a better basis for understanding and comparing wetlands on the landscape and the development and successional patterns of their vegetational communities.

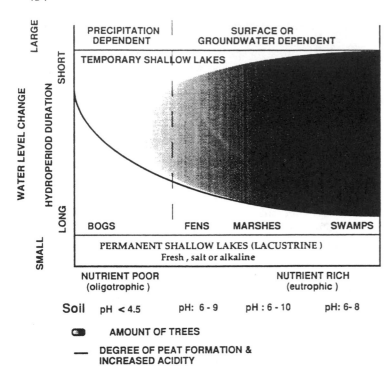

Figure 12.3. A model of wetlands found along gradients of nutrients and water regimes. Intensity of shading increases with increasing cover by tree canopies (modified from Gopal et al. 1990).

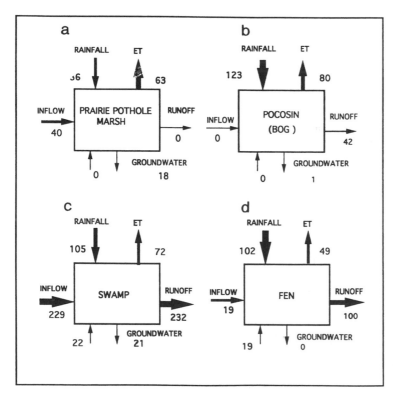

Figure 12.4. Annual water budgets for (a) a prairie pothole marsh in North Dakota, (b) a pocosin bog in North Carolina, (c) an alluvial cypress swamp in southern Illinois, and (d) a rich fen in North Wales, UK. All water flows are reported in centimeters per year (from Richardson (1994, 1995 and Mitsch and Gosselink 1993).

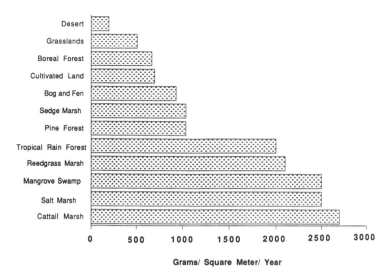

Figure 12.5. Net primary plant productivity (above- and belowground, g m^{-2} yr^{-1}) for selected wetland and terrestrial ecosystems (from Richardson (1979) and Maltby 1986).

Hydrology. The hydrologic condition of wetlands on the landscape is determined by type (e.g., bog, fen, or swamp), topographical position (i.e., proximity to large or small streams or lakes), overall size, and connection to groundwater. Many wetlands, including bogs, pocosins, peatlands, and marshes, function as "water pumps" on the landscape, losing sometimes two-thirds of their annual water by ET and leaving only 30% or less for annual runoff or groundwater recharge (Fig. 12.4a–d). The key factor controlling the vegetation within each wetland type is the period of time when water is at, near, or above the surface of the soil (i.e., hydroperiod). A general seasonal pattern for each wetland type shows that swamps often have high water levels but they dry out. Fens maintain a more constant water level due to groundwater inputs. A bog system often has high water after spring thaw, but peat soils can dry out during the high evapotranspiration (ET) of summer. Prairie marshes can have vastly fluctuating water levels due to extensive dry periods, depending on whether or not they are connected to groundwater. These varying water levels and hydroperiods in wetlands control the vegetation and successional patterns within each wetland type.

The proportion of water inflow compared to rainfall varies greatly with swamps, receiving most of their inputs by streamflow. They also gain and lose some water to the shallow aquifer beneath them. The stream flow out of swamps constitutes the biggest loss, although the water can be held for months and released more slowly than in areas where the vegetation is removed. Bogs receive most of their water by rain and lose almost none

to groundwater. The large portion of annual runoff from perched bogs (i.e., water table elevated above the regional water table) is in spring after winter snowmelt or after the bog peat has filled with rain due to low winter ET rates. The peat soils of these systems act like a sponge on the landscape, holding and slowly releasing vast quantities of rainfall when the peat is drier, and rapidly giving up water when the sponge is full. Most small bogs have no flowing outlets and lose almost all of their water by ET. Pocosin bogs, because of their size (often >20,000 ha), slowly lose water by surface and subsurface runoff, which is then distributed over vast areas to the surrounding lakes, rivers, and estuaries (Fig. 12.4 b). Fen systems are flow-through systems with runoff often exceeding ET, especially during ice thaw periods in the spring. The annual flow from fens connected to groundwater is very uniform throughout the year (Fig. 12.4 d).

Primary productivity. Annual primary productivity rates in wetlands are among the highest reported for any ecosystem in the world. Their rates greatly exceed grasslands, cultivated lands, and most forests (Fig. 12.5). Cattail, salt marsh, and reed grass marshes are the most productive and nearly double the aboveground and belowground annual plant biomass production stored in sedge marshes or bogs, fens, and muskegs. Grasslands and croplands of the United States produce only one-fourth the fiber of the most productive wetlands, whereas upland forests produce about one-half. Only the tropical rain forest comes close to matching the primary productivity of the mangrove swamps or cattail and salt marshes. These high productivity rates suggest an important role for wetlands in sequestering carbon and influencing greenhouse gases (Gorham 1991). Plant biomass also serves as food for a great multitude of both aquatic and terrestrial animals as well as for many species that are endemic to wetlands. The development of peat soils depends on the high carbon fixation found in many wetlands. Mammals such as moose, caribou, and muskrat graze on marsh plants. Waterfowl depend heavily on marsh plant seeds.

Decomposition. Another unique feature of wetlands compared to terrestrial ecosystems is that dead plants and animals do not decay as rapidly, especially in acid peatlands (Chamie and Richardson 1978). When leaves, small stems, and roots die, they normally are shredded by macroinvertebrates, earthworms, nematodes, and microbes and are decomposed almost completely in 1 yr or less in terrestrial systems. The decay rate in peatlands is only one-half to one-quarter the rates of the more aerobic uplands (Mitsch and Gosselink 1993). The complex organic matter, which is 45% carbon by weight, usually decays in terrestrial ecosystems to CO_2 and H_2O, and the mineral elements are released for new growth. This is not the case in wetlands. The slow decay rates in wetlands are due not only to a lack of oxygen but also to low pH, shortages of calcium, and low soil temperatures (Bridgham and Richardson 1993). These conditions eliminate most of the decomposer organisms, like filamentous fungi, and the existing anaerobic bacteria obtain energy by inefficient processes like fermentation (Faulkner and Richardson 1989). Very reduced soils (i.e., those with no NO_3 or SO_4 present) release "marsh gas" – such as methane and hydrogen, which are important greenhouse gases. Normal metabolism is so greatly reduced in wetlands that many of these ecosystems slowly build up peat (undecomposed plant organic matter and minerals) at a rate of 1–2 mm yr on average (Craft and Richardson 1993a). Thus, a distinct feature of wetlands versus uplands is the dominance of undecomposed plant material as the main soil matrix. Soils are Histosols. Peatlands store vast quantities of carbon and help to balance the global carbon cycle (Gorham 1991). Not only have the peat-forming organisms and local biota been preserved in this peat over thousands of years, but pollen and archaeological artifacts have been found buried almost intact. Analysis of pollen and macrofossils from peats all over the world have allowed scientists to establish historical vegetation changes, climatic variations, and human impacts on the landscape as far back as 275,000 yr ago (Walker 1970).

Undisturbed wetlands are thus a "time machine" of past vegetation and climate.

Biogeochemical cycling. One of the key ecological functions of wetlands on the landscape is their ability to store, transform, and cycle nutrients (Richardson 1989). Wetlands maintain among the widest range of oxidized and reduced chemical states of all ecosystems due to their periodic flooding and drying cycles. The unique shift from aerobic (free oxygen present) to anaerobic (no free oxygen) soil conditions provides these systems with the ability to process PO_4, NO_3, SO_4, and C and release gases (N_2O, S, H_2S, CH_4, CO_2) to the atmosphere, thus affecting biosphere problems such as acid rain, global warming, and greenhouse gases. Most flooded wetland soils have a thin oxidized layer at the surface caused by proximity to the air or higher dissolved oxygen in overlying floodwater. The reduced layer below dominates the soil and the lower water-column processes. This prevents upland plants, bacteria, and animals from invading these ecosystems because of their lack of oxygen for normal metabolism. The vegetation growing and reproducing successfully in wetlands is thus adapted to survive the lack of oxygen in the root zone and reduced forms (often toxic) of iron (Fe^{+2}), manganese (Mn^{+2}), nitrogen (NH_4^+), sulfur ($S^=$), and carbon (CH_4) (Gambrell and Patrick 1978). Wetland plant adaptations to these conditions include anatomical modifications to enhance the flow of O_2 (increased arenchyma tissue in roots and stems, lenticals, and adventitious roots), biochemical changes like malate storage, to prevent ethanol buildup, as well as alternate metabolic pathways, and enhanced radial oxygen loss in the roots (Koslowski 1984). Thus, wetland with a wide range of oxidation/reduction reactions and unique biogeochemical shifts in valence states for ions provide one of the greatest challenges for plant and animal growth and survival.

Community/habitat. There are a diverse number of plant habitats within each wetland based on water depth, hydroperiod, and soil conditions in each zone. For example, southern bottomland hardwood forests are comprised of numerous tree and shrub species specially adapted to seasonal flooding conditions and water depths such that only flood-tolerant species, like bald cypress (*Taxodium distichum*), are found growing in deep standing water, whereas moderately tolerant species, like American elm (*Ulmus americana*), are found in shallow water areas that dry out during a good portion of the growing season (Hook and Scholtens 1978). Moreover, rare orchids (*Habenaria* spp.) and unusual insectivorous plants such as sundews (*Drosera* spp.), Venus flytraps (*Dionea muscipula*), and pitcher plants (*Sarracenia* spp.) exist only in wetland habitats.

Wetlands are important year-round habitats for thousands of plant species and hundreds of bird species, amphibians, reptiles, and mammals, especially in the warmer climates. It has been estimated that more than 150 types of birds and 200 fish species are wetland-dependent. Wetlands also provide important breeding grounds, overwintering areas, and feeding grounds for millions of migratory waterfowl and other birds. Sixty to 70 percent of the 10–12 million ducks that breed annually use the prairie pothole region of the United States and Canada alone. Additionally, many upland animal species depend on wetlands for water and food, especially during drought periods.

MAJOR WETLANDS
OF NORTH AMERICA

Peatlands (Northern Bogs and Fens)

Peatlands (see Fig.12.6) comprise the vast majority of area of the palustrine wetland type in North America (Tiner 1984). A peatland is defined as a wetland with a waterlogged substrate and at least 30 cm of peat (Kivimen and Pakarinen 1981). Peatlands are areas that sequester carbon via peat accretion. Boreal peatlands are commonly found in the coniferous forest region of Canada, the interior of Alaska (see Chapter 2), as well as in major areas of the Great Slave/Great Bear Lake Region, the Hudson Bay lowlands, and the Glacial Lake Agassiz region (Glaser 1987). In western Canada, Alaska, and northern Minnesota, these peatlands cover large areas; in eastern Canada and Maine, they are found in regional depressions (Zoltai and Pollet 1983; Damman and French 1987). Vast acreages of peatlands also exist in the pocosin bogs of North Carolina (608,000 ha), as well as the 500,000 ha Everglade fens of southern Florida. The total area of marshes and peatlands in the United States (excluding Alaska) is 9.8 M ha (Mitsch and Gosselink 1993).

Transeau, in 1903, completed one of the first surveys of the boreal peatlands and determined their distribution from the ranges of peatland taxa such as *Drosera* (sundew), *Sarracenia* (pitcher plant), *Larix* (tamarack), *Ledum* (labrador tea), and *Chamaedaphne* (leather leaf). Although most scientists have focused on the vascular plants in the northern peatlands, it is members of the moss genus *Sphagnum* that often dominate the soil-forming process

and characterize the northern peatlands. This is true because they have amazing water-holding capacity (15 to 23 times their dry weight) (Vitt, Achuff, and Andrus 1975) and a uniquely high capacity for cation exchange that apparently aids in their ability to maintain low water acidity and slow decomposition rates (Clymo 1967). Spectacular landform patterns showing distinct tree islands, water tracks, strings (peat ridges), and flarks (pools) are found on the landscape and were discovered from aerial surveys and photographs (Fig. 12.7) (Heinselman 1963, 1970; Glaser 1987). Glaser (1992) mainly described distinctive landform features like ovoid forested islands and bogs that occur in large fen complexes in Minnesota and eastern Canada. These islands occur in western continental Canada but are lacking in oceanic areas (Nicholson and Vitt 1990). In Quebec, bogs without trees but with distinctive pools and ridge patterns are reported by Glaser and Janssens (1986). In Arctic Canada and many parts of Alaska, permafrost dominates the landscape, but farther south it is less prevalent and is usually restricted to ombrotrophic peatlands (Zoltai, Taylor, Jeglum, Mills, and Johnson 1988). Permafrost formation causes the expansion of freezing water, which raises the surface above the water table. The drier conditions created by this result in dense growth of conifers in the south and open-canopied lichen-dominated woodlands in the north (Vitt, Halsey, and Zoltai 1994). The small peat hills thrust above the peatland are called "palsas," and larger uplifted areas are referred to as "peat plateaus" (see Chapter 1, this volume).

Peatland formation (terrestrialization and paludification) and classification. Peatlands form because the balance between plant production and plant decomposition results in a net accumulation of organic matter on an annual basis. This means that plants are responsible for the development of the underlying Histosols, which have a content of organic matter >20–35% (>40% carbon), are usually acidic, and have bulk density of 0.02–0.3 g/gm^3 (Mitsch and Gosselink 1993). The positive balance between production and decomposition is usually the result of reduced decomposition of organic matter due to a lack of oxygen, low pH, and in some cases limiting nutrients (Chamie and Richardson 1978).

These peat-based systems can form directly over forest or prairie soils by the process of paludification (Cajander 1913) or fill in lakes by the process of terrestrialization (Weber 1902). Peatlands can also be thought of as vegetation types as well as landforms since peatlands are made up of

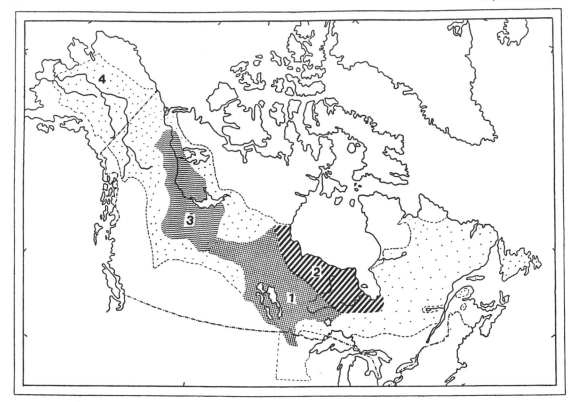

Figure 12.6. Major boreal peatlands of North America: (1) Glacial Lake Agassiz region, (2) Hudson Bay lowlands, (3) Great Bear/Great Slave Lake region, and (4) interior of Alaska, for which detailed peatland maps are not avail-able. The lightly stippled area marks the boreal region (from Viereck and Little 1972; Rowe (1972); Glaser 1987; Zoltai and Pollett 1983).

complexes of plant communities (Damman and French 1987). All peatlands include minerotrophic sites, at least at their margins. Therefore, the distinction between ombrotrophic and minerotrophic is of limited value when one is dealing with peatlands as landscape units. Damman and French (1987) suggest that a practical subdivision of peatlands be based on the nature of water that controls their development (von Post and Granlund 1926; Sjors 1948).

Climate and hydrology. Bogs are present where annual precipitation is >500 mm annually, biotemperature is <8°C, and ET/PT ratios are <1 (Vitt et al. 1994). Hydrology is the primary controlling factor that determines the type of peatland as well as plant succession (von Post and Granlund 1926; Ingram 1967, 1983; Mitsch and Gosselink 1993). Two major hypotheses explain the limits of a bog: (1) a bog is isolated from the influences of groundwater and outward expansion is blocked by minerotrophic water draining from upland or nutrient-rich adjacent fen systems; or (2) outward expansion

of a bog is blocked by discharge of groundwater at the bog margin (Siegel 1981, 1983; Boldt 1985; Siegel and Glaser 1987). Both hypotheses are based on the idea that the growth of the bog is blocked by high pH and calcium in the surface waters (Glaser 1987). An understanding of the source of water in the peatland is essential for understanding the distribution of plant species.

Water chemistry. The distribution of many circumboreal vascular plant species and cryptogams has been related to water chemistry in peatland ecosystems worldwide (Du Rietz 1949; Sjors 1946, 1948, 1952). Species have shown a sensitivity to changes in pH and the concentrations of Ca such that these species have been used as indicators of ranges of water chemistry. Heinselman (1963, 1970) applied this system to Minnesota peatlands and it allowed him to define bogs as having pH <4.2 and Ca <2.0 mg/L.

The primary cause of this acidity is open to debate, with Clymo (1963) and Clymo and Hayward (1982) presenting the argument that the cation

Figure 12.7. Patterned water track with fields of tree islands, Lost River area of northern Minnesota. Peat ridges and pools are oriented transverse to the slope (from Glaser 1987).

across a peatland. An analysis of these landforms shows that the spring-fen mound has formed over a ridge in the mineral substrate and that the raised bog has grown over a depression on the landscape (Almendinger, Almendinger, and Glaser 1986). These landforms show a striking pattern from the air (Fig. 12.7) but show almost no change in relief on the ground. Paleoecological analysis of peat cores indicates that deeply composed fen peats often first form in the peatland, followed by *Sphagnum* peat with wood (Glaser 1987). Glaser (1987) has found areas in which successional change moved from a *Sphagnum*-dominated bog or poor fen to a spring-fen in less than 2000 yr.

Most peatlands formed after the mid-Holocene when the climate was cooler and wetter. Continental bogs are generally younger (2500–5000 yr BP) than coastal maritime peatlands (6000–9000 yr BP), although there is considerable variation due to local edaphic conditions, fires, and climate (Glaser 1992; Vitt et al. 1994). Palaeoecological studies have documented that most of the present organic matter in peatlands in North America formed about 4500–5000 yr BP, with such diverse areas as the Everglades in Florida, the peatlands in Canada and the northern United States, the pocosins of North Carolina, and the Okefenokee of Georgia all showing vast expansion during this time (Gleason and Stone 1994; Glaser 1987; Daniel 1981; Cohen, Casagrande, Andrejko, and Best 1984).

The formation of *Sphagnum* mats around ponds and lakes also dates, from the mid-Holocene or earlier (Cushing 1963; Wright and Watts 1969; Lindeman 1941; Damman and French 1987), and their development is often related to rising water tables and the amount of acid coming from nutrient-poor runoff of upland pine or spruce species surrounding the depressions at this time (Jacobson 1979; Janssen 1967a, 1967b, 1968; Delcourt and Delcourt 1981; Chapter 10 this volume). A change to hardwood forests or the development of fertilized agricultural fields on uplands draining into bogs results in a significant increase in the amounts of Ca and nutrients – resulting in a demise of the native peatland vegetation, especially *Sphagnum*.

exchange process of *Sphagnum* (i.e., hydrogen pumps) is the primary cause, and Gorham, Eisenreich, Ford, and Santelmann (1984) proposing that organic acids from decomposition are the main generator of acidity. Glaser (1987) combined this information with the fact that Ca inputs arise from a mineral source either from runoff or the discharge of groundwater, or even rainfall from surrounding uplands high in Ca dust. When plotted together, pH and Ca water chemistry show that various types of peatlands form a continuum on the landscape. The high pH (7.0) and Ca concentrations >30 mg/L in extremely rich fens are maintained by a continuous discharge of groundwater from calcareous till (Glaser 1987).

Vegetation of peatlands past. There major landforms (raised bog, water track, and spring fen) can be used to separate the patterned peatlands of North America, especially those of northern Minnesota (Glaser 1987). A water track is defined as any area in which minerotrophic runoff moves

Current vegetation. The vegetation of patterned northern peatlands is usually dominated by a few plant communities with usually between 100 and 200 species of vascular plants present but seldom more than a dozen dominant plant species. Glaser's (1992), excellent cross-sectional study across central and eastern North America, has shown that vascular plant richness is limited to 81 species in bogs of eastern North America. Species numbers increase with increasing precipitation but decrease with increasing numbers of freezing days (Fig. 12.8 a, b). His study suggests that the most important variable influencing species richness is climate, but age of the bog and landform types produce regionally different patterns of species richness. In continental western Canada, Vitt, Li, and Belland (1995) found 110 species of bryophytes; 71% occurred in rich fens dominated by brown moss, and 62% were found in bogs and poor fens dominated by *Sphagnum*. Bryophyte species richness was not primarily correlated with climate but, rather, with habitat heterogeneity (r = 0.46), not temperature (r = 0.15).

The two major types of vegetation in peatlands are ombrotrophic bog and minerotrophic fen plant communities. The dominant plant community types and major species for each area, characteristic species, and environmental conditions are listed in Table 12.1. Bogs and fens usually have different species assemblages, but some species overlap. However, fen indicator species, which cannot grow in waters <pH 4.2 and Ca concentrations <2 mg/L, can be used to separate fens from bogs (Sjors 1963; Glaser 1987). Unfortunately, no species are restricted to ombrotrophic bogs, and so an absence of fen-indicator species is used to separate these types. A direct gradient analysis of the major shrubby plants and sedges found in both bogs and fens in peatlands reveals that species such as *Picea mariana* (black spruce) are often found in both the extremely rich fen and the bog, thus making them useless as indicators (Fig. 12.9 a, b). However, *Carex oligosperma* (sedge) and *Eriophorum spissum* (cottongrass) are mainly found in bogs and *C. lasiocarpa* and *Scirpus caespitosus* (deerhair bulrush) in rich fens (Fig. 12.9 a, b). As peat formers, sedges are second only to *Sphagnum* in northern peatlands.

Rare plants. One of the unique features of wetlands is the number of rare plant and animal species that are found only in these ecosystems when undisturbed. It is impossible to tabulate all those found throughout the peatland regions of North America, but a brief listing gives some idea of the importance of these habitats for many endemic species. For example, two rare sundews, *Drosera anglica* and *D. linearis*, grow only in deepwater flarks of pristine peatlands (Wheeler and Glaser 1979). Fens contain rare species like *Eleocharis rostellata* (spike rush), *Tofieldia glutinosa* (sticky false asphodel), and *Triglochin palustris* (marsh arrow grass). A number of rare bryophytes have also been noted, with the most remarkable being *Calliergon aftonianum*, which had been thought to be extinct since the Pleistocene era (Glaser 1987).

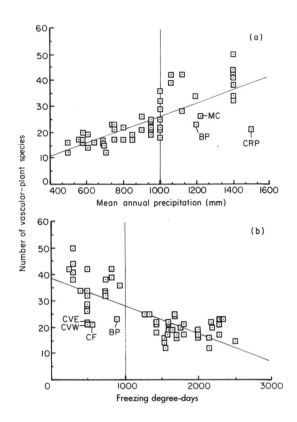

Figure 12.8. Regressions of species richness against (a) mean annual precipitation and (b) annual number of freezing degree-days for bogs in eastern North America. The vertical lines mark (a) 1000 mm precipitation and (b) 1000 freezing degree-days and the lines separate species-rich from species-poor bogs (from Glaser 1992).

Prairie Potholes (Northern Marshes and Fens)

Glaciers that once covered the central North American landscape retreated nearly 12,000 yr ago and left tens of thousands of water-filled depressions (called "prairie potholes" by the locals) on land that was once prairie grassland. These shallow depressions, mostly geologic kettle-hole formations vary in size from <0.5 ha to several hectares and usually have surface water for some period during the year (van der Valk 1989, Richardson, Arndt, and Freeland 1994). These marshes lack a well-defined outlet but often overflow during wet springs (Cowardin, Gimler, and Mechlin 1981). A small number exist as permanent lakes. The main prairie pothole region (PPR) extends from the prairie-forest line north of Edmonton, Alberta, southward toward the Wisconsin-age Des Moines lobe in Iowa (Fig. 12.10).

The boundary defining the region is partly

Table 12.1. Major plant communities in the boreal patterned peatlands of northern Minnesota

Characteristics	Bog (Ombrotrophic)		Rich Fen (Minerotrophic)		
Plant community type	Forested Bog	Open bog	Fen-flark	Fen-string	Forested Island
Dominant species	Black spruce (Picea mariana) Ericaceous shrubs- Swamp laurel (Kalmia polifolia) Bog rosemary (Andromeda glaucophylla) Labrador tea (ledum groenlandicum) Leatherleaf (Chamaedaphne calyculata) Sphagnum mosses (Sphagnum spp.)	Sedge (Carex oligosperma) Ericaceous shrubs (same as forested bog) Sphagnum mosses (Sphagnum spp.)	Sedges (Carex lasiocarpa) (C. livida) (C. limosa) Buckbean (Menyanthes trifoliata) White beak rush (Rhynchospora alba)	Bog birch (Betula pumila) Bog rosemary (Andromeda glaucophylla) Small cranberry (Vaccinium oxycoccus) Leatherleaf (Chamaedaphne calyculata)	Tamarack (Larix laricina) Black spruce (Picea mariana)
Characteristic species	Sedge (Carex trisperma) Lingberry (Vaccinium vitis-idaea) 3-leaved false Solomon's seal (Smilacina trifolia) Feathermosses (Pleurozium schreberi) (Dicranum sp.)	Sedge (Carex oligosperma)	Marsh arrow grass (Triglochin maritima) Intermediate bladderwort (Utricularia intermedia) Intermediate sundew (Drosera intermedia)	Shrubby cinquefoil (Potentilla fruitcosa) Sedge (Carex cephalantha)	Sedges (Carex pseudo-cyperus) Black chokeberry (Aronia melanocarpa) Dwarf raspberry (Rubus pubescens) Velvet honeysuckle (Lonicera villosa)
pH		very acidic (pH less than 4.2)		slightly acidic to neutral (pH greater than 5.2)	
Salt concentration		very low (e.g., Ca <2.2 mg)		moderate to high (e.g., Ca >4.3 mg)	
Species diversity		very low (9–13 plant species)		generally moderate to high (12–58 plant species)	
Associated peatland landforms		raised bogs, ovoid islands		water track features such as ribbed fens, teardrop islands, circular islands	

Sources: DNR (1984, cited in Glaser 1987).

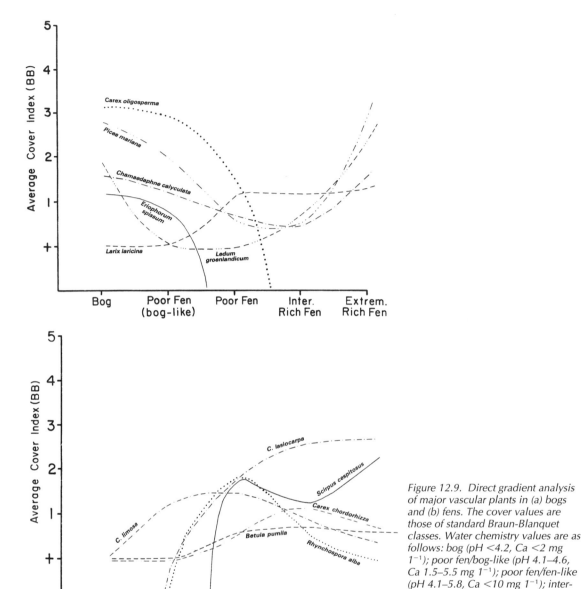

Figure 12.9. Direct gradient analysis of major vascular plants in (a) bogs and (b) fens. The cover values are those of standard Braun-Blanquet classes. Water chemistry values are as follows: bog (pH <4.2, Ca <2 mg 1^{-1}); poor fen/bog-like (pH 4.1–4.6, Ca 1.5–5.5 mg 1^{-1}); poor fen/fen-like (pH 4.1–5.8, Ca <10 mg 1^{-1}); intermediate rich fen (pH 5.8–6.7, Ca 10–32 mg 1^{-1}); extremely rich fen (pH >6.7, Ca >30 mg 1^{-1}) (from Glaser 1987).

based on geology and climate. The glaciated PPR is bounded by the Canadian Shield to the northeast, the Cordillera to the west, and the limit of the Wisconsin glacial advance to the south 14,000–9,000 BP (Winter 1989). The cool continental climate of the PPR has temperatures ranging from −40°C to +40°C annually (Winter 1989). Because of the extremely cold winters in the northern prairie, these wetlands remain frozen for nearly half a

year, many frozen solid. Droughts and pluvial cycles are the norm, and annual potential evapotranspiration generally exceeds precipitation.

It has been estimated that the PPR originally contained 8 M ha of potholes (Frayer, Manahan, Bowden, and Grayhill 1983). Nearly 50% of these wetlands have been drained, primarily for agriculture, with states like Iowa and North Dakota losing 89% and 49% of their estimated 1.6 M and 2 M ha

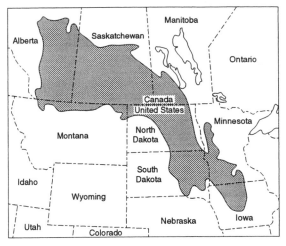

Figure 12.10. The prairie pothole region of North America (after van der Valk 1989).

of wetlands, respectively (U.S. Congress 1984; Dahl 1990). Thus, the pothole regions in many parts of the United States and Canada have been greatly altered by drainage, grazing, haying, cultivation, and pesticides from farm runoff.

Formation and classification of prairie potholes. Most of the potholes were probably formed as kettle depressions from glacial activity, but these wetlands are also found in downslope depressions in moraines, at ends of moraines, and in lake and outwash plains. The relief is also much greater in some portions of the PPR than is usually found in other glaciated areas, which gives rise to some dynamic and complex groundwater flow patterns (Winter 1989). The soils of the region are dense, silty, clayey till with high water-holding capacity and low infiltration rates. The fine-grained nature of the till significantly affects surface and groundwater flow and, in turn, the distribution of vegetation. However, the high clay content results in drying and cracking during the summer dry periods, and water can infiltrate these soils. Typically, these tills contain large amounts of dolomite and calcite derived from marine shales and Paleozoic bedrock (see J. L. Richardson et al. 1994 for a review of PPR soils). Moreover, groundwater from underlying shale bedrock in many regions is laden with high amounts of sulfate salts (Arndt and Richardson 1989). In the central and western areas (e.g., Saskatchewan), deposits of Na and Mg sulfates provide a unique geochemistry (Last and Schweyen 1983). The soils, depending on the hydrology, amount of gypsum, calcite, or organic

matter present, produce soil classes ranging from Argiaquolls to Fluvaquents and Haplaquolls to Calciaquolls; even Histosols are produced in the deeper wet areas (Arndt and Richardson 1988).

The evaporation of these leached and upwelled salts, especially at the wetlands edge, provide a unique zone of salinization or calcification that influences species composition and vegetation patterns (Richardson, et al. 1994). If carbonate is higher than calcium, then sodium carbonate soils form. If Ca is higher than $CO_3^=$ in solution after initial carbonate precipitation, then gypsiferous saline soils are formed. Under low S and Mg solute conditions, calcium carbonate salt crusts can form.

The distinct circular zonation patterns of vegetation found in potholes are based on depth of water, salinity, and disturbance history. These zones have been used to classify wetland basins into five distinct zones based on permanence of water, groundwater seepage, and salinity. Potholes are mostly palustrine systems, a few areas associated with open water being lacustrine.

Climate and hydrology. Because of a 5–10 yr cycle of periodic droughts and wet periods, potholes undergo dramatic changes in vegetation patterns due to alterations in water depth (Kantrud, Millar, and van der Valk 1989). The amount of water found in potholes depends on annual rainfall as well as spring runoff from the surrounding grasslands or agricultural fields. The rainfall pattern varies from semiarid in the west to humid in the east. Annual rainfall (20 year norms) is 340 mm in semiarid regions, 500 mm in subhumid regions, and 850 mm in humid regions (Richardson et al. 1994). Low precipitation coupled with high ET results in a negative water balance for potholes with respect to the atmosphere. Precipitation minus ET ranges from −100 mm in Iowa to −600 mm in southeast Saskatchewan (Winter 1989). A water balance for a representative pothole shows that 83% leaves as evapotranspiration but ≈20% is lost into groundwater (Fig. 12.4a).

Arndt and Richardson (1988) and Winter (1989) discovered that if we are to understand their vegetation patterns, prairie potholes should be classified into recharge, discharge, or flowthrough systems because these wetlands form a water continuum on the landscape. They also found that one needs to understand the dominant flow conditions because seasonal and climatic conditions create intermittent reversals. Recharge wetlands can be distinguished as wetlands temporarily or seasonally ponded with fresh water and leached soil profiles (Richardson et al. 1994). Discharge

Table 12.2. *Basic stable vegetation zones of prairie pothole wetlands*

Common description	Lifeform of dominant vegetation	Normal period of inundation
Wet meadow	Low herbaceous–grasses, fine sedges and forbs (shrubs in parkland)	A few weeks in the spring.
Shallow marsh	Mid-height herbaceous: grasses, coarse sedges	Spring to mid-summer or early fall.
(Emergent) deep marsh	Tall, coarse herbaceous: cattail, bulrushes	Spring through fall and frequently overwinter.
Shallow (permanent) open water	Submergent or floating aquatics	Year-round except in periods of severe drought.
Alkali (open or intermittent)	Either devoid of vegetation or containing the submergent *Ruppia maritima*	Highly variable–from a few weeks in the spring to fall or overwinter, depending on runoff and ground water conditions.

Source: Kantrud et al. (1989).

wetlands are semipermanent to permanent, have small catchment areas, are ponded by groundwater inflow, and often have saline soils with salt crusts at the edges of the wetlands. Flowthrough wetlands exhibit both discharge and recharge, have an intermediate range of salinity (i.e., brackish), are generally semipermanent in ponding duration, produce hydric soils that contain calcite and gypsum, and can be saline.

Water chemistry. The salinity factor is determined by the amount of dissolved sulfate, Na, Mg, and specific conductivity (SC reported as units of microsiemens per centimeter, µS/cm) is often used to explain the majority of variability in prairie pothole types (Richardson and Bigler 1984). Cowardin et al. (1979) proposed the following 6 salinity classes: fresh (<800 µS/cm), oligosaline (800–8000 µS/cm), mesosaline (8000–30,000 µS/cm), polysaline (30,000–45,000 µS/cm), eusaline (45,000–60,000 µS/cm), and hypersaline (>60,000 µS/cm). In general, a higher specific conductance is found in the arid regions of northwestern Canada, but a range is found in all areas. The regional PPT:ET ratio does not explain fully the higher SCs because the local soil and groundwater chemistry greatly influences ion chemistry, as noted earlier. However, the longer the recharge-discharge flowpath, the more salinity accumulates in soils. Also, evaporation at the edges of the wetland accumulates salts in this region, resulting in distinct vegetation zones (Steinwand and Richardson 1989).

Gorham, Dean, and Sanger (1983) determined that there was a good relationship between SCs and concentrations of major ions, although the pro-

portion of major ions (Ca, Mg, Na, K, HCO_3, SO_4, and Cl) changed considerably across the entire pothole region. These changes in ion composition were attributed to a shift from noncalcareous glacial drifts to calcareous drifts with increasing sulfur in the west part of pothole country. Regression analysis does show that SC is consistently related to Na, K, and SO_4 in the northern potholes. LaBaugh (1989) reports that most dilute wetlands in the north are dominated by Ca and HCO_3 but notes that the combination of Mg and HCO_3 is the most common combination. Sulfate is the most common anion in the northern prairies. For a more detailed analysis of the waters of the prairie wetlands, see LaBaugh (1989).

Vegetation. The flora of prairie potholes is controlled primarily by water regime and secondarily by salinity, water and soil ion chemistry and disturbance. The zonation patterns found in the wetlands are often dominated by a single species adapted to a narrow range of water depth and salinity (Kantrud et al. 1989; Table 12.2). In general, the number of species decreases as salinity increases. The highly alkali systems contain only species of *Ruppia*. The distribution of dominant plant species in saturated water and semipermanently flooded water regimes was compiled by Kantrud et al. (1989) and shows a dominance of *Typha latifolia, Phragmites australis*, and *Scirpus validus* as emergents in the deeper water regimes (Table 12.3). Sedges and grasses comprise the other dominants, with flowering plants becoming more important in less saturated soils. The dominant species of floating aquatics, mosses, rooted vascular

Table 12.3. *Dominant (D) and indicator species of persistent vegetation in mixosaline palustrine emergent wetland with saturated water regime in the northern plains and prairies*

Typha latifolia (D)
Phragmites australis (D)
Scirpus validus (D)   ↑   Wetter
Cicuta maculata
Scirpus atrovirens (D)
Carex aquatilis (D)
Eleocharis calva
Glyceria striata (D)
Asclepias incarnata
Triglochin maritima
Carex rostrata (D)
Eupatorium maculatum
Scutellaria epilobiifolia
Deschampsia caespitosa (D)
Carex aurea
Parnassia palustris
Ranunculus septentrionalis
Lysimachia thrysiflora
Eriophorum angustifolium
Aster junciformis
Juncus torreyi
Calamagrostis inexpansa (D)
Muhlenbergia glomerata
Lobelia kalmii
Carex sartwellii
Carex interior
Carex lanuginosa (D)
Viola nephrophylla
Epilobium glandulosum   ↓   Drier
Solidago graminifolia
Helianthus nuttallii

Source: Kantrud et al. (1989).

plants, and algae, in relationship to salinity classes, are shown in Table 12.4. These are based on field work by Smeins (1967) and Disrud (1968) and were combined by Kantrud with his unpublished data to produce these dominance patterns for each flooding regime.

It is important to remember that the dramatic shifts in wet–dry cycles noted earlier result in significant changes in vegetation and biomass found in any pothole in any particular year (van der Valk and Davis 1978). The vegetation dynamics of prairie wetlands show more extremes in plant communities than almost any other wetland type, as is demonstrated by the successional pattern of dry marsh to lake marsh that occurs in a 6–30 yr cycle in Iowa potholes (Fig. 12.11; van der Valk and Davis 1976, 1978). They found that rapid shifts in vegetation in response to changing water depths were possibly due to the presence of a seed bank for emergent, submersed, free-floating, and mudflat annuals. Shoot biomass for emergents peaked during the regeneration phase at nearly 4 metric tons/

ha but was less than 0.3 metric tons/ha in the lake marsh stage (van der Valk and Davis 1978). Total aboveground biomass also peaked at the regeneration stage (6 metric tons/ha) and dropped to near 0.5 metric tons/ha in the lake phase. The submerged and algal components increased standing crop after reflooding to nearly 2 metric tons/ha. The emergents are essentially gone in the degenerating and lake stage because their seeds cannot germinate under water. Muskrats quite often decrease standing crops of emergents like cattail significantly during the degenerating marsh phase

Table 12.4. *Dominance types of vegetation by salinity classes in aquatic bed algal (A), aquatic moss and liverwort (M), rooted vascular (R), and floating vascular (F) subclasses of palustrine wetland with semipermanently flooded water regime in the northern plains and prairies*

Water chemistry modifier: Fresh (<800 μ s/cm)
    Elodea longivaginata (R)
    Spirodela polyrhiza (F)
    Riccia fluitans (M)
    Potamogeton gramineus (R)
    Lemna trisulca (F)
    Utricularia vulgaris (R)
    Ricciocarpus natans (M)
    Potamogeton richardsonii (R)
    Ceratophyllum demersum (R)
    Myriophyllum spicatum (R)
    Drepanocladus spp. (M)
    Ranunculus subrigidus (R)
    Lemna minor (F)
    Ranunculus flabellaris (R)

Water chemistry modifier: Oligosaline (800–8000 μ s/cm)
    Hippuris vulgaris (R)
    Ranunculus gmelini (R)
    Ricciocarpus natans (F)
    Callitriche hermaphroditica (R)
    Potamogeton zosteriformis (R)
    Potamogeton pusillus (R)
    Lemna trisulca (F)
    Utricularia vulgaris (R)
    Potamogeton richardsonii (R)
    Ceratophyllum demersum (R)
    Myriophyllum spicatum (R)
    Drepanocladus spp. (M)
    Ranunculus subrigidus (R)
    Lemna minor (F)
    Zannichellia palustris (R)
    Chara spp. (A)
    Potamogeton pectinatus (R)

Water chemistry modifier: Mesosaline (8000–30,000 μ s/cm)
    Zannichellia palustris (R)
    Chara spp. (A)
    Potamogeton pectinatus (R)
    Ruppia maritima (R)

Source: Kantrud et al. (1989).

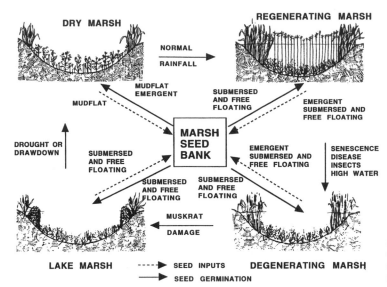

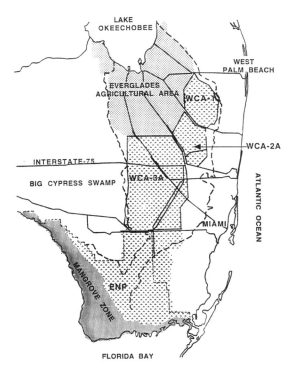

Figure 12.11. Generalized vegetation cycle in an Iowa prairie pothole marsh over a 6–30 yr drying cycle (from van der Valk and Davis 1978).

Figure 12.12. The remaining everglades area (hatched) on state land (WCA-1,-2,-3) and Everglades National Park land, Florida.

(Weller 1994). Mudflat species (2 metric tons/ha) and some emergents (2.5 metric tons/ha) grow during the dry marsh stage, but after the drought the marsh refloods. This eliminates all annuals that cannot tolerate flooding, but emergents, such as cattail and bulrush, remain (Kantrud et al. 1989). Thus, the vegetation composition and standing peak crop found in each zone of a prairie pothole marsh are also greatly dependent on what phase of the wet–dry cycle is being studied as well as soil chemistry. This shows the importance of long-term vegetation studies if we are to truly understand these systems.

California Vernal Pools

A vernal pool is a shallow depression within a mosaic of low-elevation grassland. The depression fills with water in wintertime because it is underlain by a claypan or hardpan that inhibits percolation. As it dries in the spring, various native annual plant species flower, often in concentric rings that correspond to different periods of inundation. Vernal pools alternate with "upland" hillocks at a relatively fine grain of pattern. The average area of a pool is on the order of hundreds of square meters and average depth is 40 cm. Exceptional vernal pools reach lake size (e.g., Boggs Lake in Lake County). Soils are typically old, approaching half a million years, but parent material can be diverse. Excellent reviews of pool distribution and flora have been written by Holland and Jain (1988) and Holland (1978).

Vernal pools are found throughout the Central Valley, at intermediate elevations in the North Coast Range, in Sierran foothills, and on coastal terraces in southern California. Their floristic composition varies regionally and also locally. Typically, the fruits and seeds of vernal pod species have very limited dispersal potentials. Vernal pools have been equated with oceanic islands and their biogeography studied from an evolutionary point of view (Holland and Jain 1981). Indeed, the presence of several highly speciated vernal pool genera endemic to California (e.g., *Downingia, Limnanthes, Neostapfia, Orcuttia, Pogogyne*) has been taken by Zedler (1987) as evidence of relatively recent evolution of California vernal pool flora during the past 3 M yr.

Other characteristic vernal pool genera include *Eryngium, Psilocarphus, Plagiobothrys, Callitriche, Crassula, Isoetes, Deschampsia, Alopecurus, Anagallis,* and *Myosurus*. These taxa are restricted to pools and are absent from other kinds of wetlands. In addition, pools include taxa common to other wetlands, including species of *Eleocharis, Frankenia, Juncus, Distichlis, Ranunculus,* and *Veronica*. Invasion of vernal pools by exotic weedy species has been minimal, probably because wintertime wet, anaerobic conditions shift the competitive balance to the slow-growing natives (Bauder 1989).

Although still widely distributed within California, vernal pool area has declined in this century by about 60–70%, largely due to agricultural activities and urban sprawl. Mitigation has increasingly involved the manmade creation of depressions and seeding them with topsoil of nearby pools before their destruction. The success of created pools is being actively monitored and debated. Five years of observation for a group of created pools in San Diego County, for example, showed that richness and cover by native plants are significantly lower than in natural pools, and that the invasiveness to weedy species is higher than that in natural pools (Zedler, Frazier, and Black 1993).

The Everglades (Southern Fen Peatland)

The Everglades, an alkaline fen peatland or mire, was historically a rainfall-driven, nutrient-poor (oligotrophic). Its wetland ecosystem primary vegetation, sawgrass (*Cladium jamaicense*), developed peat soils (Histosols) 0.2–4 m in depth over the past 5,000 years in southern Florida (Davis 1943; Fig. 12.12). The Everglades would be classified as a palustrien system.

The Everglades was an almost impenetrable wall of sawgrass plains and reptile-infested waters to the early explorers (Ives 1856; Lodge 1994). Its name may have come from the term "Never Glades" as first used by Vignoles (1823). Originally called Pa-hay-okee ("grassy lake") by the native Indians, it was later popularized and put forward as a threatened environment that needed federal protection by Marjory Stoneman Douglas's famous 1947 book *The Everglades: River of Grass*. It has often been referred to as the "Everglades marsh" by local biologists though it is more correctly identified as a fen peatland or mire. Its sawgrass-dominated landscape is interlaced with *Eleocharis* – periphyton sloughs and dotted with tree islands and willow heads Fig. 12.13. The plant community associations evolved under low-nutrient conditions, especially phosphorus (Davis 1943; Loveless 1959; Steward and Ornes 1975a, 1975b; Craft and Richardson 1993a). A major component of the Glades, unlike northern mires or marshes, is the periphyton, a nitrogen-fixing, blue-green algae community found in open water sloughs.

Historically, fires (controlled by seasonal wet and dry seasons) and flooding played a major role in modifying the landscape features of the Everglades. To foster urban and agriculture development during the twentieth century, more than 1500 km of drainage canals and dikes were built, altering the natural hydrologic regime, fire occurrences, and phosphorus input (Davis and Ogden 1994; Richardson and Craft 1993 Craft and Richardson 1993b). The importation and expansion of woody exotic species like *Melaleuca leucadendra* and Brazilian pepper (*Schinus terebinthifolius*), along with aquatic weeds like water hyacinth (*Eichhorina crassipes*) and hydrilla (*Hydrilla verticillata*), have also had significantly negative effects on the plant and animal communities and water balance (Bodle, Ferriter, and Thayer 1994; Gunderson and Loftus 1993).

Approximately 50% of the original 900,000 ha Everglades area has been converted to agricultural and urban development, and 350,000 ha of the original area is now under state ownership as Water Conservation Areas (WCAs) 1, 2, and 3 (Fig. 12.12), managed for "flood protection, water supply, and allied purposes of navigation and fish and wildlife protection," as mandated by the 1948 U.S. Congressional Flood Control Act. The 585,867 ha Everglades National Park (ENP) is the largest federally owned marsh in the United States and is the only subtropical wetland ecosystem in the United States that is enrolled in the Ramsar Convention of Wetlands of International Importance. Because of its size, floral and faunal diversity, geological history, and hydrological functions on the Florida landscape, it is considered by many ecologists and conservationists to be one of the most unique and important wetlands in the world.

Formation of the Everglades. The Everglades' mineral substrate formed a large basin or trough during the Pleistocene and shallow marine sediments were deposited, primarily during the Sangamon interglacial stage 125,000 yr BP (Davis 1943; Parker

Figure 12.13. Aerial photograph of tree islands surrounded by open slough communities, sawgrass marsh (photograph by the author).

and Cooke 1944; Gleason 1984). The retreat of the northern U.S. glaciers 18,000–16,000 yr BP, blockage of drainage from the Glades due to rising sea level, a change to a subtropical climate, and the concurrent increase in rainfall combined to produce the modern Everglades.

Three limestone formations underlie the Glades. The Miami Formation is found in the southern Everglades National Park region; the Anastasia Formation, comprised of sandy calcareous sandstone, is found in the northeast area; and the Fort Thompson Formation underlies the northern half to a depth of 50 m, and is mostly marine and freshwater marls, limestone, and sandstone (Enos and Perkins 1977).

The soils of the Everglades are recent Holocene Histosols and Inceptisols (Gunderson and Loftus 1993). They are primarily peats and mucks that have accumulated to a depth of nearly 4 m in the north but are less than 20 cm deep in portions of the ENP (Stephens and Johnson 1951; Richardson, unpublished data). The deepest peats in the southern Everglades are found in depressions and major water flows such as the Shark River slough. Gleason (1984) dated the basal peats and found that peat deposition began as early as 5490 ± 90 yr BP, but most peats date from 2000–4500 yr BP. How-

ever, the tree islands are more recent and date from only 1300 yr BP (Gleason and Stone 1994). The other dominant and oldest soil type is a calcitic mud, an Inceptisol, formed by cyanobacteria that reprecipitate calcium carbonate, or marl, originally derived from the limestone substrate (Browder, Gleason, and Swift 1994). It is found underlying most of the peatlands and has been dated at 6470 yr BP (Gleason and Stone 1994). It is also sometimes found in layers within the peat, indicating periods of short seasonal hydroperiod as compared to the longer period of flooding required for peat formation by macrophytes.

Climate and hydrology.   The subtropical climate of southern Florida has hot humid summers, mild winters, and a distinct wet season with 80% of the rainfall falling from mid-May through October (MacVicar and Lin 1984). The Everglades has more in common with tropical climates in that a wet/dry season is probably more important to vegetation composition than winter/summer differences in temperature. Daily temperatures average above 27°C from April through October in the northern part of the Glades and from March to November in the south. Average daily temperatures are above 10°C even in the winter, but it does drop below freezing occasionally. The key component of climate, relative to vegetation patterns and succession, is the amount of precipitation. A 71 yr analysis of south Florida (1915 to 1985) reveals that 66% of annual precipitation falls during the June to October wet season (South Florida Water Management District, unpublished data). The median annual precipitation was 1336 mm with a low of 990 mm in 1956, and a high of 1955 mm during the hurricane of 1947 (MacVicar and Lin 1984).

Hurricanes (sustained winds of 120 km hr^{-1}) are an important recurring event (usually every 3 yr) in south Florida and can produce great wind damage and significant increases in annual rainfall and storm surges (Gunderson and Loftus 1993). Evapotranspiration is an extremely important component of the Everglades; it has been estimated that 70–100% of rainfall exits the Glades this way (Dohrenwend 1977; Fennema, Neidrauer, Johnson, MacVicar, and Perkins 1994). This is due in part to the fact that with virtually no gradient and with high vegetational resistance to water flow, the Glades exhibit extremely slow surface water flow (0–1 cm/sec) (Rosendahl and Rose 1982; Romanowicz and Richardson, 1997).

Also of great ecological significance is the fact that the Everglades experiences considerable periods of drought. Additionally, during the past 50 yr Lake Okeechobee, an integral part of the northern historic Everglades waterflow system, has released nearly 50% of its water into the Atlantic Ocean and Gulf of Mexico as a result of canal construction, with only a fraction of the historic flow going to the ENP and Florida Bay (Fennema et al. 1994). Recently, the Glades has suffered further water reductions because it has received below-average rainfall 21 out of the past 30 yr, or nearly 70% of the time (Sculley 1986). Since the 1970s, most Everglades areas, with the exception of the WCA-1 (Fig. 12.12), have experienced periods of 7–10 consecutive years of drought, ending with record low rainfall in 1988 and 1989. These data, when coupled with the operation of water-control structures and changes in regulation schedules, show that both natural and manmade changes have reduced significantly the hydroperiod in the Everglades, particularly in the northern parts of the Everglades and the ENP during the past 30 yr (South Florida Water Management District 1992).

Another example of significant changes in water distribution has occurred in WCA-2A during the past 40 yr. The South Florida Water Management District (SFWMD) maintained artificially high water levels in this part of the northern Everglades during portions of the 1960s and 1970s, then reduced water levels in the early 1980s to reduce the

tree island death caused by flooding, and changes in the periphyton community caused by artificially high water levels (SFWMD 1992). The lowering of the water level in the early 1980s unfortunately coincided with an extensive drought. This resulted in dramatic reductions in the water table levels and some expansion of sawgrass communities (Worth 1988). The Florida Game and Freshwater Fish Commission also initiated a series of controlled burns in WCA-2A during the 1980–1981 dry period to reduce fuel loads (Worth 1986). This, coupled with the massive increase in phosphorus from agricultural runoff after 1979, resulted in an explosion of cattails in former sloughs and willow areas (Craft and Richardson 1993b; Davis 1994). Droughts during the last few years have also resulted in water elevations well below the surface of WCA-2A during some portions of the year (Richardson et al. 1995). These extensive shifts in water level, increased nutrient inputs, and the droughts have had a severe impact on the plant communities and have resulted in the invasion of upland species and cattail monocultures in certain portions of the Glades. The drier areas in the southern Glades have seen an expansion of sawgrass communities (39–50%) at the expense of a decrease (48–35%) in wet prairies and sloughs (Davis, Gunderson, Park, Richardson, and Mattson 1994).

Water chemistry. Historically, the Everglades survived on nutrients primarily from rainfall and recycling within the system, especially after droughts and fire (Davis 1943; Swift and Nicholas 1987). The Everglades is a P limited system that has evolved plant species that can survive under P water concentrations as low as 5 µg/L (Koch and Reddy 1992; Richardson et al. 1995; Richardson and Vaithiyanthan 1995). There is no long-term historical record of the amount of nutrients in rainfall, but an analysis of rainfall nutrient loadings over the past few decades suggests that dry-fall represents the major source of N and P into the Everglades (SFWMD 1992). The flow-weighted mean for various stations in Florida ranged from 20 to 220 µg/L of total P (SFWMD 1992). The highest values were found around Lake Okeechobee and the Everglades Agricultural Area (EAA), and the lowest values were in the ENP. The flow-weighted mean for total P in rain for south Florida averaged 62 ± 90 µg/L, but recent data show that a more reasonable estimate is nearer 30µg/L, and wetfall contributes more than half the new P input to the ecosystem. (Richardson and Vaithiyanthan, unpublished data).

Agricultural runoff from both the EAA and Lake Okeechobee contributes water with signifi-

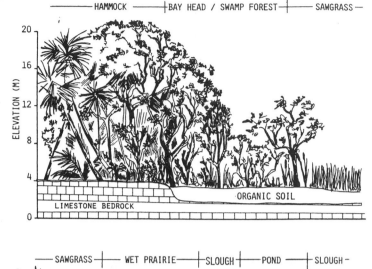

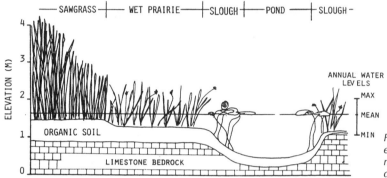

Figure 12.14. Idealized profile of everglades communities, from hummocks (a) to sloughs (b) (from Gunderson and Loftus (1993).

cantly higher concentrations of N and P than is typically found in the Everglades (Craft and Richardson 1993b; Davis and Ogden 1994). A landscape analysis of the P gradient for south Florida shows that the dairy region northeast of Lake Okeechobee has by far the highest total P concentrations and that the lake contains 75 µg/L on average (SFWMD 1992). The average P concentration in water leaving the farmland is 150 µg/L and is reduced to 115 µg/L in the canals and edges of WCA-1 (Fig. 12.12). Water flowing out of WCA-2A into WCA-3A contains 40 µg/L P. By the time surface waters reach the structures above the Everglades National Park, concentrations are <10 µg/L P, less than tap water. These data demonstrate that the Everglade plant communities to the north are under much higher P concentrations than in the past, and this in turn has had significant effects on plant communities.

Phosphorus loadings from the EAA have been implicated in the replacement of saw-grass by cattails in WCA-1 (Loxahatchee National Wildlife Refuge) and WCA-2A (Toth 1987, 1988; Belanger,

Scheidt, and Platko 1989; Urban, Davis, and Aumer 1993). Belanger et al. (1989) asserted that additions of nutrient-enriched water to WCA-2A have contributed to the invasion of a monotypic cattail community. The high P levels in vegetation, soils, and surface waters of the cattail-dominated areas of WCA-2A suggest that P may be primarily responsible for the invasion of cattails in WCA-2A (Belanger et al. 1989; Richardson and Craft 1993; Vaithiyanathan, Zakina, and Richardson 1995, 1997; Craft and Richardson 1998). Pope (1989), however, reported that both "anthropogenic nutrient enrichment of canal water and increased hydroperiod are found to diminish native sawgrass and tree island communities while enhancing expansion of cattail in the Loxahatchee" (p. 3). These data suggest that the expansion of cattail is related to alterations in the water regime as well as to increased nutrient supply.

WCA-1 is the only part of the Glades that has acidic pH waters (5.5–5.9), is truly ombrotrophic, and can be classified as a poor fen. Sodium also is quite high, which reflects the influence of Pleistocene sea water releases from beneath the Everglades and ocean-influenced rainfall (Craft and Richardson 1997). Typical sloughs, in contract, have mean pH of 8.4, are dominated by Ca and Mg chemistry, and can be classified as an alkaline-rich fen (Vymazal and Richardson 1995).

Vegetation. Six factors are responsible for the plant communities that exist within the Everglades and each may be scaled from 1 to 10, with 10 being the most important controlling factor or impact. The scaling is relative but is based in part on the percentage of the Everglades that is affected by the factor or what relative influence the factor has on plant community change from its historical base. Historically, the primary factor controlling succession in Everglades communities was climate (then rated 9–10) with fire (7–8) and hydrology (7–8) next in importance in influencing community structure. Nutrient additions, exotic species invasions, and disturbance (e.g., airboat trails, swamp buggy tracks) have all increased in importance compared to historic conditions (values of 3 now, compared to 1 in the past). The altered hydroperiod (i.e., the number of days that the wetland ecosystem has standing water at or near the surface) has been dramatically decreased in some areas and increased in others over the past 60 years (Davis and Ogden 1994) and is now scaled a 9–10 compared to a 7–8 during historic times. The flooding has decreased fire's importance from 1–2 to 2–3, caused primarily by the diking of most of the Glades. The reason for slightly decreasing the importance of climate today (down to 7–9 from 9–10) reflects the increase in the other major influences humans have put on the system. In my brief analysis of factors controlling plant communities, I have focused on climate as the primary controlling factor, as well as hydroperiod and nutrients, the latter two key factors having been significantly altered. The good news is that hydrologic increases (e.g., restored hydroperiod) and nutrient decreases (i.e., reduced runoff from agriculture) are currently mandated by the State of Florida's "Everglades Forever Act" of 1993 and the Army Corps of Engineers 1994 Restoration plan.

The major macrophyte vegetation communities of the Everglades consist of large areas of sawgrass, slough, wet prairie, and smaller areas of tree islands (Davis 1943; Loveless 1959). A typical profile of these communities with annual water levels is shown in Fig. 12.14a and b, and an aerial view is given in Fig. 12.13. Calcareous periphyton is also found in many regions of the wetlands of southern Florida, coating vegetation and bottom sediments in slough and wet prairie communities with growths of algae up to 4–7 cm thick (Loveless 1959; Van Meter 1965; Gleason and Spackman 1974; Vymazal and Richardson 1995). Calcareous periphyton develops on plants growing at variable water depths, though generally the best development of algae is in the upper 60 cm of the water column (Gleason and Spackman 1974).

Macrophytes

A complete listing of the plant species and growth forms characteristic of each community in the Everglades is shown in Table 12.5.

Sawgrass. Sawgrass (*Cladium jamaicense*) dominates the most widespread community in the everglades (Fig. 12–13). Saw-grass grows to 2–3 m in height on deep peat but only 0.5 m on shallow peat. It prefers sites with a fairly constant water depth of 10–20 cm (Toth 1987; Gunderson 1990). Its presence in the Glades is due to its ability to survive fire, low soil-nutrient content, and occasional freezing (Stewart and Ornes 1975b). It does not survive well in deep (>30 cm) water regimes (Toth 1987). The current diking and flooding in portions of WCA-2 and other parts of the Glades has resulted in the loss of this community due to deep and fluctuating water levels. Sawgrass occurs either in almost pure stands or mixed with a wide variety of other plants – for example, bulltongue (*Sagittaria lancifolia*), maidencane (*Panicum hemitomon*), pickerelweed (*Pontenderia cordata*), or cattail (*Typha* spp.) (Loveless 1959). Estimates of the extent of mixed sawgrass areas are 65–70% of the remaining Everglades fen (Kushlan 1987; Loveless 1959; Schomer and Drew 1982; Steward and Ornes 1975b; Davis 1994). Davis *et al.* (1994) estimated that pure sawgrass-dominated areas currently make up only 38% of 417,000 ha of historic sawgrass plains and sawgrass-dominated areas (Table 12.6). The sawgrass–tree island–slough mosaic has not changed much in the area.

Wet Prairie. Wet prairies are one of the important vegetation types in the northern Everglades (Table 12.6). Often referred to as "flats," they are characterized by short emergent plants and are found in the northern and central Everglades in conjunction with tree islands (Goodrick 1984; Gunderson and Loftus 1993). Wet prairies exist on both peat and marl soil; variation in species composition is shown in Table 12.5. The wet prairies in the south, on calcitic mud or marl, occur on higher and drier sites but are still wet 3–7 mo of the year (Davis 1943; Gunderson and Loftus 1993). The water depth of these areas is generally less than that of

Table 12.5. *Characteristic plant taxa in Everglades communities*

Community	Species	Growth form
Ponds, slough	(D) White water lily (*Nyphaea odorata*)	Floating aquatic
	(D) Floating heart (*Nymmphoides aqua-tica*)	Floating aquatic
	Spatterdock (*Nuphar luteum*)	Floating aquatic
	Bladderwort (*Utricularia foliosa*)	Submerged aquatic
	(*Utricularia purpurea*)	Submerged aquatic
	Water hyssop (*Bacopa caroliniana*)	Submerged aquatic
	Arrowhead (*Sagittaria lancifolia*)	Emergent aquatic
	Pickerel weed (*Pontederia cordata*)	Emergent aquatic
Sawgrass "prairie"		
Tall stature	(D) Sawgrass (*Cladium jamaicense*)	Emergent sedge
	Justicia angusta)	Emergent herb
	Spikerush (*Eleocharis cellulosa*)	Emergent rush
	Cattail (*Typha latifolia*)	Emergent aquatic
Intermediate stature	(D) Sawgrass (*Cladium jamaicense*)	Emergent sedge
	Swamp lily (*Crinum americanum*)	Emergent herb
	Arrow arum (*Peltandra virginica*)	Emergent herb
	Spider lily (*Hymenocallis latifolia*)	Emergent herb
	Shy leaf (*Aeschynomene pratensis*)	Emergent herb
	Everglades morning glory (*Ipomoea sagittata*)	Vine
Wet prairies (peat)		
Eleocharis marshes	(D) Spikerush (*Eleocharis cellulosa*)	Emergent rush
	(D) Spikerush (*Eleocharis elongata*)	Emergent rush
Rhynchospora flats	(D) Beak rush (*Rhynchospora tracyi*)	Emergent sedge
	Water rush (*Rhynchospora inundata*)	Emergent sedge
	(D) Maidencane (*Panicum hemitomon*)	Emergent grass
Maidencane flats	Water paricon (*Paspalidium gemina-tum* var. paludivagum)	Emergent grass
	Swamp lily (*Crinum americanum*)	Emergent herb
	Water hyssop (*Bacopa caroliniana*)	Submerged aquatic
	Arrowhead (*Sagittaria lancifolia*)	Emergent herb
	Water drop-wort (*Oxypolis filiformis*)	Emergent herb
Wet prairies (marl)	(D) Sawgrass (*Cladium jamaicense*)	Emergent sedge
	(D) Muhly grass (*Muhlenbergia filipes*)	Emergent grass
	Schizachyrium rhizomatum	Emergent grass
	White-top sedge (*Dichromena colorata*)	Emergent sedge
	Black-top sedge (*Schoenus nigricans*)	Emergent sedge
	Aristida purpurescens	Emergent grass
	Panicum tenerum	Emergent grass
	Rhynchospora divergens	Emergent sedge
	Rhynchospora microcarpa	Emergent sedge
Bayhead	(D) Red bay (*Persea borbonia*)	Tree
Swamp forests	(D) Sweet bay (*Magnolia virginiana*)	Tree
	(D) Dahoon holly (*Ilex cassine*)	Tree
	(D) Willow (*Salix caroliniana*)	Tree
	(D) Wax myrtle (*Myrica cerifera*)	Tree
	Cocoplum (*Chrysobalanus icaco*)	Shrub
	Swamp fern (*Blechnum serrulatum*)	Fern
	Leather fern (*Acrostichum danaei-folium*)	Fern
	Red maple (north) (*Acer rubrum*)	Tree
	Red mangrove (south) (*Rhizophora mangle*)	Tree
Pond apple	(D) Pond apple (*Annona glabra*)	Tree
forest	Pop-ash (*Fraxinus caroliniana*)	Tree
	Dew-flower (*Commelina gigas*)	Vine
	Okeechobee gourd (*Cucurbita okeechobeensis*)	Vine
	Elderberry (*Sambucus simpsonii*)	Tree
	Strangler fig (*Ficus aurea*)	Tree
	Saltbush (*Baccharis halimifolia*)	Shrub

Table 12.5. *(Cont.)*

Community	Species	Growth form
	Air plants (*Tillandsia* spp.)	Epiphyte
	Orchids (e.g. *Encyclia tampense*)	Epiphyte
Willow heads	(D) Willow (*Salix caroliniana*)	Tree
	Wax myrtle (*Myrica cerifera*)	Tree
	Buttonbush (*Cephalanthus occidentalis*)	Shrub
	Sawgrass (*Cladium jamaicense*)	Sedge
Cypress forests	(D) Cypress (*Taxodium ascendens*)	Trees
	Sawgrass (*Cladium jamaicense*)	Sedge
	Schizachrium rhizomatum	Grass
	White-top sedge (*Dichromena colorata*)	Grass
Hardwood hammocks	(D) Strangler fig (*Ficus aurea*)	Tree
	(D) Gumbo-limbo (*Bursera simarubra*)	Tree
	(D) Oak (*Quercus virginiana*)	Tree
	(D) Wild tamarind (*Lysiloma litisiliquum*)	Tree
	Cabbage palm (*Sabal palmetto*)	Tree
	Hackberry (*Celtis laevigata*)	Tree
	Mulberry (*Morus rubra*)	Tree
	Citrus (*Citrus* spp.)	Tree
	Persimmon (*Diospyros virginiana*)	Tree
	Mahogany (south) (*Swietenia mahogani*)	Tree
	Paurotis palm (*Acoelorraphe wrightii*)	Tree
	Royal palm (*Roystonea elata*)	Tree
Successional shrub		Tree
	Florida trema (*Trema micranthum*)	
	Saltbush (*Baccharis halimifolia*)	Shrub
	Wax myrtle (*Myrica cerifera*)	Tree/shrub
	Bracken fern (*Pteridium aquilinum*)	Fern

Note: Dominant taxa indicated by D. Nomenclature follows Long and Lakela (1971) and Avery and Loope (1980).
Source: Gunderson and Loftus (1993).

Table 12.6. *Aerial coverage of wetland communities in the Everglades of south Florida. Percentage loss is based on predrainage estimates and current estimates for each particular community type.*

Landscape type	Predrainage (ha)	Current (ha)	Percent loss
Swamp forest	60,000	0	100
Sawgrass plains	238,000	63,000	74
Slough/tree island/saw grass mosaic	311,000	271,000	13
Sawgrass-dominated	179,000	94,000	47
Peripheral wet prairie	117,000	0	100
Cypress strand	12,000	0	100
Southern marl-forming marshes	249,000	190,000	24
Total coverage	11666,000	618,000	49

Source: Davis et al. (1994).

sloughs but deeper than sawgrass, and thus the vegetation seldom burns. Loveless (1959) described three well-defined wet prairie associations in the northern Everglades: (1) *Rhynchospora* flats, (2) *Panicum* flats, and (3) *Eleocharis* flats. These plant associations are composed primarily of Tracey's horned rush (*Rhynchospora tracyi*), gulfcoast spikerush (*Eleocharis cellulosa*, both sedges, and the wetland panic grass, maidencane (*Panicum hemitomon*). However, many other plant species may also be present on these flats, depending on hydrological conditions, the season of the year, and soil type. Wet prairies usually dry out on an annual basis and are a transition zone between saw-grass areas and sloughs (Goodrick 1984). Wet prairies require seasonal inundation with standing water present for 6–10 mo of the year (Schomer and Drew 1982). Seasonal drying of the soil in these communities is required for seed germination and establishment of new seedlings (Dineen 1972).

Sloughs. Sloughs are open water marsh areas found primarily in the northeast and southcentral Everglades, which are dominated by floating aquatic plants with some emergent plants of low stature (Davis 1943; Loveless 1959) (Fig. 12.14). Sloughs are one of the most widespread community types in the Everglades (Table 12.6). Aquatic sloughs represent the lowest elevation of the Everglades ecosystem, except for ponds. They have deepwater levels averaging 30 cm annually and longer inundation periods than other Everglades, wetland communities (Gunderson and Loftus 1993). Sloughs occur throughout the Everglades with the largest pond-slough systems in the Everglades National Park (Shark River and Taylor sloughs) and portions of the northern Everglades (McPherson, Hendrix, Klein, and Tysus 1976). Sloughs are the narrow drainage channels that are usually water-filled, or at least wet, most of the year. The "valleys" of these channels average only a few cm to 60 cm below the elevation of adjacent marsh areas. Not as extensive as they once were, some apparently have been replaced by either sawgrass or wax myrtle and willow stands. Cattail has also filled many of the sloughs in the natural enriched areas of the northern Everglades (Rader and Richardson 1992; Urban et al. 1993; Craft and Richardson 1997). This reduction in slough areas has also been caused by artificial drainage and the resulting increase in sawgrass in the southern Everglades (Loveless 1959; Davis et al. 1994). Sloughs are easily recognized by their drainage patterns and by characteristic plant species, such as white waterlily (*Nymphaea odorata*), floating hearts (*Nymphoides peltata*), bladderworts (*Utricularia* spp.),

spikerushes (*Eleocharis* spp.), European cowlily (*Nuphar luteum*), and lemon bacopa (*Bacopa caroliniana*) (Davis 1943; Loveless 1959; Van Meter-Kasanof 1973; Gunderson and Loftus 1993) (Table 12.5).

Several factors make sloughs and wet prairies ecologically important in the landscape of the Everglades. During the dry season, sloughs serve as important feeding areas and habitats for Everglades wildlife. As the higher-elevation wet prairies dry out, sloughs provide refuge for aquatic invertebrates and fish. The high concentration of aquatic life, in turn, makes sloughs important feeding areas for Everglades populations of wading birds. When the marsh is reflooded, these areas serve to repopulate the fen as water level rises (Loveless 1959). The slough–wet prairie sawgrass mosaic covers 271,000 ha (44%) of the remaining 618,000 ha area of the Everglades (Davis et al. 1994. Table 12.6). Plant species diversity is much higher in sloughs and wet prairie communities than in pure saw-grass and cattail marsh communities (SFWMD 1992; Craft, Vymazal, and Richardson 1995). The abundance of macroinvertebrates, fish, and wading birds is also greater in sloughs (SFWMD 1992; Rader and Richardson 1992, 1994; Davis and Ogden 1994).

Ponds. Ponds are small, open water areas scattered throughout most of the Everglades. They represent the deepest water regime (Fig. 12.14.) and occur in bedrock depressions where fire has burned out the peat (Loveless 1959). Alligator activity often maintains the open water, and hence the ponds are also called "alligator holes" by the locals. They are wet except in the driest years and thus are important habitats for animals, especially birds. These holes have a border of water lilies (*Nymphaea* spp.), spatterdock (*Nuphar* spp.), pickerel weed (*Pontederia cordata*), and woody species such as willow (*Salix caroliniana*) or primrosewillow (*Ludwigia peruviana*) (Table 12.5, Gunderson and Loftus 1993).

Tree islands (bayhead–swamp forests). The broadleaf hardwood forests are called tree islands locally. The islands refer to a variety of tree clusters that stand above a matrix of shorter vegetation (Craighead 1984). They occur generally throughout the entire region but are most abundant in the central part of WCA-1 (Loveless 1959) (Fig. 12.12). Tree islands may be either bayhead (swamp forests), hammocks (upland forest), or both (Davis 1943; Gunderson and Loftus 1993). The dominant species in each type of forest are shown on Table 12.5. Red bay (*Persea borbonia*), swamp bay (*Magnolia virginiana*), dahoon holly (*Ilex cassine*), willow

(*Salix caroliniana*), and wax myrtle (*Myrica cerifera*) dominate the swamp forests. Large tree islands have a tear-drop shape, with the main axis paralleling the flow of water, whereas small islands ($\cong$100 m²) are usually round. The forests are found on the highest sites in the Glades on a peat classified as Gandy peat (Davis 1943; Loveless 1959). The sites are wet 2–6 mo yr^{-1}, but in drought conditions these systems are very susceptible to burning (Gunderson and Loftus 1993). The soil P nutrient content of tree islands is usually much higher (>1000 mg kg^{-1} vs. 500 mg kg^{-1}) than the surrounding landscape due to the fertilization of the soils by dense roosting bird populations. (Richardson, unpublished data).

Willow heads, cypress forests, pond apple forests, and hardwood upland hammocks. These forest types comprise only a small amount of area in the Everglades, but they contain distinct species (Table 12.5). The pond apple forest (*Amnona glabra*) existed primarily south of Lake Okeechobee in a band 5 km wide (Davis 1943). The land has been totally developed for agriculture and now the species exists only in small scattered stands. Willow heads exist throughout the Everglades in monotypic stands (Loveless 1959) in fire-disturbed areas, as well as around alligator holes. The upland hardwood hammocks are dominated by broadleaf hardwood trees of both temperate and tropical origin. Dominant trees include live oak (*Quercus virginiana*), gumbo limbo (*Bursera simaruba*), sabal palm (*Sabal palmetto*), and strangler fig (*Ficus aurea*). The cypress forests are found only in the southwestern Everglades and are dominant in the adjacent Big Cypress National Preserve. Pond cypress (*Taxodium ascendens*) are very short and occur as widely scattered individuals displaying very stunted growth. They are often called dwarf or hatrack cypress and seldom reach heights over 3–5 m due to extremely nutrient-poor water and soils. Cypress forests are described in Chapter 11.

Periphyton. Periphyton and algal mats are often not thought of as valuable ecological resources or even listed among vegetation community species. Several authors, however, have pointed out that components of the periphyton/algal mat (especially diatoms) are a high-quality food for some animals (Browder et al. 1981; Browder et al. 1994). Photosynthesis by the algae in sloughs can raise daytime dissolved oxygen concentrations and pH much higher (7.5–>10) than in adjacent sawgrass marshes (Belanger et al. 1989; Rader and Richardson 1992; Vaithiyanathan and Richardson 1995). The calcareous periphyton deposits marl (calcitic

mud), the second most common soil sediment type (190,000 ha) in the Everglades (Gleason and Spackman 1974; Davis et al. 1994; Table 12.6).

Periphyton covered by calcareous precipitations in the Everglades is mostly dominated by blue-green algae. *Schizothrix calcicola* and *Scytonema hofmannii* are the dominant species (Van Meter 1965; Gleason 1972; Gleason and Spackman 1974; Wood and Maynard 1974; Swift 1981, 1984; Swift and Nicholas 1987), but other genera like *Aphanothece*, *Chroococcus*, *Coccochloris*, *Gloeocapsa*, *Gomphosphaeria*, *Johannesbaptistia*, *Lyngbya*, *Microcoleus*, *Microcystis*, *Nodularia*, *Oscillatoria*, *Phormidium*, *Spirulina*, *Stigonema*, and *Tolypothrix* are frequently found in large quantities in periphyton throughout the Everglades (Wood and Maynard 1974; Gleason and Spackman 1974; Vymazal and Richardson 1995). In some areas of the Everglades, diatoms, desmids, and a few species of filamentous green algae form a significant part of the periphyton (Swift 1984; Swift and Nicholas 1987). Filamentous green algae (e.g., *Spirogyra* sp. or *Mougeotia* sp.) are found mostly in areas with elevated nutrient levels (Swift 1981, 1984; Swift and Nicholas 1987). Water quality and hydroperiod are the major factors that appear to influence the species composition and growth rates of Everglades periphyton communities (Swift 1981, 1984; Swift and Nicholas 1987; Browder et al. 1981; Flora, Walker, Scheidt, Rice, and Landers 1988; Rader and Richardson 1992).

Periphyton communities probably represent a major component of the detrital-based Everglades food web, providing organic food matter and habitat for a wide variety of grazing invertebrates and foraging fish (Craighead 1971; Carter et al. 1973; Wood and Maynard 1974; Browder et al. 1981). Periphyton photosynthesis and respiration play an important role in controlling diurnal pH, dissolved oxygen, carbon dioxide, and calcium concentration within marsh surface waters (Gleason 1972; Gleason and Spackman 1974; Wilson 1974). Algal photosynthesis accounts for a large portion of calcium precipitation within the marsh and is responsible for the formation of marl soils within the southern Everglades (Gleason 1972; Gleason and Spackman 1974; Vymazal and Richardson 1995). Gleason and Spackman (1974) concluded that the aerial extent of calcareous periphyton is impressive, being found abundantly in all of the WCAs, Everglades National Park (ENP), Big Cypress National Preserve, and inland prairies of the southern coast, the largest expanses being found in WCA-3 and the ENP.

Periphyton species composition. A total of 409 algal species representing 41 genera but only three divisions (Cyanophyta, Chrysophyta, and Chlorophyta) has been identified for the Everglades (Stevenson and Richardson 1995). The total number of diatom species found in periphyton throughout the year (31) was higher than that of blue-green algae (24), but diatom growth was very low. The most abundant species in all samples were *Amphora arcus* var. *sulcata* and *Mastogloia smithii* var. *smithii* (Vymazal and Richardson 1995).

Blue-green algae dominate in the summer and autumn whereas during winter and spring diatoms prevail. The growth of algae in winter and spring is very low and all species are then in very low quantities. Although the relative abundance of algal groups changes during the year, the total number of species remains almost constant (Vymazal and Richardson 1995). Green algae appear mostly during the summer and autumn, and their seasonal occurrence is similar to that of blue-green algae.

Succession. Succession in the Everglades has been summarized by Gunderson and Loftus (1993) (Fig. 12.15), who demonstrated that succession of Everglades communities is influenced mostly by disturbance to the hydrology and, in turn, the frequency and intensity of fire. The species noted earlier are found along an elevational gradient that translates directly into a hydrologic gradient, which controls fire intensity and frequency. The gradual buildup of peat through peat accretion (1–2 mm yr^{-1}; Craft and Richardson 1993a) results in

a gradual increase in elevation, which changes the hydroperiod. Ponds are the wettest sites, and soil accretion eventually modifies them into a wet prairie community, followed by willow heads and even sawgrass, if not severely burned. Frequent light fires have little effect on this successional sequence. Severe fires reverse this sequence and move the communities back to wetter habitats. Drought or drainage allows the invasion of upland macrophytes, scrub, and hardwood species. Alligators also change the hydrology and nutrient status of areas and can create ponds with increased plant diversity. (Kushlan 1974). The successional dynamics of the Everglades is thus controlled by the interaction of hydroperiod, fire frequency, degree of fire intensity, animal activity, and drought.

A Comparison of Dominant Marsh Vegetation Nationwide

Marsh ecosystems are frequently or continually wet and dominated by herbaceous vegetation adapted to anaerobic soil conditions. In Europe a marsh has a mineral substrate and does not accumulate peat. In North America the term "marsh" is often used generically to refer to many herb-dominated wetlands, a few having peat soils.

In the United States it was estimated by Shaw and Fredine (1956) that marshes covered nearly 7 M ha. They reported that coastal freshwater tidal marshes made up 1.5 m ha of the total and were

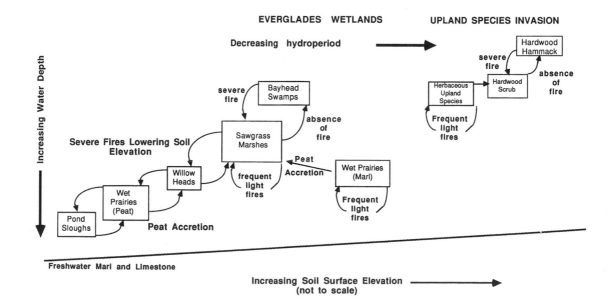

Figure 12.15. Successful relationships among everglades plant communities (modified from Gunderson and Loftus 1993).

Table 12.7. *Typical dominant emergent vegetation in different freshwater marshes*

Marsh type and location	Dominant species	Reference
Prairie glacial marsh, Iowa	*Typha latifolia* *Typha angustifolia* *Scirpus validus* *Scirpus fluviatilis* *Scirpus acutus* *Sparganium eurycarpum* *Carex* spp. *Sagittaria latifolia*	van der Valk and Davis (1978); Weller (1981)
Riverine marsh, Wisconsin	*Typha latifolia* *Scirpus fluviatilis* *Carex lacustris* *Sparganium eurycarpum* *Phalaris arundinacea*	Klopatek (1978)
Lake Erie wetlands, Ohio	*Typha angustifolia*[a, b] *Cyperus erythrorhizos*[a, b] *Scirpus validus*[a] *Leersia oryzoides*[a] *Echinochloa walteri*[a] *Ludwigia palustris*[a] *Phragmites australis*[a] *Scirpus acutus*[a] *Scirpus fluviatilis*[b] *Sparganium eurycarpum*[b] *Amaranthus tuberculatus*[b] *Nuphar advena*[b] *Nelumbo lutea*[b]	Robb (1989)
"Tule" marshes, California and Oregon	*Scirpus acutus* *Scirpus californicus* *Scirpus olneyi* *Scirpus validus* *Phragmites australis* *Cyperus* spp. *Juncus patens* *Typha latifolia*	Hofstetter (1983)
Floating freshwater coastal marsh, Louisiana	*Panicum hemitomon* *Thelypteris palustris* *Osmunda regalis* *Vigna luteola* *Polygonum sagittatum* *Sagittaria latifolia* *Decodon verticillatus*	Sasser and Gosselink (1984)

[a] Diked marshes
[b] Undiked marshes
Source: Mitsch and Gosselink (1993).

concentrated on the northern coast of the Gulf of Mexico and the south Atlantic coast. Most of these marshes would be classified as palustrine or lacustrine systems.

Most marshes dry out seasonally but may stay wet with additional sources of seepage or groundwater. The classic wet-dry cycle of marshes is demonstrated by the prairie potholes discussed earlier (Fig 12.11).

To make a comparison of the vegetation found in many different types of freshwater marshes in

the United States, the dominant species for each ecosystem were tallied by Mitsch and Gosselink (1993). This list covers marshes from Oregon, California, Louisiana, Ohio, and Wisconsin not covered in detail in this chapter. A number of common genera were found among the marshes even though they represented marshes of different latitudes, soil types, water chemistry, and varying climate conditions (Table 12.7). The most common species were *Typha* spp. (cattail), *Phragmites australis* (reed grass), *Sparganium eurycarpum* (bur reed), *Zizania aquatica* (wild rice), and *Carex* spp. (sedges). Broadleaf monocots, such as *Pontederia cordata* (pickerelweed) and *Sagittaria* spp. (arrowhead), and ferns like *Osmunda regalis* (royal fern) are also found in many marshes. Waterfowl species that migrate from northern Canadian marshes to the marshes of South America may have been dispersal vectors for many common plants.

Riparian (Bottomland Hardwood Swamps and Western Riparian Forests)

Riparian wetlands are ecosystems in which vegetation, soils, and hydroperiod are influenced by the adjacent river or stream, and they are the ecotone between aquatic and terrestrial ecosystems (Mitsch and Gosselink 1993; Fig. 12.1). The word "riparian" comes from the Latin *ripa*, which means bank or shore of a river. Ecologically it refers to the terrestrial or emergent zone immediately adjacent to the aquatic or submerged zone (Faber et al. 1989). The term has occasionally been used to describe areas adjacent to creeks and streams in both tidal and estuarine zones (e.g., tidal marshes and riverine mangroves), but it is generally restricted to freshwater systems.

The term "riparian" is often used differently in the western versus the eastern United States. Warner (1983, p. 2) defined riparian systems in California as "pertaining to the banks and other adjacent terrestrial (as opposed to aquatic) environs of freshwater bodies, watercourses, and the surface-emergent aquifers (springs, seeps, and oases) whose transported waters provide soil moisture significantly in excess of that otherwise available through precipitation." This more encompassing definition is sometimes used in the western arid regions of the western United States, but the term is usually applied in both the eastern and western states to commonly embrace the "flood lands" (say, the 100 yr flood mark) (Clark 1979). This is to be distinguished from the floodplains (i.e., 500 yr flood mark), which extend both down into the wetland and up into the upland. In contrast, the upland system on the landscape is found above the flood lands far enough not be influenced by the transported waters or watercourses on a regular basis and must depend on local rainfall for its annual water budget. The boundary between the upland and riparian system can shift in years of high or low rainfall as a result of changes in flooding conditions, dam release rates, sedimentation, and regional water table drawdown. In western states like California, Warner (1983) suggests that the riparian zone approximates the 100 yr flood zone.

Thus, the western riparian zone equals the eastern flood zone.

The delineated wetlands of the streams and rivers are often synonymous with the riparian boundaries. However, in many areas they are only a subset of the more extensive riparian zone, especially if the latter encompasses the 500 yr flood lands, which often includes a portion of the uplands. In this chapter we use the terms "riparian" and "riparian wetland" interchangeably, although legally the boundaries may be different (Mitsch and Gosselink 1993).

Riparian zones in the West often occur adjacent to low-order streams with extreme variations in fluvial conditions and geomorphic processes, and are generally narrow in width (10–100 m). In contrast, southeastern riparian zones occur on broad alluvial valleys as wide as 10 km. They are generally found as low-lying flats or extensive floodplains, with well-developed hydric soil profiles and seasonal flooding. Where streams are intermittent to ephemeral, the upland boundary is very difficult to estimate in some years. However, the presence of certain wetland species or absence of key upland species can be used to estimate the boundary. Also, the larger overall size of the plant species in the western riparian zone versus those of the upland can be used to estimate the riparian boundary (Faber et al. 1989). Riparian systems are characterized as ecotonal communities of high diversity and species density as well as extremely high productivity for the region (Fig. 12.5).

Riparian systems are most commonly recognized as bottomland hardwood and floodplain forests in the eastern and central United States and as bosque or streambank vegetation in the West (Brown, Brinson, and Lugo 1979). Abernethy and Turner (1987) estimated the total riparian forest area in the United States to be near 30 m ha in 1985, excluding Alaska, Arizona, California, Hawaii, and New Mexico. Brinson, Plantico, Swift, and Barclay (1981) estimated Alaska to have 12 m ha of forested wetlands and the other states an additional 360,000 ha. This would suggest that the entire United States had about 42 m ha of riparian wetlands in the mid-1980s, with 58% being found in the south-central and southeastern portions of the country (Fig. 12.16 from Putnam, Furnival, and McKnight 1960). No riparian area data exist for Canada, although it must be extensive, considering the acreage reported for Alaska.

Formation and classification of riparian systems. Riparian ecosystems are unique among wetland types because they are tied to a supplemental high-

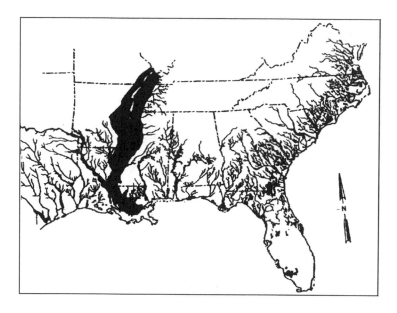

Figure 12.16. Bottomland hardwood swamps and alluvial bottoms of south-central and southeastern United States before extensive settlement, drainage, and agriculture (after Putnam et al. 1960).

energy water source from meandering streams or rivers, but most important they are open systems subjected to massive influxes or losses of nutrients and sediment during peak floods. Understanding the dynamics of the sedimentation and flooding patterns of the riparian zone is critical because these factors control vegetation and successional patterns. Two distinct features of riparian systems versus other wetland types are: (1) they are characterized by the lateral movement of pulsed water through the ecosystem with significant variations in water level tied to either daily (coastal systems), annual, seasonal, or irregular storm events (Figs. 12.1, 12.2); and (2) they have a high degree of connectedness to upstream and downstream systems as well as adjacent upland and aquatic ecosystems.

Riparian systems can be divided into three major zones along the drainage basin: (1) the erosion zone where much of the sediment is produced in the headwaters of the drainage watershed and here the riparian zone is very narrow; (2) the zone of storage and transport of sediment in the downstream or middle portion of the drainage system where the riparian zone is narrow to moderate in width; and (3) the zone of maximum deposition of sediment, which is at the lower end of the drainage system near lakes or the estuary and where the riparian system is at its maximum extent.

The zone of erosion is found in the headwaters on lower-order streams and also in the higher elevations of the mountains. Streams in this region are usually straight and narrow, and display a scoured V-shaped cross section. The steep banks have only a narrow riparian zone.

Below the erosion zone, the midorder streams widen, store coarse sediments, and form a distinctive but fairly narrow floodplain. The amount of flooding, storage of nutrients, and sediments depends on the size of the watershed, the gradient, and the rainfall. These factors all control the size of the wetland zone and the vegetation community.

The maximum zone of deposition is found on high-order, low-gradient streams on the valley floors or on coastal plains. These valley floors with gentle slope form a broader floodplain on low-gradient slow-flowing streams. Sediment deposits are much greater in these riparian areas, grading from coarse in the channel to fine at the outer limits of the floodplain (Mitsch and Gosselink 1993). The streams in this region are usually sinuous and meandering, which results in broad floodplain wetlands interlaced with braided stream channels.

The suspended sediment yield (SS) from selected streams and rivers in the United States suggests that western states, especially in northern California, have some of the highest values found in the world (approaching 4000 metric tons km^{-2} yr^{-1}) because of high-intensity rainfall, steep topography, and highly erodable rocks and soil (Faber et al. 1989). The temporal and spatial dynamics of western streams are fundamentally different from eastern streams, in that peak flooding events result in massive movement of coarse sediments, which can drastically change the successional patterns or vegetation equilibrium in the riparian depositional zone. During extensive dry periods, stream banks, sand bars in the channel, and the riparian zones revegetate. Massive storms and flooding can change the entire morphology of the stream and wetland system, resulting in a recolonization of the cleared sites. The SS yields in the central and eastern United States are much lower (100–200 metrictons km^{-2} yr^{-1}) due to the lower landscape gradients, less intense precipitation events, less erodable soil, and generally higher vegetation cover. However, intense fires and clearcutting in the watershed can increase the SS, whereas dams and channelization decrease the sediment loads downstream in the riparian zone. For example, an oak forest in the southeast was found to lose sediment via runoff at 0.11 metric ha^{-1} yr^{-1} compared to a loss of 358 metric ha^{-1} yr^{-1} for barren abandoned fields (Happ, Rittenhouse, and Dobson 1940). Moreover, wetland floodplain dynamics are tied to the river flow and sediment yield, and in turn the river flow is linked to the floodplain storage and release processes.

The development of the floodplain, the major component of the wetland system, results from both the deposition of alluvial materials (aggradation) as well as the downcutting (degradation) of the geologic materials. Dominant depositional-erosional processes such as (1) overbank deposition, (2) point bar deposition, (3) sheet and gully erosion (scour), and (4) redeposition on floodplain surfaces result in the creation of major wetland habitats along the river and stream channels (Wharton, Kitchens, Pendleton, and Swipe 1982, Fig. 12.17 a, b). The point bar is built on convex banks of meandering streams or rivers by lateral deposition, with the bulk of the sediment staying in the floodplain (Leopold and Wolman 1957). Vertical accretion by overbank deposition builds up the floodplains. As floodwaters traverse the wetland, they lose velocity and deposit the largest sediment particles first. The meandering process of swamp rivers and streams also helps to reduce the energy of the water flow, which stabilizes the system. The degree and width of meandering is a function of water volume, velocity, and density, with meanders occurring at consistent intervals of 7–15 times the width of the channel (Dury 1977). Average floodplain deposition ranges between 0.3 m–0.6 m in depth in 200–400 yr (Leopold and Wolman 1957). However, sediment accumulations as high as 46 cm have been recorded in the Atchafalaya River basin, Louisiana, during a single major flood (Wharton et al. 1982). Surface erosion, or scour, and re-deposition on the floodplain result in a mosaic of various depositional and eroded wetland habitats parallel to the stream channel (Fig. 12.17 a, b).

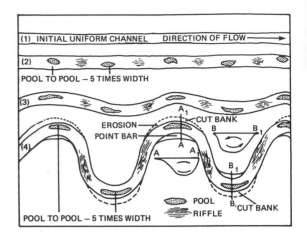

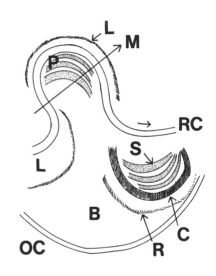

Figure 12.17. (a) Formation of point bars and meander formations in riparian floodplains. Unstable stream flow in a uniform river (1) results in the formation of pools and ripples (2), then to meanders (3), and deposition and point bar formation along the banks (4). (b) Idealized alluvial floodplain with various depositional environments: RC = river channel, M = direction of meander movement, L = natural levee, P = point bar deposit, B = backswamp, C = channel filled with deposits (a former oxbow lake), R = ridge (former natural levee), OC = overflow channel, S = swale deposit (from Wharton et al. 1982).

There is no one typical series of wetland habitats, but a number of recognized floodplain features are recognized. Natural levees are formed during periods of overbank flow as the currents abruptly slacken, and suspended sands and silt are deposited immediately adjacent and parallel to the channel (Wharton et al. 1982, Fig. 12.17 b). Levees are usually 30–100 m wide, although they can be several kilometers wide in the southeast. Backswamps, oxbow lakes, point bar deposits, hummocks (small hills), and swale ridge deposits are often formed and they provide distinctive wetland habitats. A key point is that these habitats are a product of the reworking of the sediments and that a change in topographic relief of a few centimeters results in clearly defined wetland plant communities. The National Wetlands Technical Council (NWTC) used this point to develop a framework for displaying the relationship between the geomorphologic floodplain features and bottomland plant communities (Larson et al. 1981).

An idealized floodplain sequence (Zone I–VI) proceeds from the river channel (Zone I) to the surrounding uplands along an increasing elevational gradient (Fig. 12.18 a–c). The classification system

relates to the following geomorphologic floodplain areas (Wharton et al. 1982):

Zone I: river channels, oxbow lakes, and permanently inundated backsloughs;
Zones II–V: the active floodplain including swales (II–III); flats and backswamps (IV); levees and relict levees and terraces (V);
Zone VI: the floodplain–upland transition to terrestrial ecosystems.

The presence of natural levees of varying ages, swales, and scours, as well as micosite variations in elevation, often creates a complex pattern of habitats that grade in to one another. In some locations it is not possible to discern clear zonal patterns at all, or some zones are missing. Also, the presence of a species may not reflect current hydrologic conditions but earlier water levels prior to damming of the river or old buried channels and

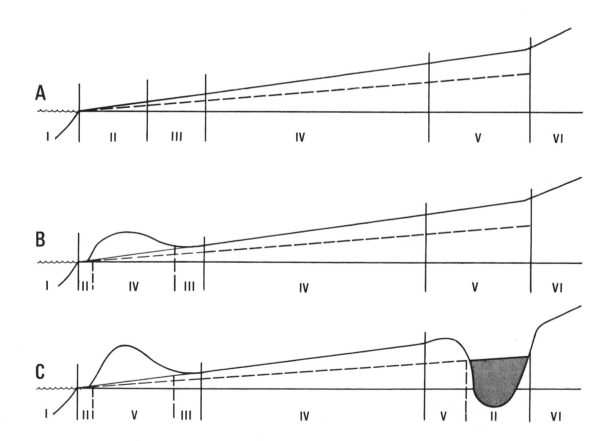

Figure 12.18. (a) Idealized sequence of bottomland hardwood zones from the river or stream to an upland along a moisture continuum; dotted line represents the depth of the water table. (b) Floodplain, from midallvial river to bluff, showing modification of zones by the intrusion of a natural levee between zones II and III. (c) Further modification by inclusion of an abandoned river channel filled with clay or peat (dark) (from Wharton et al. 1982).

so on. Thus, the floodplain of today is a complex mosaic of floral distributions that reflect individual plant species' tolerances to flooding and environmental changes, as well as manmade alterations to water flow, drainage patterns, and adjacent land use.

Riparian zones would be mostly classified as palustrine systems in the USFWS classification with a much lesser area as riverine or estuarine systems (Cowardin et al. 1979; Fig. 12.2). Finally, it should be recognized that some areas in the riparian zone may not be classified legally or ecologically as wetlands because they have water tables that recede well below 30 cm, or the rooting zone, during the growing season. Thus, there is a more restrictive definition of water dependency for wetlands than for riparian zones (Mitsch and Gosselink 1993).

Climate and hydrology. The climates of riparian systems vary greatly between the mesic east coast and arid west coast of the United States. In the West evapotranspiration (ET) greatly exceeds precipitation (PT), and the reverse is true in the East. The line of equal ET/PT is found approximately along a north to south line drawn from the eastern side of the Dakotas to eastern Texas, roughly the start of the true prairie lands (Thornthwaite 1948; see Chapter 9). These differences in rainfall and temperature greatly affect the vegetation found along the riparian zones in each region of the country. A typical water budget for an eastern riparian swamp shows that annual upstream inflows and downstream outflows dominate the hydrology of these systems (Fig. 12.4).

Two key watershed factors that influence the hydrologic patterns for rivers and in turn the acreage of wetland vegetation are the amount of drainage area and the channel slope. Water discharge volume and the duration of flooding are directly related to the drainage area upstream of the floodplain (Bedinger 1978). For example, Bedinger reported that watersheds in Arkansas with drainage basins of 116 km^2 were flooded 5–7% of the time, whereas watersheds with drainage basins in excess of 4500 km^2 were flooded 18–40% of the time. This is because small watersheds have rapid runoff and sharp peak floods, whereas large drainage areas have less severe and rapid flood peaks but longer duration of flooding. Typically small midwestern and eastern riparian watersheds or low-order streams flood for a few days to several weeks during spring thaw (Mitsch and Gosselink 1993). Zones II–V in bottomland forests in the South can flood for up to several months in the winter and spring, but Zones II remains flooded during most

of the year. By contrast, many watersheds in the arid western United States experience intense precipitation events (30–60 cm in 24 hr, Taylor 1983). Many of these riparian systems are characterized by a winter surplus of water followed by summer droughts.

These extreme differences in rainfall and runoff patterns among these riparian systems in the United States can be compared in terms of the recurrence interval (RI) of flow and discharge measured as multiples of the mean annual flood (Fig. 12.19, after Faber et al. 1989). A mean annual flood is a flood with an RI of 2.33 yr measured as a bankfull event. An RI of 50 refers to the statistical probability that a certain flow occurs every 50 yr on average, a 50 yr flood. Collectively, these flood and recurrence intervals show a distinct range in flood discharges for streams in the western compared to the eastern United States. For example, Pennsylvania rivers have a fairly flat RI (Fig. 12.19), suggesting that floods with high-magnitude recurrence intervals of 50 yr are not much larger that mean annual floods, whereas California streams have floods that exceed annual flows by more than 100 times. These flooding differences would have significant effects on the wetland species found within the wetland because it has been shown that wetland species vary greatly in their tolerance to flood duration and intensity (Whitlow and Harris 1979; Teskey and Hinkley 1980).

Water and soil chemistry. The water chemistry of riparian systems is quite variable and reflects the upstream water quality, surrounding geology, upland runoff, and groundwater inputs. Streams are usually acidic (pH 6–6.5), but if they are blackwater streams of the southeast and draining the organic soils of coastal peatlands or upstream swamps, the pH can be 5 or lower (Richardson 1981) (Table 12.8). Prolonged flooding and lower pH increase the availability of macronutrients (P, N, Mg, S) and micronutrients (Fe, Mn, Bo, Cu, and Zn) as well as the amount of dissolved organic matter (Faulkner and Richardson 1989).

The soil texture of riparian systems is generally dominated by sand, silt, silty clays, or dense clays, depending on the zone (Table 12.8). Zone II has the highest organic matter (up to 40%), which is related to its being the highest zone of continual flooding outside the stream channel (Zone I) itself. The higher levels of organic matter also result in the binding of such nutrients as Ca, N, and P (Faulkner and Richardson 1989). Alluvial floodplain soils have the lowest organic matter (Wharton et al. 1982). The soils are generally anaerobic for a portion of the year in Zones II–V, but Zones

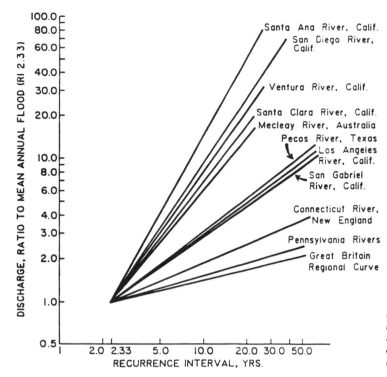

Figure 12.19. Relationship between mean annual flood and recurrence intervals for selected western and eastern rivers of the United States (adapted from Faber et al. 1989).

II, III, and IV remain much wetter during the year and are much further reduced, as shown by the gray to bluish soil color. The nutrient levels and soil redox can be directly related to the vegetation in each zone (Faulkner, Patrick, Gambrell, Parker, and Good 1991).

Vegetation

Southeastern riaprian wetlands. These wetlands comprise the largest area of forested wetlands of this type in the continental United States and are dominated by bottomland hardwood species that are adapted to the low-oxygen soil conditions of Zones II–IV and low microsites on the floodplain. The river channel (Zone I) is often dominated by a variety of submerged aquatic vascular species such as *Ceratophyllum, Nuphar, Isoetes, Lemna, Spirodela, Myriophyllum, Sagittaaria, Ludwigia,* and algae, and so on. The wettest riaparian zone is Zone II, and the least flooded area is Zone V. A recent study in the Cache River floodplain in Arkansas (Table 12.9) presented a typical southeastern floodplain vegetational sequence along a river flooding gradient (Smith 1996). Analysis of tree basal area shows that Zone II is dominated by *Nyssa aquatica* (tupelo gum) and *Taxodium distichum* (cypress), Zone III by *Quercus lyrata* and *Carya aquatica,* Zone IV by *Quer-*

cus nutalli and *Fraxinus pennsylvanica,* and Zone V by *Quercus phellos* and *Liquidambar styraciflua* (see Fig. 11.10, in this volume).

It is often quite difficult to distinguish between Zone III and Zone IV, especially if the zones do not have a distinctive dominant canopy species – that is, if they represent a transitional community. Also, many of the species found in Zones II–IV are found in each of the other zones but at different densities and basal areas. The basal area was highest in the zone nearest the river, but species richness, Shannon's diversity, and evenness were higher in less flooded zones (III–V), a trend often reported throughout the southeast (Mitsch and Gosselink 1993).

Canonical correspondence analysis indicated that species distributions were significantly correlated with flood depth and flood duration, as well as geomorphic position and soil texture (Smith 1996; Fig. 12.20). The river swamp forest (Zone II) was flooded nearly continuously, and the lower hardwood swamp forest (Zone III) was flooded up to 50% of the year. The flats or backwater swamp forests (Zones IV and V) have a higher elevation on the landscape and were flooded for approximately 30% of the year. These flooding regimes and vegetation responses closely follow those reported for swamps throughout the Southeast (Pen-

Table 12.8. *Physiochemical characteristics of floodplain soils by National Wetland Technical Council (NWTC). Zone I (permanent water courses) is excluded from this table. Data on pH include drought years.*

Characteristic	Zones				
	II	III	IV	V	VI
Degree of inundation and saturation	Intermittently exposed; nearly permanent inundation and saturation	Semipermanently inundated or saturated	Seasonally inundated or saturated	Temporarily inundated or saturated	Intermittently inundated or saturated
Timing of flooding	Year-round except during extreme droughts	Spring and summer during most of the growing season	Spring for 1–2 months of the growing season	Periodically for up to 1 month of growing season	During exceptionally high floods or extreme wet periods
Probability of annual flooding	100%	51%–100%	51%–100%	10%–50%	1%–20%
Duration of flooding	100% of the growing season	>25% of the growing season	12.5%–25% of the growing season	2%–12.5% of growing season	<2% of the growing season
Soil texture	Dominated by silty clays or sands	Dominated by dense clays	Clays dominate surface; some coarser fractions (sands) increase with depth	Clay and sandy loams dominate; sandy soils frequent	Sands to clays
Organic matter (%) Blackwater	18.0	—	7.9	—	—
Alluvial	4.5	3.4	2.8	3.8	—
Oxygenation	Moving water aerobic; stagnant anaerobic water	Anaerobic for portions of the year	Alternating anaerobic and aerobic conditions	Alternating; mostly aerobic, occasionally anaerobic	Aerobic year-round
Soil color	Gray to olive gray with greenish gray, bluish gray, and grayish green mottles	Gray with olive gray mottles	Dominantly gray on blackwater floodplains and reddish on alluvial with brownish gray and grayish brown mottles	Dominantly gray or grayish brown with brown, yellowish brown, and reddish brown mottles	Dominantly red, brown, reddish brown, yellow, yellowish red, and yellowish brown, with a wide range of mottled colors
pH Blackwater	5.0	—	5.1	—	—
Alluvial	5.0	5.3	5.5	5.6	—

Source: Zones are partially derived from Clark and Benforado (1981); modified from Wharton et al. (1982).

Table 12.9. *Tree basal area (m² ha⁻¹) for floodplain forests along the Cache River of Arkansas. Dashes indicate values <0.1 m² ha⁻¹.*

Species	Code	Dominance type Zone II	Zone III	Zone IV	Zone V
Carya tomentosa	CATO	—	—	—	0.1
Quercus falcata var. *pagodifolia*	QUPA	—	—	—	0.1
Quercus alba	QUAL	—	—	—	0.1
Quercus michauxii	QUMI	—	—	—	0.2
Quercus nigra	QUNI	—	—	—	2.7
Quercus phellos	QUPH	—	0.8	0.6	9.0
Nyssa sylvatica Marsh. var. *sylvatica*	NYSY	—	—	0.1	—
Liquidambar styraciflua	LIST	—	0.3	1.1	3.9
Carpinus caroliniana	CACA	—	—	0.6	0.5
Betula nigra	BENI	—	—	0.7	0.1
Celtis laevigata	CELA	—	—	0.4	0.1
Ulmus americana	ULAM	0.1	1.1	1.7	0.9
Quercus nutallii	QUNU	0.1	5.6	3.1	2.9
Quercus lyrata	QULY	0.7	7.2	2.3	2.7
Carya aquatica	CAAQ	0.2	3.1	2.2	0.2
Diospyros virginiana	DIVI	0.1	0.9	1.1	0.3
Fraxinus pennsylvanica	FRPE	0.3	0.3	2.1	0.1
Cornus foemina	COFO	—	—	0.1	—
Gleditsia aquatica	GLAQ	0.3	2.1	0.1	—
Populus heterophylla	POHE	0.1	0.5	—	—
Acer rubrum	ACRU	2.6	1.1	2.6	0.3
Fraxinus profunda	FRTO	3.6	1.3	1.3	—
Planera aquatica	PLAQ	1.7	0.3	0.2	—
Taxodium distichum	TADI	11.3	3.7	—	0.6
Nyssa aquatica	NYAQ	32.9	0.1	—	—
All other species		0.1	0.1	0.1	0.2
Total		54.8	29.4	21.0	25.7
Number of plots		54.0	59.0	70.0	70.0
Species richness		5.0	8.0	8.0	7.0
Shannon's Diversity (*H'*)		0.3	0.6	0.6	0.5
Evenness (*E*)		0.5	0.7	0.7	0.6

Source: Smith (1996).

found 1952; Schlesinger 1978; Conner and Day 1982; Wharton et al. 1982; Robertson, Mackenzie, and Elliot 1984; Brinson 1990; Sharitz and Mitsch 1993). An excellent review of the tolerances of vegetation to flooding by Whitlow and Harris (1979) reveals that most species on the floodplain require some period of nonflooding but that a number of species in Zones II and III can survive extensive flooding even during the growing season.

The tolerance to flooding of many species common to the swamps of the Southeast was quantified by a classic study on TVA reservoirs by Hall, Penfound, and Hess (1946). They found that a number of species like black willow (*Salix nigra*) or sweet gum (*Liquidambar styraciflua*) could tolerate 30 cm of flooding up to 2 yr but that species like loblolly pine (*Pinus taeda*) or black alder (*Alnus rugosa*) died after 1 yr of continuous flooding. Sugar

maple (*Acer saccharum*) and white oak (*Quercus alba*) could not survive 30 cm of continuous flooding for even one growing season, as was also the case for red cedar (*Juniperus virginiana*) and chestnut oak (*Quercus prinus*) (Table 12.10).

Physiological and biochemical research has shown that the species that can tolerate flooding or avoid the oxygen deficit caused by flooded soils (tupelo gum, cypress, willow, etc.) have evolved structural (e.g. arenchyma, lenticels, adventitious roots) or biochemical mechanisms (e.g. malate acid storage, alternate proton storage mechanisms and anaerobic respiration) to overcome oxygen deficiency problems (Koslowski 1984; Hook et al. 1988). Wet and dry ecotypes can exist in the same species. For example, it has been shown that wet-site loblolly pine (*Pinus taeda*) in North Carolina exhibit radial oxygen loss (i.e., ability to diffuse O_2 out

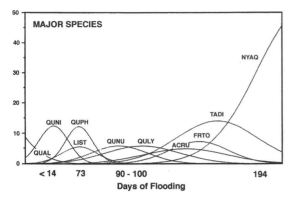

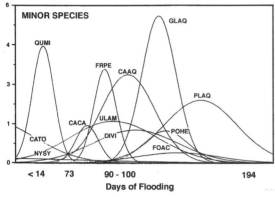

Figure 12.20. Species distributions in Arkansas riparian wetlands along the Cache River, according to days of flooding. See Table 12.9 for species codes.

of roots in flooded soils) and survive prolonged flooding, whereas loblolly pine growing in dry west Texas cannot survive flooding (Topa 1985).

The two most common swamp species reported for deep swamps and Zone II are cypress and tupelo gum. Cypress is found in bottomland riverine swamps throughout the Southeast, and as far north as Illinois and New Jersey (Conner and Day 1982). Tupelo gum is very common in all southern swamps. Because neither of these species can germinate under water and seedlings cannot survive long periods of submergence (Debell and Naylor 1972), establishment requires some period of soil drying or droughts (Wells 1942). This often results in the development of even-age stands of cypress and gum.

Indications of the flooding regime or disturbance can often be determined from the association of other species with cypress. For example, a cypress-hardwood association in a Zone II–III bottomland suggests that the area has a shorter hydroperiod than that found in a pure cypress-tupelo association. Hardwoods indicative of a shorter hydroperiod are red maple (*Acer rubrum*), American

Table 12.10. *Approximate order of tolerance of woody species to inundation in the Tennessee Valley. Tolerant: able to survive continuous flooding to a depth of 30 cm or more for up to two growing seasons. Moderately tolerant: succumb during second growing season of continuous flooding. Intolerant: unable to survive continuous flooding for one growing season.*

Common name	Scientific name
Tolerant	
Silver maple	*Acer saccharinum*
Sweet gum	*Liquidambar styracifllua*
Rattan vine	*Berchemia scandens*
Swamp rose	*Rosa palustris*
Florida vine	*Brunnichia cirrhosa*
Dogbane	*Trachelospermum difforme*
Greenbrier	*Smilax* sp.
Red maple	*Acer rubrum*
Persimmon	*Diospyros virginiana*
Green ash	*Fraxinus lanceolata*
Honey locust	*Gleditsia triacanthos*
Overcup oak	*Quercus lyrata*
Cottonwood	*Populus deltoides*
Water hickory	*Carya aquatica*
Swamp privet	*Forestiera acuminata*
Pepper vine	*Ampelopsis arborea*
Trumpet vine	*Campsis radicans*
Sandbar willow	*Salix interior*
Black willow	*S. nigra*
Buttonbush	*Cephalanthus occidentalis*
Tupelo gum	*Nyssa aquatica*
Bald cypress	*Taxodium distichum*
Moderately tolerant	
Black alder	*Alnus rugosa*
Indigo bush	*Amorpha fruticosa*
Hispid greenbrier	*Smilax hispida*
Red mulberry	*Morus rubra*
Wild grape	*Vitis* sp.
Cow oak	*Quercus michauxii*
Hackberry	*Celtis laevigata*
Winged elm	*Ulmus alata*
Hawthorn	*Crataegus* sp.
Osage orange	*Maclura pomifera*
Box elder	*Acer negundo*
Loblolly pine	*Pinus taeda*
River birch	*Betula nigra*
Water oak	*Quercus nigra*
American elm	*Ulmus americana*
Sycamore	*Platanus occidentalis*
Deciduous holly	*Ilex decidua*
Intolerant	
Post oak	*Juniperus virginiana*
Sugar maple	*Quercus stellata*
White oak	*Acer saccharum*
Yellow buckeye	*Quercus alba*
Yellow poplar	*Aesculus octandra*
Prickly ash	*Liriodendron tulipifera*
American beech	*Aralia spinosa*
Swamp hickory	*Fagus grandifolia*
Black walnut	*Carya leiodermis*
Ironwood	*Juglans nigra*
Redbud	*Carpinus caroliniana*
Red Cedar	*Cercis canadensis*

Table 12.10. *(Cont.)*

Common name	Scientific name
Scrub pine	Pinus virginiana
Shortleaf pine	P. echincata
Wild black cherry	Prunus serotina
Blackjack oak	Quercus marilandica
Basswood	Tilia sp.
Southern red oak	*Quercus falcata*
Sourwood	*Oxydendrum arboreum*
Flowering dogwood	*Cornus florida*
Sassafras	*Sassafras albidum*
Black locust	*Robinia pseudoacacia*
Snagbark hickory	*Carya ovata*
Mockernut hickory	*C. tomentosa*
Chestnut oak	*Quercus montana (=Q. prinus)*
White ash	*Fraxinus americana*

Note: Species within each group are ordered from least tolerant to most tolerant.
Source: After Hall et al. (1946).

Table 12.11. *Species tolerance to flooding in New England. Same definitions as in Table 12.10*

Common name	Scientific name
Very tolerant	
Black Willow	*Salix nigra*
Tolerant	
Red maple	*Acer rubrum*
Silver maple	*A. saccharinum*
Black alder	*Alnus glutinosa*
Slightly tolerant	
Red oak	*Quercus rubra*
Bigtooth aspen	*Populus grandidentata*
Basswood	*Tilia americana*
Ironwood	*Carpinus caroliniana*
American elm	*Ulmus americana*
Hop hornbeam	*Ostrya virginiana*
White ash	*Fraxinus americana*
Intolerant	
Sugar maple	*Acer saccharum*
Yellow birch	*Betula alleghaniensis*
Paper birch	*B. papyrifera*
White birch	*B. populifolia*
American beech	*Fagus grandifolia*
Red spruce	*Picea rubens*
White pine	*Pinus strobus*
Quaking aspen	*Populus tremuloides*
Black cherry	*Prunus serotina*
White oak	*Quercus alba*
Chinquapin oak	*Q. muehlenbergii*
Eastern hemlock	*Tsuga canadensis*

Source: Whitlow and Harris (1979).

elm (*Ulmus americana*), ash (*Fraxinus* sp.), and oaks (*Quercus* sp) (Table 12.10). Red maple and ash can also occur in areas of prolonged flooding but in reduced numbers and size. A cypress-pine association (often *Pinus taeda* in the Southeast) occurs where the area has been severely drained or the river flooding controlled by dams, water diversions, and channelization.

Northeastern riparian swamp forests. Floodplains occur on approximately 129,000 ha, or about 3% of the total land of the Northeast (U.S. Department of Agriculture 1962). Traditionally, floodplain forests in this region like most parts of the country have been harvested for timber and converted to agricultural lands. Species common to the floodplains are shown in Table 12.11. Like the southeastern floodplains, the species can be arrayed on a flood-frequency gradient that represents the physiographic features of the floodplain (e.g., flats, point bars, terraces). Sycamore (*Platanus occidentalis*) and green ash (*Fraxinus pennsylvanica*) are often found closest to the streams in areas flooded for nearly 99% of the time (Morris 1978). White oak (*Quercus alba*) and chestnut oak (*Quercus prinus*) cannot tolerate flooding conditions and are found in the upland areas.

Few data exist on the tolerance of northeastern species to flooding, but a study by McKim, Gratta, and Merry (1975) demonstrated that certain species of willow, silver maple, alder, and red maple were able to tolerate 1 yr of flooding (Table 12.11). Willow, silver maple (*Acer saccharinum*), and cottonwood (*Populus deltoides, P. balsamifera*) often dominate the wettest floodplain regions, with swamp red maple (*Acer rubrum*) and green ash (*Fraxincus pennsylvanica*) dominating many swamps and backwater flooding areas in the Northeast.

Western riparian vegetation. The plant distributions of riparian vegetation vary considerably from the northwest to the southwest because of higher rainfall in the northwestern United States (Oregon and Washington) compared to the dry southwest of California, Utah, or New Mexico. A comparative listing of flood tolerance of woody plants of the northwest suggests that willow species and red-osier dogwood (*Cornus stolonifera*) can withstand flooding in the riparian area for more than 1 yr (Table 12.12). Moreover, species like lodgepole pine (*Pinus contorta*) and box elder (*Acer negundo*) can tolerate flooding for up to 1 yr. Surprisingly, a number of alder species (*Alnus rubra* and *A. sinuata*) cannot tolerate more than a few days of flooding, nor can Douglas fir (*Pseudotsuga menziesii*) (Table 12.12). Only a few common genera

Table 12.12. *Relative flood tolerances of woody plants, North Pacific Division. Same definitions as in Table 12.10*

Common name	Scientific name
Very tolerant	
Red-osier dogwood	*Cornus stolonifera*
Narrow leaf willow	*Salix exigua*
Hooker willow	*S. hookeriana*
Pacific willow	*S. lasiandra*
Tolerant	
Box elder	*Acer negundo*
Bog laurel	*Kalmia polifolia*
Labrador tea	*Alnus glutinosa*
Lodgepole pine	*Pinus contorta*
Cottonwood	*Populus trichocarpa*
Elder	*Sambucus callicarpa*
Hardhack	*Spirea douglasii*
Western red cedar	*Thuja plicata*
Blueberry	*Vaccinium uliginosum*
Slightly tolerant	
Riverbank mugwort	*Artemesia lindleyana*
Nuttall's dogwood	*Cornus nuttallii*
Walnut	*Juglans* spp.
Apple	*Malus* spp.
Ponderosa pine	*Pinus ponderosa*
Smooth sumac	*Rhus glabra*
Western hemlock	*Tsuga heterophylla*
Intolerant	
Bigleaf maple	*Acer macrophyllum*
Alder	*Alnus rubra*
Alder	*A. sinuata*
Boxwood	*Buxus sempervirens*
Filbert	*Corylus avellana*
Hazel	*C. rostrata*
Cotoneaster	*Cotoneaster* spp.
Hawthorn	*Crataegus oxyacantha*
Holly	*Ilex aquifolium*
Mock orange	*Philadelphus gordonianus*
Bitter cherry	*Prunus emarginata*
Cherry-laurel	*P. laurocerasus*
Douglas fir	*Pseudotsuga menziesii*
Wild apple	*Pyrus rivularis*
Cascara	*Rhamnus purshiana*
Blackberry	*Rubus procerus*
Rowan tree	*Sorbus aucuparia*
Lilac	*Syringa vulgaris*

Source: Whitlow and Harris (1979).

such as *Populus* spp. and *Salix* spp. are found in the wettest zones in both the western and the eastern United States. At lower elevations, the western riparian areas support a population of willow, sycamore, alder and cottonwood, and at the higher elevations occur other species in some of the same genera. The species names may change from region to region, especially for *Salix*, but in general these vegetation patterns hold throughout the West.

Different species of *Salix* are found at different elevations from the coastal plain to the mountains. *Salix* spp. (name derived from Celtic *sal*, near, and *lis*, water) are common throughout the west and are good indicators of riparian habitat. Red willow (*Salix laevigata*) grows up to 1200 m in southern California and on Catalina Island. Yellow willow (*Salix lasiandra*) extends up to 2400 m in the mountains and is found on Santa Cruz Island. Both of these willow species occur as far east as Utah (Padgett et al. 1989). Goodings's willow (*Salix goodingii*) is found along streambanks in desert areas and the Central Valley, suggesting that it needs hot summers and groundwater (Holstein 1984). White willow or arroyo willow (*Salix lasiolepis*) is found along perennial streams at low elevations but grows up to 750 m elevation along intermittent streams, near seeps, or springs. Sandbar willow (*Salix hindsiana*) is common along coastal rivers, especially on sandbars.

Freemont cottonwood (*Populus fremontii*) is scattered throughout California along lowland streams and on Santa Cruz and Santa Catalina islands. It is also an important species for cold desert, warm desert, and Rocky Mountain riparian areas (Padgett *et al.* 1989; Chapters 3, 7, 8 of this volume). It is often confined to alluvial stream bottoms and gravely or sandy areas that stay wet. Black cottonwood (*Populus trichocarpa*) grows at elevations above Fremont cottonwood throughout California. California sycamore (*Platanus racemosa*) is abundant below 1200 m throughout cismontane California along streams, near springs or ground water supplies and on alluvial terraces. Big-leaf maple (*Acer macrophyllum*) is found along mountain springs or perennial streams in canyons. Box elder, another deciduous riparian tree (*Acer negundo*), is throughout the west. It grows in patches with alder, sycamore, and willow.

California riparian vegetation is not uniform. Roberts, Howe and Major (1977) differentiated nine regional types, including several desert types. Major nonarid types were the following:

North coast, dominated by red alder (*Alnus rubra*), big leaf maple (*Acer macrophyllum*), and black cottonwood (*Populus trichocarpa*).

South coast, dominated by sycamore (*Platanus racemosa*), Fremont cottonwood (*Populus fremontii*), white alder (*Alnus rhombifolia*), and coast live oak (*Quercus agrifolia*).

Central Valley, dominated by sycamore, Fremont cottonwood, valley oak (*Quercus lobata*), and box elder (*Acer negundo* ssp *californicum*).

Knapp (1965) referred to all these low elevation forests as "sycamore-alder bottomland woods." Holstein (1984) concluded that the Arcto-Tertiary

River Undercutting	Open Floodplain	Gravel Bar Thicket	Riparian forest	Valley Oak forest	Grassland
		overstory 2–3 m	overstory 25 m	overstory 15–30 m	
	low % cover	Salix hindsiana	Populus fremontii	Quercus lobata	Avena barbata
	high diversity	Alnus rhombifolia	Acer negundo	Platanus racemosa	Avena fatua
	Chrysopsis oregona	Acer negundo	Alnus rhombifolia	Fraxinus latifolia	Bromus diandrus
	Trichostema lanatum	Franxinus latifolia	Salix lasiolepis	Juglans hindsii	Bronus mollis
	Polypogon monspeliensis	Baccharis pilularis	SHRUB	LIANA	Carex spp.
	Verbascum blattaria	Baccharis viminea	Rubus leucodermis	Rhus diversiloba	Elymus glaucus
	Tunica prolifera	Populus fremontii	Rubus vitifolius	Aristolochia calif.	Elymus triticoides
	Bromus diandrus	HERB	Cephalanthus occidentalis	Smilax californica	Hordeum brachyantherum
		Phragmites communis	Sambucus mexicana	Clematis ligusticifolia	Hordeum geniculatum
		Artemisia douglasiana	LIANA	SHRUB	Hordeum leporinum
		Distichlis spicata	Vitis californica	Rubus vitifolius	Lolium multiflorum
		Urtica holosericea		Rosa californica	Lolium perenne
		Sorgham halepense			Stipa pulchra
		Carex senta			
		Equisetum laevigatum			
		Cynodon dactylon			

Elevation relative to river–time between major disturbances → High–long

Low–short ————

Frequency, duration, and depth of inundation

High ———→

Low ————

Figure 12.21. Idealized sequence of vegetation along major rivers in the Central Valley of California. Characteristic tree, shrub, and herb taxa are listed (from Conard et al. 1977).

geoflora origin of this forest explained the many winter-deciduous genera in common today between the Californian riparian forest and the eastern North American deciduous forest.

Much of this low-elevation forest has been destroyed by agricultural clearing. Some of the best, most expensive agricultural settings in the state are found on the well-drained natural levee soils beneath once-present riparian forests. In 1800, for example, the Central Valley was thought to contain 370,000 ha of riparian forest; major rivers were bordered by 1–5 km wide strips of forest. By 1980 only 21,500 ha of the original remained in sustainable, healthy condition (Katibah 1984). These remnants have been quantitatively summarized and classified by Conard, MacDonald, and Holland (1977), Keeler-Wolf, Lewis, and Roye (1994), Sawyer and Keeler-Wolf (1995), Ferren (1983), and Ferren Fiedler and Leidy (1994), among others.

Vegetation is zoned back from open water in the following pattern: an open herbaceous floodplain, a gravel bar scrub thicket, a riparian forest with complex physiognomy, and a valley oak–sycamore woodland (Fig. 12.21). Environmental gradients that correspond with this zonation include frequency of flooding, intensity of mechanical scouring from moving water, soil texture, and depth to water table. Apparently, valley oak and sycamore both grow best where the water table is >10 m deep, whereas cottonwood, alder, and various willow trees and shrubs grow best where the water table is shallow. The riparian forest has a four-tiered canopy architecture: an open overstory tree canopy 30 m tall and 20–80% closed, a subdominant tree canopy 5–15 m tall, a relatively continuous shrub canopy, and a patchy stratum of annual and perennial herbs. Bird and mammal diversity is also very high, possibly because of the vegetation's physiognomic complexity. The oak-sycamore woodland, farther away from and higher above the water channel, is less diverse and more open. The overstory is 15–20 m tall. Tree density is 125 ha^{-1} and the basal area is 20 m^2 ha^{-1}, about half the values of riparian forest. A continuous (perennial) herbaceous understory with scattered shrubs comprises the second canopy stratum in the woodland's two-tiered physiognomy. In both zones the overwhelming phenology is winter-deciduous.

In summary, there are significant differences in the vegetation of the western versus eastern riparian ecosystems as well as a north to south gradient on both coasts. Moreover, the species in the riparian zones change dramatically along an elevational gradient as the streams move from the mountains to the coast or valley bottoms (Padgett, Youngblood, and Winward 1989). Few species are common to both regions because of dramatic differences in climate (Brinson et al. 1981). However, generic uniformity is high; for example willow (*Salix* sp.) and cottonwood (*Populus* sp.) are found in most regions of the country and often dominate the riparian zone closest to the streams and rivers, with the exception of the cypress and gum dominance in the deep water swamps of the Southeast.

Rare and endangered species and exotic introductions. The riparian ecosystems of the north, central, and southeast support a diverse flora and fauna. A number of plant communities and species are restricted to specific regions of the country. For example, Texas and Oklahoma alone have 110 plant species, 24 plant community types, and 125 species of wildlife of special concern. The flatland hardwood community type of swamp chestnut oak (*Quercus michauxii*)–willow oak (*Q. phellos*)–laurel oak (*Q. laurifola*) is apparently unique to Texas (Marks and Harcombe 1981). The western community types dominated by cedar elm (*Ulmus crassifolia*) and sugar berry (*Celtis laevigata*) and willow oak are rare elsewhere. Today high grading of timber has resulted in the loss of most stands of large cypress and gum species. Thirty-four plant communities have been identified in the southeast (Neal and Haskins 1986).

A major concern in the riparian areas of the West is both the loss of rare species and the introduction of exotic plants. Only a few rare and endangered plant species are listed for each state. For example, California's Native Plant Society lists *Delphinium hesperium, Downingia concolor, Mahonia nevinii,* and *Monardella linoides* among others as endangered species for riparian areas. Of more importance is the fact that the riparian ecosystem itself is endangered in many parts of the developed West.

Zembal (1984) noted that approximately 30% of the species tallied for the Santa Ana River Canyon were introduced. Of particular concern throughout the West is the elimination of native riparian species by three dominant introduced species: salt cedar or tamarisk (*Tamarix* spp.), German ivy (*Senecio mikanioides*), and giant reed grass or cane (*Arundo donax*).

Succession. A key aspect of understanding the plant succession in riparian systems is that these ecosystems are subject to catastrophic flooding events, especially in the West. Also, so many of these habitats have been clearcut, farmed, or selectively cut over the past 150 years that it is difficult to determine the native species composition. The extensive alteration of riverine flooding patterns caused by dams has also greatly changed the plant community distributions for many riparian areas. Thus, it is important to understand these disturbance patterns before predicting successional pathways.

In the West succession starts near the streams with willow, cottonwood, alder and the like, and it may take only 50 to 75 yr starting from bare sand to culminate in a riparian forest (Faber et al. 1989). A more mature forest farther from the stream may require several hundred more years with no disturbance to reach maturity. This is the area where the largest sycamore, cottonwood, and oak species are found, but few such areas exist in the West due to development.

In the eastern and southern United States the diversity of species is such that each region is different. Few detailed studies exist on forest succession in riparian areas. After clearcutting or fire, Gaddy, Kohlsaat, Laurent, and Stansell (1975) noted for the Congaree Swamp in South Carolina that even-aged sweetgum stands and shade-intolerant hardwoods grow first but are replaced in ~100 years by more shade-tolerant oaks and gums. Windthrows result in gaps that release understory trees as well as shade-intolerant species like *Carpinus caroliniana* and *Ilex opaca*. In general, succession can be followed by analyzing the understory sapling and shrub density and comparing it to the overstory species composition. Pioneer successional communities in floodplain areas (III–V) are often dominated by cottonwood, black willow, or river birch but these are replaced within 100 years or less by the typical bottomland hardwood forests of the southeast. The wettest zone (II) is usually restricted to the most tolerant species like gum and cypress, although they cannot germinate or grow under continuous flooding. On sandbars and new levees, *Salix nigra* is the important pioneer, which is followed by *Populus deltoides* and *Acer saccharinum*. In the final stage these are replaced by *Quercus laurifolia* and *Carya aquatica*. Finally, a major factor controlling plant succession today along the riparian areas of the East is beaver activity (*Castor canadensis*). They have flooded tens of thousands of acres, girdled and killed native trees, and helped start the succession of *Salix, Alnus*, and *Cephlanthus occidentalis* (Hair, Hepp, Lackett, Reese, and Woodward 1979). Along the Atlantic coast infrequent but regular fires that burn to the mineral soil favor the formation of monocultures of Atlantic white cedar (*Chamaecyparis thyoi-*

des), although few stands exist today because of clear-cutting (Laderman 1989; also see Chapter 11).

AREAS FOR FUTURE RESEARCH

Wetland science is relatively young compared to other areas of ecosystem study. Unfortunately, both ecologists and the national funding agencies have not taken much interest in these unique ecosystems. Many of the outstanding wetland areas of the country, such as Carolina bays or pocosins, have never been studied in detail; only fragmented studies have been done on the hydrology, soils, biogeochemical cycles, rare and endemic species, and plant communities for many wetlands. If we are to maintain and restore many of the plant communities, we need information on the natural variation in hydroperiod and hydropattern. Information on flood tolerance or hydroperiod requirements for most species is unknown, and thus we cannot accurately place boundaries on these systems. We also lack research that integrates geology, soils, hydrology, and plant communities of wetland systems. A classic example of this is the U.S. Army Corps' recent plans to restore Everglades hydrology without any paleoecological information on the natural Everglades plant communities and their hydrological requirements.

Recent work in the role of boreal peatlands in the global warming scenario has helped our understanding of carbon storage in these systems. However, their hydrology is poorly understood, and vegetation successional patterns are virtually unknown for many areas. The hydrologic requirements and tolerance limits to changing water systems are unstudied for most peatland species.

The complexity of the prairie pothole region is just now being realized. Interdisciplinary research relating vegetation, soils, and groundwater hydrology is needed. Additional studies of the biogeochemistry of S, N, Ca, and redox chemistry are essential to our understanding of vegetation patterns in this region. No work has been done on species competition and precious little on salinity tolerances. Further work is needed on the seedbank so that we can more intelligently manage water levels in these systems to regenerate plant communities. Research on groundwater movement in complex glacial geology is also needed to unravel the relationship between wetland plants on the landscape and regional water regimes.

An understanding of factors controlling establishment and structure of Everglades macrophyte communities is critical if this ecosystem is to be protected, restored, and managed. Research on ecologically sound ways to remove the exotic species is also essential to the survival of the Glades. Work is desperately needed on the effects of changing water levels and pulsed water flow on plant community structure and species survival. The effects of nutrient inputs, such as P, have been studied intensively for the past 8 yrs (Richardson, Craft, Qualls, Stevenson, and Vaithi-yanathan 1994; Richardson et al. 1995), but carefully planned research is needed to understand the interaction of varying nutrient concentrations and changing water levels on plant successional patterns.

Between 1940 and 1980 the national loss for riparian forests, mainly bottomland hardwood forests, was approximately 2.8 M ha (Abernethy and Turner 1987). Most of these losses occurred in the south-central and southeast U.S. forested wetlands, especially along the Mississippi River, as agricultural interests converted these lands to crops. There is a pressing need to restore some of these lost habitats and to better manage the remaining forests and marshes along our river systems throughout the United States. Little research has been done on the hydrologic requirements needed to restore the habitat for most of these species.

Massive numbers of rivers have been dammed in the United States, resulting in significant changes in the hydrologic regimes for most riparian systems. The ecological effects are unknown. In the western United States over 95% of the water in rivers and major streams is allocated for irrigation, farm, or human use, and thus many rivers are now dry downstream for a significant portion of the year. This reallocation of water has severely reduced the riparian habitat and increased the endangerment of wetland flora and fauna. Research is needed to determine the effects of this water loss on vegetation. The effect of intensive cattle grazing in the riparian zone in most western states also needs study. These problems are being partially addressed in the West by the Bureau of Land Management's (BLM) "Riparian Wetland Initiative for the 90s" (U.S. Department of the Interior 1991). The BLM has proposed an $85 million restoration and maintenance program on the 9.6 M ha that the agency manages.

The problem of loss of wetland habitat resulting from drainage and development is probably the biggest cause of the extinction of wetland plant species today. The magnitude of this loss is unknown for most wetlands of North America.

REFERENCES

Abernethy, V., and R. E. Turner, 1987. U.S. forested wetlands: 1940–1980. Bioscience 37: 721–727.

Almendinger, J. C., J. E. Almendinger, and P. H. Glaser. 1986. Topographic changes across a spring fen and raised bog in the Lost River peatland, northern Minnesota. J. Ecol. 74: 393–401.

Arndt, J. L., and J. L. Richardson. 1988. Hydrology, salinity and hydric soil development in a North Dakota prairie-pothole wetland system. Wetlands 8: 93–107.

Arndt, J. L., and J. L. Richardson. 1989. Geochemical development of hydric soil salinity in a North Dakota prairie-pothole wetland system. Soil Sci. Soc. Am. J. 53: 848–855.

Avery, G., and L. Loope. 1980. Plants of Everglades National Park: a preliminary checklist of vascular plants. South Florida Research Center Rep. T-558, Everglades National Park, Homestead, Fla.

Bauder, E. T. 1989. Drought stress and competition effects on the local distribution of Pogogyne abramsii. Ecology 70: 1083–1089.

Bedinger, M. S. 1978. Relation between forest species and flooding, pp. 427–435 in P. E. Greeson, J. R. Clark, and J. E. Clark (eds.), Wetland functions and values: the state of our understanding. American Water Resources Association, Minneapolis, Minn.

Belanger, T. V., D. J. Scheidt, and J. R. Platko II. 1989. Effects of nutrient enrichment on the Florida Everglades. Lake and Reservoir Mgt. 5: 101–111.

Bodle, M. J., A. P. Ferriter, and D. D. Thayer. 1994. The biology, distribution, and ecological consequences of Melaleuca quinquenervia in the Everglades, pp. 341–355 in S. M. Davis, and J. C. Ogden (eds.), Everglades: the ecosystem and its restoration. St. Lucie Press, Delray Beach, Fla.

Bolen, E. G. 1982. Playa wetlands of the U.S. southern high plains: their wildlife values and challenges for management, pp. 9–20 in B. Gopal, R. E. Turner, R. G. Wetzel, and D. F. Whigham (eds.), Wetlands ecology and management. Proceedings of the First International Wetlands Conference, New Delhi, India, September 10–17, 1980.

Boldt, D. R. 1985. Computer simulations of groundwater flow in a raised bog system, Glacial Lake Agassiz peatlands, northern Minnesota. Master's thesis, Syracuse University, Syracuse, N.Y.

Bord na Mona. 1984. Fuel peat in developing countries. Study report prepared for the World Bank, Dublin, Ireland.

Bridgham, S. D., and C. J. Richardson. 1993. Hydrology and nutrient gradients in North Carolina peatlands. Wetlands 13: 207–218.

Brinson, M. M. 1990. Riverine forests, pp. 87–141 in Lugo, A. E., M. M. Brinson, and S. L. Brown (eds.), Forested wetlands, Ecosystems of the world 15. Elsevier, Amsterdam.

Brinson, M. M. 1993. A hydrogeomorphic classification for wetlands. Technical Report WRP-DE-4., U.S. Army Corps of Engineers. Waterways Experiment Station. Vicksburg, Miss.

Brinson, M. M., R. Plantico, B. L. Swift, and J. S. Barclay. 1981. Functions, values and management of riparian ecosystems. U.S. Fish Wildl. Serv. Biol. Serv. Program FWS/OBS-draft.

Browder, J. A., S. Black, P. Schroeder, M. Brown, M. Newman, D. Cottrell, D. Black, R. Pope and P. Pope. 1981. Perspective on the ecological causes and effects of the variable algal composition of southern Everglades periphyton. Report T-643.

South Florida Research Center, National Park Service, Everglades National Park. Homestead, Fla.

Browder, J. A., D. Cottrell, M. Brown, M. Newman, R. Edwards, J. Yuska, M. Browder, and J. Krakoski. 1982. Biomass and primary production of microphytes and macrophytes in periphyton habitats of the southern Everglades. Report T-662. South Florida Research Center, National Park Service, Everglades National Park, Homestead, Fla.

Browder, J. A., P. J. Gleason, and D. R. Swift. 1994. Periphyton in the Everglades: spatial variation, environmental correlates, and ecological implications, pp. 379–418 in S. M. Davis, and J. C. Ogden (eds.), Everglades: the ecosystem and its restoration. St. Lucie Press, Delray Beach, Fla.

Brown, S., M. M. Brinson, and A. E. Lugo. 1979. Structure and function of riparian wetlands, pp. 17–31 in R. R. Johnson and J. F. McCormick (tech. coords.), Strategies for protection and management of floodplain wetlands and other riparian ecosystems. U.S. Forest Service, Washington, D.C.

Cajander, A. K. 1913. Studien über die Moore Finnlands. Acta For. Fenn. 2: 1–208.

Carter, M. R., L. A. Burns, T. R. Cavinder, K. R. Dugger, P. L. Fore, D. B. Hicks, H. L. Revells, and T. W. Schmidt. 1973. Ecosystem analysis of the Big Cypress Swamp and estuaries. U.S. EPA 904/9-74-002, Region IV, Atlanta.

Chamie, J. P. M., and C. J. Richardson. 1978. Decomposition in northern wetlands, pp. 115–130 in R. E. Good, D. F. Whigham, and R. L. Simpson (eds.), Freshwater wetlands: ecological processes and management potential. Academic Press, New York.

Clark, J. R. 1979. Science and the conservation of riparian systems, pp. 13–16 in U.S. Department of Agriculture Forest Service. Strategies for protection and management of floodplain wetlands and other riparian ecosystems. Proceedings of the Symposium, December 11–13, 1978, Callaway Gardens, Ga.

Clark, J. R., and J. Benforado, eds. 1981. Wetlands of bottomland hardwood forests. Elsevier, Amsterdam.

Clymo, R. S. 1963. Ion exchange in Sphagnum and its relation to bog ecology. Ann. Bot (Lond.) N.S. 27: 309–324.

Clymo, R. S. 1967. Control of cation concentrations and in particular of pH in Sphagnum dominated communities, pp. 273–284 in H. L. Golterman and R. S. Clymo (eds.), Chemical environment in the aquatic habitat. North-Holland, Amsterdam.

Clymo, R. S., and P. M. Hayward. 1982. The ecology of Sphagnum, pp. 229–289 in A. J. E. Smith (ed.), Bryophyte ecology. Chapman and Hall, London.

Cohen, A. D., D. J. Casagrande, M. J. Andrejko, and G. R. Best (eds.). 1984. The Okeefenokee swamp: its natural history, geology, and geochemistry. Wetland Surveys, Los Alamos, N. Mex.

Conard, S. G., R. L. MacDonald, and R. F. Holland. 1977. Riparian vegetation and flora of the Sacramento Valley, pp. 47–55 in A. Sands (ed.), Riparian forests in California. University of California Institute of Ecology Publication 15, Davis.

Conner, W. H., and J. W. Day, Jr. 1982. The ecology of forested wetlands in the southeastern United States, pp. 69–87 in B. Gopal, R. E. Turner, R. G. Wetzel, and D. F. Whigham (eds.). National Institute of Ecology and International Scientific Publications. Jaipur, India.

Cowardin, L. M., V. Carter, F. C. Golet, and E. T. LaRoe. 1979. Classification of wetlands and deepwater habitats of the United States. Technical Report FWS/OBS/79/31. U.S. Fish and Wildlife Service, Official Bulletin Service.

Cowardin, L. M., O. S. Gimler, and L. M. Mechlin. 1981. Characteristics of central North Dakota wetlands determined from sample aerial photographs and ground study. Wildl. Soc. Bull. 9: 280–288.

Craft, C. B., and C. J. Richardson. 1993a. Peat accretion and N, P, and organic C accumulation in nutrient-enriched and unenriched Everglades peatlands. Ecol. Applic. 3(3): 446–458.

Craft, C. B., and C. J. Richardson. 1993b. Peat accretion and phosphorus accumulation along a eutrophication gradient in the northern Everglades. Biogeochem. 22: 133–156.

Craft, C. B., and C. J. Richardson. 1997. Relationships between soil nutrients and plant species composition in Everglades peatlands. J. Environ. Qual. 26: 224–232.

Craft, C. B., J. Vymazal, and C. J. Richardson. 1995. Response of Everglades plant communities to nitrogen and phosphorus additions. Wetlands 15: 258–271.

Craighead, F. C. 1971. The trees of south Florida. University of Miami Press, Coral Gables.

Craighead, F. C. Sr. 1984. Hammocks of South Florida, pp. 53–56 in P. J. Gleason (ed.), Environments of South Florida: present and past II. Miami Geol. Soc. Mem. II, pp. 53–56.

Cushing, E. J. 1963. Late-Wisconsin pollen stratigraphy in east-central Minnesota. Ph.D. dissertation, University of Minnesota, Minneapolis.

Dahl, T. E. 1990. Wetlands losses in the United States, 1780s to 1980s. U.S. Department of the Interior, Fish and Wildlife Service. Government Printing Office, Washington, D.C.

Dahl, T. E., and C. E. Johnson. 1991. Wetlands status and trends in the coterminous United States, mid-1970s to mid-1980s. U.S. Department of Interior, Fish and Wildlife Service. Government Printing Office, Washington, D.C.

Damman, A. W. H., and T. W. French. 1987. The ecology of peat bogs of the glaciated northeastern United States: a community profile. U.S. Fish Wildl. Serv. Biol. Rep. 85: 100.

Daniel, C. C. III. 1981. Hydrology, geology and soils of pocosins: a comparison of natural and altered systems, pp. 69–108 in Richardson, C. J. (ed.), Pocosin wetlands. Hutchinson Ross, Stroudsburg, Pa.

Davis, J. H. 1943. The natural features of southern Florida, especially the vegetation, and the Everglades. Florida Geol. Surv. Bull. No. 25.

Davis, S. M. 1994. Phosphorus inputs and vegetation sensitivity in the Everglades, pp. 357–378 in S. M. Davis and J. C. Ogden (eds.), Everglades: the ecosystem and its restoration. St. Lucie Press, Delray Beach, Fla.

Davis, S. M., and J. C. Ogden (eds.). 1994. Everglades: the ecosystem and its restoration. St. Lucie Press, Delray Beach, Fla.

Davis, S. M., L. H. Gunderson, W. A. Park, J. R. Richardson, and J. E. Mattson. 1994. Landscape dimension, composition, and function in a changing Everglades ecosystem, pp. 419–444 in S. M. Davis and J. C. Ogden (eds.), Everglades: the ecosystem and its restoration. St. Lucie Press, Delray Beach, Fla.

Debell, D. S., and A. W. Naylor. 1972. Some factors affecting germination of swamp tupelo seeds. Ecology 53: 504–506.

Delcourt, P. A., and H. R. Delcourt. 1981. Vegetation maps for eastern North America: 40,000 yr BP to the present, pp. 123–166 in R. Romans (ed.), Geobotany II. Plenum Press, New York.

Dineen, J. W. 1972. Life in the tenacious Everglades. In-Depth Report 1(5): 1–10, Central and Southern Flood Control District, West Palm Beach, Fla.

Disrud, D. T. 1968. Wetland vegetation of the Turtle Mountains of North Dakota. Master's thesis, North Dakota State University, Fargo.

DNR. 1984. Unpublished report. Minnesota Department of Natural Resources. Minneapolis.

Douglas, M. S. 1947. The Everglades: river of grass. Ballantine, New York.

Dohrenwend, R. E. 1977. Evapotranspiration patterns in Florida. Fla. Sci. 40: 184–192.

Du Rietz, G. E. 1949. Hvudhenter och huvudgranser I svensk myrveetation. [Summary: main units and main limits in Swedish mire vegetation.] Svensk. Bot. Tidsk. 43: 274–309.

Dury, G. H. 1977. Underfit streams: retrospect, perspect and prospect, pp. 281–293 in K. J. Gregory (ed.), River channel changes. Wiley, New York.

Enos, P., and R. D. Perkins. 1977. Quaternary sedimentation in south Florida. Geol. Soc. of Amer., Mem. 147.

Ewel, K. C., and H. T. Odum (eds.). 1984. Cypress swamps. University Presses of Florida, Gainesville.

Faber, P. M., E. Keller, A. Sands, and B. M. Massey. 1989. The ecology of riparian habitats of the southern California coastal region: a community profile. U.S. Fish Wildl. Serv. Biol. Rep. 85 (7.27).

Faulkner, S. P., and C. J. Richardson, 1989. Physical and chemical characteristics of freshwater wetland soils, pp. 41–72 in D. A. Hammer (ed.), Constructed wetlands for wastewater treatment: municipal, industrial and agricultural. Lewis, Chelsea, Mich.

Faulkner, S. P., W. H. Patrick, R. P. Gambrell, W. B. Parker, and B. J. Good. 1991. Characterization of soil processes in bottomland hardwood wetland-nonwetland transition zones in the lower Mississippi River valley. U.S. Army Corps of Engineers Contract Report WRP-91-1.

Fennema, R. J., C. J. Neidrauer, R. A. Johnson, T. K. MacVicar, and W. A. Perkins. 1994. A computer model to simulate natural everglades hydrology, pp. 249–289 in S. M. Davis and J. C. Ogden (eds.), Everglades: the ecosystem and its restoration. St. Lucie Press, Delray Beach, Fla.

Ferren, W. R. 1983. The vegetation and flora of the streams and slough. In C. P. Onuf (ed.) The proposed Corps of Engineers flood control and groundwater recharge project for the Goleta Valley, Santa Barbara County, California: inventories of the biological resources of the affected creeks and an analysis of effects on the creeks and the slough. Prepared for the U.S. Army Corps of Engineers, Santa Barbara.

Ferren, W. R. Jr., P. L. Fiedler, and R. A. Leidy. 1994. Wetlands of the central and southern California coast and coastal watersheds. U.S. Environmental Protection Agency, Region IX, San Francisco, Calif.

Flora, M. D., D. R. Walker, D. J. Scheidt, R. G. Rice, and D. H. Landers. 1988. The response of the Everglades marsh to increased nitrogen and phosphorus loading. Part I: Nutrient dosing, water chemistry and periphyton productivity. Report to the Superintendent. Everglades National Park, Homestead, Fla.

Frayer, W. E., T. J. Manahan, D. C. Bowden, and F. A. Grayhill. 1983. Status and trends of wetlands and deepwater habitats in the coterminous United States, 1950s to 1970s. Dept. For. Wood Serv., Colorado State University, Fort Collins.

Gaddy, L. L., T. H. Kohlsaat, E. A. Laurent, and K. B. Stansell. 1975. A vegetation analysis of preserve alternatives involving the Beidler Tract of the Congaree Swamp. Div. Nat. Area Acquisition and Resource Planning, S. C. Wildl. and Mar. Resour. Dept. Charleston, SC.

Gambreil, R. P., and W. H. Patrick Jr. 1978. Chemical and microbiological properties of anaerobic soils and sediments, pp. 375–423 in D. D. Hook and R. M. M. Crawford (eds.), Plant life in anaerobic environments. Ann Arbor Sci. Pub., Ann Arbor, Mich.

Glaser, P. H. 1987. The ecology of patterned boreal peatlands of Northern Minnesota: a community profile. U.S. Fish Wildl. Serv. Rep., Report 85 (7.14), Washington, D.C.

Glaser, P. H. 1992. Raised bogs in eastern North America – regional controls for species richness and floristic assemblages. J. Ecol. 80: 535–554.

Glaser, P. H., and J. A. Janssens. 1986. Raised bogs in eastern North America: transitions in landforms and gross stratigraphy. Can. J. Bot. 64: 395–415.

Glaser, P. H., G. A. Wheeler, E. Gorham, and H. E. Wright Jr. 1981. The patterned mires of the Red Lake peatland. northern Minnesota: vegetation, water chemistry, and landforms. J. Ecol. 69: 575–599.

Gleason, P. J. 1972. The origin, sedimentation, and stratigraphy of a calcite mud located in the southern freshwater Everglades. Ph.D. dissertation, Pennsylvania State University, University Park.

Gleason, P. J. 1984. The environmental significance of Holocene sediments from the Everglades and saline tidal plain, pp. 297–351 in P. J. Gleason (ed.) Environments of south Florida: present and past. Miami Geol. Soc., Miami.

Gleason, P. J., and P. Stone. 1994. Age, origin and landscape evolution of the Everglades peatland, pp. 149–197 in S. M. Davis and J. C. Ogden (eds.), Everglades: the ecosystem and its restoration. St. Lucie Press, Delray Beach, Fla.

Gleason, P. J., and W. Spackman Jr. 1974. Calcareous periphyton and water chemistry in the Everglades, pp. 146–181 in P. J. Gleason (ed.), Environments of South Florida, present and past. Miami Geolo. Soc., Miami.

Good, R., D. Whigham, and R. Simpson. 1978. Freshwater wetlands: ecological processes and management potential. Academic Press, New York.

Goodrick, R. L. 1984. The wet prairies of the northern Everglades, pp. 185–189 in P. J. Gleason (ed.), Environments of South Florida, present and past. Miami Geol. Soc., Miami.

Gopal, B., J. Kvet, H. Loffler, V. Masing, and B. C. Patten. 1990. Chapter 2, Definition and classification. vol. 1, pp. 9–15 in B. C. Patten et al. (eds.), Wetlands and shallow continental water bodies. S. P. B. Academic Publishing, The Hague.

Gore, A. J. P. (ed.). 1983. Ecosystems of the world, mires, swamp, bog, fen, and moor. Vol. 4A, General Studies. Vol. 4B, Regional Studies. Elsevier, Amsterdam.

Gorham, E. 1991. Northern peatlands: role in the carbon cycle and probable responses to climatic warming. Ecol. Applic. 1: 182–195.

Gorham, E., W. E. Dean, and J. E. Sanger. 1983. The chemical composition of lakes in the north-central United States. Limnol. and Oceanogr. 28: 287–301.

Gorham, E., S. J. Eisenreich, J. Ford, and M. V. Santelmann. 1984. The chemistry of bog waters, pp. 339–363 in W. Strum (ed.), Chemical processes in lakes. Wiley, New York.

Gosselink, J. G. 1984. The ecology of delta marshes of coastal Louisiana: a community profile. U.S. Fish Wildl. Serv. FWS/OBS-84/09.

Greeson, P. E., J. R. Clark, and J. E. Clark (eds.). 1978. Wetland functions and values: the state of our understanding. American Water Resources Association, Minneapolis, Minn.

Gunderson, L. H. 1990. Historical hydropatterns in vegetation communities of Everglades National Park, pp. 1099–1111 in R. R. Sharitz and J. W. Gibbons (eds.), Freshwater wetlands and wildlife. Ninth Annual Symposium, Savannah River Ecology Laboratory, 24–27 March 1986, Charleston, S.C.

Gunderson, L. H., and W. F. Loftus. 1993. The Everglades, pp. 199–255 in W. H. Martin, S. G. Boyce, and A. C. Echternacht (eds.), Biodiversity of the southeastern United States: lowland terrestrial communities. Wiley, New York.

Hair, J. D., G. T. Hepp, L. M. Luckett, K. P. Reese, and D. K. Woodward. 1979. Beaver pond ecosystems and their relationships to multi-use natural resource management, pp. 80–92 in U.S. Department of Agriculture Forest Service. Strategies for protection and management of floodplain wetlands and other riparian ecosystems. Proceedings of the Symposium, December 11–13, 1978, Callaway Gardens, Ga.

Hall, T. F., W. T. Penfound, and A. D. Hess. 1946. Water level relationships of plants in the Tennessee Valley with particular reference to malaria control. J. Tenn. Acad. Soil Sci. 21: 18–59.

Happ, S. C., G. Rittenhouse, and G. C. Dobson. 1940. Some principles of accelerated stream and valley sedimentation. U.S. Dept. Agric. Tech. Bull. 695. U.S. Supt. of Documents, Washington, D.C.

Heinselman, M. L. 1963. Forest sites, bog processes, and peatland types in the glacial Lake Agassiz region, Minnesota. Ecol. Monogr. 33: 327–374.

Heinselman, M. L. 1970. Landscape evolution and peatland types, and the Lake Agassiz Peatlands Natural Area, Minnesota. Ecol. Monogr. 40: 235–261.

Heinselman, M. L. 1975. Boreal peat lands in relation to environment, pp. 93–103 in A. D. Hasler (ed.), Coupling of land and water systems. Ecological Studies No. 10. Springer-Verlag, New York.

Herdendorf, C. E. 1987. The ecology of the coastal marshes of western Lake Erie: a community profile. U.S. Fish Wildl. Serv. Biol. Rep. 85 (7.9).

Hofstetter, R. H. 1983. Wetlands in the United States, pp. 201–244 in A. J. P. Gore, (ed.), Ecosystems of the world, vol. 4B, Mires: swamp, bog, fen, and moor. Elsevier, Amsterdam.

Holland, R. F. 1978. The geographic and edaphic distribution of vernal pools in the Great Central Valley. California Native Plant Society, Sacramento.

Holland, R. F., and S. K. Jain. 1981. Insular biogeography of vernal pools in the Central Valley of California. Amer. Nat. 117:24–37.

Holland, R. F. and S. K. Jain. 1988. Vernal pools, pp. 515–533 in M. G. Barbour and J. Major (eds.), Terrestrial vegetation of California, 2nd ed. California Native Plant Society, Sacramento.

Holstein, G. 1984. California riparian forests: deciduous islands in an evergreen sea, pp. 2–22 in R. E. Warner and K. M. Hendrix (eds.), California riparian systems: ecology, conservation, and productive management. University of California Press, Berkeley.

Hook, D. D. et al., eds. 1988. The ecology and management of wetlands, Vols. I and II, Croon Helm, London, and Timber Press, Portland, Ore.

Hook, D. D., and J. R. Scholtens. 1978. Adaptations and flood tolerance of tree species, pp. 299–332 in D. D. Hook and R. M. M. Crawford (eds.), Plant life in anaerobic environments. Ann Arbor Science, Ann Arbor, Mich.

Ingram, H. A. P. 1967. Problems of hydrology and plant distribution in mires. J. Ecol. 55:711–724.

Ingram, H. A. P. 1983. Hydrology in ecosystems of the world, pp. 67–158 in A. J. P. Gore (ed.), Ecosystems of the world, vol. 43, Mires, swamp, bog, fen, and moor. Elsevier, Amsterdam.

Ives, J. C. 1856. Memoir to accompany a military map of Florida south of Tampa Bay. U.S. War Department, Topographical Engineers, Washington, D.C.

Jacobson, G. L., Jr. 1979. The paleoecology of white pine (*Pinus strobus*) in Minnesota. J. Ecol. 67:697–726.

Janssen, C. R. 1967a. Stevens Pond: a postglacial pollen diagram from a small Typha swamp in northwestern Minnesota, interpreted from pollen indicators and surface samples. Ecol. Monogr. 37:145–172.

Janssen, C. R. 1967b. A floristic study of forests and bog vegetation, northwestern Minnesota. Ecology 48:751–765

Janssen, C. R. 1968. Myrtle Lake: a late-and post-glacial pollen diagram from northern Minnesota. Can. J. Bot. 46:1397–1408.

Janssens, J. A., and P. H. Glaser. 1986. The bryophyte flora and major peat-forming mosses at Red Lake peatland, Minnesota. Can. J. Bot. 64:427–442.

Johnston, C. A., N. E. Detenbeck, and G. J. Niemi. 1990. The cumulative effect of wetlands on stream water quality and quantity: a landscape approach. Biogeochemistry 10:105–141.

Kantrud, H. A., J. B. Millar, and A. G. van der Valk. 1989. Vegetation of wetlands of the prairie pothole region, pp. 132–187 in A. van der Valk (ed.), Northern prairie wetlands. Iowa State University Press, Ames.

Keeler-Wolf, T., K. Lewis, and C. Roye. 1994. The definition and location of sycamore alluvial woodland in California. California Department of Fish and Game, Sacramento.

Katibah, E. F. 1984. A brief history of riparian forests in the Central Valley of California, pp. 23–29 in R. E. Warner and K. M. Hendrix (eds.), California riparian systems. University of California Press, Berkeley.

Kivimen, E., and P. Pakarinen. 1981. Geographical distribution of peat resources and major peatland complex types in the world. Annals Acad. Sciencia Fennicae, Series A, Geology Geography 132:1–28.

Klopatek, J. M. 1978. Nutrient dynamics of freshwater riverine marshes and the role of emergent macrophytes, pp. 195–216 in R. E. Good, D. F. Whigham, and R. L. Simpson (eds.), Freshwater wetlands: ecological processes and management potential, Academic Press, New York.

Knapp, R. 1965. Die vegetation von nord und mittelamerika und der Hawaii-inseln. Gustav Fischer, Stuttgart.

Koch, M. S., and K. R. Reddy. 1992. Distribution of soil and plant nutrients along a trophic gradient in the Florida Everglades. Soil Sci. Soc. Am. J. 56:1492–1499.

Koslowski, T. J. 1984. Plant responses to flooding of soil. Bioscience 34:162–167.

Kushlan, J. A. 1974. Observations on the role of the American alligator (*Alligator mississippiensis*) in the southern Florida wetlands. Copeia 1974:993–996.

Kushlan, J. A. 1987. External threats and internal management: the hydrologic regulation of the Everglades, FL, USA. Environ. Mgt. 11:109–119.

LaBaugh, J. W. 1989. Chemical characteristics of water in northern prairie wetlands, pp. 56–90 in A. van der Valk (ed.), Northern prairie wetlands. Iowa State University Press, Ames.

Laderman, A. D. (ed.). 1987. Atlantic white cedar wetlands. Westview Press, Boulder and London.

Laderman, A. D. 1989. The ecology of Atlantic white cedar wetlands: a community profile. U.S. Fish Wildl. Serv. Biol. Rep. 85 (7.21).

Larson, J. S., M. S. Bedinger, C. F. Bryan, S. Brown, R. T. Huffman, E. L. Miller, D. G. Rhodes, and B. A. Touchet. 1981. Transition from wetlands to uplands in southeastern bottomland hardwood forests, pp. 225–273 in J. R. Clark and J. Benforado, eds. Wetlands of bottomland hardwood forests. Proceedings of a workshop on bottomland hardwood forest wetlands of the Southeastern United States, Lake Lanier, Ga., June 1–5, 1980. Developments in Agricultural and Managed-forest Ecology, Vol. 11. Elsevier, New York.

Last, W. M., and T. H. Schweyen. 1983. Sedimentology and geochemistry of saline lakes of the Great Plains. Hydrobiology 105:245–263.

Leopold, L. B., and M. G. Wolman. 1957. River channel patterns: braided, meandering and straight, pp. 39–85 in Geological Survey Professional Paper 282-B. Government Printing Office, Washington, D.C.

Lindeman, R. L. 1941. The developmental history of Cedar Creek Lake, Minnesota. Am. Midl. Nat. 25:101–112.

Lodge, T. E. 1994. The Everglades handbook: understanding the ecosystem. St. Lucie Press, Delray Beach, Fla.

Logan, T., A. C. Eller, Jr., R. Morrell, D. Ruffner, and J. Sewell. 1993. Florida panther habitat preservation plan, South Florida population. Prepared by the U.S. Fish and Wildlife Service, Florida Game and Fresh Water Fish Commission, Florida Department of Environmental Protection, and National Park Service for the Florida Panther Interagency Committee, Tallahassee, Fla.

Long, R. W., and O. Lakela. 1971. A flora of tropical Florida. A manual of the seed plants and ferns of southern peninsular Florida. University of Miami Press, Coral Gables.

Loveless, C. M. 1959. A study of the vegetation of the Florida Everglades. Ecology 40:1–9.

Lugo, A. E., M. M. Brinson, and S. L. Brown (eds.). 1990. Forested wetlands, Ecosystems of the world 15. Elsevier, Amsterdam.

MacVicar, T. K., and S. S. T. Lin. 1984. Historical rainfall activity in central and southern Florida: average, return periods estimates and selected extremes, pp. 477–509 in P. J. Gleason (ed.), Environments of South Florida: present and past II. Miami Geological Society, Memoir II.

Maltby, E. 1986. Waterlogged wealth. Earthscan, London.

Maltby, E., and R. E. Turner. 1983. Wetlands of the world. Geog. Mag. 55:12–17.

Marks, P. L., and P. A. Harcombe. 1981. Forest vegetation of the Big Thicket, southeast Texas. Ecol. Monogr. 51:287–305.

Matthews, E. 1990. Global distribution of forested wetlands. Addendum to Forested wetlands, A. E. Lugo, M. Brinson, and S. Brown, (eds.). Elsevier, Amsterdam.

Matthews, E., and I. Fung. 1987. Methane emissions from natural wetlands: global distribution, area, and environmental characteristics of sources. Global Biogeochem. Cycles, 1:61–86.

McKim, H. L., L. W. Gratta, and C. J. Merry. 1975. Inundation damage to vegetation at selected New England flood control reservoirs. Cold Regions Research and Engineering Laboratory, Hanover, N.H. U.S. Army Engineer Division, New England, Waltham, Mass.

McPherson, B. F., G. Y. Hendrix, H. Klein, and H. M. Tysus. 1976. The environment of South Florida: a summary report. U.S.G.S. Professional Paper 1011. U.S. Geological Survey. Government Printing Office, Washington, D.C.

Mitsch, W. J., and J. G. Gosselink. 1993. Wetlands, 2nd ed. Van Nostrand Reinhold, New York.

Morris, L. A. 1978. Evaluation, classification and management of the floodplain forests of south-central New York. Master's thesis, SUNY College of Environmental Science and Forestry, Syracuse, N.Y.

Neal, J. A., and J. W. Haskins. 1986. Bottomland hardwoods: ecology, management, and preservation, in D. L. Kulhavy, and R. N. Conner (eds.), Wilderness and natural areas in the eastern United States: a management challenge. Center for Applied Studies, School of Forestry, Stephen F. Austin State University, Nacogdoches, Texas.

Nicholson, B. J., and D. H. Vitt, 1990. The paleoecology of a peatland complex in continental western Canada. Can. J. Bot. 68:121–138.

Novitzki, R. P. 1979. Hydrologic characteristics of Wisconsin's wetlands and their influence on floods, stream flow, and sediment, pp. 377–388 in P. E. Greeson, J. R. Clark, and J. E. Clark (eds.), Wetland functions and values: the state of our understanding. American Water Resource Association, Minneapolis, Minn.

Padgett, W. G., A. P. Youngblood, and A. H. Winward. 1989. Riparian community type classification of Utah and Southeastern Idaho, Intermountain Region, U.S. Department of Agriculture Forest Service, R4-Ecol-89-01.

Parker, G. G., and C. W. Cooke. 1944. Late Cenozoic geology of southern Florida with a discussion of groundwater. Florida Geol. Surv. Bull. No. 27.

Penfound, W. T. 1952. Southern swamps and marshes. Bot. Rev. 18:413–446.

Pope, K. R. 1989. Vegetation in relation to water quality

and hydroperiod on the Loxahatchee National Wildlife Refuge. Master's thesis, University of Florida. Gainesville.

Putnam, J. A., G. M. Furnival, and J. S. McKnight. 1960. Management and inventory of southern hardwoods. U.S. Department of Agriculture Forest Service, Washington, D. C. Agric. Handbook No. 181.

Rader, R. B., and C. J. Richardson. 1992. The effects of nutrient enrichment on algae and macroinvertebrates in the Everglades: a review. Wetlands 12:121–135.

Rader, R. B., and C. J. Richardson. 1994. Response of macroinvertebrates and small fish to nutrient enrichment in the northern Everglades. Wetlands 14:134–146.

Richardson, C. J. 1979. Primary productivity values in freshwater wetlands, pp. 131–145 in P. E. Greeson, J. R. Clark, and J. E. Clark (eds.), Wetland functions and values: the state of our understanding. American Water Resource Association, Minneapolis, Minn.

Richardson, C. J. 1981. Pocosin wetlands. Hutchinson Ross, Stroudsburg, Pa.

Richardson, C. J. 1985. Mechanisms controlling phosphorus retention capacity in freshwater wetlands. Science 228:1424–1427.

Richardson, C. J. 1989. Freshwater wetlands: transformers, filters, or sinks? pp. 25–46 in R. R. Sharitz and J. W. Gibbons (eds.), Freshwater wetlands and wildlife. CONF-8603101, DOE Symposium Series No. 61, USDOE.

Richardson, C. J. 1993. Pocosins, Carolina bays, and mountain bogs, pp. 257–310 in W. H. Martin, S. G. Boyce, and A. C. Echternacht (eds.), Biodiversity of the Southeastern United States/lowland terrestrial communities. Wiley, New York.

Richardson, C. J. 1994. Ecological functions and human values in wetlands: a framework for assessing impacts. Wetlands 14:1–9.

Richardson, C. J. 1995. Wetlands ecology, pp. 535–550 in Encyclopedia of environmental biology, Vol. 3. Academic Press, New York.

Richardson, J. L., J. L. Arndt, and J. Freeland. 1994. Wetland soils of the prairie potholes. Adv. in Agron. 52:121–171.

Richardson, J. L., and R. J. Bigler. 1984. Principal component analysis of prairie pothole soils in North Dakota. Soil Sci. Soc. Amer. J. 48:1350–1355.

Richardson, C. J., and C. B. Craft. 1993. Effective phosphorus retention in wetlands: fact or fiction? pp. 271–282 in G. A. Moshiri (ed.), Constructed wetlands for water quality improvement. Lewis Publishers, Boca Raton, Fla.

Richardson, C. J., and P. Vaithiyanathan. 1995. Phosphorus sorption characteristics of the Everglades soils along a eutrophication gradient. Soil Sci. Soc. of Amer. J. 59:1782–1788.

Richardson, C. J., C. B. Craft, R. G. Qualls, J. Stevenson, and P. Vaithiyanathan. 1994. Annual Report: Effects of nutrient loadings and hydroperiod alterations on control of cattail expansion, community structure and nutrient retention in the water conservation areas of south Florida. Duke Wetland Center publication 94–08. School of the Environment, Duke University, Durham, N.C.

Richardson, C. J., C. B. Craft, R. G. Qualls, J. Stevenson, P. Vaithiyanathan, M. Bush, and J. Zahina. 1995. Effects of phosphorus and hydroperiod alterations

on ecosystem structure and function in the Everglades. Duke Wetland Center publication 95–05. Nicholas School of the Environment, Duke University, Durham, N.C.

Robb, D. M. 1989. Diked and undiked freshwater coastal marshes of western Lake Erie. Master's thesis. Ohio State University, Columbus.

Roberts, W. G., J. G. Howe, and J. Major. 1977. A survey of riparian forest flora and fauna in California, pp 3–19 in A. Sands (ed.), Riparian forests in California. University of California Institute of Ecology Publication 15, Davis.

Robertson, P. A., M. D. MacKenzie, L. F. Elliot. 1984. Gradient analysis and classification of the woody vegetation for four sites in southern Illinois and adjacent Missouri. Vegetation 58:87–104.

Romanowicz, E. R., and C. J. Richardson. 1997. Field measurements of hydrologic conditions in WCA-2A, pp. 167–190 in C. J. Richardson, Effects of phosphorus and hydroperiod alterations on Everglades structure and function in the Everglades. Duke Wetland Center Pub. No. 97–08. Nicholas School of the Environment. Duke University, Durham.

Rosendahl, P. C., and P. W. Rose. 1982. Freshwater flow rates and distribution within the Everglades marsh, pp. 385–401 in R. D. Cross and D. L. Williams (eds.), Proceedings of the National Symposium on Freshwater Inflow to Estuaries, Coastal Ecosystems Project. Washington, D.C.: U.S. Fish and Wildlife Service.

Rowe, J. S. 1972. Forest regions of Canada. Can. For. Serv. Publ. No. 1300.

Sasser, C. E., and J. G. Gosselink. 1984. Vegetation and primary production in a freshwater marsh in Louisiana. Aquatic Bot. 41:317–331.

Sawyer, J. O., and T. Keeler-Wolf. 1995. A manual of California vegetation. Calif. Native Plant Soc., Sacramento.

Sharitz, R. R., and W. J. Mitsch. 1993. Southern floodplain forests, pp. 311–372 in W. H. Martin, S. G. Boyce, and A. C. Echternacht (eds.), Biodiversity of the southeastern United States: lowland terrestrial communities. Wiley, New York.

Shaw, S. P., and C. G. Fredine. 1956. Wetlands of the United States, their extent, and their value for waterfowl and other wildlife. U.S. Department of Interior, Fish and Wildlife Service, Circular 39, Washington, D.C.

Schlesinger, W. H. 1978. Community structure, dynamics and nutrient cycling in the Okefenokee cypress swamp forest. Ecol. Monogr. 48:43–65.

Schomer, N. S., and R. D. Drew. 1982. An ecological characterization of the lower Everglades, Florida Bay, and the Florida Keys. FWS/OBS-82/58.1 U.S. Fish and Wildlife Service, Office of Biological Services, Washington, D.C.

Sculley, S. P. 1986. Frequency analysis of SFWMD rainfall. South Florida Water Management District, Technical publication 86–6.

Siegel, D. I. 1981. Hydrogeologic setting of the Glacial Lake Agassiz peatlands, northern Minnesota. U.S. Geol. Surv. Water Resour. Invest. 81–24.

Siegel, D. I. 1983. Ground water and the evolution of patterned mires, Glacial Lake Agassiz peatlands, northern Minnesota. J. Ecol. 71:913–921.

Siegel, D. I., and P. H. Glaser. 1987. Groundwater flow in a spring fen, raised bog complex, Lost River peatland, northern Minnesota. J. Ecol. 71:913–921.

Sjors, H. 1946. Myrvegetationen I ovre Langanomradet I Jamtland. Ark. Bot. 33A: 1–96.

Sjors, H. 1948. Myrvegetatinen I Bergslagen. Acta Phytogeogr. Suec. 21:1–299.

Sjors, H. 1952. On the relation between vegetation and electrolytes in North Swedish mire waters. Oikos 2 (1950):241–258.

Sjors, H. 1963. Bogs and fens on Attawapiskat River, northern Ontario. Mus. Canada Bull., Contrib. Bot. 186:45–133.

Smeins, F. E. 1967. The wetland vegetation of the Red River Valley and drift prairie regions of Minnesota, North Dakota and Manitoba. Ph.D. dissertation, University of Saskatchewan, Saskatoon.

Smith, R. D. 1996. Composition, structure, and distribution of woody vegetation on the Cache River floodplain, Arkansas. Wetlands 16(3):264–278.

South Florida Water Management District (SFWMD). 1992. Surface water improvement and management plan for the Everglades. Vol. III. Technical report, West Palm Beach, Fla.

Steinwand, A. L., and J. L. Richardson. 1989. Gypsum occurrence in soils on the margin of semipermanent prairie pothole wetlands. Soil Sci. Soc. Amer. 53:836–842.

Stephens, J. C., and L. Johnson. 1951. Subsidence of organic soils in the upper Everglades region of Florida. Proc. Soil Sci. Florida 9:191–237.

Stevenson, R. J., and C. J. Richardson. 1995. Interhabitat variability in Everglades algae communities, pp. 107–184 in C. J. Richardson (ed.), Effects of phosphorus and hydroperiod alterations on ecosystem structure and function in the Everglades. Duke University Wetlands Center Pub. No. 95–05. Nicholas School of the Environment. Duke University, Durham.

Steward, K. K., and W. H. Ornes. 1975a. Assessing a marsh environment for wastewater renovation. J. Water Poll. Contr. Fed. 47:1880–1891.

Steward, K. K., and W. H. Ornes. 1975b. The autecology of sawgrass in the Florida Everglades. Ecology 56: 162–171.

Stewart, R. E., and H. A. Kantrud, 1972. Vegetation of prairie potholes, North Dakota, in relation to quality of water and other environmental factors. Professional Paper 585-D. U.S. Geological Survey.

Swift, D. R. 1981. Preliminary investigations of periphyton and water quality in the Everglades Water Conservation Areas. Technical Publication 81–05. South Florida Water Management District, West Palm Beach, Fla.

Swift, D. R. 1984. Periphyton and water quality relationships in Everglades Water Conservation Areas, pp. 97–117 in P. J. Gleason (ed.), Environments of South Florida, present and past. Miami Geological Society, Miami, Fla.

Swift, D. R., and R. B. Nicholas. 1987. Periphyton and water quality relationships in the Everglades water conservation areas: 1978–1982. Technical Publication 87–2. South Florida Water Management District, West Palm Beach, Fla.

Taylor, B. D. 1983. Sediment yields in coastal Southern California. Amer. Soc. Civil Eng. J. Hydrol. Eng. 109(1):71–85.

Teskey, R. O., and T. M. Hinkley. 1980. Impact of water level changes on woody riparian and wetland communities. Vol. 1. Plant and soil responses to flooding. U.S. Fish Wildl. Serv. Off. Biol. Serv. FWS/OBS-77/58.

Thornthwaite, C. W. 1948. An approach toward a rational classification of climate. Geogr. Rev. 38:55–94.

Tiner 1984. Wetlands of the United States: current status and recent trends. U.S. Fish and Wildlife Service. Government Printing Office, Washington, D.C.

Topa, M. A. 1985. The effects of radial oxygen loss on phosphorus uptake in four pine species grown under flooded soil conditions. Ph.D. dissertation, Duke University, Durham.

Toth, L. A. 1987. Effects of hydrologic regimes on lifetime production and nutrient dynamics of sawgrass. Technical publication #87-6. South Florida Water Management District, West Palm Beach, Fla.

Toth, L. A. 1988. Effects of hydrologic regimes on lifetime production and nutrient dynamics of cattail. Technical publication #88-6. South Florida Water Management District, West Palm Beach, Fla.

Transeau, E. N. 1903. On the geographic distribution and ecological relations of the bog plant societies of northern North America. Bot. Gaz. 36:401–420.

U.S. Army Corps of Engineers. 1987. Corps of Engineers wetlands delineation manual, Technical Report Y-87-1. U.S. Army Engineer Waterways Experiment Station, Vicksburg, Miss.

U.S. Congress. Office of Technology Assessment. 1984. Wetlands: their use and regulation. Government Printing Office, Washington, D.C.

U.S. Department of Agriculture, Conservation Needs Inventory Committee. 1962. Basic statistics of the national inventory of soil and water conservation needs. U.S. Dep. Agr. Statistical Bull. 317. Government Printing Office, Washington, D.C.

U.S. Department of the Interior; Bureau of Land Management. 1991. Riparian Wetland Initiative for the 1990s. BLM/WO/GI-91/001+4340. Government Printing Office, Washington, D.C.

Urban, N. H., S. M. Davis, and N. G. Aumen. 1993. Fluctuations in sawgrass and cattail densities in Everglades Water Conservation Area 2A under varying nutrient, hydrologic and fire regimes. Aquatic Bot. 46:203–223.

Vaithiyanathan, P., J. Zahina, and C. J. Richardson. 1995. Gradient study: Macrophyte species changes along the phosphorus gradient, pp. 273–297 in C. J. Richardson, C. B. Craft, R. G. Qualls, J. Stevenson, P. Vaithiyanathan, M. Bush, and J. Zahina. Effects of phosphorus and hydroperiod alterations on ecosystem structure and function in the Everglades. Duke Wetland Center publication 95–05. Nicholas School of the Environment, Duke University, Durham.

Vaithiyanathan, P., J. Zahina, and C. J. Richardson. 1998. Macrophyte species changes in the Everglades: examination along the eutrophication gradient. J of Environment Quality (in press).

van der Valk, A. G. (ed.). 1989. Northern prairie wetlands. Iowa State University Press, Ames.

van der Valk, A. G., and C. B. Davis. 1976. The seed banks of prairie glacial marshes. Can. J. Bot. 54: 1832–38.

van der Valk, A. G., and C. B. Davis. 1978. The role of seed banks in the vegetation dynamics of prairie glacial marshes. Ecol. 59:322–35.

Van Meter, N. 1965. Some quantitative and qualitative aspects of periphyton in the Everglades. Master's thesis, University of Miami, Coral Gables, Fla.

Van Meter-Kasanof, N. N. 1973. Ecology of the micro-algae of the Florida Everglades. Part 1. Environment and some aspects of freshwater periphyton, 1959–1963. Nowa Hedwigia 24:619–664.

Viereck, L. A., and E. L. Little, Jr.. 1972. Alaska trees and shrubs. U.S. Department of Agriculture Handbook 410.

Vignoles, C. B. 1823. Observations upon Florida. Bliss and E. White, New York.

Vitt, D. H. and W.-L. Chee. 1990. The relationships of vegetation to surface water chemistry and peat chemistry in fens of Alberta, Canada. Vegetation 89: 87–106.

Vitt, D. H. H., P. Achuff, and R. E. Andrus. 1975. The vegetation and chemical properties of patterned fens in the Swan Hills, north central Alberta. Can. J. Bot. 53:2776–2795.

Vitt, D. H., L. A. Halsey, and S. C. Zoltai. 1994. The bog landforms of continental western Canada in relation to climate and permafrost patterns. Arc. and Alp. Res. 26:1:1–13.

Vitt, D. H., Y. Li, and R. Belland. 1995. Patterns of bryophyte diversity in peatlands of continental western Canada. Bryologist 98(2), 218–227.

Von Post, L., and E. Granlund. 1926. Sodra Sveriges torvtillgangar I. Sver. Geol. Unders. C 335:1–127.

Vymazal, J., and C. J. Richardson. 1992. Determination of phosphorus dosing threshold concentrations altering Everglades slough communities: baseline sampling of macrophytes and periphyton, pp. 96–157 in C. J. Richardson, C. B. Craft, R. R. Johnson, R. G. Qualls, R. B. Rader, L. Sutter, and J. Vymazal. Effects of nutrient loading and hydroperiod alterations on control of cattail expansion, community structure and nutrient retention in the Water Conservation Areas of South Florida. Annual Report to the Everglades Agricultural Area Environmental Protection District. Publication no. 92–11. Duke Wetland Center, School of the Environment, Duke University, Durham.

Vymazal, J., and C. J. Richardson. 1995. Species composition, biomass, and nutrient content of periphyton in the Florida Everglades. J. Phycology. 31:343–354.

Walker, D. 1970. Direction and rate in some British postglacial hydroseres, pp. 117–139 in D. Walker and R. G. West (eds.), Studies in the vegetational history of the British Isles. Cambridge University Press, Cambridge.

Warner, R. 1983. Riparian resources of the Central Valley and California Desert: a report on their nature, history, status, and future (Draft). California Department of Fish and Game, Sacramento.

Weber, C. A. 1902. Uber die Vegetation and Entstehung des Hochmoors von Augstumal im Memeldelta mit vergleichenden Ausblicken auf andere Hochmoore der Erde. Paul Parey, Berlin.

Weller, M. W. 1981. Freshwater marshes. University of Minnesota Press, Minneapolis.

Weller, M. W. 1994. Freshwater marshes – ecology and wildlife management, 3rd ed. University of Minnesota Press, Minneapolis.

Wells, B. W. 1942. Ecological problems of the south-eastern United States coastal plain. Bot. Rev. 8:533–561.

Wharton, C. H., W. M. Kitchens, E. C. Pendleton, and T. W. Swipe. 1982. The ecology of bottomland hardwood swamps of the southeast: a community profile. U.S. Fish and Wildlife Service, Biological Services Program FWS/OBS-81/37.

Wheeler, G. A., and P. H. Glaser. 1979. Notable vascular plants of the Red Lake peatland, northern Minnesota. Mich. Bot. 18: 137–142.

Whitlow, T. H., and R. W. Harris. 1979. Flood tolerance in plants: a state-of-the-art review. U.S. Army Engineers Waterways Experiment Stn. Tech. Rep. E-79-2. Vicksburg, Miss.

Wilson, S. U. 1974. Metabolism and biology of a blue-green algal mat. Master's thesis, University of Miami, Fla.

Winter, T. C. 1989. Hydrologic studies of wetlands in the northern prairie, pp. 16–55 in A. van der Valk (ed.), Northern prairie wetlands. Iowa State University Press, Ames.

Wood, E. J. F., and N. G. Maynard. 1974. Ecology of the micro-algae of the Florida Everglades, pp. 123–145 P. J. Gleason (ed.). *Environments of South Florida: present and past.* Miami Geolo. Soc., Miami, Fla.

Worth, D. F. 1986. Preliminary environmental responses to marsh dewatering and reduction in water regulation schedule in Water Conservation Area-2A. South Florida Water Management District, Technical report no. 83-6.

Worth, D. F. 1988. Environmental response of Water Conservation Area-2A to reduction in regulation schedule and marsh drawdown. South Florida Water Management District, Technical report no. 87–5.

Wright, H. E., Jr., and W. A. Watts. 1969. Glacial and vegetational history of northeastern Minnesota. Minn. Geol. Survey, spec. paper sp-11: 1–59.

Zedler, P. H., C. K. Frazier, and C. Black. 1993. Habitat creation as a strategy in ecosystem preservation: an example from vernal pools, pp. 239–247 in J. E. Keeley (ed.), Interface between ecology and land development in California. Southern California Academy of Sciences, Los Angleles.

Zembal, R. 1984. Survey of vegetative and vertebrate fauna in the Prado Basin and the Santa Ana River Canyon, Calif. U.S. Army Corps of Engineers, Los Angeles.

Zoltai, S. C. 1979. An outline of the wetland regions of Canada, pp. 1–8 in C. D. A. Rubec, and F. C. Pollett (eds.), Proceedings of a workshop on Canadian wetlands. Environment Canada, Lands Directorate, Ecological Land Classification Series, No. 12, Saskatoon, Saskatchewan.

Zoltai, S. C., and F. C. Pollet. 1983. Wetlands in Canada: their classification, distribution, and use, pp. 245–268 in A. J. P. Gore (ed.), Mires: swamp, bog, fen and moor. Elsevier, Amsterdam.

Zoltai, S. C., S. Taylor, J. K. Jeglum, G. G. Mills, and J. D. Johnson. 1988. Wetlands of boreal Canada, pp. 97–154 in C. D. A. (coord.). Wetlands of Canada. Polyscience, Montreal.

Zedler, P. H. 1987. The Ecology of Southern California Vernol Pools: A Community Profile, Biological Report 85 (7.11), U.S Fish and Wildlife Service. Washington, D.C. NTIS.

Chapter
13

Saltmarshes and Mangroves

IRVING A. MENDELSSOHN   KAREN L. MCKEE

INTRODUCTION

The land-sea interface is a dynamic environment occupied by a unique assemblage of macrophytes. This habitat is characterized by extremes in physical and chemical factors that would prevent survival of most higher plant species. Tidal action results in periodic inundation by seawater, and the plants growing in these conditions must cope with high salt concentrations, unstable sediments, frequent submergence, and anaerobic soils. Saltmarshes and mangroves are dominant plant communities inhabiting the intertidal zone along the world's coastlines where wave energy is low and sediment is plentiful. In North America, saltmarsh vegetation dominates temperate shorelines from northern Alaska and Canada to Florida and Texas. In tropical and subtropical regions of North America, coastal saltmarshes are generally replaced by mangrove-dominated communities. However, well-developed saltmarsh communities may be found in tropical North America – for example, Belize (Costa and Davy 1992).

Once considered to be unimportant "wastelands" to be exploited or converted to other uses, saltmarshes and mangroves are now widely valued for their ecological uniqueness and linkages to estuarine food webs (Mitsch and Gosselink 1993). Several studies have documented the high primary productivity of these systems and the export of carbon to adjacent coastal waters where it is hypothesized to support secondary production. Although a cause and effect relationship has not been conclusively established, high fishery yields in temperate and tropical coastal waters are often attributed to the presence of adjacent saltmarsh and mangrove communities, respectively (Macnae 1968; Lugo and Snedaker 1974; Turner 1977; Odum, McIvor, and Smith 1982). In addition, saltmarsh and mangrove vegetation contribute to soil formation, thus helping to stabilize coastlines. They also act as buffers against storms and hurricanes; as filters for runoff of freshwater, sediment, and nutrients, preventing their direct introduction into marine systems; and as nurseries and places of refuge for commercially important finfish and shellfish.

The worldwide distribution of saltmarshes and mangroves and their ecological and societal values have attracted the attention of researchers for decades. Early work, which centered on floristics and structural characteristics, produced a large body of descriptive literature detailing various aspects of the vegetation and component species (Penfound and Hathaway 1938; Davis 1940; Chapman 1976; Tomlinson 1986). Work conducted during the past 20 yr has increasingly focused on functional attributes such as primary productivity, carbon flow, nutrient cycling, and ecophysiology (see Mitsch and Gosselink 1993). For saltmarshes in particular, this work has resulted in a relatively comprehensive understanding of the structure and function of this system. A full understanding of some important processes in mangrove forests, however, has lagged behind that in saltmarshes. A thorough grasp of the ecology of both saltmarshes and mangroves is essential to their sound management and the ability to predict effects of future global climate change on these valuable ecosystems.

This chapter is organized into 11 major topics. In many cases, the respective discussions of the two vegetation types are not balanced simply because of differences in the amount of information available for each. In other instances, the similarity of the two systems required a combined discussion to avoid unnecessary repetition. Information about floristics, geographical distribution and areal extent, development, physical and chemical features, community structure and physiognomy, zonation, succession, productivity, regeneration strategies, and disturbance is presented. Finally, areas for future research are suggested.

TERMINOLOGY AND FLORISTICS

Saltmarsh

Coastal saltmarshes are low-energy, generally intertidal habitats – dominated by halophytic forbs, graminoids, and shrubs – that periodically flood with seawater as a result of lunar (tidal) and meteorological (primarily wind) forces. Like other wetlands, saltmarshes are characterized by a pronounced hydrology, hydric soils, and vegetation adapted to saturated soil conditions. Additionally, saltmarsh plants have anatomical, morphological, and physiological adaptations that promote the avoidance and/or tolerance of elevated salinities.

The floristics of the saltmarsh habitat can vary greatly depending on the minimum salinity required to define a saltmarsh; both spatial and temporal variability in salinity make the latter point particularly problematic. An ecological definition of a halophyte is a plant that completes its life cycle in a saline environment. However, as discussed by Adam (1990), some authors have restricted the definition of halophytes to those plant species that occur only in saltmarshes, whereas others include all

We would like to thank Walt Glooschenko for information about Canadian saltmarshes and Denise Seliskar and Carole McIvor for their reviews of the chapter.

Table 13.1. *Dominant saltmarsh families and representative species for North America and Latin America*

Family	Species	Duncan 1974	Costa & Davy 1992
Aizoaceae	*Sesuvium portulacastrum*	X	X
Amaranthaceae	*Blutaparon portulacoides*		X
Amaryllidaceae	*Crinum sp.*		X
Apocynaceae	*Rhabdadenia biflora*	X	
Asclepiadaceae	*Cynanchium angustifolium*	X	
Asteraceae	*Borrichia frutescens*	X	X
Bataceae (=Batidaceae)	*Batis maritima*	X	X
Boraginaceae	*Heliotropium curassavicum*	X	X
Caryophyllaceae	*Spergularia marina*	X	
Celastraceae	*Mayternus phyllanthoides*		X
Chenopodiaceae	*Salicornia virginica*	X	X
Convolvulaceae	*Cuscuta salina*		X
Cyperaceae	*Scirpus maritimus*	X	X
Frankeniaceae	*Frankenia grandifolia*		X
Hippuridaceae	*Hippuris tetraphylla*	X	
Juncaceae	*Juncus roemerianus*	X	X
Juncaginaceae	*Triglochin maritima*	X	X
Lythraceae	*Ammania teres*	X	X
Plantaginaceae	*Plantago maritima*	X	
Plumbaginaceae	*Limonium carolinianum*	X	X
Gramineae	*Spartina alterniflora*	X	X
Polygonaceae	*Polygonum maritimum*	X	X
Portulacaceae	*Portulaca oleracea*	X	
Primulaceae	*Glaux maritima*	X	
Pteridaceae	*Acrostichum spp.*	X	X
Scrophulariaceae	*Agalinis maritima*	X	X
Solanaceae	*Lycium carolinianum*	X	
Verbenaceae	*Phylla nodiflora*	X	

Source: Duncan (1974), and Costa and Davy (1992).

plant species that have halophytic associations or grow in what could be a halophytic habitat. Barbour's 1970 review concluded that most halophytes can grow in nonsaline conditions. That is, few species are obligate halophytes.

Duncan's important review identified 75 plant families (177 genera and 347 species) that contain halophytes along the Atlantic and Gulf coasts of North America (north of Mexico). However, this enumeration of halophytes included species that would not usually be considered as part of the saltmarsh habitat, such as seagrasses, mangroves, sand dune vegetation, and low-salinity marsh dominants. When only those families that contain species more typical of saltmarshes are included, the 75 families identified by Duncan (1974) can be reduced to approximately 20 (Table 13.1). This is quantitatively similar to the "more than 18 angiosperm families" that Dawes (1981) cites for saltmarshes in general and the 22 families that Costa and Davy (1992) invoke for the saltmarshes of Latin America. The major plant families represented in saltmarshes are (1) Poaceae, (2) Chenopodiaceae, (3) Juncaceae, (4) Cyperaceae, (5) Plumbaginaceae (6) Asteraceae, (7) Plantaginaceae and (8) Frankeniaceae (Duncan 1974; Adam 1990).

Paul Adam (1990) has developed a biogeographic classification of the world's saltmarshes based on a mixture of floristic and vegetation data. This classification, which builds on that developed by Chapman (1974), consists of six major groups and assumes that saltmarsh communities and taxa in more than one region are, in general, geographical or ecological equivalents (Table 13.2). Thus, groups within this classification integrate across geographical boundaries and are similar in such characteristics as position in marsh zonation, taxonomic affinities, physiognomy, and associated species. Broad regional groupings are possible, although the boundaries between major groups are diffuse and occur over broad latitudinal zones in response to climatic gradients as well as mesoscale and local variation in environmental conditions (Fig. 13.1). Adam's (1990) six biogeographical classes of saltmarshes are the following:

1. The Arctic Type. In North America extends from northern to southern Canada and Alaska,

Table 13.2. *Biogeographical classification of saltmarshes and representative plant species of low and high marshes*

Saltmarsh type	Low marsh species	High marsh species
Arctic	*Puccinellia phryganoides*	*Carex glareosa*
Boreal	*P. phryganoides* *Triglochin maritima*	*Triglochin maritima* *Plantago maritima* *Puccinellia maritima* *Glaux maritima* *Festuca rubra* *Calamagrostis neglecta* *Salicornia europaea*
Temperate Western North America	*Scirpus maritimus*	*Juncus balticus*
Washington/Oregon	*Spergularia marina* *Triglochin maritima* *Salicornia virginica*	*Potentilla pacifica* *Distichlis spicata* *Carex lyngbyei* *Deschampsia caespitosa* *Jaumea carnosa*
California	*Spartina foliosa* *Salicornia virginica*	*Distichlis spicata* *Atriplex patula* *Grindelia humilis* *Limonium californicum*
West Atlantic Northern subtype	*Spartina alterniflora*	*Spartina patens*
Coastal plain subtype	*Spartina alterniflora*	*Juncus roemerianus*
Dry Coast Southern California/Baja	*Spartina foliosa* *Salicornia virginica*	*Jaumea carnosa* *Monanthochloe littoralis* *Salicornia subterminalis* *Frankenia grandifolia*
Tropical	*Spartina alterniflora*	*Spartina spartinae* *Sporobolus virginicus* *Distichlis spicata* *Batis maritima* *Suaeda spp.*

Source: Adam 1990.

is floristically uniform and species-poor, and consists of only a few different plant communities.

2. The Boreal Type. Found over a broad latitudinal range. Overlaps extensively with the Arctic Type and forms a wide ecotone between the Arctic Type to the north and the Temperate Type farther to the south. Vegetatively, the Boreal and Arctic Types can be differentiated by the presence of *Triglochin maritima* and *Salicornia europaea* in the former.

3. The Temperate Type. Contains the largest diversity of flora and vegetative communities compared to the other types, and Adam (1990) divided it into five subgroups (Europe, Western North America, Japan, Australia, and South Africa). Temperate marshes can be distinguished from Boreal Types by the presence of *Limonium* spp. in the former.

4. The West Atlantic Type. Characterized by the dominance of *Spartina alterniflora* in the low marsh. Adam (1990) recognizes two major subtypes – Northern and Coastal Plain – and he suggests that the Northern subtype may be further subdivided as more information becomes available (similar to Chapman's distinction between the saltmarshes of the Bay of Fundy and those of New England).

5. The Dry Coast Type. Found in subtropical and tropical coastal regions that permanently or seasonally experience a dry climate. Soil salinities in this saltmarsh type are often hypersaline, and vegetation is generally open and dominated by succulent-stemmed shrubs of low stature. Species richness within this saltmarsh type is greatest in regions that are seasonally wet, such as southern California, and it decreases sharply in permanently arid coasts like Baja California.

Figure 13.1. Geographic distribution of North American saltmarsh vegetation showing location of classification types described by Adam (1990). At tropical and subtropical latitudes, saltmarsh distribution overlaps with that of mangroves.

6. The Tropical Type. Occurs in the tropics and sometimes can be extensive, as in Belize, either in association with mangrove swamps or in isolation. These saltmarshes are generally species-poor and consist solely of halophytes.

Mangrove

Mangrove is an ecological term used to describe salt-and flood-tolerant trees and shrubs that inhabit the intertidal zone. The word "mangrove" is thought to be derived from "mangue," the Portuguese word for tree and "grove," the English word for a stand of trees (Dawes 1981). It is typically used to refer to an individual plant or species. Synonymous terms describing the entire assemblage include mangrove community, mangrove ecosystem, mangrove swamp, mangrove forest, mangal, tidal forest, and tidal swamp forest.

Mangroves may not be closely related taxonomically, but they all exhibit various morphological and physiological modifications to allow avoidance

or tolerance of toxic ions and anoxic soil conditions characteristic of the mangrove habitat. Several mangrove species are viviparous (the condition in which the embryo germinates while still attached to the parent), a strategy that circumvents the problem of germination under adverse conditions. In addition to these specialized adaptations, Tomlinson (1986) proposed that true mangrove species were distinguished by their complete fidelity to the mangrove habitat (i.e., they are never found in terrestrial communities) and by their taxonomic isolation (i.e., separated from terrestrial relatives at least at the generic level but often at the subfamily or family level). A further distinction was made by dividing mangrove species into major and minor elements, depending on their role in forest structure and ability to form pure stands. The major elements are species that dominate the structure of the mangrove ecosystem. The minor elements are most often found at the ecotone between mangrove and terrestrial habitats, exhibit transitional characteristics, and rarely form pure stands. Mangrove associates are plants that invade mangrove environments from adjacent communities and constitute a potentially large pool of species.

Although there are more than forty mangrove species worldwide, only nine species occur in North America (Table 13.3) They represent five genera in four families: the Rhizophoraceae, Avicenniaceae, Combretaceae, and Pellicieraceae. The Rhizophoraceae is a small, pantropical family of 16 genera and is represented by three species: *Rhizophora mangle, R. xharrisonii,* and *R. racemosa.* The Avicenniaceae, which is also pantropical, is represented in North America by three species (*Avicennia germinans, A. schaueriana,* and *A.s.* var. *bicolor*). The white mangrove (*Laguncularia racemosa*) and buttonwood (*Conocarpus erectus*) are two widespread species in tropical America that both belong to a relatively large, woody family, the Combretaceae. *Pelliciera rhizophoreae,* which has a limited distribution in North America, belongs to a monotypic family, the Pellicieraceae.

GEOGRAPHICAL DISTRIBUTION AND AREAL EXTENT

Saltmarsh

North American saltmarshes occur from northern Alaska and Canada to the southern boundary of Central America but exhibit their greatest development in the temperate United States (Fig. 13.1). Although mangroves dominate tropical Mexico and Central America, saltmarsh habitats are still well represented in the tropics and subtropics, such as the extensive saltmarshes along the coast of Belize (West 1977). When found in association with mangroves, saltmarsh vegetation is reported to occur on mudflats or sandflats fronting open coasts, at the inner edge or within the mangal in canopy gaps, or as a fringe community along estuaries and tidal channels (West 1977; Lopez-Portillo and Ezcurra 1989; Patterson and Mendelssohn 1991). Where mangroves and saltmarshes co-occur, the greater canopy development of the mangal enables mangrove species to out-compete saltmarsh vegetation for light. The areal extent of saltmarshes in North America is given in Table 13.4.

About 14% of the total land area (127.2 × 10⁶ ha) of Canada is wetland (Glooschenko, Tarnocal, Zoltai, and Glooschenko 1993). Although the exact area of saltmarsh is unknown, it is estimated at 44,100 ha or 0.03% of the total Canadian wetland area (Glooschenko, personal communication) and excludes brackish marshes. Because the U.S. estimates include brackish wetlands, they are not comparable with Canada's. In Canada, saltmarshes occur in Arctic, Low and High Subarctic, Boreal, Atlantic Boreal, and Pacific Oceanic regions.

Saltmarshes in the Arctic region of Canada are primarily coastal, in contrast to estuarine, and they occupy ca. 60 km² on the Arctic islands, the western and eastern Canadian Arctic, Hudson Bay James Bay, and Labrador (Table 13.4) (Glooschenko 1982). *Puccinellia phryganoides,* a turf-forming grass species that is most common in the regularly flooded low marsh, dominates these Arctic saltmarshes that also include a number of *Carex* species (e.g., *Carex subspathacea* that can occur in the low marsh where there are inputs of freshwater lower salinity). Jefferies (1977) noted that *P. phryganoides, Stellaria humifusa,* and *Cochlearia officinalis* form the pioneer plant community on open coasts, whereas species such as *Arctophila fulva, Dupontia fisheri, Hippirus tetraphylla,* and *Carex ramenskii* dominate sheltered bays and inlets.

The saltmarshes of the Low Subarctic and High Subarctic Wetland regions have similar vegetation (Glooschenko et al. 1993). The Low Subarctic extends from approximately the Ontario/Quebec border at southern James Bay north to Cape Henrietta Maria on the Ontario coast where the Hudson and James bays come together. The High Subarctic extends along the Hudson Bay shoreline north from Cape Henrietta Maria to the Manitoba–Northwest Territory border. The most extensive Subartic, as well as Arctic, saltmarsh development occurs along the western coasts of the Hudson and James bays (Glooschenko 1982). This shoreline area has a high rate of isostatic rebound, and the presence of barrier ridges and spits allows for sediment accumulation with subsequent colonization by vegetation. In the northern Hudson Bay, *Puccinellia phryganoides* colonizes tidal flats, and *Carex subspathacea, Cochlearia officinalis* var. *groenlandica,* and *Potentilla egedii* are also common (Glooschenko and Martini 1981). To the south, an estimated 85–90% of the 1100 km shoreline of the Ontario portion of the Hudson and James bays has salt and brackish marsh development (Glooschenko 1982). These saltmarshes are dominated by species such as *P. phryganoides, P. lucida, Salicornia europaea, Glaux maritima, Scirpus maritimus, Plantago maritima, Festuca rubra,* and others. In contrast, the Quebec shoreline has little saltmarsh development due to the occurrence of the Canadian Shield at the coast and absence of extensive tidal flat development.

Saltmarshes occur in Canada along the Atlantic coast in the Atlantic Boreal and Atlantic Oceanic Wetland regions (Glooschenko et al. 1993). The Atlantic provinces of Canada, including Quebec,

Table 13.3. *North American mangrove species, their northern latitudinal limits on the Atlantic/Gulf of Mexico and Pacific continental coasts, and presence/absence (+/−) in the Caribbean Islands. NP = not present*

Species by family	Continental coasts		Caribbean Islands
	Atlantic and Gulf of Mexico	Pacific	
Rhizophoraceae			
Rhizophora mangle	29°10' (Florida, USA)	30°15' (Mexico)	+
R. racemosa	13° (Nicaragua)	10° (Panama)	−
R. x harrisonii	13° (Nicaragua)	15°15' (Mexico)	−
Avicenniaceae			
Avicennia germinans	30° (Florida, USA)	30°15' (Mexico)	+
A. schaueriana	NP	NP	+
A. bicolor	NP	9–10° (Costa Rica to Panama)	−
Combretaceae			
Laguncularia racemosa	29°08' (Florida, USA)	29°17' (Mexico)	+
Conocarpus erectus	28°50' (Florida, USA)	29°17' (Mexico)	+
Pellicieriaceae			
Pelliciera rhizophorae	13° (Nicaragua)	9°30' (Costa Rica)	−

Source: Tomlinson (1986).

Table 13.4. *Areal extent of saltmarshes in Canada and the United States. Brackish marshes are excluded for Canada but included for the United States, therefore direct comparisons between countries is not possible.*

Location	Area (ha)	Percentage of country total	Source
Canada			
Prince Edward Island	5,000	11.3	Wells & Hirvoven (1988)
Nova Scotia	12,400	28.1	"
New Brunswick	2,400	5.4	"
Quebec (St. Lawrence Estuary)	8,600	19.5	"
New Foundland & Labrador	5,000	11.3	Glooschenko (pers. comm.)
British Columbia	4,700	10.7	Glooschenko et al. (1993)
Hudson and James Bays	5,000	11.3	Glooschenko (pers. comm.)
Arctic Islands	1,000	2.3	"
Subtotal	44,100	100	
United States			
New England (ME–CT)	39,142	2.0	Alexander et al. (1986)
Mid-Atlantic (NY–VA)	258,318	13.3	"
Southeast Atlantic (NC–FL)	404,236	21.0	"
Gulf of Mexico (AL–TX)	1,072,445	55.3	"
Southern Pacific (CA)	8,736	0.5	"
Northwestern Pacific (OR–WA)	17,234	0.9	"
Alaska	138,156	7.0	Hall (1991)
Subtotal	1,938,267	100	

Source: Data from Latin America were not available, although Scott and Carbonell (1986) provide some data for various wetlands in the neotropics.

New Brunswick, Nova Scotia, Prince Edward Island, Labrador, and Newfoundland, comprise about 33,400 ha of saltmarsh (Table 13.4) characterized by *Spartina alterniflora* and *S. patens*. *Spartina alterniflora* reaches its northern limit within this region. In the Bay of Fundy, the low marsh is characterized by *S. alterniflora*, and the irregularly flooded high marsh is species rich and includes *S. patens, Limonium carolinianum, Salicornia europaea, Suaeda maritima, Atriplex patula, Plantago maritima, Puccinellia lucida, Triglochin maritima, Glaux maritima*, and *Hordeum jubatum*. In the saltmarshes along the St. Lawrence River in Quebec, the low marsh is dominated by *S. alterniflora*, and the high marsh contains an *S. patens* zone and a *Juncus balticus–J. gerardii* zone. The saltmarshes along the coastlines of New Brunswick and Prince Edward Island have developed in the lee of barrier islands and are characterized by sandy sediments. The saltmarshes of Newfoundland and Labrador are similar to those described for other parts of the Canadian Atlantic coast.

The Pacific-Oceanic Wetland region, which includes all of the Queen Charlotte Islands, the northern coast of British Columbia, and the northern and western coasts of Vancouver Island, has sparse saltmarsh development – a result of the mountainous coastline (Glooschenko et al. 1993) – of only 4700 ha (Table 13.4). The most extensive saltmarsh complex of British Columbia extends from the Fraser River Delta to Point Roberts Peninsula and Boundary Bay, on the U.S. border. Saltmarshes exist primarily where no influence of the Fraser River occurs. *Triglochin maritima, Salicornia virginica*, and *Distichlis spicata* are important species in these saline marshes. The Queen Charlotte Islands contain saltmarshes fronted by shingle beaches or mudflats. Dominant species include *Deschampsia caespitosa, Hordeum branchyantherum, Festuca rubra, Triglochin maritima, Carex lyngbyei, Plantago macrocarpa*, and *Stellaria humifusa*. Saltmarshes also form in this region on tidal flats that develop at the heads of fjords – for example, the Squamish River Delta, and they are dominated by *Carex lyngbyei*.

The United States contains over 1,900,000 ha of saltmarsh (including brackish) with most being along the Gulf of Mexico and least along the West Coast (Table 13.4). The New England marshes extend from Maine south through Connecticut and account for 2% of the total U.S. coastal saltmarshes (Table 13.4; Alexander, Broutman, and Field 1986; Hall 1991). New England marshes are characterized by pure stands of *Spartina alterniflora* along the channels of the tidal creeks, *S. patens* at higher elevations, and *Juncus gerardii*, and *J. balticus* even higher in the intertidal zone.

The saltmarshes of the mid-Atlantic region account for about 13% of the U.S. total (Table 13.4). These marshes are also characterized by *S. alterniflora* in the low marsh and *S. patens* in the high marsh, although *S. patens* and *Distichlis spicata* often dominate the majority of New Jersey–Delaware marshes (Reimold 1977). To the south, the majority of saltmarshes in Maryland and Virginia are associated with the Chesapeake Bay (Fig. 13.2c). Again, *S. alterniflora* dominates the low marsh with *S. patens* and especially *D. spicata* prevalent at higher elevations. *Juncus roemerianus* replaces *Juncus gerardii* in Virginia.

The saltmarshes of the southeast Atlantic from North Carolina through Florida are extensive and account for 21% of the total and approximately 58% of the U.S. Atlantic coast saltmarshes (Table 13.4, Fig. 13.2d) (Alexander et al. 1986). Many of these marshes form where major rivers deposit sediment into relatively shallow estuaries. In general, *S. alterniflora* dominates the low marsh, and *J. roemerianus* dominates the high marsh. From Cape Henry, Virginia, to Cape Hatteras, North Carolina, there are extensive areas of microtidal, brackish marshes dominated by *J. roemerianus*. *Spartina patens* can also gain dominance in these infrequently flooded marshes (Cooper and Waits 1973). From Cape Hatteras to northern Florida, *S. alterniflora* marshes reach their greatest development. It is here where the three height forms of this species are most apparent (Mendelssohn, McKee, and Postek 1982). The tall form occurs on creekbanks and can grow to 3 m in height. Along a complex gradient landward from the creekbanks, the height and productivity of this species gradually decrease to form the medium and short height forms. *Juncus roemerianus* often occurs in discrete patches within an *S. alterniflora* matrix or as extensive stands landward of the *S. alterniflora* zone. At higher highwater, salt pannes (sparsely vegetated, hypersaline areas) may develop due to infrequent tidal inundation and a concentration of salts by evapotranspiration. Although saltmarshes and mangroves can co-occur in northern Florida, mangrove swamps dominate the southern end of the state.

The Gulf of Mexico region contains the largest area of saltmarshes, 55% of the U.S. total (Table 13.4). The saltmarshes of the eastern Gulf of Mexico (west Florida, Alabama, and Mississippi) are primarily irregularly flooded marshes dominated by *Juncus roemerianus*. Twenty-eight percent of the U.S. *J. roemerianus* marshes occur in the eastern Gulf of Mexico, a region containing only 8% of the U.S. marshland (Stout 1984). A narrow fringe of *S. alterniflora* may occur seaward of the *Juncus* and zones of *D. spicata* and *S. patens* may occur at

Figure 13.2. Examples of saltmarsh communities in different geographic regions: (A) Pacific Northwest; (B) New England; (C) Virginia; (D) Georgia; and (E) Louisiana.

higher elevations landward of the *Juncus*. In the marshes south of Cedar Key, Florida, the black mangrove, *Avicennia germinans*, co-dominates with *J. roemerianus*. However, west of the Pearl River at the Mississippi-Louisiana border, *J. roemerianus* loses its dominance in the low-lying deltaic marshes of Louisiana. Here, saltmarshes are dominated by *S. alterniflora* (Fig. 13.2e). *Juncus roemerianus*, as well as *S. patens* and *D. spicata*, can be subordinate species, depending on local topography and marsh elevation. Saltmarshes in northern Texas are also dominated by *Spartina alterniflora*. However, in the hypersaline marshes of south Texas the grasses are replaced by succulents such as *Batis maritima, Borrichia frutescens, Suaeda maritima, Sesuvium portulacastrum*, and others.

California contains only 0.5% of the total U.S. saltmarsh area (Table 13.4). In northern California, *Spartina foliosa* first makes its appearance as an initial colonizer on low-elevation mudflats (Macdonald 1977). *Salicornia virginica* and *D. spicata*, along with a number of subordinate species, dominate the high marsh. The San Francisco Bay complex, where saline and brackish marshes occur, contains the most extensive saltmarshes of California (Josselyn 1983). In south San Francisco Bay, *Spartina foliosa* again colonizes the low marsh, with *S. virginica* dominating the high marsh (Hinde 1954). Species such as *Cuscuta salina, D. spicata, Frankenia grandifolia, Jaumea carnosa*, and to a lesser extent *Limonium californicum* and *Triglochin concinna*, occur scattered among the *Salicornia*. In southern California, relatively large tidal saltmarshes occur at Mugu Lagoon, Newport Bay, Mission Bay, San Diego Bay, and Tijuana Slough (Zedler 1982; Onuf 1987; Zedler, Nordby, and Kus 1992). In San Diego Bay, *Spartina foliosa* colonizes protected areas of the low marsh, and *Salicornia bigelovii* dominates on exposed sites. With increasing elevation, *S. virginica, Batis maritima, F. grandifolia*, and *Triglochin maritima* dominate at approximately mean higher highwater. *Salicornia virginica, Suaeda californica, Limonium californicum, Distichlis spicata*, and *Jaumea carnosa* are found at even higher elevations, and at the extreme highwater line *Salicornia subterminalis, Monanthochloe littoralis*, and *Atriplex watsonii* are prevalent.

The northwestern Pacific contains approximately 1% of the total saltmarsh area in the U.S. (Table 13.4, Fig. 13.2a). Dominant saltmarsh types in this region are the following:

1. Low sandy marshes dominated by *S. virginica* or *Scirpus americanus*, located on the landward side of bay-mouth sand spits or sandy bay islands.

2. Low marshes on rapidly accreting silt or mud substrates and colonized by *Triglochin maritima*.
3. Sedge marshes (primarily *Carex lyngbyei*) on silty substrates between low silty marshes and higher, more mature marshes.
4. Immature high marshes on silty substrates rich in organic matter and dominated by a mixture of *Deschampsia caespitosa* and *D. spicata*.
5. Mature high marshes dominated by *D. caespitosa, Juncus lesueurii*, and *Agrostis alba* at lower elevations, and forbs such as *Grindelia integrifolia* and *Potentilla pacifica* at higher elevations, on highly organic substrates underlain by older clay deposits (Macdonald 1977; Seliskar and Gallagher 1983).

The Alaskan saltmarshes comprise 7% of the U.S. total (Table 13.4), but well-developed saltmarsh communities are virtually absent along the Alaskan Arctic coastline due to its unstable and erosion-prone shoreline (Macdonald 1977). Saltmarsh communities occur as mosaics whose extent and age are determined by ice action. *Puccinellia phryganoides* is the primary colonizer on the open Arctic coast where the saltmarshes may be only a few meters in extent. At higher elevations, species such as *Potentilla pacifica, Triglochin maritima, Carex glareosa*, and others can be found. East of the Alaskan Peninsula, clear zonation replaces the irregular mosaic communities more typical of the Arctic and western Alaskan coasts. Low marshes are colonized by *P. phryganoides, P. triflora*, and *Triglochin maritima*; high marshes contain the same species, but *T. maritima* becomes dominant. *Carex lyngbyei* is common in channel bank communities subject to tidal flooding (Macdonald 1977).

Olmsted (1993) estimates the extent of the major wetlands of Mexico, including coastal lagoons, fresh, brackish, and saltmarshes, mangrove swamps, freshwater lakes and riverine forests, at 3,318,500 ha. At this time, an accurate estimate of saltmarsh area for Mexico is not available, but saltmarshes per se likely comprise only a small percentage of the total, which includes both freshwater and saline wetland types. The largest continuous wetland is located in Tabasco and Campeche and is approximately 1,400,000 ha. Other large wetlands are located in Quintana Roo and the Yucatan (335,000 and 184,000 ha, respectively). These wetlands, together with those in Tabasco, Veracruz, Campeche, and Chiapas, make southeastern Mexico the most significant wetland region of Mexico (Olmsted 1993). Although saltmarshes in tropical latitudes are often outcompeted and replaced by mangroves, West (1977) has cited three environmental situations in which saltmarsh species may exist, usually on the margins or within mangrove woodlands: (1) colonizing recently formed mudflats that fringe mangrove woodlands, (2) occupying saline soils on the inner edge or within the mangrove woodland, and (3) colonizing disturbed areas within a mangrove woodland.

In northern Mexico, saltmarshes are more abundant along the northern Pacific Coast than along the tropical Gulf or Caribbean shorelines and are dominated by *Spartina* spp., *Sporobolus virginicus, Distichlis spicata, Monanthochloe littoralis,* and *Uniola* spp. (Olmsted 1993). In more tropical Mexico, brackish and saline marshes are commonly found in association with mangroves along the Pacific, Gulf, and Caribbean Coasts, especially near coastal lagoons or near river deltas with low sediment load. Infrequently inundated hypersaline salt flats, although small in areal extent, also occur along the coastlines of Mexico. Salt flats on the northern Gulf Coast of Mexico contain three associations: (1) *Suaeda nigra* and *Salicornia ambigua*; (2) *Batis maritima, Borrichia frutescens, Clappia suaedifolia,* and *Maytenus phyllanthoides*; and (3) *D. spicata* and *M. littoralis* (Olmsted 1993). On the Yucatan Peninsula, dominant plant species on salt flats include *Salicornia* spp., *B. maritima, Suaeda linearis,* and *Sesuvium portulacastrum.* For the Gulf of California, species include *Salicornia pacifica, M. littoralis, B. maritima, S. europaea,* and *Frankenia grandifolia* (Johnston 1924).

Saltmarshes occur throughout Central America and the Caribbean, although the dominant coastal habitat is mangrove. This fact may explain the absence of available information on the areal extent of saltmarshes for these regions. Costa and Davy (1992) have suggested that saltmarshes may be maintained in Latin America, where mangroves usually dominate, due either to the unavailability of mangrove propagules in specific areas or, more important, the development of hypersaline conditions resulting from infrequent flooding and evapotranspiration. Additionally, disturbed mangrove habitats, (e.g., caused by clear-cutting) are often replaced by herbaceous species such as *Paspalum vaginatum, Sesuvium portulacastrum,* and *Blutaparon vermiculare.*

Costa and Davy (1992) divided the saltmarsh species of Central and South America into six groups, segregated primarily on the basis of geography. The northern widespread species group is characterized by species that are more frequently found in the northern, tropical part of Latin America but that are widely distributed and extend considerably southward. Species such as *Sesuvium portulacastrum, Batis maritima, Spartina alterniflora,* and *Sporobolus virginicus* are included in this group. These species are often found in association with, or fringing, nearby mangrove communities. The southern widespread species group is characterized by species frequently found in temperate and subtropical zones of South America but which extend northward into the tropics. This group includes *Spartina densiflora, Salicornia fructicosa, S. virginica, S. ambigua,* and *Juncus acutus.* The Northern Hemisphere species group is restricted to the Northern Hemisphere and includes species of the genus *Spartina, Salicornia,* and *Suaeda.* This group inhabits mangrove communities and hypersaline lagoons on the Atlantic Coast and hypersaline lagoons associated with the Baja California Peninsula on the Pacific Coast. A small, heterogeneous group that is locally abundant in the tropics includes *Blutaparon vermiculare,* which occurs in the Caribbean region, and *Spartina* spp. found in Belize. The final two groups, brackish-water species of midlatitudes and species of high latitudes, are outside the scope of this review.

Mangrove

Mangrove populations are generally limited to tropical and subtropical regions between 32°N and 28°S latitudes. Latitudinal limits vary in the Northern Hemisphere from 24°N–32°N because of local variation in air and water temperatures. Three species, *Rhizophora mangle* (red mangrove), *Avicennia germinans* (black mangrove), and *Laguncularia racemosa* (white mangrove), have the widest distribution, occurring in Florida, the Caribbean, Mexico, and Central America. Although populations of *R. mangle* and *A. germinans* occur in Bermuda (32°N latitude), they are restricted to lower latitudes on the mainland of North America where their northern limits fluctuate with the frequency, duration, and/or severity of cold winter temperatures (Table 13.3). Of the three species, *A. germinans* is apparently the most tolerant of cold, with persistent populations reported as far north as 30°N latitude on the east coast of Florida (Savage 1972) and 29°18'N on the northern Gulf of Mexico coast in Louisiana and Texas (Sherrod and McMillan 1985). The northern limit on the Pacific Coast is reported to be 31°15'N near Puerto Lobos in Mexico. The ability of *A. germinans* to coppice (stump sprout) explains its survival in areas periodically subjected to freezing temperatures, for the species exists at the northern edge of its range only in a stunted, shrub-like growth form. Populations of *R. mangle* and *L. racemosa* have been reported as far north as 29°10' N latitude in Florida at Cedar Key on the west coast (Rehm 1976), and just north of

Table 13.5. *Mangrove areas by country for North America*

Country	Area (ha)	Percentage of total	Source
Continental			
Mexico	488,367	22.1	Loza (1994)
USA	280,594	12.7	Cont. Shelf Assoc. (1991)
Panama	170,827	7.7	Osorio (1994)
Nicaragua	155,000	7.0	Garcia & Camacho (1994)
Honduras	145,800	6.6	Oyuela (1994)
Belize	78,062	3.5	Zisman (1992)
Costa Rica	41,002	1.9	Pizarro & Angulo (1994)
El Salvador	26,772	1.2	Funes (1994)
Guatemala	16,035	<0.01	Aragon de Rendon et al. (1994)
Subtotal	1,402,459	63.6	
Island			
Cuba	532,400	24.1	Carrera & Santander (1994)
Bahamas	141,957	6.4	Bacon (1993)
Dominican Republic	42,401	1.9	Alvarez (1994)
Antilles	24,571	1.1	Bacon (1993)
Haiti	18,000	<0.01	Saenger et al. (1983)
Jamaica	10,624	<0.01	Bacon (1993)
Guadeloupe	8,000	<0.01	Saenger et al. (1983)
Cayman Islands	7,268	<0.01	Bacon (1993)
Trinidad & Tobago	7,150	<0.01	Bacon (1993)
Puerto Rico	9,296	<0.01	Martinez (1994)
Martinique	1,900	<0.01	Saenger et al. (1983)
Bermuda	20	<0.01	Ellison (1993)
Subtotal	803,587	36.4	
Total	2,206,046	100	

Ponce de Leon Inlet on the east coast (Teas 1977). Small, mixed stands of *R. mangle, A. germinans,* and *L. racemosa* can be found in the Gulf of California near Isla Tiburon, Punta Perla (29°12' N, 113°36' W) and at Estero Sargento tidal channel (29°17' N, 112°19' W) (Flores-Verdugo, 1992). A fourth species, *Conocarpus erectus* L., which is considered by some authorities (e.g., Tomlinson 1986) to be a mangrove associate, is found as far north as 28°50' N in Florida and 29°17' N at Estero Sargento, Mexico fronting the Pacific Ocean.

Six additional mangrove species, which are mainly South American, have ranges that extend into Central America and the Caribbean Islands. Populations of *R. racemosa* and *R. x harrisonii* have been reported on the Pacific Coast of Costa Rica and Nicaragua, but statements about the northern latitudinal limits of these species are difficult to make due to taxonomic uncertainties (Cintron-Molero and Schaeffer-Novelli 1992). Also on the Pacific Coast are small populations of *A. bicolor* and *A. tonduzii* (Lacerda et al. 1993). *Avicennia schaueriana* occurs mainly in South America, although its range extends into the Lesser Antilles (17°30' N latitude) (Bacon 1993). *Pelliciera rhizophorae* is mainly found on the Pacific Coasts of Panama and Costa

Rica (Jimenez 1992), but Atlantic Coast populations are reported as far north as 13° in Nicaragua (Polania and Mainardi 1993).

A complete listing of mangrove associates and their distributions is beyond the scope of this chapter, but species that may make an important contribution to community structure are described by Tomlinson (1986) and Jimenez (1992). As discussed in the previous section, common saltmarsh species such as *Batis maritima, Sesuvium portulacastrum, Salicornia ambigua*, and grasses such as *Sporobolus virginicus, Paspalum vaginatum, Spartina alterniflora, S. patens*, and *S. spartinae* may also be found in association with mangroves.

The total areal extent of mangroves in North and Central America and the Caribbean exceeds 2 m ha (Table 13.5). Estimates of world mangrove cover range from 15 to 30 m ha (Saenger, Hegerl, and Davie 1983), but a recent accounting places the total at 14,197,635 (Lacerda et al. 1993). Thus, the forests occurring in North America account for about 16% of the world's mangrove cover.

Approximately 64% of the North American mangrove area occurs in continental countries. Mangrove forests are well developed along the southwest coast of Florida, but large areas have been lost to development (Patterson 1986). Estimates reported by Continental Shelf Associates (1991) place total extant mangrove area in Florida at 274,857 ha and 5737 ha in Texas, Louisiana, Mississippi, and Alabama combined. Extensive mangrove forests occur on both coasts of Mexico and together total more than 488,000 ha. Well-developed forests occupying more than 130,000 ha can be found on the Gulf of Mexico in the Laguna de Terminos, Campeche (Flores-Verdugo, Gonzalez-Farias, Zamorano, and Ramirez-Garcia 1992). On the Pacific Coast, mangrove areal extent is lower, but major forests occur in the Teacapan–Agua Brava–Marismas Nacionales (113,238 ha) and the Chantuto–Teculapa–Panzacola (35,000 ha) estuaries (Flores-Verdugo et al. 1992). Panama, Nicaragua, and Honduras follow Mexico in mangrove cover with about 146,000–171,000 ha each. The remaining countries of Belize, Guatemala, El Salvador, and Costa Rica together account for about 162,000 ha of mangrove forest area. Mangrove forests generally occupy 5% or less of the land surface in these continental countries (range is from 3.4% in Belize to 0.02% in the United States).

The Caribbean Islands contain about 37% of the total mangrove forest area in North America, with the largest tracts occurring in Cuba and the Bahamas (Table 13.5). For these insular countries, the percentage of land area occupied by mangroves may be much higher than for continental countries (e.g., 27.6% in the Cayman Islands and 10.2% in the Bahamas). Some of the continental countries, however, have extensive forests occurring on offshore islands. Belize, for example, has 22,291 ha of insular mangroves, approximately 40% of the total mangrove area (Zisman 1992).

The values given here are only estimates, for some areas have not been accurately mapped and wetlands are being rapidly cleared in many countries to make way for urban, industrial, and agricultural activities. Changes in the world's climate and sea level will also have an impact on future areal extent of wetlands in North America. Thus, saltmarsh and mangrove acreage in this region is in a constant state of flux and can be expected to change in the years to come.

SALTMARSH AND MANGROVE DEVELOPMENT

Salt marsh and mangrove development is closely tied to fluctuations in sea level. Sea level has changed dramatically over the past 700,000 yr during the Quaternary period as a result of as many as nine glacial and ten interglacial events (Shackleton and Opdyke 1973). During the most recent glacial maximum – 18,000 yr BP) the Wisconsin sea level was about 100 m below its present level (Donn et al. 1962). Sea level, however, rose rapidly during the late Pleistocene and the early Holocene, from approximately 17,000–7000 yr BP, as the Wisconsin glaciers melted. Consequently, existing coastal wetlands were submerged and the formation of new wetlands was prevented, with three possible exceptions during relatively short periods of sea level standstill (Fig. 13.3). Modern coastal wetlands began to develop between 8000–4000 YBP when the rise in sea level slowed (Aubrey and Emery 1993). Prior to about 6000 yr, the rate of sea level rise was 6–12 mm yr^{-1}, whereas the rate since then has been only about 1.0–1.4 mm yr^{-1}. During the mid-to late Holocene, the rate of sediment accretion apparently equaled, if not exceeded, the rate of sea level rise and allowed for widespread coastal wetland formation. Hence, the majority of today's coastal wetlands are composed of Holocene sediments overlying a Pleistocene base. Although sea level has been relatively constant during the late Holocene, determinations of present rates of sea level rise average 2.4 mm yr^{-1} (Peltier and Tushingham 1989), about twice that of earlier estimates. Accelerated sea level rise predicted to result from global warming (e.g., a 30-cm increase by 2050; Titus 1988) may usher in a new period of wetland submergence and loss.

Both physical and biological processes are important in controlling the rates of sedimentation and vertical accretion of saltmarshes and mangroves and thus in maintaining their intertidal position during periods of rise in sea level. Tides and storm-driven water are the main physical processes controlling the introduction and distribution of sediments into coastal wetland systems. The longer the sediment-laden tidal water remains over the marsh surface, the greater the potential for the sediment to settle out; however, once the sediment load of the water column has been depleted, a new flooding event must occur to supply more sediment to the wetland.

Biological processes may also enhance sedimentation rates. For example, plant stems dampen the force of wind-generated waves and reduce current flow, allowing sediment to accumulate more readily on the soil surface (Gleason et al. 1979). Additionally, plant roots aid in binding the sediment and thereby reduce the potential for resuspension and erosion. However, the most important contribution of vegetation to vertical marsh accretion is the production of organic matter, primarily belowground, that adds mass to the soil and controls the volumetric development of the soil. Animals can also influence sedimentation rates. Mussels, for ex-

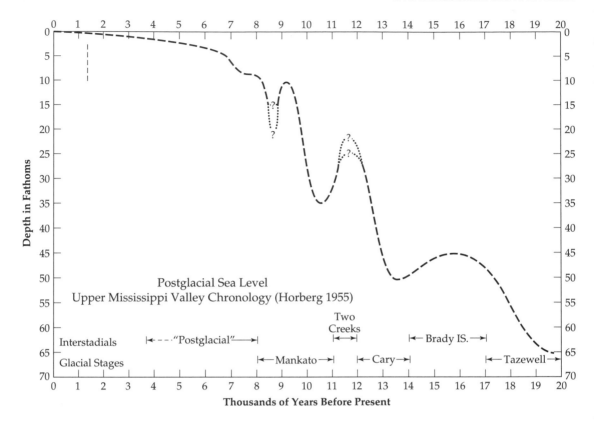

Figure 13.3. Changes in sea level over the past 20,000 years (Shepard 1960), reprinted from Tolley 1993). Exten-

sive formation of saltmarshes and mangroves has occurred during periods of slow rises or standstills in sea level.

ample, produce byssal threads that not only anchor these bivalves in place but also enhance sediment adhesion (Pestrong 1972). Invertebrate filter feeders intake large quantities of sediments and generate feces and pseudofeces that accumulate on the marsh surface (Kraeuter 1976). All these factors, biotic and physical, influence sedimentation rates and, hence, vertical accretion, a process critical to the persistence of coastal wetlands during periods of rise in sea level.

PHYSICAL AND CHEMICAL FEATURES

The distribution and development of saltmarsh and mangrove ecosystems are controlled by several physical and chemical factors, including climate, physiography, hydrology, and soil chemistry. Saltmarsh and mangrove communities both develop in the intertidal zone and are, therefore, shaped by similar forces. The intertidal habitat is characterized by anaerobic, saline soils and periodic tidal inundation. A major factor determining

the relative distributions of saltmarshes and mangroves is climate.

Climate

Saltmarshes are typically dominated by long-lived, perennial species that have underground, perennating organs capable of surviving winter temperatures and dry periods. Annuals, which avoid adverse conditions during the seed stage, also occur in saltmarshes in specific zones (e.g., the upper marsh along driftlines). Thus, life history strategies of saltmarsh species promote their survival in cold climates and explains their distribution into northern Canada and Alaska. High temperature per se is not a direct constraint on the distribution of saltmarsh vegetation but, as discussed previously, allows for the development of mangroves that outcompete saltmarsh species and thereby prevent saltmarsh dominance.

Mangroves, on the other hand, are tropical species that cannot survive in climates where the average annual temperature falls below 19° C (Waisel

1972). Rapid fluctuations in temperature ($\pm$ 10° C) or short-term exposure to freezing temperatures may be sufficient to affect viability of mangroves. Tolerance to cold, however, varies among and within species. *Avicennia germinans* seems to be the most cold-tolerant mangrove, and it can resprout if the top has been killed by frost. Although propagules of *Rhizophora mangle* are frequently dispersed from Mexico to south Texas and become established, winter freezes eventually eliminate them (Sherrod, Hockaday, and McMillan 1986).

In addition to temperature limitations on mangrove distribution, forest development is enhanced in moist or wet climates where rainfall exceeds 2000 mm yr^{-1} (Macnae 1968). In some areas of Costa Rica and Panama with high average rainfall (2100–6400 mm yr^{-1}), canopy height may reach 35 m and aboveground biomass may be 280 metric tons ha^{-1} (Golley, McGinnis, Clements, Child, and Duever 1969). The less extensive mangrove development on the Pacific Coast of Mexico can be partly attributed to the drier, colder climate there. The arid climate (<500 mm yr^{-1}) along the Gulf of California has contributed to the lower biomass and structural simplicity of mangrove forests there. Rainfall in the Caribbean islands varies from 500 to more than 5000 mm yr^{-1} and contributes to substantial variation in mangrove forest structure in that region.

Physiography and Hydrology

Physiography reflects the geology of a region and creates the physical and chemical framework on which marsh and mangrove vegetation develops. The physiographic setting of saltmarshes and mangrove swamps is similar in that development occurs primarily in areas where there is sufficient protection from wave action – for example, in protected shallow bays and estuaries, in lagoons, and behind offshore islands. Excessive wave action prevents establishment of seedlings and propagules, exposes the shallow root systems, and limits deposition of fine sediments that promote plant growth. Saltmarshes and mangroves are more extensive along low-relief coastlines where tidal intrusion reaches far inland and where there is abundant availability and accumulation of fine silty and clayey sediments.

The hydrologic regime exerts a tremendous influence on the structure and function of wetlands. Hydrology affects abiotic factors such as salinity, soil moisture, soil oxygen, and nutrient availability, as well as biotic factors such as dispersal of seeds and propagules. These factors, in turn, influence the distribution and relative abundance of plant species and ecosystem productivity. The tides constitute both a stress and a subsidy (*sensu* Odum and Fanning 1973) for saltmarsh or mangrove development. Inundation by the tides leads to soil anaerobiosis and, depending on a species' flood tolerance, may inhibit survival, growth, and spread. For saltmarsh species, effects of low oxygen may limit vegetative spread and/or seed germination. Although some mangrove species are viviparous (i.e., they germinate while still attached to the parent tree), they are still vulnerable to flooding effects during the seedling stage (McKee 1995b). Tides also import high concentrations of potentially toxic ions such as Na^+ and Cl^-. Tidal fluctuation, however, acts as a subsidy to saltmarsh and mangrove systems by importing nutrients, aerating the soil water, flushing out accumulated salts and reduced compounds that are phytotoxic, and dispersing propagules. The positive effect of this tidal subsidy can be seen in Fig. 13.4, in which saltmarsh grass and mangrove trees growing along hydrodynamically active creekbanks are taller and more productive than in the interior where tidal fluctuation is minimized.

Geology and Soils

The substrates in saltmarsh and mangrove systems may be referred to as soils or sediments. The term "soil" is used to describe substrates exhibiting pedologic structures and identifiable horizons, but it also implies a relationship with the underlying parent material from which it was derived and an ability to support vegetation. Sediment can be autochthonous (i.e., produced *in situ*) or allochthonous (i.e., originating outside the system). The words connote no relationship with an underlying bedrock or other parent material. Thus, the term that is employed depends on the nature of the deposits and the characteristics under discussion.

Although saltmarshes and mangroves achieve best development on fine-grained sediments, they can grow on a variety of substrates, including sands and volcanic lava. Terrigenous sediments may be carried by rivers from inland areas to be deposited along the sea or may originate from adjacent eroding shorelines. Fine silts and clays contain abundant exchangeable ions that fertilize and enhance the productivity of the plants. Marshes and mangroves may also develop on sandy substrata, particularly in stable, sheltered areas where the sand mixes with silt or organic matter (Chapman 1976). In the case of autochthonous deposits, the marsh or mangrove vegetation itself contrib-

Figure 13.4. Saltmarsh (A) and mangrove communities (B) both show the effect of the tidal energy subsidy that is, taller, more productive vegetation along shorelines and creekbanks where tidal fluctuation and soil drainage are maximal (saltmarsh photo by J. L. Gallagher).

utes to sedimentation through production of organic matter, primarily belowground. The organic matter content of soils may vary from <10 to >90%, depending on the relative contributions of organic versus mineral deposits. High rates of root production combined with slow decomposition rates in the anaerobic environment may promote large accumulations of peat. Other biogenic deposits include the carbonate skeletons of calcareous algae (e.g., *Halimeda* spp.), which are the major source of sand in the Caribbean region, and shells of oysters and other invertebrates, which can be important constituents of saltmarsh sediments.

Salinity

Saltmarsh and mangrove soils are typically saline, but salinity varies depending on freshwater input, rainfall to evapotranspiration potential, and hydrology. Although frequency of inundation with seawater decreases with elevation of the soil surface, there is not a simple or consistent relationship between salinity and tidal elevation. In the lower saltmarsh or mangrove forest areas that are regularly flushed by ocean tides, salinities are typically maintained near that of the flooding seawater –33–38%. At higher elevations, the interaction between flooding and climate results in substantial varia-

bility in soil salinity. During periods of high rainfall or in regions receiving freshwater runoff, salinities may be reduced between tidal flooding events. Riverine mangrove forests, for example, may experience wide fluctuations in porewater salinity, varying from less than 1–25%. Areas with high evapotranspiration rates and irregular tidal flushing develop hypersaline conditions with porewater salinities sometimes exceeding 70% in both mangrove and saltmarsh habitats. Mangrove and saltmarsh plants are able to survive and grow at elevated salinities due to several adaptations, including salt exclusion and excretion through glands in the leaves. Localized freshwater discharges in seasonally dry regions may also prevent hypersaline conditions and promote vegetative development, as reported for some Mexican and Central American mangroves (Flores-Verdugo et al. 1992; Jimenez 1992). However, along some arid tropical and subtropical coasts – for example, the Laguna Madre of southern Texas – extended periods of hypersaline conditions may prevent the survival of perennial vegetation altogether.

Oxygen

Inundation of saltmarsh and mangrove soils with water leads to anaerobic conditions due to a

10,000-times slower diffusion rate of oxygen in aqueous solution compared to air (Gambrell and Patrick 1978). Once oxygen is depleted by soil and plant root respiration, it is not quickly replaced and anaerobic conditions prevail. In the absence of oxygen, soil microorganisms use alternate oxidants (NO_3-, Mn^{+4}, Fe^{+3}, SO_4^{-2}) as electron acceptors. This process results in low redox potentials (Eh), variation in availability and form of plant nutrients, and a buildup of toxic, reduced compounds in the soil. Soil Eh is a measure of the intensity of soil reduction, and low (≤ -100 mV) values are characteristic of strongly reducing conditions. Values ranging from $+300--250$ mV are typical of flooded soils and vary depending on soil texture, availability of oxidants, and flooding regime. Several studies have also shown that the oxidation-reduction status of marsh and mangrove soils is influenced by the presence of roots (Mendelssohn and Postek 1982; Thibodeau and Nickerson 1986; McKee, Mendelssohn, Hester 1988; McKee 1993). Leakage of oxygen from the plant roots into the surrounding soil apparently creates an oxidized rhizosphere in which redox potentials are higher than in the bulk soil. Thus, the growth of saltmarsh and mangrove vegetation is influenced by the oxygenless condition of the soil substrate, but the plants themselves also modify the oxidation-reduction status of the soil.

Nutrients

Tidal regimes influence nutritional status by distributing mineral sediments and affecting the redox status of the substrate, which in turn controls nutrient transformations and form or the availability of inorganic nutrients. Nitrogen is considered to be the primary limiting nutrient in saltmarshes and mangroves. Availability of phosphorus in anaerobic sediments typically exceeds that of ammonium, the dominant nitrogen form (Mendelssohn 1979). Numerous fertilization experiments in saltmarshes have consistently demonstrated that nitrogen is the primary limiting nutrient (see Mendelssohn et al. 1982 and references therein), although phosphorus can limit plant growth in sandy environments where phosphorus availability is low (Broome et al. 1975). Recent work in Belize and Florida has also demonstrated that stunted mangroves (<2 m in height) are phosphorus-limited (McKee and Feller 1994b; Feller 1995).

Phytotoxins

Plants growing in an anaerobic soil environment may be damaged by the accumulation of soil phytotoxins. In the marine environment, a major phytotoxin that accumulates under anaerobic conditions is sulfide. Sulfate is the second most abundant anion in seawater and begins to be reduced under anaerobic conditions after NO_{3-}, Mn^{+4}, and Fe^{+3} have been reduced. The reduction of sulfate is carried out by true anaerobes (e.g., *Desulfovibrio*) and is thus dependent on anoxic conditions (Postgate 1959).

Numerous studies have demonstrated the role of sulfide in controlling the growth, productivity, and distribution of saltmarsh species (King, Klug, Weigert, and Chalmers 1982; Koch and Mendelssohn 1989; Bradley and Morris 1990; Koch, Mendelssohn, and McKee 1990). A phytotoxic effect of sulfide was also postulated as a factor controlling distribution of mangrove species based on correlative data (Nickerson and Thibodeau 1985). The spatial association of *A. germinans* with more oxidized, low H_2S areas suggested that this species was capable of oxidizing its rhizosphere and of colonizing areas with high concentrations of H_2S. Subsequent work that examined sulfide concentrations near adult *R. mangle* root systems, however, demonstrated a similar oxidizing ability (McKee et al. 1988). Experiments comparing seedling sensitivity to soil anoxia and sulfide further indicate that during early growth stages *R. mangle* is more tolerant of reducing soil conditions than are *A. germinans* and *L. racemosa* (McKee 1993; McKee 1995b).

COMMUNITY STRUCTURE AND PHYSIOGNOMY

Saltmarsh

Saltmarsh communities are relatively species-poor, and in some localities of North America (e.g., the south Atlantic and northern Gulf of Mexico coastlines), *Spartina alterniflora* may form extensive, monospecific stands. Although as many as 93 species can occur in coastal freshwater marshes, species richness in coastal saltmarshes does not exceed 17, and, as mentioned previously, most individual saltmarshes contain far fewer species (Chabreck 1972). Figure 13.5 illustrates the relationship between salinity and species richness in coastal Louisiana.

Three categories of saltmarsh plant communities are (1) communities dominated by graminoid plants such as grasses, sedges, and rushes, (2) herbaceous communities, and (3) dwarf-shrub communities common along arid and semiarid tropical and subtropical coasts (Adam 1990). Growth forms include forbs, grasses, succulents, and shrubs. Unlike forests, which contain a number of strata, the

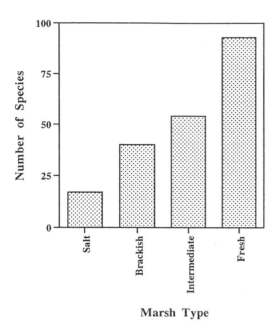

Figure 13.5. Species richness in Louisiana's coastal marshes decreases with increasing salinity (based on data reported in Chabreck, 1972).

vertical structure of saltmarshes is relatively simple. However, some saltmarshes in which the vegetative canopy is relatively sparse can allow for enough light penetration to support benthic and periphytic algal communities. The presence of both graminoids and forbs also increases the vertical structure of saltmarshes.

Saltmarshes are frequently separated into two physiographic zones. The low marsh, sometimes termed "the regularly flooded marsh," is inundated by each tidal event, once or twice a day depending on whether the tides are diurnal or semidiurnal, respectively. The high marsh, sometimes referred to as "the irregularly flooded marsh," is higher in elevation than the low marsh and thus is flooded less frequently, sometimes only during spring tides. The highest elevations of the saltmarsh can develop into hypersaline areas called "salt pannes." The salt panne is inundated only by the highest tides, and as a result, salt accumulates to lethal levels due to evapotranspiration in the absence of tidal dilution and leaching. Salt pannes are often devoid of vegetation or are characterized by stunted halophytes.

Species richness tends to increase along the elevational gradient from the sea to the marsh/terrestrial ecotone. The low marsh has low species richness, sometimes with only one species present, whereas the high marsh often exhibits a much

greater number of species, especially where freshwater runoff from adjacent uplands occurs. In addition, the deposition of marsh wrack (dead aboveground marsh vegetation) on the high marsh can open up gaps in the vegetative canopy and facilitate species recruitment and establishment (Ellison 1987; Bertness 1992). The exception to species richness that increases with elevation occurs where salt pannes are present.

Mangrove

As was true for coastal saltmarshes, the combined stresses of flooding and salinity in mangrove forests have resulted in floristically depauperate systems characterized by a relatively simple stand structure, especially in comparison with other tropical forests. Many mangrove forests throughout North and Central America and the Caribbean are comprised of only two or three species present in the canopy and few, if any, mangrove associates in the subcanopy. In some cases, extensive monospecific stands dominated by red mangrove may occur along creekbanks and shorelines, presenting a monotonous, impenetrable tangle of prop roots overtopped by a dense canopy in which 95% of incident light is intercepted (Fig. 13.6a). In such situations, the combination of edaphic stresses and low light prevents the development of an understory. Forests with abundant rainfall and/or freshwater runoff, however, may have lianas, orchids, and other epiphytes attached to branches or prop roots and grasses or ferns growing on the forest floor. Other forests (e.g., basin forests dominated by *Avicennia* spp.) may exhibit a woodland appearance with widely spaced, straight boles, dense carpets of pneumatophores (aerial roots), and an open canopy structure (Fig. 13.6b). Scrub or dwarf forests display yet another type of structure in which sparse, stunted trees create an open, low-canopied stand (Fig. 13.6c).

Descriptions of vertical structure in mangrove forests typically indicate only one major stratum, which is the main canopy. A vertical profile through a shoreline or creekbank, however, reveals three major strata: the supratidal (above the high tide mark), the intertidal (between high and low tide), and the subtidal (below the low tide mark). Each stratum is characterized by a unique assemblage of plants and animals adapted to the conditions prevalent there. The supratidal stratum is comprised of the arboreal portions of the forest, whereas the intertidal and subtidal strata include the root systems (both aerial and subaerial). This type of vertical stratification is particularly well developed in fringe forests and overwash islands.

Figure 13.6. Mangrove forests exhibit different growth forms depending on hydrology, dominant species, and climate: (A) Creekbank dominated by Rhizophora mangle *with a nearly impenetrable tangle of prop roots and overtopped by a closed canopy. (B) Basin forest dominated by* Avicennia germinans *with tall, straight-boled trees, and (C) Dwarf forest comprised of stunted* Rhizophora mangle *trees (<1.5 m in height and 40–50 yr old).*

Mangrove forests are often classified according to a scheme that relates forest physiognomy to hydrology and geomorphology. Lugo and Snedaker (1974) first classified mangrove forests into six types (overwash, fringe, basin, riverine, dwarf, and hammock), each differing in tidal characteristics, hydroperiod, and structural and functional attributes. Lugo, Brinson, and Brown (1989) have proposed a simpler classification scheme in which some of the original types were considered special cases within three basic categories: riverine; fringe, including overwash islands; and basin, including dwarf and hammock forests. Specific characteristics of selected forest types are described below and summarized in Fig. 13.7.

Riverine forests occur along tidal rivers and creeks where high inputs of freshwater, sediment, and nutrients promote maximal mangrove growth and productivity (Fig. 13.7). Trees exhibiting luxuriant growth and canopies often more than 20 m

in height typify these forests. The high productivity and dynamic hydrologic regime lead to high rates of organic matter export. Salinity may vary but is generally lower than that found in other forest types. Frequent flushing with freshwater during wet seasons can lead to salinities below 10%.

Fringe mangrove forests develop on the seaward edge of protected shorelines, adjacent to nonriverine tidal channels and on small islands (Fig. 13.7). This forest type has an open exchange with the sea and consequently is well flushed by daily tides; it experiences salinities similar to that of seawater and nutrient inputs lower than in riverine forests. Exposure of fringe forests to storm surges and strong winds is maximal because of their open shorelines. Primary productivity is lower than in riverine forests, and canopy height rarely exceeds 10 m. Overwash islands or spits are considered to be special cases of fringe forests, characterized by higher tidal velocities that "overwash" the entire

		RIVERINE	FRINGE	BASIN	SCRUB/DWARF
FOREST PHYSIOGNOMY & PRODUCTIVITY	PROFILE	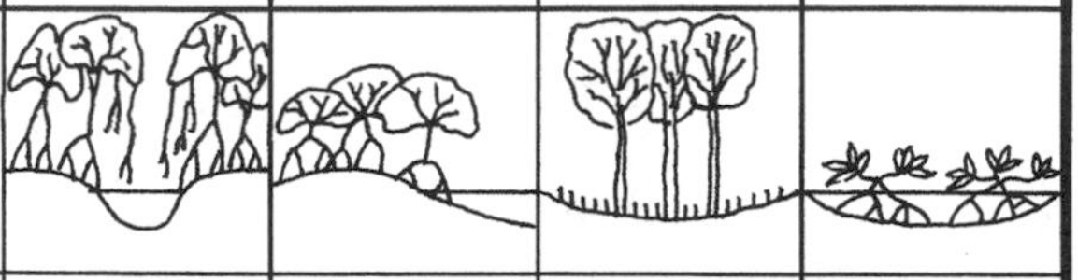			
	CANOPY HEIGHT (m)	12.64 ± 1.43	7.65 ± 0.94	12.14 ± 1.29	0.83 ± 0.09
	LITTER PRODUCTION (MT/ha/yr)	12 ± 1.0	9.0 ± 0.7	6.6 ± 0.7	1.9 ± 0.6
HYDROLOGY	WATER SOURCE	ocean tides & stream flow	ocean tides	ocean tides & saline groundwater	ocean tides & saline groundwater
	HYDROPERIOD — DURATION	hours-days	hours	days-months	perennial
	HYDROPERIOD — FREQUENCY	daily or seasonal	daily	seasonal	continuous
	HYDROPERIOD — DEPTH	shallow-deep	shallow	shallow	shallow
SOIL CHEMISTRY	SALINITY (‰)	0/26	33/38	25/60	33/46
	REDOX POTENTIAL (mV)	-48/+116	-96/+103	+87/+279	-244/-105
	SULFIDE (mM)	0.0/0.2	0.1/0.3	0.1/0.2	0.9/2.2

Figure 13.7. Comparison of four of six mangrove forest types described by Lugo and Snedaker (1974). The values for canopy height and litter production are means ± standard errors reported in Pool et al. (1975), Twilley (1986), McKee and Feller (1994b), and Feller (1995). The hydrologic characteristics are modified from Zack and Roman-Mas (1988). Minimum/maximum values for soil chemistry are based on measurements conducted in Florida and Belize (McKee et al. 1988; McKee 1993; McKee 1995b; McKee unpublished data).

land mass and limit accumulation of organic debris. The overwash island type is abundant in the Ten Thousand Island region of Florida, the southern coast of Puerto Rico, and in the barrier reef system in Belize.

Basin mangrove forests occur in topographic depressions, often inland of fringing or riverine forests (Fig. 13.7). Water movement is lower than in other forest types, with some basins experiencing infrequent tidal inundation (spring or storm tides). Once inundated by tides or freshwater, however, basin forests typically remain flooded for extended periods due to restricted drainage. Basin forests are characterized by high ratios of precipitation to evapotranspiration and annual mean litterfall averaging 7 metric tons ha^{-1} yr^{-1}. A special type of basin forest is the hammock mangrove, which occurs as isolated tree islands on slightly raised platforms. These elevated islands, or "hammocks," occur as a result of peat buildup in what was originally a depression, and they are abundant throughout the coastal fringe of the Florida Everglades.

Dwarf or scrub mangrove forests are dominated by small (<1.5 m in height), scattered trees and occur in areas characterized by environmental extremes (Fig. 13.7). The stunted trees are mature individuals, estimated, on the basis of leaf scar counts and leaf production rates, to be >40 yr old

(Feller 1995). Hydrologic energy is low in this type of forest, and restricted water movement and continuous flooding are characteristic. The resultant stagnation leads to reducing soil conditions, accumulation of sulfide, and in some cases hypersalinity. Although productivity per unit area is lower than that measured in other forest types, these low-stature forests can account for a large proportion of total mangrove area (e.g., in Belize) (Zisman 1992).

ZONATION

Zonation of species is a frequently observed characteristic of plant communities in habitats with strong physical or chemical gradients. In wetlands, spatial segregation of species often occurs in conjunction with elevational gradients that determine depth and duration of flooding and edaphic conditions influencing plant growth (Pielou and Routledge 1976; Vince and Snow 1984). Much work has centered on the role of abiotic factors as determinants of plant growth and distribution. Because the potential distributions (fundamental niches) of wetland plant species often overlap, biotic factors such as dispersal, competition, and herbivory may refine distributions so that boundaries between vegetation zones are relatively distinct.

Zonation of saltmarshes and mangrove forests

has been described for many regions, and although specific patterns vary with local conditions, relatively distinct and recurring patterns of species distribution are characteristic of these communities.

Saltmarsh

Zonation is a ubiquitous feature of most saltmarshes, a major exception being the saltmarshes of the delta plain of Louisiana where the elevational gradient is so shallow that zonation occurs over kilometers rather than meters. Because the species composition of saltmarshes varies with geographical location, the species dominating the various zones in the saltmarsh can differ greatly from one location in North America to another (Fig. 13.8). Saltmarsh vegetation zones occur along the elevational gradient from the seaward limit of the wetland to the terrestrial border. This elevational gradient is a complex gradient composed of multiple environmental factors that vary in time and space and not always in concert. The two most important abiotic factors that control zonation along this gradient are inundation and salinity. Saltmarsh species exhibit differential tolerances to these stressors. For example, *Spartina alterniflora*, a low-marsh dominant, is more flood-tolerant than *S. patens*, a high-marsh dominant (Gleason and Zieman 1981). However, the species's tolerance limits to both inundation and salinity overlap considerably so that, for example, where inundation and salinity stresses are minimal, many of these species could theoretically coexist. Thus, abiotic factors alone cannot completely explain the observed zonation in saltmarshes.

Pielou and Routledge (1976) proposed, based on a theoretical analysis, that competition controls species zonation at the generally less stressful landward boundary of the wetland and that abiotic factors control species pattern along the more stressful seaward end of the elevational gradient. This hypothesis has been rigorously tested for saltmarshes in New England where Bertness and Ellison (1987) have demonstrated that the distribution of *Spartina alterniflora* and *S. patens* is controlled by both environmental tolerances and competition. *Spartina patens* does not occur at the most seaward limit of the marsh because it cannot tolerate the inundation conditions. In contrast, *S. alterniflora* could exist in the higher elevation *S. patens* zone except that competition with *S. patens* excludes this species. As a result, competitive subordinates are displaced to the more stressful zones of the gradient, whereas competitive dominants occupy the more benign areas. Similar conclusions were drawn from a study in an Alaskan saltmarsh

with different species (Snow and Vince 1984). In this case, the fundamental niche of a number of salt marsh species was restricted by competitive interactions.

Disturbance in the form of wrack deposition or herbivory can also influence zonation patterns in saltmarsh. Bertness and Ellison (1987), for example, found that the pattern of species occurrence in a New England high marsh was generated by tidal deposition of large mats of dead plant material (wrack), causing differential plant mortality. *Spartina alterniflora* and *Distichlis spicata* are more tolerant of wrack burial than other marsh plants, and their relative abundance increases in disturbed areas. When the disturbance is more severe and of longer duration, all the underlying vegetation can be killed by the wrack and bare patches are generated. *Distichlis spicata*, *Salicornia europaea*, and *Spartina alterniflora* rapidly colonize these patches and dominate compared to adjacent nondisturbed areas. However, over time, these disturbance communities are out-competed and replaced by the surrounding communities of *Spartina patens* and *Juncus gerardii*. Furthermore, the success of *Salicornia europaea* in the bare matches is modulated by insect herbivory (Ellison 1987). Although *S. europea* is restricted to these disturbance-generated patches, the beetle *Erynephala maritima* reduces survivorship of *S. europea* to a greater extent in the patches than beneath the perennial canopy, thus affecting the dominance of this species. However, frequent wrack disturbance, especially in spring and early summer, allows this pattern mosaic to persist.

Mangrove

Zonation patterns in mangrove forests are similar to those in saltmarshes in that they vary with local conditions and species composition and exhibit recurring patterns (Fig. 13.9, p. 524). In any particular forest, the low tidal elevations will be dominated by a species different from that in the high intertidal zone. An idealized pattern was described by Davis (1940) in which the seaward position was dominated by red mangrove followed by a black mangrove–saltmarsh association and a buttonwood (*C. erectus*) transitional zone to a tropical, upland forest (Fig. 13.9a). Depiction of mangrove zones as monospecific bands of vegetation lying parallel to the shoreline is somewhat simplistic, however. Zonation patterns observed in different geographic regions support the idealized pattern that the shoreline or creekbank position is typically dominated by *Rhizophora*, but landward zones exhibit various patterns (Fig. 13.9b–e). In some in-

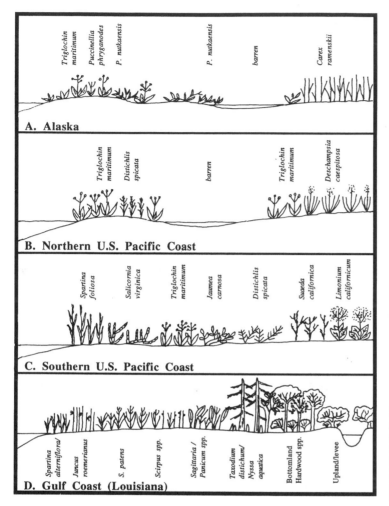

Figure 13.8. Examples of saltmarsh zonation patterns in different regions of North America. Profiles were re-drawn from Snow and Vince (1984), Mendelssohn and Marcellus (1976), Niering and Warren (1980), Dawes (1981), Pomeroy and Wiegert (1981), Stout (1984), and Knox (1986).

stances interior vegetation may be comprised of a mixture of co-occurring species, but in others the landward zones are occupied by monospecific stands. Monospecific stands of *A. germinans* are sometimes encountered landward of an *R. mangle*–dominated fringe (Fig. 13.9c). In other situations, extensive stands of scrub or dwarf *R. mangle* trees may occur in the landward position (Figs. 13.9d and e). These different patterns suggest that zonation in mangrove forests is probably best viewed as a mosaic.

In addition to differential species tolerance of environmental stress factors (McKee 1995a; McKee 1995b) and competition (Ball 1980), zonation of mangroves has been attributed to tidal sorting of propagules that differ in their size and buoyancy (Rabinowitz 1978). Work in Panama indicated that the small size and buoyancy of *A. germinans*, *A. bicolor*, and *L. racemosa* propagules should cause

them to be carried farther inland where they would be stranded at higher elevations. Conversely, larger and less buoyant propagules (*R. mangle*, *R. harrisonii*, and *P. rhizophorae*) would be expected to strand lower in the intertidal zone. Although the dispersal properties of these neotropical species matched the variation in intertidal dominance by the adult vegetation, the mechanism hypothesized to explain this relationship was never experimentally tested in Panama, and many exceptions can be found in other regions.

Seed predation, which is an important process in many plant communities, has been suggested as a major factor determining mangrove zonation patterns. Smith and coworkers (Smith 1987; Smith, Chan, McIvor, and Roblee 1989) reported an inverse relationship between propagule predation rates and conspecific dominance in some mangrove forests. Not all mangroves conform to this

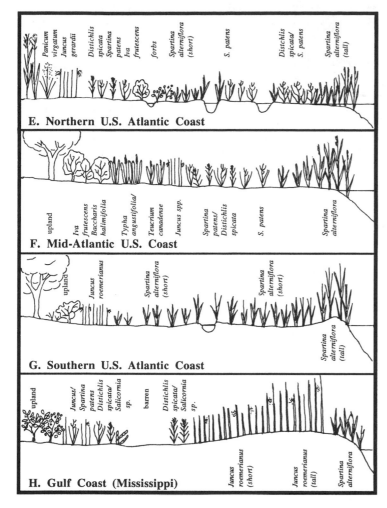

E. Northern U.S. Atlantic Coast

F. Mid-Atlantic U.S. Coast

G. Southern U.S. Atlantic Coast

H. Gulf Coast (Mississippi)

Figure 13.8 (cont'd)

predation-dominance hypothesis, however, and *Rhizophora mangle* in particular fails to support this model, suggesting that the relative importance of predators in structuring mangrove forests in North America is less than in other regions (McKee 1995c).

In summary, several biotic and abiotic factors have been identified that contribute to zonation of saltmarshes and mangroves. Although there is still some disagreement among researchers as to the relative role of various biotic and abiotic factors in determining specific zonation patterns, the consensus is that four basic processes may be involved: (1) differential dispersal and/or establishment of seeds or propagules; (2) differential susceptibility to predators;(3) differential abilities of species to tolerate stress factors that vary across the gradient; and (4) species differences in relative competitive ability. Dispersal properties, predation, and estab-

lishment abilities determine initial species distributions. After establishment, differential tolerance of stress factors such as salinity and flooding and competitive ability further refine vegetation zones.

SUCCESSION

Saltmarsh

Proponents of the Clementsian model of succession emphasized the unidirectional and autogenic nature of the process and used successional diagrams that culminated with a terrestrial forest climax (Clements 1916). Succession in marshes was often inferred from zonation patterns (Chapman 1974). For example, the zonation pattern shown in Figure 13.8d. suggested to early workers that the sequence of succession proceeds from salt to brackish, to

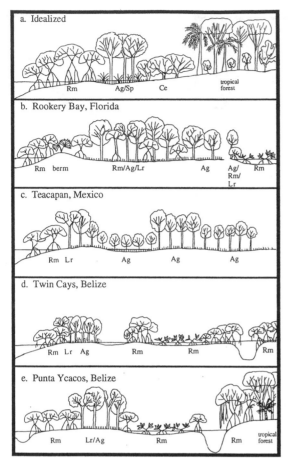

Figure 13.9. Examples of mangrove zonation patterns in tropical North America. The top panel shows an idealized pattern (redrawn from Davis 1940), and the other profiles represent patterns observed in different geographic locations (Flores-Verdugo et al. 1992; McKee 1993; and McKee unpublished data). Species depicted are Rhizophora mangle (Rm), Avicennia germinans (Ag), Laguncularia racemosa (Lr), Conocarpus erectus (Ce), and Spartina sp. (Sp).

as subsidence and salt water intrusion occur, the community succeeds to brackish, then saltmarsh communities. Also, there is little documentation that higher marsh communities have developed through a slow, autogenic-induced change from mudflat to high-marsh community (but see Ranwell 1972).

Thus, saltmarsh succession is best viewed as a Gleasonian process by which changes in vegetation occur as individual species respond over time to specific changes in environmental conditions (Gleason 1926). Vegetational change is not necessarily unidirectional nor controlled by the vegetation (although some autogenic change may occur), nor does it result in a terrestrial climax. Saltmarsh zonation is best interpreted as species sorting themselves out along an environmental gradient based on their environmental tolerance limits, competitive ability, and response to disturbance.

Mangrove

Mangrove zonation was also initially viewed as representing an autogenic succession in which the mangrove community was initiated by a pioneer species that built land in a seaward direction and was proceeding toward a terrestrial climax forest (Davis 1940; Chapman 1976). Other work provided evidence that mangroves were not "land builders" but instead responded to, rather than caused, coastal progradation (Egler 1952; Spackman, Dolsen, and Riegol 1966; Wanless 1974). Seaward migration of mangrove margins occurs in areas receiving large inputs of sediment, whereas retreat of shorelines may be observed in other locations. Succession does occur in mangrove forests, but it is not necessarily autogenic, unidirectional, or predictable. Mangrove forests are similar to saltmarshes in that they are steady-state, cyclical systems in which change is controlled by the balance between autogenic (e.g., peat accumulation) and allogenic (e.g., sediment transport, sea level change, disturbance) processes (Lugo 1980).

ABOVEGROUND AND BELOWGROUND BIOMASS AND PRODUCTIVITY

Saltmarsh

Coastal saltmarshes are one of the most productive ecosystems in the world, with rates of net primary productivity as high as 1.4–2.5 kg C m^{-2} yr^{-1} (Dring 1982). Primary producers include the emergent vascular vegetation, benthic and epiphytic algae, and phytoplankton. Few studies have quan-

fresh marsh, and finally to forested communities in Louisiana (Penfound and Hathaway 1938).

However, there are a number of exceptions to this model. Saltmarshes in particular, and wetlands in general, are relatively stable habitats that seldom, if ever, change over time to a terrestrial climax. Only a major environmental modification (e.g., a change in sea level, subsidence, glacial rebound, or tectonic uplifting) might precipitate such a change. In the example given above, we now know based on long-term studies and corings that marsh succession in the Mississippi River Delta plain occurs in the direction opposite that suggested by the zonation pattern (Fig. 13.8d). The delta is first colonized by freshwater species, and

Table 13.6. *Annual aboveground and belowground productivity* $(g \ C \ m^{-2} \ yr^{-1})$ *in Canadian and U.S. saltmarshes. N = number of observations; SE = standard error; $CI_{.05}$ = 95% confidence interval.*

Region	n	Aboveground Mean ± SE	$CI_{.05}$	n	Belowground Mean ± SE	$CI_{.05}$
Canadian Arctic[1]	1	4 ± 0	—	—	—	—
Canadian Subarctic[2]	4	67 ± 29	0–159	—	—	—
Canadian Boreal[3]	1	284 ± 0	—	—	—	—
North Atlantic[4]	18	505 ± 147	195–815	5	616 ± 224	0–1,238
Mid-Atlantic[4]	76	444 ± 51	343–545	29	1158 ± 146	859–1,457
South Atlantic[4]	38	707 ± 93	519–895	11	728 ± 198	287–1,169
Central Gulf[4]	43	812 ± 88	634–990	5	1,813 ± 423	639–2,987
Western Gulf[4]	10	391 ± 55	267–515	2	699 ± 260	0–4,002
Southern CA[4]	8	433 ± 119	152–714	—	—	—
Central CA[4]	12	448 ± 67	301–595	1	647 ± 0	—
North Pacific[4]	28	440 ± 40	358–522	—	—	—
Gulf of Alaska[4]	6	60 ± 26	0–127	4	71 ± 24	0–147
Beaufort Sea[4]	3	30 ± 15	0–95	3	41 ± 16	0–110

Sources: [1] Jefferies (1977); [2] Cargill and Jefferies (1984) and Glooschenko and Harper (1982), [3] Hatcher and Mann (1975), [4] Continental Shelf Associates, Inc. (1991). Conversion of grams dry weight to grams carbon was as noted in Continental Shelf Associates, Inc. (1991). North Atlantic = Maine to Massachusetts; Mid-Atlantic = Connecticut to mid-North Carolina; South Atlantic = mid–North Carolina to Florida; Central Gulf of Mexico = Alabama to Louisiana; Western Gulf of Mexico = Texas; North Pacific = Oregon to Washington.

tified the relative contribution of each of these sources of primary productivity. One investigation for the Duplin River marsh in Georgia demonstrated that the emergent vegetation (*Spartina alterniflora*) contributed 84% of the total primary productivity, the benthic algae 10%, and phytoplankton 6% (Pomeroy et al. 1981). However, algal productivity in some saltmarshes (e.g., hypersaline marshes of southern California where light penetration to the soil surface is relatively great because of a sparse emergent plant canopy) can account for 40–60% of the total marsh primary productivity (Zedler 1980). Most research, however, has concentrated on quantifying the productivity of the emergent vascular vegetation.

A comparison of aboveground and belowground net annual primary productivities of coastal saltmarshes in the United States and Canada is provided in Table 13.6 (Continental Shelf Associates 1991 and references therein). Although rates of primary productivity may vary depending on the methodology, plant species, environmental conditions, latitude, grazing pressure, or annual variation, a latitudinal gradient in primary productivity is apparent. Arctic saltmarshes have extremely low aboveground primary productivities (e.g., 10 g m^{-2} y^{-1}) (Jefferies 1977). The primary productivities of subarctic saltmarshes in Alaska are, on average, only marginally greater (76–150 g m^{-2} y^{-1}). Farther south along the western coast of

North America, saltmarsh aboveground productivity overlaps considerably from Washington to southern California (1100–1363 g m^{-2} y^{-1}). However, aboveground productivity increases by an order of magnitude from Alaska. Although considerable overlap is also seen along the eastern coast of North America, productivities generally increase in a southward direction with the exception of the western Gulf of Mexico (1221–1976 g m^{-2} y^{-1}). Belowground rates of production are more highly variable, possibly as a result of smaller sample sizes, than for aboveground estimates as well as greater inherent variation in the data (Table 13.6). Regardless, two points can be made: Belowground productivities are as high or higher than aboveground, and no clear latitudinal gradient is apparent, although belowground productivity in Alaska is an order of magnitude lower than elsewhere (Table 13.6).

Factors controlling productivity of coastal saltmarshes dominated by *Spartina alterniflora*, the dominant intertidal saltmarsh plant along the Atlantic and Gulf coasts of North America, have been extensively reviewed (Mendelssohn et al. 1982; Smart 1982; Howes, Dacey, and Goehringer 1986). In general, the primary factors determining the growth of this species are salinity and soil waterlogging, both of which affect use and allocation of nitrogen by plants. Prolonged flooding results in soil anoxia, biochemically reduced soil conditions, and the accumulation of hydrogen sulfide in coastal saltmarshes (DeLaune et al. 1983; Mendelssohn and McKee 1988). Soil anaerobiosis and phytotoxin accumulation inhibit the uptake of ammonium-nitrogen, the primary nutrient limiting plant growth in these systems (Morris 1984; DeLaune, Smith, and Tolley 1984; Koch et al. 1990). Additionally, the roots may become deficient in oxygen and exhibit limited aerobic respiration and reduced energy for nutrient uptake (Mendelssohn, McKee, and Patrick 1981; Koch et al. 1990). Hydrogen sulfide accumulation further inhibits root energy production and, hence, exacerbates plant nitrogen deficiencies (Koch et al. 1990). Elevated salinities can also negatively affect the growth of S. alterniflora by competitively inhibiting ammonium uptake (Morris 1984). In addition, nitrogen-containing plant osmotica that aid in maintaining plant water status are synthesized at elevated salinities; the allocation of nitrogen to osmotica production decreases the amount available for growth (Cavalieri and Huang 1979). Considerably less is known about the factors controlling the production of other saltmarsh plant species. However, since hydrology and salinity are recognized as primary forcing functions in coastal saltmarsh systems (Mitsch and Gosselink 1993), the primary productivity of other marsh species are likely controlled, at least qualitatively, by similar factors.

In addition to environmental controls, biotic factors may also influence saltmarsh productivity. Bertness (1984, 1985) has shown that both fiddler crabs and mussels can enhance the production of Spartina alterniflora, the former by aerating the soil and the latter by increasing soil fertility through the production of feces and pseudo feces. Moderate grazing by snow geese in Canadian saltmarshes also stimulates vascular plant primary production via the input of nitrogen from feces (Hik and Jefferies 1990). There is also evidence for autogenic control of plant productivity. The accumulation of peat by the vegetation over time can inhibit plant growth, possibly as a result of the increased hardness of the substrate and/or the lower fertility of a peaty soil (Bertness 1988). These biotic controls on saltmarsh plant production have generally been overlooked and require verification in other marsh types.

Mangrove

Accurate determination of standing biomass and net primary production is exceedingly difficult in mangrove forests for obvious reasons. Nonetheless, attempts have been made to estimate these values in various forests using both direct and in-

direct methods. Aboveground biomass may range from 8 metric tons ha^{-1} in dwarf stands (Lugo and Snedaker 1974) to >279 metric tons ha^{-1} in moist, hurricane-free forests in Darien, Panama (Golley, McGinniss, Clements, Child, and Duever 1975). Production estimates are derived from four methods: litterfall rates, gas exchange rates, changes in tree diameter, and harvest of trees of known age. The most comprehensive data set is based on litterfall rates, which vary from 2 (dwarf forest) to 13 (riverine forest) metric tons ha^{-1} yr^{-1}; these data suggest that maximum production occurs in the 0–20° latitudes and decreases toward the subtropics. Worldwide rates of net aboveground production range from near zero in dwarf forests to >45 metric tons ha^{-1} in riverine forests. Net primary production values for North American mangrove forests estimated from litterfall data (Table 13.7) show R. mangle to be the highest net producer, followed by A. germinans and L. racemosa (Lugo, Evink, Brinson, Brace, and Snedaker 1975). This comparison, however, reflects differences in environmental constraints on growth present in the areas dominated by each species in addition to interspecific differences in inherent production potential. Another consideration is that litterfall rates are not necessarily accurate reflections of net primary production.

Limited data exist for belowground biomass, and no accurate information is available for belowground production. Estimates of belowground biomass range from 8 metric tons ha^{-1} in Florida (Lugo and Snedaker 1974) to 190 metric tons ha^{-1} in Panama (Golley et al. 1975) and average 66 ± 43 metric tons ha^{-1} (n = 4). Aboveground to belowground biomass ratios range from 0.6–1.5, but insufficient information prevents any broad conclusions regarding biomass allocation patterns.

REGENERATION STRATEGIES

Saltmarsh

Unlike mangroves, true vivipary has not been found for saltmarsh species. Adam (1990) reported that in some saltmarsh species germination may occur in the infructescence, but this is not common. Plantlets may develop in place of spikelets in a number of saltmarsh grasses, although the phenomenon is rare and is not characteristic of many saltmarsh species. Thus, seeds of saltmarsh plants, like seeds of most plant species, must be dispersed from the parent plant, occupy a favorable site, survive a dormancy period, and germinate before establishment may proceed.

Seeds of saltmarsh species, which remain viable

Table 13.7. *Annual litterfall rates and net aboveground productivity for mangrove vegetation in North America*

Region	n	Litterfall (g OM m^{-2} yr^{-1})		Net primary production (g C m^{-2} yr^{-1})	
		Mean ± SE	CI$_{.05}$	Mean ± SE	CI$_{.05}$
Florida	29	832 ± 77	672–1000	909 ± 85	736–1081
Mexico	10	1179 ± 66	1033–1325	1291 ± 72	1130–1449
Puerto Rico	3	1026 ± 226	51–2004	1125 ± 249	55–2191
Cuba*	1	1059	—	1161	—
El Salvador*	1	993	—	1087	—
Panama*	?	(902–2000)	—	987–2183	—
Total	46	956 ± 62	832–1077	1045 ± 66	911–1178

Notes: [a]n = number of observations.
[b]* = based on a single value only.
Sources: For litterfall include Pool et al. (1975), Continental Shelf Associates (1991), Flores-Verdugo et al. (1992), D'Croz (1993), and Lacerda et al. (1993). Net aboveground primary production was estimated by assuming a litterfall:NPP ratio of 0.53 and a litter carbon content of 58%.

for extended periods in seawater, are probably most commonly dispersed by tides and currents. In addition, birds may carry seeds either externally or internally (Proctor 1968). Although seedling establishment is typically rare in established saltmarshes because of the relatively closed nature of the canopy, recruitment on unvegetated sandflats and mudflats is generally by seedling establishment. Vegetative fragments may also serve an important role in establishment on unvegetated flats. Storm events, herbivore action, ice rafting, and other events may cause fragments of marsh vegetation or whole sections of the marsh to be detached and transported by water to unvegetated shorelines where they may become rooted.

In contrast to fresh marshes, which generally have extensive soil seed banks, the seed bank of saltmarsh soils is often depauperate, especially in *Spartina alterniflora* marshes. However, the limited seed banks in these marshes can be important in marsh recovery from disturbance. For example, the seed bank of a salt marsh in California had much higher seed densities in a disturbed site compared to an undisturbed location (Hopkins and Parker 1984). The longevity of seeds in the seed bank is undoubtedly species-specific but requires further study.

In most saltmarshes, clonal growth is the primary mode of plant lateral expansion and regeneration within the marsh. Seedling establishment, even if germination occurs, is unlikely within the relatively closed canopy of many saltmarshes. For example, although seedlings of *S. alterniflora* were common in a mature marsh in New England, they did not survive the growing season (Metcalfe, El-lison, and Bertness 1986). Clonal growth provides an advantage of physiological integration between ramets whereby carbohydrates, for example, can be translocated from parent to daughter ramets and provide an energy source for growth. Additionally, the ability of the daughter ramets to invade more stressful sites within the marsh may be enhanced by the translocation of growth-limiting resources (e.g., water and nutrients) from parent to daughter ramets (Evans 1988; Hester et al. 1994). Clonal integration may also aid in allowing species to overcome such short-term fluctuations in environmental condition as transient variations in salinity and inundation.

Mangrove

As discussed previously, the stumps of some mangrove species have the capacity to sprout if the shoot is killed or removed. However, this process simply replaces that individual tree and generally does not constitute a means of propagation. In addition, coppiced shoots do not always survive, particularly where stress factors such as high sulfide concentrations prevail (McKee unpublished data). Mangroves are, therefore, primarily dependent on seedling recruitment for forest maintenance and spread. Conditions in the mangrove habitat (e.g., low light, high sedimentation rates, high salinity, and low oxygen) would normally inhibit seed germination. Some mangroves have circumvented this limitation through vivipary, in which development is continuous from fertilization through germination and there is no resting stage represented by the seed in other species. This strategy is exemplified by *R. mangle* in which germination occurs on the tree, and the seedling axis elongates through growth of the hypocotyl over a 4–6 mo period. At maturity, the seedling detaches and falls into the water, where it is ready to take root as soon as it comes into contact with the soil. Mangroves belonging to the genus *Avicennia* are considered to be cryptoviviparous – that is the embryo emerges from the seed coat but not from the fruit before detachment from the parent. Other species such as *L. racemosa* exhibit precocious germination in which emergence of the radicle often occurs during dispersal. The dispersal unit, which is typically referred to as a propagule, is buoyant and dispersed by tidal action. In order to become established, a mangrove propagule must strand in a location where it is undisturbed long enough to allow rooting to anchor it in place (Rabinowitz 1978; McKee 1995b).

DISTURBANCE

Saltmarsh

Disturbance in the form of herbivory, hurricane and storm erosion, ice scouring, and wrack deposition can result in bare patches in the marsh and allow for species recruitment into the saltmarsh. For example, bare patches created by wrack deposition in New England high marshes rapidly become hypersaline due to direct exposure to the sun, soil water evaporation, and infrequent tidal leaching of accumulated salts in the soil (Bertness 1992). Few species can colonize these hypersaline zones; however, *Salicornia europaea* and *Distichlis spicata* are highly salt-tolerant and do establish in these areas, *Salicornia* from seeds attached to wrack (Ellison 1987) and *Distichlis* from runners originating from outside the bare patches. These species shade the soil, reduce evaporation, and ameliorate salinity. As salinity decreases, *Juncus gerardii* and *Spartina patens* can invade, out-compete the initial colonizers, and dominate the patches after 2–4 yr. By reducing soil salinities in the disturbed areas, the initial colonizers facilitate succession to more competitive dominants. Because disturbance by wrack recurs periodically, the patch mosaic of vegetative communities characteristic of these marshes generally persists. Biotic disturbances can also modify marsh structure. Grazing by snow geese, for example, affected the species composition of a Canadian Arctic saltmarsh by promoting the dominance of *Puccinellia phryganoides* (Bazely and Jefferies 1986). When grazing was prevented, the vegetative structure rapidly returned to that characteristic of ungrazed marshes.

Larger-scale physical disturbances from storms

and hurricanes can also have dramatic effects. Fires, natural or human-induced (Fig. 13.10a), can retard plant succession and modify species composition at least in the short term (Kirby, Lewis, and Sexton 1988). Human disturbances have had the greatest negative impacts on wetlands in general and on saltmarshes in particular. In many cases, the saltmarsh habitat has been completely eliminated – for example, through dredge and fill operations (Fig. 13.10b). Between the mid-1970s and the mid-1980s, approximately 28,700 ha of intertidal estuarine emergent vegetation in the conterminous United States have been lost (Dahl, Johnson, and Frayer 1991). California has lost 91% of its original wetlands during the past 200 years (Dahl 1990). In Louisiana, which has 41% of the coastal wetlands in the conterminous United States, the dredging of canals for oil- and gas-related activities had modified the coastal landscape and resulted in direct conversion of wetland to open water (Turner 1990). Hurricane Andrew, which made landfall in Louisiana on August 24, 1992, literally carried away sections of saltmarsh and deposited them in shallow ponds and bays (Fig. 13.10c).

Mangrove

Although the scale and frequency of disturbance certainly influence patterns of mangrove forest structure and productivity, little information currently exists demonstrating this relationship. Disturbance in mangrove forests may be caused by large-scale events such as hurricanes, frost damage, or clear-cutting but also by small-scale events such as lightning strikes or attack by wood-boring beetles, which cause death of individual trees or small groups of trees (Lugo and Patterson-Zucca 1977; Feller 1994; Smith, Robblee, Wanless, and Doyle 1994) (Fig. 13.11). The relative importance of these different types of disturbance varies geographically, with some forests being more susceptible to frost, xylovores (wood-feeding insects), or lightning strikes than others. Mangrove forests in Florida were devastated by Hurricane Andrew in 1992, and tree mortality varied spatially and among mangrove species (Smith et al. 1994; Baldwin, Platt, Gathen, Lessmann, and Rauch 1995) (Fig. 13.11a). Recovery from large-scale disturbance may be slow and probably varies depending on species composition and intensity of stress factors subsequent to the disturbance event. Long-term monitoring of a clear-cut mangrove forest in Belize demonstrated that removal of the vegetation was followed by changes in physicochemical conditions—for example, increases in irradiance and soil temperature, decreases in soil redox potential, and

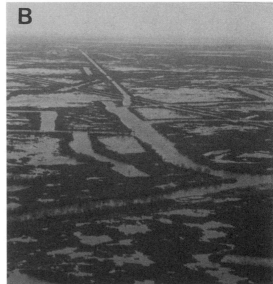

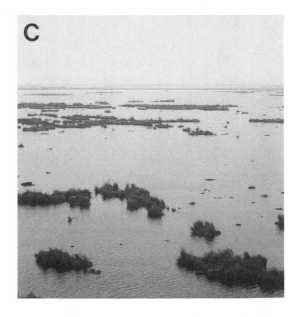

Figure 13.10. Examples of disturbance in Louisiana salt-marshes. (A) Fire may occur naturally but is more often set by humans; (B) Canals dredged for navigation and oil exploration cause major disruptions in marsh hydrology; and (C) Sections of saltmarsh were uprooted and moved by Hurricane Andrew.

increases in salinity and sulfide concentrations that inhibited regeneration (McKee and Feller, 1994a) (Fig. 13.11b).

Biotic agents of disturbance were unrecognized until recently when Feller (1994) reported that up to 30% of the canopy in a red mangrove forest in Belize was removed by the activities of wood-boring beetles that kill branches and even entire trees. The larvae of the beetles are active in the phloem and outer xylem, creating extensive feeding galleries that are still evident in the standing deadwood and fallen litter for many years. The activity of a single larva can ultimately girdle a branch or bole, resulting in a slow thinning of the canopy and terminating in the creation of a light

gap and standing deadwood that is secondarily invaded by other xylovores (Fig. 13.11c).

AREAS FOR FUTURE RESEARCH

Although thousands of publications representing a variety of descriptive and experimental studies of saltmarshes and mangroves exist, much remains to be understood about these intertidal systems. The following discussion of research topics is not meant to be exhaustive but only suggestive of the variety of areas that require further work.

Considerable empirical data have given us an understanding of the controls on species zonation and the role of stress and disturbance in New En-

Figure 13.11. Examples of different types and scales of disturbance in mangrove forests. (A) a mangrove forest near Everglades City, Florida damaged by Hurricane Andrew;

(B) a Belize island forest clear-cut for a tourist resort; (C) a red mangrove tree killed by a wood-boring beetle.

gland saltmarshes (Bertness 1992). Such research should be expanded to other marsh types as well as to mangroves to test the generality of mechanisms causing patterns in these systems. An understanding of pattern development requires more information about regeneration strategies and relationships to resource availability and nonresource stress factors such as salinity and flooding. Interactions of predicted global change – for example, elevated CO_2, altered precipitation patterns, or sea level rise – must be empirically examined to generate more accurate predictions of future impacts on community structure and succession. Studies relating climatic data to structural attributes would also be particularly informative in understanding geographic variation and in predicting response to specific changes in climate. Areas dom-

inated by *S. alterniflora* or *R. mangle* have been emphasized in productivity studies, but marsh and mangrove communities dominated by other species should also be investigated. Also, the contributions of algae and other primary producers to total primary production have been generally neglected and require further investigation. Most productivity studies have concentrated on aboveground production, but belowground production is at least equally important and needs further quantification, especially in the context of biomass partitioning and response to environmental stressors.

Quantifying the effects of plants on biotic and abiotic conditions in saltmarshes and mangroves requires additional work. The importance of saltmarsh and mangrove ecosystems as sinks and sources for greenhouse gases is also of significance. The factors that control the differential value of these systems for support of food chain and habitat need clarification for a variety of saltmarsh and mangrove community types (e.g., high vs. low marsh; riverine vs. scrub forest). Work on plant–animal interactions is limited but suggests that they may have large impacts on the structure and function of these systems. Direct herbivory of saltmarsh or mangrove tissues is assumed to be low and relatively unimportant, but a few studies have shown that specialized insects (e.g., bud moths and phloem feeders) can have potentially large effects on production or nutrient cycling by causing leaf abortion, deformed leaves, or premature leaf abscission. Regeneration rates and patterns may also be influenced by herbivores – for example, by stem-boring insects that kill mangrove seedlings or pollen-feeding insects that have an impact on seed production by *S. alterniflora* (Bertness and Shumway 1992). The role of herbivores, particularly xylovores, and their contribution to heterogeneity in mangrove forests through creation of standing deadwood is a topic that has great potential. Quantitative information about frequency and patterns of disturbance and impacts on saltmarsh and mangrove vegetation is needed. Detailed information about changes in environmental factors following disturbance would contribute to an understanding of mechanisms controlling regeneration rates and patterns. Data on gap dynamics or shade and sun tolerance characteristics of mangrove species in North America are also needed.

A final important point is that the answers to questions such as these will require information from more than one geographic region. Research carried out in a diversity of locations is essential to the achievement of a more comprehensive and representative body of data and to the conservation of these systems. The protection and proper management of saltmarshes and mangroves require basic information from each country or geographic region about areal extent; inventories of flora and fauna; soils, hydrology, and physicochemical factors; structure, biomass distribution, and net primary productivity; and natural and anthropogenic impacts. Multinational or multiregional projects that promote interactions among scientists, managers, and policymakers will lead to a synthesis of information with a broad scope of inference. Equally important are environmental education programs that stimulate public interest in and support of conservation efforts. These approaches will help to ensure the continued existence of these unique and valuable ecosystems.

REFERENCES

Adam, P. 1990. Saltmarsh ecology. Cambridge University Press, Cambridge.

Alexander, C. E., M. A. Broutman, and D. W. Field. 1986. An inventory of coastal wetlands of the USA. National Oceanic and Atmospheric Administration Washington, D.C.

Alvarez, V. 1994. Los manglares de la Republica Dominicana, pp. 209–217 in D. O. Suman (ed.), El ecosystema de manglar en America Latina y la cuenca del Caribe: su manejo y conservacion. University of Miami, Fla.

Aragon de Rendon, B. B., A. E. Barrios A., and L. M. De Leon Gamboa. 1994. Los manglares de Guatemala, pp. 125–132 in D. O. Suman (ed.).

Aubrey, D. G., and K. O. Emery. 1993. Recent global sea levels and land changes, pp. 45–46 in R. A. Warrick, E. M. Barrow and T. M. L. Wigley (eds.), Climate and sea level change: observations, projections and implications. Cambridge University Press, Cambridge.

Bacon, P. R. 1993. Mangroves in the Lesser Antilles, Jamaica, Trinidad & Tobago, pp. 155–209 in L. D. Lacerda (ed.), Conservation and sustainable utilization of mangrove forests in Latin America and Africa regions. International Society for Mangrove Ecosystems, Okinawa.

Baldwin, A. H., W. J. Platt, K. L. Gathen, J. M. Lessmann, and T. J. Rauch. 1995. Hurricane damage and regeneration in fringe mangrove forests of southeast Florida USA. J. Coastal Res. SI 21:169–183.

Ball, M. C. 1980. Patterns of secondary succession in a mangrove forest of southern Florida. Oecologia 44: 226–235.

Ball, M. C. 1988. Ecophysiology of mangroves. Trees 2: 129–142.

Barbour, M. G. 1970. In any angiosperm an obligate halophyte? Amer. Midl, Natur. 84:105–120.

Bazely, D. R., and R. L. Jefferies. 1986. Changes in the composition and standing crop of salt-marsh communities in response to the removal of a grazer. J. Ecol. 74:693–706.

Bertness, M. D. 1984. Ribbed mussels and *Spartina alterniflora* production in a New England salt marsh. Ecology 65:1794–1807.

Bertness, M. D. 1985. Fiddler crab regulation of *Spartina alterniflora* production in a New England salt marsh. Ecology 66:1042–1055.

Bertness, M. D. 1988. Peat accumulation and the success of marsh plants. Ecology 69: 703–713.

Bertness, M. D. 1992. The ecology of a New England salt marsh. Am. Sci. 80:260–268.

Bertness, M. D., and A. M. Ellison. 1987. Determinants of pattern in a New England salt marsh plant community. Ecol. Monogr. 57:129–147.

Bertness, M. D., and S. W. Shumway. 1992. Consumer driven pollen limitation of seed production in marsh grasses. Am. J. Bot. 79:288–293.

Bradley, P. M., and J. T. Morris. 1990. Influence of oxygen and sulfide concentration on nitrogen uptake kinetics in *Spartina alterniflora*. Ecology 71:282–287.

Broome, S. W., W. W. Woodhouse, Jr., and E. D. Seneca. 1975. The relationship of mineral nutrients to growth of *Spartina alterniflora* in North Carolina. II. The effects of N, P, and Fe fertilizers. Soil Sci. Soc. Amer. Proc. 39:301–307.

Cargill, S. M., and R. L. Jefferies. 1984. Nutrient limitation of primary production in a sub-arctic salt marsh. J. Appl. Ecol. 21:657–668.

Carrera, L. M., and A. P. Santander. 1994. Los manglares de Cuba: ecologia, pp. 64–75 in D. O. Suman (ed.), El ecosystema de manglar en America Latina y la cuenca del Caribe: su manejo y conservacion. University of Miami, Fla.

Cavalieri, A. J., and A. H. C. Huang. 1979. Accumulation of proline and glycinebetaine in Spartina alterniflora Loisel. in response to NaCl and nitrogen in the marsh. Oecologia 49:224–228.

Chabreck. 1972. Vegetation, water and soil characteristics of the Louisiana coastal region. Bulletin No. 664, Agricultural Experiment Station, Louisiana State University.

Chapman, V. J. 1974. Salt marshes and salt deserts of the world. 2nd ed. Cramer, Lehre.

Chapman, V. J. 1976. Coastal vegetation. Pergamon, Oxford.

Cintron-Molero, G., and Y. Schaeffer-Novelli. 1992. Ecology and management of New World mangroves, pp. 233–258 in U. Seeliger (ed.), Coastal plant communities of Latin America. Academic Press, San Diego.

Clements, F. E. 1916. Plant succession: analysis of the development of vegetation. Carnegie Institute of Washington, Washington, D.C.

Continental Shelf Associates. 1991. A comparison of marine productivity among outer continental shelf planning areas. A final report for the U.S. Department of the Interior, Minerals Management Service, Herndon, VA. OCS Study MMS 91-0001.

Cooper, A. W., and E. D. Waits. 1973. Vegetation types in an irregularly flooded salt marsh on the North Carolina outer banks. J. Elisha Mitchell Sci. Soc. 89: 78–91.

Costa, C. S. B., and A. J. Davy. 1992. Coastal saltmarsh communities of Latin America, pp. 179–199 in U. Seeliger (ed.), Coastal plant communities of Latin America. Academic Press, San Diego.

Dahl, T. E. 1990. Wetlands losses in the United States, 1780's to 1980's. U.S. Department of Interior, Fish and Wildlife Service, Government Printing Office, Washington, D.C.

Dahl, T. E., C. E. Johnson, and W. E. Frayer. 1991. Status and trends of wetlands in the conterminous United States, mid-1970's to mid-1980's. U.S. Department of Interior, Fish and Wildlife Service. Government Printing Office, Washington, D.C.

Davis, J. H. 1940. The ecology and geologic role of mangroves in Florida. Publication 517. Carnegie Institute, Washington, D.C.

Dawes, C. J. 1981. Marine botany. Wiley, New York.

D'Croz, L. 1993. Status and uses of mangroves in the Republic of Panama, pp. 115–127 in L. D. Lacerda (ed.), Conservation and sustainable utilization of mangrove forests in Latin America and Africa regions. International Society for Mangrove Ecosystems, Okinawa.

DeLaune, R. D., C. J. Smith, and W. H. Patrick, Jr. 1983. Relationships of marsh elevation, redox potential, and sulfide to Spartina alterniflora productivity. Soil Sci. Soc. Am. J. 47:930–935.

DeLaune, R. D., C. J. Smith, and M. D. Tolley. 1984. The effect of sediment redox potential on nitrogen uptake, anaerobic root respiration and growth of Spartina alterniflora Loisel. Aquat. Bot. 18:223–230.

Donn, W. L., W. R. Farrand, and M. Ewing. 1962. Pleistocene ice volumes and sea level lowering. J. Geol. 70:206–214.

Dring, M. J. 1982. The biology of marine plants. Thomson Litho Ltd., East Kilbride, Scotland.

Duncan, W. H. 1974. Vascular halophytes of the Atlantic and Gulf Coasts of North America North of Mexico, pp. 23–50 in R. J. Reimold and W. H. Queen (eds.), Ecology of halophytes. Academic Press, New York.

Egler, F. A. 1952. Southeast saline Everglades vegetation, Florida, and its management. Vegetatio 3:213–265.

Ellison, A. M. 1987. Effects of competition, disturbance, and herbivory on Salicornia europaea. Ecology 68:576–586.

Ellison, J. C. 1993. Mangrove retreat with rising sea-level, Bermuda. Estuarine, Coastal Shelf Sci. 37:75–87.

Evans, J. P. 1988. Nitrogen translocation in a clonal dune perennial Hydrocotyle bonariensis. Oecologia 77: 64–68.

Feller, I. C. 1994. Role of wood-boring insects in mangroves. Second Coastal Wetland Ecology and Management Symposium, Key Largo, Fla., USA (abstract).

Feller, I. C. 1995. Effects of nutrient enrichment on growth and herbivory of dwarf red mangrove (Rhizophora mangle). Ecol. Monogr. 65(4):477–505.

Flores-Verdugo, F., F. Gonzalez-Farias, D. S. Zamorano, and P. Ramirez-Garcia. 1992. Mangrove ecosystems of the Pacific coast of Mexico: distribution, structure, litterfall, and detritus dynamics, pp. 269–288, in U. Seeliger (ed.), Coastal plant communities of Latin America. Academic Press, San Diego.

Funes, C. A. 1994. Situacion de los bosques salados en El Salvador, pp. 115–124 in D. O. Suman (ed.), El ecosystema de manglar en America Latina y la cuenca del Caribe: su manejo y conservacion. University of Miami, Fla.

Gambrell, R. P., and W. H. Patrick, Jr. 1978. Chemical and microbiological properties of anaerobic soils and sediments, pp. 375–423, in D. D. Hook and R. M. M. Crawford, (eds.), Plant life in anaerobic environments. Ann Arbor Sci. Pub. Inc., Ann Arbor, Mich.

Garcia, N. H., and J. J. Camacho. 1994. Informe sobre manglares de Nicaragua, Central America, pp. 160–167 in D. O. Suman (ed.), El ecosystema de manglar en America Latina y la cuenca del Caribe: su manejo y conservacion. University of Miami, Fla.

Gleason, H. A. 1926. The individualistic concept of the plant association. Bull. Torr. Bot. Club 53:7–26.

Gleason, M. L., D. A. Elmer, and N. C. Pien. 1979. Effects of stem density upon sediment retention by salt marsh grass, Spartina alterniflora Loisel. Estuaries 2:271–273.

Gleason, M. L., and J. C. Zieman. 1981. Influence of tidal inundation on internal oxygen supply of Spartina alterniflora and Spartina patens. Estuarine Coastal Shelf Sci. 13: 47–57.

Glooschenko, W. A. 1982. Salt marshes of Canada, pp. 1–8 in R. E. Turner, B. Gopal, R. G. Wetzel, and D. F. Whigham (eds.), Wetlands ecology and management. International Scientific Publications, Jaipur, India.

Glooschenko, W. A., and N. S. Harper. 1982. Net aerial

primary production of a James Bay, Ontario, salt marsh. Can. J. Bot. 60: 1060–1067.

Glooschenko, W. A., and I. P. Martini. 1981. Salt marshes of the Ontario coast of Hudson Bay, Canada. Wetlands 1: 9–18.

Glooschenko, W. A., C. Tarnocal, S. Zoltai, and V. Glooschenko. 1993. Wetlands of Canada and Greenland, pp. 415–514 in D. Dykyjova, D. F. Whigham, and S. Hejny (eds.), Wetlands of the world I: Inventory, ecology and management. Kluwer Academic, Dordrecht, The Netherlands.

Golley, F. B., J. T. McGinnis, R. G. Clements, G. I. Child, and M. J. Duever. 1969. The structure of tropical forests in Panama and Colombia. Bioscience 19: 693–698.

Golley, F. B., J. T. McGinnis, R. G. Clements, G. I. Child, and M. I. Duever. 1975. Mineral cycling in a tropical moist forest ecosystem. University of Georgia Press, Athens.

Hall, J. V. 1991. Alaska coastal wetlands survey, pp. 1–15 in H. S. Bolton (ed.), Coastal wetlands. American Society of Civil Engineers, New York.

Hatcher, B. G., and K. H. Mann. 1975. Above-ground production of marsh cordgrass (*Spartina alterniflora*) near the northern end of its range. J. Fish. Res. Bd. Can. 32: 83–87.

Hester, M. W., K. L. McKee, D. M. Burdick, F. M. Flynn, M. S. Koch, S. Patterson, and I. A. Mendelssohn. 1994. Clonal integration in *Spartina patens* across a nitrogen and salinity gradient. Can. J. Bot. 72: 767–770.

Hik, D. S., and R. L. Jefferies. 1990. Increases in the net above-ground primary production of a salt-marsh forage grass: a test of the predictions of the herbivore-optimization model. J. Ecol. 78: 180–185.

Hinde, H. P. 1954. Vertical distribution of salt marsh phanerogams in relation to tidal levels. Ecolog. Monogr. 24: 209–225.

Hopkins, D. R., and V. T. Parker. 1984. The study of a seed bank of a salt marsh in northern San Francisco Bay. Am. J. Bot. 71: 348–355.

Howes, B. L., J. W. H. Dacey, and D. D. Goehringer. 1986. Factors controlling the growth form of *Spartina alterniflora*: feedbacks between above-ground production, sediment oxidation, nitrogen and salinity. J. Ecol. 74: 881–898.

Jefferies, R. L. 1977. The vegetation of salt marshes at some coastal sites in Arctic North America. J. Ecol. 65: 661–672.

Jimenez, J. A. 1992. Mangrove forests of the Pacific coast of Central America, pp. 259–267 in U. Seeliger (ed.), Coastal plant communities of Latin America. Academic Press, San Diego.

Josselyn, M. 1983. Ecology of San Francisco Bay tidal marshes: a community profile. U.S. Department of Interior, Fish and Wildlife Service USFWS-085-83-23. Washington, D.C.

Johnston, I. M. 1924. Expedition of the California Academy of Sciences to the Gulf of California in 1921, The Botany (The Vascular Plants). Proc. Calif. Acad. Sci., 4th Ser., 12: 951–1118.

King, G. M., M. J. Klug, R. G. Weigert, and A. G. Chalmers. 1982. Relation of soil water movement and sulfide concentration to *Spartina alterniflora* production in a Georgia, U.S.A., salt marsh. Science 218: 61–63.

Kirby, R. E., S. J. Lewis, and T. N. Sexson. 1988. Fire in North American wetland ecosystems and fire –

wildlife relations: an annotated bibliography. U.S. Department of Interior Fish and Wildlife Service, Biol. Rep. 88(1) Washington, D.C.

Knox, G. A. 1986. Estuarine ecosystems: A systems approach. CRC Press, Boca Raton, Fla.

Koch, M. S., and I. A. Mendelssohn. 1989. Sulfide as a soil phytotoxin: Differential responses in two marsh species. J. Ecol. 77: 565–578.

Koch, M. S., I. A. Mendelssohn, and K. L. McKee. 1990. Mechanism for the hydrogen sulfide – induced growth limitation in wetland plants. Limnol. and Oceanogr. 35: 399–408.

Kraeuter, J. 1976. Biodeposition by salt-marsh invertebrates. Mar. Biol. 35: 215–223.

Lacerda, L. D., J. E. Conde, P. R. Bacon, C. Alarcon, L. D'Croz, B. Kjerfve, J. Polania, and M. Vannucci. 1993. Mangrove ecosystems of Latin America and the Caribbean: a summary, pp. 1–42 in L. D. Lacerda (ed.), Conservation and sustainable utilization of mangrove forests in Latin America and Africa regions. International Society for Mangrove Ecosystems, Okinawa.

Lopez-Portillo, J., and E. Ezcurra. 1989. Zonation in mangrove and salt marsh vegetation at Laguna de Micoacan, México. Biotropica 21: 107–114.

Loza, E. L. 1994. Los manglares de México: sinopsis general para su manejo, pp. 144–151 in D. O. Suman (ed.), El ecosistema de manglar en America Latina y la cuenca del Caribe: su manejo y conservacion. University of Miami, Fla.

Lugo, A. E. 1980. Mangrove ecosystems: successional or steady state? Biotropica 12: 65–73.

Lugo, A. E., M. M. Brinson, and S. Brown. 1989. Forested wetlands. Ecosystems of the world. Elsevier, Amsterdam.

Lugo, A. E., G. Evink, M. M. Brinson, A. Broce, and S. C. Snedaker. 1975. Diurnal rates of photosynthesis, respiration and transpiration in mangrove forests of south Florida, pp. 335–350 in F. B. Golley and E. Medina (eds.), Tropical ecological systems. Springer-Verlag, New York.

Lugo, A. E., and C. Patterson-Zucca. 1977. The impact of low temperature stress on mangrove structure and growth. Trop. Ecol. 18: 149–161.

Lugo, A. E., and S. C. Snedaker. 1974. The ecology of mangroves, pp. 39–64 in R. F. Johnston, P. W. Frank, and C. D. Michener (eds.), Annual review of ecology and systematics. Annual Reviews Inc., Palo Alto.

Macnae, W. 1968. A general account of the flora and fauna of mangrove swamps in the Indio-west Pacific region. Adv. Mar. Biol. 6: 73–270.

Macdonald, K. B. 1977. Plant and animal communities of Pacific North American salt marshes, pp. 167–192 in V. J. Chapman (ed.), Wet coastal ecosystems. Ecosystems of the world. Elsevier, Amsterdam.

Martinez, R. F. 1994. Status del manejo y reglamentacion de los manglares en Puerto Rico, pp. 194–208 in D. O. Suman (ed.), El ecosistema de manglar en America Latina y la cuenca del Caribe: su manejo y conservacion. University of Miami, Fla.

McKee, K. L. 1993. Soil physicochemical patterns and mangrove species distribution: reciprocal effects? J. Ecol. 81: 477–487.

McKee, K. L. 1995a. Interspecific variation in growth, biomass partitioning, and defensive characteristics of neotropical mangrove seedlings: response to availability of light and nutrients. Amer. J. Bot. 82(3): 299–307.

McKee, K. L. 1995b. Seedling recruitment patterns in a Belizean mangrove forest: effects of establishment ability and physico-chemical factors. Oecologia 101: 448–460.

McKee, K. L. 1995c. Mangrove species distribution patterns and propagule predation in Belize: an exception to the dominance-predation hypothesis. Biotropica 27: 334–345.

McKee, K. L., and I. C. Feller. 1994a. Effects of nutrients and shading on growth and architecture of mangroves in Belize. Ecol. Soc. Amer. Bull. Suppl. 75: 149 (abstract).

McKee K. L., and I. C. Feller. 1994b. Interactions among nutrients, chemical and structural defense, and herbivory in mangroves at Rookery Bay, Florida. Technical report to Office of Ocean and Coastal Resource Management/NOAA, Silver Spring, Md.

McKee, K. L., I. A. Mendelssohn, and M. W. Hester. 1988. Reexamination of pore water sulfide concentrations and redox potentials near the aerial roots of *Rhizophora mangle* and *Avicennia germinans*. Amer. J. Bot. 75: 1352–1359.

Mendelssohn, I. A. 1979. Nitrogen metabolism in the height forms of *Spartina alterniflora* in North Carolina. Ecology 60: 574–584.

Mendelssohn, I. A., and K. L. Marcellus. 1976. Angiosperm productivity of three Virginia marshes in various salinity and soil nutrient regimes. Chesapeake Sci. 17: 15–23.

Mendelssohn, I. A., and K. L. McKee. 1988. *Spartina alterniflora* die-back in Louisiana: time–course investigation of soil waterlogging effects. J. Ecol. 76: 509–521.

Mendelssohn, I. A., K. L. McKee, and W. H. Patrick, Jr. 1981. Oxygen deficiency in *Spartina alterniflora* roots: metabolic adaptation to anoxia. Science 214:439–441.

Mendelssohn, I. A., K. L. McKee, and M. T. Postek. 1982. Sublethal stresses controlling *Spartina alterniflora* productivity, pp. 223–242 in R. E. Turner, B. Gopal, R. G. Wetzel, D. F. Whigham (eds.), Wetlands ecology and management. International Scientific Publications, Jaipur, India.

Mendelssohn, I. A., and M. T. Postek. 1982. Elemental analysis of deposits on the roots of *Spartina alterniflora*, Loisel. Am. J. Bot. 69: 904–912.

Metcalfe, W. S., A. M. Ellison, and M. C. Bertness. 1986. Survivorship and spatial development of *Spartina alterniflora* Loisel. (Gramineae) seedlings in a New England salt marsh. Annals Bot. 58: 249–258.

Mitsch, W. J., and J. G. Gosselink. 1993. Wetlands, 2nd ed. Van Nostrand Reinhold, New York.

Morris, J. T. 1984. Effects of oxygen and salinity on ammonium uptake by *Spartina alterniflora* Loisel. and *Spartina patens* (Ait.) Muhl. J. Exp. Mar. Biol. 78: 87–98.

Nickerson, N. H., and F. R. Thibodeau. 1985. Association between pore water sulfide concentrations and the distribution of mangroves. Biogeochemistry 1: 183–192.

Niering, W. A., and R. S. Warren. 1980. Vegetation patterns and processes in New England salt marshes. BioScience 30: 301–307.

Odum, E. P., and M. E. Fanning. 1973. Comparison of the productivity of *Spartina alterniflora* and *Spartina*

cynosuroides in Georgia coastal marshes. Bull. Georgia Acad. Sci. 31: 1–12.

Odum, W. E., C. C. McIvor, and T. J. Smith, III. 1982. The ecology of mangroves of South Florida: a community profile. U.S. Department of Interior, Fish and Wildlife Service (FWS/OBS-81/24.) Washington, D.C.

Olmsted, I. 1993. Wetlands of Mexico, pp. 637–677 in D. Dykyjova D. F. Whigham, and S. Hejny (eds.), Wetlands of the world. I. Inventory, ecology and management. Kluwer Academic, Dordrecht, The Netherlands.

Onuf, C. 1987. The ecology of Mugu Lagoon, California: an estuarine profile U.S. Department of Interior, Fish and Wildlife Service Biological Report 85 (7.15). Washington, D.C.

Osorio, O. O. 1994. Situacion de los manglares de Panama, pp. 176–193 in D. O. Suman (ed.), El ecosystema de manglar en America Latina y la cuenca del Caribe: su manejo y conservacion. University of Miami, Fla.

Oyuela, O. 1994. Los manglares del Golfo de Fonseca-Honduras, pp. 133–143 in D. O. Suman (ed.), El ecosystema de manglar en America Latina y la cuenca del Caribe: su manejo y conservacion. University of Miami, Fla.

Patterson, S. G. 1986. Mangrove community boundary interpretation and detection of areal changes on Marco Island, Florida: application of digital image processing and remote sensing techniques. U.S. Department of Interior, Fish and Wildlife Service Biol. Rep. 86(10).

Patterson, C. S., and I. A. Mendelssohn. 1991. A comparison of physicochemical variables across plant zones in a mangal/salt marsh community in Louisiana. Wetlands 11:139–161.

Peltier, W. R., and A. M. Tushingham. 1989. Global sea level rise and the greenhouse effect: might they be connected? Science 244: 806–810.

Penfound, W. T., and E. S. Hathaway. 1938. Plant communities in the marshlands of southeastern Louisiana. Ecol. Monogr. 8: 1–56.

Pestrong, R. 1972. Tidal flat sedimentation at Cooley Landing, southwest San Francisco Bay. Sed. Geol. 8: 251–288.

Pielou, E. C., and R. D. Routledge. 1976. Salt marsh vegetation: latitudinal gradients in the zonation patterns. Oecologia 24: 311–321.

Pizarro, F., and H. Angulo. 1994. Diagnostico de los manglares de la Costa Pacifica de Costa Rica, pp. 34–63 in D. O. Suman (ed.), El ecosystema de manglar en America Latina y la cuenca del Caribe: su manejo y conservacion. University of Miami, Fla.

Polania, J., and V. Mainardi. 1993. Mangrove forests of Nicaragua, pp. 139–145 in L. D. Lacerda (ed.), Conservation and sustainable utilization of mangrove forests in Latin America and Africa regions. International Society for Mangrove Ecosystems, Okinawa.

Pomeroy, L. R., W. M. Darley, E. L. Dunn, J. L. Gallagher, E. B. Haines, and D. M. Whitney. 1981. Primary productivity, pp. 39–67 in L. R. Pomeroy and R. G. Wiegert (eds.), The ecology of a salt marsh. Springer-Verlag, New York.

Pomeroy, L. R., and R. G. Wiegert. (eds.). 1981. The ecology of a salt marsh. Springer-Verlag, New York.

Pool, D. J., A. E. Lugo, and S. C. Snedaker. 1975. Litter production in mangrove forests of southern Florida and Puerto Rico, pp. 213–237 in G. E. Walsh, S. C. Snedaker, and H. J. Teas (eds.), Proceedings of the International Symposium on Biology and Management of Mangroves. University of Florida, Gainesville.

Postgate, J. 1959. Sulphate reduction by bacteria. Ann. Rev. Microbiol. 13: 505–520.

Proctor, V. W. 1968. Long-distance dispersal of seeds by retention in the digestive tract of birds. Science 160: 321–322.

Rabinowitz, D. 1978. Dispersal properties of mangrove propagules. Biotropica 10: 47–57.

Ranwell, D. S. 1972. Ecology of salt marshes and sand dunes. Chapman and Hall, London.

Rehm, A. 1976. The effects of the wood-boring isopod Sphaeroma terebrans on the mangrove communities of Florida. Environ. Conserv. 3:47–57.

Reimold, R. J. 1977. Mangals and salt marshes of eastern United States, pp. 157–166 in V. J. Chapman (ed.), Ecosystems of the world. I. Wet coastal ecosystems. Elsevier, Amsterdam.

Saenger, P., E. J. Hegerl, and J. D. S. Davie. 1983. Global status of mangrove ecosystems. Environmentalist 3: 1–88.

Savage, T. 1972. Florida mangroves as shoreline stabilizer. Florida Dept. Nat. Resour. Prof. Pap. 19: 1–46

Scott, D. A., and M. Carbonell. 1986. A directory of neotropical wetlands. IUCN, Cambridge, and IWRB, Slimbridge.

Seliskar, D., and J. L. Gallagher. 1983. The ecology of tidal marshes of the Pacific Northwest coast: a community profile U.S. Department of Interior, Fish and Wildlife Service (USFWS-085-82-32). Washington, D.C.

Shackleton, N. J., and N. D. Opdyke. 1973. Oxygen isotope paleomagnetic stratigraphy of Equatorial Pacific core V-28-238, oxygen isotope temperatures and ice volumes on a 105 year and 106 year scale. Quat. Res. 3: 39–55.

Shepard, E. P. 1960. Rise of sea level along Northwest Gulf of Mexico, pp. 110–122 in F. B. Phleger and T. H. van Andel F. P. Shepard (eds.), Recent sediments: northwest Gulf of Mexico. American Association of Petroleum Geologists, Tulsa.

Sherrod, C. L., D. L. Hockaday, and C. McMillan. 1986. Survival of red mangrove Rhizophora mangle, on the Gulf of Mexico coast of Texas. Contributions in Marine Science 29: 27–36.

Sherrod, C. L., and C. McMillan. 1985. The distributional history and ecology of mangrove vegetation along the northern Gulf of Mexico coastal region. Contr. Mar. Sci. 28: 129–140.

Smart, R. M. 1982. Distribution and environmental control of productivity and growth of Spartina alterniflora (Loisel.), pp. 127–142 in D. N. Sen and K. S. Rajpurohit (eds.), Contributions to the ecology of halophytes. Junk, The Hague.

Smith, T. J., III. 1987. Seed predation in relation to tree dominance and distribution in mangrove forests. Ecology 68: 266–273

Smith, T. J., III. 1992. Forest structure, pp. 101–136 in A. I. Robertson and D. M. Alongi (eds.), Tropical mangrove ecosystems. American Geophysical Union, Washington, D.C.

Smith, T. J., III., H. T. Chan, C. C. McIvor, and M. B.

Roblee. 1989. Comparisons of seed predation in tropical, tidal forests from three continents. Ecology 70: 146–151.

Smith, T. J., III, M. B. Robblee, H. R. Wanless, and T. W. Doyle. 1994. Mangroves, hurricanes, and lightning strikes. BioScience 44: 256–262.

Snow, A. A., and S. W. Vince. 1984. Plant zonation in an Alaskan salt marsh. II. An experimental study of the role of edaphic conditions. J. Ecol. 72: 669–684.

Spackman, W., C. P. Dolsen, and W. Riegal. 1966. Phytogenic organic sediments and sedimentary environments in the Everglades–mangrove complex. I. Evidence of a transgressing sea and its effect on environments of the Shark River area of southwest Florida. Paleotropica 117: 135–152.

Stout, J. P. 1984. The ecology of irregularly flooded salt marshes of the northeastern Gulf of Mexico: a community profile. U.S. Department of Interior, Fish and Wildlife Service Biol. Rep. 85(7.1) Washington, D.C.

Teas, H. 1977. Ecology and restoration of mangrove shorelines in Florida. Environ. Conserv. 4: 51–57.

Thibodeau, F. R., and N. H. Nickerson. 1986. Differential oxidation of mangrove substrate by Avicennia germinans and Rhizophora mangle. Amer. J. Bot. 73: 512–516.

Titus, J. G. 1988. Greenhouse effect, sea level rise and coastal wetlands. Environmental Protection Agency, EPA-230-05-86-013 Washington, D.C.

Tomlinson, P. B. 1986. The botany of mangroves. Cambridge University, Cambridge.

Tolley, M. J. 1993. Long term changes in eustatic sea level, pp. 81–107 in R. A. Warrick, E. M. Barrow, and T. M. L. Wigley (eds.), Climate and sea level change: observations, projections and implications. Cambridge University Press, Cambridge.

Turner, R. E. 1977. Intertidal vegetation and commercial yields of penaid shrimp. Trans. Am. Fish. Soc. 106: 411–416.

Turner, R. E. 1990. Landscape development and coastal wetland losses in the northern Gulf of Mexico. Amer. Zool. 30: 89–105.

Twilley, R. R., A. E. Lugo, and C. Patterson-Zucca. 1986. Litter production and turnover in basin mangrove forests in southwest Florida. Ecology 67: 670–683.

Vince, S. W., and A. Snow. 1984. Plant zonation in an Alaskan salt marsh. I. Distribution, abundance, and environmental factors. J. Ecol. 72: 651–667.

Waisel, Y. 1972. Biology of halophytes. Academic Press, New York.

Wanless, H. 1974. Mangrove sedimentation in geological perspective. Miami Geol. Soc. 2: 190–200.

Wells, E. D., and H. E. Hirvonen. 1988. Wetlands of Atlantic Canada, pp. 249–303 in National wetlands working group (eds.), Wetlands of Canada. Ecol. Land Classif. Ser. No. 24. Polyscience Publ., Montreal.

West, R. C. 1977. Tidal salt-marsh and mangal formations of Middle and South America, pp. 157–166 in V. J. Chapman (ed.), Ecosystems of the world, I. Wet coastal ecosystems. Elsevier, Amsterdam.

Zack, A., and A. Roman-Mas. 1988. Hydrology of the Caribbean Island wetlands. Acta Cientifica 2: 65–73.

Zedler, J. B. 1980. Algal mat productivity: comparisons in a salt marsh. Estuaries 5: 122–131.

Zedler, J. B. 1982. Ecology of southern California coastal salt marshes: a community profile. U.S. Department of Interior, Fish and Wildlife Service (USFWS-OBS-81-54) Washington, D.C.

Zedler, J. B., C. S. Nordby, and B. E. Kus. 1992. Ecology of Tijuana estuary, California: a national estuarine research reserve. NOAA – OCRM, Sanctuaries and Reserves Division, Washington, D.C.

Zisman, S. 1992. Mangroves in Belize: their characteristics, use and conservation. Technical Report to the Government of Belize FPMP/Forest Dept. Belmopan, Belize.

Chapter
14

Alpine Vegetation

WILLIAM DWIGHT BILLINGS

INTRODUCTION

The alpine biome, one of the smaller in area of the major North American ecosystem complexes, occupies high mountain summits, slopes, and ridges above the upper elevational limits of forests. In an environment of intense solar radiation, wind, cold, snow, and ice, alpine vegetation is close to the ground and consists mainly of dwarf shrubs and perennial herbaceous plants less than a few decimeters tall. Vegetational types vary from lawn-like meadows in the moister and wetter sites to relatively dry rocky fellfields on windward slopes with nothing more than scattered cushion plants and lichens on the ridges. Places with persistent or long-lasting snowdrifts, screes, and cliffs are almost plantless.

Except for isolated summits – for example, Mt. Washington and Mt. Katahdin in the northern Appalachians, the Long Range of Newfoundland, and the Torngat Mountains of Labrador – alpine vegetation in North America exists today only on the high mountains in the western third of the continent. There, as can be seen on the map at the beginning of the chapter, alpine vegetation occurs from the arctic Brooks Range in northern Alaska southward through the Alaska Range, the Wrangells, and the Chugach Mountains of coastal Alaska.

From southern Alaska, alpine plants occur still farther to the south on hundreds of montane "islands" rising along two great cordilleras: (1) the Coastal Cordillera from the Alaska Range to the Sierra Nevada of California far to the south, and (2) the Rocky Mountain Cordillera from the Brooks Range to the higher peaks of northern New Mexico and Arizona. South of latitude 35°N, there is a gap of more than 1500 km (ca. 950 mi) to a few alpine summits in the Sierra Madre Oriental of Mexico. Deeper into the tropics, the very high volcanic peaks rise above the Mexican Plateau. On these mountains, vascular plants extend upward to at least 4570 m (15,000 ft) according to Webster (1961). The southernmost mountains in North America with vegetation that can be called "North American alpine" are in Guatemala. The richest alpine floras in Guatemala are on the oldest mountains, not on the young volcanos, and are related to the mountain floras of the southern Rocky Mountains and Mexico (Steyermark 1950). Southward, the high mountain floras on Cerro de la Muerte and Cerro Chirripó (Costa Rica) and Volcan de Chiriquí (Panama) form "páramos" in the sense of Weber (1958) that are more closely related to the paramo floras and vegetation of the northern Andes of Venezuela and Colombia than they are to alpine vegetation farther north.

On a map of this small scale, it is not possible to indicate the extent of permanent ice and snow that exists under present climatic conditions. Therefore, in the mountains of coastal Alaska, considerable glacial ice is included in the alpine ecosystems as mapped. The same is true of the ice caps on Ellesmere, Devon Island, Baffin, and Greenland in the eastern Arctic. In Greenland, more than 95% of the land is covered with a truly continental ice cap. Although glaciers exist in the mountains of the middle latitudes in western North America, they are now too small to affect the areal distribution of alpine ecosystems as depicted here. Glacial ice in alpine situations was much more extensive at times in the past. With the postulated climatic warming resulting from the CO_2 effect and other environmental changes of the twenty-first century, montane glaciers may become less extensive in the future.

There is a tendency among some ecologists to speak of "alpine tundra" and to equate this with the true tundra of the Arctic. It is tempting to accept this equivalence from the standpoint of environment, flora, and vegetation, but it is only partly true. As shown in Figure 14.1, environmentally, only low daily mean temperatures during the growing season are held in common by the arctic tundra of the North and the alpine ecosystems of the Rocky Mountains near 40°N latitude. All else in their respective physical environments differs in considerable degree: intensity and wavelengths of solar radiation, ultraviolet radiation, day length, wind, soils, snow cover, and topography (Billings 1973, 1979; Caldwell et al. 1980).

However, with changing climates during the Pleistocene and Holocene, there was considerable north and south migration of plant species along the Cordillera of the Rocky Mountains. Also, in eastern North America, migration southward of arctic species in front of the Wisconsin ice and back northward after the melting of the ice sheets resulted in the alpine floras of the New England mountains having a strong arctic component. On Mt. Washington, this amounts to about 70% of the alpine flora, using the 63 alpine vascular plant species mentioned by Bliss (1963) as a sample. Similarly, about 50% of the alpine flora at the species level in the Beartooth Mountains in the central Rocky Mountains also occurs in the Arctic (Johnson and Billings 1962; Billings 1978). However, the younger Sierra Nevada of California, with a richer and more diverse alpine flora, has only about a 20% floristic relationship with the flora of the Arctic (Billings 1978).

Alpine vegetation is also discussed in Chapters 15 (Mexico, p. 584), 17 (Mesoamerica, p. 644), and 18 (Hawaiian Islands, pp. 673–676).

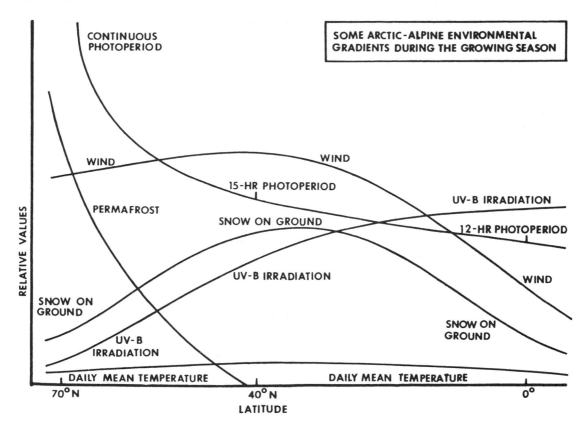

Figure 14.1. Relative values for six principal environmental factors in Arctic and alpine ecosystems along a latitudinal gradient. (From Billings 1979.)

ALPINE ENVIRONMENTS, TIMBERLINES, AND ELEVATIONAL MIGRATIONS

In alpine regions, the physical aspects of the mountain massif dominate: bare rocky crests, relatively low atmospheric pressure, low temperatures, wind, blowing snow, long-lasting drifts, and intense solar radiation. However, these are not separable from the principal biological factor: the presence of forest cover below timberline and its absence above. The sharpest environmental gradient on the mountain is that crossing timberline from protection within the forest to the elemental dangers on the exposed alpine ridges. Even scattered patches or ribbons of trees and scrubby krummholz reduce the intensity of solar radiation at ground level, capture drifting snow, provide protection from wind, and ameliorate day and night temperatures (Billings 1969). Timberline, in turn, is controlled largely by the physical aspects of the mountain: weather, temperature, wind, snowdrifts, rocks, and soils (Arno 1984). As these factors change through long periods, trees die or become established higher yet, so that timberline retreats or advances. The forest gives way and alpine vegetation replaces it, or coversely the new forest shades out the fellfield plants as tree seedlings establish upward during warmer times.

The slow upward movement of timberlines in the western North American cordilleras could result eventually in smaller and less continuous "islands" on mountain peaks and high ridges, which would decrease the opportunities for migration of the alpine either north or south. In colder climates of the past, these cordilleran routes were open and available to many cold-tolerant species of plants and animals, a situation that has already been reduced in the past two or three centuries. Some species will become extinct, locally at least, but others migrating up from below may take their places and change the composition and functioning of the ecosystem.

As an example, Hofer (1992) has tabulated lists of vascular plant species on 14 peaks in the Bernina region of the Swiss Alps between 1905 and 1985. On 12 out of the 14 peaks, in the subnival and nival zones, significant immigration of species from below has occurred in that period of 80 yr. All the adventive species are typically alpine or arctic taxa with seeds mostly distributed by wind. The average number of species per peak has increased from 16 to 28. Hofer attributes this increase to the rise in temperature on these mountains in this century and to the retreat of glacial ice. The rise in temperature between 1911 and 1990 in the higher zones of the Alps is about 0.6°–0.8°C. Rübel (1912), reporting on his 1905 survey that is the initial basis of Hofer's comparison, listed 70 species of vascular plants above timberline on 20 peaks in the Bernina region. Hofer, using only 14 of these 20 peaks, now has found 108 species. Only one peak, Piz Languard at 3261 m, has fewer species now than found there by Rübel in his 1905 census. Apparently, in the Bernina Alps, immigration from below has dominated considerably over extinction during this twentieth century.

Geological and Geomorphological Components

At the risk of oversimplification, one can recognize four large mountain systems in North America north of Mexico:

1. In the East, the Appalachians trend northeast from Alabama to Newfoundland, a distance of almost 3000 km.
2. Across the Straits of Belle Isle from the Long Range of Newfoundland, the mountains of the eastern rim of the Canadian Shield extend into the American Arctic for 3600 km from Labrador to Ellesmere and Axel Heiberg islands, and to within 750 km of the North Pole.
3. The longest and most massive of the mountain systems is the Cordillera of the western Continental Divide, the Rocky Mountains. These mountains stretch for more than 5100 km from southern New Mexico to the western end of the Brooks Range in northern Alaska. The Mexican extension of the Cordillera includes the Sierra Madre Occidental and Oriental.
4. The Western or Coastal Cordillera is almost as long (4000 km) as the Rocky Mountains. With the highest summits on the continent, it extends from its highest peak, Denali, in southwestern Alaska, along the Pacific Coast through the Cascades and the Sierra Nevada to southern California and into Baja California.

South of the desert gap in northern Mexico, the two Sierra Madres frame the Mexican Plateau until they meet the transverse belt of high volcanos south of 20°N latitude near Mexico City (see Chapter 15). Beyond the Tehuantepec Isthmus, the North America mountains continue in a southeast-erly direction from Guatemala to Panama with much physical and biological diversity.

In general, among the four main mountain systems, their tectonic ages decrease markedly from east to west. The Appalachians are very old (Cambrian to Pennsylvanian), the eastern mountains of the Canadian Arctic are not quite as old, the Rocky Mountain Cordillera much less so (Upper Cretaceous to Oligocene), and the various members of the Coastal Cordillera from the Alaska Range to the Sierra Nevada are relatively young, having risen to their present heights during the Pliocene, the Pleistocene, and including the present. Not all ranges that are components in each system are equally old, however.

The present elevations of the highest peaks in each mountain system also increase from east to west. Mt. Mitchell in the Black Mountains of western North Carolina is the highest peak in the Appalachians at 2037 m and quite close to the western limits of this old mountain system. Under present climatic conditions, Mitchell is capped to the top with spruce-fir forest. It probably had a timberline and alpine vegetation when the continental ice of the Wisconsin stage reached its terminal moraines only 375 km to the north of this mountain. The same would be true of the other peaks over 2000 m in elevation in the southern Appalachians: LeConte, Clingman's Dome, and Guyot, all in the Great Smoky Mountains of North Carolina and Tennessee. None of these peaks was glaciated. Conversely, the higher peaks of the northern Appalachians, such as Mt. Washington (1909 m), were covered by the continental ice sheet. The present climate of these northern mountains is such that timberlines there are considerably below the summits, which are capped with low alpine vegetation.

The Rocky Mountains and the Sierra Nevada have several peaks over 4300 m in elevation, with the highest being Mt. Whitney in the Sierra at 4418 m. The Rocky Mountains are geologically complex and consist of folded strata and uplifted blocks. The Sierra Nevada are a large fault block of granitoid rocks tilted to the west and capped, particularly in the north, by younger andesitic flows. Between the Rocky Mountains and the Sierra are many mountain ranges and isolated peaks. In the Great Basin, these ranges trend north-south, with some peaks being over 4000 m and as high as 4341 m. The general crest level of the Cascade Mountains is modest (1500 to 1800 m), but this ridge mountain system is punctuated from northern California to northwestern Washington by high volcanoes between 3200 and 4400 m in elevation.

The highest mountains in North America are in Alaska and along the Yukon–Alaska boundary near the Pacific Ocean. There is a sizable suite of peaks in the Alaska Range, the Wrangells, and the St. Elias Mountains over 5000 m in elevation. These include Denali (Mt. McKinley) in the Alaska Range at 6187 m and Mt. Logan in the St. Elias Range at 6050 m that are the highest. Several others are almost as high. Only Pico de Orizaba (5700 m) and Popocatépetl (5452 m) in central Mexico are in this class. The high Alaskan mountains, being close to the Pacific Ocean, are laden with alpine ice and valley glaciers extending in some cases all the way to the sea.

All the higher mountains in North America north of Mexico were glacially carved during the Pleistocene and Holocene with the exception of the southern Appalachians. Alpine glaciers and ice fields still characterize many of these mountains, especially those near the Pacific from northern California to Alaska. Smaller glaciers occur as far inland as Colorado, Wyoming, and Montana. Under present climatic conditions, many of these smaller glaciers are wasting away. There are extensive glaciers and ice caps on the Canadian Rockies, however, and also on the mountains of the eastern Canadian Arctic from Baffin to the northern tip of Ellesmere.

Weather and Climate

The common characteristic of high mountains anywhere is cold weather. This holds for the tropical mountains as well as those in arctic and subarctic Alaska. Absolute alpine temperatures should also be considered relative to those mountains in the adjacent lowlands and those along latitudinal gradients on the mountain crests.

Because high mountains rise into thin, clear atmosphere, solar radiation received at any given latitude will be more intense at high elevations than in the valleys. This is true not only across the visible spectrum but especially so in the ultraviolet. Caldwell, Robberecht, and Billings (1980) measured global UV–B (280–320 nm) in the arctic–alpine life zone from sea level on the Alaskan North Slope at 70°28' N latitude to Hidden Peak (3352 m; 40°32'N lat.) above Snowbird, Utah, and thence to Laguna Mucubají (3,560 m) at 8°50'N latitude in the Venezuelan Andes, and still farther south to the Peruvian Andes (4200 m) above Tarma and Huancayo at 11°30'S latitude. Their results show that the integrated effective UV–B irradiance can differ by a full order of magnitude along this arctic to alpine gradient. Robberecht et al. (1980) found that the leaves of alpine plant species along this same gradient attenuate the high UV–B irradiation mainly by absorption in the epidermis and

that the leaves of certain species reflect at least some of the UV radiation.

The normal temperature lapse rate in the atmosphere with increasing altitude assures that temperatures at higher elevations will be cooler at midday than those on the lower mountain slopes and in the valleys. However, on a micro scale, the absorption of intense solar radiation by rocks, soils, and plants above timberline allows air temperatures near the ground to heat up during the day. Alpine air temperatures, therefore, are much higher near the soil surface than they are in the wind at 1 m above. At timberline and above, small plants are at an advantage as compared to trees.

If one takes long-term weather data from standard stations at or above timberline, the general climatic background of alpine vegetation in middle-latitude North America can be seen in Figure 14.2. These Walter types of climate diagrams are representative of alpine sites on Mt. Washington in the northern Appalachians, Niwot Ridge in the Rocky Mountains of Colorado (as photographed in Fig. 14.3), and at Twin Lakes (now Caples Lake) near timberline in the northern Sierra Nevada of California. The Appalachian and Rocky Mountain alpine zones have much in common in regard to temperature. The mean temperature of the warmest month of the year in each of those two zones is less than 10° C, and the season during which the mean air temperature is above 0° C is short: 4–5 mo at most. On the other hand, the sunny Sierra warms up more during its longer summer after snowmelt, with the mean air temperature of the warmest month reaching 13° C. In the high radiation alpine environment of the Sierra, the climate near the ground can be quite warm and dry as compared to foggy and damp Mt. Washington; Niwot Ridge is intermediate.

From the standpoint of precipitation, Mt. Washington receives a lot of snow and rain rather evenly throughout the year. Fog is common and drought is rare there. Niwot Ridge receives only about half as much precipitation as Mt. Washington, but most of that occurs as snow in winter; its summer thundershowers keep the Niwot alpine vegetation generally green until the gentle drought of late summer brings the brown of dormancy. In contrast, the High Sierra, with a semimediterranean mountain climate, receives 1.25 m of precipitation during the year (25% more than Niwot) but almost all of it as snow in the winter and spring months. The summers are dry on its alpine ridges; moist meadows are confined to meltwater slopes below long-lasting snowdrifts. On many rocky ridges and peaks, Sierran alpine vegetation has a desert-like aspect. There are clear floristic relationships with

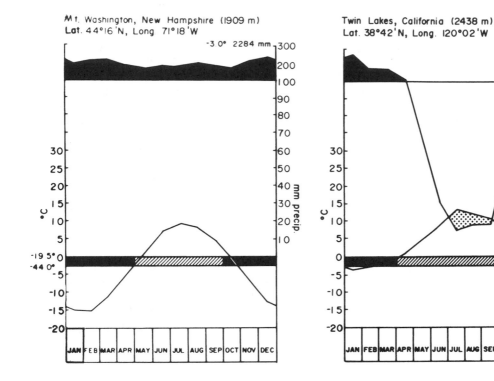

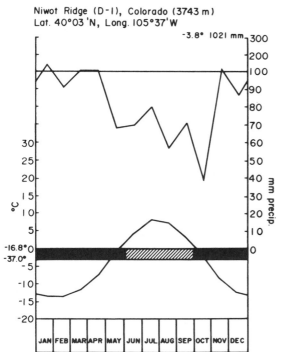

Figure 14.2. Walter-type climate diagrams for three kinds
of alpine climates: northern Appalachians (Mt. Washington), Rocky Mountains (Niwot Ridge), and Sierra Nevada
(Twin Lakes). For explanation of data and axes, see Figure
11.2 in Chapter 11 of this volume.

Figure 14.3. Aerial photograph looking southwest over Niwot Ridge alpine area in the Indian Peaks area of the Colorado Front Range. (Photograph by Patrick J. Webber.)

the arid ecosystems of the Great Basin at the eastern base of the steep Sierran escarpment. Notably absent in the higher Sierra are many arctic plant species so characteristic of the Rocky Mountains and the northern Appalachians.

Wind is a strong modifying factor in alpine environments: It chills during the day, it warms at night, and it moves water around, especially in the form of snow and also in clouds both high and low. There are few quiet days on the alpine scene, particularly during the winter and in storms. The highest straight-line wind speed so far recorded at the surface of the earth was in the alpine zone at the summit of Mt. Washington, New Hampshire, in the northern Appalachians, on April 12, 1934: 231 mph.

Strong wind action and its movement of snow off the peaks and ridges results in topographic moisture gradients in alpine regions outside the tropics. Such a mesotopographic gradient (Billings 1973) is illustrated in Figure 14.4. On any high mountain, such repeatable wind–snow gradients determine to a large extent the patterning of alpine vegetational gradients from dry ridges to the wet meadows and bogs below large snowdrifts. This vegetational patterning results from complex combinations of thickness of snowpack, time of snowmelt, speed and direction of wind, steepness of slopes, and the diversity of the extant flora and vegetation. In the high mountains of the North American middle latitudes, the most common wind direction is from the southwest. This causes the uppermost slopes that face west or southwest to be relatively snow-free, or at least to have a shallower snowpack than the great drifts on eastern or northeastern slopes in the lee of the relatively dry ridges. Spring and summer come early where snow lies thinly or not at all. Drought also comes early on these upper windward slopes and bare ridges.

Such steep local gradients of snow cover and soil moisture result in similarly steep vegetational gradients. Within the cold alpine zone, it is often the availability of water that governs ecosystem productivity and the distribution of species. But long-lasting snowbanks shorten the growing season so much that only a few species can tolerate the aridity and low temperatures of late summer after 10–11 mo of relative darkness under the snow. So productivity of alpine vegetation that lies under big drifts is quite

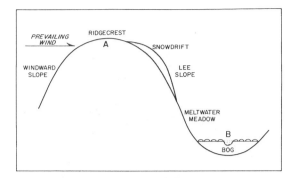

Figure 14.4. Diagram of a typical alpine mesotopographic gradient. (Adapted from Billings 1973.)

low compared to intermediate sites. Also, only a few plant species can tolerate the very short growing season of late summer before freeze-up. Late-appearing populations of *Oxyria digyna* that come after late snowmelt consist of dwarf individuals with dark purple leaves rich in anthocyanins This adaptation could be ecotypic. These late-appearing populations are also able to survive for several years without being exposed by timely melting of the overlying snowdrift.

By measuring leaf-water potentials and leaf conductances within the array of species along an alpine mesotopographic gradient throughout the short growing season, one can evaluate the roles of soil water and soil drought on the growth and local distributions of the different species. Oberbauer and Billings (1981) did this across an alpine ridge at 3300 m in the Medicine Bow Mountains of Wyoming (see Table 14.1). As expected, leaf-water potentials were generally lowest on the ridgetop and highest in the meltwater wet meadow. Highest leaf conductances were in the wet meadow and lowest on the upper windward slope. All plants growing on the ridgetop had very small, tightly overlapping leaves that were too small for the porometer. Each of the 29 species measured at frequent intervals had its own individualistic distribution along the gradient. Some were very localized. For example, *Paronychia pulvinata* occurred only on the upper windward slope and ridgetop, whereas *Caltha leptosepala* was present only in the wet meadow. A few species, such as *Geum rossii* and *Trifolium parryi*, were present almost universally along the gradient. At each site, plants of *Geum rossii* showed diurnal and seasonal courses of leaf-water potentials typical of the other species characteristic of that particular site. Whether or not this striking similarity in water potentials is caused by phenotypic acclimation in *Geum* or by local ecotypes is not known, but both are probably involved. Certainly, rooting depth has considerable effect. For example, the deep-

rooted *Trifolium parryi* maintained steady leaf-water potentials at midday of about −2.00 MPa, no matter what the site and no matter that some of its more shallowly rooted neighbors were showing very low potentials near −3.5 to 4.0 MPa at the same time.

These data on the seasonal courses of plant water potentials and stomatal conductances translate well into alpine vegetational patterning along a topographic gradient. The floristic composition along this same vegetational transect in the Medicine Bow Mountains is shown in Table 14.2. Although the vegetational gradient shows rather gradual floristic shifts from the lower windward slope to the lee slope, there are some fairly abrupt spatial changes in species composition. The most conspicuous of these occur just to the lee of the ridgetop and also near the bottom of the lee slope where the large rocks abut the level wet meadow and temporary pond. Also, the ground under the longest-lasting snowbanks is rather barren and is occupied only by an open vegetation of distinct floristic composition that includes *Sibbaldia procumbens*, *Juncus drummondii*, and *Deschampsia caespitosa*. So it is possible to recognize, in an arbitrary way, certain plant communities along the transect:

Windward slope from −150 m to ridgetop at 0 m: open, rocky fellfield.
Upper lee slope from 0–40 m: modified fellfield.
Upper lee slope from 40–60 m: transitional from fellfield to early snowbed community.
Middle lee slope from 60–90 m: early snowbed community.
Middle lee slope from 90–120 m: late snowbed community.
Middle lee slope from 120–150 m: moist meadow.
Lower lee slope from 150–260 m: mostly late snowbed community.
Bottom of lee slope from 260–300 m: very wet meadow.

This complex gradient of plant communities is depicted in Figure 14.5, which accompanies Table 14.2. From these data, the effects of water distribution on vegetational composition across a typical alpine rocky ridge is obvious. An ordination by May and Webber (1982) shows a strikingly similar set of vegetational noda on Niwot Ridge in the Front Range of Colorado only about 150 km south of the Medicine Bow site. The Medicine Bow transect lacks only one node of the Niwot set: the *Kobresia* meadow. A notable vegetational-environmental study on Niwot Ridge Isard (1986), comes to the same conclusion: that the alpine plant communities there are controlled primarily by snow cover and soil moisture.

Table 14.1. *Lowest mean midday (P_{mm}) and lowest mean predawn (P_{max}) leaf-water potentials (MPa) for the season. Change in mean predawn water potential from the fifth to the sixth week following light rain (dP_{max}). Maximum leaf conductance (C_{max}) for the season. Genera appear in Table 14.2.*

Site	P_{min}	P_{max}	dP_{max}	C_{max}	Site	P_{min}	P_{max}	dP_{max}	C_{max}
Windward (−60 m)					**Snowdrift (140m)**				
M. obtusiloba	−3.10	−1.75	+0.79	—	T. parryi	−1.92	−0.99	+0.04	0.81
H. grandiflora	−3.72	−2.90	+1.43	—	G. rossii	−2.04	−0.89	+0.19	—
A. scopulorum	−4.00	−2.33	+1.17	—	A. scopulorum	−3.15	−1.83	+0.50	—
C. jamesii	−3.42	−2.07	+0.95	0.92	B. bistortoides	−1.65	−0.82	−0.03	1.44
T. parryi	−2.18	−1.57	−0.06	0.81	P. diversifolia	−2.81	−0.78	+0.17	0.73
T. dasyphyllum	−2.54	−1.46	−0.14	—	C. rupestris	−2.92	−1.96	+0.57	—
G. rossii	−2.43	−1.82	+0.86	—	C. jamesii	−2.84	−1.37	+0.38	0.56
Ridge (0 m)					**Rocks (190 m)**				
M. obtusiloba	−4.00	−2.81	+1.29	—	G. rossii	−1.30	−0.60	—	—
T. dasyphyllum	−2.96	−1.65	−0.04	—	E. peregrinus	−1.38	−0.62	—	0.51
G. rossii	−3.12	−2.15	+0.61	—	S. dimorphophyllus	−1.08	−0.56	—	0.83
P. pulvinata	−4.00	−3.70	+2.14	—	S. procumbens	−1.60	−0.52	—	0.63
A. scribneri	−4.00	−4.00	+2.82	—	A. mollis	−1.41	−0.32	—	0.81
					R. montigenum	−1.89	−0.54	—	0.67
Lee (60 m)					S. caerulea	−1.65	−0.96	—	0.42
T. parryi	−2.00	−0.86	+0.03	0.89	P. engelmannii	−1.50	−0.59	—	—
G. rossii	−2.76	−1.69	−0.18	—					
A. scopulorum	−3.28	−1.79	+0.46	—	**Pond (230 m)**				
B. bistortoides	−2.28	−1.46	−0.05	1.37	B. bistortoides	−1.11	−0.42	—	1.20
T. alpestre	—	—	—	—	D. caespitosa	−1.89	−0.28	—	—
P. rupicola	−4.00	−2.30	−0.02	—	S. dimorphophyllus	−1.36	−0.59	—	0.99
P. diversifolia	−2.84	−1.72	−0.20	0.90	S. brachycarpa	−1.21	−0.54	—	1.85
A. lanulosa	—	—	—	—	S. rhomboidea	−1.10	−0.30	—	0.87
					S. rodanthum	−0.88	−0.34	—	—
Snowdrift (100 m)									
T. parryi	−2.34	−1.04	−0.29	0.78	**Meadow (280 m)**				
A. scopulorum	−4.00	−1.95	+0.43	—	G. rossii	−1.41	−0.51	−0.08	—
B. bistortoides	−2.02	−0.77	−0.10	1.86	P. diversifolia	−1.78	−0.36	−0.17	0.98
T. alpestre	−4.00	−4.00	+0.07	—	B. bistortoides	−1.10	−0.28	−0.04	2.17
P. diversifolia	−2.68	−1.33	+0.02	1.28	A. scopulorum	−1.95	−0.68	+0.02	—
P. alpinum	−4.00	−1.64	+0.09	1.06	C. leptosepalla	−1.35	−0.38	−0.28	1.36
D. caespitosa	−3.25	−1.35	−0.14	—	D. caespitosa	−2.10	−0.48	−0.06	—
E. peregrinus	−3.39	−1.33	+0.04	0.91	C. chalciolepsis	−1.76	−0.25	−0.12	1.29

Source: Oberbauer and Billings (1981).

Soil Frost Action

Alpine soils are cold, and they are often wet from snowmelt. These environmental circumstances, when combined with strong diurnal cycling of soil surface temperatures, lead to frequent freezing and thawing of the upper soil. Because these soils are shallow and overlie bedrock, rocky glacial till, or permafrost, the consequent formation of ice pushes boulders, rocks, and gravel toward the surface and thence outward along the surface. The result is an array of patterned ground made up of sorted polygons of various types: stone nets, frost boils, or sorted circles, sorted stripes, sorted steps, and hummocks, as described by Washburn (1956) and Billings and Mooney (1959).

Such frost-churned soils provide an unstable soil environment through much, but not all, of the

alpine zone. The principal exceptions are the rocky outcrops on ridgetops, glacially scoured rocks, cliffs, and wet meadows on deep peat. The most active sites of cryopedogenic activity are barren gravelly places in which the water table fluctuates between just above and just below the surface. Such an active substratum is not easily invaded by plants. Active polygon centers are similarly quite barren and plants are restricted to the rocky borders of such polygons. It is only after the water table falls that plants can invade the polygon centers and participate in the production of peat and that succession can proceed. The peat acts as insulation to temperature changes, ice lenses appear in the peat, and the permafrost table may rise (Billings and Mooney 1959; Johnson and Billings 1962).

In wetter and colder glacial times, on ridges above the Rocky Mountain alpine glaciers, stone

Table 14.2. *Floristic composition of alpine vegetational gradient at 3300 m in the Medicine Bow Mountains, Wyoming. The figures are percentage cover (p = <1%) for the most important taxa at 10 m intervals, from the lowermost windward slope at −140 m across the ridge at 0 m to the wet meadow at +280 m. From −140 through +60 m, data are from late June, 1978; the remainder of the data are from August of that year. Data are averages of three plots, each 20 × 50 cm.*

Species	−140	−130	−120	−110	−100	−90	−80	−70	−60	−50	−40
Geum rossii	2[a]	2	7	5	8	9	5	13	8	8	7
Minuarta obtusiloba	2	4	1	1	5	4	16	6	13	4	10
Potentilla diversifolia	1	p	2	2	9	1	—	p	—	—	—
Trifolium parryi	1	p	p	1	7	2	8	2	6	5	7
Carex elynoides	1	3	1	3	1	p	1	1	1	1	1
Silene acaulis	1	2	9	3	1	p	4	1	—	1	—
Trifolium dasyphyllum	1	2	2	3	1	16	2	7	p	p	5
Artemisia scopulorum	p[b]	—	—	—	—	—	—	—	p	—	1
Bistorta bistortoides	—	1	—	1	—	—	—	—	—	1	1
Selaginella densa	—	1	1	1	p	—	p	3	1	4	2
Phlox pulvinata	—	3	3	2	—	8	1	1	p	1	2
Saxifraga rhomboidea	—	1	—	—	—	—	—	—	—	—	—
Poa alpina	—	—	1	—	—	—	—	—	—	—	—
Paronychia pulvinata	—	—	1	—	—	—	p	5	3	1	6
Lichens	—	1	—	1	—	p	1	p	—	5	4
Calamagrostis purpurascens	—	—	—	—	—	—	—	—	—	—	—
Achillea lanulosa	—	—	—	—	—	—	—	—	—	—	—
Juncus drummondii	—	—	—	—	—	—	—	—	—	—	—
Deschampsia caespitosa	—	—	—	—	—	—	—	—	—	—	—
Sibbaldia procumbens	—	—	—	—	—	—	—	—	—	—	—
Senecio dimorphophyllus	—	—	—	—	—	—	—	—	—	—	—
Caltha leptosepala	—	—	—	—	—	—	—	—	—	—	—
Carex scopulorum	—	—	—	—	—	—	—	—	—	—	—
Sedum rhodanthum	—	—	—	—	—	—	—	—	—	—	—
Dead and litter	79	49	27	44	14	16	9	16	10	14	12
Rock	8	20	15	20	16	20	35	28	30	43	26
Bare soil	0	1	22	1	37	20	10	17	25	11	15

Species	−30	−20	−10	0	+10	+20	+30	+40	+50	+60	+70
Geum rossii	5	9	14	6	7	9	2	8	7	p	2
Minuarta obtusiloba	5	13	8	6	5	10	11	4	2	1	3
Potentilla diversifolia	—	—	—	—	—	—	—	1	1	p	3
Trifolium parryi	3	—	—	—	—	7	8	—	8	3	28
Carex elynoides	p	1	p	p	1	1	1	1	—	—	—
Silene acaulis	—	1	p	1	1	2	5	5	3	2	—
Trifolium dasyphyllum	10	17	17	17	13	5	4	11	9	—	—
Artemisia scopulorum	—	—	—	—	—	—	—	1	4	4	13
Bistorta bistortoides	—	—	—	—	—	—	—	2	—	1	2
Selaginella densa	1	1	3	3	1	1	p	1	1	1	—
Phlox pulvinata	—	—	p	p	1	6	p	1	—	—	—
Saxifraga rhomboidea	—	—	—	—	—	—	—	—	—	—	—
Poa alpina	—	—	—	—	—	—	—	—	—	—	p
Paronychia pulvinata	5	8	6	5	2	1	1	2	1	—	—
Lichens	1	1	1	1	2	2	8	p	4	1	2
Calamagrostis purpurascens	—	—	—	1	1	1	1	—	—	1	1
Achillea lanulosa	—	—	—	—	—	—	—	—	—	p	6
Juncus drummondii	—	—	—	—	—	—	—	—	—	15	—
Deschampsia caespitosa	—	—	—	—	—	—	—	—	—	—	1
Sibbaldia procumbens	—	—	—	—	—	—	—	—	—	—	—
Senecio dimorphophyllus	—	—	—	—	—	—	—	—	—	—	—
Caltha leptosepala	—	—	—	—	—	—	—	—	—	—	—
Carex scopulorum	—	—	—	—	—	—	—	—	—	—	—
Sedum rhodanthum	—	—	—	—	—	—	—	—	—	—	—
Dead and litter	12	13	28	19	29	24	21	25	18	17	14
Rock	47	30	20	35	27	25	28	14	21	20	14
Bare soil	11	11	5	7	13	3	6	15	13	30	8

Table 14.2 (Cont.)

Species	+80	+90	+100	+110	+120	+130	+140	+150	+280
Geum rossii	5	—	—	—	—	2	7	20	6
Minuarta obtusiloba	5	—	—	—	—	—	1	1	—
Potentilla diversifolia	3	—	2	1	4	7	3	5	7
Trifolium parryi	22	12	8	3	6	6	16	19	10
Carex elynoides	—	—	—	—	—	—	—	1	—
Silene acaulis	2	—	—	—	—	—	—	—	—
Trifolium dasyphyllum	—	—	—	—	—	—	—	—	—
Artemisia scopulorum	10	2	p	—	—	1	p	—	2
Bistorta bistortoides	3	p	p	—	11	8	1	1	8
Selaginella densa	—	—	—	—	—	—	1	1	—
Phlox pulvinata	—	—	—	—	—	—	—	—	—
Saxifraga rhomboides	—	1	—	—	—	—	—	—	3
Poa alpina	—	—	1	—	—	—	—	—	5
Paronychia pulvinata	—	—	—	—	—	—	—	—	—
Lichens	—	3	—	—	—	5	1	1	—
Calamagrostis purpurascens	1	1	—	—	—	1	p	—	—
Achillea lanulosa	7	14	5	2	7	—	—	—	—
Juncus drummondii	—	—	—	—	—	—	—	—	—
Deschampsia caespitosa	—	13	6	8	29	18	2	—	35
Sibbaldia procumbens	—	—	23	8	9	5	—	—	—
Senecio dimorphophyllus	—	—	1	3	1	3	—	—	—
Caltha leptosepala	—	—	—	—	—	—	—	—	4
Carex scopulorum	—	—	—	—	—	—	—	—	4
Sedum rhodanthum	—	—	—	—	—	—	—	—	9
Dead and litter	23	20	13	10	17	18	18	8	water
Rock	5	17	24	46	5	8	8	3	water
Bare soil	5	17	24	46	5	8	8	3	water

[a] The figures are percentage coverages for the most important plant species at 10 m intervals from the lowermost windward slope at −140 m across the ridge at 0 m to the wet meadow at +280 m. From −140 m through +60 m, the data are from late June 1978. The remainder of the data (+70 m to +280 m) are from August of that year; these were snow-covered in June. Data are averages of three plots of 20 × 50 cm across the line of the transect. Compare with map and slope diagram in Fig. 14.4.
[b] p = present, but <1% cover.
Source: Unpublished data of the author.

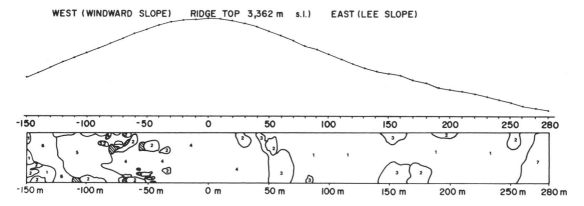

Type 1 = snowcover as of June 23, 1978
Type 2 = spruce krummholz. ⬭ = dead and flattened branches of spruce krummholz
Type 3 = early July snowmelt moist meadow (40-65% plant cover)
Type 4 = cushion plant community (35-50% plant cover)
Type 5 = fell field (30-45% plant cover)
Type 6 = open fell-field (15-30% plant cover)
Type 7 = meltwater wet meadow (75-100% plant cover)

WEST (WINDWARD SLOPE)   RIDGE TOP 3,362 m s.l.)   EAST (LEE SLOPE)

Figure 14.5. *Vegetation map and profile diagram along a mesotopographic gradient at 3300 m in the Medicine Bow Mountains of Wyoming.*

nets of large size and with very large rocks were formed. Today, these large "fossil" stone polygons with turf-covered soil in their centers are characteristic of many relatively level alpine sites in these mountains. Among such sites are Niwot Ridge, the Colorado Front Range (Fig. 14.3), and the Beartooth Plateau of northwest Wyoming.

ALPINE VEGETATION

Plant Life Forms

The vegetation of alpine environments is short in stature with a tendency toward a perennial herbaceous habit with deep root systems or occasional low or prostrate shrubs. Such shrubs may be either deciduous or evergreen. Perennial herbs dominate the alpine landscape. Most of these are graminoids and dicots. All these have much more biomass of roots and rhizomes than of shoots, leaves, and flowers. The roots and rhizomes not only function in water and nutrient absorption but also play a very important role in overwinter storage of carbohydrates (Mooney and Billings 1960; Fonda and Bliss 1966; Rochow 1969; Wallace and Harrison 1978). Annual plants, for example, *Koenigia islandica*, are rare in the vegetation and are usually only a few cm tall with a shallow root system (Reynolds 1984). Annuals constitute only about 1–2% of the flora in most alpine regions and an even smaller percentage of their vegetation.

Although most alpine plants have the perennial herbaceous life form (with preformed flower buds) or the dwarf shrub life form, other life forms do exist above timberline, including lichens, mosses, giant rosette plants with columnar stems, and even succulents such as cacti. The giant rosette plants and succulents, such as a few species of cacti, are in the Americas almost restricted to the very high tropical mountains of South America where, for example, the Andean paramos north of the Equator are characterized by a number of species of the genus *Espeletia* with their single-stemmed "trees" and shorter rosette perennials (Rundel, Smith, and Meinzer 1994). However, there are such plants in the paramos of Costa Rica in North America (see Chapter 17 in this volume, and Weber 1958). Similar plants occur above timberline on other high tropical mountains – Haleakala on Maui in Hawaii (see Chapter 18 in this volume) and the high East African volcanoes such as Mt. Kenya (Hedberg 1964; Smith and Young 1987; Rundel, 1994; Smith 1994), for example.

Table 14.3 shows a scheme that I believe is realistic for the classification of growth forms of alpine plants. It could be useful in the high moun-

Table 14.3. *Classification of alpine plant growth forms*

Perennial herbs
  Ferns
  Dicots
    Acaulescent rosettes
    Caulescent forbs
  Graminoids
    Sod-formers
    Tussocks
    Dwarf mat types
  Lycopodioids
    True lycopods
    Lycopodioid angiosperms
Perennial cushion plants
  Small
  Large (giant)
Giant rosette plants
  Acaulescent
  Tall-stemmed
Shrubs (dwarf or prostrate)
  Deciduous
  Evergreen (also wintergreen)
    Flat-leaved
    Scale-leaved, including lycopodioid angiosperms
Succulents
  Stem (cacti)
  Leaf (Sedum)
Annual herbs
  Graminoids
  Dicots
    Terminal flowering
    Lateral flowering
Lichens
  Crustose
  Thallose
  Fruticose
Mosses
  Acrocarpous
  Pleurocarpous

tains of North America, South America, and the mountainous islands of the Pacific, such as Hawaii, New Zealand, and New Guinea.

Geographic Coverage

Reference to the map of alpine regions of North America at the beginning of this chapter will convey some sense of the scattered array of alpine terrain on this continent. The vegetation and flora of these widely reparated places have been studied in a modern sense only here and there. Also, some mountaintops, especially in the Arctic and Subarctic, have emerged from glaciation only in the last few millennia and centuries of the Holocene. Others, such as the alpine slopes of the St. Elias and Alaska ranges, are still covered by ice. Existing glaciers also occupy many of the mountains of the eastern Canadian Arctic on Ellesmere, Devon, and

Baffin, and, of course, on Greenland with its massive ice sheet. Conversely, glaciation touched, but only lightly, some of the alpine areas in the American Southwest, and most of this ice has long since melted and gone.

It is my purpose here to describe briefly the alpine flora and vegetation of those mountain systems that are most prominent and ice-free. Even among these, information can be given only for certain places that seem representative of alpine vegetation for relatively large geographic regions. Unfortunately, little is known of the flora and vegetation of some large mountainous regions, particularly in the Subarctic. However, the following sections will serve as an introduction to the plant communities of many of the higher mountain ranges of the continent.

Mountains of Labrador. These low, heavily glaciated mountains are botanical bridges between the truly arctic mountains of the Canadian Archipelago and the northern Appalachians. As with so many isolated mountain ranges in the Subarctic, the floras of the mountains of Labrador are better known than the vegetation. Hustich (1962), using his own collections and those of L. A. Viereck, has compared the flora of Mt. Gerin (940 m) with that of Pallas-Ounastunturi in arctic Finland. The alpine area of Mt. Gerin is about 39 km².

Of the 151 species of plants in the alpine part of Gerin, 94 (62%) also occur on Ounastunturi, 4800 km across the Atlantic. Such "circumpolarity" in the nature of the flora is characteristic of many places in the Arctic, but it is also prominent in the vegetation of subarctic mountain ranges, as in Labrador. Because of close genetic relationship between some vicarious species, the arctic element is certainly greater than 62%. We can conclude that on Mt. Gerin and probably on other mountains of Labrador the flora is primarily of arctic derivation; one could classify it as truly "arctic-alpine" A few of the species held in common between mountainous Labrador and arctic Finland are *Poa alpina*, *Eriophorum vaginatum*, *Carex aquatilis*, *C. bigelowii*, *Oxyria digyna*, *Sibbaldia procumbens*, *Rubus chamaemorus*, *Arctostaphylos alpina*, *Diapensia lapponica*, *Bartsia alpina*, and *Phyllodoce coerulea*. Most of these species populations probably consist of ecotypes that differ between North America and northern Europe. The arctic-alpine vegetation on these Labrador mountains is typically low, dry tundra of a fellfield type.

Northern Appalachian Mountains. The vegetation and floristics above alpine timberlines were first studied in North America on the White Mountains of New Hampshire. On Mt. Washington alone, the topography of the alpine zone bears the imprint of these early nineteenth-century plant explorers: Tuckerman's Ravine, Boott Spur, Bigelow Lawn. And in the twentieth century, came the plant ecologists: Robert F. Griggs, Robert S. Monahan, Henry I. Baldwin, and Lawrence C. Bliss, among others. For the past century and a half, such people have been drawn to this rather small (1917 m) deglaciated mountain by the presence of an island of arctic plant species surrounded at lower elevations by the forests of a temperate climate. But the top of this mountain is not temperate. The weather there can be as severe as any on earth: the world's record wind speed, cloud and fog, loss of visibility, rime ice, extreme and sudden drops in temperature. Even the averages in the climate diagram (Fig. 4.2) reflect its nature. As Bob Griggs (1956) said, "It is nothing much of a mountain . . . yet . . . Mt. Washington has one of the worst of climates." The changeability and severity of the summer weather on the summit contribute to its danger for the unprepared casual visitor. There have been many fatalities even during the summer months, not to mention those during the winter.

The alpine plant communities of Mt. Washington and the other peaks in the Presidential Range of the White Mountains have been described thoroughly by Bliss (1963). The environmental relationships of the principal alpine vegetation types are shown in Figure 14.6 (Fig. 10 of Bliss, 1963) as these communities exist along gradients of snow, wind, and fog. The most exposed, snow-free sites are characterized by communities of *Diapensia lapponica*, and sedge meadows of *Carex bigelowii* occupy the snowy and foggy places. The floristic compositions of the eight main community types are listed as importance values in Table 14.4, adapted from Bliss.

The alpine vegetation of the northern Appalachian Mountains is more closely related to the vegetation of the eastern Canadian Arctic than it is to that of the Rocky Mountains or that of the Sierra Nevada, partly because of the cold and foggy climate of places such as Mt. Washington but also to some extent because of the relatively recent continental deglaciation of the New England mountains. There was lowland tundra vegetation south of the continental ice on the mid-Atlantic east coast at the last full glacial. The taxa in this vegetation were of arctic derivation. The present alpine plants of the summit of Mt. Washington are the taxa derived from this lowland tundra, and they show arctic characteristics. For example, *Oxyria digyna* on Mt. Washington have rhizomes as do arctic *Oxyria*. Plants of this species from the alpine areas of the

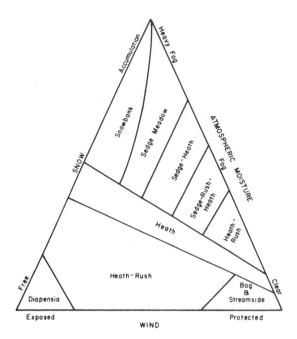

Figure 14.6. Environmental relationships of alpine plant communities of the Presidential Range of the White Mountains, New Hampshire. (Adapted from Bliss 1963.)

western mountains of the United States do not have rhizomes.

In the Presidential Range of the northern Appalachians, Bliss reports a vascular flora of about 110 species in the 19 km² above timberline (1525 m). Of these taxa, about 75 are true alpines, and 35 others are boreal species that enter the alpine zone primarily in patchy krummholz. Some of these boreal species also occur as high as 1830 m in snowbank environments.

The alpine area on Mt. Washington is small in size and almost 2000 m above the adjacent lowlands, but it was entirely covered by continental ice in the last glaciation. Therefore, an above-timberline flora of 110 vascular plant species shows a rather remarkable diversity although the species total is small compared to those of other larger alpine regions such as the Rocky Mountains. For example, in western North America, in a comparable latitude (45°N), Johnson and Billings (1962) and Billings (1978) described an alpine flora of 191 vascular plant species on the Beartooth Plateau, an alpine area of about 186 km² at elevations between 3050 and 3360 m in northwestern Wyoming. Since 1962, we have added 11 more vascular species for a total of 202 species.

Brooks Range, Northern Alaska. There are not many ecological publications on the alpine vege-

tation of the east-west mountain system of the Brooks Range (68°–69°N latitude) in northern Alaska, in contrast to the relatively abundant literature on the arctic tundra to the north of these mountains and the taiga to the south. The Brooks Range is a very real barrier to the northern limit of forest in Alaska. The spruce taiga comes up from the south to reach its limits at 675–800 m where it forms an open forest and dwarf woodland (Densmore, 1980; Billings, personal observations). Figure 14.7 diagrams a vegetational cross section of the Brooks Range near the Dietrich River Valley. There are no forests in a real sense north of these mountains in Alaska.

During the 1930s, Robert Marshall made many explorations of the Brooks Range and described beautifully the natural history, vegetation, and geomorphology of these spectacular mountains when they were pure wilderness. Marshall's book (1956) is the best general introduction to these northernmost mountains in Alaska.

In his vegetational survey of the Arctic Slope, Spetzman (1959) did visit the alpine zone of the Brooks in several places, but he reported only short lists of species and general descriptions. At 1860 m elevation near the Killik River, he listed *Luzula confusa, Potentilla elegans, Saxifraga bronchialis, Selaginella siberica,* and *Smelowskia calycina* as growing sparsely on the north face of a summit ridge of quartzite, shale, and sandstone rubble. At the time of his publication, this was the highest elevation from which seed plants had been collected on the North Slope. This kind of vegetation might be called "alpine desert." All these species also grow at lower elevations in the main alpine zone of the Brooks.

Spetzman and others have also collected around and above Anaktuvuk Pass (elevation 550 m) on the North Slope. The most widespread community in that part of the Brooks Range is the *Dryas* – lichen dry meadow only a few centimeters in height. About 90% of the plant cover is *Dryas octopetala*. Associated species are *Lupinus arcticus, Hierochloe alpina, Silene acaulis, Kobresia myosuroides,* and *Polygonum viviparum.* In wetter places, there are sedge meadows and tussock vegetation with *Carex aquatilis, C. bigelowii, C. membranacea, Dryas integrifolia, Eriophorum angustifolium, E vaginatum,* and *Salix reticulata.* Intermediate sites in regard to moisture are occupied by low shrubs with herbaceous undergrowth.

Around Lake Peters and Lake Schrader at 1070 m, Spetzman describes a complex vegetation with 150 species of vascular plants. Dry meadows are the most extensive type, with minor amounts of tussock meadow, wet sedge meadows, and wil-

Table 14.4. *Relative importance values (importance percentage scale of 0–100) of vascular plants in the alpine plant communities of the Presidential Range, New Hampshire*

Species	Sedge meadow	Sedge-dwarf shrub heath	Sedge-rush-dwarf shrub heath	Shrub heath-rush	Dwarf shrub heath	Diapensia	Snow-bank	Stream-side	Bog
Vascular plants									
Carex bigelowii	93.0	60.6	31.7	6.0	4.2	1.9	11.4	4.6	46.9
Vaccinium uliginosum incl. var. *alpinum*		2.6	3.6	11.0	18.8	8.8	11.3	9.8	8.5
Arenaria groenlandica	7.0	17.7	2.2	1.4	0.5	2.5		0.1	0.4
Juncus trifidus		3.3	23.8	25.7	1.1	21.8	1.0	0.7	
Potentilla tridentata		2.7	9.3	23.6	4.7	4.6	0.2	6.5	
Vaccinium vitis-idaea var. *minus*		12.8	19.7	21.6	12.7	0.4		4.6	8.2
Diapensia lapponica		0.2	5.4	4.6		31.3		0.1	
Agrostis borealis			0.3	1.1	0.3	5.5		1.3	
Rhododendron lapponicum			3.0	0.8		6.8		0.2	
Solidago cutleri			0.8	1.0		8.3		1.0	
Prenanthes nana				0.1	0.1	0.1	0.1	4.4	
Scirpus caespitosus var. *callosus*				0.7		0.2		10.2	11.0
Loiseleuria procumbens				0.5		7.9			0.2
Hierochloe alpina			0.1	0.6				2.1	
Carex canescens				0.5			4.0		
Betula minor				0.7					
Geum peckii				0.2				16.2	5.2
Ledum groenlandicum					15.4		1.2		0.5
Cornus canadensis					11.8		5.1		
Vaccinium angustifolium					10.7		1.0		
Maianthemum canadense					6.8		1.4		
Betula glandulosa var. *rotundifolia*					4.5				6.6
Empetrum eamesii spp. *hermaphroditum*					2.1				
Deschampsia flexuosa					2.0		14.0		
Trientalis borealis					1.5		0.3		0.8
Vaccinium caespitosum					1.5		16.3		
Lycopodium annotinum var. *pungens*					1.6		0.2		
Solidago macrophylla var. *thyrsoidea*							11.0		
Clintonia borealis							7.4		
Coptis groenlandica							4.7		
Veratrum viride							3.8		
Houstonia caerulea var. *faxonorum*							2.9		
Juncus filiformis							1.3		
Calamagrostis canadensis var. *scabra*							1.0	0.1	
Cassiope hypnoides							0.2		
Luzula parviflora var. *melanocarpa*							0.2		
Spiraea latifolia var. *septentrionalis*							0.2		
Dryopteris spinulosa							0.1		
Salix uva-ursi								16.8	
Carex scirpoidea								10.4	
Polygonum viviparum								6.2	
Campanula rotundifolia var. *arctica*								3.0	

Table 14.4. (*Cont.*)

Species	Sedge meadow	Sedge-dwarf shrub heath	Sedge-rush-dwarf shrub heath	Shrub heath-rush	Dwarf shrub heath	Diapensia	Snow-bank	Stream-side	Bog
Salix planifolia								0.8	
Carex capitata								0.4	
Poa alpigena								0.3	
Carex capillaris								0.1	
Kalmia polifolia									6.3
Vaccinium oxycoccos									5.5
Total number of species	2	7	11	17	18	13	25	23	12
Average number of species	2	5	7	9	13	10	14	14	9
Number of species restricted to community type	0	0	0	0	1	0	10	8	2

Source: Bliss (1963).

lows. Most of the vegetation > 1225 m is confined to rock crevices and the edges of small mountain streams. On north-facing slopes with late snow cover, *Cassiope tetragona* is the dominant species, but the most widespread and conspicuous genus in this elevational zone is *Saxifraga*, with the following species being important: *S. bronchialis, S. caespitosa, S. eschscholtzii, S. flagellaris, S. oppositifolia, S. punctata, S. reflexa, S. serpyllifolia, S. tricuspidata,* and *S. davurica*. The presence or absence of water plays an important role in plant distribution and community development in this part of the Brooks Range.

The most thorough and extensive vegetational study of the higher Brooks Range is that of Odasz (1983) on vegetation patterns through the tree-limit ecotone in the headwaters of the Alatna River. She sampled vegetation, soils, slope aspect, snow depths, and depth to permafrost in 247 stands, using the Braun–Blanquet system along mesotopographic gradients. Similarities in these stands were measured by Sorensen's Index. Odasz concludes that understory ecotonal vegetation is a more finely tuned indicator of climatic change than observable tree limit.

Mountains of central and southern Alaska and the Yukon. Between the Brooks Range of northern Alaska and the very high and icy Alaska, Wrangell, and St. Elias ranges, there are some relatively low mountains of about 1800 m elevation with varied geological composition, including the White Mountains and Crazy Mountains that lie between the Tanana and Yukon Rivers. Only the summits are alpine. General descriptions of the alpine vegetation and detailed lists of the floras have been provided by Gjaerevoll (1958, 1963, 1980). How-

ever, quantitative vegetational data are almost lacking. On the other hand, the ecosystem ecology of the alpine zone of these mountains has been rather thoroughly investigated around Eagle Summit by P. C. Miller and his colleagues (Miller 1982), who included many tables that describe the vegetation in the quantitative terms of physiological plant ecology. One gets a very good idea of the structure and operation of this arctic-alpine vegetation and its communities from this useful viewpoint.

The alpine vegetation of the Alaska Range and the Wrangells has not been studied thoroughly above an elevation of about 900 m. Where it is not covered by ice and snow, Viereck (1966) describes it as "low alpine tundra." Viereck's research on succession on the outwash gravels of the Muldrow Glacier in the Alaska Range at 700–760 m describes climax vegetation as low shrub–tussock–moss tundra dominated by *Eriophorum vaginatum, Sphagnum warmstorfii,* and *Vaccinium vitis-idaea.*

Rocky Mountains. The most intensively studied alpine vegetation in North America is that of the Rocky Mountains. Surprisingly, there is a rather strong coherence in floristic composition of the alpine vegetation along this latitudinal gradient at least as far south as the high mountains of northern New Mexico and northeastern Arizona. This seems to be because the Cordilleran pathway connecting the Arctic with the Continental Divide in Colorado has been there for a long time: at least from the beginning of the Cenozoic and late Cretaceous. As Billings (1974) and Axelrod and Raven (1985) have suggested, most of the upward and latitudinal migrations of plant taxa, however, have occurred during the wildly fluctuating climates of the Pleisto-

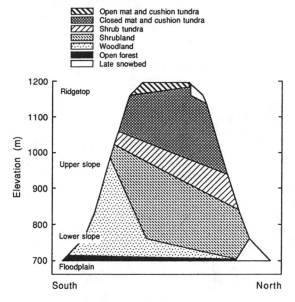

Figure 14.7. Vegetation zonation in the Dietrich River Valley, southern Brooks Range, at the northern limit of forest in the boreal region of the Rocky Mountains. (From Billings 1990, after Densmore 1980.)

cene. There have been many times in the last million years or so when cold land bridges at relatively low elevation have connected separate ranges of the Rockies and provided avenues of dispersal to and from Colorado, New Mexico, Wyoming, and the Arctic. Nor did these all have to be open simultaneously. Plants can be patient in their migratory movements.

Alpine vegetation here is a repeatable gradient from open fellfields with dwarf cushion plants on upper windward slopes, across rather barren ridges to moist graminoid-dicot meadows, and to wet meadows and bogs at the bases of the lee slopes. From north to south, these differ very little in floristic composition.

Summer precipitation increases toward the south, particularly in the form of thunderstorms that reach maximum frequency and precipitation amounts in southern Colorado and northern New Mexico. This summer rain, hail, and snow have relatively little effect on the moist and wet meadows near the lee bases of these mesotopographic gradients but I believe that it is a strong factor in maintaining certain arctic species on the ridges and fellfields.

Because environment plays such an important role in the present distribution and composition of the Rocky Mountain alpine vegetation types, it is not surprising that there has been emphasis in the

past 40 yr on the effects of environmental factors on these communities. Bliss (1956), for example, compared plant growth at four alpine sites in the Medicine Bow Mountains of Wyoming with similar sites in the Arctic in regard to temperature, water, and soils. Scott and Billings (1964) carried the temperature-water approach even further with their studies of alpine primary productivity and net photosynthesis, both in the field and in the laboratory.

The nature and activity of soils and geologic substrata have also been subjects of attention in regard to vegetation-environment interactions above timberline in the Rocky Mountains. For example, from much previous work in the mountains of Europe, it has been known for a long time that alpine plant distribution and communities are influenced to a great extent by limestone and other calcareous substrata. Bamberg and Major (1968) found that the effects of calcareous parent materials persist a long time even in cold, frost-churned soil. Such cryopedogenic activity and its effects on the nature and dynamics of Rocky Mountain alpine vegetation have been the subject of research by Billings and Mooney (1959), Johnson and Billings (1962), and Bryant and Scheinberg (1970).

The most thorough research on alpine vegetation in the Rocky Mountains has been done in the Indian Peaks area of the Front Range above Boulder, Colorado by Marr (1961) and Komárkova (1979, 1980) using standard Braun-Blanquet methods plus ordination. Niwot Ridge (Fig. 4.3) was the site of most of this work, and it has resulted in a detailed vegetation map in color at a scale of 1:10,000 (Komárkova and Webber 1978) that details the distribution of 21 associations, alliances, orders, and classes of vegetation occurring on Niwot Ridge above timberline.

In the central and southern Rocky Mountains, several alpine vegetation types appear over and over again along dry to wet moisture gradients. Among these gradients are relatively dry fellfields (Fig. 14.8), moist meadows (Fig. 14.9), wet meltwater meadows (Fig. 14.10), and rather barren areas beneath long-lasting snowbanks. These appear also in Figure 4.5 and in the data of Table 14.2 earlier in this chapter. Stand data from the most productive of the gradients (moist meadow = *Geum* turf; wet meadow = *Deschampsia* meadow) have been published by Johnson and Billings (1962) and are included here in Tables 14.5 and 14.6 as representative of the vegetational structure of such alpine communities of the Rocky Mountains.

Mountains of the Great Basin and the Southwest. From the Wasatch Mountains of central

Figure 14.8. Alpine fellfield vegetation on glacially deposited quartzite at 3300 m in the Medicine Bow Mountains, Wyoming. The prominent plant in flower is Hymenoxys grandiflora. (Photo by W. D. Billings.)

Figure 14.9. Moist alpine meadow at 3275 m in the Beartooth Mountains, Wyoming. The prominent plant in flower is Bistorta bistortoides. In the middleground is a semipermanent snowbank, still present near the end of summer and surrounded by barren gravels recently exposed by snowmelt. (Photo by W. D. Billings.)

Figure 14.10. Meltwater pond and hummocky wet meadow of sedges at 3675 m near Independence Pass, Colorado. (Photo by W. D. Billings.)

Utah to the Sierra Nevada, the rivers drain inward to the old lake basins of the Pleistocene and the early Holocene, now mostly dry. Rimming the long desert valleys of Nevada and western Utah there are almost 200 high but narrow mountain ranges. These montane "islands" trend north and south above the cold desert valleys, or "seas" (as in Billings 1978). Many of the Great Basin mountain ranges are capped with alpine vegetation beyond timberline that extends far above the forest zones on the mountain slopes and the deserts that cover the valleys, as illustrated in Figure 14.11. Southward, other montane islands rise above the warm deserts of Arizona (Billings 1980; Gehlbach 1981). One of the latter mountains is of particular importance as an alpine refugium: San Francisco Mountain (3857 m) in northern Arizona. The classic description of the life zones of this high and isolated peak and its surroundings is the pioneer work of Merriam (1890). This quiescent Pleistocene volcano rises above the pines and junipers of the Coconino Plateau, with no other alpine peaks visible even in the distance. Baldy Peak (3533 m) in the White Mountains, 260 km to the southeast, and Navajo Mountain (3175 m), 185 km to the north, are the nearest candidates. But Baldy is marginally alpine at best; it is essentially a subalpine "bald." Navajo Mountain is in the same category. Charleston Mountain (3630 m) in southern Nevada, 380 km to the northwest, has only a small alpine flora on limestone.

The remarkable thing about the top of San Francisco Mountain is that within an alpine area of only 5.18 km² is an aggregation of alpine plant species with a close relationship to the circumpolar arctic flora and to the alpine floras of the Rocky Mountains, as carefully checked in herbaria by Merriam (1890). Much later, Little (1941) listed these species, described the communities, and speculated on the geographic origins of this particular alpine flora. Little's figures show 41% arctic-alpine, 49% Rocky Mountain alpine, 6% southwestern North American, and 4% endemic to the mountain. The total alpine flora on this mountain, according to Little, was 49 vascular plant species at the time of his 1941 publication.

Only two associations were recognized by Little: alpine rock field and alpine meadow. The rock field has nearly all the species that grow above timberline on this mountain. On my visit to the sum-

Table 14.5. *Cover data (as percentages) for taxa in Geum turf.*

	Stands																					
Species	24	36	23	38	31	6	15	14	32	5	13	17	18	19	37	33	12	11	34	10	4	Mean
Geum rossii	21	26	23	26	22	20	16	12	12	10	15	15	5	6	13	7	10	5	5	3		11.7
Carex drummondiana		8	9	14	13	10	10	7	8	8	14	15	5	8	17	11	18	11	19	20	16	16.1
Phlox caespitosa	4		5	8	9	6			3	3		5	4	3	5		10	5	7	3	3	4.0
Artemisia scopulorum	10	13	13	4	7	6	6	4														2.8
Trifolium parryi	8	13	5	5																		1.5
Luzula spicata	2	4		5		4	4	3	2	3									3	2		1.5
Polygonum bistortoides	4		6			4		4		3	5	3	3	10	5	4	4		3			3.0
Potentilla diversifolia	2	3	4	3			8															1.0
Deschampsia caespitosa	2	3	4	3			8		3								2					1.5
Mosses	4	5	5				6	2		2												0.9
Carex scopulorum	3							5														0.4
Festuca ovina	3	3	3				2	4	4	3		10		4			2	2	2		2	2.1
Trifolium nanum		2	2		8				25				6	20					4			3.3
Erigeron simplex		2	2	2					7	2	2		7		2		2		5	4	7	2.0
Poa species	5	6	6	2									2	3				2				1.2
Senecio fuscatus	2																					0.1
Gentiana algida						4																0.2
Polemonium viscosum						2						2		3								0.3
Cerastium beeringianum						2						3										0.2
Mertensia alpina						2											2					0.2
Bupleurum americanum						4			3	3		10		5				3				1.3
Antennaria species					6							3								2		0.5
Sedum rosea							2															0.1
Carex elynoides					3																	0.1
Lupinus monticola												3								3		0.4
Dodecatheon radicatum												2										0.2
Smelowskia calycina				2	4				5	4			2				4	4	5	2	3	0.5
Silene acaulis				4		2	2		3	4	4		11	12			14	7				2.1
Arenaria obtusiloba				2		4	2		6	6		7			6			8	4	6	4	1.9
Trisetum spicatum		3				2									6			3		3	7	1.9
Lomatium montanum			2	2		4			5									8	4	4		1.5
Selaginella densa				2				2						3	3	5	2	3	5			1.3
Eritrichium elongatum													2	2		2		3	5		5	0.9
Oxytropis species													2	2			4	5	5	2	5	1.0
Potentilla nivea																			2	5	2	0.1
Senecio canus																				2	4	0.2
Lichens	2	2	2	2		10	4	8		15		6	18	9	20			5		14	19	4.5
Litter	6	4	2	4	34	14	16	18	6	10	35	6	8	9	7	61	60	19	6	20	10	14.6
Bare soil or rock	10	18	8	14	1	4	14	11	8	9		6	8	9			6	10	6	6	6	7.4

Source: Johnson and Billings (1962).

Table 14.6. *Cover data (as percentages) for taxa in* Deschampsia *meadow*

Species	16	35	25	27	26	3	40	42	41	30	28	29	43	Mean
Deschampsia caespitosa	11	10	44	33	43		13		2	10	4	6		13.5
Geum rossii	5		11	9	9	11								3.5
Trifolium parryi	31		3											2.6
Luzula spicata	2	2			2									0.5
Trisetum spicatum		2			2									0.3
Arenaria obtusiloba	3													0.2
Salix nivalis	12	5												1.3
Salix planifolia		19												1.5
Calamagrostis purpurascens		7												0.5
Polygonum viviparum		5												0.4
Artemisia scopulorum	11	2	4	5										1.7
Potentilla diversifolia		2	2	3	2	9						8		2.0
Festuca ovina		2		3	4	5							2	1.2
Erigeron simplex			4			7								0.9
Mertensia alpina			3											0.2
Cerastium beeringianum			3	4	5									0.9
Polygonum bistortoides	7		6	5					4		2	9		2.5
Gentiana algida			2											0.2
Lichens	3	4						2	11					1.5
Poa species			2		22				3	3	3			2.5
Smelowskia calycina						5								0.4
Mosses		3		13		2		4	8	4	8	11	3	4.2
Carex scopulorum		3			44	44	37	26	32	25	7	3		16.8
Antennaria species							5	9	20	3				2.8
Carex species		2		2			4		4	19	2	8	5	3.5
Caltha leptosepala		5		5						9	24	24		5.2
Sibbaldia procumbens							5			3	4	14		2.0
Senecio cymbalarioides										3	6	4		1.0
Juncus drummondii										2	7		2	0.8
Arabis species													3	0.2
Water										2	10	7		1.5
Litter	3	6	15	7	10		12	4	5	2	3			5.2
Bare soil or rock	8	22	5	2		9	14	40	12	5			81	15.2

Source: Johnson and Billings (1962).

mit during the 1960s, I found the soil surface to be generally covered with volcanic rocks of various sizes but with much fine material beneath the dry rocks. The fine, dark material is surprisingly moist in summer, with water derived from the frequent thundershowers that gather over the peak in the afternoons. My data on the principal species growing in this moist fine "soil" are *Geum rossii, Silene acaulis, Cerastium beeringianum, Festuca brachyphylla, Poa rupicola, Trisetum spicatum, Primula parryi,* and *Oxyria digyna.* These all have deep, well-developed root systems in the loose porous soil. The alpine meadow community also is dominated by *Geum rossii.* Present there, in addition, are *Sibbaldia procumbens, Silene acaulis, Trisetum spicatum, Carex albonegra, Pedicularis parryi,* and *Luzula spicata.*

Schaak (1983) has restudied the alpine vegetation of this mountain. With additional collections, he now lists the alpine flora as consisting of 80 taxa of vascular plants. Most of these are rare or uncommon, but their presence is significant phytogeographically. This is particularly true of the arctic saxifrages *Saxifraga cespitosa* and *S. flagellaris,* which reach their southernmost limits on this continent on San Francisco Peak. I concur with Schaak's hypothesis that the dispersal pathway for these taxa to this particular isolated mountain was from the Rocky Mountains by way of the mountains of western New Mexico, the White Mountains of Arizona, and the nearby Mogollon Rim. Such migrations would have had to take place during the colder climates of the late Pleistocene. In sum, San Francisco Peak has a remarkable assemblage of arctic-alpine plants, most of them at their southern and most isolated limits in North America. Merriam was right more than a century ago when he wrote, "Many of the plants found on the high rocky summit of San Francisco Mountain . . . in-

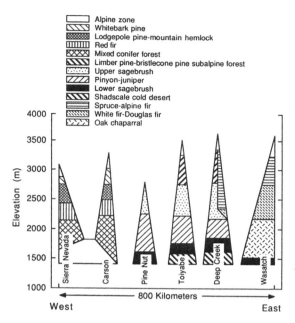

Legend:
- ☐ Alpine zone
- ⧖ Whitebark pine
- ▨ Lodgepole pine-mountain hemlock
- ▥ Red fir
- ▧ Mixed conifer forest
- ⧗ Limber pine-bristlecone pine subalpine forest
- ▦ Upper sagebrush
- ▧ Pinyon-juniper
- ▉ Lower sagebrush
- ⧕ Shadscale cold desert
- ▤ Spruce-alpine fir
- ▨ White fir-Douglas fir
- ▧ Oak chaparral

Figure 14.11. Vegetation zonation across the Basin and Range from the west face of the Wasatch Mountains to the east face of the Sierra Nevada at approximately 39° N latitude. Not all intermediate mountain ranges are shown. (Drawing by the author.)

habit the arctic regions of the globe and extend far south on the summits of the higher mountain ranges.''

Relatively little is known about the alpine flora and vegetation of the mountains of the Great Basin per se. There are some notable exceptions. Two of these are the Deep Creek Mountains in western Utah and the Ruby Mountains in northeastern Nevada. McMillan (1948) listed 80 species as being present above timberline in the relatively tall and narrow Deep Creeks. In Loope's (1969) classic work on the Ruby Mountains, a much larger range than the Deep Creeks, 189 alpine species are reported. In both cases, the alpine floras have a distinctly Rocky Mountain affinity. This is brought out in Table 14.7 (from Loope 1969) that lists the cover of plant species in six alpine turf communities as compared with four dry alpine sites. I do not know of similar studies on Wheeler Peak in the Snake Range of eastern Nevada, but in that higher and more extensive alpine area there should also be strong floristic ties with the Rocky Mountains.

There have been scattered studies on the mountain ranges of central Nevada, but as yet there has been nothing as definitive in regard to alpine vegetation and floristics as those of McMillan's or Loope's earlier research above timberline. The Toquima, Toiyabe, Grant, and other high ranges

await more alpine work. However, there has been an extensive and intensive piece of research on the subalpine conifer woodlands and timberlines of all the high mountains of the Great Basin by David Charlet (1995, 1996), who has detailed his hundreds of collections of all the conifer species and their general ecology throughout the Great Basin with special reference to timberlines.

The alpine floras and vegetation of some of the high mountains in the western Great Basin are fairly well known. These mountains are all west of the Pleistocene Lake Lahontan shoreline. The alpine zone on Mt. Grant (3426 m) in the Wassuk Range has an area of only 2.6 km² but has a flora of 70 species (Bell and Johnson 1980). The affinities of this flora are mainly with that of the Sierra Nevada to the west and with mountainous western North America in general. However, 12 species are widespread arctic-alpine. The same authors (Bell Hunter and Johnson 1983) also studied the larger alpine zone of the Sweetwater Mountains (3558 m) about 45 km southwest of Mt. Grant and only 33 km east of the crest of the Sierra. The alpine area of the Sweetwaters is about 16 km². The alpine flora there consists of 173 species of vascular plants. Of these, 94% also occur in the Sierra, 75% in the Great Basin, 52% in the southern Rocky Mountains, and 18% in the Arctic. The Sweetwater Mountains alpine flora is thus clearly Sierran in derivation.

The White Mountains of eastern California and western Nevada stand tall at the very western edge of the Great Basin, a figurative stone's throw east of the Sierra Nevada. The two big mountain ranges are 40 km apart from crest to crest and separated only by the deep Owens Valley. White Mountain Peak reaches 4341 m; much of the range is above its timberline of *Pinus longaeva*. Of all the Great Basin mountains, the White Mountains are probably the best known ecologically.

Lloyd and Mitchell (1973) have made a thorough study of the flora of this whole range in which they discuss its phytogeography and comparative floristics. Half of the endemic plant taxa in the White Mountains occur in its alpine zone. The flora of the whole alpine zone in this range is about 200 taxa, of which 125 are characteristically alpine. About 62% of these also occur in the nearby Sierra Nevada, and only 28% are also in the Rocky Mountains.

Mooney (1973) described the plant communities of the White Mountains. The alpine zone extends from about 3500 m to the summit at 4341 m. This mountain range is dry and very cold so that the vegetation around White Mountain Peak itself is extremely sparse. Even so, three species are restricted to the high alpine zone above 4000 m:

Table 14.7. Cover data (as percentages) for taxa in alpine turf (columns 1–6) and dry alpine communities (columns 7–10) of the North Ruby Mountains. Sites 1, 2, and 7 are at 3047 m, south of Lamoilie Lake; site 3 is 3100 m, right fork of Lamoille Canyon; sites 4–6 and 8–10 are at 3100 m, Island Lake cirque.

Species	1[a]	2	3	4	5	6	7	8	9	10
Erigeron peregrinus	18.9	25.3	13.2	13.9	12.7	3.1	4.5	1.1	—	0.4
Salix arctica	12.0	23.9	7.1	1.9	22.5	33.2	—	5.6	—	—
Carex elynoides	—	8.5	5.4	21.3	—	42.7	—	—	—	—
Caltha leptosepala	10.2	22.0	—	10.4	10.5	—	—	—	—	—
Sibbaldia procumbens	0.7	2.1	6.4	2.7	4.7	8.4	—	2.2	18.6	4.5
Polygonum bistortoides	6.9	3.9	8.9	10.1	3.3	—	—	2.6	3.6	0.7
Pedicularis groenlandica	1.1	1.1	2.5	6.6	5.1	—	—	—	—	—
Festuca brachyphylla	2.2	0.7	4.6	0.8	3.3	—	—	—	—	—
Geum rossii	3.3	—	—	—	—	1.1	76.5	5.6	14.7	5.2
Epilobium alpinum	2.2	—	0.4	—	—	—	—	—	—	—
Epilobium latifolium	1.8	—	—	—	—	—	1.5	—	—	—
Veronica wormskjoldii	2.2	—	6.4	2.3	1.1	—	—	—	—	—
Astragalus alpinus	5.1	6.7	5.0	0.4	0.4	2.3	—	0.7	—	—
Antennaria umbrinella	4.4	—	2.1	—	—	1.1	—	—	—	—
Potentilla diversifolia	4.4	0.7	—	2.3	7.6	—	—	0.7	1.1	0.7
Senecio cymbalarioides	0.4	—	—	16.3	13.4	1.1	7.1	0.7	0.7	0.4
Viola adunca	2.5	—	1.4	—	—	—	—	—	—	—
Gentiana calycosa	1.5	—	0.4	1.2	2.9	—	—	—	—	—
Carex pseudoscirpoidea	3.3	—	—	—	—	—	—	4.8	—	4.5
Carex festivella	0.4	—	—	—	—	—	—	—	—	—
Ranunculus eschscholtzii	—	1.1	—	—	—	4.2	—	—	0.4	—
Thalictrum alpinum	—	—	6.1	1.9	—	—	—	—	—	—
Taraxacum officinale	—	—	2.1	—	—	—	—	—	—	—
Dodecatheon alpinum	—	—	0.7	—	—	—	—	—	—	—
Kalmia polifolia	—	—	—	0.4	1.8	—	—	—	0.4	—
Mimulus primuloides	—	—	1.8	1.9	—	—	—	—	—	—
Juncus mertensianus	—	—	—	—	0.4	—	—	—	—	—
Vaccinium caespitosum	—	—	1.4	0.8	—	—	—	27.4	—	23.6
Parnassia fimbriata	—	—	3.9	—	1.1	—	—	—	—	—
Potentilla fruticosa	—	—	0.4	—	—	—	—	—	—	—
Castilleja linariaefolia	—	—	—	—	—	—	1.9	—	—	—
Mertensia ciliata	—	—	—	—	—	—	0.7	—	—	—
Hackelia jessicae	—	—	—	—	—	—	0.4	—	—	—
Festuca ovina	—	—	—	—	—	0.4	—	5.2	36.5	—
Phleum alpinum	—	—	0.4	—	0.4	—	—	—	—	—
Juncus drummondii	—	—	1.1	—	—	0.4	—	1.5	3.6	1.4
Deschampsia caespitosa	—	—	—	—	—	—	—	—	3.6	—
Salix orestera	—	—	—	1.9	—	—	—	—	—	—
Penstemon procerus	—	—	—	—	—	—	—	0.4	—	1.1
Trifolium monanthum	—	—	—	—	—	—	—	1.1	—	—
Antennaria rosea	—	—	—	—	—	—	—	5.6	4.6	3.7
Agrostis rossae	—	—	—	—	—	—	—	—	0.4	—
Trisetum spicatum	—	—	—	—	—	—	—	—	—	1.4
Carex nova	—	—	5.0	—	4.4	—	—	—	—	—
Carex species	—	—	—	—	—	—	—	—	1.1	—
Poa palustris	—	—	—	—	—	—	1.5	—	—	—
Poa epilis	—	—	—	—	—	—	—	3.8	0.7	4.5
Poa fendleriana	—	—	—	—	—	—	—	—	—	4.1
Soil	17.1	1.1	6.8	2.7	2.5	—	4.1	14.8	1.4	26.2
Litter	—	2.1	6.4	—	2.2	1.9	—	1.9	4.6	9.7
Rock	—	1.1	—	—	—	—	1.9	—	3.8	5.2

[a] Locations of sampling sites: 1, 2, and 7, 9750 ft, south of Lamoilie Lake (7/30/68); 3, 10,000 ft, right fork of Lamoille Canyon (8/4/68); 4–6, 8–10, 10,000 ft, Island Lake cirque (8/22/68) and (8/23/68).
Source: Loope (1969).

Phoenicaulis eurycarpa, Polemonium chartaceum, and *Erigeron vagus*. There is a strong relationship to geologic substratum. The dolomites have a rather barren appearance and low plant cover except for their open and very old forests of *Pinus longaeva* as compared to the granites and quartzites. Table 14.8 compares the plant cover at three sites in the alpine zone. Additional data on the alpine vegetation and flora in relation to active soil frost features are provided by Mitchell, La Marche, and Lloyd (1966).

Billings (1978) has calculated floristic similarity (Sørensen' Index) between alpine floras across the Great Basin along two transects from the Rocky Mountains to the Sierra Nevada. The northern transect extends from the Beartooth Mountains on the Wyoming–Montana state line to Piute Pass in the Sierra. The southern transect is from San Francisco Peak to Olancha Peak in the Sierra. These percentages are shown in Table 14.9. These data demonstrate that the alpine floras east of a hypothetical line extending southward from near Elko, Nevada, to Las Vegas, Nevada, are strongly related to those of the Rocky Mountains. West of that line, alpine floras of the Great Basin are unique. However, in the White Mountains there is some floristic relationship of the alpine flora to that of the Sierra, but vegetationally the two mountain ranges are distinct.

The Sierra Nevada. This large mountain range in eastern California and western Nevada extends almost 650 km from northwest to southeast. Its base is a large granitic batholith tilted toward the west and capped by andesitic flows in its northern reaches. In the south, Mt. Whitney (4418 m) is the highest peak in the lower 48 states. The Sierra is a young range uplifted mainly during the Pliocene and Pleistocene, particularly the northern Sierra in the region of Lake Tahoe. Axelrod (1962, 1976) estimates that in early to middle Pliocene the main ridge of the ancestral Sierra Nevada was near 915 m in elevation in the Tahoe region as compared to its present elevation of more than 3000 m. The relatively low Pliocene elevations allowed the migrations of *Sequoidendron*, the giant Sequoia, westward across the Sierra from the western Great Basin where its Pliocene fossils now occur. The present geographical range of *Sequoidendron* is now restricted to the western slopes of the Sierra Nevada.

Timberline in the present northern Sierra Nevada ranges from 3300 m to 2550 m. The timberline tree species in the northern Sierra is whitebark pine, *Pinus albicaulis* (see Chapter 5 of this volume). At windy timberline, it forms a krummholz scarcely a meter high. However, in protected snowy places, it forms a multistemmed tree as tall as 10–12 m. During full glacial in the northern Sierra, whitebark pine ranged eastward to the western shores of Pleistocene (Pluvial) Lake Lahontan at an elevation of 1380 m. Nowak, Nowak, Tausch, and Wigand (1994) found macrofossils of *Pinus albicaulis* C-dated at about 23,000 yr BP in packrat middens within 2–3 km of the ancient shoreline of the Pleistocene lake. Such a fossil location is about 1100 m below whitebark pine's present elevation in the Sierra 50 km to the southwest.

The Sierran alpine flora is the largest and richest such flora on the continent. Neither the alpine flora nor its vegetation is as well known as they could be in a definitive sense. However, Major and Taylor (1988) have done an excellent job in synthesizing the available information on the alpine vegetation. They present long and detailed tables on the compositions of the main alpine communities. In their paper are details particularly concerning the close relationships between the high Sierran geologic substrata and their vegetative cover.

Chabot and Billings (1972) studied a series of elevational sites on the steep eastern escarpment of the Sierra from the desert near Bishop at 1400 m to above Piute Pass at 3540 m on the glacially scoured granites of the Sierran crest (see Fig. 14.12). Emphasis was on the question, "How does an alpine flora originate?" Much of the research in the field was devoted to measuring the microenvironments along this steep transect in regard to the distribution of different plant species along the gradient. Physiological measurements, including water relations, photosynthesis, respiration, storage products, and acclimation, were made on these species under controlled conditions in the Duke University Phytotron.

From a phytogeographical standpoint, only 19% of the alpine species on the granites above Piute Pass also occur in the Arctic, in strong contrast to the almost 50% in the Beartooth Mountains of Montana and Wyoming. Those alpine species near Piute Pass that are endemic to the Sierra constitute 17% of the Sierran alpine flora. These are in genera commonly present in California or Great Basin floras at lower elevations. A number of the remaining alpine species near Piute Pass also have populations in the desert below. Desert plant taxa certainly have contributed to the Sierran alpine gene pool. Most of the alpine flora consists of perennial herbaceous species, but there are more annual species in this alpine region than in other alpine floras.

The alpine vegetation on these granites is particularly sparse (Table 14.10). This very open vegetation is characterized by low plants of *Phlox caespitosa, Pentstemon davidsonii, Ivesia pygmaea, Ivesia lycopodioides, Draba lemmonii. Arenaria nuttallii, Carex hellerii, Eriogonum ochrocephalum*, and *Potentilla brewerii*.

Table 14.8. *Plant cover (%) in three alpine communities of the White Mountains, California*

Species	White Mt. Peak pyramid, 4176 m	Dolomite barrens, 3597 m	Fellfield granite, 3871 m
Herbs			
Erigeron vagus	0.40	—	0.52
Festuca brachyphylla	0.56	—	—
Calyptridium umbellatum	0.24	—	—
Polemonium chartaceum	0.28	—	—
Eriogonum gracilipes	—	1.32	—
Poa rupicola	—	2.00	—
Sitanion hystrix	—	0.36	0.96
Phlox covillei	—	6.52	—
Erigeron pygmaeus	—	0.92	—
Draba sierrae	—	0.28	0.48
Arenaria kingii	—	0.64	—
Castilleja nana	—	0.20	—
Eriogonum ovalifolium	—	—	6.16
Carex helleri	—	—	0.24
Trifolium monoense	—	—	27.64
Selaginella watsoni	—	—	0.68
Haplopappus apargioides	—	—	1.04
Koeleria cristata	—	—	11.12
Lewisia pygmaea	—	—	0.08
Potentilla pennsylvanica	—	—	0.20
Total cover, all plants	1.48	12.24	50.08

Source: Mooney (1973).

Table 14.9. *Sørensen's index of floristic similarity (%) for all combinations of selected alpine areas. Numbers in parentheses are distances (km) between the alpine regions.*

	Northern transect						Southern transect		
	Bear-tooth	Deep Creeks	Ruby Mts.	Toiyabe Range	Pellisier Flats (White Mts.)	Piute Pass (Sierra)	Olancha Pk. group (Sierra)	Spring Mts.	San Francisco Pks.
Beartooth	—	24	33	14	14	9	14	8	23
Deep Creeks	(676)	—	39	22	19	10	15	13	31
Ruby Mts.	(692)	(153)	—	20	14	11	14	8	19
Toiyabe Range	(917)	(322)	(257)	—	21	16	13	18	21
Pellisier Flats (White Mts.)	(1102)	(451)	(418)	(145)	—	34	36	16	19
Piute Pass (Sierra)	(1167)	(515)	(483)	(217)	(72)	—	32	10	4
Olancha Pk. group (Sierra)	(1223)	(547)	(539)	(290)	(169)	(129)	—	13	13
Spring Mts.	(1110)	(434)	(475)	(325)	(298)	(306)	(225)	—	14
San Francisco Pks.	(1086)	(531)	(668)	(636)	(660)	(668)	(603)	(378)	—

Source: Billings (1978).

At slightly higher elevations on the same granitoid substratum, *Oxyria digyna*, *Polemonium eximium*, and *Hulsia algida* are constant members of the alpine vegetation wherever soil has accumulated in the granite. Extremely arid sites of this type have almost no vascular plants except for small individuals of *Calyptri-* *dium umbellatum* and *Polygonum minimum*; the latter is a miniature annual.

Carson Pass, about 27 km south of Lake Tahoe, is the northern limit for many alpine species in the High Sierra according to Major and Taylor (1988). Here, in the northern Sierra, the old granodiorites

Figure 14.12. Very open fellfield on glacially scoured
granite at 3540 m above Piute Pass in the Sierra Nevada.
Scattered plants are primarily Lupinus breweri, Potentilla
breweri, Antennaria rosea, and Calyptridium umbellatum.
Mt. Humphreys appears in the background. (Photo by
W. D. Billings.)

are overlain by volcanic flows, mainly andesites.
The granitic rocks have a better representation of
truly alpine plant species, particularly on north-
facing slopes. Even on the andesites, however, in
places where snow stays late into the summer,
there are alpine plants such as the rare but ubiquitous Oxyria digyna. However, where the andeites
are swept clear of winter snow on ridgetops, there
are communities of plant species more typical of
the low-elevation cold deserts of nearby western
Nevada along the eastern base of the Sierra: Sitanion hystrix, Balsamorrhiza sagittata, Viola beckwithii,
Wyethia mollis, Eriogonum umbellatum, E. ovalifolium,
Crepis occidentalis, Leptodactylon pungens, and Lygo-
desmia spinosa (Billings, unpublished data). From
these ridgetop alpine semideserts down the lee
slopes of Meiss Ridge between Carson Pass and
Lake Tahoe there is a classic mesotopographic gradient from ridge to snowbank to wet meadow.

The Cascades Range. The most conspicuous alpine mountains of the Cascades Range are the high
ice-covered volcanoes that are strung like beads
along the crest of the Cascades from Mt. Lassen
and Mt. Shasta in northern California to Mt. Baker
in northwestern Washington. Above timberline on
high ridges, but below the ice, these "island"
mountains support small, open communities of
herbaceous plants in the alpine zone. These show
considerable relationship with similar vegetation
on ridges in the Sierra Nevada.

However, in northern Washington, the northern
Cascades provide a rugged set of young mountains, geologically complex and heavily glaciated
in the past. The alpine vegetation is well developed
in this snowy, cool environment. Douglas and Bliss
(1977) sampled 128 stands of these alpine communities and related the alpine vegetation to soils
and microenvironments by using methods of ordination. Their results show a large array of plant
community types across these mountains from
west to east.

The alpine flora of the northern Cascades has
floristic affinities with many mountain regions in
western North America. Unlike the alpine flora of
the Sierra, the strongest relationships are with the
arctic and cordilleran regions to the north and east.
A smaller part of the flora is restricted to the

Table 14.10. *Composition and coverage of the alpine plant community on granite above Piute Pass, 3540 m, central Sierra Nevada. P indicates <1% cover.*

Species	Coverage (%)
Lupinus breweri	7.0
Antennaria rosea	6.5
Carex helleri	5.1
Carex phaeocephala	2.1
Sedum rosea ssp. integrifolium	1.7
Ivesia pygmaea	1.5
Solidago multiradiata	1.2
Selaginella watsoni	1.1
Lewisia pygmaea	0.5
Phlox covillei	0.3
Agrostis variabilis	0.3
Gentiana newberryi	0.2
Draba lemmonii	0.2
Trisetum spicatum	0.1
Calyptridium umbellatum	0.1
Castilleja nana	p
Antennaria alpina var. media	p
Arenaria nuttallii ssp. gracilis	p
Eriogonum ovalifolium	p
Dodecatheon jeffreyi	p
Potentilla breweri	p
Poa hansenii	p
Sitanion hystrix	p
Polygonum minimum	p
Penstemon davidsonii	p
Poa nervosa	p
Carex nigricans	p
Total plant cover	27.9
Rock	40.7
Soil	31.4

Source: Chabot and Billings (1972).

mountains from British Columbia to California. The number of Cascadian endemic alpine species in the North Cascades is very few.

The Olympic Mountains. The Olympic Mountains constitute a unique and rather isolated massif in the middle of the Olympic Peninsula in the northwestern part of the State of Washington. These mountains are not particularly high; their western slopes and peaks, such as Mt. Olympus and several others, are in the neighborhood of 2400 m. They are the wettest mountains in the lower 48 states. The higher peaks are very snowy and heavily glaciated at present. Their mean annual precipitation ranges from 400–500 cm. The lower slopes and western valleys are densely forested with very large and old trees of evergreen conifers such as *Pseudotsuga menziesii, Picea sitchensis,* and *Tsuga heterophylla* (see Chapter 5 in this book). Such forests receive mean annual rainfalls of 300–375 cm in this

relatively mild maritime climate. The northeastern slopes of these mountains between the Straits of Juan de Fuca and Puget Sound lie in a rainshadow of the higher peaks to the southwest and receive much lower precipitation of 45–60 cm.

Between the forests and the glacial ice, there are productive graminoid and dicot subalpine meadows (Kuramoto and Bliss, 1970), and, finally upslope, an alpine zone with dwarf vegetation of low shrubs and small perennial plants. These subalpine and alpine communities have been described by Bliss, Fonda, and Kuramoto (1969). According to their classifications, the following alpine communities are the most common:

1. Cushion plant vegetation on ridgetops and slopes facing south or west.
2. Vegetation stripes on unstable substrates.
3. Alpine turf communities occurring where there is protection from wind on lee slopes where some snow accumulates.
4. Snowbed communities of diverse natures depending on time of melt.

The principal dominant plant species in these four community types are the following:

1. Cushion plant type: *Phlox diffusa, Lupinus lepidus, Synthyris lanuginosa,* and *Erigeron compositus.*
2. Vegetation stripes: *Phlox diffusa, Arenaria obtusiloba, Draba lonchocarpa, Potentilla diversifolia,* and *Lupinus lepidus.*
3. Alpine turf type: *Carex phaeocepha, Potentilla, diversifolia, Phlox diffusa,* and *Solidago spathulata.*
4. Snowbed communities:
   a) Early melt: *Antennaria lanata, Carex albonigra, Polygonum bistortoides,* and *Aster alpigenus.*
   b) Late melt: *Carex nigricans, Veronica cusickii,* and *Epilobium alpinum.*
   c) Steep north-and east-facing slopes with various melt times: *Douglasia laevigata, Ranunculus eschscholtzii, Saxifraga cespitosa,* and *Oxyria digyna* (personal observations and collections by W. D. Billings).

The High Mountains of Mexico. The alpine floras of subtropical and tropical mountains are highly specialized, and very few arctic-alpine species are represented in them (see Chapter 15). However, some of the genera are present in the northern latitudes. Many other genera and species are endemic, and still others are more characteristic of the Southern Hemisphere. Beaman and Andresen (1966) collected specimens of 81 species of vascular plants in the few square kilometers of subalpine and alpine habitats > 3500 m on Cerro Potosí (ca.

3800 m) at 24°53' N in the Sierra Madre Oriental. From the standpoint of genera, the flora of Cerro Potosí is northern, but in a sample of 64 species from the summit area, 27 (42.2%) are endemic to the Sierra Madre Oriental, and of these 13 are known only from Cerro Potosí – a high percentage of endemics from a single peak!

Webster (1961) listed the upper elevational limits of alpine plants on the higher mountains of the earth, including Citlaltepetl (Píco de Orizaba) at 19°N latitude whose summit elevation in central Mexico is about 5700 m. The highest vascular plant species there is apparently *Castilleja tolucensis* at 4570 m. As Webster notes, vascular plants in the Himalayas and the southern Andes reach their upper limits almost 1500 m higher. Why should plants not go higher in Mexico? In my opinion, this could possibly be a result of the relatively young age of these high Mexican volcanoes and their isolated geographic position between the Rocky Mountains to the north and the Andes to the south. Also, arctic taxa at the species level apparently have had difficulties getting to these high mountains.

The same thing could be said for the members of the antarctic element, some of which do come up the Andes to Colombia and Venezuela. One of these, *Colobanthus quitensis*, does reach the high mountains of central Mexico occurring at 3900 m on Píco de Orizaba. Its nearest occurrence to the south is on Pichincha, the volcano overlooking Quito, Ecuador. How did this species make such a migratory jump, if indeed it did? No one has yet collected one of its associated antarctic species, the arctic-antarctic *Saxifraga cespitosa* on the Mexican volcanoes. The gap in this saxifrage's distribution is wide: San Francisco Mountain, Arizona, to Pichincha is an airline distance of 5500 km (Mulroy 1979). Why isn't it in Mexico? From Pichincha, the distribution of *S. cespitosa* extends down the high Andes to the Beagle Channel at 55°S latitude. *Saxifraga cespitosa* also has a number of geographic ecotypes: I have seen this species at elevations of over 5000 m in the Peruvian Andes, where it forms a flat mat 10 cm across and 2 cm high with a light purple color caused by considerable amounts of ultraviolet-screening anthocyanin in its leaves.

Beaman and Andresen (1966) also described alpine meadow vegetation from the summit of Cerro Potosí (ca. 3650 m) at 24°53'N latitude in the Sierra Nevada Oriental. In this alpine meadow, the presence and percentage of plant cover by species were recorded on 100 quadrats that were 50 cm on a side along a line 1 km long extending from the north to the south end of the meadow. The results are shown in Table 14.11. *Potentilla leonina*, a species of *Arenaria*, and *Bidens muelleri* together comprise almost half of total plant cover. These dominant plant species are all perennial herbs with a depressed or prostrate habit and small or finely divided leaves. In the sixth and seventh places are the familiar *Linum lewisii* and *Trisetum spicatum*, the former is a Rocky Mountain species, and the latter is a true arctic-alpine taxon. Table 4.11 shows that other species also add to the feeling that there is an influence from the north on the alpine summit of the Cerro.

ALPINE VEGETATIONAL SUCCESSION

Vegetational change and soil formation are very slow in any cold climate. These processes are even slower on glacially polished rock than on glacial till or alluvium. For example, such succession is extremely slow on the Medicine Bow Peak quartzite in Wyoming at elevations of over 3000 m. Not only is this rock extremely hard, but it has been subjected to heavy and polishing glaciation, and the present climate is quite cold. The same properties hold true for the Sierran granites west of Lake Tahoe in Desolation Valley and on the granites above Piute Pass in the Sierra farther south. Volcanic rocks in the Sierra, however, are more readily covered by alpine vegetation and soils. Pockets of till or alluvium in the lower parts of the topographic gradient are invaded rather readily; given enough time they become moist or wet alpine meadows. Flow diagrams for succession in alpine ecosystems have been published; most of these are hypothetical. It would be difficult to use them for prediction except in a general way.

Two successional processes, however, are worthy of note; one is biological, and the other is physical. Griggs (1956) presented interesting evidence for the role of plant competition in regard to invasion and succession on alpine fellfields in the Front Range of Colorado. He found that cushion plants such as *Silene acaulis*, *Minuartia obtusiloba*, *Paronychia pulvinata*, *Trifolium nanum*, and *Trifolium dasyphyllum* are readily invaded by a number of species. As many as 33 other species were observed invading *Silene acaulis* alone. From such observations, Griggs constructed a tentative competitive "ladder," near the top of which are *Geum rossii* and *Trifolium parryi*. Table 14.2 shows that these two species do prevail also in the alpine zone of the Medicine Bow Mountains of Wyoming, but their success there is primarily in the moister parts of the gradient rather than in the dry or wet sections. Surprisingly, Griggs's ideas about alpine suc-

Table 14.11. *Frequency, cover, importance value (scale 0–200), and life form for 38 taxa encountered in an alpine meadow of Cerro Potosí, Mexico. Ch = chamaephyte, H = hemicryptophyte, N = nanophanerophyte, G = geophyte.*

Species	(%) Frequency	(%) Cover	Importance value	Life form
Potentilla leonina	81	8.63	33.54	Ch
Arenaria species (Beaman 2664)	82	4.84	23.14	Ch
Bidens muelleri	54	4.31	18.36	H
Astragalus purpusii	72	1.36	12.31	Ch
Lupinus cacuminis	32	2.76	11.45	H
Festuca hephaestophila	55	1.28	10.79	H
Linum lewisii	69	0.93	10.77	H
Trisetum spicatum	62	0.83	9.65	H
Grindelia inuloides	38	1.74	9.34	H
Pinus culminicola	11	2.35	7.82	N
Astranthium beamanii	42	0.55	6.39	H
Thlaspi mexicanum	41	0.43	6.05	H
Senecio loratifolius	11	1.59	5.71	H
Draba helleriana	32	0.35	4.77	H
Senecio scalaris	26	0.36	4.08	H
Phacelia platycarpa	26	0.31	3.94	H
Hymenoxys insignis	7	0.56	2.38	H
Senecio sanguisorbae	7	0.50	2.22	H
Castilleia bella	13	0.20	2.10	H
Juniperus monticola	2	0.60	1.90	N
Villadia species (Beaman 4461)	13	0.03	1.62	H
Senecio carnerensis	7	0.25	1.52	Ch
Trifolium species	9	0.15	1.47	H
Tauschia madrensis	4	0.30	1.30	H
Sedum species (Beaman 4462)	8	0.09	1.20	Ch
Campanula rotundifolia	8	0.08	1.17	H
Achillea lanulosa	7	0.09	1.08	H
Delphinium valens	1	0.30	0.95	H
Sisyrinchium species (Schneider 939)	5	0.05	0.73	H
Stellaria cuspidata	3	0.06	0.52	H
Blepharoneuron tricholepis	3	0.06	0.52	H
Erysimun species (Beaman 2648)	3	0.03	0.39	H
Gnaphalium species (Beaman 2655)	2	0.02	0.30	H
Penstemon leonensis	2	0.02	0.30	H
Smilacina stellata	2	0.02	0.30	G
Allium species	1	0.01	0.15	G
Bromus frondosus	1	0.01	0.15	H
Cerastium brachypodum	1	0.01	0.15	H

Source: Beaman and Andreson (1966).

cession have not been avidly pursued; they deserve better.

Not all succession in alpine ecosystems is linear as in the biological example just mentioned. Wherever the physical process of soil permafrost polygonization is concerned, succession is likely to be cyclic, with no endpoint of stability. This is true on a large scale, as in the case of the thaw-lake cycle in the arctic wet tundra (Billings and Peterson 1980), but smaller-scale cyclic successions can be seen in the alpine zone.

Billings and Mooney (1959) studied such a cycle in the alpine meadows of the Medicine Bow Mountains of Wyoming (Fig. 14.13). As sedge peat hum-

mocks build up in the meadow, they eventually get too high for complete protection by winter snow cover. Strong winter winds carrying sharp snow and ice crystals blast the tops off of the higher hummocks, allowing frost action in the soil to thrust rocks up from the geologic substratum. The hummocks degrade into a stone net at or below the water table, with the stones being pushed outward. In this way, the rocks form the boundaries between the cells of the net that are themselves filled with barren alluvium covered with shallow water. In time, mosses and vascular plants get established on the border rocks and eventually form embryonic hummocks that grow higher than the

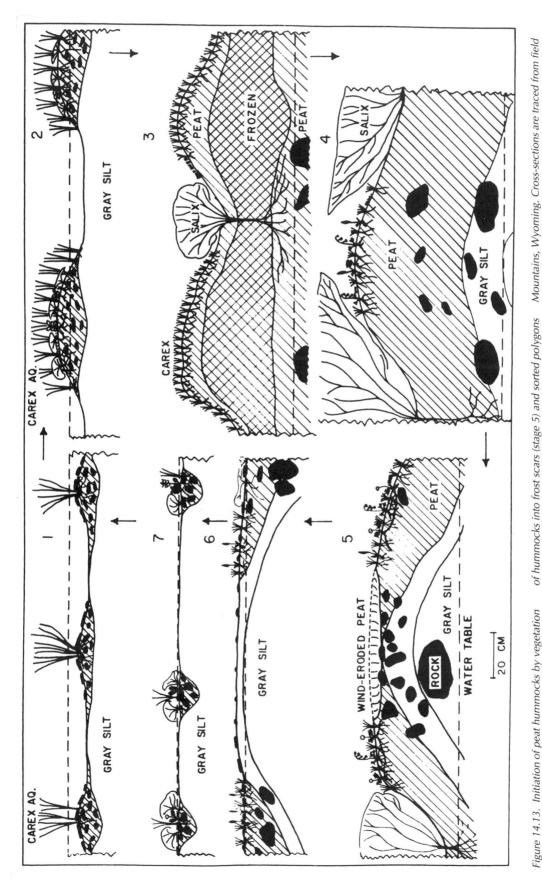

Figure 14.13. Initiation of peat hummocks by vegetation and frost action (stages 1–4) and subsequent disintegration of hummocks into frost scars (stage 5) and sorted polygons (stages 6 and 7). Alpine meadow of the Medicine Bow Mountains, Wyoming. Cross-sections are traced from field drawings by the author; see also Billings and Mooney (1959).

water table, form peat, and the whole process is renewed.

SOME EFFECTS OF CLIMATE CHANGES ON ALPINE VEGETATION

Increases in Ultraviolet-B Radiation

The high mountains of the middle latitudes and the tropics have always had strong fluxes of incoming ultraviolet radiation. Some leaves reflect it, whereas others screen it out by compounds such as flavonoids in the leaf epidermis.

Although the hazard of ultraviolet irradiation has been present as long as there have been organisms in high mountains, the UV-B gradient between the polar mountains and those in the tropics increases curvilinearly toward the tropics. The Andes, the Himalayas, the Hawaiian volcanoes, and the equatorial mountains of East Africa are examples of the steepness of the gradient as it reaches its maximum at high elevations near the equator. The increase in the UV-B dose is caused by a combination of increased elevation of the alpine vegetation, a thinner ozone layer over the tropics as compared to that over the polar regions, and the steeper angle of incidence of solar radiation as received at the surface of the earth (Caldwell, Robberecht, and Billings 1980).

The twentieth century has seen drastic increases in UV-B irradiation not only in the mountains but also at sea level. It is now known that the stratospheric ozone screen is breaking down, not just over the tropics but also over the polar regions of Antarctica and the Arctic. The cause is the continued accidental release of a product of the mid-twentieth century, the man-made chlorofluorocarbon gases (CFCs, or Freons) as summarized by Rowland (1989).

However, Caldwell's (1968) pioneer research on UV-B and alpine plants on Niwot Ridge, Colorado, set the stage for what we do know at present about the effects of such irradiation on plants. Billings (1984) summarized such information as it stood at that time. Also, Caldwell et al. (1982), in regard to arctic-alpine plants, experimentally found that UV-B inhibits photosynthesis in several ecotypic pairs of species along the North and South American Cordillera. These taxa included arctic and alpine ecotypes of *Oxyria digyna*, the arctic *Taraxacum lateritium*, and a *Taraxacum* sect. *Mexicanum* from 4000 m elevation in the Peruvian Andes, and *Lupinus arcticus* from northern Alaska versus *Lupinus meridanus* from the Venezuelan Andes at 3000 m elevation. They found that the arctic ecotypes of *Oxyria*, the arctic *Lupinus*, and the arctic *Taraxacum*

were consistently and significantly more sensitive to UV-B irradiation than were their alpine relatives and ecotypes in the southern Rocky Mountains, the Sierra Nevada, and the Andes.

Another example of the relative tolerance of certain alpine plants to high irradiation in tropical mountains is apparent in the very high UV-B environment on the alpine summit of Haleakala (3024 m) on the island of Maui in Hawaii. The light-colored pubescent leaves of *Geranium tridens* reflect about 20% of the ultraviolet irradiance across the UV spectrum from 290–400 nm (Robberecht et al., 1980). Its neighboring species in that barren, open alpine landscape, *Argyroxiphium sandwicense* (silversword, or "ahinahina"), reflects about 40% of the UV across the same spectral gradient from its silvery appressed pubescence. Modification of epidermal UV transmittance either by reflectance or absorbance that results in lower UV-B irradiance at the level of the mesophyll may represent one mechanism toward acclimation by plants to UV-B radiation. It appears that alpine plants growing in a naturally high UV-B environment have evolved genetic ways that maintain reproductive phenology and carbon uptake in spite of increased UV-B. This probably is a matter of centuries of evolution and is not a solution to one of our serious present-day environmental problems, but in the past it has been achieved under natural high UV-B conditions.

Essentially, we can conclude that middle-latitude and tropical alpine plants have evolved in the presence of high UV-B irradiation and can reflect or screen out such irradiance epidermally. However, if the polar stratospheric ozone layer breaks down as a result of CFCs, tundra plants in polar regions may be endangered physiologically or reproductively as a result of artificially increased UV-B.

Possible Changes in Alpine Snowfall, Drifting, and Runoff

It is difficult enough to predict what the climates of lowlands will be in the next century, with their thousands of first-class and long-term weather observation stations. But, as Barry (1992) reminds us, mountain weather stations are relatively few and usually far between, with records mostly spanning only the last century or less. In the western North American mountains, there are very few observation stations at all, and only two or three are equipped with first-rate instruments and personnel. Switzerland has many more and has had them for a longer time. In the United States, the better mountain weather stations are the one on Mt. Washington, New Hampshire, which was established in 1932, and the station at the Institute of Arctic and Alpine Research (INSTAAR) of the University of Colorado on Niwot Ridge in the Front Range, which was established by Dr. John Marr in 1952.

Precipitation types and regimes of the next century are more difficult to predict than are atmospheric CO_2 concentrations or changes in temperature. Such predictions are even more difficult in mountains than elsewhere because of the complexity of the topographic, geographic, and atmospheric factors involved. Certainly, whether or not precipitation, particularly snowfall, increases or decreases in the western North American mountains, there will be considerable impact on alpine ecosystems because of changes in snowdrift accumulations along mesotopographic gradients, which in turn control runoff of streams and groundwater both locally and regionally. A century ago, this became apparent to the great mountain scientist–classicist Dr. James E. Church of the University of Nevada at Reno; the result was his invention of winter snow survey techniques, now used worldwide.

Long-term studies of interactions between alpine snow and vegetation that use a hierarchic geographic information system (GIS) has been pioneered by Donald and Marilyn Walker of the Institute of Arctic and Alpine Research of the University of Colorado at Boulder. They have quantitatively brought alpine snow and vegetation together species by species and by vegetation type in the Front Range by using GIS. In regard to alpine vegetation and snow, Walker, Halfpenny, Walker, and Wessman (1993) and Walker, Billings, and De Molenaar (in press) are particularly useful.

Warming Climates, Alpine Migrations, and Extinctions

Environmental gradients up the sides of mountains are steeper than are latitudinal gradients. In general, floristic gradients become less rich with increasing elevation, especially when increasing elevation is combined with increasing latitude (Breckle 1974; Billings 1987).

As mountain climates get warmer, and perhaps get drier in the lower and middle elevations largely as a result of the greenhouse gases (CO_2, CO, CH_4), plant populations will migrate vertically up-mountain and latitudinally along the Cordillera – or possibly become extinct. In this regard, interglacial refugia are fully as important as glacial refugia, especially under present climates. Also, alpine floras and populations are evolving and consolidating their genetic adaptations after catastrophic selec-

tion during the present interglacial. Examples of such interglacial refugia are mountain peaks, certain rock types, glacial moraines, alpine bogs, and foggy cliffs with suitable crevices.

If the climate warms in the western North American mountains, timberline vegetation and its associated species, both plant and animal, will move up the mountains but rather slowly. For example, Carrara, Trimble, and Rubin (1991) have quantified Holocene timberline fluctuations in the San Juan Mountains of southwestern Colorado. Between 9600 and 5400 yr BP, treeline in the northern San Juans was at an elevation at least 80 m higher than at present. A large fragment of spruce wood with complacent annual rings and a radiocarbon date of about 8000 yr BP suggests that timberline may have been at least 140 m higher than at present. During the early and middle Holocene, mean July temperatures in the higher San Juans probably were at least 0.5°–0.9°C higher than at present.

Recently, Grabherr (1995), in the Austrian and eastern Swiss Alps, using 26 alpine summits exceeding elevations of 3000 m, found a similar situation of plant species migrating upward. He measured vascular plant cover, abundance, and species richness on those peaks and summits in 1992 as compared to historical records of similar data earlier in this century. From the very precise historical records, as compared to the present distributions of nine typical nival plant species, Grabherr calculated that maximum rates of upward migration on these 26 peaks were close to 4 m of elevation per decade. However, most rates were near 1 m per decade. In the twentieth century, the Austrian meteorological stations have shown that the mean annual air temperature of this alpine region has increased by 0.7°C. Grabherr and his colleagues ascribe the upward migration of these plant species to this moderate warming trend. They agree with my hypothesis and with the statement earlier in this chapter that if this warming continues, there could be extinction of many alpine species on the higher peaks, not only in the Alps but also on mountain ranges of the Northern Hemisphere.

PEOPLE AND ALPINE VEGETATION

In Europe, it has been traditional for centuries to use alpine and subalpine ecosystems for summer grazing and the production of hay. During the past century, in a much more casual approach, these practices became standard in the American West as mountains became available for settlement. The greatest use has been as summer rangeland for sheep and cattle. Toward the end of the nineteenth century and during the first half of the twentieth, millions of sheep were driven in the summers up into alpine and subalpine meadows of the Rocky Mountains, the Basin Ranges, and the Sierra Nevada. The effects of this heavy, relatively uncontrolled grazing are still apparent in the national parks, where such grazing is not allowed now. Grazing is also closely controlled in the national forests and other public lands. The present composition of alpine vegetation reflects the impact of this grazing. Was the unpalatable *Geum rossii* always as common as it is now in the high Rocky Mountains? Or has it increased because of grazing pressures on other more palatable species?

Mining has also had a strong local impact on mountain vegetation during the past century or more, particularly in the North American West. Much of this impact is a result of the very acid and nutrient-poor spoil dumps that often have high concentrations of toxic heavy metals. Many of these dumps are still barren of plants even after more than a century. The U.S. Forest Service and also the mining companies with large modern mines have active research in progress on the uses of native plants in revegetation plans and processes.

The twentieth century has seen an ever-increasing rise in tourist use of the American mountains, including the alpine zones. Ski resorts and recreational skiing have increased in the western North American high mountains just as they have in the European Alps. Millions of people are involved, with heavy concentrations in and around the national parks and national forests. The result is heavy impact of foot traffic on trails, with much backpacking and camping in the high country.

As hikers increase in numbers, casual trails through alpine and subalpine vegetation have proliferated into extensive anastomosing networks as vegetation is damaged or destroyed. Hartley (1976) made a thorough study of these impacts on the alpine and subalpine herbaceous vegetation of Glacier National Park, Montana. Using a "sensitivity index," he found that 35 alpine plant species decreased in abundance as the trail system grew and only 7 species increased. Some of the latter were semiweedy species from lower elevations. Hartley's research suggests some of the reasons for decreased vigor of these plants. Using cross-trail transects and carbohydrate analyses, he found that in late summer, total usable stored carbohydrates decreased 20–50% within 0.5 m of the trail as compared to the amounts stored by plants of the same species >2 m from the trail.

During his research, Hartley constructed a useful model of the effects of trampling on the physiological status of alpine vegetation under the stress of such trampling. Billings (1983) modified Hartley's model somewhat. Primarily, the model starts through physical damage to the plants and the loss of photosynthetic tissue. Secondarily, energy capture by leaves and subsequent translocation to carbohydrate storage in roots, rhizomes, and bulbs are severely hampered or stopped. Since these organ systems are the perennating parts of subalpine herbaceous plants and dwarf shrubs, lack of regrowth and reproduction can cause local extinction of susceptible species. Examples include *Phyllodoce empetriformis*, *Senecio resedifolius*, and to some extent, the bulbiferous *Erythronium grandifolium*.

Grabherr (1982) has studied similar trampling by tourists in the Tyrolean Alps of Austria, where the most sensitive plants are fruticose lichens. On the other hand, *Carex curvula* tolerates very heavy trampling. He concludes that low tolerance to trampling in those mountains is accompanied by low and slow regeneration of the plant populations.

There are other ecosystemic impacts that are not physical. An example is the accidental introduction of *Giardia*, a protozoan from Europe, into even isolated alpine streams and lakes by campers and hikers. This contamination causes the disease giardiasis, which produces diarrhea. Such water supplies in the wild are no longer safe.

SUGGESTIONS FOR FUTURE RESEARCH

The key word is "change" for future research on alpine ecosystems in an unstable biosphere. The complexities of environmental and biological changes through time that will involve new combinations of plant and animal species are almost limitless and will no doubt be filled with surprises.

One cannot view the biosphere provincially either in space or time. There are no crosswalls or real barriers to biotic migrations or environmental changes on this Earth. This is true also, of course, of regional ecosystems, including high mountains. Factors limiting the successful growth and reproduction of alpine and subalpine systems exist and are unique to specific ecotypic populations. Within such ecosystems, there is an eventual carrying capacity for people, plants, and animals that must not be exceeded.

I do not wish to restrict the scope of future research on alpine ecosystems by focusing on only certain aspects and procedures. There are enough ecological problems mentioned in this chapter to get ecologists and the general public thinking imaginatively about preservation and management. It is apparent that there is much to learn from vegetation and ecosystems at their upper limits. Within the foreseeable future, I am confident that we will know a great deal more about how such ecosystems are constructed and how they operate in a rapidly changing biosphere.

REFERENCES

Arno, S. F. 1984. Timberline: mountain and arctic forest frontiers. The Mountaineers, Seattle, Wash.

Axelrod, D. I. 1962. A Pliocene Sequoiadendron forest from western Nevada. University of California Publications in Geological Sciences, Vol. 39, No. 3, pp. 195–268, plates 38–50, plus 3 figures.

Axelrod, D. I. 1976. History of the coniferous forest, California and Nevada. University of California Publications in Botany 70: 1–62.

Axelrod, D. I., and P. H. Raven. 1985. Origins of the Cordilleran flora. J. Biogeog. 12: 21–47.

Bamberg, S. A., and J. Major. 1968. Ecology of the vegetation and soils associated with calcareous parent materials in three alpine regions of Montana. Ecolog. Monog. 38: 127–167.

Barry, R. G. 1992. Climate change in the mountains, pp. 359–380 in P. B. Stone (ed.), The state of the world's mountains. Zed Books, London.

Beaman, J. H., and J. W. Andresen. 1966. The vegetation, floristics, and phytogeography of the summit of Cerro Potosí, México. Amer. Midl. Natur. 75: 1–33.

Bell, K. L., and R. E. Johnson. 1980. Alpine flora of the Wassuk Range, Mineral County, Nevada. Madroño 27: 25–35.

Bell Hunter, K. L., and R. E, Johnson. 1983. Alpine flora of the Sweetwater Mountains, Mono County, California. Madroño 30:89–105.

Billings, W. D. 1969. Vegetational pattern near alpine timberline as affected by fire–snowdrift interactions. Vegetatio 19:192–207.

Billings, W. D. 1973. Arctic and alpine vegetations: similarities, differences, and susceptibility to disturbance. BioScience 23:697–704.

Billings, W. D. 1974. Adaptations and origins of alpine plants. Arctic and Alpine Res. 6:129–142.

Billings, W. D. 1978. Alpine phytogeography across the Great Basin. Great Basin Natur. Mem. 2:105–117.

Billings, W. D. 1979. Alpine ecosystems of western North America, pp. 6–21 in D. A. Johnson (ed.), Special management needs of alpine ecosystems. Society for Range Managemen. Denver, Colo.

Billings, W. D. 1980. American deserts and their mountains: an ecological frontier. Address of the Past President, Ecological Society of America, Tucson, Arizona.. Bull. Ecolog. Soc. Amer. 61:203–209.

Billings, W. D. 1983. Man's influence on ecosystem structure, operation, and ecophysiological processes, Vol. 12D, pp. 527–548, In O. L. Lange, P. S. Nobel, C. B. Osmond, N. H. Ziegler (eds.), Encyclopedia of plant physiology, New Series. Springer-Verlag, New York.

Billings, W. D. 1984. Effects of solar ultraviolet–B radiation on plants and vegetation as ecosystem components, pp. 206–217 in Causes and effects of changes in stratospheric ozone: update 1983. National Academy Press, Washington, D.C.

Billings, W. D. 1987. Constraints to plant growth, reproduction, and establishment in arctic environments. Arctic and Alpine Res. 19:357–365.

Billings, W. D. 1990. The mountain forests of North America and their environments, pp. 47–86 plus 8 color plates in C. B. Osmond, L. F. Pitelka, and G. M. Hidy (eds.), Plant biology of the basin and range. Ecological Studies Vol. 80. Springer-Verlag, New York.

Billings, W. D. 1992. Phytogeographic and evolutionary potential of the arctic flora and vegetation in a changing climate, pp. 91–109 in F. S. Chapin, III, R. L. Jefferies, J. F. Reynolds, G. R. Shaver, and J. Svoboda (eds.), Arctic ecosystems in a changing climate. Academic Press, San Diego, Calif.

Billings, W. D., and H. A. Mooney. 1959. An apparent frost hummock-sorted polygon cycle in the alpine tundra of Wyoming. Ecology 40:16–20.

Billings, W. D., and K. M. Peterson. 1980. Vegetational change and ice-wedge polygons through the thaw-lake cycle in arctic Alaska. Arctic and Alpine Res. 12:413–432.

Bliss, L. C. 1956. A comparison of plant development in microenvironments of arctic and alpine tundras. Ecolog. Monog. 26:303–337.

Bliss, L. C. 1963. Alpine plant communities of the Presidential Range, New Hampshire. Ecology 44: 678–697.

Bliss, L. C., R. W. Fonda, and R. T. Kuramoto. 1969. Guide to the Olympic Mountain field trip. XI International Botanical Congress. Seattle, Wash.

Breckle, Sigmar-W. 1974. Notes on alpine and nival flora of the Hindu Kush, East Afghanistan. Botaniska Notiser 127:280–284.

Bryant, J. P., and E. Scheinberg. 1970. Vegetation and frost activity in an alpine fellfield on the summit of Plateau Mountain, Alberta. Can. J. Bot. 48:751–771.

Caldwell, M. M. 1968. Solar ultraviolet radiation as an ecological factor for alpine plants. Ecolog. Monogr. 38:243–268.

Caldwell, M. M., R. Robberecht, and W. D. Billings. 1980. A steep latitudinal gradient of solar ultraviolet-B radiation in the arctic-alpine life zone. Ecology 61: 600–611.

Caldwell, M. M., R. Robberecht, R. S. Nowak, and W. D. Billings. 1982. Differential photosynthetic inhibition by ultraviolet radiation in species from the arctic-alpine life zone. Arctic and Alpine Res. 14: 195–202.

Carrara, P. E., D. A. Trimble, and M. Rubin. 1991. Holocene treeline fluctuations in the northern San Juan Mountains, Colorado, U.S.A., as indicated by radiocarbon-dated conifer wood. Arctic and Alpine Res. 23: 233–246.

Chabot, B. F., and W. D. Billings. 1972. Origins and ecology of the Sierran alpine flora and vegetation. Ecolog. Monog. 42:163–199.

Charlet, David. 1995. Great Basin conifer diversity: dispersal or extinction. Ph.D. dissertation, University of Nevada, Reno.

Charlet, David. 1996. Atlas of Nevada conifers. University of Nevada, Reno.

Densmore, D. 1980. Vegetation and forest dynamics of the upper Dietrich River valley, Alaska. Master's thesis, North Carolina State University, Raleigh.

Douglas, G. W., and L. C. Bliss. 1977. Alpine and high subalpine plant communities of the North Cascades Range, Washington and British Columbia. Ecolog. Monog. 47:113–150.

Fonda, R. W., and L. C. Bliss. 1966. Annual carbohydrate cycle of alpine plants on Mt. Washington, New Hampshire. Bull. Torrey Bot. Club 93:268–277.

Gehlbach, F. R. 1981. Mountain islands and desert seas: a natural history of the U.S.–Mexican borderlands. Texas A. and M. University Press, College Station.

Gjaerevoll, O. 1958. Botanical Investgations in central Alaska, especially in the White Mountains. Part I: Pteridophytes and Monocotyledons. Det Kgl. Norske Videnskabers Selskabs Skrifter, NR 5 Trondheim.

Gjaerevoll, O. 1963. Botanical investigations in central Alaska, especially in the White Mountains. Part II: Dicotyledons: Salicaceae-Umbelliferae. Det Kgl. Norske Videnskabers Selskabs Skrifter NR 4. Trondheim.

Gjaerevoll, O. 1980. A comparison between the alpine plant communities of Alaska and Scandinavia, pp. 83–88 in E. Sjogren (ed.), Acta Phytogeographica Suecica 68. Studies in plant biology dedicated to Hugo Sjors.

Grabherr, G. 1982. The impact of trampling by tourists on a high altudinal grassland in the Tyrolean Alps, Austria. Vegetatio 48:209–217.

Grabherr, G. 1995. Patterns and current changes in alpine plant diversity, pp. 167–181 in F. S. Chapin and C. Koerner (eds.), Arctic and alpine biodiversity, Springer-Verlag, New York.

Griggs, R. F. 1956. Competition and succession on a Rocky Mountain fellfield. Ecology 37: 8–20.

Hartley, E. A. 1976. Man's effects on the stability of alpine and subalpine vegetation in Glacier National Park, Montana. Ph.D. dissertation, Duke University, Durham.

Hedberg, O. 1964. Features of Afroalpine plant ecology. Acta Phytogeographica Suecica 49.

Hofer, H. R. 1992. Veranderungen in der Vegetation von 14 Gipfeln des Berninagebietes zwischen 1905 und 1985. Berichte des Geobotanischen Institutes der ETH, Stiftung Rubel 58: 39–54.

Hustich, I. 1962. A comparison of the floras of subarctic mountains of Labrador and in Finnish Lapland. Acta Geograph. 17: 1–24.

Isard, S. A. 1986. Factors influencing soil moisture and plant community distributions on Niwot Ridge, Front Range, Colorado, U.S.A. Arctic and Alpine Res. 18:83–96.

Johnson, P. L., and W. D. Billings. 1962. The alpine vegetation of the Beartooth Plateau in relation to cryopedogenic processes and patterns. Ecolog. Monogr. 32: 105–135.

Komárkova, V. 1979. Alpine vegetation of the Indian Peaks area, Front Range, Colorado Rocky Mountains. Flora et Vegetatio Mundi, Band VII. 2 vol. J. Cramer, Vaduz, Liechtenstein.

Komárkova, V. 1980. Classification and ordination in the Indian Peaks area, Colorado Rocky Mountains. Vegetatio 42: 149–163.

Komárkova, V., and P. J. Webber. 1978. An alpine vegetation map of Niwot Ridge, Colorado. Arctic and Alpine Res. 10: 1–29.

Kuramoto, R. T., and L. C. Bliss. 1970. Ecology of subalpine meadows in the Olympic Mountains, Washington. Ecolog. Monog. 40:317–347.

Little, E. L., Jr. 1941. Alpine flora of San Francisco Mountain, Arizona. Madroño 6: 65–96.

Lloyd, R. M., and R. S. Mitchell. 1973. A flora of the White Mountains, California and Nevada. University of California Press, Berkeley.

Loope, L. L. 1969. Subalpine and alpine vegetation of northeastern Nevada. Ph.D. dissertation, Duke University, Durham.

McMillan, C. 1948. A taxonomic and ecological study of the flora of the Deep Creek Mountains of central western Utah. Master's thesis, University of Utah, Salt Lake City.

Major, J., and D. W. Taylor. 1988. Alpine, pp. 601–675 in M. G. Barbour and J. Major (eds.), Terrestrial vegetation of California, 2nd ed. California Native Plant Society.

Marr, J. W. 1961. Ecosystems of the East Slope of the Front Range in Colorado. University of Colorado Studies, Series in Biology No. 8. Boulder, Colo.

Marshall, Robert. 1956. Arctic wilderness: exploring the central Brooks Range. George Marshall (ed.). University of California Press, Berkeley.

May, D. E., and P. J. Webber. 1982. Spatial and temporal variation of the vegetation and its productivity, Niwot Ridge, Colorado, pp. 35–62 in J. Halfpenny (ed.), Ecological studies in the Colorado Alpine: a festschrift for John W. Marr. Institute of Arctic and Alpine Research, University of Colorado, Occasional Paper No. 37. Boulder, Colo.

Merriam, C. Hart. 1890. Results of a biological survey of the San Francisco mountain region and desert of the Little Colorado in Arizona. North American Fauna No. 3. U.S. Dept of Agriculture. Government Printing Office, Washington, D.C.

Miller, P. C. (ed.). 1982. The availability and utilization of resources in tundra ecosystems. Holarctic Ecol. 5: 81–220.

Mitchell, R. S., V. C. LaMarche, Jr., and R. M. Lloyd. 1966. Alpine vegetation and active frost features of Pellisier Flats, White Mountains, California. Amer. Midl. Nat. 75: 516–625.

Mooney, H. A. 1973. Plant communities and vegetation, pp. 7–17 R. M. Lloyd, and R. S. Mitchell (eds.), A flora of the White Mountains, California and Nevada. University of California Press, Berkeley.

Mooney, H. A., and W. D. Billings. 1960. The annual carbohydrate cycle of alpine plants as related to growth. Amer. J. Bot. 47: 594–598.

Mulroy, J. C. 1979. Contributions to the ecology and biogeography of the *Saxifraga cespitosa* L. complex in the Americas. Ph.D. dissertation, Duke University, Durham.

Nowak, Cheryl L., R. S. Nowak, R. J. Tausch, and P. E. Wigand. 1994. A 30,000 year record of vegetation dynamics at a semi-arid locale in the Great Basin. J. Vegeta. Sci. 5: 579–590.

Oberbauer, S. F., and W. D. Billings. 1981. Drought tolerance and water use by plants along an alpine topographic gradient. Oecologia 50:325–331.

Odasz, Ann Marie. 1983. Vegetation patterns at the treelimit ecotone in the upper Alatna River drainage of the Central Brooks Range, Alaska. Ph.D. dissertation, University of Colorado, Boulder.

Reynolds, D. N. 1984. Alpine annual plants: phenology, germination, photosynthesis, and growth of three Rocky Mountain species. Ecology 65: 759–766.

Robberecht, R., M. M. Caldwell, and W. D. Billings. 1980. Leaf ultraviolet optical properties along a latitudinal gradient in the arctic-alpine life zone. Ecology 61:612–619.

Rochow, T. F. 1969. Growth, caloric content, and sugars

in *Caltha leptosepala* in relation to alpine snowmelt. Bull. Torrey Bot. Club 96:689–698.

Rowland, F. S. 1989. Chlorofluorocarbons and the depletion of stratospheric ozone. Amer. Sci. 77:36–45.

Rübel, E. 1912. Pflanzengeographische Monographie des Berninagebietes. Engelmann, Leipzig.

Rundel, P. W., A. P. Smith, and F. C. Meinzer. 1994. Tropical alpine environments: plant form and function. Cambridge University Press, Cambridge.

Schaak, C. G. 1983. The alpine vascular flora of Arizona. Madroño 30: 79–88.

Scott, D., and W. D. Billings. 1964. Effects of environmental factors on standing crop and productivity of an alpine tundra. Ecolog. Monog. 34: 243–270.

Smith, A. P. 1994. Introduction to tropical alpine vegetation, pp. 1–19 in P. W. Rundel, A. P. Smith, and F. C. Meinzer (eds.), Tropical alpine environments: plant form and function. Cambridge University Press, Cambridge.

Smith, A. P., and T. Young. 1987. Tropical alpine ecology. Ann. Rev. of Ecol. and System. 18:137–158.

Spetzman, L. A. 1959. Vegetation of the Arctic Slope of Alaska. U.S. Geological Survey Professional Paper 302-B. Government Printing Office, Washington, D.C.

Steyermark, J. A. 1950. Flora of Guatemala. Ecology 31: 368–372.

Viereck, L. A. 1966. Plant succession and soil development on gravel outwash of the Muldrow Glacier, Alaska. Ecolog. Monogr. 36:181–199.

Walker, D. A., W. D. Billings, and J. G. Molenaar. In press. Snow–vegetation interactions in arctic and alpine environments, in H. G. Jones, R. W. Hoham, J. W. Pomeroy, and D. A Walker (eds.) The ecology of snow. Cambridge University Press, Cambridge.

Walker, D. A., J. C. Halfpenny, Marilyn D. Walker, and Carol A. Wessman. 1993. Long-term studies of snow–vegetation interactions. BioScience 43:287–301.

Wallace, L. L., and A. T. Harrison. 1978. Carbohydrate mobilization and movement in alpine plants. Amer. J. Bot. 65:1035–1040.

Washburn, A. L. 1956. Classification of patterned ground and review of suggested origins. Bull. Geolog. Soc. Amer. 67:823–865.

Weber, H. 1958. Die Paramos von Costa Rica und ihre pflanzengeographische Verkettung mit den Hochanden Sudameikas, pp. 123–194 in Mathematisch-Natuwissenschaftlichen Klasse. Jahrgange NR 3. Akademie der Wissenschaften und der Literatur. Mainz, Germany.

Webster, G. L. 1961. The altitudinal limits of vascular plants. Ecology 42: 587–590.

Chapter
15

Mexican Temperate Vegetation

ALEJANDRO VELÁZQUEZ
VICTOR MANUEL TOLEDO   ISOLDA LUNA

DIVERSITY AND NEARCTIC AFFINITIES

Mexico has a high diversity of ecosystems (Ramamoorthy, Bye, Lot, and Fa 1993). Mexico includes six of the ten major terrestrial biomes of the world – extra-dry vegetation, mediterranean, temperate forest, temperate grassland, montane, and tropical rain forest (Cox and Moore 1993) – and it is one of the ten most megadiverse countries of the world (Mittermeier 1988), harboring 10–12% of the world's vascular species (Toledo and Ordoñez 1993).

Mexico's wide elevation range (0–5000 m), its location astride the Tropic of Cancer, and the influence of two oceans across its relatively narrow continental mass probably are determining factors for the most significant features of Mexico's climatic diversity. The Tropic of Cancer is a significant thermal demarcation and also delimits the transition between arid and semiarid climates: arid anticyclone high pressures toward the north versus humid and semihumid trade winds and cyclones in the south. The complex physiography, together with the differences determined by latitude and altitude, result in a climatic mosaic with a great number of variations (García 1981). Maximum average temperatures (28–30° C) are recorded in the low-lying regions of the Balsas Depression, whereas adjacent zones at the top of Pico de Orizaba in Veracruz have the lowest average temperatures (−6° C). Some mountains have glaciers and permanent snow. Apart from these two extremes, the range of temperatures most frequently recorded varies from 10 to 28° C. Precipitation also presents notable contrasts: from <50 mm annually and no wet season (as in parts of Baja California) to >5500 mm annually and almost no dry season (as in parts of Tabasco and Chiapas).

As a consequence of this climatic diversity, Mexico has a large variety of vegetation types, comparable only to India or Peru. Although detailed studies have distinguished up to 70 different units of vegetation, based on physiognomy and floristic composition, it is possible to differentiate fewer principal types of vegetation in Mexico at the biome category (e.g., West 1971; Rzedowski 1978, 1993; Flores 1993). At such a scale, it is apparent

Thanks to Martha Gual and R. M. Fonseca for support on floristics. Comments by Jorge Llorente, Javier Madrigal, and Richard Minnich on an early version, and by Michael Barbour on a later version, are acknowledged. J. Rzedowski fully encouraged the preparation of this chapter, and it was further supported by a grant-in-aid from DGAPA–UNAM grant IN-209094, and FUO's University of Amsterdam.

that Mexico may be divided relatively easily. Figure 15.1 shows the distribution of the main vegetation types of Mexico as a function of two climatic attributes: precipitation and temperature (the latter is represented by elevation in the figure).

Although many Cosmopolitan and Paleoarctic taxa are present in Mexico, its geographic location has favored the establishment of biotic elements characteristic of two main regions, Nearctic and Neotropical. Mexico is situated on a transitional gradient from Neotropical to Nearctic environments.

A large part of Mexico is dominated by ecosystems of northern affiliation (Beard 1944; Rzedowski 1978). Two main historical events may explain the present dominance of arctic biota: (1) most of Mexico's northern territory has been linked permanently to the rest of North America, and (2) the last glaciation (ca. 18,000 yr ago) promoted the movement south of many northern taxa within Mexico's present political borders (Ferrusquía-Villafranca 1993; Velázquez 1993).

Temperate vegetation types on long north-south mountain chains typify Mexico's arctic affinities. The Madrean Region from northern Mexico south through the Neovolcanic Transversal Belt to Las Cañadas de Chiapas is dominated by oak, alder, pine, and fir species. The absolute dominance of species of Nearctic origin in the tree layer and a large number of species of Neotropical origin in the shrub and herb layers is commonly observed (Gadow 1930). This complex vegetation pattern becomes more dominated by the Neotropical to the south and by the Nearctic to the north. The Neovolcanic Transversal Belt forms the heart of the gradient, and it contains a large number of endemic taxa (Fa 1989; Rzedowski 1993).

Data from a large number of botanical and zoological expeditions throughout the present century provide strong evidence of the great affinity of montane regions of Mexico with the rest of North America (Beard 1944; Goldman and Moore 1945; Smith 1940; Troll 1952; Wagner 1964). Structural classifications of Mexican vegetation types equated temperate ecosystems with mountanious regions (Sanders 1921; Shelford 1926). Temperate montane vegetation types have variously been called (Table 15.1) montane rain forest, high mountain forest, and páramo (Beard 1944; Beaman 1962); pine-oak forest (Leopold 1950); low evergreen forest, conifer-oak forest, oak forest, and páramo (Miranda and Hernández-X. 1963); and most recently cloud montane forest, coniferous forest, oak forest, and grassland (Rzedowski 1978). Research that defines the relationship of these vegetation types to ecological factors such as humidity, soil suitability,

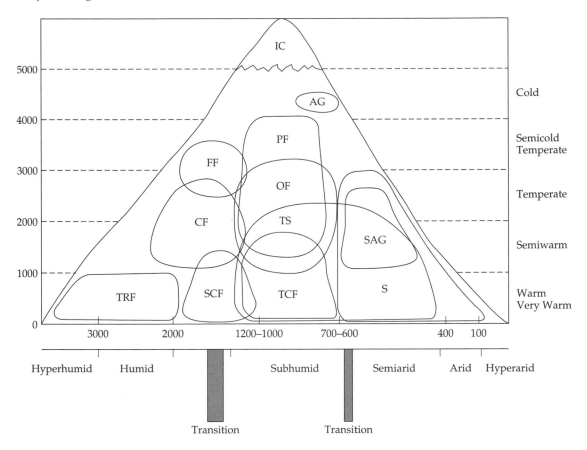

Figure 15.1. *Major Mexican vegetation types ordinated along gradients of temperature and precipitation (annual rainfall in millimeters). Abbreviations: tropical rain forest (TRF), subtropical caducifolius forest (SCF), tropical caducifolius forest (TCF), shrubland (S), subalpine grassland (SAG), thornshrub (TS), oak forest (OF), cloud forest (CF), fir forest (FF), pine forest (PF), and alpine grassland (AG). The nival zone is IC. (Modified from Toledo & Rzedowski, 1995.)*

Table 15.1. *Equivalent names of Mexican temperate vegetation types given by various authors*

Beard (1955)	Miranda & Hernandez-X. (1963)	Rzedowski (1978)	Toledo & Ordoñez (1993)
Montane rain forest	Deciduous forest	Cloud forest	Humid temperate
Montane forest	Pine-fir forest	Conifer forest	Subhumid temperate
Páramo	Alpine bunchgrassland	"Zacatonal"	Cool temperate

Source: Rzedowski, (1978).

fire, number of days with temperatures below zero, and mean annual precipitation is, on the whole, scanty or restricted to specific places. General definitions of temperate Mexican environments, however, have been made (Rzedowski 1978; Toledo and Ordoñez 1993). Major temperate biogeoclimatic ecoregions are the humid temperate, the subhumid temperate, and the cool (or alpine) (Fig. 15.2).

The humid temperate region is characterized by cloud forests dominated by oaks. The forests' floristic composition includes both boreal and tropical elements. This region occupies very restricted sites of 600–3200 m elevation, mainly on slopes facing the Gulf of Mexico from Tamaulipas to Chiapas. Distributed in 21 states, it covers an area of approximately 10,000 km².

The subhumid temperate region covers the

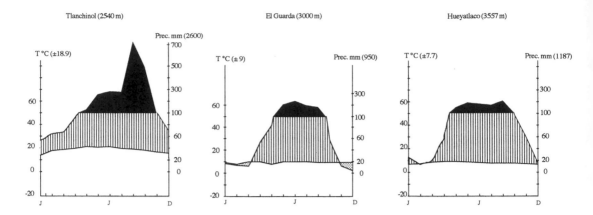

Figure 15.2. Climatograms representative of three main elevation belts of mountain environments: temperate humid (Tlanchinol), temperate subhumid (El Guarda), and temperate cool (Hueyatlaco). Months (January–December) are arranged along the horizontal axis, mean monthly temperature along the left vertical axis, and monthly precipitation along the right vertical axis. Vertical shading represents soil moisture recharge; solid shading represents precipitation beyond soil storage capacity, and dotted shading represents soil moisture deficit. Parenthetical information is elevation, mean annual temperature, and annual precipitation. Climatic data are averages of 1980–1990.

greatest part of the mountains of Mexico at elevations of 2800–4000 m. The characteristic vegetation is forest of fir, pine, oak, or mixtures. It is distributed through 20 states (principally Chihuahua, Michoacán, Durango, and Oaxaca) and covers an area of approximately 333,000 km².

The cool (or alpine) region is located above timberline (>4000 m) on the 12 highest mountains of Mexico. It is dominated by alpine bunchgrasses, or zacatonales.

GEOGRAPHICAL DISTRIBUTION OF TEMPERATE VEGETATION

Mexico is a very mountainous country, with over half of its territory >1000 m in elevation. Arid and semiarid vegetation dominates the high plateau of central and northern Mexico, whereas temperate vegetation (as defined in this chapter) covers the steep and more humid areas above 1000 m. Temperate vegetation thus covers 22% of the Mexican territory. Mexico is also a country with evidence of past volcanic activity. The most spectacular volcanic area is the great Neovolcanic Transversal Belt, which crosses Mexico from west to east at the latitude of Mexico City (19–20°). Its landscape is characterized by thousands of old cinder cones and dozens of tall volcanic peaks (Fig. 15.3). Volcanism continues today, with many active or temporarily dormant volcanoes.

Earthquake activity is common, mostly along the Pacific Coast and the Gulf of California. Earthquakes are also frequent in the Neovolcanic Trans-

versal Belt, often causing considerable damage in this heavily populated region.

Mexico can be divided into five general realms: extratropical drylands, tropical highlands, tropical lowlands, extratropical highlands, and subhumid lowlands (West 1971). These realms match most Mexican physiographic regions – for example, the Baja California and Buried Ranges of northwest Mexico, the western and eastern Sierra Madres, the Neovolcanic Transversal Belt, and the Highlands of southern Mexico. Four regions are considered temperate: the western Sierra Madre, the eastern Sierra Madre, the Neovolcanic Transversal Belt, and the southern Sierra Madre.

The western and eastern Sierra Madres form dissected borders of the western and eastern edges of the central plateau. The broad crest of the western Sierra Madre rises up to 3000 m. The upper portion of the range is covered with thick layers of lava. The western slope of the range forms rugged canyons and narrow ridges dropping down to the Pacific coastal plain. The eastern Sierra Madre rises to a sharper crest on the eastern rim of the central plateau, with elevations up to 4000 m. In the North, the Sierra is comprised of several irregular ridges separated by basins descending gradually to the Gulf coastal plain.

The Neovolcanic Transversal Belt forms a major geological break with the central plateau. The belt is bordered on the north by a series of high basins; on the south the land drops sharply into the deep Balsas Depression. Included in this volcanic area are Mexico's highest and best known peaks: Pico

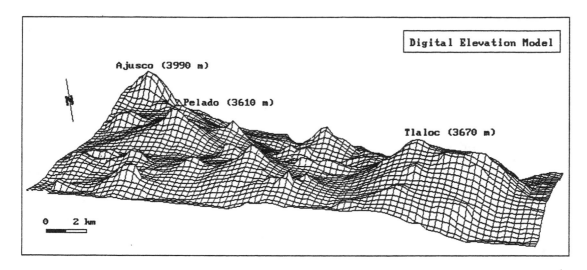

Figure 15.3. Digital elevation model of southern portion of the Valley of México (vertical exaggeration 5x). Within this *area of 800 km² are >200 volcanic cones. (From Veláz-quez 1993.)*

de Orizaba (5700 m), Popocatépetl (5452 m), Ixtaccíhuatl (5285 m), and Nevado de Toluca (4392 m).

The Highlands of southern Mexico are a geologically complex region separated into two sections by the Isthmus of Tehuantepec: the southern Sierra Madre in the west and the Chiapas Highlands in the east. The southern Sierra Madre is a highly dissected mountain system with narrow valleys, a discontinuous Pacific coastal plain, and a few highland basins. The southeastern highlands are dominated by the altiplano of Chiapas, a plateau rising up to an elevation of 2500 m.

These regions occur in Chihuahua, Durango, Coahuila, Nuevo Leon, Tamaulipas, Zacatecs, San Luis Potosí, Jalisco, Guanajuato, Hidalgo, Mexico, Distrito Federal, Puebla, Veracruz, Morelos, Michoacan, Oaxaca, and Chiapas. Some other scattered regions in the country also harbor temperate species: high elevations of Sierra de San Lazaro in Baja California Sur, Cerro Potosí in Tamaulipas, Sierra Fria in Aquascalientes, and high elevations of Sierra Lacandona and Las Canadas in Chiapas.

MAJOR TEMPERATE VEGETATION TYPES

The four major temperate vegetation types of Mexico are comprised of many plant communities (*sensu* Rzedowski 1978) or associations (*sensu* Beard 1944 and Braun-Blanquet 1951). Most Mexican classifications are purely based on the canopy layer and at the generic level (Table 15.2). This emphasis masks considerable variation at the species level. Fir forest (*Abies*), for instance, occurs

Table 15.2. Characteristic genera of the four main temperate vegetation types of Mexico. These genera may vary from slope to slope and from sierra to sierra, so that no single genus can be found throughout a single temperate vegetation type.

Genera	Cloud forest	Pine forest	Fir forest	Alpine grassland
Fagus	X			
Weinmannia	X			
Engelhardtia	X			
Celtis	X			
Liquidambar	X			
Acer	X			
Tilia	X			
Muhlenbergia		X		
Ribes		X		
Arbutus		X		
Baccharis		X		
Stevia		X		
Halenia		X		
Geranium		X		
Roldana			X	
Buddleia			X	
Thuidium			X	
Sibthorpia			X	
Salix			X	
Cinna			X	
Symphoricarpos			X	
Calamagrostis				X
Agrostis				X
Trisetum				X
Umbilicaria				X
Bryoerythrophyllum				X
Stereocaulon				X
Cladonia				X

throughout Mexico, but co-dominant shrub species and fir species change from north to south (Veláz-quez and Cleef 1993; Islebe, Cleef, and Velázquez 1995). Only a few phytosociological studies have been conducted in central Mexico, sufficient to de-fine associations and alliances based on fine-scale differences in species. Another limitation to our ability to summarize the vegetation is that a large part of the territory where temperate vegetation types are distributed has not been surveyed in de-tail. This lack of homogeneous information does not permit us to provide a thorough description of all communities. Thus a detailed description of temperate vegetation is beyond the scope this chapter.

Our objective is to provide an overview of ma-jor vegetation types and to emphasize those that have been best studied. We also intend to outline the conservation possibilities for temperate vege-tation in Mexico. Additional details on the vege-tation of Mexican forests, grasslands, deserts, al-pine, and wetlands are found in Chapters 3, 8, 9, 13 and 14 of this volume.

Cloud Forest (Humid Temperate Forests)

Large biological heterogeneity typifies a cloud for-est, which is a mix of northern, southern, and en-demic taxa, and of low- and upper-elevation taxa. Because of its diversity, various names have been given to this vegetation type: bosque mesófilo de montaña (Miranda 1947), caducifolious forest (Mir-anda and Hernández-X. 1963), temperate decidu-ous forest (Rzedowski 1963).

Five environmental requirements seem to gov-ern the presence of cloud forest in Mexico: high relative humidity, montane environments, irregu-lar topography, deep litter layer, and temperate cli-mate. Cloud forest covers at most 1% of the total Mexican surface, but it includes about 3000 vas-cular species (Rzedowski 1993), which is 12% of the country's vascular flora (Toledo and Ordoñez 1993). Currently, there are only a few large pre-serves of cloud forest, but they are scattered throughout the range of the type.

Along an elevation gradient, the structural com-plexity of cloud forest decreases toward high ele-vations and varies from slope to slope. Elevation ranges from 600–3200 m, though the vegetation is best developed at 1000–1750 m. Precipitation is 1800–5800 mm yr[1], and cloudiness is common throughout the year. Freezing temperatures are rare. Major temperature changes are seasonal, in contrast to large daily changes in alpine environ-ments at higher elevations.

Physiognomically, cloud forest is dense, 15–40 m high, and multilayered (Fig. 15.4). Some of the tree genera that reach more than 40 m are *Engel-hardia* and *Platanus*. The upper tree layer is domi-nated by caducifolious (deciduous) taxa, the lower tree layer by perennifolious (evergreen) ones. The most diagnostic species is *Liquidambar macrophylla* (see Fig. 15.3), and the most common associated boreal elements are *Carpinus caroliniana, Cornus dis-ciflora, Tilia mexicana, Alnus firmifolia*, and *Quercus candicans* (Miranda 1947; Puig 1970). Some species shared with the eastern deciduous forest of North America are *Acer negundo, Carpinus caroliniana, Carya ovata, Cornus florida, Fagus mexicana, Illicium floridanum, Liquidambar macrophylla, Nyssa sylvatica, Ostrya virginiana, Prunus serotina, Tilia floridana*, and *Taxus globosa*. Species shared with western (of-ten riparian) forests are *Arbutus xalapensis, Celtis pallida, C. reticulata*, and *Sambucus mexicana*. Meso-american taxa frequently found in cloud forest are *Clethra* spp., *Weinmannia* spp., *Arctostaphylos arguta, Ilex discolor, Litsea glaucescens, Magnolia schiedeana, Pinus montezumae, P. pseudostrobus, Prunus brachy-botrya*, and *Ulmus mexicana*. Among the endemic taxa are *Carya ovata* var. *mexicana, Ilex pringlei, Jug-lans mollis*, and *Platanus mexicana* (Rzedowski 1978).

Commonly, there are two shrub layers, both with Neotropical affinities. The families Composi-tae, Gesneriaceae, Clusiaceae, Labiatae, Legumi-nosae, Malvaceae, Melastomataceae, Myrsinaceae, Piperaceae, and Rubiaceae variously dominate, de-pending on elevation, latitude, and humidity.

The herb layer increases cover and diversity when the overstory is disturbed. Arboreal ferns (e.g., *Cyathaea*) are common, as well as herbaceous species. Mosses are abundant. Flowering plants are in the families Asclepiadeacea, Begoniaceae, Bro-meliaceae (especially the genus *Tillandsia*), Cyper-aceae, Compositae, Convolvulaceae, Cucurbita-ceae, Dioscoriaceae, Equisetaceae, Gramineae, Liliaceae, Lycopodiaceae, Orchidaceae, Piperaceae, Solanaceae, Urticaceae and Verbenaceae. This layer has a Neotropical affinity, with only a few mosses and mushrooms being of boreal affinity (Crum 1951; Guzmán 1973; Delgadillo 1979).

Pine Forest (Subhumid Temperate Forests)

Mexico contains about half of the world's pine spe-cies (Critchfield and Little 1966; Styles 1993). In most areas pine forests are co-dominants with other broadleaf trees (*Alnus* and *Quercus*) and other conifer species (*Abies* and *Juniperus*). Collectively they cover 15% of Mexico: on sandy soils of coastal

Figure 15.4. Aspect of cloud forest at Teocelo area (2600 m). (Courtesy of J. L. Contreras.)

plains (*Pinus caribaea, P. oocarpa*), on lava flows (*P. michoacanus, P. montezumae*), and at high elevations (*P. hartwegii*), (Rzedowski 1978). Temperature (ca. 15° C ± 10) and precipitation (800 ± 150 mm) do not seem to be limiting factors for the establishment of pine species. Acid soils, however, are associated with pine forests (Aguilera, Dow, and Hernández 1962).

Rzedowski and McVaugh (1966) state that fire disturbance in pine forest favors the establishment of conifers, but Sanchez and Huguet (1959) have shown that fire, logging, and grazing induce succession toward pine–alder–bunchgrass. More study is needed to document the effect of fire and grazing on pine forest (Velázquez 1994).

Major types of pine forest include alder-pine, ponderosa pine, pine-oak, Hartweg pine, and mixed pine. Each is described in the following sections.

Alder-pine forest. Alder-pine forest is made up of four structural layers: (1) coniferous tree layer (50% cover, maximum height 22 m) of *Pinus*: (2) shrub layer (4 m height) of *Alnus firmifolia, Senecio cinerarioides*, and *Symphoricarpos microphyllus*; (3) dense bunchgrass layer dominated mainly by *Muhlenbergia macroura* and *Festuca tolucensis*; and (4) ground layer composed mainly of *Alchemilla procumbens* and *Arenaria lycopodioides*. The bunchgrass layer is characterized by compact bunchgrasses with

Figure 15.5. Aspect of pine forest at Tláloc volcano (3300 m).

broad, long, tough grass leaves that seem to be well adapted to fire and browsing (Fig. 15.5). The forest is restricted to very dissected, rolling-to-steep slopes, and lava flows at 2700–3500 m elevation. Soils are shallow with gravelly sandy loam texture.

Constant diagnostic species include *Alnus firmifolia, Arbutus glandulosa, Buddleia paryiflora, Eryngium carlinae, Penstemon gentianoides, P. campanulatus, Pinus montezumae, Quercus laurina, Sicyos parviflorus, Stellaria cuspidata,* and *Stevia monardifolia.*

Cervantes (1980) and González (1982) described this type of mixed forest, and Rzedowski (1954, 1978) referred to it as a mosaic of *Alnus firmifolia* forest and *Muhlenbergia quadridentata* grassland and suggested that repeated burning and grazing are the main causes of the bunchgrass. Disturbed conifer forest in some parts of central Mexico is replaced by alder forest or by subalpine coarse bunchgrassland of *Muhlenbergia* and *Festuca.*

Velázquez and Cleef (1993) described four associations within alder-pine forest in central Mexico: *Trisetum altijugum–Alnus firmifolia, Pinus–Alnus firmifolia, Eryngium carlinae–Alnus firmifolia,* and *Pinus montezumae–Alnus firmifolia.*

Pinus ponderosa *forest.* Ponderosa pine is chiefly found in the western Sierra Madre on granitic or volcanic steep slopes or plains at 400–1500 m elevation. Fires are frequent. Ponderosa pine dominates large stands, which are often mixed with *Abies concolor.* Studies of this forest's distribution and dynamics are scanty in contrast to many publications about it in the United States (Styles 1993).

Pine-oak Forest (Pinus oocarpa–Quercus laurina). Most landscapes at high elevations (2300–2400 m) of Chiapas are covered by pine-oak forest in some stage of recovery from disturbance (Wagner 1964; González-Espinosa et al. 1991). Only a few patches of old-growth forest remain. There are no freezing days during the year. Soils are moderately deep (40 cm) calcareous or clayey loams (Breedlove 1981).

Three seral stages – early, middle, and mature forest – are apparent. At early stages, *Pinus oocarpa, P. oaxacana, Quercus laurina, Q. crassifolia,* and *Q. rugosa* prevail. Below the overstory canopy is a second, lower tree layer dominated by *Rapanea juergensenii* and *Prunus serotina. Solanum nigricans* and

Litsea glaucescens are abundant on the ground layer. Midstage stands share all the species just mentioned, but pines are less important and oaks are more dominant. In addition, *Oreopanax xalapensis* and *Symplocos limoncillo* are abundant. The mature stage is an oak forest dominated by *Quercus laurina* and *Q. crassifolia*. Pines are present, but no single species prevails. *Cleyera theaoides* is an abundant subcanopy tree. Shrubs are chiefly absent (González-Espinosa et al. 1991). This pine-oak seral phase is one of the best examples of a canopy layer dominated by Holarctic species with a Neotropical understory (though the degree to which this holds depends strongly on geographic location).

Intensive grazing, cropping, and logging are common in the Cañadas of Chiapas. Slash and burn agriculture for corn production is more limited, mainly conducted by indigenous groups.

Pinus hartwegii *forest.* Rzedowski (1954, 1978) described this vegetation as being rather species-poor. It is mainly restricted to the upper part of the pedregal lava slopes below volcanic cones or to rather flat undulating slopes forming plateaus at 3350–3570 m elevation. The relatively shallow soils (1 m) are loamy clays with a thin litter layer (5 cm).

The physiognomy of this forest consists of three layers: (1) an overstory tree layer up to 20 m high; (2) a herb-bunchgrass layer with *Lupinus, Muhlenbergia, Festuca,* and *Calamagrostis* grass species; and (3) a ground layer dominated by *Alchemilla* and *Arenaria lycopodioides.* The main diagnostic species of this forest are *Pinus hartwegii* and *Muhlenbergia quadridentata*; important associated species include *Muhlenbergia macroura, Arenaria lycopodioides, Festuca tolucensis,* and *Calamagrostis tolucensis.* The latter two bunchgrass species are also diagnostic for alpine grassland (Almeida, Cleef, Herrera, Velásquez, and Luna 1994).

The forest lacks well-defined shrub and herb layers, unlike the *Pinus hartwegii* forest on Tláloc and Pelado volcanoes in central Mexico, which always have a dense herb layer. *Pinus hartwegii* forest is considered by most to be a climax community; Ern (1973), however, suggested that *Pinus hartwegii* was a successional species in severely burned *Abies religiosa* forest, a forest restricted to very steep subalpine slopes at 2900–4000 m elevation, and which grows in a matrix with *Alnus firmifolia, Pinus montezumae,* and *Quercus forest* (Rzedowski 1954, 1978; Anaya, Hernandez, and Madrigal 1980). The least boreal distribution of pines reaches high mountains of Central America and are best represented in Guatemala's mountains (Islebe et al. 1995). Several plant communities have been identified associated with bunchgrassland dominated mainly by *Muhlenbergia quadridentata* or *Festuca tolucensis.* Rzedowski (1978) referred to these bunchgrass species, together with *Calamagrostis tolucensis,* as very abundant associates of *Pinus hartwegii* forest.

Evidence of recent burning and grazing (mainly by sheep) is found in most areas of this community. Wood extraction in central Mexico is practically absent, though in Guatemala only a few patches of this forest remain, due to the intensity of fuelwood cutting (Islebe 1993).

Mixed pine forest. There are a number of large forested areas where no single species of pine seems to be dominant. Such a complex pine unit is apparently promoted by intensive logging practices and by fire. It occurs in very heterogeneous landscapes that include flats and steep slopes. Soils are shallow, acidic, and sandy. Often, there is recently deposited volcanic ash or lava (e.g., in the Paricutín volcano area and other portions of the Transversal Neovolcanic Belt). The lava flows have an irregular topography, which produces a diversity of microenvironmental situations with many endemic taxa.

Within the 40 m tall overstory are *Pinus michoacana, P. montezumae, P. leiophyla, P. pseudostrobus, P. rudis, P. teocote,* and *P. hartwegii. Alnus* and *Quercus* species are present but with less cover than *Pinus.* The tree canopy is open, more like that of a woodland than a forest. Below is an open shrub layer with *Senecio, Buddleia, Ribes,* and *Rubus* as characteristic genera. The lowest understory layer contains *Satureja, Stevia, Eupatorium, Salvia,* and cushion-like species such as *Arenaria bioides, Geranium seemanii,* and *Alchemilla procumbens.* A definite bunchgrass layer is absent.

Fir Forest (*Abies*)

The most boreal Mexican vegetation type is *Abies* forest, variously and locally described throughout Mexico by Leopold (1950), Rzedowski (1954), Beaman (1965), Anaya (1962), Madrigal (1967), and Anaya et al. (1980). Rzedowski (1978) has given a complete description on a national scale. Fir forests typically occur below *Pinus hartwegii* forest on high volcanoes, along escarpments, and in glens between lava flows. They prefer canyons or other steep slopes protected from direct sunlight and strong winds. Soils are rich in organic matter and ash.

Three fir species occur in Mexico: *Abies concolor, A. religiosa,* and *A. guatemalensis.* They usually co-occur in the overstory with *Pinus, Quercus, Pseu-*

dotsuga, and *Cupressus* species. A lower tree layer is comprised chiefly of *Alnus, Arbutus, Salix, Prunus,* and *Garrya* species. In rare undisturbed stands, there is a ground layer of mosses and cushion plants.

The *Abies religiosa* forest is mainly restricted to the Transversal Neovolcanic Belt. It occurs on steep to moderate (10–30°) outer slopes of volcanic cones at 3000–3500 m elevation. Soils are deep and there is a thick layer of litter on the surface. Some evidence of disturbance from grazing, burning, logging, and tree harvest exists (Madrigal 1967; Rzedowski 1978).

The forest is dense and tall, reaching 30 m in height (Fig. 15.6). The overstory layer is dominated by *A. religiosa*. Below is a layer of shrubs and tall herbs (0.5–3 m), dominated by *Senecio angulifolius* and *Roldana barba-johannis*, and a ground layer (5 cm tall) of rosaceous herbs (e.g., *Alchemilla procumbens*) and mosses (e.g., *Polytrichum juniperinum*).

Other diagnostic and common species of this plant community group are *Senecio toluccanus, S. callosus, S. platanifolius, Sibthorpia repens, Salix oxylepis, Festuca amplissima, Alchemilla procumbens, Thuidium delicatulum, Acaena elongata, Stachys* species, *Lolium* species, *Galium aschenbornii, Cinna poaeformis, Pernettya prostrata, Dydimaea alsinoides,* and *Buddleia sessiliflora.* At lower elevations this forest mixes with *Muhlenbergia* and *Calamagrostis* grassland and *Alnus firmifolia* forest (Velázquez and Cleef 1993).

The *Abies guatemalensis* forest is mainly restricted to the southernmost part of Mexico and into the Guatemalan mountains (Islebe et al. 1995). It occurs on very steep slopes at 2800–3400 m on deep soils rich in organic matter (Islebe and Velazques 1994; Islebe et al. 1995). It is a dense forest with three layers: an overstory fir layer 30 m tall; a shrub layer dominated by *Roldana barba-johannis* and *Tetragyron orizabensis*; and a moss layer dominated by *Thuidium delicatulum*. Other species present in this forest include *Fuchsia microphylla, Senecio callosus, Trifolium amabile, Sabazia pinetorum,* and *Pinus ayacahuite.*

In Mexico, there are no communities where *Abies concolor*, the other fir species, is dominant. It always occurs mixed with ponderosa pine. At Sierra San Pedro Mártir, in Baja California Sur, *Pinus jeffreyi* coexists with *Abies concolor* as part of other mixed forest communities (see chapter 5 of this volume).

Alpine Bunchgrassland

Mexican alpine bunchgrassland is dominated by tussock grasses restricted to steep volcanic slopes at elevations above timberline (ca. 3800 m; Fig. 15.7). It has been studied by a large number of researchers throughout this century (Standley 1936; Beaman 1962, 1965; Cruz 1969; Delgadillo 1987). Beaman (1962, and 1965) called this vegetation "alpine prairie," whereas Miranda and Hernández-X. (1963) related it to Andean ecosystems and called it "high páramo." Almeida et al. (1994), however, believed that the unique composition of Mexican alpine vegetation made it different from the actual páramo. Rzedowski (1978), in agreement, treated it as a separate vegetation type, which he called "alpine zacatonal."

Alpine bunchgrassland communities occur in six main high-mountain formations: Cerro Potosí, Nevado de Colima, Nevado de Toluca, Sierra Nevada, Malinche volcano, and Pico de Orizaba volcano. No systematic survey has described and compared these communities. Recent surveys conducted at Popocatépetl (Almeida et al. 1994), Iztaccíhuatl, and Nevado de Colima volcanoes provide a limited summary of Mexican alpine ecosystems. Near the lower limit, in the vicinity of *Pinus hart-wegii* forest, *Lupinus montanus, Festuca tolucensis, Calamagrostis tolucensis, Penstemon gentianiodes,* and *Descurainia impatiens* are the most common species. These are considered the diagnostic species of zonal alpine communities. *Arenaria bryoides* and *Juniperus monticola* typify the azonal alpine communities. Near the upper nival border, mosses and lichens dominate (*Bartramia* and *Bryoerythrophyllum*). Intensive grazing and fires are fast depleting these alpine ecosystems. This vegetation also harbors a large number of endemic taxa. Despite the small area covered by Mexican alpine bunchgrassland (0.02% of the whole country), five zonal and two azonal plant associations (*sensu* Braun-Blanquet 1951) have been described by Almeida et al. (1994).

In the southern mountains of Mexico, at the border with Guatemala, there are small patches of alpine grassland dominated by *Lupinus montanus* and tussocks of *Calamagrostis vulcanica* up to 1 m high. There is also a ground layer with mosses such as *Breutelia* and *Leptodontium*. Other common species are *Luzula racemosa, Agrostis tolucensis, Draba vulcanica, Arenaria bryoides, Gnaphalium salicifolium,* and *Potentilla heterosepala.* On rocky outcrops, *Racomitrium crispulum* is dominant. This southern alpine vegetation grows on gentle, wind-protected slopes with regosols (Islebe and Velázquez 1994).

Fire and grazing are the major causes of degradation of alpine bunchgrassland ecosystems. When fires are frequent, *Lupinus montanus* becomes dominant. Hiking paths significantly fragment this vegetation (Almeida et al. 1994).

Figure 15.6. Aspect of fir forest at Ajusco volcano (3100 m).

Figure 15.7. Aspect of alpine bunchgrassland at Iztaccíhuatl volcano (4100 m)

Subalpine Bunchgrassland (Festuca tolucensis)

This community is mainly restricted to the flat valley bottoms within volcanic craters at 3500–3550 m elevation. Soils are very deep and have a thick surface layer of litter. The community consists of a dense layer of bunchgrasses 50 cm tall, dominated by *Festuca tolucensis* and *Calamagrostis tolucensis*, and an open ground layer dominated by *Alchemilla procumbens*. Other diagnostic species are *Poa annua*, *Trisetum spicatum*, *Pinus montezumae*, *Pinus hartwegii*, *Muhlenbergia quadridentata*, *Muhlenbergia* aff. *pusilla*, *Oxalis* spp., *Sicyos parviflorus*, *Potentilla staminea*, *Pedicularis orizabae*, *Draba jorullensis*, and *Arenaria bryoides*.

Beaman (1965), Cruz (1969), Rzedowski (1978), and Almeida et al. (1994) have observed this community in volcanoes along the Transversal Neovolcanic Belt where continuous burning and grazing disturbances take place. This is the reason these authors considered this grassland to be a seral community (Velázquez 1994). Woodcutting could promote this type of vegetation.

Less Common Communities

Megarosettes of *Furcraea bedinghausii* indicate a vegetation type restricted to the rolling, dissected, rocky lower slopes of a few volcanoes in central Mexico, such as Pelado (3090–3340 m) and Tláloc. Soils are shallow, gravelly, loamy clays (pH 5.0–6.5). Half-meter-high monocaulescent agavaceous megarosettes of the endemic *Furcraea bedinghausii* characterize this community. The maximum height ever measured for *Furcraea* is 5.3 m. A floristically rich but relatively open herb layer is common. Other characteristic species include *Senecio angulifolius*, *Stipa ichu*, *Symphoricarpos microphyllus*, *Conyza schiedeana*, *Muhlenbergia macroura*, *M. quadridentata*, *Geranium potentillaefolium*, *Gnaphalium oxyphyllum*, *Alchemilla procumbens*, *Sibthorpia repens*, and *Festuca amplissima*.

Stipa ichu meadow is commonly known as a "pradera" of *Potentilla candicans*. It occurs on poorly drained soils (Cruz 1969) on flats surrounding volcanic cones at 3000–3300 m elevation. It is restricted to the Valley of Mexico (Rzedowski and Rzedowski 1978; Rzedowski 1981). Soils are deep sandy loams of pH 5.0–6.2. Vegetation is a single ground layer (0.15 m height), consisting of low forbs and grasses. Diagnostic and associated species include *Stipa ichu*, *Potentilla candicans*, *Astragalus micranthus*, *Reseda luteola*, *Bidens triplinervia*, *Hedeoma piperitum*, *Commelina alpestris*, *Vulpia myuros*, *Alchemilla procumbens*, *Gnaphalium seemannii*, and *Salvia* species. This vegetation is significantly disturbed by hikers and campers.

An alpine scrub of *Juniperus monticola*, co-dominated by *Tortula andicola*, *Eryngium proteiflorum*, and *Senecio mairetianus*, is restricted to rocky, wet places above timberline on most high Mexican volcanoes, such as Popocatépetl, and it is also present in Guatemala (Islebe and Velázquez 1994).

A narrow ecotonal *Cupressus lusitanica* forest used to be widely common between fir forest and cloud forest at about 2600 m elevation. This transition has largely been deforested and transformed into farmland. In the few remaining patches, *Abies* is sometimes co-dominant in the overstory. *Senecio platanifolius*, *S. cinerarioides*, *Fuchsia microphyla*, and species of *Rubus* and *Ribes* are common in an understory shrub layer.

BIOGEOGRAPHIC HYPOTHESES

Mexico's large biodiversity has been attributed to two main hypotheses. First, its geographic location – where the Neartic and the Neotropical zones overlap – necessarily mixes temperate and tropical elements. This is the dispersal hypothesis. Additionally, a substantial proportion of the total flora of this region is of autochthonous origin, forming a heterogeneous mosaic of species (e.g., Rzedowski 1978). Second, its role as a Pleistocene refuge could make palaeoclimatic events of paramount importance in governing present vegetation (Toledo 1982).

The dispersal hypothesis imagines that the major mountain ranges served as bridges connecting Mexican and North American floras. As a consequence, Holarctic taxa reached Mexico across these mountain bridges, which provide a continuously similar climatic condition across latitude (e.g., Graham 1972; Rzedowski 1978). The arrival period of most boreal elements is controversial: Some authors believe that it was in the Tertiary and Early Quaternary (Graham 1972), whereas Martin and Harrell (1957), among others, suggest a more recent arrival. The close affinity between Mexican and North American temperate taxa has been commented on by several biogeographers (e.g., Islebe and Velázquez 1994). Rzedowski (1993) estimated the affinity at 95% at the generic level and named it the "megamexico phytogeographic unit."

According to Toledo (1976), two areas of Mexico functioned as Pleistocene refugia: the Lacondona and Soconusco regions, both of which still harbor a large number of endemic species and subspecies. Toledo also hypothesized that additional areas in Guatemala and Belize served as refuges and that collectively they played this role several times during the Quaternary. The expansion and contraction of glaciers may additionally have modified the present distribution of temperate humid montane vegetation.

There are also alternative hypotheses. Croizat (1982), for example, finds a close correspondence between biologic and geologic histories, suggesting convergent evolution and vicariant patterns that extend back across longer periods. The panbiogeograpy/vicariance ideas of Espinosa and Llorente (1993) extend back to Laurasian time.

DYNAMICS OF VEGETATION

Indigenous Indians, contact-period Spanish settlers, and current mixed populations of many cultures all have modified the temperate Mexican landscape. There is very little published speculation about reconstructing maps or descriptions of pre-contact landscapes. Largely we have had to rely on experienced botanists (e.g., Miranda, Hernandez-X., Rzedowski, Madrigal) who infer general successional pathways, and on the few remaining Indian groups that still manage vegetation and maintain seral stages for cultural reasons (e.g., Tarahumaras, Purépechas, Tzotziles) (Toledo 1994). There is some evidence from pollen analysis that suggests that there was deforestation prior to colonial times (Metcalfe et al. 1991). The deforestation could perhaps have been caused by large stand-replacing fires promoted by drought periods (Islebe et al. 1995).

By the time of the arrival of the Spanish (ca. 1500), most areas in central Mexico were densely covered by forest in different successional stages. When woodcutting took place to fulfill European demands (± 1700–1800), most temperate forests were clear-cut on plateaus, valleys, around lakes, and near large human settlements. Remaining, to some extent, was montane temperate vegetation. However, logging in montane areas has been increasing over the past 50 yr of the twentieth century, explaining the present patchiness of montane forest except in the highest, most inaccessible areas. Livestock foraging in the nineteenth century extended into these mountane environments. With human population increase, more meat was needed, and burning in the dry season became a popular method to increase forage. This fire-grazing practice is still implemented on a yearly basis throughout most of the temperate region (Toledo 1988). Although the fires are intended to promote regrowth in the understory vegetation, accidental canopy fires frequently take place (Velázquez 1992).

Most researchers (Miranda and Hernández-X. 1963; Madrigal 1967; Rzedowski 1978; Velázqu͏ and Cleef 1993) agree that fir forest is the mes͏ climax type of Mexican temperate ecosystems (Fig. 15.8). *Abies* communities are favored by soils rich in organic matter, humid terrains, and middle to high elevations (Madrigal 1967). Clear-cutting for paper production transforms these forests into subalpine bunchgrasslands dominated by *Muhlenbergia, Calamagrostis*, and *Festuca*; selective cutting transforms them into mixed oak forest with many sympatric *Quercus* species. Fire and grazing may then degrade subalpine bunchgrasslands into a scrubland of *Senecio* and *Ribes*, or into meadows of *Potentilla* and *Stipa* where soils are poorly drained.

Scrub and meadow may regenerate into conifer forest if there is not strong human interference (Fig. 15.8). Mixed oak forest can develop into mixed alder-oak forest and then into either cloud forest (given sufficient moisture) or into mixed pine-alder forest (where soils are sandy and acidic). Mixed pine-alder forest is the most widely distributed Mexican temperate ecosystem at this time. Fire and grazing favor pine species, transforming this vegetation type into pure stands of pine forest (*Pinus hartwegii* at high elevations). Mixed pine-alder forest may, under very limited circumstances, develop back into fir forest (Fig. 15.8). This happens, for instance, in very humid canyons with a thick litter layer and an absence of fire and grazing. Pure pine forest can also be replaced by fir forest but only at elevations where fir species are better adapted than pines.

High-elevation fir and pine forests, when harvested, revert to alpine bunchgrassland dominated by *Calamagrostis, Trisetum, Agrostis*, and *Festuca*. Mexican alpine bunchgrassland, as well as the tropical alpine grasslands of páramo and puna, expand where deforestation takes place (Balslev and Luteyn 1992). Forest regeneration is suppressed by fire- and grazing activities.

Nearly all temperate vegetation types of Mexico are in some stage of regression or progression. Fire, grazing, wind, herbivory, avalanches, landslides, volcanism, and human disturbances are the causes of this seral landscape. Natural fire and arson may be the most common disturbances as evidenced by the charcoal that is found in most soils throughout Mexico's mountains. At return intervals of 20 yr, fire seems to be a suitable tool for forest management, but in central Mexico fires recur every 1–5 yr. Only a few places have remained unburned >5 yr. The reason for frequent burning is that the forage becomes less palatable for domesticated animals over time. A large amount of oxalates and silicates, which accumulate in the leaves of grasses, may be the cause of low consumption of forage

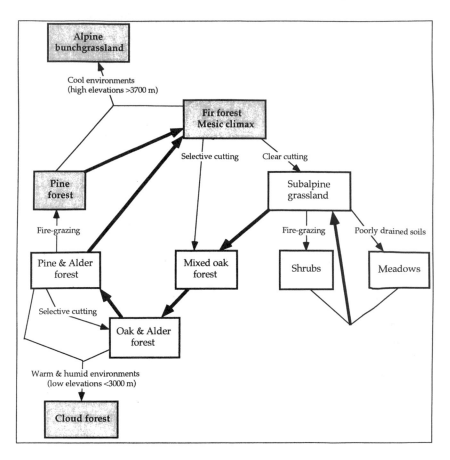

Figure 15.8. Schematic summary of successional relationships among Mexican temperate vegetation types. Mesic climax types are shaded darker than seral stages. Thick arrows represent natural, progressive succession, whereas thin arrows represent retrogressive succession caused by human interference or by azonal environmental conditions.

(Velázquez 1988). Peasants who set fires, however, are ignorant of the fact that many nutrients released by the fire are leached and eroded away.

Those few maps of current Mexican vegetation that have been published (e.g., Leopold 1950; Miranda and Hernández-X. 1963; Flores, Jimenéz, Madrigal, Moncayo, and Takaki 1971; Rzedowski 1978) rapidly became out of date as extensive human modification of the landscape continued. Consequently, these maps should be considered as depicting potential vegetation rather than actual vegetation (Velázquez and Cleef 1993). According to Masera, Ordonez, and Dirzo (1992), temperate forest deforestation (excluding cloud forest) has been estimated at 163,000 ha yr^{-1}, equivalent to 0.01% of the total Mexican surface being deforested every year. In contrast, reforestation has been attempted on only 13,000 ha, and not all the attempts have been successful.

Mexican temperate and tropical forests have been more impacted, fragmented, and depleted than any other vegetation type (Toledo 1988; Masera et al. 1992). Recent estimations by Masera et al. (1992) are that fires (49%), livestock production (28%), and agriculture (16%) are the main causes of the depletion of temperate forests. However, these estimates are based on only general field observations and anecdotal comments from rural people. Intensive logging activities were common throughout the country 30 yr ago. These activities became regulated by law in the 1950s, although actual implementation of the regulations was limited to central Mexico. The western and eastern Sierra Madres are still being clear-cut where the original forests remain. Clear-cutting in Mexico is not followed by reforestation; consequently, the negative effects of soil erosion and soil productivity are astonishing.

In contrast, less intensive fuelwood gathering, as practiced for local consumption by rural people, favors both regeneration and development of adult trees. Minnich et al. (1994) provide evidence from repeat aerial photographs that land use may not have had a severe impact on pine-oak forest in Aguascalientes. The Sierra Fría, located in the State of Aguascalientes, is dominated by pine-oak forests (*Quercus potosina, Q. laeta, Q. eduardii, Q. sideroxyla, Q. rugosa, Pinus teocote, P. leiophylla, Juniperus deppeana*) that form contiguous forests on steep slopes of barrancas, and open savannas on mesas (Fig. 15.9). Chaparral dominated by *Arctostaphylos pungens*, with scattered *Arbutus glandulosa, A. halopensis, Garrya* spp. and *Comerostaphylos polifolia*, is abundant on steep slopes. Technologies for forest exploitation for timber and charcoal production, using rudimentary ground kilns, were limited until the 1920s due to the inaccessibility of the range. Repeat aerial photographs taken between 1942 and 1993 reveal that the range experienced a distinct pulse of woodcutting with the introduction of gasoline sawmills after 1920. This land use ended in 1950 when the urban demand for fuelwood ceased with the introduction of natural gas pipelines into the city of Aguascalientes. Oaks were still common in exploited forests because most species resprout from pollarding. Since 1942 there has also been broad-scale thickening of pine-oak forests caused by rapid increases of *Juniperus deppeana* and slow establishment of *Quercus* spp. There has been little change in the distribution of pines except for local declines in *Pinus leiophylla* and *P. teocote* from insect attack during an El Niño–related drought in 1984 (Siquéiros-Delgado 1989). *Arctostaphylos pungens* chaparral experienced little change in spite of large fires between 1920 and 1950. Most species appear to resprout or to establish numerous postfire seedlings from seedbanks (*A. pungens*), similar to congeners in California. Stand-thickening coincided with open-range cattle grazing.

Recent directional vegetation changes may be caused by anomalously infrequent fire as a natural disturbance in the sierra. A dense herbaceous understory in pine-oak forests probably supported frequent fire prior to livestock grazing. The various species of *Quercus* are adapted to survive ground fires with moderately thick bark, tall canopies, and an ability to resprout from rootcrowns. Surface fires may have selectively eliminated young pines and encouraged open old-growth forests, as was characteristic in yellow pine forests in the western United States (see Chapter 5). Recurrent fires may reduce the extent of *Juniperus deppeana* and *Arctostaphylos pungens* because they are nonsprouters

and do not establish abundant postfire seedlings from seedbanks. *J. deppeana* may have survived in fire-protected canyons where old-growth stands now occur.

Urbanization is perhaps the most threatening human activity to temperate forests in central Mexico. Many of the main urban concentrations of the country, including Mexico City, are located in (or near) temperate vegetation. According to the National Census of Population of 1990, the area covered by temperate vegetation is inhabited by about 19 m people, or one quarter of the total Mexican population (Toledo and Rzedowski 1995). Of these "temperate" people, 62% live in cities and 38% in the countryside. It is this latter nonurban population of approximately 7.3 million who live most intimately with temperate vegetation. We estimate that the precontact population within temperate vegetation was – in contrast – only about 2 m indigenous people belonging to 40 different ethnic groups.

CONSERVATION IMPORTANCE AND PROBLEMS

The three ecological regions that constitute the temperate vegetation area of Mexico are of great importance from a biodiversity point of view. Despite its relatively small area, the humid temperate cloud forest is biologically very rich. It harbors a large number of endemic plant species, especially orchids, ferns, and mosses. Given the small area and the large number of species, it is floristically the richest zone in Mexico by unit area (Rzedowski 1993; Styles 1993). The zone is notable as well for its large number of endemic mammals, amphibians, reptiles, and butterflies (Flores 1993; Flores and Gerez 1994) to such an extent that it is one of the principal centers of autochthonous species. Areas above timberline (>4000 m) are also of notable biological and biogeographical importance.

Overall, the temperate zone covers the greatest part of the mountainous areas of Mexico. Of special importance is the Transversal Neovolcanic Belt because it harbors one of the highest concentrations of species diversity and endemism presently known (Fa 1989). Rzedowski (1993) estimates that there are 7000 species of flowering plants, of which 4900 (ca. 75%) are endemic.

About 7.3% of the Mexican territory is under some policy of protection (Flores and Gerez 1994). The criteria for the selection of protected areas and their boundaries have changed from time to time. Most protected areas (79 out of 1660) are within the temperate region and harbor temperate vege-

Figure 15.9. Pine-oak forest in the Sierra Friá, Aguascalientes. (A) General aspect, showing mosaic of woodlands
and savannas. (B) Typical structure of the forest/woodland.
(Photographs Courtesy of Richard Minnich.)

Figure 15.10. Deforestation in Desierto de los Leones National Park (2900 m) near Mexico City, caused by bark beetle (Dendroctonus adjuntus). Air pollutants initially *weakened the forest; infestation by bark beetles then caused the death of 25% of the trees.*

tation types (Flores and Gerez 1994), but their cumulative area is modest, accounting for only 4% of all protected area, or only 0.32% of the Mexican surface. Most of the protected areas with temperate vegetation are very small in size, which implies that the probability of their continued existence into the future is low. Financial support for patrolling to reduce poaching and other human encroachment is limited. Furthermore, no social aspects have been taken into account to ensure their permanence and conservation (McNeely 1989; Alcorn 1994; Toledo and Ordoñez 1993).

AREAS FOR FUTURE RESEARCH

Temperate vegetation types of Mexico have enormous importance as reservoirs of biodiversity. They also have economic and human health importance as sources of timber, fuelwood, pharmaceuticals, water, erosion control, and oxygen. Current knowledge is, however, far from complete. Most studies have concentrated on describing communities in Central Mexico, leaving large gaps in information about northern and southern communities. In addition to the need for more descriptive studies, great effort should be given to documenting ecosystem processes such as vegetation dynamics and succession. These ecosystem processes are the least known aspects of temperate vegetation. The data are needed to model future distributions and compositions of these communities. If financial support were available, such data could be obtained relatively quickly. In the absence of support, the most feasible future of temperate plant communities is their progressive destruction (Fig. 15.10).

REFERENCES

Alcorn, J. B. 1994. Noble savage or noble state?: northern myths and southern realities in biodiversity conservation. Etnolecologica II:1–8.

Aguilera, N., T. M. Dow, and R. Hernández S. 1962. Suelos, problema básico en silvicultura, pp. 108–140 in Seminario y viaje de estudio de coníferas latinoamericanas. Ins. Nac. Invest. Forest Publ. Esp. 1. México, D. F.

Almeida, L., A. M. Cleef, a. Herrera, A. Velázquez, and I. Luna. 1994. 1994. El zacatonal alpino del

Volcán Popocatépetl, México y su posición en las montañas tropicales de América. Phytocoenologia 22:391–436.

Anaya, L. A. L. 1962. Estudio de las relaciones entre la vegetación, el suelo y algunos factores climáticos en seis sitios del declive occidental del Iztaccíhuatl. Tesis. Facultad de Ciencias, UNAM. México, D.F.

Anaya, L. A. L., S. R. Hernández, and S. X. Madrigal. 1980. La vegetación y los suelos de un transecto altitudinal del declive occidental del Iztaccíhuatl (México). Boletín Técnico 65. INIF, SARH. México.

Balslev, H., and J. L. Luteyn (eds.). 1992. Páramo. Academic Press, New York.

Beaman, J. H. 1962. The timberline of Iztaccíhuatl and Popocatépetl, México. Ecology 43:377–385.

Beaman, J. H. 1965. A preliminary ecological study of the alpine flora of Popocatépetl and Iztaccíhuatl. Bol. Soc. Bot. México 29:63–75.

Beard, J. S. 1944. Climax vegetation in tropical America. Ecology 25:58–125.

Beard, J. S. 1955. The classification of tropical American vegetation types. Ecology 36:89–100.

Benítez, B. G. 1988. Efectos del fuego en la vegetación herbácea de un bosque de Pinus hartwegii Lindl. de la Sierra del Ajusco, pp. 111–152 in E. H. Rappoport, y I. R. López Moreno (eds), Aportes a la ecología urbana de la ciudad de México. Editorial Limusa, México.

Braun-Blanquet, J. J. 1951. Pflanzensoziologie, Grundzüge der Vegetationskunde, 2nd ed. Springer-Verlag, New York.

Breedlove, D. E. 1981. Flora de Chiapas. Part I: Introduction. California Academy of Sciences, San Francisco.

Cervantes, F. A. 1980. Principales características biológicas del conejo de los volcanes Romerolagus diazi, Ferrari Pérez 1893 (Mammalia: Lagomorpha). Tesis de licenciatura. Facultad de Ciencias, UNAM. México, D. F.

Critchfield, W. B., and E. L. Little. 1966. Geographic distribution of the pines of the world (Pinus section Strobus). Taxon 35:647–656.

Cox, B. C., and P. D. Moore. 1993. Biogeography. An ecological and evolutionary approach, 5th ed. Blackwell Scientific Publications, Oxford.

Croizat, L. 1982. Vicaraince/vicariism, panbiogeography, "vicariance biogeography," a clarification. System. Zool. 31:291–304.

Crum, H. A. 1951. The Appalachian-Ozarkian element in the moss flora of Mexico with a checklist of all known Mexican mosses. Ph.D. dissertation, University of Michigan, Ann Arbor.

Cruz, C. R. 1969. Contribución al estudio de los pastizales en el Valle de México. Tesis. Escuela Nacional de Ciencias Biológicas, IPN. México, D. F.

Delgadillo, C. 1979. Mosses and phytogeography of the Liquidambar forest of Mexico. Bryologist 82:432–449.

Delgadillo, C. 1987. Moss distribution and phytogeographical significance of the Neovolcanic Belt of Mexico. J. Biogeo. 14:69–78.

Espinosa, D. O., and J. B. Llorente. 1993. Fundamentos de biogeografias filogenéticas. UNAM–CONABIO. México.

Ern, H. 1973. Repartición, ecología e importancia económica de los bosques de coníferas en los Estados mexicanos de Puebla y Tlaxcala. Com. Proy. Pue. Tlax. 7:21–23.

Fa, J. E. 1989. Conservation-motivated analysis of mammalian biogeography in the Trans-Mexican Neovolcanic Belt. Nat. Geog. Res. 5:296–315.

Ferrusquía-Villafranca, I. 1993. Geology of Mexico: a synopsis, pp. 3–103 in T. P. Ramamoorthy, R. Bye, A. Lot, and J. Fa. (eds.), Biological diversity of Mexico (origins and distribution). Oxford University Press, Oxford.

Flores, M. G., J. L. Jiménez, X. S. Madrigal, F. R. Moncayo, and F. T. Takaki. 1971. Memoria del mapa de tipos de vegetación de la Republica Méxicana. Secretaría de Recursos Hidráulicos. México, D. F.

Flores, O. 1993. Herpetofauna of Mexico: distribution and endemism, pp. 253–280 in T. P. Ramamoorthy, R. Bye, A. Lot, y J. Fa. (eds.), Biological diversity of Mexico. Oxford University Press, Oxford.

Flores, O., and P. Gerez. 1994. Biodiversidad y conservación en México: vertebrados, vegetación y uso del suelo. UNAM – CONABIO. México, D. F.

Gadow, H. 1930. Jorullo: The history of the volcano Jorullo and reclamation of the devastated district by animals and plants. Cambridge University Press, Cambridge.

García, E. 1981. Modificaciones al sistema de clasificación de Koeppen. 2a. Edición. Instituto de Geografía, Universidad Nacional Autónoma de México. México, D. F.

García, E., and Z. Falcón. 1986. Nuevo Atlas Porrúa de la República Méxicana. 7a. edición. Editorial Porrua, S. A. México.

Goldman, E. A., and R. T. Moore. 1945. The biotic provinces of Mexico. J. Mammal. 26:347–360.

González, J. G. 1982. El Volcán el Pelado como una reserva natural. Tesis. Facultad de Filosofía y Letras, Colegio de Geografía, UNAM. México, D. F.

González-Espinosa, M., P. F. Quintana-Ascencio, N. Ramírez-Marcial, and P. Gaytán-Guzmán. 1991. Secondary succession in disturbed Pinus-Quercus forest in the highlands of Chiapas, Mexico. J. Vegeta. Sci. 2:351–360.

Graham, A. 1972. Some aspects of Tertiary vegetation history about the Caribbean Basin. Mem. Simposium Congress Latinoamericano de Botánica, pp. 97–117.

Guzmán, G. 1973. Some distributional relationships between Mexican and United States mycofloras. Mycologia 45:1319–1330.

Islebe, A. G. 1993. Will Guatemala's juniperus-pinus forest survive? Environ. Conserv. 20:167–168.

Islebe, A. G., and A. Velázquez. 1994. Affinity among mountain ranges in Megamexico: a phytogeographic scenario. Vegetatio 115:1–9.

Islebe, A. G., A. M. Cleef, and A. Velázquez. 1995. High elevation coniferous vegetation of Guatemala. Vegetatio 116:7–23.

Jardel, P. E., and L. R. Sánchez-Velásquez. 1989. La sucesión forestal: fundamento ecológico de la silvicultura. Ciencia y desarrollo, Vol. XIV, Num. 84. CONACYT México, D. F.

Leopold, A. S. 1950. Vegetation zones of Mexico. Ecology 31:507–518.

MacNally, R. C. 1989. The relationship between habitat breadth, habitat position, and abundance in forest and woodland birds along a continental gradient. Oikos 54:44–54.

Madrigal, S. X. 1967. Contribución al conocimiento de la ecología de los bosques de oyamel (Abies religiosa H.B.K., Schl. et Cham.) en el Valle de México. Instituto Nacional de Investigaciones Forestales. Boletín Técnico No. 18. México, D. F.

Martin, P. S., and B. E. Harrell. 1957. The Pleistocene history of temperate biotas in Mexico and eastern United States. Ecology 38:468–480.

Masera, O., M. J. Ordonez, and R. Dirzo. 1992. Carbon emissions from Mexican forest: current situation and long-term scenarios. Lawrence Berkeley Laboratory Report, University of California, Berkeley.

McNeely, R. C. 1989. The relationship between habitat breadth, habitat position, and abundance in forest and woodland birds along a continental gradient. Oikos 54:44–54.

Metcalfe, S. E., F. A. Street-Perrot, R. A. Perrot, and D. D. Harkness. 1991. Paleolimnology of the upper Lerma Basin, Central Mexico: a record of climatic change and anthropogenic disturbance since 11,600 BP. J. Paleolimn. 5:197–218.

Minnich, R. A., J. Sosa Ramírez, E. Franco Vizcaíno, W. J. Barry, and M. E. Siquéiros Delgado. 1994. Reconocimiento preliminar de la vegetación y de los impactos de las actividdes humanas en la Sierra Fría, Aguascalientes, México. Investigación y Ciencia. Universidad Autónoma de Aguascalientes. Edición Cuatrimestral 12:23–29.

Miranda, F. 1947. Estudios sobre la vegetación de México. V Rasgos de la vegetación en la Cuenca del Río de las Balsas. Rev. Soc. Méx. Hist. Nat. 8:95–114.

Miranda, F., and E. Hernández-X. 1963. Los tipos de vegetación de México y su clasificación. Boletín de la Sociedad Botánica de México 28:29–179.

Mittermeier, R. A. 1988. Primate diversity and the tropical forest: case studies from Brazil and Madagascar and the importance of megadiversity countries, pp. 145–154. in E. O. Wilson (ed.), Biodiversity. National Academy Press, Washington D.C.

Puig, H. 1970. Notas acerca de la flora y la vegetación de Tamaulipas. Anales Esc. Nac. Ci. Biol. 17:37–49.

Ramamoorthy, T. P., R. Bye, A. Lot, and J. Fa, (eds.). 1993. Biological diversity of Mexico (origins and distribution). Oxford University Press, Oxford.

Rzedowski, J. 1954. Vegetación del Pedregal de San Angel, Distrito Federal, México. Anales Esc. Nac. Ci. Biol. 8:59–129.

Rzedowski, J. 1963. El extremo boreal del bosque tropical siempre verde en Norteamérica continental. Vegetatio 11:173–198.

Rzedowski, J. 1978. Vegetación de México. Editorial Limusa, México, D. F.

Rzedowski, J. 1993. Diversity and origins of the Phanerogamic flora of Mexico, pp. 129–146. in T. P. Ramamoorthy, R. Bye, A. Lot, y J. Fa. (eds.), Biological diversity of Mexico. Oxford University Press, Oxford.

Rzedowski, J., and R. McVaugh. 1996. La vegetación de Nueva Galicia. Contr. Univ. Michigan Herb. 9:1–123.

Rzedowski, J., and G. C. de Rzedowski. 1981. Flora fanerogámica del Valle de México. 3a ed. Vol. 1. Editorial CECSA. México, D. F.

Sánchez, M. N., and L. Huguet. 1959. Las coníferas méxicanas. Unasylva 3:24–35.

Sanders, E. M. 1921. The natural regions of Mexico. Geograph. Rev. 11:212–226.

Shelford, V. E. (ed). 1926. Naturalist's guide to the Americas. Williams & Wilkins, Baltimore.

Siquéiros-Delgado, M. 1989. Coníferas de Aguascalientes. Universidad Autónoma de Aguascalientes. Dirección General de Investigación Científica y Superación Académica de la SEP, mediante los convenios Núm. C86-01-0206, C87-01-0251 y C88-01-0075,

Smith, L. B. 1940. Las provincias bióticas de México, según la distribución geográfica de las lagartijas del género *Sceloporus*. Anales Esc. Nac. Ci. Biol. Ci. 2:95–110.

Standley, P. C. 1936. Las relaciones geográficas de la flora mexicana. Anales Inst. Biol. Univ. Nac. Autón. México 7:9–16.

Styles, B. T. 1993. Genus *Pinus*: a Mexican purview, pp. 397–420 in T. P. Ramamoorthy, R. Bye, A Lot, y J. Fa. (eds.), Biological diversity of Mexico. Oxford University Press, Oxford.

Toledo, V. M. 1976. Los cambios climáticos del Plesitoceno y sus efectos sobre la vegetatión tropical calida y húmeda de México. Master's thesis. UNAM. México, D. F.

Toledo, V. M. 1982. Pleisticene changes of vegetation in tropical Mexico, pp. 63–71 in G. T. Prance (ed.), Biological diversification in the tropics. Columbia University Press, New York.

Toledo, V. M. 1988. La diversidad biológica de México. Ciencia y Desarrollo 81:17–30.

Toledo, V. M. 1994. La diversidad biológica de México. Ciencias 34:43–59.

Toledo V. M., and M. J. Ordoñez. 1993. The biodiversity scenario of Mexico: a review of terrestrial habitats, pp. 757–778. in T. P. Ramamoorthy, R. Bye, A. Lot, and J. Fa. (eds.), Biological diversity of Mexico. Oxford University Press, Oxford.

Toledo, V. M., and J. Rzedowski. 1995. Mexico, in V. Heywood, and O. Herrera, (eds.), Centres of plant diversity: a guide and strategy for their conservation. IUCN, International Union for the Conservation of Nature, Geneva, Switzerland.

Toledo, V. M., J. Rzedowski, and J. Villa-Lobos. 1997. Regional overview: middle America, pp. 97–124 in S. D. Davis et al. (eds.), Centers of plant diversity, Vol. 3, The Americas. World Wildlife Fund and the International Union for the Conservation of Nature, New York.

Troll, C. 1952. Das Pflanzenkleid der Tropen in seiner Abhüngigkeit von klima, Boden und Mensch. Deutscher Geog. Frankfurt, 1951, Tagungsber. und wiss. Abh., Remagen, pp. 35–56.

van der Hammen, T. 1991. Paleoecology of the neotropics: an overview of the state of affairs. Proceedings of Global Changes in South America During the Quarternary, Boletin IG-USP, Publicacao espcial, No. 8, Universidade de São Paulo, Instituto de Ciencias, Brasil.

Velázquez, A. 1988. Especies y hábitats en peligro de extinción. El caso del conejo de los volcanes. Revista de Información Científica y Tecnológica (CONACYT) 10 (147):45–49.

Velázquez, A. 1992. Grazing and burning in grassland communities of high volcanoes in Mexico, pp. 216–231 in H. Balslev, and J. L. Luteyn, (eds.), Páramo. Academic Press, New York.

Velázquez, A. 1993. Landscape ecology of Tláloc and Pelado volcanoes, Mexico. International Institute for Aerospace Survey and Earth Sciences (ITC), Pub. No. 16, Enschede, The Netherlands. 152 p.

Velázquez, A. 1994. Multivariate analysis of the vegetation of volcanoes Tláloc and Pelado, Mexico. J. Vegetat. Sci. 5:263–270.

Velázquez, A., and A. M. Cleef. 1993. The plant communities of the valcanoes Tláloc and Pelado, Mexico. Phytocoenologia 22:145–191.

Vink, R., and V. Wijninga. 1988. The vegetation of the semi-arid region of La Herrera (Cundinamarca, Colombia). Internal Report. Hugo de Vries Laboratory. University of Amsterdam.

Wagner, P. L. 1964. Natural vegetation of Middle America pp. 216–264 in C. R. West (ed.), Handbook of Middle American Indians, Vol. 1. University of Texas Press, Austin.

West, C. R. 1971. The natural regions of Middle America, pp. 363–383 in Handbook of Middle American Indians, Vol. 2 University of Texas Press, Austin.

Chapter 16

The Caribbean

ARIEL E. LUGO
JULIO FIGUEROA COLÓN   F. N. SCATENA

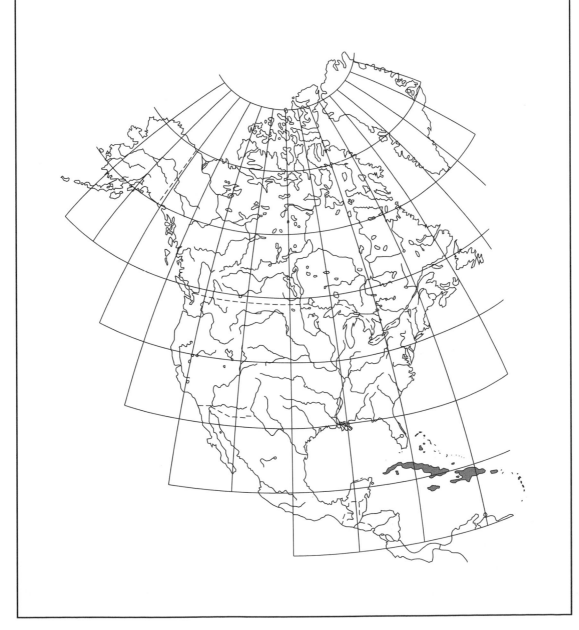

INTRODUCTION

An overview of Caribbean vegetation would take us through seagrass beds, mangroves, coastal forests, sand dunes, swamps, marshes, bogs, savannas, and lower montane and montane forests, all found in a wide variety of edaphic and climatic conditions (Fig. 16.1). In fact, the number of plant associations described for the Caribbean is staggering. Although Beard (1955) listed 28 plant formations for all of tropical America (including the Caribbean), the vegetation map of Cuba (Borhidi 1993) lists 62 plant formations that he subdivided into 27 classes and 56 orders of vegetation (Borhidi 1991). In Jamaica, Grossman, Iremonger, and Muchoney (ND) classified about 70 plant associations. Holdridge (1983) listed 30 plant associations within only 30,000 ha of public forests in Puerto Rico.

Certainly, the diversity of plant associations in the Caribbean islands reflects the wide range of environmental factors that occur at differing scales of time, space, and intensity. Other general factors that shape Caribbean vegetation are geologic and human history, climate, and natural disturbances. The effect of these factors on vegetation is further modified by aspect, topography, soils, and elevation.

Caribbean vegetation is distributed along a chain of islands with varying degrees of isolation from mainlands to the south (South America), west (Central America), and northwest (North America). This geographic position and environmental history produced a distinct flora that reached, colonized, and evolved on the islands and was further modified by at least 2000 yr of human occupation.

We focus our chapter on today's Caribbean island vegetation (especially forest vegetation) and on providing a synthesis explanation for why the vegetation is so diverse. This chapter has three parts: (1) environmental setting, (2) vegetation, and (3) research priorities.

ENVIRONMENTAL SETTING

The Caribbean (Fig. 16.2) has a notably heterogeneous environment. The region includes several thousand islands scattered over a 1900 km north-

This study was done in cooperation with the University of Puerto Rico and is part of the USDA Forest Service contribution to the Long-Term Ecological Research Program (Grant BSR-8811902) of the National Science Foundation. We thank M. Barbour, D. Billings, G. J. Breckon, S. Brown, C. Domínguez Cristobal, N. Fetcher, J. K. Francis, J. Frangi, J. Lodge, R. Myster, G. Reyes, M. Rivera, D. Schaefer, W. Silver, J. A. Torres, J. K. Zimmerman, and X. Zou for their help with the manuscript. José Colón provided the photographs of vegetation.

south distance from the Bahamas to Grenada, and a 2700 km east-west distance from Barbados to Cuba. This geographic expanse contains at least 14 Holdridge life zones (Table 16.1), complex geology, all taxonomic soil orders, and a potent group of periodic natural disturbances. Human-induced disturbances also affect vegetation both directly and indirectly through interactions with natural disturbances.

Diverse conditions and steep gradients of change are the common denominators of the six environmental influences that we discuss (geology, climate, topography, soils, natural disturbances, and human effects). This scale of diversity brings into close proximity, and allows exchange of materials and organisms among, different vegetation types.

On many Caribbean islands one can observe major gradients of environmental and vegetation change over distances of <100 km. Similar gradients occur in the continental tropics only over hundreds or even thousands of kilometers. Although most of the land surface of the Caribbean islands is <300 m in elevation (Howard 1979), individual islands can have dramatic elevation gradients. For example, the lowest (−30 m) and highest (3200 m) elevations in the region occur on the island of Hispaniola within a horizontal distance of only 75 km. The plant associations along this gradient range from thorn woodland to montane wet forest and the climatic change is the equivalent of nine Holdridge life zones (Hartshorn et al. 1981). Southeastern Cuba, between Guantánamo and Pico Bayamonesa, has a 1370 m elevation gradient over a 50 km distance, with a vegetation change from semidesert to montane wet conditions. The mean annual temperature and precipitation along this gradient range from 28° C and 390 mm at sea level to 12.5° C and 2674 mm at the montane extreme. An even more dramatic precipitation gradient occurs in the Luquillo Mountains of Puerto Rico, where within a horizontal distance of only 10 km, there is a 1000 m elevation gradient accompanied by 6° C and 4000 mm differences in mean annual temperature and rainfall. This gradient occurs on volcanic bedrock, but in other parts of the island one can find similar gradients over ultramafic (serpentine) soils, on limestone-derived soils, and mixed bedrock.

The occurrence of these environmental gradients over such short distances highlights two salient environmental aspects of the Caribbean. First, these gradients represent significant ecophysiological challenges to the vegetation. There is a notion among many ecologists that the tropics are benign environments favorable to unhindered plant

Figure 16.1. Diversity of Caribbean vegetation. (A) Freshwater marsh on Tortuguero Lagoon, north coast of Puerto Rico. The dominant species is Typha domingensis, with Sagittaria in flower just in front of it. Exotic coconut palms (Cocos nucifera) are in the background. Photograph courtesy of José Colón.

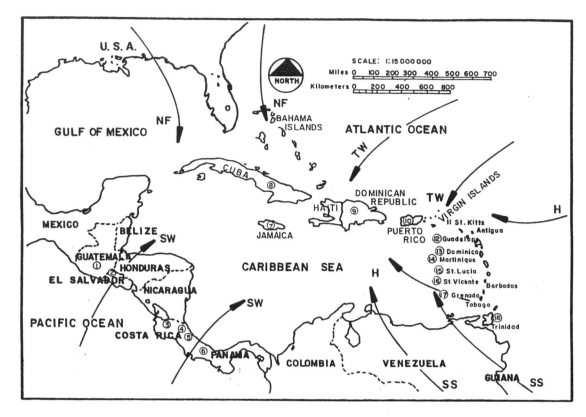

Figure 16.2. The Caribbean region, showing land masses and major climatic systems that converge on it. NF = northern fronts, TW = trade winds, H = hurricanes, SS = southern systems, SW = southwest winds. Modified from Lugo and Scatena (1992).

Table 16.1. Ecological life zones (sensu Holdridge 1967) within the Greater Antilles

Life zone	Cuba	Hispan-iola	Jamaica	Puerto Rico
T–Thorn woodland[a]	X			
T–Very dry forest	X		X	
T–Dry forest	X		X	
T–Moist forest	X		X	
T–Wet forest			X	
ST–Thorn woodland		X	X	
ST–Dry forest	X	X	X	X
ST–Moist forest	X	X	X	X
ST–Wet forest	X	X	X	X
ST–Rain forest		X		X
ST–LM–Moist forest	X	X		
ST–LM–Wet forest	X	X		X
ST–LM–Rain forest		X		X
ST–M–Wet forest	X	X		
Total	10	9	8	6

[a]T = Tropical, ST = Subtropical, LM = Lower montane, M = Montane. Although Borhidi (1991) reports Tropical life zones in Cuba, we believe these are subtropical.
Sources: Hartshorn et al. (1981), Thompson et al. (1986), Ewel and Whitmore (1973), and Borhidi (1991).

growth and high diversity of species. This is true for the Caribbean in terms of the absence of frost and subfreezing temperatures in the lowlands, which removes a significant environmental barrier to plant growth. Frost occurs in the Caribbean only at high elevations in the mountains of Cuba and Hispaniola. Annual temperature fluctuations are narrow in the region (Portig 1976). But within the benign temperature setting of the Caribbean, plants can be exposed to other dramatic environmental extremes that stress vegetation. For example, wet conditions lead to mechanical stress on vegetation, chemical leaching of nutrients, unstable terrain, and saturated anaerobic soils. Ultramafic soils can lead to toxic levels of metals in soil, physiological drought, and low Ca/Mg ratios. Limestone-derived soils, coastal habitats, and human-dominated conditions are also associated with stressful conditions for vegetation (Fig. 16.3; Borhidi 1991; Lugo and Scatena 1995; Lugo, González Liboy, Citrón, and Dugger 1978; and Medina, Cuevzs, Figuero, and Lugo 1994).

Second, Caribbean vegetation is patchy in its

Figure 16.3. Dry forest and scrub vegetation on Miocene limestone outcrops near the coast in the Guanica Forest of Puerto Rico. The two cacti shown are species of Melocactus *(foreground) and* Cephalocereus *(background). Vegetation is wind-sculpted, increasing in height back from the ocean (to the background in this photograph). Photograph courtesy of José Colón.*

distribution and can exhibit a diversity of eco-physiological strategies within relatively short distances. For example, it is common to find ever-green, deciduous, and succulent plant species growing in close proximity. They can exhibit different heights and leaf-water potentials; different rates of photosynthesis, respiration, and transpiration; and different degrees of leaf wilting (Lugo et al. 1978; Medina 1983). The roots of these plants are exposed to dramatically different conditions of available soil water, nutrients, and soil volume. Similar examples, but in response to different conditions, occur in wet montane regions where plants growing near each other may experience different wind, atmospheric, or light conditions (Shreve 1914). In coastal settings, wind, soil salinity, hydroperiod, and aeration can significantly change over short vertical or horizontal distances and have contrasting effects on coastal plant communities (Cintrón, Lugo, Pool, and Morris 1978).

Another result of the sharp environmental gradients in the region is the frequent occurrence of montane wetland conditions on steep slopes and mountain tops. Studies in the Luquillo Mountains suggest that the abundance of the palm *Prestoea montana* (Fig. 16.1b; plant nomenclature follows Liogier and Martorell 1982) and the broadleaf colorado tree, *Cyrilla racemiflora* (Fig. 16.4), is correlated with low soil oxygen resulting from high rainfall and a combination of edaphic and topographic conditions (Silver, Lugo, and Keller 1995). Frangi (1983) suggested that plant associations such as elfin cloud forests, palm brakes, palm floodplain forests, and colorado forests were all wetland forests in the Luquillo Mountains. It appears that the role of soil oxygen has been overlooked in wet tropical areas as a regulating factor of vegetation structure, composition, and function (Silver et al. 1995). To further understand the causal factors of the patchiness and diversity of Caribbean island environments, it is necessary to expand on each of the main factors that shape vegetation in the region.

Geology

The Caribbean region occupies a relatively small tectonic plate that is situated between the North and South American crustal plates. The region's geologic history is a complex sequence of events and has led to more than 100 geologic provinces (Case, Holcomb, and Martin 1984). Although there is considerable controversy concerning the actual sequence of events involved in the development of the region (Malfait and Dinkelman 1972; Pindell and Dewey 1982; Mattson 1984; Donnelly 1985; Burke, Cooper, Dewey, Mann, and Pindell 1984; Pindell and Barnett 1990), there is general agreement in several important aspects: (1) the opening of the Atlantic and the formation of the Gulf of Mexico occurred in the Jurassic; (2) North and South America were totally separated by the early Cretaceous; and (3) the subduction of the Caribbean Plate and the resulting formation of an island arc followed in the late Cretaceous. However, the movement, enlargement, and suturing of particular islands is still debated (Donnelly 1989).

Figure 16.4. Forest dominated by Cyrilla racemiflora in Luquillo Experimental Forest, Puerto Rico. This upper montane forest occurs at elevations >600 m on saturated soils. The open canopy allows light transmission and an extensive growth of ferns, vines, and epiphytes. Photograph courtesy of José Colón.

Physiographically, the Caribbean is made up of three distinct groups of islands: (1) the Bahamas (including Turks and Caicos); (2) the Greater Antilles (Cuba, Hispaniola, Jamaica, and Puerto Rico, including the Virgin Islands); and (3) the Lesser Antilles (from Anguilla to Grenada). Barbados is excluded here because it is actually an exposure of the submerged Barbados Ridge that connects to Trinidad, Tobago, and South America. The oldest rocks and geological features in the Caribbean have been found along the region's boundaries. The Lesser Antilles are of more recent Eocene origin. Jurassic to early Cretaceous deposits occur in the Bahamas and off the north coast of Cuba. Early to middle Jurassic sedimentary rocks from western Cuba are also well documented (Perfit and Williams 1989). Caribbean vegetation has been tracked back approximately 45 m yr to the middle of the Eocene (Graham and Jarzen 1969; Graham 1990).

Excluding Cuba, the oldest rocks in the Greater Antilles are primitive island arc remnants that form the foundations of the region, and upthrusted fragments of basaltic ultramafic oceanic crust. Significant exposures of ultramafic rock in the Caribbean occur on the islands of Cuba, Jamaica, Hispaniola, and Puerto Rico. Geologic evidence in the form of volcanic and clastic rocks of Eocene-Oligocene age suggest that certain parts of Cuba, Hispaniola, and Puerto Rico were never totally submerged (Mattson 1984), but Jamaica (Arden 1975) and many of the smaller islands have been totally submerged at different stages of their development.

Climate

The climate of the Caribbean is influenced by local conditions and meteorologic events that originate far from the region (Fig. 16.2). For example, Colón (1987) estimated that half of the annual rainfall of Puerto Rico originates in tropical latitudes and falls in summer under the influence of northeast trade winds. The other half originates in northern latitudes and falls in winter under the influence of northwest winds. These external climatic events generate atmospheric systems that interact with the land masses and vegetation as they pass over the Caribbean islands, creating gradients of climatic factors on both regional and local scales and also through time.

Three climatic gradients predominate in the region. One is the gradient that spans about 15° of latitude (12–27°N) between the Bahamas and Grenada. The main climatic variables here are increasing mean annual temperature, decreasing frequency and strength of cold fronts, and decreasing amplitude of daylength. The length of the dry season also varies along this gradient. Mean annual temperatures are cooler, and frost occasionally occurs in the Bahamas and south Florida but not in the islands south of them. Portig (1976) mapped the southern limits of the mean wind streamlines in the region and showed northerlies extending as an arc that passed over latitude 10°N on the south and over the northwestern tip of Hispaniola on the north. However, cold fronts do reach Puerto Rico at 18°N (Colón 1987). Annual differences in daylength decrease southward, and this amplitude can affect vegetation phenology (Negrón 1980; Cintrón and Shaeffer Novelli 1983).

The other two climatic gradients are orographic in nature. The first occurs on mountains within high relief islands. Rainfall, humidity, and wind speed increase, and air temperature decreases with elevation (Brown, Lugo, Silander, and Liegel 1983; Portig 1976). The lapse rate is about 6° C/1000 m. The other orographic gradient concerns the spatial climatic gradient within islands caused by the interaction of trade winds and topography. This interaction results in a gradient from wet and cool windward slopes to warm and dry leeward aspects (Portig 1976). Research in montane forests of Jamaica found that these factors do not influence vegetation independently. Vegetation in depressions and swales on windward slopes resembled vegetation higher up the slope on the windward side. Also, vegetation on peaks and ridges at lower elevations resembled vegetation at higher elevations (Shreve 1914).

Islands are also grouped as windward or leeward, depending on their location relative to powerful synoptic climatic systems that approach the region from the east (Fig. 16.2). Windward islands are exposed to more wind, rain, and salt spray and to heavier seas than are leeward islands. However, because of their southern location, windward islands are exposed to fewer hurricanes than are leeward islands.

Vegetation is different on leeward and windward coasts of individual islands in terms of zonation patterns, height, and degree of wind sculpturing. For example, the distribution and type of mangroves on these coastlines is dependent on geomorphology. Windward coastlines are less likely to have ocean-fringing mangroves. Instead, mangroves occur behind sand dunes or in inland basins and river floodplains. Leeward coastlines are less likely to have sand dune vegetation or riverine floodplains, but they can develop ocean-fringing mangroves and mangrove islands that grow in close proximity to coral reefs and seagrass beds (see Chapter 13).

Portig (1976) also highlighted the temporal aspects of climatic events in the Caribbean. All climatic factors exhibit periodicity at almost all temporal scales: diurnal, seasonal, El Niño events (cf. Neuman, Caso, and Jaruinen 1987; Brown et al. 1983; Colón 1987; Scatena and Larsen 1991; Lugo and Scatena 1995), episodic drought or high rainfall, and high or low hurricane incidence. All these factors are critical to vegetation function (Lugo and Scatena 1995; Scatena and Lugo 1995). Similarly, Foster (1973) found that unusually high rainfall on the Caribbean coast of Panama was a "false clue"

that caused lowland rain forest trees to bloom and bear fruit out of their normal season. This phenomenon subsequently triggered such far-reaching effects through the ecosystem as starvation and mortality among the frugivores and omnivores of the forest.

Topography

The topography of most Caribbean islands is steep (Fig. 16.5). This feature alone has considerable effect on vegetation by establishing the conditions for growth, for it affects the hydrology, water budget, and nutrient dynamics, determines the degree of exposure of vegetation to wind damage, and limits the space available for establishment and survival of plants.

The main topographical difference in the region is the distinction between low-lying (low) and mountainous (high) islands. Low islands have hot and dry climates and are mostly capped by limestone. High islands have more variability in topography, creating more variation in soil, climate, and exposure. In general, the region's high islands have a core of volcanic materials topped by low-lying clastic and carbonate rocks. The notable exception is Jamaica, whose long periods of subsidence are responsible for the majority of its high peaks being capped by limestone.

Topography and landform influence the pattern of wind damage to vegetation. Landscape patterns protect vegetation from hurricane winds (Bellingham 1991; Lugo 1995). Using spatially distributed simulation modeling, Boose, Foster, and Fluet (1994) made the following observations based on a large-scale simulation of wind effects on Caribbean forests:

Differences in topography determined what percentage of an area was classified as protected from wind damage.

Complex topography greatly modified hurricane strength and the effect of wind on forest.

Variation in wind damage to vegetation coincided with major physiographic features.

Large-scale geographic factors, such as proximity to water, presence of large land masses, and presence or absence of mountains, explained the frequency of hurricane passage.

The topographic catena from valleys to slopes to ridges shown in Fig. 16.5 is a common geomorphological feature that repeats itself over the region's landscapes at all elevations. As we explain later, this topographic sequence is associated with significant changes in the conditions for vegetation growth.

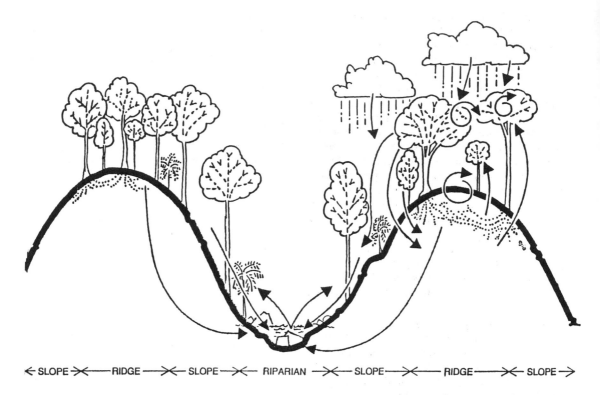

← SLOPE ⟩⟨ ——RIDGE ——⟩⟨— SLOPE—⟩⟨— RIPARIAN —⟩⟨— SLOPE——⟩⟨—— RIDGE——⟩⟨— SLOPE —⟩

Figure 16.5. The topographic catena that dominates Carib-bean forests. Arrows show fluxes of water, carbon, and nu- *trients. From Lugo and Scatena (1995). Data later shown in Figure 16.12 refer to this catena.*

Soil

The soils of the Caribbean are extremely diverse. Within 8900 km² on the island of Puerto Rico, there are nine taxonomic soil orders, 29 suborders, 38 great groups, 82 subgroups, 51 families, and 164 soil series (Beinroth 1982; Fox 1982; Bonnet Benítez 1983). Inceptisols and Ultisols are the most common soil orders on Puerto Rico. Puerto Rico is not the exception, for all of the region's larger islands have high soil diversity (Hartshorn et al. 1981, Kelly et al. 1988; Borhidi 1991).

Although soil surveys are available for most Caribbean islands, they are of limited use for descriptions of natural vegetation because they were mostly conducted for agricultural and development purposes. These types of surveys usually ignore forest soils on steep terrain or unusual soils unsuitable for agriculture. Notable among the available soil surveys are the dozens of volumes published since 1955 by the Regional Research Centre of the British Caribbean at the Imperial College of Tropical Agriculture in Trinidad. These reports contain systematic information on insular soils of the West Indies, and they include soil chemistry data. Most of the soil nutrient data have

been collected for agricultural purposes and therefore focus on the available fraction of the nutrient pool, whereas the total nutrient fraction can be 10 times higher than the available fraction.

Forest soils have been described in relation to vegetation in Jamaica (Grubb and Tanner 1976; Tanner 1977, 1986) and Puerto Rico; Lugo and Murphy 1986; Weaver, Birdsey, and Lugo 1987; Sánchez Irizarry 1989; Brown and Lugo 1990b; Johnston 1990; Lugo 1992a; Silver 1992; Medina et al. 1994; Silver, Scatena, Johnson Siccama, and Sanchez 1994; Soil Survey Staff 1995; Lugo and Scatena 1995; Lugo, Bokkestijn, and Scatena 1995). These studies included coastal dry, lower montane, and montane forests. Medina et al. (1994) reported on ultramafic soils on moist and wet lower montane climates in relation to tissue chemistry of various plant species.

Both volcanic- and limestone-derived forest soils are high in organic matter and low in bulk density, with most of the ecosystem's pool of nutrient and carbon accumulation being in the soil as opposed to in the vegetation. Soil organic matter in montane conditions is a function of climate (i.e., there is greater accumulation with increasing water availability) (Weaver, Birdsey, and Lugo 1987). Soil organic matter and earthworm species richness increased and earthworm density decreased along a successional sequence from pastures to mature forests in lower montane wet conditions (Zou and González 1997). Dry forest soils also have a high accumulation of organic matter under mature vegetation (Lugo and Murphy 1986; Brown and Lugo 1990b).

Calcareous substrates are a prominent feature of the Caribbean. The Bahamas and the southern warm temperate portion of south Florida are entirely underlain by limestone, as are the islands of the outer arc of the Lesser Antilles from the Virgin Islands to Barbados. Predominantly small in area and relief, the forests on limestone in these areas are generally xeric in composition and structure. However, the prominent topography and environmental diversity that occur on the larger islands of the Greater Antilles present conditions for the development of forests on limestone over a wider range of climates.

Limestone deposits account for 15–66% of the area of the climatically diverse Greater Antilles (Table 16.2). There is a catena of vegetation types growing on limestone that range from desert scrub on "dogtooth" rock pavement (Howard and Briggs 1953; Lugo et al. 1978) (Fig. 16.3) to montane rain forests (Kelly et al. 1988; Borhidi 1991) (Fig. 16.6). Throughout their range in the region, mesic forests on limestone are also centers of high species richness and endemism (Hartshorn et al. 1981; Borhidi and Muñoz 1983; Kelly 1988; Figueroa Colón 1995).

Ultramafic geological parent material (mainly serpentinized peridotite) is also widely distributed in the Greater Antilles (Mattson 1979; Burke et al. 1984; Figueroa Colón 1992). The largest area of outcrops occurs in Cuba; it is 7500 km² or 6% of the island's area. This is 10 times the total aggregate area of outcrops on Jamaica, Hispaniola, and Puerto Rico (Figueroa Colón 1992). Vegetation occurs on ultramafic soils along an annual precipitation gradient from <500 mm–>4500 mm, and along an elevation gradient that extends from sea level to >2000 m. Studies in Cuba (Berazain Iturralde 1976, 1981a, 1981b; Borhidi 1991), and Puerto Rico (Tschirley, Dowler, and Duke 1970; Beinroth 1982; Figueroa and Schmidt 1983; Medina et al. 1994) show correlation between ultramafic soils and the xeromorphic nature of the vegetation. When compared to forests on nonultramafic soils, these forests generally show higher values of species richness and endemism, higher sclerophylly, less structure and biomass, and a greater component of plant species that are accumulators or hyperaccumulators of heavy metals.

Regional soils have been shown to sustain ad-

Table 16.2. *Environmental ranges of calcareous substrates in the Greater Antilles*

Island	Area (km²)	Percentage of total area	Mean annual rainfall (mm)	Elevation (m)
Cuba	110,922	50	700–2000	0–930
Hispaniola	72,914	30	700–2400	0–2400
Jamaica	11,290	66	700–6000	0–1200
Puerto Rico	8,900	15	800–2000	0–560

Sources: Borhidi (1991, 1993), Hartshorn et al. (1981), Kapos (1986), Kelly (1986), Kelly et al. (1988), Chinea (1980), Monroe (1976).

equate nutrient pools. Estimates suggest the capacity of these soils to allow for the loss of nutrients in several successive harvests of vegetation (Lugo and Murphy 1986; Brown and Lugo 1990a; Lugo 1992a; Scatena, Silver, Siccama; Johnson, and Sanchez 1993). Moreover, there is evidence of strong vegetation-soil interaction in the control of soil nutrient distributions on a landscape scale (Silver 1992; Silver et al. 1994; Lugo and Scatena 1995). Nitrogen does not appear to be a limiting factor in Caribbean forests (Lugo and Murphy 1986; Lugo 1992a; Silver 1992; Scatena et al. 1993) except for vegetation growing on white sands in various locations such as the Bahamas, Cuba, and Puerto Rico, or the vegetation on sites damaged, for example, by landslides or human disturbance.

Agricultural use of forest soils can deplete reserves of nutrients and organic matter (Brown, Glubczynski, and Lugo 1984; Weaver et al. 1987; Brown and Lugo 1990b). Following abandonment of agricultural activities, secondary forests and tree plantations contribute to the rapid accumulation of nutrients and organic matter in soils depleted of these constituents by human use (Brown et al. 1984; Sanchez, and Brown 1986; Weaver et al. 1987; Brown and Lugo 1990a, 1990b; Lugo, Cuevas, and Sánchez 1990; Lugo, Wang, and Bormann 1990; Wang, Bormann, Lugo, and Bowden 1991; Lugo 1992a). Changes in land use result in temporary increases in N_2O emissions in some of these forest soils (Steudler, Mellilo, Bowden, Castro, and Lugo 1991). Dry and wet forests are carbon, nitrogen, and methane sinks (Brown et al. 1984; Lugo and Murphy 1986; Brown and Lugo 1990b; Steudler et al. 1991; Lugo 1992a), whereas soils of montane wetlands are carbon sinks but methane sources (Silver et al. 1995).

The agricultural soils of the region are generally fertile, but their area is small. Roberts (1942) classified the soils of Puerto Rico according to their fertility

Figure 16.6. Montane forest on lime-stone, Rio Abajo, Puerto Rico. (A) The absence of emergent trees is typical of Caribbean forests. Tree crowns are relatively small, and there are many canopy openings through which light penetrates to the forest floor. Distinct plant associations occur on tops, slopes, and bottoms of such limestone slopes. (B) Interior view of the forest; note limestone outcrop on the right. Average stem diameter is small in spite of high rainfall. Abundant un-derstory indicates many light gaps ex-ist. Photographs courtesy of José Co-lón.

for agricultural production using a value of 1 to rep-resent the most fertile soils and 10 for the least pro-ductive. He found that only 6% of the island soils could be classified as 1, 22% 2–5, and 72% as 6–10. These latter soils are the montane soils where most of the remaining natural vegetation occurs.

Natural Disturbances

Caribbean vegetation is exposed to many natural disturbances including hurricanes, high winds, high rainfall, earthquakes, volcanism, high-pres-sure systems accompanied by drought and high temperatures, tsunamis, extreme high or low tides, and exotic genetic materials transported by hurricanes (Lugo 1988a; Torres 1988). Frequent hurricanes are one of the major natural distur-bances, and some sectors of the Caribbean experi-ence hurricanes more often than others (Fig. 16.7). Sectors west of Cuba and the northern Bahamas have experienced about 70 hurricanes since 1871, whereas the coastal zones around the southern pe-

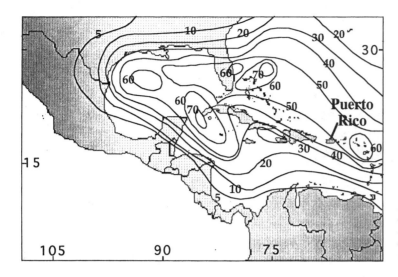

Figure 16.7. Isopleths of the number of hurricanes that have passed over the Caribbean since 1871. Analysis of original data from Neuman et al. (1978); courtesy of Geraldo Camilo.

riphery of the Caribbean Sea, have experienced <5 hurricanes over the same period. The Greater Antilles are intermediate, having experienced 30–50 since 1871.

Tectonic movements and volcanic activity shaped the region and gave rise to the high islands. Though the Greater Antilles have been volcanically senescent for a long time, the frequency of volcanic activity increases toward the geologically younger lower portions of the Lesser Antilles. Islands such as Martinique, St. Lucia, Montserrat, and St. Vincent still harbor active volcanoes (Tomblin 1981).

Earthquakes also exhibit periodicity and spatial gradients in the Caribbean. Between 1898 and 1952, >125 earthquakes, ranging in magnitude from 5.0 to 8.0+ on the Richter scale, have had epicenters within the geographic area of the Caribbean Sea (Tomblin 1981).

Landslides are associated with hurricanes, earthquakes, and low-pressure systems. Larsen and Simon (1993) described the rainfall duration and intensity threshold beyond which landslides are triggered in the Caribbean. They found that for storm events of <10 hr duration, more rainfall is required to trigger landslides in Puerto Rico than has been documented for other areas. For events of >10 hr duration, the data approach world averages. They hypothesize that world average data reflect temperate zones. Some geological formations have more landslides than others (Guariguata 1990; Larsen and Torres Sánchez 1992). Guariguata and Larsen (1989) mapped landslides and found that road construction and poor road maintenance increase the probability of landslides in the Caribbean.

Treefall gaps are not as common in the Caribbean as they are in hurricane-free tropical main-lands (Scatena and Lugo 1995). Treefall gaps occur in response to external disturbances that either blow trees down, make soils soggy and unable to support standing trees, or a combination of both factors. Many trees fall away from the direction of the prevailing winds (Pérez Viera 1986) or slopes (Lugo et al. 1995). Age, size, and topographic location of trees can modify the effects of the particular external causal forces that trigger their fall.

Human Disturbances

Humans have deforested the Caribbean, burned vegetation, channelized and dammed rivers, filled or drained wetlands, modified the topography, flooded valleys, promoted erosion, introduced hundreds of exotic plant and animal species, and generally created havoc with the natural vegetation and landscape (Westermann 1952; Ambio 1981; Gajraj 1981). The density of the human population and its use of energy per unit area is 10–100 times higher in Caribbean islands than on the mainland of the periphery of the Caribbean. These parameters correlate inversely with insular forest cover (Lugo, Schmidt, and Brown 1981). The intensity of human activity is a dominant factor in the modification of Caribbean island vegetation. The Caribbean is a harbinger of the future for the rest of the world because unchecked population growth in the mainland will eventually lead to population density values that the Caribbean has experienced for centuries. It is impossible to interpret the vegetation of the Caribbean without taking into consideration human disturbances in combination with other components of the environmental setting (Lugo et al. 1981; García Montiel and Scatena 1994).

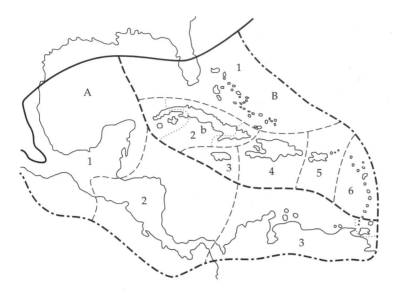

Figure 16.8. Phytogeographical units of the Caribbean region. (A) is the Central American subregion. Provinces within it are (1) Mexico; (2) Panama-Guatamala; and (3) Colombia and Venezualan coasts. (B) is the Antillean subregion. Provinces within it are (1) south Florida, the Bahamas, and Bermuda; (2) Cuba; (3) Jamaica; (4) Hispaniola; (5) Puerto Rico; and (6) the Lesser Antilles. From Borhidi (1991).

Flora

Good (1953) categorized the geographical region comprised of southern Florida, all the Caribbean, Central America, northern Colombia, and northern Venezuela as the Caribbean Province of the Neotropical Kingdom. His scheme was modified by Borhidi (1991) to highlight regional floristic variability within the Caribbean province. Because Borhidi relied heavily on bioclimatic criteria to delineate vegetation types, his resulting phytogeographic units (Fig. 16.8) are consistent with the rainfall map of the region derived independently by Portig (1976). Borhidi recognized differences between the vegetation of Caribbean islands and the vegetation of the mainland to the south and west. He recognized the following six floristic subgroups within the Caribbean islands:

1. Bahamas, plus southern Florida
2. Cuba
3. Jamaica
4. Hispaniola
5. Puerto Rico, plus the Virgin Islands
6. The Lesser Antilles.

The oldest fossil angiosperm flora from northern Latin America (Mexico–Central America–Antilles) is a megafossil assemblage of Late Cretaceous age described from Coahuila in northern Mexico (Graham 1992). Within the Caribbean proper, mega-and microfossil assemblages have been described for all the islands that make up the Greater Antilles. The oldest material described to date comes from Cuba and is of Middle Eocene age. For Jamaica, Puerto Rico, and Hispaniola, the oldest reported fossil floras are from the Eocene,

Oligocene, and Miocene, respectively (Graham 1992). Though very limited and patchy, these fossil records are beginning to reveal the paleoclimatic history of the region as well as the driving forces behind the biogeography of the area. Tertiary microfossil floras known from the northern neotropics suggest a general temperature drop in the Middle Miocene, and a slight rise in the Late Pliocene that preceded the fall caused by glacial conditions in the Pleistocene (Graham and Jarzen 1969; Graham 1990; Borhidi 1991).

Several vegetation assemblages have been identified in Eocene and Oligocene sediments of the Caribbean, including coastal, brackish, upland tropical to subtropical communities, and an arboreal cool-temperate community (Graham and Jarzen 1969; Graham 1993). Studies of pollen and macrofossil remains indicate that mangrove communities have had a progressive increase in diversity through the Cenozoic (Graham 1995). Paleovegetation studies also indicate that about 18,000 yr BP, parts of the lower Caribbean and Venezuela were more arid than at present, tropical savanna had a larger areal extent, and the altitudinal zonation on mountains was lowered by several hundred meters (Schubert 1988). A Quaternary glaciation has been suggested in the Dominican Republic (Schubert and Medina 1982). Paleoclimatic and vegetation sequences from Haitian lake sediments agree well with other Holocene sequences from Central America, the Amazon, and Africa (Hodell et al. 1991). Apparently, the early Holocene was relatively dry, whereas the mid-Holocene was relatively wet.

Given the environmental setting and history of the Caribbean, it is not surprising that the region is highly diverse floristically. The region harbors an estimated 12,000–15,000 vascular plant species (Toledo 1985), one endemic family, 200 endemic genera in 49 families, and 7500 endemic species (Howard 1979; Borhidi 1991; Gentry 1992). The native flora of the region is characterized by species that are mostly endemics or of Caribbean distribution rather than of mainland affinity. For example, 80% of the native flora of Cuba is endemic or of Caribbean distribution, whereas only 1% (65 species) occur in both Cuba and South America (Borhidi 1991). Endemic species are fewer in the Lesser Antilles, and the proportion of Neotropical elements is higher (Howard 1979). Beard (1949) and Gleason and Cook (1927) suggested that the Caribbean islands are in floristic steady state and thus are less susceptible to species invasions than are younger floras elsewhere (eg, the Hawaiian Islands; see Chapter 18).

The plant species of Caribbean islands are tightly packed per unit area. Species-area curves for Caribbean islands typically have steep slopes (Kapos 1986; Tanner 1986; Lugo 1987; Wadsworth 1987; Kelly et al. 1988; Borhidi 1991; Scatena and Lugo 1995). The slope or z values of the species-area curves (sensu MacArthur and Wilson 1967) are higher than those of the much younger Hawaiian island chain (Lugo 1994a).

Howard (1977) and Figueroa Colón (1996) discussed the threats to endangerment and extinction within the Caribbean flora and could document only a few examples of recent extinctions of plant species. In the fossil record of the Oligocene for Puerto Rico, Graham and Jarzen (1969) were able to identify 44 endemic genera of which 31 are still components of the flora. An additional 10 genera grow elsewhere in the region in habitats comparable to their original ones, and only three are presently extinct from the region. Forces that shaped Caribbean vegetation may also provide resilience to the flora and preadaptation to the onslaught of human activity (Lugo 1988c).

VEGETATION

Despite the floristic and phytogeographic diversity in the region, the vegetation of the Caribbean shares many salient features. Shreve (1914) was the first to write about these affinities, and he did so by comparing Jamaican montane forests with the mythical mainland lowland rain forest. From a functional perspective, it is difficult to find fundamental differences between Caribbean island vegetation and other tropical plant associations in other island groups or in mainland scenarios (Lugo 1987).

Major Types of Vegetation

Howard (1973) categorized the vegetation of Caribbean islands into three formations (coastal, lowland, and montane), and within these formations he listed and described a variety of plant associations. Not all these have received equal scientific attention (Table 16.3). In the coastal formation, for example, we know more about the mangroves than the other three associations recognized by Howard (i.e., beach [Fig. 16.9], strand, and rock pavement [Fig. 16.3]). In the lowland formations (thorn scrub, savanna, marsh (Fig. 16.12) or swamp, and alluvium), the most studied are the dry forests on limestone (thorn scrub, Fig. 16.1c) followed by the pterocarpus swamps. The montane formations (wet or dry forests on limestone, montane sclerophylls, palm brakes, tree fern communities, pine forests, wet forests on volcanic soils, volcanic and soufrière communities, crater lakes, and elfin thickets or cloud forests) are the most studied of all.

Mangroves occur in protected shores in dry or moist life zones. They form a variety of distinct communities depending upon tidal, hydrologic, or topographic conditions. Fringe mangroves and mangrove islands grow in close proximity to the ocean, are dominated by *Rhizophora mangle*, and may have landward zones dominated by *Avicennia germinans* or *Laguncularia racemosa*. Basin mangroves occur in inland depressions and are usually dominated by *A. germinans*. Riverine mangroves exhibit the most extensive growth and can be dominated by *R. mangle* or *L. racemosa*. Under extreme conditions of salinity or nutrient depletion *R. mangle* and *A. germinans* form scrub mangrove communities. For additional details about mangrove vegetation, see Chapter 13 of this book.

Swamps of *Pterocarpus officinalis* occur in moist or wet life zones behind mangroves, along river floodplains, and in riparian zones at up to 400 m elevation. The tree species richness increases with distance from the coast, where *Pterocarpus* grows with *Laguncularia* at soil salinities that can exceed 10 parts per 1000. Under montane conditions there is no soil salinity, trees exceed 30 m in height, and more than 10 tree species share the canopy with *Pterocarpus*.

The vegetation gradient from rock pavement to thorn scrub and evergreen and deciduous dry forest in the coastal lowlands of Puerto Rico has been described in detail (references in Table 16.3). This gradient is characterized by increasing soil volume and moisture, and increasing vegetation height, species richness, and plant productivity. This veg-

Table 16.3. *Selected references for studies that address the plant ecology and the major plant associations of the Caribbean*

Ecological topic or plant association	Sources of information[a]
Plant succession	Beard 1945, 1976; Byer and Weaver 1977; Crow 1980; Dunevitz 1985; Ewel 1971, 1980; Pérez Viera 1986; Walker et al. 1991; Weaver 1990.
Overview of vegetation and plant ecology	Asprey and Robbins 1953; Beard 1944, 1949; Borhidi 1991; Brown et al. 1983; Ewel 1977; Gleason and Cook 1927; Grubb 1971, 1977; Herrera, Méndex, Rodríguez, and García 1988; Howard 1952; Odum and Pigeon 1970; Samek 1973; Smith 1954; Stehlé 1945; Soffers 1956; Thompson, Bretting, and Humphreys 1986; Weaver 1983; Weaver and Murphy 1990.
Dry coastal forests	Borhidi 1993; Kapos 1986; Lugo et al. 1978; Lugo and Murphy 1986; Moreno Casasola 1993; Murphy and Lugo 1986a, 1986b; Murphy et al. 1995; Stoffers 1993.
Montane forests	Beard 1942.
A. Tabonuco forests	Briscoe and Wadsworth 1970; Lugo and Scatena 1995; Odum and Pigeon 1970; Smith 1970
B. Colorado forests	Weaver 1987, 1986, 1995
C. Palm forests	Frangi and Lugo 1985, 1991; Lugo and Rivera Batlle 1987; Lugo et al. 1995;
D. Wet limestone forests	Chinea 1980; Kelly 1986; Kelly et al. 1988; Proctor 1986a.
E. Elfin cloud forests	Baynton 1969; Gill 1969; Grubb and Tanner 1976; Howard 1968, 1969, 1970; Medina, Cuevas, and Weaver 1981; Nevling 1971; Russell and Miller 1977; Scatena 1995; Shreve 1914; Tanner 1977, 1986; Weaver 1972, 1976, 1995; Weaver et al. 1986.
Pterocarpus forests and other freshwater wetlands	Alvarez Lopez 1990, Bacon 1990; Borhidi, Muñiz, and Del Risco 1983; Figueroa, Totti, Lugo, and Woodbury 1983; Lugo and Brown 1988; Proctor 1986b.
Mangrove forests and other coastal vegetation	Chapman 1940, 1944; Cintrón et al. 1978; Egler 1950; Lugo and Snedaker 1974; Lugo 1980.
Savannas	Beard 1953.
Seagrass beds	Thorhaug 1981, Vicente 1992.
Pine forests	Chardón 1941.
Plantation forests	Cuevas et al. 1991; Francis 1995; Francis and Weaver 1988; Liegel 1984; Lugo 1988, 1992a; Lugo and Figueroa 1985; Lugo, Cuevas, and Sanchez, 1990; Lugo, Wang, and Bormann 1990; Marrero 1947; Wadsworth 1960; Weaver 1989b.

Note: Additional references appear in the text.

etation is very responsive to rainfall and is characterized by a diverse array of plant life forms growing in close proximity. Some examples of prominent growth forms and genera are succulent herbaceous plants such as *Sesuvium* and *Portulaca*; shrubs such as *Lantana* and *Jacquinia*; cacti such as *Melocactus* and *Cephalocerus*; trees such as *Tabebuia, Acacia, Pictetia, Plumeria, Bursera*; vines such as *Stigmaphyllon*; and epiphytes such as *Tillandsia*. Tree height ranges from <1 m in windy conditions to about 20 m in protected valleys. This vegetation usually shows strong wind sculpturing.

The palm *Prestoea montana* is common on wet and steep slopes, riparian areas, and floodplains >500 m elevation. In Puerto Rico, these forests occur on saturated soils and share dominance with endemic species such as *Calycogonium squamulosum* and *Croton poecilanthus*. *Cecropia peltata* and *Cyathea arborea* grow in these forests as successional species after disturbance (Fig. 16.10a). The latter species, a native tree fern throughout the Caribbean, forms monocultures in landslide areas where it also acts as a pioneer (Fig. 16.10b).

Cyrilla racemiflora, a swamp shrub in North

Figure 16.9. Sand dune vegetation, north coast of Puerto Rico. Foreground herb is Ipomoea; *background trees are the exotic* Casuarina equisetifolia, *planted throughout Puerto Rico to reduce erosion. Photograph courtesy of José Colón.*

America, is known as palo colorado in Puerto Rico and grows to gigantic size throughout the Caribbean (Fig. 16.4). Trees in excess of a meter in diameter are common in the upper montane forest that it dominates. Colorado forests occur above the cloud condensation level and have trees that rarely exceed 15 m in height. *Micropholis, Tabebuia, Cordia, Ocotea,* and *Eugenia* are examples of other dominant tree genera of the colorado forest. Tank bromeliad epiphytes of the genus *Guzmania* are common and they sometimes root on the forest floor.

The tallest trees in the montane formations occur in the lower montane forest dominated by *Dacryodes excelsa* (tabonuco), which can reach 30 m. This tabonuco forest is characterized by straight boles, open understory, few epiphytes, and high tree species richness. *Sloanea, Manilkara, Swartzia, Pithecellobium,* and *Sterculia* are examples of some of the dominant tree genera in the tabonuco forest.

The shortest trees of the montane formation occur in the elfin thickets on the top of mountain ridges (Fig. 16.11). These are cloud forests dominated by *Tabebuia* and *Clusia.* Epiphytes are extremely abundant in these cloud forests and include lower vascular plants as well as orchids and other flowering vascular plants. Like the coastal strand, this forest type is exposed to strong winds and exhibits pronounced wind sculpturing and xeromorphic characteristics in some of the wettest climates of the Caribbean (>5000 mm rainfall per year).

Physiognomy and Dominance

Caribbean vegetation exhibits a common canopy physiognomy throughout the region. Caribbean vegetation has smooth, wind-sculptured canopies with a conspicuous absence of emergent trees (Fig. 16.6a Shreve 1914). Both Odum (1970) and N. Brokaw (1995, personal communication) showed this to be an adaptation to prevailing winds and hurricanes. Trees in the Caribbean do not attain the heights observed in the lowlands of continents outside the trade wind and hurricane belts (Kelly et al. 1988). For example, the tallest known tree in Puerto Rico is an individual of the exotic *Casuarina equisetifolia* (J. R. Francis, personal communication). Forest canopies in the island rarely exceed 30 m.

Positive contributions of hurricanes to vegetation and adaptations of vegetation to these disturbances and to excessive moisture are listed in Lugo (1986) and Lugo and Scatena (1995). There are some indications that vegetation with a high Holdridge complexity index (i.e., the product of tree density, basal area, number of species, and height of the three tallest trees divided by 1000 for trees with dbh >10 cm in a 0.1 ha plot) (Holdridge 1967) is less likely to suffer damage from hurricanes than vegetation with low complexity indices, all other factors being equal (Lugo et al. 1983).

Because of frequent hurricanes and other disturbances, Caribbean vegetation is maintained in a youthful and vigorous condition with a mixture of

Figure 16.10. Successional forests on disturbed sites of Luquillo Experimental Forest, Puerto Rico. (A) The tree fern Cyathea arborea *forms monocultures on landslides. Dicotyledous trees are beginning to emerge through the fern overstory. (B) Wet montane forest on volcanic soil five years after Hurricane Hugo passed. The successional species* Cecropia peltata *is visible in canopy gaps in the left foreground and elsewhere. The palms are* Prestoea montana. This lower montane forest is known as tabonuco forest (after the species Dacryodes excelsa) and is characterized by a high diversity of tree species, individuals of which are typically >20 m tall, with large stem diameters, and at low density. Photographs courtesy of José Colón.

mature and successional species. Caribbean forests have been categorized as "storm forests" because the structure, physiognomy, species composition, abundance of lianas, reduced number of vegetation strata, and other characteristics are believed to be hurricane-induced (Fig. 16.10b; Beard 1945, 1976; Lugo et al. 1981; Borhidi 1991; Lugo and Scatena 1995). Borhidi (1991) added that hurricanes are dispersal agents of species and narrated how whole large trees with their full compliments of associated epiphytes, microorganisms, and animals can be moved over large land distances. Torres (1988)

documented the introduction of African insect species into the Caribbean as a result of transport by hurricanes.

An attribute of native vegetation in the Caribbean is the high proportion of biomass allocated to roots (30–50% in a gradient from wet to dry forests) and the high turnover of mass and nutrients associated with the turnover of roots and litter (Murphy and Lugo 1986a, 1986b; Cuevas, Brown, and Lugo 1991; Lugo 1992a; Scatena et al. 1993; Silver and Vogt 1993). In fact, high biomass and nutrient turnover may occur in many tropical eco-

Figure 16.11. Cloud forest, El Yunque Peak (1065 m) in the Luquillo Experimental Forest of Puerto Rico. This vegetation is typical of mountain summits where rainfall, wind, and humidity are high and temperatures, solar radiation, and transpiration are low. Epiphytes are abundant. Tree height, diameter, and species richness are low; tree density is high. Photograph courtesy of José Colón.

systems, but they have been demonstrated most convincingly in the Caribbean (Cuevas et al. 1991; Parrotta and Lodge 1991; Silver 1992; Lugo 1992a; Scatena et al. 1993; Silver and Vogt 1993).

Another salient feature of vegetation in the region is the high dominance of a few species within plant associations (Weaver 1983, 1987; Johnston 1990; Lugo 1991; Lugo and Scatena 1995; Lugo et al. 1995). Species dominance curves (*sensu* Whittaker 1965, 1970) range from geometric to log-normal for Caribbean island vegetation, in contrast to log-normal to random for mainland lowlands, excluding wetlands (see Lugo 1991; Lugo and Scatena 1995).

Missing in Caribbean plant communities are the continent's relatively large number of rare plant species. Obviously, Caribbean vegetation has a higher proportion of rare species than dominant ones, but the number of rare species and the degree of their rarity in a given plant association is not as pronounced as in the continental lowland forests of the Amazon basin. The result is a lower species richness in Caribbean islands compared to the mainland. This relatively high dominance has been attributed to disturbance events such as hurricanes or human interventions rather than to island effects (Lugo 1987, 1991; Wadsworth 1987).

Tree Mortality and Turnover Time

Tree mortality in Caribbean forests is episodic in response to catastrophic events. "Normal" or "background" tree mortality rates in this region are similar to those in other tropical regions (i.e.,

2–5% of stems or biomass/ha/year) (compare data in Jiménez, Lugo, and Cintrón 1985; Lugo et al. 1995; Lugo and Scatena 1995; Scatena and Lugo 1995; and Weaver 1983, 1986, 1987, 1994, 1995; with data for Venezuela and Costa Rica in Carey, Brown, Gillespie, and Lugo 1994, and D. Lieberman, M. Lieberman, Peralta, and Hartshorn 1985). "Catastrophic" or "sudden" tree mortality rates occur at >5% ha^{-1} yr^{-1} in response to disturbances such as hurricanes, tornadoes, landslides, windstorms, periods of high salinity in coastal mangroves, or chronic inundation in the lowlands (Lugo, Applefield, Pool, and McDonald 1983; Jiménez et al. 1985; Weaver 1989a, 1994, 1995; Walker, Lodge, Brokaw, and Waide 1991; Lugo and Scatena 1995; Scatena and Lugo 1995).

The rate of tree mortality is a function of area sampled and the magnitude of the disturbance event. Small sampling areas can inflate results by overestimating rates of tree mortality as well as values of stand biomass and species richness. The reason is the patchy distribution of tree mortality, species, and tree biomass. Some events cause high mortality in small areas (landslides), others cause large scale mortality with high spatial variability (hurricanes), while still others cause patches of high mortality scattered over large areas with lower rates of mortality (gap-forming treefalls).

Episodes of high tree mortality lead to cohorts of regeneration that grow and mature as a group. This in turn leads to size and age class dominance in Caribbean forests (Lugo and Rivera Batlle 1987; Weaver 1986, 1987, 1989b, 1995; Johnston 1990; Lugo and Scatena 1995; Lugo et al. 1995). If a disturbance fails to occur by the time these dominant age classes reach maturity, age-induced mortality may lead to pulses of treefall gaps.

In spite of the high instantaneous rates of episodic tree mortality and regeneration in Caribbean forests, Scatena and Lugo (1995) found that over 100 yr intervals, background tree mortality turned over more biomass and individuals (turnover time, TOT, of 55 yr for both parameters) than any other disturbance in lower montane forests in Puerto Rico. Hurricanes (TOT of 105 and 220 yr, respectively), followed by treefall gaps (TOT of 150 and 380 yr, respectively), and landslides (TOT of 3350 and 3300 yr, respectively) were the most significant disturbance-related causes of mortality in these forests. In summary, background tree mortality is a low-intensity, high-frequency, and large-scale mechanism that over a period of a century is responsible for most of the turnover of vegetation. It does not open the canopy and does not involve dramatic environmental change. In contrast, sudden-death tree mortality exhibits higher rates of turnover on smaller scales and shorter periods. This mechanism of vegetation turnover is associated with disturbances, which over periods of centuries turn over less mass, nutrients, and individuals than background mortality. Yet, because these events open the canopy and are recurrent over many centuries (Fig. 16.4), they help shape vegetation and create opportunity for changes in species composition and succession.

Morphological Plasticity

Lugo and Scatena (1995) noted the morphological plasticity of trees in the Luquillo Mountains in Puerto Rico and gave many examples. This plasticity, observable along the whole elevation gradient on the Luquillo Mountains, included the following: formation of tree unions due to extensive root grafting (Basnet 1992; Basnet, Scatena, Likens, and Lugo 1993); abundance of clonal reproduction (Nevling 1971); coalescence of smaller trees into larger individuals (Weaver 1986, 1994); ability to keep leaning stems upright through production of support roots (Weaver 1987, 1994, 1995); ability for downed stems to sprout and regain canopy position (Frangi and Lugo 1981); persistence of live tree stumps for decades; and adventitious root structures (see Gill 1969; Frangi and Ponce 1985; Torres 1994; Lugo and Scatena 1995; and Weaver 1986, 1994). Adventitious root structures include roots with negative geotropism, canopy roots, roots invading humus formed in cavities of the parent tree, tabular roots, and roots with specialized mechanisms to enhance gas exchange with or without pneumatorhizae.

Collectively, these morphological responses of vegetation allow species to maintain dominance on certain favorable sites (e.g., well-aerated ridges by *Dacryodes excelsa*), conduct gas exchange in saturated soils (e.g., the palm *Prestoea montana*), respond rapidly to disturbances after trees are knocked down (*Cyrilla racemiflora*), cycle nutrients efficiently, modify the microtopography to favor tree growth (*D. excelsa*), and facilitate nutrient capture from cloud mist (Gill 1969; Weaver 1972; Asbury, McDowell, Trinidad Pizarro, and Berrias 1994).

Spatial Gradients

The structural and functional attributes of Caribbean vegetation parallel environmental gradients. For example, the height of mangrove trees follows a latitudinal gradient associated with decreasing temperatures (Cintrón and Shaeffer Novelli 1983). Shreve (1914), Beard (1944, 1949, 1955), and Odum (1970) related changes in the physiognomy of vegetation to elevation gradients of temperature, rainfall, wind, and other environmental variables such as humidity, soil waterlogging, or atmospheric saturation deficit. In landslides, vegetation distribution follows gradients of light, nutrients, and mychorrizae (Fernández and Myster 1995; Myster and Fernández 1995). Weaver and Murphy (1990) summarized changes in structure and productivity along an elevation gradient in Puerto Rico. Shreve (1914), Gleason and Cook (1927), Beard (1944, 1949), and Borhidi (1991) described many examples of changes in vegetation structure, composition, and physiognomy along gradients of water availability in several islands in the Caribbean.

Research in the Luquillo Mountains of Puerto Rico is documenting vegetation responses to the topographic catena that includes valley, slope, and ridges (Weaver 1983, 1987, 1989a, 1995; Basnet, Likens, Scatena, and Lugo 1992; Scatena et al. 1993; Silver et al. 1994; Lugo 1995) (Figs. 16.3, 16.5, and 16.6). Vegetation change along such catenas are associated with changes in soil chemistry, soil atmosphere (Silver et al. 1994; Silver et al. 1995), and hydrological conditions (Scatena 1989) (Fig. 16.12). Both the catena and vegetation have reciprocal interactions with disturbance events such as landslides and treefalls (Scatena and Lugo 1995). For example, the location along the catena influences rates of treefall gap formation (more in slopes and valleys), occurrence of landslides (mostly on slopes), and degree of hurricane impact on vegetation (ridges are more vulnerable).

In mangroves, the catena results from centimeter-level differences in elevation that influence hydroperiod and soil salinity. Mangrove species

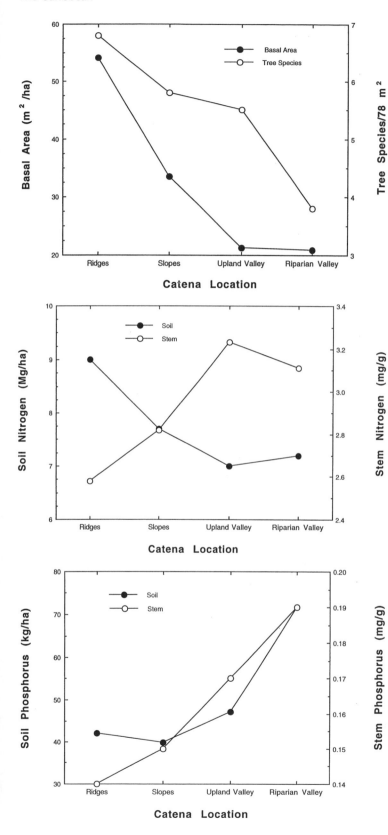

Figure 16.12. Biotic and abiotic gradients along the topographic catena diagrammed in Figure 16.5, Luquillo Mountains of Puerto Rico. (A) Tree species richness and basal area both decline from ridge tops to valley bottoms. From Scatena and Lugo (1995). (B) Soil and stem nitrogen exhibit opposite patterns. Mg is equivalent to metric tons. From Scatena et al. (1993) and Silver et al. (1994). (C) Soil and stem phosphorus both increase from valley bottoms to ridge tops. From Scatena et al. (1993) and Silver, Scatena, Johnson, et al. et al. (1994).

Table 16.4. *Number of rare and endangered species in selected plant families in selected Caribbean islands*

| Family | Number of species | | | |
	Cuba	Hispaniola	Jamaica	Puerto Rico
Asteraceae	69	21	30	6
Euphorbiaceae	58	7	18	3
Melastomataceae	38	6	11	8
Myrtaceae	87	3	37	18
Orchidaceae	18	110	41	6
Rubiaceae	77	11	43	11
Percentage[a]	36	58	66	33

[a] Note: Percentage refers to the percentage of the total flora of rare and endangered taxa for that island.
Sources: Borhidi and Muñiz (1983), Hartshorn et al. (1981), Kelly (1988), and Figueroa and Woodbury (1996).

are very sensitive to these differences (Cintrón et al. 1978; Twilley, Lugo, and Patterson Zucca 1986; see also Chapter 13 of this volume). Structural development and rates of primary productivity and organic matter export increase with decreasing soil salinity and reductions in the hydroperiod.

Along the topographic or climatic gradients that we have described so far, one also encounters gradients in the numbers of species (Fig. 16.12a). For example, the number of fern species increases with elevation amplitude among islands, and they increase with rainfall in any particular island (Tryon 1979). Murphy and Lugo (1986a) reported the following gradient of tree species richness along a rainfall gradient on various substrates in Puerto Rico's southwest coast: 169, 218, and 265 species at 860, 1413, and 2550 mm of annual rainfall, respectively. A similar analysis along a rainfall gradient over limestone-derived soils in Jamaica resulted in a peak of tree species richness at an intermediate rainfall regime (Kelly et al. 1988): 129 and 81, 247 and 135, and 280 and 118, plant and tree species respectively at an annual rainfall of 1000, 1600, and 4000 mm.

Maps of the distribution of endemic species in Cuba show steep gradients in the number of species per area (Borhidi 1991). Similar observations are available for Jamaica (Thompson et al. 1986) and Puerto Rico (Figueroa Colón 1996). In Puerto Rico, montane sites averaged three times more endemic species than nonmontane sites (Figueroa Colón 1996). Concentrations of endemics are associated with high elevation and extreme conditions, such as very wet or dry sites, and particular edaphic conditions, such as ultramafic soils or oligotrophic white sands (cf. Gentry 1992). Endemic

species are also more susceptible to endangerment and extinction (Howard 1977). Rare and endangered species are also more common in particular plant families (Table 16.4): Up to 66% of the total rare and endangered plants of a given island's native vegetation in the Greater Antilles is found in just six families.

Anthropogenic influences can exacerbate what is a naturally patchy distribution of native species. Type of land use and the uses of trees influence the species composition of secondary forests. Gradients of human disturbance are accompanied by the increased number of exotic species that mix with native and naturalized species (Birdsey and Weaver 1982). In locations with maximum human modification, the species composition is sometimes exclusively exotic. For example, forests of the exotic nitrogen fixer *Albizia procera* are now developing in Puerto Rico on the highly compacted soils of pastures and roadsides (Chinea 1992).

Succession

The predominant type of succession in the Caribbean is what Egler (1950) termed cyclic. This is plant succession driven by external (allogenic) factors as opposed to internal (autogenic) vegetation factors. Long-term observations of coastal mangroves (Cintrón et al. 1978), montane forests (Crow 1980; Weaver 1987, 1980a, 1990, 1992, 1995; Johnston 1990; Lugo et al. 1995; Taylor, Silander, Waide, and Pfeiffer 1995; Trevin, Rodney, Glasgow, and Weekes 1993), and lowland dry forests (Murphy, Lugo, Murphy, and Nepstad 1995) in the Caribbean all document this kind of succession.

Secondary succession seres are typically rapid. Forest maturity is reached within 60–100 yr following harvests (Dunevitz 1985; Ewel 1971, 1977), landslides (Guariguata 1990; Zarin 1993), exposure to gamma radiation (Odum and Pigeon 1970, Crow 1980), agricultural use (Johnston 1990), and hurricanes (Weaver 1987, 1995; Walker et al. 1991; Lugo 1992a; Silver 1992; Lugo et al. 1995; Taylor et al. 1995), However, maturity cannot be construed as carbon steady-state because forests continue to accumulate carbon (Odum and Pigeon 1970; Lugo 1992; Lugo and Murphy 1986).

Some situations lead to slower succession: for example, where fire recurs; on intensively used pastures (Aide, Zimmerman, Herrara, Rosario, and Serrano 1995); after disturbances of elfin cloud forest on steep, wet, and exposed locations (Byer and Weaver 1977; Weaver 1990, 1995); and on highly compacted dry-forest soils (S. Molina, personal communication, 1990). When humans are the cause of delayed succession, the "human effect" remains even after hurricanes have visited the same sites (Zimmerman, Aide, Herrara, et al. 1995). The recovery of species on human-damaged sites is slower than the recovery of functional parameters (S. Molina, personal communication 1990; Zou et al. 1995)

Seres tend to have fewer stages and to be increasingly autogenic as environmental conditions become harsher (Lugo and Scatena 1995). For example, conditions of high salinity (Lugo 1980), high rainfall with soil saturation (Weaver 1990: Lugo et al. 1995), and extreme drought (Murphy et al. 1995. Under less challenging conditions, studies show rapid substitution of species and a larger number of successional stages (Crow 1980; and Weaver 1987, 1995; Lugo and Scatena 1995; Lugo et al. 1995).

Studies of primary succession following landslides (Guariguata 1990; Zarin 1993) and volcanic explosions (Beard 1945, 1976) show surprisingly rapid vegetation colonization following the catastrophic event. However, arrested successions can occur at any of several critical stages in locations where edaphic conditions are extreme.

The diversity of conditions created by frequent disturbances leads to the development of a flora that is both adapted to colonization and to survival in mature stands. Many studies address the successional status of particular species that are common to both pioneer and mature stages of succession. Examples are *R. mangle*, *P. montana*, and *C. racemiflora* (Beard 1949, 1955, 1976; Bannister 1970; Lugo 1980; Lugo et al. 1995; McCormick 1995; Weaver 1987). Smith (1970) and Weaver (1992) categorized dozens of tree species from the Luquillo Mountains into a continuum between pioneer and mature stages of succession. Zimmerman et al. (1994) related life history characteristics of tree species to hurricane damage. Data on how species behave over several decades of observation on 1-ha patches are available in Crow (1980), Johnston (1990), Lugo et al. (1995), Lugo and Rivera Batlle (1987), and Weaver (1983, 1986, 1987, 1989a, 1990, 1992, 1994, 1995).

Response to hurricanes. Figure 16.13 is a model of the changes that occur after the passage of a Category 4 or 5 hurricane over a 1 ha patch of Caribbean island lower montane forest. At this scale, one can observe changes in the presence or absence of species that do not occur at larger scales. The model is based on data for trees with dbh >4 cm. A single event that lasts less than 10 hr (Hurricane Hugo lasted 4 hr over the Luquillo Forest in Puerto Rico [Scatena and Larsen 1991]) can set in motion vegetation changes that continue throughout the

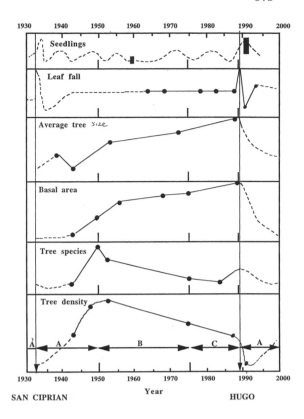

Figure 16.13. Seventy years of vegetation change, from 1930 to 2000, in the lower montane forest of the Luquillo Mountains in Puerto Rico. The two vertical lines indicate passage of type 4–5 hurricanes in 1932 and 1989. Dots represent years for which data are available. Dotted lines represent future projected pathways. Time segments A, B, and C are periods of recovery (see text). The vertical axis is scale-less because absolute values can change depending on the size of sample plots, intensity and duration of events, and local conditions. From Crow (1980), Guzman, Grajales, and Walker (1991), Johnston (1990), Lugo and Scatena (1995), Odum (1970), Smith (1970), Weaver (1994), and unpublished data of Scatena.

model's 60 yr period between hurricane events. Only functional attributes such as leaf fall and the chemical and physical processes of soil appear to reach steady state in the inter-hurricane period. All other indices of vegetation structure continue to change without reaching a steady state, with the exception of low biomass compartments such as canopy leaves or ground litter.

We can arbitrarily break the interhurricane time interval into three distinct periods of vegetation response to the hurricane event. Table 16.5 lists some of the major changes in vegetation that typify each period of response. We also describe in Table 16.5 events during the hurricane and how the forest appears after the passage of the hurricane. Hurricane effects on both plants and animals are also sum-

Table 16.5. *Events within lower montane forests of Puerto Rico during, and following, a hurricane*

Period and duration	Events
During the hurricane; hours	Trees can lose branches and leaves; be uprooted, snapped, or inclined to various degrees; many plants crushed by falling debris from above; massive transfer of mass and nutrients from the canopy to the forest floor; export of organic matter and nutrients off-site and to other downhill locations; many landslides and canopy gaps.
Forest condition immediately after the hurricane; days	Canopy is destroyed, leaving large open areas; forest floor is a tangle of fallen stems and piles of leaves many meters thick; high level of illumination, high temperature and low humidity near the forest floor; dark, humid, and cool inside the tangle of fallen vegetation; absence of green foliage on the canopy, except in protected areas or in short forest floor vegetation in areas without overhead canopy before the hurricane (i.e., forest gaps and landslide areas); massive root mortality possible.
Period of rapid change yr 0 to 20	Begin germination of herbaceous plants and lianas, including exotic species; explosion of herbivorous activity on herbs; many trees begin to resprout; epicormic sprouting common due to loss of apical dominance; downed trees also sprout; herbaceous plants and lianas form a new forest canopy very close to the forest floor; limp, chlorophyll-free, xanthophyll-rich leaves are common; sequence of massive flowering by herbaceous plants followed by lianas; tree ferns and palms begin production of new fronds
A—Reorganization phase; yr 0 to 10	and leaves; epicormic leaf production abundant in trees; delayed mortality of injured trees; massive seedling germination; continued wood fall; root biomass recovers; canopy begins to move upward as successional species elongate stems and surviving trees produce new leaves and branches; intense competition for space and light; massive mortality of herbaceous vegetation and lianas; litter and wood decomposition peaks; leaf fall of successional species increases, and leaf litter fall reaches prehurricane values; decomposition of downed wood begins to have a significant role in nutrient availability; fine wood fall lags behind leaf fall; opportunity for establishment of successional species; seedling populations decline in density;
B—Aggrading phase; yr 10 to 20	average tree diameter increases afer a slight dip; canopy is closed and reaches its normal height; tree density reaches its peak; tree species richness peaks; basal area and biomass increase rapidly; apical dominance reutrns as epicormic leaves and branches die off;
Period of transition; yr 20 to 45	mortality of successional species increases; tree density steadily declines; successional species become less abundant; basal area and biomass increase slowly; total species richness decreases; average tree diameter increases steadily; functional processes are at steady state; forest regains the closed, smooth canopy physiognomy typical of the region.
Period of maturity; yr 45 to 60	Tree density continues to decline; tree species richness may increase due to treefall gap formation; basal area and biomass increase sharply; average tree size increases at a faster rate; functional processes remain at steady state; ground litter at steady state.

Sources: Crow (1980), Fernandez and Fetcher (1991), Frangi and Lugo (1991), Guzman Granjales and Walker (1991), Johnston (1990), Lodge Scatena, Asbury, and Sánchez (1991), Lugo (1995), Lugo et al. (1983; 1995), Lugo and Rivera Batlle (1987), Lugo and Scatena (1995), Lugo and Waide (1993), Odum (1970), Scatena and Larsen (1991), Scatena and Lugo (1995), Scatena et al. (1993), Silver (1992), Smith (1970), Torres (1992), Wever (1983, 1989a, 1994), Zimmerman, Everham, and Waide, et al. (1994) and Zimmerman, Pulliam, and Lodge, et al. (1995).

marized by Lugo and Waide (1993) and Walker et al. (1991).

The first 20 yr period is when all structural and functional parameters are in flux (Fig. 16.13). This period can be subdivided into two 10-yr phases. The first 10-yr phase is one of reorganization, and the second 10-yr phase is of aggradation, or rapid growth. The transition from reorganization to aggradation ends with the closure of the canopy and the leveling of leaf litterfall rates at normal values. The first 20-yr period of rapid change ends when the number of tree species that invade open areas of the forest reaches a peak value.

Next follows a 25-yr transition period during which tree density decreases steadily as a result of self-thinning (Fig. 16.13). Many of the successional tree species that had entered the patch after the hurricane now die off. New species continue to enter the patch, but the net effect is a reduction in the number of tree species.

Forty-five years after the hurricane, the forest reaches maturity when the complement of species, density, and cover regain their predisturbance values. Tree basal area increases very rapidly because dominant trees have room to grow and less competition. Average tree size, which had increased steadily after the hurricane, increases rapidly during this maturity stage (Fig. 16.13). Seedling populations decrease and return to exhibiting periodic cycles of abundance and absence from the forest floor.

Rehabilitation and restoration. The particular economic and sociopolitical history of Puerto Rico offers a landscape-level example of how native tropical vegetation can be restored or rehabilitated following almost complete deforestation. The island, presumed at one point to have been almost 100% covered by forests (Wadsworth 1950), lost 90% of its forest cover and 99% of its primary forests by the 1950s (Birdsey and Weaver 1982, 1987). Over the next 40 yr there was massive land abandonment due to the collapse of agriculture and population migration to urban centers. This was followed by forest regeneration initially dominated by those exotic or native species that had been used by people for a variety of purposes (i.e., timber, fuelwood, fruit gathering, coffee shade, etc.) (Birdsey and Weaver 1982).

Today, underneath the canopy of exotic, naturalized, and native successional tree species, one finds an understory composed exclusively of native or naturalized species in quantities that suggest adequate stocking for wood production (Wadsworth and Birdsey 1983). This phenomenon has also been documented in the understory of plantations of exotic trees established for wood production purposes (Lugo 1988b, 1992a, 1992b; Lugo, Parrotta, and Brown 1993; Parrotta 1992, 1993; Weaver 1989b). These natural and managed examples of the rehabilitation of damaged lands, and the restoration of native forests following extensive human activities provide hope for the future rehabilitation of Caribbean vegetation.

The composition of the future vegetation of the region will be different from pre-Colombian vegetation and also different from today's vegetation. The tree flora of Puerto Rico (547 native species), for example, has been enriched by 203 introductions according to Little, Woodbury, and Wadsworth (1974). Forty-five exotic tree species have been naturalized within this century alone (Francis and Liogier 1991). These trends are true for other growth forms and for other islands (Liogier 1990), and they are bound to continue. In Jamaica, for example, of the 70 plant associations that were identified by Grossman, Iremonger, and Muchoney (ND), over half of them were human-altered associations, new to the island. The human effects are felt in almost every square meter of the landscape, and it is not appropriate for ecologists to ignore the new associations and ecosystems that characterize our environs and that will sustain us into the 21st century.

AREAS FOR FUTURE RESEARCH

Caribbean vegetation has been subjected to extensive and intensive scientific study, beginning with taxonomic and natural history surveys (e.g., Sloane 1707, 1725; Britton 1919; Chapman 1940; Proctor 1989; Liogier 1995). The oldest research plots for the continuous measuring of tree growth and mortality and vegetation change in the neotropics are also located in the region (Wadsworth 1995). Moreover, the environmental setting of the Caribbean is fairly well understood.

We have already noted that most of the plant associations in the region are yet to be adequately described both functionally and compositionally. New plant associations being formed as a result of human activity also require scientific attention. The paleoclimate, biogeography, paleoecology, and long-term dynamics of Caribbean vegetation are areas of research currently receiving increasing attention, and they are of clear importance to plant ecology and to the interpretation of global change (Lugo and Scatena 1992).

Vegetation disturbance and recovery provide a sound unifying focus for vegetation research in the region (Waide and Lugo 1992). However, what remains as the main gap in our knowledge of the region's vegetation is understanding the connection between vegetation and environment along explicit spatial and temporal scales, and establishing cause-effect relationships that can be used for managing ecosystems of all types. Management and conservation of Caribbean vegetation offer challenges that are different from those on continental mainlands (Lugo 1994b). Plant ecologists must take account of the unique environmental setting of the Caribbean when managing, conserving, and preserving its diverse vegetation.

REFERENCES

Aide, T. M., J. K. Zimmerman, L. Herrera, M. Rosario, and M. Serrano. 1995. Forest recovery in abandoned tropical pastures in Puerto Rico. Forest Ecology and Management 77:77–86

Alvarez López, M. 1990. Ecology of *Pterocarpus officinalis* forested wetlands in Puerto Rico, pp. 251–265 in A. E. Lugo, M. M. Brinson, and S. Brown (eds.), Forested wetlands. Elsevier, Amsterdam.

Ambio. 1981. The Caribbean 10:274–346.

Arden, D. D., Jr. 1975. The geology of Jamaica and the Nicaraguan rise, pp. 617–661 in A. E. M. Nairn and F. G. Stchli (eds.), Ocean basins and margins, 3. Gulf coast, Mexico and the Caribbean. Plenum Press, New York.

Asbury, C. E., W. H. McDowell, R. Trinidad Pizarro, and S. Berrios. 1994. Solute deposition from cloud water to the canopy of a Puerto Rican montane forest. Atmos. Environ. 28:1773–1780.

Asprey, G. F. and R. G. Robbins. 1953. The vegetation of Jamaica. Ecolog. Monogr. 23:359–412.

Bacon, P. R. 1990. 1990. Ecology and management of swamp forests in the Guianas and Caribbean region, pp. 213–250 in A. E. Lugo, M. M. Brinson, and S. Brown (eds.), Forested wetlands. Elsevier, Amsterdam.

Bannister, B. A. 1970. Ecological life cycle of *Euterpe globosa* Gaertn, Chap. B-18 in H. T. Odum and R. F. Pigeon (eds.), A tropical rain forest. National Technical Information Service, Springfield, Va.

Basnet, K. 1992. Effect of topography on the pattern of trees in tabonuco (*Dacryodes excelsa*) dominated rain forests of Puerto Rico. Biotropica 24:31–42.

Basnet, K., G. E. Likens, F. N. Scatena, and A. E. Lugo. 1992. Hurricane Hugo: damage to a tropical rain forest in Puerto Rico. J. Trop. Ecol. 8:47–55.

Basnet, K., F. N. Scatena, G. E. Likens, and A. E. Lugo. 1993. Ecological consequences of root grafting in tabonuco (*Dacryodes excelsa*) trees in the Luquillo Experimental Forest, Puerto Rico. Biotropica 25:28–35.

Baynton, H. W. 1969. The ecology of an elfin forest in Puerto Rico, 3. Hilltop and forest influences on the microclimate of Pico del Oeste. J. Arnold Arbor. 50: 80–92.

Beard, J. S. 1942. Montane vegetation in the Antilles. Carib. For. 3:61–74.

Beard, J. S. 1944. Climax vegetation in tropical America. Ecology 25:127–158.

Beard, J. S. 1945. The progress of plant succession on the Soufriere of St. Vincent. J. Ecol. 33:1–9.

Beard, J. S. 1949. Natural vegetation of the windward and leeward islands. Oxford For. Mem. 21:1–192.

Beard, J. S. 1953. The savanna vegetation of northern tropical America. Ecolog. Monogr. 23:149–215.

Beard, J. S. 1955. The classification of tropical American vegetation types. Ecology 36:89–100.

Beard, J. S. 1976. The progress of plant succession on the Soufriere of St. Vincent: observations in 1972. Vegetatio 31:69–77.

Beinroth, F. H. 1982. Some highly weathered soils of Puerto Rico, 1. Morphology, formation and classification. Geoderma 27:1–74.

Bellingham, P. J. 1991. Landforms influence patterns of hurricane damage: evidence from Jamaican montane forests. Biotropica 23:427–433.

Berazain Iturralde, R. 1976. Estudio preliminar de la flora serpentinicola de Cuba. Ciencias 10:11–26.

Berazain Iturralde, R. 1981a. Sobre el endemismo de la flora serpentinicola de 'Lomas de Galindo' Canasi, Habana. Revista Jardín Botánico Nacional de Cuba 2:29–47.

Berazain Iturralde, R. 1981b. Reporte preliminar de plantas serpentinicolas acumuladoras e hiperacumuladoras de algunos elementos. Revista del Jardín Botánico Nacional de Cuba 2:48–59.

Birdsey, R. A., and P. L. Weaver. 1982. The forest resources of Puerto Rico. Resource Bulletin SO-85. USDA Forest Service, Southern Forest Experiment Station, New Orleans.

Birdsey, R. A., and P. L. Weaver. 1987. Forest area trends in Puerto Rico. Research Note SO-331. USDA Forest Service, Southern Forest Experiment Station, New Orleans.

Bonnet Benítez, J. A. 1983. Evaluación de la nueva clasificación taxonómica de los suelos de Puerto Rico.

University of Puerto Rico, Agriculture Experiment Station, Publication 147, Río Piedras.

Boose, E. R., D. R. Foster, and M. Fluet. 1994. Hurricane impacts to tropical and temperate forest landscapes. Ecolog. Monogr. 64:369–400.

Borhidi, A. 1991. Phytogeography and vegetation ecology of Cuba. Akademiai Kiado, Budapest.

Borhidi, A. 1993. Dry coastal ecosystems of Cuba, pp. 423–452 in E. van der Maarel (ed.), Dry coastal ecosystems Africa, America, Asia, and Oceania. Elsevier, Amsterdam.

Borhidi, A., and O. Muñiz. 1983. Catálogo de plantas cubanas amenazadas o extinguidas. Academia de Ciencias de Cuba. Instituto de Botánica. Editorial Academia, Habana.

Borhidi, A., O. Muñiz, and E. Del Risco. 1983. Plant communities of Cuba. I. Fresh and saltwater, swamp and coastal vegetation. Acta Bot. Hungarica 29:337–376.

Briscoe, C. B., and F. H Wadsworth. 1970. Stand structure and yield in the tabonuco forest of Puerto Rico, Chap. B-6 in H. T. Odum and R. F. Pigeon (eds.), A tropical rain forest. National Technical Information Service, Springfield, Va.

Britton, N. L. 1919. History of the survey. Scientific survey of Porto Rico and the Virgin Islands 1:1–10. New York Academy of Sciences, New York.

Brown, S., A. Glubczynski, and A. E. Lugo. 1984. Effects of land use and climate on the organic carbon content of tropical forest soils in Puerto Rico, pp. 204–209 in New forests for a changing world. Proc. of the Conv. of the Soc. of Ameri. Fore. Portland, Ore. Society of American Foresters, Washington, D.C.

Brown, S., and A. E. Lugo. 1990a. Tropical secondary forests. J. Trop. Ecol. 6:1–32.

Brown, S., and A. E. Lugo. 1990b. Effects of forest clearing and succession on the carbon and nitrogen content of soils in Puerto Rico and the U.S. Virgin Islands. Plant and Soil 124:53–64.

Brown, S., and A. E. Lugo. 1994. Rehabilitation of tropical lands: a key to sustaining development. Restora. Ecol. 2:97–111.

Brown, S., A. E. Lugo, S. Silander, and L. H. Liegel. 1983. Research history and opportunities in the Luquillo Experimental Forest. General Technical Report SO-44. USDA Forest Service, Southern Forest Experiment Station, New Orleans.

Burke, K., C. Cooper, J. F. Dewey, J. P. Mann, and J. Pindell. 1984. Caribbean tectonics and relative plate movements. Mem. Geolog. Soc. Amer. 162:31–64.

Byer, M. D. and P. L. Weaver. 1977. Early secondary succession in an elfin woodland in the Luquillo Mountains of Puerto Rico. Biotropica 9:35–47.

Carey, E. V., S. Brown, A. J. R. Gillespie, and A. E. Lugo. 1994. Tree mortality in mature lowland tropical moist and tropical lower montane moist forests of Venezuela. Biotropica 26:255–265.

Case, J. E., T. L. Holcombe, and R. G. Martin. 1984. Map of geologic provinces in the Caribbean region. Mem. Geolog. Soc. Amer. 162:1–30.

Chapman, V. J. 1940. 1939 Cambridge University expedition to Jamaica. Geogr. J. 96:305–328.

Chapman, V. J. 1940. The botany of the Jamaican shoreline. Bot. J. Linnean Soc. 52:407–447.

Chardón, C. E. 1941. Los pinares de la República Dominicana. Carib. For. 2:120–131.

Chinea, J. D. 1980. The forest vegetation of the limestone hills of northern Puerto Rico. Master's thesis. Cornell University, Ithaca, N.Y.

Chinea, J. D. 1992. Invasion dynamics of the exotic legume tree *Albizia procera* (Roxb.) Benth., in Puerto Rico. Ph.D. dissertation. Cornell University, Ithaca, N.Y.

Cintrón, G., A. E. Lugo, D. J. Pool, and G. Morris. 1978. Mangroves of arid environments in Puerto Rico and adjacent islands. Biotropica 10:110–121.

Cintrón, G., and Y. Shaeffer Novelli. 1983. Introducción a la ecología del manglar. Oficina Regional de Ciencia y Tecnología de la Unesco para América Latina. Montevideo, Uruguay.

Colón, J. A. 1987. Algunos aspectos de la climatología de Puerto Rico. Acta Científica 1:55–63.

Crow, T. R. 1980. A rainforest chronicle: a 30-year record of change in structure and composition at El Verde, Puerto Rico. Biotropica 12:42–55.

Cuevas E., S. Brown, and A. E. Lugo. 1991. Above-and belowground organic matter storage and production in a tropical pine plantation and a paired broadleaf secondary forest. Plant and Soil 135:257–268.

Donnelly, T. W. 1985. Mesozoic and Cenozoic plate evolution of the Caribbean region, pp. 89–121 in F. G. Stehli and S. D. Web (eds.), The great American biotic interchange. Plenum Press, New York.

Donnelly, T. W. 1989. History of marine barriers and terrestrial connections: Caribbean paleogeographic inference from pelagic sediment analysis, pp. 103–118 in C. A. Woods (ed.), Biogeography of the West Indies: past, present, and future. Sandhill Crane Press, Gainesville, Fla.

Dunevitz, V. L. 1985. Regrowth of clearcut subtropical dry forests: mechanisms of recovery and quantification of resiliency. Master's thesis, Michigan State University, East Lansing.

Egler, F. E. 1950. Southeast saline everglades vegetation, Florida, and its management. Vegetatio 3:213–265.

Ewel, J. J. 1971. Experiments in arresting succession with cutting and herbicides in five tropical environments. Ph.D. dissertation. University of North Carolina, Chapel Hill.

Ewel, J. J. 1977. Differences between wet and dry successional tropical ecosystems. Geo-Eco-Trop 1:103–117.

Ewel, J. J. 1980. Tropical succession: manyfold routes to maturity. Biotropica 12 (supplement):2–7.

Ewel, J. J., and J. L. Whitmore. 1973. The ecological life zones of Puerto Rico and the U.S. Virgin Islands. USDA Forest Service Research Paper ITF-18. Institute of Tropical Forestry, Río Piedras.

Fernández, D. S., and N. Fetcher. 1991. Changes in light availability following Hurricane Hugo in a subtropical montane forest in Puerto Rico. Biotropica 23: 393–399.

Fernández, D. S., and R. N. Myster. 1995. Temporal variation and frequency distribution of photosynthetic photon flux densities on two landslides in Puerto Rico. Tropical Ecology, 36:73–87.

Figueroa Colón, J. 1992. La ecología de los suelos ultramaficos. Acta Científica 6:49–58.

Figueroa Colón, J. 1996. Phytogeographical trends, centers of high species richness and endemism, and the question of species extinctions in the native flora of Puerto Rico, pp. 89–102 in J. Figueroa Colón (ed.), The scientific survey of Puerto Rico and the Virgin Islands: an eighty year reassessment of the Islands' natural history. Annals of the New York Academy of Sciences, 776 New York.

Figueroa, J., and R. Schmidt. 1983. Caracterización y estructura de dos bosques en bosque pluvial sobre suelos serpentiniticos, Maricao, Puerto Rico, pp. 27–42 in C. A. Abrahamson, J. Vivaldi, D. Folch, and B. Cintrón (eds.), Compendio de ponencias presentadas en el octavo simposio de recursos naturales. Departamento de Recursos Naturales, San Juan.

Figueroa, J. C., L. Totti, A. E. Lugo, and R. Woodbury. 1983. Structure and composition of moist coastal forests in Dorado, Puerto Rico. USDA Forest Service, Southern Forest Experiment Station Research Paper SO-202, New Orleans.

Figueroa Colón, J., and R. O. Woodbury. 1996. Rare and endangered plant species of Puerto Rico and the Virgin Islands: an annotated checklist, pp. 65–72 in J. Figueroa Colón (ed.), The scientific survey of Puerto Rico and the Virgin Islands: an eighty year reassessment of the Islands' natural history. Ann. of the New York Acad. of Sci., 776, New York.

Foster, R. B. 1973. Seasonality of fruit production and seed fall in a tropical forest ecosystem in Panama. Ph.D. dissertation, Duke University, Durham.

Fox, R. L. 1982. Some highly weathered soils of Puerto Rico, 3. Chemical properties. Geoderma 27:139–176.

Francis, J. K. 1995. Forest plantations in Puerto Rico, pp. 210–223 in A. E. Lugo and C. Lowe (eds.), Tropical forests: management and ecology. Springer-Verlag, New York.

Francis, J. K., and H. A. Liogier. 1991. Naturalized exotic tree species in Puerto Rico. USDA Forest Service, Southern Forest Experiment Station, General Technical Report SO-82, New Orleans.

Francis, J. K., and P. L. Weaver. 1988. Performance of *Hibiscus elatus* in Puerto Rico. Com. For. Rev. 67: 327–338.

Frangi, J. L. 1983. Las tierras pantanosas de la montaña Puertorriqueña, pp. 233–247 in A. E. Lugo (ed.), Los bosques de Puerto Rico. USDA Forest Service, Institute of Tropical Forestry, and Puerto Rico Department of Natural Resources, San Juan.

Frangi, J. L. and A. E. Lugo. 1985. Ecosystem dynamics of a subtropical floodplain forest. Ecolog. Monogr. 55:351–369.

Frangi, J. L., and A. E. Lugo. 1991. Hurricane damage to a floodplain forest in the Luquillo Mountains of Puerto Rico. Biotropica 23:324–335.

Frangi, J. L., and M. Ponce. 1985. The root system of *Prestoea montana* and its ecological significance. Principes 29:13–19.

Gajraj, A. M. 1981. Threats to the terrestrial resources of the Caribbean. Ambio 10:307–311.

García Montiel, D., and F. N. Scatena. 1994. The effect of human activity on the structure and composition of a tropical forest in Puerto Rico. For. Ecol. and Mgt. 63:57–78.

Gentry, A. H. 1992. Tropical forest biodiversity: distributional patterns and their conservation significance. Oikos 63:19–28.

Gill, A. M. 1969. The ecology of an elfin forest in Puerto Rico, 6. Aerial roots. J. Arnold Arbor. 50:197–209.

Gleason, H. A., and M. T. Cook. 1927. Plant ecology of Porto Rico. Scientific Survey of Porto Rico and the Virgin Islands 7:1–173. Ann. of the New York Acad. of Sci., New York.

Good, R. 1953. The geography of flowering plants. Longmans, Green, London.

Graham, A. 1990. A late Tertiary microfossil flora from the Republic of Haiti. Amer. J. Bot. 77:911–926.

Graham, A. 1992. The current status of the legume fossil record in the Caribbean region, pp. 161–167 in P. S. Herendeen and D. L. Dilcher (eds.), Advances in legume systematics. Part 4. The fossil record. Royal Botanic Gardens, Kew, London.

Graham, A. 1993. Contribution toward a Tertiary palynostratigraphy for Jamaica: the status of Tertiary paleobotanical studies in northern Latin America and a preliminary analysis of the Guys Hill Member (Chapelton Formation, middle Eocene) of Jamaica. Mem. Geolog. Soc. Amer. 182:443–461.

Graham, A. 1995. Diversification of Gulf/Caribbean mangrove communities through Cenozoic time. Biotropica 27:20–27.

Graham, A., and D. M. Jarzen. 1969. Studies in Neotropical paleobotany. I. The Oligocene communities of Puerto Rico. Ann. Missouri Bot. Gard. 56:308–357.

Grossman, D. H., S. Iremonger, and D. M. Muchoney. ND. Jamaica: a rapid ecological assessment. Nature Conservancy, Arlington, Va.

Grubb, P. J. 1971. Interpretation of the "Massenerhebung Effect" on tropical mountains. Nature 229:44–45.

Grubb, P. J. 1977. Control of forest growth and distribution on wet tropical mountains with special reference to mineral nutrition. Ann. Rev. Ecol. and System. 8:83–107.

Grubb, P. J., and E. V. J. Tanner. 1976. The montane forests and soils of Jamaica: a reassessment. J. Arnold Arbor. 57:313–368.

Guariguata, M. R. 1990. Landslide disturbances and forest regeneration in the upper Luquillo mountains of Puerto Rico. J. Ecol. 78:814–832.

Guariguata, M. R., and M. C. Larsen. 1989. Preliminary map showing locations of landslides in El Yunque Quadrangle, Puerto Rico. U.S. Department of Interior, Geological Survey Open File Report 89–257. Government Printing Office, Washington, D.C.

Guzmán Grajales, S. M., and L. R. Walker. 1991. Differential seedling responses to litter after Hurricane Hugo in the Luquillo Experimental forest of Puerto Rico. Biotropica 23:407–413.

Hartshorn, G., G. Antonini, R. Du Bois, D. Harcharik, S. Heckadon, H. Newton, C. Quesada, J. Shores, and G. Staples. 1981. The Dominican Republic country environmental profile: a field study. AID Contract AID/SOD/PDC-C-0247. JRB Asso., McLean, Va.

Herrera, R. A., L. Méndez, M. E. Rodríguez, and E. E. García. 1988. Ecología de los bosques siempreverdes de la Sierra del Rosario, Cuba. Instituto de Ecología y Sistemática, Academia de Ciencias, Habana.

Hodell, D. A., J. H. Curtis, G. A. Jones, A. Higuera Gundy, M. Brenner, M. W. Binford, and K. T. Dorsey. 1991. Reconstruction of Caribbean climate change over the past 10,500 years. Nature 352:790–793.

Holdridge, L. R. 1967. Life zone ecology. Tropical Science Center. San José, Costa Rica.

Holdridge, L. R. 1983. Puerto Rican public forest land associations, pp. 35–40 in A. E. Lugo (ed.), Los bosques de Puerto Rico. USDA Forest Service, Institute of Tropical Forestry, and Puerto Rico Department of Natural Resources, San Juan.

Howard, R. A. 1952. The vegetation of the Grenadines, Windward Isles, British West Indies. Contribution 174, Gray Herbarium, Harvard University, Cambridge, Mass.

Howard, R. A. 1968. The ecology of an elfin forest in Puerto Rico, 1. Introduction and composition studies. J. Arnold Arbor. 49:381–418.

Howard, R. A. 1969. The ecology of an elfin forest in Puerto Rico, 8. Studies of stem growth and form and of leaf structure. J. Arnold Arbor. 50:225–261.

Howard, R. A. 1970. The summit forest of Pico del Oeste, Puerto Rico, Chap. B-20 in H. T. Odum and R. F. Pigeon (eds.), A tropical rain forest. National Technical Information Service, Springfield, Va.

Howard, R. A. 1973. The vegetation of the Antilles, pp. 1–38 in A. Graham (ed.), Vegetation and vegetational history of northern Latin America. Elsevier, Amsterdam.

Howard, R. A. 1977. Conservation and the endangered species of plants in the Caribbean islands, pp. 105–114 in G. T. Prance and T. S. Elias (eds.), Extinction is forever. New York Botanical Gardens, New York.

Howard, R. A. 1979. Flora of the West Indies, pp. 239–250 in K. Larsen and L. B. Holm Nielsen (eds.), Tropical botany. Academic Press, New York.

Howard, R. A., and W. R. Briggs. 1953. The vegetation on coastal dogtooth limestone in southern Cuba. J. Arnold Arbor. 34:88–95.

Jiménez, J. A., A. E. Lugo, and G. Cintrón. 1985. Tree mortality in mangrove forests. Biotropica 17:177–185.

Johnston, M. H. 1990. Successional change and species/site relationships in a Puerto Rican tropical forest. Ph.D. dissertation, State University of New York, College of Environmental Science and Forestry, Syracuse.

Kapos, V. 1986. Dry limestone forests of Jamaica, pp. 49–58 in D. A. Thompson, P. K. Bretting, and M. Humphreys (eds.), Forests of Jamaica. Jamaican Society of Scientists and Technologists. Kingston, Jamaica.

Kelly, D. L. 1986. Native forests on wet limestone in Northeastern Jamaica, pp. 31–42 in D. A. Thompson, P. K. Bretting, and M. Humphreys (eds.), Forests of Jamaica. Jamaican Society of Scientists and Technologists. Kingston, Jamaica.

Kelly, D. L. 1988. The threatened flowering plants of Jamaica. Biolog. Cons. 46:201–216.

Kelly, D. L., E. V. J. Tanner, V. Kapos, T. A. Dickinson, G. A. Goodfriend, and P. Fairbairn. 1988. Jamaican limestone forests: floristics, structure and environment of three examples along a rainfall gradient. J. Trop. Ecol. 4:121–156.

Larsen, M. C., and A. Simon. 1993. A rainfall intensity-duration threshold for landslides in a humid-tropical environment, Puerto Rico. Geografiska Annaler 75A:13–23.

Larsen, M. C., and A. J. Torres Sánchez. 1992. Landslides triggered by Hurricane Hugo in eastern Puerto Rico, September 1989. Carib. J. Sci. 28:113–120.

Lieberman, D., M. Lieberman, R. Peralta, and G. S. Hartshorn. 1985. Mortality patterns and stand turnover rates in wet tropical forest in Costa Rica. J. Ecol. 73:915–924.

Liegel, L. H. 1984. Status, growth, and development of unthinned Honduras pine plantations in Puerto Rico. Turrialba 34:313–324.

Liogier, H. A. 1990. The plants introduced in the West

Indies after the discovery of America and their impact on the ecology. Boletín de la Comición Puertorriqueña para la Celebración del Quinto Aniversario del Descubrimiento de America y Puerto Rico. San Juan.

Liogier, H. A. 1996. Botany and botanists in Puerto Rico, pp. 41–54 in J. Figueroa Colón (ed.), The scientific survey of Puerto Rico and the Virgin Islands: an eighty year reassessment of the Island's natural history. Ann. New York Acad. Sci., 776, New York.

Liogier, H. A., and L. F. Martorell. 1982. Flora of Puerto Rico and adjacent islands: a systematic synopsis. Editorial de la Universidad de Puerto Rico, Río Piedras.

Little, E. L., R. O. Woodbury, and F. H. Wadsworth. 1974. Trees of Puerto Rico and the Virgin Islands. USDA Forest Service Agriculture Handbook 449, Washington, D.C.

Lodge, D. J., F. N. Scatena, C. E. Asbury, and M. J. Sánchez. 1991. Fine litterfall and related nutrient inputs resulting from Hurricane Hugo in subtropical wet and lower montane rain forests of Puerto Rico. Biotropica 23:336–342.

Lugo, A. E. 1980. Mangrove ecosystems: successional or steady state? Biotropica 12 (supplement):65–72.

Lugo, A. E. 1986. Water and the ecosystems of the Luquillo Experimental Forest. General Technical Report SO-63. USDA Forest Service, Southern Forest Experiment Station, New Orleans.

Lugo, A. E. 1987. Are island ecosystems different from continental ecosystems? Acta Científica 1:48–54.

Lugo, A. E. 1988a. Ecological aspects of catastrophes in Caribbean islands. Acta Científica 2:24–31.

Lugo, A. E. 1988b. The future of the forest: ecosystem rehabilitation in the tropics. Environment 30:17–20, 41–45.

Lugo, A. E. 1988c. Estimating reductions in the diversity of tropical forest species, pp. 58–70 in E. O. Wilson and F. M. Peter (eds.), Biodiversity. National Academy Press, Washington, D.C.

Lugo, A. E. 1991. Dominancia y diversidad de plantas en Isla de Mona. Acta Científica 5:65–71.

Lugo, A. E. 1992a. Comparison of tropical tree plantations with secondary forests of similar age. Ecolog. Monogr. 62:1–37.

Lugo, A. E. 1992b. Tree plantations for rehabilitating damaged forest lands in the tropics, pp. 247–255 in M. K. Waly (ed.), Ecosystem rehabilitation, Vol. 2: ecosystem analysis and synthesis. Academic Press The Hague.

Lugo, A. E. 1994a. Maintaining an open mind on exotic species, pp. 218–220 in G. K. Meffe and R. C. Carroll (eds.), Principles of Conserva. Biol. Sinauer Associates, Sunderland, Mass.

Lugo, A. E. 1994b. Preservation of primary forests in the Luquillo Mountains, Puerto Rico. Conserva. Biol. 8:1122–1131.

Lugo, A. E. 1995. Reconstructing hurricane passages over forests: a tool for understanding multiple scale responses to disturbance. TREE 10:98–99.

Lugo, A. E., M. Applefield, D. J. Pool, and R. B. McDonald. 1983. The impact of Hurricane David on the forests of Dominica. Cana. J. For. Res. 13:201–211.

Lugo, A. E., A. Bokkestijn, and F. N. Scatena. 1995. Structure, succession, and soil chemistry of palm forests in the Luquillo Experimental Forest, pp. 142–177 in A. E. Lugo and C. Lowe (eds.), Tropical forests: management and ecology. Springer-Verlag, New York.

Lugo, A. E., and S. Brown. 1988. The wetlands of Caribbean islands. Acta Científica 2:48–61.

Lugo, A. E., E. Cuevas, and M. J. Sánchez. 1990. Nutrients and mass in litter and top soil of ten tropical tree plantations. Plant and Soil 125:263–286.

Lugo, A. E., and J. Figueroa. 1985. Performance of *Anthocephalus chinensis* in Puerto Rico. Can. J. of For. Res. 15:577–585.

Lugo, A. E., J. A. González Liboy, B. Cintrón, and K. Dugger. 1978. Structure, productivity, and transpiration of a subtropical dry forest. Biotropica 10:278–291.

Lugo, A. E., and P. G. Murphy. 1986. Nutrient dynamics of a Puerto Rican subtropical dry forest. J. Trop. Ecol. 1:55–72.

Lugo, A. E., J. A. Parrotta, and S. Brown. 1993. Loss in species caused by tropical deforestation and their recovery through management. Ambio 22:106–109.

Lugo, A. E., and C. T. Rivera Batlle. 1987. Leaf production, growth rate, and age of the palm *Prestoea montana* in the Luquillo Experimental Forest. J. Trop. Ecol. 3:151–161.

Lugo, A. E., M. J. Sánchez, and S. Brown. 1986. Land use and organic carbon content of some subtropical soils. Plant and Soil 96:185–196.

Lugo, A. E., and F. N. Scatena. 1992. Epiphytes and climate change research in the Caribbean. Selbyana 13:123–130.

Lugo, A. E., and F. N. Scatena. 1995. Ecosystem-level properties of the Luquillo Experimental Forest with emphasis on the tabonuco forest, pp. 59–108 in A. E. Lugo and C. Lowe (eds.)., Tropical forests: management and ecology. Springer-Verlag, New York.

Lugo, A. E., R. Schmidt, and S. Brown. 1981. Tropical forests in the Caribbean. Ambio 10:318–324.

Lugo, A. E., and S. C. Snedaker. 1974. The ecology of mangroves. Ann. Rev. Ecol. and System. 5:39–64.

Lugo, A. E., and R. B. Waide. 1993. Catastrophic and background disturbance of tropical ecosystems at the Luquillo Experimental Forest. J. Biosci. 18:475–481.

Lugo, A. E., D. Wang, and F. H. Bormann. 1990. A comparative analysis of biomass production in five tropical tree species. For. Ecol. and Mgt. 31:153–166.

MacArthur, R. H., and E. O. Wilson. 1967. The theory of island biogeography. Princeton University Press, Princeton, N.J.

Malfait, B. T., and M. G. Dinkleman. 1972. Circum-Caribbean tectonic and igneous activity and the evolution of the Caribbean plate. Bull. Geologi. Soc. of Ameri. 83:251–272.

Marrero, J. 1947. A survey of forest plantations in the Caribbean National Forest. Master's thesis. School of Forestry and conservation, University of Michigan, Ann. Arbor.

Mattson, P. H. 1979. Subduction, buoyant braking, flipping, and strikeship in the northern Caribbean. J. Geol. 87:293–304.

Mattson, P. H. 1984. Caribbean structural breaks and plate movements. Mem. Geolog. Soci. Amer. 162:131–152.

McCormick, J. F. 1995. A review of the population dynamics of selected tree species in the Luquillo Experimental Forest, Puerto Rico, pp. 224–257 in A. E.

Lugo and C. Lowe (eds.), Tropical forest management and ecology. Springer-Verlag, New York.

Medina, E. 1983. Adaptation of tropical trees to moisture stress, pp. 225–237 in F. B. Golley (ed.), Tropical rain forest ecosystems: structure and function. Elsevier, Amsterdam.

Medina, E., E. Cuevas, J. Figueroa, and A. E. Lugo. 1994. Mineral content of leaves from trees growing on serpentine soils under contrasting rainfall regimes in Puerto Rico. Plant and Soil 158:13–21.

Medina, E., E. Cuevas, and P. L. Weaver. 1981. Composición foliar y transpiración de especies leñosas de Pico del Este, Sierra de Luquillo, Puerto Rico. Acta Científica Venezolana 32:159–165.

Monroe, W. T. 1976. The karst landforms of Puerto Rico. Geological Survey Professional Paper 899. Government Printing Office. Washington, D.C.

Moreno Casasola, P. 1993. Dry coastal ecosystems of the Atlantic coasts of Mexico and Central America, pp. 389–405 in E. van der Maarel (ed.), Dry coastal ecosystems Africa, America, Asia, and Oceania. Elsevier, Amsterdam.

Murphy, P. G., A. E. Lugo, A. J. Murphy, and D. C. Nepstad. 1995. The dry forests of Puerto Rico's south coast, pp. 178–209 in A. E. Lugo and C. Lowe (eds.), Tropical forests: management and ecology. Springer-Verlag, New York.

Murphy, P. G., and A. E. Lugo. 1986a. Ecology of tropical dry forest. Ann. Rev. Ecol. System. 17:67–88.

Murphy, P. G., and A. E. Lugo. 1986b. Structure and biomass of a subtropical dry forest in Puerto Rico. Biotropica 18:89–96.

Myster, R. W., and D. S. Fernández. 1995. Spatial gradients and structure on two Puerto Rican landslides. Biotropica 27:149–159.

Negrón, L. N. 1980. La producción de hojarasca en el manglar ribereño del Espíritu Santo. Master's thesis. University of Puerto Rico, Río Piedras.

Neuman, C. J., G. W. Caso, and B. R. Jaruinen. 1978. Tropical cyclones of the north Atlantic Ocean, 1871–1978. U.S. Department of Commerce, National Oceanic and Atmospheric Administration, National Climatic Center. Asheville, N. C.

Nevling, L. I. Jr. 1971. The ecology of an elfin forest in Puerto Rico, 16. The flowering cycle and an interpretation of its seasonality. J. Arnold Arbor. 52:586–613.

Odum, H. T. 1970. An emerging view of the ecological system at El Verde, pp. I-191–I – 289 in H. T. Odum and R. F. Pigeon (eds.), A tropical rain forest. National Technical Information Service, Springfield, Va.

Odum, H. T., and R. F. Pigeon, eds. 1970. A tropical rain forest. National Technical Information Service, Springfield, Va.

Parrotta, J. A. 1992. The role of plantation forests in rehabilitating degraded tropical ecosystems. Agric., Ecosys. and Environ. 41:115–133.

Parrotta, J. A. 1993. Secondary forest regeneration on degraded tropical lands. The role of plantations as "foster ecosystems," pp. 63–73 in H. Lieth and M. Lohmann (eds.), Restoration of tropical forest ecosystems. Kluwer Academic Publishers, Dordrecht, The Netherlands.

Parrotta, J. A., and D. J. Lodge. 1991. Fine root dynamics in a subtropical wet forest following hurricane disturbance in Puerto Rico. Biotropica 23:343–347.

Pérez Viera, I. E. 1986. Tree regeneration in two tropical rain forests. Master's thesis. Department of Biology, University of Puerto Rico, Río Piedras.

Perfit, M. R., and E. E. Williams. 1989. Geological constraints and biological retrodictions in the evolution of the Caribbean Sea and its islands, pp. 77–102 in C. A. Woods (ed.), Biogeography of the West Indies: past, present, and future. Sandhill Crane Press, Gainesville, Fla.

Pindell, J. L., and S. F. Barnett. 1990. Geological evolution of the Caribbean region: a plate tectonics perspective, pp. 405–472 in G. Dengo and J. E. Case (eds.), The Caribbean region. The geology of North America, vol. H. Geological Society of America, Boulder, Colo.

Pindell, J. L., and J. F. Dewey. 1982. Permo-triassic reconstruction of western Pangea and the evolution of the Gulf of Mexico/Caribbean region. Tectonics 1:179–212.

Portig, W. H. 1976. The climate of Central America, pp. 405–454 in W. Schwerdtfeger (ed.), Climates of Central and South America. Elsevier, Amsterdam.

Proctor, G. R. 1986a. Cockpit country and its vegetation, pp. 43–47 in D. A. Thompson, P. K. Bretting, and M. Humphreys (eds.), Forests of Jamaica. Jamaican Society of Scientists and Technologists. Kingston, Jamaica.

Proctor, G. R. 1986b. Vegetation of the Black River morass, pp. 59–65 in D. A. Thompson, P. K. Bretting, and M. Humphreys (eds.), Forests of Jamaica. Jamaican Society of Scientists and Technologists. Kingston, Jamaica.

Proctor, G. R. 1989. Ferns of Puerto Rico and the Virgin Islands. Mem. New York Bot. Gard. 53:1–389.

Roberts, R. C. 1942. Soil survey of Puerto Rico. USDA Series 1936, No. 8. Government Printing Office. Washington, D.C.

Russel, K. W., and H. A. Miller. 1977. The ecology of an elfin forest in Puerto Rico, 17. Epiphytic mossy vegetation of Pico del Oeste. J. Arnold Arbor. 58:1–24.

Samek, V. 1973. Regiones fitogeográficas de Cuba. Serie Forestal No. 15. Academia de Ciencias de Cuba. Habana.

Sánchez Irizarry, M. J. 1989. Estudio comparativo de algunas propiedades químicas y físicas de suelos de bosque bajo uso natural y plantaciones silvestres. Río Piedras? Master's thesis. University of Puerto Rico, Mayagüez.

Scatena, F. N. 1989. An introduction to the physiography and history of the Bisley Experimental Watersheds in the Luquillo Mountains of Puerto Rico. USDA Forest Service, Southern Forest Experiment Station General Technical Report SO-72.

Scatena, F. N. 1995. Management of the Luquillo elfin cloud forest ecosystems: Irreversible decisions in a nonsustainable ecosystem, pp. 296–308 in L. S. Hamilton, J. O. Juvik, and F. N. Scatena (eds.), Tropical montane cloud forests. Springer-Verlag, New York.

Scatena, F. N., and M. C. Larsen. 1991. Physical aspects of Hurricane Hugo in Puerto Rico. Biotropica 23: 317–323.

Scatena, F. N., and A. E. Lugo. 1995. Geomorphology, disturbance, and the soil and vegetation of two subtropical wet steepland watersheds of Puerto Rico. Geomorphology 13:199–213.

Scatena, F. N., W. Silver, T. Siccama, A. Johnson, and M. J. Sánchez. 1993. Biomass and nutrient content of the Bisley Experimental Watersheds, Luquillo

Experimental Forest, Puerto Rico, after Hurricane Hugo, 1989. Biotropica 25:15–27.

Schubert, C. 1988. Climatic changes during the last glacial maximum in northern South America and the Caribbean: a review. Interciencia 13:128–137.

Schubert, C., and E. Medina. 1982. Evidence of Quaternary glaciation in the Dominican Republic: some implications for Caribbean paleoclimatology. Paleoecology 39:281–294.

Shreve, F. 1914. A montane rain forest. A contribution to the physiological plant geography of Jamaica. Publ. of the Carnegie Institu. 199:1–110.

Silver, W. 1992. Effects of small-scale and catastrophic disturbance on carbon and nutrient cycling in a lower montane subtropical wet forest in Puerto Rico. Ph.D. dissertation. Yale School of Forestry and the Environment. New Haven.

Silver, W. L., A. E. Lugo, and M. Keller. 1995. Soil oxygen links tropical rain forest biodiversity with climate. Manuscript submitted for publication.

Silver, W. L., F. N. Scatena, A. H. Johnson, T. G. Siccama, and M. J. Sánchez. 1994. Nutrient availability in a montane wet tropical forest in Puerto Rico: spatial patterns and methodological considerations. Plant and Soil 164:129–145.

Silver, W. L., and K. A. Vogt. 1993. Fine root dynamics following single and multiple disturbances in a subtropical wet forest ecosystem. J. Ecol. 81:729–738.

Sloane, H. 1707, 1725. A voyage to the islands Madera, Barbados, Nieves, S. Christophers and Jamaica. Two volumes. London.

Smith, E. E. 1954. The forests of Cuba. Maria Moors Cabot Foundation. Publication No. 2. Cienfuegos, Cuba.

Smith, R. F. 1970. The vegetation structure of a Puerto Rican rain forest before and after short-term gamma irradiation, Chap. D-3 in H. T. Odum and R. F. Pigeon (eds.), A tropical rain forest. National Technical Information Services. Springfield, Va.

Soil Survey Staff. 1995. Order 1 Soil Survey of the Luquillo long-term ecological research grid, Puerto Rico. USDA Natural Resources Conservation Service, Lincoln, Nebr.

Stehlé, H. 1945. Forest types of the Caribbean islands. Carib. For. 6 (supplement):273–408.

Steudler, P. A., J. M. Mellilo, R. D. Bowden, M. S. Castro, and A. E. Lugo. 1991. The effects of natural and human disturbances on soil nitrogen dynamics and trace gas fluxes in a Puerto Rican wet forest. Biotropica 23:356–363.

Stoffers, A. L. 1956. The vegetation of the Netherland Antilles, studies on the flora of Curacao and other Caribbean islands, Vol. 1. Martinus Nijhoff, The Hague.

Stoffers, A. L. 1993. Dry coastal ecosystems of the West Indies, pp. 407–421 in E. van der Maarel (ed.), Dry coastal ecosystems Africa, America, Asia, and Oceania. Elsevier, Amsterdam.

Tanner, E. V. J. 1977. Four montane forests of Jamaica: a quantitative characterization of the floristics, the soils, and the foliar mineral levels, and a discussion of the interrelations. J. Ecol. 65:883–918.

Tanner, E. V. J. 1986. Forests of the Blue Mountains and the Port Royal Mountains of Jamaica, pp. 15–30 in D. A. Thompson, P. K. Bretting, and M. Humphreys (eds.), Forests of Jamaica. Jamaican Society of Scientists and Technologists. Kingston, Jamaica.

Taylor, C. M., S. Silander, R. B. Waide, and W. J. Pfeiffer. 1995. Recovery of a tropical forest after gamma irradiation: a 23-year chronicle, pp. 258–285 in A. E. Lugo and C. Lowe (eds.), Tropical forest management and ecology. Springer-Verlag, New York.

Thompson, D. A., P. K. Bretting, and M. Humphreys (eds.). 1986. Forests of Jamaica. Jamaican Society of Scientists and Technologists. Kingston, Jamaica.

Thorhaug, A. 1981. Biology and management of seagrass in the Caribbean. Ambio 10:295–298.

Toledo, V. M. 1985. A critical evaluation of the floristic knowledge in Latin America and the Caribbean. A report presented to the Nature Conservancy International Program. Washington, D.C.

Tomblin, J. 1981. Earthquakes, volcanoes and hurricanes: a review of natural hazards and vulnerability in the West Indies. Ambio 10:340–345.

Torres, J. A. 1988. Tropical cyclone effects on insect colonization and abundance in Puerto Rico. Acta Científica 2:40–44.

Torres, J. A. 1992. Lepidoptera outbreaks in response to successional changes after the passage of Hurricane Hugo in Puerto Rico. J. Trop. Ecol. 8:285–298.

Torres, J. A. 1994. Wood decomposition of *Cyrilla racemiflora* in a tropical montane forest. Biotropica 26:124–140.

Trevin, J. O., K. P. Rodney, A. Glasgow, and N. Weekes. 1993. Forest change in a subtropical moist forest of St. Vincent, West Indies: the King's Hill Forest Reserve, 1945–1990. Com. For. Rev. 72:187–192.

Tryon, R. 1979. Biogeography of the antillean fern flora, pp. 55–68 in D. Bramwell (ed.) Plants and islands. Academic Press, New York.

Tschirley, F. H., C. C. Dowler, and J. A. Duke. 1970. Species diversity in two plant communities of Puerto Rico, Chap. B-7 in H. T. Odum and R. F. Pigeon (eds.), A tropical rain forest. National Technical Information Service, Springfield, Va.

Twilley, R. R., A. E. Lugo, and C. Patterson Zucca. 1986. Litter production and turnover in basin mangrove forests in southwest Florida. Ecology 67:670–683.

Vicente, V. P. 1992. A summary of ecological information on the seagrass beds of Puerto Rico, pp. 123–133 in U. Seeliger (ed.), Coastal plant communities of Latin America. Academic Press, New York.

Wadsworth, F. H. 1950. Notes on the climax forests of Puerto Rico and their destruction and conservation prior to 1900. Carib. For. 11(1):38–47.

Wadsworth, F. H. 1960. Records of forest plantation growth in Mexico, the West Indies, and Central and South America. Carib. For. 21:i – A10.

Wadsworth, F. H. 1987. Composition of trees in insular forest ecosystems. Acta Científica 1:77–80.

Wadsworth, F. H. 1995. A forest research institution in the West Indies: the first 50 years, pp. 33–56 in A. E. Lugo and C. Lowe (eds.), Tropical forests: management and ecology. Springer-Verlag, New York.

Wadsworth, F. H., and R. A. Birdsey. 1983. Un nuevo enfoque de los bosques de Puerto Rico, pages 12–27 in Puerto Rico Department of Natural Resources ninth symposium on natural resources. Puerto Rico Department of Natural Resources, San Juan.

Waide, R. B., and A. E. Lugo. 1992. A research perspective on disturbance and recovery of a tropical montane forest, pp. 173–190 in J. G. Goldammer (ed.), Tropical forests in transition. Birkhäuser Verlag, Basel, Switzerland.

Walker, L. R., D. J. Lodge, N. V. L. Brokaw, and R. B. Waide. 1991. Special issue: ecosystem, plant, and animal responses to hurricanes in the Caribbean. Biotropica 23:313–521.

Wang, D., F. H. Bormann, A. E. Lugo, and R. D. Bowden. 1991. Comparison of nutrient-use efficiency and biomass production in five tropical tree taxa. For. Ecol. and Manage. 46:1–21.

Weaver, P. L. 1972. Cloud moisture interception in the Luquillo Mountains of Puerto Rico. Carib. J. of Sci. 12:129–144.

Weaver, P. L. 1976. Transpiration rates in the dwarf forest of the Luquillo Mountains of Puerto Rico. Carib. J. Sci. 15:21–30.

Weaver, P. L. 1983. Tree growth and stand changes in the subtropical life zones of the Luquillo Mountains of Puerto Rico. Research Paper SO-190. USDA Forest Service, Southern Forest Experiment Station, New Orleans.

Weaver, P. L. 1986. Growth and age of *Cyrilla racemiflora* L. in montane forests of Puerto Rico. Interciencia 11:47–58.

Weaver, P. L. 1987. Structure and dynamics in the colorado forests of the Luquillo Mountains of Puerto Rico. Ph.D. dissertation. Michigan State University, East Lansing.

Weaver, P. L. 1989a. Forest changes after hurricanes in Puerto Rico's Luquillo Mountains. Interciencia 14: 181–192.

Weaver, P. L. 1989b. Taungya plantings in Puerto Rico. J. For. 87:37–39, 41.

Weaver, P. L. 1990. Succession in the elfin woodland of the Luquillo Mountains of Puerto Rico. Biotropica 22:83–89.

Weaver, P. L. 1992. An ecological comparison of canopy trees in the montane rain forest of Puerto Rico's Luquillo Mountains. Carib. J. Sci. 28:62–69.

Weaver, P. L. 1994. Baño de Oro natural area Luquillo Mountains, Puerto Rico. USDA Forest Service, Southern Forest Experiment Station, General Technical Report SO-111. New Orleans.

Weaver, P. L. 1995. The colorado and dwarf forests of Puerto Rico's Luquillo Mountains, pp. 109–141 in A. E. Lugo and C. Lowe (eds.), Tropical forests: management and ecology. Springer-Verlag, New York.

Weaver, P. L., R. A. Birdsey, and A. E. Lugo. 1987. Soil organic matter in secondary forests of Puerto Rico. Biotropica 19:17–23.

Weaver, P. L., E. Medina, D. Pool, K. Dugger, J. González Liboy, and E. Cuevas. 1986. Ecological observations in the dwarf cloud forest of the Luquillo Mountains. Biotropica 18:79–85.

Weaver, P. L., and P. G. Murphy. 1990. Forest structure and productivity in Puerto Rico's Luquillo Mountains. Biotropica 22:69–82.

Westermann, J. H. 1952. Conservation in the Caribbean. Foundation for Scientific Research in Surinam and the Netherlands Antilles, No. 7. Utrecht.

Whittaker, R. H. 1965. Dominance and diversity in land plant communities. Science 147:250–260.

Whittaker, R. H. 1970. Communities and ecosystems. Macmillan, London.

Zarin, D. J. 1993. Nutrient accumulation during succession in subtropical lower montane wet forests, Puerto Rico. Ph.D. dissertation. University of Pennsylvania, College Station.

Zimmerman, J. K., T. M. Aide, J. Herrera, M. A. Rosario, M. Serrano, and M. I. Serrano. 1995. Effects of land management and a recent hurricane on forest structure and composition in the Luquillo Experimental Forest, Puerto Rico. For. Ecol. and Mgt. 77: 65–76.

Zimmerman, J. K., E. M. Everham III, R. B. Waide, D. J. Lodge, C. M. Taylor, and N. Brokaw. 1994. Responses of tree species to hurricane winds in subtropical wet forest in Puerto Rico: implications for tropical tree life histories. J. Ecol. 82:911–922.

Zimmerman, J. K., W. M. Pulliam, D. J. Lodge, V. Quiñones Orfila, N. Fetcher, S. Guzmán Grajales, J. A. Parrotta, C. E. Asbury, L. R. Walker, and R. B. Waide. 1995. Nitrogen immobilization by decomposing woody debris and the recovery of tropical wet forest from hurricane damage. Oikos 72:316–322.

Zou, X., and G. González. 1997. Changes in earthworm density and community structure during secondary succession in abandoned tropical pastures. Soil Biology and Biochemistry 29:627–629.

Zou, X., C. P. Zucca, R. B. Waide, and W. H. McDowell. 1995. Long-term influence of deforestation on tree species composition and litter dynamics of a tropical rain forest in Puerto Rico. For. Ecol. and Mgt. 78:147–157.

Chapter
17

Tropical and Subtropical Vegetation of Mesoamerica

GARY S. HARTSHORN

SETTING

Mesoamerica is a convenient, nonpolitical term for the region between South America and North America, but where do we draw the southern and northern limits for purposes of this chapter? It makes no sense ecologically to separate the Darién forests of eastern Panama from the very similar forests of the Colombian Chocó. A more reasonable geographical boundary is the isthmus of Panama, preferably east of the Panama Canal, if for no other reason than that the renowned Barro Colorado Island (BCI) has been the site of much research on tropical vegetation and plants.

The northern ecological limit of Mesoamerica is not so easily defined. One possibility is the Tropic of Cancer (23°30'N), that is, the astronomic northern limit of the tropics. The narrow isthmus of Tehuántepec (225 km wide) in southern Mexico is an attractive candidate for the northern limit, but that would exclude the lowland subtropical forests of Veracruz (cf. Wendt 1989, 1993). A more meaningful ecological boundary for Mesoamerica is the northern limit of mangrove forests at 26–27°N (Chapter 13). However, the preponderance of temperate vegetation in the Mexican highlands makes it impossible to draw a northern latitudinal boundary to Mesoamerica (Chapter 15). This chapter focuses on the major types of tropical and subtropical vegetation prevalent in Mesoamerica and their conservation status, without attempting to define precise latitudinal or altitudinal limits. Vegetation of the Caribbean islands is described in Chapter 16.

The division of Mesoamerica into tropical and subtropical regions occurs at approximately 12–13°N latitude (Holdridge 1967). This latitudinal division corresponds to the inner tropics and outer tropics of geographers. Others prefer to consider the entire region as tropical. Less understandable is the designation of all of Mesoamerica as subtropical (Encyclopaedia Britannica 1977). Tropical Mesoamerica and subtropical Mesoamerica differ in several ways. Seasonal climate is more pronounced farther from the equator. Tropical storms and hurricanes often strike subtropical Mesoamerica, whereas tropical Mesoamerica is usually south of hurricane paths. More important for this chapter, forest structure and composition differ between subtropical and tropical Mesoamerica.

This chapter is dedicated to Dr. L. R. Holdridge. Early drafts were improved by the comments and suggestions of Lynne Hartshorn, Les Holdridge, David Janos, Diana Lieberman, and Milton Lieberman. I thank Jim Barborak, Steve Cornelius and Richard Margoluis for assisting with the compilation of Table 17.18.

Physical Features

Physiography. The remarkable diversity of vegetation types highlighted in this chapter is due not only to Mesoamerica's low latitude but also to the fundamental influences that mountains and volcanoes have on the region's climate, geology, and soils. Mountain ranges are dominant physiographic features in every country of the region, ranging from the 5700 m Mount Orizaba in Mexico (the highest North American peak south of Alaska) to the low Maya Mountains in Belize. Several active volcanoes (e.g., Mexico's El Chichón, Guatemala's Fuego, Nicaragua's Masaya, Costa Rica's Arena) are part of the Pacific "ring of fire" that forms the spectacularly beautiful backbone of Mesoamerica.

The rugged Sierra Madre Occidental and Sierra Madre Oriental dominate the physiography of much of Mexico, with vast, warm-temperate tablelands between these two parallel ranges. The transversal Cordillera Neo-Volcánica, which forms the southern rim of the Mexico City basin, is one of several west-east mountain ranges prominent in southern Mexico and in Guatemala, a phytogeographic unit termed "Megamexico" by Islebe and Velásquez (1994). A dominant mountain range is absent east of Guatemala; nevertheless, rugged highlands occupy 65% of Honduras. These highlands extend into northern and central Nicaragua, finally disappearing in the Nicaraguan Depression (Lake Managua, Lake Nicaragua, and the Río San Juan). Costa Rica differs from the pattern of parallel mountain ranges in northern Mesoamerica in that its four major ranges are aligned northwest to southeast. The Cordillera de Talamanca includes the highest peaks of Costa Rica as well as of Panama.

Mexico's Yucatán Peninsula is part of an extensive limestone platform that also underlies northern Belize and the Guatemalan Petén. Another prominent physiographic feature along the Caribbean coast is the Honduran-Nicaraguan bulge terminating at Cape Gracias de Diós. The Pacific coast of Mesoamerica has prominent peninsulas (Nicoya, Osa, Azuero) and gulfs (Tehuántepec, Fonseca, Nicoya, Dulce, Chiriquí, Montijo, Panamá). The coastal plain is more extensive on the Caribbean side than on the Pacific side of Mesoamerica.

Climate. The Mesoamerican climate is characterized by predictable temperature regimes and unpredictable rainfall patterns. The daily variation in temperature (on a sunny day) exceeds the difference between average temperatures for the coolest and warmest months. Because of minimal variation in temperature patterns during the year, seasonal differences are based on rainfall. When the "heat equator," or Intertropical Convergence Zone (ITCZ), is south of the region, weather depends on the northeast tradewinds that blow incessantly from November–December to March–April. Picking up moisture in their passage over the Caribbean Sea, the tradewinds are the source of the famous cloud forests characteristic of the windward mountain slopes in many parts of the circum-Caribbean region. Stripped of moisture by the mountainous cloud forests and rapidly heated as they descend the hot Pacific slopes, the northeast tradewinds have an opposite, rainshadow effect on the leeward side of the mountains. Most of the Pacific lowlands and slopes, as well as many of the intermountain valleys, have a monsoon type of climate, with a very pronounced dry season of up to 6 mo without significant rainfall.

The wet season is associated with the presence of the ITCZ over the southern part of Mesoamerica. Late in the dry season, the northeast tradewinds diminish some 6–8 wk before the abrupt start of the rains. Because of proximity to large bodies of water along both coasts, the region's climate is strongly maritime, with onshore breezes usually bringing afternoon rains. Convectional air currents produce abundant thunderstorms, with typically heavy rainfall. The predictable pattern of daily rains associated with the ITCZ is overlain by the unpredictable occurrence of storms caused by northern polar air masses penetrating far to the south. These northers occur between November and March, bringing rainy and cooler than normal weather to the highlands and Caribbean lowlands during the normal dry season.

Tropical cyclones also bring heavy rains to the region, usually during the traditional rainy season. Pacific storms often arrive late in the rainy season (August to October), not only resulting in very heavy rainfall but also producing a modest rainshadow effect on the Caribbean side. Tropical cyclones known as hurricanes occasionally come across the Caribbean to strike northern Mesoamerica, often with devastating consequences (Brokaw and Walker 1991; Whigham, Olmsted, Cabrero Cano, and Harmon 1991; Yih, Boucher, Vandermeer, and Zamora 1991; Although hurricane paths are usually absent from the inner tropics (Coen 1983), high winds and heavy rains do affect the eastern lowlands of Nicaragua (cf. Boucher, Vandermeer, Mallona, Zamona, and Perfecto 1994) and Costa Rica.

Geology and soils. The geological history of Mesoamerica has been characterized by episodes of mountain building and oceanic submergence that first provided islands in the region and finally the present land bridge between continents (Raven and Axelrod 1975). The Quaternary has been characterized by intense volcanic activity throughout much of the region. Even in Belize, which has no geological vestiges of volcanoes and lies far from known volcanoes, some of the soils have volcanic ash among marine sediments (Wright, Romney, Arbuckle, and Vial 1959). Volcanic eruptions have blanketed extensive areas with ash; the Central Valley of Costa Rica has multiple layers of volcanic ash as deep as 50 m.

In addition to lava flows, volcanoes eject two principal types of ash: andesitic and rhyolitic. Andesitic ash is mostly plagioclase feldspar, whereas rhyolitic ash is high in silica. As parent material, these two different types of volcanic ash have influenced both soil genesis and vegetation, as well as land-use patterns. Andesitic ash is the parent material for a well-known group of soils called "andepts" that characterize the major coffee-growing areas of the world. Andepts are common in southern Mexico, western Guatemala, El Salvador, Costa Rica, and western Panama; they also characterize other regions supporting high population densities, such as highland Colombia, Kenya, Java, and Bali. In contrast to the fertile andepts, rhyolitic ash and pumice lava flows are parent materials of relatively infertile soils (e.g., Honduran highlands).

The combination of rugged topography and high rainfall results in appreciable erosion of soils, often exacerbated by inappropriate land use. The high sediment loads carried by rivers are deposited on the active floodplains. Because most of the region's mountainous areas are geologically young, frequent floods enrich the lowlands with minerals from the eroded slopes. These extensive alluvial floodplains have long been prized for intensive agriculture, such as banana plantations in eastern Costa Rica and northern Honduras, or sugarcane fields in seasonally dry valleys of the Pacific lowlands.

The Yucatán Peninsula is very different geologically and ecologically from the rest of Mesoamerica because of the preponderance of limestone and soils derived from marine sediments. Where limestone rock is exposed, the topography tends to be abrupt, and drainage is internal through the porous limestone. Although soil fertility is generally good, seasonal drought has a strong influence on natural vegetation and often limits land-use options. Other locally important soil types are mentioned in the treatments of vegetation that follow.

Vegetation Classification Systems

Though many classification systems have been developed to categorize vegetation, it is beyond the scope of this overview of Mesoamerican vegetation to evaluate or compare them. Rather, a few major classification systems relevant to Mesoamerican vegetation will be mentioned (Shimwell 1971; Mueller-Dombois and Ellenberg 1974). The vegetation classification systems used in Mesoamerica can be grouped into three types: floristic, physiognomic (structural), and bioclimatic.

Most of the early descriptions of Mesoamerican vegetation were based on floristic criteria (Stevenson 1928; Lundell 1937; Standley 1937; Allen 1956; Gómez-Pompa, Hernandez, and Sousa 1964; Wagner 1964; Rzedowski 1978). The descriptions often included substantial lists of plant species, but there were few specifics concerning abundance, stature, or dominance. Such species lists mean very little to the reader unfamiliar with the flora. The great species richness of tropical forests, the wealth of life forms, and the frequent lack of dominance make it exceedingly difficult to classify tropical vegetation solely on the basis of floristics.

The classic work of Davis and Richards (1933–1934) was the first to include physiognomic criteria in the classification of neotropical vegetation. These authors pioneered the use of profile diagrams to illustrate the complex vertical structures of tropical forests. Key physiognomic features include the number of layers of tree crowns, the canopy height, the proportion of deciduous species in the canopy, the abundance of leaf forms, and the types of synusiae, or life forms.

Beard's (1944) physiognomic classification of Mesoamerican vegetation distinguished six formations: rain; seasonal; dry evergreen; montane; swamp forests; and marsh or seasonal swamp. The UNESCO physiognomic classification system (Mueller-Dombois and Ellenberg 1974) uses a confusing plethora of adjectives (e.g., deciduous, drought-deciduous, evergreen, ombrophilous, seasonal, semideciduous) to define vegetation formations. Except for a few superficial trials (Kuchler and Montoya Maquín 1971), the UNESCO system has not been used in Mesoamerica.

A fundamental criterion for physiognomic classification of vegetation is that it be applied to climax communities. Because of natural dynamic processes such as tree-fall gaps (Hartshorn 1980), the validity of the "climatic climax" concept for tropical forests has been seriously questioned (Aubréville 1938; Hewetson 1956). Where a dry season permits burning, slash-and-burn agriculture has profoundly and perhaps permanently altered the

natural vegetation (Gómez-Pompa, Vásquez-Yanes, and Guevara 1972). The use of fire by indigenous cultures for hunting game or shifting cultivation of subsistence crops may have affected the composition and structure of present forests (Sauer 1958; Budowski 1959).

Holdridge (1947) devised a bioclimatic classification system using precipitation and temperature, plus the ratio of potential evapotranspiration (PET) to precipitation. These parameters are arranged logarithmically in an isogonal diagram (Fig. 17.1). Each hexagon is called a life zone, and the small triangles around the periphery of a hexagon are called transitions. Because of the geometry of the life-zone diagram, tropical transitional areas can occur in the subtropical latitudinal region. Some examples occur in the lower Aguán Valley of Honduras (cool transition of tropical moist forest life zone) and in southeastern Belize (cool transition of tropical wet forest life zone). Holdridge's life-zone classification system has been used extensively in Mesoamerica to map natural vegetation (e.g., Tosi 1969, 1971; Holdridge 1975a; De la Cruz 1976; Tosi and Hartshorn 1978; González, Ramirez, and Peralta 1983; Hartshorn et al. 1984).

In the life-zone diagram (Fig. 17.1), note that a latitudinal basal belt does not have a corresponding altitudinal belt. For example, the subtropical latitudinal region does not have a premontane altitudinal belt; rather, altitudinal belts in the subtropics start with the lower montane. The lowest row of complete hexagons (Fig. 17.1) is divided by a critical-temperature line into subtropical and warm-temperate latitudinal regions or premontane and lower-montane altitudinal belts. The arrangement of latitudinal regions and altitudinal belts is clearer in the third dimension (Fig. 17.2). At the equator (right margin of Fig. 17.2), the theoretical limits of altitudinal belts are shown (e.g., 3000–4000 m for tropical montane life zone). Moving poleward, the altitudinal limits decrease; in Costa Rica (8–11°N), the change from tropical basal belt to tropical premontane altitudinal belt occurs at only 500–700 m elevation.

MAJOR VEGETATION TYPES

Mesoamerica is exceptionally rich in plant species that occur in an impressive number of plant communities (Gentry 1990, 1992). No comprehensive survey of Mesoamerican vegetation has been done, and given the extensive deforestation that has occurred, it is difficult to classify the region's natural vegetation. Excluding Mexico, Mesoamerica is estimated to have about 18,000 species of vascular plants (D'Arcy 1977). Even tiny Belize lists about

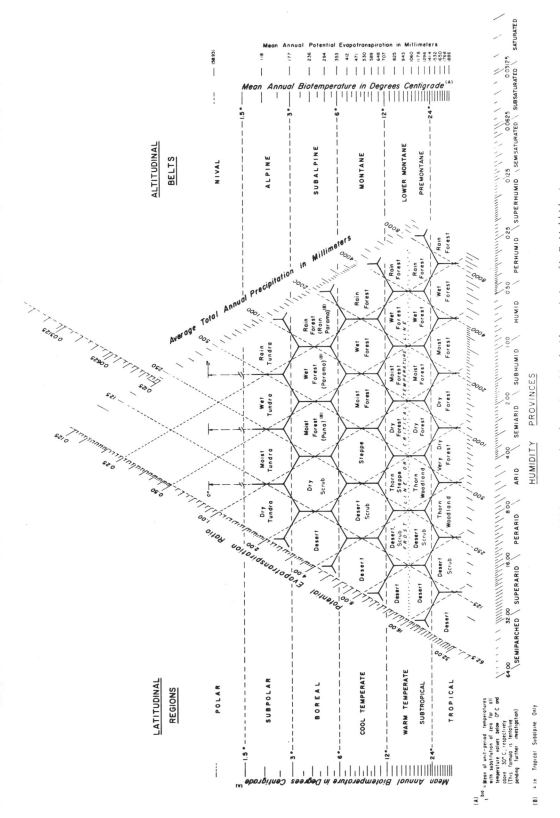

Figure 17.1. Diagram for the Holdridge classification of world life zones or plant formation. (Used with permission of L. R. Holdridge.)

4000 species of higher plants (Spellman, Dwyer, and Davidse 1975; Dwyer and Spellman 1981). The rich tree flora in some lowland sites (Table 17.1) is indicative of the plant species richness in local areas of Mesoamerica, as well as altitudinal patterns.

Protected Mesoamerica forests are major attrac-

tions for research on tropical vegetation (Leigh, Rand, and Windsor 1982; Estrada and Coates-Estrada 1983; Janzen 1983a; Clark, Dirzo, and Fetcher 1987; Leigh, Rand, and Windsor 1990; McDade, Bawa, Hespenheide, and Hartshorn 1994). It is no surprise that much of the information

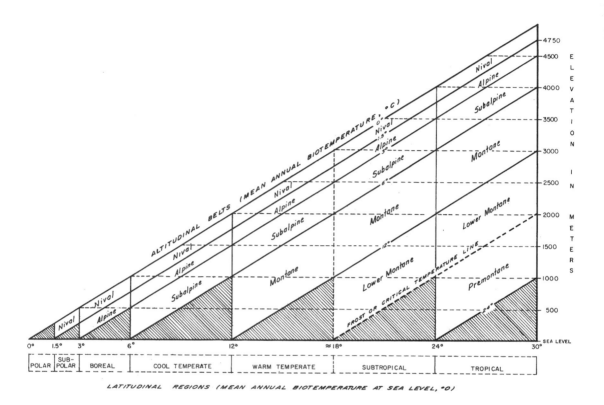

Figure 17.2. Approximate guideline positions of latitudinal regions and altitudinal belts of Holdridge's world life zone system. Based on lapse rates of 6°C per 1000 m elevation. Shaded areas represent basal belt positions. (Used with permission of L. R. Holdridge.)

Table 17.1. Tree flora richness in Mesoamerican lowland forests. Data on "species per hectare" column are based on inventories of trees >10 cm dbh on one or more hectares; data on "species" column are based on trees >5 m tall or >10 cm dbh. Legumes have been divided into three families.

Site and country	Life zone	Area (ha)	Species per ha	Species	Genera	Families	Reference
Deininger National Park, El Salvador	Subtrop. moist	732	45	143	105	48	Witsberger et al. (1982)
Palo Verde Wildlife Refuge, Costa Rica	Tropical dry	4,757	52	152	116	51	Hartshorn and Poveda (1983)
Santa Rosa National Park, Costa Rica	Tropical dry	10,700	—	206	162	60	Hartshorn and Poveda (1983)
Barro Colorado Island, Panama	Tropical moist	1,560	93	362	221	67	Croat (1978)
La Selva Biological Station, Costa Rica	Tropical wet	1,330	98	448	224	65	Hartshorn and Hammel (1982)

and many of the references in this chapter are from a few research sites: La Selva, Monteverde, Palo Verde, and Santa Rosa in Costa Rica; Barro Colorado Island (BCI) in Panama; and lowland Veracruz in Mexico. Increasing use of the Mexican biological stations at Chamela and Los Tuxtlas is contributing new information on different types of vegetation.

Even though Mexico has not been mapped using Holdridge's life-zone classification system, it can be assumed that all the subtropical life zones represented in Guatemala (which have been mapped) also occur in Mexico. Thus, approximately 25 life zones occur in the tropical and subtropical regions of Mesoamerica. For comparison, the eastern United States and and the central United States have 11 life zones (Sawyer and Lindsey 1964). Rather than treat each life zone of Mesoamerica in this short chapter, major vegetation types (Table 17.2) are grouped into three humidity regimes – perhumid (including superhumid), humid, and subhumid (including semiarid) – within three altitudinal zones: lowlands (0–500 m), low mountains (500–2000 m), and high mountains (>2000 m). These three altitudinal zones are used strictly as a framework to assist the reader rather than as an indication of specific altitudinal limits for any vegetation type. I agree with Standley (1937) and Holdridge (1967) that it is impossible to assign regional altitudinal limits to vegetation types.

Lowlands

For our purposes, lowlands are arbitrarily defined as occurring from sea level to roughly 500 m in elevation, a zone that includes tropical and subtropical basal latitudinal regions (Fig. 17.2), plus some of the lower-elevation tropical premontane altitudinal belt.

Mangrove forests. Nothing is more typical of the tropics than mangroves along the coasts of many low-latitude regions of the world. (See Chapter 13 for a broad overview of this pantropical type of vegetation.) Mangroves typify low-energy coasts, deltas, and estuaries, but they do not occur along beaches with high wave action (Lugo and Snedaker 1974). Because of much higher daily fluctuations of tides on the Pacific coast of Mesoamerica and the low base flow of rivers during the long dry season, mangrove forests extend inland for several kilometers and form extensive stands in the major Pacific coast deltas (Golfo de San Miguel, Río Grande de Térraba, Río Grande de Tárcoles, Golfo de Nicoya, Golfo de Fonseca, Bahía de Jiquilisco).

In contrast, the Caribbean coast has lower daily tidal fluctuations, much greater input of fresh water, and a less severe dry season. Hence, on the Caribbean coast, mangrove forests occur in a narrower band and penetrate only a few hundred meters inland along rivers. As rainfall decreases and seasonality increases, there is a marked decrease in mangrove height (Lot-Helgueras, Vásquez-Yanes, and Menéndez 1975; Pool, Snedaker, and Lugo 1977; Hartshorn et al. 1984). The tallest (35–40 m) mangrove forests occur in the tropical wet forest life zone, with rainfall greater than 4000 mm yr.$^{-1}$

Mesoamerica mangroves, characterized by eight tree species in five genera and four families, are not nearly as species-rich as are Old World mangroves (Chapman 1975). Best known is the red mangrove, *Rhizophora mangle* (Rhizophoraceae), with its impressive arching stilt roots forming an almost impenetrable barrier at the sea edge of mangrove forests. On reaching the substrate, the aerial roots branch profusely; up to 50% of the lower root volume is gas space (Gill and Tomlinson 1977), which facilitates root aeration in the anaerobic mud (Scholander, Van Damn, and Scholander 1955).

Rhizophora mangle, Avicennia germinans (Avicenniaceae), *Pelliciera rhizophorae* (Theaceae), *Conocarpus erecta* and *Laguncularia racemosa* (Combretaceae) usually form distinct zones along a salinity gradient (Cintron, Lugo, Pool, and Morris 1978; Ellison and Farnsworth 1993; Molina Lara and Esmeralda Esquivel 1993). Ecologists generally interpret mangrove zonation as representing a successional sequence, but Lugo (1980) argues that mangroves are steady-state ecosystems that undergo cyclic succession. Rabinowitz (1978) found that the dispersal properties of mangrove propagules correlate well with the spatial distribution of adult mangroves. Mangrove seedlings grow equally well or better when planted under different mangrove species than when under conspecifics (Rabinowitz 1975); thus, tidal sorting of available propagules may influence mangrove zonation. In the Tivives mangrove forest on the central Pacific coast of Costa Rica, all four mangrove species fruit during the rainy season when water levels within the forest are maximum (Jiménez 1988).

Freshwater swamp forests. In the wet lowlands, freshwater swamp forests often occur inland of mangrove forests, where salinity no longer restricts floristic composition. The frequency, depth, and duration of flooding appear to be important determinants of species composition and dominance in swamp forests (Budowski 1966; Holdridge, Grenke, Hatheway, Liang, and Tosi 1971). As with mangrove forests, some freshwater swamp forests

Table 17.2. *Relation of Mesoamerican vegetation types to Holdridge life zones. Standard acroynyms appear in parentheses following each vegetation type's name.*

This chapter	Holdridge life zones
Lowlands (0–500 m)	
mangrove forests (Lmf)	Tropical dry forest
	Tropical moist forest
	Tropical wet forest
	Subtropical dry forest
	Subtropical moist forest
	Subtropical wet forest
freshwater swamp forests (Lfsf)	Tropical moist forest
	Tropical wet forest
	Tropical premontane wet forest
	Subtropical wet forest
perhumid forests (Lpf)	Tropical wet forest
	Tropical premontane wet forest
	Subtropical wet forest
humid forests (Lhf)	Tropical moist forest
	Tropical premontane moist forest
	Subtropical moist forest
subhumid forests (Lsf)	Tropical dry forest
	Subtropical dry forest
	Subtropical thorn woodland
savannas (Ls)	Tropical dry forest
	Tropical moist forest
	Tropical premontane moist forest
	Subtropical moist forest
Low mountains (500–2000 m)	
subhumid forests (LMsf)	Subtropical thorn woodland
	Subtropical dry forest
	Subtropical lower-montane dry forest
humid forests (LMhf)	Tropical premontane moist forest
	Tropical lower-montane moist forest
	Subtropical moist forest
	Subtropical lower-montane moist forest
prehumid forests (LMpf)	Tropical premontane wet forest
	Tropical premontane rain forest
	Tropical lower-montane wet forest
	Tropical lower-montane rain forest
	Subtropical lower-montane wet forest
	Subtropical lower-montane rain forest
High mountains (>2000 m)	
humid forests (HMhf)	Tropical montane moist forest
	Subtropical montane moist forest
	Subtropical subalpine moist forest
perhumid forests (HMpf)	Tropical montane wet forest
	Tropical montane rain forest
	Subtropical montane wet forest
páramo (HMp)	Tropical subalpine rain páramo

are nearly pure stands of one tree species (Table 17.3). Two characteristic swamp trees (Fig. 17.3), *Mora oleifera* (Caesalpiniaceae) and *Pterocarpus officinalis* (Fabaceae), also occur occasionally in mangrove forests. *P. officinalis* may form a pure stand of large trees, such as those near the mouth of the Río Llorona in Costa Rica's Corcovado National Park (Janzen 1978). It is unclear how the Llorona stand became established, but it probably was not induced by humans.

Swamp forests of cativo, *Prioria copaifera* (Caesalpiniaceae), are truly impressive because of the

Table 17.3. *Lowland freshwater swamp forest, site 8B. Total absolute basal area (BA) = 7.0 m² for the 0.2 ha sample; total absolute density (D) = 47 trees (defined as >10 cm dbh); canopy height = 26 m. Seven tree species present. Importance value percentage (IV) = (relative BA + relative D + relative frequency)/3. Osa, Costa Rica.*

Tree species	%BA	%D	%F	%IV
Mora oleifera	95	70	48	71
Amphitecna latifolia	1	15	19	12
Avicennia germinans	2	4	10	5
Tabebuia rosea	1	4	10	5
Luehea seemannii	0.2	2	5	2
Subtotal top five species	100	96	90	95

Source: Holdridge et al. 1971.

abundance of large trees (Table 17.4). Unfortunately, all pure cativo stands have been logged for plywood core stock (Mayo Meléndez 1965). Cativo seems to do best where there are strong seasonal pulses of flooding, with substantial intervals of moderately good drainage. Interestingly, cativo is abundant (27–29 stems ha⁻¹ >10 mm dbh [diameter at breast height]) in undisturbed forest on BCI (Condit, Hubbell, and Foster 1993). Along the Caribbean, cativo occurs from Colombia to Nicaragua, whereas it is restricted to Panama and Colombia on the Pacific coast. Cativo is surprisingly absent from southwestern Costa Rica.

Even more restricted in occurrence is the oré tree, *Campnosperma panamensis* (Anacardiaceae), which forms pure stands in the Laguna de Chiriquí region of Panama's Bocas del Toro province (Holdridge and Budowski 1956), as well as along coastal Costa Rica and Nicaragua. The oré swamp forests have exceptionally high volume, averaging 382 m³ ha⁻¹ for stems greater than 40 cm dbh (Falla 1978).

Palms dominate some swamp forests on the coastal plains (Fig. 17.4). The yolillo palm, *Raphia taedigera* (Palmae), is common in southwestern and northeastern Costa Rica (Table 17.5). In a detailed study of the Tortuguero swamp forests, Myers (1981) showed that *Manicaria saccifera* (Palmae) dominates the more interior palm swamps. In northeastern Costa Rica, *Raphia* lines the natural levees along slow-moving rivers. It also occurs in natural sloughs farther inland (e.g., in the Sarapiquí region of Costa Rica) that are inundated during heavy rains with water backed up by flooding rivers (Anderson and Mori 1967).

More heterogeneous swamp forests occur farther inland or peripheral to the pure stands described earlier. At the La Selva Biological Station in northeastern Costa Rica, the primary swamp forest is dominated by *Pentaclethra macroloba* (Mimosaceae) and *Carapa nicaraguensis* (Meliaceae). A 1 ha plot had 79 species of trees and lianas (≥ 10 cm dbh), which is only about 20 species ha⁻¹ fewer than in nearby firm-ground forest (Table 17.1). In the same La Selva swamp forest there are dramatic reductions in stem density and species richness on sites with poorer drainage because of the exclusion of tree species intolerant of waterlogged soils (Lieberman, Lieberman, Hartshorn, and Peralta 1985). Several typical swamp tree species also occur in lower densities in nonswamp habitats. The absence of an effective dry season may permit swamp species like *Pentaclethra* to grow on well-drained ridges (Hartshorn 1983b). The occurrence of cativo on seasonally dry BCI (Croat 1978) is more difficult to interpret.

Most of the dominant tree species in swamp forests have large seeds that float (Janzen 1983b; McHargue and Hartshorn 1983). Many large seeds or fruits are common drift debris on tropical beaches (Gunn and Dennis 1976). The coconut palm, *Cocos nucifera*, is so widely dispersed along tropical coasts that it is difficult to determine its geographical origin (Gruezo and Harries 1984). The large cotyledonary reserves of swamp species' seeds not only enhance viability while floating but also may facilitate better anchoring of the seedling in soft soil or raise the first leaves above the typical level of floodwaters.

Perhumid forests. Lowland perhumid forests are often called tropical rain forest (*sensu* Richards 1952). A tropical rain forest life zone (*sensu* Holdridge 1967) does not occur in Mesoamerica; hence, our lowland perhumid forests fall mostly in the tropical wet forest, tropical premontane wet forest, and subtropical wet forest life zones. These lowland perhumid life zones are extensive along the Caribbean coast of western Panama, northeastern Costa Rica, much of southeastern Nicaragua, northern Honduras, from the lower Motagua Valley of Guatemala northwest to the Transversal, and the Caribbean lowlands of Veracruz, Mexico. Perhumid life zones also occur in the Pacific lowlands of Guatemala and southern Costa Rica. These two isolated perhumid areas on the Pacific side of Mesoamerica may be caused by tall mountains (>3000 m) partially blocking the northeast tradewinds and creating a vortex that brings moisture-laden air inland off the Pacific Ocean.

Lowland perhumid forests (Fig. 17.5) are the most species-rich plant communities in Mesoamerica (Table 17.1). The highest diversity of trees occurs at 300–500 m on the northern slopes of Volcán Barva (Lieberman, Lieberman, Peralta, and

Figure 17.3. Freshwater swamp forest in the Darien of Panama. The buttressed trees are Pterocarpus officinalis, *and the nonbuttressed trees in the foreground are* Prioria copaifera.

Hartshorn 1996). Even where there is moderate dominance by one species (Table 17.6), the heterogeneous forest averages about 100 tree species ha⁻¹ (Lieberman et al. 1985). Palms, which may be subcanopy trees (Fig. 17.6), understory treelets, or dwarf species only a meter tall, are a conspicuous component of lowland perhumid forests. Seven palm species constitute 25.5% of the stems (≥10 cm dbh) on 12.4 ha of permanent inventory plots at La Selva (Lieberman et al. 1985).

Lowland perhumid forests are generally considered to have three tree strata – canopy, subcanopy, and understory – though it is often difficult or impossible to recognize precisely distinct layers. Even if discrete tree strata do exist, the functional significance of layers for light interception, pollination, seed dispersal, and other processes has not yet been studied. Some canopy tree species are facultatively deciduous for a few weeks in response to a short-term moisture deficit (Borchert 1980; Reich and Borchert 1982). A few canopy species like *Ceiba pentandra* (Bombacaceae) and *Tabebuia guayacan* (Bignoniaceae) are deciduous for a few months, during which time massive flowering occurs for a few days.

Dependent life forms such as epiphytes (Fig.

Table 17.4. Lowland freshwater swamp forest, site 19C. Total absolute basal area (BA) = 16.5 m² for the 0.3 ha sample; total absolute density (D) = 87 trees; canopy height = 47 m. Five tree species present. Importance value percentage (IV) = (relative BA + relative D + relative frequency)/3. Osa, Costa Rica.

Tree species	%BA	%D	%F	%IV
Prioria copaifera	95	80	50	75
Stemmadenia donnell-smithii	1	13	29	14
Pithecellobium latifolium	3	3	11	6
Grias fendleri	0.1	2	7	3
Ixora finlaysoniana	0.1	1	4	2
Subtotal top five species	100	100	100	100

Source: Holdridge et al. 1971.

17.7), woody climbers (lianas), herbaceous vines, and stranglers are abundant in lowland perhumid forests. Epiphytic orchids, ferns, bromeliads, gesneriads, and so forth are most prevalent on the branches of canopy trees (Perry 1978). Even cacti occur as epiphytes in lowland perhumid forests. Lianas are well represented in the families Apocynaceae, Bignoniaceae, Dilleniaceae, and Marcgraviaceae. The largest may attain 35–40 cm dbh and be more than 100 m long as they spread from crown to crown through the canopy. Climbing vines are epitomized by the aroids, with dozens of species of *Anthurium, Monstera, Philodendron,* and *Syngonium* (Strong and Ray 1975; Ray 1983a, 1983b).

Though some stranglers belong to the genus *Clusia* (Clusiaceae), most are figs (*Ficus,* Moraceae). A strangler begins life as an epiphyte whose aerial roots clasp the host tree and eventually form a stout lattice of coalesced aerial roots that can stand as an independent tree long after the host tree has died and decomposed. The popular notion of "strangling" has never been proved scientifically; rather, I have noted that stranglers usually have as hosts shade-intolerant gap species. Hence, shading by the strangler crown may cause death of the shade-intolerant host tree.

Tree-fall gaps in the primary forest canopy are essential for successful regeneration of up to 50% of the tree species (Hartshorn 1980). Shade intolerance is more common (63%) among canopy tree species at La Selva, decreasing in subcanopy (43%) and understory (38%) tree species. Of 68 woody species at Los Tuxtlas classified for gap depend-

Table 17.5. *Lowland freshwater swamp forest, Tortuguero, Costa Rica. Total absolute basal area (BA) = 103.2 m² for the 0.3 ha sample; total absolute density (D) = 191 trees; canopy height = 15.3 m. Nine tree species present. Importance value percentage (IV) = (relative BA + relative D + relative frequency)/3.*

Tree species	%BA	%D	%F	%IV
Raphia taedigera	98	84	65	82
Pentaclethra macroloba	1	4	9	5
Grias fendleri	0.3	4	9	4
Crudia acuminata	0.1	3	7	3
Prioria copaifera	0.1	1	4	2
Subtotal top five species	99	96	93	96

Source: Myers (1981).

ence, 18 are obligate gap species and 32 are gap-dependent (Popma, Bongers, and Werger 1992). Detailed studies of gap-dependent species indicate that most of their regeneration must undergo a series of canopy openings and closings before successfully reaching the canopy (Denslow 1987; Clark and Clark 1993; Denslow and Hartshorn 1994).

Gaps (and mature trees in them) may serve as

Figure 17.4. Freshwater swamp forest in Corcovado National Park, Costa Rica. The dominant palm is Raphia taedigera.

Figure 17.5. Lowland perhumid forest at La Selva, Costa Rica. The emergent crown in the center of the photograph is Dipteryx panamensis.

Table 17.6. *Lowland perhumid forest, site 11. Total absolute basal area (BA) = 41.1 m² for the 0.8 ha sample; total absolute density (D) = 328 trees; canopy height = 46 m. Sixty-five tree species present. Importance value percentage (IV) = (relative BA + relative D + relative frequency)/3. Las Selva, Costa Rica, 60 m elevation.*

Tree species	%BA	%D	%F	%IV
Pentaclethra macroloba	28	16	11	19
Welfia georgii	3	13	11	9
Protium species	4	6	7	6
Guarea species	2	7	4	4
Carapa nicaraguensis	9	1	2	4
Subtotal top five species	45	44	34	41

Source: Holdridge et al. 1971.

a keystone habitat for frugivorous birds (Levey 1990). Existing gaps are often enhanced by later treefalls where canopy trees with asymmetrical crowns tend to fall on their heavy side (Young and Hubbell 1991; Young and Perkocha 1994). Using a long-term study of gap occurrence in three permanent inventory plots as a basis, Hartshorn (1978) calculated an average turnover rate of 118 yr for the La Selva primary forest. Gap-phase dynamics appear to be important in the maintenance of species-rich perhumid forests in Mesoamerica (Hartshorn 1980; Brokaw 1982, 1985; Lang and Knight 1983; Pickett 1983; Hubbell and Foster 1985a), as in other tropical areas (Whitmore 1984). Basing their conclusion on a 13 yr study of three permanent 1 ha plots in Panama, Milton, Laca, and Demment (1994) suggest that gap-neutral species replace gap-positive species in late-successional time.

Humid forests. Lowland humid forests in the tropical moist forest, tropical premontane moist forest, and subtropical moist forest life zones (Fig. 17.1) were originally the most extensive vegetation type in Mesoamerica. Large areas of these life zones occur in eastern, central, and western Panama, northwestern and northern Costa Rica, northeastern Nicaragua, eastern Honduras, northern Petén in Guatemala, northern Belize, and the southern Yucatán Peninsula in Mexico.

Tropical humid forests are nearly as tall and impressive as tropical wet forests (Table 17.7), whereas subtropical humid forests are shorter, with only two tree strata. Lowland humid forests have an appreciable proportion (25–50%) of the canopy deciduous during the dry season. Epiphytic orchids and gesneriads (Gesneriaceae) are less abundant than in lowland perhumid forests; however, supple, thin, woody vines are more abundant than in lowland humid forests (Holdridge et al. 1971). Putz (1984) counted 1597 climbing lianas on 1 ha of old-growth forest on BCI. Liana recruitment to the forest canopy is usually on the edge of tree-fall gaps and by vegetative sprouting (Peñalosa 1984; Putz 1984). Robust palms, especially *Scheelea rostrata*, are common subcanopy trees. Because adults are fire-tolerant and juveniles build a substantial trunk below ground, *Scheelea* is an aggressive colonizer of deforested land (Sousa 1964). *S. rostrata* is a conspicuous component of pastures in western Panama, Costa Rica's El General Valley and Nicoya Peninsula, the northern foothills of Honduras, and southeastern Mexico.

The seasonality of rainfall and drought not only determines many community functions and ecosystem processes in lowland humid forests but also

Figure 17.6. Stilt roots supporting the subcanopy palm
Socratea durissima at La Selva, Costa Rica.

has profound effects on faunal populations (Leigh et al. 1982). Particularly striking is the strong synchronization of massive flowering with the dry season (Janzen 1967; Frankie, Baker, and Opler 1974; Croat 1975; Opler, Frankie, and Baker 1976; Foster 1982a). In several seasonal tropical forests, leaf and flower production have evolved to coincide with seasonal peaks of irradiance (Wright and Schaik 1994). Dry-season flowering while tree crowns are leafless facilitates pollination by large bees (Janzen 1971; Frankie 1975; Frankie, Opler, and Bawa 1976) and by wind (Bullock 1994). Borchert (1983) found that seasonal moisture stress is the proximal cause of dry-season flowering. The lessening of xylem tension in stems by leaf shedding is a prerequisite for flowering and bud break (Borchert 1994a). Dry-season stem water content was lowest in heavy hardwoods and highest in light hardwoods (Borchert 1994b), but all rehydrate very rapidly in the wet season (Borchert 1994c). An unusually wet 1970 dry season on BCI limited the normal pulse of tree flowering at the start of the rainy season, causing a threefold decrease in rainy-season fruiting and a severe famine among frugivores (Foster 1982b).

The pioneering studies on spatial distributions of Mesoamerican tree species by Hubbell (1979) and Hubbell and Foster (1983, 1985a, 1985b) indicate that very few species have the classic hyper-dispersion so often suggested in the literature (e.g., Richards 1952). Many tree species have a random

Figure 17.7. Tremendous epiphyte load on an emergent Tabebuia guayacan at La Selva, Costa Rica. The 45 m tall tree is decidous during the brief dry season.

Table 17.7. *Lowland humid forest, site 2A. Total absolute basal area (BA) = 28 m² for the 0.6 ha sample; total absolute density (D) = 304 trees; canopy height = 45 m. Forty-nine tree species present. Importance value percentage (IV) = (relative BA + relative D + relative frequency)/3. Barranca, Costa Rica, 40 m elevation.*

Tree species	%BA	%D	%F	%IV
Scheelea rostrata	35	33	23	30
Luehea seemannii	22	11	10	14
Bravaisia integerrima	3	8	7	6
Enterolobium cyclocarpum	8	1	1	3
Spondias mombin	1	3	4	3
Subtotal top five species	70	56	44	57

Source: Holdridge et al. 1971.

Figure 17.8. Lowland subhumid forest at Palo Verde, CostaRica, at the beginning of the dry season. The trees *with light-colored crowns are* Calycophyllum candidissimum *in flower.*

dispersion of adults, and some species are clumped. The dispersion patterns of tree species have important consequences for pollination syndromes (Janzen 1971; Frankie 1975; Feinsinger 1976; Heithaus 1979; Bawa, Bullock, Perry, Colville, and Grayum 1985), tree breeding systems (Dobzhansky 1950; Baker 1970; Bawa 1974, 1980; Bawa and Opler 1975, 1977; Frankie et al. 1976; Opler and Bawa 1978; Bawa and Beach 1983; Bawa, Perry and Beach 1985), herbivory (Janzen 1969, 1970a), and the gap-phase regeneration of shade-intolerant trees (Hartshorn 1978, 1980; Denslow 1980, 1987; Brokaw 1982, 1985; Augspurger 1984).

Mahogany, *Swietenia macrophylla* (Meliaceae), is largely restricted to lowland humid forests, where it is more abundant in the subtropical moist forest life zone than in the tropical moist forest life zone. South of Nueva Guinea, Nicaragua, mahogany drops out of natural forests as the life zone changes to tropical wet forest. Mahogany was once quite common in southern Yucatán and northern Guatemala and Belize, where it was the principal object of logging efforts for more than two centuries (Lamb 1966; Hartshorn et al. 1984). Consistent with its dependence on large-scale disturbance for successful natural regeneration (Hartshorn 1992; Ro-

dan, Newton, and Verrissimo 1992; Snook 1994), mahogany regeneration improves with increasing opening of the forest canopy in Quintana Roo, Mexico (Negreros-Castillo and Mize 1993).

The Mayan civilization that developed in the lowland humid areas of Belize, Guatemala, Honduras, and Mexico is believed to have favored useful trees such as mahogany, zapote (*Manilkara zapota,* Sapotaceae), and ramón (*Brosimum alicastrum,* Moraceae) (Lundell 1933; Lambert and Arnason 1978). The demise of the classic Mayan civilization occurred about 900 A.D. (Deevey et al. 1979; Turner and Harrison 1983). Thus, Mayan fields have been abandoned for more than a millennium. Recent studies of tropical forest dynamics suggest that 1000 yr should be ample time for succession to have attained climax status in the region. I have observed no major differences in mahogany abundance or stand structure between mature subtropical moist forests of northern Belize and those on the Beni Plain at the northern base of the Bolivian Andes, which suggests that the effects of Mayan agriculture have long since disappeared. Future demographic studies such as those done by Hartshorn in 1975 and Sarukhán in 1978 in the protected forests at the Tikal World Heritage Site could clarify the successional status of mahogany in these humid forests.

Subhumid forests. Lowland subhumid forests and woodland occur in the tropical dry forest (Fig. 17.8), subtropical dry forest, and subtropical thorn woodland life zones. This vegetation type was once common in a wide arc bordering a large area: Panama's Bahía de Parita; Costa Rica's lower Tempisque Valley; north and west of Lake Managua; extensive areas around the Golfo de Fonseca shared by Nicaragua, Honduras, and El Salvador; much of El Salvador's La Unión department, eastern San Vicente and Cabañas departments, and the upper Río Lempa Valley in Chalatenango and Santa Ana departments; substantial areas in Guatemala's Baja Verapaz, El Progreso, Chiquimula, and Zacapa departments; Mexico's northern Yucatán Peninsula; and along the central Pacific lowlands. Because of a long history of extensive grazing, frequent burning, and some agriculture, nearly all subhumid forests have been eliminated or severely degraded. Costa Rica's Santa Rosa National Park, Palo Verde Wildlife Refuge and National Park, and Mexico's Chamela Biological Station are the only three protected areas with substantial representative vegetation of lowland subhumid forests in Mesoamerica. Tree species richness is substantially lower in these forests than in lowland humid and perhumid forests (Tables 17.1 and 17.8).

Lowland subhumid forests are seasonally deciduous (Fig. 17.9), with only two tree strata, the overstory of which seldom exceeds 20 m in height (e.g., Bullock and Solís-Magallanes 1990). The shrub layer is often dense with thorny stems. Thin, woody vines are common, but epiphytes are sparse. Columnar cacti (e.g., *Lemaireocereus aragonii*) are occasional on xeric sites. *Jacquinia pungens*

Table 17.8. *Lowland subhumid forest. Total absolute basal area (BA) = 79.8 m² for this 4.0 ha sample; total absolute density (D) = 691 trees; canopy height = 20 m. Sixty-eight tree species present. Importance value percentage (IV) = (relative BA + relative D + relative frequency)/3. Palo Verde, Costa Rica, 10 m elevation.*

Tree species	%BA	%D	%F	%IV
Calycophyllum candidissimum	21	6	6	11
Licania arborea	7	6	6	6
Brosimum alicastrum	9	4	3	5
Spondias mombin	5	5	5	5
Guazuma ulmifolia	3	6	5	5
Subtotal top five species	46	28	24	33

Source: Hartshorn (1983a).

(Theophrastaceae) is a small understory tree that is uniquely deciduous during the rainy season, but in the dry season it takes advantage of sunlight passing through the barren overstory to store appreciable photosynthate, as well as to flower and fruit (Janzen 1970b). While leafless during the rainy season, an individual can lose up to 50% of the starch stored in the stem (Janzen and Wilson 1974).

Taller, evergreen forests occur along streams or in limestone basins where groundwater is available. Subhumid riparian forests have several species (e.g., *Anacardium excelsum*) from nonriparian habitats in wetter life zones (Table 17.9). Some characteristic tree species of hillside subhumid forests have the opposite habitat pattern: *Tabebuia rosea* (Bignoniaceae) and *Bursera simaruba* (Burseraceae) are three times larger on alluvial soils in perhumid forests (where they attain 100 cm dbh and 40 m in height) than in subhumid forests.

In contrast to the monospecific stands in some freshwater swamps mentioned earlier, pure stands in subhumid and some humid lowlands are usually attributable to edaphic factors, such as pumice soil or montmorillonite clay. The calabash tree (*Crescentia alata*, Bignoniaceae), *Erythrina fusca* (Fabaceae), palo verde (*Parkinsonia aculeata*, Caesalpiniaceae), and a white oak (*Quercus oleoides*, Fagaceae) often form pure stands or dominate forests on restrictive sites (Holdridge et al. 1971; Hartshorn 1983a). *Quercus oleoides* is the only lowland oak in Mesoamerica; it occurs in a series of disjunct populations that extend from Guanacaste, Costa Rica, to Tamaulipas, Mexico (Montoya Maquín 1966). In northwestern Costa Rica it once dominated (Table 17.10) soils derived from pumice, but most of these forests have been converted to rangeland with the naturalized African savanna grass *Hyparrhenia rufa* (Daubenmire 1972). *Q. oleoides* has thick bark that effectively protects it from dry-season fires that sweep the rangeland. Seedlings can also survive fire because the seed reserves are translocated to a tuberous radicle extending 4–8 cm below the soil surface (Boucher 1983).

Savannas. Mesoamerican savanna habitats are most prevalent in the tropical dry forest, tropical moist forest, subtropical dry forest, and subtropical moist forest life zones. By definition, savanna has a continuous grass cover (Beard 1953). According to the UNESCO classification system, trees may form up to 30% of the plant cover in savannas (Mueller-Dombois and Ellenberg 1974: 479). Considerable areas of Mesoamerican vegetation meet these criteria for savanna; however, it is not clear if they all are forest or woodland degraded through repetitive burning by humans (Cook 1909;

Figure 17.9. Interior of a lowland subhumid forest at Palo Verde, Costa Rica, late in the dry season. Most trees and shrubs are deciduous. Note the accumulation of litter on the forest floor.

Budowski 1956; Sauer 1958). Regardless of origin, Mesoamerican savannas are neither as vast nor as ecologically important as the Colombian-Venezuelan llanos and the pampas of the Bolivian Beni.

The rangelands in the Pacific lowlands of Mesoamerica appear to be derived by annual burning to remove the rank old growth of grass and to suppress woody invasion. Many of these derived savannas are dotted with fire-resistant tree species such as *Acrocomia vinifera* (Palmae), *Byrsonima crassifolia* (Malpighiaceae), *Crescentia alata*, and *Curatella americana* (Dilleniaceae). The first three can be found occasionally in remnant subhumid forests, but *Curatella* is restricted to xeric savanna habitats where the soil is very shallow over pumice or a plinthic hardpan.

Lowland pine (*Pinus caribaea* var. *hondurensis*) savannas occur in northeastern Nicaragua, much of Honduras, part of the Guatemalan Petén, and east-central Belize. In contrast to the derived savannas of the Pacific lowlands, pine savannas appear to be natural associations maintained by frequent fire. The flat pine savannas of Belize occur on sandy soils that usually flood during the rainy season

(Wright et al. 1959). The seasonal aridity of sandy soils make the vegetation highly flammable. Palmetto, *Paurotis wrightii* (Palmae), is a conspicuous component of these savannas (Anderson and Fralish 1975).

Inland from the Miskito coast of Nicaragua and adjoining Honduras, extensive pine savannas occur on gravelly outwash and finer quartz sediments (Parsons 1955; Taylor 1962, 1963; Alexander 1973). The infertile soil, in conjunction with frequent burning, delimits pine savanna in northeastern Nicaragua (Alexander 1973). The presence of pine savanna and broadleaved savanna in humid and subhumid life zones indicates that climate is not the primary determinant of Mesoamerican savannas. Rather, edaphic factors such as soil infertility and shallowness, coupled with frequent burning, are the important factors. Kellman (1984) suggests that vegetation on infertile soils is more susceptible to fire than on fertile soils, because slower growth delays canopy closure, thus permitting persistence of herbaceous grasses that carry fire. Dicot tree species in savannas are important to building the nutrient pool in woody biomass during fire-free intervals (Kellman 1989).

Table 17.9. *Riparian lowland subhumid forest, site 1F. Total absolute basal area (BA) = 14 m² for this 0.4 ha sample; total absolute density (D) = 67 trees; canopy height = 33 m. Eighteen tree species present. Importance value percentage (IV) = (relative BA + relative D + relative frequency)/3. Taboga, Costa Rica, 10 m elevation.*

Tree species	%BA	%D	%F	%IV
Luehea seemannii	22	15	16	18
Guarea excelsa	10	21	15	15
Scheelea rostrata	11	16	15	14
Terminalia oblonga	5	10	11	9
Anacardium excelsum	13	4	5	8
Subtotal top five species	61	67	62	63

Source: Holdridge et al. 1971.

Table 17.10. *Lowland subhumid oak forest. Total absolute basal area (BA) = 50.7 m² for this 4.0 ha sample; total absolute density (D) = 814 trees; canopy height = 17 m. Forty-four tree species present. Importance value percentage (IV) = (relative BA + relative D + relative frequency)/3. Bagaces, Costa Rica.*

Tree species	%BA	%D	%F	%IV
Quercus oleoides	58	34	18	37
Byrsonima crassifolia	6	12	11	10
Apeiba tibourbou	6	11	11	9
Spondias mombin	7	8	8	7
Cordia alliodora	2	5	7	5
Subtotal top five species	79	70	55	68

Source: Hartshorn (1983a).

Low Mountains

For purposes of this chapter, low mountains range from 500 m to 2000 m. It is important to remember that each pair of life zones in subtropical and warm temperate latitudinal regions, or in premontane and lower-montane altitudinal belts (Fig. 17.1), has similar vegetation physiognomy. For example, subtropical wet forest and subtropical lower-montane wet forest life zones have similar stand structures but differ floristically. The critical-temperature line (Fig. 17.1) separates species by their tolerance to cool temperatures. The upper limit for coffee plantations in Mesoamerica coincides well with the critical-temperature line separating the tropical premontane belt from the lower-montane altitudinal belt, and the subtropical basal belt from the subtropical lower-montane altitudinal belt (Holdridge 1967).

The similarity in vegetation physiognomy and the high species richness within a hexagon make it difficult to detect the change between two life zones separated by the critical-temperature line (Fig. 17.1). In northern Mesoamerica, the presence of such temperate forest genera as *Almus, Carpinus, Cornus, Nyssa, Ostrya, Platanus,* and *Ulmus* is a good indication of the subtropical lower-montane altitudinal belt (i.e., above the critical temperature elevation). In southern Mesoamerica, several tree genera have a few closely related species that are geographically separated by the critical-temperature line: *Billia* (Hippocastanaceae); *Brunellia* (Brunelliaceae); *Byrsonima* (Malpighiaceae); *Cedrela* (Meliaceae); *Didymopanax* (Araliaceae); *Hedyosmum* (Chloranthaceae); *Pithecellobium* (Mimosaceae); *Rapanea* (Myrsinaceae); *Symplocos* (Symplocaceae); and *Turpinia* (Staphylaeaceae). An example is *Billia colombiana* which occurs from about 300 m to 2000 m elevation, whereas *B. hippocastanum* occurs above 2000 m on Volcán Barva.

Subhumid forests. This vegetation type was once extensive on the lower-to middle-elevation tablelands and valleys that occur over much of northern Mesoamerica. Intermountain valleys usually are in the subtropical thorn woodland or subtropical dry forest life zones, whereas the higher slopes are in the subtropical lower-montane dry forest life zone. These cooler tablelands have been inhabited by humans for centuries and have been severely degraded by overgrazing and fire. The relatively open forests are generally of short stature (<10 m tall), with an abundance of spiny, mimosoid legumes such as *Pithecellobium dulce, P. flexicaule,* and *Prosopis laevigata.* In Mexico, this vegetation type is called "low, spiny, evergreen forest" or "low, spiny, deciduous forest" (Miranda and Hernández 1963; see also Chapter 15). *Cercidium praecox* is a characteristic tree of the subtropical thorn woodland life zone in Guatemala's Zacapa Valley (Holdridge 1956). On upper slopes around Chimaltenango, Guatemala, *Pinus montezumae* is common (Holdridge 1957b).

Humid forests. Tropical premontane moist forest, subtropical moist forest, and subtropical lower-montane moist forest life zones are widespread on the low mountains of Mesoamerica. In the northern region there is a complex mosaic of pine forests on the upper slopes or tablelands and broadleaved forests in the valleys. On the middle slopes of the Sierra Madre Oriental of Mexico, the latter vegetation type is called "deciduous forest" (Miranda and Hernández 1963). It has a mixture of temperate

Table 17.11. *Low-montane humid forest, site 18. Total absolute basal area (BA) = 16.8 m² for this 0.5 ha sample; total absolute density (D) = 245 trees; canopy height = 23 m. Forty-four tree species present. Importance value percentage (IV) = (relative BA + relative D + relative frequency)/3. Alajuela, Costa Rica, 800 m elevation.*

Tree species	BA	D	F	IV
Luehea candida	29	15	9	18
Phoebe mexicana	6	11	11	9
Stemmadenia obovata	3	12	8	8
Roupala complicata	4	10	5	7
Anacardium excelsum	8	4	5	6
Subtotal top five species	50	52	39	47

Source: Holdridge et al. 1971.

Table 17.12. *Low-montane humid forest, site 3. Total absolute basal area (BA) = 17.7 m² for this 0.4 ha sample; total absolute density (D) = 236 trees; canopy height = 40 m. Seventy-one tree species present. Importance value percentage (IV) = (relative BA + relative D + relative frequency)/3. Turrialba, Costa Rica, 590 m elevation.*

Tree species	%BA	%D	%F	%IV
Brosimum alicastrum	25	17	11	18
Guarea aligera	5	9	7	7
Hymenolobium pulcherrimum	9	6	6	7
Simarouba amara	4	8	7	6
Anacardium excelsum	16	1	1	6
Subtotal top five species	58	41	32	44

Source: Holdridge et al. 1971.

trees (mentioned earlier) with tropical trees such as *Brunellia* (Brunelliaceae), *Beilschmiedia* and *Phoebe* (both Lauraceae), and *Alchornea* (Euphorbiaceae) (Pennington and Sarukhán 1968).

Low-mountain pine forests of Belize, Guatemala, Honduras, and Nicaragua usually are relatively pure stands of *Pinus oocarpa*, although some stands include *P. caribaea* (Denevan 1961; Johnson, Chaffey, and Birchall 1973). An active fire-control program by the Belize Forest Department has successfully established excellent 20–30-yr-old pine forests in the Mountain Pine Ridge (Hartshorn et al. 1984). Because of fire suppression, there is a vigorous understory of broadleaved trees, confirming the importance of fire in maintaining pine forests on these poor soils. Kellman (1979) reports that some broadleaved trees accumulate nutrients on poor sites in the pine forest and suggests that these locally enriched sites may facilitate invasion by broadleaved species typical of more fertile soils.

Low-mountain humid forests in southern Mesoamerica are restricted to intermountain valleys and Pacific slopes. This vegetation type has a relatively tall (<40 m), open, partially deciduous canopy (Tables 17.11 and 17.12). Because of accessibility and hospitable climate, virtually all of this vegetation type has been deforested, mostly for cattle pastures (Hartshorn et al. 1982; Heckadon Moreno 1983).

Perhumid forests. Because of the complex interactions of mountainous terrain and rainfall patterns, low-mountain perhumid forests occur in several life zones: tropical premontane wet forest, tropical premontane rain forest, tropical lower-montane wet forest, tropical lower-montane rain forest, subtropical lower-montane wet forest, and subtropical lower-montane rain forest life zones. Low-mountain perhumid forests occur in a fairly broad band along the "front ranges" (Fig. 17.10) exposed to the northeast tradewinds, as well as ringing most of the volcanoes, and on some of the Pacific middle slopes (e.g., Costa Rican Talamancas). Where the physiographic position is more or less perpendicular to the northeast tradewinds, luxuriant cloud forest usually occurs (Vogelmann 1973; Cruz and Erazo Peña 1977; Reyna Vásquez 1979; Lawton and Dryer 1980).

Low-mountain perhumid forests have a well-developed canopy 30–45 m tall (Tables 17.13 and 17.14), with occasional emergents 50–55 m tall, including *Ulmus mexicana* and *Quercus copeyensis*. In the Sierra Negra of Honduras, Vogel (1954) measured a *Pinus pseudostrobus* as having a diameter of 231 cm dbh and a total height of 60 m. Canopy trees are mostly evergreen, although a few are briefly deciduous. Several species of oaks and other temperate genera (mentioned earlier) are conspicuous components of the lower-montane canopy. These perhumid forests are characterized by a profusion of epiphytic bryophytes and vascular plants (Ingram and Nadkarni 1993; Bohlman, Matelson, and Nadkarni 1995). Trunks and branches are festooned with mosses, leaves are covered with epiphylls, and even epiphytic trees and shrubs are abundant. There is evidence that nitrogen fixed by free-living epiphyllous microorganisms is absorbed through the host's leaves (Bentley 1984).

The variations in vegetation structure and floristic composition on tropical mountains have long attracted the attention of plant ecologists (Shreve 1914; Beard 1949; Holdridge et al. 1971). Nevertheless, considerable controversy still exists over the physical factors controlling the distribution of forest types (Leigh 1975; Grubb and Tanner 1976; Grubb 1977). In a study of low-mountain perhumid forest in Jamaica, Tanner and Kapos (1982; Kapos and Tanner 1985) found that leaves are smaller and thicker in the cloud forest than in the lowlands, but the xeromorphic cloud forest leaves do not seem to be an adaptation to moisture limitation (Kapos and Tanner 1985).

A detailed study of tropical forest structure and composition (Lieberman et al. 1996) along a large-scale altitudinal gratient (La Selva-Volcán Barva, Costa Rica) shows that canopy height is greatest at 300 m, decreasing both above and below that elevation and reaching a minimum near the summit (2600 m). Trees of large diameter were least abundant at middle elevations, whereas basal area was greatest near the summit. Tree diversity was also highest at 300 m. Species composition varied continuously with elevation, with no discontinuities or evidence of discrete floristic zones.

Table 17.14. *Low-montane perhumid forest, site 22E. Total absolute basal area (BA)* = 13.1 m² *for this 0.4 ha sample; total absolute density (D)* = 281 *trees; canopy height* = 38 m. *Sixty-six tree species present. Importance value percentage (IV)* = *(relative BA + relative D + relative frequency)/3. Valle Escondido, Costa Rica, 1100 m elevation.*

Tree species	%BA	%D	%F	%IV
Euterpe macrospadix	6	17	9	11
Oreomunnea mexicana	18	4	4	8
Inga species	6	9	6	7
Chrysophyllum mexicanum	9	5	5	6
Clusia pithecobia	7	5	4	5
Subtotal top five species	46	40	29	38

Source: Holdridge et al. 1971.

Figure 17.10. *Lower montane perhumid forest in the rugged mountains of Braulio Carrillo National Park, Costa Rica. Quaternary lava flows are deeply dissected by streams, such as the Rio Puerto Viejo and its tributaries shown here.*

Table 17.13. *Low-montane perhumid forest, site 7. Total absolute basal area (BA)* = 22.7 m² *for this 1.0 ha sample; total absolute density (D)* = 503 *trees; canopy height* = 44 m. *Sixty-seven tree species present. Importance value percentage (IV)* = *(relative BA + relative D + relative frequency)/3. Volcan, Costa Rica, 620 m elevation.*

Tree species	%BA	%D	%F	%IV
Vantanea barbourii	38	13	12	21
Socratea durissima	11	36	13	20
Brosimum utile	20	3	3	8
Dendropanax arboreus	2	5	7	5
Conostegia globulifera	1	4	5	3
Subtotal top five species	71	61	41	58

Source: Holdridge et al. 1971.

In the Monteverde Cloud Forest Reserve of Costa Rica (Hartshorn 1983a), the location, structure, and floristic composition of forest types are controlled mainly by degree of exposure to the northeast tradewinds (Lawton and Dryer 1980). The northeast tradewinds roar up the Peñas Blancas Valley to spill over the low (1500–1600 m) continental divide. Lawton and Dryer (1980) describe six forest types for the Monteverde Reserve: cove, leeward cloud, oak ridge, windward cloud, elfin, and swamp communities. Increasing exposure to the tradewinds increases rainfall and wind shear, which result in luxuriant epiphyte loads, shorter forest stature, and a more open canopy. The smoothly sculptured canopy of the elfin forest (Fig. 17.11) is an exception to the pattern of less continuous canopy with increasing exposure to the trade-

Figure 17.11. Elfin forest on the continental divide in the Monteverde Cloud Forest Reserve, Costa Rica. Wind shear by northeast tradewinds creates the smoothly sculptured forest canopy.

winds; gusts regularly exceed 100 km across the Brillante gap (Lawton and Dryer 1980).

The abundance of brownish epiphytic mosses (Fig. 17.12) gives such a distinctive appearance to lower-montane wet forest and rain forest canopies that it is possible to recognize them from a low-flying airplane. Also notable is the abundance of ericaceous epiphytic shrubs (*Cavendishia, Gaultheria, Gonocalyx, Macleania, Psammisia, Satyria*). The profusion of epiphytes in cloud forests results in a considerable amount of organic matter on the branches. Breakage of heavily laden branches is a common source of canopy gaps (Lawton and Dryer 1980). Some host trees produce adventitious roots in the organic debris associated with epiphytes (Nadkarni 1984), thus partially short-circuiting the nutrient cycle.

High Mountains

High-mountain vegetation types generally occur above 2000–2500 m in Mesoamerica. The tropical and subtropical montane altitudinal belt is restricted to south-central Mexico, western Guatemala, Costa Rica's Cordillera Volcánica Central, and the high Talamancas of Costa Rica and Pan-

ama. High mountain forests tend to have the canopy dominated by Holarctic genera (e.g., conifers, oaks, alder, laurels), whereas the understory is a mixture of Holarctic and Neotropical elements (Luna-Vega et al. 1989; Quintana-Ascencio and González-Espinosa 1993; see also Chapter 15). The Talamancan tree flora has a closer affinity with Colombia than with Mexico (Kappelle, Cleef, and Chaverri 1992). Guatemala's highlands are the low-latitude center of diversity for conifers: *Abies, Cupressus, Juniperus* (two species), *Pinus* (nine), *Podocarpus* (two), *Taxodium*, and *Taxus* (Holdridge 1957a, 1975b; Veblen 1976). Guatemala's volcanic history (Williams 1960), rugged terrain, and altitudinal variation have contributed to the richness of pine species. Tropical and subtropical subalpine vegetation occurs only on the high peaks of southern Mexico, western Guatemala, and the Costa Rican Talamancas.

Humid forests. The high-mountain humid forests of northern Mesoamerica are characterized by coniferous trees not native to southern Mesoamerica. Subtropical montane humid forests are fairly open stands, usually dominated by *Juniperus standleyi, Pinus ayacahuite,* or *P. rudis* (Holdridge 1956). Human uses of natural resources have destroyed or degraded most native forests through logging, slash-and-burn agriculture for maize, and overgrazing by livestock. An interesting exception in the Guatemalan highlands is the Totonicapán furniture industry based on *Pinus ayacahuite* and dating to the sixteenth century (Veblen 1978). The remaining white pine forests are held communally and are vigorously protected by the communities of woodcutters and carpenters dependent on the forests for their livelihood.

The tropical montane moist forest life zone is restricted to small rainshadow areas in the high mountains of Costa Rica. These areas are mostly in agriculture, with remnant trees suggesting that the original forests were dominated by oaks (Table 17.15).

Pinus hartwegii, P. rudis, and *Juniperus standleyi* occur up to 4100 m elevation in northern Mesoamerica (Holdridge 1956; Veblen 1976). The occurrence of these high-mountain coniferous forests in the subtropical subalpine altitudinal belt (Fig. 17.2) indicates that the treeline is higher in northern than in southern Mesoamerica. The treeline in the Talamanca Mountains of Costa Rica occurs at about 3300–3500 m, above which is subalpine páramo. Longer days during summer apparently permit tree growth at higher elevations in extra-tropical regions.

Perhumid forests. *Abies guatemalensis* and *A. religiosa* often dominate the high-mountain perhumid forests of western Guatemala and south-central Mexico, respectively (Holdridge 1956; see Chapter 15). In the latitudinal transition zone between the Holarctic and Neotropical floristic regions, Velásquez (1994) found that altitude and soil moisture are the most important determinants of the major vegetation types dominated by *Pinus montezumae P. hartwegii, Alnus firmifolia,* and *Abies religiosa* on the Mexican volcanoes Tláloc and Pelado. The oyamel fir (*A. religiosa*) is an important source of timber and provides critical winter habitat for the migratory monarch butterfly (Conrad and Salas 1993). Also present in northern Mesoamerican perhumid forests are *Cupressus lusitanica, Pinus pseudostrobus,* and *Podocarpus oleifolius* (Veblen 1976), as well as mixed oak-pine forests.

The tropical montane wet forest and tropical montane rain forest life zones of the high Talamancas in Costa Rica are dominated by several species of oaks (Table 17.16). The oak forests have impressive structure, with a canopy (40–55 m) dominated by *Quercus copeyensis* and *Q. costaricensis* (timber volumes of 713 m³ ha⁻¹ and 573 m³ ha⁻¹ respectively), and the understory dominated by

Figure 17.12. Moss-laden branches in the Monteverde Cloud Forest Reserve of Costa Rica.

Table 17.15. High-montane humid forest, site 17. Total absolute basal area (BA) = 11.9 m² for this 0.3 ha sample; total absolute density (D) = 37 trees; canopy height = 33 m. Thirty-seven tree species present. Importance value percentage (IV) = (relative BA + relative D + relative frequency)/3. Irazu, Costa Rica, 2360 m elevation.

Tree species	%BA	%D	%F	%IV
Quercus copeyensis	66	32	19	39
Rapanea guianensis	4	11	12	9
Citharexylum lankesteri	6	10	10	9
Freziera candicans	5	8	10	8
Ocotea seibertii	3	8	9	7
Subtotal top five species	84	69	61	72

Source: Holdridge et al. 1971.

bamboos, *Chusquea talamancensis* and *C. tomentosa* (Blaser and Camacho 1991; Orozco Vílchez 1991). Large trees are heavily laden with epiphytes, particularly mosses, large tank bromeliads, and shrubby ericads. Many of these oak forests apparently are not regenerating, and the bamboo is hypothesized to be interfering with oak regeneration.

Páramo. In the Western Hemisphere, páramo is restricted to tropical subalpine regions of the Talamanca Mountains of Costa Rica and the northern Andes of South America (Salgado-Labouriau 1979). The Andean páramo is characterized by several *Espeletia* species (Compositae) absent from the Costa Rican páramo, but the floristic and physiognomic similarities are sufficient to make the Costa Rican páramo an outlier of the Andean páramo (Weber 1959). The rosette life form (both stem rosettes and ground rosettes) is a striking feature of these high, cold landscapes (Rundel et al. 1994).

Tropical subalpine rain páramo life zone of Costa Rica is a low scrub with occasional, taller trees. In the Cerro de la Muerte region, fires have lowered the treeline by 300–400 m (Fig. 17.13). Vegetative regeneration following fire is quite slow (Janzen 1973). Horn and Sanford (1992) dated Holocene charcoal in Cerro Chirripó soils to 2430 BP and 1110–1180 BP; interestingly, soil charcoal from La Selva (lowland perhumid forest) clusters around the same dates. During Quaternary glacials, the treeline was 700–900 m lower in the Talamanca Mountains (Martin 1964).

Several *Hypericum* species (Hypericaceae) and a miniature bamboo, *Swallenochloa subtessellata* (Gramineae), dominate the shrubby páramo. Where drainage is poor, bogs have an abundance of the cycad-like fern, *Blechnum buchtienii*, plus a species of the Andean fern *Jamesonia* and the terrestrial bromeliad *Puya dasylirioides*. High Talamancan vegetation also has strong floristic affinities with that of North America, with the presence of temperate genera such as *Alchemilla* (Rosaceae), *Castilleja* (Scrophulariaceae), *Cirsium* (Compositae), *Vaccinium* (Ericaceae).

Additional information about alpine vegetation in Mesoamerica can be found in Chapter 14.

CONSERVATION

Mesoamerica has been a key land bridge for biotic migrations between the North American and South American continents. Jaguars used to roam the southwestern United States, and white-tailed deer are common in that region as well as in South America. The sweet gum tree (*Liquidambar styraciflua*, Hamamelidaceae), so characteristic of the eastern deciduous forests, is a striking canopy component (to 150 cm dbh and 70 m tall) in the lowland humid forests of northern Mesoamerica. *Liquidambar* resin was an important Aztec tribute (Peterson and Peterson 1992). As mentioned ear-

Table 17.16. *High-montane perhumid forest, site 6. Total absolute basal area (BA) = 17.9 m² for this 0.4 ha sample; total absolute density (D) = 245 trees; canopy height = 30 m. Twenty-two tree species present. Importance value percentage (IV) = (relative BA + relative D + relative frequency)/3. Villa Mills, Costa Rica, 3080 m elevation.*

Tree species	%BA	%D	%F	%IV
Quercus costaricensis	61	41	20	41
Miconia biperulifera	10	16	15	14
Vaccinium consanguineum	4	10	12	9
Weinmannia pinnata	10	7	9	8
Didymopanax pittieri	5	4	6	5
Subtotal top five species	89	78	62	77

Source: Holdridge et al. 1971.

lier, the highlands of northern Mesoamerica are centers of species richness for temperate conifers and oaks. Pines occur naturally as far south as central Nicaragua. The southern hemisphere conifer *Podocarpus* is common in the highlands of southern Mesoamerica, extending north to the lowland perhumid forests of Mexico.

In addition to interesting assemblages of tropical and temperate species in Mesoamerica, the region's forests are critical habitats for numerous species that migrate either altitudinally (Stiles 1988; Loiselle and Blake 1991; Powell and Bjork 1995) or latitudinally (Terborgh 1989; Hagan and Johnston 1992). Though eastern songbirds (e.g., many warbler species) are well-known latitudinal migrants, several species migrate altitudinally to track seasonal food availability and/or appropriate climate. For example, the endangered resplendent quetzal (*Pharomachrus mocinno*, Trogonidae) migrates from its breeding range in or near the Monteverde Cloud Forest Reserve, Costa Rica, to spend 3–5 months in lower forests on the Caribbean slope, as much as 50 km from Monteverde. The striking three-wattled bellbird (*Procnias tricarunculata*, Cotingidae) migrates from Monteverde to Caribbean lowland forests of Costa Rica, Nicaragua, and Panama.

The species-rich communities of Mesoamerica, their importance to past and present migrations, and their conversion for economic activities all present enormous challenges for conservation. Though several national parks and equivalent reserves were established in the 1960s and 1970s, many were little more than "paper parks" that existed on a map but lacked protection on the ground (Hartshorn 1983c). Only in the past 15 yr has each of the Mesoamerican countries established a national system of protected areas (Tables 17.17,

Figure 17.13. Treeline near the Cerro de la Muerte, Costa Rica. The treeline has been lowered 300–400 m by re- *peated burning of the subalpine paramo. The forest is dominated by oaks (Quercus spp.).*

Table 17.17. *Status of forests and conservation areas in tropical and subtropical Mesoamerica. For Mexico, remaining forests and protected areas are from the subtropics only. Forest areas and loss (all in km²) are from WRI (1994).*

Country	Area	Remaining forest in 1990 (%)	Annual forest loss (%)	Protected area (%)	Number of conservation units
Belize	22,965	19,960 (87)	50 (0.2)	5,642 (25)	22
Costa Rica	50,900	14,280 (28)	500 (2.6)	14,481 (28)	41
El Salvador	21,395	1,230 (6)	30 (2.0)	62 (0.3)	3
Guatemala	108,890	42,250 (39)	810 (1.6)	19,278 (18)	21
Honduras	112,085	46,050 (41)	1,120 (2.0)	28,917 (26)	58
Mexico	1,972,545	485,860 (25)	6,780 (1.2)	30,032 (5)	48
Nicaragua	148,000	60,130 (41)	1,240 (1.7)	9,045 (6)	60
Panama	78,515	31,170 (40)	640 (1.7)	17,894 (23)	26
Central America	542,750	215,070 (40)	4,390 (1.7)	95,319 (18)	231
TOTAL	2,515,295	700,930 (28)	11,170 (1.4)	125,351 (11)	279

Source: Protected areas data are from IUCN (1994).

17.18). Costa Rica (28% in protected areas), Honduras (26%), Belize (25%), Panama (23%), and Guatemala (18%) all have national systems that greatly exceed the world average of about 4% in protected areas. Furthermore, each of these countries has a fairly high number of conservation units that ap-

pear to give reasonably comprehensive coverage of the major vegetation types in the region (Table 17.18). Costa Rica is pioneering a more holistic approach to conservation by broadening the traditional focus on protected units to more inclusive regional conservation areas (Boza 1993). Also of

Table 17.18. *National parks and equivalent reserves in tropical and subtropical Mesoamerica*

Conservation unit	Area (ha)	Vegetation type
Belize		
Aguas Turbias National Park	3,622	Lhf
Bermudian Landing Community Baboon Sanctuary	777	Lhf
Bladen Nature Reserve	39,256	Lpf
Blue Hole National Park	233	Lhf
Burdon Canal Nature Reserve	2,416	Lmf/Lhf
Caracol Archaeological Reserve	2,000	Lhf
Chiquibul Forest Reserve	77,348	Lhf/LMhf
Chiquibul National Park	107,607	Lhf/LMhf
Cockscomb Basin Wildlife Sanctuary	102,400	Lpf/LMpf
Crooked Tree Wildlife Sanctuary	1,470	Lsf
Freshwater Creek Forest Reserve	?	Lsf
Guanacaste National Park	21	Lhf
Half Moon Caye National Monument	3,925	Lmf/Lsf
Hol Chan Marine Reserve	411	Lsf
Laughing Bird Cay National Park	1	Lmf/Lsf
Maya Mountains Forest Reserves	111,325	LMhf/LMpf
Mountain Pine Ridge Forest Reserve	51,282	LMhf
Paynes Creek National Park	11,331	Lpf
Río Bravo Private Conservation Area	?	Lhf
Río Grande Nature Reserve	2,340	Lpf
Sarstoon/Temash National Park	16,592	Lmf/Lpf
Shipstern Private Nature Reserve	9,022	Lsf
Society Hall Nature Reserve	2,792	?
Costa Rica		
Arenal Conservation Area	63,000	LMhf/LMpf
Arenal National Park	2,000	LMpf
Ballena Marino National Park	4,200	Lpf
Barra del Colorado Wildlife Refuge	92,000	Lmf/Lfsf/Lpf
Barra Honda National Park	2,295	Lsf
Braulio Carillo National Park	45,899	Lpf/LMpf/HMpf
Cabo Blanco Biological Reserve	1,172	Lhf
Cahuita National Park	1,067	Lmf/Lhf
Caño Negro Wildlife Refuge	9,969	Lfsf/Lhf
Carara Biological Reserve	4,700	Lhf
Chirripó National Park	50,150	LMpf/HMpf/Hmp
Corcovado National Park	54,539	Lpf/LMpf
Cordillera Volcánica Central Biosphere Reserve	144,363	Lpf/LMpf/HMpf
Curú Wildlife Refuge	84	Lsf
Diriá National Forest	2,400	Lsf/LMhf
Gandoca/Manzanillo Wildlife Refuge	9,449	Lmf/Lfsf/Lpf
Golfito Wildlife Refuge	2,810	Lmf/Lpf
Guanacaste National Park	32,512	Ls/Lsf/LMpf
Guayabo National Monument	217	LMhf
Hitoy–Cerere Biological Reserve	9,154	Lpf
Isla Bolaños Wildlife Refuge	25	Lsf
Isla del Caño Biological Reserve	200	Lpf
Isla del Coco National Park	2,400	Lpf
Islas Guayabo, Negritos y Pájaros Biological Reserve	143	Lsf
Juan Castro Blanco National Park	14,258	LMhf/LMpf
La Amistad National Park	193,929	LMpf/HMpf/HMp
La Amistad World Heritage Site	584,592	LMpf/HMpf/HMp
La Selva Biological Station	1,534	Lpf
Las Baulas National Marine Park	445	Lmf/Lsf
Lomas Barbudal Biological Reserve	2,279	Lsf
Manuel Antonio National Park	690	Lpf
Monteverde Cloud Forest Reserve	11,931	LMhf/LMpf

Table 17.18. *(Cont.)*

Conservation unit	Area (ha)	Vegetation type
Ostional Wildlife Refuge	248	Lmf/Lsf
Verde Wildlife Refuge & National Park	16,804	Lsf
Peñas Blancas Wildlife Refuge	2,400	Lsf/LMhf
Rincón de la Vieja National Park	14,083	Lmhf/Lmpf
Santa Rosa National Park	37,217	Lmf/Lhf/Lsf/Ls
Tapantí National Park	6,080	LMpf
Tortuguero National Park	18,946	Lmf/Lfsf/Lpf
Volcán Irazú National Park	2,309	Hmpf
Volcán Poás National Park	5,600	Hmpf
El Salvador		
El Imposible National Park	3,222	?
Laguna Jocotal Wildlife Refuge	1,000	?
Montecristo National Park	2,000	LMhf/LMpf
Guatemala		
Aguateca Cultural Monument	1,709	?
Bahía de Santo Tomás National Park	1,000	?
Ceibal Cultural Monument	2,100	?
Cerro Cahui (III)	750	?
Chocón–Machacas Biotope	6,265	?
Dos Pilas Cultural Monument	3,166	?
El Tigre National Park	350,000	Lhf
Lacandón National Park	200,000	Lhf
Laguna Lachua National Park	10,000	?
Machaquilla Cultural Monument	2,000	?
Mario Dary Rivera Quetzal Biotope	1,173	?
Maya Biosphere Reserve	1,000,000	Lhf/Lpf
Mirador/Dos Lagunas/Río Azul National Park	2,000	?
Monterrico (VIII)	2,800	?
Quetzal Conservation Biotope	1,153	LMpf
San Miguel/El Zotz Biotope	42,000	?
Sierra de las Minas Biosphere Reserve	236,300	LMhf/LMpf/HMpf
Sipacate/Naranjo National Park	2,000	/
Tikal National Park	57,400	Lhf
Trifinio National Park	4,000	?
Volcán de Pacaya National Park	2,000	LMpf
Honduras		
Bosque Carías–Bermudez (VIII)	5,500	?
Capiro–Calentura National Park	5,500	?
Celaque National Park	27,000	?
Cerro Azul de Copán National Park	15,500	?
Cerro Azul Meambar National Park	20,000	?
Cuero y Salado Wildlife Refuge	13,225	?
El Armado Wildlife Refuge	3,500	?
El Cajón Forest Reserve (VIII)	33,696	?
El Chiflador (IV)	500	?
El Chile Biological Reserve	12,000	?
El Guisayote Biological Reserve	7,000	?
El Pital Biological Reserve	3,800	?
El Uyuca Biological Reserve	1,138	?
Embalse El Coyolar (VIII)	18,600	?
Erapuca Wildlife Refuge	6,500	?
Guajiquiro Biological Reserve	7,000	?
Guanacaure (VIII)	2,119	?
Isla de Exposición (II)	239	?
Lago de Yojoa (VIII)	34,628	?
Laguna de Caratasca Wildlife Refuge	120,000	?
Laguna de Guaymoreto Wildlife Refuge	5,000	?
La Muralla National Park	24,850	?
La Tigra National Park	23,821	Lhf

Table 17.18. (Cont.)

Conservation unit	Area (ha)	Vegetation type
Lancetilla Botanical Garden Biological Reserve	1,681	Lpf
Las Iguanas Wildlife Refuge	1,426	?
Manglar del Golfo de Fonseca (VIII)	71,000	Lmf/Lsf
Misoco (IV)	4,000	?
Mixcure Wildlife Refuge	8,000	?
Montaña de Comayagua National Park	18,000	?
Montaña de Corralitos Wildlife Refuge	5,500	?
Montaña de Cusuco National Park	18,400	Lhf/LMhf
Montaña de la Flor (VII)	3,199	?
Montaña de Yoro National Park	30,000	?
Montaña Verde Wildlife Refuge	8,300	?
Misoco Biological Reserve	4,000	?
Montecillos Biological Reserve	12,500	?
Montecristo–Trifinio National Park	5,400	LMhf/LMpf
Olancho (VIII)	1,000,000	LMsf/LMhf
Opalaca Biological Reserve	14,500	?
Pico Bonito National Park	112,500	?
Pico Pijol National Park	11,400	?
Puca Wildlife Refuge	4,900	?
Pinares de Guanaja (VIII)	5,400	Lsf
Punta Condega Wildlife Refuge	3,900	?
Punta Isopo Wildlife Refuge	11,200	?
Punta Sal National Park	78,162	?
Río Kruta Biological Reserve	50,000	?
Río Negro Biological Reserve	60,000	?
Río Plátano Biosphere Reserve	350,000	Lhf/LMhf/LMpf
Río Plátano World Heritage Site	500,000	Lhf/LMhf/LMpf
San Bernardo Wildlife Refuge	2,600	?
Sandy Bay Marine Reserve	420	Lmf/Lsf
Santa Bárbara National Park	13,000	?
Sierra de Agalta National Park	65,500	?
Texiguat Wildlife Refuge	10,000	?
Volcán Pacayita Bilogical Reserve	9,700	?
Yerba Buena Bilogical Reserve	3,600	?
Yuscarán Biological Reserve	2,360	?
Mexico		
Benito Juárez National Park	2,737	?
Bonampak Natural Monument	4,357	Lhf
Bosencheve National Park	15,000	?
Calakmul Biosphere Reserve	723,185	Lsf
Cañón de Río Blanco National Park	55,690	LMhf
Cañón del Sumidero National Park	21,789	?
Cascadas de Agua Azul Special B. R.	2,580	?
Cerro de la Estrella National Park	1,100	?
Chan-Kin Protection Area	12,184	Lsf
Chichinautzin Bilogical Corridor	37,302	?
El Chico National Park	2,739	?
El Cielo Biosphere Reserve	144,530	?
El Cimatario National Park	2,447	?
El Gogorrón National Park	25,000	?
El Potosí National Park	2,000	?
El Tepozteco National Park	24,000	?
El Triunfo Biosphere Reserve	119,177	LMhf/HMhf
El Veledero National Park	3,159	?
Insurgente José María Morelos y Pavón National Park	1,813	?
Insurgente Miguel Hidalgo y Costilla National Park	1,750	?
Isla Contoy Special Biosphere Res.	176	?
Isla Isabela National Park	194	?
Iztaccihuatl-Popocatepétl Nat. Park	25,679	?

Table 17.18. *(Cont.)*

Conservation unit	Area (ha)	Vegetation type
La Encrucijada Biotope	30,000	?
La Malinche National Park	45,711	?
Lacan-Tun Biosphere Reserve	6,833	Lhf
Lagunas de Chacahua National Park	14,187	?
Lagunas de Montebello National Park	6,022	?
Lagunas de Zempoala National Park	4,669	?
Los Mármoles National Park	23,150	?
Mariposa Monarca Special Biosphere Reserve	16,100	HMhf
Montes Azules Biosphere Resrve	331,200	LMhf/LMpf
Nevados de Toluca National Park	51,000	?
Omiltemi Park	3,600	?
Palenque National Park	1,772	Lsf
Pantanos de Centla Biosphere Res.	302,706	Lmf/Lfsf/L?
Pico de Orizaba National Park	19,750	?
Pico de Tancitaro National Park	29,316	?
Ría Celestún Special Biosphere Reserve	59,130	?
Ría Lagartos Special Biosphere Reserve	47,840	?
Selva El Ocote Special Biosphere Reserve	48,140	?
Sian Ka'an Biosphere Reserve	528,147	Lmf/Lfsf/Lsf
Sierra de Manantlán Biosphere Reserve	139,577	LMhf
Sierra de Santa Martha Special Biosphere Reserve	20,000	?
Volcán de San Martlín Special B. R.	1,500	?
Volcán Nevado de Colima Nat. Park	22,200	?
Yaxchilán Natural Monument	2,621	Lhf
Zoquiapán y Anexas National Park	19,418	?
Nicaragua		
Alamikamba Nature Reserve	2,100	?
Apante Nature Reserve	1,230	?
Archipiélago de Solentiname National Monument	18,930	?
Archipiélago Zapatero Nat. Park	5,227	?
Cabo Viejo Nature Reserve	5,800	?
Cayos Miskitos Biological Reserve	50,000	?
Cerro Bana Cruz Nature Reserve	10,130	?
Cerro Cola Blanca Nature Reserve	22,200	?
Cerro Cumaica/Cerro Alegre Nat. Res.	5,000	?
Cerro Datanli/El Diablo Nature Res.	2,216	?
Cerro Kilambe Nature Reserve	10,128	?
Cerro Kuskawas Nature Reserve	4,760	?
Cerro Musún Nature Reserve	4,142	?
Cerro Quiabuc Nature Reserve	3,630	?
Cerro Tisey/Estanzuela Nature Res.	6,400	?
Complejo Volcánico Momotombo y Momotombito Nature Reserve	8,500	?
Complejo Volcánico Pilas/El Hoyo Nature Reserve	7,422	?
Complejo Volcánico San Cristóbal NR.	17,950	?
Complejo Volcánico Telica/Rota N.R.	9,088	?
Cordillera de Yolaina Nature Res.	40,000	?
Cordillera Dipilto y Jalapa Nat. R.	41,200	?
Delta del Estero Real Nature Reserve	55,000	?
Estero Padre Ramos Nature Reserve	8,800	?
Fila Cerro Frío/La Cumplida Nature Reserve	1,761	?
Fila Masigüe Nature Reserve	5,580	?
Guabule Nature Reserve	1,100	?
Isla Juan Venando Nature Reserve	4,600	?
Kligna Nature Reserve	1,000	?
Laguna Bismuna-Raya Nature Reserve	11,800	?
Laguna de Apoyo Nature Reserve	3,500	?
Laguna de Mecatepe Nature Reserve	1,200	?
Laguna de Pahara Nature Reserve	10,200	?
Laguna de Tisma Nature Reserve	10,295	?

Table 17.18. *(Cont.)*

Conservation unit	Area (ha)	Vegetation type
Laguna Kukalaya Nature Reserve	3,500	?
Laguna Layasica Nature Reserve	1,800	?
Laguna Tala-Sulamas Nature Reserve	31,400	?
Laguna Yulu Karata Nature Reserve	25,300	?
Limbaika Nature Reserve	1,800	?
Llanos de Karawala Nature Reserve	2,000	?
Los Guatusos Wildlife Refuge	43,750	?
Macizos de Peñas Blancas Nature Reserve	11,308	Lsf
Makantaka Nature Reserve	2,000	?
Mesas de Monopotente Nature Reserve	7,500	?
Península de Chiltepe Nature Reserve	1,800	?
Río Escalante/Chococente Wildlife Refuge	4,800	?
Río Indio Maíz Biological Reserve	295,000	Lpf
Río Manares Nature Reserve	1,100	?
Saslaya National Park	15,000	Lhf/LMhf
Sierra Amerrisque Nature Reserve	12,073	?
Sierra Kiragua Nature Reserve	9,097	?
Tepesomoto/Pataste Nature Reserve	8,700	?
Volcán Concepción Nature Reserve	2,200	?
Volcán Cosigüina Nature Reserve	12,420	?
Volcán Madera Nature Reserve	4,100	?
Volcán Masaya National Park	5,100	Lhf
Volcán Mombacho Nature Reserve	2,487	?
Volcán Yali Nature Reserve	3,500	?
Yucul Genetic Reserve	4,826	?
Yulu Nature Reserve	1,000	?
Panama		
Altos de Campana National Park	4,816	Lsf/LMhf
Barro Colorado Nature Monument	5,400	Lhf
Camino de Cruces National Park	4,000	Lsf
Canglón Forest Reserve (VIII)	31,650	?
Cerro Hoya National Park	32,557	?
Chagres National Park	129,000	Lhf/LMhf/LMpf
Chepigana (VIII)	146,000	?
Ciénaga del Mangle (III)	776	Lmf
Coiba National Park	270,125	Lhf
Darién National Park	579,000	Lhf/Lpf/LMhf/LMpf
Isla Bastimientos National Park	13,226	Lmf/Lpf
Isla Iguana (IV)	53	Lsf
Islas Taboga y Urabá Wildlife Refuge	258	Lmf/Lsf
La Amistand National Park	207,000	Lpf/LMpf/HMhf/HMpf
LaFortuna (VIII)	15,000	LMhf
La Tronosa (VIII)	22,000	?
LaYeguada (VIII)	3,000	?
Lago Gatún (V)	348	Lhf
Metropolitano (V)	265	Lsf
Montuoso (VIII)	10,000	?
Palo Seco (VIII)	244,000	?
Peñón de la Onda Wildlife Refuge	2,000	?
Portobelo National Park	34,846	Lpf
Sarigua National Park	8,000	?
Soberanía National Park	22,104	Lhf/Lpf
Volcán Barú National Park	14,000	HMhf/HMpf

Source: IUCN (1992, 1994) and Ugalde and Godoy (1994).

note are international efforts to link protected areas through transborder collaboration, such as the Selva Maya region (e.g., Medellín 1994) and the Paseo Pantera project in the Caribbean lowlands of Mesoamerica.

Deforestation has long been the traditional cause of vegetation and habitat losses in Mesoamerica (e.g., Parsons 1976; Boyer et al. 1980; Hartshorn et al. 1982; Campanella et al. 1982; Hartshorn et al. 1984; Leonard 1987; Dirzo and Garcia 1992). In the 1970s and early 1980s, some of the countries had the highest relative rates of deforestation in the world. The rampant deforestation was primarily to establish cattle pastures for the beef export market. Though there has been some abandonment of pastures following the collapse of this market, the combination of population pressures, inequitable land tenure, government development policies, land speculation, and civil wars have prevented the reestablishment of young secondary forests. Through the past decade, relative deforestation rates declined to below 2% yr^{-1}, except for Costa Rica (2.6% yr^{-1}). Between the extremes of Belize (87%) and El Salvador (6%), the rest of the countries had 25–41% of their tropical or subtropical forests present as of 1990 (Table 17.17).

Economic difficulties hinder or threaten the survival of national parks and equivalent reserves in Mesoamerica. A recent review of protected areas in the region indicates that many of the individual units, as well as the national agency responsible for administering the protected areas system, are worse off with declining budgets, inadequately trained staff, and more responsibilities than a decade ago (Ugalde and Godoy 1994). Even though the region (especially Costa Rica) has received substantial foreign support for conservation, international assistance is even more critical to the survival of the region's protected areas. Without public appreciation and support, few conservation units in Mesoamerica will survive well into the twenty-first century.

Belize

In less than 10 years, Belize has created a comprehensive national system of protected areas that includes several outstanding terrestrial units such as the Bladen Nature Reserve, Chiquibul National Park, and the Cockscomb Basin Wildlife Sanctuary. Belize also has an impressive system of protected forest reserves associated with the extensive Maya Mountains. Private conservation efforts are also important, ranging from the small Community Baboon Sanctuary to the large Río Bravo Conservation and Management Area. Belizean scientists and

officials are actively participating in the trinational Selva Maya project that promotes coordinated initiatives in such areas as integrating conservation and development, multidisciplinary collaboration, and transborder connections of protected areas (such as the Aguas Turbias, Río Bravo, and Chiquibul units with Guatemalan homologs).

Costa Rica

Costa Rica has long been recognized for its world-class national system of protected areas (Boza and Mendoza 1981), despite having one of the world's highest rates of deforestation. For nearly 25 years, Costa Rica's leadership in conservation has attracted very significant outside funding (e.g., Abramovitz 1989, 1991) from international conservation organizations, U.S. foundations, and bilateral donors (e.g., USAID, Canada, Sweden, the Netherlands). More recently, Costa Rica's conservation leaders (including top government officials) have continued to attract outside interest and support with innovative approaches such as the broader conservation area concept, the private Neotrópica Foundation, the National Biodiversity Institute (INBio), and the proposed All-Taxa Inventory of the national biota.

The national system of protected areas is complemented by a long tradition of private lands dedicated to conservation. Although some private reserves (e.g., Cabo Blanco) became founding units of the National Park Service, several other private reserves are maintained as independent conservation units. Two of the latter, the Organization for Tropical Studies' La Selva Biological Station and the Tropical Science Center's Monteverde Cloud Forest Reserve, are world-class research and ecotourist sites, respectively. It is important to note that there are literally dozens of smaller, less well-known groups as well as private individuals who have purchased forest lands for the express purpose of conservation. For example, private landholdings now protect forests from the southern border of Corcovado National Park to the tip of the Osa Peninsula (M. Ramírez, personal communication). Conservation commitments by local communities, local environment groups (e.g., Monteverde Conservation League), and individual landowners probably extend the total area of forests protected in the Cordillera de Tilarán to more than 50,000 ha (G. Powell, personal communication).

El Salvador

Despite the highest population density in the Mesoamerican region, a devastating civil war, and the region's worst deforestation, El Salvador has embarked upon an ambitious plan to develop a national system of protected areas (Ugalde and Godoy 1994). Though most of the protected areas are small, Laguna Jocotal is an important wetlands area for migratory waterfowl, and Montecristo is part of a trinational set of protected areas in the mountainous border area with Guatemala and Honduras.

Guatemala

Guatemala has significantly improved its national system of protected areas in the past decade. Of particular significance are the recently created, very large Maya Biosphere Reserve (that includes about half of the Petén), El Tigre National Park, Sierra de las Minas Biosphere Reserve, and Lacandón National Park. However, Guatemala still has serious problems with consolidating the protection status of many of the designated areas. The University of San Carlos' Center for Conservation Studies has been instrumental in protecting several small biotopes. National and international focus on the Petén highlights the importance of the region for testing sustainable development models, integrating mahogany logging into multiple-use areas, involving indigenous peoples in conservation, and promoting greater economic returns to local communities from traditional nontimber forest products (xate palm leaves, chicle, and allspice).

Honduras

The number of designated protected areas has increased 10-fold in the past 15 yr. Many of these new areas are small (<10,000 ha) and of uncertain protection status. Best known is the Río Plátano Biosphere Reserve and World Heritage Site in eastern Honduras. It is one of the rare instances in Mesoamerica where an entire watershed and river basin have been designated for protection. Unfortunately, this protected area seems to be under almost constant threat from agricultural colonists who slash and burn the forest and from government-promoted roads built through the wilderness area. Some official forest reserves still have significant conservation potential, such as the Gulf of Fonseca mangroves, the vast Olancho pine forests, and the critical watershed for the El Cajón hydroelectric generating station.

Mexico

Mexico has an impressive array of conservation areas in the subtropical part of Mesoamerica. Best known are the Calakmul and Sian Ka'an Biosphere Reserves in the Yucatán. The Chiapas highlands include the important Montes Azules and El Triunfo Biosphere Reserves. Of special note are the Sierra de Manantlán Biosphere Reserve where perennial maize (Zea diploperennis) occurs, and the Monarch Butterfly Special Biological Reserve in the coniferous forests of the south-central highlands.

Nicaragua

On the basis of total number of conservation units, Nicaragua has the most comprehensive national system of protected areas (Table 17.18). All but two of the present 59 protected areas were created post-1980. Two regionally important conservation areas are the vast Bosawas multiple-use area in northeastern Nicaragua that could be connected to Honduras' Río Plátano unit, and the Río Indio Maíz Biological Reserve in extreme southeastern Nicaragua that is a key part of the proposed binational Si-A-Paz conservation complex along the Río San Juan border with Costa Rica.

Panama

Panama has made significant progress in creating a national system of protected areas. The world-class La Amistad and Darién national parks are solid "bookends" for several smaller conservation areas near the center of the country. Particularly noteworthy is Panama's substantial progress in creating adjoining protected areas on both sides of the Panama Canal as U.S. military holdings revert to Panama. The Smithsonian's world-famous research facility on Barro Colorado Island is part of the larger Barro Colorado Nature Monument. The series of protected areas on the east side of the Panama Canal extend right into Panama City (Metropolitan National Park). The proposed Interoceanic National Park on the west side of the Panama Canal is in limbo due to an interagency jurisdictional dispute. Several forest reserves have considerable conservation potential for Panama's biodiversity, hence could serve as corridors between the large, transborder protected areas and some of the smaller units.

AREAS FOR FUTURE RESEARCH

The scientific research needs of developing countries have been the focus of many studies (NRC 1980a, 1982). It would be easy to list many intriguing questions about tropical vegetation (cf. Meffe and Carroll 1994; Primack 1993). Rather, I prefer to stress the urgent need for conservation of tropical and subtropical vegetation. We are destroying the world's tropical and subtropical forests at alarming rates (NRC 1980b; FAO 1981; WRI 1994). In the moist tropics, nowhere is deforestation occurring more rapidly than in Mesoamerica (Heckadon Moreno and McKay 1982; Hartshorn 1983c; Nations and Komer 1983).

Excluding Belize, no country in the region has more than 40% of its territory still in natural, broadleaved forests, and the remaining primary forests are being cut down at rates of 1–3% yr^{-1}. It seems inevitable that unprotected broadleaved forests of Mesoamerica will have been destroyed by early in the next century. The consequences of tropical deforestation for global climate, food crops, pharmaceuticals, species extinction, soil erosion, hydrologic regimes and projects, desertification, and so forth have received considerable attention (Parsons 1976; Prance and Elias 1977; Myers 1979, 1984; Ehrlich and Ehrlich 1981; Caufield 1985), but they have not stopped the destruction of tropical and subtropical forests.

Basic biological inventories of tropical and subtropical forests are woefully incomplete. Expeditions to remote areas easily turn up dozens of species new to science (Pringle et al. 1984). Without basic inventories, it is impossible to know what species are adequately protected in conservation units or to legitimately list endangered or threatened species. The intensive biochemical prospecting by Costa Rica's National Biodiversity Institute (INBio) is generating much interest as a possible mechanism to finance conservation. Even more significantly, INBio is conducting the first comprehensive biodiversity inventories of the major conservation areas.

Particularly relevant to conservation is the role of gap-phase dynamics in the regeneration and maintenance of species in primary tropical forests. The stochastic nature of gap occurrence maintains dozens of ecologically similar, shade-intolerant species in complex tropical forests (Brandani, Hartshorn, and Orians 1988). If tropical forests are nonequilibrium communities with hundreds of species sharing few ecological guilds determined by unpredictable disturbances, we need to modify our thinking on how to conserve and manage tropical forests. We know far too little about managing tropical forests for particular species to be able to maximize community diversity or even to produce timber on a sustained yield basis.

REFERENCES

Abramovitz, J. N. 1989. A survey of U.S.-based efforts to research and conserve biological diversity in developing countries. World Resources Institute, Washington D.C.

Abramovitz, J. N. 1991. Investing in biological diversity: U.S. research and conservation efforts in developing countries. World Resources Institute, Washington, D.C.

Alexander, E. B. 1973. A comparison of forest and savanna soils in northeastern Nicaragua. Turrialba 23: 181–191.

Allen, P. H. 1956. The rain forests of Golfo Dulce. University of Florida Press, Gainesville.

Anderson, R. C., and J. S. Fralish. 1975. An investigation of palmetto, *Paurotis wrightii* (Griseb. & Wendl.) Britt., communities in Belize, Central America. Turrialba 25:37–44.

Anderson, R. C., and S. Mori. 1967. A preliminary investigation of *Raphia* palm swamps, Puerto Viejo, Costa Rica. Turrialba 25:37–44.

Aubréville, A. 1938. La forêt coloniale; les forêts de l'Afrique occidentale française. Ann. Acad. Sci. Colón 9:1–245.

Augspurger, C. K. 1984. Light requirements of neotropical tree seedlings: a comparative study of growth and survival. J. Ecol. 72:777–795.

Baker, H. G. 1970. Evolution in the tropics. Biotropica 2: 101–111.

Bawa, K. S. 1974. Breeding systems of tree species of a lowland tropical community. Evolution 28:85–92.

Bawa, K. S. 1980. Evolution of dioecy in flowering plants. Ann. Rev. Ecol. Syst. 11:15–39.

Bawa, K. S., and J. H. Beach. 1983. Self-incompatibility systems in the Rubiaceae of a tropical lowland wet forest. Amer. J. Bot. 70:1281–1288.

Bawa, K. S., S. H. Bullock, D. R. Perry, R. E. Colville, and M. H. Grayum. 1985. Reproductive biology of tropical lowland rain forest trees, II, Pollination systems. Amer. J. Bot. 72:346–356.

Bawa, K. S., and P. A. Opler. 1975. Dioecism in tropical forest trees. Evolution 29:167–79.

Bawa, K. S., and P. A. Opler. 1977. Spatial relationships between staminate and pistillate plants of dioecious tropical forest trees. Evolution 31:64–68.

Bawa, K. S., D. R. Perry, and J. H. Beach. 1985. Reproductive biology of tropical lowland rain forest trees, I, Sexual systems and incompatibility mechanisms. Amer. J. Bot. 72:331–345.

Beard, J. S. 1944. Climax vegetation in tropical America. Ecology 25:127–158.

Beard, J. S. 1949. The natural vegetation of the Windward and Leeward Islands. Oxford For. Mem. no. 21.

Beard, J. S. 1953. The savanna vegetation of northern tropical America. Ecol. Monogr. 23:149–215.

Bentley, B. L. 1984. Direct transfer of newly fixed nitrogen from free-living epiphyllous microorganisms to their host plant. Oecologia 63:52–56.

Blaser, J., and M. Camacho. 1991. Estructura, composición y aspectos silviculturales de un bosque de robles (*Quercus* spp.) del piso montano en Costa Rica. CATIE Serie Técnica: Colección Silvicultura y Manejo de Bosques Naturales, Publ. No. 1.

Bohlman, S. A., T. J. Matelson, and N. M. Nadkarni. 1995. Moisture and temperature patterns of canopy humus and forest floor soil of a montane cloud forest, Costa Rica. Biotropica 27:13–19.

Borchert, R. 1980. Phenology and ecophysiology of tropical trees: *Erythrina poeppigiana* O. F. Cook. Ecology 61:1065–1074.

Borchert, R. 1983. Phenology and control of flowering in tropical trees. Biotropica 15:81–89.

Borchert, R. 1994a. Water status and development of tropical trees during seasonal drought. Trees: Struct. and Func. 8:115–125.

Borchert, R. 1994b. Electric resistance as a measure of tree water status during seasonal drought in a tropical dry forest in Costa Rica. Tree Physiol. 14:299–312.

Borchert, R. 1994c. Induction of rehydration and bud break by irrigation or rain in deciduous trees of a tropical dry forest in Costa Rica. Trees: Struct. and Func. 8:198–204.

Boucher, D. H. 1983. *Quercus oleoides* (Roble Encino, Oak), pp. 319–320 in D. H. Janzen (ed.), Costa Rican natural history. University of Chicago Press, Chicago.

Boucher, D. H., J. H. Vandermeer, M. A. Mallona, N. Zamora, and I. Perfecto. 1994. Resistance and resilience in a directly regenerating rainforest: Nicaraguan trees of the Vochysiaceae after Hurricane Joan. For. Ecol. Mgt. 68:127–136.

Boyer, J., R. DuBois, G. Hartshorn, S. Heckadon, E. Ossio, F. Zadroga, and G. Schuerholtz. 1980. Panama country environmental profile: a field study. ISTI, Washington, D.C.

Boza, M. 1993. Conservation in action: Past, present, and future of the national park system of Costa Rica. Cons. Biol. 7:239–247.

Boza, M. A., and R. Mendoza. 1981. The national parks of Costa Rica. INCAFO, Madrid.

Brandani, A., G. S. Hartshorn, and G. H. Orians. 1988. Internal heterogeneity of gaps and species richness in Costa Rican tropical wet forest. J. Trop. Ecol. 4: 99–119.

Brokaw, N. V. L. 1982. Treefalls: frequency, timing, and consequences, pp. 101–108 in E. G. Leigh, Jr., A. S. Rand, and D. M. Windsor (eds.), The ecology of a tropical forest: seasonal rhythms and long-term changes. Smithsonian Institution Press, Washington, D.C.

Brokaw, N. V. L. 1985. Gap-phase regeneration in a tropical forest. Ecology 66:682–687.

Brokaw, N. V. L., and L. R. Walker. 1991. Summary of the effects of Caribbean hurricanes on vegetation. Biotropica 23:442–447.

Budowski, G. 1956. Tropical savannas, a sequence of forest felling and repeated burnings. Turrialba 6:23–33.

Budowski, G. 1959. Algunas relaciones entre la presente vegetación y antiguas actividades del hombre en el trópico americano, pp. 259–263 in Actas 33 Congreso Internacional de Americanistas. vol. 1. Lehmann, San José, Costa Rica.

Budowski, G. 1966. Los bosques de los trópicos húmedos. Turrialba 16:278–285.

Bullock, S. H. 1994. Wind pollination of neotropical dioecious trees. Biotropica 26(2):172–179.

Bullock, S. H., and J. A. Solís-Magallanes. 1990. Phenology of canopy trees of a tropical deciduous forest in Mexico. Biotropica 22:22–35.

Campanella, P., J. Dickinson, R. DuBois, P. Dulin, D. Glick, A. Merkel, D. Pool, R. Rios, D. Skillman, and J. Talbot. 1982. Honduras country environmental profile: A field study. JRB Associates, McLean, Va.

Caufield, C. 1985. In the rainforest. Knopf, New York.

Chapman, V. J. 1975. Mangrove biogeography, pp. 3–22 in G. E. Walsh, S. C. Snedaker, and H. J. Teas (eds.), Proceedings of the international symposium on biology and management of mangroves. University of Florida Institute of Food and Agricultural Sciences, Gainesville.

Cintron, G., A. E. Lugo, D. J. Pool, and G. Morris. 1978. Mangroves of arid environments in Puerto Rico and adjacent islands. Biotropica 10:110–121.

Clark, D. B., and D. A. Clark. 1993. Comparative analysis of microhabitat utilization by saplings of nine tree species in neotropical rain forest. Biotropica 25(4):397–407.

Clark, D., R. Dirzo, and N. Fetcher (eds.). 1987. Simposio sobre plantas mesoamericanas. Rev. Biol. Trop. (Costa Rica) Vol. 35 [Suppl.]

Coen, E. 1983. Climate, pp. 35–46 in D. H. Janzen (ed.), Costa Rican natural history. University of Chicago Press, Chicago.

Condit, R., S. P. Hubbell, and R. B. Foster. 1993. Mortality and growth of a commercial hardwood 'el cativo,' *Prioria copaifera*, in Panama. For. Ecol. Mgt. 62: 107–122.

Conrad, J. M., and G. Salas. 1993. Economic strategies for coevolution: timber and butterflies in Mexico. Land Econ. 69:404–415.

Cook, O. F. 1909. Vegetation affected by agriculture in Central America. USDA Bureau of Plant Science Industry, bull. no. 145. Washington, D.C.

Croat, T. B. 1975. Phenological behavior of habit and habitat classes on Barro Colorado Island (Panama Canal Zone). Biotropica 7:270–277.

Croat, T. B. 1978. Flora of Barro Colorado Island. Stanford University Press, Stanford, Cal.

Cruz, G. A., and M. Erazo Peña. 1977. Análisis de la vegetación del bosque nebuloso "La Tigra" (Reserva Forestal San Juancito). Ceiba (Tegucigalpa, Honduras) 21:19–60.

D'Arcy, W. D. 1977. Endangered landscapes in Panama and Central America: the threat to plant species, pp. 89–104 in G. T. Prance and T. S. Elias (eds.), Extinction is forever: threatened and endangered species of plants in the Americas and their significance in ecosystems today and in the future. New York Botanical Garden, New York.

Daubenmire, R. 1972. Ecology of *Hyparrhenia rufa* (Ness) in derived savanna in north-western Costa Rica. J. Appl. Ecol. 9:11–23.

Davis, T. A. W., and P. W. Richards. 1933–1934. The vegetation of Moraballi Creek, British Guiana: an ecological study of a limited area of tropical rain forest, Parts I & II. J. Ecol. 21:350–384; 22:106–155.

Deevey, E. S., D. S. Rice, P. M. Rice, H. H. Vaughan, M. Brenner, and M. S. Flannery. 1979. Mayan urbanism: impact on a tropical karst environment. Science 206:298–306.

De la Cruz, J. R. 1976. Mapa de zonas de vida de Guatemala. INAFOR, Guatemala.

Denevan, W. M. 1961. The upland pine forests of Nicaragua: a study in cultural plant geography. Univ. California Pub. Geogr. 12:251–320.

Denslow, J. S. 1980. Gap partitioning among tropical rainforest trees. Biotropica [Suppl.] 12:47–55.

Denslow, J. S. 1987. Tropical rain forest gaps and tree species diversity. Ann. Rev. Ecol. Syst. 18:431–451.

Denslow, J. S., and G. S. Hartshorn. 1994. Tree-fall gap environments and forest dynamic processes, pp. 120–127 in L. A. McDade, K. S. Bawa, H. A. Hespenheide, and G. S. Hartshorn (eds.), La Selva: ecology and natural history of a neotropical rain forest. University of Chicago Press, Chicago.

Dirzo, R., and M. C. Garcia. 1992. Rates of deforestation in Los Tuxtlas, a neotropical area in southeast Mexico. Cons. Biol. 6:84–90.

Dobzhansky, T. 1950. Evolution in the tropics. Amer. Sci. 38:209–221.

Dwyer, J. D., and D. L. Spellman. 1981. A list of the Dicotyledoneae of Belize. Rhodora 83:161–236.

Ehrlich, P., and A. Ehrlich. 1981. Extinction: the causes and consequences of the disappearance of species. Random House, New York.

Ellison, A. M., and E. J. Farnsworth. 1993. Seedling survivorship, growth, and response to disturbance in Belizean mangal. Am. J. Bot. 80:1137–1145.

Encyclopaedia Britannica. 1977. Jungles and rain forests, pp. 336–346 in Encyclopaedia Britannica, vol. 10, 15th ed. Encyclopaedia Britannica, Chicago.

Estrada, A., and R. Coates-Estrada. 1983. Rain forest in Mexico: research and conservation at Los Tuxtlas. Oryx 17:201–204.

Falla R., A. 1978. Plan de desarrollo forestal, parte II, Estudio de las perspectivas del desarrollo forestal en Panamá. PCT/6/PAN/01/1, no. 2, FAO, Panama.

FAO. 1981. Tropical forest resources assessment project (GC-MS): tropical America. FAO/UNEP, Rome.

Feinsinger, P. 1976. Organization of a tropical guild of nectarivorous birds. Ecol. Monogr. 46:257–291.

Foster, R. B. 1982a. The seasonal rhythm of fruitfall on Barro Colorado Island, pp. 151–172 in E. G. Leigh, Jr., A. S. Rand, and D. M. Windsor (eds.), The ecology of a tropical forest: seasonal rhythms and long-term changes. Smithsonian Institution Press, Washington, D.C.

Foster, R. B. 1982b. Famine on Barro Colorado Island, pp. 201–212 in E. G. Leigh, Jr., A. S. Rand, and D. M. Windsor (eds.), The ecology of a tropical forest: seasonal rhythms and long-term changes. Smithsonian Institution Press, Washington, D.C.

Frankie, G. W. 1975. Tropical forest phenology and pollinator plant coevolution, pp. 192–209 in L. E. Gilbert and P. H. Raven (eds.), Coevolution of animals and plants. University of Texas Press, Austin.

Frankie, G. W., H. G. Baker, and P. A. Opler. 1974. Comparative phenological studies of trees in tropical wet and dry forests in the lowlands of Costa Rica. J. Ecol. 62:881–919.

Frankie, G. W., P. A. Opler, and K. S. Bawa. 1976. Foraging behaviour of solitary bees: implications for outcrossing of a neotropical forest tree species. J. Ecol. 64:1049–1057.

Gentry, A. H. (ed.). 1990. Four neotropical forests. Yale University Press, New Haven.

Gentry, A. H. 1992. Tropical forest biodiversity: distributional patterns and their conservational significance. Oikos 63:19–28.

Gill, A. M., and P. B. Tomlinson. 1977. Studies on the growth of red mangrove (*Rhizophora mangle* L.), 4, The adult root system. Biotropica 9:145–155.

Gómez-Pompa, A., P. L. Hernandez, and S. M. Sousa. 1964. Estudio fitoecológico de la cuenca intermedia del Río Papaloapán. INIF (México) Publ. Esp. 3:37–90.

Gómez-Pompa, A., C. Vásquez-Yanes, and S. Guevara. 1972. The tropical rainforest: a non-renewable resource. Science 177:762–765.

González, L., M. Ramírez, and R. Peralta. 1983. Estudio ecológico y dendrológico: zonas de vida y vegetación. ACDI-COHDEFOR, Tegucigalpa, Honduras.

Grubb, P. J. 1977. Control of forest growth and distribution on wet tropical mountains. Ann. Rev. Ecol. Syst. 8:83–107.

Grubb, P. J., and E. V. J. Tanner. 1976. The montane forests and soils of Jamaica: a reassessment. J. Arnold Arboretum 57:313–368.

Gruezo, W. S., and H. C. Harries. 1984. Self-sown, wild-type coconuts in the Philippines. Biotropica 16:140–147.

Gunn, C. R., and J. V. Dennis. 1976. World guide to tropical drift seeds and fruit. Quadrangle, New York.

Hagan, J. M., III, and D. W. Johnston (eds.) 1992. Ecology and conservation of neotropical migrant landbirds. Smithsonian Institution Press, Washington D.C.

Hartshorn, G. S. 1975. A matrix model of tree population dynamics, pp. 41–51 in F. B. Golley and E. Medina (eds.), Tropical ecological systems: trends in terrestrial and aquatic research. Springer-Verlag, Berlin.

Hartshorn, G. S. 1978. Tree falls and tropical forest dynamics, pp. 617–638 in P. B. Tomlinson and M. H. Zimmermann (eds.), Tropical trees as living systems. Cambridge University Press, Cambridge.

Hartshorn, G. S. 1980. Neotropical forest dynamics. Biotropica [Suppl.] 12:23–30.

Hartshorn, G. S. 1983a. Plants: introduction, pp. 118–157 in D. H. Janzen (ed.), Costa Rican natural history. University of Chicago Press, Chicago.

Hartshorn, G. S. 1983b. *Pentaclethra macroloba* (Gavilán), pp. 301–303 in D. H. Janzen (ed.), Costa Rican natural history. University of Chicago Press, Chicago.

Hartshorn, G. S. 1983c. Wildlands conservation in Central America, pp. 423–444 in S. L. Sutton, T. C. Whitmore, and A. C. Chadwick (eds.), Tropical rain forest: ecology and management. Blackwell, London.

Hartshorn, G. S. 1992. Mahogany workshop: Review and implications of CITES. Trop. Forest Fdn., Alexandria, Va.

Hartshorn, G. S., and B. Hammel. 1982. Trees of La Selva. Organization for Tropical Studies mimeograph, San José, Costa Rica.

Hartshorn, G., L. Hartshorn, A. Atmella, L. D. Gomez, A. Mata, L. Mata, R. Morales, R. Ocampo, D. Pool, C. Quesada, C. Solera, R. Solórzano, G. Stiles, J. Tosi, Jr., A. Umaña, C. Villalobos, and R. Wells. 1982. Costa Rica country environmental profile: a field study. Tropical Science Center, San José.

Hartshorn, G., L. Nicolait, L. Hartshorn, G. Bevier, R. Brightman, J. Cal, A. Cawich, W. Davidson, R. DuBois, C. Dyer, J. Gibson, W. Hawley, J. Leonard, R. Nicolait, D. Weyer, H. White, and C. Wright. 1984. Belize country environmental profile: a field study. R. Nicolait & Associates, Belize City.

Hartshorn, G. S., and L. J. Poveda. 1983. Plants: checklist of trees, pp. 158–183 in D. H. Janzen (ed.), Costa Rican natural history. University of Chicago Press, Chicago.

Heckadon Moreno, S. 1983. Cuando se acaban los montes: los campesinos Santeños y la colonización de Tonosi. University of Panama and STRI, Panama.

Heckadon Moreno, S., and A. McKay (eds.). 1982. Colonización y destrucción de bosques en Panamá: ensayos sobre un grave problema ecológico. Asociación Panameña de Antropología, Panamá.

Heithaus, E. R. 1979. Community structure of neotropical flower visiting bees and wasps: diversity and phenology. Ecology 60:190–202.

Hewetson, C. E. 1956. A discussion on the "climax" concept in relation to the tropical rain and deciduous forest. Empire For. Rev. 35:274–291.

Holdridge, L. R. 1947. Determination of world plant formations from simple climatic data. Science 105: 367–368.

Holdridge, L. R. 1956. Middle America, pp. 183–200 in S. Haden-Guest, J. K. Wright, and E. M. Teclaff (eds.), A world geography of forest resources. Ronald Press, New York.

Holdridge, L. R. 1957a. The vegetation of mainland Middle America. Eighth Pacific Sci. Congr. Proc. IV: 148–161.

Holdridge, L. R. 1957b. Pine and other conifers, pp. 332–338 in Tropical silviculture. FAO Forestry and Forest Products Studies no. 13, vol. 2. FAO, Rome.

Holdridge, L. R. 1967. Life zone ecology, rev. ed. Tropical Science Center, San José, Costa Rica.

Holdridge, L. R. 1975a. Zonas de vida de El Salvador. PNUD/FAO/ELS/73/004, no. 6, San Salvador.

Holdridge, L. R. 1975b. Las coníferas de Guatemala PUND/FAO/FO:DP/GUA/72/006, no. 1, Guatemala City.

Holdridge, L. R., and G. Budowski. 1956. Report of an ecological survey of the Republic of Panama. Carib. For. 17:92–110.

Holdridge, L. R., W. C. Grenke, W. H. Hatheway, T. Liang, and J. A. Tosi, Jr. 1971. Forest environments in tropical life zones: a pilot study. Pergamon Press, Elmsford, N.Y.

Horn, S. P., and R. L. Sanford, Jr. 1992. Holocene fires in Costa Rica. Biotropica 24:354–361.

Hubbell, S. P. 1979. Tree dispersion, abundance, and diversity in a tropical dry forest. Science 203:1299–1309.

Hubbell, S. P., and R. B. Foster. 1983. Diversity of canopy trees in a neotropical forest and implications for conservation, pp. 25–41 in S. L. Sutton, T. C. Whitmore, and A. C. Chadwick (eds.), Tropical rain forest: ecology and management. Blackwell, London.

Hubbell, S. P., and R. B. Foster. 1985a. Canopy gaps and the dynamics of a neotropical forest, in M. J. Crawley (ed.), Plant ecology. Blackwell, London.

Hubbell, S. P., and R. B. Foster. 1985b. Biology, chance, and history, and the structure of tropical tree communities, in J. M. Diamond and T. J. Case (eds.), Community ecology. Harper & Row, New York.

Ingram, S. W., and N. M. Nadkarni. 1993. Composition and distribution of epiphytic organic matter in a neotropical cloud forest, Costa Rica. Biotropica 25: 370–383.

Islebe, G. A., and A. Velásquez. 1994. Affinity among mountain ranges in Megamexico: a phytogeographical scenario. Vegetatio 115(1):1–9.

IUCN. 1992. Protected areas of the world: a review of national systems, volume 4, nearctic and neotropical. IUCN, Gland, Switzerland.

IUCN. 1994. 1993 United Nations list of national parks and protected areas. IUCN, Gland, Switzerland.

Janzen, D. H. 1967. Synchronization of sexual reproduction of trees within the dry season in Central America. Evolution 21:620–637.

Janzen, D. H. 1969. Seed-eaters versus seed size, number, toxicity and dispersal. Evolution 23:1–27.

Janzen, D. H. 1970a. Herbivores and the number of tree species in tropical forests. Ameri. Nat. 104:501–528.

Janzen, D. H. 1970b. Jacquinia pungens, a heliophile from the understory of tropical deciduous forest. Biotropica 2:112–119.

Janzen, D. H. 1971. Euglossine bees as long-distance pollinators of tropical plants. Science 171:203–205.

Janzen, D. H. 1973. Rate of regeneration after a tropical high elevation fire. Biotropica 5:117–122.

Janzen, D. H. 1978. Description of a Pterocarpus officinalis (Leguminosae) monoculture in Corcovado National Park, Costa Rica. Brenesia 14–15:305–309.

Janzen, D. H. (ed.). 1983a. Costa Rican natural history. University of Chicago Press, Chicago.

Janzen, D. H. 1983b. Mora megistosperma (Alcornoque, Mora), pp. 280–281 in D. H. Janzen (ed.), Costa Rican natural history. University of Chicago Press, Chicago.

Janzen, D. H., and D. E. Wilson. 1974. The cost of being dormant in the tropics. Biotropica 6:260–262.

Jiménez, J. A. 1988. Floral and fruiting phenology of trees in a mangrove forest on the dry Pacific coast of Costa Rica. Brenesia 29:33–50.

Johnson, M. S., D. R. Chaffey, and C. J. Birchall. 1973. A forest inventory of part of the mountain pine ridge, Belize. ODA Land Resources Division, land resource study no. 13, Surrey, England.

Kapos, V., and E. V. J. Tanner. 1985. Water relations of Jamaican upper montane rain forest trees. Ecology 66:241–250.

Kappelle, M., A. M. Cleef, and A. Chaverri. 1992. Phytogeography of Talamanca montane Quercus forests, Costa Rica. J. Biogeogr. 19:299–315.

Kellman, M. 1979. Soil enrichment by neotropical savanna trees. J. Ecol. 67:565–577.

Kellman, M. 1984. Synergistic relationships between fire and low soil fertility in neotropical savannas: a hypothesis. Biotropica 16:158–160.

Kellman, M. 1989. Mineral nutrient dynamics during savanna – forest transformation in Central America, pp. 137–151 in J. Proctor (ed.), Mineral nutrients in tropical forest and savanna ecosystems. Blackwell, Oxford.

Kuchler, A. W., and J. M. Montoya Maquín. 1971. The UNESCO classification of vegetation: some tests in the tropics. Turrialba 21:98–109.

Lamb, F. B. 1966. Mahogany of tropical America: its ecology and management. University of Michigan Press, Ann Arbor.

Lambert, J. D. H., and T. Arnason. 1978. Distribution of vegetation on Maya ruins and its relationship to ancient land-use at Lamanai, Belize. Turrialba 28:33–41.

Lang, G. E., and D. H. Knight. 1983. Tree growth, mortality, recruitment, and canopy gap formation during a 10-year period in a tropical moist forest. Ecology 64:1075–1080.

Lawton, R., and V. Dryer. 1980. The vegetation of the Monteverde Cloud Forest Reserve. Brenesia 18:101–116.

Leigh, E. G., Jr. 1975. Structure and climate in tropical rain forest. Ann. Rev. Ecol. Syst. 6:67–86.

Leigh, E. G., Jr., A. S. Rand, and D. M. Windsor (eds.). 1982. The ecology of a tropical forest: seasonal rhythms and long-term changes. Smithsonian Institution Press, Washington, D.C.

Leigh, E. G., Jr., A. S. Rand, and D. M. Windsor (eds.). 1990. Ecología de un bosque tropical: ciclos estacionales y cambios a largo plazo. Smithsonian Tropical Research Institute, Balboa, Panama.

Leonard, H. J. 1987. Natural resources and economic development in Central America. IIED, Washington, D.C.

Levey, D. J. 1990. Habitat-dependent fruiting behavior of an understory tree, *Miconia centrodesma*, and tropical treefall gaps as keystone habitats for frugivores in Costa Rica. J. Trop. Ecol. 6:409–420.

Lieberman, M., D. Lieberman, G. Hartshorn, and R. Peralta. 1985. Small-scale altitudinal variation in lowland wet tropical vegetation. J. Ecol. 73:505–516.

Lieberman, D., M. Lieberman, R. Peralta, and G. S. Hartshorn. 1996. Tropical forest structure and composition on a large-scale altitudinal gradient in Costa Rica. J. Ecol. 84:137–152.

Loiselle, B. A., and J. G. Blake. 1991. Temporal variation in birds and fruits along an elevation gradient in Costa Rica. Ecology 72:180–193.

Lot-Helgueras, A., C. Vásquez-Yanes, and L. F. Menéndez. 1975. Physiognomic and floristic changes near the northern limit of mangroves in the Gulf Coast of Mexico, pp. 52–61 in G. E. Walsh, S. C. Snedaker, and H. J. Teas (eds.), Proc. of the international symposium on biology and management of mangroves. University of Florida Institute of Food and Agricultural Sciences, Gainesville.

Lugo, A. E. 1980. Mangrove ecosystems: successional or steady state? Biotropica [Suppl.] 12:65–72.

Lugo, A. E., R. Schmidt, and S. Brown. 1981. Tropical forests in the Caribbean. Ambio 10:318–324.

Lugo, A. E., and S. C. Snedaker. 1974. The ecology of mangroves. Ann. Rev. Ecol. Syst. 5:39–64.

Luna-Vega, I., L. Almeida-Leñero, and J. Llorente-Bousquets. 1989. Florística y aspectos fitogeográficos del bosque mesófilo de montaña de las cañadas de Ocuilan, estados de Morelos y México. Anales del Inst. Biol., Serie Botánica 59:63–87.

Lundell, C. L. 1933. The agriculture of the Maya. Southwest Rev. 19:65–77.

Lundell, C. L. 1937. The vegetation of Petén. Carnegie Institution Publication no. 478, Washington, D.C.

McDade, L. A., K. S. Bawa, H. A. Hespenheide, and G. S. Hartshorn (eds.). 1994. La Selva: ecology and natural history of a neotropical rain forest. University of Chicago Press, Chicago.

McHargue, L. A., and G. S. Hartshorn. 1983. Seed and seedling ecology of *Carapa guianensis*. Turrialba 33: 399–404.

Martin, P. S. 1964. Paleoclimatology and a tropical pollen profile. Vl Int. Quat. Congr. (Warsaw) Rept. 2: 319–323.

Mayo Meléndez, E. 1965. Algunas caracteristicas ecológicas de los bosques inundables de Darién, Panamá, con miras a su posible utilización. Turrialba 15:336–347.

Medellín, R. A. 1994. Mammal diversity and conversation in the Selva Lacandona, Chiapas, Mexico. Cons. Biol. 8:780–799.

Meffe, G. K., and C. R. Carroll. 1994. Principles of conservation biology. Sinauer Associates, Sunderland, Mass.

Milne, R., and J. Waugh. 1994. North America, pp. 281–299 in J. A. McNeely, J. Harrison, and P. Dingwall (eds.), Protecting nature: regional reviews of protected areas. IUCN, Gland, Switzerland.

Milton, K., E. A. Laca, and M. W. Demment. 1994. Successional patterns of mortality and growth of large

trees in a Panamanian lowland forest. J. Ecol. 82:79–87.

Miranda, F., and X. E. Hernández. 1963. Los tipos de vegetación de México y su clasificación. Bol. Soc. Bot. Méx. 28:29–179.

Molina Lara, O. A., and R. Esmeralda Esquivel. 1993. Asociaciones vegetales en el manglar de la Barra de Santiago, Ahuachapán, El Salvador. Rev. Biol. Trop. 41:37–46.

Montoya Maquín, J. M. 1966. Notas fitogeográficas sobre el *Quercus oleoides* Cham. Schlecht. Turrialba 16: 57–66.

Mueller-Dombois, D., and H. Ellenberg. 1974. Aims and methods of vegetation ecology. Wiley, New York.

Myers, N. 1979. The sinking ark: a new look at the problem of disappearing species. Pergamon Press, Elmsford, N.Y.

Myers, N. 1984. The primary source: tropical forests and our future. Norton, New York.

Myers, R. L. 1981. The ecology of low diversity palm swamps near Tortuguero, Costa Rica. Ph.D. dissertation, University of Florida, Gainesville.

Nadkarni, N. M. 1984. Epiphyte biomass and nutrient capital of a neotropical elfin forest. Biotropica 16: 249–256.

Nations, J. D., and D. I. Komer. 1983. Central America's tropical rainforests: positive steps for survival. Ambio 12:232–238.

Negreros-Castillo, P., and C. Mize. 1993. Effects of partial overstory removal on the natural regeneration of a tropical forest in Quintana Roo, Mexico. For. Ecol. Mgt. 58:259–272.

NRC (National Research Council). 1980a. Research priorities in tropical biology. National Academy of Science, Washington, D.C.

NRC. 1980b. Conversion of tropical moist forests. National Academy of Science, Washington, D.C.

NRC. 1982. Ecological aspects of development in the humid tropics. National Academy Press, Washington, D.C.

Opler, P. A., and K. S. Bawa. 1978. Sex ratios in tropical forest trees. Evolution 32:812–821.

Opler, P. A., G. W. Frankie, and H. G. Baker. 1976. Rainfall as a factor in the synchronization, release, and timing of anthesis by tropical trees and shrubs. J. Biogeog. 3:231–236.

Orozco Vílchez, L. 1991. Estudio ecológico y de estructura horizontal de seis comunidades boscosas en la Cordillera de Talamanca, Costa Rica. CATIE Ser. Tec., Colección Silv. Manejo de Bosques Nat., Publ. No. 2.

Parsons, J. J. 1955. The Miskito pine savanna of Nicaragua and Honduras. Ann. Assoc. Amer. Geogr. 45: 36–63.

Parsons, J. J. 1976. Forest to pasture: development or destruction? Rev. Biol. Trop. [Suppl.] 24:121–138.

Peñalosa, J. 1984. Basal branching and vegetative spread in two tropical rain forest lianas. Biotropica 16:1–9.

Pennington, T. D., and J. Sarukhán. 1968. Arboles tropicales de México. INIF, México.

Perry, D. R. 1978. Factors influencing arboreal epiphytic phytosociology in Central America. Biotropica 10: 235–237.

Peterson, A. A., and A. T. Peterson. 1992. Aztec exploitation of cloud forests: tributes of liquidambar resin and quetzal feathers. Global Ecol. Biogeogr. Let. 2: 165–173.

Pickett, S. T. A. 1983. Differential adaptation of tropical

tree species to canopy gaps and its role in community dynamics. Trop. Ecol. 24:68–84.

Pool, D. J., S. C. Snedaker, and A. E. Lugo. 1977. Structure of mangrove forests in Florida, Puerto Rico, Mexico, and Costa Rica. Biotropica 9:195–212.

Popma, J., F. Bongers, and M. J. A. Werger. 1992. Gap-dependence and leaf characteristics of trees in a tropical lowland rain forest in Mexico. Oikos 63(2): 207–214.

Powell, G. V. N., and R. Bjork. 1995. Implications of intratropical migration on reserve design: a case study using *Pharomachrus mocinno*. Cons. Biol. 9:354–362.

Prance, G. T., and T. S. Elias (eds.). 1977. Extinction is forever: the status of threatened and endangered plants of the Americas. New York Botanical Garden, New York.

Primack, R. B. 1993. Essentials of conservation biology. Sinauer Assoc., Sunderland, Mass.

Pringle, C., I. Chacón, M. Grayum, H. Greene, G. Hartshorn, G. Schatz, G. Stiles, C. Gómez, and M. Rodriguez. 1984. Natural history observations and ecological evaluation of the La Selva Protection Zone, Costa Rica. Brenesia 22:189–206.

Putz, F. E. 1984. The natural history of lianas on Barro Colorado Island, Panama. Ecology 65:1713–1724.

Quintana-Ascencio, P. F., and M. González-Espinosa. 1993. Afinidad fitogeográfica y papel sucesional de la flora leñosa de los bosques de pino-encino Los Altos de Chiapas, México. Acta Bot. Méx. 21:43–57.

Rabinowitz, D. 1975. Planting experiments in mangrove swamps of Panama, pp. 385–393 in G. E. Walsh, S. C. Snedaker, and H. J. Teas (eds.), Proc. of the international symposium on biology and management of mangroves. University of Florida Institute of Food and Agricultural Sciences, Gainesville.

Rabinowitz, D. 1978. Dispersal properties of mangrove propagules. Biotropica 10:47–57.

Raven, P. H., and D. I. Axelrod. 1975. History of the flora and fauna of Latin America. Amer. Sci. 63:420–429.

Ray, T. 1983a. *Monstera tenuis* (Chirravaca, Mano de Tigre, Monstera), pp. 278–280 in D. H. Janzen (ed.), Costa Rican natural history. University of Chicago Press, Chicago.

Ray, T. 1983b. *Syngonium triphyllum* (Mano de Tigre), pp. 333–335 in D. H. Janzen (ed.), Costa Rican natural history. University of Chicago Press, Chicago.

Reich, P. B., and R. Borchert. 1982. Phenology and eco-physiology of the tropical tree, *Tabebuia neochrysantha* (Bignoniaceae). Ecology 63:294–299.

Reyna Vásquez, M. L. 1979. Vegetación arborea del bosque nebuloso de Montecristo. Master's thesis, University of El Salvador, San Salvador.

Richards, P. W. 1952. The tropical rainforest: an ecological study. Cambridge University Press, Cambridge.

Rodan, B. D., A. C. Newton, and A. Verissimo. 1992. Mahogany conservation: status and policy initiatives. Env. Cons. 19:331–342.

Rundel, P. W., A. P. Smith, and F. C. Meinzer. 1994. Tropical alpine environments: plant form and function. Cambridge University Press, Cambridge.

Rzedowski, J. 1978. Vegetación de México. Editorial Limusa, Mexico, D. F.

Salgado-Labouriau, M. L. (ed.). 1979. El medio ambiente páramo. Ediciones Centro de Estudios Avanzados, IVIC, Caracas.

Sarukhán, J. 1978. Studies on the demography of tropi-

cal trees, pp. 163–184 in P. B. Tomlinson and M. H. Zimmermann (eds.), Tropical trees as living systems. Cambridge University Press, Cambridge.

Sauer, C. O. 1958. Man in the ecology of tropical America, pp. 104–110 in Ninth Pacific science congress proceedings (Bangkok).

Sawyer, J. O., and A. A. Lindsey. 1964. The Holdridge bioclimatic formations of the eastern and central United States. Indiana Acad. Sci. Proc. 73:105–112.

Scholander, P. F., L. Van Dam, and S. I. Scholander. 1955. Gas exchange in the roots of mangroves. Amer. J. Bot. 42:92–98.

Shimwell, D. W. 1971. The description and classification of vegetation. University of Washington Press, Seattle.

Shreve, F. 1914. A montane rain-forest: a contribution to the physiological plant geography of Jamaica. Carnegie Institution Publication no. 199, Washington, D.C.

Snook, L. K. 1994. Mahogany: ecology, exploitation, trade and implications for CITES. WWF, Washington, D.C.

Sousa S., M. 1964. Estudio de la vegetación secundaria en la región de Túxtepec, Oax. INIF (México) Publ. Esp. 3:91–105.

Spellman, D. L., J. D. Dwyer, and G. Davidse. 1975. A list of the Monocotyledoneae of Belize: including a historical introduction to plant collecting in Belize. Rhodora 77:105–140.

Standley, P. C. 1937. Flora of Costa Rica. Field Museum of Natural History, Chicago.

Stevenson, D. 1928. Types of forest growth in British Honduras. Trop. Woods 14:20–25.

Stiles, F. G. 1988. Altitudinal movement of birds on the Caribbean slope of Costa Rica: implications for conservation, pp. 243–258 in F. Almeda and C. M. Pringle (eds.), Tropical rain forest: diversity and conservation. California Academy of Sciences, San Francisco.

Strong, D. R., Jr., and T. S. Ray, Jr. 1975. Host tree location behavior of a tropical vine, *Monstera gigantea*, by skototropism. Science 190:804–806.

Tanner, E. V. J., and V. Kapos. 1982. Leaf structure of Jamaican upper montane rain-forest trees. Biotropica 14:16–24.

Taylor, B. W. 1962. The status and development of the Nicaraguan pine savannas. Carib. For. 23:21–26.

Taylor, B. W. 1963. An outline of the vegetation of Nicaragua. J. Ecol. 51:27–54.

Terborgh, J. 1989. Where have all the birds gone? Essays on the biology and conservation of birds that migrate to the American tropics. Princeton University Press, Princeton, N.J.

Tosi, J. A., Jr. 1969. Mapa Ecológico, República de Costa Rica. Centro Científico Tropical, San José.

Tosi, J. A., Jr. 1971. Zonas de vida: una base ecológica para investigaciones silvícolas é inventariación forestal en la República de Panamá. FAO/FO:SF/PAN/6, no. 2, Panama.

Tosi, J., Jr., and G. Hartshorn. 1978. Mapa ecológico de El Salvador. MAG & CATIE, San Salvador.

Turner, B. L., II, and P. D. Harrison (eds.). 1983. Pulltrouser Swamp: ancient Maya habitat, agriculture, and settlement in northern Belize. University of Texas Press, Austin.

Ugalde, A., and J. C. Godoy. 1994. Central America, pp. 305–322 in J. A. McNeely, J. Harrison, and P.

Dingwall (eds.), Protecting nature: regional reviews of protected areas. IUCN, Gland, Switzerland.

Veblen, T. T. 1976. The urgent need for forest conservation in highland Guatemala. Biol. Conserv. 9:141–154.

Veblen, T. T. 1978. Forest preservation in the western highlands of Guatemala. Geogr. Rev. 68:417–434.

Velásquez, A. 1994. Multivariate analysis of the vegetation of the volcanoes Tláloc and Pelado, Mexico. J. Veg. Science 5:263–270.

Vogel, F. H. 1954. Los bosques de Honduras. Ceiba 4:85–121.

Vogelmann, H. W. 1973. Fog precipitation in the cloud forests of eastern Mexico. BioScience 23:96–100.

Wagner, P. L. 1964. Natural vegetation of Middle America, pp. 216–264 in R. C. West (ed.), Natural environment and early cultures. University of Texas Press, Austin.

Weber, H. 1959. Los Páramos de Costa Rica y su concatenación fitogeográfica con los Andes Suramericanos. Instituto Geográfico Nacional, San José.

Wendt, T. 1989. Las selvas de Uxpanapa, Veracruz-Oaxaca, México: evidencia de refugios florísticos cenozoicos. Anales Inst. Biol., Ser Bot. 58:29–54.

Wendt, T. 1993. Composition, floristic affinities, and origins of the canopy tree flora of the Mexican Atlantic slope rain forest, pp. 595–680 in T. P. Ramamoorthy, R. Bye, A. Lot, and J. Fa (eds), Biological diversity of Mexico: origins and distribution. Oxford University Press, New York.

Whigham, D. F., I. Olmsted, E. Cabrera Cano, and M. E. Harmon. 1991. The impact of Hurricane Gilbert on trees, litterfall, and woody debris in a dry tropical forest in the northeastern Yucatán Península. Biotropica 23:434–441.

Whitmore, T. C. 1984. Tropical rain forests of the Far East, 2nd ed. Oxford University Press, Oxford.

Williams, H. 1960. Volcanic history of the Guatemalan highlands. University of California Publications in Geological Science, no. 38, Berkeley.

Witsberger, D., D. Current, and E. Archer. 1982. Arboles del Parque Deininger. Ministerio de Educación, San Salvador.

WRI. 1994. World resources 1994–1995: a guide to the global environment. Oxford University Press, New York.

Wright, A. C. S., D. H. Romney, R. H. Arbuckle, and V. E. Vial. 1959. Land in British Honduras. Colonial Research Publication no. 24, London.

Wright, S. J., and C. P. van Schaik. 1994. Light and the phenology of tropical trees. Am. Nat. 143:192–199.

Yih, K., D. H. Boucher, J. H. Vandermeer, and N. Zamora. 1991. Recovery of the rain forest of southeastern Nicaragua after destruction by Hurricane Joan. Biotropica 23:106–113.

Young, T. P., and S. P. Hubbell. 1991. Crown asymmetry, treefalls, and repeat disturbance of broadleaved forest gaps. Ecology 72:1464–1471.

Young, T. P., and V. Perkocha. 1994. Treefalls, crown asymmetry, and buttresses. J. Ecol. 82:319–324.

Chapter
18

*Vegetation of
the Hawaiian Islands*

LLOYD L. LOOPE

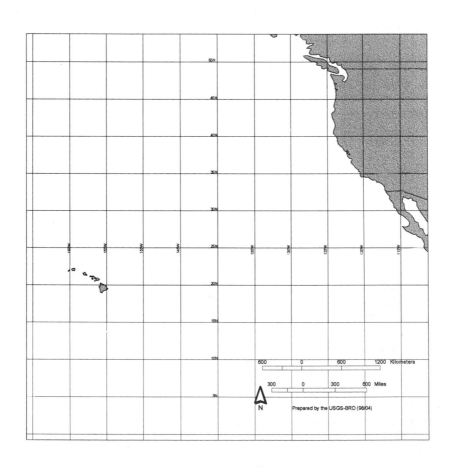

Prepared by the USGS-BRD (98/04)

INTRODUCTION

The Hawaiian archipelago, the most isolated island group of comparable size and topographic diversity on earth, is located about 4000 km from the nearest continent and 3200 km from the nearest high-island group (the Marquesas Islands of French Polynesia). The State of Hawaii consists of 132 islands, reefs, and shoals stretching 2400 km in a northwest-southeast direction between latitudes 28°N to 19°N (Armstrong 1983). The eight major high islands (Fig. 18.1), located at the southeast end of the chain, comprise over 99% of the total land area. The youngest island, Hawaii, with over 10,000 km² (63% of the total area of the state) and with elevations to more than 4000 m, is comprised of five volcanoes, two of which are highly active.

These islands are part of a longer chain that was produced by the northwestward movement of the Pacific Plate over a hot spot in the Earth's mantle over a 70-m-yr period. The oldest rocks of the eight high islands range in age from 420,000 yr for the island of Hawaii to about 5 m yr for Kauai (Macdonald, Abbott, and Peterson 1983). Islands of the chain extending to the northwest are eroded and submerged remnants of what were once high islands.

Climate

The Hawaiian Islands have a remarkable diversity of climates. Over the open ocean near Hawaii, annual rainfall averages 600–750 mm. However, certain locations on the islands receive up to 15 times this amount as a result of orographic lifting of moist northeast trade-wind air (Price 1983). In contrast, areas on the leeward side of islands receive annual precipitation as low as 200 mm.

As the northeast trade winds rise over the islands, zones of often dense cloud formation occupy northern and northeastern exposures, creating a cloud forest zone, delineated on East Maui by Kitayama and Mueller-Dombois (1994) as the zone between the lower cloud limit (i.e., the lifting condensation level) at about 1000 m elevation and the upper cloud limit set by the trade wind inversion at about 1900 m elevation. Maximum rainfall is normally near the base of this zone (Fig. 18.2). Within the cloud forest zone, fog may add significant amounts of precipitation to a very wet environment (Juvik and Ekern 1978).

Hawaii's precipitation is often sharply reduced during El Niño/Southern Oscillation (ENSO) events (Chu 1989), and rainforests which normally have an annual precipitation of > 5000 mm may

be subjected to substantial water stress (Medeiros, Loope, and Mobdy 1993).

Temperature decreases by about 0.55°C per 100 m increase in elevation (Price 1983). Whereas mean annual temperature at sea level is about 24°C, mean annual temperature in the montane cloud forest on windward east Maui at 1000–1900 m elevation is about 13.5–18.5°C (Fig. 18.2). Mean annual temperature at the 4200-m summit of Mauna Kea volcano on the island of Hawaii is about 4°C and glaciation occurred there during the late Pleistocene (Porter 1979). The mean temperature of the warmest month and the coldest month differ by only about 4°C (Price 1983).

The upper rainforest limit at about 1900 m roughly coincides with a major microclimatic discontinuity: the zone of the trade wind inversion (Leuschner and Schulte 1991; Giambelluca and Nullet 1991) a phenomenon associated with subtropical high-pressure cells. The inversion zone normally limits upward movement of clouds; above the inversion, the climate becomes substantially drier because of lower rainfall, lower humidity, and higher solar radiation (Fig. 18.2).

Tropical cyclonic storms are seasonally frequent in the vicinity of the Hawaiian Islands, but in recent times hurricanes have struck only the island of Kauai, with major forest damage from Hurricanes Iwa in 1982 (Fig. 18.3) and Iniki in 1991.

Biota

The Hawaiian biota started to evolve as much as 70 m yr ago in nearly total isolation – with successful colonization occurring through long-distance dispersal and establishment for major taxonomic groups occurring at very infrequent intervals. The few colonizing species that reached Hawaii over thousands of kilometers of open ocean by wind, flotation, or attachment to storm-driven birds had remarkably diverse potential habitats. Many groups of organisms common on continents were unable to make the journey successfully. Hawaii lacks native representatives of ants, coniferous trees, and most bird families, for example, and has only one native terrestrial mammal (a bat). The low number of colonizers has been partially offset by enrichment of biological diversity through evolution after establishment.

The percentage of endemism (species found nowhere else in the world than in Hawaii) is very high. The total known macroscopic terrestrial biota of Hawaii (including liverworts, mosses, fungi, and lichens, in addition to groups enumerated later) is believed to have evolved from roughly 2000 colo-

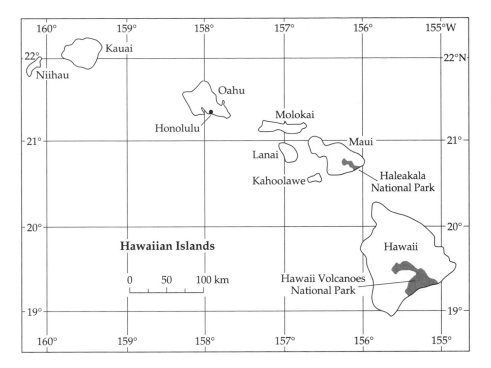

Figure 18.1. Map of the main Hawaiian Islands.

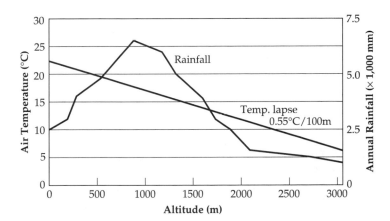

Figure 18.2. Variation of mean annual air temperature and precipitation along an elevational gradient of the windward slopes of Haleakala volcano, Maui. (From Kitayama and Mueller-Dombois.)

nizing ancestral species (i.e., no more on average than one successful immigrant per 35,000 years). A flora of about 960 species of flowering plants (Wagner, Herbst, and Sohmer 1990) is believed to have evolved from about 270 colonizing ancestors (Fosberg 1948); 168 species of ferns and fern allies evolved from about 135 original immigrants (Fosberg 1948); a native arthropod biota of 6000–10,000 species evolved from about 300–400 ancestral immigrant species (Gagné and Christiansen 1985; Hardy 1983); a native land snail biota of about 1200 species evolved from as few as 22–24 long-distance immigrants, probably carried by birds (Zimmer-

man 1948); and about 115 known species of endemic land birds (including species known only as fossils) evolved from as few as 20 ancestral immigrants (Olson and James 1982; James 1995).

The differing dispersal and establishment modes of various groups of organisms as colonizers of Hawaii have resulted in evolutionary adaptive radiation. Beginning with a single ancestral population, certain groups have undergone spectacular adaptive radiation, resulting in a diverse assemblage of closely related species occupying a wide range of habitats. As pointed out by Wagner et al. (1990) for flowering plants, adaptive radiation

Figure 18.3. Downed trunks of Metrosideros polymorpha, *Alakai Swamp, Kauai. Photograph taken September 1983 after a hurricane in 1982. Such severe hurricane effects on forests have occurred recently only on the island of Kauai; resulting forest openings have accelerated invasion by nonnative plant species. (Photography courtesy of A. C. Medeiros.)*

actually involves only a small percentage of the original colonists. Nearly 50% (over 450) of the flowering plant species are derived from only 20 genera representing only 26–32 original colonists (10–12% of the total). Presumably these genera represent evolutionarily plastic groups that arrived relatively early in the evolution of the island biota. Examples include the silversword alliance in the Asteraceae (with the endemic genera *Argyroxiphium, Dubautia,* and *Wilkesia*) and the lobelioids in the Campanulaceae (with the endemic genera *Brighamia, Clermontia, Cyanea, Rollandia,* and *Trematolobelia*) (Wagner et al. 1990).

Prehistoric Human Impact

Until about 15 yr ago, the conventional wisdom was that indigenous human Pacific populations had little impact on island ecosystems and that the dramatic deterioration of island environments was almost entirely a result of effects brought by western civilization. In fact, Polynesians had been drastically altering their environment for thousands of years prior to western contact, through cutting, burning, and the introduction of nonnative plants for agriculture (Kirch 1982). The most dramatic evidence for environmental modification comes from bird bones in archaeological sites, lava tubes, and sand deposits.

The Hawaiian Islands were reached by colonizing Polynesians, probably from the Marquesas, in about the Fourth century A.D., somewhat later than most other Polynesian islands. A peak human population of about 200,000 was attained before western contact in 1778. Landscapes were modified through the use of shifting cultivation and fire and through creation of sizeable wetlands for aquaculture. Throughout the Hawaiian Islands, most land below 600-m elevation with even moderately good soils was cultivated by the Hawaiians from the thirteenth through the eighteenth centuries (Kirch 1982). Polynesians introduced pigs (*Sus scrofa*), jungle fowl (*Gallus gallus*), dogs (*Canis familiaris*), Polynesian rats (*Rattus exulans*), and various stowaway geckos, skinks, and snails (Kirch 1982) as well as at least 32 plant species, including major food plants such as taro (*Colocasia esculenta*) and sweet potato (*Ipomoea batatas*), and also species providing sources for the manufacture of cloth, cordage, musical instruments, and so on (Nagata 1985). The kukui, or candlenut, tree (*Aleurites moluccana*), brought by Polynesians for lamp oil and various other uses (Abbott 1992), persists today as a local dominant in lowland vegetation.

Approximately 50% of the native land bird species were eliminated prior to observations by western scientists; this figure is continually being revised upward as more new species of extinct birds are discovered in lava tubes and sand dune deposits (James 1995). Based on the known pollen record, plant species may have been more resistant to prehistoric extinction than animals were (James 1995). A dramatic exception to this generalization has recently been noted, however, and more exceptions may be expected, because few pollen cores from low elevations have yet been analyzed (Athens and Ward 1993).

Historic Human Impact

Damage to native ecosystems accelerated after the arrival of Captain James Cook's ships in the Hawaiian Islands in 1778. The greatest early impacts were from grazing and browsing animals – especially feral cattle, goats, and sheep. These impacts have continued until the present, but feral pigs are now causing the most destruction to remaining native ecosystems. Commercial exploitation of sandalwood and firewood for whaling ships, sugar and pineapple production, logging of koa and 'ohi'a, ranching, and real estate development have also progressively destroyed native habitats over the past two centuries (Cuddihy and Stone 1990). Although there are still quite significant direct conflicts in Hawaii between economic development and preservation of native biota (especially conflicts for water resources), the greater conflicts are indirect, through continued introduction of invasive nonnative species, intentionally and inadvertently, and by modern transportation and commerce.

Botanical Loss

Not surprisingly, in view of the vulnerability of island biota and the drastic changes brought about by humans through habitat modification and species introduction, the native Hawaiian biota has suffered substantial extinction and endangerment. Much has been lost, with land snails and birds having been more decimated than other groups, but rich native biological diversity remains in the Hawaiian Islands in relatively intact ecosystems at upper elevations. Biological invasions, from exotic organisms already present and those that may arrive in the next decades, present the greatest threat to sustained survival of diverse native ecosystems.

The known Hawaiian vascular flora includes about 1302 taxa (including species, subspecies, and varieties), of which 1158 are endemic to Hawaii. About 106 (8%) of these taxa are now extinct. An additional 373 taxa (29%) are considered at some risk of becoming extinct in the near future. As of March, 1995, Hawaii had 199 taxa federally listed as endangered or threatened under the Endangered Species Act; this is 38% of the total such listed endangered and threatened vascular plants nationwide. The act mandates that the U.S. Fish and Wildlife Service is to provide for recovery of all listed species – an extremely difficult task, for 104 taxa have fewer than 20 known individuals remaining in the wild (L. Mehrhoff, U.S. Fish and Wildlife Service, personal communication).

VEGETATION

The standard treatment of native Hawaiian vegetation is that of Gagné and Cuddihy (1990), summarized by Cuddihy and Stone (1990), based on a classification developed by Jacobi (1989, 1990). These treatments use five elevational zones: coastal (0–50 m, extending sometimes as high as 300 m);

lowland (extending up to 500–1000 m; montane (500–1000 m to 2000 m); subalpine (2000 to 2800 m); and alpine (above 2800 m). Within each zone, vegetation is categorized by moisture regime: dry (<1250 mm of mean annual precipitation); moist or mesic (1250 mm to 2500 mm); and wet (>2500 mm). Additionally, five physiognomic types are used, based on the dominant life form: grassland, shrubland, forest, open forest, and parkland. A much simplified overview of the vegetation is given here.

Gagné and Cuddihy (1990) estimate that only 20% of the land area of the Hawaiian Islands is currently occupied by native and nonnative plant communities, with the remaining 80% of the land area occupied by cattle pasture; sugar, pineapple and other agriculture, and urban areas. Coastal and lowland vegetation has been largely obliterated in most areas, whereas substantial areas of native vegetation survive in the uplands.

Coastal Vegetation

A very narrow belt of strand vegetation occurs around each island. On the main Hawaiian Islands, these communities have been severely damaged, especially by development and by grazing mammals. Some largely intact coastal habitats persist, however, the most notable of which is at Mo'omomi on the north coast of Molokai. Many native coastal plant species generally persist in rocky shoreline habitats. Relatively few coastal species have been lost to extinction, largely because of steep coastal habitats. Restoration of coastal habitats has much potential. Excellent small-scale restoration examples occur at Wailea Point on Maui, Kilauea Point on Kauai, and Kaena Point on Oahu.

Common and widespread species of the native, low-growing coastal strand include *Scaevola sericea* (most abundant), *Ipomoea pescaprae*, *Jacquemontia ovalifolia*, *Sesuvium portulacastrum*, and *Sporobolus virginicus*. Locally important species include *Heliotropium anomalum*, *Lipochaeta* spp., *Myoporum sandwicense*, and *Sida fallax*. Less common or localized elements include *Boerhavia repens*, *Capparis sandwichiana*, *Chamaesyce celastroides*, *Fimbristylis cymosa*, *Gossypium tomentosum*, *Heliotropium curassavicum*, *Lycium sandwicense*, *Nama sandwicensis*, *Sesbania tomentosa*, *Vitex rotundifolia*, and *Wikstroemia* spp. Coastal forests dominated by *Pandanus tectorius* occur on some windward shores.

The northwestern Hawaiian Islands (politically part of the State of Hawaii except for Midway, which belongs to the U.S. Navy) extend northwest of Kauai for 1210 km from Nihoa to Kure Atoll. About 40 emergent islands comprise a total of about 10.1 km² of dry land. They vary in maximum elevation from 3–277 m. The strand vegetation that covers most of these islands is generally more intact than the strand of the main islands because of less disturbance, although disturbance to individual islands (e.g., Laysan and Lisianski, where rabbits were introduced) has been dramatic. Whereas the islands have only eight endemic and 42 indigenous plant species, there are 114 naturalized nonnative species. The nonnative species appear largely in balance with the rest of the flora, although close monitoring is warranted (Herbst and Wagner 1992).

The Hawaiian Islands notably lack native mangrove species. *Rhizophora mangle* (Neotropical) and *Brugiera gymnorrhiza* (from Asia) were introduced after 1900 to combat erosion. Both introduced species are spreading, with *Rhizophora* invading aggressively in some locations.

Lowland Vegetation

Most land below 600 m elevation with even moderately good soil was cultivated by the Hawaiians from the thirteenth through the eighteenth centuries (Kirch 1982; Cuddihy and Stone 1990). The little knowledge that we have of pristine lowland vegetation of the Hawaiian Islands comes from a few remnant lowland areas dominated by native species and from a recent pollen analysis from Oahu.

Two low-elevation cores from Oahu, one from a windward area (Kawainui Marsh) and one from a leeward area (Ft. Shafter flats), contain a well-dated pollen record from 1200 B.C. to about 1500–1600 A.D.: The dominant types, collectively comprising 50–70% of the total pollen in the lower part of the record, include *Pritchardia* palm pollen (two species) and pollen of a leguminous type that could not be matched to an extant species (Athens, Ward, and Wickler 1992). *Pritchardia* was much reduced by the end of the pollen record in about 1565 A.D.; the leguminous pollen type disappeared from the pollen column by 1200 A.D.. At Kawainui, the early dominant pollen types were replaced later in the record by Chenopodiaceae/ Amaranthaceae, grasses, and sedges. At Ft. Shafter, the replacement was more complex.

All 19 species of *Pritchardia* in Hawaii have been uncommon throughout historic times. The unidentified pollen type has since been matched with a newly discovered legume genus and species (*Kanaloa kahoolawensis*), known only from two individuals from a cliff face on the island of Kahoolawe (Lorence and Wood 1994). Anthropogenic clearing for agriculture in lowland Oahu is extremely likely

to have been responsible for this vegetation conversion but would have been unlikely to almost completely eliminate *Kanaloa* prior to western contact. Predation by the Polynesian rat is a likely contributing factor.

The implication of the Kawainui and Ft. Shafter pollen cores reported by Athens et al. (1992) is that *Pritchardia* palm forests dominated both windward and leeward forests of lowland Oahu prior to arrival of Polynesians. On windward Oahu, a *Dodonaea* shrubland may have been present on lower elevation slopes, with a mixed wet to mesic forest similar to the native upland vegetation persisting today on higher slopes. On leeward Oahu, there was no *Dodonaea* shrubland, but a diverse dryland to mixed mesic forest rich in species was present.

Natural lowland rainforests may still be seen in relatively few areas with rocky substrates or steep terrain on most islands (Cuddihy and Stone 1990). Such forests are dominated by *Metrosideros polymorpha*, often with a lower stratum of native trees such as *Psychotria* spp. and *Antidesma platyphyllum*, tree ferns (*Cibotium* spp.), and the vine *Freycinetia arborea*.

Most remaining native dry and mesic lowland forests are dominated by *Metrosideros polymorpha*, *Diospyros sandwicensis*, or *Erythrina sandwicensis*, but a large number of tree species are locally present, including *Canthium odoratum*, *Colubrina oppositifolia*, *Nestegis sandwicensis*, *Nothocestrum* spp., *Rauvolfia sandwicensis*, and *Reynoldsia sandwicensis*.

Lowland leeward forests were considered by Rock (1913) to be the richest of all Hawaiian forests in numbers of tree species, but dryland forests have been largely reduced to small remnants (Fig. 18.4). Direct habitat destruction by humans has been a factor (especially on Oahu and Lanai, and in the Kona District of Hawaii), but even more important has been the impact of browsing animals, invasion of nonnative grasses, and fire. Spectacular "museum-piece forests" remain in sites such as Auwahi, Maui, and Puuwaawaa, Hawaii, where rich assemblages of tree species persist, but with only very old and decadent individuals, no reproduction for many years, and conditions that appear to make reproduction impossible without heroic restoration efforts (Medeiros, Loope, and Holt 1986).

The understory of the remnant native forest at Auwahi, Maui (which extends upslope to 1200 m and thus into the montane zone), is covered by a dense mat (30–40 cm thick) of kikuyu grass (*Pennisetum clandestinum*), introduced as a pasture grass from East Africa, which smothers any reproduction. Dryland forests of Puuwaawaa and other sites on the island of Hawaii are being rapidly

Figure 18.4. Dryland forest. (Top) Relatively intact stand in Kanaio Natural Area Reserve, East Maui, 800 m elevation. Dominant trees are Diospyros sandwicensis and Pleomele auwahiensis. (Bottom) Isolated remnant individual Reynoldsia sandwicensis, same area. (Photographs courtesy of A. C. Medeiros.)

(over 1–2 decades) degraded by fires carried by nonnative fountain grass *Pennisetum setaceaum* from North Africa. Attack by nonnative termites (*Coptotermes formosanus*) may be another significant factor hastening the demise of native dryland trees (Lai et al. 1983; A. C. Medeiros, personal observation). Protection of dryland forest sites poses one of the greatest challenges to conservation managers.

Montane Vegetation

Montane forest. Although most rainforests at low elevations were destroyed long ago, rainforests still cover relatively large expanses on the islands of Maui (Fig. 18.5A) and Hawaii and also occur on the steep windward slopes, ridges, and peaks of Kauai, Oahu, and Molokai (Cuddihy and Stone 1990). Jacobi and Scott (1985) reported the presence

Figure 18.5. Montane wet zone. (A) Kipahulu Valley in Haleakala National Park has extensive, relatively pristine rainforest vegetation. (B) Metrosideros polymorpha *is the dominant tree in most Hawaiian rainforests. Waikamoi Preserve, East Maui. (C) Extensive monospecific understory stands of uluhe* (Dicranopteris linearis) *occur on all islands. (Photographs courtesy of A. C. Medeiros.)*

of about 140,000 ha of wet forest dominated by native species on the island of Hawaii. Cuddihy and Stone (1990) estimate a similar area of wet forest for the other four largest islands combined. Most of these forests are dominated by 'ohi'a (*Metrosideros polymorpha*) (Fig. 18.5B), with a closed canopy and a well-developed subcanopy layer of mixed native tree species, shrubs, and tree ferns. An important understory dominant in open *Metrosideros* stands on all islands is the native fern uluhe, *Dicranopteris linearis* (Fig. 18.5C). Koa (*Acacia koa*) is locally dominant or co-dominant with 'ohi'a.

A good overview of island-wide mesic and wet forest vegetation and flora is provided by Kitayama and Mueller-Dombois (1995), who sampled forest stands at eight mesic sites (mean annual rainfall ca. 2500 mm) and eight wet sites (mean annual rainfall ca. 4000 mm) at 1200 m elevation along a substrate age gradient from Hawaii (Big Island) to Kauai. Five nearly adjacent 20 m × 20 m plots were used at each site. Floristic and vegetation differences between mesic and wet sites were not dramatic. *Metrosideros* was overwhelmingly dominant at all sites; *Acacia koa* was important (29% cover) at only one site (mesic site at Waikamoi, Maui). The greatest number of species was found for mesic sites at Waikamoi, Maui (93 taxa) and for wet sites on west Maui (74 taxa). Genera present, with number of species per genus, are shown in Table 18.1.

Hawaiian wet forests lack obligately epiphytic species; however, many species grow both terrestrially and epiphytically. Two types of sites are favored for epiphytic growth (Medeiros et al. 1993): (1) the bryophyte mats that form on the branches and trunks of forest trees; and (2) trunks of tree ferns.

The long-term prognosis for ecological integrity of unmanaged Hawaiian rainforests is not good, given the recent invasion of feral pigs. Rainforests of the two major national parks, some reserves managed by The Nature Conservancy, and a few State Natural Area Reserves, are receiving protection from pigs; most other rainforest areas are being degraded by feral pigs. As a result, although high-elevation Hawaiian rainforests are the most intact extensive native ecosystems remaining in Hawaii, their ecological integrity is being gradually eroded. This process is difficult to reverse after degradation reaches a certain stage.

Mesic forests at higher elevations are generally dominated by *Metrosideros polymorpha* or *Acacia koa*. Mesic forests differ from wet forests in the relative scarcity of tree ferns (*Cibotium* spp.) and epiphytes, the abundance of shrubs such as *Styphelia tameiameiae* in the understory, and a different comple-

Table 18.1. *Genera and number of species per genus present in 16 samples of montane forests, 1200 m elevation, in the Hawaiian Islands. Each sample was 0.2 ha; eight were at mesic localities and eight were at wet localities.*

Tree genera (no. of spp./family)
Acacia (1 sp./Fabaceae)
Cheirodendron (2 spp./Araliaceae)
Ilex (1 sp./Aquifoliaceae)
Metrosideros (1 sp./Myrtaceae)
Myrsine (2 spp./Myrsinaceae)
Psychotria (1 sp./Rubiaceae)
Syzygium (1 sp./Myrtaceae)

Shrub and tree fern genera (no. of spp./family)
Broussaisia (1 sp./Hydrangeaceae)
Cibotium (3 spp./Cyatheaceae)
Clermontia (7 spp./Campanulaceae)
Coprosma (6 spp./Rubiaceae)
Cryptocarya (1 sp./Lauraceae)
Cyrtandra (5 spp./Gesneriaceae)
Dubautia (3 spp./Asteraceae)
Elaeocarpus (1 sp./Elaeocarpaceae)
Eurya (1 sp./Theaceae)
Hedytois (3 spp./Rubiaceae)
Labordia (3 spp./Loganiaceae)
Melicope (9 spp./Rutaceae)
Myrsine (2 spp./Myrsinaceae)
Perotettia (1 sp./Celastraceae)
Pipturus (1 sp./Urticaceae)
Pittosporum (2 spp./Pittosporaceae)
Psychotria (1 sp./Rubiaceae)
Rubus (1 sp./Rosaceae)
Styphelia (1 sp./Epacridaceae)
Tetraplasandra (1 sp./Araliaceae)
Vaccinium (2 spp./Ericaceae)
Wikstroemia (2 spp./Thymelaeaceae)

Pteridophyte genera (no. of spp./family)
Adenophorus (6 spp./Grammitidaceae)
Asplenium (10 spp./Aspleniaceae)
Athyrium (2 spp./Dryopteridaceae)
Callistopteris (1 sp./Hymenophyllaceae)
Coniogramme (1 sp./Pteridaceae)
Ctenitis (1 sp./Dryopteridaceae)
Dicranopteris (1 sp./Gleicheniaceae)
Diplopterygium (1 sp./Gleicheniaceae)
Dryopteris (8 spp./Dryopteridaceae)
Elaphoglossum (4 spp./Dryopteridaceae)
Grammitis (2 spp./Grammitidaceae)
Huperzia (2 spp./Lycopodiaceae)
Hypolepis (1 sp./Dennstaedtiaceae)
Lycopodium (2 spp./Lycopodiaceae)
Marattia (1 sp./Marattiaceae)
Mecodium (1 sp./Hymenophyllaceae)
Microlepia (1 sp./Dennstaedtiaceae)
Nephrolepis (2 spp./Dryopteridaceae)
Odontosoria (1 sp./Lindsaeaceae)
Ophioglossum (1 sp./Ophioglossaceae)
Pleopeltis (1 sp./Polypodiaceae)
Polypodium (1 sp./Polypodiaceae)
Psilotum (2 spp./Psilotaceae)
Pteris (1 sp./Pteridaceae)
Sadleria (3 spp./Blechnaceae)
Schizea (1 sp./Schizaeaceae)
Sphaerocionium (2 spp./Hymenophyllaceae)

Table 18.1. *(Cont.)*

Sticherus (1 sp./Gleicheniaceae)
Thelypteris (4 spp./Thelypteridaceae)
Vandenboschia (1 sp./Thelypteridaceae)
Xiphopteris (1 sp./Grammitidaceae)

Low shrub genera (no. of spp./family)
Cyanea (4 spp./Campanulaceae)
Cyrtandra (4 spp./Gesneriaceae)
Labordia (2 spp./Loganiaceae)
Viola (2 spp./Violaceae)

Liana genera (no. of spp./family)
Alyxia (1 sp./Apocynaceae)
Embelia (1 sp./Myrsinaceae)
Freycinetia (1 sp./Pandanaceae)
Phyllostegia (1 sp./Lamiaceae)
Smilax (1 sp./Smilaceae)
Stenogyne (5 spp./Lamiaceae)

Herbaceous genera (no. of spp./family)
Astelia (2 spp./Liliaceae)
Carex (1 sp./Cyperaceae)
Deschampsia (1 sp./Poaceae)
Dianella (1 sp./Liliaceae)
Gahnia (1 sp./Cyperaceae)
Isachne (1 sp./Poaceae)
Liparis (1 sp./Orchidaceae)
Luzula (1 sp./Juncaceae)
Machaerina (1 sp./Cyperaceae)
Nertera (1 sp./Rubiaceae)
Peperomia (8 spp./Piperaceae)
Uncinia (1 sp./Cyperaceae)
Viola (1 sp./Violaceae)

Parasitic genera (no. of spp./family)
Korthalsella (3 spp./Viscaceae)

Source: Kitayama and Mueller-Dombois (1995).

ment of native ferns in the ground cover (Cuddihy and Stone 1990). About 50,000 ha of mesic forests occur on the island of Hawaii. Maui and Kauai have only a few thousand hectares each. Mesic forests have often been degraded by browsing goats and cattle, as is the case in Kahikinui, Maui (Fig. 18.6), but excellent potential for recovery exists.

Montane parkland. Open woodlands ("mountain parklands") of *Acacia koa* and *Sophora chrysophylla* often occur at the transition of wet forest to the subalpine zone on the island of Hawaii (Cuddihy and Stone 1990). *Sophora* woodland at 1800—3000 m elevation on Mauna Kea declined with heavy browsing pressure by sheep and cattle (Scowcroft 1983; Scowcroft and Giffin 1983; Scowcroft and Sakai 1983) but began to recover well when these feral herbivores were removed (Scowcroft and Conrad 1992). *Acacia koa* parkland within Hawaii Volcanoes National Park at 1500–2000 m elevation was damaged by years of cattle browsing (Baldwin

and Fagerlund 1943; Cuddihy 1984) but has recovered dramatically after 20 years of protection from cattle (Tunison, Markiewicz, McKinney 1995).

Montane wetland. Hawaiian bogs occur, mostly at higher elevations, as openings in cloud forest on Kauai, Oahu, Molokai, Maui, and Hawaii. The vegetation of these areas contains taxa thought to be derived from ancestors in bogs, wet places, and alpine habitats elsewhere in the world and from Hawaiian wet forests (Carlquist 1980). At the generic level, affinities are particularly strong with the bogs of Malaysia, New Zealand, and southern Chile (Godley 1978; Canfield 1986).

The largest wetland complex in the Hawaiian Islands occurs on the Alakai Plateau of Kauai. Intense orographic precipitation falls on the Alakai Plateau, reaching the highest recorded precipitation on earth at Mt. Waialale, which received a record 16.9 m in 1982 and has a recorded annual mean precipitation of 11.3 m (Canfield 1986). Canfield distinguished three bog communities based on life form, species richness, structure, and floristics: prostrate bog, dwarf woody bog, and shrubby bog/bog margin. Important taxa included *Metrosideros polymorpha* vars. *incana* and *glabrifolia*, *Oreobolus furcatus*, *Rhynchospora spicaeformis*, and *Dicranopteris linearis*. These communities intergrade slightly and include, in world bog terminology, four mire types: poor fen, mixed sedge mire, lawn, and raised bog (Canfield 1986).

Montane bogs occur on northeastern Haleakala (East Maui) volcano at 1450–2270 m elevation, with vegetation dominated by the sedges *Carex alligata*, *Carex echinata*, *Carex montis-eeka*, *Oreobolus furcatus*, and *Rhynchospora rugosa* subsp. *lavarum*, the grass *Deschampsia nubigena*, and the moss *Rhacomitrium lanuginosum* (Loope, Medeiros, and Gagné 1991a) (Fig. 18.7). Although these treeless areas, surrounded by cloud forest, occupy fewer than 30 ha on East Maui, 15 endemic plant species are largely confined to them. A mostly water-impervious substrate layer and a very high annual rainfall create nearly permanently flooded conditions.

Because of their remote locations and extremely wet climates, most Hawaiian montane bogs have until recently been little disturbed. However, feral pigs had finally reached bogs on all islands by the 1980s. An example of dramatic, progressive pig damage to an area pristine until recently has occurred in the montane bogs of east Maui. Feral pigs arrived in bogs dominated by sedge (*Oreobolus furcatus* and *Carex echinata*) in the upper Hana rainforest of Haleakala National Park in the early 1970s. The pigs were apparently attracted to bog habitats by the availability of non-

Figure 18.6. Impact of goats and cattle on mesic forest in Maui. (A) Stand severely degraded by browsing goats and/or cattle. (B) Goat/cattle exclosure demonstrates that Aca-cia koa *reproduction becomes abundant in the absence of goats and/or cattle. Healani exclosure, 1280 m elevation.*

Figure 18.7. Pristine bog vegetation, East Maui, 1860 m elevation, prior to impact by feral pigs. Dominant species are Oreobolus furcatus and Carex echinata sedges. (Photograph taken by J. D. Jacobi in 1973.)

native earthworms, a probably important protein source (Loope et al. 1991a). In extreme situations, removal of plant cover in these bogs by pigs can approach 100% (Loope, Madeiros, and Gagné 1991b).

In the early 1980s, digging by pigs increased in bogs at 1650–1660 m elevation, and nonnative plant invasion was underway. A series of sampling plots was established to determine the trends. In sites dominated by *Carex echinata*, cover of nonnative species increased from 6% in 1982 to 30% in 1988, at which time the habitat was fenced to protect it from pigs. Since fencing, however, cover by native species has not increased because the invasion of nonnative plant species inhibited reproduc-

tion of rare endemic plants such as the Haleakala greensword (*Argyroxiphium grayanum*) (Medeiros, Loope, and Gagné 1991).

Montane grassland. Karpa and Vitousek (1994) present evidence that montane grasslands dominated by *Deschampsia nubigena* develop successionally from woody plant–dominated vegetation on pahoehoe lava flows on Mauna Loa at 1700–1950 m elevation. Woody species increase with substrate age initially (through 1300 yr of succession) but are eliminated by the age of 3000 years. On pahoehoe, woody species persist only below 1700 m elevation and above 1950 m. In contrast, on 'a'a lava, woody vegetation continues to increase at all elevations.

The mechanisms causing the shift from forest and shrubland to grassland vegetation on Mauna Loa are not known. However, impediment of drainage by accumulation of organic matter in the few cracks in pahoehoe lava, with consequent elimination of trees by waterlogging, appears to be a possibility (Karpa and Vitousek 1994).

Subalpine/Alpine Vegetation

Closed forest extends to only about 1900–2000 m, a limit that roughly coincides with the average level of the trade wind inversion (Leuschner and Schulte 1991) (Fig. 18.8a). Shrublands or open parkland with small trees (*Sophora/Myoporum* on Mauna Kea) occur above this elevation on volcanoes of Maui and Hawaii islands. Extensive high-elevations on these volcanoes is dominated by *Styphelia tameiameiae*, *Vaccinium reticulatum*, *Sophora chrysophylla*, *Dodonea viscosa*, *Coprosma montana*, *Geranium cuneatum*, and *Dubautia* spp. (Fig. 18.8b). Shrubland extends to lower elevation (montane zone) in mesic and dry sites with relatively young substrates.

A distinctive area of grassland dominated by the endemic bunchgrass *Deschampsia nubigena* occurs at 2200–2500 m elevation on the northeast outer slope of Haleakala, east Maui (Vogl 1971) (Fig. 18.8c). Grassland and subalpine shrubland vegetation of Haleakala National Park was highly impacted by pigs prior to their elimination in the 1980s. Nonnative gosmore (*Hypochoeris radicata*) leaves and roots, and bracken fern (*Pteridium aquilinum*) rhizomes were thought to be the preferred foods of pigs in subalpine grasslands and shrublands (Jacobi 1981). Chronic pig digging in such habitats gradually removes native vegetation and results in a progressive increase in nonnative grasses and forbs.

Subalpine and alpine communities were probably very little affected by Hawaiians and even today remain largely undeveloped, except for astronomical observatories with associated buildings and roads. Goats, sheep, and cattle have recently damaged vegetation of these communities.

One localized, monospecific community that has undergone a remarkable decline and recovery over the past century is that of the Haleakala silversword (*Argyroxiphium sandwicense* subsp. *macrocephalum*) (Fig. 18.8). Endemic to a 1000-ha area at 2100–3000 m elevation in the crater and outer slopes of Haleakala volcano, this distinctive globed-shaped rosette plant has rigid (swordlike) succulent leaves densely covered by silver hairs. The silversword grows in an environment that has one of the highest solar ultraviolet-B irradiation fluxes measured anywhere on earth (Caldwell, Robberecht, and Billings 1980). It is no surprise that this plant's leaves have one of the highest reflectances of both visible light and ultraviolet-B radiation (Robberecht et al. 1980).

Silversword is perhaps the most famous member of the endemic Hawaiian silversword alliance, a premier example of evolutionary adaptive radiation in plants. This morphologically diverse group is comprised of 28 species of herbs, vines, shrubs, trees, and rosette plants in three genera that evolved in the Hawaiian islands from a North American tarweed (Asteraceae: Madiinae) ancestor (Robichaux, Carr, Liebman, and Pearey 1990; Baldwin, Kyhos, Dvorak, and Carr 1991). The monocarpic (flowers only once, at the end of its lifetime) silversword matures from seed to its flowering stage in about 15–50 yr. The plant remains a compact rosette until it sends up an erect, central flowering stalk, sets seed, and dies.

Numbers of the Haleakala silversword were so depleted (probably to only 2000–4000 plants, although Otto Degener estimated fewer than 100 plants in 1927) by human vandalism and browsing by goats and cattle that the Maui Chamber of Commerce sent a petition to Washington, D.C., requesting that a serious effort be made to save the species within the newly established national park. Human vandalism was curtailed in the 1930s by the national park staff; cattle were eliminated from the park in the 1930s, feral goats in the 1980s. With protection, the silversword population has increased to more than 60,000 plants. More than 6000 flowered in a single year in 1991 (Loope and Medeiros 1994, 1995).

Alpine Vegetation

With increasingly severe climatic conditions at higher elevations and with proximity to a mountain's summit, vegetation becomes more sparse and smaller in stature. Vegetation at the uppermost limits of plant growth consists of prostrate shrubs (*Styphelia tameiameiae*, *Vaccinium reticulatum*), graminoids (*Deschampsia nubigena*, *Trisetum glomeratum*, *Agrostis sandwicensis*, *Luzula hawaiiensis*, *Carex* spp.), and ferns (*Pteridium aquilinum*, *Asplenium* spp.) (Hartt and Neal 1940; Fosberg 1959; Mueller-Dombois and Krajina 1968; Whiteaker 1983). The fresh lava substrates on Mauna Loa (4170 m) result in much less development of soil and vegetation there than on older Mauna Kea (4207 m) and Haleakala (3056 m), both of which have extensive outcrops of cinder and ash deposits (Fosberg 1959).

Figure 18.8. Subalpine communities on East Maui. (A) Abrupt transition of montane rainforest to subalpine shrubland and grassland at 2000 m, a limit roughly coinciding with the average elevation of trade wind inversion. (B) High-elevation shrubland, 2500 m, with Styphelia tameiameiae, Vaccinium reticulatum, Sophora chrysophylla, Co-

C

D

prosma montana, and Geranium cuneatum var. tridens. *(C) High-elevation grassland, 2200–2500 m, dominated by the endemic bunchgrass* Deschampsia nubigena. *(D) Haleakala silversword* (Argyroxiphium sandwicense *subsp.* macrocephalum) *on north slope of cinder cone within Haleakala Crater. (Photographs A, B, C courtesy of A. C. Medeiros; D. from R. P. Pharis.)*

Snow falls only occasionally on upper Haleakala volcano, occurring as low as 2500 m, but winter snow is frequent on Mauna Kea and Mauna Loa. Freeze–thaw processes result in a variety of stone stripes and other active patterned ground phenomena near Haleakala's summit (Noguchi, Tabuchi, and Hasegawa 1987). Haleakala occupies a marginal periglacial environment, whereas permanent ice exists < 1 m below the surface in the cinder of the summit cones of Mauna Kea (Woodcock 1976).

ECOSYSTEM DYNAMICS

Succession

With two active volcanoes, Kilauea and Mauna Loa, the island of Hawaii provides a prime site for examination of primary succession. For example, Smathers and Mueller-Dombois (1974) described the first nine years of succession on bare volcanic substrate of Kilauea Iki, a December 1959 eruption that devastated an existing montane rain and seasonal forest, covering an area of about 500 ha, leaving pahoehoe lava and a variable fallout of cinder. It was found that plants moved into the crater floor within the first year, with invasion related to favorable microsites (e.g., the bases of snags where moisture was enhanced), distance from seed sources, and a substrate-heat gradient. The increase in plant cover was much more rapid on habitats with remaining vegetation than on those without. There was recovery of some Metrosideros polymorpha trees that had been buried to depths of cinder of 2.5 m. Several native shrubs recovered after their entire shoot systems had been buried. The best herbaceous survivors in areas with light cinder deposition were those with underground storage organs, which included both native and nonnative species. On substrates without vegetation remnants, the sequence of life form establishment was algae first, then mosses and ferns, then lichens, then native woody seed plants, and finally nonnative woody and herbaceous seed plants. Native plants were in general considered better adapted to these pioneer environments (Smathers and Mueller-Dombois 1974).

In the past, numerous studies attempted to characterize succession on volcanic substrates of the island of Hawaii over a longer time scale, with very limited success (e.g., Forbes 1912; MacCaughey 1917; Skottsberg 1941; Atkinson 1970; and Eggler 1971). After studying vegetation on 16 young lava flows of Mauna Loa and Kilauea, Eggler (1971) was unable to find a detectable pattern.

Recent precise dating of lava flows, made possible by obtaining datable charcoal from beneath flows (Lockwood and Lipman 1980), has been done extensively by the U.S. Geological Survey with the primary objective of determining volcanic hazards (Lockwood, Lipman, Petersen, and Warshauer 1988). A byproduct of this effort has been the acquisition of knowledge about primary plant succession and ecosystem development, especially on the east flank of Mauna Loa volcano. On eastern Mauna Loa, environmental factors vary across a remarkably broad but well-defined range of conditions of rainfall, temperature, and substrate age, while the organisms and the chemical composition of parent material remain virtually constant. As a result, ecosystem scientists are provided with a "model system for ecosystem studies" (Vitousek 1995).

Drake and Mueller-Dombois (1993) examined forest structure on 'a'a lava flows at 1200 m elevation and mean annual precipitation of about 4000 mm. The eight flows ranged in age from 47 to 3400 yr. They found that (1) the biomass and height of the dominant Metrosideros polymorpha increased consistently with age, most rapidly in the early stages, and (2) colonizing individuals of Metrosideros with pubescent leaves were replaced in later stages by Metrosideros with glabrous leaves, confirming a phenomenon reported earlier by Stemmermann (1983).

Kitayama, Mueller-Dombois, and Vitousek (1995) studied community development flows aged 8–9000 yr in the same area where Drake and Mueller-Dombois (1993) worked, including some 'a'a flows partially covered with younger ash deposits. They found that the presence of ash increases nitrogen availability and standing biomass but does not appreciably alter the course of succession. Lower layers of the forest are dominated initially by a matted fern Dicranopteris linearis up to 300 yr and subsequently by tree ferns (Cibotium spp.) to 9000 yr (Fig. 18.9). The cover of Cibotium declined slightly after 3000 yr, but other native herb and shrub species increased. A climax stage was not reached on the observed age gradient because the sere changed continuously in biomass and species.

Aplet and Vitousek (1994) were able to extend these analyses by evaluating primary succession on a remarkable age-elevation matrix in the same general area of eastern Mauna Loa volcano, examining 'a'a flows 5–3400 yr old at six elevations (915–2400 m). A major objective was to determine how climate (which varies along the elevation gradient) affects the rate and pathway of primary succession. They found that standing plant biomass increased with substrate age and ranged from 0 to 36,000 g

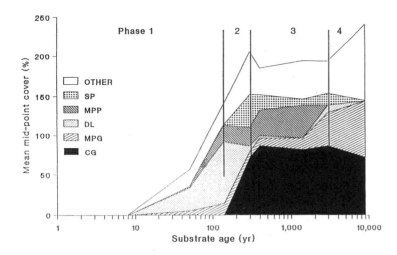

Figure 18.9. Four phases in the primary successional sere of Hawaiian montane rainforest. The vertical axis indicates cumulative cover values of the following successional dominants: CG = Cibotium glaucum, MPG = Metrosideros polymorpha var. glaberrima, DL = Dicranopteris linearis, MPP = Metrosideros polymorpha var. polymorpha, and SP = Sadleria pollida. Other = all other taxa combined. (From Kitayama and Mueller-Dombois 1995.)

Table 18.2. Aboveground live biomass (g m⁻²) on 29 sites representing six elevations and five 'a'a substrate ages. Eastern Mauna Loa volcano.

Altitude (m)	Flow age				
	5 yr	47 yr	137 yr	300 yr	3400 yr
914	0.462 (8)	653.0 (17)	1880.0 (12)	—(−)	31100 (30)
1219	0.028 (5)	462.0 (17)	1730.0 (17)	9180.0 (25)	11000 (24)
1524	0 (0)	72.9 (9)	323.0 (11)	5480.0 (18)	36400 (23)
1829	0 (0)	2.11 (3)	90.8 (13)	715.0 (12)	11100 (19)
2134	0 (0)	0.242 (6)	2.80 (7)	45.1 (6)	1370 (14)
2438	0 (0)	0 (0)	0 (0)	2.86 (3)	121 (9)

Source: Aplet and Vitousek (1994).

m⁻² across the matrix (Table 18.2). Biomass generally increased as elevation decreased (i.e., as climate became warmer and wetter), although the maximum value was observed at intermediate elevation on the oldest flow. Species richness ranged from 0 to 30 species plot⁻¹, with the maximum at the lowest elevation on the oldest flow.

Kitayama and Mueller-Dombois (1995) report on investigations along a 4.1 m yr soil age gradient at 1200 m elevation from Hawaii (Big Island) to Kauai. Eight "mesic" forest sites with annual rainfall of about 2500 mm and with substrate ages of 300, 2100, 5000, 20,000, 150,000 yr (all on the island of Hawaii), 410,000 yr (east Maui), 1.4 m yr (Molokai), and 4.1 m yr (Kauai) were chosen. In addition, eight "wet" forest sites with annual rainfall of about 4000 mm and with substrate ages of 400, 1400, 5000, 9000 yr (all on the island of Hawaii), 410,000 yr (east Maui), 1.3 m yr (west Maui), 1.4 m yr (Molokai), and 4.1 m yr (Kauai) were chosen.

A single tree species, *Metrosideros polymorpha*, dominates the forest canopies at all these sites. Mean height and diameter at breast height (dbh) of canopy *Metrosideros* trees increased from the

youngest site to peak values at the 2100, 5000, and 9000 yr sites and successively declined at older sites. Maximum standing biomass on intermediate-aged soils is consistent with findings of high availability of phosphorus in Hawaiian soils of intermediate age (Crews et al. in press) and high concentrations of the nutrients phosphorus and nitrogen in leaves of trees on intermediate-aged soils (Vitousek, Gerrish, Turner, Walker, Mueller-Dambois). The crucial element phosphorus becomes less and less available to plant roots as Hawaiian soils age and lose available phosphorus to insoluble and physically protected forms and (to a lesser extent) through leaching.

The pattern of plant successional replacement of "pioneer" species was similar in the wet and mesic sites, with the replacement occurring more rapidly at the wetter sites, consistent with the concept that fertility declines more rapidly in soils of wetter forests because of more rapid leaching. Species richness on both mesic and wet gradients peaked at East Maui sites (410,000-yr-old substrate, well past maximum phosphorus availability), indicating that species richness is not determined by soil fertility alone. The authors conclude: "Since the size of canopy trees and the turnover of associated plant species vary as predicted from the established pattern of nutrient availability, we suggest that the underlying resource gradient (i.e., availability of phosphorus, primarily) is a principal cause for the observed vegetation pattern" (Kitayama and Mueller-Dombois 1995).

Whereas succession in wet forest sites is becoming well understood, little is known about dry forest succession, partly because this type of vegetation has suffered so severely from degradation by nonnative grazing animals, fire, land conversion, and invading plant species that natural succession is only rarely actively occurring and is difficult to conceptualize. Nevertheless, work by Stemmermann and Ihsle (1993), using dated lava flows up to 4000 years in age, has immensely advanced knowledge of dry forest succession on the island of Hawaii. They found that *Metrosideros polymorpha* consistently dominates in early succession on lava flows in dry areas (as it does in wet areas) but is replaced by other tree taxa (e.g. *Diospyros, Sophora, Myoporum, Dodonaea*) in later seral stages. Tree density is greater on older lava flows, and the dominant species there exhibit significantly lower midday plant water potentials than *Metrosideros* growing on adjacent younger flows. This pattern suggests that the greater water stress to which the later seral trees are subjected may be due to competition for limited water, and this may restrict *Metrosideros* from these sites (Stemmermann and Ihsle 1993). In contrast, of course, *Metrosideros* persists as the dominant tree on wet sites where water stress is only rarely a problem.

Succession and Ecosystem Function

The environmental matrix on eastern Mauna Loa has been further used to examine parameters of ecosystem function. Annual net primary production (ANPP) was examined in sites dominated by *Metrosideros polymorpha* and *Dicranopteris linearis*, along an elevation gradient of 290–1660 m with a parallel temperature gradient of 22–13° C. Raich, Russell, and Vitousek (1997) and Vitousek et al. (1994) found that annual net primary productivity of the vegetation decreased linearly by 35 g/m^2/y for every 100 m increase in elevation. Maximum productivity was about 600 g/m^2/y.

Vitousek et al. (1994) determined rates of decomposition at four elevations on the same flows, based on measurement of mass loss over two years of *Metrosideros* leaf litter collected at each site, of litter collected at a single site and distributed to all four elevations, and of wooden dowels. Decom-

position increased exponentially with decreasing elevation (increasing temperature) for all substrates used (Fig. 18.9). Vitousek, Turner, Parton, and Sanford (1994) conclude that increasing temperature increases both ANPP and decomposition but that temperature affects decomposition more than it does production at these sites. Since decomposition slows with decreasing temperature, nutrient availability to plants may decline with decreasing temperature, and nutrient supply may be reduced to a greater extent than nutrient demand. Consistent with this deduction, concentrations of nitrogen (N) and phosphorus (P) in *Metrosideros* foliage were found to decrease significantly with increasing elevation on the east flank of Mauna Loa (Vitousek et al. 1992).

Whereas decreasing temperature slows decomposition more than it slows production, decreased precipitation reduces nutrient demand by plants more than it reduces supply from decomposers, so that nutrient availability is relatively greater on dry sites than on wet sites of comparable age and elevation on eastern Mauna Loa. Concentrations of both N and P in foliage are higher in dry sites, and the decomposability of tissue produced there is substantially greater (Vitousek, Aplet, Turner, and Lockwood 1992; Vitousek et al. 1994).

Accumulation of biologically available N and P occurs in developing ecosystems on lava flows up to several thousand years old, as nutrients are added from the atmosphere and from mineral weathering (Vitousek, Matson, and Turner 1988; Vitousek et al. 1992). Biologically available P decreases substantially on older substrates in the Hawaiian Islands (Crews et al. 1995; Vitousek, Turner, and Kitayama 1995).

'Ohi'a (*Metrosideros*) Dieback

An understanding of the soil resource gradient is likely to be important in understanding the phenomenon known as 'ohi'a (*Metrosideros*) dieback. Because *Metrosideros* dominates 80% of the relatively intact remaining forest in the Hawaiian Islands, 'ohi'a dieback is an extremely important phenomenon. An added significance is that aggressive nonnative plants can establish on a large scale during a dieback if seed sources are present in the area, resulting in relatively irreversible native forest deterioration.

In the 1970s, concern arose because of major stand-level dieback of *Metrosideros* on the island of Hawaii. Death of virtually all canopy trees in entire stands had moved through and affected about half of an 80,000 ha area of rainforest on the eastern slopes of Mauna Loa and Mauna Kea from the

1950s through the 1970s (Hodges, Adee, Stein, Wood, and Doty 1986). The phenomenon was initially thought to be caused by a pathogen, but a decade of disease research led to the conclusion that the trees were dying of some other cause (Mueller-Dombois 1985; Hodges et al. 1986).

Suggested environmental causes of dieback at first included flooding of *Metrosideros* root systems after periods of above-normal rainfall, drought, sulfur dioxide fumes, and damage by feral ungulates, but on closer examination none of these factors appeared related to dieback patterns. Mueller-Dombois (1983) concluded that many dieback stands of *Metrosideros* are in a senescing life stage and that less noticeable climatic perturbations had triggered dieback over the large area.

Another possibility is that this dominant species faces reduced competition in its island environment and thus invades sites over an extremely broad ecological spectrum, on some of which it is not well adapted (e.g., soils with toxic aluminum levels or waterlogged soils). As a consequence, it dies periodically over part of its range in response to environmental triggers (Mueller-Dombois 1986). But since no native species is capable of replacing it as a dominant, *Metrosideros* seedlings reinvade.

Jacobi, Gerrish, and Mueller-Dombois (1983) have determined, through analysis of permanent plots, that forest dieback is indeed followed in many instances by vigorous regeneration of *Metrosideros*. The predisposing cause for stand-level dieback appears to be the cohort (even-aged) nature of most *Metrosideros* forest stands. After canopy closure of the first-generation forest, *Metrosideros* regeneration is confined to small seedlings only, which remain small and turn over under the canopy without graduating into the sapling stage. *Metrosideros* seedlings and saplings are recruited into the canopy only after the canopy collapses. Such synchronous stand dieback is to be expected only in forests with a canopy composed of a single species and thus only (at least in the tropics or subtropics) on islands or mountains where biogeographic isolation has strongly limited colonization by other canopy species (Mueller-Dombois 1987).

Fire and Hawaiian Vegetation

Fire does not appear to have played an important evolutionary role in most native ecosystems of the Hawaiian Islands, and relatively few Hawaiian endemic plant species have adaptations to fire (Mueller-Dombois 1981b). Lightning is uncommon on oceanic islands because a small island's land mass is not conducive to convective buildup of thunderheads. Many native Hawaiian ecosystems may have lacked adequate fuel to carry fires ignited by lightning or volcanism (Mueller-Dombois 1981b). Pollen cores from lowland Oahu (Athens et al. 1992) lack charcoal and suggest infrequency of fires either before or after Polynesian settlement in Hawaii. Fires in modern Hawaii are mostly human-caused, are fueled primarily by nonnative grasses, and are generally highly destructive to native plant species.

Opportunistic invasive nonnative plant species, on the other hand, typically spread rapidly following fire or other disturbance. In natural areas of Hawaii, fire is therefore usually considered a negative influence that must be suppressed to the extent possible. Exceptions may involve the cautious use of prescribed fire, with accompanying ecological research, in restoration of *Acacia koa* woodlands or *Dodonea* and *Sophora* shrublands.

Polynesians may have used fire to manipulate vegetation in the lowlands in a slash-and-burn form of agriculture (Kirch 1982; Cuddihy and Stone 1990). Although fire frequency and size have generally been low since European contact, large fires have occurred occasionally, and their incidence has greatly increased in recent decades because of spread of nonnative grasses (Smith and Tunison 1992).

Hawaii Volcanoes National Park is the only natural area in Hawaii with a substantial fire history continuously recorded over the past 60 years. Most of the fires (73%) in the park have been human-caused; lava flows ignited 25% and lightning ignited 2% (Smith and Tunison 1992). Fire frequency and size in Hawaii Volcanoes National Park have increased dramatically since goat control, which began in the early 1970s, resulted in dramatically increased accumulation of fine fuels. The protection of Hawaii Volcanoes park and other natural areas can be enhanced by aggressive fire suppression and presuppresion planning, including construction of fuel breaks, preparation of fuels distribution maps, strengthening of interagency fire agreements, and fire prevention (Smith and Tunison 1992).

Mechanisms driving loss of native species through fires fueled by nonnative grasses are being explored. The major grasses fueling fires in Hawaii Volcanoes National Park are *Schizachyrium condensatum* and *Melinis minutiflora*. The initial invader of otherwise undisturbed native ecosystems is *Schizachyrium*, which adds enough fine fuel to previously fire-free sites to carry fire (Hughes, Vitousek, and Tunison 1991). Most native species are eliminated by fire (if not by the first, then by later ones), whereas *Schizachyrium* recovers rapidly after fire and the even more flammable *Melinis* invades. The increased dominance of grasses enhances flammability and subsequent fires lead to nearly mono-specific stands of *Melinis* (Hughes et al. 1991; Hughes and Vitousek 1993).

CONTINUING EFFECTS OF INVASIVE PLANTS AND ANIMALS

Oceanic islands are extremely vulnerable to certain types of human-related biological invasions because the biota of such islands has evolved for such a long time in isolation from the continual challenge of some of the selective forces that have shaped the evolution of continental organisms. These forces include browsing and trampling by herbivorous mammals, predation by ants, virulent disease, and frequent and intense fires (Loope and Mueller-Dombois 1989). Taxonomic "disharmony," the absence or relative absence of certain important groups of organisms, which is a major cause of evolutionary adaptive radiation of successful immigrants over time, is another cause of vulnerability to invasion. And the inherent vulnerability of islands to invasion and extinction has clearly been exacerbated by exceptionally heavy exploitation by human populations.

It was stated earlier that the known native macroscopic biota of the Hawaiian Islands can be accounted for by one successful immigrant every 35,000 years over a 70-M-yr period. With the arrival of the Polynesians in the fourth century A.D., this rate of immigration began to accelerate to about 3–4 per century for about 1400 years (Kirch 1982; Nagata 1985). In modern times, Beardsley (1979) found that 15–20 species of immigrant insects alone become established in Hawaii each year. The Hawaiian archipelago has more than 8000 introduced plant species or cultivars (Yee and Gagné 1992), an average of 40 introductions per year when averaged over the past two centuries; 861 (11%) of these are now naturalized (Wagner et al. 1990). Smith (1985) listed 86 invasive nonnative plant species present in Hawaii that pose threats to native Hawaiian ecosystems, but this list is undergoing reevaluation because new invaders and trends have become apparent within the past decade.

The following accounts sketch the biology and effects of some of the most destructive invading species in Hawaii. Efforts to control these species for conservation purposes have met with varying levels of success. In addition, species already present may increase their destructiveness in the future, and new species will invade. Efforts to prevent new invasive species from establishing in Hawaii are extremely important (Nature Conservancy of Hawaii and Natural Resources Defense Council 1992; U.S. Congress, Office of Technology

Assessment 1993). Active management is needed to protect native Hawaiian ecosystems from being eventually overwhelmed by forces that began to act 1500 yr ago with the arrival of Polynesians on Hawaiian shores.

Feral Pigs (*Sus scrofa*)

Feral pigs are currently the primary modifiers of remaining Hawaiian rainforest and have substantial effects on other ecosystems. Although pigs were brought to the Hawaiian Islands by Polynesians as early as the fourth century A.D., the currently severe environmental damage done by pigs apparently began much more recently and seems to have resulted entirely from release of domestic, non-Polynesian genotypes (Diong 1982). There is little or no evidence that pigs were a major factor degrading Hawaiian ecosystems prior to the twentieth century. In the early 1900s, the damage by feral European pigs in native rainforests was recognized; the Hawaii Territorial Board of Agriculture and Forestry started a feral pig eradication project that lasted until 1958 and removed 170,000 pigs from forests statewide (Diong 1982).

Perhaps the best case for the recentness of pig impacts comes from the island of Maui. Although goats, cattle, and wild dogs were reported by nineteenth century explorers of the island, the explorers made no mention of seeing pigs (Diong 1982). There were no feral pigs in west Maui until someone introduced them in the 1960s; although they have spread throughout the northeastern portion of the west Maui Mountains, the remainder of high-elevation west Maui has never been degraded by pigs (R. W. Hobdy, Hawaii Division of Forestry and Wildlife, personal communication). Feral pigs were first seen in high-elevation east Maui (Haleakala Crater) in the 1930s at Paliku (elevation 1920 m). These pigs were probably derived from runaway domestic breeds in lowland areas. Perhaps aided by a seasonally abundant and expanding carbohydrate source from the invasion of nonnative strawberry guava and by an enhanced protein source from abundant nonnative earthworms, truly feral pig populations developed and spread into adjacent pristine forest. By 1945, pigs had moved into the upper Kipahulu Valley of east Maui, though an expedition in that year found the valley pig-free between 610 and 1375 m elevation. By 1967, pig damage could be found for the first time throughout Kipahulu Valley – although damage at that time was still moderate (Warner 1968). After 1967, pig densities greatly increased. By 1979–1981, pig densities in Kipahulu Valley ranged from 5 to 31 per km² (13–80 per mi²) (Diong 1982). A similar pattern has taken place across the entire north and northeastern slope of east Maui during the past 30 years (R. W. Hobdy, personal communication).

Pigs are omnivores and opportunistic feeders. Diong (1982) found the following contents in pig stomachs collected throughout the year in koa forests of Kipahulu Valley: native tree fern (*Cibotium*) core – 43%; strawberry guava fruits – 28%; other plant material (leaves, bark, roots) – 19%; earthworms – 6%. Analysis of stomachs collected in koa forests during the September-November fruiting season yielded a 78% volume of strawberry guava fruits; seed germination of strawberry guava was enhanced by passing through pig digestive tracts. Native tree ferns provided an abundant, high-energy food source available throughout the year in Kipahulu Valley and may have determined the carrying capacity of the habitat for feral pigs.

In addition to their common food items, pigs selectively seek out certain currently rare plant species for food – such as the fern *Marattia*, the lily-like plant *Astelia*, the vine *Freycinetia*, and lobelias. Some plants with particularly fragile stems and leaves (*Cyrtandra*, mints, orchids) have drastically declined due to trampling by pigs. The ground cover of small ferns (*Vandenboschia, Adenophorus, Grammitis, Xiphopteris, Sphaerocionium, Mecodium, Callistopteris*, etc.) has probably been altered most, based on observations made in the few remaining pig-free areas of the Hawaiian Islands – areas protected by surrounding cliffs such as Lihau on West Maui, Olokui on Molokai, and a few kipukas (areas with soil and vegetation surrounded with recent lava flows) on Hawaii. These mosses and ferns have, for the most part, not been totally eliminated from the native vegetation since they survive as epiphytes on tree trunks (especially trunks of native tree ferns) and on downed logs. In rainforests of east Maui, tree ferns (*Cibotium*) – originally the dominant subcanopy species at elevations up to 1500 m – are rapidly being depleted. As mature tree ferns become further reduced in abundance, fewer individuals of other rare plant species will be able to survive as epiphytes.

Opportunistic plant species, often nonnatives, occupy the habitats left by removal of native species by feral pigs. Seeds of nonnative plants are carried on pigs' coats or in their digestive tracts; the seeds thrive on germination on the forest floor where pig-digging has exposed mineral soil. Once aggressive plant invaders have obtained a new foothold in the forest, they spread opportunistically, aided by pigs and nonnative birds. The spread of nonnative plants has been better documented for Kipahulu Valley than for most remote areas in Hawaii (Yoshinaga 1980; Anderson, Stone, and Higashino 1992). Removal of feral pigs from the valley in the late 1980s (Anderson and Stone 1993) has substantially slowed the rate of nonnative plant invasion and has already resulted in substantial recovery of the forest understory.

Feral Goats (*Capra hircus*)

The negative impact of goats on vegetation is well known worldwide. The Hawaiian situation is worse than elsewhere because Hawaiian plant species evolved in the absence of mammalian grazing and browsing pressure. Goats were introduced to the eight major Hawaiian islands before 1800 and within a few decades had reached remote areas (Cuddihy and Stone 1990). Goats have reduced or eliminated whole populations of native plants, have facilitated nonnative plant invasion, and have hastened soil erosion (Yocum 1967; Baker and Reeser 1972; Spatz and Mueller-Dombois 1973; Mueller-Dombois and Spatz 1975). Sustained locally high populations of goats have resulted in obliteration of even the most unpalatable native plant species. Indeed, some plant species survive only on ledges and other sites that are inaccessible to goats, or as trees that are apparently older than the period during which goats have occupied their habitats (Stone and Loope 1987).

Goats have been eradicated on Niihau, Lanai (1981), and Kahoolawe (1990). Eradication of feral goats on uninhabited Kahoolawe (a former bombing range, 115 km² or 45 mi² in area) was conducted by the U.S. Navy prior to returning the island to the State of Hawaii; unfortunately, most native vegetation had already been destroyed. Feral goats have also been virtually eliminated in Hawaii's two major national parks, Hawaii Volcanoes and Haleakala, resulting in dramatic changes in vegetation dynamics (e.g., Mueller-Dombois 1981a; Stone, Cuddihy, and Tunison 1992; Tunison et al. 1995) and at least partial recovery of rare plant populations (e.g., Loope and Medeiros 1994).

Cattle (*Bos taurus*)

Feral cattle were abundant in the Hawaiian Islands in the 1800s and early 1900s (e.g., Tomich 1986). Cattle have been a major contributor to the decline of many plant species (cf. Rock 1913). Most cattle grazing is now on private, managed lands, but wild cattle persist locally, numbering in the thousands on the island of Hawaii (P. M. Vitousek, Stanford University, personal communication), in forested areas. Relatively small numbers of feral or domestic cattle can do appreciable damage in areas with native vegetation. Cattle are a continuing significant threat to the Hawaiian flora, especially in certain coastal and lowland leeward habitats.

Rats and Mice (*Rattus rattus, Rattus exulans,* and *Mus musculus*)

Of four rodent species introduced to the Hawaiian Islands, the arboreal black rat (*Rattus rattus*) probably has had the greatest impact on native fauna and flora (Stone and Loope 1987). Rodents feed on the fleshy fruits and flowers of Hawaiian plants and girdle and strip tender branches (Scowcroft and Sakai 1984; Cuddihy and Stone 1990, Sugihara 1997).

Fountain Grass (*Pennisetum setaceum*)

This large bunchgrass from northern Africa has spread aggressively, with high densities, throughout the North Kona District of the island of Hawaii since its introduction there in the early 1900s (Tunison 1992). It is a fire-stimulated grass infamous for carrying intense fires through formerly open lava flow. Scattered native vegetation and many rare plants not jeopardized by fire in the days before fountain grass are now killed. Concerted effort at Hawaii Volcanoes National Park has successfully kept it contained there for more than 15 yr (Tunison 1992). Fountain grass has been present on Maui for 30 yr only in a small area of the sandhills of Wailuku. However, it poses a serious threat to rangelands and remnant dryland forest of southern east Maui with lightly vegetated, young volcanic substrates, as well as to the largely barren, relatively undisturbed ecosystems of upper Haleakala (Loope, Nagata, and Medeiros 1992). Based on its occurrence as high as 2900 m on Mauna Kea (Jacobi and Warshauer 1992), it must be regarded as a potential invader of Haleakala Crater.

Banana Poka (*Passiflora mollissima*)

This passionflower vine, locally known as banana poka, was introduced to the Hawaiian Islands as an ornamental about 1900. It has since become established in more than 520 km² of native forest on the islands of Hawaii and Kauai. In some areas, it has become so dense that the vines drape from tree to tree, smothering large tracts of native forest. It is established on a relatively small area of Maui. It potentially occupies elevations of 500–2500 m (Jacobi and Warshauer 1992) and thrives where mean annual precipitation is 1000–5000 mm. Seeds are spread by feral pigs, birds, and humans (Jacobi and Warshauer 1992; La Rosa 1992). Several insects have been released for biological control of banana poka (Markin, Lai, and Funasaki, 1992), but no negative effects on the health of plants in the wild have yet been noted.

Strawberry Guava (*Psidium cattleianum*)

A shrub, small tree, or large tree, depending on density of stocking and habitat conditions, strawberry guava establishes very dense stands from primarily pig-dispersed seed (Diong 1982) and shades out native species. Introduced to the islands in the 1800s, it is now very widespread but still spreading in rainforests of all islands. On the island of Hawaii, its elevational range is up to 1300 m, and it occupies sites with mean annual rainfall ranging from <1250 mm to >7000 mm (Jacobi and Warshauer 1992). Its elevational range in Haleakala National Park's Kipahulu Valley, Maui, is 90–1190 m, where it grows in dense, "dog-hair" thickets at 460–760 m. Attainment of local dominance is facilitated by clonal reproduction (Huenneke and Vitousek 1990).

The rate of spread of strawberry guava appears to have been reduced dramatically since most pigs were removed from Kipahulu Valley in the late 1980s (S. Anderson, pers. comm.). However, the species is also dispersed by birds. Huenneke and Vitousek (1990) found that seedling germination of *P. cattleianum* in the field appears to be independent of soil disturbance; naturally occurring seedlings were found disproportionately on bryophyte mats and other undisturbed substrates – the same substrates that support germination of native forest plant species.

Strawberry guava potentially threatens numerous middle- to low-elevation rainforest plant species with local extirpation through displacement of reproduction. Species of east Maui particularly threatened in this way include *Antidesma platyphyllum, Claoxylon sandwicense, Joinvillea ascendens, Nothocestrum longifolium, Psychotria mariniana, Sicyos cucumerinus,* and *Strongylodon ruber* (Loope et al. 1992).

Australian Tree Fern (*Cyathea cooperi*)

Cyathea tree ferns have been in cultivation in the Hawaiian Islands at least since the 1960s as ornamentals around homes and in botanical gardens. The widely cultivated species, *Cyathea cooperi,* is native to Queensland and New South Wales in eastern Australia. It is widely planted in Hawaii because it is a hardy, attractive species and is faster growing than native Hawaiian tree ferns (*Cibotium* spp.).

It has been recently (late 1980s) discovered that populations of *C. cooperi* are invasive in 'ohi'a (*Metrosideros polymorpha*) and koa (*Acacia koa*) rainforests on Kauai and in the Kipahulu Valley of Haleakala Park on Maui (Medeiros et al.

1992). Given the early stage of the invasion, the ultimate outcome is difficult to predict. Based on early observations, the greatest threat posed by *C. cooperi* to Hawaiian forests is its displacement of native species where the fern has achieved high densities and local dominance. Unlike native *Cibotium* tree ferns, *Cyathea* does not support the dense colonies of epiphytic native plant species that often colonize the trunks of tree ferns (Medeiros et al. 1993). Where *Cyathea* forms dense stands in Kipahulu, the understory is conspicuously open and lacking many characteristic native species normally found there. This is apparently due to exclusion of other species by the thick layering of fibrous roots that forms at the soil surface surrounding a growing tree fern. On large tree ferns of this species, this dense layer of near-surface roots may extend out over a diameter of 3–4 m, effectively excluding most other vegetation (Medeiros et al. 1992).

Miconia calvescens

Miconia calvescens (Melastomataceae), native to neotropical forests at 300–1800 m elevation, is now known to be an unusually aggressive invader of moist island habitats. Introduced to Tahiti in 1937, dense thickets of *M. calvescens* had by the 1980s replaced the native forest over most of the island, with dramatic reduction of biological diversity (Meyer 1996). After the late F. R. Fosberg saw this species in Tahiti in 1971, he reported that "it is the one plant that could really destroy the native Hawaiian forest." Because of its attractive purple and green foliage, it was inadvertently introduced to Hawaii as an ornamental in the 1970s. After its detection on Maui by conservation agencies in 1990, an alarm was raised. Nearly 20,000 individuals of *M. calvescens* were removed from private lands by agency staff and volunteers from 1991–1993, and control appeared feasible. However, in September 1993, an aerial vegetation survey discovered a previously undetected *Miconia* population on state land – far larger (> 100 ha) than all previously known populations on Maui. An interagency working group developed and began (January 1994) implementation of a containment strategy, initially involving helicopter herbicide (Garlon 4) spraying of individual emergent *Miconia* trees and monitoring of results. Efforts to mobilize a control effort commensurate with the task began in early 1995 (Conant, Medeiros, and Loope 1997).

Clidemia hirta

This densely branching shrub (to 4 m tall) is native to the neotropics (southern Mexico and West Indies

to Argentina). It has become an aggressive invader in many parts of India, Southeast Asia, and the Pacific islands, including the Hawaiian Islands, where it was introduced about 1940. In Hawaii, *Clidemia* forms dense, monotypic stands in mesic to wet environments. The infestation is particularly severe on Oahu, where it expanded its area, probably by bird-dispersal, from less than 100 ha in 1952 to an estimated 31,000–38,000 ha in the late 1970s; more recently, the area occupied has approached the habitat available and is estimated at 100,000 ha (Smith 1992). It has now spread to Hawaii (1972), Molokai (1973), Maui (1976), Kauai (1982), and Lanai (1988). Its elevation range is from just above sea level to about 1220 m. The primary mode of interisland dispersal is believed to be in mud on boots. Several biological control agents show promise of limiting the further invasion of *Clidemia* on islands other than Oahu (Smith 1992; Nakahara, Burkhart, and Funasaki 1992).

Faya Tree (*Myrica faya*)

One of the worst invaders in Hawaii Volcanoes National Park, this small tree (4–6 m tall) from the Azores, Madeira, and the Canary Islands is an actinorhizal nitrogen fixer with great potential for massive alteration of early successional ecosystems through nutrient enrichment. Rapidly spreading, aided by efficient dispersal by the nonnative bird *Zosterops japonica*, seedlings of *Myrica* generally appear under *Metrosideros* perch trees (Woodward et al. 1990). Planted seeds of *Myrica* also germinate and establish better under *Metrosideros*; diameter growth of *Myrica* is 15-fold greater than that of *Metrosideros* (Vitousek and Walker 1989). The rate of nitrogen fixation in a dense stand of *Myrica* was measured at 18 kg ha^{-1} yr^{-1} (Vitousek and Walker 1989) the maximum rate in native forest – where the major substrates were lichens, liverworts, decaying leaf litter, and decaying wood – was only 2.8 kg ha^{-1} yr^{-1} (Vitousek 1994). In response to nitrogen-fixing *Myrica*, nonnative earthworm biomass in a high-density *Myrica* stand was over three times that in nearby native forest (Aplet 1990).

Myrica often forms dense stands that shade out native competitors. This species was brought to Hawaii in the 1920s for reforestation. It has invaded about 34,000 ha statewide, of which 29,000 ha are on the island of Hawaii (Whiteaker and Gardner 1992). It is also spreading on Kauai, Oahu, Lanai, and Maui.

AREAS FOR FUTURE RESEARCH

Much has been written about the tragic dismemberment of the Hawaiian biota. The extent of the losses is unequaled in any other region of the United States. What is not generally appreciated is that much of Hawaii's unique biological heritage remains and can be protected with careful management. Large tracts of near-pristine ecosystems remain at high elevation on Hawaii, Maui, Molokai, and Kauai. Much remains to be lost. Careful management of the flow of species into Hawaii from the continental United States and from foreign countries is crucial to long-term protection of Hawaii's natural heritage.

If the Hawaiian Islands are to continue as a prime research site for evolutionary studies, conservation efforts, supported by basic and applied research, must flourish. Continuing applied research is needed: (1) to understand the biology and impact of invasive species; (2) to provide the tools needed to manage the most destructive invasive species; and (3) to provide the tools for ecological restoration.

Hawaii's surviving natural heritage is a unique national treasure. As a result of human introduction of invasive nonnative species, this natural heritage is in serious jeopardy. Continuing ecological research and refinement of management strategies have important roles. However, biologists know enough to confidently predict that much can be saved in the long run if the political will exists to implement appropriate management. It seems that the crucial factor limiting conservation of biological diversity in Hawaii is not a dearth of research but of public understanding and support, at both the state and national levels. Ironically, protection of terrestrial biodiversity in Hawaii is threatened less by economic conflicts than by ignorance and apathy.

REFERENCES

Abbott, I. A. 1992. La'au Hawai'i: traditional Hawaiian uses of plants. Bishop Museum Press. Honolulu, Hawaii.

Anderson, S. J., and C. P. Stone. 1993. Snaring to control feral pigs *Sus scrofa* in a remote Hawaiian rain forest. Biological Conservation 63:195–201.

Anderson, S. J., C. P. Stone, and P. K. Higashino. 1992. Distribution and spread of alien plants in Kipahulu Valley, Haleakala National Park, above 2300 ft elevation, pp. 300–338 in C. P. Stone, C. W. Smith, and J. T. Tunison (eds.), Alien plant invasions in native ecosystems of Hawaii: management and research. University of Hawaii Cooperative National Park Resources Studies Unit, Honolulu.

Aplet, G. H. 1990. Alteration of earthworm community biomass by the alien *Myrica faya* in Hawai'i. Oecologia 82:414–416.

Aplet, G. H., and P. M. Vitousek. 1994. An age-altitude matrix analysis of Hawaiian rain-forest succession. J. Ecol. 82:137–147.

Armstrong, R. W. (ed.). 1983. Atlas of Hawaii. University Press of Hawaii, Honolulu.

Athens, J. S., and J. V. Ward. 1993. Environmental change and prehistoric Polynesian settlement in Hawai'i. Asian Perspect. 32:205–223.

Athens, J. S., J. V. Ward, and S. Wickler. 1992. New Zealand J. Archeol. 14:9–34.

Atkinson, I. A. E. 1970. Successional trends in the coastal and lowland forest of Mauna Loa and Kilauea Volcanoes, Hawaii. Pacific Sci. 24:387–400.

Baker, J. K., and D. W. Reeser. 1972. Goat management problems in Hawaii Volcanoes National Park: a history, analysis, and management plan. National Park Service Natural Resources Rept. 2. U.S. Department of the Interior, Washington, D.C.

Baldwin, B. G., D. W. Kyhos, J. Dvorak, and G. D. Carr. 1991. Chloroplast DNA evidence for a North American origin of the Hawaiian silversword alliance. Proceedings of the National Academy of Sci. (USA) 88:1840–1843.

Baldwin, P. H., and G. O. Fagerlund. 1943. The effect of cattle grazing on koa reproduction in Hawaii National Park. Ecology 24:118–122.

Beardsley, J. W. 1979. New immigrant insects in Hawaii: 1962 through 1976. Proceedings of the Hawaiian Entomological Society 23:35–44.

Caldwell, M. M., R. Robberecht, and W. D. Billings. 1980. A steep latitudinal gradient of solar ultraviolet-B radiation in the arctic-alpine life zone. Ecology 61:600–611.

Canfield, J. E. 1986. The role of edaphic factors and plant water relations in plant distribution in the bog/wet forest complex of Alaka'i Swamp, Kaua'i, Hawaii. Ph.D. dissertations, University of Hawaii, Honolulu.

Carlquist, S. 1980. Hawaii: a natural history. 2nd ed. Pacific Tropical Botanical Garden, Lawai, Kauai, Hawaii.

Chu, P. 1989. Hawaiian drought and the Southern Oscillation. Intl. Jour. Climatol. 9:619–631.

Conant, P., A. C. Medeiros, and L. L. Loope. 1997. A multi-agency containment program for miconia (*Miconia calvescens*), an invasive tree in Hawaiian rain forests, pp. 249–254 in J. Luken and J. Thieret (eds.), Assessment and management of invasive plants. Springer-Verlag, New York.

Crews, T. E., K. Kitayama, J. Fownes, D. Herbert, D. Mueller-Dombois, R. Riley, and P. M. Vitousek. 1995. Changes in soil phosphorus and ecosystem dynamics across a long soil chronosequence in Hawaii. Ecology 76:1407–1424.

Cuddihy, L. W. 1984. Effects of cattle grazing on the mountain parkland ecosystem, Manna Loa, Hawaii. University of Hawaii, Cooperative National Park Resources Studies Unit. Tech. Rept. 51. Honolulu.

Cuddihy, L. W., and C. P. Stone. 1990. Alteration of native Hawaiian vegetation: effects of humans, their activities and introductions. University of Hawaii, Cooperative National Park Resources Studies Unit, Honolulu.

Diong, C. H. 1982. Population biology and management of the feral pig (*Sus scrofa* L.) in Kipahulu Valley, Maui. Ph.D. dissertation, University of Hawaii, Honolulu.

Drake, D. R., and D. Mueller-Dombois. 1993. Population development of rain forest trees on a chronosequence of Hawaiian lava flows. Ecology 74:1012–1019.

Eggler, W. 1971. Quantitative studies of vegetation on sixteen young lava flows on the island of Hawaii. Trop. Ecol. 12:66–100.

Forbes, C. N. 1912. Preliminary observations concerning the plant invasion in some of the lava flows of Mauna Loa, Hawaii. Occasional Papers of the Bishop Museum 5:15–23.

Fosberg, F. R. 1948. Derivation of the flora of the Hawaiian Islands, pp. 107–119 in E. C. Zimmerman (ed.), Insects of Hawaii, Vol. 1. University of Hawaii Press, Honolulu.

Fosberg, F. R. 1959. The upper limits of vegetation on Mauna Loa. Ecology 40:144–146.

Gagné, W. C., and C. C. Christensen. 1985. Conservation status of native terrestrial invertebrates in Hawaii, pp. 105–126 in C. P. Stone and J. M. Scott (eds.), Hawaii's terrestrial ecosystems: preservation and management. University of Hawaii Cooperative National Park Resources Studies Unit, Honolulu.

Gagné, W. C., and L. W. Cuddihy. 1990. Vegetation, pp. 45–114 in W. L. Wagner, D. R. Herbst, and S. H. Sohmer (eds.), Manual of the flowering plants of Hawaii. Bishop Museum and University of Hawaii Presses, Honolulu.

Giambelluca, T. W., and D. Nullet. 1991. Influence of the trade-wind inversion on the climate of a leeward mountain slope in Hawaii. Climate Res. 1:207–216.

Godley, E. J. 1978. Cushion bogs, pp. 141–158 in C. Troll and W. Lauer (eds.), Geological relations between the Southern Temperate Zone and the Tropical Mountains. Proceedings of the Symposium of International Geographical Union Commission on High-Altitude Geoecology, 1974. Franz Steiner Verlag, Wiesbaden.

Hardy, C. E. 1983. Insects, pp. 80–82 in R. W. Armstrong (ed.), Atlas of Hawaii. University Press of Hawaii, Honolulu.

Hartt, C. E., and M. Neal. 1940. The plant ecology of Mauna Kea, Hawaii. Ecology 21:237–266.

Herbst, D. R., and W. L. Wagner. 1992. Alien plants on the northwestern Hawaiian Islands, pp. 189–224 in C. P. Stone, C. W. Smith, and J. T. Tunison (eds.), Alien plant invasions in native ecosystems of Hawaii: management and research. University of Hawaii Cooperative National Park Resources Studies Unit, Honolulu.

Hodges, C. S., K. T. Adee, J. D. Stein, H. B. Wood, and R. D. Doty. 1986. Decline of 'ohi'a (Metrosideros polymorpha) in Hawaii: a review. U.S. Forest Service, Pacific Southwest Forest and Range Experiment Station, General Technical Report PSW-86, Berkeley, Calif.

Huenneke, L. F., and P. M. Vitousek. 1990. Seedling and clonal recruitment of the invasive tree Psidium cattleianum: implications for management of native Hawaiian forests. Biolog. Conserva. 53:199–211.

Hughes, R. F., P. M. Vitousek, and T. Tunison. 1991. Alien grass invasion and fire in the seasonal submontane zone of Hawai'i. Ecology 72:743–746.

Hughes, R. F., and P. M. Vitousek. 1993. Barriers to shrub reestablishment following fire in the seasonal submontane zone of Hawai'i. Oecologia 93 (4):557–563.

Jacobi, J. D. 1981. Vegetation changes in a subalpine grassland in Hawaii following disturbance by feral pigs. University of Hawaii, Department of Botany,

Cooperative National Park Resources Studies Unit, Tech. Rept. 41, Honolulu.

Jacobi, J. D. 1989. Vegetation maps of the upland plant communities on the islands of Hawaii, Maui, Molokai, and Lanai. University of Hawaii, Department of Botany, Cooperative National Park Resources Studies Unit, Tech. Rept. 68, Honolulu.

Jacobi, J. D. 1990. Distribution maps, ecological relationships and status of native plant communities on the island of Hawaii. Ph.D. dissertation, Botany Department, University of Hawaii at Manoa, Honolulu.

Jacobi, J. D., G. Gerrish, and D. Mueller-Dombois. 1983. 'Ohi'a dieback in Hawaii: vegetation changes in permanent plots. Pacific Sci. 37:327–337.

Jacobi, J. D., and J. M. Scott. 1985. An assessment of the current status of native upland habitats and associated endangered species on the island of Hawaii, pp. 3–22 in C. P. Stone and J. M. Scott (eds.), Hawaii's terrestrial ecosystems: preservation and management. University of Hawaii, Cooperative National Park Resources Studies Unit, Honolulu.

Jacobi, J. D., and F. R. Warshauer. 1992. Distribution of six alien plant species in upland habitats on the island of Hawaii, pp. 155–188 in C. P. Stone, C. W. Smith and J. T. Tunison (eds.), Alien plant invasions in native ecosystems of Hawaii: management and research. University of Hawaii, Cooperative National Park Resources Studies Unit, Honolulu.

James, H. F. 1995. Prehistoric extinctions and ecological changes on oceanic islands, pp. 87–101 in P. Vitousek, L. Loope, and H. Adersen (eds.), Biological diversity and ecosystem function on islands. Springer-Verlag, Heidelberg.

Juvik, J. O., and P. C. Ekern. 1978. A climatology of mountain fog on Mauna Loa, Hawaii Island. Tech. Rept. 118. Water Resources Research Center. University of Hawaii, Honolulu.

Karpa, D. M., and P. M. Vitousek. 1994. Successional development of a Hawaiian montane grassland. Biotropica 26:2–11.

Kirch, P. V. 1982. The impact of prehistoric Polynesians on the Hawaiian ecosystem. Pacific Sci. 36:1–14.

Kitayama, K., and D. Mueller-Dombois. 1994. An altitudinal transect analysis of the windward vegetation on Haleakala, a Hawaiian island mountain: (1) climate and soils. Phytocoenologia 24:111–133.

Kitayama, K., and D. Mueller-Dombois. 1995. Vegetation changes along gradients of long-term soil development in the Hawaiian montane rainforest zone. Vegetatio. 120:1–20.

Kitayama, K., D. Mueller-Dombois, and P. M. Vitousek. 1995. Primary succession of the Hawaiian montane rainforest on a chronosequence of eight lava flows. J. Vege. Sci. 6:211–222.

Lai, P. Y., M. Tamashiro, J. R. Yates, N. Y. Su, J. K. Fujii, and R. H. Ebesu. 1983. Living plants in Hawaii attacked by Coptotermes formosanus. Proceedings, Hawaiian Entomological Society 24:283–286.

La Rosa, A. M. 1992. The status of banana poka in Hawaii, pp. 271–299 in C. P. Stone, C. W. Smith, and J. T. Tunison (eds.), Alien plant invasions in native ecosystems of Hawaii: management and research. University of Hawaii, Cooperative National Park Resources Studies Unit, Honolulu.

Leuschner, C., and M. Schulte. 1991. Microclimatological investigations in the tropical alpine scrub of

Maui, Hawaii: evidence for a drought-induced alpine timberline. Pacific Sci. 45:152–168.

Lockwood, J. P., and P. W. Lipman. 1980. Recovery of datable charcoal beneath young lavas: lessons from Hawaii. Bull. Volcanologique 43:609–615.

Lockwood, J. P., P. W. Lipman, L. D. Petersen, and F. R. Warshauer. 1988. Generalized ages of surface lava flows of Mauna Loa Volcano, Hawaii. U.S. Geological Survey, Misc. Investigations Series Map I-1908. U.S. Department of the Interior, Washington, D.C.

Loope, L. L., and A. C. Medeiros. 1994. Impacts of biological invasions, management needs, and recovery efforts for rare plant species in Haleakala National Park, Maui, Hawaiian Islands, pp. 143–158 in M. Bowles and C. J. Whelan (eds.), Restoration of endangered species. Cambridge University Press, Cambridge.

Loope, L. L., and A. C. Medeiros. 1995. Haleakala silversword (*Argyroxiphium sandwicense* DC. ssp. *macrocephalum*), pp. 363–364 in E. T. LaRoe, G. S. Farris, C. E. Puckett, P. D. Doran, and M. J. Mac (eds.). Our living resources: a report to the nation on the distribution, abundance, and health of U.S. plants, animals, and ecosystems. U.S. Department of the Interior, National Biological Service, Washington, D.C.

Loope, L. L., A. C. Medeiros, and B. H. Gagné. 1991a. Aspects of the history and biology of the montane bogs of Haleakala National Park. University of Hawaii, Department of Botany, Cooperative National Park Resources Studies Unit, Tech. Rept. 76, Honolulu.

Loope, L. L., A. C. Medeiros, and B. H. Gagné. 1991b. Recovery of vegetation of a montane bog in Haleakala National Park following protection from feral pig rooting. University of Hawaii, Department of Botany, Cooperative National Park Resources Studies Unit, Tech. Rept. 77, Honolulu.

Loope, L. L., and D. Mueller-Dombois. 1989. Characteristics of invaded islands, pp. 257–280 in H. A. Mooney et al. (eds.), Ecology of biological invasions: a global synthesis. Wiley, Chichester.

Loope, L. L., R. J. Nagata, and A. C. Medeiros. 1992. Introduced plants in Haleakala National Park, pp. 551–576 in C. P. Stone, C. W. Smith, and J. T. Tunison (eds.), Alien plant invasions in native ecosystems of Hawaii: management and research. University of Hawaii, Cooperative National Park Resources Studies Unit, Honolulu.

Lorence, D. H., and K. R. Wood. 1994. *Kanaloa*, a new genus of Fabaceae (Mimosoideae) from Hawaii. Novon 4:137–145.

MacCaughey, V. 1917. Vegetation of Hawaiian lava flows. Botanical Gazette 64:386–420.

Macdonald, G. A., A. T. Abbott, and F. L. Petersen. 1983. Volcanoes in the sea: the geology of Hawaii. University of Hawaii Press, Honolulu.

Markin, G. P., P. Y. Lai, and G. Y. Funasaki. 1992. Status of biological control of weeds in Hawaii and implications for managing native ecosystems, pp. 466–482 in C. P. Stone, C. W. Smith, and J. T. Tunison (eds.), Alien plant invasions in native ecosystems of Hawaii: management and research. University of Hawaii, Cooperative National Park Resources Studies Unit, Honolulu.

Medeiros, A. C., L. L. Loope, and S. Anderson. 1993. Differential colonization by epiphytes on native (*Ci-*

botium spp.) and alien (*Cyathea cooperi*) tree ferns in a Hawaiian rain forest. Selbyana 14:71–74.

Medeiros, A. C., L. L. Loope, T. Flynn, L. Cuddihy, K. A. Wilson, and S. Anderson. 1992. The naturalization of an Australian tree fern (*Cyathea cooperi*) in Hawaiian rain forests. Amer. Fern J. 82(1):27–33.

Medeiros, A. C., L. L. Loope, and B. H. Gagné. 1991. Degradation of vegetation in two montane bogs in Haleakala National Park: 1982–1988. University of Hawaii, Department of Botany, Cooperative National Park Resources Studies Unit, Tech. Rept. 78, Honolulu.

Medeiros, A. C., L. L. Loope, and R. Hobdy. 1993. Conservation of cloud forests in Maui County (Maui, Moloka'i, and Lana'i), Hawaii, pp. 142–148 in L. S. Hamilton, J. O. Juvik, and F. N. Scatena (eds.), Tropical montane cloud forests. Proceedings of an international symposium. East–West Center, Honolulu.

Medeiros, A. C., L. L. Loope, P. Conant, and S. McElvaney. 1997. Status, ecology, and management of the invasive tree *Miconia calvescens* DC (Melastomataceae) in the Hawaiian Islands, pp. 23–35 in Records of the the Hawaii Biological Survey for 1996, N. L. Evenhuis and S. E. Miller (eds.), Bishop Museum Occasional Papers No. 48.

Medeiros, A. C., L. L. Loope, and R. A. Holt. 1986. Status of native flowering plant species on the south slope of Haleakala, East Maui. University of Hawaii, Department of Botany, Cooperative National Park Resources Studies Unit, Tech. Rept. 59, Honolulu.

Meyer, J.–Y. 1996. *Miconia calvescens*, the major plant pest in the Society Islands (French Polynesia). Pacific Sci. 50:66–76.

Mueller-Dombois, D. 1981a. Vegetation dynamics in a coastal grassland of Hawaii. Vegetatio 46:131–140.

Mueller-Dombois, D. 1981b. Fire in tropical ecosystems, pp. 137–176 in H. A. Mooney, T. M. Bonnicksen, N. L. Christensen, J. E. Lotan, and W. A. Reiners (eds.), Fire regimes and ecosystem properties. Proc. Conference Dec. 11–15, 1978, Honolulu. U.S. Dept. Agriculture, Forest Service Gen. Tech. Rept. WO-26. Washington, D.C.

Mueller-Dombois, D. 1983. Canopy dieback and successional processes in Pacific forests. Pacific Sci. 37:317–325.

Mueller-Dombois, D. 1985. 'Ohi'a dieback in Hawaii: 1984 synthesis and evaluation. Pacific Sci. 39:150–170.

Mueller-Dombois, D. 1986. Perspectives for an etiology of stand-level dieback. Ann. Rev. Ecol. and System. 17:221–243.

Mueller-Dombois, D. 1987. Forest dynamics in Hawaii. Trends in Ecol. and Evol. 2:216–220.

Mueller-Dombois, D., and V. J. Krajina. 1968. Comparison of east-flank vegetations on Mauna Loa and Mauna Kea, Hawaii, pp. 502–520 in R. Misra and B. Copal (eds.), Proceedings, Symposium on recent advances in tropical ecology. Intl. Soc. for Tropical Ecology, Varanasi, India.

Mueller-Dombois, D., and G. Spatz. 1975. The influence of feral goats on the lowland vegetation in Hawaii Volcanoes National Park. Phytocoenologica 3:1–29.

Nagata, K. M. 1985. Early plant introductions in Hawaii. Hawaiian J. History. 19:35–61.

Nakahara, L. M., R. M. Burkhart, and G. Y. Funasaki.

1992. Review and status of biological control of clidemia in Hawaii, pp. 52–465 in C. P. Stone, C. W.
Smith, and J. T. Tunison (eds.), Alien plant invasions in native ecosystems of Hawaii: management
and research. University of Hawaii, Cooperative
National Park Resources Studies Unit, Honolulu.

Nature Conservancy of Hawaii and the Natural Resources Defense Council. 1992. The alien pest species invasion in Hawaii: Background study and rec-
ommendations for interagency planning. Honolulu.

Noguchi, Y., H. Tabuchi, and H. Hasegawa. 1987. Physical factors controlling the formation of patterned
ground on Haleakala, Maui. Geografiska Annaler
69A:329–342.

Olson, S. L., and H. F. James. 1982. Prodromus of the
fossil avifauna of the Hawaiian Islands. Smithsonian Contributions in Zoology 365:1–59.

Porter, S. C. 1979. Hawaiian glacial ages. Quat. Res. 12:
161–187.

Price, S. 1983. Climate, pp. 53–62 in R. W. Armstrong
(ed.), Atlas of Hawaii. University Press of Hawaii.
Honolulu.

Raich, J. W., A. E. Russell, and P. M. Vitousek. 1997.
Primary production and ecosystem development
along elevational and age gradients in Hawaii.
Ecology, 707–721.

Robberecht, R., M. M. Caldwell, and W. D. Billings.
1980. Leaf ultraviolet optical properties along a latitudinal gradient in the arctic-alpine life zone. Ecology 61:612–619.

Robichaux, R. H., G. D. Carr, M. Liebman, and R. W.
Pearcy. 1990. Adaptive radiation of the silversword
alliance (Compositae: Madiinae): ecological, morphological, and physiological diversity. Ann. Missouri Bot. Gard. 77:64–72.

Rock, J. F. 1913. The indigenous trees of the Hawaiian
Islands. Charles E. Tuttle, Rutland, Vt. [Reprinted
1974, with annotations, by Pacific Tropical Botanical Garden, Lawai, Kauai, Hawaii.]

Scowcroft, P. G. 1983. Tree cover changes in mamane
(*Sophora chrysophylla*) forests grazed by sheep and
cattle. Pacific Sci. 37:109–119.

Scowcroft, P. G., and C. E. Conrad. 1992. Alien and native plant response to release from feral sheep
browsing on Mauna Kea, pp. 625–665 in C. P.
Stone, C. W. Smith, and J. T. Tunison (eds.), Alien
plant invasions in native ecosystems of Hawaii:
management and research. University of Hawaii,
Cooperative National Park Resources Studies Unit,
Honolulu.

Scowcroft, P. G., and J. G. Griffin. 1983. Feral herbivores suppress mamane and other browse species
on Mauna Kea, Hawaii. J. Range Manage. 36:638–
645.

Scowcroft, P. G., and R. Hobdy. 1986. Recovery of montane koa parkland vegetation protected from feral
goats. Biotropica 19:208–215.

Scowcroft, P. G., and H. F. Sakai. 1983. Impact of feral
herbivores on mamane forests of Mauna Kea, Hawaii: bark stripping and diameter class structure. J.
Range Manage. 36:495–498.

Scowcroft, P. G., and H. F. Sakai. 1984. Stripping of
Acacia koa bark by rats on Hawaii and Maui. Pacific
Sci. 38:80–86.

Skottsberg, C. 1941. Plant succession on recent lava
flows in the island of Hawaii. Goteborgs Kungl.
Vetenskapsoch Vitterhetssamhalles Handlingar.
Sjatte Folden, Series B, Band 1, No. 8.

Smathers, G. A., and D. Mueller-Dombois. 1974. Invasion and recovery of vegetation after a volcanic
eruption in Hawaii. National Park Service Scientific
Monograph Series, No. 5. U.S. Dept. of the Interior,
Washington, D.C.

Smith, C. W. 1985. Impacts of alien plants on Hawaii's
native biota, pp. 180–250 in C. P. Stone and J. M.
Scott (eds.), Hawaii's terrestrial ecosystems: preservation and management. University of Hawaii, Cooperative National Park Resources Studies Unit,
Honolulu.

Smith, C. W. 1992. Distribution, status, phenology, rate
of spread, and management of *Clidemia* in Hawaii,
pp. 241–253 in C. P. Stone, C. W. Smith, and J. T.
Tunison (eds.), Alien plant invasions in native ecosystems of Hawaii: management and research. University of Hawaii, Cooperative National Park Resources Studies Unit, Honolulu.

Smith, C. W., and J. T. Tunison. 1992. Fire and alien
plants in Hawaii: research and management implications for native ecosystems, pp. 394–408 in C. P.
Stone, C. W. Smith, and J. T. Tunison (eds.), Alien
plant invasions in native ecosystems of Hawaii:
management and research. University of Hawaii,
Cooperative National Park Resources Studies Unit,
Honolulu.

Spatz, G., and D. Mueller-Dombois. 1973. The influence
of feral goats on koa tree reproduction in Hawaii
Volcanoes National Park. Ecology 54:870–876.

Stemmermann, L. 1983. Ecological studies of Hawaiian
Metrosideros in a successional context. Pacific Sci. 37:
361–373.

Stemmermann, L., and T. Ihsle. 1993. Replacement of
Metrosideros polymorpha, 'ohi'a, in Hawaiian dry forest succession. Biotropica 25:36–45.

Stone, C. P., L. W. Cuddihy, and J. T. Tunison. 1992.
Responses of Hawaiian ecosystems to removal of
feral pigs and goats, pp. 666–704 in C. P. Stone,
C. W. Smith, and J. T. Tunison (eds.), Alien plant
invasions in native ecosystems of Hawaii: management and research. University of Hawaii, Cooperative National Park Resources Studies Unit, Honolulu.

Stone, C. P., and L. L. Loope. 1987. Reducing negative
effects of introduced animals on native biotas in
Hawaii: what is being done, what needs doing, and
the role of national parks. Environ. Conserva. 14:
245–258.

Sugihara, R. T. 1997. Abundance and diets of rats in two
native Hawaiian forests. Pacific Science 51:189–199.

Tomich, P. Q. 1986. Mammals in Hawaii: a synopsis
and notational bibliography, 2nd ed. B. P. Bishop
Museum Special Publ. 76. Bishop Museum Press,
Honolulu.

Tunison, J. T. 1992. Fountain grass control in Hawaii
Volcanoes National Park: management considerations and strategies, pp. 376–393 in C. P. Stone,
C. W. Smith, and J. T. Tunison (eds.), Alien plant
invasions in native ecosystems of Hawaii: management and research. University of Hawaii, Cooperative National Park Resources Studies Unit, Honolulu.

Tunison, J. T., W. L. Markiewicz, and A. A. McKinney.
1995. The expansion of koa forest after cattle and
goat removal, Hawaii Volcanoes National Park.
University of Hawaii, Department of Botany, Cooperative National Park Resources Studies Unit, Tech.
Rept., Honolulu.

U.S. Congress, Office of Technology Assessment. 1993. Harmful non-indigenous species in the United States. OTA-F-565. Government Printing Office, Washington, D.C.

Vitousek, P. M. 1994. Potential nitrogen fixation during primary succession in Hawai'i Volcanoes National Park. Biotropica 26:234–240.

Vitousek, P. M. 1995. The Hawaiian Islands as a model system for ecosystem studies. Pacific Sci. 49: 2–16.

Vitousek, P. M., G. H. Aplet, D. R. Turner, and J. P. Lockwood. 1992. The Mauna Loa environmental matrix: foliar and soil nutrients. Oecologia 89:372–382.

Vitousek, P. M., G. Gerrish, D. R. Turner, L. R. Walker, D. Mueller-Dombois. 1995. Litterfall and nutrient cycling in four Hawaiian montane rain forests. J. Trop. Ecol. 11:189–203.

Vitousek, P. M., P. A. Matson, and D. R. Turner. 1988. Elevation and age gradients in Hawaiian montane rainforest: foliar and soil nutrients. Oecologia 77: 565–570.

Vitousek, P. M., D. R. Turner, and K. Kitayama. 1995. Foliar nutrients during long-term soil development in Hawaiian montane rain forest. Ecology 76:712–720.

Vitousek, P. M., D. R. Turner, W. J. Parton, and R. L. Sanford. 1994. Litter decomposition on the Mauna Loa environmental matrix, Hawaii: patterns, mechanisms, and models. Ecology 75:418–429.

Vitousek, P. M., K. van Cleve, N. Balakrishnan, and D. Mueller-Dombois. 1983. Soil development and nitrogen turnover on recent volcanic substrates in Hawaii. Biotropica 15:268–274.

Vitousek, P. M. and L. R. Walker. 1989. Biological invasion by *Myrica faya* in Hawai'i: plant demography, nitrogen fixation, ecosystem effects. Ecol. Monog. 59:247–265.

Vogl, R. J. 1971. General ecology of the northeast outer slopes of Haleakala Crater, East Maui, Hawaii. Contributions from The Nature Conservancy, No. 6.

Wagner, W. L., D. R. Herbst, and S. H. Sohmer. 1990. Manual of the flowering plants of Hawaii. Bishop Museum and University of Hawaii Presses, Honolulu.

Warner, R. E. 1968. Scientific report of the Kipahula Valley expedition, Maui, Hawaii. Sponsored by The Nature Conservancy, Arlington, Va.

Whiteaker, L. D. 1983. The vegetation and environment in the Crater District of Haleakala National Park. Pacific Sci. 37:1–24.

Whiteaker, L. D., and D. E. Gardner. 1992. Firetree (*Myrica faya*) distribution in Hawaii, pp. 225–240 in C. P. Stone, C. W. Smith, and J. T. Tunison (eds.), Alien plant invasions in native ecosystems of Hawaii, Cooperative National Park Resources Studies Unit, Honolulu.

Woodcock, A. H. 1976. Permafrost and climatology of a Hawaii volcano crater. Arctic and Alpine Res. 6:49–62.

Woodward, S. A., P. M. Vitousek, K. A. Matson, R. F. Hughès, K. Benvenuto, and P. A. Matson. 1990. Avian use of the introduced nitrogen-fixer *Myrica faya* in Hawaii Volcanoes National Park. Pacific Sci. 44:88–93.

Yee, R. S. N., and W. C. Gagné. 1992. Activities and needs of the horticulture industry in relation to alien plant problems in Hawaii, pp. 712–725 in C. P. Stone, C. W. Smith, and J. T. Tunison (eds.), Alien plant invasions in native ecosystems of Hawaii: management and research. University of Hawaii, Cooperative National Park Resources Studies Unit, Honolulu.

Yocum, C. F. 1967. Ecology of feral goats in Haleakala National Park, Maui, Hawaii. Am. Midl. Nat. 77: 418–451.

Yoshinaga, A. Y. 1980. Upper Kipahulu Valley weed survey. University of Hawaii, Department of Botany, Cooperative National Park Resources Studies Unit, Tech. Rept. 33, Honolulu.

Zimmerman, E. C. 1948. Insects of Hawaii. Vol. 1. Introduction. University of Hawaii Press, Honolulu.

Subject Index

Entries in this index are limited to countries, regions, major topographic features, and recurring text topics or themes. Apart from "Alaska" and the "Californian region," states are not mentioned. Many cross-references are included. Entries in the species index are limited to binomials. Genera and families, for example, are not included. Common plant names do not appear in this index even if used in the text.

These two indices were prepared with the vital assistance of Spencer Graff, Allan Shanfield, and Mandy Tu.

Species Index

All binomials mentioned in the text are listed, but subspecific trinomials are not included.